PRINCIPLES OF OPTICS

FIFTH EDITION

To the Memory of

SIR ERNEST OPPENHEIMER

Principles of Optics

Electromagnetic Theory of Propagation, Interference and Diffraction of Light

by

MAX BORN
M.A., Dr.Phil., F.R.S.
Nobel Laureate
Formerly Professor at the Universities of Göttingen and Edinburgh

and

EMIL WOLF
Ph.D., D.Sc.
Professor of Physics, University of Rochester, N.Y.

with contributions by

A. B. Bhatia, P. C. Clemmow, D. Gabor, A. R. Stokes, A. M. Taylor, P. A. Wayman and W. L. Wilcock

FIFTH EDITION
1975

PERGAMON PRESS
OXFORD · NEW YORK · TORONTO · SYDNEY · PARIS · FRANKFURT

U.K.	Pergamon Press Ltd., Headington Hill Hall, Oxford OX3 0BW, England
U.S.A.	Pergamon Press Inc., Maxwell House, Fairview Park, Elmsford, New York 10523, U.S.A.
CANADA	Pergamon of Canada Ltd., 75 The East Mall, Toronto, Ontario, Canada
AUSTRALIA	Pergamon Press (Aust.) Pty. Ltd., 19a Boundary Street, Rushcutters Bay, N.S.W. 2011, Australia
FRANCE	Pergamon Press SARL, 24 rue des Ecoles, 75240 Paris, Cedex 05, France
WEST GERMANY	Pergamon Press GmbH, 6242 Kronberg-Taunus, Pferdstrasse 1, West Germany

First edition 1959
Second (revised) edition 1964
Third (revised) edition 1965
Fourth (revised) edition 1970
Fifth (revised) edition 1975
Reprinted 1975, 1977

Library of Congress Cataloging in Publication Data

Born, Max, 1882–1970.
Principles of optics.
Includes bibliographical references
1. Optics. 2. Electromagnetic theory.
I. Wolf, Emil, joint author. II. Title
QC355.2.B67 1975 535 74-8136
ISBN 0-08-018018 3

Printed in Great Britain by Page Bros (Norwich) Ltd.

PREFACE TO THE FIRST EDITION

THE idea of writing this book was a result of frequent enquiries about the possibility of publishing in the English language a book on optics written by one of us* more than twenty-five years ago. A preliminary survey of the literature showed that numerous researches on almost every aspect of optics have been carried out in the intervening years, so that the book no longer gives a comprehensive and balanced picture of the field. In consequence it was felt that a translation was hardly appropriate; instead a substantially new book was prepared, which we are now placing before the reader. In planning this book it soon became apparent that even if only the most important developments which took place since the publication of *Optik* were incorporated, the book would become impracticably large. It was, therefore, deemed necessary to restrict its scope to a narrower field. *Optik* itself did not treat the whole of optics. The optics of moving media, optics of X-rays and γ rays, the theory of spectra and the full connection between optics and atomic physics were not discussed; nor did the old book consider the effects of light on our visual sense organ—the eye. These subjects can be treated more appropriately in connection with other fields such as relativity, quantum mechanics, atomic and nuclear physics, and physiology. In this book not only are these subjects excluded, but also the classical molecular optics which was the subject-matter of almost half of the German book. Thus our discussion is restricted to those optical phenomena which may be treated in terms of MAXWELL's phenomenological theory. This includes all situations in which the atomistic structure of matter plays no decisive part. The connection with atomic physics, quantum mechanics, and physiology is indicated only by short references wherever necessary. The fact that, even after this limitation, the book is much larger than *Optik*, gives some indication about the extent of the researches that have been carried out in classical optics in recent times.

We have aimed at giving, within the framework just outlined, a reasonably complete picture of our present knowledge. We have attempted to present the theory in such a way that practically all the results can be traced back to the basic equations of MAXWELL's electromagnetic theory, from which our whole consideration starts.

In Chapter I the main properties of the electromagnetic field are discussed and the effect of matter on the propagation of the electromagnetic disturbance is described formally, in terms of the usual material constants. A more physical approach to the question of influence of matter is developed in Chapter II: it is shown that in the presence of an external incident field, each volume element of a material medium may be assumed to give rise to a secondary (scattered) wavelet and that the combination of these wavelets leads to the observable, macroscopic field. This approach is of considerable physical significance and its power is illustrated in a later chapter (Chapter XII) in connection with the diffraction of light by ultrasonic waves, first treated in this way by A. B. BHATIA and W. J. NOBLE; Chapter XII was contributed by Prof. BHATIA himself.

A considerable part of Chapter III is devoted to showing how geometrical optics follows from MAXWELL's wave theory as a limiting case of short wavelengths. In addition to discussing the main properties of rays and wave-fronts, the vectorial

---

* MAX BORN, *Optik* (Berlin, Springer, 1933).

aspects of the problem (propagation of the directions of the field vectors) are also considered. A detailed discussion of the foundations of geometrical optics seemed to us desirable in view of the important developments made in recent years in the related field of microwave optics (optics of short radio waves). These developments were often stimulated by the close analogy between the two fields and have provided new experimental techniques for testing the predictions of the theory. We found it convenient to separate the mathematical apparatus of geometrical optics—the calculus of variations—from the main text; an appendix on this subject (Appendix I) is based in the main part on unpublished lectures given by D. HILBERT at Göttingen University in the early years of this century. The following appendix (Appendix II), contributed by Prof. D. GABOR, shows the close formal analogy that exists between geometrical optics, classical mechanics, and electron optics, when these subjects are presented in the language of the calculus of variations.

We make no apology for basing our treatment of geometrical theory of imaging (Chapter IV) on HAMILTON's classical methods of characteristic functions. Though these methods have found little favour in connection with the design of optical instruments, they represent nevertheless an essential tool for presenting in a unified manner the many diverse aspects of the subject. It is, of course, possible to derive some of the results more simply from *ad hoc* assumptions; but, however valuable such an approach may be for the solution of individual problems, it cannot have more than illustrative value in a book concerned with a systematic development of a theory from a few simple postulates.

The defect of optical images (the influence of aberrations) may be studied either by geometrical optics (appropriate when the aberrations are large), or by diffraction theory (when they are sufficiently small). Since one usually proceeds from quite different starting points in the two methods of treatments, a comparison of results has in the past not always been easy. We have attempted to develop a more unified treatment, based on the concept of the deformation of wave-fronts. In the geometrical analysis of aberrations (Chapter V) we have found it possible and advantageous to follow, after a slight modification of his eikonal, the old method of K. SCHWARZSCHILD. The chapter on diffraction theory of aberrations (Chapter IX) gives an account of the NIJBOER-ZERNIKE theory and also includes an introductory section on the imaging of extended objects, in coherent and in incoherent illumination, based on the techniques of FOURIER transforms.

Chapter VI, contributed by Dr. P. A. WAYMAN, gives a brief description of the main image-forming optical systems. Its purpose is to provide a framework for those parts of the book which deal with the theory of image formation.

Chapter VII is concerned with the elements of the theory of interference and with interferometers. Some of the theoretical sections have their nucleus in the corresponding sections of *Optik*, but the chapter has been completely re-written by Dr. W. L. WILCOCK, who has also considerably broadened its scope.

Chapter VIII is mainly concerned with the FRESNEL-KIRCHHOFF diffraction theory and with some of its applications. In addition to the usual topics, the chapter includes a detailed discussion of the central problem of optical image formation—the analysis of the three-dimensional light distribution near the geometrical focus. An account is also given of a less familiar alternative approach to diffraction, based on the notion of the boundary diffraction wave of T. YOUNG.

The chapters so far referred to are mainly concerned with perfectly monochromatic (and therefore completely coherent) light, produced by point sources. Chapter X deals with the more realistic case of light produced by sources of finite extension and covering a finite frequency range. This is the subject of partial coherence, where considerable progress has been made in recent years. In fact, a systematic theory of interference and diffraction with partially coherent light has now been developed. This chapter also includes an account of the closely related subject of partial polarization, from the standpoint of coherence theory.

Chapter XI deals with rigorous diffraction theory, a field that has witnessed a tremendous development over the period of the last twenty years,* stimulated largely by advances in the ultra-shortwave radio techniques. This chapter was contributed by Dr. P. C. CLEMMOW who also prepared Appendix III, which deals with the mathematical methods of steepest descent and stationary phase.

The last two chapters, Optics of Metals (Chapter XIII) and Optics of Crystals (Chapter XIV) are based largely on the corresponding chapters of *Optik*, but were revised and extended with the help of Prof. A. M. TAYLOR and Dr. A. R. STOKES respectively. These two subjects are perhaps discussed in less detail than might seem appropriate. However, the optics of metals can only be treated adequately with the help of quantum mechanics of electrons, which is outside the scope of this book. In crystal optics the centre of interest has gradually shifted from visible radiation to X-rays, and the progress made in recent years has been of a technical rather than theoretical nature.

Though we have aimed at producing a book which in its methods of presentation and general approach would be similar to *Optik*, it will be evident that the present book is neither a translation of *Optik*, nor entirely a compilation of known data. As regards our own share in its production, the elder co-author (M. B.) has contributed that material from *Optik* which has been used as a basis for some of the chapters in the present treatise, and has taken an active part in the general planning of the book and in numerous discussions concerning disputable points, presentation, etc. Most of the compiling, writing, and checking of the text was done by the younger co-author (E. W.).

Naturally we have tried to use systematic notation throughout the book. But in a book that covers such a wide field, the number of letters in available alphabets is far too limited. We have, therefore, not always been able to use the most elegant notation but we hope that we have succeeded, at least, in avoiding the use in any one section of the same symbol for different quantities.

In general we use vector notation as customary in Great Britain. After much reflection we rejected the use of the nabla operator alone and employed also the customary "div", "grad", and "curl". Also, we did not adopt the modern electrotechnical units, as their main advantage lies in connection with purely electromagnetic measurements, and these play a negligible part in our discussions; moreover, we hope, that if ever a second volume (*Molecular and Atomic Optics*) and perhaps a third volume (*Quantum Optics*) is written, the C.G.S. system, as used in Theoretical Physics, will have returned to favour. Although, in this system of units, the magnetic permeability μ of most substances differs inappreciably from unity at optical frequencies, we have retained it in some of the equations. This has the advantage of greater symmetry and makes it possible to derive "dual" results by making use of the symmetry properties of MAXWELL's equations. For time periodic fields we have used, in complex representation, the factor $\exp(-i\omega t)$ throughout.

We have not attempted the task of referring to all the relevant publications. The

---

* The important review article by C. J. BOUWKAMP, *Rep. Progr. Phys.* (London, Physical Society), **17** (1954), 35, records more than 500 papers published in the period 1940–1954.

references that are given, and which, we hope, include the most important papers, are to help the reader to gain some orientation in the literature; an omission of any particular reference should not be interpreted as due to our lack of regard for its merit.

In conclusion it is a pleasure to thank many friends and colleagues for advice and help. In the first place we wish to record our gratitude to Professor D. GABOR for useful advice and assistance in the early stages of this project, as well as for providing a draft concerning his ingenious method of reconstructed wave-fronts (§ 8.10). We are also greatly indebted to Dr. F. ABELÈS, who prepared a draft, which is the backbone of § 1.6, on the propagation of electromagnetic waves through stratified media, a field to which he himself has made a substantial contribution. We have also benefited by advice on this subject from Dr. B. H. BILLINGS.

We are much indebted to Dr. H. H. HOPKINS, Dr. R. A. SILVERMAN, Dr. W. T. WELFORD and Dr. G. WYLLIE for critical comments and valuable advice, and to them and also to Dr. G. BLACK, Dr. H. J. J. BRADDICK, Dr. N. CHAKO, Dr. F. D. KAHN, Mr. A. NISBET, Dr. M. ROSS and Mr. R. M. SILLITTO for scrutinizing various sections of the manuscript. We are obliged to Polaroid Corporation for information concerning dichroic materials. Dr. F. D. KAHN helped with proof-reading and Dr. P. ROMAN and Mrs. M. PODOLANSKI with the preparation of the author index.

The main part of the writing was done at the Universities of Edinburgh and Manchester. The last stages were completed whilst one of the authors (E. W.) was a guest at the Institute of Mathematical Sciences, New York University. We are grateful to Professor M. KLINE, Head of its Division of Electromagnetic Research, for his helpful interest and for placing at our disposal some of the technical facilities of the Institute.

We gratefully acknowledge the loan of original photographs by Professor M. FRANÇON and Dr. M. CAGNET (Figs. 7.4, 7.26, 7.28, 7.60, 14.24, 14.26), Professor H. LIPSON and his co-workers at the Manchester College of Science and Technology (Figs. 8.10, 8.12, 8.15), Dr. O. W. RICHARDS (Figs. 8.34, 8.35), and Professor F. ZERNIKE and Dr. K. NIENHUIS (Figs. 9.4, 9.5, 9.8, 9.10, 9.11). Figure 7.66 is reproduced by courtesy of the Director of the Mount Wilson and Palomar Observatories. The blocks of Fig. 7.42 were kindly loaned by Messrs. Hilger and Watts, Ltd., and those of Figs. 7.64 and 7.65 by Dr. K. W. MEISSNER.

Financial assistance was provided by Messrs. Industrial Distributors Ltd., London, and we wish to acknowledge the generosity of the late Sir ERNEST OPPENHEIMER, its former head.

Finally, it is a pleasure to thank our publishers and in particular Mr. E. J. BUCKLEY, Mr. D. M. LOWE and also Dr. P. ROSBAUD, who as a former Director of Pergamon Press was closely associated with this project in its early stages, for the great care they have taken in the production of the book. It is a pleasure to pay tribute also to the printers, Pitman Press of Bath, for the excellence of their printing.

Bad Pyrmont and Manchester
January 1959

MAX BORN
EMIL WOLF

PREFACE TO THE SECOND EDITION

ADVANTAGE has been taken in the preparation of a new edition of this work to make a number of corrections of errors and misprints, to make a few minor additions and to include some new references.

Since the appearance of the first edition almost exactly three years ago, the first optical masers (lasers) have been developed. By means of these devices very intense and highly coherent light beams may be produced. Whilst it is evident that optical masers will prove of considerable value not only for optics but also for other sciences and for technology, no account of them is given in this new edition. For the basic principles of maser action have roots outside the domain of classical electromagnetic theory on which considerations of this book are based. We have, however, included a few references to recent researches in which light generated by optical masers was utilized or which have been stimulated by the potentialities of these new optical devices.

We wish to acknowledge our gratitude to a number of readers who drew our attention to errors and misprints. We are also obliged to Dr. B. KARCZEWSKI and Mr. C. L. MEHTA for assistance with the revisions.

Bad Pyrmont and Rochester
November 1962

M. B.
E. W.

PREFACE TO THE FIFTH EDITION

SOME further errors and misprints that were found in the earlier editions of this work have been corrected, the text in several sections has been improved and a number of references to recent publications have been added. More extensive changes have been made in §§13.1–13.3, dealing with the optical properties of metals. It is well known that a purely classical theory is inadequate to describe the interaction of an electromagnetic field with a metal in the optical range of the spectrum. Nevertheless, it is possible to indicate some of the main features of this process by means of a classical model, provided that the frequency dependence of the conductivity is properly taken into account and the role that the free, as well as the bound, electrons play in the response of the metal to an external electromagnetic field is understood, at least in qualitative terms. The changes in §§13.1–13.3 concern mainly these aspects of the theory and the revised sections are believed to be free of misleading statements and inaccuracies that were present in this connection in the earlier editions of this work and which can also be commonly found in many other optical texts.

I am grateful to some of our readers for informing me about misprints and errors. I wish to specifically acknowledge my indebtedness to Prof. A. D. BUCKINGHAM, Dr. D. CANALS FRAU and, once again, Dr. E. W. MARCHAND, who supplied me with detailed lists of corrections and to Dr. É. LALOR and Dr. G. C. SHERMAN for having drawn my attention to the need for making more substantial changes in Chapter XIII. I am also much obliged to Mr. J. T. FOLEY for assistance with the revisions.

Rochester
January 1974

E. W.

CONTENTS

HISTORICAL INTRODUCTION

THE physical principles underlying the optical phenomena with which we are concerned in this treatise were substantially formulated before 1900. Since that year, optics, like the rest of physics, has undergone a thorough revolution by the discovery of the quantum of energy. While this discovery has profoundly affected our views about the nature of light, it has not made the earlier theories and techniques superfluous; rather, it has brought out their limitations and defined their range of validity. The extension of the older principles and methods and their applications to very many diverse situations has continued, and is continuing with undiminished intensity.

In attempting to present in an orderly way the knowledge acquired over a period of several centuries in such a vast field it is impossible to follow the historical development, with its numerous false starts and detours. It is therefore deemed necessary to record separately, in this preliminary section, the main landmarks in the evolution of ideas concerning the nature of light.*

The philosophers of antiquity speculated about the nature of light, being familiar with burning glasses, with the rectilinear propagation of light, and with refraction and reflection. The first systematic writings on optics of which we have any definite knowledge are due to the Greek philosophers and mathematicians [EMPEDOCLES (*c.* 490–430 B.C.), EUCLID (*c.* 300 B.C.)].

Amongst the founders of the new philosophy, RENÉ DESCARTES (1596–1650) may be singled out for mention as having formulated views on the nature of light on the basis of his metaphysical ideas.† DESCARTES considered light to be essentially a pressure transmitted through a perfectly elastic medium (the aether) which fills all space, and he attributed the diversity of colours to rotary motions with different velocities of the particles in this medium. But it was only after GALILEO GALILEI (1564–1642) had, by his development of mechanics, demonstrated the power of the experimental method that optics was put on a firm foundation. The law of reflection was known to the Greeks; the law of refraction was discovered experimentally in 1621 by WILLEBRORD SNELL‡ (SNELLIUS, *c.* 1580–1626). In 1657 PIERRE DE FERMAT (1601–1665) enunciated the celebrated *Principle of Least Time*§ in the form "Nature

* For a more extensive account of the history of optics, reference may be made to: J. PRIESTLEY, *History and Present State of Discoveries relating to Vision, Light and Colours* (2 Vols., London, 1772); THOMAS YOUNG, *A Course of Lectures on Natural Philosophy and the Mechanical Arts* (London, 1845, Vol. 1, pp. 374–385); E. WILDE, *Geschichte der Optik vom Ursprung dieser Wissenschaft bis auf die gegenwärtige Zeit* (2 Vols., Berlin, 1838, 1843); ERNST MACH, *The Principles of Physical Optics*, A historical and philosophical treatment (First German edition 1913. English translation 1926, reprinted by Dover Publications, New York, 1953); E. HOPPE, *Geschichte der Optik* (Leipzig, Weber, 1926); V. RONCHI, *Storia della Luce* (Bologna: Zanichelli, 2nd. Ed., 1952). A comprehensive historical account up to recent times is E. T. WHITTAKER's *A History of the Theories of Aether and Electricity*, Vol. I (The Classical Theories), revised and enlarged edition 1952; Vol. II (The Modern Theories 1900–1926), 1953, published by T. Nelson and Sons, London and Edinburgh. The first volume was used as the chief source for this introductory section.

† R. DESCARTES, *Dioptrique, Météores* [published (anonymously) in Leyden in 1637 with prefactory essay "Discours de la méthode"]. *Principia Philosophiae* (Amsterdam, 1644).

‡ SNELL died in 1626 without making his discoveries public. The law was first published by DESCARTES in his *Dioptrique* without an acknowledgement to SNELL, though it is generally believed that DESCARTES had seen SNELL's manuscript on this subject.

§ In a letter to CUREAU DE LA CHAMBRE. It is published in *Oeuvres de Fermat* (Paris, 1891, **2**, 354).

always acts by the shortest course". According to this principle, light always follows that path which brings it to its destination in the shortest time, and from this, in turn, and from the assumption of varying "resistance" in different media, the law of refraction follows. This principle is of great philosophical significance, and because it seems to imply a teleological manner of explanation, foreign to natural science, it has raised a great deal of controversy.

The first phenomenon of interference, the colours exhibited by thin films now known as "Newton's rings", was discovered independently by ROBERT BOYLE* (1627–1691) and ROBERT HOOKE† (1635–1703). HOOKE also observed the presence of light in the geometrical shadow, the "diffraction" of light but this phenomenon had been noted previously by FRANCESCO MARIA GRIMALDI‡ (1618–1663). HOOKE was the first to advocate the view that light consists of rapid vibrations propagated instantaneously, or with a very great speed, over any distance, and believed that in an homogeneous medium every vibration will generate a sphere which will grow steadily.§ By means of these ideas HOOKE attempted an explanation of the phenomena of refraction, and an interpretation of colours. But the basic quality of colour was revealed only when ISAAC NEWTON (1642–1727) discovered|| in 1666 that white light could be split up into component colours by means of a prism, and found that each pure colour is characterized by a specific refrangibility. The difficulties which the wave theory encountered in connection with the rectilinear propagation of light and of polarization (discovered by HUYGENS¶) seemed to NEWTON so decisive that he devoted himself to the development of an emission (or corpuscular) theory, according to which light is propagated from a luminous body in the form of minute particles.

At the time of the publication of NEWTON's theory of colour it was not known whether light was propagated instantaneously or not. The discovery of the finite speed of light was made in 1675 by OLAF RÖMER (1644–1710) from the observations of the eclipses of Jupiter's satellites.**

The wave theory of light which, as we saw, had HOOKE amongst its first champions was greatly improved and extended by CHRISTIAN HUYGENS¶ (1629–1695). He enunciated the principle, subsequently named after him, according to which every point of the "aether" upon which the luminous disturbance falls may be regarded as the centre of a new disturbance propagated in the form of spherical waves; these secondary waves combine in such a manner that their envelope determines the wave-front at any later time. With the aid of this principle he succeeded in deriving the laws of reflection and refraction. He was also able to interpret the double refraction of calc-spar (discovered in 1669 by ERASMUS BARTHOLINUS (1625–1698)) by assuming that in the crystal there is, in addition to a primary spherical wave, a secondary ellipsoidal wave. It was in the course of this investigation that HUYGENS made the fundamental discovery of polarization: each of the two rays arising from refraction by calc-spar may be extinguished by passing it through a second crystal of the same material if the latter crystal be rotated about the direction of the ray. It was, however, left to NEWTON to interpret these phenomena; he assumed that rays have "sides";

---

* *The Philosophical Works of* ROBERT BOYLE (abridged by P. SHAW), Second Ed. (London, 1738), Vol. II, p. 70.

† R. HOOKE, *Micrographia* (1665), 47.

‡ F. M. GRIMALDI, *Physico-Mathesis de lumine, coloribus, et iride* (Bologna, 1665).

§ The early wave theories of HOOKE and HUYGENS operate with single "pulses" rather than with wave trains of definite wavelengths.

|| I. NEWTON, *Phil. Trans.* No. 80 (Feb. 1672), 3075.

¶ CHR. HUYGENS, *Traité de la lumière* (completed in 1678, published in Leyden in 1690).

** OLAF RÖMER, *Mém. de l'Acad. Sci. Paris,* **10** (1666–1699), 575; *J. de Sav.* (1676), 223.

and indeed this "transversality" seemed to him an insuperable objection to the acceptance of the wave theory, since at that time scientists were familiar only with longitudinal waves (from the propagation of sound).

The rejection of the wave theory on the authority of NEWTON led to its abeyance for nearly a century. But it still found an occasional supporter, such as the great mathematician LEONHARD EULER (1707–1783).*

It was not until the beginning of the 19th century that the decisive discoveries were made which led to general acceptance of the wave theory. The first step towards this was the enunciation in 1801 by THOMAS YOUNG (1773–1829) of the principle of interference and the explanation of the colours of thin films.† But as YOUNG's views were expressed largely in a qualitative manner they did not gain general recognition.

About this time polarization of light by reflection was discovered by ÉTIENNE LOUIS MALUS‡ (1775–1812). Apparently, one evening, in 1808, he observed the reflection of the sun from a window pane through a calc-spar crystal, and found that the two images obtained by double refraction varied in relative intensities as the crystal was rotated about the line of sight. But MALUS did not attempt an interpretation of this phenomenon, being of the opinion that current theories were incapable of providing an explanation.

In the meantime the emission theory had been developed further by PIERRE SIMON DE LAPLACE (1749–1827) and JEAN-BAPTISTE BIOT (1774–1862). Its supporters proposed the subject of diffraction for the prize question set by the Paris Academy for 1818, in the expectation that a treatment of this subject would lead to the crowning triumph of the emission theory. But their hopes were disappointed, for, in spite of strong opposition, the prize was awarded to AUGUSTIN JEAN FRESNEL (1788–1827), whose treatment§ was based on the wave theory, and was the first of a succession of investigations which, in the course of a few years, were to discredit the corpuscular theory completely. The substance of his memoir consisted of a synthesis of HUYGENS's Envelope Construction with YOUNG's Principle of Interference. This, as FRESNEL showed, was sufficient to explain not only the "rectilinear propagation" of light but also the minute deviations from it—diffraction phenomena. FRESNEL calculated the diffraction caused by straight edges, small apertures, and screens; particularly impressive was the experimental confirmation by ARAGO of a prediction, deduced by POISSON from FRESNEL's theory, that in the centre of the shadow of a small circular disc there should appear a bright spot.

In the same year (1818) FRESNEL also investigated the important problem of the influence of the earth's motion on the propagation of light, the question being whether there was any difference between the light from stellar and terrestrial sources. DOMINIQUE FRANÇOIS ARAGO (1786–1853) found from experiment that (apart from aberration) there was no difference. On the basis of these findings FRESNEL developed his theory of the partial convection of the luminiferous aether by matter, a theory confirmed in 1851, by direct experiment carried out by ARMAND HYPOLITE LOUIS FIZEAU (1819–1896). Together with ARAGO, FRESNEL investigated the interference of polarized rays of light and found (in 1816) that two rays polarized at right angles to each other never interfere. This fact could not be reconciled with the assumption of

* *L. Euleri Opuscula varii argumenti*, Berlin (1746), 169.

† TH. YOUNG, *Phil. Trans. Roy. Soc., London* xcii (1802) 12, 387. *Young's Works*, Vol. 1, p. 202.

‡ É. L. MALUS, *Nouveau Bull. d. Sci., par la Soc. Philomatique*, Vol. 1 (1809), 266. *Mém. de la Soc. d'Arcueil*, Vol. 2 (1809).

§ A. FRESNEL, *Ann. Chim. et Phys.*, (2), **1** (1816) 239; *Oeuvres*, Vol. 1, 89, 129.

longitudinal waves, which had hitherto been taken for granted. YOUNG, who had heard of this discovery from ARAGO, found in 1817 the key to the solution when he assumed that the vibrations were transverse.

FRESNEL at once grasped the full significance of this hypothesis, which he sought to put on a more secure dynamical basis* and from which he drew numerous conclusions. For, since only longitudinal oscillations in a fluid are possible, the aether must behave like a solid body; but at that time a theory of elastic waves in solids had not yet been formulated. Instead of developing such a theory and deducing the optical consequences from it, FRESNEL proceeded by inference, and sought to deduce the properties of the luminiferous aether from the observations. The peculiar laws of light propagation in crystals were FRESNEL's starting point; the elucidation of these laws and their reduction to a few simple assumptions about the nature of elementary waves represents one of the greatest achievements of natural science. In 1832, WILLIAM ROWAN HAMILTON† (1805–1865), who himself made important contributions to the development of optics, drew attention to an important deduction from FRESNEL's construction, by predicting the so-called conical refraction, whose existence was confirmed experimentally shortly afterwards by HUMPHREY LLOYD‡ (1800–1881).

It was also FRESNEL who (in 1821) gave the first indication of the cause of dispersion, by taking into account the molecular structure of matter§, a suggestion elaborated later by CAUCHY.

Dynamical models of the mechanism of aether vibrations led FRESNEL to deduce the laws which now bear his name, governing the intensity and polarization of light rays produced by reflection and refraction.||

FRESNEL's work had put the wave theory on such a secure foundation that it seemed almost superfluous when in 1850 FOUCAULT¶ and FIZEAU and BREGUET** undertook a crucial experiment first suggested by ARAGO. The corpuscular theory explains refraction in terms of the attraction of the light-corpuscles at the boundary towards the optically denser medium, and this implies a greater velocity in the denser medium; on the other hand the wave theory demands, according to HUYGENS's construction, that a smaller velocity obtains in the optically denser medium. The direct measurement of the velocity of light in air and water decided unambiguously in favour of the wave theory.

The decades that followed witnessed the development of the elastic aether theory. The first step was the formulation of a theory of the elasticity of solid bodies. CLAUDE LOUIS MARIE HENRI NAVIER (1785–1836) developed such a theory††, discerning that matter consists of countless particles (mass points, atoms) exerting on each other forces along the lines joining them. The now customary derivation of the equations of elasticity by means of the continuum concept is due to AUGUSTINE LOUIS CAUCHY‡‡ (1789–1857). Of other scientists who participated in the development of optical theory mention must be made of SIMÉON DENIS POISSON§§ (1781–1840), GEORGE

---

* A. FRESNEL, *Oeuvres*, **2**, 261, 479.

† W. R. HAMILTON, *Trans. Roy. Irish Acad.*, **17** (1833), 1. Also *Hamilton's Mathematical Papers*, edited by J. L. SYNGE and W. CONWAY (Cambridge University Press, 1931), Vol. 1, p. 285.

‡ H. LLOYD, *Trans. Roy. Irish Acad.*, **17** (1833), 145.

§ A. FRESNEL, *Oeuvres*, **2**, 438.

|| A. FRESNEL, *Mém. de l'Acad.*, **11** (1832), 393; *Oeuvres*, **1**, 767.

¶ L. FOUCAULT, *Compt. Rend. Acad. Sci. Paris*, **30** (1850), 551.

** H. FIZEAU and L. BREGUET, *Compt. Rend. Acad. Sci. Paris*, **30** (1850), 562, 771.

†† C. L. M. H. NAVIER, *Mém. de l'Acad.*, **7**, 375 (submitted in 1821, published in 1827).

‡‡ A. L. CAUCHY, *Exercise de Mathématiques*, **3** (1828), 160.

§§ S. D. POISSON, *Mém. de l'Acad.*, Vol. 8 (1828), 623.

GREEN* (1793–1841), JAMES MACCULLAGH† (1809–1847) and FRANZ NEUMANN‡ (1798–1895). To-day it is no longer relevant to enter into the details of these theories or into the difficulties which they encountered; for the difficulties were all caused by the requirement that optical processes should be explicable in mechanical terms, a condition which has long since been abandoned. The following indication will suffice. Consider two contiguous elastic media, and assume that in the first a transverse wave is propagated towards their common boundary. In the second medium the wave will be resolved, in accordance with the laws of mechanics, into longitudinal and transverse waves. But, according to ARAGO's and FRESNEL's experiments, elastic longitudinal waves must be ruled out and must therefore be eliminated somehow. This, however, is not possible without violating the laws of mechanics expressed by the boundary conditions for strains and stresses. The various theories put forward by the authors mentioned above differ in regard to the assumed boundary conditions, which always conflicted in some way with the laws of mechanics.

An obvious objection to regarding the aether as an elastic solid is expressed in the following query: How is one to imagine planets travelling through such a medium at enormous speeds without any appreciable resistance? GEORGE GABRIEL STOKES (1819-1903) thought that this objection could be met on the grounds that the planetary speeds are very small compared to the speeds of the aetherial particles in the vibrations constituting light; for it is known that bodies like pitch or sealing wax are capable of rapid vibrations but yield completely to stresses applied over a long period. Such controversies seem superfluous today since we no longer consider it necessary to have mechanical pictures of all natural phenomena.

A first step away from the concept of an elastic aether was taken by MACCULLAGH, § who postulated a medium with properties not possessed by ordinary bodies. The latter store up energy when the volume elements change shape, but not during rotation. In MACCULLAGH's aether the inverse conditions prevail. The laws of propagation of waves in such a medium show a close similarity to MAXWELL's equations of electromagnetic waves which are the basis of modern optics.

In spite of the many difficulties, the theory of an elastic aether persisted for a long time and most of the great physicists of the 19th century contributed to it. In addition to those already named, mention must be made of WILLIAM THOMSON|| (Lord KELVIN, 1824–1908), CARL NEUMANN¶ (1832–1925), JOHN WILLIAM STRUTT** (Lord RAYLEIGH, 1842–1919), and GUSTAV KIRCHHOFF†† (1832–1887). During this period many optical problems were solved, but the foundations of optics remained in an unsatisfactory state.

In the meantime researches in electricity and magnetism had developed almost independently of optics, culminating in the discoveries of MICHAEL FARADAY‡‡ (1791–1867). JAMES CLERK MAXWELL§§ (1831–1879) succeeded in summing up all previous experiences in this field in a system of equations, the most important

* G. GREEN, *Trans. Camb. Phil. Soc.* (1838); *Math. Papers*, 245.

† J. MACCULLAGH, *Phil. Mag.* (3), **10** (1837), 42, 382; *Proc. Roy. Irish Acad.*, **18** (1837).

‡ F. NEUMANN, *Abh. Berl. Akad., Math. Kl.* (1835), 1.

§ J. MACCULLAGH, *Trans. Roy. Irish Acad.*, **21**, Coll. Works, Dublin (1880), 145.

|| W. THOMSON, *Phil. Mag.*, (5), **26** (1888), 414. Baltimore Lectures (London, 1904).

¶ C. NEUMANN, *Math. Ann.*, **1** (1869), 325, **2** (1870), 182.

** J. W. STRUTT, (Lord Rayleigh), *Phil. Mag.*, (4) **41** (1871), 519; **42** (1871), 81.

†† G. KIRCHHOFF, *Berl. Abh. Physik., Abteilg.* **2** (1876), 57; *Ges. Abh.* 352; *Berl. Ber.* (1882), 641; *Pogg. Ann. Physik. u. Chem.* (2), **18** (1883), 663; *Ges. Abh., Nachtrag.* 22.

‡‡ M. FARADAY, *Experimental Researches in Electricity* (London, 1839).

§§ J. C. MAXWELL, *A Treatise on Electricity and Magnetism*, 2 Vols. (Oxford, 1873).

consequence of which was to establish the possibility of electromagnetic waves, propagated with a velocity which could be calculated from the results of purely electrical measurements. When RUDOLPH KOHLRAUSCH (1809–1858) and WILHELM WEBER* (1804–1891) carried out these measurements, the velocity turned out to be that of light. This led MAXWELL to conjecture that light waves are electromagnetic waves; a conjecture verified by direct experiment in 1888 by HEINRICH HERTZ† (1857–1894). In spite of this, MAXWELL's electromagnetic theory had a long struggle to gain general acceptance. It seems to be a characteristic of the human mind that familiar concepts are abandoned only with the greatest reluctance, especially when a concrete picture of phenomena has to be sacrificed. MAXWELL himself, and his followers, tried for a long time to describe the electromagnetic field with the aid of mechanical models. It was only gradually, as MAXWELL's concepts became more familiar, that the search for an "explanation" of his equations in terms of mechanical models was abandoned; today there is no conceptual difficulty in regarding MAXWELL's field as something which cannot be reduced to anything simpler.

But even the electromagnetic theory of light has attained the limits of its serviceability. It is capable of explaining, in their main features, all phenomena connected with the propagation of light. However, it fails to elucidate the processes of emission and absorption, in which the finer features of the interaction between matter and the optical field are manifested.

The laws underlying these processes are the proper object of modern optics, indeed of modern physics. Their story begins with the discovery of certain regularities in spectra. The first step was JOSEF FRAUNHOFER's (1787–1826) discovery‡ (1814–1817) of the dark lines in the solar spectrum, since named after him; and their interpretation as absorption lines given in 1861 on the basis of experiments by ROBERT WILHELM BUNSEN (1811–1899) and GUSTAV KIRCHHOFF§ (1824–1887). The light of the continuous spectrum of the body of the sun, passing the cooler gases of the sun's atmosphere, loses by absorption just those wavelengths which are emitted by the gases. This discovery was the beginning of spectrum analysis, which is based on the recognition that every gaseous chemical element possesses a characteristic line spectrum. The investigation of these spectra has been a major object of physical research up to and including the present; and since light is its subject and optical methods are employed, it is often considered as a part of optics. The problem of how light is produced or destroyed in atoms is, however, not exclusively of an optical nature, as it involves equally the mechanics of the atom itself; and the laws of spectral lines reveal not so much the nature of light as the structure of the emitting particles. Thus, from being a part of optics, spectroscopy has gradually evolved into a separate discipline which provides the empirical foundations for atomic and molecular physics. This field is, however, beyond the scope of this book.

Concerning methods, it has become apparent that classical mechanics is inadequate for a proper description of events occurring within the atom and must be replaced by

---

* R. KOHLRAUSCH and W. WEBER, *Pogg. Ann. Physik u. Chem.* (2), **99** (1856), 10.

† H. HERTZ, *Sitzb. Berl. Akad. Wiss.*, Feb. 2, 1888; *Wiedem. Ann.* **34** (1888), 551; English translation in his *Electric Waves* (London, Macmillan, 1893), 107.

‡ J. FRAUNHOFER, *Gilberts Ann.*, **56** (1817), 264. W. H. WOLLASTON (1766–1828) observed these lines in 1802 (*Phil. Trans. Roy. Soc.*, London, 1802, 365) but had not appreciated his discovery and interpreted them incorrectly.

§ R. BUNSEN and G. KIRCHHOFF, *Untersuchungen über das Sonnenspektrum und die Spektren der Chemischen Elemente.*, *Abh. kgl. Akad. Wiss.*, Berlin, 1861, 1863.

the quantum theory, originated in 1900 by MAX PLANCK* (1858–1947). Its application to the atomic structure, led in 1913, to the explanation by NIELS BOHR† (born in 1885) of the simple laws of line spectra of gases. From these beginnings and from the ever increasing experimental material, modern quantum mechanics developed (HEISENBERG, BORN, JORDAN, DE BROGLIE, SCHRÖDINGER, DIRAC).‡ By its means considerable insight has been obtained into the structure of atoms and molecules.

However, our concept of the nature of light has also been greatly influenced by quantum theory. Even in its first form due to PLANCK there appears a proposition which is directly opposed to classical ideas, namely that an oscillating electric system does not impart its energy to the electromagnetic field in a continuous manner but in finite amounts, or "quanta" $\varepsilon = h\nu$, proportional to the frequency ν of the light, where $h = 6{\cdot}55 \times 10^{-27}$ erg/sec is PLANCK's constant. We may say that the occurrence of the constant h is the feature which distinguishes modern physics from the old.

It was only gradually that the paradoxical, almost irrational, character of PLANCK's equation $\varepsilon = h\nu$ was fully realized by physicists. This was brought about mainly by the work of EINSTEIN and BOHR. On the basis of PLANCK's theory, EINSTEIN (1879–1955) revived in 1905 the corpuscular theory of light in a new form§ by assuming that PLANCK's energy quanta exist as real light-particles, called "light quanta" or "photons". He thereby succeeded in explaining some phenomena which had been discovered more recently in connection with the transformation of light into corpuscular energy, phenomena which were inexplicable by the wave theory. Chief among these are the so-called photo-electric effect and the fundamentals of photochemistry. In phenomena of this type light does not impart to a detached particle an energy proportional to its intensity, as demanded by the wave theory, but behaves rather like a hail of small shots. The energy imparted to the secondary particles is independent of the intensity, and depends only on the frequency of the light (according to the law $\varepsilon = h\nu$). The number of observations confirming this property of light increased year by year and the situation arose that the simultaneous validity of both wave and corpuscular theories had to be recognized, the former being exemplified by the phenomena of interference, the latter by the photo-electric effect. It is only in more recent years that the development of quantum mechanics has led to a partial elucidation of this paradoxical state of affairs, but this has entailed giving up a fundamental principle of the older physics, namely the principle of deterministic causality.

The detailed theory of the interaction between field and matter required the extension of the methods of quantum mechanics (field quantization). For the electromagnetic radiation field this was first carried out by DIRAC¶ and these investigations form the basis of quantum optics. A discussion of these topics is, however, outside the scope of this book.

Another branch of optics, not touched upon in this work, is the optics of moving bodies. Like the quantum theory it has grown into a vast independent field of study.

---

* M. PLANCK, *Verh. d. deutsch phys. Ges.*, **2**, 1900, 202, 237. *Ann. d. Physik*, (4), **4** (1901), 553.

† N. BOHR, *Phil. Mag.*, (6), **26** (1913), 1, 476, 857.

‡ W. HEISENBERG, *Z.f. Phys.*, **33** (1925), 879; M. BORN and P. JORDAN, *ibid.*, **34** (1925), 858; M. BORN, W. HEISENBERG, and P. JORDAN, *ibid.*, **35** (1926), 557; L. DE BROGLIE, *Thèse*, Paris, 1924; *Ann. de Physique* (10), **3** (1925), 22; E. SCHRÖDINGER, *Ann. d. Physik*, (4), **79** (1926), 361, 489 and 734; **80** (1926), 437; **81** (1926), 109. English translation: "Collected Papers on Wave Mechanics" by E. SCHRÖDINGER (London and Glasgow, Blackie, 1928); P. A. M. DIRAC, *Proc. Roy. Soc.* A, **109** (1925), 642; *ibid.*, **110** (1926), 561.

§ A. EINSTEIN, *Ann. d. Physik*, (4), **17** (1905), 132; **20** (1906), 199.

¶ P. A. M. DIRAC, *Proc. Roy. Soc.*, A**114** (1927), 243, 710.

The first observed phenomenon in this field, recorded in 1728 by JAMES BRADLEY* (1692–1762), was the aberration of "fixed stars", i.e. the observation of slightly different angular positions of the stars according to the motion of the earth relative to the direction of the light ray. BRADLEY correctly interpreted this phenomenon as being due to the finite velocity of light and thus was able to determine the velocity. We have already mentioned other phenomena belonging to optics of moving media: FRESNEL was the first to enquire into the convection of light by moving bodies and to show that it behaved as if the luminiferous aether participated in the movement only with a fraction of the speed of the moving bodies; FIZEAU then demonstrated this partial convection experimentally with the aid of flowing water. The effect of the motion of the light source or of the observer was investigated by CHRISTIAN DOPPLER† (1803–1853) who formulated the well-known principle named after him. So long as the elastic theory of light held the field and the precision of measurements was sufficiently limited, FRESNEL's ideas on partial convection sufficed for a satisfactory explanation of all the phenomena. But the electromagnetic theory of light encountered difficulties of a fundamental nature. HERTZ was the first to attempt to generalize MAXWELL's laws to moving bodies. His formulae were, however, in conflict with some electromagnetic and optical experiments. Of great importance was the theory of HENDRIK ANTOON LORENTZ (1853–1928) who assumed an "aether in a state of absolute rest" to be the carrier of the electromagnetic field and deduced the properties of material bodies from the interaction of elementary electric particles—the electrons. He was able to show that FRESNEL's coefficient of convection could be obtained correctly from his theory and that in general all phenomena known at the time (1895) lent themselves to explanation by this hypothesis.‡ But the enormous increase of precision in the determination of optical paths by means of the interferometer of ALBERT ABRAHAM MICHELSON (1852–1931) led to a new anomaly: it proved impossible to demonstrate the existence of an "aether drift" required by the theory of the "stationary aether".§ The anomaly was resolved by ALBERT EINSTEIN|| in 1905 in his special theory of relativity. The theory is founded on a critique of the concepts of time and space and leads to the abandonment of euclidian geometry and the intuitive conception of simultaneity. Its further development into the so-called general theory of relativity¶ led to a completely new conception of gravitational phenomena by a "geometrization" of the space-time manifold. The application of this theory involves the use of special mathematical and physical methods which, although relevant to optics in many cases, may easily be considered separately from it. The number of optical phenomena in which the motion of bodies (e.g. light sources) plays a significant part is rather small.

* J. BRADLEY, *Phil. Trans.*, **35** (1728), 637.

† CHR. DOPPLER, *Abh. Königl. böhm. Geselsch*, **2** (1842), 466; *Pogg. Ann.* **68** (1847), 1.

‡ H. A. LORENTZ, *Versuch einer Theorie der electrischen und optischen Erscheinungen in bewegten Körpern* (Leiden, E. J. Brill 1895; reprinted by Teubner, Leipzig, 1906).

§ A. A. MICHELSON, *Amer. Jour. Sci.*, (3), **22** (1881), 20; A. A. MICHELSON and E. W. MORLEY, *Amer. Jour. Sci.*, (3), **34** (1887), 333; *Phil. Mag.*, **24** (1887), 449.

|| A. EINSTEIN, *Ann. d. Physik* (4), **17** (1905), 891.

¶ A. EINSTEIN, *Berl. Sitz.* (1915), 778, 799, 831, 844. *Ann. d. Physik*, (4) **49** (1916), 769

CHAPTER 1

BASIC PROPERTIES OF THE ELECTROMAGNETIC FIELD

1.1 THE ELECTROMAGNETIC FIELD

1.1.1 Maxwell's equations

THE state of excitation which is established in space by the presence of electric charges is said to constitute an *electromagnetic field*. It is represented by two vectors, $\boldsymbol{E}$ and $\boldsymbol{B}$, called the *electric vector* and the *magnetic induction* respectively.*

To describe the effect of the field on material objects, it is necessary to introduce a second set of vectors, viz. *the electric current density* $\boldsymbol{j}$, *the electric displacement* $\boldsymbol{D}$, and *the magnetic vector* $\boldsymbol{H}$.

The space and time derivatives of the five vectors are related by *Maxwell's equations*, which hold at every point in whose neighbourhood the physical properties of the medium are continuous:†

$$\operatorname{curl} \boldsymbol{H} - \frac{1}{c}\dot{\boldsymbol{D}} = \frac{4\pi}{c}\boldsymbol{j}, \tag{1}$$

$$\operatorname{curl} \boldsymbol{E} + \frac{1}{c}\dot{\boldsymbol{B}} = 0, \tag{2}$$

the dot denoting differentiation with respect to time.

They are supplemented by two scalar relations:

$$\operatorname{div} \boldsymbol{D} = 4\pi\rho, \tag{3}$$

$$\operatorname{div} \boldsymbol{B} = 0. \tag{4}$$

Eq. (3) may be regarded as a defining equation for the electric charge density ρ and (4) may be said to imply that no free magnetic poles exist.

---

* In elementary considerations $\boldsymbol{E}$ and $\boldsymbol{H}$ are, for historical reasons, usually regarded as the basic field vectors, and $\boldsymbol{D}$ and $\boldsymbol{B}$ as describing the influence of matter. In general theory, however, the present interpretation is compulsory for reasons connected with the electrodynamics of moving media.

The four MAXWELL equations (1)–(4) can be divided into two sets of equations, one consisting of two homogeneous equations (right-hand side zero), containing $\boldsymbol{E}$ and $\boldsymbol{B}$, the other of two non-homogeneous equations (right-hand side different from zero), containing $\boldsymbol{D}$ and $\boldsymbol{H}$. If a co-ordinate transformation of space and time (relativistic LORENTZ transformation) is carried out, the equations of each group transform together, the equations remaining unaltered in form if $\boldsymbol{j}/c$ and ρ are transformed as a four-vector, and each of the pairs $\boldsymbol{E}$, $\boldsymbol{B}$ and $\boldsymbol{D}$, $\boldsymbol{H}$ as a six-vector (antisymmetric tensor of the second order). Since the nonhomogeneous set contains charges and currents (which represent influence of matter), one has to attribute the corresponding pair ($\boldsymbol{D}$, $\boldsymbol{H}$) to the influence of matter. It is, however, customary to refer to $\boldsymbol{H}$ and not to $\boldsymbol{B}$ as the *magnetic field vector*; we shall conform to this terminology when there is no risk of confusion.

† The so-called Gaussian system of units is used here, i.e. the electrical quantities ($\boldsymbol{E}$, $\boldsymbol{D}$, $\boldsymbol{j}$ and ρ) are measured in electrostatic units, and the magnetic quantities ($\boldsymbol{H}$ and $\boldsymbol{B}$) in electromagnetic units. The constant c in (1) and (2) relates the unit of charge in the two systems; it is the velocity of light in the vacuum and is approximately equal to 3×10^{10} cm/sec. (A more accurate value is given in § 1.2.)

From (1) it follows (since div curl ≡ 0) that

$$\operatorname{div} \boldsymbol{j} = -\frac{1}{4\pi} \operatorname{div} \dot{\boldsymbol{D}},$$

or, using (3),

$$\frac{\partial \rho}{\partial t} + \operatorname{div} \boldsymbol{j} = 0. \tag{5}$$

By analogy with a similar relation encountered in hydrodynamics, (5) is called *equation of continuity*. It expresses the fact that the charge is conserved in the neighbourhood of any point. For if one integrates (5) over any region of space, one obtains, with the help of GAUSS' theorem,

$$\frac{d}{dt}\int \rho \, dV + \int \boldsymbol{j} \, . \, \boldsymbol{n} \, dS = 0, \tag{6}$$

the second integral being taken over the surface bounding the region and the first throughout the volume, $\boldsymbol{n}$ denoting the unit outward normal. This equation implies that the total charge

$$e = \int \rho \, dV \tag{7}$$

contained within the domain can only increase on account of the flow of electric current

$$\mathscr{J} = \int \boldsymbol{j} \, . \, \boldsymbol{n} \, dS. \tag{8}$$

If all the field quantities are independent of time, and if, moreover, there are no currents ($\boldsymbol{j} = 0$), the field is said to be *static*. If all the field quantities are time independent, but currents are present ($\boldsymbol{j} \neq 0$), one speaks of a *stationary field*. In optical fields the field vectors are very rapidly varying functions of time; but the sources of the field are usually such that, when averages over any macroscopic time interval are considered rather than the instantaneous values, the properties of the field are found to be independent of the instant of time at which the average is taken. The word *stationary* is often used in a wider sense to describe a field of this type. An example is a field constituted by the steady flux of radiation (say from a distant star) through an optical system.

1.1.2 Material equations

The MAXWELL equations (1)–(4) connect the five basic quantities $\boldsymbol{E}$, $\boldsymbol{H}$, $\boldsymbol{B}$, $\boldsymbol{D}$ and $\boldsymbol{j}$. To allow a unique determination of the field vectors from a given distribution of currents and charges, these equations must be supplemented by relations which describe the behaviour of substances under the influence of the field. These relations are known as *material equations** (or *constitutive relations*). In general they are rather complicated, but if the bodies are at rest (or in very slow motion) relative to each other and if the material is *isotropic* (i.e. when its physical properties at each point are independent of direction) they take usually the relatively simple form†

* There is an alternative way of describing the behaviour of matter. Instead of the quantities $\varepsilon = D/E$, $\mu = B/H$ one considers the differences $\boldsymbol{D} - \boldsymbol{E}$ and $\boldsymbol{B} - \boldsymbol{H}$; these have a simpler physical significance and will be discussed in Chapter II.

† The more general relations, applicable also to moving bodies, are studied in the theory of relativity. We shall only need the following result from the more general theory: that in the case of moving charges there is also, in addition to the conduction current $\sigma\boldsymbol{E}$, a convection current $\rho\boldsymbol{v}$, where $\boldsymbol{v}$ is the velocity of the moving charges and ρ the charge density (cf. p. 8).

$$j = \sigma E, \tag{9}$$

$$D = \varepsilon E, \tag{10}$$

$$B = \mu H. \tag{11}$$

Here σ is called the *specific conductivity*, ε is known as the *dielectric constant* (or permittivity), and μ is called the *magnetic permeability*.

Eq. (9) is the differential form of OHM's law. Substances for which $\sigma \neq 0$ (or more precisely is not negligibly small; the precise meaning of this cannot however be discussed here) are called *conductors*. Metals are very good conductors, but there are other classes of good conducting materials such as ionic solutions in liquids and also in solids. In metals the conductivity decreases with increasing temperature. However, in other classes of materials, known as *semiconductors* (e.g. germanium), conductivity increases with temperature over a wide range.

Substances for which σ is negligibly small are called *insulators* or *dielectrics*. Their electric and magnetic properties are then completely determined by ε and μ. For most substances the magnetic permeability μ is practically unity. If this is not the case, i.e. if μ differs appreciably from unity, we say that the substance is *magnetic*. In particular, if $\mu > 1$, the substance is said to be *paramagnetic* (e.g. platinum, oxygen, nitrogen dioxide), while if $\mu < 1$ it is said to be *diamagnetic* (e.g. bismuth, copper, hydrogen, water).

If the fields are exceptionally strong, such as are obtained, for example, by focusing light that is generated by an optical maser, the right-hand sides of the material equations may have to be supplemented by terms involving components of the field vectors in powers higher than the first.*

In many cases the quantities σ, ε and μ will be independent of the field strengths; in other cases, however, the behaviour of the material cannot be described in such a simple way. Thus, for example, in a gas of free ions the current, which is determined by the mean speed of the ions, depends, at any moment, not on the instantaneous value of $\boldsymbol{E}$, but on all its previous values. Again, in so-called *ferromagnetic* substances (substances which are very highly magnetic, e.g. iron, cobalt and nickel) the value of the magnetic induction $\boldsymbol{B}$ is determined by the past history of the field $\boldsymbol{H}$ rather than by its instantaneous value. The substance is then said to exhibit *hysteresis*. A similar history-dependence will be found for the electric displacement in certain dielectric materials. Fortunately hysteretic effects are rarely significant for the high-frequency field encountered in optics.

In the main part of this book we shall study the propagation in substances which light can penetrate without appreciable weakening (e.g. air, glass). Such substances are said to be *transparent* and must be electrical nonconductors ($\sigma = 0$), since conduction implies the evolution of Joule heat (cf. §1.1.4) and therefore loss of electromagnetic energy. Optical properties of conducting media will be discussed in Chapter XIII.

---

* Non-linear relationship between the displacement vector $\boldsymbol{D}$ and the electric field $\boldsymbol{E}$ was first demonstrated in this way by P. A. FRANKEN, A. E. HILL, C. W. PETERS and G. WEINRICH, *Phys. Rev. Let.*, **7** (1961), 118. The theory of propagation, of refraction and reflection of electromagnetic waves under conditions when such non-linear effects have to be taken into account has been developed by J. A. ARMSTRONG, N. BLOEMBERGEN, J. DUCUING and P. S. PERSHAN, *Phys. Rev.*, **127** (1962), 1918 and N. BLOEMBERGEN and P. S. PERSHAN, *ibid*, **128** (1962), 606.

For systematic treatments of non-linear effects see N. BLOEMBERGEN, *Nonlinear Optics* (New York, W. A. Benjamin, Inc., 1965) and P. S. PERSHAN, *Progress in Optics*, Vol. 5, ed. E. WOLF (Amsterdam, North Holland Publishing Company and New York, J. Wiley and Sons, 1965), p. 83. See also P. A. FRANKEN and J. F. WARD, *Rev. Mod. Phys.*, **35** (1963), 23.

1.1.3 Boundary conditions at a surface of discontinuity

MAXWELL's equations were only stated for regions of space throughout which the physical properties of the medium (characterized by ε and μ) are continuous. In optics one often deals with situations in which the properties change abruptly across one or more surfaces. The vectors $\boldsymbol{E}, \boldsymbol{H}, \boldsymbol{B}$ and $\boldsymbol{D}$ may then be expected also to become discontinuous, while ρ and $\boldsymbol{j}$ will degenerate into corresponding surface quantities. We shall derive relations describing the transition across such a discontinuity surface.

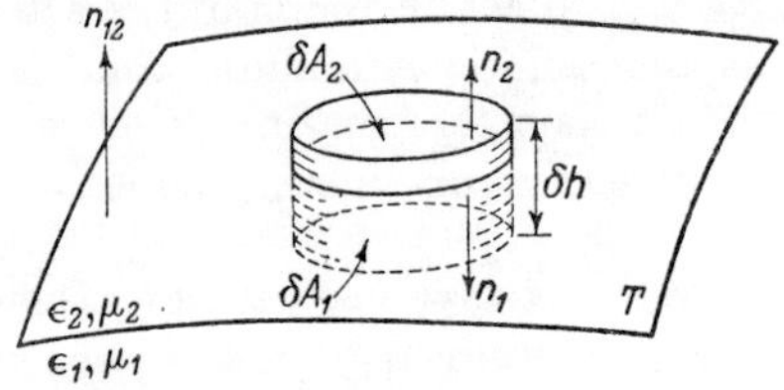

Fig. 1.1. Derivation of boundary conditions for the normal components of $\boldsymbol{B}$ and $\boldsymbol{D}$.

Let us replace the sharp discontinuity surface T by a thin transition layer within which ε and μ vary rapidly but continuously from their values near T on one side to their value near T on the other. Within this layer we construct a small near-cylinder, bounded by a stockade of normals to T, roofed and floored by small areas δA_1 and δA_2 on each side of T, at constant distance from it, measured along their common normal (Fig. 1.1). Since $\boldsymbol{B}$ and its derivatives may be assumed to be continuous throughout this cylinder, we may apply GAUSS' theorem to the integral of div $\boldsymbol{B}$ taken throughout the volume of the cylinder and obtain, from (4),

$$\int \operatorname{div} \boldsymbol{B}\, dV = \int \boldsymbol{B} \cdot \boldsymbol{n}\, dS = 0; \tag{12}$$

the second integral is taken over the surface of the cylinder, and $\boldsymbol{n}$ is the unit outward normal.

Since the areas δA_1 and δA_2 are assumed to be small, $\boldsymbol{B}$ may be considered to have constants values $\boldsymbol{B}^{(1)}$ and $\boldsymbol{B}^{(2)}$ on δA_1 and δA_2, and (12) may then be replaced by

$$\boldsymbol{B}^{(1)} \cdot \boldsymbol{n}_1\, \delta A_1 + \boldsymbol{B}^{(2)} \cdot \boldsymbol{n}_2\, \delta A_2 + \text{contribution from walls} = 0. \tag{13}$$

If the height δh of the cylinder decreases towards zero, the transition layer shrinks into the surface and the contribution from the walls of the cylinder tends to zero, provided that there is no surface flux of magnetic induction. Such flux never occurs, and consequently in the limit,

$$(\boldsymbol{B}^{(1)} \cdot \boldsymbol{n}_1 + \boldsymbol{B}^{(2)} \cdot \boldsymbol{n}_2)\delta A = 0, \tag{14}$$

δA being the area in which the cylinder intersects T. If $\boldsymbol{n}_{12}$ is the unit normal pointing from the first into the second medium, then $\boldsymbol{n}_1 = -\boldsymbol{n}_{12}$, $\boldsymbol{n}_2 = \boldsymbol{n}_{12}$ and (14) gives

$$\boldsymbol{n}_{12} \cdot (\boldsymbol{B}^{(2)} - \boldsymbol{B}^{(1)}) = 0, \tag{15}$$

i.e. *the normal component of the magnetic induction is continuous across the surface of discontinuity.*

The electric displacement $\boldsymbol{D}$ may be treated in a similar way, but there will be an additional term if charges are present. In place of (12) we now have from (3)

$$\int \operatorname{div} \boldsymbol{D}\, dV = \int \boldsymbol{D} \cdot \boldsymbol{n}\, dS = 4\pi \int \rho\, dV. \tag{16}$$

As the areas δA_1 and δA_2 shrink together, the total charge remains finite, so that the volume density becomes infinite. Instead of the volume charge density ρ the concept of *surface charge density* $\hat{\rho}$ must then be used. It is defined by*

$$\lim_{\delta h \to 0} \int \rho \, dV = \int \hat{\rho} \, dA. \tag{17}$$

We shall also need later the concept of *surface current density* $\hat{\boldsymbol{j}}$, defined in a similar way:

$$\lim_{\delta h \to 0} \int \boldsymbol{j} \, dV = \int \hat{\boldsymbol{j}} \, dA. \tag{18}$$

If the area δA and the height δh are taken sufficiently small, (16) gives

$$\boldsymbol{D}^{(1)} \cdot \boldsymbol{n}_1 \, \delta A_1 + \boldsymbol{D}^{(2)} \cdot \boldsymbol{n}_2 \, \delta A_2 + \text{contribution from walls} = 4\pi\hat{\rho} \, \delta A.$$

The contribution from the walls tends to zero with δh, and we therefore obtain in the limit as $\delta h \to 0$,

$$\boldsymbol{n}_{12} \cdot (\boldsymbol{D}^{(2)} - \boldsymbol{D}^{(1)}) = 4\pi\hat{\rho}, \tag{19}$$

i.e. *in the presence of a layer of surface charge density* $\hat{\rho}$ *on the surface, the normal component of the electric displacement changes abruptly across the surface, by an amount equal to* $4\pi\hat{\rho}$.

Next, we examine the behaviour of the tangential components. Let us replace the sharp discontinuity surface by a continuous transition layer. We also replace the

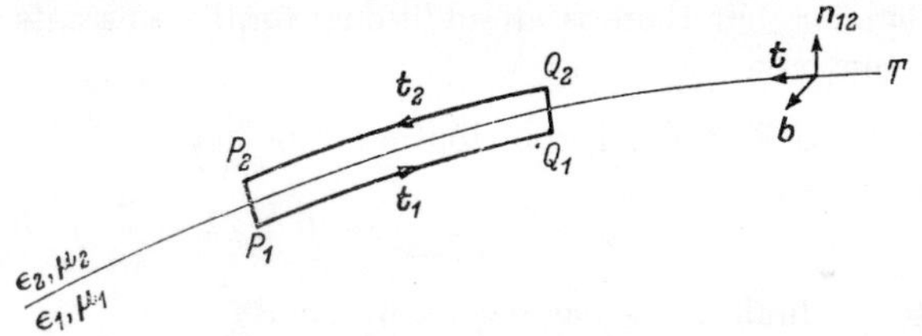

Fig. 1.2. Derivation of the boundary conditions for the tangentia components of $\boldsymbol{E}$ and $\boldsymbol{H}$.

cylinder of Fig. 1.1 by a "rectangular" area with sides parallel and perpendicular to T (Fig. 1.2).

Let $\boldsymbol{b}$ be the unit vector perpendicular to the plane of the rectangle. Then it follows from (2) and from STOKES' theorem that

$$\int \operatorname{curl} \boldsymbol{E} \cdot \boldsymbol{b} \, dS = \int \boldsymbol{E} \cdot d\boldsymbol{r} = -\frac{1}{c} \int \dot{\boldsymbol{B}} \cdot \boldsymbol{b} \, dS, \tag{20}$$

---

* For later purposes we note a representation of the surface charge density and the surface current density in terms of the DIRAC delta function (see Appendix IV). If the equation of the surface of discontinuity is $F(x,y,z) = 0$, then

$$\rho = \hat{\rho} \, |\text{grad } F| \delta(F), \tag{17a}$$

$$\boldsymbol{j} = \hat{\boldsymbol{j}} \, |\text{grad } F| \delta(F). \tag{18a}$$

These relations can immediately be verified by substituting from (17a) and (18a) into (17) and (18) and using the relation $dF = |\text{grad } F| dh$ and the sifting property of the δ function.

the first and third integrals being taken throughout the area of the rectangle, and the second along its boundary. If the lengths P_1Q_1 ($= \delta s_1$), and P_2Q_2 ($= \delta s_2$) are small, $\boldsymbol{E}$ may be replaced by constant values $\boldsymbol{E}^{(1)}$ and $\boldsymbol{E}^{(2)}$ along each of these segments. Similarly $\dot{\boldsymbol{B}}$ may be replaced by a constant value. (20) then gives

$$\boldsymbol{E}^{(1)} \cdot \boldsymbol{t}_1 \, \delta s_1 + \boldsymbol{E}^{(2)} \cdot \boldsymbol{t}_2 \, \delta s_2 + \text{contribution from ends} = -\frac{1}{c} \dot{\boldsymbol{B}} \cdot \boldsymbol{b} \, \delta s \delta h, \tag{21}$$

where δs is the line element in which the rectangle intersects the surface. If now the height of the rectangle is gradually decreased, the contribution from the ends P_1P_2 and Q_1Q_2 will tend to zero, provided that $\boldsymbol{E}$ does not in the limit acquire sufficiently sharp singularities; this possibility will be excluded. Assuming also $\dot{\boldsymbol{B}}$ to remain finite, we obtain in the limit as $\delta h \to 0$,

$$(\boldsymbol{E}^{(1)} \cdot \boldsymbol{t}_1 + \boldsymbol{E}^{(2)} \cdot \boldsymbol{t}_2)\delta s = 0. \tag{22}$$

If $\boldsymbol{t}$ is the unit tangent along the surface, then (see Fig. 1.2) $\boldsymbol{t}_1 = -\boldsymbol{t} = -\boldsymbol{b} \wedge \boldsymbol{n}_{12}$, $\boldsymbol{t}_2 = \boldsymbol{t} = \boldsymbol{b} \wedge \boldsymbol{n}_{12}$, and (22) gives

$$\boldsymbol{b} \cdot [\boldsymbol{n}_{12} \wedge (\boldsymbol{E}^{(2)} - \boldsymbol{E}^{(1)})] = 0.$$

Since the orientation of the rectangle and consequently that of the unit vector $\boldsymbol{b}$ is arbitrary, it follows that

$$\boldsymbol{n}_{12} \wedge (\boldsymbol{E}^{(2)} - \boldsymbol{E}^{(1)}) = 0, \tag{23}$$

i.e. *the tangential component of the electric vector is continuous across the surface.*

Finally consider the behaviour of the tangential component of the magnetic vector. The analysis is similar, but there is an additional term if currents are present. In place of (21) we now have

$$\boldsymbol{H}^{(1)} \cdot \boldsymbol{t}_1 \, \delta s_1 + \boldsymbol{H}^{(2)} \cdot \boldsymbol{t}_2 \, \delta s_2 + \text{contribution from ends}$$
$$= \frac{1}{c} \dot{\boldsymbol{D}} \cdot \boldsymbol{b} \, \delta s \, \delta h + \frac{4\pi}{c} \hat{\boldsymbol{j}} \cdot \boldsymbol{b} \, \delta s. \tag{24}$$

On proceeding to the limit $\delta h \to 0$ as before, we obtain

$$\boldsymbol{n}_{12} \wedge (\boldsymbol{H}^{(2)} - \boldsymbol{H}^{(1)}) = \frac{4\pi}{c} \hat{\boldsymbol{j}}. \tag{25}$$

From (25) it follows that *in the presence of a surface current of density $\hat{\boldsymbol{j}}$, the tangential component (considered as a vector quantity) of the magnetic vector changes abruptly, its discontinuity being $\frac{4\pi}{c} \hat{\boldsymbol{j}} \wedge \boldsymbol{n}_{12}$.*

Apart from discontinuities due to the abrupt changes in the physical properties of the medium, the field vectors may also be discontinuous because of the presence of a source which begins to radiate at a particular instant of time $t = t_0$. The disturbance then spreads into the surrounding space, and at any later instant $t_1 > t_0$ will have filled a well-defined region. Across the (moving) boundary of this region, the field vectors will change abruptly from finite values on the boundary to the value zero outside it.

The various cases of discontinuity may be covered by rewriting MAXWELL's equations in an integral form.* The general discontinuity conditions may also be

* See for example A. SOMMERFELD, *Electrodynamics* (New York, Academic Press, 1952), p. 11; or J. A. STRATTON, *Electromagnetic Theory* (New York, McGraw-Hill, 1941), p. 6.

written in the form of simple difference equations; a derivation of these equations is given in Appendix VI.

1.1.4 The energy law of the electromagnetic field

Electromagnetic theory interprets the light intensity as the energy flux of the field. It is therefore necessary to recall the energy law of MAXWELL's theory.

From (1) and (2) it follows that

$$\boldsymbol{E} \cdot \operatorname{curl} \boldsymbol{H} - \boldsymbol{H} \cdot \operatorname{curl} \boldsymbol{E} = \frac{4\pi}{c} \boldsymbol{j} \cdot \boldsymbol{E} + \frac{1}{c} \boldsymbol{E} \cdot \dot{\boldsymbol{D}} + \frac{1}{c} \boldsymbol{H} \cdot \dot{\boldsymbol{B}}. \tag{26}$$

Also, by a well-known vector identity, the term on the left may be expressed as the divergence of the vector product of $\boldsymbol{H}$ and $\boldsymbol{E}$:

$$\boldsymbol{E} \cdot \operatorname{curl} \boldsymbol{H} - \boldsymbol{H} \cdot \operatorname{curl} \boldsymbol{E} = -\operatorname{div} (\boldsymbol{E} \wedge \boldsymbol{H}). \tag{27}$$

From (26) and (27) we have that

$$\frac{1}{c} (\boldsymbol{E} \cdot \dot{\boldsymbol{D}} + \boldsymbol{H} \cdot \dot{\boldsymbol{B}}) + \frac{4\pi}{c} \boldsymbol{j} \cdot \boldsymbol{E} + \operatorname{div} (\boldsymbol{E} \wedge \boldsymbol{H}) = 0. \tag{28}$$

When we multiply this equation by $c/4\pi$, integrate throughout an arbitrary volume and apply GAUSS's theorem, this gives

$$\frac{1}{4\pi} \int (\boldsymbol{E} \cdot \dot{\boldsymbol{D}} + \boldsymbol{H} \cdot \dot{\boldsymbol{B}})\, dV + \int \boldsymbol{j} \cdot \boldsymbol{E}\, dV + \frac{c}{4\pi} \int (\boldsymbol{E} \wedge \boldsymbol{H}) \cdot \boldsymbol{n}\, dS = 0, \tag{29}$$

where the last integral is taken over the boundary of the volume, $\boldsymbol{n}$ being the unit outward normal.

The relation (29) is a direct consequence of MAXWELL's equations and is therefore valid whether or not the material equations (9)–(11) hold. It represents, as will be seen, the *energy law* of an electromagnetic field. We shall discuss it here only for the case where the material equations (9)–(11) are satisfied. Generalizations to anisotropic media, where the material equations are of a more complicated form, will be considered later (Chapter XIV).

We have, on using the material equations,

$$\left.\begin{aligned} \frac{1}{4\pi} (\boldsymbol{E} \cdot \dot{\boldsymbol{D}}) &= \frac{1}{4\pi} \boldsymbol{E} \cdot \frac{\partial}{\partial t} (\varepsilon \boldsymbol{E}) = \frac{1}{8\pi} \frac{\partial}{\partial t} (\varepsilon \boldsymbol{E}^2) = \frac{1}{8\pi} \frac{\partial}{\partial t} (\boldsymbol{E} \cdot \boldsymbol{D}), \\ \frac{1}{4\pi} (\boldsymbol{H} \cdot \dot{\boldsymbol{B}}) &= \frac{1}{4\pi} \boldsymbol{H} \cdot \frac{\partial}{\partial t} (\mu \boldsymbol{H}) = \frac{1}{8\pi} \frac{\partial}{\partial t} (\mu \boldsymbol{H}^2) = \frac{1}{8\pi} \frac{\partial}{\partial t} (\boldsymbol{H} \cdot \boldsymbol{B}). \end{aligned}\right\} \tag{30}$$

Setting

$$w_e = \frac{1}{8\pi} \boldsymbol{E} \cdot \boldsymbol{D}, \qquad w_m = \frac{1}{8\pi} \boldsymbol{H} \cdot \boldsymbol{B}, \tag{31}$$

and

$$W = \int (w_e + w_m)\, dV, \tag{32}$$

(29) becomes,

$$\frac{dW}{dt} + \int \boldsymbol{j} \cdot \boldsymbol{E}\, dV + \frac{c}{4\pi} \int (\boldsymbol{E} \wedge \boldsymbol{H}) \cdot \boldsymbol{n}\, dS = 0. \tag{33}$$

We shall show that W represents the total energy contained within the volume, so that w_e may be identified with the *electric energy density* and w_m with the *magnetic energy density* of the field.*

To justify the interpretation of W as the total energy we have to show that, for a closed system (i.e. one in which the field on the boundary surface may be neglected), the change in W as defined above is due to the work done by the field on the material charged bodies which are embedded in it. It suffices to do this for slow motion of the material bodies, which themselves may be assumed to be so small that they can be regarded as point charges e_k ($k = 1, 2, \ldots$). Let the velocity of the charge e_k be $\mathbf{v}_k$ ($v_k \ll c$).

The force exerted by a field ($\boldsymbol{E}$, $\boldsymbol{B}$) on a charge e moving with velocity $\mathbf{v}$ is given by the so-called *Lorentz law*,

$$\boldsymbol{F} = e\left(\boldsymbol{E} + \frac{1}{c}\mathbf{v} \wedge \boldsymbol{B}\right), \tag{34}$$

which is based on experience. It follows that if all the charges e_k are displaced by $\delta\boldsymbol{x}_k$ ($k = 1, 2, \ldots$) in time δt, the total work done is

$$\begin{aligned}\delta A &= \sum_k \boldsymbol{F}_k \cdot \delta\boldsymbol{x}_k = \sum_k e_k \left(\boldsymbol{E}_k + \frac{1}{c}\mathbf{v}_k \wedge \boldsymbol{B}\right) \cdot \delta\boldsymbol{x}_k \\ &= \sum_k e_k\, \boldsymbol{E}_k \cdot \delta\boldsymbol{x}_k = \sum_k e_k\, \boldsymbol{E}_k \cdot \mathbf{v}_k\, \delta t,\end{aligned}$$

since $\delta\boldsymbol{x}_k = \mathbf{v}_k\, \delta t$. If the number of the charged particles is large, we can consider the distribution to be continuous. We introduce the charge density ρ (i.e. total charge per unit volume) and the last equation becomes

$$\delta A = \delta t \int \rho\, \mathbf{v} \cdot \boldsymbol{E}\, dV, \tag{35}$$

the integration being carried throughout an arbitrary volume. Now the velocity $\mathbf{v}$ does not appear explicitly in MAXWELL's equations, but it may be introduced by using an experimental result found by RÖNTGEN†, according to which a *convection current* (i.e. a set of moving charges) has the same electromagnetic effect as a *conduction current* in a wire. Hence the current density $\boldsymbol{j}$ appearing in MAXWELL's equations can be split into two parts

$$\boldsymbol{j} = \boldsymbol{j}_c + \boldsymbol{j}_v, \tag{36}$$

where

$$\boldsymbol{j}_c = \sigma \mathbf{E}$$

is the conduction current density, and

$$\boldsymbol{j}_v = \rho \mathbf{v}$$

* In the general case the densities are defined by the expressions

$$w_e = \frac{1}{4\pi}\int \boldsymbol{E} \cdot d\boldsymbol{D}, \qquad w_m = \frac{1}{4\pi}\int \boldsymbol{H} \cdot d\boldsymbol{B}$$

When the relationship between $\boldsymbol{E}$ and $\boldsymbol{D}$ and between $\boldsymbol{H}$ and $\boldsymbol{B}$ is linear, as here assumed, these expressions reduce to (31).

† W. C. RÖNTGEN, *Ann. d. Physik.*, **35** (1888), 264; **40** (1890), 93.

represents the convection current density. (35) may therefore be written as

$$\delta A = \delta t \int \boldsymbol{j}_v \, . \, \boldsymbol{E} \, dV. \tag{37}$$

Let us now define a vector $\boldsymbol{S}$ and a scalar Q by the relations

$$\boldsymbol{S} = \frac{c}{4\pi} (\boldsymbol{E} \wedge \boldsymbol{H}), \tag{38}$$

$$Q = \int \boldsymbol{j}_c \, . \, \boldsymbol{E} \, dV = \int \sigma E^2 \, dV. \tag{39}$$

Then by (35) and (36)

$$\int \boldsymbol{j} \, . \, \boldsymbol{E} \, dV = Q + \int \boldsymbol{j}_v \, . \, \boldsymbol{E} \, dV$$

$$= Q + \frac{\delta A}{\delta t}, \tag{40}$$

where the second function is not, of course, a total derivative of a space time function. (33) now takes the form

$$\frac{dW}{dt} = -\frac{\delta A}{\delta t} - Q - \int \boldsymbol{S} \, . \, \boldsymbol{n} \, dS. \tag{41}$$

For a nonconductor ($\sigma = 0$) we have that $Q = 0$. Assume also that the boundary surface is so far away that we can neglect the field on it, due to the electromagnetic processes inside; then $\int \boldsymbol{S} \, . \, \boldsymbol{n} \, dS = 0$, and integration of (41) gives

$$W + A = \text{constant}. \tag{42}$$

Hence, for an isolated system, the increase of W per unit time is due to the work done on the system during this time. This result justifies our definition of electromagnetic energy by means of (32).

The term Q represents the resistive dissipation of energy (called *Joule's heat*) in a conductor ($\sigma \neq 0$). According to (41) there is a further decrease in energy if the field extends to the boundary surface. The surface integral must therefore represent the flow of energy across this boundary surface. The vector $\boldsymbol{S}$ is known as the *Poynting vector* and represents the amount of energy which crosses per second a unit area normal to the directions of $\boldsymbol{E}$ and $\boldsymbol{H}$.

It should be noted that the interpretation of $\boldsymbol{S}$ as energy flow (more precisely as the density of the flow) is an abstraction which introduces a certain degree of arbitrariness. For the quantity which is physically significant is, according to (41), not $\boldsymbol{S}$ itself, but the integral of $\boldsymbol{S} \, . \, \boldsymbol{n}$ taken over a closed surface. Clearly, from the value of the integral, no unambiguous conclusion can be drawn about the detailed distribution of $\boldsymbol{S}$, and alternative definitions of the energy flux density are therefore possible. One can always add to $\boldsymbol{S}$ the curl of an arbitrary vector, since such a term will not contribute to the surface integral as can be seen from GAUSS' theorem and the identity div curl $\equiv 0$.* However, when the definition has been applied cautiously, in particular

* According to modern theories of fields the arbitrariness is even greater, allowing for alternative expressions for both the energy density and the energy flux, but consistent with the change of the Lagrangian density of the field by the addition of a four-divergence.

for averages over small but finite regions of space or time, no contradictions with experiments have been found. We shall therefore accept the above definition in terms of the Poynting vector of the density of the energy flow.

Finally we note that in a nonconducting medium ($\sigma = 0$) where no mechanical work is done ($A = 0$), the energy law may be written in the form of a hydrodynamical continuity equation for noncompressible fluids:

$$\frac{\partial w}{\partial t} + \text{div}\, \boldsymbol{S} = 0, \qquad (w = w_e + w_m). \tag{43}$$

A description of propagation of light in terms of a hydrodynamical model is often helpful, particularly in the domain of geometrical optics and in connection with scalar diffraction fields, as it gives a picture of the energy transport in a simple and graphic manner. In optics, the (averaged) Poynting vector is the chief quantity of interest. The magnitude of the Poynting vector is a measure of the light intensity, and its direction represents the direction of propagation of the light.

1.2 THE WAVE EQUATION AND THE VELOCITY OF LIGHT

MAXWELL's equations relate the field vectors by means of simultaneous differential equations. On elimination we obtain differential equations which each of the vectors must separately satisfy. We shall confine our attention to that part of the field which contains no charges or currents, i.e. where $\boldsymbol{j} = 0$ and $\rho = 0$.

We substitute for $\boldsymbol{B}$ from the material equation § 1.1 (11) into the second Maxwell equation § 1.1 (2), divide both sides by μ and apply the operator curl. This gives

$$\text{curl}\left(\frac{1}{\mu}\,\text{curl}\, \boldsymbol{E}\right) + \frac{1}{c}\,\text{curl}\, \dot{\boldsymbol{H}} = 0. \tag{1}$$

Next we differentiate the first Maxwell equation § 1.1 (1) with respect to time, use the material equation §1.1 (10) for $\boldsymbol{D}$, and eliminate curl $\dot{\boldsymbol{H}}$ between the resulting equation and (1); this gives

$$\text{curl}\left(\frac{1}{\mu}\,\text{curl}\, \boldsymbol{E}\right) + \frac{\varepsilon}{c^2}\,\ddot{\boldsymbol{E}} = 0. \tag{2}$$

If we use the identities curl $u\mathbf{v} = u$ curl $\mathbf{v}$ + (grad u) $\wedge$ $\mathbf{v}$ and curl curl = grad div $- \nabla^2$, (2) becomes

$$\nabla^2 \boldsymbol{E} - \frac{\varepsilon\mu}{c^2}\ddot{\boldsymbol{E}} + (\text{grad} \log \mu) \wedge \text{curl}\, \boldsymbol{E} - \text{grad div}\, \boldsymbol{E} = 0. \tag{3}$$

Also from §1.1 (3), using again the material equation for $\boldsymbol{D}$ and applying the identity div $u\mathbf{v} = u$ div $\mathbf{v} + \mathbf{v}\,.$ grad u we find

$$\varepsilon\, \text{div}\, \boldsymbol{E} + \boldsymbol{E}\, .\, \text{grad}\, \varepsilon = 0. \tag{4}$$

Hence (3) may be written in the form

$$\nabla^2 \boldsymbol{E} - \frac{\varepsilon\mu}{c^2}\ddot{\boldsymbol{E}} + (\text{grad} \log \mu) \wedge \text{curl}\, \boldsymbol{E} + \text{grad}\, (\boldsymbol{E}\, .\, \text{grad} \log \varepsilon) = 0. \tag{5}$$

In a similar way we obtain an equation for $\boldsymbol{H}$ alone:

$$\nabla^2 \boldsymbol{H} - \frac{\varepsilon\mu}{c^2}\ddot{\boldsymbol{H}} + (\text{grad} \log \varepsilon) \wedge \text{curl}\, \boldsymbol{H} + \text{grad}\, (\boldsymbol{H}\, .\, \text{grad} \log \mu) = 0. \tag{6}$$

In particular, if the medium is homogeneous, grad log ε = grad log $\mu = 0$, and (5) and (6) reduce to

$$\nabla^2 \boldsymbol{E} - \frac{\varepsilon\mu}{c^2}\ddot{\boldsymbol{E}} = 0, \qquad \nabla^2 \boldsymbol{H} - \frac{\varepsilon\mu}{c^2}\ddot{\boldsymbol{H}} = 0. \tag{7}$$

These are standard equations of wave motion and suggest the existence of electromagnetic waves propagated with a velocity*

$$v = c/\sqrt{\varepsilon\mu}. \tag{8}$$

The constant c was first determined by R. KOHLRAUSCH and W. WEBER in 1856 from ratio of the values of the capacity of a condenser measured in electrostatic and electromagnetic units, and it was found to be identical with the velocity of light in free space. Using this result, MAXWELL developed his electromagnetic theory of light, predicting the existence of electromagnetic waves; the correctness of this prediction was confirmed by the celebrated experiments of H. HERTZ (cf. Historical Introduction).

As in all wave theories of light, the elementary process which produces the optical impression is regarded as being a harmonic wave in space-time (studied in its simplest form in § 1.3 and § 1.4). If its frequency is in the range from 4×10^{14} sec^{-1} to $7{\cdot}5 \times 10^{14}$ sec^{-1} (approximately) it gives rise to the psychological impression of a definite colour. (The opposite however is not true: coloured light of a certain subjective quality may be a composition of harmonic waves of very different frequency distributions.) The actual connection between colour and frequency is very involved and will not be studied in this book.†

The first determination of the velocity of light‡ was made by RÖMER in 1675 from observations of the eclipses of the first satellite of Jupiter and later in a different way (from aberration of fixed stars) by BRADLEY (1728).

The first measurements of the velocity of light from terrestrial sources were carried out by FIZEAU in 1849. It is necessary to employ a modulator, which marks off a portion of the beam§ and for this purpose FIZEAU used a rotating wheel. Later methods employed rotating mirrors or electronic shutters. The rotating mirror method was suggested by WHEATSTONE in 1834 and was used by FOUCAULT in 1860. It was later systematically developed over a period of many years by MICHELSON. The average value based on about 200 measurements by MICHELSON gave c as 299,796 km/sec. An optical shutter method employing a KERR cell was developed by KAROLUS and MITTELSTAEDT (1928), ANDERSON (1937), and HÜTTEL (1940). The values of c obtained from these measurements are in excellent agreement with those based on

---

* The concept of a velocity of an electromagnetic wave has actually an unambiguous meaning only in connection with waves of very simple kind, e.g. plane waves. That v does not represent the velocity of propagation of an arbitrary solution of (7) is obvious if we bear in mind that these equations also admit standing waves as solutions.

In this introductory section it is assumed that the reader is familiar with the concept of a plane wave, and we regard v as the velocity with which such a wave advances. The mathematical representation of a plane wave will be discussed in § 1.3 and § 1.4.

† The sensitivity of the human eye to different colours is, however, briefly discussed in § 4.8.1.

‡ For description of the methods used for determination of the velocity of light, see for example, E. BERGSTRAND, *Encyclopedia of Physics*, edit. S. FLÜGGE, (Berlin, Springer), **24** (1956), 1. Detailed analysis of the results obtained by different methods is also given by R. T. BIRGE in *Rep. Progr. Phys.* (London, The Physical Society), **8** (1941), p. 90.

§ Such determinations give essentially the group velocity (see § 1.3.4). The difference between the group velocity and the phase velocity in air at standard temperature and pressure is about 1 part in 50,000.

indirect methods, such as determinations from the ratio of an electric charge measured in electrostatic and electromagnetic units; for example ROSA and DORSEY (1907) found in this way c as 299,784 km/sec. Measurements of the velocity of electromagnetic waves on wires carried out by MERCIER (1923) gave the value of c equal to 299,782 km/sec. The value adopted by R. T. BIRGE (1941, *loc. cit.*) from a careful analysis of all the available data is

$$c = 299{,}776 \pm 4 \text{ km/sec.} \tag{9}$$

The close agreement between the values of c obtained from measurements of very different kinds (and in some cases using radiation whose frequencies differ by a factor

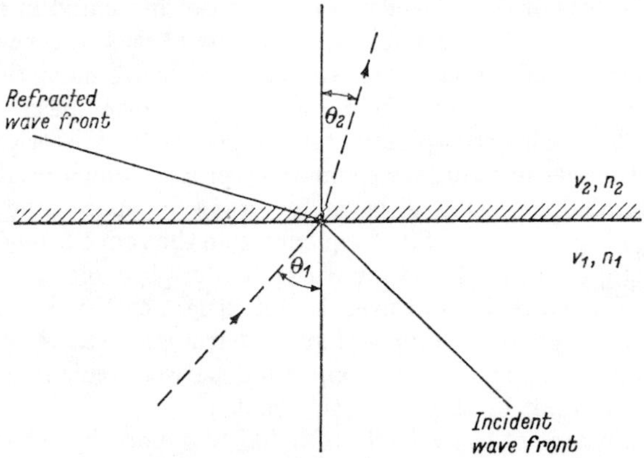

Fig. 1.3. Illustrating the refraction of a plane wave.

of hundreds of thousands from those used in the optical measurements) gives a striking confirmation of MAXWELL's theory.

The dielectric constant ε is usually greater than unity, and μ is practically equal to unity for transparent substances, so that the velocity v is then according to (8) smaller than the vacuum velocity c. This conclusion was first demonstrated experimentally for propagation of light in water in 1850 by FOUCAULT and FIZEAU.

The value of v is not as a rule determined directly, but only relative to c, with the help of the law of refraction. According to this law, if a plane electromagnetic wave falls on to a plane boundary between two homogeneous media, the sine of the angle θ_1 between the normal to the incident wave and the normal to the surface bears a constant ratio to the sine of the angle θ_2 between the normal of the refracted wave and the surface normal (Fig. 1.3), this constant ratio being equal to the ratio of the velocities v_1 and v_2 of propagation in the two media:

$$\frac{\sin \theta_1}{\sin \theta_2} = \frac{v_1}{v_2}. \tag{10}$$

This result will be derived in § 1.5. Here we only note that it is equivalent to the assumption that the wave-front, though it has a kink at the boundary, is continuous, so that the line of intersection between the incident wave and the boundary travels at the same speed (v', say) as the line of intersection between the refracted wave and the boundary. We then have

$$v_1 = v' \sin \theta_1, \qquad v_2 = v' \sin \theta_2, \tag{11}$$

from which, on elimination of v', (10) follows. This argument, in a slightly more elaborate form, is often given as an illustration of HUYGENS' construction (cf. § 3.3).

The value of the constant ratio in (10) is usually denoted by n_{12} and is called *the refractive index*, for refraction from the first into the second medium. We also define an "*absolute refractive index*" n of a medium; it is the refractive index for refraction from vacuum into that medium,

$$n = \frac{c}{v}. \tag{12}$$

If n_1 and n_2 are the absolute refractive indices of two media, the (relative) refractive index n_{12} for refraction from the first into the second medium then is

$$n_{12} = \frac{n_2}{n_1} = \frac{v_1}{v_2}. \tag{13}$$

Comparison of (12) and (8) gives MAXWELL's formula:

$$n = \sqrt{\varepsilon\mu}. \tag{14}$$

Since for all substances with which we shall be concerned, μ is effectively unity (nonmagnetic substances), the refractive index should then be equal to the square root of the dielectric constant, which has been assumed to be a constant of the material. On the other hand, well-known experiments on prismatic colours, first carried out by NEWTON, show that the index of refraction depends on the colour, i.e. on the frequency of the light. If we are to retain MAXWELL's formula, it must be supposed that ε is not a constant characteristic of the material, but is a function of the frequency of the field. The dependence of ε on frequency can only be treated by taking into account the atomic structure of matter, and will be briefly discussed in § 2.3.

MAXWELL's formula (with ε equal to the static dielectric constant) gives a good approximation for such substances as gases with a simple chemical structure which do not disperse light substantially, i.e. for those whose optical properties do not strongly depend on the colour of the light. Results of some early measurements for such gases, carried out by L. BOLTZMANN* are given in Table I. (14) also gives a good approximation for liquid hydrocarbons; for example benzene C_6H_6 has $n = 1{\cdot}482$ for yellow light whilst $\sqrt{\varepsilon} = 1{\cdot}489$.

TABLE I

	n (yellow light)	$\sqrt{\varepsilon}$
Air	1·000294	1·000295
Hydrogen H_2 . .	1·000138	1·000132
Carbon dioxide CO_2 .	1·000449	1·000473
Carbon monoxide CO .	1·000340	1·000345

On the other hand, there is a strong deviation from the formula for many solid bodies (e.g. glasses), and for some liquids, as illustrated in Table II.

* L. BOLTZMANN, *Wien. Ber.*, **69** (1874), 795; *Pogg. Ann.*, **155** (1875), 403; *Wiss. Abh. Physik-techn. Reichsanst.*, **1**, Nr. 26, 537.

TABLE II

	n (yellow light)	$\sqrt{\varepsilon}$
Methyl alcohol CH_3OH .	1·34	5·7
Ethyl alcohol C_2H_5OH .	1·36	5.0
Water H_2O . .	1·33	9·0

1.3 SCALAR WAVES

In an homogeneous medium in regions free of currents and charges, each rectangular component $V(\mathbf{r}, t)$ of the field vectors satisfies, according to § 1.2 (7), the homogeneous wave equation

$$\nabla^2 V - \frac{1}{v^2}\frac{\partial^2 V}{\partial t^2} = 0. \tag{1}$$

We shall now briefly examine the simplest solution of this equation.

1.3.1 Plane waves

Let $\mathbf{r}(x,y,z)$ be a position vector of a point P in space and $\mathbf{s}(s_x,s_y,s_z)$ a unit vector in a fixed direction. Any solution of (1) of the form

$$V = V(\mathbf{r} \cdot \mathbf{s}, t) \tag{2}$$

is said to represent a *plane wave*, since at each instant of time V is constant over each of the planes

$$\mathbf{r} \cdot \mathbf{s} = \text{constant}$$

which are perpendicular to the unit vector $\mathbf{s}$.

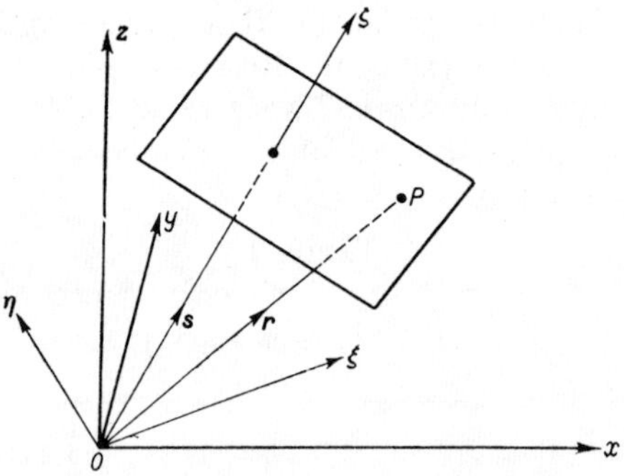

Fig. 1.4. Propagation of a plane wave.

It will be convenient to choose a new set of Cartesian axes $O\xi$, $O\eta$, $O\zeta$ with $O\zeta$ in the direction of $\mathbf{s}$. Then (see Fig. 1.4)

$$\mathbf{r} \cdot \mathbf{s} = \zeta, \tag{3}$$

and one has

$$\frac{\partial}{\partial x} = s_x \frac{\partial}{\partial \zeta}, \quad \frac{\partial}{\partial y} = s_y \frac{\partial}{\partial \zeta}, \quad \frac{\partial}{\partial z} = s_z \frac{\partial}{\partial \zeta}.$$

From these relations one easily finds that

$$\nabla^2 V = \frac{\partial^2 V}{\partial \zeta^2}, \tag{4}$$

so that (1) becomes

$$\frac{\partial^2 V}{\partial \zeta^2} - \frac{1}{v^2}\frac{\partial^2 V}{\partial t^2} = 0. \tag{5}$$

If we set

$$\zeta - vt = p, \qquad \zeta + vt = q, \tag{6}$$

(5) takes the form

$$\frac{\partial^2 V}{\partial p \partial q} = 0. \tag{7}$$

The general solution of this equation is

$$\begin{aligned} V &= V_1(p) + V_2(q) \\ &= V_1(\mathbf{r} \cdot \mathbf{s} - vt) + V_2(\mathbf{r} \cdot \mathbf{s} + vt), \end{aligned} \tag{8}$$

where V_1 and V_2 are arbitrary functions.

We see that the argument of V_1 is unchanged when (ζ,t) is replaced by $(\zeta + v\tau, t + \tau)$, where τ is arbitrary. Hence V_1 represents a disturbance which is propagated with velocity v in the positive ζ direction. Similarly $V_2(\zeta + vt)$ represents a disturbance which is propagated with velocity v in the negative ζ direction.

1.3.2 Spherical waves

Next we consider solutions representing spherical waves, i.e. solutions of the form

$$V = V(r,t) \tag{9}$$

where $r = |\mathbf{r}| = \sqrt{x^2 + y^2 + z^2}$.

Using the relations $\frac{\partial}{\partial x} = \frac{\partial r}{\partial x}\frac{\partial}{\partial r} = \frac{x}{r}\frac{\partial}{\partial r}$, etc., one finds after a straightforward calculation that

$$\nabla^2 V = \frac{1}{r}\frac{\partial^2}{\partial r^2}(rV), \tag{10}$$

so that the wave equation (1) now becomes

$$\frac{\partial^2}{\partial r^2}(rV) - \frac{1}{v^2}\frac{\partial^2}{\partial t^2}(rV) = 0. \tag{11}$$

Now this equation is identical with (5), if ζ is replaced in the latter by r and V by rV. Hence the solution of (11) can immediately be written down from (8):

$$V = \frac{V_1(r - vt)}{r} + \frac{V_2(r + vt)}{r}, \tag{12}$$

V_1 and V_2 being again arbitrary functions. The first term on the right hand side of (12) represents a spherical wave diverging from the origin, the second a spherical wave converging towards the origin, the velocity of propagation being v in both cases.

1.3.3 Harmonic waves. The phase velocity

At a point $\boldsymbol{r}_0$ in space the wave disturbance is a function of time only:

$$V(\boldsymbol{r}_0,t) = F(t). \tag{13}$$

As will be evident from our earlier remarks about colour, the case when F is periodic is of particular interest. Accordingly we consider the case when F has the form

$$F(t) = a \cos [\omega t + \delta]. \tag{14}$$

Here a (> 0) is called the *amplitude*, and the argument $\omega t + \delta$ of the cosine term is called the *phase*. The quantity

$$\nu = \frac{\omega}{2\pi} = \frac{1}{T} \tag{15}$$

is called the *frequency* and represents the number of vibrations per second. ω is called the *angular* frequency and gives the number of vibrations in 2π seconds. Since F remains unchanged when t is replaced by $t + T$, T is the *period* of the vibrations. Wave functions (i.e. solutions of the wave equation) of the form (14) are said to be *harmonic* with respect to time.

Let us first consider a wave function which represents a *harmonic plane wave* propagated in the direction specified by a unit vector $\boldsymbol{s}$. According to § 1.3.1 it is obtained on replacing t by $t - \boldsymbol{r} \cdot \boldsymbol{s}/v$ in (14):

$$V(\boldsymbol{r},t) = a \cos \left[\omega \left(t - \frac{\boldsymbol{r} \cdot \boldsymbol{s}}{v}\right) + \delta\right]. \tag{16}$$

Eq. (16) remains unchanged when $\boldsymbol{r} \cdot \boldsymbol{s}$ is replaced by $\boldsymbol{r} \cdot \boldsymbol{s} + \lambda$, where

$$\lambda = v \frac{2\pi}{\omega} = vT. \tag{17}$$

The length λ is called the *wavelength*. It is also useful to define a *reduced wavelength* λ_0 as

$$\lambda_0 = cT = n\lambda; \tag{18}$$

this is the wavelength which corresponds to a harmonic wave of the same frequency propagated *in vacuo*. In spectroscopy one uses also the concept of a *wave number** κ, which is defined as the number of wavelengths *in vacuo*, per unit of length (cm):

$$\kappa = \frac{1}{\lambda_0} = \frac{\nu}{c}. \tag{19}$$

It is also convenient to define vectors $\boldsymbol{k}_0$ and $\boldsymbol{k}$ in the direction $\boldsymbol{s}$ of propagation, whose lengths are respectively

$$k_0 = 2\pi\kappa = \frac{2\pi}{\lambda_0} = \frac{\omega}{c}, \tag{20}$$

and

$$k = nk_0 = \frac{2\pi}{\lambda} = \frac{n\omega}{c} = \frac{\omega}{v}. \tag{21}$$

---

* We shall refer to κ as the "spectroscopic wave number" and reserve the term "wave number" for k_0 or k, defined by (20) and (21), as customary in optics.

The vector $\boldsymbol{k} = k\boldsymbol{s}$ is called the *wave vector* or the *propagation vector* in the medium, $\boldsymbol{k}_0 = k_0\boldsymbol{s}$ being the corresponding vector in the vacuum.

Instead of the constant δ one also uses the concept of *path length* l, which is the distance through which a wave-front recedes when the phase increases by δ:

$$l = \frac{v}{\omega}\,\delta = \frac{\lambda}{2\pi}\,\delta = \frac{\lambda_0}{2\pi n}\,\delta. \tag{22}$$

Let us now consider time harmonic waves of more complicated form. A general time harmonic real scalar wave of frequency ω may be defined as a real solution of the wave equation, of the form

$$V(\boldsymbol{r},t) = a(\boldsymbol{r}) \cos\,[\omega t - g(\boldsymbol{r})], \tag{23}$$

a (> 0) and g being real scalar functions of positions. The surfaces

$$g(\boldsymbol{r}) = \text{constant} \tag{24}$$

are called *co-phasal surfaces* or *wave surfaces*. In contrast with the previous case, the surfaces of constant amplitude of the wave (23) do not, in general, coincide with the surfaces of constant phase. Such a wave is said to be *inhomogeneous*.

Calculations with harmonic waves are simplified by the use of exponential instead of trigonometric functions. Eq. (23) may be written as

$$V(\boldsymbol{r},t) = \mathscr{R}\{U(\boldsymbol{r})e^{-i\omega t}\} \tag{25}$$

where

$$U(\boldsymbol{r}) = a(\boldsymbol{r})e^{ig(\boldsymbol{r})}, \tag{26}$$

and $\mathscr{R}$ denotes the real part. On substitution from (26) into the wave equation (1), one finds that U must satisfy the equation

$$\nabla^2 U + n^2 k_0{}^2 U = 0. \tag{27}$$

U is called the *complex amplitude** of the wave. In particular, if the wave is plane, one has

$$g(\boldsymbol{r}) = \omega\left(\frac{\boldsymbol{r}\,.\,\boldsymbol{s}}{v}\right) - \delta = k(\boldsymbol{r}\,.\,\boldsymbol{s}) - \delta = \boldsymbol{k}\,.\,\boldsymbol{r} - \delta. \tag{28}$$

If the operations on V are linear, one may drop the symbol $\mathscr{R}$ in (25) and operate directly with the complex function, the real part of the final expression being then understood to represent the physical quantity in question. However, when dealing with expressions which involve non-linear operations such as squaring, etc. (e.g. in calculations of the electric or magnetic energy densities), one must in general take the real parts first and operate with these alone.†

Unlike a plane harmonic wave, the more general wave (25) is not periodic with respect to space. The phase $\omega t - g(\boldsymbol{r})$ is, however, seen to be the same for $(\boldsymbol{r},t)$ and $(\boldsymbol{r} + d\boldsymbol{r}, t + dt)$, provided that

$$\omega\, dt - (\text{grad}\, g)\,.\,d\boldsymbol{r} = 0. \tag{29}$$

---

* In the case of plane waves, one often separates the constant factor $e^{-i\delta}$ and implies by complex amplitude only the variable part $ae^{i\boldsymbol{k}.\boldsymbol{r}}$.

† This is not necessary when only a time average of a quadratic expression is required [cf. § 1.4, eq. (54)–(56)].

If we denote by $\boldsymbol{q}$ the unit vector in the direction of $d\boldsymbol{r}$, and write $d\boldsymbol{r} = \boldsymbol{q}ds$, then (29) gives

$$\frac{ds}{dt} = \frac{\omega}{\boldsymbol{q} \cdot \text{grad } g}. \tag{30}$$

This expression will be numerically smallest when $\boldsymbol{q}$ is the normal to the co-phasal surface, i.e. when $\boldsymbol{q} = \text{grad } g/|\text{grad } g|$, the value then being

$$v^{(p)}(\boldsymbol{r}) = \frac{\omega}{|\text{grad } g|}. \tag{31}$$

$v^{(p)}(\boldsymbol{r})$ is called the *phase velocity* and is the speed with which each of the co-phasal surfaces advances. For a plane electromagnetic wave one has from (28) that grad $g = \boldsymbol{k}$, and therefore

$$v^{(p)} = \frac{\omega}{k} = \frac{c}{\sqrt{\varepsilon\mu}},$$

because of (21). For waves of more complicated form, the phase velocity $v^{(p)}$ will in general differ from $c/\sqrt{\varepsilon\mu}$ and will vary from point to point even in a homogeneous medium. However, it will be seen later (§ 3.1.2) that, when the frequency is sufficiently large, the phase velocity is *approximately* equal to $c/\sqrt{\varepsilon\mu}$, even for waves whose co-phasal surfaces are not plane.

It must be noted that the expression for ds/dt given by (30) is not the resolute of the phase velocity in the $\boldsymbol{q}$-direction, i.e. the phase velocity does not behave as a vector. On the other hand its reciprocal, i.e. the quantity

$$\frac{dt}{ds} = \frac{\boldsymbol{q} \cdot \text{grad } g}{\omega}, \tag{32}$$

is seen to be the component of the vector $(\text{grad } g)/\omega$ in the $\boldsymbol{q}$ direction. The vector $(\text{grad } g)/\omega$ is sometimes called *phase slowness*.

The phase velocity may in certain cases be greater than c. For plane waves this will be so when $n = \sqrt{\varepsilon\mu}$ is smaller than unity, as in the case of dispersing media in regions of the so-called anomalous dispersion* (cf. §2.3.4). Now according to the theory of relativity, signals can never travel faster than c. This implies that the phase velocity cannot correspond to a velocity with which a signal is propagated. It is, in fact, easy to see that the phase velocity cannot be determined experimentally and must therefore be considered to be void of any direct physical significance. For in order to measure this velocity, it would be necessary to affix a mark to the infinitely extended smooth wave and to measure the velocity of the mark. This would, however, mean the replacement of the infinite harmonic wave train by another function of space and time.

1.3.4 Wave packets. The group velocity

The monochromatic waves considered in the preceding section are idealizations never strictly realized in practice. It follows from FOURIER's theorem that any wave $V(\boldsymbol{r},t)$

* The problem of propagation of electromagnetic signals in dispersive media has been very fully investigated by A. SOMMERFELD, *Ann. d. Physik*, **44** (1914), 177 and by L. BRILLOUIN, *ibid.*, **44** (1914), 203. English translations of these papers are included in L. BRILLOUIN, *Wave Propagation and Group Velocity* (New York, Academic Press, 1960), 17, 43. See also H. G. BAERWALD, *ibid.* (5), **6** (1930), 295; **7** (1930), 731; **8** (1931), 565.

(provided it satisfies certain very general conditions), may be regarded as a superposition of monochromatic waves of different frequencies:

$$V(\mathbf{r},t) = \int_0^\infty a_\omega(\mathbf{r}) \cos [\omega t - g_\omega(\mathbf{r})]\, d\omega. \tag{33}$$

It will again be convenient to use a complex representation, in which V is regarded as the real part of an associated complex wave:*

$$V(\mathbf{r},t) = \mathscr{R} \int_0^\infty a_\omega(\mathbf{r}) e^{-i[\omega t - g_\omega(\mathbf{r})]}\, d\omega. \tag{33a}$$

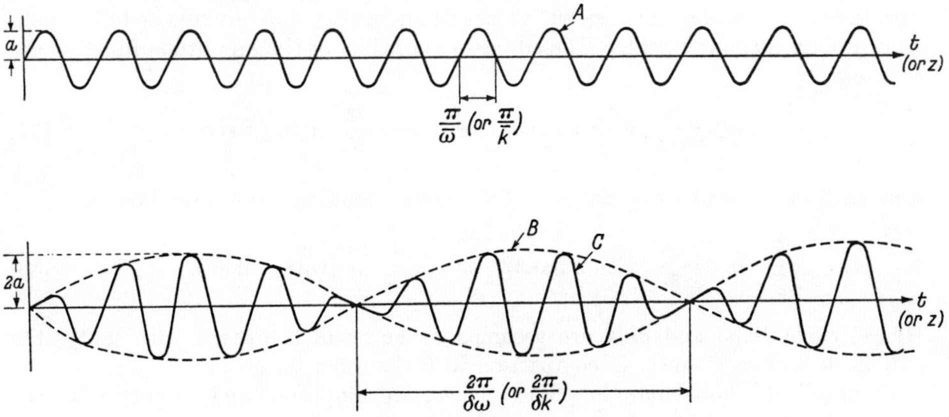

Fig. 1.5. A simple wave group.

(A) The wave $a \cos(\bar{\omega}t - \bar{k}z)$.

(B) The wave $2a \cos [\frac{1}{2}(t\delta\omega - z\delta k)]$.

(C) The wave group $2a \cos [\frac{1}{2}(t\delta\omega - z\delta k)] \cos (\bar{\omega}t - \bar{k}z)$.

The ordinate represents one of the two independent variables (t or z) whilst the other is kept constant.

A wave may be said to be "almost monochromatic," if the Fourier amplitudes a_ω differ appreciably from zero only within a narrow range

$$\bar{\omega} - \tfrac{1}{2}\Delta\omega \leqslant \omega \leqslant \bar{\omega} + \tfrac{1}{2}\Delta\omega \qquad (\Delta\omega/\bar{\omega} \ll 1)$$

around a mean frequency $\bar{\omega}$. In such a case one usually speaks of a *wave group* or a *wave packet*.†

To illustrate some of the main properties of a wave group, consider first a wave formed by the superposition of two plane monochromatic waves of the same amplitudes and slightly different frequencies and wave numbers, propagated in the direction of the z-axis:

$$V(z,t) = ae^{-i(\omega t - kz)} + ae^{-i[(\omega+\delta\omega)t - (k+\delta k)z]}. \tag{34}$$

* For a fuller discussion of the complex representation of real polychromatic waves see §10.2.

† Strictly speaking, in order that V should exhibit properties commonly attributed to a wave group, one should also assume that over the effective frequency range the phase function g_ω can be approximated to by a linear function of ω.

The symbol $\mathscr{R}$ is omitted here in accordance with the convention explained earlier. Eq. (34) may be written in the form

$$V(z,t) = a\{e^{\frac{1}{2}i(t\delta\omega - z\delta k)} + e^{-\frac{1}{2}i(t\delta\omega - z\delta k)}\}e^{-i(\bar{\omega}t - \bar{k}z)}$$

$$= 2a \cos [\tfrac{1}{2}(t\delta\omega - z\delta k)]e^{-i(\bar{\omega}t - \bar{k}z)}, \tag{35}$$

where

$$\bar{\omega} = \omega + \tfrac{1}{2}\delta\omega, \qquad \bar{k} = k + \tfrac{1}{2}\delta k, \tag{36}$$

are the mean frequency and the mean wave number respectively. (35) may be interpreted as representing a plane wave of frequency $\bar{\omega}$ and wavelength $2\pi/\bar{k}$ propagated in the z direction. The amplitude of this wave is, however, not constant, but varies with time and position, between the values $2a$ and 0 (Fig. 1.5), giving rise to the well-known phenomenon of beats. The successive maxima of the amplitude function are at intervals

$$\delta t = \frac{4\pi}{\delta\omega} \text{ (with } z \text{ fixed)} \quad \text{or} \quad \delta z = \frac{4\pi}{\delta k} \text{ (with } t \text{ fixed)} \tag{37}$$

from each other, whilst the maxima of the phase function are at intervals

$$\delta t = \frac{2\pi}{\bar{\omega}} \text{ (with } z \text{ fixed)} \quad \text{or} \quad \delta z = \frac{2\pi}{\bar{k}} \text{ (with } t \text{ fixed)}. \tag{38}$$

Hence, since $\delta\omega/\bar{\omega}$ and $\delta k/\bar{k}$ are assumed to be small compared with unity, the amplitude will vary slowly in comparison with the other term.

From (35) it follows that the planes of constant amplitude and, in particular, the maxima of the amplitude are propagated with the velocity

$$v^{(g)} = \frac{\delta\omega}{\delta k}, \tag{39}$$

whilst the planes of constant phase are propagated with the velocity

$$v^{(p)} = \frac{\bar{\omega}}{\bar{k}}. \tag{40}$$

$v^{(g)}$ is called the *group velocity* of the wave. Since V obeys the wave equation, the frequency ω and the wave number k are related; in a medium of refractive index n, one has (see (21))

$$k = n(\omega)\frac{\omega}{c}, \tag{41}$$

where n is the refractive index function. Equation (41) expresses the *dispersion* of the wave. In a non-dispersive medium, n is independent of ω; the phase velocity $v^{(p)}$ and the group velocity $v^{(g)}$ are then both equal to c/n. In a dispersive medium, however, the two velocities will, in general, be different.

Since $\delta\omega$ is assumed to be small, $\delta\omega/\delta k$ may be replaced by the differential coefficient $d\omega/dk$, so that the expression for the group velocity may be written as

$$v^{(g)} = \frac{d\omega}{dk}. \tag{42}$$

This relation is, in fact, valid under more general conditions, as we shall now show. Consider a one-dimensional wave group

$$V(z,t) = \int_{(\Delta\omega)} a_\omega e^{-i(\omega t - kz)}\, d\omega, \tag{43}$$

where $\Delta\omega$ denotes the small interval around a mean frequency $\bar{\omega}$ ($\Delta\omega/\bar{\omega} \ll 1$) for which a_ω differs appreciably from zero. Let $\bar{k} = n(\bar{\omega})\bar{\omega}/c$ be the corresponding wave number. Then (43) may be expressed in the form

$$V(z,t) = A(z,t)e^{-i(\bar{\omega}t - \bar{k}z)}, \tag{44}$$

where

$$A(z,t) = \int_{(\Delta\omega)} a_\omega e^{-i[(\omega-\bar{\omega})t-(k-\bar{k})z]}\, d\omega \sim \int_{(\Delta\omega)} a_\omega e^{-i\left[(\omega-\bar{\omega})\left\{t-\left(\frac{dk}{d\omega}\right)_{\bar{\omega}} z\right\}\right]}\, d\omega, \tag{45}$$

if $\Delta\omega$ is sufficiently small. V may again be interpreted as a plane wave with variable amplitude, of frequency $\bar{\omega}$ and wave number $\bar{k}$, propagated in the z direction. The amplitude $A(z,t)$ is represented as a superposition of harmonic waves of frequencies $\omega - \bar{\omega}$. Since $\Delta\omega/\bar{\omega}$ is assumed to be small compared with unity, A will vary slowly in comparison with the other term. In general A is complex, so that there is a contribution ($\arg A$) to the phase $\bar{\omega}t - \bar{k}z$. The surfaces

$$t = \left(\frac{dk}{d\omega}\right)_{\bar{\omega}} z \tag{46}$$

are seen to play a special role: on each such surface $A(z,t)$ is constant. Hence the velocity of advance of a definite value of A and also of the maximum of $|A|$ is given by the *group velocity*

$$v^{(g)} = \left(\frac{d\omega}{dk}\right)_{\bar{k}} \tag{47}$$

as before.

The following relations are seen to hold between the group velocity and the phase velocity:

$$v^{(g)} = \frac{d}{dk}(v^{(p)}k) = v^{(p)} + k\frac{dv^{(p)}}{dk} = v^{(p)} - \lambda\frac{dv^{(p)}}{d\lambda}, \tag{48}$$

all the quantities here referring to mean frequency $\bar{\omega}$.

Finally, let us consider the general three-dimensional group

$$V(\mathbf{r},t) = \mathscr{R}\int_{(\Delta\omega)} a_\omega(\mathbf{r})e^{-i[\omega t - g_\omega(\mathbf{r})]}\, d\omega. \tag{49}$$

By analogy with (43), we separate a term corresponding to the mean frequency $\bar{\omega}$, and write

$$V(\mathbf{r},t) = A(\mathbf{r},t)e^{-i[\bar{\omega}t - g_{\bar{\omega}}(\mathbf{r})]}, \tag{50}$$

where

$$A(\mathbf{r},t) = \int_{(\Delta\omega)} a_\omega(\mathbf{r})e^{-i\{(\omega-\bar{\omega})t-[g_\omega(\mathbf{r})-g_{\bar{\omega}}(\mathbf{r})]\}}\, d\omega$$

$$\sim \int_{(\Delta\omega)} a_\omega(\mathbf{r})e^{-i\left\{(\omega-\bar{\omega})\left[t-\left(\frac{\partial g(\mathbf{r})}{\partial\omega}\right)_{\bar{\omega}}\right]\right\}}\, d\omega, \tag{51}$$

if $\Delta\omega$ is sufficiently small. Eq. (50) represents a wave of frequency $\bar{\omega}$ whose (generally complex) amplitude $A(\mathbf{r},t)$ varies both in space and time, this variation again being slow in comparison with the other term. By analogy with (46) the surface

$$t = \left[\frac{\partial g(\mathbf{r})}{\partial \omega}\right]_{\bar{\omega}} \tag{52}$$

may be expected to play a special part. However, the amplitude function A is now not necessarily constant over each such surface, since the Fourier amplitudes a_ω are now functions not only of the frequency but also of the position. The significance of (52) is seen by considering the absolute amplitude $M = |A|$. We have

$$M^2(\mathbf{r},t) = A(\mathbf{r},t)A^\star(\mathbf{r},t) = \int_{(\Delta\omega)}\int_{(\Delta\omega)} a_\omega(\mathbf{r})a_{\omega'}(\mathbf{r})e^{-i\left\{(\omega-\omega')\left[t-\left(\frac{\partial g(\mathbf{r})}{\partial\omega}\right)_{\bar{\omega}}\right]\right\}}\, d\omega d\omega'. \tag{53}$$

Obviously the imaginary part of the double integral vanishes since M^2 is real. (Formally this may be verified by interchanging the independent variables ω and ω' and noting that the imaginary part of the integrand then changes sign.) Hence

$$M^2(\mathbf{r},t) = \int_{(\Delta\omega)}\int_{(\Delta\omega)} a_\omega(\mathbf{r})a_{\omega'}(\mathbf{r}) \cos\left\{(\omega-\omega')\left[t-\left(\frac{\partial g(\mathbf{r})}{\partial\omega}\right)_{\bar{\omega}}\right]\right\} d\omega d\omega'. \tag{54}$$

Fixing our attention on any particular point $\mathbf{r} = \mathbf{r}_0$ and remembering that a_ω is either positive or zero, we see that $M^2(\mathbf{r}_0,t)$ attains its maximum when the argument of the cosine term is zero, i.e. when $t = \left(\dfrac{\partial g(\mathbf{r}_0)}{\partial\omega}\right)_{\bar{\omega}}$. Thus (52) represents the surfaces on which the absolute amplitude attains its maximum at time t in the sense just explained. It is therefore appropriate to define the group velocity of the general three-dimensional wave group as the velocity with which these surfaces advance. Considering a small displacement $\delta\mathbf{r} = \mathbf{q}\delta s$, where $\mathbf{q}$ is a unit vector in the direction normal to the surface, we have from (52) that the corresponding change δt is given by

$$\delta t = \delta s \left| \operatorname{grad}\left(\frac{\partial g(\mathbf{r})}{\partial\omega}\right)_{\bar{\omega}}\right|, \tag{55}$$

so that the group velocity of the general three-dimensional group is given by

$$v^{(g)} = \frac{1}{\left|\operatorname{grad}\left(\dfrac{\partial g}{\partial\omega}\right)_{\bar{\omega}}\right|}. \tag{56}$$

This expression should be compared with the expression (31)

$$v^{(p)} = \frac{1}{\left|\operatorname{grad}\dfrac{g}{\omega}\right|} \tag{57}$$

for the phase velocity of a general harmonic wave. In the special case of a group of plane waves propagated in the z direction, $g_\omega = kz$, and (56) reduces to (47).

It is evident from the preceding discussion that the effective frequency range $\Delta\omega$ is an important parameter relating to a wave group; it is this quantity which substantially determines the rate of fluctuation of the amplitude and the phase. If the medium is not strongly dispersive, a wave group will travel a considerable distance without appreciable "diffusion". In such circumstances, the group velocity, which may be considered as the velocity of the propagation of the group as a whole, will also represent the velocity at which the energy is propagated.* This, however, is not true in general. In particular, in regions of anomalous dispersion (cf. § 2.3.4) the group velocity may exceed the velocity of light or become negative, and in such cases it has no longer any appreciable physical significance.

1.4 VECTOR WAVES

1.4.1 The general electromagnetic plane wave

The simplest electromagnetic field is that of a plane wave; then each Cartesian component of the field vectors and consequently $\boldsymbol{E}$ and $\boldsymbol{H}$ are, according to §1.3.1, functions of the variable $u = \boldsymbol{r} . \boldsymbol{s} - vt$ only:

$$\boldsymbol{E} = \boldsymbol{E}(\boldsymbol{r} . \boldsymbol{s} - vt), \qquad \boldsymbol{H} = \boldsymbol{H}(\boldsymbol{r} . \boldsymbol{s} - vt), \tag{1}$$

$\boldsymbol{s}$ denoting as before a unit vector in the direction of propagation.

Denoting by a dot differentiation with respect to t, and by a prime differentiation with respect to the variable u, we have

$$\left.\begin{aligned} \dot{\boldsymbol{E}} &= -v\boldsymbol{E}' \\ (\text{curl } \boldsymbol{E})_x &= \frac{\partial E_z}{\partial y} - \frac{\partial E_y}{\partial z} = E_z' s_y - E_y' s_z = (\boldsymbol{s} \wedge \boldsymbol{E}')_x . \end{aligned}\right\} \tag{2}$$

Substituting these expressions into MAXWELL's equations § 1.1 (1), (2) with $\boldsymbol{j} = 0$, and using the material equations § 1.1 (10), (11) we obtain

$$\left.\begin{aligned} \boldsymbol{s} \wedge \boldsymbol{H}' + \frac{\varepsilon v}{c} \boldsymbol{E}' &= 0, \\ \boldsymbol{s} \wedge \boldsymbol{E}' - \frac{\mu v}{c} \boldsymbol{H}' &= 0. \end{aligned}\right\} \tag{3}$$

If we set the additive constants of integration equal to zero (i.e. neglect a field constant in space) and set, as before, $v/c = 1/\sqrt{\varepsilon\mu}$, (3) gives, on integration,

$$\left.\begin{aligned} \boldsymbol{E} &= -\sqrt{\frac{\mu}{\varepsilon}}\, \boldsymbol{s} \wedge \boldsymbol{H}, \\ \boldsymbol{H} &= \sqrt{\frac{\varepsilon}{\mu}}\, \boldsymbol{s} \wedge \boldsymbol{E}. \end{aligned}\right\} \tag{4}$$

Scalar multiplication with $\boldsymbol{s}$ gives

$$\boldsymbol{E} . \boldsymbol{s} = \boldsymbol{H} . \boldsymbol{s} = 0. \tag{5}$$

This relation expresses the "transversality" of the field, i.e. it shows that the electric and magnetic field vectors lie in planes normal to the direction of propagation.

* See, for example, F. BORGNIS, *Z. f. Phys.*, **117** (1941), 642; L. J. F. BROER, *Appl. Sci. Res.*, **A2** (1951), 329.

From (4) and (5) it is seen that $\boldsymbol{E}$, $\boldsymbol{H}$ and $\boldsymbol{s}$ form a right-handed orthogonal triad of vectors. We also have from (4) that

$$\sqrt{\mu}H = \sqrt{\varepsilon}E. \tag{6}$$

where $E = |\boldsymbol{E}|$, $H = |\boldsymbol{H}|$.

Let us now consider the amount of energy which crosses, in unit time, an element of area perpendicular to the direction of propagation. Imagine a cylinder whose axis is parallel to $\boldsymbol{s}$ and whose cross-sectional area is unity. The amount of energy which crosses the base of the cylinder in unit time is then equal to the energy which was contained in the portion of the cylinder of length v. Hence the energy flux is equal to vw where w is the energy density. According to (6) and § 1.1 (31), the energy density is given by

$$w = \frac{\varepsilon}{4\pi} E^2 = \frac{\mu}{4\pi} H^2. \tag{7}$$

The Poynting vector, on the other hand, is according to § 1.1 (38), given by

$$\boldsymbol{S} = \frac{c}{4\pi} EH\boldsymbol{s} = \frac{c}{4\pi}\sqrt{\frac{\varepsilon}{\mu}}E^2\boldsymbol{s} = \frac{c}{4\pi}\sqrt{\frac{\mu}{\varepsilon}}H^2\boldsymbol{s}. \tag{8}$$

Comparison of (7) and (8) shows that

$$\boldsymbol{S} = \frac{c}{\sqrt{\varepsilon\mu}} w\boldsymbol{s} = vw\boldsymbol{s}. \tag{9}$$

We see that, in agreement with § 1.1.4, the Poynting vector represents the flow of energy both with regard to its magnitude and direction of propagation.

1.4.2 The harmonic electromagnetic plane wave

Of particular interest is the case when the plane wave is time harmonic, i.e. when each Cartesian component of $\boldsymbol{E}$ and $\boldsymbol{H}$ is of the form

$$a \cos(\tau + \delta) = \mathscr{R}\{ae^{-i(\tau+\delta)}\}. \qquad (a > 0). \tag{10}$$

Here τ denotes the variable part of the phase factor, i.e.

$$\tau = \omega\left(t - \frac{\boldsymbol{r}\,.\,\boldsymbol{s}}{v}\right) = \omega t - \boldsymbol{k}\,.\,\boldsymbol{r}. \tag{11}$$

We choose the z-axis in the $\boldsymbol{s}$ direction. Then, since according to (5) the field is transversal, only the x and y components of $\boldsymbol{E}$ and $\boldsymbol{H}$ are different from zero. We shall now consider the nature of the curve which the end point of the electric vector describes at a typical point in space; this curve is the locus of the points whose coordinates (E_x, E_y) are

$$\left.\begin{aligned} E_x &= a_1 \cos(\tau + \delta_1), \\ E_y &= a_2 \cos(\tau + \delta_2), \\ E_z &= 0. \end{aligned}\right\} \tag{12}$$

(a) *Elliptic polarization*

In order to eliminate τ between the first two equations of (12), we re-write them in the form

$$\left.\begin{aligned} \frac{E_x}{a_1} &= \cos\tau\cos\delta_1 - \sin\tau\sin\delta_1, \\ \frac{E_y}{a_2} &= \cos\tau\cos\delta_2 - \sin\tau\sin\delta_2. \end{aligned}\right\} \tag{13}$$

Hence

$$\left.\begin{aligned} \frac{E_x}{a_1}\sin\delta_2 - \frac{E_y}{a_2}\sin\delta_1 &= \cos\tau\sin(\delta_2-\delta_1), \\ \frac{E_x}{a_1}\cos\delta_2 - \frac{E_y}{a_2}\cos\delta_1 &= \sin\tau\sin(\delta_2-\delta_1). \end{aligned}\right\} \tag{14}$$

Squaring and adding gives

$$\left(\frac{E_x}{a_1}\right)^2 + \left(\frac{E_y}{a_2}\right)^2 - 2\frac{E_x}{a_1}\frac{E_y}{a_2}\cos\delta = \sin^2\delta, \tag{15}$$

where

$$\delta = \delta_2 - \delta_1. \tag{16}$$

(15) is the equation of a conic. It is an ellipse, since the associated determinant is not negative:

$$\begin{vmatrix} \dfrac{1}{a_1^2} & -\dfrac{\cos\delta}{a_1a_2} \\ -\dfrac{\cos\delta}{a_1a_2} & \dfrac{1}{a_2^2} \end{vmatrix} = \frac{1}{a_1^2a_2^2}[1-\cos^2\delta] = \frac{\sin^2\delta}{a_1^2a_2^2} \geqslant 0.$$

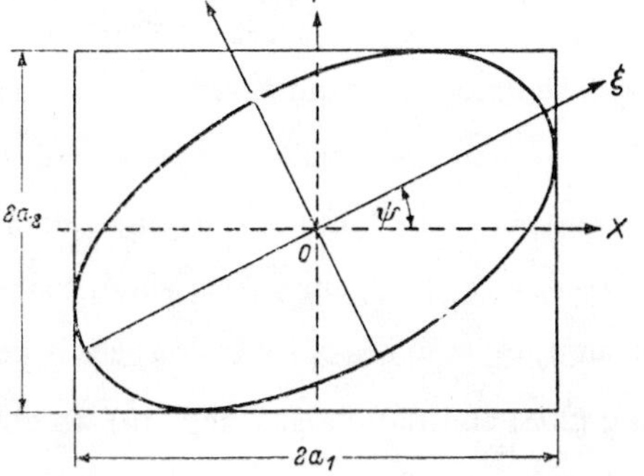

Fig. 1.6. Elliptically polarized wave. The vibrational ellipse for the electric vector.

The ellipse is inscribed into a rectangle whose sides are parallel to the co-ordinate axes and whose lengths are $2a_1$ and $2a_2$ (Fig. 1.6). The ellipse touches the sides at the points $(\pm a_1, \pm a_2\cos\delta)$ and $(\pm a_1\cos\delta, \pm a_2)$.

The wave (12) is then said to be *elliptically polarized.* It is easily seen that the wave associated with the magnetic vector is also elliptically polarized. For by (5) and (12),

$$\left.\begin{aligned} H_x &= -\sqrt{\frac{\varepsilon}{\mu}}\, E_y = -\sqrt{\frac{\varepsilon}{\mu}}\, a_2 \cos(\tau + \delta_2), \\ H_y &= \sqrt{\frac{\varepsilon}{\mu}}\, E_x \quad = \sqrt{\frac{\varepsilon}{\mu}}\, a_1 \cos(\tau + \delta_1), \\ H_z &= 0. \end{aligned}\right\} \tag{17}$$

The end point of the magnetic vector therefore describes an ellipse which is inscribed into a rectangle whose sides are parallel to the x and y directions and whose lengths are $2\sqrt{\frac{\varepsilon}{\mu}}\, a_2$, $2\sqrt{\frac{\varepsilon}{\mu}}\, a_1$.

In general the axes of the ellipse are not in the Ox and Oy directions. Let $O\xi$, $O\eta$ be a new set of axes along the axes of the ellipse and let ψ $(0 \leqslant \psi < \pi)$ be the angle between Ox and the direction $O\xi$ of the major axis (Fig. 1.6). Then the components E_ξ and E_η are related to E_x and E_y by

$$\left.\begin{aligned} E_\xi &= \quad E_x \cos\psi + E_y \sin\psi, \\ E_\eta &= -E_x \sin\psi + E_y \cos\psi. \end{aligned}\right\} \tag{18}$$

If $2a$ and $2b$ $(a \geqslant b)$ are the lengths of the axes of the ellipse, the equation of the ellipse referred to $O\xi$, $O\eta$ is:

$$\left.\begin{aligned} E_\xi &= \quad a \cos(\tau + \delta_0), \\ E_\eta &= \pm b \sin(\tau + \delta_0). \end{aligned}\right\} \tag{19}$$

The two signs distinguish the two possible senses in which the end point of the electric vector may describe the ellipse.

To determine a and b we compare (18) and (19) and use (13):

$$\begin{aligned} a(\cos\tau\cos\delta_0 - \sin\tau\sin\delta_0) &= a_1[\cos\tau\cos\delta_1 - \sin\tau\sin\delta_1]\cos\psi \\ &\quad + a_2[\cos\tau\cos\delta_2 - \sin\tau\sin\delta_2]\sin\psi. \\ \pm b(\sin\tau\cos\delta_0 + \cos\tau\sin\delta_0) &= -a_1[\cos\tau\cos\delta_1 - \sin\tau\sin\delta_1]\sin\psi \\ &\quad + a_2[\cos\tau\cos\delta_2 - \sin\tau\sin\delta_2]\cos\psi. \end{aligned}$$

Next we equate the coefficients of $\cos\tau$ and $\sin\tau$:

$$a\cos\delta_0 = a_1\cos\delta_1\cos\psi + a_2\cos\delta_2\sin\psi, \tag{20a}$$

$$a\sin\delta_0 = a_1\sin\delta_1\cos\psi + a_2\sin\delta_2\sin\psi, \tag{20b}$$

$$\pm b\cos\delta_0 = a_1\sin\delta_1\sin\psi - a_2\sin\delta_2\cos\psi, \tag{21a}$$

$$\pm b\sin\delta_0 = -a_1\cos\delta_1\sin\psi + a_2\cos\delta_2\cos\psi. \tag{21b}$$

On squaring and adding (20a) and (20b) and using (16) we obtain

$$a^2 = a_1^2\cos^2\psi + a_2^2\sin^2\psi + 2a_1a_2\cos\psi\sin\psi\cos\delta,$$

and similarly from (21a) and (21b)

$$b^2 = a_1^2\sin^2\psi + a_2^2\cos^2\psi - 2a_1a_2\cos\psi\sin\psi\cos\delta. \tag{22}$$

Hence

$$a^2 + b^2 = a_1^2 + a_2^2. \tag{23}$$

Next we multiply (20a) by (21a), (20b) by (21b) and add. This gives

$$\mp ab = a_1 a_2 \sin \delta. \tag{24}$$

Further on dividing (21a) by (20a) and (21b) by (20b) we obtain

$$\pm \frac{b}{a} = \frac{a_1 \sin \delta_1 \sin \psi - a_2 \sin \delta_2 \cos \psi}{a_1 \cos \delta_1 \cos \psi + a_2 \cos \delta_2 \sin \psi} = \frac{-a_1 \cos \delta_1 \sin \psi + a_2 \cos \delta_2 \cos \psi}{a_1 \sin \delta_1 \cos \psi + a_2 \sin \delta_2 \sin \psi},$$

and these relations give the following equation for ψ:

$$(a_1^2 - a_2^2) \sin 2\psi = 2a_1 a_2 \cos \delta \cos 2\psi.$$

It will be convenient to introduce an auxiliary angle α $(0 \leqslant \alpha \leqslant \pi/2)$, such that

$$\frac{a_2}{a_1} = \tan \alpha. \tag{25}$$

The preceding equation then becomes

$$\tan 2\psi = \frac{2a_1 a_2}{a_1^2 - a_2^2} \cos \delta = \frac{2 \tan \alpha}{1 - \tan^2 \alpha} \cos \delta,$$

i.e.

$$\tan 2\psi = (\tan 2\alpha) \cos \delta. \tag{26}$$

Now from (23) and (24) we also have

$$\mp \frac{2ab}{a^2 + b^2} = \frac{2a_1 a_2}{a_1^2 + a_2^2} \sin \delta = (\sin 2\alpha) \sin \delta. \tag{27}$$

Let χ $(-\pi/4 \leqslant \chi \leqslant \pi/4)$ be another auxiliary angle, such that

$$\pm \frac{b}{a} = \tan \chi. \tag{28}$$

The numerical value of $\tan \chi$ represents the ratio of the axes of the ellipse and the sign of χ distinguishes the two senses in which the ellipse may be described. Eq. (27) may be written in the form

$$\sin 2\chi = -(\sin 2\alpha) \sin \delta. \tag{29}$$

It will be useful to summarize the results. If a_1, a_2 and the phase difference δ are given, referred to an arbitrary set of axes, and if α $(0 \leqslant \alpha \leqslant \pi/2)$ denotes an angle such that

$$\tan \alpha = \frac{a_2}{a_1}, \tag{30}$$

then the principal semi-axes a and b of the ellipse and the angle ψ $(0 \leqslant \psi < \pi)$ which the major axis makes with Ox are specified by the formulae

$$a^2 + b^2 = a_1^2 + a_2^2, \tag{31a}$$

$$\tan 2\psi = (\tan 2\alpha) \cos \delta, \tag{31b}$$

$$\sin 2\chi = -(\sin 2\alpha) \sin \delta, \tag{31c}$$

where χ $(-\pi/4 < \chi \leqslant \pi/4)$ is an auxiliary angle which specifies the shape and orientation of the vibrational ellipse:

$$\tan \chi = \pm b/a. \tag{32}$$

Conversely, if the lengths a and b of the axes and the orientation of the ellipse is known (i.e. a, b and ψ given) these formulae enable the determination of the amplitudes a_1, a_2 and the phase difference δ. In Chapter XIV, instruments will be described by means of which these quantities may be directly determined.

Before discussing some important special cases, we must say a few words about the terminology. We distinguish two cases of polarization, according to the sense in which the end point of the electric vector describes the ellipse. It seems natural to call the polarization right-handed or left-handed according as to whether the rotation of $\boldsymbol{E}$ and the direction of propagation form a right-handed or left-handed screw. But the traditional terminology is just the opposite—being based on the apparent behaviour of $\boldsymbol{E}$ when "viewed" face on by the observer. We shall conform throughout this book to this customary usage. Thus we say that the polarization is *right-handed* when to an observer looking in the direction from which the light is coming, the end-point of the electric vector would appear to describe the ellipse in the clockwise sense. If we consider the values of (12) for two time instants separated by a quarter of a period, we see that in this case $\sin \delta > 0$, or by (29), $0 < \chi \leqslant \pi/4$. For *left-handed* polarization the opposite is the case, i.e. to an observer looking in the direction from which the light is propagated, the electric vector would appear to describe the ellipse anticlockwise; in this case $\sin \delta < 0$, so that $-\pi/4 \leqslant \chi < 0$.

For reasons connected with the historical development of optics, the direction of the magnetic vector is often called the *direction of polarization* and the plane containing the magnetic vector and the direction of propagation is known as the *plane of polarization.* This terminology is, however, not used by all writers; some define these quantities with respect to the electric rather than the magnetic vector. This lack of uniformity arises partly from the fact that there is no single physical entity which could be described without ambiguity as "the light vector". When particular attention is paid to the physical effect of the field vectors, there would actually be some grounds for regarding $\boldsymbol{E}$ as the light vector. For every action is a consequence of the motion of elementary charged particles (electrons, nuclei) set into motion by the electromagnetic field. The mechanical force $\boldsymbol{F}$ of the field on the particle is then given by LORENTZ' law, § 1.1 (34).

$$\boldsymbol{F} = e\left(\boldsymbol{E} + \frac{\mu}{c}\,\mathbf{v} \wedge \boldsymbol{H}\right),$$

e being the charge and $\mathbf{v}$ the velocity of the particle. Hence the electric vector is seen to act even when the particle is at rest. On the other hand, the magnetic vector plays a part only when the particle is in motion; however, since v/c is usually very small compared to unity this effect may often be neglected. Nevertheless the "direction of polarization" and the "plane of polarization" are usually associated with the magnetic vector. The reason for this nomenclature will become apparent in the next section when polarization on reflection is discussed.

To avoid confusion we shall, in accordance with more recent practice, not use the terms "direction of polarization" and "plane of polarization"; instead we shall speak of *direction of vibration* and *plane of vibration* to denote the direction of a field vector and the plane containing the field vector and the direction of propagation, the vector in question being specified in each case.

(b) *Linear and circular polarization*

Two special cases are of particular importance, namely when the polarization ellipse degenerates into a straight line or a circle.

According to (12) the ellipse will reduce to a straight line, when

$$\delta = \delta_2 - \delta_1 = m\pi \qquad (m = 0, \pm 1, \pm 2, \ldots). \tag{33}$$

Then

$$\frac{E_y}{E_x} = (-1)^m \frac{a_2}{a_1}, \tag{34}$$

and we say that $\boldsymbol{E}$ is *linearly polarized.** One of the coordinate axes, x say, may be chosen along this line. Then only one component (E_x) remains. Moreover, since the electric and magnetic vectors are orthogonal and lie in the plane perpendicular to the z direction, the component H_x then vanishes, so that $\boldsymbol{H}$ is linearly polarized in the y direction.

The other special case of importance is that of a *circularly polarized* wave, the ellipse then degenerating into a circle. Clearly a necessary condition for this is that the circumscribed rectangle shall become a square:

$$a_1 = a_2 = a. \tag{35}$$

Also, one of the $\boldsymbol{E}$ components must be zero when the other has an extreme value; this demands that

$$\delta = \delta_2 - \delta_1 = m\pi/2 \qquad (m = \pm 1, \pm 3, \pm 5 \ldots), \tag{36}$$

and (15) then reduces to the equation of the circle

$$E_x^2 + E_y^2 = a^2. \tag{37}$$

When the polarization is *right-handed* $\sin \delta > 0$, so that

$$\delta = \frac{\pi}{2} + 2m\pi \qquad (m = 0, \pm 1, \pm 2, \ldots) \tag{38}$$

$$\left.\begin{aligned} E_x &= a \cos(\tau + \delta_1) \\ E_y &= a \cos(\tau + \delta_1 + \pi/2) = -a \sin(\tau + \delta_1). \end{aligned}\right\} \tag{39}$$

For *left-handed polarization* $\sin \delta < 0$, so that

$$\delta = -\frac{\pi}{2} + 2m\pi \qquad (m = 0, \pm 1, \pm 2, \ldots) \tag{40}$$

$$\left.\begin{aligned} E_x &= a \cos(\tau + \delta_1), \\ E_y &= a \cos(\tau + \delta_1 - \pi/2) = a \sin(\tau + \delta_1). \end{aligned}\right\} \tag{41}$$

If, instead of the real representation, the complex one is used (i.e. if the exponential instead of the cosine function is written in (12)), then

$$\frac{E_y}{E_x} = \frac{a_2}{a_1} e^{i(\delta_1 - \delta_2)} = \frac{a_2}{a_1} e^{-i\delta}, \tag{42}$$

and one can immediately determine from the value of this ratio the nature of the polarization. One has for

(a) *Linearly polarized electric wave* ($\delta = m\pi$, $m = 0, \pm 1, \pm 2, \ldots$):

$$\frac{E_y}{E_x} = (-1)^m \frac{a_2}{a_1}.$$

* The less appropriate term *plane polarized* is also used.

(b) *Right-handed circularly polarized electric wave** ($a_1 = a_2$, $\delta = \pi/2$):

$$\frac{E_y}{E_x} = e^{-i\pi/2} = -i.$$

(c) *Left-handed circularly polarized electric wave* ($a_1 = a_2$, $\delta = -\pi/2$):

$$\frac{E_y}{E_x} = e^{i\pi/2} = i.$$

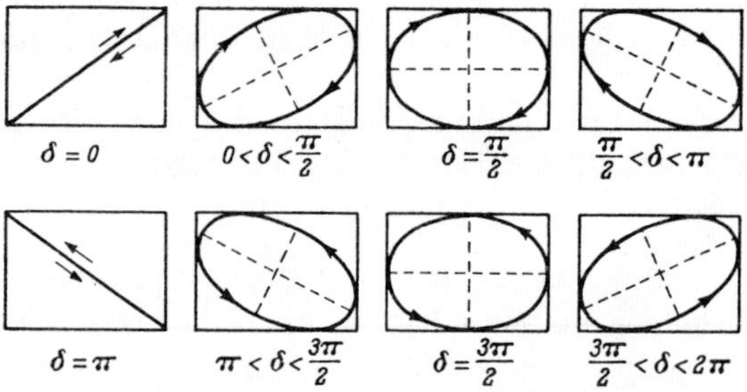

Fig. 1.7. Elliptical polarization with various values of the phase difference δ.

More generally it may be shown that for right-handed elliptical polarization, the ratio E_y/E_x has a positive imaginary part, whereas for left-handed elliptical polarization the imaginary part is negative.

Fig. 1.7 illustrates how the polarization ellipse changes with varying δ.

(c) *Characterization of the state of polarization by Stokes parameters*

To characterize the polarization ellipse three independent quantities are necessary, e.g. the amplitudes a_1 and a_2 and the phase difference δ, or the major and minor axes a, b and the angle χ which specifies the orientation of the ellipse. For practical purposes it is convenient to characterize the state of polarization by certain parameters which are all of the same physical dimensions, and which were introduced by G. G. STOKES in 1852, in his investigations relating to partially polarized light. We shall define them later (§ 10.8.3) in their full generality. We will also show there, that for any given wave these parameters may be determined from simple experiments.

The *Stokes parameters* of a plane monochromatic wave are the four quantities

$$\left.\begin{aligned} s_0 &= a_1^2 + a_2^2, \\ s_1 &= a_1^2 - a_2^2, \\ s_2 &= 2a_1a_2 \cos \delta, \\ s_3 &= 2a_1a_2 \sin \delta. \end{aligned}\right\} \tag{43}$$

Only three of them are independent since they are related by the identity

$$s_0^2 = s_1^2 + s_2^2 + s_3^2. \tag{44}$$

The parameter s_0 is evidently proportional to the intensity of the wave. The parameters s_1, s_2 and s_3 are related in a simple way to the angle ψ $(0 \leqslant \psi < \pi)$ which specifies the orientation of the ellipse and the angle χ $(-\pi/4 \leqslant \chi \leqslant \pi/4)$

* N.B. This is right-handed according to the traditional, not the natural nomenclature.

which characterizes the ellipticity and the sense in which the ellipse is being described. In fact the following relations hold:

$$s_1 = s_0 \cos 2\chi \cos 2\psi, \tag{45a}$$

$$s_2 = s_0 \cos 2\chi \sin 2\psi, \tag{45b}$$

$$s_3 = s_0 \sin 2\chi. \tag{45c}$$

The relation (45c) follows from (25) and (29). To derive the other two relations we note that according to the equation preceding (26),

$$s_2 = s_1 \tan 2\psi. \tag{46}$$

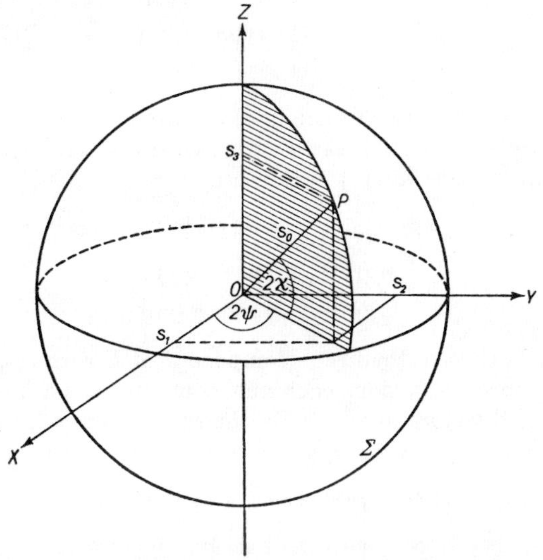

Fig. 1.8. POINCARÉ's representation of the state of polarization of a monochromatic wave. (The POINCARÉ sphere.)

The relation (45a) follows on substitution from (46) and from (45c) into (44). Finally, (45b) is obtained on substitution from (45a) into (46).

The relations (45) indicate a simple geometrical representation of all the different states of polarization: s_1, s_2 and s_3 may be regarded as the Cartesian coordinates of a point P on a sphere Σ of radius s_0, such that 2χ and 2ψ are the spherical angular coordinates of this point (see Fig. 1.8). Thus *to every possible state of polarization of a plane monochromatic wave of a given intensity* ($s_0 =$ *constant*), *there corresponds one point on* Σ *and vice versa.* Since χ is positive or negative according as the polarization is right-handed or left-handed, it follows from (45c) that right-handed polarization is represented by points on Σ which lie above the equatorial plane (xy-plane), and left-handed polarization by points on Σ which lie below this plane. Further, for linearly polarized light the phase difference δ is zero or an integral multiple of π; according to (43) the STOKES parameter s_3 is then zero, so that linear polarization is represented by points in the equatorial plane. For circular polarization $a_1 = a_2$ and $\delta = \pi/2$ or $-\pi/2$ according as the polarization is right- or left-handed; hence right-handed circular polarization is represented by the north pole ($s_1 = s_2 = 0$, $s_3 = s_0$) and left-handed circular polarization by the south pole ($s_1 = s_2 = 0$,

$s_3 = - s_0$). This geometrical representation of different states of polarization by points on a sphere is due to POINCARÉ,* and is very useful in crystal optics for determining the effect of crystalline media on the state of polarization of light traversing it.† The sphere Σ is called *the Poincaré sphere.*

1.4.3 Harmonic vector waves of arbitrary form

The main results of the preceding sections can easily be extended to time-harmonic waves of more complicated form.

A general time-harmonic real vector wave $\boldsymbol{V}(\boldsymbol{r},t)$ is a solution of the vector wave equation, whose Cartesian components V_x, V_y, V_z are represented by expressions of the form § 1.3 (23):

$$\left.\begin{aligned} V_x(\boldsymbol{r},t) &= a_1(\boldsymbol{r}) \cos [\omega t - g_1(\boldsymbol{r})], \\ V_y(\boldsymbol{r},t) &= a_2(\boldsymbol{r}) \cos [\omega t - g_2(\boldsymbol{r})], \\ V_z(\boldsymbol{r},t) &= a_3(\boldsymbol{r}) \cos [\omega t - g_3(\boldsymbol{r})]. \end{aligned}\right\} \tag{47}$$

where a_s and g_s ($s = 1,2,3$) are real functions of position. For the plane harmonic wave considered in the preceding section the a's were constant, and $g_s(\boldsymbol{r}) = \boldsymbol{k} \cdot \boldsymbol{r} - \delta_s$.

It will be convenient to write (47) in the form

$$V_x(\boldsymbol{r},t) = p_x(\boldsymbol{r}) \cos \omega t + q_x(\boldsymbol{r}) \sin \omega t \qquad \text{etc.,} \tag{48}$$

where

$$\left.\begin{aligned} p_x(\boldsymbol{r}) &= a_1(\boldsymbol{r}) \cos g_1(\boldsymbol{r}), \\ q_x(\boldsymbol{r}) &= a_1(\boldsymbol{r}) \sin g_1(\boldsymbol{r}). \end{aligned}\right\} \tag{49}$$

If a new set of axes is chosen, then each component of $\boldsymbol{V}$ with respect to the new set will again be of the form (47), since each new component is a linear combination of the old ones and can therefore involve only the sum of terms in $\cos \omega t$ and $\sin \omega t$.

We may regard (p_x,p_y,p_z) and (q_x,q_y,q_z) as components of two real vectors $\boldsymbol{p}$ and $\boldsymbol{q}$. Then

$$\boldsymbol{V}(\boldsymbol{r},t) = \boldsymbol{p}(\boldsymbol{r}) \cos \omega t + \boldsymbol{q}(\boldsymbol{r}) \sin \omega t. \tag{50}$$

By Fourier analysis, an arbitrary vector wave may be expressed as superposition of waves of this type.

As in the case of scalar waves, it is often convenient to use a complex representation. We write (50) in the form

$$\boldsymbol{V}(\boldsymbol{r},t) = \mathscr{R}\{\boldsymbol{U}(\boldsymbol{r})e^{-i\omega t}\}, \tag{51}$$

where $\boldsymbol{U}$ is the complex vector

$$\boldsymbol{U}(\boldsymbol{r}) = \boldsymbol{p}(\boldsymbol{r}) + i\boldsymbol{q}(\boldsymbol{r}), \tag{52}$$

$\mathscr{R}$ denoting the real part. When the operations on $\boldsymbol{V}$ are linear one can, as in the corresponding scalar case, operate directly on the complex quantity, omitting the symbol $\mathscr{R}$ altogether. The real part of the final expression is then understood to represent the physical quantity in question.

Operation with complex vectors follow the usual rules of vector algebra and of algebra of complex numbers. For example, the conjugate of $\boldsymbol{U}$ is the vector

$$\boldsymbol{U}^\star = \boldsymbol{p} - i\boldsymbol{q}.$$

---

* H. POINCARÉ, *Théorie Mathématique de la Lumière* (Paris, Georges Carré), Vol. 2, (1892) Chap. 12.

† Examples illustrating the method and references to the relevant literature are given in H. G. JERRARD, *J. Opt. Soc. Amer.*, **44** (1954), 634. See also the paper by M. J. WALKER, referred to in § 10.8.3 and S. PANCHARATNAM, *Proc. Ind. Acad. Sci.*, A, **44** (1956), 247.

Similarly

$$U^2 = U \cdot U = p^2 - q^2 + 2ip \cdot q$$

$$U \cdot U^\star = (p + iq) \cdot (p - iq) = p^2 + q^2,$$

etc.

To illustrate calculations with complex vectors we shall derive formulae needed later for the energy densities and the Poynting vector in a time-harmonic electromagnetic field. The electric and magnetic vectors then are of the form

$$\left.\begin{aligned} E(r,t) &= \mathscr{R}\{E_0(r)e^{-i\omega t}\} = \tfrac{1}{2}[E_0(r)e^{-i\omega t} + E_0^\star(r)e^{i\omega t}], \\ H(r,t) &= \mathscr{R}\{H_0(r)e^{-i\omega t}\} = \tfrac{1}{2}[H_0(r)e^{-i\omega t} + H_0^\star(r)e^{i\omega t}], \end{aligned}\right\} \tag{53}$$

E_0 and H_0 being complex vector functions of position. Since the optical frequencies are very large (ω is of order 10^{15} sec^{-1}), one cannot observe the instantaneous values of any of the rapidly oscillating quantities, but only their time average taken over a time interval (say $-T' \leqslant t \leqslant T'$) which is large compared to the fundamental period $T = 2\pi/\omega$. In particular, the time-averaged electric energy density is

$$\langle w_e \rangle = \frac{1}{2T'}\int_{-T}^{T} \frac{\varepsilon}{8\pi} E^2\, dt = \frac{\varepsilon}{8\pi}\frac{1}{2T'}\int_{-T'}^{T'} \tfrac{1}{4}[E_0{}^2 e^{-2i\omega t} + 2E_0 \cdot E_0^\star + E_0^{\star 2} e^{2i\omega t}]dt.$$

Now

$$\frac{1}{2T'}\int_{-T}^{T} e^{-2i\omega t}\, dt = -\frac{1}{4i\omega T'}\left[e^{-2i\omega t}\right]_{-T'}^{T'} = \frac{1}{4\pi}\frac{T}{T'}\sin 2\omega T'.$$

Since T' is assumed to be large compared to T, T/T' will be small compared to unity, so that the integral involving $e^{-2i\omega t}$ may be neglected. Similarly the integral involving $e^{2i\omega t}$ may also be neglected, and we finally obtain

$$\langle w_e \rangle = \frac{\varepsilon}{16\pi} E_0 \cdot E_0^\star. \tag{54}$$

In a similar way the time average of the magnetic energy density is seen to be

$$\langle w_m \rangle = \frac{\mu}{16\pi} H_0 \cdot H_0^\star. \tag{55}$$

The average of the Poynting vector is given by

$$\begin{aligned} \langle S \rangle &= \frac{1}{2T'}\int_{-T'}^{T'} \frac{c}{4\pi}(E \wedge H)dt \\ &= \frac{c}{4\pi}\frac{1}{2T'}\int_{-T'}^{T'} \tfrac{1}{4}[E_0 \wedge H_0 e^{-2i\omega t} + E_0 \wedge H_0^\star + E_0^\star \wedge H_0 + E_0^\star \wedge H_0^\star e^{2i\omega t}]dt \\ &\simeq \frac{c}{16\pi}(E_0 \wedge H_0^\star + E_0^\star \wedge H_0) \\ &= \frac{c}{8\pi}\mathscr{R}(E_0 \wedge H_0^\star). \end{aligned} \tag{56}$$

The law of conservation of energy also takes a simple form. For a nonconducting medium ($\sigma = 0$) where no mechanical work is done, we find, on taking the time average of eq. (43) of § 1.1 that

$$\operatorname{div}\langle S \rangle = 0. \tag{57}$$

If we integrate (57) throughout an arbitrary volume which contains no radiator or absorber of energy, and apply GAUSS' theorem, it follows that

$$\int \langle \boldsymbol{S} \rangle \, . \, \boldsymbol{n} \, dS = 0, \tag{58}$$

$\boldsymbol{n}$ being the outward normal to the surface; the integration is taken over the boundary. Thus the averaged total flux of energy through any closed surface is zero.

We now return to the general time-harmonic vector wave (50), and consider the behaviour of $\boldsymbol{V}$ at a point $\boldsymbol{r} = \boldsymbol{r}_0$ in space. In general, as time varies, the end point of $\boldsymbol{V}$ describes an ellipse, so that, like the plane wave, the wave (50) is in general also elliptically polarized. To see this we note first that, with varying time, the end point describes a curve in the plane specified by $\boldsymbol{p}\ (\boldsymbol{r}_0)$ and $\boldsymbol{q}\ (\boldsymbol{r}_0)$. Since $\boldsymbol{V}$ is periodic, the curve must be closed. Now we may set

$$(\boldsymbol{p} + i\boldsymbol{q}) = (\boldsymbol{a} + i\boldsymbol{b})e^{i\varepsilon}, \tag{59}$$

where ε is any scalar. In terms of $\boldsymbol{p}$, $\boldsymbol{q}$ and ε,

$$\left.\begin{aligned} \boldsymbol{a} &= \quad \boldsymbol{p} \cos \varepsilon + \boldsymbol{q} \sin \varepsilon, \\ \boldsymbol{b} &= -\boldsymbol{p} \sin \varepsilon + \boldsymbol{q} \cos \varepsilon. \end{aligned}\right\} \tag{60}$$

Let us choose ε so that the vectors $\boldsymbol{a}$ and $\boldsymbol{b}$ are perpendicular to each other and let $|\boldsymbol{a}| \geqslant |\boldsymbol{b}|$. For $\boldsymbol{a}$ and $\boldsymbol{b}$ to be orthogonal, ε must satisfy the equation

$$(\boldsymbol{p} \cos \varepsilon + \boldsymbol{q} \sin \varepsilon) \, . \, (-\boldsymbol{p} \sin \varepsilon + \boldsymbol{q} \cos \varepsilon) = 0, \tag{61}$$

i.e.

$$\tan 2\varepsilon = \frac{2\boldsymbol{p} \, . \, \boldsymbol{q}}{\boldsymbol{p}^2 - \boldsymbol{q}^2}. \tag{62}$$

We now consider as parameters specifying the wave the 5 independent components of the orthogonal vectors $\boldsymbol{a}$ and $\boldsymbol{b}$ and the corresponding phase factor ε, instead of the 6 rectangular components of $\boldsymbol{p}$ and $\boldsymbol{q}$. Then from (51), (52) and (59),

$$\begin{aligned} \boldsymbol{V} &= \mathscr{R}\{[\boldsymbol{a} + i\boldsymbol{b}]e^{-i(\omega t - \varepsilon)}\} \\ &= \boldsymbol{a} \cos (\omega t - \varepsilon) + \boldsymbol{b} \sin (\omega t - \varepsilon). \end{aligned} \tag{63}$$

If we take Cartesian axes with origin at $\boldsymbol{r}_0$ and with the x and y directions along $\boldsymbol{a}$ and $\boldsymbol{b}$, then

$$V_x = a \cos (\omega t - \varepsilon), \qquad V_y = b \sin (\omega t - \varepsilon), \qquad V_z = 0. \tag{64}$$

This represents an *ellipse* (the *polarization* ellipse)

$$\frac{V_x^{\,2}}{a^2} + \frac{V_y^{\,2}}{b^2} = 1, \tag{65}$$

with semi-axes of lengths a and b along the x and y co-ordinate axes. By elementary geometry it may be shown that $\boldsymbol{p}$ and $\boldsymbol{q}$ are a pair of conjugate semi-diameters of the ellipse.

As in the case of plane waves, the ellipse may be described in two possible senses, corresponding to left- and right-handed polarization; these are distinguished by the sign of the scalar triple product $[\boldsymbol{a},\boldsymbol{b},\nabla\varepsilon] = [\boldsymbol{p},\boldsymbol{q},\nabla\varepsilon]$.

The lengths of the semi-axes of the polarization ellipse are easily obtained from (60) and (62). From (60)

$$a^2 = p^2 \cos^2 \varepsilon + q^2 \sin^2 \varepsilon + 2\boldsymbol{p} \cdot \boldsymbol{q} \sin \varepsilon \cos \varepsilon$$
$$= \tfrac{1}{2}(p^2 + q^2) + \tfrac{1}{2}(p^2 - q^2) \cos 2\varepsilon + \boldsymbol{p} \cdot \boldsymbol{q} \sin 2\varepsilon.$$

From (62)

$$\sin 2\varepsilon = \frac{2\boldsymbol{p} \cdot \boldsymbol{q}}{\sqrt{(p^2 - q^2)^2 + 4(\boldsymbol{p} \cdot \boldsymbol{q})^2}}, \qquad \cos 2\varepsilon = \frac{p^2 - q^2}{\sqrt{(p^2 - q^2)^2 + 4(\boldsymbol{p} \cdot \boldsymbol{q})^2}}.$$

Hence

Similarly one finds

$$\left.\begin{aligned} a^2 &= \tfrac{1}{2}[p^2 + q^2 + \sqrt{(p^2 - q^2)^2 + 4(\boldsymbol{p} \cdot \boldsymbol{q})^2}] \\ b^2 &= \tfrac{1}{2}[p^2 + q^2 - \sqrt{(p^2 - q^2)^2 + 4(\boldsymbol{p} \cdot \boldsymbol{q})^2}]. \end{aligned}\right\} \tag{66}$$

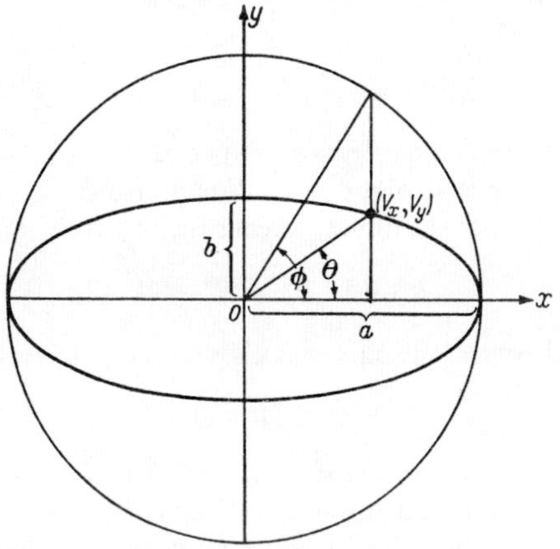

Fig. 1.9. Illustrating the notation relating to eqs. (67) and (68).

To find an expression for the angle between $\boldsymbol{a}$ and $\boldsymbol{p}$ we express the equation of the ellipse in parametric form:

$$V_x = a \cos \phi, \qquad V_y = b \sin \phi, \tag{67}$$

where ϕ is the eccentric angle (see Fig. 1.9). From elementary geometry we learn that this angle is related to the polar angle θ of the point (V_x, V_y) by

$$\tan \theta = \frac{b}{a} \tan \phi. \tag{68}$$

Comparison of (64) and (67) shows that in the present case $\phi = \omega t - \varepsilon$. Now according to (50), $\boldsymbol{V} = \boldsymbol{p}$ when $t = 0$, so that the eccentric angle of $\boldsymbol{p}$ is $-\varepsilon$. Hence the angle ψ between $\boldsymbol{p}$ and $\boldsymbol{a}$ is given by

$$\tan \psi = \frac{b}{a} \tan \varepsilon. \tag{69}$$

If γ denotes the angle between $\boldsymbol{p}$ and $\boldsymbol{q}$, and we introduce an auxiliary angle β such that

$$\frac{q}{p} = \tan \beta, \tag{70}$$

then (62) becomes

$$\tan 2\varepsilon = \frac{2pq}{p^2 - q^2} \cos \gamma$$
$$= \tan 2\beta \cos \gamma. \tag{71}$$

Let us again summarize the results: If $\boldsymbol{p}$ and $\boldsymbol{q}$ are given, γ denotes the angle between these vectors and β denotes the auxiliary angle defined by (70), then the principal semi-axes of the ellipse and the angle ψ which the major axis makes with $\boldsymbol{p}$ are given by

$$\left.\begin{aligned} a^2 &= \tfrac{1}{2}[p^2 + q^2 + \sqrt{(p^2 - q^2)^2 + 4p^2q^2 \cos^2 \gamma}],\\ b^2 &= \tfrac{1}{2}[p^2 + q^2 - \sqrt{(p^2 - q^2)^2 + 4p^2q^2 \cos^2 \gamma}],\\ \tan \psi &= \frac{b}{a} \tan \varepsilon, \end{aligned}\right\} \tag{72}$$

where

$$\tan 2\varepsilon = \tan 2\beta \cos \gamma. \tag{73}$$

As for plane waves, there are two cases of special interest, namely when the ellipse degenerates into a circle or a straight line. For a *circularly polarized* wave, $\boldsymbol{a}$ and $\boldsymbol{b}$ and consequently ε are not determined. According to (62), for this to be the case,

$$\boldsymbol{p} \cdot \boldsymbol{q} = \boldsymbol{p}^2 - \boldsymbol{q}^2 = 0. \tag{74}$$

For a *linearly polarized* wave, the minor axis vanishes ($b^2 = 0$) and (66) then gives

$$p^2q^2 = (\boldsymbol{p} \cdot \boldsymbol{q})^2. \tag{75}$$

Finally we stress that the term *polarization* refers to the behaviour at a particular *point* in the field, and that the state of polarization will therefore in general be different at different points of the field. Thus a wave may be linearly or circularly polarized at some points and elliptically at others.* Only in special cases, as for example for the homogeneous plane wave, will the state of polarization be the same at every point in the field.

1.5 REFLECTION AND REFRACTION OF A PLANE WAVE

In § 1.1.3 relations were derived which the field vectors must satisfy across surfaces at which the physical properties of the medium are discontinuous. These formulae will now be applied to the study of the propagation of a plane wave incident on a plane boundary between two homogeneous isotropic media.

1.5.1 The laws of reflection and refraction

When a plane wave falls on to a boundary between two homogeneous media of different optical properties, it is split into two waves: a transmitted wave proceeding into the second medium and a reflected wave propagated back into the first medium. The existence of these two waves can be demonstrated from the boundary conditions, since it is easily seen that these conditions cannot be satisfied without postulating both the transmitted and the reflected wave. We shall tentatively assume that these waves are

* General properties of time-harmonic electromagnetic waves of arbitrary form but with at least one of the field vectors linearly polarized have been investigated by A. Nisbet and E. Wolf, *Proc. Camb. Phil. Soc.*, **50** (1954), 614.

also plane, and derive expressions for their directions of propagation and their amplitudes.

A plane wave propagated in the direction specified by the unit vector* $\mathbf{s}^{(i)}$ is completely determined when the time behaviour at one particular point in space is known. For if $F(t)$ represents the time behaviour at any one point, the time behaviour at another point, whose position vector relative to the first point is $\mathbf{r}$, is given by $F[t - (\mathbf{r} \cdot \mathbf{s})/v]$. At the boundary between the two media, the time variation of the secondary fields will be the same as that of the incident primary field. Hence, if $\mathbf{s}^{(r)}$ and $\mathbf{s}^{(t)}$ denote unit vectors in the direction of propagation of the reflected and transmitted wave, one has, on equating the arguments of the three wave functions at a point $\mathbf{r}$ on the boundary plane $z = 0$:

$$t - \frac{\mathbf{r} \cdot \mathbf{s}^{(i)}}{v_1} = t - \frac{\mathbf{r} \cdot \mathbf{s}^{(r)}}{v_1} = t - \frac{\mathbf{r} \cdot \mathbf{s}^{(t)}}{v_2}, \tag{1}$$

v_1 and v_2 being the velocities of propagation in the two media. Written out more explicitly, one has, with $\mathbf{r} \equiv x, y, 0$:

$$\frac{x s_x^{(i)} + y s_y^{(i)}}{v_1} = \frac{x s_x^{(r)} + y s_y^{(r)}}{v_1} = \frac{x s_x^{(t)} + y s_y^{(t)}}{v_2}. \tag{2}$$

Since (2) must hold for all values x and y on the boundary,

$$\frac{s_x^{(i)}}{v_1} = \frac{s_x^{(r)}}{v_1} = \frac{s_x^{(t)}}{v_2}, \qquad \frac{s_y^{(i)}}{v_1} = \frac{s_y^{(r)}}{v_1} = \frac{s_y^{(t)}}{v_2}. \tag{3}$$

The plane specified by $\mathbf{s}^{(i)}$ and the normal to the boundary is called the *plane of incidence.* (3) shows that both $\mathbf{s}^{(t)}$ and $\mathbf{s}^{(r)}$ lie in this plane.

Taking the plane of incidence as the xz-plane and denoting by θ_i, θ_r and θ_t the angle which $\mathbf{s}^{(i)}$, $\mathbf{s}^{(r)}$ and $\mathbf{s}^{(t)}$ make with Oz, one has (see Fig. 1.10)

$$\left.\begin{aligned} s_x^{(i)} &= \sin\theta_i, & s_y^{(i)} &= 0, & s_z^{(i)} &= \cos\theta_i, \\ s_x^{(r)} &= \sin\theta_r, & s_y^{(r)} &= 0, & s_z^{(r)} &= \cos\theta_r, \\ s_x^{(t)} &= \sin\theta_t, & s_y^{(t)} &= 0, & s_z^{(t)} &= \cos\theta_t. \end{aligned}\right\} \tag{4}$$

For waves propagated from the first into the second medium, the z components of the $\mathbf{s}$ vectors are positive; for those propagated in the opposite sense, they are negative:

$$s_z^{(i)} = \cos\theta_i \geqslant 0, \qquad s_z^{(r)} = \cos\theta_r \leqslant 0, \qquad s_z^{(t)} = \cos\theta_t \geqslant 0. \tag{5}$$

The first set in (3) gives, on substituting from (4)

$$\frac{\sin\theta_i}{v_1} = \frac{\sin\theta_r}{v_1} = \frac{\sin\theta_t}{v_2}. \tag{6}$$

Hence, $\sin\theta_r = \sin\theta_i$, and we find, also using (5), that $\cos\theta_r = -\cos\theta_i$, so that

$$\theta_r = \pi - \theta_i. \tag{7}$$

This relation, together with the statement that the reflected wave normal $\mathbf{s}^{(r)}$ is in the plane of incidence, constitute the *law of reflection.*

---

* The suffixes i, r and t refer throughout to the incident, reflected and transmitted (refracted) waves respectively.

Also from (6), using MAXWELL's relation § 1.2 (14) connecting the refractive index and the dielectric constant,

$$\frac{\sin \theta_i}{\sin \theta_t} = \frac{v_1}{v_2} = \sqrt{\frac{\varepsilon_2 \mu_2}{\varepsilon_1 \mu_1}} = \frac{n_2}{n_1} = n_{12}. \tag{8}$$

The relation $\sin \theta_i/\sin \theta_t = n_2/n_1$, together with the statement that the refracted wave normal $\mathbf{s}^{(t)}$ is in the plane of incidence, constitute the *law of refraction* (or SNELL's law).

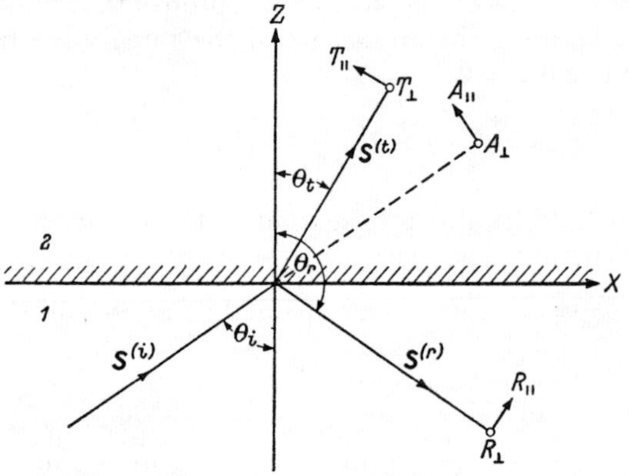

Fig. 1.10. Refraction and reflection of a plane wave. Plane of incidence.

When $n_2 > n_1$, then $n_{12} > 1$, and one says that the second medium is optically denser than the first medium. In this case, by (8),

$$\sin \theta_t = \frac{1}{n_{12}} \sin \theta_i < \sin \theta_i, \tag{9}$$

so that there is a real angle θ_t of refraction for every angle of incidence. If, however, the second medium is optically less dense than the first medium (i.e. if $n_{12} < 1$), one obtains a real value for θ_t only for those incident angles θ_i for which $\sin \theta_i \leqslant n_{12}$. For larger values of the angle of incidence, so-called *total reflection* takes place. It will be considered separately in § 1.5.4.

1.5.2 Fresnel formulae

Next we consider the amplitudes of the reflected and the transmitted waves. We shall assume that the two (homogeneous and isotropic) media are both of zero conductivity and consequently perfectly transparent; their magnetic permeabilities will then in fact differ from unity by negligible amounts, and accordingly we take $\mu_1 = \mu_2 = 1$.

Let A be the amplitude of the electric vector of the incident field. We take A to be complex, with its phase equal to the constant part of the argument of the wave function; the variable part is

$$\tau_i = \omega \left(t - \frac{\mathbf{r} \cdot \mathbf{s}^{(i)}}{v_1} \right) = \omega \left(t - \frac{x \sin \theta_i + z \cos \theta_i}{v_1} \right). \tag{10}$$

We resolve each vector into components parallel (denoted by subscript $\parallel$) and perpendicular (subscript $\perp$) to the plane of incidence. The choice of the positive directions for the parallel components is indicated in Fig. 1.10. The perpendicular components must be visualized at right angles to the plane of the figure.

The components of the electric vector of the incident field then are

$$E_x^{(i)} = -A_{\parallel} \cos \theta_i e^{-i\tau_i}, \qquad E_y^{(i)} = A_{\perp} e^{-i\tau_i}, \qquad E_z^{(i)} = A_{\parallel} \sin \theta_i e^{-i\tau_i}. \tag{11}$$

The components of the magnetic vector are immediately obtained by using the relation 1.4 (4) (with $\mu = 1$):

$$\boldsymbol{H} = \sqrt{\varepsilon}\, \boldsymbol{s} \wedge \boldsymbol{E}. \tag{12}$$

This gives

$$H_x^{(i)} = -A_{\perp} \cos \theta_i \sqrt{\varepsilon_1} e^{-i\tau_i}, \; H_y^{(i)} = -A_{\parallel} \sqrt{\varepsilon_1} e^{-i\tau_i}, \; H_z^{(i)} = A_{\perp} \sin \theta_i \sqrt{\varepsilon_1} e^{-i\tau_i}. \tag{13}$$

Similarly if T and R are the complex amplitudes of the transmitted and reflected waves, the corresponding components of the electric and magnetic vectors are:

Transmitted field

$$\left.\begin{aligned} E_x^{(t)} &= -T_{\parallel} \cos \theta_t e^{-i\tau_t}, & E_y^{(t)} &= T_{\perp} e^{-i\tau_t}, & E_z^{(t)} &= T_{\parallel} \sin \theta_t e^{-i\tau_t}, \\ H_x^{(t)} &= -T_{\perp} \cos \theta_t \sqrt{\varepsilon_2} e^{-i\tau_t}, & H_y^{(t)} &= -T_{\parallel} \sqrt{\varepsilon_2} e^{-i\tau_t}, & H_z^{(t)} &= T_{\perp} \sin \theta_t \sqrt{\varepsilon_2} e^{-i\tau_t}, \end{aligned}\right\} \tag{14}$$

with

$$\tau_t = \omega \left(t - \frac{\boldsymbol{r} \cdot \boldsymbol{s}^{(t)}}{v_2} \right) = \omega \left(t - \frac{x \sin \theta_t + z \cos \theta_t}{v_2} \right). \tag{15}$$

Reflected field

$$\left.\begin{aligned} E_x^{(r)} &= -R_{\parallel} \cos \theta_r e^{-i\tau_r}, & E_y^{(r)} &= R_{\perp} e^{-i\tau_r}, & E_z^{(r)} &= R_{\parallel} \sin \theta_r e^{-i\tau_r}, \\ H_x^{(r)} &= -R_{\perp} \cos \theta_r \sqrt{\varepsilon_1} e^{-i\tau_r}, & H_y^{(r)} &= -R_{\parallel} \sqrt{\varepsilon_1} e^{-i\tau_r}, & H_z^{(r)} &= R_{\perp} \sin \theta_r \sqrt{\varepsilon_1} e^{-i\tau_r}, \end{aligned}\right\} \tag{16}$$

with

$$\tau_r = \omega \left(t - \frac{\boldsymbol{r} \cdot \boldsymbol{s}^{(r)}}{v_1} \right) = \omega \left(t - \frac{x \sin \theta_r + z \cos \theta_r}{v_1} \right). \tag{17}$$

The boundary conditions § 1.1 (23), (25) demand that across the boundary the tangential components of $\boldsymbol{E}$ and $\boldsymbol{H}$ should be continuous. Hence we must have

$$\left.\begin{aligned} E_x^{(i)} + E_x^{(r)} &= E_x^{(t)}, & E_y^{(i)} + E_y^{(r)} &= E_y^{(t)}, \\ H_x^{(i)} + H_x^{(r)} &= H_x^{(t)}, & H_y^{(i)} + H_y^{(r)} &= H_y^{(t)}, \end{aligned}\right\} \tag{18}$$

the conditions § 1.1 (15), and (19), being then automatically fulfilled for the normal components of $\boldsymbol{B}$ and $\boldsymbol{D}$. On substituting into (18) for all the components, and using the fact that $\cos \theta_r = \cos (\pi - \theta_i) = -\cos \theta_i$, we obtain the four relations

$$\left.\begin{aligned} \cos \theta_i (A_{\parallel} - R_{\parallel}) &= \cos \theta_t T_{\parallel}, \\ A_{\perp} + R_{\perp} &= T_{\perp}, \\ \sqrt{\varepsilon_1} \cos \theta_i (A_{\perp} - R_{\perp}) &= \sqrt{\varepsilon_2} \cos \theta_t T_{\perp}, \\ \sqrt{\varepsilon_1} (A_{\parallel} + R_{\parallel}) &= \sqrt{\varepsilon_2}\, T_{\parallel}. \end{aligned}\right\} \tag{19}$$

We note that the equations (19) fall into two groups, one of which contains only the components parallel to the plane of incidence, whilst the other contains only those which are perpendicular to the plane of incidence. *These two kinds of waves are, therefore, independent of one another.*

We can solve (19) for the components of the reflected and transmitted waves in terms of those of the incident wave, giving (using again the Maxwell relation $n = \sqrt{\varepsilon}$)

$$\left.\begin{aligned} T_\parallel &= \frac{2n_1 \cos\theta_i}{n_2\cos\theta_i + n_1\cos\theta_t} A_\parallel, \\ T_\perp &= \frac{2n_1\cos\theta_i}{n_1\cos\theta_i + n_2\cos\theta_t} A_\perp, \end{aligned}\right\} \tag{20}$$

$$\left.\begin{aligned} R_\parallel &= \frac{n_2\cos\theta_i - n_1\cos\theta_t}{n_2\cos\theta_i + n_1\cos\theta_t} A_\parallel, \\ R_\perp &= \frac{n_1\cos\theta_i - n_2\cos\theta_t}{n_1\cos\theta_i + n_2\cos\theta_t} A_\perp. \end{aligned}\right\} \tag{21}$$

Eqs. (20) and (21) are called *Fresnel formulae*, having first been derived in a slightly less general form by FRESNEL in 1823, on the basis of his elastic theory of light. They are usually written in the following alternative form, which may be obtained from (20) and (21) by using the law of refraction (8):

$$\left.\begin{aligned} T_\parallel &= \frac{2\sin\theta_t\cos\theta_i}{\sin(\theta_i+\theta_t)\cos(\theta_i-\theta_t)} A_\parallel, \\ T_\perp &= \frac{2\sin\theta_t\cos\theta_i}{\sin(\theta_i+\theta_t)} A_\perp, \end{aligned}\right\} \tag{20a}$$

$$\left.\begin{aligned} R_\parallel &= \frac{\tan(\theta_i-\theta_t)}{\tan(\theta_i+\theta_t)} A_\parallel, \\ R_\perp &= -\frac{\sin(\theta_i-\theta_t)}{\sin(\theta_i+\theta_t)} A_\perp. \end{aligned}\right\} \tag{21a}$$

Since θ_i and θ_t are real (the case of total reflection being excluded for the present), the trigonometrical factors on the right-hand side of (20a) and (21a) will also be real. Consequently the phase of each component of the reflected or transmitted wave is either equal to the phase of the corresponding component of the incident wave or differs from it by π. Since $T_\parallel$ and $T_\perp$ have the same signs as $A_\parallel$ and $A_\perp$, the phase of the transmitted wave is actually equal to that of the incident wave. In the case of the reflected wave, the phase will, however, depend on the relative magnitudes of θ_i and θ_t. For, if the second medium is optically denser than the first ($\varepsilon_2 > \varepsilon_1$), then $\theta_t < \theta_i$; according to (21), the signs of $R_\perp$ and $A_\perp$ are different and the phases therefore differ* by π. Under the same circumstances $\tan(\theta_i - \theta_t)$ is positive, but the denominator $\tan(\theta_i + \theta_t)$ becomes negative for $\theta_i + \theta_t > \pi/2$, and the phase of $R_\parallel$ and $A_\parallel$ then differ by π. Similar considerations apply when the second medium is optically less dense than the first.

For *normal incidence*, $\theta_i = 0$ and consequently $\theta_t = 0$, and (20) and (21) reduce to

* From (11) and (16) it follows that in the plane $z = 0$

$$\frac{E_y^{(r)}}{E_y^{(i)}} = \frac{R_\perp}{A_\perp}, \qquad \frac{E_x^{(i)}}{E_x^{(i)}} = -\frac{R_\parallel}{A_\parallel}.$$

This result implies, that in the plane $z = 0$, the phases of $E_y^{(r)}$ and $E_y^{(i)}$ differ by π, whereas the phases of $E_x^{(r)}$ and $E_x^{(i)}$ are equal to each other. This difference in the behaviour of the phases of the y- and x-components is rather formal, arising from the way in which the angle of reflection θ_r was defined (see Fig. 1.10).

$$\left.\begin{aligned} T_{\parallel} &= \frac{2}{n+1} A_{\parallel}, \\ T_{\perp} &= \frac{2}{n+1} A_{\perp}, \end{aligned}\right\} \tag{22}$$

$$\left.\begin{aligned} R_{\parallel} &= \frac{n-1}{n+1} A_{\parallel}, \\ R_{\perp} &= -\frac{n-1}{n+1} A_{\perp}, \end{aligned}\right\} \tag{23}$$

where $n = n_2/n_1$.

1.5.3 The reflectivity and transmissivity; polarization on reflection and refraction

We shall now examine how the energy of the incident field is divided between the two secondary fields.

According to § 1.4 (8), the light intensity is given (again assuming $\mu = 1$) by

$$S = \frac{c}{4\pi} \sqrt{\varepsilon}\, E^2 = \frac{cn}{4\pi} E^2. \tag{24}$$

The amount of energy in the primary wave which is incident on a unit area of the boundary per second is therefore

$$J^{(i)} = S^{(i)} \cos\theta_i = \frac{cn_1}{4\pi} |A|^2 \cos\theta_i, \tag{25}$$

and the energies of the reflected and transmitted wave leaving a unit area of the boundary per second are given by similar expressions:

$$\left.\begin{aligned} J^{(r)} &= S^{(r)} \cos\theta_i = \frac{cn_1}{4\pi} |R|^2 \cos\theta_i, \\ J^{(t)} &= S^{(t)} \cos\theta_t = \frac{cn_2}{4\pi} |T|^2 \cos\theta_t. \end{aligned}\right\} \tag{26}$$

The ratios

$$\mathscr{R} = \frac{J^{(r)}}{J^{(i)}} = \frac{|R|^2}{|A|^2} \quad \text{and} \quad \mathscr{T} = \frac{J^{(t)}}{J^{(i)}} = \frac{n_2 \cos\theta_t}{n_1 \cos\theta_i} \frac{|T|^2}{|A|^2} \tag{27}$$

are called the *reflectivity* and *transmissivity* respectively.* It can easily be verified that, in agreement with the law of conservation of energy,

$$\mathscr{R} + \mathscr{T} = 1. \tag{28}$$

The reflectivity and transmissivity depend on the polarization of the incident wave. They may be expressed in terms of the reflectivity and transmissivity associated with polarizations in the parallel and perpendicular directions respectively.

Let α_i be the angle which the $\boldsymbol{E}$ vector of the incident wave makes with the plane of incidence. Then

$$A_{\parallel} = A \cos\alpha_i, \qquad A_{\perp} = A \sin\alpha_i. \tag{29}$$

---

* If $\mu \neq 1$, the factor $\frac{n_2}{n_1}$ in the expression for $\mathscr{T}$ must be replaced by $\sqrt{\frac{\varepsilon_2}{\mu_2}}\Big/\sqrt{\frac{\varepsilon_1}{\mu_1}}$, as is immediately evident from § 1.4 (8).

Let

$$\left.\begin{aligned} J_{\parallel}^{(i)} &= \frac{cn_1}{4\pi}|A_{\parallel}|^2 \cos\theta_i = J^{(i)} \cos^2 \alpha_i, \\ J_{\perp}^{(i)} &= \frac{cn_1}{4\pi}|A_{\perp}|^2 \cos\theta_i = J^{(i)} \sin^2 \alpha_i, \end{aligned}\right\} \tag{30}$$

and

$$\left.\begin{aligned} J_{\parallel}^{(r)} &= \frac{cn_1}{4\pi}|R_{\parallel}|^2 \cos\theta_i, \\ J_{\perp}^{(r)} &= \frac{cn_1}{4\pi}|R_{\perp}|^2 \cos\theta_i. \end{aligned}\right\} \tag{31}$$

Then

$$\begin{aligned} \mathscr{R} = \frac{J^{(r)}}{J^{(i)}} &= \frac{J_{\parallel}^{(r)} + J_{\perp}^{(r)}}{J^{(i)}} = \frac{J_{\parallel}^{(r)}}{J_{\parallel}^{(i)}} \cos^2\alpha_i + \frac{J_{\perp}^{(r)}}{J_{\perp}^{(i)}} \sin^2\alpha_i \\ &= \mathscr{R}_{\parallel} \cos^2\alpha_i + \mathscr{R}_{\perp} \sin^2\alpha_i, \end{aligned} \tag{32}$$

where

$$\left.\begin{aligned} \mathscr{R}_{\parallel} &= \frac{J_{\parallel}^{(r)}}{J_{\parallel}^{(i)}} = \frac{|R_{\parallel}|^2}{|A_{\parallel}|^2} = \frac{\tan^2(\theta_i - \theta_t)}{\tan^2(\theta_i + \theta_t)}, \\ \mathscr{R}_{\perp} &= \frac{J_{\perp}^{(r)}}{J_{\perp}^{(i)}} = \frac{|R_{\perp}|^2}{|A_{\perp}|^2} = \frac{\sin^2(\theta_i - \theta_t)}{\sin^2(\theta_i + \theta_t)}. \end{aligned}\right\} \tag{33}$$

In a similar way, we obtain

$$\mathscr{T} = \frac{J^{(t)}}{J^{(i)}} = \mathscr{T}_{\parallel} \cos^2\alpha_i + \mathscr{T}_{\perp} \sin^2\alpha_i, \tag{34}$$

with

$$\left.\begin{aligned} \mathscr{T}_{\parallel} &= \frac{J_{\parallel}^{(t)}}{J_{\parallel}^{(i)}} = \frac{n_2}{n_1}\frac{\cos\theta_t}{\cos\theta_i}\frac{|T_{\parallel}|^2}{|A_{\parallel}|^2} = \frac{\sin 2\theta_i \sin 2\theta_t}{\sin^2(\theta_i + \theta_t)\cos^2(\theta_i - \theta_t)}, \\ \mathscr{T}_{\perp} &= \frac{J_{\perp}^{(t)}}{J_{\perp}^{(i)}} = \frac{n_2}{n_1}\frac{\cos\theta_t}{\cos\theta_i}\frac{|T_{\perp}|^2}{|A_{\perp}|^2} = \frac{\sin 2\theta_i \sin 2\theta_t}{\sin^2(\theta_i + \theta_t)}. \end{aligned}\right\} \tag{35}$$

Again we may verify that

$$\mathscr{R}_{\parallel} + \mathscr{T}_{\parallel} = 1, \qquad \mathscr{R}_{\perp} + \mathscr{T}_{\perp} = 1. \tag{36}$$

For *normal incidence* the distinction between the parallel and perpendicular components disappears, and one has from (22), (23) and (27)

$$\left.\begin{aligned} \mathscr{R} &= \left(\frac{n-1}{n+1}\right)^2, \\ \mathscr{T} &= \frac{4n}{(n+1)^2}. \end{aligned}\right\} \tag{37}$$

It is seen from (37) that

$$\lim_{n\to 1} \mathscr{R} = 0, \qquad \lim_{n\to 1} \mathscr{T} = 1. \tag{38}$$

Similar results also hold for the limiting values of $\mathscr{R}_{\parallel}$, $\mathscr{T}_{\parallel}$, and $\mathscr{R}_{\perp}$, $\mathscr{T}_{\perp}$, as can easily be seen from (33) and (35), making use of the fact that, according to the law of refraction, $\theta_t \to \theta_i$ as $n \to 1$. Hence the smaller the difference in the optical densities of the two media, the less energy is carried away by the reflected wave.

The denominators in (33) and (35) are finite, except when $\theta_i + \theta_t = \pi/2$. Then $\tan(\theta_i + \theta_t) = \infty$ and consequently $\mathscr{R}_\parallel = 0$. In this case (see Fig. 1.11) the reflected and transmitted rays are perpendicular to each other, and it follows from the law of refraction (since now $\sin\theta_t = \sin\{(\pi/2) - \theta_i\} = \cos\theta_i$) that

$$\tan\theta_i = n. \tag{39}$$

The angle θ_i given by (39) is called the *polarizing* or *Brewster angle*; its significance was noted first in 1815 by DAVID BREWSTER (1781–1868): *If light is incident under this angle, the electric vector of the reflected light has no component in the plane of incidence.*

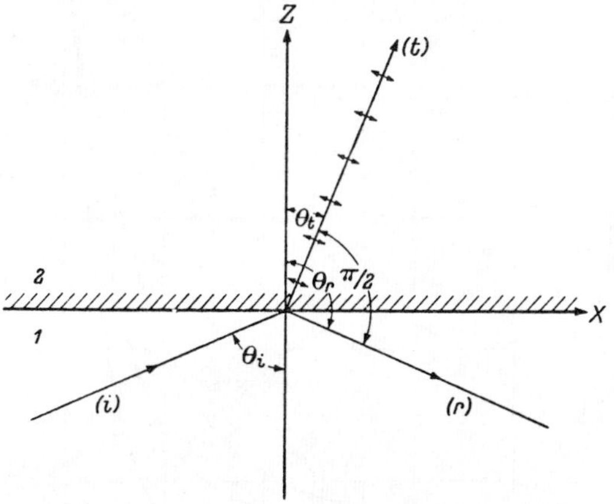

Fig. 1.11. Illustrating the polarizing (BREWSTER'S) angle.

One usually says that the light is then polarized "in the plane of incidence". According to this traditional terminology the plane of polarization is therefore the plane which contains the magnetic vector and the direction of propagation. For reasons already explained in § 1.4.2 it is, however, better not to use this term.

The above result, often called *Brewster's law*, can also be explained by the following, more direct argument: The incident field gives rise to vibrations of electrons in the atoms of the second medium. These vibrations are in the direction of the electric vector of the transmitted wave. The vibrating electrons give rise to another wave, the reflected wave, which will be propagated back into the first medium. Now a linearly vibrating electron radiates *transversally* (cf. § 2.2.3) so that there is no flux of radiant energy in the direction of vibration, and it follows that when the reflected and transmitted rays are at right angles to each other, the reflected ray does not receive any energy for oscillations in the plane of incidence.

In Fig. 1.12 the reflectivity for glass of refractive index 1·52 are plotted against the angle of incidence θ_i. The numbers along the upper horizontal refer to the angle of refraction θ_t. The zero value of the curve (a) (for $\mathscr{R}_\parallel$) corresponds to the polarizing angle $\tan^{-1} 1{\cdot}52 = 56° \, 40'$.

The refractive indices with respect to air are usually of the order of 1·5 at optical wavelengths, but at radio wavelengths they are much larger, there being a corresponding increase in the polarizing angle. For example, at optical wavelengths the refractive index of water is about 1·3 and the polarizing angle 53°. For radio wavelengths its value is about 9, the polarizing angle then being in the neighbourhood of 84°.

According to (32), the curve (b) in Fig. 1.12 is seen to correspond to $\alpha = 45°$. The same curve, as will now be shown, represents also the reflectivity $\overline{\mathscr{R}}$ for natural light, i.e. for light obtained from a body which is made to glow, by raising its temperature. The directions of vibration of natural light vary rapidly in a random, irregular manner. The corresponding reflectivity $\overline{\mathscr{R}}$ may be obtained by averaging over all

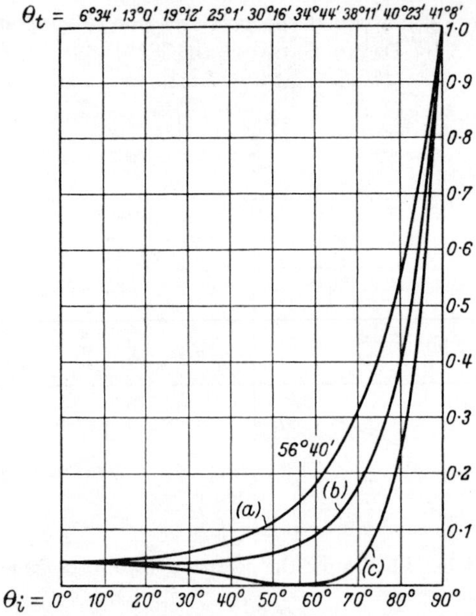

Fig. 1.12. Intensity of reflected light as a function of the angle of incidence.

(a) $\mathscr{R}_\perp$; (b) $\frac{1}{2}(\mathscr{R}_\parallel + \mathscr{R}_\perp)$; (c) $\mathscr{R}_\parallel$.

(After O. D. Chwolson, *Lehrb. d. Physik* (Braunschweig, Vieweg), Bd. II. 2, 2 Aufl. (1922), 716.)

directions. Since the averages of $\sin^2\alpha$ and $\cos^2\alpha$ are $\frac{1}{2}$, we obtain for the average values of $J_\parallel^{(i)}$ and $J_\perp^{(i)}$ the relations

$$\overline{J}_\parallel^{(i)} = \overline{J}_\perp^{(i)} = \tfrac{1}{2}J^{(i)}. \tag{40}$$

For the reflected light, however, the two components will in general differ from each other. For, using (40), one has

$$\left.\begin{aligned} \overline{J}_\parallel^{(r)} &= \frac{1}{2}\frac{\overline{J}_\parallel^{(r)}}{\overline{J}_\parallel^{(i)}}J^{(i)} = \tfrac{1}{2}\mathscr{R}_\parallel J^{(i)}, \\ \overline{J}_\perp^{(r)} &= \frac{1}{2}\frac{\overline{J}_\perp^{(r)}}{\overline{J}_\perp^{(i)}}J^{(i)} = \tfrac{1}{2}\mathscr{R}_\perp J^{(i)}. \end{aligned}\right\} \tag{41}$$

The reflected light is then said to be partially polarized and its *degree of polarization* P may be defined to be*

$$P = \left|\frac{\mathscr{R}_\parallel - \mathscr{R}_\perp}{\mathscr{R}_\parallel + \mathscr{R}_\perp}\right|. \tag{42}$$

* A more general definition of the *degree of polarization* is given in § 10.8.2, where its physical significance is also discussed.

The reflectivity $\overline{\mathscr{R}}$ is now given by

$$\overline{\mathscr{R}} = \frac{\overline{J}^{(r)}}{\overline{J}^{(i)}} = \frac{\overline{J}_{\parallel}^{(r)} + \overline{J}_{\perp}^{(r)}}{\overline{J}^{(i)}} = \tfrac{1}{2}(\mathscr{R}_{\parallel} + \mathscr{R}_{\perp}) \tag{43}$$

and is therefore again represented by the curve (b) in Fig. 1.12. The degree of polarization may now be expressed in the form

$$P = \frac{1}{\overline{\mathscr{R}}}\tfrac{1}{2}\{|\mathscr{R}_{\parallel} - \mathscr{R}_{\perp}|\};$$

the quantity in the brace brackets is sometimes called the *polarized proportion* of the light reflected.

Similar results hold for the transmitted light. We also have for natural light

$$\overline{\mathscr{R}} + \overline{\mathscr{T}} = 1. \tag{44}$$

Returning to the case where the incident light is linearly polarized we see that this is true also for the reflected and for the transmitted light, since the phases change only by 0 or π. The directions of vibration in the reflected and transmitted light as compared with the incident light are, however, turned in opposite directions. This can be seen from the following:

The angle which we denoted by α, i.e. the angle between the plane of vibration and the plane of incidence, may be called *the azimuth* of the vibration, and we shall regard it as positive when the plane of vibration turns clockwise around the direction of propagation (Fig. 1.13). It may be assumed that the azimuthal angle is in the range $-\pi/2$ to $\pi/2$. We have for the incident, reflected and transmitted electric wave

$$\tan \alpha_i = \frac{A_{\perp}}{A_{\parallel}}, \qquad \tan \alpha_r = \frac{R_{\perp}}{R_{\parallel}}, \qquad \tan \alpha_t = \frac{T_{\perp}}{T_{\parallel}}. \tag{45}$$

Using the FRESNEL formulae (20) and (21),

$$\tan \alpha_r = -\frac{\cos(\theta_i - \theta_t)}{\cos(\theta_i + \theta_t)}\tan \alpha_i, \tag{46}$$

$$\tan \alpha_t = \cos(\theta_i - \theta_t) \tan \alpha_i. \tag{47}$$

Since $0 \leqslant \theta_i \leqslant \pi/2, \quad 0 < \theta_t < \pi/2$,

$$|\tan \alpha_r| \geqslant |\tan \alpha_i|, \tag{48}$$

$$|\tan \alpha_t| \leqslant |\tan \alpha_i|. \tag{49}$$

In (48) the equality sign holds only for normal or tangential incidence ($\theta_i = \theta_t = 0$ or $\theta_i = \pi/2$); in (49) it holds only for normal incidence. The two inequalities imply that on reflection the plane of vibration is turned away from the plane of incidence, whereas on refraction it is turned towards it. In Fig. 1.14 the behaviour of α_r and α_t is illustrated for $n = 1{\cdot}52$ and $\alpha_i = 45°$. We see that, when θ_i is equal to the polarizing angle 56° 40′, $\alpha_r = 90°$. In fact, according to (46), $\tan \alpha_r = \infty$ (i.e. $\alpha_r = \pi/2$), for $\theta_i + \theta_t = \pi/2$, whatever the value of α_i may be.

It follows from BREWSTER's law that polarized light may be produced simply by allowing reflection to take place at the polarizing angle. One of the oldest instruments based on this principle is the so-called *Nörrenberg's Reflecting Polariscope* (after

NÖRRENBERG, 1787–1862). It consists essentially of two glassplates (Fig. 1.15) which the rays meet at the polarizing angle. The first plate plays the role of a *polarizer*, i.e. of an instrument producing linearly polarized light from unpolarized light; the second plays the role of an *analyzer*, i.e. an arrangement which detects linearly polarized light. This instrument has however several disadvantages, chiefly that the fraction of the light reflected at the polarizing angle is comparatively small, and the

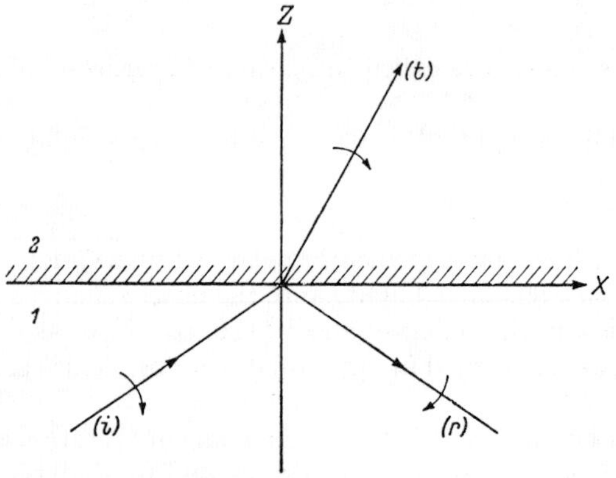

Fig. 1.13. Illustrating the signs of the azimuthal angles.

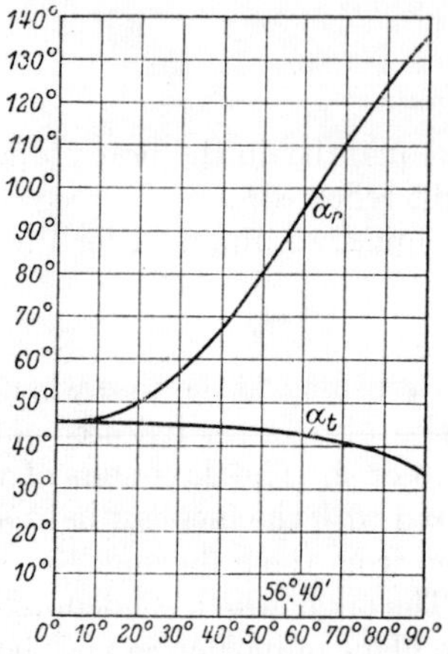

Fig. 1.14. Azimuthal angles as functions of the angle of incidence. (After O. D. CHWOLSON, *Lehrb. d. Physik*, (Braunschweig, Vieweg) Bd. II. 2, 2 Aufl. (1922), 716.)

path of the ray through the instrument is rather complicated. It is preferable to employ instruments which polarize the incident light without changing its direction of propagation. This can be done, for example, by using a *pile of thin plane-parallel plates*. If a beam of unpolarized light is incident on the pile, it is partially polarized on each refraction, and one can achieve a reasonably high degree of polarization even

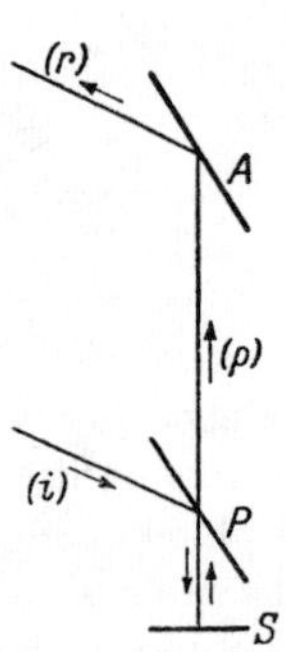

Fig. 1.15. Principle of the *Nörrenberg's Reflecting Polariscope.*

P = polarizing glass plate
S = reflecting mirror
A = analyser
i = incident beam
p = polarized beam
r = beam reflected from A.

with a small number of plates. After passing through both boundaries of a plate the intensities of the two components will be in the ratio

$$\left(\frac{\mathcal{T}_\perp}{\mathcal{T}_\parallel}\right)^2 = \cos^4(\theta_i - \theta_t) < 1, \tag{50}$$

a result obtained by applying (35) twice in succession. This shows that, on emerging from the plate, the parallel component is stronger than the perpendicular component, the degree of polarization being greater, the greater θ_i is. If θ_i is equal to the polarizing angle, $\theta_i + \theta_t = \pi/2$, $\tan\theta_i = n$, and we then have

$$\left(\frac{\mathscr{T}_\perp}{\mathscr{T}_\parallel}\right)^2 = \sin^4 2\theta_i = \left(\frac{2n}{1+n^2}\right)^4. \tag{51}$$

For $n = 1{\cdot}5$ this expression has the value 0·73. Hence, if the light passes through five plates, for example, we obtain the ratio $0{\cdot}73^5 \simeq 0{\cdot}2$.

In the past, polarized light was as a rule produced by double refraction in crystals of calcite or quartz, as described in § 14.4.1. To-day the most convenient method is by the use of so-called *polaroid* films. Their action is based on a property known as *dichroism*. By this we mean the property shown by certain materials of having different absorption coefficients for light polarized in different directions. Polyvinyl alcohol films impregnated by iodine, for example, can be made which transmit nearly 80 per cent of light polarized in one plane, and less than 1 per cent of light polarized at right angles to this plane. The theory of this effect will be briefly discussed in § 14.6.3.

1.5.4 Total reflection

So far we have excluded the case when the law of refraction

$$\sin\theta_t = \frac{\sin\theta_i}{n_{12}} \tag{52}$$

does not give a real value for the angle of refraction θ_t. We shall now examine this case. It occurs when light is propagated from an optically denser medium into one which is optically less dense, i.e. when

$$n_{12} = \frac{n_2}{n_1} = \sqrt{\frac{\varepsilon_2\mu_2}{\varepsilon_1\mu_1}} < 1,$$

provided that the angle of incidence θ_i exceeds the critical value $\bar{\theta}_i$ given by

$$\sin\bar{\theta}_i = n_{12}. \tag{53}$$

When $\theta_i = \bar{\theta}_i$, $\sin\theta_t = 1$, i.e. $\theta_t = 90°$, so that the light emerges in a direction tangent to the boundary. If θ_i exceeds the limiting value $\bar{\theta}_i$, no light enters the second medium. All the incident light is then reflected back into the first medium and we speak of *total reflection.*

Nevertheless the electromagnetic field in the second medium does not disappear, only there is no longer a flow of energy across the boundary. For, if (omitting the subscript 12 on n_{12}) we set

$$\sin\theta_t = \frac{\sin\theta_i}{n}, \qquad \cos\theta_t = \pm i\sqrt{\frac{\sin^2\theta_i}{n^2} - 1}, \tag{54}$$

in the phase factor (15) of the transmitted wave, we have

$$e^{-i\tau_t} = e^{-i\omega\left(t - \frac{x\sin\theta_i}{nv_2}\right)} e^{\mp\frac{\omega z}{v_2}\sqrt{\frac{\sin^2\theta_i}{n^2}-1}}. \tag{55}$$

(55) represents an inhomogeneous wave which is propagated along the boundary in the plane of incidence (i.e. in the x direction), and which varies exponentially with the distance z from the boundary. Naturally only the negative sign in front of the square

root in (55) corresponds to the physical situation, since otherwise the amplitude would tend to infinity with increasing distance. The amplitude is seen to decrease very rapidly with the depth z of penetration, the effective depth of penetration being of the order of $v_2/\omega = \lambda_2/2\pi$, i.e. of the order of a wavelength. The wave is not transversal since, as will be shown below, the component of the electric vector in the direction of propagation does not vanish.

Experimental verification of the disturbance in the second (less dense) medium is somewhat troublesome, since any arrangement used for its detection will perturb the boundary conditions. A rough confirmation may be obtained by bringing up a second refracting medium within about a quarter of a wavelength of the interface at which total reflection is taking place, and observing the entry of the radiation into the second medium.*

To apply the FRESNEL formulae (21a) to the case of total reflection, we rewrite them in the form

$$\left.\begin{aligned} R_{\parallel} &= \frac{\sin\theta_i\cos\theta_i - \sin\theta_t\cos\theta_t}{\sin\theta_i\cos\theta_i + \sin\theta_t\cos\theta_t} A_{\parallel}, \\ R_{\perp} &= -\frac{\sin\theta_i\cos\theta_t - \sin\theta_t\cos\theta_i}{\sin\theta_i\cos\theta_t + \sin\theta_t\cos\theta_i} A_{\perp}, \end{aligned}\right\} \qquad (56)$$

and substitute into these expressions from (54), remembering that the upper sign is to be taken in front of the square root. We then obtain

$$\left.\begin{aligned} R_{\parallel} &= \frac{n^2\cos\theta_i - i\sqrt{\sin^2\theta_i - n^2}}{n^2\cos\theta_i + i\sqrt{\sin^2\theta_i - n^2}} A_{\parallel}, \\ R_{\perp} &= \frac{\cos\theta_i - i\sqrt{\sin^2\theta_i - n^2}}{\cos\theta_i + i\sqrt{\sin^2\theta_i - n^2}} A_{\perp}. \end{aligned}\right\} \qquad (57)$$

Hence

$$|R_{\parallel}| = |A_{\parallel}|, \qquad |R_{\perp}| = |A_{\perp}|, \qquad (58)$$

i.e. for each component, the intensity of the light which is totally reflected is equal to the intensity of the incident light.

Although there is a field in the second medium, it is easy to see that no energy flows across the boundary. More precisely it will be shown that, although the component of the Poynting vector in the direction normal to the boundary is in general finite, its time average vanishes; this implies that the energy flows to and fro, but that there is no lasting flow into the second medium.

We write down the x and y components of the transmitted field, for $z = 0$, and make use of (54). (Real expressions must now be used for $\boldsymbol{E}$ and $\boldsymbol{H}$ since the energy flow is a quadratic function of the components). Denoting the conjugate of a complex quantity by an asterisk, we have from (14),

$$\begin{aligned} E_x^{(t)} &= -\tfrac{1}{2}[T_{\parallel}\cos\theta_t\, e^{-i\tau_t^o} + T_{\parallel}^{\star}\cos^{\star}\theta_t\, e^{+i\tau_t^o}] \\ &= -\frac{i}{2}\sqrt{\frac{\sin^2\theta_i}{n^2} - 1}\left[T_{\parallel}e^{-i\tau_t^o} - T_{\parallel}^{\star}e^{+i\tau_t^o}\right], \\ E_y^{(t)} &= \tfrac{1}{2}[T_{\perp}e^{-i\tau_t^o} + T_{\perp}^{\star}e^{+i\tau_t^o}], \end{aligned}$$

* An elegant way of doing this was described by W. CULSHAW and D. S. JONES, *Proc. Phys. Soc.*, B, **66** (1954), 859, using electrically generated waves of wavelength 1·25 cm.

$$H_x^{(t)} = -\tfrac{1}{2}[T_\perp \cos\theta_t \sqrt{\varepsilon_2}\, e^{-i\tau_t^o} + T_\perp^\star \cos^\star\theta_t \sqrt{\varepsilon_2}\, e^{+i\tau_t^o}]$$
$$= -\frac{i}{2}\sqrt{\varepsilon_2}\sqrt{\frac{\sin^2\theta_i}{n^2}-1}\left[T_\perp e^{-i\tau_t^o} - T_\perp^\star e^{+i\tau_t^o}\right],$$
$$H_y^{(t)} = -\tfrac{1}{2}\sqrt{\varepsilon_2}[T_\parallel e^{-i\tau_t^o} + T_\parallel^\star e^{+i\tau_t^o}],$$

where

$$\tau_t^o = \omega\left[t - \frac{x\sin\theta_i}{nv_2}\right].$$

If we now form the time average of

$$S_z^{(t)} = \frac{c}{4\pi}[E_x^{(t)}H_y^{(t)} - E_y^{(t)}H_x^{(t)}]$$

over an interval $-t' \leqslant t \leqslant t'$, where t' is large compared with the period $\mathrm{T} = 2\pi/\omega$, then both terms disappear for $z = 0$; for one contains the factor

$$\frac{1}{2t'}\int_{-t'}^{t'}[T_\parallel^2 e^{-2i\tau_t^o} - T_\parallel^{\star 2} e^{+2i\tau_t^o}]dt = [T_\parallel^2 e^{+\frac{2i\omega x\sin\theta_i}{nv_2}} - T_\parallel^{\star 2} e^{-\frac{2i\omega x\sin\theta_i}{nv_2}}]\,O\left(\frac{\mathrm{T}}{t'}\right),$$

which is negligibly small when $t' \gg \mathrm{T}$; the other contains a similar factor with $T_\perp$ in place of $T_\parallel$.

On the other hand, if the other two components of $\mathbf{S}^{(t)}$ for $z = 0$ are calculated, namely $S_x^{(t)}$ and $S_y^{(t)}$, their time averaged values are in general found to be finite. Energy therefore does not penetrate into the second medium, but flows along the boundary in the plane of incidence.

The preceding analysis applies to a stationary state, and is based on the assumption that the boundary surface and the wave-fronts are of infinite extent. It does not explain how the energy initially entered the second medium. In an actual experiment, the incident wave will be bounded both in space and time*; at the beginning of the process a small amount of energy will penetrate into the second medium and will give rise to a field there.

Finally, we determine the changes in the phases of the components of the reflected and the incident wave. On account of (58) we may set

$$\frac{R_\parallel}{A_\parallel} = e^{i\delta_\parallel}, \qquad \frac{R_\perp}{A_\perp} = e^{i\delta_\perp}. \tag{59}$$

Now according to (57), $\dfrac{R_\parallel}{A_\parallel}$ and $\dfrac{R_\perp}{A_\perp}$ are each of the form $z(z^\star)^{-1}$. Hence if α is the argument of z (i.e. $z = ae^{i\alpha}$, with a, α both real), then

$$e^{i\delta} = z(z^\star)^{-1} = e^{2i\alpha}, \qquad \text{i.e.}\quad \tan\frac{\delta}{2} = \tan\alpha,$$

and therefore

$$\left.\begin{aligned}\tan\frac{\delta_\parallel}{2} &= -\frac{\sqrt{\sin^2\theta_i - n^2}}{n^2\cos\theta_i},\\ \tan\frac{\delta_\perp}{2} &= -\frac{\sqrt{\sin^2\theta_i - n^2}}{\cos\theta_i}.\end{aligned}\right\} \tag{60}$$

* The total reflection of a beam of light of finite cross-section has been treated by J. Picht, *Ann. d. Physik* (5), **3** (1929), 433.

The two components are seen to undergo phase jumps of different amounts. Linearly polarized light will in consequence become elliptically polarized on total reflection.

One can also immediately write down an expression for the relative phase difference $\delta = \delta_\perp - \delta_\parallel$:

$$\tan\frac{\delta}{2} = \frac{\tan\frac{\delta_\perp}{2} - \tan\frac{\delta_\parallel}{2}}{1 + \tan\frac{\delta_\perp}{2}\tan\frac{\delta_\parallel}{2}} = \frac{\left(\frac{1}{n^2} - 1\right)\frac{\sqrt{\sin^2\theta_i - n^2}}{\cos\theta_i}}{1 + \frac{\sin^2\theta_i - n^2}{n^2\cos^2\theta_i}},$$

i.e.

$$\tan\frac{\delta}{2} = \frac{\cos\theta_i\sqrt{\sin^2\theta_i - n^2}}{\sin^2\theta_i}. \tag{61}$$

This expression vanishes for grazing incidence ($\theta_i = \pi/2$), and for incidence at the critical angle $\bar{\theta}_i$ ($\sin\bar{\theta}_i = n$). Between these two values there lies the maximum value of the relative phase difference; it is determined from the equation

$$\frac{d}{d\theta_i}(\tan\delta/2) = \frac{2n^2 - (1 + n^2)\sin^2\theta_i}{\sin^3\theta_i\sqrt{\sin^2\theta_i - n^2}} = 0.$$

This is satisfied when

$$\sin^2\theta_i = \frac{2n^2}{1 + n^2}. \tag{62}$$

On substituting from (62) into (61), we obtain for the maximum δ_m of the relative phase difference δ, the expression

$$\tan\frac{\delta_m}{2} = \frac{1 - n^2}{2n}. \tag{63}$$

From (61) it is seen that, with n given, there are two values of the angle θ_i of incidence for every value of δ.

The phase change which takes place on total reflection may be used (as shown already by FRESNEL) to produce circularly polarized light from light which is linearly polarized. The amplitude components of the incident light are made equal ($|A_\parallel| = |A_\perp|$), by taking the incident wave to be polarized in a direction which makes an angle of 45° with the normal to the plane of incidence (i.e. $\alpha_i = 45°$). Then, by (58), $|R_\parallel| = |R_\perp|$. Further, n and θ_i are chosen in such a way that the relative phase difference δ is equal to 90°. To attain this value of δ by a single reflection, it would be necessary, according to (63), that

$$\tan\frac{\pi}{4} = 1 < \frac{1 - n^2}{2n},$$

i.e.

$$n = n_{12} < \sqrt{2} - 1 = 0{\cdot}414.$$

This means that the refractive index $n_{21} = 1/n$ of the denser with respect to the less dense medium would have to be at least 2·41. This value is rather large, although there are non-absorbing substances whose refractive index comes close to, and even exceeds, this value. FRESNEL made use of two total reflections on glass. When $n_{21} = 1{\cdot}51$, one obtains, according to (62) and (63), a maximum relative phase difference $\delta_m = 45°\,56'$ when the angle of incidence θ_i equals to 51° 20′. It is therefore just possible to attain the value $\delta = 45°$, namely, with either of the following angles of incidence:

$$\theta_i = 48°\,37', \qquad \theta_i = 54°\,37'.$$

A phase difference of 90° may therefore be obtained by means of two successive total reflections at either of these angles. For this purpose a glass block is used, of the form shown in Fig. 1.16, known as *Fresnel's rhomb.*

FRESNEL's rhomb may, of course, be also used to produce elliptically polarized light; the azimuth of the incident (linearly polarized) light must then be taken

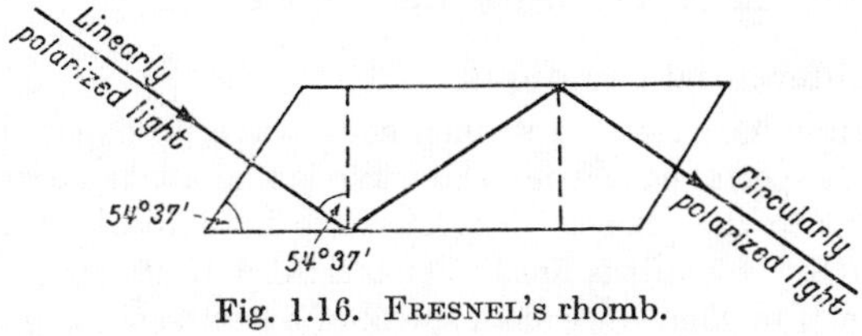

Fig. 1.16. FRESNEL's rhomb.

different from 45°. One may also invert the procedure and produce, by means of FRESNEL's rhomb, linearly polarized light from elliptically polarized light.

Measurement of the critical angle $\bar{\theta}_i$ gives a convenient and accurate way of determining the index of refraction $n = \sin \bar{\theta}_i$. Instruments used for this purpose are called *refractometers.*

1.6 WAVE PROPAGATION IN A STRATIFIED MEDIUM. THEORY OF DIELECTRIC FILMS

A medium whose properties are constant throughout each plane perpendicular to a fixed direction is called a *stratified medium.* If the z-axis of a Cartesian reference system is taken along this special direction, then

$$\varepsilon = \varepsilon(z), \qquad \mu = \mu(z). \tag{1}$$

We shall consider the propagation of a plane, time-harmonic electromagnetic wave through such a medium; this is a natural generalization of the simple case treated in the previous section.

The theory of stratified media is of considerable importance in optics, in connection with *multilayers,* i.e. a succession of thin plane-parallel films. Such films may be produced with the help of high-vacuum evaporation techniques, and their thickness may be controlled with very high accuracy. They have many useful applications. For example, as will be demonstrated later, they may be employed as *antireflection films,* i.e. as coatings which reduce the reflectivity of a given surface. On the other hand thin films will, under appropriate conditions, *enhance* the reflectivity so that when deposited on a glass surface they may be used as beam-splitters, i.e. arrangements employed in interferometry for the division of an incident beam into two parts. Under appropriate conditions a multilayer may also be employed as a filter which transmits (or reflects) only selected regions of the spectrum. Multilayers may also be used as polarizers.

The subject of dielectric and metallic films has been very extensively treated in the scientific literature and many schemes for the computation of the optical effects of multilayers have been proposed. We shall give an outline of the general theory as developed in elegant and important investigations by F. ABELÈS,* and consider in

* F. ABELÈS, *Ann. de Physique,* **5** (1950), 596–640 and 706–782. For a detailed treatment of the subject of thin films see a more specialized treatise e.g. H. MAYER, *Physik dünner Schichten* (Stuttgart, Wissenschaftliche Verlagsgessellschaft, 1950); S. METHFESSEL, *Dünne Schichten* (Halle (Saale), VEB Wilhelm Knapp Verlag, 1953); or O. S. HEAVENS, *Optical Properties of Thin Solid Films* (London, Butterworths Scientific Publications, 1955).

detail some special cases of particular interest. For the treatment of problems involving only a small number of films it is naturally not necessary to use the general theory, and accordingly we shall later (§ 7.6) describe an alternative and older method based on the concept of multiple reflections.

Only dielectric stratified media will be treated in this section. The extension of the analysis to conducting media will be described in § 13.4.

1.6.1 The basic differential equations

Consider a plane, time-harmonic electromagnetic wave propagated through a stratified medium. In the special case when the wave is linearly polarized with its electric vector perpendicular to the plane of incidence we shall speak of a *transverse electric wave* (denoted by TE); when it is linearly polarized with its magnetic vector perpendicular to the plane of incidence we shall speak of a *transverse magnetic wave* (denoted by TM).* Any arbitrarily polarized plane wave may be resolved into two waves, one of which is a TE wave and the other a TM wave. Since according to § 1.5 the boundary conditions at a discontinuity surface for the perpendicular and parallel components are independent of each other, these two waves will also be mutually independent. Moreover, MAXWELL's equations remain unchanged when E and H and simultaneously ε and $-\mu$ are interchanged. Thus any theorem relating to TM waves may immediately be deduced from the corresponding result for TE waves by making this change. It will, therefore, be sufficient to study in detail the TE waves only.

We take the plane of incidence to be the yz-plane,† z being the direction of stratification. For a TE wave, $E_y = E_z = 0$ and MAXWELL's equations reduce to the following six scalar equations [time dependence $\exp(-i\omega t)$ being assumed]:

$$\frac{\partial H_z}{\partial y} - \frac{\partial H_y}{\partial z} + \frac{i\varepsilon\omega}{c} E_x = 0, \quad (1a) \qquad \frac{i\omega\mu}{c} H_x = 0, \quad (2a)$$

$$\frac{\partial H_x}{\partial z} - \frac{\partial H_z}{\partial x} = 0, \quad (1b) \qquad \frac{\partial E_x}{\partial z} - \frac{i\omega\mu}{c} H_y = 0, \quad (2b)$$

$$\frac{\partial H_y}{\partial x} - \frac{\partial H_x}{\partial y} = 0, \quad (1c) \qquad \frac{\partial E_x}{\partial y} + \frac{i\omega\mu}{c} H_z = 0. \quad (2c)$$

These equations show that H_y, H_z and E_x are functions of y and z only. Eliminating H_y and H_z between (1a), (2b) and (2c) [or by taking the x component of the wave equation § 1.2 (5) for **E**] it follows that

$$\frac{\partial^2 E_x}{\partial y^2} + \frac{\partial^2 E_x}{\partial z^2} + n^2 k_0^2 E_x = \frac{d(\log\mu)}{dz}\frac{\partial E_x}{\partial z}, \quad (3)$$

where

$$n^2 = \varepsilon\mu, \qquad k_0 = \frac{\omega}{c} = \frac{2\pi}{\lambda_0}. \quad (4)$$

To solve (3) we take, as a trial solution, a product of two functions, one involving y only and the other involving z only:

$$E_x(y, z) = Y(y)U(z). \quad (5)$$

---

* The terms "E-polarized" and "H-polarized" are also used (cf. § 11.4.1). It should be mentioned that the terms "transverse electric wave" and "transverse magnetic wave" have different meanings in the theory of wave guides.

† Not the xz-plane as in the previous section.

Eq. (3) then becomes

$$\frac{1}{Y}\frac{d^2 Y}{dy^2} = -\frac{1}{U}\frac{d^2 U}{dz^2} - n^2 k_0^2 + \frac{d\,(\log \mu)}{dz}\frac{1}{U}\frac{dU}{dz}. \tag{6}$$

Now the term on the left is a function of y only whilst the terms on the right depend only on z. Hence (6) can only hold if each side is equal to a constant ($-K^2$ say):

$$\frac{1}{Y}\frac{d^2 Y}{dy^2} = -K^2, \tag{7}$$

$$\frac{d^2 U}{dz^2} - \frac{d\,(\log \mu)}{dz}\frac{dU}{dz} + n^2 k_0^2 U = K^2 U. \tag{8}$$

It will be convenient to set

$$K^2 = k_0^2 \alpha^2. \tag{9}$$

Then (7) gives

$$Y = \text{const.}\, e^{ik_0 \alpha y},$$

and consequently E_x is of the form

$$E_x = U(z) e^{i(k_0 \alpha y - \omega t)}, \tag{10}$$

where $U(z)$ is a (possibly complex) function of z. From (2b) and (2c) we see that H_y and H_z are given by expressions of the same form:

$$H_y = V(z) e^{i(k_0 \alpha y - \omega t)}, \tag{11}$$

$$H_z = W(z) e^{i(k_0 \alpha y - \omega t)}. \tag{12}$$

On account of (1a), (2b) and (2c), the amplitude functions U, V and W are related by the following equations:

$$V' = ik_0[\alpha W + \varepsilon U], \tag{13a}$$

$$U' = ik_0 \mu V, \tag{13b}$$

$$\alpha U + \mu W = 0, \tag{13c}$$

the prime denoting differentiation with respect to z. Substituting for W from (13c) into (13a) we have, together with (13b), a pair of simultaneous first-order differential equations* for U and V:

$$\left.\begin{aligned} U' &= ik_0 \mu V, \\ V' &= ik_0 \left(\varepsilon - \frac{\alpha^2}{\mu}\right) U. \end{aligned}\right\} \tag{14}$$

* Eqs. (14) are of the same form as the equations of an electric transmission line, i.e.

$$\frac{dV}{dz} = -ZI, \qquad \frac{dI}{dz} = -YV,$$

where V is the voltage across the line, I is the current in the line, Z is the series impedance, and Y the shunt admittance. The theory of stratified media may therefore be developed in a strict analogy with the theory of electric transmission lines, as has been done by several authors, for example, R. B. Muchmore, *J. Opt. Soc. Amer.*, **38** (1948), 20; K. Schuster, *Ann. d. Physik* (6), **4** (1949), 352; R. Kronig, R. S. Blaisse and J. J. v.v. Sande, *J. Appl. Sci. Res.*, **B,1** (1947), 63.

Elimination between these equations finally gives the following second-order linear differential equations for U and V:

$$\frac{d^2U}{dz^2} - \frac{d\,(\log \mu)}{dz}\frac{dU}{dz} + k_0^2(n^2 - \alpha^2)U = 0, \tag{15}$$

$$\frac{d^2V}{dz^2} - \frac{d\left[\log\left(\varepsilon - \frac{\alpha^2}{\mu}\right)\right]}{dz}\frac{dV}{dz} + k_0^2(n^2 - \alpha^2)V = 0. \tag{16}$$

According to the substitution rule which is a consequence of the symmetry of MAXWELL's equations, it immediately follows that *for the TM wave* ($H_y = H_z = 0$), the non-vanishing components of the field vectors are of the form:

$$H_x = \quad U(z)e^{i(k_0\alpha y - \omega t)}, \tag{17}$$

$$E_y = -\ V(z)e^{i(k_0\alpha y - \omega t)}, \tag{18}$$

$$E_z = -\ W(z)e^{i(k_0\alpha y - \omega t)} \tag{19}$$

where

$$\left.\begin{aligned} U' &= ik_0\varepsilon V, \\ V' &= ik_0\left(\mu - \frac{\alpha^2}{\varepsilon}\right)U, \end{aligned}\right\} \tag{20}$$

and W is related to U by means of the equation

$$\alpha U + \varepsilon W = 0. \tag{21}$$

U and V now satisfy the following second-order linear differential equations:

$$\frac{d^2U}{dz^2} - \frac{d\{\log \varepsilon\}}{dz}\frac{dU}{dz} + k_0^2(n^2 - \alpha^2)U = 0, \tag{22}$$

$$\frac{d^2V}{dz^2} - \frac{d\left\{\log\left(\mu - \frac{\alpha^2}{\varepsilon}\right)\right\}}{dz}\frac{dV}{dz} + k_0^2(n^2 - \alpha^2)V = 0. \tag{23}$$

U, V and W are in general complex functions of z. The surfaces of constant amplitude of E_x are given by

$$|U(z)| = \text{constant},$$

whilst the surfaces of constant phase have the equation

$$\phi(z) + k_0\alpha y = \text{constant},$$

where $\phi(z)$ is the phase of U. The two sets of surfaces do not in general coincide so that E_x (and similarly H_y and H_z) is an inhomogeneous wave. For a small displacement (dy, dz) along a co-phasal surface, $\phi'(z)dz + k_0\alpha dy = 0$; hence if θ denotes the angle which the normal to the co-phasal surface makes with OZ, then

$$\tan\theta = -\frac{dz}{dy} = \frac{k_0\alpha}{\phi'(z)}.$$

In the special case when the wave is an homogeneous plane wave,

$$\phi(z) = k_0 nz \cos\theta, \qquad \alpha = n \sin\theta. \tag{24}$$

Hence the relation

$$\alpha = \text{constant}$$

imposed by (9) may be regarded as a generalization of *Snell's law of refraction* to stratified media.

1.6.2 The characteristic matrix of a stratified medium

The solutions, subject to appropriate boundary conditions, of the differential equations which we have just derived, and various theorems relating to stratified media, can most conveniently be expressed in terms of matrices. We shall therefore give a brief account of the main definitions relating to matrices before discussing the consequences of our equations.

I. By a matrix one understands a system of real or complex numbers, arranged in a rectangular or a square array:

$$\begin{bmatrix} a_{11} & a_{12} & \cdots & a_{1n} \\ a_{21} & a_{22} & \cdots & a_{2n} \\ \cdot & \cdot & \cdots & \cdot \\ \cdot & \cdot & \cdots & \cdot \\ a_{m1} & a_{m2} & \cdots & a_{mn} \end{bmatrix},$$

a_{ij} denoting the element in the ith row and the jth column. The matrix is denoted symbolically by $\mathbf{A}$ or $[a_{ij}]$, and is said to be an m by n matrix (or $m \times n$ matrix), since it contains m rows and n columns. In the special case when $m = n$, $\mathbf{A}$ is said to be a *square matrix* of order m. If $\mathbf{A}$ is a square matrix, the determinant whose elements are the same, and are in the same positions as the elements of $\mathbf{A}$, is said to be the *determinant of the matrix* $\mathbf{A}$; it is denoted by $|\mathbf{A}|$ or $|a_{ij}|$. If $|\mathbf{A}| = 1$, $\mathbf{A}$ is said to be *unimodular*.

By definition two matrices are *equal* only if they have the same number of rows (m) and the same number of columns (n), and if their corresponding elements are equal. If $\mathbf{A} = [a_{ij}]$ and $\mathbf{B} = [b_{ij}]$ are two matrices with the same number of rows and the same number of columns, then their *sum* $\mathbf{A} + \mathbf{B}$ is defined as the matrix $\mathbf{C}$ whose elements are $c_{ij} = a_{ij} + b_{ij}$. Similarly their *difference* $\mathbf{A} - \mathbf{B}$ is defined as the matrix $\mathbf{D}$ with elements $d_{ij} = a_{ij} - b_{ij}$.

A matrix having every element zero is called a *null matrix*. The square matrix with elements $a_{ij} = 0$ when $i \neq j$ and $a_{ii} = 1$ for every value of i is called *unit matrix* and will be denoted by $\mathbf{I}$.

The product of a matrix $\mathbf{A}$ and a number λ (real or complex) is defined as the matrix $\mathbf{B}$ with elements $b_{ij} = \lambda a_{ij}$.

The product $\mathbf{AB}$ of two matrices is defined only when the number of columns in $\mathbf{A}$ is equal to the number of rows in $\mathbf{B}$. If $\mathbf{A}$ is a $m \times p$ matrix and $\mathbf{B}$ is a $p \times n$ matrix the product is then by definition the $m \times n$ matrix with elements

$$c_{ij} = \sum_{k=1}^{p} a_{ik} b_{kj}.$$

The process of multiplication of two matrices is thus analogous to the row-by-column rule for multiplication of determinants of equal orders. In general $\mathbf{AB} \neq \mathbf{BA}$. For example

$$\begin{bmatrix} 0 & 1 \\ 1 & 0 \end{bmatrix} \begin{bmatrix} 1 & 0 \\ 0 & -1 \end{bmatrix} = \begin{bmatrix} 0 & -1 \\ 1 & 0 \end{bmatrix},$$

whilst

$$\begin{bmatrix} 1 & 0 \\ 0 & -1 \end{bmatrix} \begin{bmatrix} 0 & 1 \\ 1 & 0 \end{bmatrix} = \begin{bmatrix} 0 & 1 \\ -1 & 0 \end{bmatrix}.$$

In the special case when $\mathbf{AB} = \mathbf{BA}$, the matrices $\mathbf{A}$ and $\mathbf{B}$ are said to *commute.*

The above definitions and properties of matrices are the only ones necessary for our purposes, and we can, therefore, now return to our discussion of propagation of electromagnetic waves through a stratified medium.

II. Since the functions $U(z)$ and $V(z)$ of § 1.6.1 each satisfy a second-order linear differential equation [(15) and (16)], it follows that U and V may each be expressed as a linear combination of two particular solutions, say U_1, U_2 and V_1, V_2. These particular solutions cannot be arbitrary; they must be coupled by the first order differential equations (14):

$$\left.\begin{aligned} U_1' &= ik_0\mu V_1, \\ V_1' &= ik_0\left(\varepsilon - \frac{\alpha^2}{\mu}\right) U_1, \end{aligned}\right\} \qquad \left.\begin{aligned} U_2' &= ik_0\mu V_2, \\ V_2' &= ik_0\left(\varepsilon - \frac{\alpha^2}{\mu}\right) U_2. \end{aligned}\right\} \tag{25}$$

From these relations it follows that

$$V_1 U_2' - U_1' V_2 = 0, \qquad U_1 V_2' - V_1' U_2 = 0,$$

so that

$$\frac{d}{dz}(U_1 V_2 - U_2 V_1) = 0.$$

This relation implies that *the determinant*

$$D = \begin{vmatrix} U_1 & V_1 \\ U_2 & V_2 \end{vmatrix} \tag{26}$$

associated with any two arbitrary solutions of (14) is a constant, i.e. that D is an invariant of our system of equations.*

For our purposes the most convenient choice of the particular solutions is

$$\left.\begin{aligned} U_1 &= f(z), \qquad & U_2 &= F(z), \\ V_1 &= g(z), \qquad & V_2 &= G(z), \end{aligned}\right\} \tag{27}$$

such that

$$f(0) = G(0) = 0 \quad \text{and} \quad F(0) = g(0) = 1. \tag{28}$$

* This also follows from a well-known property of a Wronskian of second-order linear differential equations. Moreover it may also be shown that, if U_1 is known, U_2 may be obtained by integration from the relation

$$U_2 = ikDU_1 \int \frac{\mu}{U_1^{\,2}}\, dz.$$

Cf. F. Abelès, *Ann. de Physique*, **5** (1950), 603.

Then the solutions with

$$U(0) = U_0, \qquad V(0) = V_0, \tag{29}$$

may be expressed in the form

$$\left.\begin{aligned} U &= FU_0 + fV_0, \\ V &= GU_0 + gV_0, \end{aligned}\right\}$$

or, in matrix notation,

$$\mathbf{Q} = \mathbf{N}\mathbf{Q}_0, \tag{30}$$

where

$$\mathbf{Q} = \begin{bmatrix} U(z) \\ V(z) \end{bmatrix}, \quad \mathbf{Q}_0 = \begin{bmatrix} U_0 \\ V_0 \end{bmatrix}, \quad \mathbf{N} = \begin{bmatrix} F(z) & f(z) \\ G(z) & g(z) \end{bmatrix}. \tag{31}$$

On account of the relation $D =$ constant, the determinant of the square matrix $\mathbf{N}$ is a constant. The value of this constant may immediately be found by taking $z = 0$, giving

$$|\mathbf{N}| = Fg - fG = 1.$$

It is usually more convenient to express U_0 and V_0 as functions of $U(z)$ and $V(z)$. Solving for U_0 and V_0, we obtain

$$\mathbf{Q}_0 = \mathbf{M}\mathbf{Q}, \tag{32}$$

where

$$\mathbf{M} = \begin{bmatrix} g(z) & -f(z) \\ -G(z) & F(z) \end{bmatrix}. \tag{33}$$

This matrix is also unimodular,

$$|\mathbf{M}| = 1. \tag{34}$$

The significance of $\mathbf{M}$ is clear: it relates the x and y components of the electric (or magnetic) vectors in the plane $z = 0$, to the components in an arbitrary plane $z =$ constant. Now we saw that the knowledge of U and V is sufficient for the complete specification of the field. Hence *for the purposes of determining the propagation of a plane monochromatic wave through a stratified medium, the medium only need be specified by an appropriate two by two unimodular matrix* $\mathbf{M}$. For this reason we shall call $\mathbf{M}$ the *characteristic matrix* of the stratified medium. The constancy of the determinant $|\mathbf{M}|$ may be shown to imply the conservation of energy.*

We shall now consider the form of the characteristic matrix for cases of particular interest.

(a) *A homogeneous dielectric film*

In this case ε, μ and $n = \sqrt{\varepsilon\mu}$ are constants. If θ denotes the angle which the normal to the wave makes with the z-axis, we have by (24),

$$\alpha = n \sin \theta,$$

For a TE wave, we have according to (15) and (16),

$$\left.\begin{aligned} \frac{d^2U}{dz^2} + (k_0^2 n^2 \cos^2 \theta) U = 0, \\ \frac{d^2V}{dz^2} + (k_0^2 n^2 \cos^2 \theta) V = 0. \end{aligned}\right\} \tag{35}$$

---

* To show this, one evaluates the reflectivity and transmissivity (51) in terms of the matrix elements. If further one uses the fact that for a non-absorbing medium the characteristic matrix is of the form indicated by (45), it follows that the conservation law $\mathscr{R} + \mathscr{T} = 1$ will be satisfied provided that $|\mathbf{M}| = 1$.

The solutions of these equations, subject to the relations (14), are easily seen to be

$$\left.\begin{aligned} U(z) &= A\cos(k_0 nz\cos\theta) + B\sin(k_0 nz\cos\theta), \\ V(z) &= \frac{1}{i}\sqrt{\frac{\varepsilon}{\mu}}\cos\theta\{B\cos(k_0 nz\cos\theta) - A\sin(k_0 nz\cos\theta)\}. \end{aligned}\right\} \tag{36}$$

Hence the particular solutions (27) which satisfy the boundary conditions (28) are

$$\left.\begin{aligned} U_1 = f(z) &= \frac{i}{\cos\theta}\sqrt{\frac{\mu}{\varepsilon}}\sin(k_0 nz\cos\theta), \\ V_1 = g(z) &= \cos(k_0 nz\cos\theta), \\ U_2 = F(z) &= \cos(k_0 nz\cos\theta), \\ V_2 = G(z) &= i\sqrt{\frac{\varepsilon}{\mu}}\cos\theta\sin(k_0 nz\cos\theta). \end{aligned}\right\} \tag{37}$$

If we set

$$p = \sqrt{\frac{\varepsilon}{\mu}}\cos\theta, \tag{38}$$

the characteristic matrix is seen to be

$$\mathbf{M}(z) = \begin{bmatrix} \cos(k_0 nz\cos\theta) & -\frac{i}{p}\sin(k_0 nz\cos\theta) \\ -ip\sin(k_0 nz\cos\theta) & \cos(k_0 nz\cos\theta) \end{bmatrix}. \tag{39}$$

For a *TM* wave, the same equations hold, with p replaced by

$$q = \sqrt{\frac{\mu}{\varepsilon}}\cos\theta. \tag{40}$$

(b) *A stratified medium as a pile of thin homogeneous films*

Consider two adjacent stratified media, the first one extending from $z = 0$ to $z = z_1$, and the second from $z = z_1$ to $z = z_2$. If $\mathbf{M}_1(z)$ and $\mathbf{M}_2(z)$ are the characteristic matrices of the two media, then

$$\mathbf{Q}_0 = \mathbf{M}_1(z_1)\mathbf{Q}(z_1), \qquad \mathbf{Q}(z_1) = \mathbf{M}_2(z_2 - z_1)\mathbf{Q}(z_2),$$

so that

$$\mathbf{Q}_0 = \mathbf{M}(z_2)\mathbf{Q}(z_2),$$

where

$$\mathbf{M}(z_2) = \mathbf{M}_1(z_1)\mathbf{M}_2(z_2 - z_1).$$

This result may immediately be generalized to the case of a succession of stratified media extending from $0 \leqslant z \leqslant z_1$, $z_1 \leqslant z \leqslant z_2$, . . . $z_{N-1} \leqslant z \leqslant z_N$. If the characteristic matrices are $\mathbf{M}_1$, $\mathbf{M}_2$, . . . $\mathbf{M}_N$, then

$$\left.\begin{aligned} &\mathbf{Q}_0 = \mathbf{M}(z_N)\mathbf{Q}(z_N), \\ &\text{where} \\ &\mathbf{M}(z_N) = \mathbf{M}_1(z_1)\mathbf{M}_2(z_2-z_1)\ldots\mathbf{M}_N(z_N-z_{N-1}). \end{aligned}\right\} \tag{41}$$

With the help of (41) an approximate expression for the characteristic matrix of any stratified medium may easily be derived*: we regard the medium as consisting of a very large number of thin films of thickness δz_1, δz_2, δz_3, . . . δz_n. If the maximum

* For a fuller treatment of stratified media of continuously varying refractive index see R. JACOBSSON, *Progress in Optics*, Vol. 5, ed. E. WOLF (Amsterdam, North Holland Publishing Company and New York, J. Wiley and Sons, 1965), p. 247.

thickness is sufficiently small, it is permissible to regard ε, μ and n to be constant throughout each film. From (39) it is seen that the characteristic matrix of the jth film is then approximately given by

$$\mathsf{M}_j = \begin{bmatrix} 1 & -\dfrac{i}{p_j} k_0 n_j \delta z_j \cos\theta_j \\ -ip_j k_0 n_j \delta z_j \cos\theta_j & 1 \end{bmatrix}.$$

Hence the characteristic matrix of the whole medium, considered as a pile of thin films, is approximately equal to (again retaining terms up to the first power in δz only):

$$\mathsf{M} = \prod_{j=1}^{N} \mathsf{M}_j = \begin{bmatrix} 1 & -ik_0 B \\ -ik_0 A & 1 \end{bmatrix}, \tag{42}$$

where

$$A = \sum_{j=1}^{N} p_j n_j \delta z_j \cos\theta_j = \sum_{j=1}^{N} \left(\varepsilon_j - \frac{\alpha^2}{\mu_j}\right) \delta z_j,$$

$$B = \sum_{j=1}^{N} \frac{n_j}{p_j} \delta z_j \cos\theta_j = \sum_{j=1}^{N} \mu_j \delta z_j.$$

Proceeding to the limit as $N \to \infty$ in such a way that $\max|\delta z_j| \to 0$, we obtain

$$\mathsf{M} = \begin{bmatrix} 1 & -ik_0 \mathscr{B} \\ -ik_0 \mathscr{A} & 1 \end{bmatrix}, \tag{43}$$

where

$$\mathscr{A} = \int \left(\varepsilon - \frac{\alpha^2}{\mu}\right) dz, \qquad \mathscr{B} = \int \mu dz, \tag{44}$$

the integration being taken throughout the whole z range. Eq. (43) gives a first approximation to the characteristic matrix of an arbitrary stratified medium. Improved approximations can be obtained by retaining higher-order terms* in the expansions of $\cos(k_0 n \delta z \cos\theta)$ and $\sin(k_0 n \delta z \cos\theta)$ and in the product (42).

Since, for a non-absorbing medium, ε and μ are real, it is also seen that *the characteristic matrix of a non-absorbing stratified medium has the form*

$$\mathsf{M} = \begin{bmatrix} a & ib \\ ic & d \end{bmatrix}, \tag{45}$$

where a, b, c and d are real.

1.6.3 The reflection and transmission coefficients

Consider a plane wave incident upon a stratified medium that extends from $z = 0$ to $z = z_1$ and that is bounded on each side by a homogeneous, semi-infinite medium. We shall derive expressions for the amplitudes and intensities of the reflected and transmitted waves.†

Let A, R and T denote as before the amplitudes (possibly complex) of the electric vectors of the incident, reflected and transmitted waves. Further, let ε_1, μ_1 and ε_l, μ_l be the dielectric constant and the magnetic permeability of the first and the last

* This is discussed fully in the paper by F. Abelès, *Ann. d Physique*, **5** (1950), p. 611.

† We consider the amplitudes of the electric vectors when studying a *TE* wave and those of the magnetic vectors when studying a *TM* wave.

medium, and let θ_1 and θ_l be the angles which the normals to the incident and the transmitted waves make with the z-direction (direction of stratification).

The boundary conditions of § 1.1 demand that the tangential components of $\boldsymbol{E}$ and $\boldsymbol{H}$ shall be continuous across each of the two boundaries of the stratified medium. This gives, if the relation § 1.4 (4)

$$\boldsymbol{H} = \sqrt{\frac{\varepsilon}{\mu}}\, \boldsymbol{s} \wedge \boldsymbol{E}$$

is also used, the following relations for a *TE* wave:

$$\left.\begin{aligned} U_0 &= A + R, \qquad & U(z_1) &= T, \\ V_0 &= p_1(A - R), \qquad & V(z_1) &= p_l T, \end{aligned}\right\} \tag{46}$$

where

$$p_1 = \sqrt{\frac{\varepsilon_1}{\mu_1}} \cos\theta_1, \qquad p_l = \sqrt{\frac{\varepsilon_l}{\mu_l}} \cos\theta_l. \tag{47}$$

The four quantities U_0, V_0, U and V given by (46) are connected by the basic relation (32); hence

$$\left.\begin{aligned} A + R &= (m'_{11} + m'_{12} p_l) T, \\ p_1(A - R) &= (m'_{21} + m'_{22} p_l) T, \end{aligned}\right\} \tag{48}$$

m'_{ij} being the elements of the characteristic matrix of the medium, evaluated for $z = z_1$.

From (48) we obtain the reflection and transmission coefficients of the film:

$$r = \frac{R}{A} = \frac{(m'_{11} + m'_{12} p_l) p_1 - (m'_{21} + m'_{22} p_l)}{(m'_{11} + m'_{12} p_l) p_1 + (m'_{21} + m'_{22} p_l)}, \tag{49}$$

$$t = \frac{T}{A} = \frac{2 p_1}{(m'_{11} + m'_{12} p_l) p_1 + (m'_{21} + m'_{22} p_l)}. \tag{50}$$

In terms of r and t, the *reflectivity* and *transmissivity* are

$$\mathscr{R} = |r|^2, \qquad \mathscr{T} = \frac{p_l}{p_1} |t|^2. \tag{51}$$

The phase δ_r of r may be called *the phase change on reflection* and the phase δ_t of t *the phase change on transmission*. The phase change δ_r is referred to the first surface of discontinuity, whilst the phase change δ_t is referred to the plane boundary between the stratified medium and the last semi-infinite medium.

The corresponding formulae for a *TM* wave are immediately obtained from (49)–(51) on replacing the quantities p_1 and p_l by

$$q_1 = \sqrt{\frac{\mu_1}{\varepsilon_1}} \cos\theta_1, \qquad q_l = \sqrt{\frac{\mu_l}{\varepsilon_l}} \cos\theta_l. \tag{52}$$

r and t are then the ratios of the amplitudes of the magnetic and not the electric vectors.

1.6.4 A homogeneous dielectric film*

The properties of an homogeneous dielectric film situated between two homogeneous media is of particular interest in optics, and we shall, therefore, study this case more fully. We assume all the media to be non-magnetic ($\mu = 1$).

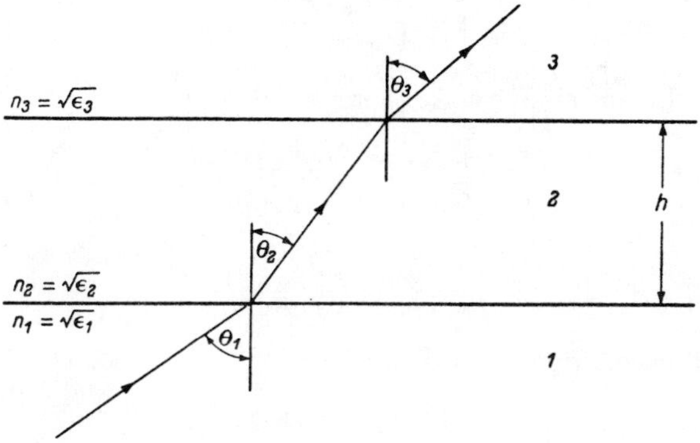

Fig. 1.17. Propagation of an electromagnetic wave through a homogeneous film.

The characteristic matrix of a homogeneous dielectric film is given by (39). Denoting by subscripts 1, 2 and 3 quantities which refer to the three media (see Fig. 1.17), and by h the thickness of the film, we have

$$m'_{11} = m'_{22} = \cos\beta, \qquad m'_{12} = -\frac{i}{p_2}\sin\beta, \qquad m'_{21} = -ip_2\sin\beta, \tag{53}$$

where

$$\beta = \frac{2\pi}{\lambda_0} n_2 h \cos\theta_2$$

and

$$p_j = n_j \cos\theta_j, \qquad (j = 1, 2, 3). \tag{54}$$

The reflection and transmission coefficients r and t may be obtained by substituting these expressions into (49) and (50), with $l = 3$. The resulting formulae may be conveniently expressed in terms of the corresponding coefficients r_{12}, t_{12} and r_{23}, t_{23} associated with the reflection and transmission at the first and the second surface respectively. According to the FRESNEL formulae § 1.5 (20) and (21) we have for a TE wave,

$$r_{12} = \frac{n_1\cos\theta_1 - n_2\cos\theta_2}{n_1\cos\theta_1 + n_2\cos\theta_2} = \frac{p_1 - p_2}{p_1 + p_2}, \tag{55}$$

$$t_{12} = \frac{2n_1\cos\theta_1}{n_1\cos\theta_1 + n_2\cos\theta_2} = \frac{2p_1}{p_1 + p_2}, \tag{56}$$

---

* An alternative derivation of the main formulae relating to the properties of a single dielectric film will be found in § 7.6.1. The formulae may, of course, also be derived directly by applying the boundary conditions of § 1.1.3 at each boundary of the film (cf. M. BORN, *Optik* (Berlin, Springer, 1933), p. 125; or H. MAYER, *Physik dünner Schichten* (Stuttgart, Wissenschaftliche Verlagsgesellschaft 1950), p. 145.

with analogous expressions for r_{23} and t_{23}. In terms of these expressions, the formulae for r and t become*

$$r = \frac{r_{12} + r_{23}e^{2i\beta}}{1 + r_{12}r_{23}e^{2i\beta}}, \tag{57}$$

$$t = \frac{t_{12}t_{23}e^{i\beta}}{1 + r_{12}r_{23}e^{2i\beta}}; \tag{58}$$

the reflectivity and transmissivity are therefore given by

$$\mathscr{R} = |r|^2 = \frac{r_{12}^2 + r_{23}^2 + 2r_{12}r_{23}\cos 2\beta}{1 + r_{12}^2 r_{23}^2 + 2r_{12}r_{23}\cos 2\beta}; \tag{59}$$

and

$$\mathscr{T} = \frac{p_3}{p_1}|t|^2 = \frac{n_3\cos\theta_3}{n_1\cos\theta_1}\frac{t_{12}^2 t_{23}^2}{1 + r_{12}^2 r_{23}^2 + 2r_{12}r_{23}\cos 2\beta}. \tag{60}$$

A straightforward calculation gives, as expected,

$$\mathscr{R} + \mathscr{T} = 1.$$

The phase changes can also easily be calculated from (57) and (58), and are found to be given by

$$\tan\delta_r = \tan(\arg r) = \frac{r_{23}(1 - r_{12}^2)\sin 2\beta}{r_{12}(1 + r_{23}^2) + r_{23}(1 + r_{12}^2)\cos 2\beta}, \tag{61}$$

$$\tan\delta_t = \tan(\arg t) = \frac{1 - r_{12}r_{23}}{1 + r_{12}r_{23}}\tan\beta. \tag{62}$$

Let us now briefly consider the implications of these formulae. We first note that (59) and (60) remain unchanged when β is replaced by $\beta + \pi$, i.e. when h is replaced by $h + \Delta h$, where

$$\Delta h = \frac{\lambda_0}{2n_2\cos\theta_2}. \tag{63}$$

Hence *the reflectivity and transmissivity of dielectric films which differ in thickness by an integral multiple of* $\lambda_0/2n_2\cos\theta_2$ *are the same.*

Next we determine the optical thickness for which the reflection coefficient has a maximum or a minimum. If we set

$$H = n_2 h, \tag{64}$$

we find from (59) that

$$\frac{d\mathscr{R}}{dH} = 0 \qquad \text{when} \qquad \sin 2\beta = 0,$$

i.e. when

$$H = \frac{m\lambda_0}{4\cos\theta_2}, \qquad (m = 0, 1, 2, \ldots).$$

We must distinguish two cases:

(1) When *m is odd*, i.e. when H has any of the values

$$H = \frac{\lambda_0}{4\cos\theta_2}, \quad \frac{3\lambda_0}{4\cos\theta_2}, \quad \frac{5\lambda_0}{4\cos\theta_2}, \ldots$$

* These formulae were first derived in a different manner by G. B. AIRY, *Phil. Mag.*, **2** (1833), 20; also *Ann. Phys. und Chem.* (Ed. Poggendorf), **41** (1837), 512.

then $\cos 2\beta = -1$ and (59) reduces to

$$\mathscr{R} = \left(\frac{r_{12} - r_{23}}{1 - r_{12}r_{23}}\right)^2. \tag{65}$$

In particular for *normal incidence*, one has from (55)

$$r_{12} = \frac{n_1 - n_2}{n_1 + n_2}, \qquad r_{23} = \frac{n_2 - n_3}{n_2 + n_3}, \tag{66}$$

and (65) becomes

$$\mathscr{R} = \left(\frac{n_1 n_3 - n_2^2}{n_1 n_3 + n_2^2}\right)^2. \tag{67}$$

(2) When *m is even*, i.e. when the optical thickness has any of the values

$$H = \frac{\lambda_0}{2\cos\theta_2}, \qquad \frac{2\lambda_0}{2\cos\theta_2}, \qquad \frac{3\lambda_0}{2\cos\theta_2}, \ldots$$

then $\cos 2\beta = 1$ and (59) reduces to

$$\mathscr{R} = \left(\frac{r_{12} + r_{23}}{1 + r_{12}r_{23}}\right)^2. \tag{68}$$

In particular, for *normal incidence*, this becomes

$$\mathscr{R} = \left(\frac{n_1 - n_3}{n_1 + n_3}\right)^2, \tag{69}$$

and is seen to be independent of n_2. Now the only difference in the case of oblique incidence is the replacement of n_j by $n_j \cos\theta_j$ $(j = 1, 2, 3)$ in all the formulae; hence *a plate whose optical thickness is* $m\lambda_0/2\cos\theta_2$ $(m = 1, 2, 3 \ldots)$ *has no influence on the intensity of the reflected* (*or transmitted*) *radiation.*

Next we must determine the nature of these extreme values. After a straightforward calculation we find that when $H = m\lambda_0/4\cos\theta_2$ $(m = 1, 2, \ldots)$

$$\left.\begin{aligned} &\left(\frac{d^2\mathscr{R}}{dH^2}\right) \gtrless 0 \\ \text{according as} \quad &(-1)^m r_{12}r_{23}[1 + r_{12}^2 r_{23}^2 - r_{12}^2 - r_{23}^2] \lesseqgtr 0, \end{aligned}\right\} \tag{70}$$

so that with the upper sign there is a minimum and with the lower sign a maximum. In particular, for *normal incidence*, r_{12} and r_{23} are given by (66) and we have

$$\left.\begin{aligned} &\textit{maximum}, \text{ if } \quad (-1)^m(n_1 - n_2)(n_2 - n_3) > 0, \\ &\textit{minimum}, \text{ if } \quad (-1)^m(n_1 - n_2)(n_2 - n_3) < 0. \end{aligned}\right\} \tag{71}$$

Usually the first medium is air $(n_1 \sim 1)$ and we see that *with a film whose optical thickness has any of the values* $\lambda_0/4$, $3\lambda_0/4$, $5\lambda_0/4, \ldots$ *the reflectivity is then a maximum or a minimum according to whether the refractive index of the film is greater or smaller than the refractive index of the last medium; for a film whose optical thickness has any of the values* $\lambda_0/2$, $2\lambda_0/2$, $3\lambda_0/2, \ldots$ *the opposite is the case.*

These results, which are illustrated in Fig. 1.18, are found to be in good agreement with experiment.*

It is evident from the preceding analysis that a plate whose optical thickness is a quarter of the wavelength and whose refractive index is low enough may be used as an *antireflection film*, i.e. a film by means of which the reflectivity of a surface is reduced. (The surface is then said to be "bloomed".) The two substances most commonly used for this purpose are cryolite ($n \sim 1{\cdot}35$) and magnesium fluoride (MgF_2, $n \sim 1{\cdot}38$)†:

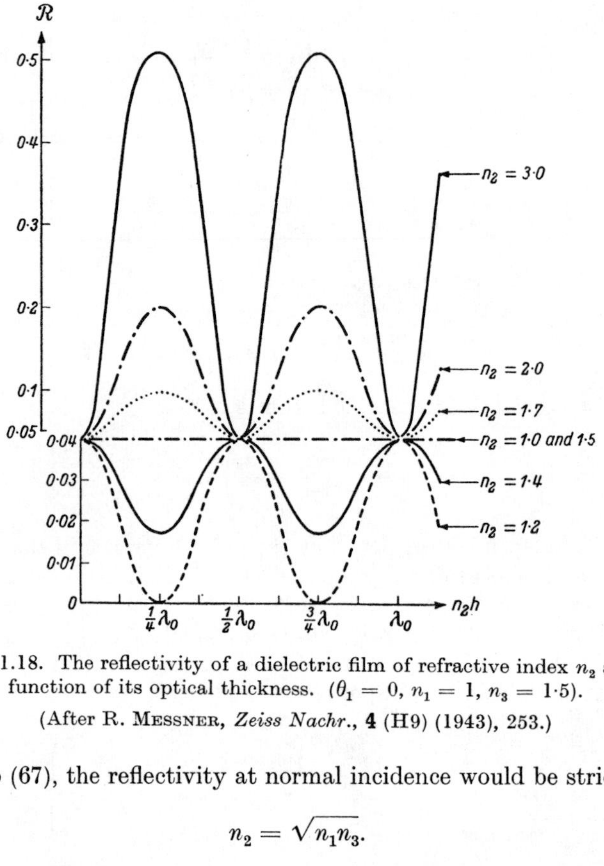

Fig. 1.18. The reflectivity of a dielectric film of refractive index n_2 as a function of its optical thickness. ($\theta_1 = 0$, $n_1 = 1$, $n_3 = 1{\cdot}5$).
(After R. MESSNER, *Zeiss Nachr.*, **4** (H9) (1943), 253.)

According to (67), the reflectivity at normal incidence would be strictly zero if

$$n_2 = \sqrt{n_1 n_3}. \tag{72}$$

With $n_1 = 1$, $n_3 = 1{\cdot}5$ this demands $n_2 \sim 1{\cdot}22$, a condition which cannot be satisfied in practice. A fuller analysis of (59) shows, however, that with oblique incidence it is possible to have zero reflectivity for a TM wave (electric vector parallel to the plane of incidence) but not for a TE wave (electric vector perpendicular to the plane of incidence), i.e. under favourable conditions one can have simultaneously $\mathscr{R}_\parallel = 0, \mathscr{R}_\perp \neq 0$. Hence a thin film of a suitable dielectric material may also be used as a *polarizer*, working by reflection. Such a polarizer may be regarded as a generalization of the

* See, for example, K. HAMMER, *Z. tech. Phys.*, **24** (1943), 169.

† The design and performance of multilayer antireflection films is discussed by A. MUSSET and A. THELEN in *Progress in Optics*, vol. 8, ed. E. WOLF (Amsterdam, North-Holland Publishing Company and New York, American Elsevier Publishing Company, 1970), p. 201.

simple arrangement discussed earlier in connection with BREWSTER's angle. To obtain a large value for $\mathscr{R}_\perp$ (with $\mathscr{R}_\parallel = 0$), the refractive index n_2 of the film must be as large as possible.* For example, with $n_1 = 1$, $n_2 = 2{\cdot}5$, $n_3 = 1{\cdot}53$ one obtains $\mathscr{R}_\parallel = 0$, $\mathscr{R}_\perp = 0{\cdot}79$ when $\theta_1 = 74°\,30'$.

If a glass surface is coated with a material of sufficiently high refractive index, the reflectivity of the surface will, according to the preceding analysis, be greatly enhanced (see Figs. 1.18 and 1.19). The surface will then act as a good beam splitter. Coatings of titanium dioxide (TiO_2, $n \sim 2{\cdot}45$) or zinc sulphide (ZnS, $n \sim 2{\cdot}3$) are very suitable for this purpose, giving a maximum reflectivity of about 0·3. There are other substances which have high refractive indices, but they absorb some of the incident light. For example, with a coating of stibnite (Sb_2S_3, $n \sim 2{\cdot}8$) one can attain the values $\mathscr{R} = \mathscr{T} = 0{\cdot}46$, but 8 per cent of the incident light is then absorbed by the film.

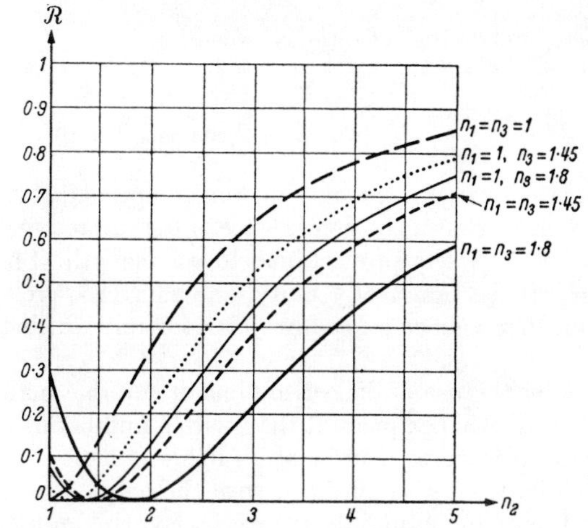

Fig. 1.19. The reflectivity at normal incidence of a quarter-wave film ($n_2 h = \lambda_0/4$) as function of the refractive index n_2 of the film.
(After K. HAMMER, *Z. tech. Phys.*, **24** (1943), 169.)

It is also of interest to examine the case when total reflection takes place on the first boundary. In this case

$$n_1 \sin\theta_1 > n_2, \qquad n_1 \sin\theta_1 < n_3,$$

and (cf. § 1.5, eq. 54),

$$n_2 \cos\theta_2 = i\sqrt{n_1^2 \sin^2\theta_1 - n_2^2}. \tag{73}$$

The coefficients for reflection at the two boundaries now are (cf. § 1.5 (21))

$$\left.\begin{aligned} r_{12} &= \frac{n_1\cos\theta_1 - i\sqrt{n_1^2\sin^2\theta_1 - n_2^2}}{n_1\cos\theta_1 + i\sqrt{n_1^2\sin^2\theta_1 - n_2^2}}, \\ r_{23} &= \frac{i\sqrt{n_1^2\sin^2\theta_1 - n_2^2} - n_3\cos\theta_3}{i\sqrt{n_1^2\sin^2\theta_1 - n_2^2} + n_3\cos\theta_3}. \end{aligned}\right\} \tag{74}$$

If we set

$$k_0 n_2 h \cos\theta_2 = ib, \tag{75}$$

* Cf. H. SCHRÖDER, *Optik*, **3** (1948), 499.

where, according to (73),

$$b = \frac{2\pi}{\lambda_0} h\sqrt{n_1^2 \sin^2 \theta_1 - n_2^2}, \tag{76}$$

we obtain for the reflection coefficients the following expression, in place of (57):

$$r = \frac{r_{12} + r_{23}e^{-2b}}{1 + r_{12}r_{23}e^{-2b}}. \tag{77}$$

Since $|r_{12}| = |r_{23}| = 1$, r_{12} and r_{23} are of the form

$$r_{12} = e^{i\phi_{12}}, \qquad r_{23} = e^{i\phi_{23}}, \tag{78}$$

where the ϕ's are real; hence the reflectivity now is

$$\mathscr{R} = |r|^2 = \frac{e^{2b} + e^{-2b} + 2\cos(\phi_{12} - \phi_{23})}{e^{2b} + e^{-2b} + 2\cos(\phi_{12} + \phi_{23})}. \tag{79}$$

In contrast with the previous case, $\mathscr{R}$ is now no longer a periodic function of the thickness of the film. (76) shows that if the dependence of the refractive index on the wavelength is neglected, b is inversely proportional to wavelength. Since for sufficiently large values of b, $\mathscr{R}$ will be practically unity, the shorter wavelengths will not be transmitted; the film then acts as a *low-pass filter*, i.e. one which transmits the long wavelengths only.

We have seen that by the use of dielectric films of suitable material, many useful effects can be attained. It will be apparent that, with a number of such films arranged in succession, the desired features may be still further enhanced. The characteristic matrix of such a *multilayer* may be obtained with the help of the theorem expressed by (41).* We shall discuss in detail only the case when the multilayer is periodic.

1.6.5 Periodically stratified media

A stratified periodic medium with period h is characterized by a dielectric constant ε and a magnetic permeability μ which are functions of z only and are such that

$$\varepsilon(z + jh) = \varepsilon(z), \qquad \mu(z + jh) = \mu(z),$$

being any integer in some fixed range $1 \leqslant j \leqslant N$.

Let $\mathbf{M}(h)$ be the characteristic matrix corresponding to one period and write†

$$\mathbf{M}(h) = \begin{bmatrix} m_{11} & m_{12} \\ m_{21} & m_{22} \end{bmatrix}. \tag{80}$$

According to (41) we then have, on account of the periodicity,

$$\mathbf{M}(Nh) = \underbrace{\mathbf{M}(h) \,.\, \mathbf{M}(h) \,.\,.\,.\, \mathbf{M}(h)}_{N \text{ times}} = (\mathbf{M}(h))^N. \tag{81}$$

---

* Formulae relating to multilayers have been given by many writers, e.g. R. L. Mooney, *J. Opt. Soc. Amer.*, **36** (1946), 256; W. Weinstein, *ibid*, **37** (1947), 576.

† We now omit the prime on the matrix elements.

To evaluate the elements of the matrix $\mathbf{M}(Nh)$ we use a result from the theory of matrices, according to which the Nth power of a unimodular matrix $\mathbf{M}(h)$ is*

$$[\mathbf{M}(h)]^N = \begin{bmatrix} m_{11}\mathscr{U}_{N-1}(a) - \mathscr{U}_{N-2}(a) & m_{12}\mathscr{U}_{N-1}(a) \\ m_{21}\mathscr{U}_{N-1}(a) & m_{22}\mathscr{U}_{N-1}(a) - \mathscr{U}_{N-2}(a) \end{bmatrix}, \tag{82}$$

where

$$a = \tfrac{1}{2}(m_{11} + m_{22}), \tag{83}$$

and $\mathscr{U}_N$ are the *Chebyshev Polynomials* of the second kind†:

$$\mathscr{U}_N(x) = \frac{\sin[(N+1)\cos^{-1}x]}{\sqrt{1-x^2}}. \tag{84}$$

A multilayer usually consists of a succession of homogeneous layers of alternately low and high refractive indices n_2 and n_3 and of thickness h_2 and h_3, placed between two homogeneous media of refractive indices n_1 and n_l. (See Fig. 1.20.) We again assume the media to be non-magnetic ($\mu = 1$) and set

$$\left.\begin{aligned} \beta_2 &= \frac{2\pi}{\lambda_0} n_2 h_2 \cos\theta_2, \qquad & \beta_3 &= \frac{2\pi}{\lambda_0} n_3 h_3 \cos\theta_3, \\ p_2 &= n_2\cos\theta_2, & p_3 &= n_3\cos\theta_3. \\ & & h = h_2 + h_3. \end{aligned}\right\} \tag{85}$$

The characteristic matrix $\mathbf{M}_2(h)$ of one period then is, according to (39) and (41),

$$\mathbf{M}_2(h) = \begin{bmatrix} \cos\beta_2 & -\dfrac{i}{p_2}\sin\beta_2 \\ -ip_2\sin\beta_2 & \cos\beta_2 \end{bmatrix} \begin{bmatrix} \cos\beta_3 & -\dfrac{i}{p_3}\sin\beta_3 \\ -ip_3\sin\beta_3 & \cos\beta_3 \end{bmatrix}$$

$$= \begin{bmatrix} \cos\beta_2\cos\beta_3 - \dfrac{p_3}{p_2}\sin\beta_2\sin\beta_3 & -\dfrac{i}{p_3}\cos\beta_2\sin\beta_3 - \dfrac{i}{p_2}\sin\beta_2\cos\beta_3 \\ -ip_2\sin\beta_2\cos\beta_3 - ip_3\cos\beta_2\sin\beta_3 & \cos\beta_2\cos\beta_3 - \dfrac{p_2}{p_3}\sin\beta_2\sin\beta_3 \end{bmatrix}. \tag{86}$$

---

* The correctness of this result may be verified by induction, using the recurrence relation

$$\mathscr{U}_j(x) = 2x\mathscr{U}_{j-1}(x) - \mathscr{U}_{j-2}(x),$$

which follows as an identity from the definition of the Chebyshev Polynomials.

A direct proof based on the theory of matrices was given by F. Abelès, *Ann. de Physique*, **5** (1950), 777.

† These polynomials satisfy the following orthogonality and normalizing conditions:

$$\begin{aligned} \int_{-1}^{+1} \mathscr{U}_m(x)\mathscr{U}_n(x)\sqrt{1-x^2}\,dx &= 0 \text{ when } n \neq m \\ &= \frac{\pi}{2} \text{ when } n = m. \end{aligned}$$

For convenience we note the explicit expressions of the first six polynomials:

$$\begin{aligned} \mathscr{U}_0(x) &= 1, & \mathscr{U}_3(x) &= 8x^3 - 4x, \\ \mathscr{U}_1(x) &= 2x, & \mathscr{U}_4(x) &= 16x^4 - 12x^2 + 1, \\ \mathscr{U}_2(x) &= 4x^2 - 1, & \mathscr{U}_5(x) &= 32x^5 - 32x^3 + 6x. \end{aligned}$$

Tables of Chebyshev polynomials have been published by the National Bureau of Standards Washington (Applied Mathematics Series 9 (1952)), where the main properties of the polynomials are also summarized. See also *Higher Transcendental Functions* (Bateman Manuscript Project, New York, McGraw-Hill, Vol. **2** (1953), p. 183).

Hence according to (81), the characteristic matrix $\mathbf{M}_{2N}(Nh)$ of the multilayer (with $2N$ films in all) is given by the following formula due to ABELÈS:

$$\mathbf{M}_{2N}(Nh) = \begin{bmatrix} \mathscr{M}_{11} & \mathscr{M}_{12} \\ \mathscr{M}_{21} & \mathscr{M}_{22} \end{bmatrix}, \tag{87}$$

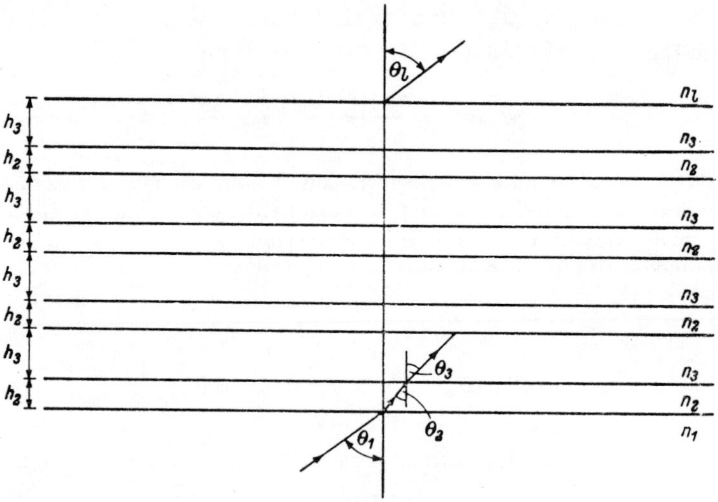

Fig. 1.20. A periodic multilayer.

where

$$\left.\begin{aligned}
\mathscr{M}_{11} &= \left[\cos \beta_2 \cos \beta_3 - \frac{p_3}{p_2} \sin \beta_2 \sin \beta_3\right] \mathscr{U}_{N-1}(a) - \mathscr{U}_{N-2}(a), \\
\mathscr{M}_{12} &= -i \left[\frac{1}{p_3} \cos \beta_2 \sin \beta_3 + \frac{1}{p_2} \sin \beta_2 \cos \beta_3\right] \mathscr{U}_{N-1}(a), \\
\mathscr{M}_{21} &= -i[p_2 \sin \beta_2 \cos \beta_3 + p_3 \cos \beta_2 \sin \beta_3] \mathscr{U}_{N-1}(a), \\
\mathscr{M}_{22} &= \left[\cos \beta_2 \cos \beta_3 - \frac{p_2}{p_3} \sin \beta_2 \sin \beta_3\right] \mathscr{U}_{N-1}(a) - \mathscr{U}_{N-2}(a),
\end{aligned}\right\} \tag{88}$$

and

$$a = \cos \beta_2 \cos \beta_3 - \tfrac{1}{2} \left(\frac{p_2}{p_3} + \frac{p_3}{p_2}\right) \sin \beta_2 \sin \beta_3. \tag{89}$$

The reflection and transmission coefficients of the multilayer are immediately obtained by substituting these expressions into (49) and (50).

Of particular interest is the case when the two basic layers are of the same optical thickness (usually $\lambda_0/4$), i.e. when

$$n_2 h_2 = n_3 h_3, \tag{90}$$

and the incidence is normal ($\theta_1 = 0$). Then

$$\beta_2 = \beta_3 = \frac{2\pi}{\lambda_0} n_2 h_2 = \frac{2\pi}{\lambda_0} n_3 h_3, \tag{91}$$

and if we denote this common value by β, the argument of the CHEBYSHEV polynomials reduces to

$$a = \cos^2 \beta - \tfrac{1}{2}\left(\frac{n_2}{n_3} + \frac{n_3}{n_2}\right) \sin^2 \beta. \tag{92}$$

It is seen that a cannot exceed unity, but that for some values of β it may become smaller than -1. Then $\cos^{-1} a$ will be imaginary and consequently, since for any χ,

$$\sin i\chi = i \sinh \chi = i\, \frac{e^{\chi} - e^{-\chi}}{2}$$

$\mathscr{U}_N$ will have exponential behaviour. It follows that the reflectivity of such a multilayer will increase rapidly with the number of the periods.

With *quarter-wave films* ($n_2 h_2 = n_3 h_3 = \lambda_0/4$) at normal incidence (again assuming non-magnetic media),

$$\beta = \pi/2, \qquad p_2 = n_2, \qquad p_3 = n_3, \tag{93}$$

and (86) reduces to

$$\mathbf{M}_2(h) = \begin{bmatrix} -\dfrac{n_3}{n_2} & 0 \\ 0 & -\dfrac{n_2}{n_3} \end{bmatrix}. \tag{94}$$

The characteristic matrix (87) of the multilayer whose basic period is such a double layer is, as can be directly verified by multiplying (94) N times by itself,

$$\mathbf{M}_{2N}(Nh) = \begin{bmatrix} \left(-\dfrac{n_3}{n_2}\right)^N & 0 \\ 0 & \left(-\dfrac{n_2}{n_3}\right)^N \end{bmatrix}. \tag{95}$$

According to (49) and (51) the reflectivity is

$$\mathscr{R}_{2N} = \left(\frac{1 - \dfrac{n_l}{n_1}\left(\dfrac{n_2}{n_3}\right)^{2N}}{1 + \dfrac{n_l}{n_1}\left(\dfrac{n_2}{n_3}\right)^{2N}}\right)^2. \tag{96}$$

This shows that for a fixed number N of the double layers,* $\mathscr{R}_{2N}$ increases when the ratio n_2/n_3 is increased, and that if this ratio is fixed $\mathscr{R}_{2N}$ increases with N.

Sometimes, for example, for plate coatings of the FABRY–PEROT Interferometer (cf. § 7.6) the layers are arranged in succession characterized by the sequence $n_2, n_3, n_2, n_3, \ldots n_2, n_3, n_2$ of refractive indices. The characteristic matrix of this multilayer is

$$\mathbf{M}_{2N+1} = \mathbf{M}_{2N} \,.\, \mathbf{M}, \tag{97}$$

* A thorough discussion of the properties of a system of double layers will be found in a paper by C. DUFOUR and A. HERPIN, *Rev. Opt.*, **32** (1953), 321.

TABLE III

Reflectivity $\mathscr{R}_{2N+1}$ of multilayers formed by a periodic succession of quarter-wave films of zinc sulphide and cryolite at normal incidence ($n_1 = 1$, $n_2 = 2{\cdot}3$, $n_3 = 1{\cdot}35$, $n_l = 1{\cdot}52$, $n_2 h_2 = n_3 h_3 = \lambda_0/4$, $\lambda_0 = 5460$ Å, $\theta_1 = 0$).

The values in brackets are experimental results obtained by P. GIACOMO, *Compt. Rend. Acad. Sci., Paris*, **235** (1952), 1627.

N	$\mathscr{R}_{2N+1}$
0	0·306
1	0·672
2	0·872 (0·865)
3	0·954 (0·945)
4	0·984 (0·97)

where $\mathbf{M}_{2N}$ is given by (87) and $\mathbf{M}$ is the characteristic matrix of the last film in the sequence. In particular with quarter-wave films at normal incidence, $\mathbf{M}_{2N}$ reduces to (95), $\beta_2 = \pi/2$, and (97) then becomes

$$\mathbf{M}_{2N+1} = \begin{bmatrix} 0 & -\dfrac{i}{n_2}\left(-\dfrac{n_3}{n_2}\right)^N \\ -in_2\left(-\dfrac{n_2}{n_3}\right)^N & 0 \end{bmatrix}. \tag{98}$$

Substitution into (49) and (51) gives the required reflectivity:

$$\mathscr{R}_{2N+1} = \left(\frac{1 - \left(\dfrac{n_2}{n_1}\right)\left(\dfrac{n_2}{n_l}\right)\left(\dfrac{n_2}{n_3}\right)^{2N}}{1 + \left(\dfrac{n_2}{n_1}\right)\left(\dfrac{n_2}{n_l}\right)\left(\dfrac{n_2}{n_3}\right)^{2N}}\right)^2. \tag{99}$$

The reflectivity is seen to increase rapidly with the ratio n_2/n_3 and with N (see Table III).

CHAPTER II

ELECTROMAGNETIC POTENTIALS AND POLARIZATION

In the preceding chapter the effect of matter on an electromagnetic field was expressed in terms of a number of macroscopic constants. These have only a limited range of validity and are in fact inadequate to describe certain processes, such as the emission, absorption and dispersion of light. A full account of these phenomena would involve an extensive study of the atomistic theory and lies therefore outside the scope of this book.

It is possible, however, to describe the interaction of field and matter by means of a simple model which is entirely adequate for most branches of optics. For this purpose each of the vectors $\boldsymbol{D}$ and $\boldsymbol{B}$ is expressed as the sum of two terms.* Of these one is taken to be the vacuum field and the other is regarded as arising from the influence of matter. Thus one is led to the introduction of two new vectors for describing the effects of matter: the *electric polarization* ($\boldsymbol{P}$) and the *magnetic polarization* or *magnetization* ($\boldsymbol{M}$). Instead of the material equations (10) and (11) in § 1.1 connecting $\boldsymbol{D}$ and $\boldsymbol{B}$ with $\boldsymbol{E}$ and $\boldsymbol{H}$, we now have equations connecting $\boldsymbol{P}$ and $\boldsymbol{M}$ with $\boldsymbol{E}$ and $\boldsymbol{H}$. These new equations have a more direct physical meaning and lead to the following conception of the propagation of an electromagnetic field in matter:

An electromagnetic field produces at a given volume element certain amounts of polarization $\boldsymbol{P}$ and $\boldsymbol{M}$ which, in the first approximation, are proportional to the field, the constant of proportionality being a measure of the reaction of the field. Each volume element then becomes a source of a new secondary or scattered wavelet, whose strength is related in a simple way to $\boldsymbol{P}$ and $\boldsymbol{M}$. All the secondary wavelets combine with each other and with the incident field and form the total field, this being the one from which the whole consideration proceeds. By expressing this identity formally, we obtain two integral equations,† which may be shown to be equivalent to Maxwell's differential equations, but which describe the propagation of the electromagnetic field in a manner more clearly related to the atomic constitution of matter.

The two main results which will be obtained from the theory are: (1) The Lorentz–Lorenz formula, which relates the macroscopic-optical properties of the medium to the number and the properties of the scattering particles, and (2) the so-called extinction theorem of Ewald and Oseen, which shows how an external electromagnetic disturbance travelling with the velocity of the light in vacuum is exactly cancelled out and replaced in the substance by the secondary disturbance travelling with an appropriately smaller velocity.

* According to the remarks on p. 1, it would be more appropriate to write $\boldsymbol{H}$ in place of $\boldsymbol{B}$. This departure from the general theory is made here for the sake of the trivial convenience of having the positive signs on the right-hand side of eq. (2) in § 2.2, as is customary.

† In the case of non-magnetic substances there is only one integral equation. The second integral equation reduces to a relatively simple expression for the magnetic field, and may be evaluated when the solution of the first equation has been determined.

The "material" integral equations discussed here must be distinguished from the "geometrical" integral equations used in the treatment of certain diffraction problems (cf. Chapter XI).

The theory provides an alternative mathematical approach to the treatment of some problems of electromagnetic theory. This will be illustrated by the derivation of the laws of refraction and reflection and the FRESNEL formulae, and will be further illustrated by the treatment of a more complex problem in Chapter XII.

In the derivation of these results some new mathematical apparatus will be required. Accordingly, we shall first discuss the representation of an electromagnetic field in terms of so-called retarded potentials, these being generalizations of the well-known static potentials. The expressions of the potentials in terms of the polarization vectors lead to another set of auxiliary quantities, known as HERTZ vectors. In § 2.1 and § 2.2 we discuss these mathematical preliminaries and in § 2.3 we briefly explain the underlying physical concepts. The integral equations and the two basic theorems already referred to will be derived in § 2.4.

2.1 THE ELECTRODYNAMIC POTENTIALS IN THE VACUUM

2.1.1 The vector and scalar potentials

Let us consider an electromagnetic field in vacuum due to a given distribution of charges $\rho(\boldsymbol{r}, t)$ and currents $\boldsymbol{j}(\boldsymbol{r}, t)$. It obeys MAXWELL's equations § 1.1 (1)–(4) which may be written (since in vacuum $\boldsymbol{D} = \boldsymbol{E}$ and $\boldsymbol{B} = \boldsymbol{H}$) in the form:

$$\text{curl}\, \boldsymbol{B} - \frac{1}{c}\dot{\boldsymbol{E}} = \frac{4\pi}{c}\boldsymbol{j}, \tag{1}$$

$$\text{curl}\, \boldsymbol{E} + \frac{1}{c}\dot{\boldsymbol{B}} = 0, \tag{2}$$

$$\text{div}\, \boldsymbol{E} = 4\pi\rho, \tag{3}$$

$$\text{div}\, \boldsymbol{B} = 0. \tag{4}$$

Since the divergence of the curl of any vector is zero, (4) will be satisfied if we set

$$\boldsymbol{B} = \text{curl}\, \boldsymbol{A}, \tag{5}$$

$\boldsymbol{A}$ being an arbitrary vector function of position and time. If we substitute from (5) into MAXWELL's second equation, we have

$$\text{curl}\left(\boldsymbol{E} + \frac{1}{c}\dot{\boldsymbol{A}}\right) = 0. \tag{6}$$

(6) will hold if

$$\boldsymbol{E} = -\frac{1}{c}\dot{\boldsymbol{A}} - \text{grad}\, \phi, \tag{7}$$

ϕ being an arbitrary scalar function. $\boldsymbol{A}$ and ϕ must now be determined in such a way as to satisfy the remaining Maxwell equations.

Substituting from (5) and (7) into (1) and (3), and using the identities curl curl $\equiv$ grad div $- \nabla^2$ and div grad $\equiv \nabla^2$, we obtain

$$\nabla^2\boldsymbol{A} - \frac{1}{c^2}\ddot{\boldsymbol{A}} - \text{grad}\left(\text{div}\, \boldsymbol{A} + \frac{1}{c}\dot{\phi}\right) = -\frac{4\pi}{c}\boldsymbol{j}, \tag{8}$$

and

$$\nabla^2\phi - \frac{1}{c^2}\ddot{\phi} + \frac{1}{c}\frac{\partial}{\partial t}\left(\text{div}\, \boldsymbol{A} + \frac{1}{c}\dot{\phi}\right) = -4\pi\rho. \tag{9}$$

If we prescribe between A and ϕ the relation

$$\operatorname{div} \boldsymbol{A} + \frac{1}{c}\dot{\phi} = 0, \tag{10}$$

then (8) and (9) reduce to the inhomogeneous wave equations

$$\nabla^2 \boldsymbol{A} - \frac{1}{c^2}\ddot{\boldsymbol{A}} = -\frac{4\pi}{c}\boldsymbol{j}, \tag{11}$$

and

$$\nabla^2 \phi - \frac{1}{c^2}\ddot{\phi} = -4\pi\rho. \tag{12}$$

The two functions $\boldsymbol{A}$ and ϕ, from which $\boldsymbol{B}$ and $\boldsymbol{E}$ may be determined by means of the relations (5) and (7), are known as the *magnetic vector potential* and the *electric scalar potential* respectively. The relation (10), which couples the two potentials, is called the *Lorentz condition*. We note that it is consistent with the continuity equation § 1.1 (5)

$$\dot{\rho} + \operatorname{div} \boldsymbol{j} = 0. \tag{13}$$

It is to be observed that (11), (12) and (10) do not define the two potentials uniquely. For if we add to $\boldsymbol{A}$ the vector grad χ, where χ is arbitrary, $\boldsymbol{B}$ will remain unchanged, and if, in addition, ϕ is replaced by $\phi - \dot{\chi}/c$, $\boldsymbol{E}$ also will remain unchanged. In other words $\boldsymbol{B}$ and $\boldsymbol{E}$ are invariant under the transformation

$$\boldsymbol{A}' = \boldsymbol{A} + \operatorname{grad} \chi, \tag{14a}$$

$$\phi' = \phi - \frac{1}{c}\dot{\chi}. \tag{14b}$$

Now from (10) and (14),

$$\operatorname{div} \boldsymbol{A}' + \frac{1}{c}\dot{\phi}' - \left(\nabla^2\chi - \frac{1}{c^2}\ddot{\chi}\right) = 0. \tag{15}$$

Hence $\boldsymbol{A}'$ and ϕ' will satisfy the LORENTZ relation, if one imposes on χ the condition

$$\nabla^2\chi - \frac{1}{c^2}\ddot{\chi} = 0. \tag{16}$$

Eqs. (14), subject to the condition (16), express the so-called *gauge transformation*. The gauge transformation may be used to simplify the expressions for the field vectors. For example in a region where the charge density ρ is zero, ϕ satisfies the homogeneous wave equation

$$\nabla^2\phi - \frac{1}{c^2}\ddot{\phi} = 0 \tag{17}$$

and χ may then be so chosen, that the scalar potential vanishes. According to (14b) and (16) it is only necessary to take

$$\chi = c\int \phi \, dt. \tag{18}$$

The field can then be derived from the vector potential alone* by means of the relations (dropping the prime on A)

$$B = \text{curl } A, \qquad E = -\frac{1}{c}\dot{A}, \tag{19}$$

whilst the LORENTZ condition reduces to

$$\text{div } A = 0. \tag{20}$$

2.1.2 Retarded potentials

We now consider the solution (subject to the relation (10)) of the inhomogeneous wave equations (11) and (12) for the vector and scalar potentials, and show first that these equations are satisfied by the following functions:

$$A(r,t) = \frac{1}{c}\int \frac{j(r', t - R/c)}{R}\, dV', \tag{21}$$

$$\phi(r,t) = \int \frac{\rho(r', t - R/c)}{R}\, dV'. \tag{22}$$

Here

$$R = |r - r'| = \sqrt{(x - x')^2 + (y - y')^2 + (z - z')^2} \tag{23}$$

is the distance from the point $r(x, y, z)$ to the volume element dV' at $r'(x', y', z')$, the integration being carried throughout the whole space.

To verify that (22) satisfies the inhomogeneous wave equation for the scalar potential, we imagine that the point r is surrounded by a sphere of radius a, centred on that point, and divide (22) into two parts:

$$\phi = \phi_1 + \phi_2, \tag{24}$$

where ϕ_1 represents the contribution to the integral from the interior of the sphere and ϕ_2 the contribution from the rest of space. Since $R \neq 0$ for every point $r'(x', y', z')$ outside the sphere it will be permissible to differentiate ϕ_2 under the integral sign. We can then verify by a straightforward calculation that ϕ_2 satisfies the homogeneous wave equation

$$\nabla^2\phi_2 - \frac{1}{c^2}\ddot{\phi}_2 = 0, \tag{25}$$

consisting, simply as it does, of a set of superposed spherical waves (cf. eq. (12), § 1.3). In the case of ϕ_1 we must, however, proceed differently, since the integrand has a singularity at the centre $R = 0$.

We note that by making the radius of the sphere sufficiently small, we can secure (assuming ρ to be a continuous function of r and t) that for all points r' inside the sphere $\rho(r', t - R/c)$ differs from $\rho(r, t)$ by an amount smaller than any prescribed value. Hence, as the radius a tends to zero, $\nabla^2\phi_1$ will approach more and more closely the value of the electrostatic potential of a homogeneously charged sphere, with charge density ρ; i.e. for a sufficiently small a,

$$\nabla^2\phi_1 = -4\pi\rho(r,t). \tag{26}$$

---

* Eq. (19) expresses the field in a charge-free region in vacuum in terms of three scalar wave functions (Cartesian components of A). On account of (20) they are, however, not independent. It may be shown that in such a region the field can in fact be derived from two real scalar wave functions alone (see, for example, S. A. SCHELKUNOFF, *Electromagnetic Waves*, New York, Van Nostrand, 1943, 382; H. S. GREEN and E. WOLF, *Proc. Phys. Soc.*, A, **66** (1953), 1129; A. NISBET, *Proc. Roy. Soc.*, A, **231** (1955) 251).

Also, as $a \to 0$, $\ddot{\phi}_1 \to 0$. For if a is sufficiently small we may write

$$\begin{aligned}\ddot{\phi}_1 &= \ddot{\rho}(\boldsymbol{r},t)\int_{R\leqslant a}\frac{dV'}{R}\\ &= 4\pi\ddot{\rho}\int_0^a R dr\\ &= 2\pi a^2\ddot{\rho},\end{aligned} \tag{27}$$

and this tends to zero with a. Hence it follows from (24)–(26) that as $a \to 0$,

$$\nabla^2\phi - \frac{1}{c^2}\ddot{\phi} = \nabla^2(\phi_1 + \phi_2) - \frac{1}{c^2}(\ddot{\phi}_1 + \ddot{\phi}_2) \to -4\pi\rho(\boldsymbol{r},t), \tag{28}$$

so that (22) satisfies the inhomogeneous wave equation for the scalar potential. In a strictly analogous manner we can show that each Cartesian component of (21) is a solution of the corresponding scalar wave equation, whose inhomogeneous term contains, in place of ρ, the appropriate component of $\boldsymbol{j}/c$. Hence (21) satisfies the inhomogeneous wave equation for the vector potential. Moreover, on account of the continuity relation (13), these solutions also satisfy the Lorentz relation (10).

The expressions (21) and (22) have a simple physical interpretation. They show that one may regard $\boldsymbol{A}$ and ϕ as arising from contributions from each volume element of space, a typical element dV' contributing the amounts

$$\frac{1}{c}\frac{\boldsymbol{j}(\boldsymbol{r}',t-R/c)}{R} \quad \text{and} \quad \frac{\rho(\boldsymbol{r}',t-R/c)}{R}$$

to $\boldsymbol{A}$ and ϕ respectively. Now R/c is precisely the time needed for light to travel from the point $\boldsymbol{r}'$ to $\boldsymbol{r}$, so that each contribution has to leave the element at such preceding time as to reach the point of observation at the required time t. For this reason (21) and (22) are called *retarded potentials*.*

(21) and (22) represent a special solution of the wave equations (11) and (12), namely that which arises from the given charges and currents. The general solution is obtained by adding to this the general solutions (again subject to the Lorentz condition) of the homogeneous wave equations

$$\nabla^2\boldsymbol{A} - \frac{1}{c^2}\ddot{\boldsymbol{A}} = 0, \tag{29}$$

$$\nabla^2\phi - \frac{1}{c^2}\ddot{\phi} = 0. \tag{30}$$

---

* It is also possible to construct solutions in the form of *advanced potentials* (with $t + R/c$ in place of $t - R/c$). These represent the effect of incoming spherical waves, whereas the retarded potentials represent the effect of outgoing spherical waves.

At the cost of some formal asymmetry, we may restrict our attention to solutions in terms of retarded potentials only. When the interaction of radiation and matter is considered thermodynamically, then there is an asymmetry in the physical situation which reinforces this choice (cf. M. Born, *Natural Philosophy of Cause and Chance*, Oxford, Clarendon Press, 1949; Dover Publications, New York, 1964; especially p. 26).

2.2 POLARIZATION AND MAGNETIZATION

2.2.1 The potentials in terms of polarization and magnetization

In Chapter I the field was studied with the help of the material relations $\boldsymbol{D} = \varepsilon\boldsymbol{E}$, $\boldsymbol{B} = \mu\boldsymbol{H}$; these imply that the fields $\boldsymbol{E}$ and $\boldsymbol{H}$ are regarded as producing at each point of the substance certain displacements $\boldsymbol{D}$ and $\boldsymbol{B}$ which are proportional to them. Instead of these "multiplicative relations" we shall now describe the interaction of matter and field by means of "additive relations",

$$\boldsymbol{D} = \boldsymbol{E} + 4\pi\boldsymbol{P}, \tag{1}$$

$$\boldsymbol{B} = \boldsymbol{H} + 4\pi\boldsymbol{M}. \tag{2}$$

Here $\boldsymbol{P}$ is called the *electric polarization* and $\boldsymbol{M}$ the *magnetic polarization* or *magnetization.* The physical meaning of these quantities will become clear later; here we only observe that $\boldsymbol{P}$ and $\boldsymbol{M}$ both vanish in vacuum so that these quantities represent the influence of matter on the field in a simple intuitive way.

Assume the material to be nonconducting ($\sigma = 0$) and consider the field in a region where the current and charge densities are zero ($\rho = \boldsymbol{j} = 0$). Eliminating $\boldsymbol{D}$ and $\boldsymbol{H}$ from MAXWELL's equations § 1.1 (1)–(4) by means of (1) and (2), we obtain

$$\operatorname{curl} \boldsymbol{B} - \frac{1}{c}\dot{\boldsymbol{E}} = \frac{4\pi}{c}\tilde{\boldsymbol{j}}, \tag{3}$$

$$\operatorname{curl} \boldsymbol{E} + \frac{1}{c}\dot{\boldsymbol{B}} = 0, \tag{4}$$

$$\operatorname{div} \boldsymbol{E} = 4\pi\tilde{\rho}, \tag{5}$$

$$\operatorname{div} \boldsymbol{B} = 0, \tag{6}$$

where the *free current density* $\tilde{\boldsymbol{j}}$ and the *free charge density* $\tilde{\rho}$ are defined by the relations

$$\tilde{\boldsymbol{j}} = \dot{\boldsymbol{P}} + c \operatorname{curl} \boldsymbol{M}, \tag{7}$$

$$\tilde{\rho} = - \operatorname{div} \boldsymbol{P}. \tag{8}$$

Eqs. (3)–(6) *are formally identical with the equations for the field in vacuum considered* in the previous section. Hence, as in the vacuum case, we may introduce a vector potential $\boldsymbol{A}$ and a scalar potential ϕ such that

$$\boldsymbol{B} = \operatorname{curl} \boldsymbol{A}, \tag{9}$$

$$\boldsymbol{E} = -\frac{1}{c}\dot{\boldsymbol{A}} - \operatorname{grad}\phi. \tag{10}$$

Moreover, if we impose, as before, the LORENTZ condition

$$\operatorname{div} \boldsymbol{A} + \frac{1}{c}\dot{\phi} = 0, \tag{11}$$

we have, by analogy with (11) and (12) in § 2.1, the following equations for $\boldsymbol{A}$ and ϕ:

$$\nabla^2\boldsymbol{A} - \frac{1}{c^2}\ddot{\boldsymbol{A}} = -\frac{4\pi}{c}\tilde{\boldsymbol{j}}, \tag{12}$$

$$\nabla^2\phi - \frac{1}{c^2}\ddot{\phi} = -4\pi\tilde{\rho}. \tag{13}$$

Now condition (11) is consistent with (12) and (13) provided that

$$\frac{\partial \tilde{\rho}}{\partial t} + \operatorname{div} \tilde{\boldsymbol{j}} = 0; \tag{14}$$

this relation is, in fact, satisfied identically, as can be immediately seen from (7) and (8), on using the vector identity div curl $\equiv 0$.

In terms of the polarization and magnetization, the solutions of (12) and (13) may, according to (21) and (22) in § 2.1, and by (7) and (8), be expressed in the form

$$\phi = -\int \frac{1}{R} [\operatorname{div}' \boldsymbol{P}] \, dV', \tag{15}$$

$$\boldsymbol{A} = \int \frac{1}{R} \left[\operatorname{curl}' \boldsymbol{M} + \frac{1}{c} \dot{\boldsymbol{P}}\right] dV'. \tag{16}$$

Here the differential operators div$'$ and curl$'$ are taken with respect to the coordinates (x', y', z') of the variable point of integration $\boldsymbol{r}'$ at which the volume element dV' is situated, and the square brackets denote *retarded values*, i.e. they imply that the argument t is replaced by $t - R/c$ inside each bracket.

By a straightforward calculation the following identities can be verified:

$$\operatorname{div}' [\boldsymbol{P}] = [\operatorname{div}' \boldsymbol{P}] + \frac{1}{cR} \boldsymbol{R} \cdot [\dot{\boldsymbol{P}}], \tag{17a}$$

$$\operatorname{curl}' [\boldsymbol{M}] = [\operatorname{curl}' \boldsymbol{M}] + \frac{1}{cR} \boldsymbol{R} \wedge [\dot{\boldsymbol{M}}], \tag{17b}$$

where

$$\boldsymbol{R} = \boldsymbol{r} - \boldsymbol{r}'. \tag{18}$$

Hence (15) and (16) may be written as

$$\phi = -\int \left\{\frac{1}{R} \operatorname{div}' [\boldsymbol{P}] - \frac{1}{cR^2} \boldsymbol{R} \cdot [\dot{\boldsymbol{P}}]\right\} dV', \tag{19}$$

$$\boldsymbol{A} = \int \left\{\frac{1}{R} \operatorname{curl}' [\boldsymbol{M}] - \frac{1}{cR^2} \boldsymbol{R} \wedge [\dot{\boldsymbol{M}}] + \frac{1}{cR} [\dot{\boldsymbol{P}}]\right\} dV'. \tag{20}$$

From the first term in each of these integrals we can separate an integral over the boundary surface. To do this we use the vector identities

$$\operatorname{div} \left(\frac{1}{R} \boldsymbol{P}\right) = \frac{1}{R} \operatorname{div} \boldsymbol{P} + \boldsymbol{P} \cdot \operatorname{grad} \frac{1}{R}, \tag{21a}$$

$$\operatorname{curl} \left(\frac{1}{R} \boldsymbol{M}\right) = \frac{1}{R} \operatorname{curl} \boldsymbol{M} - \boldsymbol{M} \wedge \operatorname{grad} \frac{1}{R}. \tag{21b}$$

Integrating these relations throughout an arbitrary finite domain, we obtain, on using the Gauss theorem,

$$\int \frac{1}{R} (\boldsymbol{n} \cdot \boldsymbol{P}) dS = \int \left\{\frac{1}{R} \operatorname{div} \boldsymbol{P} + \boldsymbol{P} \cdot \operatorname{grad} \frac{1}{R}\right\} dV, \tag{22a}$$

$$\int \frac{1}{R} (\boldsymbol{n} \wedge \boldsymbol{M}) dS = \int \left\{\frac{1}{R} \operatorname{curl} \boldsymbol{M} - \boldsymbol{M} \wedge \operatorname{grad} \frac{1}{R}\right\} dV, \tag{22b}$$

the integrals on the left extending over the surface bounding the domain, $\boldsymbol{n}$ being the unit outward normal to the surface.

Assume now that the substance (i.e. the region of space where $\boldsymbol{P}$ and $\boldsymbol{M}$ differ from zero) remains within a finite closed surface. If then the integrals (19) and (20) are extended over the volume inside this surface, the surface integrals in (22) vanish and (19) and (20) may be expressed in the form

$$\phi = \int\left\{[\boldsymbol{P}] \cdot \text{grad}' \frac{1}{R} + \frac{1}{cR^2} \boldsymbol{R} \cdot [\dot{\boldsymbol{P}}]\right\} dV', \tag{23}$$

$$\boldsymbol{A} = \int\left\{[\boldsymbol{M}] \wedge \text{grad}' \frac{1}{R} - \frac{1}{cR^2} \boldsymbol{R} \wedge [\dot{\boldsymbol{M}}] + \frac{1}{cR}[\dot{\boldsymbol{P}}]\right\} dV'. \tag{24}$$

From (23) and (24) the physical significance of $\boldsymbol{P}$ and $\boldsymbol{M}$ can be appreciated. For this purpose consider the case when $\boldsymbol{P}$ and $\boldsymbol{M}$ are zero everywhere except within a

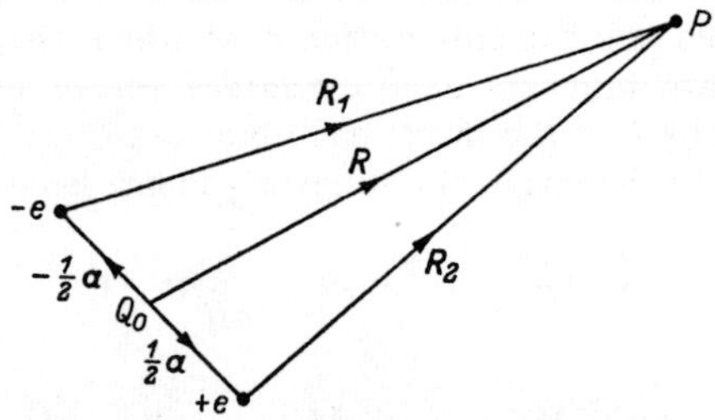

Fig. 2.1. Notation used in calculating the potentials of an electric dipole.

vanishingly small volume element at $\boldsymbol{r}_0(x_0, y_0, z_0)$. We may express this assumption formally with the help of Dirac's delta function (see Appendix IV):

$$\boldsymbol{P}(\boldsymbol{r}',t) = \boldsymbol{p}(t)\delta(\boldsymbol{r}' - \boldsymbol{r}_0), \tag{25}$$

$$\boldsymbol{M}(\boldsymbol{r}',t) = \boldsymbol{m}(t)\delta(\boldsymbol{r}' - \boldsymbol{r}_0). \tag{26}$$

(23) and (24) then become

$$\phi = [\boldsymbol{p}] \cdot \text{grad}_0 \frac{1}{R} + \frac{1}{cR^2} \boldsymbol{R} \cdot [\dot{\boldsymbol{p}}], \tag{27}$$

$$\boldsymbol{A} = [\boldsymbol{m}] \wedge \text{grad}_0 \frac{1}{R} - \frac{1}{cR^2} \boldsymbol{R} \wedge [\dot{\boldsymbol{m}}] + \frac{1}{cR}[\dot{\boldsymbol{p}}], \tag{28}$$

where now

$$\boldsymbol{R} = \boldsymbol{r} - \boldsymbol{r}_0, \qquad R = |\boldsymbol{r} - \boldsymbol{r}_0|, \tag{29}$$

and grad_0 implies that the operator is taken with respect to the coordinates x_0, y_0, z_0 of the point $\boldsymbol{r}_0$.

Eqs. (27) and (28) have a simple interpretation. Consider the electrostatic potential of two fixed electric charges $-e$ and $+e$ situated at points whose position vectors with respect to a fixed point $Q_0(\boldsymbol{r}_0)$ are $-\frac{1}{2}\boldsymbol{a}$ and $\frac{1}{2}\boldsymbol{a}$ respectively. The Coulomb potential $\phi(\boldsymbol{R})$ due to the charges is (see Fig. 2.1):

$$\phi = \frac{e}{R_2} - \frac{e}{R_1}. \tag{30}$$

If a is sufficiently small, we may expand $1/R_2$ in powers of the components of $\boldsymbol{a}$ and obtain

$$\frac{1}{R_2} = \frac{1}{R_1} + \boldsymbol{a} \cdot \operatorname{grad} \frac{1}{R_1} + \ldots, \tag{31}$$

where the operator grad is taken with respect to the coordinates of the point at which the charge $-e$ is situated. If we neglect the higher-order terms in (31), (30) becomes

$$\phi = e\boldsymbol{a} \cdot \operatorname{grad} \frac{1}{R_1} = e\,\frac{\boldsymbol{a} \cdot \boldsymbol{R}_1}{R_1{}^3}. \tag{32}$$

Suppose now that a is gradually decreased whilst e is increased without limits in such a way that $e\boldsymbol{a}$ approaches a finite value $\boldsymbol{p}$:

$$\lim_{a\to 0} e\boldsymbol{a} = \boldsymbol{p}. \tag{33}$$

Then R_1 approaches R and (32) becomes in the limit

$$\phi = \boldsymbol{p} \cdot \operatorname{grad}_0 \frac{1}{R}. \tag{34}$$

This expression is identical with (27) in the special case when $\boldsymbol{p}$ is independent of time.

If the charges depend on time but at each instant of time differ only in sign, we have, in place of (30),

$$\phi = \frac{e(t - R_2/c)}{R_2} - \frac{e(t - R_1/c)}{R_1} \tag{35}$$

and obtain by the same limiting process

$$\phi = [\boldsymbol{p}] \cdot \operatorname{grad}_0 \frac{1}{R} + \frac{1}{cR^2}\boldsymbol{R} \cdot [\dot{\boldsymbol{p}}],$$

which is (27). Thus (23) may be interpreted as the scalar potential of a *distribution of electric dipoles of moment* $\boldsymbol{P}$ *per unit volume.* The last term in (28) may be shown to be the magnetic potential arising from these dipoles.

In a similar way (28) may be shown to be the vector potential of a magnetic dipole of moment $\boldsymbol{m}(t)$, such a dipole being equivalent, of course, to an infinitesimal closed circuit of area A normal to $\boldsymbol{m}$, carrying the current* $c\boldsymbol{m}/A$. Hence the first two terms in (24) may be interpreted as the vector potential of a *distribution of magnetic dipoles of moment* $\boldsymbol{M}$ *per unit volume.*

2.2.2 Hertz vectors

Instead of using the potentials $\boldsymbol{A}$ and ϕ one can represent the field in terms of another pair of potential functions $\boldsymbol{\Pi}_e$ and $\boldsymbol{\Pi}_m$ which depend on $\boldsymbol{P}$ and $\boldsymbol{M}$ in a much simpler way. They are known as the *Hertz vectors* or *Polarization Potentials*† and may be introduced by means of the relations

---

* For a discussion of AMPÈRE's infinitesimal currents see any standard textbook on electricity, or MAX BORN, *Optik* (Berlin, Springer, 1933), pp. 305–306.

† $\boldsymbol{\Pi}_e$ and $\boldsymbol{\Pi}_m$ have similar transformation properties as $\boldsymbol{E}$ and $\boldsymbol{B}$ (or $\boldsymbol{D}$ and $\boldsymbol{H}$), i.e. they form a *six vector* (antisymmetric tensor of the second order). They are generalizations of a certain potential function introduced for the electromagnetic field of an oscillating dipole by H. HERTZ, *Ann. d. Physik.*, **36** (1889), 1. The vector character of the HERTZ potential was noted by A. RIGHI, *Nuovo Cimento*, **2** (1901), 104, who introduced the corresponding magnetic potential. A general theory of the HERTZ vectors and the associated gauge transformation is due to F. KOTTLER, *Ann. d. Physik*, (4), **71** (1923), 462 and A. NISBET, *Proc. Roy. Soc.*, A **231** (1955), 250. See also W. H. MCCREA, *ibid.*, **240** (1957), 447.

$$A = \frac{1}{c}\dot{\boldsymbol{\Pi}}_e + \operatorname{curl} \boldsymbol{\Pi}_m, \tag{36}$$

$$\phi = -\operatorname{div} \boldsymbol{\Pi}_e. \tag{37}$$

$\boldsymbol{\Pi}_e$ and $\boldsymbol{\Pi}_m$ are seen to bear the same mathematical relationship to A and ϕ as the polarizations $\boldsymbol{P}$ and $\boldsymbol{M}$ bear to $\tilde{j}/c$ and $\tilde{\rho}$.

The LORENTZ condition is now automatically satisfied and the eqs. (12) and (13) for A and ϕ will also be satisfied if $\boldsymbol{\Pi}_e$ and $\boldsymbol{\Pi}_m$ are solutions of the inhomogeneous wave equations

$$\nabla^2 \boldsymbol{\Pi}_e - \frac{1}{c^2}\ddot{\boldsymbol{\Pi}}_e = -4\pi \boldsymbol{P}, \tag{38}$$

$$\nabla^2 \boldsymbol{\Pi}_m - \frac{1}{c^2}\ddot{\boldsymbol{\Pi}}_m = -4\pi \boldsymbol{M}. \tag{39}$$

According to § 2.1, particular solutions of these equations may be expressed in the form of retarded potentials

$$\boldsymbol{\Pi}_e = \int \frac{[\boldsymbol{P}]}{R}\, dV', \tag{40}$$

$$\boldsymbol{\Pi}_m = \int \frac{[\boldsymbol{M}]}{R}\, dV'. \tag{41}$$

The field vectors may be expressed directly in terms of $\boldsymbol{\Pi}_e$ and $\boldsymbol{\Pi}_m$ by simple differentiation. We have from (9) and (36),

$$\boldsymbol{B} = \operatorname{curl}\left(\frac{1}{c}\dot{\boldsymbol{\Pi}}_e + \operatorname{curl} \boldsymbol{\Pi}_m\right), \tag{42}$$

and from (10), (36) and (37),

$$\begin{aligned} \boldsymbol{E} &= -\frac{1}{c}\frac{\partial}{\partial t}\left(\frac{1}{c}\dot{\boldsymbol{\Pi}}_e + \operatorname{curl} \boldsymbol{\Pi}_m\right) + \operatorname{grad}\operatorname{div} \boldsymbol{\Pi}_e \\ &= \operatorname{curl}\left(-\frac{1}{c}\dot{\boldsymbol{\Pi}}_m + \operatorname{curl} \boldsymbol{\Pi}_e\right) + \left(\nabla^2 \boldsymbol{\Pi}_e - \frac{1}{c^2}\ddot{\boldsymbol{\Pi}}_e\right), \end{aligned} \tag{43}$$

where the identity grad div $\equiv$ curl curl $+ \nabla^2$ has been used. According to (38) the term in the second bracket is equal to $-4\pi \boldsymbol{P}$; hence, using (1),

$$\boldsymbol{D} = \operatorname{curl}\left(-\frac{1}{c}\dot{\boldsymbol{\Pi}}_m + \operatorname{curl} \boldsymbol{\Pi}_e\right). \tag{44}$$

Finally, from (2), (42) and (39)

$$\begin{aligned} \boldsymbol{H} &= \operatorname{curl}\left(\frac{1}{c}\dot{\boldsymbol{\Pi}}_e + \operatorname{curl} \boldsymbol{\Pi}_m\right) + \left(\nabla^2 \boldsymbol{\Pi}_m - \frac{1}{c^2}\ddot{\boldsymbol{\Pi}}_m\right) \\ &= \frac{1}{c}\frac{\partial}{\partial t}\left(-\frac{1}{c}\dot{\boldsymbol{\Pi}}_m + \operatorname{curl} \boldsymbol{\Pi}_e\right) + \operatorname{grad}\operatorname{div} \boldsymbol{\Pi}_m. \end{aligned} \tag{45}$$

It was shown in § 2.1 that the potentials A and ϕ associated with a given field are not unique; any other pair (A',ϕ') which can be derived from a possible (A,ϕ) by means of a gauge transformation § 2.1 (14) will represent the same field. The HERTZ vectors too are not unique. They may be subjected to the following transformation, which leaves the field vectors invariant:

$$\left.\begin{aligned}\boldsymbol{\Pi}_e' &= \boldsymbol{\Pi}_e + \operatorname{curl} \boldsymbol{F} - \operatorname{grad} G,\\ \boldsymbol{\Pi}_m' &= \boldsymbol{\Pi}_m - \frac{1}{c}\dot{\boldsymbol{F}},\end{aligned}\right\} \tag{46}$$

where the vector function $\boldsymbol{F}$ and the scalar function G are any solutions of the homogeneous wave equations

$$\left.\begin{aligned}\nabla^2 \boldsymbol{F} - \frac{1}{c^2}\ddot{\boldsymbol{F}} &= 0,\\ \nabla^2 G - \frac{1}{c^2}\ddot{G} &= 0.\end{aligned}\right\} \tag{47}$$

The invariance under this transformation is readily verified by substituting from (46) into (36) and (37). $\boldsymbol{A}$ and ϕ are then transformed into $\boldsymbol{A}'$ and ϕ' according to (14), § 2.1, with

$$\chi = -\frac{1}{c}\dot{G}.$$

Moreover, substitution from (46) into (38) and (39) shows that $\boldsymbol{\Pi}_e'$ and $\boldsymbol{\Pi}_m'$ also satisfy the inhomogeneous wave equations for the HERTZ vectors.

2.2.3 The field of a linear electric dipole

For the purpose of later applications it will be useful to write down the full explicit solutions of the field equations for a field in vacuum produced by a linear electric dipole, situated at a point $\boldsymbol{r}_0$ and vibrating in a fixed direction specified by a unit vector $\boldsymbol{n}$. Such a dipole is characterized by an electric polarization

$$\boldsymbol{P}(\boldsymbol{r},t) = p(t)\delta(\boldsymbol{r} - \boldsymbol{r}_0)\boldsymbol{n}, \tag{48}$$

where p is a function of time and δ the DIRAC delta function. A general dipole whose orientation $\boldsymbol{n}$ is also a function of time is equivalent to three linear dipoles, with their moment vectors oriented along three mutually perpendicular directions.

According to (40) the electric HERTZ vector associated with (48) is

$$\boldsymbol{\Pi}_e = \frac{p(t - R/c)}{R}\boldsymbol{n}, \tag{49}$$

where R is the distance of the point $\boldsymbol{r}$ from the point $\boldsymbol{r}_0$.

Since $\boldsymbol{\Pi}_m = 0$ and $\boldsymbol{\Pi}_e$ satisfies (except at the origin) the homogeneous wave equation for vacuum, the eqs. (42)–(45) become

$$\boldsymbol{E} = \boldsymbol{D} = \operatorname{curl}\operatorname{curl}\boldsymbol{\Pi}_e, \tag{50}$$

$$\boldsymbol{B} = \boldsymbol{H} = \frac{1}{c}\operatorname{curl}\dot{\boldsymbol{\Pi}}_e. \tag{51}$$

Using the identity $\operatorname{curl}\operatorname{curl} \equiv \operatorname{grad}\operatorname{div} - \nabla^2$ and the relation $\nabla^2\boldsymbol{\Pi}_e = \frac{1}{c^2}\ddot{\boldsymbol{\Pi}}_e$, (50) may be re-written in the form

$$\boldsymbol{E} = \boldsymbol{D} = \operatorname{grad}\operatorname{div}\boldsymbol{\Pi}_e - \frac{1}{c^2}\ddot{\boldsymbol{\Pi}}_e. \tag{52}$$

From (49) we obtain, after a simple calculation,

$$\operatorname{div} \mathbf{\Pi}_e = -\left\{\frac{[p]}{R^3} + \frac{[\dot{p}]}{cR^2}\right\} (\boldsymbol{n} \cdot \boldsymbol{R}),$$

$$\operatorname{grad} \operatorname{div} \mathbf{\Pi}_e = \left\{\frac{3[p]}{R^5} + \frac{3[\dot{p}]}{cR^4} + \frac{[\ddot{p}]}{c^2R^3}\right\} (\boldsymbol{n} \cdot \boldsymbol{R})\boldsymbol{R} - \left\{\frac{[p]}{R^3} + \frac{[\dot{p}]}{cR^2}\right\} \boldsymbol{n},$$

$$\operatorname{curl} \mathbf{\Pi}_e = \left\{\frac{[p]}{R^3} + \frac{[\dot{p}]}{cR^2}\right\} (\boldsymbol{n} \wedge \boldsymbol{R}),$$

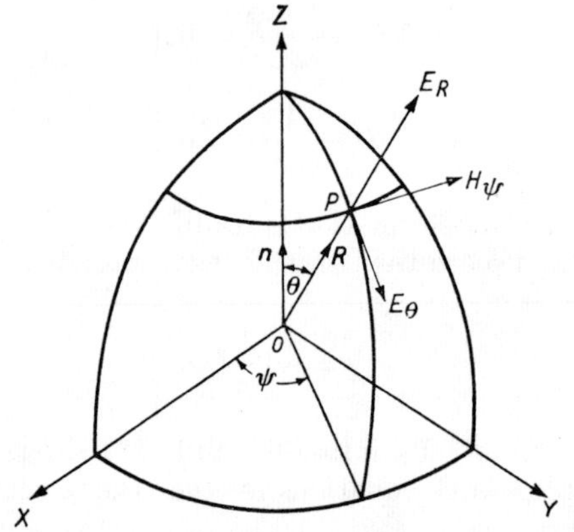

Fig. 2.2. Calculation of the field of a linear electric dipole with moment-vector along the z-axis.

where, as before, the square brackets denote retarded values. Substituting into (52) and (51) we obtain the required formulae for the field vectors:

$$\boldsymbol{E} = \boldsymbol{D} = \left\{\frac{3[p]}{R^5} + \frac{3[\dot{p}]}{cR^4} + \frac{[\ddot{p}]}{c^2R^3}\right\} (\boldsymbol{n} \cdot \boldsymbol{R})\boldsymbol{R} - \left\{\frac{[p]}{R^3} + \frac{[\dot{p}]}{cR^2} + \frac{[\ddot{p}]}{c^2R}\right\} \boldsymbol{n}, \tag{53}$$

$$\boldsymbol{B} = \boldsymbol{H} = \left\{\frac{[\dot{p}]}{cR^3} + \frac{[\ddot{p}]}{c^2R^2}\right\} (\boldsymbol{n} \wedge \boldsymbol{R}). \tag{54}$$

We shall also later need expressions for $\boldsymbol{E}$ and $\boldsymbol{H}$ in terms of spherical polar coordinates. Taking the $\boldsymbol{n}$ direction as the z-axis (see Fig. 2.2), and denoting by $\boldsymbol{i}_R, \boldsymbol{i}_\theta$ and $\boldsymbol{i}_\psi$ the unit vectors in the direction of increasing R, θ and ψ we have

$$\boldsymbol{R} = R\boldsymbol{i}_R, \qquad \boldsymbol{n} = (\cos\theta)\boldsymbol{i}_R - (\sin\theta)\boldsymbol{i}_\theta,$$
$$(\boldsymbol{n} \cdot \boldsymbol{R}) = R\cos\theta, \qquad (\boldsymbol{n} \wedge \boldsymbol{R}) = (R\sin\theta)\boldsymbol{i}_\psi, \tag{55}$$

and (53) and (54) give

$$\boldsymbol{E} = E_R\boldsymbol{i}_R + E_\theta\boldsymbol{i}_\theta, \qquad \boldsymbol{H} = H_\psi\boldsymbol{i}_\psi, \tag{56}$$

where

$$\left.\begin{aligned} E_R &= 2\left(\frac{[p]}{R^3} + \frac{[\dot{p}]}{cR^2}\right)\cos\theta, \\ E_\theta &= \left(\frac{[p]}{R^3} + \frac{[\dot{p}]}{cR^2} + \frac{[\ddot{p}]}{c^2R}\right)\sin\theta, \\ H_\psi &= \left(\frac{[\dot{p}]}{cR^2} + \frac{[\ddot{p}]}{c^2R}\right)\sin\theta. \end{aligned}\right\} \tag{57}$$

Thus the electric vector lies in the meridional plane through the axis of the dipole, and the magnetic vector is perpendicular to this plane.

Of particular interest is the field in regions which are so far away from the dipole that all terms but those in $1/R$ may be neglected in the above expressions. This *wave-* (or *radiation-*) *zone* is characterized by

$$R \gg c\left|\frac{p}{\dot{p}}\right|, \qquad R \gg c\left|\frac{\dot{p}}{\ddot{p}}\right|. \tag{58}$$

In this region

$$E_\theta \sim H_\varphi \sim \frac{[\ddot{p}]}{c^2 R}\sin\theta, \tag{59}$$

and the other components are negligible. Hence, in the wave zone, $\boldsymbol{E}$ and $\boldsymbol{H}$ are of equal magnitude and perpendicular to each other and to the radius vector $\boldsymbol{R}$, which now coincides with the direction of the Poynting vector. In this zone the structure of the field of a linear electric dipole is therefore similar to that of a plane wave. However, on each "wave surface" (sphere centred on $\boldsymbol{r}_0$) the field vectors now vary from point to point, diminishing in magnitude from the equator to the poles, being zero on the axis of the oscillator. The dipole therefore does not radiate energy in the direction of its axis.

Let us now calculate the amount of energy radiated per second across each of the spherical wave surfaces. It is given by the integral of the magnitude S of the Poynting vector taken over this surface. In the wave zone,

$$S = \frac{c}{4\pi}|\boldsymbol{E}\wedge\boldsymbol{H}| = \frac{c}{4\pi}|E_\theta H_\varphi| = \frac{[\ddot{p}]^2}{4\pi c^3 R^2}\sin^2\theta. \tag{60}$$

Hence the total amount of energy which is radiated across the spherical surface per second is ($d\sigma$ denoting element of the surface)

$$\begin{aligned}\int S\,d\sigma &= \frac{[\ddot{p}]^2}{4\pi c^3 R^2}\int_0^\pi \sin^2\theta\,.\,2\pi R^2\sin\theta\,d\theta \\ &= \frac{2[\ddot{p}]^2}{3c^3}.\end{aligned} \tag{61}$$

Consider the special case when $p(t)$ is periodic with angular frequency ω:

$$p(t) = p_0 e^{-i\omega t}, \tag{62}$$

where p_0 is a complex constant, and the real part of (62) is understood to represent p. The two conditions (58) now reduce to the single condition

$$R \gg \lambda/2\pi, \qquad (\lambda = 2\pi c/\omega) \tag{63}$$

and (59) gives, for the non-vanishing components of the field in the wave zone,

$$E_\theta = H_\varphi \sim -\left(\frac{\omega}{c}\right)^2 p_0 \sin\theta\,\frac{e^{-i\omega(t-R/c)}}{R}, \tag{64}$$

it being understood that the real part of the expression on the right-hand side of (64) is taken. The amount of energy crossing per second a unit area of the spherical surface in the wave zone is

$$S = \frac{c}{4\pi}\left(\frac{\omega}{c}\right)^4\frac{|p_0|^2}{R^2}\sin^2\theta\cos^2[\omega(t-R/c)-\alpha], \tag{65}$$

α being the phase of p_0.

On taking the time average over an interval of time which is large compared with the period $T = 2\pi/\omega$, (65) reduces to

$$\langle S \rangle = \frac{|p_0|^2}{8\pi R^2} \frac{\omega^4}{c^3} \sin^2 \theta. \tag{66}$$

The (time averaged) energy crossing the whole surface per second is therefore

$$\int \langle S \rangle \, \delta\sigma = \frac{1}{3} \frac{|p_0|^2 \omega^4}{c^3}. \tag{67}$$

2.3 THE LORENTZ–LORENZ FORMULA AND ELEMENTARY DISPERSION THEORY*

2.3.1 The dielectric and magnetic susceptibilities

The interpretation of the potentials in terms of polarization and magnetization is of fundamental importance in the atomic theory of matter. From the standpoint of such a theory the additive relations § 2.2 (1) and (2) connecting $\boldsymbol{E}$ with $\boldsymbol{D}$ and $\boldsymbol{P}$ and $\boldsymbol{H}$ with $\boldsymbol{B}$ and $\boldsymbol{M}$ have a more direct physical meaning than the multiplicative relations (the material equations (10) and (11), in § 1.1).

In the atomic theory one regards matter as composed of interacting particles (atoms and molecules) embedded in the vacuum. These entities produce a field which has large local variations in the interior of the matter. This internal field is modified by any field which is applied externally; the properties of the matter are then derived by averaging over the total field within it. As long as the region over which the average is taken is large compared with the linear dimensions of the particles, the electromagnetic properties of each can be simply described by an electric and a magnetic dipole; the secondary field is then the field due to these dipoles (with retardation). This is, in fact, exactly what we have just described by regarding matter as a continuous distribution interacting with the field: it corresponds to the first approximation (for slow variation in space) of the atomic theory. In this approximation for sufficiently weak fields† one can assume $\boldsymbol{P}$ and $\boldsymbol{M}$ to be proportional to $\boldsymbol{E}$ and $\boldsymbol{H}$ respectively:

$$\boldsymbol{P} = \eta \boldsymbol{E}, \qquad \boldsymbol{M} = \chi \boldsymbol{H}. \tag{1}$$

The factor η is called the *dielectric susceptibility* and χ the *magnetic susceptibility*, the latter being a concept current in practical magnetism. From (1) and from eq. (1) and (2) in § 2.2 it follows, on comparison with the material equations (10) and (11) in § 1.1, that the dielectric constant ε and the magnetic permeability μ are related to the dielectric and magnetic susceptibilities by the formulae

$$\varepsilon = 1 + 4\pi\eta, \qquad \mu = 1 + 4\pi\chi. \tag{2}$$

---

* In this section we depart from our deductive treatment in order to obtain a clearer understanding of the phenomena before formulating the integral equations which will be introduced in § 2.4 in place of MAXWELL's differential equations. The considerations of this section are more illustrative than rigorous, since a rigorous treatment would involve most of the atomistic theory of matter and take us far beyond the reasonable bounds of this book.

† A field which is considered strong on the atomic level is one in which the electrostatic energy of an electron changes by an appreciable fraction of the ionization potential of the atom in a distance equal to the atomic diameter, i.e. a field of millions of volts per centimetre. Electrical breakdowns in real materials actually occur at much smaller fields.

By means of (2) the complete formal equivalence of the "multiplicative" and "additive" treatments is secured. But such formal comparison of the constants makes no use of any of the consequences of the atomistic structure. While a detailed discussion involving the results of the atomistic theory is outside the scope of this book, it will be useful to introduce some of the relevant concepts and formulae. With their help we shall obtain a clearer understanding of the physical content of the integral equations which will be introduced in § 2.4 in place of the usual differential equations of MAXWELL's theory.

The simplest assumption which leads one step into the atomistic theory is to regard matter as being made up of certain physical entities—the molecules—which are polarizable, so that under the influence of an external field they show electric and magnetic moments. In the first approximation it may be assumed that the components of these moments are linear functions of the components of the field; in general the direction of the moment vector does not coincide with the direction of the field. This assumption has many consequences, which can only be briefly studied here. We shall confine our attention to isotropic non-magnetic substances, and examine first the dependence of the electric constants on the density in a substance composed of similar molecules. We shall also study the dependence of the refractive index on frequency. Only somewhat oversimplified arguments can be put forward here, but a rigorous, though a more formal, derivation of the main result (the LORENTZ–LORENZ formula) will be given in § 2.4.

2.3.2 The effective field

To begin with we must distinguish between the *effective field* $\boldsymbol{E}'$, $\boldsymbol{H}'$ acting on a molecule and the *mean* or *observed* field $\boldsymbol{E}$, $\boldsymbol{H}$ obtained by averaging over a region which contains a great number of molecules. The difference between the two fields is due to the gaps between the molecules and depends on the number of molecules per unit volume (their number density N).

To estimate the difference $\boldsymbol{E}' - \boldsymbol{E}$, consider a particular molecule and imagine it to be surrounded by a small sphere whose radius is nevertheless large compared with its linear dimensions. We shall consider separately the effects on the central molecule produced by the matter outside and inside this sphere.

In determining the effect of the matter *outside* the sphere we may clearly neglect the molecular structure and treat the substance as continuous. We may then assume that, outside the sphere, the polarization $\boldsymbol{P}$ produced by the mean electric field is constant. Concerning the effect of the molecules *inside* the sphere we shall assume, as may be shown for a number of important special cases, including that of random distribution, that they do not produce any resulting field at the central molecule. Hence we may regard the molecule as being situated in a spherical region, inside which there is vacuum and outside which there is a homogeneously polarized medium. We must then determine the potential ϕ of this configuration, i.e. of the free charges on the spherical discontinuity surface at which $\boldsymbol{P}$ changes from the interior value zero to a constant exterior value.

For this purpose consider the potential $\tilde{\phi}$ of the "complementary" configuration, namely of a homogeneously polarized sphere surrounded by vacuum. The superposition of these two configurations is a homogeneously polarized substance with no boundary. Hence the potential due to the boundary is zero and we have

$$\phi + \tilde{\phi} = 0. \tag{3}$$

Now the potential $\tilde{\phi}$ can be immediately obtained from § 2.2 (23) by taking $\boldsymbol{P}$ to be constant. This gives

$$\phi = -\tilde{\phi} = -\boldsymbol{P}\,.\int \text{grad}'\,\frac{1}{R}\,dV'. \tag{4}$$

Since

$$R = \sqrt{(x-x')^2 + (y-y')^2 + (z-z')^2},$$

we may replace grad' by — grad in (4), and write

$$\phi = -\tilde{\phi} = \boldsymbol{P}\,.\,\text{grad}\int \frac{dV'}{R} = -\boldsymbol{P}\,.\,\text{grad}\,\phi_0, \tag{5}$$

where

$$\phi_0 = -\int \frac{dV'}{R}. \tag{6}$$

(6) may be interpreted as the potential of a uniformly charged sphere of charge density — 1. Hence it satisfies the Poisson equation

$$\nabla^2\phi_0 = 4\pi. \tag{7}$$

Now the components of the field strengths associated with the potential (5) are

$$\begin{aligned} -\frac{\partial\phi}{\partial x} &= \frac{\partial}{\partial x}\left[P_x\frac{\partial\phi_0}{\partial x} + P_y\frac{\partial\phi_0}{\partial y} + P_z\frac{\partial\phi_0}{\partial z}\right] \\ &= P_x\frac{\partial^2\phi_0}{\partial x^2} + P_y\frac{\partial^2\phi_0}{\partial x\partial y} + P_z\frac{\partial^2\phi_0}{\partial x\partial z} \end{aligned} \tag{8}$$

etc. At the centre of the sphere we have, by symmetry,

$$\frac{\partial^2\phi_0}{\partial x\partial y} = \frac{\partial^2\phi_0}{\partial y\partial z} = \frac{\partial^2\phi_0}{\partial z\partial x} = 0, \tag{9}$$

and

$$\frac{\partial^2\phi_0}{\partial x^2} = \frac{\partial^2\phi_0}{\partial y^2} = \frac{\partial^2\phi_0}{\partial z^2}. \tag{10}$$

Using (7) it follows that each term in (10) is equal to $4\pi/3$; (8) then shows that the required contribution to the effective field strength is

$$-\nabla\phi = \frac{4\pi}{3}\boldsymbol{P}. \tag{11}$$

The total field inside the sphere, which is the effective field acting on the central molecule, is obtained by adding to this the mean field $\boldsymbol{E}$, giving

$$\boldsymbol{E}' = \boldsymbol{E} + \frac{4\pi}{3}\boldsymbol{P}. \tag{12}$$

For magnetizable substances there is a corresponding relation between $\boldsymbol{H}'$, $\boldsymbol{H}$ and $\boldsymbol{M}$. As we shall however treat only non-magnetic substances we shall always have $\boldsymbol{H}' = \boldsymbol{H}$.

Next we must connect the polarization $\boldsymbol{P}$ with the density.

2.3.3 The mean polarizability: the Lorentz-Lorenz formula

As already explained, it will be assumed that, for each molecule, the electric dipole moment $\boldsymbol{p}$ established under the influence of the field is proportional to the effective field* $\boldsymbol{E}'$

$$\boldsymbol{p} = \alpha \boldsymbol{E}'. \tag{13}$$

The molecule is here assumed to be isotropic;† but as we are interested only in the average effect over all possible orientations of the molecule, it will not be necessary to assume that each *individual* molecule is isotropic. α will then be regarded as representing the *mean polarizability*, and since $\boldsymbol{p}$ is of dimensions $[el]$ and $\boldsymbol{E}'$ of dimensions $[el^{-2}]$ (e = charge, l = length), α is seen to have the dimensions $[l^3]$, i.e. those of a volume.

If, as before, N is the number of molecules per unit volume, the total *electric moment* $\boldsymbol{P}$ *per unit volume* is given by

$$\boldsymbol{P} = N\boldsymbol{p} = N\alpha\boldsymbol{E}'. \tag{14}$$

On eliminating $\boldsymbol{E}'$ between (12) and (14) we obtain the first formulae (1), with the following explicit expression for the dielectric susceptibility:

$$\eta = \frac{N\alpha}{1 - \frac{4\pi}{3}N\alpha}. \tag{15}$$

If now we substitute for η into the first formula in (2) we obtain the following expression for the dielectric constant:

$$\varepsilon = \frac{1 + \frac{8\pi}{3}N\alpha}{1 - \frac{4\pi}{3}N\alpha}. \tag{16}$$

Conversely (16) gives information about the dependence of the mean polarizability on ε and N, or (if MAXWELL's relation $\varepsilon = n^2$ is used) on n and N:

$$\alpha = \frac{3}{4\pi N}\frac{\varepsilon - 1}{\varepsilon + 2} = \frac{3}{4\pi N}\frac{n^2 - 1}{n^2 + 2}. \tag{17}$$

By a remarkable coincidence, the relation (17) was discovered independently and practically at the same time by two scientists of almost identical names,‡ LORENTZ and LORENZ, and is accordingly called the *Lorentz–Lorenz formula*. It is seen to be a

---

* We are restricting our discussion to so-called non-polar molecules, having no permanent dipole moment in the absence of the field.

† In general the polarizability α is a tensor of the second rank.

‡ H. A. LORENTZ, *Wiedem. Ann.*, **9** (1880), 641; L. LORENZ, *Wiedem. Ann.*, **11** (1881), 70.
An analogous formula for statical fields was derived earlier by R. CLAUSIUS, *Mechanische Wärmetheorie*, **2** (Braunschweig, 2nd ed. 1879), p. 62), and by O. F. MOSSOTTI, *Mem. Soc. Sci. Modena*, **14** (1850), p. 49. These authors tried to explain the dielectric properties of insulators on the assumption that the atoms are small conducting spheres whose mutual distances are large compared to their diameters, and derived the following expression for the number density g of the spheres in terms of the dielectric constant:

$$g = \frac{\varepsilon - 1}{\varepsilon + 2}.$$

bridge which connects MAXWELL's phenomenological theory with the atomistic theory of matter.

Instead of the mean polarizability α one often uses another quantity A, called *molar refractivity*. (In the case of monatomic substances it is also called *atomic refractivity*). This is essentially the total polarizability of a mole of the substance,* being defined as

$$A = \frac{4\pi}{3} N_m \alpha. \tag{18}$$

Here $N_m = 6{\cdot}02 \,.\, 10^{23}$ is the AVOGADRO number,† namely the number of molecules in a mole. If W is the molecular weight, ρ the density, p the pressure and T the absolute temperature, then the molar volume is

$$\frac{N_m}{N} = \frac{W}{\rho} = \frac{RT}{p}, \tag{19}$$

where BOYLE's law was used. Hence the molar refractivity may be written in the alternative forms

$$A = \frac{W}{\rho}\frac{n^2-1}{n^2+2} = \frac{RT}{p}\frac{n^2-1}{n^2+2}. \tag{20}$$

(20) shows that A has the dimensions and the order of magnitude of a molar volume.

For a gas, n^2 is nearly unity, so that

$$\alpha \sim \frac{n^2-1}{4\pi N}, \qquad A \sim \frac{W}{\rho}\frac{n^2-1}{3} = \frac{RT}{p}\frac{n^2-1}{3}. \tag{21}$$

The formula (20) gives the explicit dependence of the refractive index, for light of any one particular colour, on the density, and it should hold when the density is changed provided that isotropy is preserved. In the case of *gases* we actually find that $n^2 - 1$ and ρ are very nearly proportional to each other as demanded by the second formula in (21). But even at high pressure, when n will differ appreciably from unity, the molar refractivity remains substantially constant, as illustrated in Table IV. The molar refractivity is also found to remain practically constant when the gas is condensed into a liquid (Table V).

TABLE IV

The molar refractivity A of air at different pressures and at about 14·5°C, *for sodium D light*

Pressure in atm	n	A
1·00	1·0002929	4·606
42·13	1·01241	4·661
96·16	1·02842	4·713
136·21	1·04027	4·743
176·27	1·05213	4·772

* A mole or a gram-molecule is that amount of the substance whose weight, in grams, is equal to the molecular weight of the substance.

† This is the name internationally accepted; however, in the German literature the term *Loschmidt's number* is used, after the physicist who first gave a numerical estimate for it from experimental data.

TABLE V

Molar refractivity A of different compounds for sodium D light; in the columns giving n and A the first entry refers to liquid state, the second to vapour. The densities of the vapours are not stated but were calculated assuming perfect gas law from the known density of hydrogen

Substance	Formula	W	n	ρ	A
Oxygen . . .	O_2	32	1·221 1·000271	1·124	4·00 4·05
Hydrochloric acid .	HCl	36·5	1·245 1·000447	0·95	5·95 6·68
Water . . .	H_2O	18	1·334 1·000249	1·00	3·71 3·72
Carbon disulphide .	CS_2	76	1·628 1·00147	1·264	21·33 21·99
Acetone . . .	C_3H_6O	58	1·3589 1·00108	0·791	16·14 16·16

From (20) it can be shown that, to a good approximation, the molar refractivity of a mixture of two substances is equal to the sum of the contributions due to each substance. Thus, if two liquids of refractivities A_1 and A_2 are mixed, and if a unit volume of the first liquid contains N_1 molecules and of the second N_2 molecules, then the molar refractivity of the mixture will be

$$A = \frac{N_1 A_1 + N_2 A_2}{N_1 + N_2}. \tag{22}$$

This result is also found to be in reasonably good agreement with experiment [cf. Table VI].

TABLE VI

Molar refractivity for sodium D light of mixtures of water and sulphuric acid at 15°C

Weight per cent	$\frac{N_1}{N_1+N_2}$	W	n	ρ	A (interpolated from (22))	A (calculated from (20))
0	0	18·00	1·3336	0·9991	3·72	3·72
19·98	0·044	21·52	1·3578	1·1381	4·19	4·15
39·76	0·109	26·72	1·3817	1·2936	4·80	4·81
59·98	0·216	35·28	1·4065	1·4803	5·86	5·86
80·10	0·425	52·00	1·4308	1·6955	7·95	7·93
100	1	98	1·4277	1·8417	13·68	13·68

Finally, we consider the dependence of the molar refractivity A of a compound on the atomic refractivities of its constituent atoms. If the molecule consists of $\mathcal{N}_1$ atoms of refractivity A_1 and of $\mathcal{N}_2$ atoms of refractivity A_2, etc., then obviously

$$A = \mathcal{N}_1 A_1 + \mathcal{N}_2 A_2. \tag{23}$$

This formula too agrees fairly well with experiment, as is seen from Table VIII. The refractivities of the constituent atoms are given in Table VII.

TABLE VII

Atomic refractivities of various elements for sodium D light

Element	Formula	Atomic Weight	A
Hydrogen . .	H	1	1·02
Carbon . . .	C	12	2·11
Oxygen . . .	O	16	2·01
Sulphur . .	S	32	8·23
Chlorine . .	Cl	35·5	5·72

TABLE VIII

The molar refractivities of various compounds for sodium D light

Substance	Formula	A (calculated from (20))	A (calculated from (23))
Hydrochloric acid .	HCl	6·68	6·74
Water . . .	H_2O	3·72	4·05
Carbon disulphide .	CS_2	21·97	18·57
Acetone . .	C_3H_6O	16·14	14·46

The good agreement obtained in this case is really better than one could reasonably expect, on the basis of our simple considerations. Refinements are in fact necessary in some cases; for example in the case of nitrogen one has to attribute different atomic polarizabilities to the nitrogen atom according to the compounds which it enters. With the necessary modifications, the polarizability α is quite a well defined atomic parameter which may be used also in the study of other phenomena, such as the VAN DER WAALS cohesion forces, adsorption phenomena on surfaces, etc. These can only be satisfactorily treated, however, with the help of quantum mechanics.

2.3.4 Elementary theory of dispersion

In § 1.2 it was pointed out that the phase velocity and consequently the refractive index cannot be constants of the medium as suggested by our earlier formal considerations, but that these quantities must in fact depend on frequency. The variation of the refractive index with frequency constitutes the phenomenon of *dispersion*. For an adequate treatment of dispersion, it would be necessary to go deeply into the atomic theory of matter, but it is possible to give a simplified model of a dispersing medium (essentially due to H. A. LORENTZ), by making use of one or two basic results concerning the structure of molecules.

A molecule consists of a number of heavy particles (the atomic nuclei of the atoms forming the molecule) around which light particles (electrons) revolve. The electrons carry negative and the nuclei positive charges. In neutral molecules the charges of the electrons just compensate those of the nuclei. However, the centres of the positive (nuclear) charges and that of the negative (electronic) charges may not coincide; this system has then an electric dipole and is called *polar*. For the sake of simplicity we shall here exclude polar molecules, although these play a great part in many physical and chemical phenomena.

If a non-polar molecule is subjected to an electric field, the electrons and nuclei are displaced and a dipole moment is generated. The vectorial sum of all the dipole moments of the molecules in the unit of volume is essentially the polarization vector $\boldsymbol{P}$ introduced formally in the preceding section.

In order to determine the dependence of the polarization and of the refractive index on the frequency of the field we must first find the displacement $\boldsymbol{r}$ of each charged particle from its equilibrium position. We may assume that each electron is acted on by the LORENTZ force $\boldsymbol{F}$ (see eq. (34) in § 1.1):

$$\boldsymbol{F} = e\left(\boldsymbol{E}' + \frac{\boldsymbol{v}}{c} \wedge \boldsymbol{B}'\right),$$

where e is the charge of the electron and $\boldsymbol{v}$ its velocity. It will be assumed that the velocity of the electron is small in comparison with the velocity c of light in vacuum, so that the contribution of the magnetic field may be neglected in the expression for the LORENTZ force. The rigorous determination of the effective displacement of the nuclei and electrons under the action of the electric force is a complicated problem of quantum mechanics. It is however plausible, and actually confirmed by the rigorous theory, that the electron behaves, to a good approximation, as if it were bound to an equilibrium position by a quasi-elastic restoring force

$$\boldsymbol{Q} = -q\boldsymbol{r}.$$

Hence if m denotes the mass of the electron, its equation of motion is

$$m\ddot{\boldsymbol{r}} + q\boldsymbol{r} = e\boldsymbol{E}'. \tag{24}$$

Let ω be the angular frequency of the incident field,

$$\boldsymbol{E}' = \boldsymbol{E}_0' e^{-i\omega t}, \tag{25}$$

and take as trial solution of (24) the expression

$$\boldsymbol{r} = \boldsymbol{r}_0 e^{-i\omega t}. \tag{26}$$

(24) then gives the required stationary solution

$$\boldsymbol{r} = \frac{e\boldsymbol{E}'}{m(\omega_0^2 - \omega^2)}, \tag{27}$$

where

$$\omega_0 = \sqrt{\frac{q}{m}} \tag{28}$$

is called the *resonance* (or *absorption*) *frequency*. According to (27) the electron oscillates with the frequency of the incident field.

Each electron contributes to the polarization a moment $\boldsymbol{p} = e\boldsymbol{r}$. There will also be contributions from the nuclei; but since the nuclear masses are heavy in comparison with the masses of the electrons, their contributions may in the first approximation be neglected. Also, assuming for the moment that there is only one effective electron in a molecule with resonance frequency ω_0, we obtain for the total polarization $\boldsymbol{P}$ the expression

$$\boldsymbol{P} = N\boldsymbol{p} = Ne\boldsymbol{r} = N\frac{e^2}{m}\frac{\boldsymbol{E}'}{(\omega_0^2 - \omega^2)}. \tag{29}$$

Comparison of (29) with (14) gives

$$N\alpha = N\frac{e^2}{m(\omega_0^2 - \omega^2)}, \tag{30}$$

expressing the "density of polarizability" in terms of atomic parameters. Thus it turns out that the quantity α is not a constant, as it ought to be if the ε in (17) denotes the static dielectric constant. It is convenient to introduce the notion of a frequency-dependent dielectric constant $\varepsilon(\omega)$ defined by MAXWELL's relation $\varepsilon = n^2$, where n is the refractive index, a function $n(\omega)$ of ω. The static dielectric constant is then the value $\varepsilon(0) = n^2(0)$; it corresponds, according to (17), to the limiting value $N\alpha(0)$ which is found from (30) to be

$$N\alpha(0) = \frac{Ne^2}{m\omega_0^2};$$

if this is introduced into (16), one obtains the static constant $\varepsilon(0)$.

For $\omega \neq 0$, the function $N\alpha(\omega)$ is, according to (30), monotonically increasing with ω, but has an infinity (resonance point) for $\omega = \omega_0$; for $\omega > \omega_0$, it approaches with increasing ω the value 0 from the negative side. Substituting from (30) into (17), we find the explicit dependence of the refractive index on frequency:

$$\frac{n^2 - 1}{n^2 + 2} = \frac{4\pi}{3}\frac{Ne^2}{m(\omega_0^2 - \omega^2)}. \tag{31}$$

For a gas, n is close to unity so that we may set $n^2 + 2 \sim 3$ in the denominator on the left-hand side and obtain

$$n^2 - 1 \sim 4\pi N\alpha = \frac{4\pi Ne^2}{m(\omega_0^2 - \omega^2)}. \tag{32}$$

It is seen that n is an increasing function of the frequency. The dispersion is then said to be *normal*. Further $n \gtrless 1$ according as $\omega \lessgtr \omega_0$, and n approaches unity with increasing ω (see Fig. 2.3).

At the resonance frequency ($\omega = \omega_0$) n and α are actually not infinite as our formula suggests. The singularity arises only formally, because we neglected the effect of damping. Damping is in fact an essential factor in the whole process as the vibrating electrons emit electromagnetic waves which carry away energy. But there are also other reasons for the dissipation of energy (e.g. because of collisions between the atoms). Formally damping may be taken into account by adding in the equation of motion (24) a term $g\dot{\boldsymbol{r}}$ representing a resisting force:

$$m\ddot{\boldsymbol{r}} + g\dot{\boldsymbol{r}} + q\boldsymbol{r} = e\boldsymbol{E}'. \tag{33}$$

Instead of (27) we now have

$$\mathbf{r} = \frac{eE'}{m(\omega_0^2 - \omega^2) - i\omega g}. \tag{34}$$

The polarization, and hence also $N\alpha$, becomes a complex quantity. It can be shown (cf. the similar situation in the theory of metals where a complex refractive index is used—Chapter XIII) that the modulus of this complex function is—apart from a small factor, due to absorption—the actual "density of polarizability" and this is shown by the dotted curve in Fig. 2.3. The curve has a sharp maximum at a value of ω which is slightly smaller than ω_0, and a sharp minimum at a value slightly greater than ω_0. Between the maximum and minimum the function decreases with increasing frequency

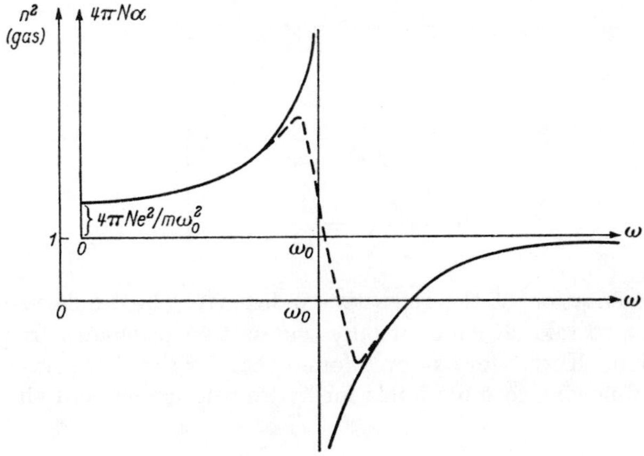

Fig. 2.3. Dispersion curves.
Full curve: effect of damping neglected.
Broken curve: effect of damping taken into account. (The ordinates now represent the real parts of n^2 and of $4\pi N\alpha$.)

and we speak of a region of *anomalous dispersion*. The rays of shorter wavelength are then refracted less than those of longer wavelength, and this results in the reversal of the usual sequence of prismatic colours. For our purposes the region of anomalous dispersion is of little importance, since the absorption frequencies of free atoms lie almost exclusively in the ultra-violet region of the spectrum. The refractive index for visible light is then always greater than unity.

So far we have assumed that the system has only one resonance frequency. In general there will be many such frequencies, even in systems with the same kind of molecules; (31) and (32) must then be replaced by more general expressions. Neglecting for the moment again the motion of the nuclei, one has, in place of (31),

$$\frac{4\pi}{3} N\alpha = \frac{n^2 - 1}{n^2 + 2} = \frac{4\pi}{3} N \frac{e^2}{m} \sum_k \frac{f_k}{\omega_k^2 - \omega^2}, \tag{35}$$

where Nf_k is the number of electrons corresponding to the resonance frequency ω_k. For gases ($n \sim 1$) we may rewrite (35) in the form

$$n^2 - 1 = 4\pi N\alpha = \sum_k \frac{\rho_k}{\nu_k^2 - \nu^2} = \sum_k \frac{\rho_k}{c^2} \frac{\lambda^2 \lambda_k^2}{\lambda^2 - \lambda_k^2}, \tag{36}$$

where

$$\left.\begin{aligned}\rho_k &= N\frac{e^2}{\pi m}f_k,\\ \nu_k &= \frac{\omega_k}{2\pi} = \frac{c}{\lambda_k}, \qquad \nu = \frac{\omega}{2\pi} = \frac{c}{\lambda}.\end{aligned}\right\} \tag{37}$$

Using the identity

$$\frac{\lambda^2}{\lambda^2 - \lambda_k^2} = 1 + \frac{\lambda_k^2}{\lambda^2 - \lambda_k^2},$$

(36) becomes

$$n^2 - 1 = a + \sum_k \frac{b_k}{\lambda^2 - \lambda_k^2}, \tag{38}$$

with

$$\left.\begin{aligned}a &= \sum_k \frac{\rho_k}{\nu_k^2} = \frac{1}{c^2}\sum_k \rho_k\lambda_k^2,\\ b_k &= \frac{c^2\rho_k}{\nu_k^4} = \frac{1}{c^2}\rho_k\lambda_k^4.\end{aligned}\right\} \tag{39}$$

For the representation of the refractive index over the whole visible range it is usually sufficient to take account of only one or two resonance frequencies in the ultra-violet region. Koch* for example found that for the range from $\lambda = 0{\cdot}436\mu$ to $\lambda = 8{\cdot}68\mu$, the following formula holds for hydrogen, oxygen and air:

$$n^2 - 1 = a + \frac{b}{\lambda^2 - \lambda_0^2}, \tag{40}$$

where the constants a, b and λ_0 are shown in Table IX.

Table IX

Dispersion constants for oxygen, hydrogen and air between $\lambda = 0{\cdot}436\ \mu$ and $\lambda = 8{\cdot}68\ \mu$ at 0°C and 760 mm Hg

Gas	$a \,.\, 10^8$	$b \,.\, 10^8$	λ_0^2 in 10^{-8} cm²	ν_0 in 10^{15} sec⁻¹
Hydrogen	27,216	211·2	0·007760	3·40
Oxygen	52,842	369·9	0·007000	3·55
Air	57,642	327·7	0·005685	3·98

In a spectral range which contains no resonance frequencies, (38) may with a good accuracy be replaced by a simpler formula. For all substances which appear transparent to the eye, the *visible* domain of the spectrum is such a range. Denoting by ν_v the absorption frequencies which are on the short wavelength side (violet), and by ν_r

* J. Koch, *Nov. Act. Soc. Ups.* (4), 2 (1909), p. 61.

those on the long wavelength side (red), the dispersion formula (36) becomes, on expanding into power series with respect to ν and λ respectively:

$$\left.\begin{aligned} n^2 - 1 &= A + B\nu^2 + C\nu^4 + \dots \\ &\quad - \frac{B'}{\nu^2} - \frac{C'}{\nu^4} - \dots \\ &= A + \frac{Bc^2}{\lambda^2} + \frac{Cc^4}{\lambda^4} + \dots \\ &\quad - \frac{B'\lambda^2}{c^2} - \frac{C'\lambda^4}{c^4} - \dots \end{aligned}\right\} \tag{41}$$

where

$$\left.\begin{aligned} &A = \sum_v \frac{\rho_v}{\nu_v^2}, \quad B = \sum_v \frac{\rho_v}{\nu_v^4}, \quad C = \sum_v \frac{\rho_v}{\nu_v^6}, \dots \\ &B' = \sum_r \rho_r, \quad C' = \sum_r \rho_r \nu_r^2, \dots \end{aligned}\right\} \tag{42}$$

In such an absorption-free range the value of n for gases differs so little from unity that one may replace $n^2 - 1$ by $2(n - 1)$. Moreover the terms B', C', . . . arising from the ultra-violet resonance frequencies are usually without appreciable influence. Then, if only terms down to $1/\lambda^2$ are retained, (41) reduces to *Cauchy's formula**

$$n - 1 = A_1 \left(1 + \frac{B_1}{\lambda^2}\right). \tag{43}$$

where

$$A_1 = \frac{A}{2}, \qquad B_1 = \frac{Bc^2}{A}. \tag{44}$$

In Table X the values of A_1 and B_1 for the more important gases are given. Also, to illustrate the accuracy of the formula, the values of the refractive index for air as given by CAUCHY's formula and the observed values are compared with each other in Table XI.

TABLE X

The constants A_1 and B_1 of Cauchy's dispersion formula for various gases

Gas	$A_1 \cdot 10^5$	$B_1 \cdot 10^5$
Argon . .	27·92	5·6
Nitrogen . .	29·19	7·7
Helium . .	3·48	2·3
Hydrogen .	13·6	7·7
Oxygen . .	26·63	5·07
Air . . .	28·79	5·67
Ethane . .	73·65	9·08
Methane . .	42·6	14·41

* L. CAUCHY, *Bull. des. sc. math.*, **14** (1830), 9; Sur la dispersion de la lumière (*Nouv. exerc. de math.*, 1836).

TABLE XI

Observed values of the refractive index for air and the values given by Cauchy's dispersion formula (43)

$\lambda \cdot 10^5$ cm	$(n-1) \cdot 10^4$ observed	$(n-1) \cdot 10^4$ calculated	Difference
7·594	2·905	2·907	0·002
6·563	2·916	2·917	0·001
5·896	2·926	2·926	0·000
5·378	2·935	2·935	0·000
5·184	2·940	2·940	0·000
4·861	2·948	2·948	0·000
4·677	2·951	2·954	0·003
4·308	2·966	2·967	0·001
3·969	2·983	2·983	0·000
3·728	2·995	2·996	0·001
3·441	3·016	3·017	0·001
3·180	3·040	3·041	0·001
3·021	3·056	3·058	0·002
2·948	3·065	3·067	0·002

In the case of substances of high density, i.e. liquids or solids, it is no longer permissible to replace n by unity in the denominator of the second term in (35). Nevertheless, it is possible to reduce (35) to the same form as before. Since

$$\frac{n^2-1}{n^2+2} = \frac{4\pi}{3} N\alpha,$$

it follows that

$$n^2 - 1 = \frac{12\pi N\alpha}{3 - 4\pi N\alpha}, \tag{45}$$

where

$$4\pi N\alpha = \frac{Ne^2}{\pi m} \sum_k \frac{f_k}{\nu_k^2 - \nu^2} = \sum_k \frac{\rho_k}{\nu_k^2 - \nu^2}. \tag{46}$$

It will, as a rule, be sufficient to take into account only a finite number of the absorption frequencies, and it follows that $n^2 - 1$, given by (45), represents a rational function in ν^2, and may therefore be decomposed into partial fractions. For this purpose one must find the zero values of the denominator, i.e. the roots of the equation

$$3 - 4\pi N\alpha = 3 - \sum_k \frac{\rho_k}{\nu_k^2 - \nu^2} = 0. \tag{47}$$

If the roots are denoted by $\bar{\nu}_k$, then (45) may be reduced to the form

$$n^2 - 1 = \sum_k \frac{\bar{\rho}_k}{\bar{\nu}_k^2 - \nu^2}, \tag{48}$$

which is identical in form with the formula (36) for gases.

For example if there is only a single absorption frequency, $\bar{\nu}_1$, this frequency will be the root of the equation

$$3 - \frac{\rho_1}{\nu_1^2 - \nu^2} = 0, \tag{49}$$

whence

$$\left.\begin{aligned} \bar{\nu}_1^2 &= \nu_1^2 - \tfrac{1}{3}\rho_1, \\ \bar{\rho}_1 &= \rho_1. \end{aligned}\right\} \tag{50}$$

Eq. (48), which may also be written in the form (38), is known as *Sellmeir's dispersion formula.*

So far the effect of the motion of the nuclei has been neglected. It is, in fact, only of importance at very long wavelengths (infra-red). The reason for this can easily be understood. The electric moment and the mean polarizability may be approximately divided into two parts, one associated with the electrons and the other with the nuclei,

$$\boldsymbol{p} = \boldsymbol{p}_e + \boldsymbol{p}_n, \tag{51}$$

where

$$\boldsymbol{p}_e = \alpha_e \boldsymbol{E}', \qquad \boldsymbol{p}_n = \alpha_n \boldsymbol{E}'. \tag{52}$$

The electrons will follow the field almost instantly, up to rather high frequencies, including in many cases those of the whole visible spectrum. The nuclear masses, on the other hand, are so heavy that they cannot follow the field in the high-frequency region, i.e. $\alpha_n \sim 0$ for visible light. This can also be seen from the dispersion formula (41) where $n^2 - 1$ is represented by two sets of terms, one due to the high-frequency (violet) vibration ν_v, the other due to the low-frequency (red) vibrations ν_r. It is reasonable to assume, and it follows rigorously from quantum mechanics, that the quasi-elastic forces binding the nuclei and electrons are all of the same order of magnitude; hence the order of magnitude of the frequencies will be related by

$$m\nu_e^2 \sim M\nu_n^2, \tag{53}$$

where m is the electronic and M the nuclear mass, and ν_e and ν_n are the frequencies of vibrations of the electrons and nuclei respectively. In order to extend the formulae (41) and (42) to take into account also the motion of the nuclei, we may simply assume that the violet vibrations are due to the electrons, and the red vibrations due to the nuclei. Then we have also to distinguish two types of factors ρ_k as given by (37), those due to the electronic and those due to the nuclear vibrations, and we have the order of magnitude relation

$$m\rho_e \sim M\rho_n. \tag{54}$$

The coefficients (42) of the dispersion formulae are therefore of the following orders of magnitude:

$$\left.\begin{aligned} B &\sim \frac{A}{\nu_e^2}, & C &\sim \frac{A}{\nu_e^4}, \\ B' &\sim \frac{m}{M} A\nu_e^2, & C' &\sim \frac{m}{M} A\nu_e^4, \end{aligned}\right\} \tag{55}$$

and we have

$$n^2 - 1 \sim A\left\{1 + a\frac{\nu^2}{\nu_e^2} + b\frac{\nu^4}{\nu_e^4} + \dots \right. \\ \left. - \frac{m}{M}\left[a'\frac{\nu_e^2}{\nu^2} + b'\frac{\nu_e^4}{\nu^4} + \dots\right]\right\}, \tag{56}$$

where $a,b, \ldots a',b', \ldots$ are numerical constants of the order of unity. The terms in the first line correspond to the electronic part α_e of the polarizability, in the second line to its nuclear part α_n; since m/M is small ($\sim 1/1840$) we see that the latter terms will be negligible, provided that the frequency ν is not a very small fraction of the violet resonance frequencies ν_e.

Hence for an optical frequency (i.e. one corresponding to visible light), the polarization is given essentially by α_e alone (as assumed in our preceding calculations), while for a static field it is given by $\alpha_e + \alpha_n$.

The considerations of the present section were entirely based on classical mechanics. When similar calculations are carried on the basis of quantum mechanics, the interaction of the field and matter may still be described in terms of virtual oscillators, but the number of the oscillators, even when only a single electron is present, is found to be infinite. Eq. (35) still applies, but the strength factors f_k no longer specify the number of electrons of a particular type, but rather the number of virtual oscillators belonging to one electron or to a set of electrons. In most cases only a finite number of the f_k factors have an appreciable value, the others may be neglected. In fact the whole formal theory is hardly affected by the introduction of quantum mechanics, but with its help one can calculate the f_k factors for a given electronic system.

2.4 PROPAGATION OF ELECTROMAGNETIC WAVES TREATED BY INTEGRAL EQUATIONS*

At the beginning of this chapter a method was outlined for the treatment of propagation of electromagnetic waves in terms of integral equations, and in the sections which followed, auxiliary quantities needed for this purpose were introduced. We shall now formulate these integral equations and discuss some of their consequences. This may be done entirely within the framework of the macroscopic theory as dealt with throughout this book. Within this framework the integral equations are equivalent to MAXWELL's equations and represent the mathematical description of the electromagnetic phenomena in terms of mutual interactions over finite distances (which of course need time for transfer). A definite advantage of this treatment, which in some cases is more powerful than the more customary treatment based on differential equations, is that it connects macroscopic phenomena with the molecular ones, discussed in the previous section.

As already explained, a substance may be considered to consist of molecules which react to the incident waves like dipoles. All the emitted dipole waves then give rise to the effective force acting on any other dipole and also determine the average measurable field. It will be assumed that the dipoles are evenly distributed throughout the medium, and the average value $\mathbf{P}$ of their electric moment per unit volume will be considered as the basic quantity. Actually the molecules are never quite evenly distributed throughout a medium (i.e. there are density fluctuations) and consequently the electric moment of the individual particles exhibits fluctuations about the average

---

* The theory here described was developed chiefly by W. ESMARCH, *Ann. d. Physik*, **42** (1913), 1257; C. W. OSEEN, *ibid.*, **48** (1915), 1; and W. BOETHE, Beiträge zur Theorie der Brechung und Reflexion (Dissertation, Berlin, 1914), and *Ann. d. Physik*, **64** (1921), 693. A full account of the theory was given by R. LUNDBLAD, *Univ. Årsskrift, Upsala* (1920). See also C. G. DARWIN, *Trans. Camb. Phil. Soc.*, **23** (1924), 137, and H. HOEK, Algemeene theorie der optische activiteit van isotrope media (Thesis, Leiden, 1939).

Some generalizations of this theory which take in to account non-linear behaviour of the medium (arising when the fields are exceptionally strong) were discussed by N. BLOEMBERGEN and P. S. PERSHAN, *Phys. Rev.*, **128** (1962), 619.

value. The resulting phenomena may be accounted for in the present theory by carrying the calculations one step further, i.e. one calculates not only the averages but also their mean square deviations. Such calculations are of importance in some problems, e.g. for the explanation of the blue colour of the sky, as first given by RAYLEIGH.* But this extension of the theory cannot be given here.†

2.4.1 The basic integral equation

We consider the propagation of an electromagnetic wave in an homogeneous, isotropic non-magnetic medium. The electric and magnetic fields $\boldsymbol{E}'_j$ and $\boldsymbol{H}'_j$ which act on the jth dipole in the interior of the medium can be divided into the incident fields $\boldsymbol{E}^{(i)}, \boldsymbol{H}^{(i)}$ (propagated with the vacuum velocity c) and the contribution arising from all the dipoles:

$$\left.\begin{aligned} \boldsymbol{E}'_j &= \boldsymbol{E}^{(i)} + \sum_l{}' \boldsymbol{E}_{jl}, \\ \boldsymbol{H}'_j &= \boldsymbol{H}^{(i)} + \sum_l{}' \boldsymbol{H}_{jl}, \end{aligned}\right\} \tag{1}$$

the summation extending over all the dipoles except the jth one. At the point $\boldsymbol{r}_j$ where the jth dipole is situated, the field of the lth dipole is obtained from § 2.2 (49)–(51):

$$\left.\begin{aligned} \boldsymbol{E}_{jl} &= \operatorname{curl}\operatorname{curl}\frac{\boldsymbol{p}_l(t - R_{jl}/c)}{R_{jl}}, \\ \boldsymbol{H}_{jl} &= \frac{1}{c}\operatorname{curl}\frac{\dot{\boldsymbol{p}}_l(t - R_{jl}/c)}{R_{jl}}, \end{aligned}\right\} \tag{2}$$

$\boldsymbol{p}_l(t)$ denoting the moment of the lth dipole, $R_{jl} = |\boldsymbol{r}_j - \boldsymbol{r}_l|$, and the operation curl is taken with respect to the coordinates x_j, y_j, z_j of the jth dipole.

As already explained, the distribution may to a good approximation be treated as continuous, i.e. the moment of the dipoles may be considered to be a continuous function of position (and of time): $\boldsymbol{p} = \boldsymbol{p}(\boldsymbol{r}, t)$. Likewise the number density N will be assumed to be a continuous function $N(\boldsymbol{r})$ of position. The total electric dipole $\boldsymbol{P}$ per unit volume is then given by (14), § 2.3:

$$\boldsymbol{P} = N\boldsymbol{p} = N\alpha\boldsymbol{E}'. \tag{3}$$

For reasons explained in § 2.3, the contribution of the magnetic force has been neglected in (3). Since the material is also assumed to be non-magnetic (i.e. $\boldsymbol{M} = 0$), the effective field $\boldsymbol{H}'$ will not enter into the conditions for the dynamical equilibrium to be established.

If we substitute from (2) into (1) and go over to a continuous distribution, using (3), we obtain‡

---

* We shall refer to RAYLEIGH's theory again in a later section (§ 13.5) which deals with the scattering of spherical particles embedded in vacuum. If the particles are small, the dependence of the intensity of the scattered radiation on the wavelength is the same as in the present case of spontaneous fluctuations of density in a homogeneous medium (namely inversely proportional to the fourth power of the wavelength). The fluctuation theory, however, gives in addition to this result also the dependence of the scattered intensity on the density fluctuation.

† It is discussed, for example, in M. BORN, *Optik* (Berlin, Springer, 1933), § 81.

‡ One must sum the fields and not the potentials, i.e. in (4) the operator curl curl must be under the integration sign and not outside it; for the field acting on the molecule as expressed by (3) is obtained by summing all the individual fields. The singularities of the fields at the sources lead to a difference between the sum of the fields and the field associated with the total potential (sum of potentials associated with the individual fields). Cf. (15) below.

$$\boldsymbol{E}'(\boldsymbol{r}, t) = \boldsymbol{E}^{(i)} + \int \text{curl curl } N\alpha \frac{\boldsymbol{E}'(\boldsymbol{r}', t - R/c)}{R} dV', \tag{4}$$

$$\boldsymbol{H}'(\boldsymbol{r}, t) = \boldsymbol{H}^{(i)} + \frac{1}{c} \int \text{curl } N\alpha \frac{\dot{\boldsymbol{E}}'(\boldsymbol{r}', t - R/c)}{R} dV', \tag{5}$$

where

$$R = |\boldsymbol{r} - \boldsymbol{r}'|.$$

If the point of observation $\boldsymbol{r}$ is outside the medium, the integral is taken throughout the whole medium. If it is inside the medium, a small domain occupied by the atom must be first excluded; we consider it to be a small sphere σ of radius a. Eventually we will proceed to the limit $a \to 0$ in the usual way.

Eq. (4) is an integro-differential equation for $\boldsymbol{E}'$. When it is solved, $\boldsymbol{H}'$ may be obtained from (5). The two equations are essentially equivalent to MAXWELL's equations for isotropic non-magnetic substances. Generalization to magnetic substances may be obtained with the help of the second HERTZ vector.

2.4.2 The Ewald-Oseen extinction theorem and a rigorous derivation of the Lorentz-Lorenz formula

Eq. (4) connects the effective electric field with the incident electric field in a rather complicated manner and only in special cases can this equation be solved explicitly. Nevertheless it is possible to derive from it some basic results, such as the LORENTZ–LORENZ formula, the laws of refraction and reflection, and the FRESNEL formulae. Before showing this we shall establish an important general consequence of the solution.

Let Σ denote the boundary of the medium. For points of observation inside the medium the basic equation (4) may be written in the form

$$\boldsymbol{P} = N\alpha(\boldsymbol{E}^{(i)} + \boldsymbol{E}^{(d)}), \tag{6}$$

where $\boldsymbol{E}^{(d)}$ denotes the contribution from the dipoles:

$$\boldsymbol{E}^{(d)} = \int_{\sigma}^{\Sigma} \text{curl curl} \frac{\boldsymbol{P}(\boldsymbol{r}', t - R/c)}{R} dV'. \tag{7}$$

Here we have indicated explicitly the boundary of the volume over which the integration is carried out.

The incident field $\boldsymbol{E}^{(i)}$ will be assumed to be monochromatic, with angular frequency ω,

$$\boldsymbol{E}^{(i)} = \boldsymbol{A}^{(i)}(\boldsymbol{r})e^{-i\omega t}, \tag{8}$$

and we take, as a trial solution for $\boldsymbol{P}$, a wave which is also monochromatic and of the same frequency, but which has a different velocity of propagation (say c/n),

$$\boldsymbol{P} = (n^2 - 1)k_0^2\boldsymbol{Q}(\boldsymbol{r})e^{-i\omega t}, \tag{9}$$

where, as before, $k_0 = \omega/c$ and

$$\nabla^2\boldsymbol{Q} + n^2k_0^2\boldsymbol{Q} = 0. \tag{10}$$

In (9) the constant factor $(n^2 - 1)k_0^2$ has been introduced to simplify later formulae. The constant n must be regarded as an unknown quantity, its determination being one of the chief aims of the present analysis.

We also assume that $\boldsymbol{Q}$ has no sources in the medium, i.e. that inside the medium

$$\text{div } \boldsymbol{Q} = 0. \tag{11}$$

The possibility of solving the basic integro-differential equation (6) in the form represented by our trial solution may appear somewhat strange at first sight, since $\boldsymbol{E}^{(i)}$ represents a wave propagated with the vacuum velocity c, whereas $\boldsymbol{P}$ is assumed to be propagated with the velocity c/n. It will however be shown that the dipole field $\boldsymbol{E}^{(d)}$ may be expressed as the sum of two terms, one of which obeys the wave equation in the vacuum and cancels out exactly the incident wave, whereas the other satisfies the wave equation for propagation with velocity c/n. The incident wave may therefore be regarded as extinguished at any point within the medium by interference with the dipole field and replaced by another wave with a different velocity (and generally also with a different direction) of propagation. This result is known as the *extinction theorem* and was established first for crystalline media by EWALD* and for isotropic media by OSEEN.† In more recent times it has been shown that an analogous extinction theorem can also be formulated within the framework of Maxwell's electromagnetic theory.‡

To prove the extinction theorem we rewrite (6) in its time-free form

$$\boldsymbol{Q} = N\alpha\left\{\frac{1}{(n^2-1)k_0^2}\boldsymbol{A}^{(i)} + \boldsymbol{A}^{(d)}\right\}, \tag{12}$$

where

$$\left.\begin{aligned} \boldsymbol{E}^{(d)} &= (n^2-1)k_0^2\,\boldsymbol{A}^{(d)}e^{-i\omega t}, \\ \boldsymbol{A}^{(d)} &= \int_\sigma^\Sigma \operatorname{curl}\operatorname{curl}\boldsymbol{Q}(\boldsymbol{r}')G(R)dV', \end{aligned}\right\} \tag{13}$$

and

$$G = \frac{e^{ik_0R}}{R}. \tag{14}$$

It is shown in Appendix V that, when the radius of the sphere σ is sufficiently small,

$$\int_\sigma^\Sigma \operatorname{curl}\operatorname{curl}\boldsymbol{Q}(\boldsymbol{r}')G(R)dV' = \operatorname{curl}\operatorname{curl}\int_\sigma^\Sigma \boldsymbol{Q}(\boldsymbol{r}')G(R)dV' - \frac{8\pi}{3}\boldsymbol{Q}(\boldsymbol{r}). \tag{15}$$

Now the function G represents a spherical wave in the vacuum and therefore satisfies the wave equation

$$\nabla^2 G + k_0^2 G = 0. \tag{16}$$

Hence it follows from (10) and (16) that the term under the integral sign on the right of (15) may be expressed in the form

$$\boldsymbol{Q}G = \frac{1}{(n^2-1)k_0^2}(\boldsymbol{Q}\nabla^2 G - G\nabla^2\boldsymbol{Q}), \tag{17}$$

giving on integration with the help of GREEN's theorem,

$$\int_\sigma^\Sigma \boldsymbol{Q}(\boldsymbol{r}')G(R)dV' = \frac{1}{(n^2-1)k_0^2}\left[\int_\Sigma \left\{\boldsymbol{Q}\frac{\partial G}{\partial \nu'} - G\frac{\partial \boldsymbol{Q}}{\partial \nu'}\right\}dS' - \int_\sigma \left\{\boldsymbol{Q}\frac{dG}{dR} - G\frac{\partial \boldsymbol{Q}}{\partial R}\right\}dS'\right], \tag{18}$$

where $\partial/\partial\nu'$ denotes differentiation along the outward normal to the boundary Σ. A

---

* P. P. EWALD (Dissertation, München, 1912); *Ann. d. Physik*, **49** (1916), 1.

† C. W. OSEEN, *Ann. d. Physik*, **48** (1915), 1.

‡ J. J. SEIN, *Opt. Comm.*, **2** (1970). 170; J. DE GOEDE and P. MAZUR, *Physica*, **58** (1972), 568; D. N. PATTANAYAK and E. WOLF, *Opt. Comm.*, **6** (1972), 217; E. WOLF in *Coherence and Quantum Optics*, ed. L. MANDEL and E. WOLF (New York, Plenum Press, 1973), 339..

straightforward calculation gives (cf. § 8.3.1) for the limit of the surface integral over σ as $a \to 0$ the value $-4\pi Q(\mathbf{r})$, so that

$$\int_{\sigma}^{\Sigma} Q(\mathbf{r}')G(R)dV' \to \frac{1}{(n^2-1)k_0^2}\left\{\int_{\Sigma}\left(Q\frac{\partial G}{\partial \nu'} - G\frac{\partial Q}{\partial \nu'}\right)dS' + 4\pi Q\right\}. \qquad (19)$$

From (13), (15) and (19) we obtain, in the limit,

$$A^{(d)} = \int_{\sigma}^{\Sigma} \text{curl curl } Q(\mathbf{r}')G(R)dV' =$$

$$\frac{1}{(n^2-1)k_0^2}\left[4\pi \text{ curl curl } Q + \text{curl curl}\int_{\Sigma}\left\{Q\frac{\partial G}{\partial \nu'} - G\frac{\partial Q}{\partial \nu'}\right\}dS'\right] - \frac{8\pi}{3}Q(\mathbf{r}). \qquad (20)$$

Next we use the identity

$$\text{curl curl } Q \equiv \text{grad div } Q - \nabla^2 Q.$$

The first term on the right vanishes on account of (11), and the second is by (10) equal to $n^2k_0^2Q$. Hence, on substituting into (20), we finally obtain the following expression for the dipole field:

$$A^{(d)} = \frac{4\pi}{3}\left(\frac{n^2+2}{n^2-1}\right)Q + \frac{1}{(n^2-1)k_0^2}\text{ curl curl}\int_{\Sigma}\left\{Q\frac{\partial G}{\partial \nu'} - G\frac{\partial Q}{\partial \nu'}\right\}dS'. \qquad (21)$$

According to (10) the first term on the right represents a wave propagated with the velocity c/n, whereas the second is seen to represent a wave which, like G, is propagated with the vacuum velocity c. Hence, as anticipated, the basic equation (12) splits into two groups of terms, each of which represents a wave propagated with a different velocity of propagation. This is only possible if each group vanishes separately, so that

$$\frac{4\pi}{3}N\alpha = \frac{n^2-1}{n^2+2}, \qquad (22)$$

and

$$A^{(i)} + \text{curl curl}\int_{\Sigma}\left\{Q\frac{\partial G}{\partial \nu'} - G\frac{\partial Q}{\partial \nu'}\right\}dS' = 0. \qquad (23)$$

The relation (23) expresses the *extinction* of the incident wave

$$E^{(i)} = A^{(i)}e^{-i\omega t}$$

at any point inside the medium, by interference with part of the dipole field. The incident wave is replaced by another wave, namely by

$$E' = \frac{1}{N\alpha}P = \frac{1}{N\alpha}(n^2-1)k_0^2Qe^{-i\omega t} = \frac{4\pi}{3}(n^2+2)k_0^2Qe^{-i\omega t}, \qquad (24)$$

which is propagated inside the medium with velocity c/n. n is given by (22) in terms of the number density N and the polarizability α; this relation will be recognized as the *Lorentz–Lorenz formula*, tentatively introduced in Section 2.3.3.

The general solution of the integro-differential equation (23) may, in principle, be obtained by reducing it to a system of two coupled FREDHOLM integral equations for Q and $\partial Q/\partial \nu'$ on the surface. Q at any point inside the medium may then be obtained by solving (10) subject to the condition that it takes these values on the boundary. This may be done by standard methods, using GREEN's functions, or by applying the HELMHOLTZ–KIRCHHOFF integral formula (§ 8.3, eq. (7)).

We must now connect the effective field $\boldsymbol{E}'$ with the ordinary field $\boldsymbol{E}$ of MAXWELL's theory. To do this, we write down the two definitions of the electric displacement $\boldsymbol{D}$:

$$\boldsymbol{D} = \varepsilon \boldsymbol{E} = n^2 \boldsymbol{E} \tag{25a}$$

and

$$\boldsymbol{D} = \boldsymbol{E} + 4\pi \boldsymbol{P}. \tag{25b}$$

Elimination of $\boldsymbol{D}$ between these two equations gives

$$\boldsymbol{E} = \frac{4\pi}{n^2 - 1} \boldsymbol{P}. \tag{26}$$

Now from (3) and (22),

$$\boldsymbol{E}' = \frac{1}{N\alpha} \boldsymbol{P} = \frac{4\pi}{3} \frac{n^2 + 2}{n^2 - 1} \boldsymbol{P},$$

so that, using (26), we obtain

$$\boldsymbol{E}' = \boldsymbol{E} + \frac{4\pi}{3} \boldsymbol{P}, \tag{27}$$

in agreement with (12), § 2.3. Finally from (26) and (9) it is seen that, in terms of $\boldsymbol{Q}$, the field $\boldsymbol{E}$ *inside the medium* ($n \neq 1$) is given by

$$\boldsymbol{E} = 4\pi k_0^2 \boldsymbol{Q} e^{-i\omega t}. \tag{28}$$

To determine the electric field *outside the medium* we must return to (4). The integrals occurring in these equations now extend over the whole medium, so that the operator curl curl may be taken outside the respective integrals. Further, since outside the medium $\boldsymbol{D} = \boldsymbol{E}$, we have $\boldsymbol{P} = 0$ there, and consequently, by (27), we have, formally, $\boldsymbol{E}' = \boldsymbol{E}$. In place of (6) we now obtain

$$\left.\begin{aligned} \boldsymbol{E} &= \boldsymbol{E}^{(i)} + \boldsymbol{E}^{(r)}, \\ \text{where}\qquad & \\ \boldsymbol{E}^{(r)} &= \text{curl curl} \int \frac{\boldsymbol{P}(\boldsymbol{r}', t - R/c)}{R} \, dV'. \end{aligned}\right\} \tag{29}$$

For the time harmonic case here considered, we have, on substituting from (9), using (17) and GREEN's theorem,

$$\left.\begin{aligned} \boldsymbol{E}^{(r)} &= \boldsymbol{A}^{(r)} e^{-i\omega t}, \\ \text{where}\qquad & \\ \boldsymbol{A}^{(r)} &= \text{curl curl} \int_{\Sigma} \left\{ \boldsymbol{Q} \frac{\partial G}{\partial \nu'} - G \frac{\partial \boldsymbol{Q}}{\partial \nu'} \right\} dS', \end{aligned}\right\} \tag{30}$$

and $\boldsymbol{Q}$ is the same function on the surface Σ as before.

So far we have considered the electric field only. To determine the magnetic field we have only to substitute for $\boldsymbol{E}'$ into (5) and evaluate the integral. This can be done by a similar calculation as above, but the evaluation is somewhat simpler now, since the operator curl in (5) may be taken outside the integral sign, as is evident from discussion of Appendix V. On comparing the resulting expressions for $\boldsymbol{H}$ with those for $\boldsymbol{E}$ we find that the solutions are consistent with MAXWELL's equations. Since these calculations are straightforward and do not bring out any important new features, they will not be given here.

It is of very great interest that an approach based on the physical remark that the field in matter is more naturally characterized by the polarization than by the displacement vector, should lead so elegantly, via the integro-differential equation (4), to a strict derivation of the LORENTZ–LORENZ formula and the extinction theorem. This powerful method has so far been little used in treating more specific problems,* but an example of its application will be given in Chapter XIII.

2.4.3 Refraction and reflection of a plane wave, treated with the help of the Ewald-Oseen extinction theorem

The EWALD–OSEEN extinction theorem, expressed by the formula (23), will now be applied to the case of a plane monochromatic wave entering a homogeneous medium which fills the half-space $z < 0$. It will be shown that it leads to the laws of refraction and reflection and to the FRESNEL formulae.

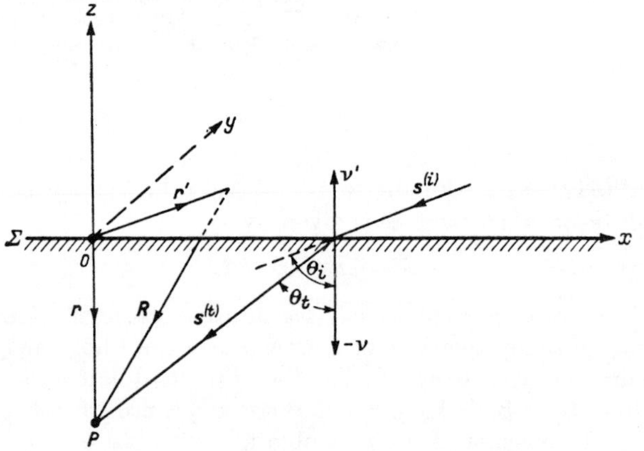

Fig. 2.4. Penetration of a wave into a homogeneous medium which is considered as a system of dipoles.

The incident wave will be written in the form

$$\boldsymbol{E}^{(i)} = \boldsymbol{A}^{(i)}(\boldsymbol{r})e^{-i\omega t} = \boldsymbol{A}_0^{(i)}e^{i[k_0(\boldsymbol{r}.\boldsymbol{s}^{(i)})-\omega t]}, \tag{31}$$

where $k_0 = \omega/c$, $\boldsymbol{A}_0^{(i)}$ is a constant vector and $\boldsymbol{s}^{(i)}$ the unit vector in the direction of propagation.

We choose the z-axis so that it passes through the point of observation P, assumed first to be inside the medium at a distance r from the boundary Σ (see Fig. 2.4). The x-axis will be chosen so that $\boldsymbol{s}^{(i)}$ lies in the xz-plane. Hence, if θ_i denotes the angle of incidence, we have

$$s_x^{(i)} = -\sin\theta_i, \qquad s_y^{(i)} = 0, \qquad s_z^{(i)} = -\cos\theta_i. \tag{32}$$

In accordance with the analysis of the preceding section, the transmitted wave will be assumed to have the same frequency as the incident wave, but a different velocity c/n, n being given in terms of the polarizability and the density by the LORENTZ–LORENZ formula. As our trial solution for the transmitted wave we take a plane wave propagated in the direction of a unit vector $\boldsymbol{s}^{(t)}$ which is assumed to lie in the xz-plane:

$$s_x^{(t)} = -\sin\theta_t, \qquad s_y^{(t)} = 0, \qquad s_z^{(t)} = -\cos\theta_t. \tag{33}$$

* The method has been applied to the propagation of electromagnetic waves through a stratified medium by D. R. HARTREE, *Proc. Cambr. Phil. Soc.*, **25** (1929), 97.

(28) then becomes

$$E = 4\pi k_0^2 Q e^{-i\omega t} = 4\pi k_0^2 Q_0 e^{i[k_0 n(\boldsymbol{r}.\boldsymbol{s}^{(t)}) - \omega t]}, \tag{34}$$

Q_0 being a constant vector which, on account of (11), is orthogonal to $\boldsymbol{s}^{(t)}$.

The solution of the integro-differential equation can easily be obtained if we only consider points of observation P whose distance from the boundary $z = 0$ is large compared to the wavelength, i.e. if

$$\frac{2\pi r}{\lambda} = k_0 r \gg 1. \tag{35}$$

This condition is merely imposed to simplify the calculations and does not correspond to a limitation on the physical validity of the integro-differential equation.*

The derivatives occurring in the integral over Σ are now

$$\frac{\partial Q(\boldsymbol{r}')}{\partial \nu'} = \frac{\partial}{\partial \nu'} Q_0 e^{ink_0(\boldsymbol{r}'.\boldsymbol{s}^{(t)})} = ink_0 \left(\frac{\partial \boldsymbol{r}'}{\partial \nu'} . \boldsymbol{s}^{(t)}\right) Q_0 e^{ink_0(\boldsymbol{r}'.\boldsymbol{s}^{(t)})}, \tag{36a}$$

$$\frac{\partial G(R)}{\partial \nu'} = \frac{\partial}{\partial \nu'} \frac{e^{ik_0 R}}{R} = ik_0 \frac{\partial R}{\partial \nu'} G \left(1 + \frac{i}{k_0 R}\right). \tag{36b}$$

The last term in (36b) may be neglected because of (35), and the integral occurring in (23) becomes

$$J = \int_\Sigma \left(Q \frac{\partial G}{\partial \nu'} - G \frac{\partial Q}{\partial \nu'}\right) dS' = ik_0 Q_0 \int_\Sigma \left\{\frac{\partial R}{\partial \nu'} - n \left(\frac{\partial \boldsymbol{r}'}{\partial \nu'} . \boldsymbol{s}^{(t)}\right)\right\} \frac{e^{ik_0[R + n(\boldsymbol{r}'.\boldsymbol{s}^{(t)})]}}{R} dS'. \tag{37}$$

The components of the vectors $\boldsymbol{r}$, $\boldsymbol{r}'$ and $\boldsymbol{R}$ are (see Fig. 2.4),

$$\begin{array}{llll} \boldsymbol{r}: & 0, & 0, & -r, \\ \boldsymbol{r}': & x', & y', & 0, \\ \boldsymbol{R}: & -x', & -y', & -r, \end{array}$$

so that

$$\left.\begin{array}{ll} R = \sqrt{x'^2 + y'^2 + r^2}, & \boldsymbol{r}' . \boldsymbol{s}^{(t)} = -x' \sin\theta_t, \\ \dfrac{\partial R}{\partial \nu'} = \dfrac{r}{R}, & \dfrac{\partial \boldsymbol{r}'}{\partial \nu'} . \boldsymbol{s}^{(t)} = s_z^{(t)} = -\cos\theta_t. \end{array}\right\} \tag{38}$$

(37) becomes on substitution

$$J = ik_0 Q_0 \int_{-\infty}^{+\infty}\int_{-\infty}^{+\infty} \frac{1}{R}\left(\frac{r}{R} + n\cos\theta_t\right) e^{ik_0(R - nx'\sin\theta_t)}\, dx'dy'. \tag{39}$$

It will be convenient to introduce an angle ϕ and an associated unit vector $\boldsymbol{s}$, defined by

$$n \sin\theta_t = \sin\phi, \tag{40}$$

$$s_x = -\sin\phi, \qquad s_y = 0, \qquad s_z = -\cos\phi. \tag{41}$$

Since $k_0 r$ was assumed to be large compared with unity, the exponent in (39) will also be

* The exact solution of the integro-differential equation for the problem of refraction and reflection was given by É. Lalor and E. Wolf, *J. Opt. Soc. Amer.*, **62**, 1165 (1972).

large compared with unity and the exponential term will oscillate rapidly and change sign many times as the point (x', y') explores the domain of integration. Under these conditions, a good approximation to the value of J is obtained by the application of the following formula, which is a consequence of the principle of stationary phase (see Appendix III, eq. (20)):

$$\iint_D g(x', y')e^{ik_0 f(x',y')}dx'dy' \sim \frac{2\pi i}{k_0} \sum_j \frac{\sigma_j}{\sqrt{|\alpha_j\beta_j - \gamma_j^2|}} g(x_j', y_j')e^{ik_0 f(x_j',y_j')}. \tag{42}$$

Here (x_j', y_j') are the points in the domain D of integration at which f is stationary and

$$\alpha_j = \left(\frac{\partial^2 f}{\partial x'^2}\right)_{x_j',y_j'}, \qquad \beta_j = \left(\frac{\partial^2 f}{\partial y'^2}\right)_{x_j',y_j'}, \qquad \gamma_j = \left(\frac{\partial^2 f}{\partial x_j' \partial y_j'}\right)_{x',y_j'}. \tag{43a}$$

and

$$\sigma_j = \left.\begin{array}{rl} +1 & \text{when} \quad \alpha_j\beta_j > \gamma_j^2, \quad \alpha_j > 0, \\ -1 & \text{when} \quad \alpha_j\beta_j > \gamma_j^2, \quad \alpha_j < 0, \\ -i & \text{when} \quad \alpha_j\beta_j < \gamma_j^2. \end{array}\right\} \tag{43b}$$

In the present case, we have, if we also use (40) and (41),

$$g = \frac{1}{R}\left(\frac{r}{R} + \frac{\sin\phi}{\sin\theta_t}\cos\theta_t\right), \qquad f = R - x'\sin\phi, \tag{44}$$

and consequently

$$\frac{\partial f}{\partial x'} = \frac{x'}{\sqrt{x'^2 + y'^2 + r^2}} - \sin\phi, \qquad \frac{\partial f}{\partial y'} = \frac{y'}{\sqrt{x'^2 + y'^2 + r^2}}.$$

For f to be stationary,

$$\frac{\partial f}{\partial x'} = \frac{\partial f}{\partial y'} = 0.$$

Hence (noting that the positive square root is to be taken in evaluating the partial derivatives), f is seen to be stationary only when

$$x' = x_1', \qquad y' = y_1',$$

where

$$x_1' = r\tan\phi, \qquad y_1' = 0. \tag{45}$$

The expressions in (43) and (44) take the following values at this point:

$$\alpha_1 = \left(\frac{\partial^2 f}{\partial x'^2}\right)_{x_1',y_1'} = \frac{\cos^3\phi}{r}, \qquad \beta_1 = \left(\frac{\partial^2 f}{\partial y'^2}\right)_{x_1',y_1'} = \frac{\cos\phi}{r}, \qquad \gamma_1 = 0, \tag{46a}$$

$$\sigma_1 = 1, \tag{46b}$$

$$R_1 = (R)_{x_1',y_1'} = \frac{r}{\cos\phi}, \tag{46c}$$

$$f(x_1', y_1') = r\cos\phi = \mathbf{r}\cdot\mathbf{s}, \tag{46d}$$

$$g(x_1', y_1') = \frac{\cos\phi}{r}\,\frac{\sin(\phi + \theta_t)}{\sin\theta_t}. \tag{46e}$$

The formula (42) now gives

$$\int_{-\infty}^{+\infty}\int_{-\infty}^{+\infty}\frac{1}{R}\left(\frac{r}{R}+n\cos\theta_t\right)e^{ik_0(R-nx'\sin\theta_t)}\,dx'dy'=\frac{2\pi i}{k_0}\frac{\sin(\phi+\theta_t)}{\cos\phi\sin\theta_t}e^{ik_0(\boldsymbol{r}.\boldsymbol{s})},$$

so that (39) becomes

$$J=\int_\Sigma\left\{Q\frac{\partial G}{\partial\nu'}-G\frac{\partial Q}{\partial\nu'}\right\}dS'=-2\pi\frac{\sin(\phi+\theta_t)}{\cos\phi\sin\theta_t}Q_0e^{ik_0(\boldsymbol{r}.\boldsymbol{s})}. \tag{47}$$

We now substitute this expression into the integro-differential equation (23), which expresses the extinction of the incident wave. We have

$$\text{curl}\ \boldsymbol{Q}_0\,e^{ik_0(\boldsymbol{r}.\boldsymbol{s})}=ik_0\,(\boldsymbol{s}\wedge\boldsymbol{Q}_0)e^{ik_0(\boldsymbol{r}.\boldsymbol{s})},$$

$$\text{curl curl}\ \boldsymbol{Q}_0\,e^{ik_0(\boldsymbol{r}.\boldsymbol{s})}=-k_0^2\,\boldsymbol{s}\wedge(\boldsymbol{s}\wedge\boldsymbol{Q}_0)e^{ik_0(\boldsymbol{r}.\boldsymbol{s})}.$$

Hence (23) reduces to

$$\boldsymbol{A}_0^{(i)}\,e^{ik_0(\boldsymbol{r}.\boldsymbol{s}^{(i)})}=-2\pi k_0^2\frac{\sin(\phi+\theta_t)}{\cos\phi\sin\theta_t}\boldsymbol{s}\wedge(\boldsymbol{s}\wedge\boldsymbol{Q}_0)e^{ik_0(\boldsymbol{r}.\boldsymbol{s})}. \tag{48}$$

Now this must be an identity for all points $\boldsymbol{r}$ on the boundary. Hence

$$\boldsymbol{s}=\boldsymbol{s}^{(i)}, \tag{49}$$

and

$$\boldsymbol{A}_0^{(i)}=-2\pi k_0^2\frac{\sin(\phi+\theta_t)}{\cos\phi\sin\theta_t}\{\boldsymbol{s}^{(i)}\wedge(\boldsymbol{s}^{(i)}\wedge\boldsymbol{Q}_0)\}. \tag{50}$$

Eq. (49) expresses the *law of refraction* § 1.5 (8). For it implies that

$$\phi=\theta_i, \tag{51}$$

or, by (40),

$$n\sin\theta_t=\sin\theta_i. \tag{52}$$

Moreover (41) and (32) show that the vector $\boldsymbol{s}^{(t)}$ lies in the plane specified by $\boldsymbol{s}^{(i)}$ and the normal to Σ.

Eq. (50) connects the amplitudes of the incident wave (31) and the transmitted wave (34). Denoting the (vector) amplitude of the transmitted wave by $\boldsymbol{T}_0$,

$$\boldsymbol{T}_0=4\pi k_0^2\boldsymbol{Q}_0, \tag{53}$$

and expanding the vector triple product, (50) becomes

$$\boldsymbol{A}_0^{(i)}=\frac{1}{2}\frac{\sin(\theta_i+\theta_t)}{\cos\theta_i\sin\theta_t}[\boldsymbol{T}_0-\boldsymbol{s}^{(i)}(\boldsymbol{s}^{(i)}\,.\,\boldsymbol{T}_0)]. \tag{54}$$

Let $A_\perp$, $A_\parallel$ and $T_\perp$, $T_\parallel$ be the resolutes of $\boldsymbol{A}_0^{(i)}$ and of $\boldsymbol{T}_0$ in directions perpendicular and parallel to the plane of incidence. Then, remembering that $\boldsymbol{A}_0^{(i)}$ is orthogonal to $\boldsymbol{s}^{(i)}$ and $\boldsymbol{T}_0$ is orthogonal to $\boldsymbol{s}^{(t)}$, and that $\boldsymbol{s}^{(i)}$ and $\boldsymbol{s}^{(t)}$ are inclined to each other at an angle $(\theta_i-\theta_t)$, it follows from (54) that

$$A_\perp=\frac{1}{2}\frac{\sin(\theta_i+\theta_t)}{\cos\theta_i\sin\theta_t}T_\perp, \tag{55a}$$

$$A_\parallel=\frac{1}{2}\frac{\sin(\theta_i+\theta_t)\cos(\theta_i-\theta_t)}{\cos\theta_i\sin\theta_t}T_\parallel. \tag{55b}$$

These relations will be recognized as the *Fresnel formulae* § 1.5 (20a) *for refraction*.

Finally we consider the case when the point of observation is outside the medium ($z > 0$). The calculations are strictly analogous, only $z' = -r$ must now be replaced by $z' = r$ in the appropriate formulae. This is equivalent to replacing s_z by $-s_z$ in (46d), i.e. to replacing $\phi = \theta_i$ by $\phi = \theta_r$, where

$$\theta_r = \pi - \theta_i. \tag{56}$$

In place of the unit vector $\mathbf{s} = \mathbf{s}^{(i)}$ we now introduce the unit vector $\mathbf{s}^{(r)}$ with components

$$s_x^{(r)} = -\sin\theta_r = -\sin\theta_i, \quad s_y^{(r)} = 0, \quad s_z^{(r)} = -\cos\theta_r = \cos\theta_i, \tag{57}$$

and obtain instead of (47),

$$\mathbf{J}' = \int_\Sigma \left\{ \mathbf{Q}\frac{\partial G}{\partial \nu'} - G\frac{\partial \mathbf{Q}}{\partial \nu'} \right\} dS' = -2\pi \frac{\sin(\theta_i - \theta_t)}{\cos\theta_i \sin\theta_t} \mathbf{Q}_0\, e^{ik_0(\mathbf{r}.\mathbf{s}^{(r)})}. \tag{58}$$

(30) now gives for the time-free part of the reflected wave

$$\mathbf{A}_0^{(r)} e^{ik_0(\mathbf{r}.\mathbf{s}^{(r)})} = 2\pi k_0^2 \frac{\sin(\theta_i - \theta_t)}{\cos\theta_i \sin\theta_t} \{\mathbf{s}^{(r)} \wedge (\mathbf{s}^{(r)} \wedge \mathbf{Q}_0)\}\, e^{ik_0(\mathbf{r}.\mathbf{s}^{(r)})}. \tag{59}$$

(59) represents a plane wave propagated in the direction specified by the unit vector $\mathbf{s}^{(r)}$, this direction being related to the direction $\mathbf{s}^{(i)}$ of the incident wave by (56); relation (56) expresses the *law of reflection*, in agreement with § 1.5 (7).

In terms of the amplitude of the transmitted wave, the amplitude $\mathbf{A}_0^{(r)}$ of the reflected wave is

$$\mathbf{A}_0^{(r)} = -\frac{1}{2}\frac{\sin(\theta_i - \theta_t)}{\cos\theta_i \sin\theta_t}[\mathbf{T}_0 - \mathbf{s}^{(r)}(\mathbf{s}^{(r)}.\mathbf{T}_0)]. \tag{60}$$

Denoting by $R_\perp$ and $R_\parallel$ the resolutes of $\mathbf{A}_0^{(r)}$ in directions perpendicular and parallel to the plane of incidence, and using (55), we obtain

$$R_\perp = -\frac{1}{2}\frac{\sin(\theta_i - \theta_t)}{\cos\theta_i \sin\theta_t} T_\perp = -\frac{\sin(\theta_i - \theta_t)}{\sin(\theta_i + \theta_t)} A_\perp, \tag{61a}$$

and

$$R_\parallel = -\frac{1}{2}\frac{\sin(\theta_i - \theta_t)\cos(\theta_r - \theta_t)}{\cos\theta_i \sin\theta_t} T_\parallel = \frac{\tan(\theta_i - \theta_t)}{\tan(\theta_i + \theta_t)} A_\parallel. \tag{61b}$$

Relations (61) are the *Fresnel formulae* § 1.5 (21a) *for reflection.*

CHAPTER III

FOUNDATIONS OF GEOMETRICAL OPTICS

3.1 APPROXIMATION FOR VERY SHORT WAVELENGTHS

THE electromagnetic field associated with the propagation of visible light is characterized by very rapid oscillations (frequencies of the order of 10^{14} sec^{-1}) or, what amounts to the same thing, by the smallness of the wavelength (of order 10^{-5} cm). It may therefore be expected that a good first approximation to the propagation laws in such cases may be obtained by a complete neglect of the finiteness of the wavelength. It is found that for many optical problems such a procedure is entirely adequate; in fact, phenomena which can be attributed to departures from this approximate theory (so-called diffraction phenomena, studied in Chapter VIII) can only be demonstrated by means of carefully conducted experiments.

The branch of optics which is characterized by the neglect of the wavelength, i.e. that corresponding to the limiting case $\lambda_0 \rightarrow 0$, is known as *geometrical optics,** since in this approximation the optical laws may be formulated in the language of geometry. The energy may then be regarded as being transported along certain curves (light rays). A physical model of a pencil of rays may be obtained by allowing the light from a source of negligible extension to pass through a very small opening in an opaque screen. The light which reaches the space behind the screen will fill a region the boundary of which (the edge of the pencil) will, at first sight, appear to be sharp. A more careful examination will reveal, however, that the light intensity near the boundary varies rapidly but continuously from darkness in the shadow to lightness in the illuminated region, and that the variation is not monotonic but is of an oscillatory character, manifested by the appearance of bright and dark bands, called diffraction fringes. The region in which this rapid variation takes place is only of the order of magnitude of the wavelength. Hence, as long as this magnitude is neglected in comparison with the dimensions of the opening, we may speak of a sharply bounded pencil of rays.† On reducing the size of the opening down to the dimensions of the wavelength phenomena appear which need more refined study. If, however, one considers only the limiting case of negligible wavelengths, no restriction on the size of the opening is imposed, and we may say that an opening of vanishingly small dimensions defines an infinitely thin pencil—the light ray. It will be shown that the variation in the cross-section of a pencil of rays is a measure of the variation of the intensity of the light. Moreover it will be seen that it is possible to associate a state of polarization with each ray, and to study its variation along the ray.

---

* The historical development of geometrical optics is described by M. HERZBERGER, *Strahlenoptik* (Berlin, Springer, 1931), 179; *Z. Instrumentenkunde*, **52** (1932), 429–435, 485–493, 534–542, C. CARATHÉODORY, *Geometrische Optik* (Berlin, Springer, 1937) and E. MACH, *The Principles of Physical Optics, A Historical and Philosophical Treatment* (First German edition 1913, English translation: London, Methuen, 1926; reprinted by Dover Publications, New York, 1953).

† That the boundary becomes sharp in the limit as $\lambda_0 \rightarrow 0$ was first shown by G. KIRCHHOFF, *Vorlesungen ü. Math. Phys.*, **2** (*Mathematische Optik*), (Leipzig, Teubner, 1891), p. 33. See also B. B. BAKER and E. T. COPSON, *The Mathematical Theory of Huygens' Principle* (Oxford, Clarendon Press, 2nd edition, 1950), p. 79, and A. SOMMERFELD, *Optics* (New York, Academic Press, 1954), § 35.

Further it will be seen that for small wavelengths the field has the same general character as that of a plane wave, and, moreover, that within the approximation of geometrical optics the laws of refraction and reflection established for plane waves incident upon a plane boundary remain valid under more general conditions. Hence if a light ray falls on a sharp boundary (e.g. the surface of a lens) it is split into a reflected ray and a transmitted ray, and the changes in the state of polarization as well as the reflectivity and transmissivity may be calculated from the corresponding formulae for plane waves.

The preceding remarks imply that, when the wavelength is small enough, the sum total of optical phenomena may be deduced from geometrical considerations, by determining the paths of the light rays and calculating the associated intensity and polarization. We shall now formulate the appropriate laws by considering the implications of MAXWELL's equations when $\lambda_0 \to 0$.*

3.1.1 Derivation of the eikonal equation

We consider a general time-harmonic field

$$\left.\begin{aligned} \boldsymbol{E}(\boldsymbol{r}, t) &= \boldsymbol{E}_0(\boldsymbol{r})e^{-i\omega t}, \\ \boldsymbol{H}(\boldsymbol{r}, t) &= \boldsymbol{H}_0(\boldsymbol{r})e^{-i\omega t}, \end{aligned}\right\} \tag{1}$$

in a non-conducting isotropic medium. $\boldsymbol{E}_0$ and $\boldsymbol{H}_0$ denote complex vector functions of positions, and, as explained in § 1.4.3, the real parts of the expressions on the right-hand side of (1) are understood to represent the fields.

The complex vectors $\boldsymbol{E}_0$ and $\boldsymbol{H}_0$ will satisfy MAXWELL's equations in their time-free form, obtained on substituting (1) into (1)–(4) of § 1.1. In regions free of currents and charges ($\boldsymbol{j} = \rho = 0$), these equations are

$$\text{curl } \boldsymbol{H}_0 + ik_0\varepsilon\boldsymbol{E}_0 = 0, \tag{2}$$

$$\text{curl } \boldsymbol{E}_0 - ik_0\mu\boldsymbol{H}_0 = 0, \tag{3}$$

$$\text{div } \varepsilon\boldsymbol{E}_0 = 0, \tag{4}$$

$$\text{div } \mu\boldsymbol{H}_0 = 0. \tag{5}$$

Here the material relations $\boldsymbol{D} = \varepsilon\boldsymbol{E}$, $\boldsymbol{B} = \mu\boldsymbol{H}$ have been used and, as before, $k_0 = \omega/c = 2\pi/\lambda_0$, λ_0 being the vacuum wavelength.

We have seen that a homogeneous plane wave in a medium of refractive index $n = \sqrt{\varepsilon\mu}$, propagated in the direction specified by the unit vector $\boldsymbol{s}$, is represented by

$$\boldsymbol{E}_0 = \boldsymbol{e}e^{ik_0n(\boldsymbol{s}.\boldsymbol{r})}, \qquad \boldsymbol{H}_0 = \boldsymbol{h}e^{ik_0n(\boldsymbol{s}.\boldsymbol{r})}, \tag{6}$$

where $\boldsymbol{e}$ and $\boldsymbol{h}$ are constant, generally complex vectors. For a (monochromatic) electric dipole field in the vacuum we found (cf. § 2.2) that

$$\boldsymbol{E}_0 = \boldsymbol{e}e^{ik_0r}, \qquad \boldsymbol{H}_0 = \boldsymbol{h}e^{ik_0r}, \tag{7}$$

---

* It was first shown by A. SOMMERFELD and J. RUNGE, *Ann. d. Physik*, **35** (1911), 289, using a suggestion of P. DEBYE, that the basic equation of geometrical optics (the eikonal equation (15b)) may be derived from the (scalar) wave equation in the limiting case $\lambda_0 \to 0$. Generalizations which take into account the vectorial character of the electromagnetic field are due to W. IGNATOWSKY, *Trans. State Opt. Institute* (*Petrograd*), **1** (1919), III; V. A. FOCK, *ibid.*, **3** (1924), 3; S. M. RYTOV, *Compt. Rend.* (*Doklady*) *Acad. Sci. URSS*, **18** (1938), 263; N. ARLEY, *Det. Kgl. Danske Videns Selsk.*, **22** (1945), No. 8; F. G. FRIEDLANDER, *Proc. Cambr. Phil. Soc.*, **43** (1947), 284; K. SUCHY, *Ann. d. Physik.*, **11** (1952), 113, *ibid.*, **12** (1953), 423, and *ibid.*, **13** (1953), 178; R. S. INGARDEN and A. KRZYWICKI, *Acta Phys. Polonica*, **14** (1955), 255.

r being the distance from the dipole. Here $\boldsymbol{e}$ and $\boldsymbol{h}$ are no longer constant vectors, but at distances sufficiently far away from the dipole ($r \gg \lambda_0$) these vectors are, with suitable normalization of the dipole moment, independent of k_0.

These examples suggest that in regions which are many wavelengths away from the sources we may represent more general types of fields in the form

$$\boldsymbol{E}_0 = \boldsymbol{e}(\boldsymbol{r})e^{ik_0\mathscr{S}(\boldsymbol{r})}, \qquad \boldsymbol{H}_0 = \boldsymbol{h}(\boldsymbol{r})e^{ik_0\mathscr{S}(\boldsymbol{r})}, \tag{8}$$

where $\mathscr{S}(\boldsymbol{r})$, "the optical path", is a real scalar function of position, and $\boldsymbol{e}(\boldsymbol{r})$ and $\boldsymbol{h}(\boldsymbol{r})$ are vector functions of position, which may in general be complex.* With (8) as trial solution, MAXWELL's equations lead to a set of relations between $\boldsymbol{e}$, $\boldsymbol{h}$ and $\mathscr{S}$. It will be shown that for large k_0 (small λ_0) these relations demand that $\mathscr{S}$ should satisfy a certain differential equation, which is independent of the amplitude vectors $\boldsymbol{e}$ and $\boldsymbol{h}$.

From (8), using well-known vector identities,

$$\text{curl}\, \boldsymbol{H}_0 = (\text{curl}\, \boldsymbol{h} + ik_0 \,\text{grad}\, \mathscr{S} \wedge \boldsymbol{h})e^{ik_0\mathscr{S}}, \tag{9}$$

$$\text{div}\, \mu\boldsymbol{H}_0 = (\mu \,\text{div}\, \boldsymbol{h} + \boldsymbol{h} \,.\, \text{grad}\, \mu + ik_0\mu\boldsymbol{h} \,.\, \text{grad}\, \mathscr{S})e^{ik_0\mathscr{S}}, \tag{10}$$

with similar expressions for curl $\boldsymbol{E}_0$ and div $\varepsilon\boldsymbol{E}_0$. Hence (2)–(5) become

$$\text{grad}\, \mathscr{S} \wedge \boldsymbol{h} + \varepsilon\boldsymbol{e} = -\frac{1}{ik_0}\,\text{curl}\, \boldsymbol{h}, \tag{11}$$

$$\text{grad}\, \mathscr{S} \wedge \boldsymbol{e} - \mu\boldsymbol{h} = -\frac{1}{ik_0}\,\text{curl}\, \boldsymbol{e}, \tag{12}$$

$$\boldsymbol{e} \,.\, \text{grad}\, \mathscr{S} = -\frac{1}{ik_0}(\boldsymbol{e} \,.\, \text{grad}\log\varepsilon + \text{div}\, \boldsymbol{e}), \tag{13}$$

$$\boldsymbol{h} \,.\, \text{grad}\, \mathscr{S} = -\frac{1}{ik_0}(\boldsymbol{h} \,.\, \text{grad}\log\mu + \text{div}\, \boldsymbol{h}). \tag{14}$$

We are interested in the solution for very large values of k_0. Hence as long as the multiplicative factors of $1/ik_0$ on the right-hand side are not exceptionally large they may be neglected, and the equations then reduce to

$$\text{grad}\, \mathscr{S} \wedge \boldsymbol{h} + \varepsilon\boldsymbol{e} = 0, \tag{11a}$$

$$\text{grad}\, \mathscr{S} \wedge \boldsymbol{e} - \mu\boldsymbol{h} = 0, \tag{12a}$$

$$\boldsymbol{e} \,.\, \text{grad}\, \mathscr{S} = 0, \tag{13a}$$

$$\boldsymbol{h} \,.\, \text{grad}\, \mathscr{S} = 0. \tag{14a}$$

We can confine our attention to equations (11a) and (12a) alone, since (13a) and (14a) follow from them on scalar multiplication with grad $\mathscr{S}$. Now (11a) and (12a) may be regarded as a set of six simultaneous homogeneous linear scalar equations for the Cartesian components e_x, h_x, . . . of $\boldsymbol{e}$ and $\boldsymbol{h}$. These simultaneous equations have non-trivial solutions only if a consistency condition (the vanishing of the associated

---

* Complex $\boldsymbol{e}$ and $\boldsymbol{h}$ are necessary, if all possible states of polarization are to be included. According to § 1.4 (75) real $\boldsymbol{e}$ and $\boldsymbol{h}$ correspond to fields which are linearly polarized.

determinant) is satisfied. This condition may be obtained simply by eliminating $\boldsymbol{e}$ or $\boldsymbol{h}$ between (11a) and (12a). Substituting for $\boldsymbol{h}$ from (12a), (11a) becomes

$$\frac{1}{\mu}[(\boldsymbol{e}\,.\,\operatorname{grad}\mathscr{S})\operatorname{grad}\mathscr{S}-\boldsymbol{e}(\operatorname{grad}\mathscr{S})^2]+\varepsilon\boldsymbol{e}=0.$$

The first term vanishes on account of (13a), and the equation then reduces, since $\boldsymbol{e}$ does not vanish everywhere, to

$$(\operatorname{grad}\mathscr{S})^2=n^2, \tag{15a}$$

or, written explicitly,

$$\left(\frac{\partial\mathscr{S}}{\partial x}\right)^2+\left(\frac{\partial\mathscr{S}}{\partial y}\right)^2+\left(\frac{\partial\mathscr{S}}{\partial z}\right)^2=n^2(x,y,z), \tag{15b}$$

where as before $n=\sqrt{\varepsilon\mu}$ denotes the refractive index. The function $\mathscr{S}$ is often called the *eikonal** and (15b) is known as the *eikonal equation*; it is the basic equation of geometrical optics.† The surfaces

$$\mathscr{S}(\boldsymbol{r})=\text{constant}$$

may be called the *geometrical wave surfaces* or the *geometrical wave-fronts*.‡

The eikonal equation was derived here by using the first-order MAXWELL's equations, but it may also be derived from the second-order wave equations for the electric or magnetic field vectors. To show this one substitutes from (1) and (8) into the wave equation § 1.2 (5) and obtains, after a straightforward calculation,

$$\mathbf{K}(\boldsymbol{e},\mathscr{S},n)+\frac{1}{ik_0}\mathbf{L}(\boldsymbol{e},\mathscr{S},n,\mu)+\frac{1}{(ik_0)^2}\mathbf{M}(\boldsymbol{e},\varepsilon,\mu)=0, \tag{16}$$

where

$$\mathbf{K}(\boldsymbol{e},\mathscr{S},n)=\{n^2-(\operatorname{grad}\mathscr{S})^2\}\boldsymbol{e},$$

$$\mathbf{L}(\boldsymbol{e},\mathscr{S},n,\mu)=\{\operatorname{grad}\mathscr{S}\,.\,\operatorname{grad}\log\mu-\nabla^2\mathscr{S}\}\boldsymbol{e}-2\{\boldsymbol{e}\,.\,\operatorname{grad}\log n\}\operatorname{grad}\mathscr{S}$$
$$-2\{\operatorname{grad}\mathscr{S}\,.\,\operatorname{grad}\}\boldsymbol{e},$$

$$\mathbf{M}(\boldsymbol{e},\varepsilon,\mu)=\operatorname{curl}\boldsymbol{e}\wedge\operatorname{grad}\log\mu-\nabla^2\boldsymbol{e}-\operatorname{grad}(\boldsymbol{e}\,.\,\operatorname{grad}\log\varepsilon).$$

The corresponding equation involving $\boldsymbol{h}$, obtained on substitution into the wave equation (6) in § 1.2 for $\boldsymbol{H}$ (or more simply by using the fact that MAXWELL's equations remain unchanged when $\boldsymbol{E}$ and $\boldsymbol{H}$ and simultaneously ε and $-\mu$ are interchanged), is

$$\mathbf{K}(\boldsymbol{h},\mathscr{S},n)+\frac{1}{(ik_0)}\mathbf{L}(\boldsymbol{h},\mathscr{S},n,\varepsilon)+\frac{1}{(ik_0)^2}\mathbf{M}(\boldsymbol{h},\mu,\varepsilon)=0. \tag{17}$$

---

* The term *eikonal* (from Greek εἰκῶν = image) was introduced in 1895 by H. BRUNS to describe certain related functions (cf. p. 133), but has come to be used in a wider sense.

† The eikonal equation may also be regarded as the equation of the characteristics of the wave equations (5) and (6), in § 1.2, for $\boldsymbol{E}$ and $\boldsymbol{H}$, and describes rigorously the propagation of discontinuities of the solutions of these equations. In geometrical optics we are, however, not concerned with the propagation of discontinuities but with time-harmonic (or nearly time-harmonic) solutions. The formal equivalence of the two interpretations is demonstrated in Appendix VI.

The eikonal equation will also be recognized as the HAMILTON–JACOBI equation of the variational problem $\delta\int n\,ds=0$, the optical counterpart of which goes back to FERMAT (cf. § 3.3.2 and Appendix I).

‡ In future we shall drop the adjective "geometrical" when there is no risk of confusion.

For sufficiently large k_0 the second and third terms may in general be neglected; then $K = 0$, giving again the eikonal equation. It will be seen later that the terms in the first power of $1/(ik_0)$ in (16) and (17) also possess a physical interpretation.

It may be shown that in many cases of importance the spatial parts $\mathbf{E}_0$ and $\mathbf{H}_0$ of the field vectors may be developed into asymptotic series of the form*

$$\mathbf{E}_0 = e^{ik_0\mathcal{S}} \sum_{m\geqslant 0} \frac{\mathbf{e}^{(m)}}{(ik_0)^m}, \qquad \mathbf{H}_0 = e^{ik_0\mathcal{S}} \sum_{m\geqslant 0} \frac{\mathbf{h}^{(m)}}{(ik_0)^m}, \tag{18}$$

where $\mathbf{e}^{(m)}$ and $\mathbf{h}^{(m)}$ are functions of position, and $\mathcal{S}$ is the same function as before.† Geometrical optics corresponds to the leading terms of these expansions.

3.1.2 The light rays and the intensity law of geometrical optics

From (8), and from (54) and (55) in § 1.4, it follows that the time averages of the electric and magnetic energy densities $\langle w_e\rangle$ and $\langle w_m\rangle$ are given by

$$\langle w_e\rangle = \frac{\varepsilon}{16\pi}\,\mathbf{e}\,.\,\mathbf{e}^\star, \qquad \langle w_m\rangle = \frac{\mu}{16\pi}\,\mathbf{h}\,.\,\mathbf{h}^\star. \tag{19}$$

Substitution for $\mathbf{e}^\star$ from (11a) and for $\mathbf{h}$ from (12a) gives

$$\langle w_e\rangle = \langle w_m\rangle = \frac{1}{16\pi}[\mathbf{e}, \mathbf{h}^\star, \text{grad}\,\mathcal{S}], \tag{20}$$

the square bracket denoting the scalar triple product. Hence, *within the accuracy of geometrical optics, the time-averaged electric and magnetic energy densities are equal.*

The time average of the Poynting vector is obtained from (8) and § 1.4 (52):

$$\langle \mathbf{S}\rangle = \frac{c}{8\pi}\mathcal{R}(\mathbf{e} \wedge \mathbf{h}^\star).$$

Using (12a), we obtain

$$\langle \mathbf{S}\rangle = \frac{c}{8\pi\mu}\{(\mathbf{e}\,.\,\mathbf{e}^\star)\,\text{grad}\,\mathcal{S} - (\mathbf{e}\,.\,\text{grad}\,\mathcal{S})\mathbf{e}^\star\}.$$

The last term vanishes on account of (13a) so that we have, if use is made of the expression for $\langle w_e\rangle$ and of MAXWELL's relation $\varepsilon\mu = n^2$,

$$\langle \mathbf{S}\rangle = \frac{2c}{n^2}\langle w_e\rangle\,\text{grad}\,\mathcal{S}. \tag{21}$$

---

* We assume here that only one geometrical wave-front passes through each point. In some cases, for example when reflection takes place at obstacles present in the medium, several wave-fronts may pass through each point. The resulting field is then represented by the addition of series of the above type.

† The theory of such asymptotic expansions has its origin chiefly in the work of R. K. LUNEBURG, *Propagation of Electromagnetic Waves* (mimeographed lecture notes, New York University, 1947–1948). See also M. KLINE, *Comm. Pure and Appl. Math.*, **4** (1951), 225; *ibid.*, **8** (1955), 595 and W. BRAUNBECK, *Z. Naturforsch.*, **6** (1951), 672. A comprehensive account of the theory is given in M. KLINE and I. W. KAY, *Electromagnetic Theory and Geometrical Optics* [New York, Interscience Publishers, (1965)].

Since $\langle w_e \rangle = \langle w_m \rangle$, the term $2\langle w_e \rangle$ represents the time average $\langle w \rangle$ of the total energy density (i.e. $\langle w \rangle = \langle w_e \rangle + \langle w_m \rangle$). Also, on account of the eikonal equation, $(\text{grad}\, \mathscr{S})/n$ is a unit vector ($\mathbf{s}$ say),

$$\mathbf{s} = \frac{\text{grad}\, \mathscr{S}}{n} = \frac{\text{grad}\, \mathscr{S}}{|\text{grad}\, \mathscr{S}|}, \tag{22}$$

and (21) shows that $\mathbf{s}$ is in the direction of the average Poynting vector. If, as before, we set $c/n = v$, (21) becomes

$$\langle \mathbf{S} \rangle = v\langle w \rangle \mathbf{s}. \tag{23}$$

Hence *the average Poynting vector is in the direction of the normal to the geometrical wave-front, and its magnitude is equal to the product of the average energy density and the velocity* $v = c/n$. This result is analogous to the relation (9) in § 1.4 for plane waves,

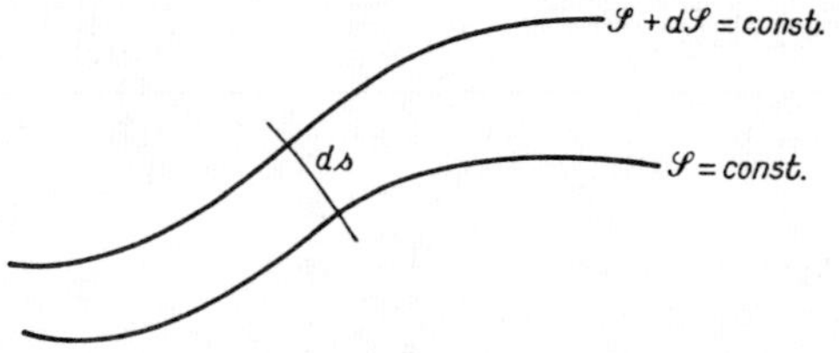

Fig. 3.1. Illustrating the meaning of the relation $n\mathbf{s} = \text{grad}\, \mathscr{S}$.

and shows that *within the accuracy of geometrical optics the average energy density is propagated with the velocity* $v = c/n$.

The *geometrical light rays* may now be defined as the orthogonal trajectories to the geometrical wave-fronts $\mathscr{S}$ = constants. We shall regard them as oriented curves whose direction coincides everywhere with the direction of the average Poynting vector.* If $\mathbf{r}(s)$ denotes the position vector of a point P on a ray, considered as a function of the length of arc s of the ray, then $d\mathbf{r}/ds = \mathbf{s}$, and the equation of the ray may be written as

$$n\frac{d\mathbf{r}}{ds} = \text{grad}\, \mathscr{S}. \tag{24}$$

From (13a) and (14a) it is seen that *the electric and magnetic vectors are at every point orthogonal to the ray.*

The meaning of (24) may be made clearer from the following remarks. Consider two neighbouring wave-fronts $\mathscr{S}$ = constant and $\mathscr{S} + d\mathscr{S}$ = constant (Fig. 3.1). Then

$$\frac{d\mathscr{S}}{ds} = \frac{d\mathbf{r}}{ds} \cdot \text{grad}\, \mathscr{S} = n. \tag{25}$$

Hence the distance ds between points on the opposite ends of a normal cutting the two wave-fronts is inversely proportional to the refractive index, i.e. directly proportional to v.

* This definition of light rays is appropriate for isotropic media only. We shall see later (Chapter XIV) that in an anisotropic medium the direction of the wave-front normal does not, in general, coincide with the direction of the Poynting vector.

The integral $\int_C nds$ taken along a curve C is known as the *optical length* of the curve. Denoting by square brackets the optical length of the *ray* which joins points P_1 and P_2, we have

$$[P_1P_2] = \int_{P_1}^{P_2} nds = \mathscr{S}(P_2) - \mathscr{S}(P_1). \tag{26}$$

Since, as we have seen, the average energy density is propagated with the velocity $v = c/n$ along the ray,

$$nds = \frac{c}{v}\, ds = cdt,$$

where dt is the time needed for the energy to travel the distance ds along the ray; hence

$$[P_1P_2] = c\int_{P_1}^{P_2} dt, \tag{27}$$

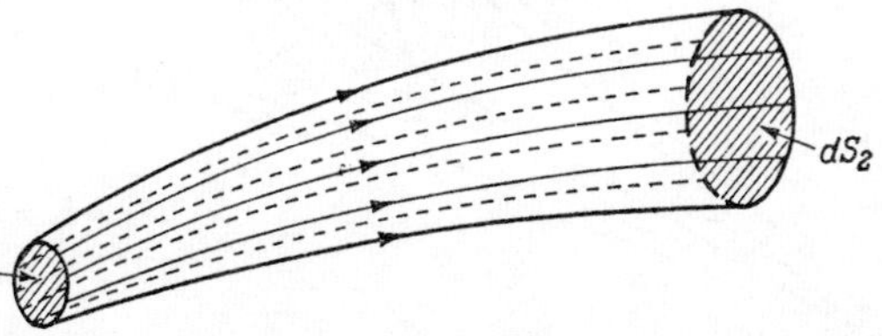

Fig. 3.2. Illustrating the intensity law of geometrical optics.

i.e. *the optical length* $[P_1P_2]$ *is equal to the product of the vacuum velocity of light and the time needed for light to travel from* P_1 *to* P_2.

The intensity of light I was defined as the absolute value of the time average of the Poynting vector. We therefore have from (23),

$$I = |\langle \mathbf{S}\rangle| = v\langle w\rangle, \tag{28}$$

and the conservation law § 1.4 (57) gives

$$\text{div}\,(I\mathbf{s}) = 0. \tag{29}$$

To see the implications of this relation we take a narrow tube formed by all the rays proceeding from an element dS_1 of a wave-front $\mathscr{S}(\mathbf{r}) = a_1$ (a_1 being a constant), and denote by dS_2 the corresponding element in which these rays intersect another wave-front $\mathscr{S}(\mathbf{r}) = a_2$ (Fig. 3.2). Integrating (29) throughout the tube and applying Gauss' theorem we obtain

$$\int I\mathbf{s}.\boldsymbol{\nu}\, dS = 0, \tag{30}$$

$\boldsymbol{\nu}$ denoting the outward normal to the tube. Now

$$\begin{aligned} \mathbf{s}\,.\,\boldsymbol{\nu} &= 1 \text{ on } dS_2, \\ &= -1 \text{ on } dS_1, \\ &= 0 \text{ elsewhere,} \end{aligned}$$

so that (30) reduces to

$$I_1 dS_1 = I_2 dS_2, \tag{31}$$

I_1 and I_2 denoting the intensity on dS_1 and on dS_2 respectively. Hence *IdS remains constant along a tube of rays.* This result expresses *the intensity law of geometrical optics.*

We shall see later that in a homogeneous medium the rays are straight lines. The intensity law may then be expressed in a somewhat different form. Assume first that dS_1, and consequently also dS_2, are bounded by segments of lines of curvature (see Fig. 3.3). If R_1 and R_1' are the principal radii of curvature (cf. § 4.6.1) of the segments A_1B_1 and B_1C_1, then

$$A_1B_1 = R_1 d\theta, \qquad B_1C_1 = R_1' d\phi,$$

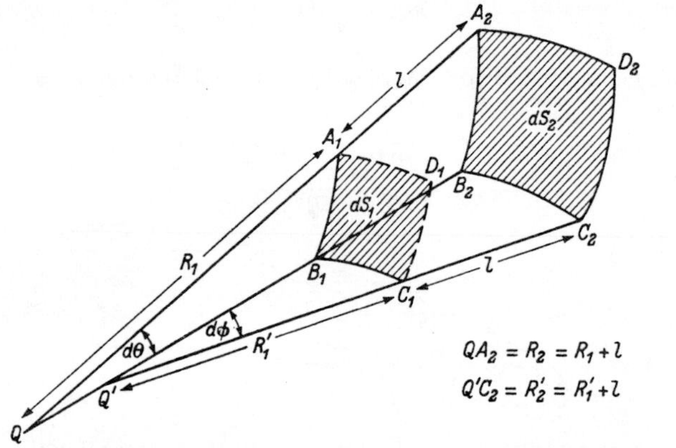

Fig. 3.3. Illustrating the intensity law of geometrical optics for rectilinear rays.

where $d\theta$ and $d\phi$ are the angles which A_1B_1 and B_1C_1 subtend at the respective centres of curvature Q and Q'. Hence

$$dS_1 = A_1B_1 \,.\, B_1C_1 = R_1R_1' d\theta d\phi; \tag{32}$$

similarly for an element dS_2 in which the bundle of rays through dS_1 meets another wave-front of the family,

$$dS_2 = A_2B_2 \,.\, B_2C_2 = R_2R_2' d\theta d\phi. \tag{33}$$

If l is the distance between dS_1 and dS_2 measured along the rays, then

$$R_2 = R_1 + l, \qquad R_2' = R_1' + l,$$

and it follows that

$$\frac{I_2}{I_1} = \frac{dS_1}{dS_2} = \frac{R_1R_1'}{R_2R_2'} = \frac{R_1R_1'}{(R_1 + l)(R_1' + l)}. \tag{34}$$

If the areas dS_1 and dS_2 are bounded by arbitrary curves, (34) still holds. This can immediately be seen if we regard them as made up of a number of elements bounded by lines of curvature, and sum their contributions.

If $R_1 \ll l$, $R_1' \ll l$, (34) reduces to

$$\frac{I_2}{I_1} = \frac{R_1R_1'}{l^2}. \tag{35}$$

This formula is sometimes used in connection with the scattering of radiation.

The reciprocal $1/RR'$ of the product of the two principal radii of curvature is called the *Gaussian curvature* (or second curvature) of the surface. (34) shows that *the intensity at any point of a rectilinear ray is proportional to the Gaussian curvature of the wave-front which passes through that point.* In particular if all the (rectilinear) rays have a point in common, the wave-fronts are spheres centred at that point; then $R_1 = R_1'$, $R_2 = R_2'$ and we obtain (dropping the suffixes) the *inverse square law*

$$I = \frac{\text{constant}}{R^2}. \tag{36}$$

Returning to the general case of an arbitrary pencil of rays (curved or straight), we can write down an explicit expression in terms of the $\mathscr{S}$ function for the variation of the intensity along each ray. Substituting for $\mathbf{s}$ from (22) into (29), and using the identities div $u\mathbf{v} = u$ div $\mathbf{v} + \mathbf{v}$. grad u, and div grad $= \nabla^2$, we obtain

$$\frac{I}{n}\nabla^2\mathscr{S} + \text{grad}\,\mathscr{S}\,.\,\text{grad}\,\frac{I}{n} = 0.$$

This may also be written as

$$\nabla^2\mathscr{S} + \text{grad}\,\mathscr{S}\,.\,\text{grad}\,\log\frac{I}{n} = 0. \tag{37}$$

Let us now introduce the operator

$$\frac{\partial}{\partial\tau} = \text{grad}\,\mathscr{S}\,.\,\text{grad}, \tag{38}$$

where τ is a parameter which specifies position along the ray. Then (37) may be written as

$$\frac{\partial}{\partial\tau}\log\frac{I}{n} = -\nabla^2\mathscr{S}$$

whence, on integration,

$$I = ne^{-\int^{\tau}\nabla^2\mathscr{S}d\tau}$$

But by (38), (15), and (25),

$$d\tau = \frac{d\mathscr{S}}{(\text{grad}\,\mathscr{S})^2} = \frac{1}{n^2}\,d\mathscr{S} = \frac{1}{n}\,ds, \tag{39}$$

so that we finally obtain the following expressions for the ratio of the intensities at any two points of a ray:

$$\frac{I_2}{I_1} = \frac{n_2}{n_1}e^{-\int_{\mathscr{S}_1}^{\mathscr{S}_2}\frac{\nabla^2\mathscr{S}}{n^2}d\mathscr{S}} = \frac{n_2}{n_1}e^{-\int_{s_1}^{s_2}\frac{\nabla^2\mathscr{S}}{n}ds}, \tag{40}$$

the integrals being taken along the ray*.

3.1.3 Propagation of the amplitude vectors

We have seen that, when the wavelength is sufficiently small, the transport of energy may be represented by means of a simple hydrodynamical model which may be completely described in terms of the real scalar function $\mathscr{S}$, this function being a solution of the eikonal equation (15). According to traditional terminology, one understands by geometrical optics this approximate picture of energy propagation, using the concept of rays and wave-fronts. In other words polarization properties

* It has been shown by M. Kline, *Comm. Pure and Appl. Maths.*, **14** (1961), 473 that the intensity ratio (40) may be expressed in terms of an integral which involves the principal radii of curvature of the associated wavefronts. Kline's formula is a natural generalization, to inhomogeneous media, of the formula (34). See also M. Kline and I. W. Kay, *Electromagnetic Theory and Geometrical Optics* (New York, Interscience Publishers, 1965), p. 184.

are excluded. The reason for this restriction is undoubtedly due to the fact that the simple laws of geometrical optics concerning rays and wave-fronts were known from experiments long before the electromagnetic theory of light was established. It is however possible, and from our point of view quite natural, to extend the meaning of geometrical optics to embrace also certain geometrical laws relating to the propagation of the "amplitude vectors" $\boldsymbol{e}$ and $\boldsymbol{h}$. These laws may be easily deduced from the wave equations (16)–(17).

Since $\mathscr{S}$ satisfies the eikonal equation, it follows that $\mathbf{K} = 0$, and we see that when k_0 is sufficiently large (λ_0 small enough), only the $\mathbf{L}$-terms need to be retained in (16) and (17). Hence, in the present approximation, the amplitude vectors and the eikonal are connected by the relations $\mathbf{L} = 0$. If we use again the operator $\partial/\partial\tau$ introduced by (38), the equations $\mathbf{L} = 0$ become

$$\frac{\partial \boldsymbol{e}}{\partial\tau} + \frac{1}{2}\left(\nabla^2\mathscr{S} - \frac{\partial \log \mu}{\partial\tau}\right)\boldsymbol{e} + (\boldsymbol{e} \,.\, \text{grad} \log n) \,\text{grad}\, \mathscr{S} = 0, \tag{41}$$

$$\frac{\partial \boldsymbol{h}}{\partial\tau} + \frac{1}{2}\left(\nabla^2\mathscr{S} - \frac{\partial \log \varepsilon}{\partial\tau}\right)\boldsymbol{h} + (\boldsymbol{h} \,.\, \text{grad} \log n) \,\text{grad}\, \mathscr{S} = 0. \tag{42}$$

These are the required *transport equations* for the variation of $\boldsymbol{e}$ and $\boldsymbol{h}$ along each ray. The implications of these equations can best be understood by examining separately the variation of the magnitude and of the direction of these vectors.

We multiply (41) scalarly by $\boldsymbol{e}^\star$ and add to the resulting equation the corresponding equation obtained by taking the complex conjugate. This gives

$$\frac{\partial}{\partial\tau}(\boldsymbol{e} \,.\, \boldsymbol{e}^\star) + \left(\nabla^2\mathscr{S} - \frac{\partial \log \mu}{\partial\tau}\right)\boldsymbol{e} \,.\, \boldsymbol{e}^\star = 0. \tag{43}$$

On account of the identity $\text{div}\, u\mathbf{v} = u \,\text{div}\, \mathbf{v} + \mathbf{v} \,.\, \text{grad}\, u$, the second and third term may be combined as follows:

$$\nabla^2\mathscr{S} - \frac{\partial \log \mu}{\partial\tau} = \nabla^2\mathscr{S} - \text{grad}\,\mathscr{S} \,.\, \text{grad} \log \mu = \mu \,\text{div}\left(\frac{1}{\mu}\,\text{grad}\,\mathscr{S}\right). \tag{44}$$

Integrating (43) along a ray, the following expression for the ratio of $\boldsymbol{e} \,.\, \boldsymbol{e}^\star$ at any two points of the ray is obtained:*

$$\frac{(\boldsymbol{e} \,.\, \boldsymbol{e}^\star)_2}{(\boldsymbol{e} \,.\, \boldsymbol{e}^\star)_1} = e^{-\int_{\tau_1}^{\tau_2} \mu\, \text{div}\left(\frac{1}{\mu}\text{grad}\,\mathscr{S}\right) d\tau} = e^{-\int_{s_1}^{s_2} \sqrt{\frac{\mu}{\varepsilon}}\, \text{div}\left(\frac{1}{\mu}\text{grad}\,\mathscr{S}\right) ds} \tag{45}$$

---

* This relation may also be written in the alternative form

$$\left(\frac{\boldsymbol{e} \,.\, \boldsymbol{e}^\star}{\mu}\right)_2 = \left(\frac{\boldsymbol{e} \,.\, \boldsymbol{e}^\star}{\mu}\right)_1 e^{-\int_{s_1}^{s_2} \frac{\nabla^2\mathscr{S}}{n} ds} \tag{45a}$$

which follows when (43) is re-written in the form

$$\frac{\partial}{\partial\tau}\left[\log\left(\frac{\boldsymbol{e} \,.\, \boldsymbol{e}^\star}{\mu}\right)\right] = -\nabla^2\mathscr{S},$$

and the integral is taken along a ray. (45a) is in fact only another way of expressing the relation (40) for the variation of intensity, and follows from it when the relation

$$I = \frac{2c}{n}\langle w_e\rangle = \frac{c\varepsilon}{8\pi n}(\boldsymbol{e} \,.\, \boldsymbol{e}^\star)$$

and the MAXWELL formula $\varepsilon\mu = n^2$ are used.

Similarly

$$\frac{(\boldsymbol{h}\,.\,\boldsymbol{h}^\star)_2}{(\boldsymbol{h}\,.\,\boldsymbol{h}^\star)_1} = e^{-\int_{s_1}^{s_2}\sqrt{\frac{\varepsilon}{\mu}}\,\mathrm{div}\left(\frac{1}{\varepsilon}\,\mathrm{grad}\,\mathscr{S}\right)ds}. \tag{46}$$

Next consider the variation of the complex unit vectors

$$\boldsymbol{u} = \frac{\boldsymbol{e}}{\sqrt{\boldsymbol{e}\,.\,\boldsymbol{e}^\star}}, \qquad \mathbf{v} = \frac{\boldsymbol{h}}{\sqrt{\boldsymbol{h}\,.\,\boldsymbol{h}^\star}}, \tag{47}$$

along each ray. Substitution into (41) gives

$$\frac{\partial \boldsymbol{u}}{\partial \tau} + \frac{1}{2}\left[\frac{\partial \log (\boldsymbol{e}\,.\,\boldsymbol{e}^\star)}{\partial \tau} + \nabla^2\mathscr{S} - \frac{\partial \log \mu}{\partial \tau}\right]\boldsymbol{u} + (\boldsymbol{u}\,.\,\mathrm{grad}\,\log n)\,\mathrm{grad}\,\mathscr{S} = 0.$$

The second, third and fourth terms vanish on account of (43), and it follows that

$$\frac{d\boldsymbol{u}}{d\tau} \equiv n\frac{d\boldsymbol{u}}{ds} = -\,(\boldsymbol{u}\,.\,\mathrm{grad}\,\log n)\,\mathrm{grad}\,\mathscr{S}, \tag{48}$$

and similarly

$$\frac{d\mathbf{v}}{d\tau} \equiv n\frac{d\mathbf{v}}{ds} = -\,(\mathbf{v}\,.\,\mathrm{grad}\,\log n)\,\mathrm{grad}\,\mathscr{S}. \tag{49}$$

This is the required law for the variation of $\boldsymbol{u}$ and $\mathbf{v}$ along each ray.* In particular, *for a homogeneous medium* (n = constant) (48) and (49) reduce to $d\boldsymbol{u}/ds = d\mathbf{v}/ds = 0$ *so that* $\boldsymbol{u}$ *and* $\mathbf{v}$ *then remain constant along each ray.*

Finally we note that for a time-harmonic homogeneous plane wave in a homogeneous medium, $\mathscr{S} = n\mathbf{s}.\mathbf{r}$ and $\boldsymbol{e}$, $\boldsymbol{h}$, ε and μ are all constants, and consequently $\mathbf{K} = \mathbf{L} = \mathbf{M} \equiv 0$ in (16). Such a wave (whatever its frequency) therefore obeys rigorously the laws of geometrical optics.

3.1.4 Generalizations and the limits of validity of geometrical optics

The considerations of the preceding sections apply to a strictly monochromatic field. Such a field, which may be regarded as a typical Fourier component of an arbitrary field, is produced by a harmonic oscillator, or by a set of such oscillators of the same frequency.

In optics one usually deals with a source which emits light within a narrow, but nevertheless finite, frequency range. The source may then be regarded as arising from a large number of harmonic oscillators whose frequencies fall within this range. To obtain the intensity at a typical field point P one has to sum the individual fields produced by each oscillator (element of the source):

$$\boldsymbol{E} = \sum_n \boldsymbol{E}_n, \qquad \boldsymbol{H} = \sum_n \boldsymbol{H}_n. \tag{50}$$

---

* The relations (48) and (49) have an interesting interpretation in terms of non-Euclidean geometry. If we consider the associated non-Euclidean space whose line element is given by

$$ds' = nds = n\sqrt{dx^2 + dy^2 + dz^2},$$

then the geometrical light rays correspond to geodesics in this space, and (48) and (49) may be shown to imply that each of the two vectors $\boldsymbol{u}$ and $\mathbf{v}$ is transferred parallel to itself (in the sense of Levi-Civita parallelism) along each ray. Cf. F. Bortolotti, *Rend. R. Acc. Naz. Linc.*, 6a, **4** (1926), 552; R. K. Luneburg, *Mathematical Theory of Optics* (mimeographed lecture notes, Brown University, Providence, R.I., 1944, p. 55–59; printed version published by University of California Press, Berkeley and Los Angeles, 1964, p. 51–55); M. Kline and I. W. Kay, *Electromagnetic Theory and Geometrical Optics* (New York, Interscience Publishers, 1965), p. 180–183.

The intensity is then given by (using real representation)

$$I(P) = |\langle \mathbf{S} \rangle| = \frac{c}{4\pi} |\langle \mathbf{E} \wedge \mathbf{H} \rangle| = \frac{c}{4\pi} |\sum_{n,m} \langle \mathbf{E}_n \wedge \mathbf{H}_m \rangle|$$

$$= \frac{c}{4\pi} |\sum_{n} \langle \mathbf{E}_n \wedge \mathbf{H}_n \rangle + \sum_{n \neq m} \langle \mathbf{E}_n \wedge \mathbf{H}_m \rangle|. \qquad (51)$$

In many optical problems it is usually permissible to assume that the second sum in (51) vanishes (the fields are then said to be *incoherent*), so that

$$I(P) = \frac{c}{4\pi} |\sum_{n} \langle \mathbf{E}_n \wedge \mathbf{H}_n \rangle| = |\sum_{n} \langle \mathbf{S}_n \rangle|, \qquad (52)$$

$\mathbf{S}_n$ denoting the Poynting vector due to the nth element of the source. It is not possible to discuss at this stage the conditions under which the neglect of the second

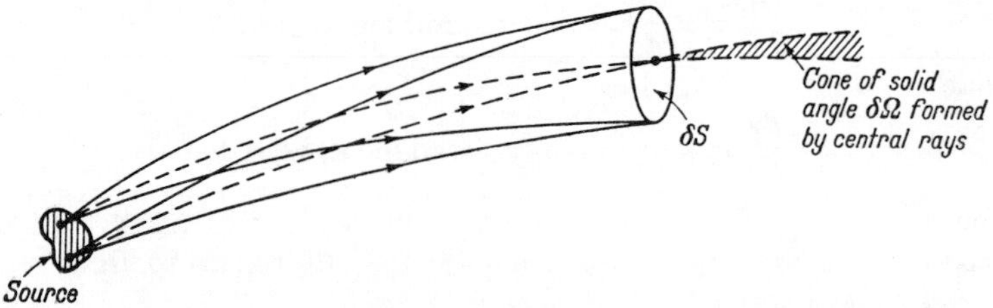

Fig. 3.4. Illustrating the intensity law of geometrical optics for an extended incoherent source.

term in (51) is justified, but this point will be considered fully later, in connection with partial coherence (Chapter X).

Let δS be a small portion of a wave-front associated with one particular element of the source. Every element of the source sends through δS a tube of rays, and the central rays of these tubes fill a cone of solid angle $\delta\Omega$ (Fig. 3.4). If the semi-vertical angle of this cone is small enough, we may neglect the variation of $\mathbf{S}_n$ with direction, and (52) may then be replaced by

$$I(P) = \sum_{n} |\langle \mathbf{S}_n \rangle| = \sum_{n} I_n. \qquad (53)$$

Now the number of elements (oscillators) may be regarded as being so large that no appreciable error is introduced by treating the distribution as continuous. The contribution due to each element is then infinitesimal, but the total effect is finite. The sum (integral) is then proportional to $\delta\Omega$:

$$I(P) = B\delta\Omega,$$

and the total (time-averaged) energy flux δF which crosses the element δS per unit time is given by

$$\delta F = B\delta\Omega\delta S. \qquad (54)$$

This formula is of importance in photometry, and will be used later.

We must now briefly consider the limits of validity of geometrical optics. The eikonal equation was derived on the assumption that the terms on the right-hand sides of (11) and (12) may be neglected. If the dimensionless quantities ε, μ and

$|\text{grad}\,\mathscr{S}|$ are assumed to be of order unity, we see that this neglect will be justified provided that the magnitudes of the changes in $\boldsymbol{e}$ and $\boldsymbol{h}$ are small compared with the magnitudes of $\boldsymbol{e}$ and $\boldsymbol{h}$ over domains whose linear dimensions are of the order of a wavelength. This condition is violated, for example, at boundaries of shadows, for across such boundaries the intensity (and therefore also $\boldsymbol{e}$ and $\boldsymbol{h}$) changes rapidly. In the neighbourhood of points where the intensity distribution has a very sharp maximum (e.g. at a focus, see § 8.8), geometrical optics likewise cannot be expected to describe correctly the behaviour of the field.

The transport equations (41) and (42) for the complex amplitude vector $\boldsymbol{e}$ and $\boldsymbol{h}$ were obtained on the assumption that $\mathscr{S}$ satisfies the eikonal equation, and that the terms $\lambda_0|\boldsymbol{M}(\boldsymbol{e}, \varepsilon, \mu)|$ and $\lambda_0|\boldsymbol{M}(\boldsymbol{h}, \mu, \varepsilon)|$ are small compared with $|\boldsymbol{L}(\boldsymbol{e}, \mathscr{S}, n, \mu)|$ and $|\boldsymbol{L}(\boldsymbol{h}, \mathscr{S}, n, \varepsilon)|$ respectively. This imposes certain additional restrictions on, not only the first, but also the second derivatives of $\boldsymbol{e}$ and $\boldsymbol{h}$. These conditions are rather complicated and will not be studied here.

It is, of course, possible to obtain improved approximations by retaining some of the higher-order terms in the expansions (18) for the field vectors.* In problems of instrumental optics, the practical advantage of such a procedure is, however, doubtful, since the closer the special regions are approached the more terms have to be included, and the expansions usually break down completely at points of particular interest (e.g. at a focus or at a caustic surface). A more powerful approach to the study of the intensity distribution in such regions is offered by methods which will be discussed in the chapters on diffraction.

Finally we stress that the simplicity of the geometrical optics model arises essentially from the fact that, in general, the field behaves *locally* as a plane wave. At optical wavelengths, the regions for which this simple geometrical model is inadequate are an exception rather than a rule; in fact for most optical problems geometrical optics furnishes at least a good starting point for more refined investigations.

3.2 GENERAL PROPERTIES OF RAYS

3.2.1 The differential equation of light rays

The light rays have been defined as the orthogonal trajectories to the geometrical wave-fronts $\mathscr{S}(x, y, z) = \text{constant}$ and we have seen that, if $\boldsymbol{r}$ is a position vector of a typical point on a ray and s the length of the ray measured from a fixed point on it, then

$$n\frac{d\boldsymbol{r}}{ds} = \text{grad}\,\mathscr{S}. \tag{1}$$

---

* It has been suggested by J. B. KELLER [*J. Appl. Phys.*, **28** (1957), 426; also *Calculus of Variations and its Application*, ed. L. M. GRAVES (New York, McGraw-Hill, 1958), 27] that the behaviour of the contributions represented by the higher-order terms may be studied by means of a model which is an extension of ordinary geometrical optics. In this theory the concept of a *diffracted ray* is introduced, which obeys a generalized FERMAT's principle. With each such ray an appropriate field is associated and is assumed to satisfy the same propagation laws as the geometrical optics field. Some applications of the theory were described by J. B. KELLER, *Trans. Inst. Radio Eng.*, A.P.—**4** (1956), 312 and J. B. KELLER, R. M. LEWIS and B. D. SECKLER, *J. Appl. Phys.*, **28** (1957), 570. See also M. KLINE and I. W. KAY, *loc. cit.*

This equation specifies the rays by means of the function $\mathscr{S}$, but one can easily derive from it a differential equation which specifies the rays directly in terms of the refractive index function $n(\mathbf{r})$.

Differentiating (1) with respect to s we obtain

$$\begin{aligned}\frac{d}{ds}\left(n\frac{d\mathbf{r}}{ds}\right) &= \frac{d}{ds}(\text{grad } \mathscr{S})\\ &= \frac{d\mathbf{r}}{ds}\cdot\text{grad (grad } \mathscr{S})\\ &= \frac{1}{n}\text{grad } \mathscr{S}\cdot\text{grad (grad } \mathscr{S}) \qquad \text{(by (1))}\\ &= \frac{1}{2n}\text{grad }[(\text{grad } \mathscr{S})^2]\\ &= \frac{1}{2n}\text{grad } n^2 \qquad \text{(by § 3.1 (15))}\end{aligned}$$

i.e.

$$\frac{d}{ds}\left(n\frac{d\mathbf{r}}{ds}\right) = \text{grad } n. \tag{2}$$

This is the vector form of the differential equations of the light rays. In particular, *in a homogeneous medium* $n =$ constant and (2) then reduces to

$$\frac{d^2\mathbf{r}}{ds^2} = 0,$$

whence

$$\mathbf{r} = s\mathbf{a} + \mathbf{b}, \tag{3}$$

$\mathbf{a}$ and $\mathbf{b}$ being constant vectors. (3) is a vector equation of a straight line in the direction of the vector $\mathbf{a}$, passing through the point $\mathbf{r} = \mathbf{b}$. Hence *in a homogeneous medium the light rays have the form of straight lines.*

As an example of some interest, let us consider rays in a medium which has spherical symmetry, i.e. where the refractive index depends only on the distance r from a fixed point O:

$$n = n(r). \tag{4}$$

This case is approximately realized by the earth's atmosphere, when the curvature of the earth is taken into account.

Consider the variation of the vector $\mathbf{r} \wedge [n(r)\mathbf{s}]$ along the ray. We have

$$\frac{d}{ds}(\mathbf{r} \wedge n\mathbf{s}) = \frac{d\mathbf{r}}{ds} \wedge n\mathbf{s} + \mathbf{r} \wedge \frac{d}{ds}(n\mathbf{s}). \tag{5}$$

Since $d\mathbf{r}/ds = \mathbf{s}$, the first term on the right vanishes. The second term may, on account of (2), be written as $\mathbf{r} \wedge \text{grad } n$. Now from (4)

$$\text{grad } n = \frac{\mathbf{r}}{r}\frac{dn}{dr},$$

so that the second term on the right-hand side of (5) also vanishes. Hence

$$\mathbf{r} \wedge n\mathbf{s} = \text{constant}. \tag{6}$$

This relation implies that all the rays are plane curves, situated in a plane through the origin, and that along each ray

$$nr \sin \phi = \text{constant}, \tag{7}$$

where ϕ is the angle between the position vector $\boldsymbol{r}$ and the tangent at the point $\boldsymbol{r}$ on the ray (see Fig. 3.5). Since $r \sin \phi$ represents the perpendicular distance d from the origin to the tangent, (7) may also be written as

$$nd = \text{constant}. \tag{8}$$

This relation is sometimes called the *formula of Bouguer* and is the analogue of a well-known formula in dynamics, which expresses the conservation of angular momentum of a particle moving under the action of a central force.

To obtain an explicit expression for the rays in a spherically symmetrical medium, we recall from elementary geometry that, if (r,θ) are the polar coordinates of a

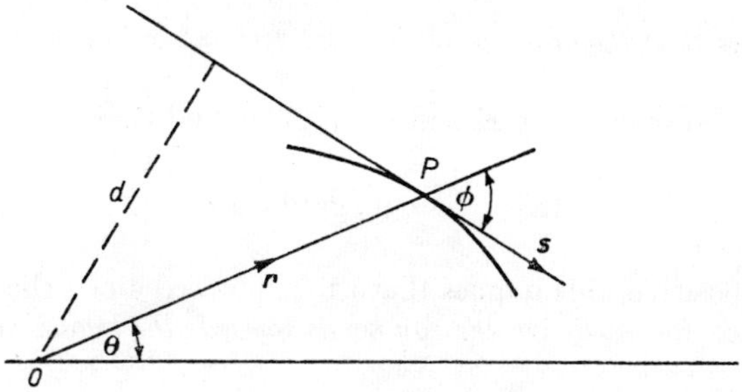

Fig. 3.5. Illustrating BOUGUER's formula nd = constant, for rays in a medium with spherical symmetry.

plane curve, then the angle ϕ between the radius vector to a point P on the curve and the tangent at P is given by*

$$\sin \phi = \frac{r(\theta)}{\sqrt{r^2(\theta) + \left(\dfrac{dr}{d\theta}\right)^2}}. \tag{9}$$

From (7) and (9)

$$\frac{dr}{d\theta} = \frac{r}{c} \sqrt{n^2 r^2 - c^2}, \tag{10}$$

c being a constant. The equation of rays in a medium with spherical symmetry may therefore be written in the form

$$\theta = c \int^{r} \frac{dr}{r\sqrt{n^2 r^2 - c^2}}. \tag{11}$$

Let us now return to the general case and consider the *curvature vector* of a ray, i.e. the vector

$$\boldsymbol{K} = \frac{d\boldsymbol{s}}{ds} = \frac{1}{\rho} \boldsymbol{\nu}, \tag{12}$$

* See, for example, R. COURANT, *Differential and Integral Calculus*, Vol. I (Glasgow, Blackie, 2nd edition, 1942), p. 265.

whose magnitude $1/\rho$ is the reciprocal of the radius of curvature; $\boldsymbol{\nu}$ is the unit principal normal at a typical point of the ray.

From (2) and (12) it follows that

$$n\mathbf{K} = \text{grad } n - \frac{dn}{ds}\mathbf{s}. \tag{13}$$

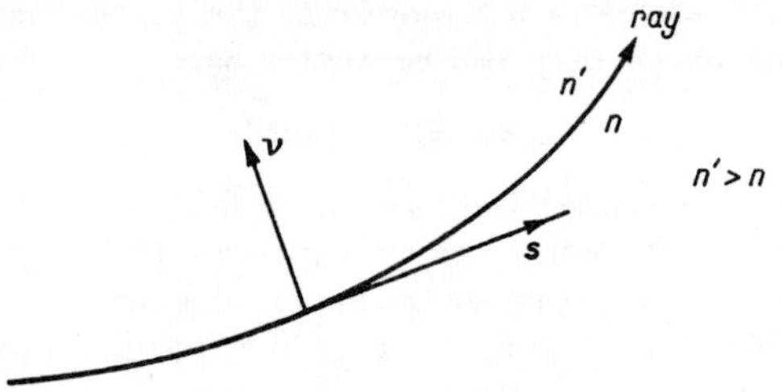

Fig. 3.6. Bending of a ray in a heterogeneous medium.

This relation shows that *the gradient of the refractive index lies in the osculating plane of the ray.*

If we multiply (13) scalarly by $\mathbf{K}$ and use (12) we find that

$$|\mathbf{K}| = \frac{1}{\rho} = \boldsymbol{\nu} \,.\, \text{grad log } n. \tag{14}$$

Since ρ is always positive, this implies that as we proceed along the principal normal the refractive index increases i.e. *the ray bends towards the region of higher refractive index* (Fig. 3.6).

3.2.2 The laws of refraction and reflection

So far it has been assumed that the refractive index function n is continuous. We must now discuss the behaviour of rays when they cross a surface separating two homogeneous media of different refractive indices. It has been shown by SOMMERFELD

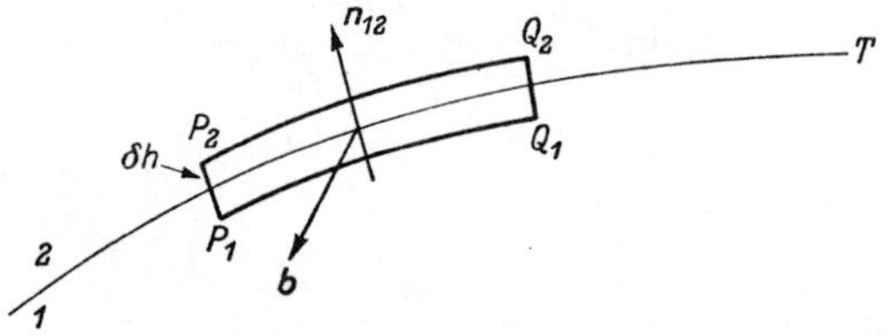

Fig. 3.7. Derivation of the laws of refraction and reflection.

and RUNGE (*loc. cit.*) that the behaviour can easily be determined by an argument similar to that used in deriving the conditions relating to the changes in the field vectors across a surface discontinuity (cf. § 1.1.3).

It follows from (1), on account of the identity curl grad $\equiv 0$, that the vector $n\mathbf{s} = n d\mathbf{r}/ds$, called sometimes the *ray vector*, satisfies the relation

$$\text{curl } n\mathbf{s} = 0. \tag{15}$$

As in § 1.1.3 we replace the discontinuity surface T by a transition layer throughout which ε, μ and n change rapidly but continuously from their values near T on one side to their values near T on the other. Next we take a plane element of area with its sides P_1Q_1 and P_2Q_2 parallel and with P_1P_2 and Q_1Q_2 perpendicular to T (Fig. 3.7).

If $\boldsymbol{b}$ denotes the unit normal to this area, then we have from (15), on integrating throughout the area and applying STOKES' theorem,

$$\int (\text{curl } n\boldsymbol{s}) \cdot \boldsymbol{b} \, dS = \int n\boldsymbol{s} \cdot d\boldsymbol{r} = 0, \tag{16}$$

the second integral being taken along the boundary curve $P_1Q_1Q_2P_2$. Proceeding to the limit as the height $\delta h \to 0$, in a strictly similar manner as in the derivation of § 1.1 (23), we obtain

$$\boldsymbol{n}_{12} \wedge (n_2\boldsymbol{s}_2 - n_1\boldsymbol{s}_1) = 0, \tag{17}$$

where $\boldsymbol{n}_{12}$ is the unit normal to the boundary surface pointing from the first into the second medium. (17) implies that *the tangential component of the ray vector $n\boldsymbol{s}$ is*

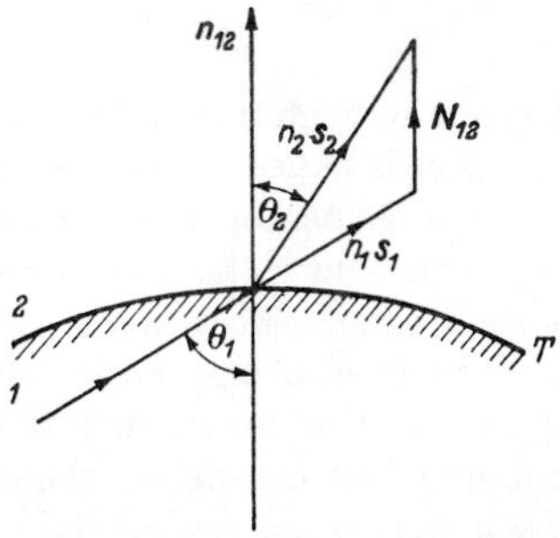

Fig. 3.8 (a). Illustrating the law of refraction.

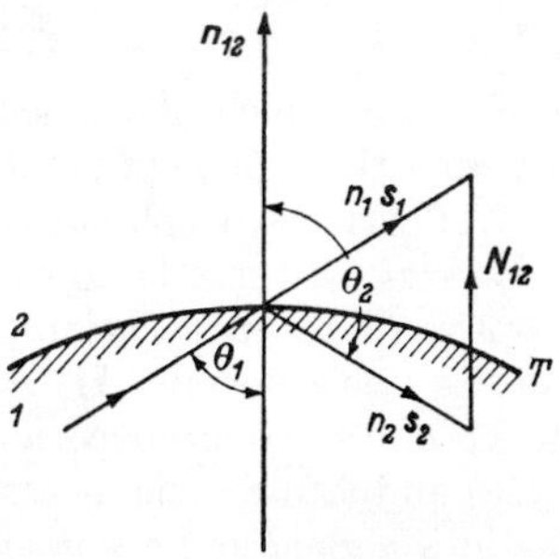

Fig. 3.8 (b). Illustrating the law of reflection.

continuous across the surface or, what amounts to the same thing, *the vector $\boldsymbol{N} = n_2\boldsymbol{s}_2 - n_1\boldsymbol{s}_1$ is normal to the surface.*

Let θ_1 and θ_2 be the angles which the incident ray and the refracted ray make with the normal $\boldsymbol{n}_{12}$ to the surface (see Fig. 3.8a). Then it follows from (17) that

$$n_2(\boldsymbol{n}_{12} \wedge \boldsymbol{s}_2) = n_1(\boldsymbol{n}_{12} \wedge \boldsymbol{s}_1), \tag{18}$$

so that

$$n_2 \sin \theta_2 = n_1 \sin \theta_1. \tag{19}$$

(18) implies that *the refracted ray lies in the same plane as the incident ray and the normal to the surface* (*the plane of incidence*) and (19) shows that *the ratio of the sine of the angle of refraction to the sine of the angle of incidence is equal to the ratio n_1/n_2 of the refractive indices*. These two results express *the law of refraction* (*Snell's law*). This law has already been derived in § 1.5 for the special case of plane waves. But whilst the earlier discussion concerned a plane wave of *arbitrary* wavelength falling upon a plane refracting surface, the present analysis applies to waves and refracting surfaces of more general form, provided that the wavelength is sufficiently small ($\lambda_0 \to 0$). This condition means, in practice, that the radii of curvature of the incident wave and of the boundary surface must be large compared to the wavelength of the incident light.

As in the case treated in § 1.5 we must expect that there will be another wave, the reflected wave, propagated back into the first medium. Setting $n_2 = n_1$ in (18) and (19) (see Fig. 3.8b) it follows that *the reflected ray lies in the plane of incidence* and that $\sin \theta_2 = \sin \theta_1$; hence

$$\theta_2 = \pi - \theta_1. \tag{20}$$

The last two results express *the law of reflection.*

3.2.3 Ray congruences and their focal properties

The relation (15), namely

$$\text{curl } n\mathbf{s} = 0, \tag{21}$$

characterizes all the ray systems which can be realized in an isotropic medium and distinguishes them from more general families of curves. In a homogeneous isotropic medium n is constant, and (21) then reduces to

$$\text{curl } \mathbf{s} = 0. \tag{22}$$

Rays in a heterogeneous isotropic medium can also be characterized by a relation independent of n. It may be obtained by applying to (21) the identity curl $n\mathbf{s} = n$ curl $\mathbf{s}$ + (grad n) $\wedge$ $\mathbf{s}$ and taking the scalar product with $\mathbf{s}$. It then follows that a system of rays in any *isotropic* medium must satisfy the relation

$$\mathbf{s} \,.\, \text{curl } \mathbf{s} = 0. \tag{23}$$

A system of curves which fills a portion of space in such a way that in general a single curve passes through each point of the region is called a *congruence*. If there exists a family of surfaces which cut each of the curves orthogonally the congruence is said to be *normal*; if there is no such family, it is said to be *skew*. For ordinary geometrical optics (light propagation) only normal congruences are of interest, but in electron optics (see Appendix II) skew congruences also play an important part.

If each curve of the congruence is a straight line the congruence is said to be *rectilinear*; (23) and (22) are the necessary and sufficient conditions that the curves should represent a *normal* and a *normal rectilinear congruence* respectively.*

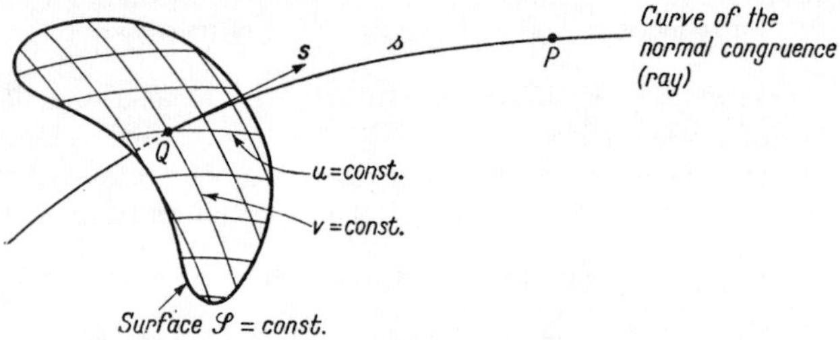

Fig 3.9. Notation relating to a normal congruence.

Let us choose a set of curvilinear coordinate lines u, v on one of the orthogonal surfaces $\mathcal{S}(x, y, z)$ = constant. To every point $Q(u, v)$ of this surface there will then correspond one curve of the congruence, namely that curve which meets $\mathcal{S}$ in Q. Let $\mathbf{r}$ denote the position vector of a point P on the curve. $\mathbf{r}$ may then be regarded as a function of the coordinates (u, v) and of the length of arc s between Q and P, measured along the curve (Fig. 3.9).

Consider two neighbouring curves of the congruence passing through the points (u, v) and $(u + du, v + dv)$ on $\mathcal{S}$, and let us examine whether there are points on these curves such that the distance between them is of the second or higher order (one says that the curves cut to first order at such points). Points with this property are called *foci* and must satisfy the equation

$$\mathbf{r}(u, v, s) = \mathbf{r}(u + du, v + dv, s + ds) \tag{24}$$

to the first order.

* For a more detailed discussion of congruences of curves see, for example, C. E. Weatherburn, *Differential Geometry of Three Dimensions* (Cambridge University Press), Vol. I (1927), Chapter X; Vol. II (1930), Chapter XIII.

Expanding (24) we obtain

$$\boldsymbol{r}_u du + \boldsymbol{r}_v dv + \boldsymbol{s} ds = 0, \tag{25}$$

where $\boldsymbol{r}_u$, $\boldsymbol{r}_v$ are the partial derivatives with respect to u and v. Condition (25) implies that $\boldsymbol{r}_u$, $\boldsymbol{r}_v$ and $\boldsymbol{s}$ are coplanar. This is equivalent to saying that the scalar triple product of the three vectors vanishes, i.e.

$$[\boldsymbol{r}_u, \boldsymbol{r}_v, \boldsymbol{s}] = 0. \tag{26}$$

The number of foci on a given curve (u, v) depends on the number of values of s which satisfy (26). If $\boldsymbol{r}$ is a polynomial in s of degree m, then since $\boldsymbol{s} = d\boldsymbol{r}/ds$, it is seen that (26) is an equation of degree $3m - 1$ in s. In particular, if the congruence is rectilinear, $\boldsymbol{r}$ is a linear function of s ($m = 1$), showing that *there are two foci on each ray of a rectilinear congruence.*

If u and v take on all possible values, the foci will describe a surface, represented by (26), known as the *focal surface*; in optics it is called the *caustic surface.* Any curve of the congruence is tangent to the focal surface at each focus of the curve. The tangent plane at any point of the focal surface is known as the *focal plane.*

We shall mainly be concerned with rays in a homogeneous medium, i.e. with rectilinear congruences. Some further properties of such congruences will be discussed in § 4.6, in connection with astigmatic pencils of rays.

3.3 OTHER BASIC THEOREMS OF GEOMETRICAL OPTICS

With the help of the relations established in the preceding sections, we shall now derive a number of theorems concerning rays and wave-fronts.

3.3.1 Lagrange's integral invariant

Assume first that the refractive index n is a continuous function of position. Then as in § 3.2 (16) it follows on applying STOKES' theorem to the integral, taken over any open surface, of the normal component of curl $n\boldsymbol{s}$, that

$$\oint n\boldsymbol{s} \,.\, d\boldsymbol{r} = 0. \tag{1}$$

The integral extends over the closed boundary curve C of the surface. (1) is known as *Lagrange's integral invariant** and implies that *the integral*

$$\int_{P_1}^{P_2} n\boldsymbol{s} \,.\, d\boldsymbol{r} \tag{2}$$

taken between any two points P_1 and P_2 in the field, is independent of the path of integration.

---

* Sometimes called *Poincaré's invariant.* In fact it is only a special one-dimensional case of much more general integral invariants discussed by J. H. POINCARÉ in his *Les Méthodes Nouvelles de la Mécanique Céleste,* **3** (Paris, Gauthier-Villars, 1899). Cf. E. CARTAN, *Leçons sur les Invariants Integraux* (Paris, Hermann, 1922). See also our Appendix I, eq. (85).

With the help of the law of refraction it is easily seen that (1) also holds when the curve C intersects a surface which separates two homogeneous media of different refractive indices. To show this, let C_1 and C_2 be the portions of C on each side of the refracting surface T (Fig. 3.10), and let the points of intersection of C with the surface T be joined by another curve K in the surface. On taking (1) along each of the loops C_1K and C_2K and on adding the equations, we obtain

$$\int_{C_1} n_1\mathbf{s}_1 \cdot d\mathbf{r} + \int_{C_2} n_2\mathbf{s}_2 \cdot d\mathbf{r} + \int_K (n_2\mathbf{s}_2 - n_1\mathbf{s}_1) \cdot d\mathbf{r} = 0. \tag{3}$$

The integral over K vanishes, since according to the law of refraction the vector $\mathbf{N} = n_1\mathbf{s}_1 - n_2\mathbf{s}_2$ is at each point of K perpendicular to the surface, and consequently (3) reduces to (1).

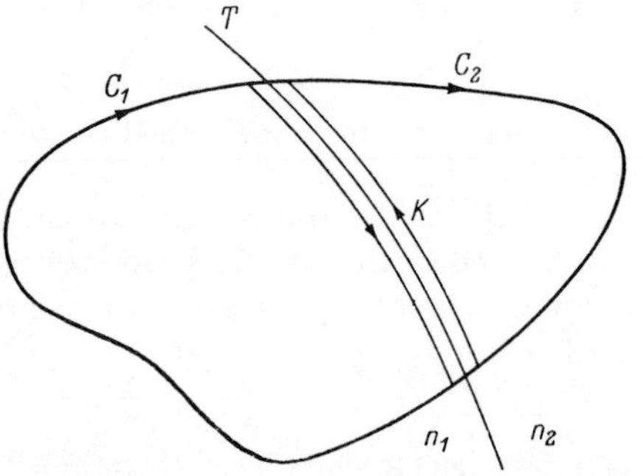

Fig. 3.10. Derivation of the LAGRANGE's integral invariant in the presence of a surface of discontinuity of the refractive index.

3.3.2 The principle of Fermat

The *principle of Fermat*, known also as the principle of the *shortest optical path** asserts that *the optical length*

$$\int_{P}^{P_2} nds \tag{4}$$

of an actual ray between any two points P_1 and P_2 is shorter than the optical length of any other curve which joins these points and which lies in a certain regular neighbourhood of it. By a regular neighbourhood we mean one that may be covered by rays in such a way that one (and only one) ray passes through each point of it. Such a covering is exhibited, for example, by rays from a point source P_1 in that domain around P_1 where the rays on account of refraction or reflection or on account of their curvature do not intersect each other.

Before proving this theorem it may be mentioned that it is possible to formulate FERMAT's principle in a form which is weaker but which has a wider range of validity. According to this formulation the actual ray is distinguished from other curves (no

* Since by § 3.1 (27)

$$\int_{P_1}^{P_2} nds = c\int_{P_1}^{P_2} dt$$

it is also known as the *principle of least time*.

longer restricted to lie in a regular neighbourhood) by a *stationary value* of the integral.*

To prove FERMAT's principle, we take a pencil of rays and compare a segment P_1P_2 of a ray $\bar{C}$ with an arbitrary curve C joining P_1 and P_2 (Fig. 3.11). Let two neighbouring orthogonal trajectories (wave-fronts) of the pencil intersect C in Q_1 and Q_2 and $\bar{C}$ in $\bar{Q}_1$ and $\bar{Q}_2$. Further let Q_2' be the point of intersection of the trajectory $Q_2\bar{Q}_2$ with the ray $\bar{C}'$ which passes through Q_1.

Applying LAGRANGE's integral relation to the small triangle $Q_1Q_2Q_2'$, we have

$$(n\mathbf{s} \,.\, d\mathbf{r})_{Q_1Q_2} + (n\mathbf{s} \,.\, d\mathbf{r})_{Q_2Q_2'} - (nds)_{Q_1Q_2} = 0. \tag{5}$$

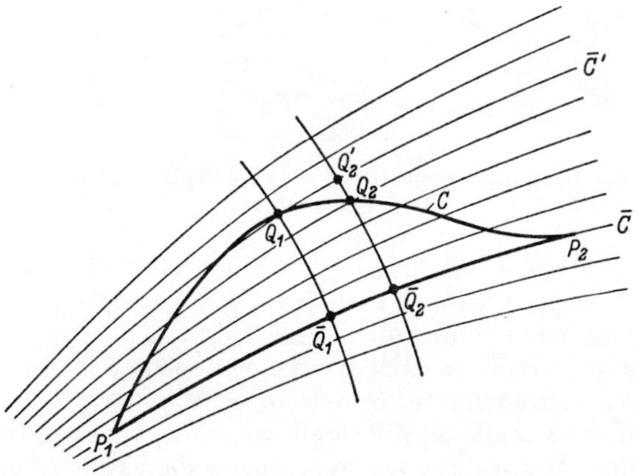

Fig. 3.11. Illustrating FERMAT's principle.

Now from the definition of the scalar product

$$(n\mathbf{s} \,.\, d\mathbf{r})_{Q_1Q_2} \leqslant (nds)_{Q_1Q_2}.$$

Further, $\mathbf{s}$ is orthogonal to $d\mathbf{r}$ on the wave-front, so that

$$(n\mathbf{s} \,.\, d\mathbf{r})_{Q_2Q_2'} = 0.$$

Also from § 3.1 (25), since Q_1, Q_2' and $\bar{Q}_1, \bar{Q}_2$ are corresponding points on the two wave-fronts,

$$(nds)_{Q_1Q_2'} = (nds)_{\bar{Q}_1\bar{Q}_2}.$$

On substituting from the last three relations into (5) we find that

$$(nds)_{\bar{Q}_1\bar{Q}_2} \leqslant (nds)_{Q_1Q_2}, \tag{6}$$

whence, on integration,

$$\int_{\bar{C}} nds \leqslant \int_{C} nds. \tag{7}$$

* To find the curves for which the integral has a stationary value we must apply in general the methods of the variational calculus, described in Appendix I. It is shown there that such curves satisfy the EULER Differential Equations AI (7). In the present case these are nothing but the equations § 3.2. (2) of the rays as shown in section 11 of Appendix I.

It has been stressed by C. CARATHÉODORY (*loc. cit.*) that the stationary value is never a true maximum. In the weaker formulation of FERMAT's principle it is therefore appropriate to speak of a stationary value rather than of an extremal value. The minimal formulation on the other hand corresponds to a "strong minimum" in the sense of JACOBI (Appendix I, section 10).

Moreover, the equality sign could only hold if the directions of $\boldsymbol{s}$ and $\boldsymbol{dr}$ were coincident at every point of C, i.e. if the comparison curve was an actual ray. This case is excluded by our assumption that not more than one ray passes through any point of the neighbourhood. Hence the optical length of the ray is smaller than the optical length of the comparison curve, which is FERMAT's principle.

It can easily be seen that, when the regularity condition is not fulfilled, the optical length of the ray may no longer be a minimum. Consider for example a field of rays

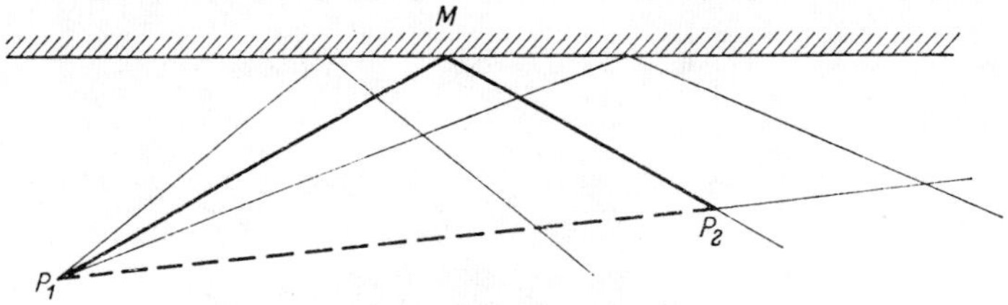

Fig. 3.12. Field of rays obtained by reflection of light from a point source on a plane mirror.

from a point source P_1 in a homogeneous medium, reflected by a plane mirror (Fig. 3.12). Two rays then pass through each point P_2; the optical length of the direct ray P_1P_2 is an absolute minimum but the reflected ray P_1MP_2 gives a minimum only relative to curves in a certain restricted neighbourhood of it. In general when rays from a point source P_1 are refracted or reflected at boundaries between homogeneous media, the regular neighbourhood will terminate on the envelope (caustic) formed by the rays. The point P_1' at which a ray from a point source at P_1 touches the envelope is called the *conjugate* of P_1 on the particular ray. For the optical length of a ray P_1P_2

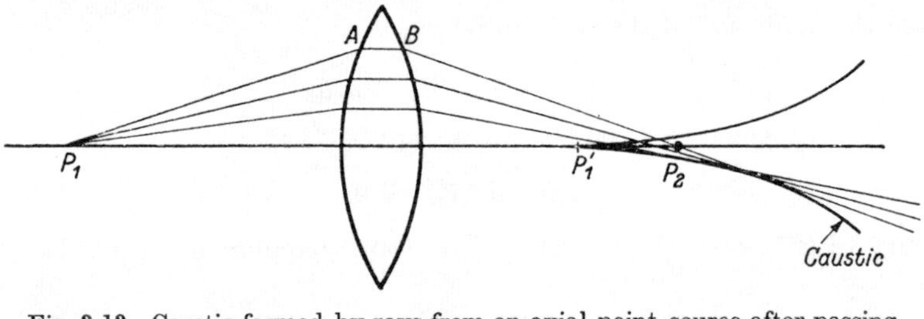

Fig. 3.13. Caustic formed by rays from an axial point source after passing through a lens.

to be a minimum, P_2 must lie between P_1 and P_1', i.e. P_1 and P_2 must lie on the same side of the caustic. For example, in the case of an uncorrected lens (Fig. 3.13) the central ray from P_1 has a minimal optical length only up to the tip (P_1') of the caustic (the Gaussian image of P_1). For any point P_2 which lies behind the envelope the optical length of the direct path $P_1P_1'P_2$ exceeds that of the broken path P_1ABP_2.

3.3.3 The theorem of Malus and Dupin and some related theorems

The light rays have been defined as the orthogonal trajectories of the wave surfaces $\mathscr{S}(x, y, z) =$ constant, $\mathscr{S}$ being a solution of the eikonal equation (15) in § 3.1. This is a natural way of introducing the light rays when the laws of geometrical optics are to be deduced from MAXWELL's equations. Historically, however, geometrical optics developed as the theory of light rays which were defined differently, namely as curves for which the line integral $\int nds$ has a stationary value. Formulated this way geometrical optics may then be developed purely along the lines of calculus of variations.*

Variational considerations are of considerable importance as they often reveal analogies between different branches of physics. In particular there is a close analogy between geometrical optics and the mechanics of a moving particle; this was brought out very clearly by the celebrated investigations of Sir W. R. HAMILTON, whose approach became of great value in modern physics, especially in applications to DE BROGLIE's wave mechanics. In order not to interrupt the optical considerations, an account of the relevant parts of the calculus of variation and of the Hamiltonian analogy are given in separate sections (Appendix I and II). Here we shall only show

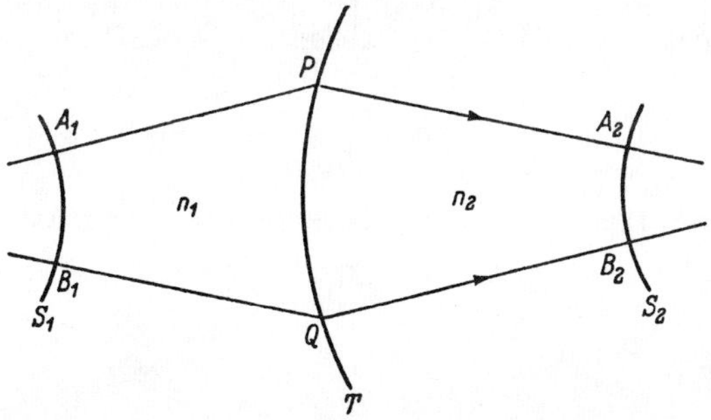

Fig. 3.14. Illustrating the theorem of MALUS and DUPIN.

how several theorems, which played an important part in the development of geometrical optics, may be derived from LAGRANGE's integral invariant.

Consider rays in a homogeneous medium: if they all have a point in common, e.g. when they then proceed from a point source, they are said to form a *homocentric pencil*. Such a pencil forms a normal congruence, since every ray of the pencil is cut orthogonally by spheres centred on the mutual point of intersection of the rays. In 1808 MALUS† showed that, if a homocentric pencil of rectilinear rays is refracted or reflected at a surface, the resulting pencil (in general no longer homocentric) will again form a normal congruence. Later DUPIN (1816), QUETELET (1825), and GERGONNE (1825) generalized MALUS's result. These investigations lead to the following theorem, known sometimes as *the theorem of Malus and Dupin: A normal rectilinear congruence remains normal after any number of refractions or reflections.*‡

It will be sufficient to establish the theorem for a single refraction. Consider a normal rectilinear congruence of rays in a homogeneous medium of refractive index n_1 and assume that the rays undergo a refraction at a surface T which separates this medium from another homogeneous medium of refractive index n_2 (Fig. 3.14).

Let S_1 be one of the orthogonal trajectories (wave-fronts) in the first region, and let A_1 and P be the points of intersections of a typical ray in the first medium with

* A systematic treatment of this kind is given for example in C. CARATHÉODORY (*loc. cit.*).

† E. MALUS, Optique Dioptrique, *J. École polytechn.*, **7** (1808), 1–44, 84–129. Also his "Traité d'optique", *Mém. présent. à l'Institut par divers savants*, **2** (1811), 214–302. References and an account of the interesting history of the MALUS–DUPIN theorem can be found in the *Mathematical Papers of Sir William Rowan Hamilton*, **1** (Geometrical Optics), edited by A. W. CONWAY and J. L. SYNGE (Cambridge University Press, 1931), p. 463.

‡ T. LEVI-CIVITA, *Rend. R. Acc. Naz. Linc.*, **9** (1900) 237 established the converse theorem, namely that in general two normal rectilinear congruences may be transformed into each other by a single refraction or reflection.

S_1 and with T respectively, and let A_2 be any point on the refracted ray. If the point A_1 is displaced to another point B_1 on the wave-front, the point P will be displaced to another point Q on the refracting surface. Now take a point B_2, on the ray which is refracted at Q, such that the optical path from B_1 to B_2 is equal to the optical path from A_1 to A_2:

$$[A_1PA_2] = [B_1QB_2]. \tag{8}$$

As B_1 takes on all possible positions on S_1 the point B_2 describes a surface S_2. It will now be shown that the refracted ray QB_2 is perpendicular to this surface.

Applying LAGRANGE's integral invariant to the closed path $A_1PA_2B_2QB_1A_1$, it follows that

$$\int_{A_1PA_2} nds + \int_{A_2B_2} n\boldsymbol{s} \, . \, d\boldsymbol{r} + \int_{B_2QB_1} nds + \int_{B_1A_1} n\boldsymbol{s} \, . \, d\boldsymbol{r} = 0. \tag{9}$$

Now by (8),

$$\int_{A_1PA_2} nds + \int_{B_2QB_1} nds = 0. \tag{10}$$

Moreover, since on S_1 the unit vector $\boldsymbol{s}$ is everywhere orthogonal to S_1,

$$\int_{B_1A_1} n\boldsymbol{s} \, . \, d\boldsymbol{r} = 0, \tag{11}$$

so that (9) reduces to

$$\int_{A_2B_2} n\boldsymbol{s} \, . \, d\boldsymbol{r} = 0. \tag{12}$$

This relation must hold for every curve on S_2. This is only possible if $\boldsymbol{s} \, . \, d\boldsymbol{r} = 0$ for every linear element $d\boldsymbol{r}$ of S_2, i.e. if the refracted rays are orthogonal to the surface; in other words *if the refracted rays form a normal congruence.* The proof for reflection is strictly analogous.

Since $[A_1PA_2] = [B_1QB_2]$ it follows that *the optical path length between any two orthogonal surfaces (wave-fronts) is the same for all rays.* This result clearly remains valid when several successive refractions or reflections takes place and, as is immediately obvious from eq. (26) in § 3.1 it also applies to rays in a medium with continuously varying refractive index. This theorem is known as the *principle of equal optical path*; it implies that the orthogonal trajectories (geometrical wave-fronts) of a normal congruence of rays, or of a set of normal congruences generated by successive refractions or reflections, are "optically parallel" to each other (cf. Appendix I).

A related theorem, first put forward by HUYGENS* asserts that *each element of a wave-front may be regarded as the centre of a secondary disturbance which gives rise to spherical wavelets*; and moreover that *the position of the wave-front at any later time is the envelope of all such wavelets.* This result, sometimes called *Huygens' construction*, is essentially a rule for the construction of a set of surfaces which are "optically parallel" to each other. If the medium is homogeneous, one can use in the construction wavelets of finite radius, in other cases one has to proceed in infinitesimal steps.

HUYGENS' theorem was later extended by FRESNEL and led to the formulation of the so-called Huygens-Fresnel principle, which is of great importance in the theory of diffraction (see § 8.2), and which may be regarded as the basic postulate of the wave theory of light.

---

* *Traité de la Lumière* (Leyden, 1690); English translation (*Treatise on Light*) by S. P. THOMPSON (London, Macmillan & Co., 1912).

CHAPTER IV

GEOMETRICAL THEORY OF OPTICAL IMAGING

4.1 THE CHARACTERISTIC FUNCTIONS OF HAMILTON

IN § 3.1 it was shown that, within the approximations of geometrical optics, the field may be characterized by a single scalar function $\mathscr{S}(\mathbf{r})$. Since $\mathscr{S}(\mathbf{r})$ satisfies the eikonal equation (15) in § 3.1, this function is fully specified by the refractive index function $n(\mathbf{r})$ alone, together with the appropriate boundary conditions.

Instead of the function $\mathscr{S}(\mathbf{r})$ closely related functions known as *characteristic functions* of the medium are often used. They were introduced into optics by W. R. HAMILTON, in a series of classical papers.* Although on account of algebraic complexity it is impossible to determine the characteristic functions explicitly for all but the simplest media, HAMILTON's methods nevertheless form a very powerful tool for systematic analytical investigations of the general properties of optical systems.

In discussing the properties of these functions and their applications, an isotropic but generally heterogeneous medium will be assumed.

4.1.1 The point characteristic

Let (x_0, y_0, z_0) and (x_1, y_1, z_1) be respectively the coordinates of two points P_0 and P_1 each referred to a different set of mutually parallel, rectangular axes† (Fig. 4.1). If the two points are imagined to be joined by all possible curves, there will, in general, be some amongst them, the optical rays, which satisfy FERMAT's principle. Assume for the present that not more than one ray joins any two arbitrary points. The characteristic function V, or the *point characteristic*, is then defined as *the optical length* $[P_0P_1]$ *of the ray between the two points*, considered as a function of their coordinates,

$$V(x_0, y_0, z_0; x_1, y_1, z_1) = \int_{P_0}^{P_1} n ds. \tag{1}$$

It is important to note that this function is defined *by the medium*.

* Sir W. R. HAMILTON, *Trans. Roy. Irish Acad.*, **15** (1828), 69; *ibid.*, **16** (1830), 1; *ibid.*, **16** (1831), 93; *ibid.*, **17** (1837), 1. Reprinted in *The Mathematical Papers of Sir W. R. Hamilton*, Vol. I (*Geometrical Optics*), edited by A. W. CONWAY and J. L. SYNGE (Cambridge University Press, 1931).

Many years later BRUNS independently considered similar functions which he called *eikonals.* (H. BRUNS, *Abh. Kgl. Sächs. Ges. Wiss., math-phys. Kl.*, **21** (1895), 323.) As already mentioned on p. 112, this term has come to be used in a wider sense. The characteristic functions of HAMILTON are themselves often referred to as eikonals.

A useful introduction to HAMILTON's methods is a monograph by J. L. SYNGE, *Geometrical Optics* (Cambridge University Press, 1937). The relationship between the work of HAMILTON and BRUNS was discussed by F. KLEIN in *Z. Math. Phys.*, **46** (1901), 376, and *Ges. Math. Abh.*, **2** (1922), 603, C. CARATHÉODORY, *Geometrische Optik* (Berlin, Springer, 1937), p. 4, and in a polemic between M. HERZBERGER and J. L. SYNGE, *J. Opt. Soc. Amer.*, **27** (1937), 75, 133, 138.

† The use of two reference systems has some advantages, since P_0 and P_1 are often situated in different regions, namely, the object- and image-spaces of an optical system.

From (1) and § 3.1 (26) it follows that

$$V(x_0, y_0, z_0; x_1, y_1, z_1) = \mathscr{S}(x_1, y_1, z_1) - \mathscr{S}(x_0, y_0, z_0), \tag{2}$$

where the function $\mathscr{S}$ is now associated with any pencil of rays to which the natural ray through P_0 and P_1 belongs [for example, a pencil produced by a point source at P_0].* Then by § 3.1 (24), the unit vectors $\mathbf{s}_0$ and $\mathbf{s}_1$ at P_0 and P_1, in the direction of the ray, are given by

$$\left.\begin{aligned} \text{grad}^0\, V &= -\, n_0\mathbf{s}_0, \\ \text{grad}^1\, V &= \quad n_1\mathbf{s}_1, \end{aligned}\right\} \tag{3}$$

the superscripts 0 and 1 implying that the operator grad is taken with respect to the coordinates (x_0, y_0, z_0) and (x_1, y_1, z_1) respectively.

The vector

$$\mathbf{g} = n\mathbf{s} \tag{4}$$

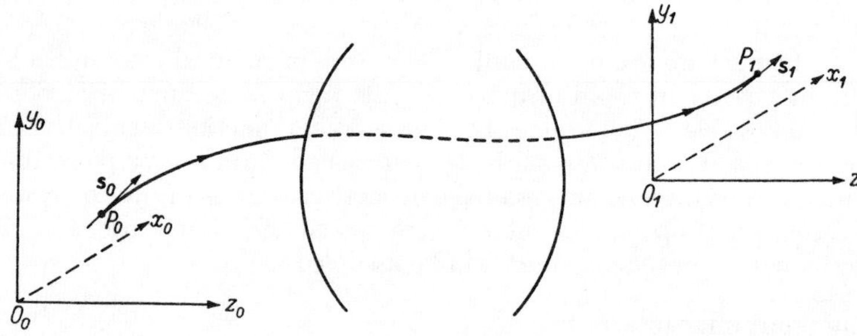

Fig. 4.1. Illustrating the definition of the point characteristic function.

is sometimes called the *ray vector*. If α, β and γ are the angles which the ray vector makes with the coordinate axes, its projections†

$$p = n \cos \alpha, \qquad q = n \cos \beta, \qquad m = n \cos \gamma, \tag{5}$$

on to the axes are called the *ray components*. On account of the identity

$$\cos^2 \alpha + \cos^2 \beta + \cos^2 \gamma = 1,$$

they satisfy the relation

$$p^2 + q^2 + m^2 = n^2. \tag{6}$$

From (3) it follows that the ray components at P_0 and P_1 are given by

$$p_0 = -\frac{\partial V}{\partial x_0}, \qquad p_1 = \frac{\partial V}{\partial x_1}, \tag{7}$$

with similar expressions for q_0, q_1, and m_0, m_1. These relations show that, from the knowledge of the point characteristic, the components of the ray which joins any two points in the medium can immediately be determined. Further it follows from (6)

---

* In the language of the calculus of variations, $\mathscr{S}$ represents the solution which involves a two-parameter family (∞^2) of extremals, of the Hamilton–Jacobi equation, associated with Fermat's variational problem. The point characteristic function V represents the general solution, which involves all (∞^4) extremals (cf. Appendix I).

† This "asymmetrical" notation is purposely chosen here to remind us that on account of the identity (6), only two of the ray components are independent.

and (7) that the point characteristic satisfies the eikonal equation in both sets of variables:

$$\left(\frac{\partial V}{\partial x_0}\right)^2 + \left(\frac{\partial V}{\partial y_0}\right)^2 + \left(\frac{\partial V}{\partial z_0}\right)^2 = n_0^2, \tag{8}$$

$$\left(\frac{\partial V}{\partial x_1}\right)^2 + \left(\frac{\partial V}{\partial y_1}\right)^2 + \left(\frac{\partial V}{\partial z_1}\right)^2 = n_1^2. \tag{9}$$

Instead of the point characteristic it is often convenient to use certain related functions (also introduced by HAMILTON), known as the *mixed characteristic* and the *angle characteristic*. They may be derived from the point characteristic by means of LEGENDRE transformations,* and are particularly useful when either P_0 or P_1 or both these points are at infinity.

4.1.2 The mixed characteristic

The mixed characteristic function W is defined by the equation†

$$W = V - \Sigma p_1 x_1, \tag{10}$$

where Σ denotes summation over the three similar terms with suffix one. (10) expresses W as a function of nine variables, but in general only six (and in a homogeneous medium only five) are independent. To show this, consider the effect of small displacements of the points P_0 and P_1. The corresponding change in W is then given by

$$\delta W = \delta V - \Sigma p_1 \delta x_1 - \Sigma x_1 \delta p_1. \tag{11}$$

Now by (7),

$$\delta V = \Sigma p_1 \delta x_1 - \Sigma p_0 \delta x_0. \tag{12}$$

From (11) and (12),

$$\delta W = -\Sigma p_0 \delta x_0 - \Sigma x_1 \delta p_1. \tag{13}$$

This relation shows that in general W can be expressed as a function of the six variables x_0, y_0, z_0, p_1, q_1 and m_1, and that, when it is so expressed,

$$p_0 = -\frac{\partial W}{\partial x_0}, \qquad x_1 = -\frac{\partial W}{\partial p_1}, \tag{14}$$

with similar expressions for q_0, y_1, m_0 and z_1. On account of (6), $W(x_0, y_0, z_0; p_1, q_1, m_1)$ satisfies the eikonal equation

$$\left(\frac{\partial W}{\partial x_0}\right)^2 + \left(\frac{\partial W}{\partial y_0}\right)^2 + \left(\frac{\partial W}{\partial z_0}\right)^2 = n_0^2. \tag{15}$$

It is to be observed (see Fig. 4.2) that the sum $\Sigma p_1 x_1$ has a simple geometrical interpretation:

$$\Sigma p_1 x_1 = n_1 d_1, \tag{16}$$

* A LEGENDRE transformation transforms in general a function $f(x,y)$ into a function $g(x,z)$ where $z = \partial f/\partial y$, in such a way that the derivative of g with respect to the new variable z is equal to the old variable y.

† To bring out more clearly its physical meaning, we follow SYNGE in defining the mixed characteristic with opposite sign to that used by HAMILTON.

One can also define a mixed characteristic

$$W' = V + \Sigma p_0 x_0,$$

the summation being taken over similar terms with suffix zero. The properties of W and W' are, of course, strictly analogous.

where $d_1 = Q_1P_1$ is the projection of O_1P_1 on to the tangent to the ray at P_1. If P_1 is situated in a homogeneous region, the portion of the ray near P_1 coincides with the line segment Q_1P_1; according to (10) and (16) *W then represents the optical length of*

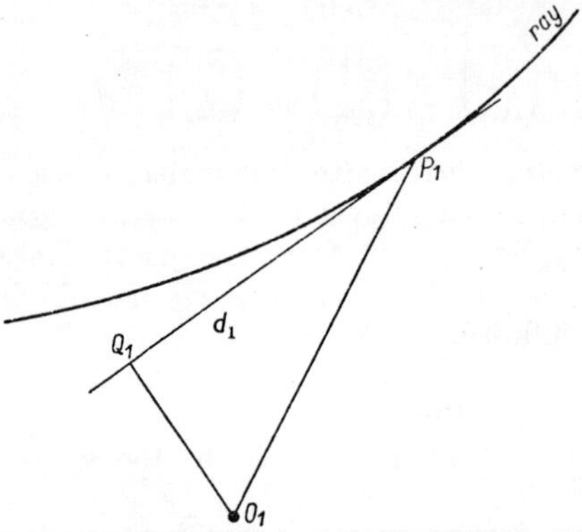

Fig. 4.2. Illustrating the meaning of the mixed characteristic function.

the ray from P_0 to the foot Q_1 of the perpendicular drawn from the origin O_1 on to the final portion of the ray (Fig. 4.3):

$$W = [P_0Q_1]. \tag{17}$$

Since in this case the refractive index of the medium around P_1 has a constant value it follows from (6) that

$$\delta m_1 = -\frac{p_1\delta p_1 + q_1\delta q_1}{m_1}, \tag{18}$$

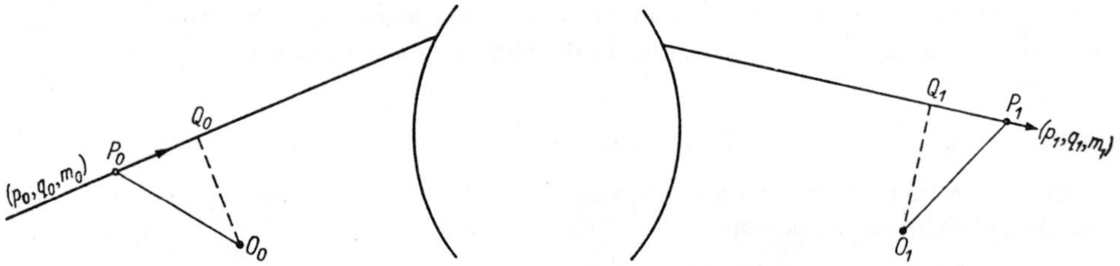

Fig. 4.3. Interpretation of HAMILTON's characteristic functions when the initial and final media are homogeneous:

$$V(x_0, y_0, z_0;\ x_1, y_1, z_1) = [P_0P_1],$$
$$W(x_0, y_0, z_0;\ p_1, q_1) = [P_0Q_1],$$
$$T(p_0, q_0;\ p_1, q_1) = [Q_0Q_1].$$

and (13) becomes on substitution from (18)*

$$\delta W = -\Sigma p_0\delta x_0 - \left(x_1 - \frac{p_1}{m_1}z_1\right)\delta p_1 - \left(y_1 - \frac{q_1}{m_1}z_1\right)\delta q_1. \tag{19}$$

---

* If a function depends on variables which are connected by subsidiary relations, such as (6), some of the variables may be eliminated, or alternatively the relations may be used to express it as a *homogeneous* function in all the variables. The alternative procedure, which is somewhat more difficult to handle, was frequently employed by HAMILTON.

Hence *when the final medium is homogeneous, the mixed characteristic is expressible as a function of five variables:*

$$W = W(x_0, y_0, z_0; p_1, q_1), \tag{20}$$

and its derivatives then satisfy the relations

$$p_0 = -\frac{\partial W}{\partial x_0}, \qquad q_0 = -\frac{\partial W}{\partial y_0}, \qquad m_0 = -\frac{\partial W}{\partial z_0}, \tag{21}$$

$$x_1 - \frac{p_1}{m_1} z_1 = -\frac{\partial W}{\partial p_1}, \qquad y_1 - \frac{q_1}{m_1} z_1 = -\frac{\partial W}{\partial q_1}. \tag{22}$$

(21) and (22) show that if a point on the ray in the initial medium and the components of the ray in the final medium are given, the ray components in the initial medium and points on the ray in the final medium may immediately be determined from the knowledge of the mixed characteristic.

4.1.3 The angle characteristic

The angle characteristic T may be defined by means of the relation

$$T = V + \Sigma p_0 x_0 - \Sigma p_1 x_1. \tag{23}$$

If P_0 and P_1 are slightly displaced, the corresponding change in T is given by

$$\delta T = \Sigma x_0 \delta p_0 - \Sigma x_1 \delta p_1, \tag{24}$$

where (12) was used. Hence T is expressible as a function of the six ray components, and when expressed in this way,

$$x_0 = \frac{\partial T}{\partial p_0}, \qquad x_1 = -\frac{\partial T}{\partial p_1}, \tag{25}$$

with similar relations for the other coordinates.

It is seen from (23) that *if the regions in which P_0 and P_1 are situated are both homogeneous, T represents the optical length of the ray between the feet Q_0 and Q_1 of perpendiculars drawn from O_0 and O_1 on to the initial and final portions of the ray* (see Fig. 4.3),

$$T = [Q_0 Q_1]. \tag{26}$$

In this case the angle characteristic may be expressed as a function of four variables only. For if we substitute for δm_1 from (18) and for δm_0 from a similar relation, (24) becomes

$$\delta T = \left(x_0 - \frac{p_0}{m_0} z_0\right) \delta p_0 + \left(y_0 - \frac{q_0}{m_0} z_0\right) \delta q_0$$

$$- \left(x_1 - \frac{p_1}{m_1} z_1\right) \delta p_1 - \left(y_1 - \frac{q_1}{m_1} z_1\right) \delta q_1. \tag{27}$$

This relation shows that *when the initial and final media are homogeneous, the angle characteristic is expressible as a function of the four variables p_0, q_0, p_1 and q_1:*

$$T = T(p_0, q_0; p_1, q_1), \tag{28}$$

and its derivatives then satisfy the relations

$$\left.\begin{aligned} x_0 - \frac{p_0}{m_0} z_0 &= \frac{\partial T}{\partial p_0}, & y_0 - \frac{q_0}{m_0} z_0 &= \frac{\partial T}{\partial q_0}, \\ x_1 - \frac{p_1}{m_1} z_1 &= -\frac{\partial T}{\partial p_1}, & y_1 - \frac{q_1}{m_1} z_1 &= -\frac{\partial T}{\partial q_1}. \end{aligned}\right\} \tag{29}$$

If the components of the initial and final portion of a ray are given, the coordinates of points on these portions may, according to (29), be determined immediately from the knowledge of the angle characteristic.

4.1.4 Approximate form of the angle characteristic of a refracting surface of revolution

Let

$$z = c_2(x^2 + y^2) + c_4(x^2 + y^2)^2 + \ldots, \tag{30}$$

where c_2, c_4, . . . are constants, be the equation of a refracting surface of revolution, referred to Cartesian axes, whose origin O coincides with the axial point (called pole) of the surface, and whose z-direction is along the axis of symmetry. If r denotes the radius of curvature at the pole of the surface (measured as positive when the surface is convex towards light incident from the negative z-direction), then

$$c_2 = \frac{1}{2r}. \tag{31}$$

For a spherical surface of radius r, $c_4 = 1/8r^3$. For a general surface of revolution we may write

$$c_4 = \frac{1}{8r^3}(1 + b); \tag{32}$$

the constant b (sometimes called the *deformation coefficient*) is a rough measure of the departure of the surface from spherical form. In terms of r and b,

$$z = \frac{x^2 + y^2}{2r} + \frac{(x^2 + y^2)^2}{8r^3}(1 + b) + \ldots \tag{33}$$

It will be assumed that the regions on either side of the surface are homogeneous and of refractive indices n_0 and n_1 respectively. The angle characteristic will be referred to systems of axes parallel to those at O and with origins at axial points $O_0(0, 0, a_0)$ and $O_1(0, 0, a_1)$ [$a_0 < 0$, $a_1 > 0$, $r > 0$ in Fig. 4.4].

If P is the point of intersection of the incident ray with the refracting surface, and if Q_0 and Q_1 are the feet of the perpendiculars drawn from O_0 and O_1 on to the incident and the refracted ray, the angle characteristic T is then, according to (26),

$$\begin{aligned} T &= [Q_0P] + [PQ_1] \\ &= \{xp_0 + yq_0 + (z - a_0)m_0\} - \{xp_1 + yq_1 + (z - a_1)m_1\}, \end{aligned} \tag{34}$$

where (x, y, z) are the coordinates of P with respect to the axes at O, and (p_0, q_0, m_0), (p_1, q_1, m_1) are the components of the ray incident and refracted at P.

The coordinates (x, y, z) may be eliminated from (34) with the help of the law of refraction. According to § 3.2.2, the law of refraction is equivalent to the assertion that the vector $\mathbf{N}(p_0 - p_1, q_0 - q_1, m_0 - m_1)$ is normal to the surface at P. Hence if (33) is written in the form

$$F(x, y, z) \equiv z - \frac{x^2 + y^2}{2r} - \frac{(x^2 + y^2)^2}{8r^3}(1 + b) - \ldots = 0, \tag{35}$$

then

$$\left.\begin{aligned} \lambda \frac{\partial F}{\partial x} &= -\lambda \left[\frac{x}{r} + \ldots\right] = p_0 - p_1, \\ \lambda \frac{\partial F}{\partial y} &= -\lambda \left[\frac{y}{r} + \ldots\right] = q_0 - q_1, \\ \lambda \frac{\partial F}{\partial z} &= \quad \lambda \qquad\qquad\;\; = m_0 - m_1. \end{aligned}\right\} \tag{36}$$

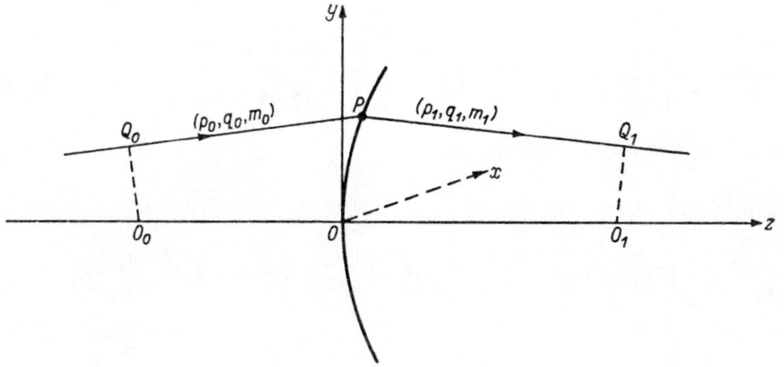

Fig. 4.4. The angle characteristic of a refracting surface of revolution. (The points O_0, O, O_1, Q_1, P, Q_0 are not necessarily coplanar.)

These equations imply that

$$\left.\begin{aligned} x &= -r\frac{p_0 - p_1}{m_0 - m_1} + \Delta x, \\ y &= -r\frac{q_0 - q_1}{m_0 - m_1} + \Delta y, \end{aligned}\right\} \tag{37}$$

where Δx and Δy are quantities of the third order in p, q, x/r, y/r. To express z in terms of the ray components, we substitute from (37) into (35), and obtain

$$\begin{aligned} z = \frac{r}{2(m_1 - m_0)^2}[(p_0 - p_1)^2 + (q_0 - q_1)^2] + \frac{1}{m_1 - m_0}[\Delta x(p_0 - p_1) + \Delta y(q_0 - q_1)] \\ + \frac{1}{8}\frac{r(1 + b)}{(m_1 - m_0)^4}[(p_0 - p_1)^2 + (q_0 - q_1)^2]^2 + \ldots \end{aligned} \tag{38}$$

To find the expansion of T up to and including the fourth-order terms it is not necessary to evaluate Δx and Δy; for when we substitute from (37) and (38) into (34), the contributions involving Δx and Δy are seen to be of order higher than the fourth and may therefore be neglected. (34) then becomes

$$T(p_0, q_0, m_0; p_1, q_1, m_1) = -m_0 a_0 + m_1 a_1 + \frac{r}{2(m_1 - m_0)}[(p_0 - p_1)^2 + (q_0 - q_1)^2]$$
$$- \frac{1}{8}\frac{r(1+b)}{(m_1 - m_0)^3}[(p_0 - p_1)^2 + (q_0 - q_1)^2]^2. \qquad (39)$$

(39) is the expansion of the angle characteristic up to the fourth order, the angle characteristic being considered as a function of all the six ray components. Two of the components may be eliminated by using the identity (6). We have from (6)

$$\left.\begin{aligned} m_0 &= n_0 - \frac{1}{2n_0}(p_0^2 + q_0^2) - \frac{1}{8n_0^3}(p_0^2 + q_0^2)^2 + \dots \\ m_1 &= n_1 - \frac{1}{2n_1}(p_1^2 + q_1^2) - \frac{1}{8n_1^3}(p_1^2 + q_1^2)^2 + \dots \end{aligned}\right\} \qquad (40)$$

so that

$$\frac{1}{m_1 - m_0} = \frac{1}{n_1 - n_0}\left[1 - \frac{1}{2n_0(n_1 - n_0)}(p_0^2 + q_0^2) + \frac{1}{2n_1(n_1 - n_0)}(p_1^2 + q_1^2) + \dots\right]. \qquad (41)$$

(39) becomes, on substitution from (41):

$$\begin{aligned} T(p_0, q_0; p_1, q_1) &= n_1 a_1 - n_0 a_0 \\ &+ \frac{r}{2(n_1 - n_0)}[(p_0 - p_1)^2 + (q_0 - q_1)^2] + \frac{a_0}{2n_0}(p_0^2 + q_0^2) - \frac{a_1}{2n_1}(p_1^2 + q_1^2) \\ &- \frac{r}{4(n_1 - n_0)^2}[(p_0 - p_1)^2 + (q_0 - q_1)^2]\left[\frac{p_0^2 + q_0^2}{n_0} - \frac{p_1^2 + q_1^2}{n_1}\right] \\ &- \frac{(1+b)r}{8(n_1 - n_0)^3}[(p_0 - p_1)^2 + (q_0 - q_1)^2]^2 \\ &+ \frac{a_0}{8n_0^3}(p_0^2 + q_0^2)^2 - \frac{a_1}{8n_1^3}(p_1^2 + q_1^2)^2 + \dots \end{aligned} \qquad (42)$$

The four variables p_0, q_0, p_1 and q_1 are seen to enter this expression only in the three combinations*

$$p_0^2 + q_0^2 = u^2, \quad p_1^2 + q_1^2 = v^2, \quad \text{and} \quad p_0 p_1 + q_0 q_1 = w^2. \qquad (43)$$

With this substitution (42) becomes, on separating into orders,

$$T(p_0, q_0; p_1, q_1) = T^{(0)} + T^{(2)} + T^{(4)} + \dots,$$

---

* It can be shown more generally that the angle characteristic of any medium which is rotationally symmetrical about the z-axis depends on the four variables only through the three combinations (43). To see this, we use a result proved in § 5.1, according to which any function $F(x_0, y_0; x_1, y_1)$ which is invariant with respect to rotation of axes about the origin in the xy-plane depends only on the three scalar products

$$\mathbf{r}_0^2 = x_0^2 + y_0^2, \quad \mathbf{r}_1^2 = x_1^2 + y_1^2, \quad \mathbf{r}_0 \cdot \mathbf{r}_1 = x_0 x_1 + y_0 y_1,$$

of the two vectors $\mathbf{r}_0(x_0, y_0)$, $\mathbf{r}_1(x_1, y_1)$. Identifying $\mathbf{r}_0$ and $\mathbf{r}_1$ with the projections of the propagation vectors $\mathbf{s}_0(p_0, q_0, m_0)$ and $\mathbf{s}_1(p_1, q_1, m_1)$ on to the xy-planes, the result follows.

where

$$\left.\begin{aligned} T^{(0)} &= n_1a_1 - n_0a_0, \\ T^{(2)} &= \mathscr{A}u^2 + \mathscr{B}v^2 + \mathscr{C}w^2, \\ T^{(4)} &= \mathscr{D}u^4 + \mathscr{E}v^4 + \mathscr{F}w^4 + \mathscr{G}u^2v^2 + \mathscr{H}u^2w^2 + \mathscr{K}v^2w, \end{aligned}\right\} \tag{44}$$

and

$$\left.\begin{aligned} \mathscr{A} &= \frac{1}{2}\left[\frac{r}{n_1 - n_0} + \frac{a_0}{n_0}\right], \\ \mathscr{B} &= \frac{1}{2}\left[\frac{r}{n_1 - n_0} - \frac{a_1}{n_1}\right], \\ \mathscr{C} &= -\frac{r}{n_1 - n_0}, \\ \mathscr{D} &= -\frac{r}{4(n_1 - n_0)^2}\left[\frac{1+b}{2(n_1 - n_0)} + \frac{1}{n_0}\right] + \frac{a_0}{8n_0^3}, \\ \mathscr{E} &= -\frac{r}{4(n_1 - n_0)^2}\left[\frac{1+b}{2(n_1 - n_0)} - \frac{1}{n_1}\right] - \frac{a_1}{8n_1^3}, \\ \mathscr{F} &= \frac{-(1+b)r}{2(n_1 - n_0)^3}, \\ \mathscr{G} &= -\frac{r}{4(n_1 - n_0)^2}\left[\frac{1+b}{n_1 - n_0} + \frac{1}{n_0} - \frac{1}{n_1}\right], \\ \mathscr{H} &= \frac{r}{2(n_1 - n_0)^2}\left[\frac{1+b}{n_1 - n_0} + \frac{1}{n_0}\right], \\ \mathscr{K} &= \frac{r}{2(n_1 - n_0)^2}\left[\frac{1+b}{n_1 - n_0} - \frac{1}{n_1}\right]. \end{aligned}\right\} \tag{45}$$

4.1.5 Approximate form of the angle characteristic of a reflecting surface of revolution

The expansion up to fourth degree for the angle characteristic associated with a reflecting surface of revolution can be derived in a similar manner. It is however not necessary to carry out the calculations in full. Using the same notation as in the preceding section (cf. Fig. 4.4 and 4.5) all the equations of § 4.1.4 up to and including (39) apply without change in the present case; hence (39) is also the *angle characteristic of a reflecting surface of revolution, when regarded as a function of all the six ray components*. However, when m_0 and m_1 are eliminated from (39) with the help of the two identities connecting the ray components, different expressions for T (as a function of four ray components) are obtained in the two cases. Denoting by n the refractive index of the medium in which the rays are situated, we have in place of (40),

$$\left.\begin{aligned} m_0 &= \quad n - \frac{1}{2n}(p_0^2 + q_0^2) - \frac{1}{8n^3}(p_0^2 + q_0^2)^2 + \dots, \\ m_1 &= -\left[n - \frac{1}{2n}(p_1^2 + q_1^2) - \frac{1}{8n^3}(p_1^2 + q_1^2)^2 + \dots\right]. \end{aligned}\right\} \tag{46}$$

In the second relation the negative square root $-\sqrt{n^2 - (p^2 + q^2)}$ has been taken, as we assume that the reflected ray returns into the region from which the light is propagated

($z < 0$); the direction cosine of the reflected ray with respect to the positive z-direction, and consequently m, is therefore negative. Since (40) reduces to (46) on setting $n_0 = -n_1 = n$, it follows that *the angle characteristic, considered as a function of the four ray components p_0, q_0, p_1 and q_1, of a reflecting surface of revolution, can be obtained*

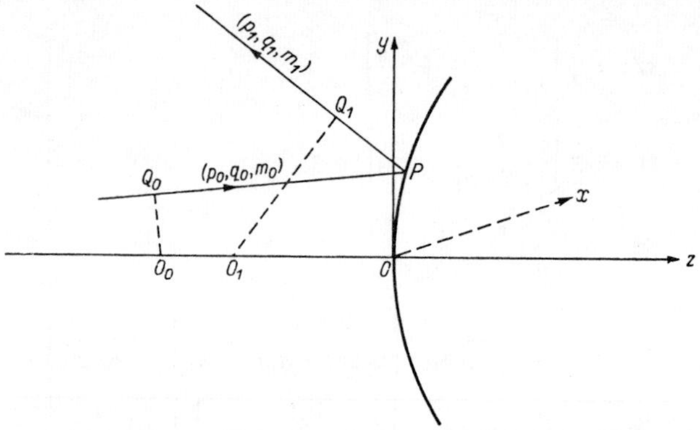

Fig. 4.5. The angle characteristic of a reflecting surface of revolution. (The points O_0, O_1, O, Q_0, P, Q_1 are not necessarily coplanar.)

from the angle characteristic of a refracting surface of revolution by setting $n_0 = -n_1 = n$. Hence, for the case of reflection, we have

$$\left.\begin{aligned} T^{(0)} &= -n(a_0 + a_1), \\ T^{(2)} &= \mathscr{A}'u^2 + \mathscr{B}'v^2 + \mathscr{C}'w^2, \\ T^{(4)} &= \mathscr{D}'u^4 + \mathscr{E}'v^4 + \mathscr{F}'w^4 + \mathscr{G}'u^2v^2 + \mathscr{H}'u^2w^2 + \mathscr{K}'v^2w^2, \end{aligned}\right\} \tag{47}$$

where

$$\left.\begin{aligned} \mathscr{A}' &= \frac{1}{2n}[-\tfrac{1}{2}r + a_0], \\ \mathscr{B}' &= \frac{1}{2n}[-\tfrac{1}{2}r + a_1], \\ \mathscr{C}' &= \frac{r}{2n}, \\ \mathscr{D}' &= -\frac{r}{16n^3}\left[-\frac{1+b}{4} + 1\right] + \frac{a_0}{8n^3}, \\ \mathscr{E}' &= -\frac{r}{16n^3}\left[-\frac{1+b}{4} + 1\right] + \frac{a_1}{8n^3}, \\ \mathscr{F}' &= \frac{(1+b)r}{16n^3}, \\ \mathscr{G}' &= \frac{r}{16n^3}\left[\frac{1+b}{2} - 2\right], \\ \mathscr{H}' &= \frac{r}{8n^3}\left[-\frac{1+b}{2} + 1\right], \\ \mathscr{K}' &= \frac{r}{8n^3}\left[-\frac{1+b}{2} + 1\right]. \end{aligned}\right\} \tag{48}$$

4.2 PERFECT IMAGING

Consider the propagation of light from a point source situated at a point P_0 in a medium specified by a refractive index function $n(x, y, z)$. An infinite number of rays will then proceed from P_0, but in general only a finite number will pass through any other point of the medium. In special cases, however, a point P_1 may be found through which an infinity of rays pass. Such a point P_1 is said to be a *stigmatic* (or a *sharp*) image of P_0.

In an ideal optical instrument every point P_0 of a three-dimensional region, called the *object space*, will give rise to a stigmatic image P_1. The totality of the image points defines the *image space*. The corresponding points in the two spaces are said to be *conjugate points*. In general not all the rays which proceed from P_0 will reach the image space; some, for example, will be excluded by the diaphragms of the instrument. Those rays which reach the image space will be said *to lie in the field of the instrument*. When P_0 describes a curve C_0 in the object space, P_1 will describe a conjugate curve C_1. The two curves will not necessarily be geometrically similar to each other. If *every* curve C_0 of the object space is geometrically similar to its image, we may say that the imaging between the two spaces is *perfect*. In a similar way we may define perfect imaging between two surfaces.

Optical instruments which are perfect in the sense just defined are of considerable interest, and accordingly we shall formulate some general theorems relating to perfect, or at least sharp, imaging of three-dimensional domains. Some results relating to sharp imaging of two-dimensional domains (surfaces) will be briefly discussed in § 4.2.3.

4.2.1 General theorems

An optical system $\mathscr{I}$ which images stigmatically a three-dimensional domain is often called an *absolute instrument*. It will be shown that *in an absolute instrument the optical length of any curve in the object space is equal to the optical length of its image.* This theorem was first put forward by MAXWELL* in 1858 for the special case when both the object and the image space are homogeneous. More rigorous proofs were later given by H. BRUNS (1895), F. KLEIN (1901) and H. LIEBMANN (1916).†

The theorem was later shown by CARATHÉODORY‡ not to be restricted to homogeneous media, but to be valid also when the media are heterogeneous and anisotropic. In proving this theorem the method of CARATHÉODORY will be used, but our discussion will be restricted to absolute instruments with isotropic (but generally heterogeneous) object and image spaces. §

---

* J. C. MAXWELL, *Quart. J. Pure Appl. Maths.*, **2** (1858), p. 233. Also his *Scientific Papers*, I (Cambridge University Press, 1890), 271.

† H. BRUNS, "Das Eikonal", *Abh. Kgl. sächs. Ges. Wiss., math-phys. Kl.*, **21** (1895), 370; F. KLEIN, *Z. Math. Phys.*, **46** (1901), 376; *Ges. Math. Abh.*, **2** (1922), 607. (See also E. T. WHITTAKER, *The Theory of Optical Instruments* (Cambridge University Press, 1907), 47. H. LIEBMANN, *Sitzgsber. bayer. Akad. Wiss., Math-naturw. Abt.* (1916), 183. An account of these researches will also be found in an article by H. BOEGEHOLD in S. CZAPSKI and O. EPPENSTEIN, *Grundzüge der Optischen Instrumente nach Abbe* (Leipzig, Barth, 3rd edition (1924), p. 213).

‡ C. CARATHÉODORY, *Sitzgsber. bayer. Akad. Wiss. Math-naturw. Abt.*, **56** (1926), 1.

§ In microwave optics the use of heterogeneous substances is common [see, for example, J. BROWN, *Microwave Lenses* (London, Methuen), 1953]. In light optics interest in them has developed more recently [cf. E. MARCHAND, *Progress in Optics*, vol. 11, ed. E. WOLF (Amsterdam, North Holland Publishing Company and New York, American Elsevier Publishing Company, 1973), p. 305].

Let A_0B_0 and A_1B_1 be ray-segments in the object and image spaces (Fig. 4.6) of a ray which lies in the field of an absolute instrument $\mathscr{I}$. Any other ray with a line-element which neither in position nor in direction departs appreciably from an element of $A_0B_0A_1B_1$, will also lie in the field of the instrument.

If each element of a curve (which is assumed to have a continuously turning tangent) coincides with an element of some ray which lies completely in the field of $\mathscr{I}$, the curve will be said to lie *tangentially* in the field of $\mathscr{I}$. If a "polygon" with a sufficient number of sides is inscribed into such a curve, each side of the polygon will coincide with an element of a ray which lies completely in the field of the instrument.

According to the principle of equal optical path (see § 3.3.3), all the rays which join A_0 to its image A_1 have the same optical length. We shall denote this optical path length by $V(A_0)$ and will show that it is in fact independent of A_0.

Let B_0 and B_1 be another pair of conjugate points. Then (see Fig. 4.6)

$$[A_1B_1] = [A_0B_0] + V(B_0) - V(A_0). \tag{1}$$

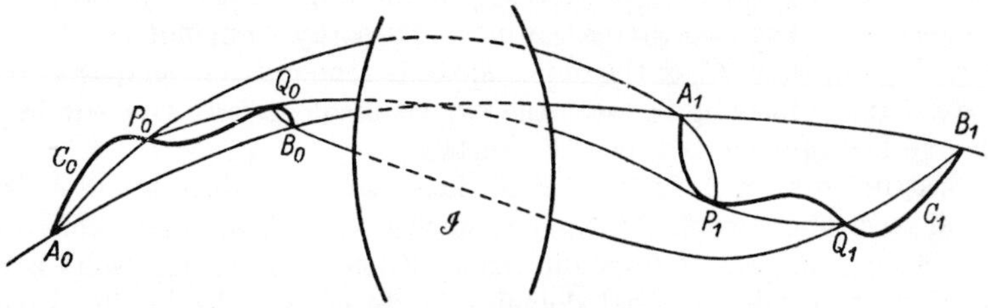

Fig. 4.6. An absolute optical instrument.

Let C_0 be a curve which joins A_0 and B_0 and which lies tangentially in the field of the instrument, and let C_1 be its image. We inscribe into C_0 a polygon $A_0P_0Q_0B_0$ and denote the image points of P_0 and Q_0 by P_1 and Q_1. Then, on applying (1) to the side A_1P_1, we have

$$[A_1P_1] = [A_0P_0] + V(P_0) - V(A_0),$$

and similarly for the other sides:

$$[P_1Q_1] = [P_0Q_0] + V(Q_0) - V(P_0),$$

and

$$[Q_1B_1] = [Q_0B_0] + V(B_0) - V(Q_0).$$

Hence

$$[A_1P_1] + [P_1Q_1] + [Q_1B_1] = [A_0P_0] + [P_0Q_0] + [Q_0B_0] + V(B_0) - V(A_0).$$

Obviously this result can be extended to a polygon of any number of sides N. Proceeding to the limit, as $N \to \infty$ in such a way that the greatest side tends to zero, we obtain the relation

$$L_1 = L_0 + V(B_0) - V(A_0), \tag{2}$$

where

$$L_0 = \int_{C_0} n_0 ds_0, \qquad L_1 = \int_{C_1} n_1 ds_1 \tag{3}$$

are the optical lengths of the curves C_0 and C_1. Next, it will be shown that $V(B_0) = V(A_0)$.

The points on the two curves are in a one-one correspondence, which may be expressed by relations of the form

$$x_1 = f(x_0, y_0, z_0), \qquad y_1 = g(x_0, y_0, z_0), \qquad z_1 = h(x_0, y_0, z_0). \tag{4}$$

An element ds_1 of C_1 is a function of the corresponding element ds_0,

$$ds_1 = \sqrt{\left(\frac{dx_1}{ds_0}\right)^2 + \left(\frac{dy_1}{ds_0}\right)^2 + \left(\frac{dz_1}{ds_0}\right)^2} ds_0. \tag{5}$$

Hence

$$L_1 = \int_{C_1} F\left(x_1, y_1, z_1, \frac{dx_1}{ds_0}, \frac{dy_1}{ds_0}, \frac{dz_1}{ds_0}\right) ds_0, \tag{6}$$

where

$$F\left(x_1, y_1, z_1, \frac{dx_1}{ds_0}, \frac{dy_1}{ds_0}, \frac{dz_1}{ds_0}\right) = n_1(x_1, y_1, z_1) \sqrt{\left(\frac{dx_1}{ds_0}\right)^2 + \left(\frac{dy_1}{ds_0}\right)^2 + \left(\frac{dz_1}{ds_0}\right)^2}$$

is a homogeneous function of the first degree in the derivatives dx_1/ds_0, dy_1/ds_0 and dz_1/ds_0; moreover F remains unchanged when $dx_1/ds_0, \ldots,$ is replaced by $-dx_1/ds_0, \ldots$. Now from (4),

$$\frac{dx_1}{ds_0} = \frac{\partial f}{\partial x_0}\frac{dx_0}{ds_0} + \frac{\partial f}{\partial y_0}\frac{dy_0}{ds_0} + \frac{\partial f}{\partial z_0}\frac{dz_0}{ds_0}, \tag{7}$$

with similar expressions for dy_1/ds_0 and dz_1/ds_0. Hence using (7) and (4), F can be expressed in the form

$$F\left(x_1, y_1, z_1, \frac{dx_1}{ds_0}, \frac{dy_1}{ds_0}, \frac{dz_1}{ds_0}\right) = \Phi\left(x_0, y_0, z_0, \frac{dx_0}{ds_0}, \frac{dy_0}{ds_0}, \frac{dz_0}{ds_0}\right), \tag{8}$$

Φ being also a homogeneous function of the first degree in $dx_0/ds_0, \ldots$; moreover, Φ remains unchanged when $dx_0/ds_0, \ldots,$ is replaced by $-dx_0/ds_0, \ldots,$

$$\Phi\left(x_0, y_0, z_0, -\frac{dx_0}{ds_0}, -\frac{dy_0}{ds_0}, -\frac{dz_0}{ds_0}\right) = \Phi\left(x_0, y_0, z_0, \frac{dx_0}{ds_0}, \frac{dy_0}{ds_0}, \frac{dz_0}{ds_0}\right). \tag{9}$$

From (2), (6), and (8) it follows that

$$\int_{C_0} (n_0 - \Phi) ds_0 = V(A_0) - V(B_0), \tag{10}$$

showing that the value of the curvilinear integral in (10) depends only on the end points A_0, B_0 and not on the choice of C_0. The curve C_0 is, however, not quite arbitrary, for it must lie tangentially in the field of the instrument. Nevertheless, it may be concluded that the expression $(n_0 - \Phi)ds_0$ must be a complete differential of some function Ψ,

$$n_0 - \Phi = \frac{\partial \Psi}{\partial x_0}\frac{dx_0}{ds_0} + \frac{\partial \Psi}{\partial y_0}\frac{dy_0}{ds_0} + \frac{\partial \Psi}{\partial z_0}\frac{dz_0}{ds_0}.$$

If now the derivatives $dx_0/ds_0, \ldots,$ are replaced by $-dx_0/ds_0$, the right-hand side will change sign, but the left-hand side will, on account of (9), remain unchanged. This is only possible if each side vanishes (Ψ = constant); hence

$$\Phi = n_0. \tag{11}$$

Equation (10) shows that $V(A_0) = V(B_0)$ and consequently (2) reduces to the relation $L_1 = L_0$. Hence for any curve, whether or not it lies tangentially in the field, provided only it has an image,

$$\int_{C_0} n_0 ds_0 = \int_{C_1} n_1 ds_1. \tag{12}$$

This is *Maxwell's theorem for an absolute instrument.**

From MAXWELL's theorem a number of interesting conclusions can immediately be drawn. Consider a small triangle whose sides are of lengths $ds_0^{(1)}$, $ds_0^{(2)}$, $ds_0^{(3)}$ and let $ds_1^{(1)}$, $ds_1^{(2)}$, $ds_1^{(3)}$ be the sides of its image formed by an absolute instrument. Further let n_0 and n_1 be the refractive indices of the regions where the triangles are situated. By MAXWELL's theorem,

$$n_0 ds_0^{(1)} = n_1 ds_1^{(1)}, \qquad n_0 ds_0^{(2)} = n_1 ds_1^{(2)}, \qquad n_0 ds_0^{(3)} = n_1 ds_1^{(3)}. \tag{13}$$

Hence the two triangles are similar to each other and corresponding sides are in the inverse ratio of the refractive indices. The angle between any two curves will therefore be preserved in the imaging, i.e. the imaging must be a *conformal transformation.* Now there is a general theorem due to LIOUVILLE† according to which a conformal transformation of a three-dimensional domain on to a three-dimensional domain can only be a projective transformation (collineation), an inversion‡ or the combination of these two transformations. We have thus established the following theorem due to

* The following less general, but very simple and elegant proof of MAXWELL's theorem was given by W. LENZ (contribution in *Probleme der Modernen Physik*, edited by P. DEBYE (Leipzig, Hirzel, 1928), p. 198.

Assume that all the rays from each object point reach the image and let (A_0, A_1) and (B_0, B_1) be two conjugate pairs. By hypothesis, the ray A_0B_0 must pass through A_1 and B_1. Likewise B_0A_0

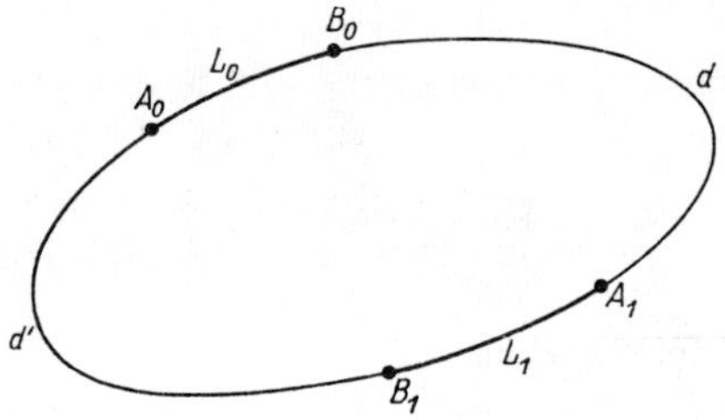

Fig. 4.7. Illustrating LENZ's proof of MAXWELL's theorem for an absolute instrument.

must pass through these points. Hence each ray must be a closed curve and by the principle of equal optical path it follows that (see Fig. 4.7)

$$[A_0A_1]_{\text{clockwise}} = [A_0A_1]_{\text{anticlockwise}}$$

and

$$[B_0B_1]_{\text{clockwise}} = [B_0B_1]_{\text{anticlockwise}}$$

Let

$$[A_0B_0] = L_0, \quad [A_1B_1] = L_1, \quad [B_0A_1] = d, \quad [B_1A_0] = d'.$$

The two equations then become

$$L_0 + d = d' + L_1,$$
$$d + L_1 = L_0 + d'.$$

and on subtraction we obtain

$$L_0 = L_1.$$

This proves MAXWELL's theorem for the special case when the curve is a portion of a ray. Generalization to an arbitrary curve may be obtained as in our main proof, by regarding the curve as a limiting form of a polygon formed by a large number of ray segments.

† See, for example, W. BLASCHKE, *Vorlesungen über Differential-Geometrie* I (Berlin, Springer, 2nd ed. (1924), p. 68; 4th ed. (1945), p. 101)).

‡ An inversion transforms each point P_0 into a point P_1 on the line joining P_0 with a fixed origin O, and the product $OP_0 \,.\, OP_1$ is constant.

CARATHÉODORY: *The imaging by an absolute instrument is either a projective transformation, an inversion, or a combination of the two.*

Let us now briefly consider the case when the imaging between the two spaces is not only stigmatic but is perfect, i.e. it is such that any figure is transformed into one which is geometrically similar to it. Clearly the imaging must be a projective transformation, since it transforms lines into lines.* It then follows from (13) that *the magnification ds_1/ds_0 between any two conjugate linear elements is equal to the ratio n_0/n_1 of the refractive indices.* In particular, if $n_0 = n_1 =$ constant, then $ds_1/ds_0 = 1$, so that *a perfect imaging between two homogeneous spaces of equal refractive indices is always trivial in the sense that it produces an image which is congruent with the object.* A plane mirror (or a combination of plane mirrors) is the only known instrument which produces such imaging.

These general considerations imply that, in order to obtain non-trivial imaging between homogeneous spaces of equal refractive indices, the requirement of exact stigmatism or of strict similarity between the object and the image must be dropped.

4.2.2 Maxwell's "fish-eye"

A simple and interesting example of an absolute instrument is presented by the medium which is characterized by the refractive index function

$$n(r) = \frac{1}{1 + (r/a)^2}\, n_0, \tag{14}$$

where r denotes the distance from a fixed point O, and n_0 and a are constants. It is known as the "fish-eye" and was first investigated by MAXWELL.†

It was shown in § 3.2 that in a medium with spherical symmetry the rays are plane curves which lie in planes through the origin, and that the equation of the rays may be written in the form (cf. § 3.2 (11))

$$\theta = c\int^{r} \frac{dr}{r\sqrt{n^2(r)r^2 - c^2}},$$

c being a constant. On substituting from (14) and setting

$$\rho = \frac{r}{a}, \qquad K = \frac{c}{an_0}, \tag{15}$$

---

* Cf. F. KLEIN, *Elementary Mathematics from an Advanced Standpoint*, Vol. II (translated from third German ed., London, Macmillan, 1939, 89; reprinted by Dover Publications, New York).

† J. C. MAXWELL, *Cambridge and Dublin Math. J.*, **8** (1854), 188; also *Scientific Papers*, I (Cambridge University Press), p. 76.

Interesting generalizations of MAXWELL's fish-eye were found by W. LENZ, contribution in *Probleme der Modernen Physik*, edited by P. DEBYE (Leipzig, Hirzel, 1928), 198 and R. STETTLER, *Optik*, **12** (1955), 529. The latter paper also includes a generalization of the so-called *Luneburg lens* which, because of its wide angle scanning capabilities, has useful applications in microwave antenna design. This lens, first considered by R. K. LUNUBURG in his *Mathematical theory of Optics* (mimeographed lecture notes, Brown University, Providence, R.I., 1944; printed version published by University of California Press, Berkeley and Los Angeles, 1964, § 29), is an inhomogeneous sphere with the refractive index function $n(r) = \sqrt{2 - r^2}$ $(0 \leqslant r \leqslant 1)$. When placed in a homogeneous medium of unit refractive index, it brings to a sharp focus every incident pencil of parallel rays. See also R. F. RINEHART, *J. Appl. Phys.*, **19** (1948), 860; A. FLETCHER, T. MURPHY and A. YOUNG, *Proc. Roy. Soc.*, A **223** (1954), 216; and G. TORALDO DI FRANCIA, *Optica Acta*, **1** (1954–1955), 157.

we obtain

$$\theta = \int^{\rho} \frac{K(1+\rho^2)d\rho}{\rho\sqrt{\rho^2 - K^2(1+\rho^2)^2}}. \tag{16}$$

It may be verified that

$$\frac{K(1+\rho^2)}{\rho\sqrt{\rho^2 - K^2(1+\rho^2)^2}} = \frac{d}{d\rho}\left[\text{arc sin}\left(\frac{K}{\sqrt{1-4K^2}}\frac{\rho^2-1}{\rho}\right)\right],$$

so that (16) becomes

$$\sin(\theta - \alpha) = \frac{c}{\sqrt{a^2 n_0^2 - 4c^2}} \frac{r^2 - a^2}{ar}, \tag{17}$$

where α is a constant of integration.

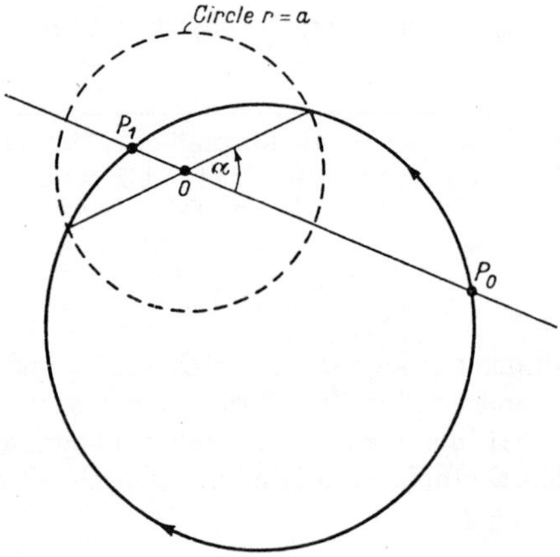

Fig. 4.8. Rays in MAXWELL's "fish-eye".

(17) is the polar equation of the rays. The one-parameter family of rays through a fixed point $P_0(r_0,\theta_0)$ is, therefore, given by

$$\frac{r^2 - a^2}{r\sin(\theta - \alpha)} = \frac{r_0^2 - a^2}{r_0 \sin(\theta_0 - \alpha)}. \tag{18}$$

It is seen that whatever the value of α, this equation is satisfied by $r = r_1$, $\theta = \theta_1$, where

$$r_1 = \frac{a^2}{r_0}, \qquad \theta_1 = \pi + \theta_0, \tag{19}$$

showing that *all the rays from an arbitrary point P_0 meet in a point P_1 on the line joining P_0 to O; P_0 and P_1 are on opposite sides of O and $OP_0 \,.\, OP_1 = a^2$.* Hence the fish-eye is an absolute instrument in which the imaging is an *inversion.*

We note that (17) is satisfied by $r = a$, $\theta = \alpha$ and $r = a$, $\theta = \pi + \alpha$; each ray therefore intersects the fixed circle $r = a$ in diametrically opposite points (see Fig. 4.8).

To obtain the equation of the rays in Cartesian coordinates, we put $x = r\cos\theta$, $y = r\sin\theta$ in (17), and find

$$y\cos\alpha - x\sin\alpha = \frac{c}{a\sqrt{a^2 n_0^2 - 4c^2}}(x^2 + y^2 - a^2),$$

or

$$(x + b \sin \alpha)^2 + (y - b \cos \alpha)^2 = a^2 + b^2 \tag{20}$$

where

$$b = \frac{a}{2c} \sqrt{a^2 n_0^{\,2} - 4c^2}.$$

(20) shows that each ray is a circle.

4.2.3 Stigmatic imaging of surfaces

So far we have been concerned only with perfect or sharp imaging of three-dimensional domains. We saw that when the object space and image space are homogeneous and of equal refractive indices, perfect imaging can only be of a trivial kind, producing a mirror image of the object. It is natural to inquire whether non-trivial imaging may be obtained, when it is required that only certain *surfaces* should be imaged perfectly (or at least sharply) by the instrument. This question has been investigated by a number of authors,* who found that *when the object space and image space are homogeneous, not more than two surfaces may in general*† *be sharply imaged by a rotationally symmetrical system.* For the proof of this theorem the papers by Boegehold and Herzberger and by Smith should be consulted. Here we shall only consider in detail a simple case of sharp imaging of a spherical surface which is of particular interest in practice.

Consider the refraction at a solid homogeneous sphere S embedded in a homogeneous medium. Let O be the centre of the sphere, r its radius, and n and n' the refractive indices of the sphere and of the surrounding medium respectively. Further, let AQ be a ray incident upon the sphere. The refracted ray QB can easily be found by means of the following construction:

Let S_0 and S_1 be two spheres whose centres are at O and whose radii are

$$r_0 = \frac{n}{n'}\, r, \qquad r_1 = \frac{n'}{n}\, r. \tag{21}$$

If P_0 is the point of intersection of AQ with S_0, and P_1 is the point at which OP_0 meets S_1, then QP_1 is the refracted ray. For one has, by construction (see Fig. 4.9)

$$\frac{OQ}{OP_0} = \frac{OP_1}{OQ} = \frac{n'}{n}. \tag{22}$$

Moreover

$$\widehat{QOP_0} = \widehat{QOP_1}. \tag{23}$$

Hence the triangles OQP_0 and OP_1Q are similar, and consequently

$$\frac{\sin \phi_0}{\sin \phi_1} = \frac{OP_0}{OQ} = \frac{n}{n'}, \tag{24}$$

* H. Boegehold and M. Herzberger, *Compositio Mathematica*, 1 (1935), 448; M. Herzberger, *Ann. New York Acad. Sci.*, **48** (1946), *Art.* 1, p. 1; T. Smith, *Proc. Phys. Soc.*, **60** (1948), 293. See also C. G. Wynne, *Proc. Phys. Soc.*, **65B** (1952), 436.

† "In general" implies here that certain degenerate cases in which the *whole* object space is imaged sharply (e.g. by reflection on a plane mirror) are excluded.

where $\phi_0 = \widehat{OQP_0}$ and $\phi_1 = \widehat{OQP_1}$ are the angles of incidence and refraction respectively. ϕ_0 and ϕ_1 satisfy the law of refraction, and consequently QP_1 is the refracted ray.

The construction implies that all rays which diverge from a point P_0 on S_0 will form a (virtual) stigmatic image at the point P_1 in which the diameter OP_0 intersects S_1. Hence *the sphere* S_1 *is a stigmatic image of* S_0, *and vice versa.*

It will be useful to express (24) in a somewhat different form. If we denote by θ_0 and θ_1 the angles which the two conjugate rays make with the line P_0P_1, i.e. $\theta_0 = \widehat{OP_0Q}$,

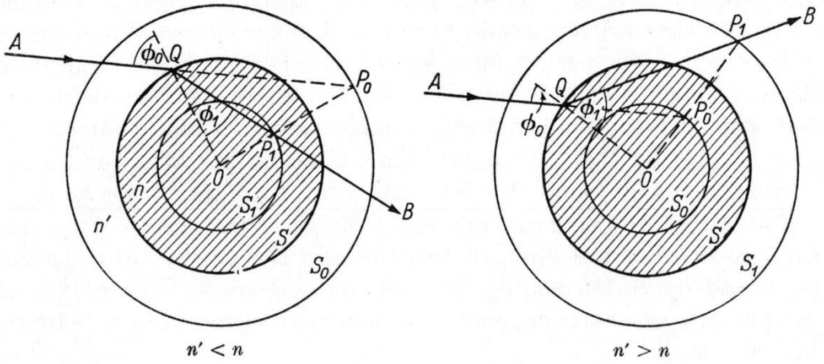

Fig. 4.9. Refraction at a spherical surface. Aplanatic points.

$\theta_1 = \widehat{OP_1Q}$, then, since the two triangles are similar, it follows that $\theta_0 = \phi_1$ and $\theta_1 = \phi_0$; hence

$$\frac{\sin\theta_1}{\sin\theta_0} = \frac{n}{n'} = \text{constant}. \tag{25}$$

Equation (25) is a special case of the so-called *sine condition* whose significance will be explained in § 4.5. In accordance with the terminology of § 4.5, P_0 and P_1 are called *aplanatic points* of the spherical surface S.

The existence of the aplanatic points for refraction at a spherical surface is made use of, as will be shown in § 6.6, in the construction of certain microscope objectives.

4.3 PROJECTIVE TRANSFORMATION (COLLINEATION) WITH AXIAL SYMMETRY

It has been shown in the previous section that perfect imaging between three-dimensional domains must necessarily be a projective transformation, since it transforms lines into lines. But even when the requirements for perfect imaging are not strictly fulfilled, the properties of projective transformations are of great importance. For, as will be seen later, the relationship between the object and the image in any optical system is, *at least to a first approximation*, a transformation of this kind. It will therefore be convenient to study the general properties of a projective transformation before deriving the laws of image formation in actual instruments. Though this preliminary discussion is essentially of a geometrical nature, it will be convenient to retain, where possible, the terminology of optics.

4.3.1 General formulae

Let (x, y, z) be the coordinates of a point P of the object space and (x', y', z') the coordinates of a point P' in the image space, both referred at present to the same set of Cartesian rectangular axes, chosen arbitrarily. A projective relationship between the two spaces is mathematically expressed by relations of the form

$$x' = \frac{F_1}{F_0}, \qquad y' = \frac{F_2}{F_0}, \qquad z' = \frac{F_3}{F_0}, \tag{1}$$

where

$$F_i = a_i x + b_i y + c_i z + d_i \qquad (i = 0, 1, 2, 3). \tag{2}$$

Pairs of points related by (1) will be said to form a *conjugate* pair.

Solving (1) for x, y, z, we obtain relations of the same form

$$x = \frac{F'_1}{F'_0}, \qquad y = \frac{F'_2}{F'_0}, \qquad z = \frac{F'_3}{F'_0}, \tag{3}$$

where

$$F'_i = a'_i x' + b'_i y' + c'_i z' + d'_i.$$

From (1) it follows that the image of any point situated in the plane $F_0 = 0$ will lie at infinity. Similarly from (3) it is seen that all object points whose images lie in the plane $F'_0 = 0$ are at infinity. The plane $F_0 = 0$ is called the *focal plane* of the *object space* and the plane $F'_0 = 0$ is known as the *focal plane* of the *image space*.* Rays which are parallel in the object space will be transformed into rays which intersect in a point on the focal plane $F'_0 = 0$. Similarly rays from a point in the focal plane $F_0 = 0$ will transform into a pencil of parallel rays. In special cases both the focal planes may lie at infinity. The transformation is then said to be *affine* or *telescopic*. Then to finite values (x, y, z) there correspond finite values of (x', y', z'), so that in a telescopic transformation one always has $F_0 \neq 0$ and $F'_0 \neq 0$. This clearly is only possible when $a_0 = b_0 = c_0 = 0$ and $a'_0 = b'_0 = c'_0 = 0$.

Of special importance for optics is the case of *axial symmetry*, since the majority of optical systems consist of surfaces of revolution with a common axis (called usually *centred systems*). It then follows from symmetry that the image of each point P_0 lies in the plane which contains P_0 and the axis; in considering the properties of the associated projective transformation we may therefore restrict our discussion to points situated in such a *meridional plane*. Let the meridional plane be the yz-plane and take the z-axis along the axis of symmetry. A point $(0, y, z)$ of the object space will then be transformed into a point $(0, y', z')$ of the image space, where

$$y' = \frac{b_2 y + c_2 z + d_2}{b_0 y + c_0 z + d_0}, \qquad z' = \frac{b_3 y + c_3 z + d_3}{b_0 y + c_0 z + d_0}. \tag{4}$$

Now it follows from symmetry that z' remains unchanged when y is changed into $-y$. This, in general, is only possible if $b_0 = b_3 = 0$. Further it follows that, if $y \to -y$, then $y' \to -y'$, which implies that $c_2 = d_2 = 0$. Hence (4) reduces to

$$y' = \frac{b_2 y}{c_0 z + d_0}, \qquad z' = \frac{c_3 z + d_3}{c_0 z + d_0}. \tag{5}$$

* The terms "focal plane" and "focal points" have here a somewhat different meaning than in connection with normal congruences (§ 3.2.3) and astigmatic pencils (§ 4.6).

These equations contain five constants but only their ratios are significant. Hence *a projective transformation with axial symmetry is characterized by four parameters.*

Solving (5) for y and z, we obtain

$$y = \frac{c_0 d_3 - c_3 d_0}{b_2} \frac{y'}{c_0 z' - c_3}, \qquad z = \frac{-d_0 z' + d_3}{c_0 z' - c_3}. \tag{6}$$

From (5) and (6) it is seen that the focal planes are given by

$$F_0 \equiv c_0 z + d_0 = 0, \qquad F_0' \equiv c_0 z' - c_3 = 0;$$

hence the focal planes intersect the axis at right angles in the points whose abscissae are

$$z = -\frac{d_0}{c_0}, \qquad z' = \frac{c_3}{c_0}. \tag{7}$$

These points are called the *principal foci* and are denoted by F and F' in Fig. 4.10.

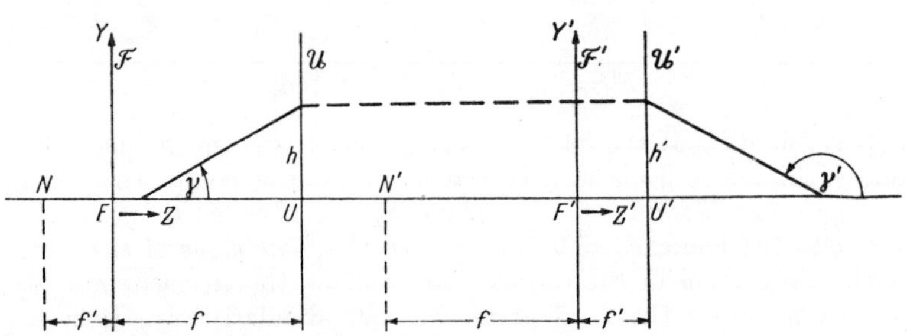

Fig. 4.10. The cardinal points and planes of an optical system: F, F' = foci; U, U' = unit (or principal) points; N, N' = nodal points; $\mathscr{F}, \mathscr{F}'$ = focal planes; $\mathscr{U}, \mathscr{U}'$ = unit (principal) planes.

It will now be convenient to introduce a separate coordinate system for each of the two spaces, measuring the z coordinates from the principal foci; i.e. we set

$$\left.\begin{aligned} y &= Y, & c_0 z + d_0 &= c_0 Z, \\ y' &= Y', & c_0 z' - c_3 &= c_0 Z'. \end{aligned}\right\} \tag{8}$$

(5) then become

$$Y' = \frac{b_2}{c_0} \frac{Y}{Z}, \qquad Z' = \frac{c_0 d_3 - c_3 d_0}{c_0{}^2 Z}.$$

Further we set

$$f = \frac{b_2}{c_0}, \qquad f' = \frac{c_0 d_3 - c_3 d_0}{b_2 c_0}. \tag{9}$$

With this substitution, the equations of the transformation take the simple form

$$\frac{Y'}{Y} = \frac{f}{Z} = \frac{Z'}{f'}. \tag{10}$$

The second relation, $ZZ' = ff'$, is usually called *Newton's equation*. The constant f is known as *the focal length of the object space and f' as the focal length of the image space.*

From (10) it follows that, for fixed object and image planes,

$$\left(\frac{dY'}{dY}\right)_{Z=\text{const.}} = \frac{Y'}{Y} = \frac{f}{Z} = \frac{Z'}{f'}. \tag{11}$$

This quantity is known as the *lateral magnification*. Further, we have independently of Y and Y', the *longitudinal magnification*

$$\frac{dZ'}{dZ} = -\frac{Z'}{Z} = -\frac{ff'}{Z^2} = -\frac{Z'^2}{ff'}. \tag{12}$$

From (11) and (12) it is seen that the two magnifications are related by the formula

$$\frac{dZ'}{dZ} = -\frac{f'}{f}\left(\frac{dY'}{dY}\right)^2_{Z=\text{const.}} \tag{13}$$

Since the lateral magnification depends on Z but not on Y it follows that an object which is situated in a plane perpendicular to the axis will be transformed into one which is geometrically similar to it.

The *lateral magnification* is equal to unity when $Z = f$ and $Z' = f'$. These planes are called the *unit* or *principal planes* and are denoted by the letters $\mathscr{U}$ and $\mathscr{U}'$ in Figs. 4.10 and 4.11. The points U and U' in which these planes intersect the axis are known as the *unit* or *principal points*.

Let h be the distance from the axis at which a ray from an axial point $(0, 0, Z)$ meets the unit plane of the object space. The conjugate ray meets the other unit plane at the same distance (h) from the axis; and the angles γ and γ' which the two rays make with the axis are given by (see Fig. 4.10)

$$\tan\gamma = \frac{h}{f - Z}, \qquad \tan\gamma' = \frac{h}{f' - Z'}.$$

The ratio

$$\frac{\tan\gamma'}{\tan\gamma} = \frac{f - Z}{f' - Z'} = -\frac{Z}{f'} = -\frac{f}{Z'} \tag{14}$$

is known as the *angular magnification* or *convergence ratio*. It is independent of h and h' and is equal to unity when $Z = -f'$ and $Z' = -f$. The two axial points N and N' specified by these expressions are known as *nodal points*. They are the two conjugate points which are characterized by the property that conjugate rays passing through them are parallel to each other. The distance between the nodal points is equal to the distance between the unit points. When $f = -f'$ the nodal points and the unit points coincide. The foci, the unit points, and the nodal points are called *cardinal* points of the transformation.

It is sometimes convenient to measure distances from the unit planes rather than from the focal planes. This means that in place of Z and Z' we use the variables

$$\zeta = Z - f, \qquad \zeta' = Z' - f'. \tag{15}$$

NEWTON's equation $ZZ' = ff'$ then becomes

$$\frac{f}{\zeta} + \frac{f'}{\zeta'} = -1. \tag{16}$$

Using the properties of the focal points and the principal planes, we can find the point P' conjugate to a given point P by a simple geometrical construction. Two lines are drawn through P, one parallel to the axis and the other through the focus F (Fig. 4.11). Let A and B be the points in which these lines intersect the unit plane $\mathscr{U}$. From the property of the unit planes it then follows that the points A' and B', which are conjugate to A and B, are the points of intersection with $\mathscr{U}'$ of the two lines through A and B drawn parallel to the axis. Moreover, since PA passes through the focus F, $P'A'$ must be parallel to the axis, and since PB is parallel to the axis, $P'B'$ must pass through the other focus F'. Hence the image point P' is the point of intersection of AA' with $B'F'$.

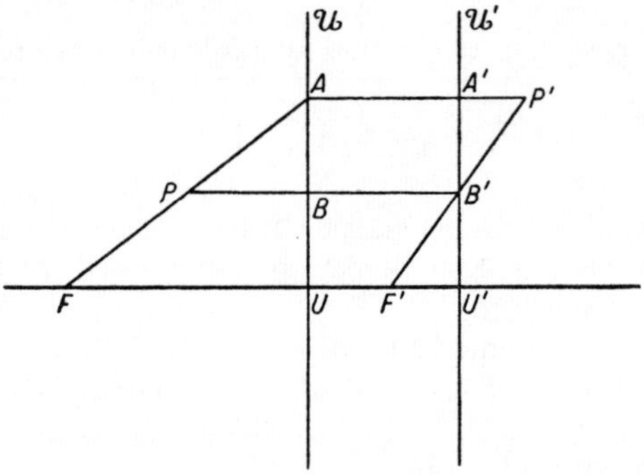

Fig. 4.11. Graphical determination of the image point.

4.3.2 The telescopic case

We shall now consider the special case of telescopic (affine) collineation. As already explained it is characterized by having both focal planes at infinity. The coefficient c_0 in (5) and (6) then vanishes, and (5) reduces to

$$y' = \frac{b_2 y}{d_0}, \qquad z' = \frac{c_3 z + d_3}{d_0}. \tag{17}$$

We shall again choose separate coordinate systems for the object space and for the image space. As origins we take any conjugate axial pair of points. Referred to the new set of axes, the equations of the transformation take the simple form

$$Y' = \alpha Y, \qquad Z' = \beta Z, \tag{18}$$

where $\alpha = b_2/d_0$ and $\beta = c_3/d_0$.

The lateral as well as the longitudinal magnification is now constant. The angular magnification is also constant; to show this, consider two conjugate rays and take the origins of the two coordinate systems at the points in which these rays intersect the axis. The equations of the two rays then are

$$Y = Z \tan \gamma, \qquad Y' = Z' \tan \gamma'. \tag{19}$$

Hence

$$\frac{\tan \gamma'}{\tan \gamma} = \frac{Y'}{Z'}\frac{Z}{Y} = \frac{\alpha}{\beta} = \frac{b_2}{c_3}. \tag{20}$$

Although in a telescopic transformation both f and f' are infinite, the ratio f'/f must be regarded as finite. For by (9), as $c_0 \to 0$,

$$\frac{f'}{f} \to -\frac{c_3 d_0}{b_2^2} = -\frac{\beta}{\alpha^2}. \tag{21}$$

4.3.3 Classification of projective transformations

Projective transformations may be classified according to the signs of the focal lengths.

When the focal lengths are of opposite signs, $ff' < 0$, and according to (12) $dZ'/dZ > 0$. This implies that if the object is displaced in a direction parallel to the axis, the image will be displaced in the same direction. It will be seen later that such imaging occurs whenever the image is produced either by refraction alone, or by an even number of reflections, or by a combination of the two. Such imaging is called *concurrent* or *dioptric*.

When the focal lengths have equal signs, $ff' > 0$, $dZ'/dZ < 0$, so that to a displacement of the object in a direction parallel to the axis, there corresponds a displacement of the image in the opposite direction. This type of imaging is produced by means of an odd number of reflections, or by a combination of an odd number of reflections with any number of refractions. Imaging of this type is called *contracurrent* or *katoptric*.

In each group two types of transformations are distinguished, according to the sign of the focal length f. It is seen from (11) that for $Z > 0$ the lateral magnification is positive or negative according as $f > 0$ or $f < 0$. Hence an object situated in the right-hand half of the object space will have an image which is similarly oriented or inverted, according as f is positive or negative. In the former case the transformation is said to be *convergent*; in the latter, it is said to be *divergent*. This terminology is derived from the fact that an incident bundle of parallel rays, after passing the unit plane of the image space, is rendered convergent in the former case and divergent in the latter. The four cases are summarized in Table XII.

TABLE XII

Classification of projective transformations (Cartesian sign-convention)

	Convergent	Divergent
Concurrent (Dioptric)	$f > 0;\ f' < 0$	$f < 0;\ f' > 0$
Contracurrent (Katoptric)	$f > 0;\ f' > 0$	$< 0;\ f' < 0$

In the special case when the transformation is telescopic, the four types are distinguished by the signs of α and β. It follows from (21) that the transformation will be concurrent when β is positive and contracurrent when it is negative. Further it is seen from (18) that it will be convergent or divergent according as to α is positive or negative.

4.3.4 Combination of projective transformations

We now consider the combination of two successive projective transformations, which are rotationally symmetrical about the same axis.

Let the subscript 0 refer to the first transformation, and subscript 1 to the second. Then the equations specifying the two transformations are:

$$\left.\begin{aligned} \frac{Y_0'}{Y_0} &= \frac{f_0}{Z_0} = \frac{Z_0'}{f_0'}, \\ \frac{Y_1'}{Y_1} &= \frac{f_1}{Z_1} = \frac{Z_1'}{f_1'}. \end{aligned}\right\} \qquad (22)$$

Let c be the distance between the focal points F'_0 and F_1. Since the image space of the first transformation coincides with the object space of the second,

$$Z_1 = Z'_0 - c, \qquad Y_1 = Y'_0. \tag{23}$$

Elimination of the coordinates of the intermediate space from (22) by means of (23) gives

$$\left.\begin{aligned} Y'_1 &= \frac{Z'_1 Y_1}{f'_1} = \frac{Z'_1 Y'_0}{f'_1} = \frac{Z'_1 f_0 Y_0}{f'_1 Z_0} = \frac{f_0 f_1 Y_0}{f_0 f'_0 - cZ_0}, \\ Z'_1 &= \frac{f_1 f'_1}{Z_1} = \frac{f_1 f'_1}{Z'_0 - c} = \frac{f_1 f'_1}{\dfrac{f_0 f'_0}{Z_0} - c} = \frac{f_1 f'_1 Z_0}{f_0 f'_0 - cZ_0}. \end{aligned}\right\} \tag{24}$$

Let

$$\left.\begin{aligned} Y &= Y_0, \qquad Z = Z_0 - \frac{f_0 f'_0}{c}, \\ Y' &= Y'_1, \qquad Z' = Z'_1 + \frac{f_1 f'_1}{c}. \end{aligned}\right\} \tag{25}$$

Equations (25) express a change of coordinates, the origins of the two systems being shifted by distances $f'_0 f_0/c$ and $-f_1 f'_1/c$ respectively in the Z-direction. In terms of these variables, the equations of the combined transformation become

$$\frac{Y'}{Y} = \frac{f}{Z} = \frac{Z'}{f'}, \tag{26}$$

where

$$f = -\frac{f_0 f_1}{c}, \qquad f' = \frac{f'_0 f'_1}{c}. \tag{27}$$

The distance between the origins of the new and the old systems of coordinates, i.e. the distances $\delta = F_0 F$ and $\delta' = F'_1 F'$ of the foci of the equivalent transformation from the foci of the individual transformations are seen from (25) to be

$$\delta = \frac{f_0 f'_0}{c}, \qquad \delta' = -\frac{f_1 f'_1}{c}. \tag{28}$$

If $c = 0$, then $f = f' = \infty$ so that the equivalent collineation is telescopic. The equations (24) then reduce to

$$\left.\begin{aligned} Y'_1 &= \frac{f_1}{f'_0} Y_0, \\ Z'_1 &= \frac{f_1 f'_1}{f_0 f'_0} Z_0; \end{aligned}\right\} \tag{29}$$

the constants α and β in (18) of the equivalent transformation, are therefore

$$\alpha = \frac{f_1}{f'_0}, \qquad \beta = \frac{f_1 f'_1}{f_0 f'_0}. \tag{30}$$

The angular magnification is now

$$\frac{\tan \gamma'}{\tan \gamma} = \frac{\alpha}{\beta} = \frac{f_0}{f'_1}. \tag{31}$$

If one or both of the transformations are telescopic, the above considerations must be somewhat modified.

4.4 GAUSSIAN OPTICS

We shall now study the elementary properties of lenses, mirrors, and their combinations. In this elementary theory only those points and rays will be considered which lie in the immediate neighbourhood of the axis; terms involving squares and higher powers of off-axis distances, or of the angles which the rays make with the axis, will be neglected. The resulting theory is known as Gaussian optics.*

4.4.1 Refracting surface of revolution

Consider a pencil of rays incident on a refracting surface of revolution which separates two homogeneous media of refractive indices n_0 and n_1. To begin with, points and rays in both media will be referred to the same Cartesian reference system, whose origin will be taken at the pole O of the surface, with the z-direction along the axis of symmetry.

Let $P_0(x_0, y_0, z_0)$ and $P_1(x_1, y_1, z_1)$ be points on the incident and on the refracted ray respectively. Neglecting terms of degree higher than first, it follows from § 4.1 (29), § 4.1 (40), and § 4.1 (44) that the coordinates of these points and the components of the two rays are connected by the relations

$$\left.\begin{aligned} x_0 - \frac{p_0}{n_0} z_0 &= \frac{\partial T^{(2)}}{\partial p_0} = 2\mathscr{A} p_0 + \mathscr{C} p_1, \\ x_1 - \frac{p_1}{n_1} z_1 &= -\frac{\partial T^{(2)}}{\partial p_1} = -2\mathscr{B} p_1 - \mathscr{C} p_0, \end{aligned}\right\} \tag{1a}$$

$$\left.\begin{aligned} y_0 - \frac{q_0}{n_0} z_0 &= \frac{\partial T^{(2)}}{\partial q_0} = 2\mathscr{A} q_0 + \mathscr{C} q_1, \\ y_1 - \frac{q_1}{n_1} z_1 &= -\frac{\partial T^{(2)}}{\partial q_1} = -2\mathscr{B} q_1 - \mathscr{C} q_0, \end{aligned}\right\} \tag{1b}$$

where, according to § 4.1 (45),

$$\mathscr{A} = \mathscr{B} = \frac{1}{2}\frac{r}{n_1 - n_0}, \qquad \mathscr{C} = -\frac{r}{n_1 - n_0}, \tag{2}$$

r being the paraxial radius of curvature of the surface.

Let us examine under what conditions all the rays from P_0 (which may be assumed to lie in the plane $x = 0$) will, after refraction, pass through P_1. The coordinates of P_1 will then depend only on the coordinates of P_0 and not on the components of the rays, so that when q_1 is eliminated from (1b), q_0 must also disappear.

Now from the first equation (1b)

$$q_1 = \frac{1}{\mathscr{C}}\left\{y_0 - q_0\left(2\mathscr{A} + \frac{1}{n_0} z_0\right)\right\}, \tag{3}$$

and substituting this into the second equation, we obtain

$$y_1 = -\left(2\mathscr{B} - \frac{1}{n_1} z_1\right)\frac{1}{\mathscr{C}} y_0 + \left[\frac{1}{\mathscr{C}}\left(2\mathscr{B} - \frac{1}{n_1} z_1\right)\left(2\mathscr{A} + \frac{1}{n_0} z_0\right) - \mathscr{C}\right] q_0. \tag{4}$$

---

* As before, the usual sign convention of analytical geometry (Cartesian sign convention) is used. The various sign conventions employed in practice are very fully discussed in a Report on the Teaching of Geometrical Optics published by the Physical Society (London) in 1934.

Hence P_1 will be a stigmatic image of P_0 if

$$\left(2\mathscr{A} + \frac{1}{n_0} z_0\right)\left(2\mathscr{B} - \frac{1}{n_1} z_1\right) = \mathscr{C}^2, \tag{5}$$

or, on substituting from (2), if

$$\left[\frac{r}{n_1 - n_0} + \frac{z_0}{n_0}\right]\left[\frac{r}{n_1 - n_0} - \frac{z_1}{n_1}\right] = \frac{r^2}{(n_1 - n_0)^2}. \tag{6}$$

(6) may be written in the form

$$n_0 \left[\frac{1}{r} - \frac{1}{z_0}\right] = n_1 \left[\frac{1}{r} - \frac{1}{z_1}\right]. \tag{7}$$

It is seen that, within the present degree of approximation, every point gives rise to a stigmatic image; and the distances of the conjugate planes from the pole O of the surface are related by (7). Moreover, equation (4), subject to (5), implies that the imaging is a *projective transformation.*

The expressions on either side of (7) are known as *Abbe's* (*refraction*) *invariant* and play an important part in the theory of optical imaging. (7) may also be written in the form

$$\frac{n_1}{z_1} - \frac{n_0}{z_0} = \frac{n_1 - n_0}{r}. \tag{8}$$

The quantity $(n_1 - n_0)/r$ is known as the *power* of the refracting surface and will be denoted by $\mathscr{P}$,

$$\mathscr{P} = \frac{n_1 - n_0}{r}. \tag{9}$$

According to (4) and (5) the lateral magnification y_1/y_0 is equal to unity when $z_1/n_1 = 2\mathscr{B} + \mathscr{C}$. But by (2), $2\mathscr{B} + \mathscr{C} = 0$. Hence the unit points U_0 and U_1 are given by $z_0 = z_1 = 0$, i.e. *the unit points coincide with the pole of the surface.* Further, from (8),

$$z_0 \to -\frac{n_0 r}{n_1 - n_0} \qquad \text{as } z_1 \to \infty$$

and

$$z_1 \to \frac{n_1 r}{n_1 - n_0} \qquad \text{as } z_0 \to -\infty,$$

so that the abscissae of the foci F_0 and F_1 are $-n_0 r/(n_1 - n_0)$ and $n_1 r/(n_1 - n_0)$ respectively. The focal lengths $f_0 = F_0U_0$, $f_1 = F_1U_1$ are therefore given by

$$\left.\begin{aligned} f_0 &= \frac{n_0 r}{n_1 - n_0}, \\ &\text{and} \\ f_1 &= -\frac{n_1 r}{n_1 - n_0}; \end{aligned}\right\} \tag{10}$$

or, in terms of the power $\mathscr{P}$ of the surface,*

$$\frac{n_0}{f_0} = -\frac{n_1}{f_1} = \mathscr{P}. \tag{11}$$

Since f_0 and f_1 have different signs, the imaging is *concurrent* (cf. § 4.3.3). If the surface is convex towards the incident light $(r > 0)$ and if $n_0 < n_1$ then $f_0 > 0$, $f_1 < 0$, and the imaging is convergent. If $r > 0$ and $n_0 > n_1$ it is *divergent* (Fig. 4.12). When the surface is concave towards the incident light the situation is reversed.

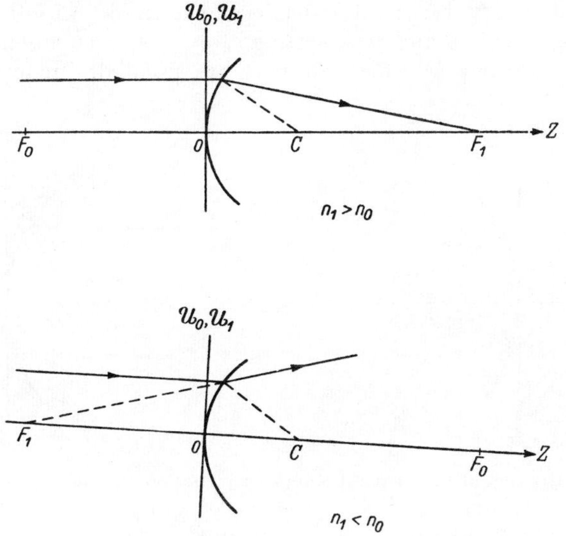

Fig. 4.12. Position of the cardinal points for refraction at a surface of revolution.

Using the expressions (10) for the focal lengths, (8) becomes

$$\frac{f_0}{z_0} + \frac{f_1}{z_1} = -1; \tag{12}$$

and the coefficients (2) may be written in the form

$$\mathscr{A} = \frac{f_0}{2n_0}, \qquad \mathscr{B} = -\frac{f_1}{2n_1}, \qquad \mathscr{C} = -\frac{f_0}{n_0} = \frac{f_1}{n_1}. \tag{13}$$

Let us introduce separate coordinate systems in the two spaces, with the origins at the foci, and with the axes parallel to those at O:

$$\begin{aligned} X_0 &= x_0, & X_1 &= x_1, \\ Y_0 &= y_0, & Y_1 &= y_1, \\ Z_0 &= z_0 + f_0, & Z_1 &= z_1 + f_1. \end{aligned}$$

---

* It will be seen later that the relation $n_0/f_0 = -n_1/f_1$ is not restricted to a single refracting surface, but holds in general for any centred system, the quantities with suffix zero referring to the object space, those with the suffix 1 to the image space. (11) may therefore be regarded as defining the *power* of a general centred system. The practical unit of power is a *dioptre*; it is the reciprocal of the focal length, when the focal length is expressed in meters. The power is positive when the system is convergent ($f_0 > 0$) and negative when it is divergent ($f_0 < 0$).

Equations (12) and (4), subject to (5), then reduce to the standard form § 4.3 (10):

$$\frac{Y_1}{Y_0} = \frac{f_0}{Z_0} = \frac{Z_1}{f_1}. \tag{14}$$

4.4.2 Reflecting surface of revolution

It is seen that with a notation strictly analogous to that used in the previous section, equations of the form (1a) and (1b) also hold when reflection takes place on a surface of revolution, the coefficients $\mathscr{A}$, $\mathscr{B}$ and $\mathscr{C}$ now being replaced by the corresponding coefficients $\mathscr{A}'$, $\mathscr{B}'$ and $\mathscr{C}'$ of § 4.1.5. It has been shown in § 4.1.5 that $\mathscr{A}'$, $\mathscr{B}'$ and $\mathscr{C}'$ can be obtained from $\mathscr{A}$, $\mathscr{B}$ and $\mathscr{C}$ on setting $n_0 = -n_1 = n$, n denoting the refractive index of the medium in which the rays are situated. Hence the appropriate

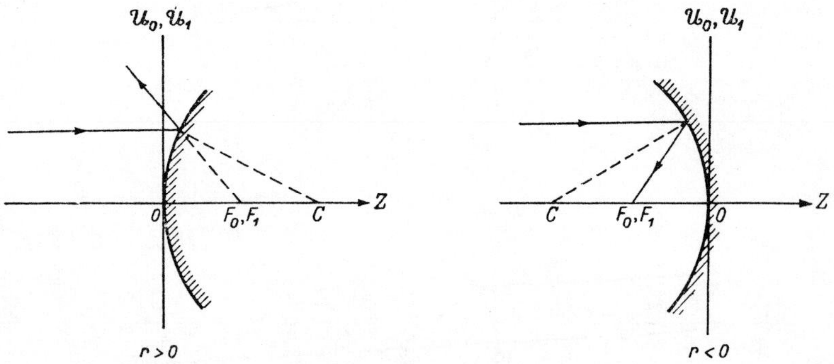

Fig. 4.13. Position of the cardinal points for reflection on a mirror of revolution.

formulae for reflection may be immediately written down by making this substitution in the preceding formulae. In particular, (7) gives

$$\frac{1}{z_0} - \frac{1}{r} = -\frac{1}{z_1} + \frac{1}{r}, \tag{15}$$

the expression on either side of (15) being *Abbe's (reflection) invariant*. (15) may also be written as

$$\frac{1}{z_0} + \frac{1}{z_1} = \frac{2}{r}. \tag{16}$$

The focal lengths f_0 and f_1 are now given by

$$f_0 = f_1 = -\frac{r}{2}, \tag{17}$$

and the power $\mathscr{P}$ is

$$\mathscr{P} = -\frac{2n}{r}. \tag{18}$$

Since $f_0 f_1 > 0$, the imaging is *contracurrent* (katoptric). When the surface is convex towards the incident light $(r > 0)$ $f_0 < 0$, and the imaging is then divergent; when it is concave towards the incident light, $(r < 0)$, $f_0 > 0$ and the imaging is convergent (Fig. 4.13).

4.4.3 The thick lens

Next we derive the Gaussian formulae relating to imaging by two surfaces which are rotationally symmetrical about the same axis.

Let n_0, n_1 and n_2 be the refractive indices of the three regions, taken in the order in which light passes through them, and let r_1 and r_2 be the radii of curvature of the surfaces at their axial points, measured positive when the surface is convex towards the incident light.

By (10), the focal lengths of the first surface are given by

$$f_0 = \frac{n_0 r_1}{n_1 - n_0}, \qquad f_0' = -\frac{n_1 r_1}{n_1 - n_0}, \tag{19}$$

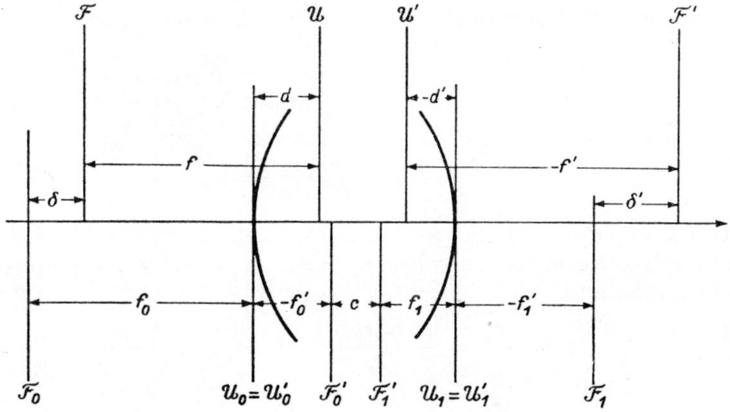

Fig. 4.14. The cardinal points of a combined system (thick lens).

and of the second surface by

$$f_1 = \frac{n_1 r_2}{n_2 - n_1}, \qquad f_1' = -\frac{n_2 r_2}{n_2 - n_1}. \tag{20}$$

According to § 4.3 (27) the focal lengths of the combination are

$$f = -\frac{f_0 f_1}{c}, \qquad f' = \frac{f_0' f_1'}{c}, \tag{21}$$

where c is the distance between the foci F_0' and F_1. Let t be the axial thickness of the lens, i.e. the distance between the poles of the two surfaces; then (see Fig. 4.14)

$$c = t + f_0' - f_1. \tag{22}$$

On substituting into (22) for f_0' and for f_1 we obtain

$$c = \frac{D}{(n_1 - n_0)(n_2 - n_1)}, \tag{23}$$

where

$$D = (n_1 - n_0)(n_2 - n_1)t - n_1[(n_2 - n_1)r_1 + (n_1 - n_0)r_2]. \tag{24}$$

The required expressions for the focal lengths of the combination are now obtained on substituting for f_0, f_1, f_0', f_1' and c into (21). This gives

$$f = -n_0 n_1 \frac{r_1 r_2}{D}; \qquad f' = n_1 n_2 \frac{r_1 r_2}{D}. \tag{25}$$

Since ff' is negative, the imaging is concurrent. The power $\mathscr{P}$ of the lens is

$$\mathscr{P} = \frac{n_0}{f} = -\frac{n_2}{f'} = -\frac{1}{n_1}\frac{D}{r_1 r_2}$$

$$= \mathscr{P}_1 + \mathscr{P}_2 - \frac{t}{n_1}\mathscr{P}_1\mathscr{P}_2, \tag{26}$$

where $\mathscr{P}_1$ and $\mathscr{P}_2$ are the powers of the two surfaces.

By § 4.3 (28), the distances $\delta = F_0F$ and $\delta' = F_1'F'$ are seen to be

$$\delta = -n_0 n_1 \frac{n_2 - n_1}{n_1 - n_0}\frac{r_1^2}{D}, \qquad \delta' = n_1 n_2 \frac{n_1 - n_0}{n_2 - n_1}\frac{r_2^2}{D}. \tag{27}$$

The distances d and d' of the principal planes $\mathscr{U}$ and $\mathscr{U}'$ from the poles of the surfaces are (see Fig. 4.14)

$$\left.\begin{aligned} d &= \delta + f - f_0 = -n_0(n_2 - n_1)\frac{r_1 t}{D}, \\ d' &= \delta' + f' - f_1' = n_2(n_1 - n_0)\frac{r_2 t}{D}. \end{aligned}\right\} \tag{28}$$

Of particular importance is the case where the media on both sides of the lens are of the same refractive index, i.e. when $n_2 = n_0$. If we set $n_1/n_0 = n_1/n_2 = n$, the formulae then reduce to

$$\left.\begin{aligned} f &= -f' = -\frac{n r_1 r_2}{\Delta}, \\ \delta &= n\frac{r_1^2}{\Delta}, \qquad \delta' = -n\frac{r_2^2}{\Delta}, \\ d &= (n-1)\frac{r_1 t}{\Delta}, \qquad d' = (n-1)\frac{r_2 t}{\Delta}, \end{aligned}\right\} \tag{29}$$

where

$$\Delta = (n-1)[n(r_1 - r_2) - (n-1)t]. \tag{30}$$

Referred to axes at the foci F and F', the abscissae of the unit points are $Z = f$ and $Z' = f'$, and the nodal points are given by $Z = -f'$, $Z = -f$. Since $f = -f'$ the unit points and the nodal points now coincide. The formula § 4.3 (16) which relates the distances ζ and ζ' of conjugate planes from the unit planes becomes

$$\frac{1}{\zeta} - \frac{1}{\zeta'} = -\frac{1}{f}. \tag{31}$$

The lens is *convergent* ($f > 0$) or *divergent* ($f < 0$) according as

$$f = -n\frac{r_1 r_2}{\Delta} \gtrless 0, \tag{32}$$

i.e. according as

$$\frac{1}{r_2} - \frac{1}{r_1} \lessgtr \frac{n-1}{n}\frac{t}{r_1 r_2}. \tag{33}$$

When $f = \infty$, we have the intermediate case of *telescopic* imaging. Then $\Delta = 0$, i.e.

$$r_1 - r_2 = \frac{n-1}{n}t. \tag{34}$$

The three cases may be illustrated by considering a double convex lens, $r_1 > 0$, $r_2 < 0$ (cf. Fig. 4.15). If, for simplicity, we assume that both radii are numerically equal, i.e. $r_1 = -r_2 = r$, the imaging will be convergent or divergent according as $t \lessgtr 2nr/(n-1)$ and telescopic when $t = 2nr/(n-1)$.

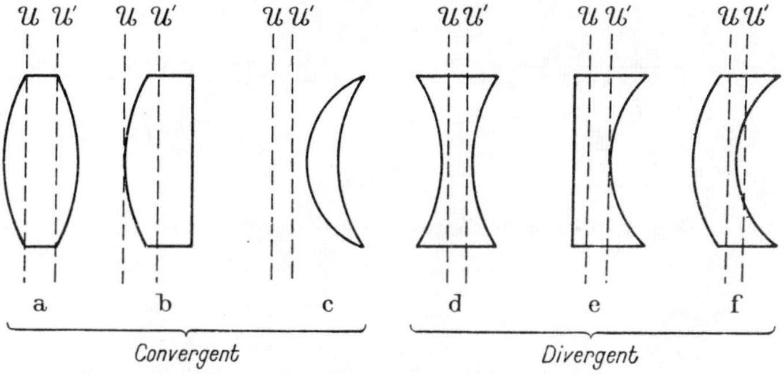

Fig. 4.15. Common types of lenses:
(a) Double-convex; (b) Plano-convex; (c) Convergent meniscus;
(d) Double-concave; (e) Plano-concave; (f) Divergent meniscus.
$\mathcal{U}$ and $\mathcal{U}'$ are the unit planes, light being assumed to be incident from the left.

4.4.4 The thin lens

The preceding formulae take a particularly simple form when the lens is so thin that the axial thickness t may be neglected. Then, according to (28), $d = d' = 0$, so that the unit planes pass through the axial point of the (infinitely thin) lens. Consequently, the rays which go through the centre of the lens will not suffer any deviation; this implies that *imaging by a thin lens is a central projection from the centre of the lens.*

From (26) it follows, on setting $t = 0$, that

$$\mathscr{P} = \mathscr{P}_1 + \mathscr{P}_2 = \frac{n_1 - n_0}{r_1} + \frac{n_2 - n_1}{r_2}, \tag{35}$$

i.e. the power of a thin lens is equal to the sum of the powers of the surfaces forming it.

If the media on the two sides of the lens are of equal refractive indices ($n_0 = n_2$), we have from (25)

$$\frac{1}{f} = -\frac{1}{f'} = (n-1)\left(\frac{1}{r_1} - \frac{1}{r_2}\right), \tag{36}$$

where as before, $n = n_1/n_0 = n_1/n_2$. Assuming that $n > 1$, as is usually the case, f is seen to be positive or negative according as the curvature $1/r_1$ of the first surface is greater or smaller than the curvature $1/r_2$ of the second surface (an appropriate sign being associated with the curvature). This implies that thin lenses whose thickness decreases from centre to the edge are convergent, and those whose thickness increases to the edge are divergent.

For later purposes we shall write down an expression for the focal lengths f and f' of a system consisting of two centred thin lenses, situated in air. According to § 4.3 (27) we have, since $f_0 = -f_0'$, $f_1 = -f_1'$,

$$\frac{1}{f} = -\frac{1}{f'} = -\frac{c}{f_0 f_1}, \tag{37}$$

c being the distance between the foci F_0' and F_1 (see Fig. 4.16). If l is the distance between the two lenses, then

$$l = f_0 + c + f_1. \tag{38}$$

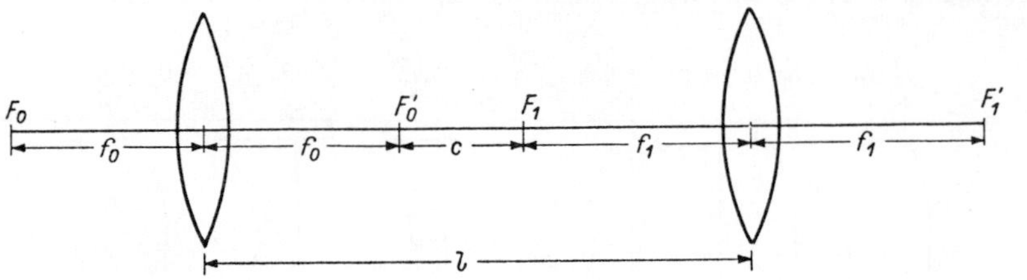

Fig. 4.16. System formed by two centred thin lenses.

Hence

$$\frac{1}{f} = -\frac{1}{f'} = \frac{1}{f_0} + \frac{1}{f_1} - \frac{l}{f_0 f_1}. \tag{39}$$

When the lenses are in contact ($l = 0$), (39) may also be written in the form, $\mathscr{P} = \mathscr{P}_1 + \mathscr{P}_2$, so that the power of the combination is then simply equal to sum of the powers of the two lenses.

4.4.5 The general centred system

Within the approximations of Gaussian theory, a refraction or a reflection at a surface of revolution was seen to give rise to a projective relationship between the object and the image spaces.* Since according to § 4.3 successive applications of projective transformations are equivalent to a single projective transformation, it follows that imaging by a centred system is, to the present degree of approximation, also a transformation of this type. The cardinal points of the equivalent transformation may be found by the application of the formulae of § 4.4.1, § 4.4.2 and § 4.3.4. We shall mainly confine our discussion to the derivation of an important invariant relation valid (within the present degree of accuracy) for any centred system.

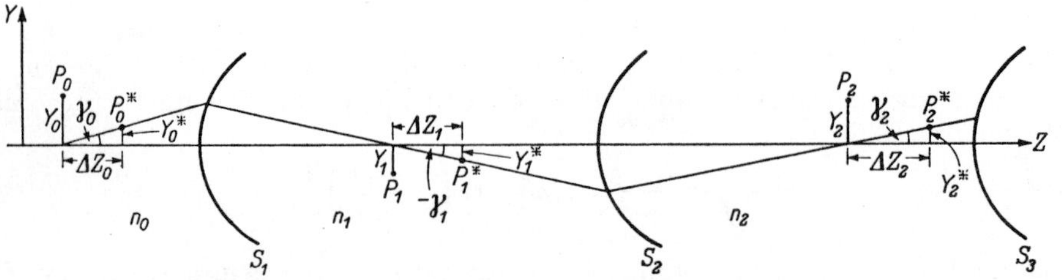

Fig. 4.17. Illustrating the SMITH–HELMHOLTZ formula.

Let $S_1, S_2, \ldots S_m$ be the successive surfaces of the system, $f_0, f_1, \ldots f_{m-1}$ the corresponding focal lengths, and $n_0, n_1, \ldots n_m$ the refractive indices of the successive spaces (Fig. 4.17). Further, let P_0, P_0^* be two points in the object space, situated in a

* As in the case of a single surface, the object and the image spaces are regarded as superposed on to each other and extending indefinitely in all directions. The part of the object space which lies in front of the first surface (counted in the order in which the light traverses the system) is said to form the real portion of the object space and the portion of the image space which follows the last surface is called the real portion of the image space. The remaining portions of the two spaces are said to be *virtual*. In a similar manner we may define the real and virtual parts of any of the intermediate spaces of the system.

meridional plane, and let $P_1, P_1^\star$, $P_2, P_2^\star$ . . ., be their images in the successive surfaces. Referred to axes at the foci of the first surface, the coordinates of $P_0, P_0^\star$ and $P_1, P_1^\star$ are related by eq. (14), viz.

$$\frac{Y_1}{Y_0} = \frac{f_0}{Z_0} = \frac{Z_1}{f_1}, \tag{40}$$

$$\frac{Y_1^\star}{Y_0^\star} = \frac{f_0}{Z_0^\star} = \frac{Z_1^\star}{f_1}. \tag{41}$$

Hence

$$\frac{Z_1^\star - Z_1}{f_1} = \frac{Y_1^\star}{Y_0^\star} - \frac{Y_1}{Y_0} = \frac{Y_0 Y_1^\star - Y_1 Y_0^\star}{Y_0 Y_0^\star}, \tag{42}$$

and

$$\frac{Z_0^\star - Z_0}{f_0} = \frac{Y_0^\star}{Y_1^\star} - \frac{Y_0}{Y_1} = \frac{Y_1 Y_0^\star - Y_0 Y_1^\star}{Y_1 Y_1^\star}. \tag{43}$$

Let

$$\left.\begin{aligned} Z_0^\star - Z_0 &= \Delta Z_0, \\ Z_1^\star - Z_1 &= \Delta Z_1. \end{aligned}\right\} \tag{44}$$

Then from (42) and (43),

$$\frac{f_0 Y_0 Y_0^\star}{\Delta Z_0} = -\frac{f_1 Y_1 Y_1^\star}{\Delta Z_1}. \tag{45}$$

Now by (10), $f_0/f_1 = -n_0/n_1$, so that this equation may be written as

$$\frac{n_0 Y_0 Y_0^\star}{\Delta Z_0} = \frac{n_1 Y_1 Y_1^\star}{\Delta Z_1}. \tag{46}$$

Similarly, for refraction at the second surface,

$$\frac{n_1 Y_1 Y_1^\star}{\Delta Z_1} = \frac{n_2 Y_2 Y_2^\star}{\Delta Z_2}, \tag{47}$$

and generally

$$\frac{n_{i-1} Y_{i-1} Y_{i-1}^\star}{\Delta Z_{i-1}} = \frac{n_i Y_i Y_i^\star}{\Delta Z_i} \qquad (1 \leqslant i \leqslant m). \tag{48}$$

Hence $n_i Y_i Y_i^\star/\Delta Z_i$ *is an invariant in the successive transformations.* This result plays an important part in the geometrical theory of image formation. If we set $Y_i^\star/\Delta Z_i = \tan \gamma_i$ (see Fig. 4.17), (48) becomes

$$n_{i-1} Y_{i-1} \tan \gamma_{i-1} = n_i Y_i \tan \gamma_i.$$

Since to the present degree of accuracy, $\tan \gamma$ and $\tan \gamma'$ may be replaced by γ and γ' respectively, we obtain the *Smith–Helmholtz formula**

$$n_{i-1} Y_{i-1} \gamma_{i-1} = n_i Y_i \gamma_i. \tag{49}$$

The quantity $n_i Y_i \gamma_i$ is known as the *Smith–Helmholtz invariant.*

From (48) and (49) a number of important conclusions may be drawn. As one is usually interested only in relations between quantities pertaining to the first and the last medium (object and image space), we shall drop the suffixes and denote quantities which refer to these two spaces by unprimed and primed symbols respectively.

* This formula is also associated with the names of LAGRANGE and CLAUSIUS. It was actually known in more restricted forms to earlier writers, e.g. HUYGENS and COTES. (Cf. Lord RAYLEIGH, *Phil. Mag.*, (5) **21** (1886), 466).

Let (Y,Z) and $(Y + \delta Y,\ Z + \delta Z)$ be two neighbouring points in the object space and (Y',Z') and $(Y' + \delta Y',\ Z' + \delta Z')$ the conjugate points. The SMITH–HELMHOLTZ formula gives, by successive application, the following relation:

$$\frac{nY(Y + \delta Y)}{\delta Z} = \frac{n'Y'(Y' + \delta Y')}{\delta Z'}. \tag{50}$$

In the limit as $\delta Y \to 0$ and $\delta Z \to 0$, this reduces to

$$\frac{dZ'}{dZ} = \frac{n'}{n}\frac{Y'^2}{Y^2}. \tag{51}$$

According to § 4.3 (11), we may write $Y'/Y = (dY'/dY)_{Z=\text{const.}}$, and (51) becomes

$$\frac{dZ'}{dZ} = \frac{n'}{n}\left(\frac{dY'}{dY}\right)^2_{Z=\text{const.}}, \tag{52}$$

known as *Maxwell's elongation formula*. It implies that the *longitudinal magnification is equal to the square of the lateral magnification multiplied by the ratio n'/n of the refractive indices.* Now in § 4.3 we derived an analogous formula [(13)] connecting the magnifications and the ratio of the focal lengths. On comparing these two formulae it is seen that

$$\frac{f'}{f} = -\frac{n'}{n}, \tag{53}$$

i.e. the *ratio of the focal lengths of the instrument is equal to the ratio n'/n of the refractive indices, taken with a negative sign.*

From the SMITH–HELMHOLTZ formula it also follows, that

$$\frac{dY'}{dY}\frac{\gamma'}{\gamma} = \frac{n}{n'}. \tag{54}$$

showing that *the product of the lateral and the angular magnification is independent of the choice of the conjugate planes.*

It has been assumed so far that the system consists of refracting surfaces only. If one of the surfaces (say the ith one) is a mirror, we obtain in place of (48),

$$\frac{Y_{i-1}Y^\star_{i-1}}{\Delta Z_{i-1}} = -\frac{Y_iY^\star_i}{\Delta Z_i},$$

the negative sign arising from the fact that for reflection $f_{i-1}/f_i = 1$, whereas for refraction $f_{i-1}/f_i = -n_{i-1}/n_i$. In consequence n' must be replaced by $-n'$ in the final formulae. More generally n' must be replaced by $-n'$ when the system contains an odd number of mirrors; when it contains an even number of mirrors, the final formulae remain unchanged.

4.5 STIGMATIC IMAGING WITH WIDE-ANGLE PENCILS

The laws of Gaussian Optics were derived under the assumption that the size of the object and the angles which the rays make with the axis are sufficiently small. Often one also has to consider systems where the object is of small linear dimensions, but where the inclination of the rays is appreciable. There are two simple criteria, known as the *sine condition** and the *Herschel condition*† relating to the stigmatic imaging in such instruments.

Let O_0 be an axial object point and P_0 any point in its neighbourhood, not necessarily on the axis. Assume that the system images these two points stigmatically, and let O_1 and P_1 be the stigmatic images.

Let (x_0, y_0, z_0) and (x_1, y_1, z_1) be the coordinates of P_0 and P_1 respectively, P_0 being referred to rectangular axes at O_0 and P_1 to parallel axes at O_1, the z-directions being taken along the axis of the system (Fig. 4.18). By the principle of equal optical path, the path lengths of all the rays joining P_0 and P_1 are the same. Hence, if V denotes the point characteristic of the medium,

$$V(x_0, y_0, z_0; x_1, y_1, z_1) - V(0, 0, 0; 0, 0, 0) = F(x_0, y_0, z_0; x_1, y_1, z_1), \qquad (1)$$

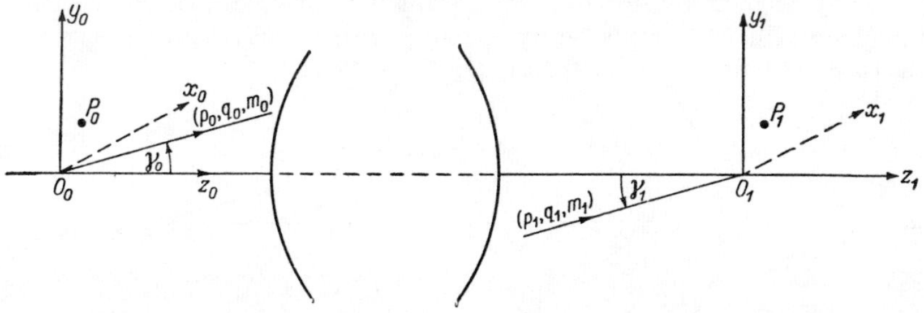

Fig. 4.18. Illustrating the sine condition and the HERSCHEL condition.

F being some function which is independent of the ray components. Using the basic relations § 4.1 (7) which express the ray component in terms of the point characteristic, we have from (1), if terms above the first power in distances are neglected,

$$(p_1^{(0)}x_1 + q_1^{(0)}y_1 + m_1^{(0)}z_1) - (p_0^{(0)}x_0 + q_0^{(0)}y_0 + m_0^{(0)}z_0) = F(x_0, y_0, z_0; x_1, y_1, z_1), \quad (2)$$

$(p_0^{(0)}, q_0^{(0)}, m_0^{(0)})$ and $(p_1^{(0)}, q_1^{(0)}, m_1^{(0)})$ being the ray components of any pair of corresponding rays through O_0 and O_1. It is to be noted that, although the points P_0 and P_1 are assumed to be in the neighbourhood of O_0 and O_1, no restriction as to the magnitude of the ray components is imposed.

Two cases are of particular interest, namely when (i) P_0 and P_1 lie in the planes $z_0 = 0$ and $z_1 = 0$ respectively, and (ii) when P_0 and P_1 are on the axis of symmetry. The two cases will be considered separately.

4.5.1 The sine condition

Without loss of generality we may again consider only points in a meridional plane $(x_0 = x_1 = 0)$. If P_0 lies in the plane $z_0 = 0$ and P_1 in the plane $z_1 = 0$, (2) becomes

$$q_1^{(0)}y_1 - q_0^{(0)}y_0 = F(0, y_0, 0; 0, y_1, 0). \qquad (3)$$

* The sine condition was first derived by R. CLAUSIUS (*Pogg. Ann.*, **121** (1864), 1) and by H. HELMHOLTZ (*Pogg. Ann. Jubelband* (1874), 557) from thermodynamical considerations. Its importance was, however, not recognized until it was rediscovered by E. ABBE (*Jenaisch. Ges. Med. Naturw.* 1879), 129, also *Carl. Repert. Phys.*, **16** (1880), 303).

The derivations given here are essentially due to C. HOCKIN, *J. Roy. Micro. Soc.* (2) **4** (1884), 337.

† J. F. W. HERSCHEL, *Phil. Trans. Roy. Soc.*, **111** (1821), 226.

This relation holds for each pair of conjugate rays. In particular it must, therefore, hold for the axial pair $p_0^{(0)} = q_0^{(0)} = 0,\ p_1^{(0)} = q_1^{(0)} = 0$. Hence

$$F(0, y_0, 0; 0, y_1, 0) = 0. \tag{4}$$

Relation (3) becomes

$$q_1^{(0)} y_1 = q_0^{(0)} y_0, \tag{5}$$

or, more explicitly,

$$n_1 y_1 \sin \gamma_1 = n_0 y_0 \sin \gamma_0, \tag{6}$$

γ_0 and γ_1 being the angles which the corresponding rays through O_0 and O_1 make with the z-axis, and n_0 and n_1 being the refractive indices of the object and image spaces. (6) is known as the *sine condition*, and is the required condition under which a small region of the object plane in the neighbourhood of the axis is imaged sharply by a pencil of any angular divergence. If the angular divergence is sufficiently small, $\sin \gamma_0$ and $\sin \gamma_1$ may be replaced by γ_0 and γ_1 respectively, and the sine condition reduces to the SMITH–HELMHOLTZ formula, § 4.4 (49).

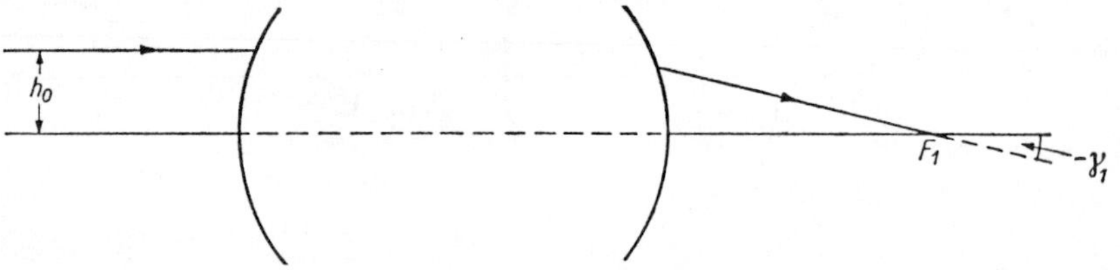

Fig. 4.19. The sine condition, when the object is at infinity.

If the object lies at infinity, the sine condition takes a different form. Assume first that the axial object point is at a great distance from the first surface. If Z_0 is the abscissa of this point referred now to axes at the first focus, and h_0 is the height above the axis at which a ray from the axial point meets the first surface, then $\sin \gamma_0 \sim -h_0/Z_0$; more precisely $Z_0 \sin \gamma_0/h_0 \to -1$ as $Z_0 \to -\infty$ whilst h_0 is kept fixed. Hence, if Z_0 is large enough, (6) may be written as

$$\frac{n_0}{n_1} h_0 = -\frac{y_1}{y_0} Z_0 \sin \gamma_1. \tag{7}$$

But by § 4.3 (10), $y_1 Z_0/y_0 = f_0$, and by § 4.4 (53) $n_0/n_1 = -f_0/f_1$, so that in the limit (6) reduces to (see Fig. 4.19)

$$\frac{h_0}{\sin \gamma_1} = f_1. \tag{8}$$

This implies that each ray which is incident in the direction parallel to the axis intersects its conjugate ray on a sphere of radius f_1, which is centred at the focus F_1.

Axial points which are stigmatic images of each other, and which, in addition, have the property that conjugate rays which pass through them satisfy the sine condition, are said to form an *aplanatic* pair. We have already encountered such point pairs when studying the refraction at a spherical surface (§ 4.2.3).

In the terminology of the theory of aberrations (cf. Chapter V), axial stigmatism implies the absence of all those terms in the expansion of the characteristic function which do not depend on the off-axis distance of the object, i.e. it implies the absence of spherical aberration of all orders. If, in addition, the sine condition is satisfied, then all terms in the characteristic function which depend on the first power of the off-axis distance must also vanish; these terms represent aberrations known as *circular coma*.

Since the sine condition gives information about the quality of the off-axis image in terms of the properties of axial pencils it is of great importance for optical design.

4.5.2 The Herschel condition

Next consider the case when P_0 and P_1 lie on the axis of the system ($x_0 = y_0 = 0$, $x_1 = y_1 = 0$). The condition (2) for sharp imaging reduces to

$$m_1{}^{(0)}z_1 - m_0{}^{(0)}z_0 = F(0, 0, z_0; 0, 0, z_1), \tag{9}$$

or, in terms of γ_0 and γ_1,

$$n_1 z_1 \cos \gamma_1 - n_0 z_0 \cos \gamma_0 = F(0, 0, z_0; 0, 0, z_1). \tag{10}$$

In particular for the axial ray this gives

$$F(0, 0, z_0; 0, 0, z_1) = n_1 z_1 - n_0 z_0. \tag{11}$$

Hence (10) may be written as

$$n_1 z_1 \sin^2(\gamma_1/2) = n_0 z_0 \sin^2(\gamma_0/2). \tag{12}$$

This is one form of the *Herschel condition*. Since the distances from the origin are assumed to be small, we have, by § 4.4 (52),

$$\frac{z_1}{z_0} = \frac{n_1}{n_0}\left(\frac{y_1}{y_0}\right)^2, \tag{13}$$

so that the HERSCHEL condition may also be written in the form

$$n_1 y_1 \sin(\gamma_1/2) = n_0 y_0 \sin(\gamma_0/2). \tag{14}$$

When the HERSCHEL condition is satisfied, an element of the axis in the immediate neighbourhood of O_0 will be imaged sharply by a pencil of rays, irrespective of the angular divergence of the pencil.

It is to be noted that the sine condition and the HERSCHEL condition cannot hold simultaneously unless $\gamma_1 = \gamma_0$. Then $y_1/y_0 = z_1/z_0 = n_0/n_1$, i.e. the longitudinal and lateral magnifications must then be equal to the ratio of the refractive indices of the object and image space.

4.6 ASTIGMATIC PENCILS OF RAYS

Rectilinear rays which have a point in common are said to form a *homocentric pencil*. The associated wave-fronts are then spherical, centred on their common point of intersection. It was with such pencils that we were concerned in the preceding sections.

In general, the homocentricity of a pencil is destroyed on refraction or reflection. It will therefore be useful to study the properties of more general pencils of rectilinear rays.

4.6.1 Focal properties of a thin pencil

Let S denote one of the orthogonal trajectories (wave-fronts) of a pencil of rectilinear rays, and let P be any point on it (Fig. 4.20). We take a plane through the ray at P and denote by C the curve in which this plane intersects S. Since the rays of the pencil are all normal to S, the centre of the circle of curvature at P will be on the ray through P.

If now the plane is gradually rotated around the ray, the curve C and consequently the radius of curvature will change continuously. When the plane has undergone a rotation of 180°, the radius of curvature will have passed through its maximum and minimum values. It can be shown by elementary geometry* that the two planes which contain the shortest and the longest radius of curvature are perpendicular to each other. These two planes are known as the *principal planes*† at P and the corresponding radii are called *principal radii of curvature.* The curves on S which have the property that at each point they are tangential to the principal planes form two mutually orthogonal families of curves, called the *lines of curvature*.

In general two normals at adjacent points of a surface do not intersect, to first order. But if they are drawn from adjacent points on a line of curvature they will intersect to this order, and their point of intersection is a focus of the congruence formed by

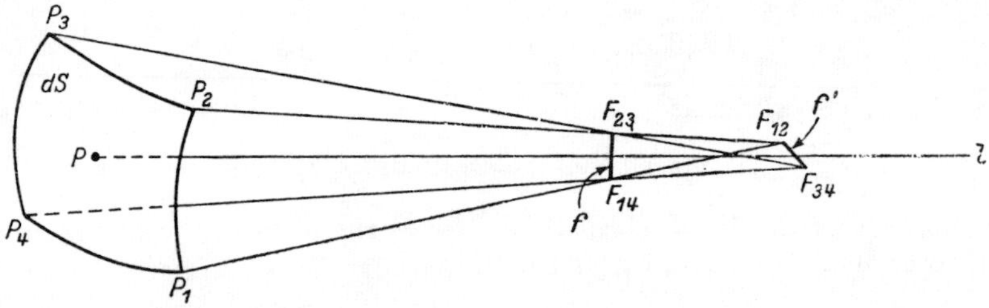

Fig. 4.20. A thin pencil of rays.

the normals (rays). Hence, in agreement with the general conclusions of § 3.2.3 there are two foci on each normal, these being the two principal centres of curvature. The caustic surface of a pencil of rectilinear rays has therefore, in general, two branches and is the *evolute* of the wave-fronts; conversely the wave-fronts are the *involutes* of the caustic surface. If the wave-fronts are surfaces of revolution, one branch of the caustic surface degenerates into a segment of the axis of revolution; the other is a surface of revolution whose meridional section is the evolute of the meridional section of the wave-front.

Let us consider a thin pencil consisting of all rays which intersect an element dS of the wave-front. It will be convenient to take as boundary of dS two pairs of lines of curvature, which may be assumed to be arcs of circles. Two of them (P_1P_2 and P_4P_3) may be taken to be vertical, and the other two (P_1P_4 and P_2P_3) to be horizontal (Fig. 4.20).

To the first order in small quantities, all the rays which pass through the arc P_1P_2 will intersect in a focus F_{12}, and those through P_3P_4 will intersect at F_{34}. The line f' joining F_{12} and F_{34} is known as a *focal line* of the pencil and is seen to be horizontal. Similarly the rays through P_1P_4 and P_2P_3 will give rise to a vertical focal line f.

If the lines of curvature are drawn through any point on dS, the corresponding two foci will lie on the two focal lines, and conversely. Hence *an approximate model of a thin pencil of rays is obtained by joining all pairs of points on two mutually orthogonal elements of lines.*

* See, for example, C. E. Weatherburn, *Differential Geometry of Three Dimensions* (Cambridge University Press, 1927), p. 185.

† The terms principal planes and focal planes have now a different meaning than in connection with projective transformations discussed in § 4.3.

The ray l through the centre point P is called the *central* (or *principal*) *ray* of the pencil, and the distance between the focal lines, measured along this ray, is called the *astigmatic focal distance* of the pencil. The two planes specified by f and l, and by f' and l, are known as the *focal planes* of the pencil, and are mutually perpendicular. It is not, however, necessarily true (as is often incorrectly asserted in the literature) that the focal lines are perpendicular to the central ray. Consider for example a family of wave-fronts which have cylindrical symmetry about a common axis (Fig. 4.21). Let dS be a surface element (assumed not to contain an axial point) of one of the wave-fronts, and let ds be the curve of intersection of dS with a plane which contains the axis. Then, clearly, the focal line f at the centre of curvature K of ds is perpendicular to this plane. The other focal line, f', coincides with a portion of the axis, namely with that portion which is bounded by the normals at the end points of ds. In general this focal line is not perpendicular to l.

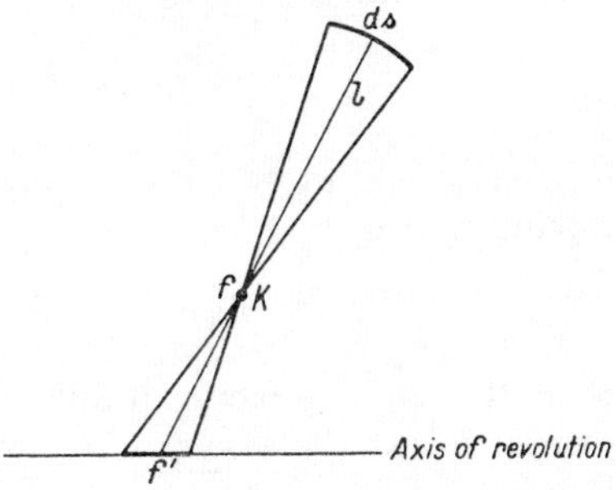

Fig. 4.21. Focal lines of a wave-front with cylindrical symmetry.

4.6.2 Refraction of a thin pencil

It was seen that a thin pencil of rays is completely specified by its central ray and its two focal lines. Suppose that a pencil specified in this way is incident on a refracting surface. It is of importance to determine the central ray and the focal lines of the refracted pencil. We shall consider the case of particular importance in practice, namely, when one of the principal planes of the incident pencil coincides with a

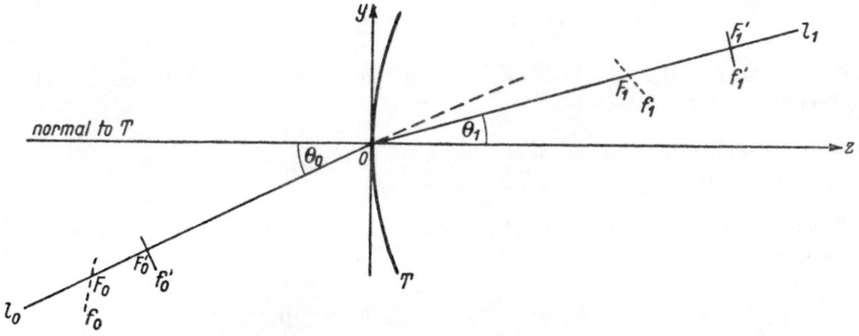

Fig. 4.22. Refraction of a thin astigmatic pencil.

principal plane of curvature of the surface at the point O at which the central ray meets it (Fig. 4.22).

Take Cartesian axes at O, with Oz along the normal to the surface T and with Ox and Oy in the direction of the principal lines of curvature of T.

Further, let θ_0 and θ_1 be the angles which the central rays l_0 and l_1 of the two pencils make with Oz. Let F_0 and F_0' be the foci of the incident pencil, situated at $z = \zeta_0$ and $z = \zeta_0'$ respectively. The focal line at F_0 will be assumed to be perpendicular to the plane of incidence; F_0 is then said to be a *primary focus*, and the corresponding focal line f_0 *the primary focal line*. The focus F_0' is called the *secondary focus*; the corresponding focal line f_0' (which lies in the plane of incidence) is known as the *secondary focal line*. In the case of a centred system, the primary and secondary focus of a pencil whose central ray lies in the meridional plane are known as the *tangential* and the *sagittal focus* respectively.

To find the focal lines of the refracted pencil it is necessary to write down first an expression for the angle characteristic of the refracting surface. If the radii of curvature of the surface in the principal directions Ox and Oy are r_x and r_y respectively, the equation of the surface is

$$z = \frac{x^2}{2r_x} + \frac{y^2}{2r_y} + \ldots \tag{1}$$

Now according to § 4.1 (34), the angle characteristic referred to a set of axes at O is (taking $a_0 = a_1 = 0$)

$$T = (p_0 - p_1)x + (q_0 - q_1)y + (m_0 - m_1)z. \tag{2}$$

From the law of refraction it follows, in a way similar to that described in § 4.1 (which corresponds to the case $r_x = r_y$), when terms of the lowest degree only are retained, that

$$\left.\begin{aligned} x &= -r_x \frac{p_0 - p_1}{m_0 - m_1}, \\ y &= -r_y \frac{q_0 - q_1}{m_0 - m_1}. \end{aligned}\right\} \tag{3}$$

Substitution into (1) gives

$$z = \frac{1}{2\mu^2}[r_x(p_0 - p_1)^2 + r_y(q_0 - q_1)^2], \tag{4}$$

with

$$\mu = m_1 - m_0 = n_1 \cos\theta_1 - n_0 \cos\theta_0. \tag{5}$$

Substitution from (3) and (4) then leads to the required expression for the angle characteristic:

$$T(p_0, q_0; p_1, q_1) = \frac{1}{2\mu}[r_x(p_0 - p_1)^2 + r_y(q_0 - q_1)^2] + \ldots \tag{6}$$

On using this expression in § 4.1 (29), the equations of the incident and refracted ray are obtained:

$$x_0 - \frac{p_0}{m_0} z_0 = \frac{1}{\mu} r_x(p_0 - p_1), \tag{7}$$

$$x_1 - \frac{p_1}{m_1} z_1 = \frac{1}{\mu} r_x(p_0 - p_1), \tag{8}$$

the corresponding equations involving the y coordinates being strictly analogous.

Consider now the changes in the various quantities as we pass from the central ray to a neighbouring ray. From (7) and (8)

$$\delta x_0 - \frac{p_0}{m_0}\delta z_0 = z_0 \delta\left(\frac{p_0}{m_0}\right) + r_x\left\{\delta\left(\frac{p_0}{\mu}\right) - \delta\left(\frac{p_1}{\mu}\right)\right\}, \tag{9}$$

$$\delta x_1 - \frac{p_1}{m_1}\delta z_1 = z_1 \delta\left(\frac{p_1}{m_1}\right) + r_x\left\{\delta\left(\frac{p_0}{\mu}\right) - \delta\left(\frac{p_1}{\mu}\right)\right\} \tag{10}$$

Now the components of the central ray of the incident pencil are

$$p_0 = 0, \qquad q_0 = n_0 \sin \theta_0, \qquad m_0 = n_0 \cos \theta_0, \tag{11}$$

so that

$$\begin{aligned} \delta \left(\frac{p_0}{m_0}\right) &= \frac{1}{m_0^{\,2}} (m_0 \delta p_0 - p_0 \delta m_0) \\ &= \frac{1}{n_0} \sec \theta_0 \delta p_0, \end{aligned} \tag{12}$$

and

$$\begin{aligned} \delta \left(\frac{q_0}{m_0}\right) &= \frac{1}{m_0^{\,2}} (m_0 \delta q_0 - q_0 \delta m_0) \\ &= \frac{1}{n_0} \sec^3 \theta_0 \delta q_0. \end{aligned} \tag{13}$$

Here use was made of the identity $m_0 \delta m_0 + p_0 \delta p_0 + q_0 \delta q_0 = 0$.

Equations (9) and (10) become

$$\delta x_0 = \frac{z_0}{n_0} \sec \theta_0 \delta p_0 - \frac{1}{\mu} r_x(\delta p_1 - \delta p_0), \tag{14}$$

$$\delta x_1 = \frac{z_1}{n_1} \sec \theta_1 \delta p_1 - \frac{1}{\mu} r_x(\delta p_1 - \delta p_0). \tag{15}$$

In deriving (15) use was also made of the fact that $p_1 = 0$; this result follows from the law of refraction and the assumption that $p_0 = 0$. In a similar way we find

$$\delta y_0 - (\tan \theta_0) \delta z_0 = \frac{z_0}{n_0} (\sec^3 \theta_0) \delta q_0 - r_y \left\{ \delta \left(\frac{q_1}{\mu}\right) - \delta \left(\frac{q_0}{\mu}\right) \right\}, \tag{16}$$

$$\delta y_1 - (\tan \theta_1) \delta z_1 = \frac{z_1}{n_1} (\sec^3 \theta_1) \delta q_1 - r_y \left\{ \delta \left(\frac{q_1}{\mu}\right) - \delta \left(\frac{q_0}{\mu}\right) \right\}. \tag{17}$$

Consider now those rays of the pencil which pass through the focus F_0. Then $z_0 = \zeta_0$, $\delta x_0 = \delta y_0 = \delta z_0 = 0$. Since all these rays also intersect the focal line f_0', $\delta p_0 = 0$. With this substitution (14) and (16) give

$$\delta p_1 = 0, \tag{14a}$$

and

$$\frac{\zeta_0}{n_0} \sec^3 \theta_0 \delta q_0 - \frac{1}{\mu} r_y(\delta q_1 - \delta q_0) = 0. \tag{16a}$$

(14a) shows that the corresponding refracted rays lie in the yz-plane. Now since all the rays from F_0 pass through the focus F_1 ($z_1 = \zeta_1$), (17) must hold with $z_1 = \zeta_1$, $\delta x_1 = \delta y_1 = \delta z_1 = 0$, whatever the value of δq_0, so that

$$\frac{\zeta_1}{n_1} \sec^3 \theta_1 \delta q_1 - \frac{1}{\mu} r_y(\delta q_1 - \delta q_0) = 0. \tag{17a}$$

Equations (16a) and (17a) can be satisfied simultaneously for an arbitrary value of δq_0 only if

$$\frac{n_0 \cos^3 \theta_0}{\zeta_0} - \frac{n_1 \cos^3 \theta_1}{\zeta_1} = \frac{n_0 \cos \theta_0 - n_1 \cos \theta_1}{r_y}. \tag{18}$$

This relation gives the position of the focus F_1 of the refracted rays. From (14a) it is seen that the focal line through F_1 is perpendicular to the yz-plane so that F_1 is a *primary focus.*

To find the position of the other focus, consider the rays which proceed from F_0'. Then $z_0 = \zeta_0'$, $\delta x_0 = \delta y_0 = \delta z_0 = 0$. Since all these rays intersect the focal line f_0, $\delta q_0 = \delta m_0 = 0$. Equations (14) and (16) now give

$$\frac{\zeta_0'}{n_0} \sec \theta_0 \delta p_0 - \frac{1}{\mu} r_x(\delta p_1 - \delta p_0) = 0, \tag{14b}$$

and

$$\delta q_1 = 0. \tag{16b}$$

(16b) shows that the refracted rays now lie in the xz-plane. All these rays will pass through the other focus $F_1'(z_1 = \zeta_1')$, so that (15) must be satisfied with $z_1 = \zeta_1'$, $\delta x_1 = \delta y_1 = \delta z_1 = 0$, whatever the value of δp_0. Hence,

$$\frac{\zeta_1'}{n_1} \sec \theta_1 \delta p_1 - \frac{1}{\mu} r_x(\delta p_1 - \delta p_0) = 0. \tag{15b}$$

Since (15b) and (14b) hold simultaneously for any arbitrary value of δp_0, it follows that

$$\frac{n_0 \cos \theta_0}{\zeta_0'} - \frac{n_1 \cos \theta_1}{\zeta_1'} = \frac{n_0 \cos \theta_0 - n_1 \cos \theta_1}{r_x}. \tag{19}$$

This relation gives the position of the *secondary focus* F_1'.

It is often convenient to specify the position of the foci by means of their distances from O rather than by means of their z coordinates. If $OF_0 = d_0^{(t)}$, $OF_0' = d_0^{(s)}$, $OF_1 = d_1^{(t)}$, $OF_1' = d_1^{(s)}$ (in Fig. 4.22 $d_0^{(t)} < 0$, $d_0^{(s)} < 0$, $d_1^{(t)} > 0$, $d_1^{(s)} > 0$), then

$$\left.\begin{aligned} \zeta_0 &= d_0^{(t)} \cos \theta_0, & \zeta_1 &= d_1^{(t)} \cos \theta_1, \\ \zeta_0' &= d_0^{(s)} \cos \theta_0, & \zeta_1' &= d_1^{(s)} \cos \theta_1, \end{aligned}\right\} \tag{20}$$

and the two relations (18) and (19) become

$$\frac{n_0 \cos^2 \theta_0}{d_0^{(t)}} - \frac{n_1 \cos^2 \theta_1}{d_1^{(t)}} = \frac{n_0 \cos \theta_0 - n_1 \cos \theta_1}{r_y}, \tag{21}$$

and

$$\frac{n_0}{d_0^{(s)}} - \frac{n_1}{d_1^{(s)}} = \frac{n_0 \cos \theta_0 - n_1 \cos \theta_1}{r_x}. \tag{22}$$

The corresponding relations for reflection may be obtained by setting $n_1 = -n_0$.

4.7 CHROMATIC ABERRATION. DISPERSION BY A PRISM

In Chapter II it was shown that the refractive index is not a material constant but depends on colour, i.e. on the wavelength of light. We shall now discuss some elementary consequences of this result in relation to the performance of lenses and prisms.

4.7.1 Chromatic aberration

If a ray of polychromatic light is incident upon a refracting surface, it is split into a set of rays, each of which is associated with a different wavelength. In traversing an optical system, light of different wavelengths will therefore, after the first refraction, follow slightly different paths. In consequence, the image will not be sharp and the system is said to suffer from *chromatic aberration.*

We shall again confine our attention to points and rays in the immediate neighbourhood of the axis, i.e. it will be assumed that the imaging in each wavelength obeys the laws of Gaussian optics. The chromatic aberration is then said to be of the first order, or primary. If Q_α and Q_β are the images of a point P in two different wavelengths (Fig. 4.23), the projections of $Q_\alpha Q_\beta$ in the directions parallel and perpendicular to the axis are known as *longitudinal* and *lateral* chromatic aberration respectively.

Consider the change δf in the focal length of a thin lens, due to a change δn in the refractive index. According to § 4.4 (36) the quantity $(n - 1)f$ will, for a given lens, be independent of the wavelength. Hence

$$\frac{\delta f}{f} + \frac{\delta n}{n-1} = 0. \tag{1}$$

The quantity

$$\Delta = \frac{n_F - n_C}{n_D - 1}, \tag{2}$$

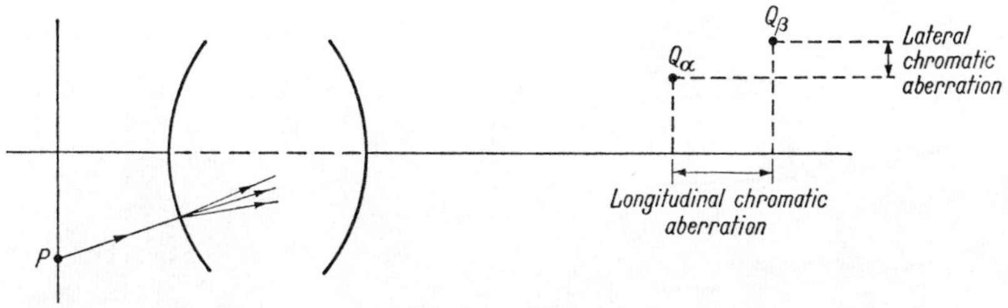

Fig. 4.23. The longitudinal and lateral chromatic aberration.

where n_F, n_D and n_C are the refractive indices for the Fraunhofer F, D and C lines ($\lambda = 4861$ Å, 5893 Å, 6563 Å respectively) is a rough measure of the dispersive properties of the glass, and is called the *dispersive power*. From (1) it is seen that it is approximately equal to the distance between the red and blue image divided by the focal length of the lens, when the object is at infinity. The variation with wavelength, of the refractive index of the usual types of glass employed in optical systems is shown in Fig. 4.24. The corresponding values of Δ lie between about 1/60 and 1/30.

To obtain an image of good quality, the monochromatic as well as the chromatic aberrations must be small. Usually a compromise has to be made, since in general it is impossible to eliminate all the aberrations simultaneously. Often it is sufficient to eliminate the chromatic aberration for two selected wavelengths only. The choice of these wavelengths will naturally depend on the purpose for which the system is designed; for example, since the ordinary photographic plate is more sensitive to the blue region than is the human eye, photographic objectives are usually "achromatized" for colours nearer to the blue end of the spectrum than is the case in visual instruments. Achromatization with respect to two wavelengths does, of course, not secure a complete removal of the colour error. The remaining chromatic aberration is known as *the secondary spectrum*.

Let us now examine under what conditions two thin lenses will form an achromatic combination with respect to their focal lengths. According to § 4.4 (39) the reciprocal of the focal length of a combination of two thin lenses separated by a distance l is given by

$$\frac{1}{f} = \frac{1}{f_1} + \frac{1}{f_2} - \frac{l}{f_1 f_2}. \tag{3}$$

It is seen that $\delta f = 0$ when

$$\frac{\delta f_1}{f_1{}^2} + \frac{\delta f_2}{f_2{}^2} - \frac{l}{f_1 f_2}\left[\frac{\delta f_1}{f_1} + \frac{\delta f_2}{f_2}\right] = 0. \tag{4}$$

If the achromatization is made for the C and F lines, we have, using (1) and (2),

$$l = \frac{\Delta_1 f_2 + \Delta_2 f_1}{\Delta_1 + \Delta_2}, \tag{5}$$

where Δ_1 and Δ_2 are the dispersive powers of the two lenses.

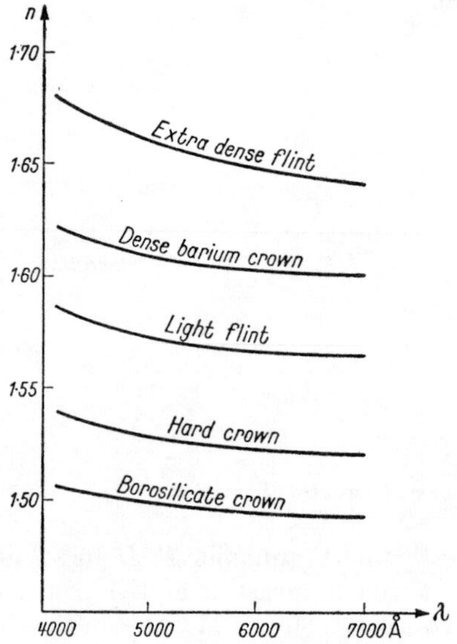

Fig. 4.24. Typical dispersion curves of various types of glass.

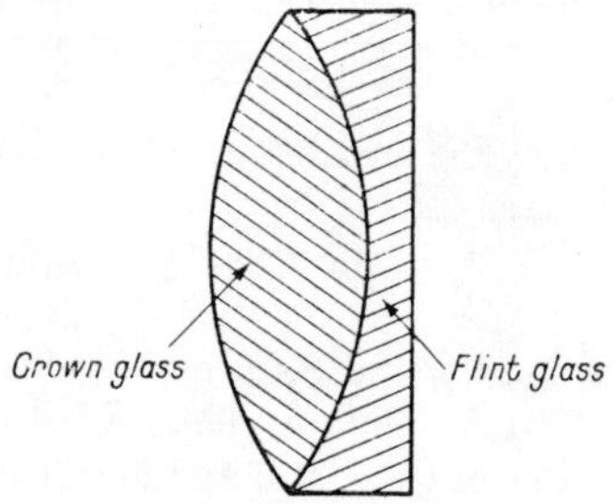

Fig. 4.25. An achromatic doublet.

One method of reducing the chromatic aberration is to employ two thin lenses in contact (Fig. 4.25), one made of crown glass, and the other of flint glass. In this case, since $l = 0$, we have, according to (5),

$$\frac{\Delta_1}{f_1} + \frac{\Delta_2}{f_2} = 0, \tag{6}$$

or, using (3),

$$\frac{1}{f_1} = \frac{1}{f}\frac{\Delta_2}{\Delta_2 - \Delta_1}, \qquad \frac{1}{f_2} = -\frac{1}{f}\frac{\Delta_1}{\Delta_2 - \Delta_1}. \tag{7}$$

Now for given glass, and with a fixed value of the focal length f, (7) specifies f_1 and f_2 uniquely. But f_1 and f_2 depend on three radii of curvature; hence one of the radii may be chosen arbitrarily. This degree of freedom is sometimes used to make the spherical aberration as small as possible.

Another method of obtaining an achromatic system is to employ two thin lenses made of the same glass ($\Delta_1 = \Delta_2$) and separated by a distance equal to half of the sum of their focal lengths

$$l = \tfrac{1}{2}(f_1 + f_2). \tag{8}$$

That such a combination is achromatic follows immediately from (5).

An instrument consisting of several components cannot, in general, be made achromatic with respect to both position and magnification, unless each component is itself achromatic in this sense. We shall prove this for the case of two centred thin lenses separated by a distance l. Since according to § 4.4.4 the imaging by a thin lens is a central projection from the centre of the lens, we have (see Fig. 4.26)

$$\frac{Y_1'}{Y_1} = -\frac{\zeta_1'}{\zeta_1}, \qquad \frac{Y_2'}{Y_2} = -\frac{\zeta_2'}{\zeta_2}. \tag{9}$$

Since $Y_2 = Y_1'$, the magnification is

$$\frac{Y_2'}{Y_1} = \frac{\zeta_1'}{\zeta_1} \cdot \frac{\zeta_2'}{\zeta_2}. \tag{10}$$

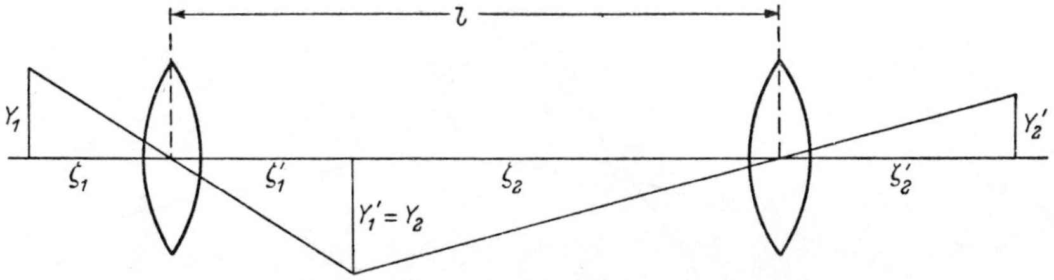

Fig. 4.26. Achromatization of two thin lenses.

If now the wavelength is altered, ζ_1 will remain unchanged, and if the position of the image is assumed to be achromatized, ζ_2' will also remain unchanged. Hence the condition for achromatization of the magnification may be expressed by the formula

$$\delta\left(\frac{\zeta_1'}{\zeta_2}\right) = \frac{1}{\zeta_2^{\,2}}(\zeta_2\delta\zeta_1' - \zeta_1'\delta\zeta_2) = 0. \tag{11}$$

Since $\zeta_1' + \zeta_2 = l$, $\delta\zeta_1' = -\delta\zeta_2$, and it is seen that (11) can only be satisfied if $\delta\zeta_1' = \delta\zeta_2 = 0$, i.e. if each of the lenses is achromatized.

So far we have only considered the primary chromatic aberration of a thin lens and of a combination of two such lenses. Expressions for the primary chromatic aberration of a general centred system will be derived in Chapter V.

4.7.2 Dispersion by a prism

We shall now briefly discuss the passage of light through a prism.

Let α be the angle between the two faces of the prism. It is assumed that the edge A in which the two faces meet is perpendicular to the plane which contains the incident, transmitted, and emergent rays (Fig. 4.27). To begin with the light will be assumed to be strictly monochromatic.

Let B_1 and B_2 be the points of intersection of the incident and the emergent ray with the two faces, ϕ_1 and ψ_1 the angles of incidence and refraction at B_1, and ψ_2 and ϕ_2 the inner and outer angles at B_2 (i.e. the angles which the ray B_1B_2 and the emergent ray make with the normal at B_2). Further let C be the point of intersection of the normals to the prism at B_1 and B_2, and D the point of intersection of the incident and the emergent rays, when these are prolonged sufficiently far.

If ε is the *angle of deviation*, i.e. the angle which the emergent ray makes with the incident ray, then

$$\phi_1 + \phi_2 = \varepsilon + \alpha, \tag{12}$$

$$\psi_1 + \psi_2 = \alpha. \tag{13}$$

Further, by the law of refraction,

$$\left.\begin{aligned} \sin\phi_1 &= n \sin\psi_1, \\ \sin\phi_2 &= n \sin\psi_2, \end{aligned}\right\} \tag{14}$$

where n is the refractive index of the glass with respect to the surrounding air. The deviation ε will have an extremum when

$$\frac{d\varepsilon}{d\phi_1} = 0. \tag{15}$$

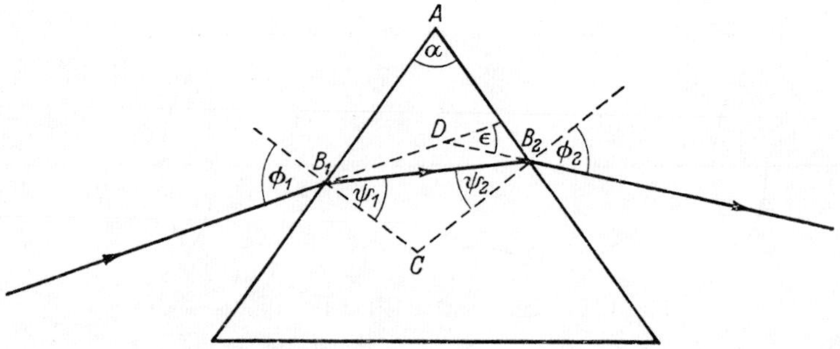

Fig. 4.27. Passage of a ray through a prism.

Using (12) this implies that

$$\left(\frac{d\phi_2}{d\phi_1}\right)_{\text{extr.}} = -1. \tag{16}$$

Now we have from (13) and (14),

$$\left.\begin{aligned} \frac{d\psi_1}{d\phi_1} &= -\frac{d\psi_2}{d\phi_1}, \\ \cos\phi_1 &= n\cos\psi_1 \frac{d\psi_1}{d\phi_1}, \\ \cos\phi_2 \frac{d\phi_2}{d\phi_1} &= n\cos\psi_2 \frac{d\psi_2}{d\phi_1}, \end{aligned}\right\} \tag{17}$$

and hence, on elimination

$$\frac{d\phi_2}{d\phi_1} = -\frac{\cos\phi_1\cos\psi_2}{\cos\psi_1\cos\phi_2}. \tag{18}$$

From (16) and (18) it follows, that for an extremum,

$$\frac{\cos\phi_1\cos\psi_2}{\cos\psi_1\cos\phi_2} = 1, \tag{19}$$

whence, on squaring and using (14)

$$\frac{1-\sin^2\phi_1}{n^2-\sin^2\phi_1} = \frac{1-\sin^2\phi_2}{n^2-\sin^2\phi_2}. \tag{20}$$

This equation is satisfied by

then
$$\left.\begin{aligned}\phi_1 &= \phi_2;\\ \psi_1 &= \psi_2.\end{aligned}\right\} \tag{21}$$

To determine the nature of the extremum we must evaluate $d^2\varepsilon/d\phi_1^2$. From (12) and (18),

$$\begin{aligned}\frac{d^2\varepsilon}{d\phi_1^2} = \frac{d^2\phi_2}{d\phi_1^2} &= \frac{d\phi_2}{d\phi_1}\frac{d}{d\phi_1}\left[\log\left(-\frac{d\phi_2}{d\phi_1}\right)\right]\\ &= \frac{d\phi_2}{d\phi_1}\left[-\tan\phi_1 - \tan\psi_2\frac{d\psi_2}{d\phi_1} + \tan\psi_1\frac{d\psi_1}{d\phi_1} + \tan\phi_2\frac{d\phi_2}{d\phi_1}\right].\end{aligned} \tag{22}$$

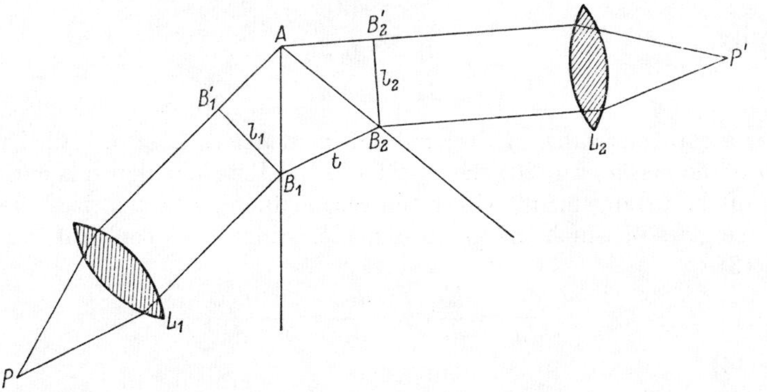

Fig. 4.28. Dispersion by a prism.

When $\phi_1 = \phi_2$, $\psi_1 = \psi_2$, this becomes with the help of (16), (17) and (14),

$$\left(\frac{d^2\varepsilon}{d\phi_1^2}\right)_{\text{extr.}} = 2\tan\phi_1 - 2\tan\psi_1\frac{\cos\phi_1}{n\cos\psi_1} = 2\tan\phi_1\left(1 - \frac{\tan^2\psi_1}{\tan^2\phi_1}\right). \tag{23}$$

Since $n > 1$, $\phi_1 > \psi_1$; also since $0 < \phi_1 < \pi/2$, $\tan\phi_1 > 0$. Hence $(d^2\varepsilon/d\phi_1^2) > 0$, so that *the deviation is a minimum.* According to (21) it takes place when the passage of the rays through the prism is *symmetrical.* The minimal value of the deviation then is

$$\varepsilon_{\min} = 2\phi_1 - \alpha. \tag{24}$$

In terms of $\varepsilon_{\min}$ and α, the angle of incidence and the angle of refraction at the first face of the prism are

$$\phi_1 = \tfrac{1}{2}(\varepsilon_{\min} + \alpha), \qquad \psi_1 = \tfrac{1}{2}\alpha, \tag{25}$$

so that

$$n = \frac{\sin\phi_1}{\sin\psi_1} = \frac{\sin\left[\frac{1}{2}(\varepsilon_{\min} + \alpha)\right]}{\sin\left(\frac{1}{2}\alpha\right)}. \tag{26}$$

This formula is often used in determinations of the refractive index of glass. One measures $\varepsilon_{\min}$ and α with the help of a spectrometer and evaluates n from (26).

Instead of a single ray, let us now consider the passage of a pencil of parallel rays through the prism, for example from a point source P placed in the focal plane of a lens L_1 (Fig. 4.28), the light still being assumed to be monochromatic. Let B_1' and B_2' be the feet of the perpendiculars dropped from B_1 and B_2 on to the rays which pass through the edge A. Then B_1B_1' and B_2B_2' are the lines of intersection of two wave-fronts with the plane of incidence (plane of the figure). These two lines are inclined to each other at the angle of deviation ε. We set

$$B_1B_1' = l_1, \quad B_2B_2' = l_2', \quad B_1B_2 = t. \tag{27}$$

Consider now a parallel beam of polychromatic instead of monochromatic light. If the lens L_1 is corrected for chromatic aberration, B_1B_1' will still be in the wave front of the incident pencil. On the other hand the line B_2B_2' will no longer be unique, but will depend on the wavelength λ. For the refractive index of the prism is a function of the wavelength,

$$n = n(\lambda), \tag{28}$$

and consequently the deviation ε also depends on λ:

$$\varepsilon = \varepsilon(\lambda). \tag{29}$$

The quantity

$$\frac{d\varepsilon}{d\lambda} = \frac{d\varepsilon}{dn}\frac{dn}{d\lambda} \tag{30}$$

formed for a constant value of the angle of incidence ϕ_1 is often called the *angular dispersion of the prism*. In (30) the first factor on the right depends entirely on the geometry of the arrangement, whilst the second factor characterizes the dispersive power of the glass of which the prism is made. Since ϕ_1 = constant, we have from (12) and (13)

$$\frac{d\varepsilon}{dn} = \frac{d\phi_2}{dn}, \qquad \frac{d\psi_1}{dn} = -\frac{d\psi_2}{dn}, \tag{31}$$

and from (14)

$$\sin\psi_1 + n\cos\psi_1 \frac{d\psi_1}{dn} = 0,$$

$$\cos\phi_2 \frac{d\phi_2}{dn} = \sin\psi_2 + n\cos\psi_2 \frac{d\psi_2}{dn}, \tag{32}$$

whence on elimination

$$\frac{d\varepsilon}{dn} = \frac{\sin(\psi_1 + \psi_2)}{\cos\phi_2\cos\psi_1} = \frac{\sin\alpha}{\cos\phi_2\cos\psi_1}. \tag{33}$$

From the triangle AB_1B_2, we have, by the sine theorem,

$$AB_2 = \frac{\cos\psi_1}{\sin\alpha}t, \tag{34}$$

and from the triangle AB_2B_2',

$$AB_2 = l_2\sec\phi_2. \tag{35}$$

Using (34) and (35), (33) gives

$$\frac{d\varepsilon}{d\lambda} = \frac{t}{l_2}\frac{dn}{d\lambda}. \tag{36}$$

In the position of minimum deviation, we have by symmetry $l_1 = l_2$. If, moreover, the lenses are so large that the pencil completely fills the prism, then t will be equal to the length b of the base of the prism. Equation (36) then gives for the *angular dispersion* $\delta\varepsilon$, i.e. for the angle by which the emergent wave-front is rotated when the wavelength is changed from λ to $\lambda + \delta\lambda$, the following expression:

$$\delta\varepsilon = \frac{b}{l_1}\frac{dn}{d\lambda}\delta\lambda. \tag{37}$$

4.8 PHOTOMETRY AND APERTURES

The branch of optics concerned with the measurement of light is called *photometry*. Strictly, photometry is not part of geometrical optics, but it seems appropriate to include a short account of it in the present chapter, since in many practical applications the approximate geometrical picture of an optical field forms an adequate basis for photometric investigations. We shall, therefore, take over the geometrical model according to which light is regarded as the flow of luminous energy along the geometrical rays, subject to the geometrical law of conservation of energy. This states (cf. § 3.1, eq. (31)), that the energy which is transmitted in unit time through any cross-section of a tube of rays is constant.

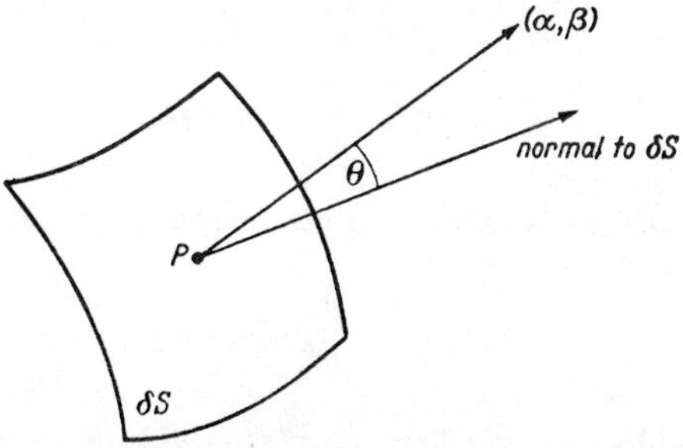

Fig. 4.29. Energy transmission from a surface element.

4.8.1 Basic concepts of photometry*

In photometry we are concerned essentially with the light energy emerging from a portion of a surface S. This surface may be fictitious, or it may coincide with the actual radiating surface of a source, or with an illuminated surface of a solid. If the latter is opaque, it is the reflected light which is considered; if it is transparent or semi-transparent (in which case the light is partly absorbed or scattered), it is the transmitted light which is usually measured.

Let $P(\xi, \eta)$ be a typical point on S referred to any convenient set of curvilinear coordinates on the surface. The (time averaged) amount of energy which emerges per unit time from the element δS of the surface at P and which falls within an element $\delta\Omega$ of the solid angle around a direction specified by the polar angles (α,β), may be expressed in the form

$$\delta F = B \cos \theta \delta S \delta \Omega. \tag{1}$$

Here θ is the angle which the direction (α, β) makes with the normal to the surface element (see Fig. 4.29), and B is a factor which in general depends on (ξ, η) and (α, β),

$$B = B(\xi, \eta; \alpha, \beta). \tag{2}$$

The factor $\cos \theta$ is introduced in (1) since it is the projection of δS on to a plane normal to the direction (α, β), rather than δS itself, which is the physically significant quantity. B is called the *photometric brightness* at the point (ξ, η) in the direction (α, β). It must be distinguished from the visual sensation of brightness, from which

* Only the fundamental notions of photometry will be discussed. For further information and for the description of instruments used for measuring light see, for example, J. W. T. WALSH, *Photometry* (London, Constable, 2nd ed., 1953) or *Measurement of Radiant Energy* (McGraw-Hill, New York and London, 1937), edited by W. E. FORSYTHE.

it will in general differ, because the eye is not equally sensitive to all colours;* this point will be discussed more fully later.

δF is usually decomposed in two different ways, to show the explicit dependence on $\delta\Omega$ and δS:

$$\delta F = \delta I \delta\Omega = \delta E \delta S. \tag{3}$$

Comparison of (1) and (3) gives

$$\delta I = \frac{\delta F}{\delta\Omega} = B \cos\theta \delta S, \tag{4}$$

$$\delta E = \frac{\delta F}{\delta S} = B \cos\theta \delta\Omega. \tag{5}$$

The integral

$$I(\alpha, \beta) = \int B \cos\theta dS \tag{6}$$

taken over a piece of surface is called the *photometric intensity*† *in the direction* (α,β), and the integral

$$E(\xi, \eta) = \int B \cos\theta d\Omega \tag{7}$$

taken throughout a solid angle is called the *photometric illumination at the point* (ξ, η).

The variation of B with direction will depend on the nature of the surface, especially on whether it is rough or smooth, whether it is self-luminous, or whether it transmits or reflects other light. Often it is permissible to assume that, to a good approximation, B is independent of the direction. The radiation is then said to be *isotropic*. If the radiation is isotropic and if the radiating surface is plane, (6) reduces to

$$I(\alpha,\beta) = I_0 \cos\theta, \tag{8}$$

where

$$I_0 = \int B\, dS.$$

The photometric intensity in any direction then varies as the cosine of the angle between that direction and the normal to the surface. (8) is known as *Lambert's (cosine) law*, and when satisfied one speaks of *diffuse emission* or *diffuse reflection*, according as to whether the surface is emitting or reflecting.

The measurement of the quantities F, B, I and E involves the determination of a time interval, an area, a direction, a solid angle and an energy. The averages involved are often small and consequently sensitive instruments have to be used. They are essentially of two kinds. First, those which react to the heat developed in an absorbing medium (e.g. bolometer, thermo-couple, etc.) and are mainly used in studying heat

* We use the adjective "photometric" when we wish to stress that a particular quantity is evaluated with regard to its true physical, rather than its visual, effect.

† In Chapter I the light intensity was defined as the time average of the amount of energy which crosses per second a unit area perpendicular to the direction of the flow. This quantity must not be confused with the photometric intensity as defined by (6). It is unfortunate that the same word is used to denote two different quantities. Except in the present section we shall always understand by "intensity" the quantity introduced in Chapter I. If the surface element δS at P is orthogonal to the Poynting vector, the intensity (in the sense of Chapter I) is equal to the photometric illumination δE at P.

radiation (infra-red); secondly, those which are based on the (surface) photo-electric effect, i.e. on the phenomenon of emission of electrons from a metal, caused by the incidence of light on the surface of the metal (in monochromatic light the number of emitted electrons, i.e. the current produced, is proportional to the energy of the incident light). Instruments of this kind are used, for instance, as exposure meters in photography.

In technical photometry, however, indirect methods are used. One defines and constructs a standard source of light, and expresses its photometric data in absolute energy units. Measurements are then made relative to this source, often using the eye as a null indicator, namely as an indicator of equal brightness. The comparison is based on a simple law which holds for the illumination due to a very small source* Q:

Let δS be a surface element at P and let $QP = r$. If θ is the angle which QP makes with the normal to δS (see Fig. 4.30), then the energy which the source sends through δS per unit time is $I\delta\Omega$, where I is the photometric intensity of the source in the

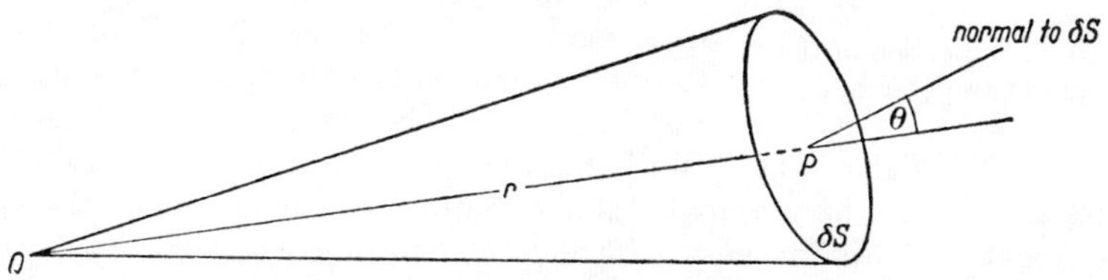

Fig. 4.30. Illumination from a point source.

direction QP, and $\delta\Omega$ is the solid angle which δS subtends at Q. Now by elementary geometry,

$$\cos\theta\delta S = r^2\delta\Omega. \tag{9}$$

Hence, using (3), we have

$$E = \frac{I\cos\theta}{r^2}. \tag{10}$$

(10) is the basic equation of practical photometry. It expresses the so-called *cosine law of illumination* (E proportional to $\cos\theta$) and the *inverse square law* (E inversely proportional to r^2), and enables a comparison of the intensities of sources with the help of simple geometry. If a surface element δS is illuminated by two point sources Q_1 and Q_2 of photometric intensities I_1 and I_2, and if the lines connecting δS with the sources make angles θ_1 and θ_2 with the normal to δS (Fig. 4.31) then, when the photometric illuminations are equal, we have

$$I_1 : I_2 = \frac{r_1^{\,2}}{\cos\theta_1} : \frac{r_2^{\,2}}{\cos\theta_2}. \tag{11}$$

With the help of (11), the strength of a given source may be determined by a comparison with a standard source.

---

* The photometric intensity I of a point source is defined by a procedure often used in connection with limiting concepts. Let us suppose that the area of the surface decreases towards zero, while at the same time B increases towards infinity, in such a way that the integral (6) remains finite. Then I is a function of α and β and of the position of the point source.

In calculating the photometric illumination, the finite extension of the source is usually neglected when its linear dimensions are less than about 1/20 of its distance from the illuminated surface.

The equality of the illumination produced by two sources can be detected either by physical means, or directly, by the eye. The comparison by the eye is relatively easy when the light from the two sources is monochromatic and of the same frequency, but in general one has to compare sources which emit light of different spectral compositions. This simple procedure can then no longer be used, since the eye is not equally sensitive to light of all wavelengths. Filters, each of which transmits light in a narrow range around a known wavelength of the spectrum, must be used instead; the determination of equality of illumination for different colours is then reduced to the determination of the relative energy values, taking into account the *relative visibility curve* of the eye. This is the curve obtained by plotting the reciprocal K_λ of that amount of flux which produces the same sensation of brightness, against the wavelength. The curve depends to some extent on the strength of the illumination. For bright light it has a maximum near 550 mμ. (See Fig. 4.32 and Table XIII.) With decreasing strength of illumination the curve retains its shape but the maximum shifts towards the blue end of the spectrum, being at about 507 mμ for very faint light. This phenomenon is known as the *Purkinje effect*.

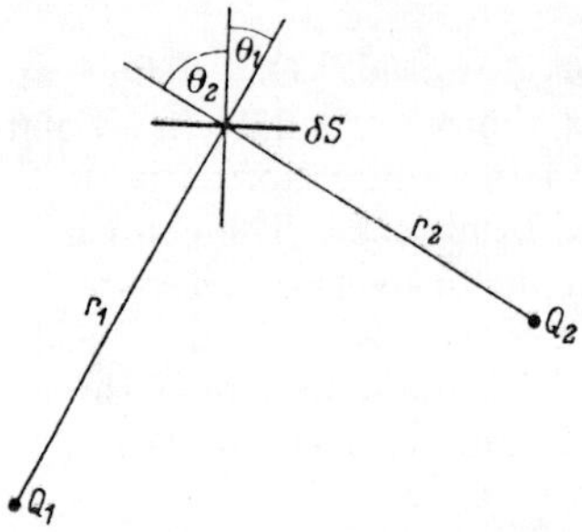

Fig. 4.31. Comparison of the intensities of two point sources.

If the flux of energy is evaluated with respect to the visual sensation which it

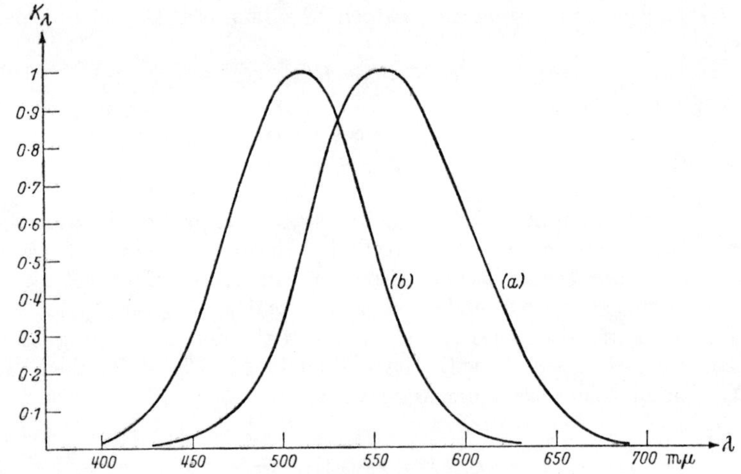

Fig. 4.32. The relative visibility curve K_λ of the average human eye: (a) for bright light and (b) for feeble light.

produces, rather than with regard to its true physical magnitude, one speaks of *the luminous energy* F':

$$F' = \frac{\int K_\lambda F_\lambda d\lambda}{\int K_\lambda d\lambda}, \tag{12}$$

$F_\lambda d\lambda$ being the amount of energy in the range $(\lambda, \lambda + d\lambda)$, K_λ the relative visibility and the integration being taken throughout the spectral range. The quantities B', I' and E', which bear the same relations to B, I and E as F' bears to F, are usually called the *luminance*, the *luminous intensity* (or *candle power*) and the *illumination* respectively. These concepts are extensively used in visual photometry.

There are practical units for each of the four quantities F', B', I' and E'. Since it is easier to maintain a standard of luminous intensity rather than of luminous flux, the unit of luminous intensity is usually considered to be the basic photometric unit, and those for F', B' and E' are expressed in terms of it. The adopted standard of luminous intensity was at one time the *international candle*, a standard preserved by

TABLE XIII

The Relative Visibility Factor K_λ of an average human eye (bright light)

λ (in mμ)	K_λ	λ (in mμ)	K_λ
400	0·0004	600	0·631
410	0·0012	610	0·503
420	0·0040	620	0·381
430	0·0116	630	0·265
440	0·023	640	0·175
450	0·038	650	0·107
460	0·060	660	0·061
470	0·091	670	0·032
480	0·139	680	0·017
490	0·208	690	0·0082
500	0·323	700	0·0041
510	0·503	710	0·0021
520	0·710	720	0·00105
530	0·862	730	0·00052
540	0·954	740	0·00025
550	0·995	750	0·00012
560	0·995	760	0·00006
570	0·952		
580	0·870		
590	0·757		

a number of carbon lamps kept at various national laboratories. In more recent times it has been replaced by a new standard called the *candela* (cd); this is defined as one-sixtieth of the luminous intensity per square centimetre of a black body radiator at the temperature of solidification of platinum (2042°K approx.). The value of the luminous intensity of light of a different spectral composition must be evaluated by the procedure already explained, taking into account the relative visibility curve.

The unit of luminous flux is called the *lumen*. It is the luminous flux emitted within a unit solid angle by a uniform point source of luminous intensity 1 candela.

The unit of illumination depends on the unit of length employed. The metric unit of illumination is the *lux* (lx), sometimes also called the *metre-candle*; it is the illumination of a surface area of one square metre receiving a luminous flux of one lumen. The British unit is the lumen per square foot, formerly called the *foot-candle* (f.c.).

The unit of luminance is the candela per square centimetre, termed the *stilb* (sb), and the candela per square metre termed the *nit*. In the British system, the units used are the candela per square inch or candela per square foot.

Other units are also used, but for their definitions and their relation to the units here discussed we must refer to books on photometry.

4.8.2 Stops and pupils*

The amount of light which reaches the image space of an optical system depends not only on the brightness of the object but also on the dimensions of the optical elements (lenses, mirrors) and of the stops. A stop (or diaphragm) is an opening, usually circular, in an opaque screen. The opaque parts of the screen prevent some of the severely aberrated rays from reaching the image. For the purposes of the present

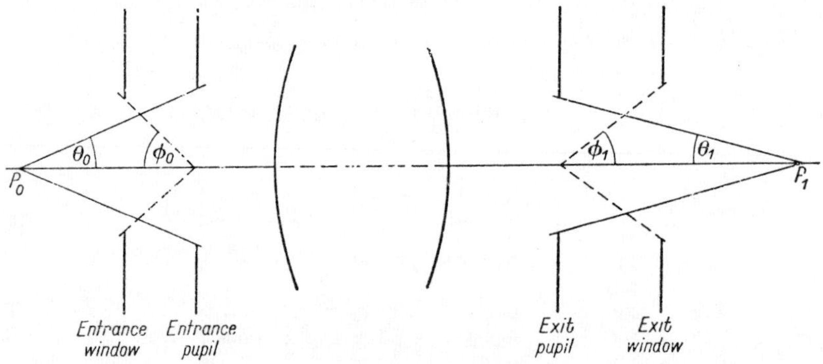

Fig. 4.33. Stops and pupils.

discussion it will be convenient to include the edges of the lenses and mirrors as well as diaphragms in the term "stop".

Consider all the rays from the axial object point P_0. The stop which determines the cross-section of the image-forming pencil is called the *aperture stop*. To determine its position, the Gaussian image of each stop must be found in the part of the system which precedes it; the image which subtends the smallest angle at P_0 is called the *entrance pupil*. The physical stop which gives rise to the entrance pupil is the aperture stop. (If it lies in front of the first surface it is identical with the entrance pupil.) The angle $2\theta_0$ which the diameter of the entrance pupil subtends at P_0 is called the *angular aperture on the object side*, or simply *angular aperture* (Fig. 4.33).

The image of the aperture stop formed by the part of the system which follows it (also the image of the entry pupil by the whole system) is known as the *exit pupil*; the angle $2\theta_1$ which its diameter subtends at the image P_1 may be called the *angular aperture on the image side* (sometimes also the *projection angle*).

In the pencil of rays which proceeds from each object point, there will be a ray which passes through the centre of the entrance pupil. This special ray is known as the *principal ray* (also the *chief* or the *reference ray*) of the pencil, and is of particular importance in the theory of aberrations. In the absence of aberrations, the principal ray will also pass through the centre of the aperture stop and through the centre of the exit pupil.

If the aperture stop is situated in the back focal plane of that part of the system which precedes it, then the entrance pupil will be at infinity and all the principal rays

* The theory of stops was formulated by E. Abbe, *Jena Z. Naturwiss.*, **6** (1871), 263, and was extended by M. von Rohr, *Zentr. Ztg. Opt. Mech.*, **41** (1920), 145, 159, 171.

in the object space will be parallel to the axis. Such a system is said to be *telecentric on the object side*. If the aperture stop is in the front focal plane of the part of the system which follows it, then the exit pupil will be at infinity and the principal rays in the image space will be parallel to the axis; the system is then said to be *telecentric on the image side*. Telecentric arrangements are useful in measurements of the size of the object.

If other parameters are kept fixed, the angular aperture is a measure of the amount of light which traverses the system. There are other quantities which are frequently used to specify the "light-gathering power" of an optical system, for example the *numerical aperture* (N.A.) of a microscope objective. This is defined as the sine of the angular semi-aperture in the object space multiplied by the refractive index of the object space,

$$N.A. = n \sin \theta_0. \tag{13}$$

When a system is designed to work with objects at a great distance, as in the case of telescopes or certain photographic lenses, a convenient measure of its light-gathering

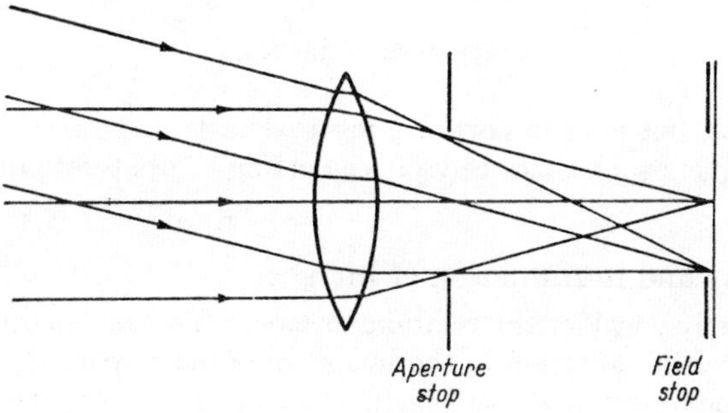

Fig. 4.34. Illustrating the distinction between the aperture stop and a field stop.

power is the so-called "*F number*" or "*nominal focal ratio*". It is the ratio of the focal length f of the system to the diameter d of the entrance pupil:

$$F = f/d. \tag{14}$$

Thus for a lens of 10 cm focal length which works with an aperture of 2 cm, $F = 5$. It is called an $f/5$ lens and is said to work at a "speed" of $f/5$.

Quantities such as the angular aperture, the numerical aperture, and the F number, which may be taken as the measure of the light-gathering power of the instrument, are often called *relative apertures*.

In addition to aperture stops, optical systems also possess *field stops*; they determine what proportion of the surface of an extended object is imaged by the instrument. The distinction between the two types of stops is illustrated in Fig. 4.34.

To determine the field stop, we again find first the image of each stop in the part of the system which precedes it. That image which subtends the smallest angle ($2\phi_0$ in Fig. 4.33) at the centre of the entrance pupil is called the *entrance window*, and the angle $2\phi_0$ is called the *field angle* or *the angular field of view*. The physical stop which corresponds to the entrance window is then the required field stop.

The image of the entrance window by the instrument (that is also the image of the field stop by the part of the system which follows it), is called the *exit window*. The angle ($2\phi_1$ in Fig. 4.33) which the diameter of the exit window subtends at the centre of the exit pupil is sometimes called *the image field angle*.

It may happen that, although the aperture stop is smaller than the lenses, some of the rays miss part of a lens entirely and parts of a lens may receive no light at all from certain regions of the object. This effect is known as *vignetting*, and is illustrated in Fig. 4.35. It is seldom encountered in systems like telescopes which have a relatively

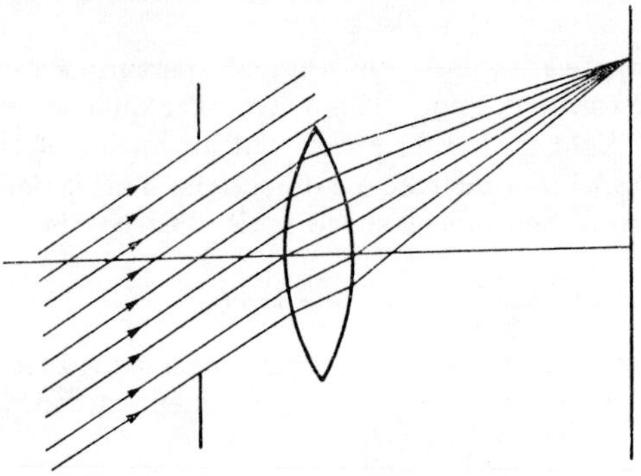

Fig. 4.35. Vignetting.

small field of view, but is of importance in other instruments, such as photographic objectives. Designers sometimes rely on vignetting to obliterate undesirable off-axis aberrations.

4.8.3 Brightness and illumination of images

We shall now briefly consider the relations between the basic photometric quantities which characterize the radiation in the image and object space.

Assume that the object is a small plane element of area δS_0, perpendicular to the axis and radiating in accordance with LAMBERT's law. The photometric brightness B_0 is then independent of direction. The amount of energy δF_0 which falls per unit time on to an annular element of the entry pupil centred on the axis is

$$\delta F_0 = B_0 \cos \gamma_0 \delta S_0 \delta \Omega_0, \tag{15}$$

where

$$\delta \Omega_0 = 2\pi \sin \gamma_0 \delta \gamma_0, \tag{16}$$

γ_0 being the angle which a typical ray passing through the annulus makes with the axis. Hence if θ_0 denotes, as before, the angular semi-aperture on the object side, the total flux of energy which falls on to the entrance pupil per unit time is

$$\begin{aligned} F_0 &= 2\pi B_0 \delta S_0 \int_0^{\theta_0} \sin \gamma_0 \cos \gamma_0 d\gamma_0 \\ &= \pi B_0 \delta S_0 \sin^2 \theta_0. \end{aligned} \tag{17}$$

The energy flux F_1 emerging from the exit pupil may be expressed in a similar form:

$$F_1 = \pi B_1 \delta S_1 \sin^2 \theta_1. \tag{18}$$

F_1 cannot exceed F_0 and can only be equal to it if there are no losses due to reflection, absorption or scattering within the system; hence

$$B_1 \sin^2 \theta_1 \delta S_1 \leqslant B_0 \sin^2 \theta_0 \delta S_0. \tag{19}$$

Now the ratio $\delta S_1/\delta S_0$ is equal to the square of the lateral magnification M:

$$\frac{\delta S_1}{\delta S_0} = M^2. \tag{20}$$

If further it is assumed that the system obeys the sine condition (§ 4.5, eq. (6)),

$$\frac{n_0 \sin \theta_0}{n_1 \sin \theta_1} = M. \tag{21}$$

On substituting from (20) and (21) into (19) it follows that

$$B_1 \leqslant \left(\frac{n_1}{n_0}\right)^2 B_0. \tag{22}$$

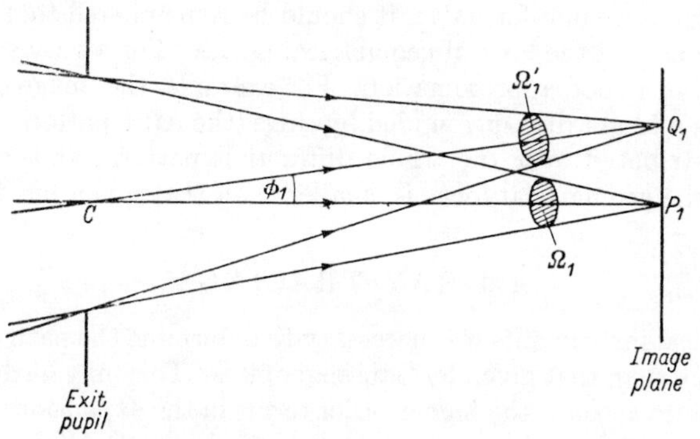

Fig. 4.36. Illumination at an off-axis image point.

In particular, *if the refractive indices of the object and image spaces are equal*, then according to (22), *the photometric brightness of the image cannot exceed that of the object, and can only be equal to it if the losses of light within the system are negligible.*

From (18) and (22) it follows, assuming the losses to be negligible, that

$$F_1 = \pi \left(\frac{n_1}{n_0}\right)^2 B_0 \delta S_1 \sin^2 \theta_1, \tag{23}$$

so that the photometric illumination $E_1 = F_1/\delta S_1$ at the axial point P_1 of the image is

$$E_1 = \pi \left(\frac{n_1}{n_0}\right)^2 B_0 \sin^2 \theta_1. \tag{24}$$

If θ_1 is small, the solid angle Ω_1, which the exit pupil subtends at P_1, is approximately equal to $\pi \sin^2 \theta_1$, so that (24) may then be written as

$$E_1 = \left(\frac{n_1}{n_0}\right)^2 B_0 \Omega_1. \tag{25}$$

This relation applies to the axial image, but the off-axis image may be treated in a similar way. If ϕ_1 is the angle which the principal ray CQ_1 makes with the axis (see Fig. 4.36), then we have in place of (25),

$$E_1 = \left(\frac{n_1}{n_0}\right)^2 B_0 \Omega_1' \cos \phi_1, \tag{26}$$

Ω_1' being the solid angle which the exit pupil subtends at Q_1. With the help of (9) it follows that

$$\frac{\Omega_1'}{\Omega_1} = \cos\phi_1 \left(\frac{CP_1}{CQ_1}\right)^2 = \cos^3\phi_1, \tag{27}$$

so that

$$E_1 = \left(\frac{n_1}{n_0}\right)^2 B_0\Omega_1 \cos^4\phi_1. \tag{28}$$

This formula shows that *the illumination in the image decreases as the fourth power of the cosine of the angle which the principal ray through the image point makes with the axis*, it being assumed that the object radiates according to LAMBERT's law, that there are no losses of light within the system, and that the angular semi-aperture θ_1 is small.

In applying the preceding formulae, it should be remembered that they have been derived with the help of the laws of geometrical optics. For a very small source they may no longer give a good approximation. For example, the image of a point source is not a point but a bright disc surrounded by rings (the AIRY pattern, see § 8.5.2.); the light is then distributed over the whole diffraction pattern, and consequently the illumination at the geometrical focus is smaller than that given by (24).

4.9 RAY TRACING*

In designing optical instruments it is necessary to determine the path of the light with a greater accuracy than that given by Gaussian optics. This may be done by algebraic analysis, taking into account the higher-order terms in the expansion of the characteristic function, a procedure discussed in the next chapter. Alternatively one may determine the path of the light rays accurately with the help of elementary geometry, by successive application of the law of refraction (or reflection); this method, which will now be briefly described, is known as *ray tracing* and is extensively employed in practice.

4.9.1 Oblique meridional rays†

We consider first the tracing of an oblique meridional ray, i.e. a meridional ray from an extra-axial object point. Let A be the pole of the first surface of the system. The surface will be assumed to be a spherical refracting surface of radius r centred at a point C, and separating media of refractive indices n and n'. An incident ray OP (see Fig. 4.37) in the meridional plane is specified by the angle U which it makes with the axis, and by the distance $L = AB$ between the pole of A and the point B at which it meets the axis. Let I be the angle between the incident ray and the normal

---

* For a fuller discussion of ray tracing, see for example A. E. CONRADY, *Applied Optics and Optical Design*, Part I (Oxford University Press, 1929; reprinted by Dover Publications, Inc., New York, 1957); M. VON ROHR, *The Geometrical Investigation of Formation of Images in Optical Systems*, translated from German by R. KANTHACK (London, H.M. Stationery Office, 1920), and M. HERZBERGER, *Modern Geometrical Optics* (New York, Interscience Publishers, 1958). A method for the tracing of rays with the help of electronic computing machines was described by G. BLACK, *Proc. Phys. Soc.* B, **68** (1954), 569. A method for the tracing of rays through non-spherical surfaces was proposed by T. SMITH, *Proc. Phys. Soc.*, **57** (1945), 286; see also W. WEINSTEIN, *Proc. Phys. Soc.* B, **65** (1952), 731.

† In §4.9.1, §4.9.2, and §4.10 the notation and the sign convention usually employed in a meridional ray trace is used. The sign convention for angles differs from the Cartesian sign convention used throughout the rest of the book; to revert to the Cartesian sign convention set $U = -\gamma$, $U' = -\gamma'$.

PC. The corresponding quantities relating to the refracted ray are denoted by primed symbols.

The following sign convention is used: The quantities r, L and L' are taken to be positive when C, B and B' are to the right of A, the light being assumed to be incident from the left. The angles U and U' are considered to be positive if the axis can be brought into coincidence with the rays PB and PB' by a clockwise rotation of less than 90° about B or B' respectively. The angles I and I' are taken to be positive if the incident and the refracted rays may be made to coincide with the normal PC by a clockwise rotation of less than 90° about the point P of incidence.

The quantities L and U which specify the incident ray may be assumed to be known.

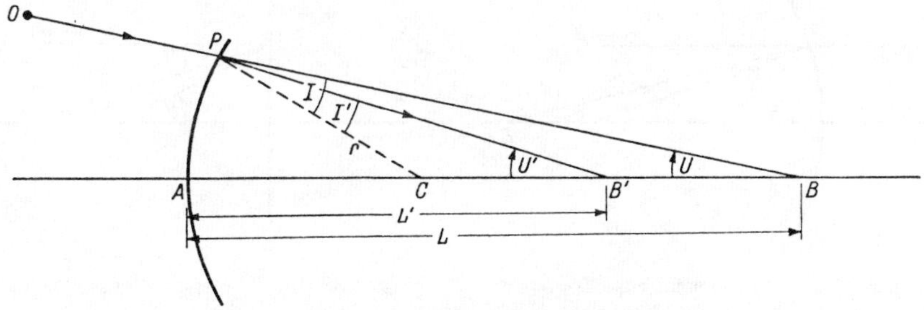

Fig. 4.37. Notation used in an oblique meridional ray trace.

It is then necessary to calculate L' and U'. Assuming also for the moment that both L and r are finite, we have, from the triangle PCB:

$$\sin I = \frac{L - r}{r} \sin U. \tag{1}$$

By the law of refraction

$$\sin I' = \frac{n}{n'} \sin I. \tag{2}$$

Also from the figure

$$U' = U + I - I'. \tag{3}$$

Finally, from the triangle PCB',

$$L' = \frac{\sin I'}{\sin U'} r + r. \tag{4}$$

By successive application of the *refraction equations* (1)–(4), the quantities L' and U', which specify the refracted ray PB', are obtained.

The refracted ray PB' now becomes the incident ray for the second surface. Writing L_1' in place of L' and U_1' in place of U', and denoting by L_2, U_2 the corresponding values, with L_2 referred to the pole of the second surface, we have the *transfer equations*

$$L_2 = L_1' - d, \tag{5}$$

$$U_2 = U_1', \tag{6}$$

where $d > 0$ is the distance between the poles of the two surfaces.

Next the "incident values" given by (5) and (6) are substituted into the refraction equations (1)–(4). Solving for the primed quantities, the ray is then traced through the second surface. In this way, by the repeated application of the refraction and the transfer equations, the values L' and U', relating to the ray in the image space, are obtained. The point of intersection of this ray with the image plane may then be determined. In practice, one would naturally trace not a single ray, but a number of suitably selected rays through the system; their intersection points with the image plane then give a rough estimate of the performance of the system.

If one of the surfaces (say the kth) is a mirror, the appropriate formulae to be used may be formally deduced from the preceding ones by setting $n'_k = -n_k$. Then d_k must be considered to be negative. Moreover all the remaining refractive indices and

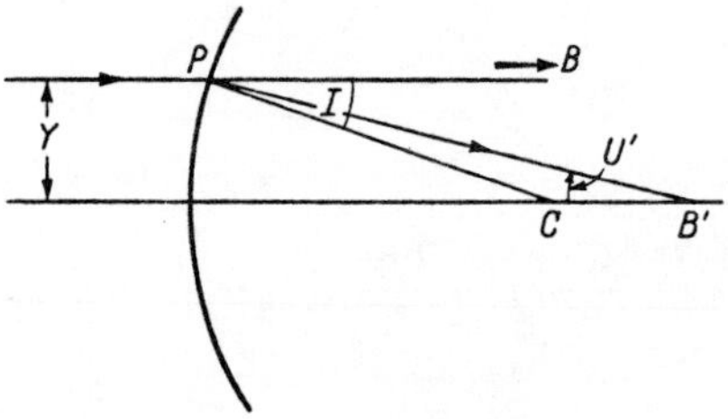

Fig. 4.38 (a). The special case $L = \infty$.

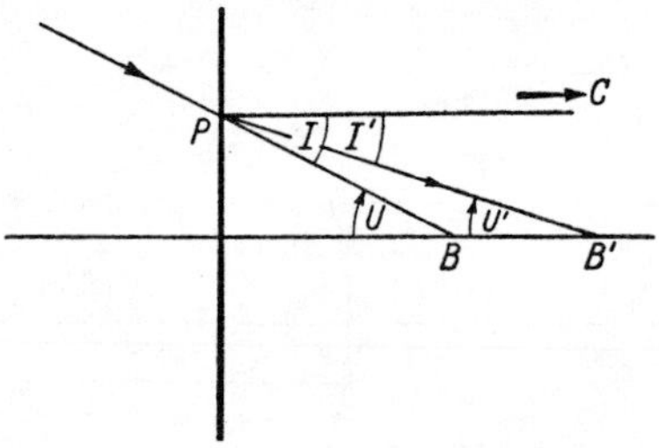

Fig. 4.38 (b). The special case $r = \infty$.

the subsequent d values must also be considered to be negative, unless a second reflection takes place, when they revert to positive signs.

Next consider the two special cases which were so far excluded. If the incident ray is parallel to the axis ($L = \infty$) the equation

$$\sin I = \frac{Y}{r}, \tag{7}$$

is used in place of (1), where Y is the distance of the ray from the axis (see Fig. 4.38(a)).

If the surface is plane ($r = \infty$) we have, in place of (1)–(4), the following set of equations (cf. Fig. 4.38(b)):

$$I = -U, \tag{8}$$

$$\sin U' = \frac{n}{n'} \sin U, \tag{9}$$

$$I' = -U', \tag{10}$$

$$L' = \frac{\tan U}{\tan U'} L. \tag{11}$$

On account of (9), equation (11) may also be written in the form

$$L' = \frac{n'}{n} \frac{\cos U'}{\cos U} L, \tag{11a}$$

which is more convenient for computation than (11) if the angles are small.

It is useful to determine also the coordinates (Y_k, Z_k) of the point P_k of incidence at the kth surface, and the distance $D_k = P_kP_{k+1}$ between two successive points of incidence (see Fig. 4.39).

From the figure,

$$Y_k = r_k \sin (U_k + I_k), \tag{12}$$

$$Z_k = r_k - r_k \cos (U_k + I_k) = \frac{Y_k^2}{[1 + \cos (U_k + I_k)]r_k}, \tag{13}$$

$$D_k = \{d_k + Z_{k+1} - Z_k\} \sec U'_k. \tag{14}$$

In terms of Y_k, Z_k and U'_k,

$$L'_k = Z_k + \frac{Y_k}{\tan U'_k}. \tag{15}$$

This relation may be used as a check on the value computed from (11a) or (11).

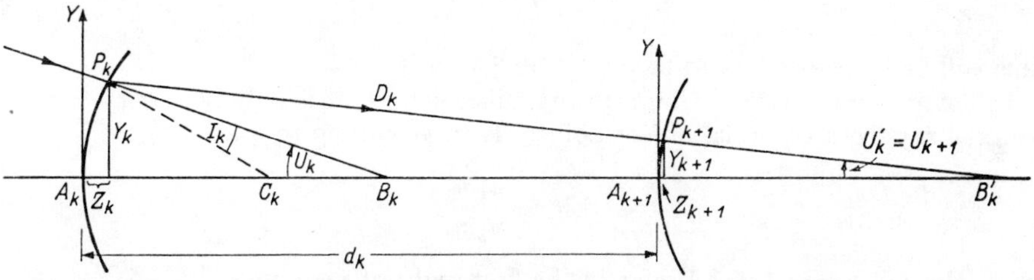

Fig. 4.39. Oblique meridional ray trace through two successive refracting surfaces.

In the special case when L_k is infinite, $U_k = 0$, and (12)–(15) still holds. If r_k is infinite, Y_k may be computed from the relation

$$Y_k = L_k \tan U_k, \tag{16}$$

Z_k then being zero.

4.9.2 Paraxial rays

If the inclination of a ray to the axis is sufficiently small, the sines of the various angles may be replaced, in the preceding formulae, by the angles themselves. The formulae then reduce to the Gaussian approximation for the path of the light. Such "*paraxial* ray-tracing formulae" are used in practice for computing the Gaussian magnification and the focal length of the system. A brief summary of these formulae will therefore be given here.

It is customary to denote quantities which refer to the paraxial region by small letters. The refraction equations (1)–(4) become

$$i = \frac{l - r}{r} u, \tag{17}$$

$$i' = \frac{n}{n'} i, \tag{18}$$

$$u' = u + i - i', \tag{19}$$

$$l' = \frac{i'}{u'} r + r. \tag{20}$$

The transfer equations (5) and (6) take the form

$$l_2 = l'_1 - d, \tag{21}$$

$$u_2 = u'_1. \tag{22}$$

In a similar way the paraxial equations for the cases $L = \infty$ and $r = \infty$ are obtained from (7)–(11a).

The paraxial equation for the incidence height, needed later, follows from (12):

$$y_k = r_k(u_k + i_k). \tag{23}$$

Although the relations (17)–(20) involve the angles which the incident ray and the refracted ray make with the axis, l' is independent of these quantities. This result, established in a different manner in § 4.4.1 follows when i, i' and u' are eliminated from the above relations. It is then found that u also disappears, and we obtain

$$n'\left(\frac{1}{r} - \frac{1}{l'}\right) = n\left(\frac{1}{r} - \frac{1}{l}\right). \tag{24}$$

This will be recognized as ABBE's relation § 4.4 (7).

To determine the lateral Gaussian magnification M, it is only necessary to trace a paraxial ray from the axial object point. Then according to § 4.4 (54),

$$M = \frac{n_1}{n_l}\frac{u_1}{u_l}, \tag{25}$$

where the subscripts 1 and l refer to the first and last medium.

The focal length f' of the system may be obtained by tracing a paraxial ray at any desired height y_1 from an infinitely distant object. The equation of the conjugate ray in the image space, referred to axes at the second focal point, is then $y_l/z_l = -u'_l$, and it follows from § 4.3 (10) that

$$f' = -\frac{y_1}{u'_l}. \tag{26}$$

4.9.3 Skew rays

So far only rays which lie in a meridional plane were considered. We shall now briefly discuss the tracing of *skew* rays, i.e. rays which are not coplanar with the axis. The tracing of such rays is much more laborious and is usually carried out only in the design of systems with very high apertures.

A ray will now be specified by its direction cosines and by the coordinates of the point at which it meets a particular surface of the system. We take Cartesian rectangular axes at the pole A_1 of the first surface, with the Z-axis along the axis of the system. Let L_1, M_1, N_1, $(L_1^2 + M_1^2 + N_1^2 = 1)$ be the direction cosines of a ray incident at the point $P_1(X_1, Y_1, Z_1)$ (see Fig. 4.40).

The first step is to calculate the cosine of the angle I_1 of incidence. If r_1 is the radius of the first surface, the direction cosines of the normal at P_1 are

$$\overline{L}_1 = -\frac{X_1}{r_1}, \qquad \overline{M}_1 = -\frac{Y_1}{r_1}, \qquad \overline{N}_1 = \frac{r_1 - Z_1}{r_1},$$

so that

$$\begin{aligned}\cos I_1 &= L_1\overline{L}_1 + M_1\overline{M}_1 + N_1\overline{N}_1 \\ &= N_1 - \frac{1}{r_1}(L_1X_1 + M_1Y_1 + N_1Z_1).\end{aligned} \tag{27}$$

The next step is to determine the direction cosines L_1', M_1', N_1' of the refracted ray. This is done in two stages: One calculates first the cosine of the angle of refraction, using the law of refraction in the form

$$n'\cos I_1' = \sqrt{n'^2 - n^2 + n^2\cos^2 I_1}. \tag{28}$$

One then uses the fact that the refracted ray lies in the plane specified by the incident ray and the surface normal. Denoting by $\mathbf{s}_1$, $\mathbf{s}_1'$ and $\bar{\mathbf{s}}_1$ the unit vectors along the incident ray, the refracted ray, and the normal, i.e. the vectors with components (L_1,M_1,N_1), (L_1',M_1',N_1') and $(\bar{L}_1,\bar{M}_1,\bar{N}_1)$, the coplanarity condition gives

$$\mathbf{s}_1' = \lambda\mathbf{s}_1 + \mu\bar{\mathbf{s}}_1, \tag{29}$$

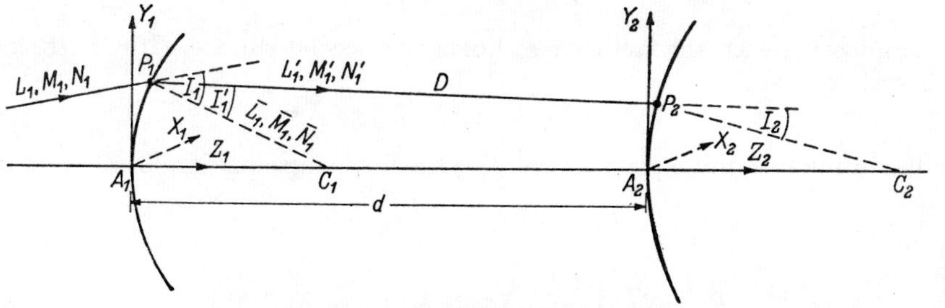

Fig. 4.40. Tracing of a skew ray.

where λ and μ are certain scalar functions. To determine λ and μ we first multiply (29) scalarly by $\mathbf{s}_1$, and use the fact that (see Fig. 4.40) $\mathbf{s}_1 \cdot \mathbf{s}_1' = \cos(I_1 - I_1')$, $\mathbf{s}_1 \cdot \bar{\mathbf{s}}_1 = \cos I_1$. This gives

$$\cos(I_1 - I_1') = \lambda + \mu\cos I_1.$$

Next we multiply (29) scalarly by $\bar{\mathbf{s}}_1$, and use the relations $\bar{\mathbf{s}}_1 \cdot \mathbf{s}_1' = \cos I_1'$, $\bar{\mathbf{s}}_1 \cdot \mathbf{s}_1 = \cos I_1$. We then obtain

$$\cos I_1' = \lambda\cos I_1 + \mu.$$

From the last two relations it follows that

$$\lambda = \frac{\sin I_1'}{\sin I_1} = \frac{n}{n'}, \qquad \mu = \frac{1}{n'}(n'\cos I_1' - n\cos I_1),$$

and (29) then gives the following three equations for the direction cosines of the refracted ray:

$$\left.\begin{aligned} n'L_1' &= nL_1 - \sigma X_1, \\ n'M_1' &= nM_1 - \sigma Y_1, \\ n'N_1' &= nN_1 - \sigma(Z_1 - r_1), \end{aligned}\right\} \tag{30}$$

where

$$\sigma = \frac{1}{r_1}(n'\cos I_1' - n\cos I_1). \tag{31}$$

This completes the ray trace through the first surface, by means of the *refraction equations* (27), (28), (30) and (31).

The refracted ray now becomes the incident ray for the second surface. Taking parallel axes at the pole A_2 of the second surface, we have the *transfer equations for the direction cosines*

$$L_2 = L_1', \qquad M_2 = M_1', \qquad N_2 = N_1'. \tag{32}$$

Let d be the distance A_1A_2 between the poles of the two surfaces. Denoting by X_1^+, Y_1^+, Z_1^+ the coordinates of the point P_1 referred to the axes at A_2, we have the *transfer equations for the coordinates*

$$X_1^+ = X_1, \qquad Y_1^+ = Y_1, \qquad Z_1^+ = Z_1 - d. \tag{33}$$

Next the coordinates (X_2, Y_2, Z_2) of the point P_2 in which the refracted ray meets the second surface must be determined. If D denotes the distance from P_1 and P_2, then

$$\left.\begin{aligned} X_2 &= X_1^+ + L_2 D, \\ Y_2 &= Y_1^+ + M_2 D, \\ Z_2 &= Z_1^+ + N_2 D. \end{aligned}\right\} \tag{34}$$

To determine D we use the fact that P_2 lies on the second surface. If r_2 is the radius of this surface, then

$$X_2^2 + Y_2^2 + Z_2^2 - 2Z_2 r_2 = 0. \tag{35}$$

On substituting into this equation from (34) the following equation for D is obtained:

$$D^2 - 2Fr_2 D + Gr_2 = 0, \tag{36}$$

where

$$\left.\begin{aligned} F &= N_2 - \frac{1}{r_2}(L_2 X_1^+ + M_2 Y_1^+ + N_2 Z_1^+), \\ G &= \frac{1}{r_2}[(X_1^+)^2 + (Y_1^+)^2 + (Z_1^+)^2] - 2Z_1^+. \end{aligned}\right\} \tag{37}$$

Equation (36) gives the following expression for D, if we also make use of the fact that $D = d$ for the axial ray:

$$D = r_2\left(F - \sqrt{F^2 - \frac{1}{r_2}G}\right). \tag{38}$$

The coordinates of P_2 are determined on substituting for D into (34).

Finally, to complete the cycle, the cosine of the angle I_2 of incidence at the second surface must be calculated. It is given by a relation strictly analogous to (27):

$$\cos I_2 = N_2 - \frac{1}{r_2}(L_2 X_2 + M_2 Y_2 + N_2 Z_2), \tag{39}$$

or, using (34) and (38)

$$\begin{aligned} \cos I_2 &= F - \frac{1}{r_2} D \\ &= \sqrt{F^2 - \frac{1}{r_2}G}. \end{aligned} \tag{40}$$

Hence the last stage consists of the evaluation of the formulae (40), (38) and (34).

Because of the labour involved in the tracing of a general skew ray, the calculations are sometimes restricted to skew rays which lie in the immediate neighbourhood of a selected principal ray. Such skew rays may be traced through the system with the help of simplified schemes,* which are similar to those used for paraxial ray trace and are adequate for determining the position of the sagittal focal surface.

* Cf. H. H. Hopkins, *Proc. Phys. Soc.*, **58** (1946), 663; also his *Wave Theory of Aberrations* (Oxford, Clarendon Press, 1950), pp. 59, 65.

4.10 DESIGN OF ASPHERIC SURFACES

In the great majority of optical systems lenses and mirrors are employed, the surfaces of which have plane, spherical or paraboloidal form. The restriction to surfaces of such simple form is mainly due to the practical difficulties encountered in the production of surfaces of more complicated shapes with the high degree of precision required in optics. The restriction to surfaces of simple form naturally imposes limitations on the ultimate performance which systems of conventional design can attain. For this reason, in spite of the difficulties of manufacture, surfaces of more complicated form, called *aspheric* surfaces, are employed in certain systems. As early as 1905, K. SCHWARZSCHILD* considered a class of telescope objectives consisting of two aspheric mirrors, and showed that such systems can be made aplanatic.

In 1930, BERNHARD SCHMIDT, an optician of Hamburg, constructed a telescope of a new type, which consisted of a spherical mirror and a suitably designed aspheric lens placed at its centre of curvature. The performance of this system (considered more fully in § 6.4) was found to be quite outstanding. By means of such a telescope it is possible to photograph on one plate a very large region of the sky, many hundred times larger than can be obtained with telescopes of conventional design. The *Schmidt camera* has since become an important tool in astronomical research. Aspheric systems which use the principle of the SCHMIDT camera are also employed in certain projection-type television receivers,† in X-ray fluorescent screen photography, and certain fast low-dispersion spectrographs. Aspheric surfaces find also useful application in microscopy (cf. § 6.6).

By making one surface of any centred system aspherical, it is possible, in general, to ensure exact axial stigmatism; by means of two aspheric surfaces any centred system may in general be made aplanatic. In this section formulae will be derived for the design of such aspheric surfaces.

4.10.1 Attainment of axial stigmatism‡

Consider the rays from an axial object point P. In the image space of the system the rays from different zones of the exit pupil will in general intersect the axis at different points. Let $S^{(0)}$ be the last surface of the system and O its axial point (Fig. 4.41). It will be shown that it is possible to modify the profile of $S^{(0)}$ in such a way as to compensate exactly for the departure from the homocentricity of the image-forming pencil; more precisely we shall show that it is possible in general to replace $S^{(0)}$ by a new surface S which will ensure that all the rays in the image space meet the axis at any prescribed point Q.

On account of symmetry only meridional rays need be considered. Each ray in the space which precedes the last surface will be specified by the following parameters: the angle U which it makes with the axis and the distance $H = OM$ at which it intersects the Y-axis (see Fig. 4.41). It will be convenient to label the rays: Let t be

* K. SCHWARZSCHILD, *Astr. Mitt. Königl. Sternwarte Göttingen* (1905). Reprinted from *Abh. Königl. Ges. Wiss. Göttingen, Math. Phys. Klasse*, **4** (1905–1906), No. 2.

Two telescopes of this type were later constructed, one with aperture of 24 in. at the University of Indiana and another, with a 12-in. aperture, at Brown University.

† See, for example, I. G. MALLOFF and D. W. EPSTEIN, *Electronics*, **17** (1944), 98.

‡ The methods here described are due to E. WOLF and W. S. PREDDY, *Proc. Phys. Soc.*, **59** (1947), 704; also E. WOLF, *Proc. Phys. Soc.*, **61** (1948), 494. Similar formulae were also given by R. K. LUNEBURG in his *Mathematical Theory of Optics* (mimeographed notes of lectures delivered at Brown University, R.I. (1944); printed version published by University of California Press, Berkeley and Los Angeles (1964), § 24.

any convenient parameter, for example the angle U_0 which the corresponding ray in the object space makes with the axis, or the height at which it meets the first surface. The pencil may then be completely specified by the two functional relations

$$U = U(t), \qquad H = H(t). \tag{1}$$

It may be assumed that $t = 0$ for the axial ray. In general the relations (1) will not be known in an explicit form; a table of values of U and H for any prescribed set of t values may, however, be obtained from a ray trace.

Let W be the orthogonal surface (wave-front) through O of the pencil (Fig. 4.41), and let N be the point in which the ray through M meets it. Assuming that a correcting surface S with the required property can be found, it follows that in the corrected

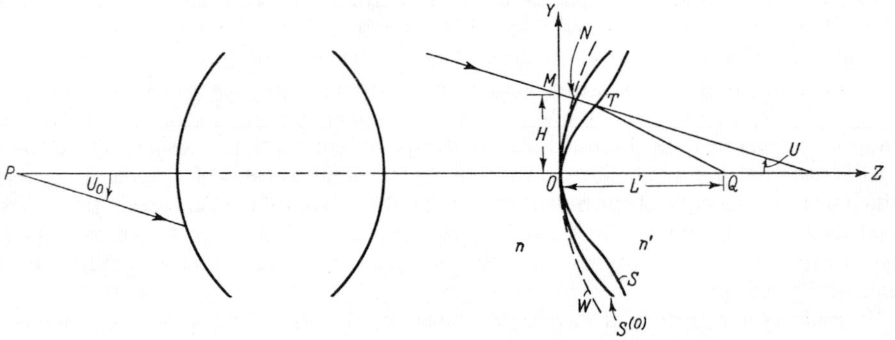

Fig. 4.41. Design of an aspheric surface to attain axial stigmatism.

system the optical path length from N to Q must be equal to the optical path length from O to Q. Hence, if $T(Y,Z)$ is the point in which the ray meets S,

$$[NT] + [TQ] = [OQ]. \tag{2}$$

If n is the refractive index of the space which precedes S, and n' the refractive index of the image space, we have*

$$\begin{aligned} [NT] &= [MT] - [MN] = nZ \sec U - [MN], \\ [TQ] &= n'\sqrt{(L' - Z)^2 + (H - Z \tan U)^2}, \\ [OQ] &= n'L', \end{aligned} \tag{3}$$

L' denoting the distance from O to Q.

The optical path $[MN]$ which occurs in the expression for $[NT]$ may also be evaluated in terms of the given quantities. By applying LAGRANGE's invariant relation §3.3 (1) to the curve formed by the segment OM of the Y-axis, the segment MN of the ray and the curve NO on the wave-front W, we have

$$\int_{OM} n\mathbf{s} \cdot d\mathbf{r} + \int_{MN} n\mathbf{s} \cdot d\mathbf{r} + \int_{NO} n\mathbf{s} \cdot d\mathbf{r} = 0, \tag{4}$$

* As in §4.9.1, the angle U is considered to be positive if the axis can be brought into coincidence with the ray by a clockwise rotation of less than 90° about the axial point.

where $\boldsymbol{s}$ is the unit vector along the ray and $d\boldsymbol{r}$ an element of the path of integration. Now from the figure,

$$\left.\begin{aligned}\int_{OM} n\boldsymbol{s}\,.\,d\boldsymbol{r} &= -n\int_0^H \sin U dH = -n\int_0^t \sin U \frac{dH}{dt}\,dt,\\ \int_{MN} n\boldsymbol{s}\,.\,d\boldsymbol{r} &= [MN],\\ \int_{NO} n\boldsymbol{s}\,.\,d\boldsymbol{r} &= 0,\end{aligned}\right\} \tag{5}$$

and hence (4) gives

$$[MN] = n\int_0^t \sin U \frac{dH}{dt}\,dt. \tag{6}$$

(6) expresses $[MN]$ as an integral which may be evaluated numerically from a table of U and H values.*

Substitution from (3) and (6) into (2) leads to the following equation for Z:

$$\mathscr{A}Z^2 + 2\mathscr{B}Z\cos U + \mathscr{C}\cos^2 U = 0, \tag{7}$$

with

$$\left.\begin{aligned}\mathscr{A} &= n'^2 - n^2,\\ \mathscr{B} &= n^2\int_0^t \sin U\frac{dH}{dt}\,dt - n'^2(L'\cos U + H\sin U) + nn'L',\\ \mathscr{C} &= n'^2H^2 - \left(n\int_0^t \sin U\frac{dH}{dt}\,dt\right)\left(n\int_0^t \sin U\frac{dH}{dt}\,dt + 2n'L'\right).\end{aligned}\right\} \tag{8}$$

Hence

$$Z = \frac{\cos U}{\mathscr{A}}\left[-\mathscr{B} \pm \sqrt{\mathscr{B}^2 - \mathscr{A}\mathscr{C}}\right]. \tag{9}$$

Also, from the figure,

$$Y = H - Z\tan U. \tag{10}$$

Since $Z = U = H = 0$ when $t = 0$, and since we assume L' to be positive (see Fig. 4.41), the positive sign must be taken in front of the square root in (9). Finally, combining (9) and (10), we obtain

$$Z + iY = \frac{-\mathscr{B} + \sqrt{\mathscr{B}^2 - \mathscr{A}\mathscr{C}}}{\mathscr{A}}\,e^{-iU} + iH. \tag{11}$$

(11) is an exact parametric equation of the aspheric surface S in terms of the free parameter t.

The special case when the focus Q is at infinity ($L' = \infty$) is also of interest. To derive the appropriate formula we note that both $\mathscr{B}$ and $\mathscr{C}$ contain L' in the first power only. Hence for sufficiently large L',

$$\begin{aligned}-\mathscr{B} + \sqrt{\mathscr{B}^2 - \mathscr{A}\mathscr{C}} &= \mathscr{B}\left[-1 + \sqrt{1 - \frac{\mathscr{A}\mathscr{C}}{\mathscr{B}^2}}\right]\\ &= -\frac{1}{2}\frac{\mathscr{A}\mathscr{C}}{\mathscr{B}} - \frac{1}{8}\frac{\mathscr{A}^2\mathscr{C}^2}{\mathscr{B}^3} - \dots\\ &= -\frac{1}{2}\frac{\mathscr{A}\mathscr{C}}{\mathscr{B}} + O\left(\frac{1}{L'}\right).\end{aligned}$$

* It is, of course, possible to evaluate $[MN]$ directly from a ray trace by making use of the property that the optical path from P to N is equal to the optical path from P to O. This gives

$$[MN] = [PO] - [PM].$$

In the limit, as $L' \to \infty$, (11) therefore reduces to

$$Z + iY = \frac{ne^{-iU}}{n - n' \cos U} \int_0^t \sin U \frac{dH}{dt}\, dt + iH. \tag{12}$$

We have only considered the case when the aspheric surface is the last surface of the system but the method may be extended to the design of an aspheric surface situated in the interior of an optical system. The computation is, however, much more laborious in such cases and will not be considered here.*

4.10.2 Attainment of aplanatism†

We have seen that by making one surface of a system aspherical, it is possible to attain exact axial stigmatism. We shall now consider the design of two aspheric

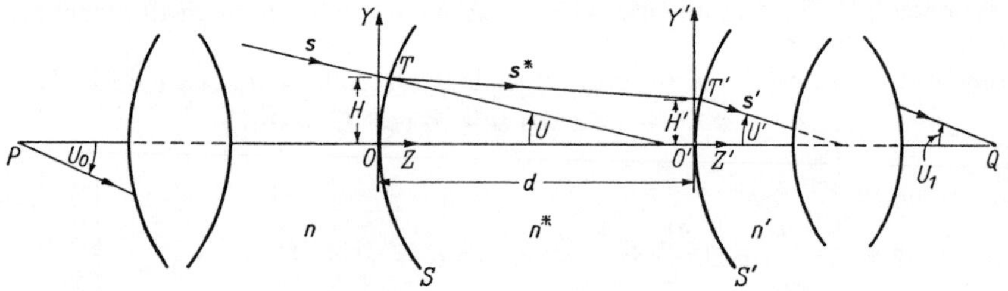

Fig. 4.42. Design of two aspheric surfaces to attain aplanatism.

surfaces which will ensure not only axial stigmatism but also the satisfaction of the sine condition.

Let S and S' be the two aspheric surfaces, the profiles of which are to be determined. It will be assumed that S and S' are neighbouring surfaces in the system.‡ They may, however, be separated from the object or image points by any number of refracting or reflecting surfaces. Again, we shall be concerned only with the final correction of the system, and assume that all the design data, save the profiles of S and S', are known.

We introduce two sets of Cartesian rectangular axes, with origins at the poles O and O' of S and S', and with the Z-axes along the axis of the system. The surface S will be referred to the axes at O, and S' to the axes at O'.

In the space which precedes S, the pencil of rays from the axial object point P will again be specified by a relation of the form (see Fig. 4.42)

$$U = U(t), \qquad H = H(t). \tag{13}$$

The pencil of rays in the space which follows S' (in the corrected system) will be specified by a similar relation:

$$U' = U'(t'), \qquad H' = H'(t'). \tag{14}$$

(14) may be obtained in a tabulated form by tracing rays backwards from the selected axial image point P'.

---

* For details see E. Wolf, *Proc. Phys. Soc.*, **61** (1948), 494. Other methods were described by M. Herzberger and H. O. Hoadley, *J. Opt. Soc. Amer.*, **36** (1946), 334; and D. S. Volosov *J. Opt. Soc. Amer.*, **37** (1947), 342.

† The methods described in this section are due to G. D. Wassermann and E. Wolf, *Proc. Phys. Soc.* B, **62** (1949), 2.

‡ A generalization to systems where S and S' are not optical neighbours was described by E. M. Vaskas, *J. Opt. Soc. Amer.*, **47** (1957), 669.

If the object and image are at finite distances we choose as the parameters t and t' the sines of the angles which the corresponding rays in the object and image spaces make with the axis of the system:*

$$t = \sin U_0, \qquad t' = \sin U_1. \tag{15}$$

If the object is at infinity, we choose as the t parameter the distance H_0 of the corresponding ray in the object space from the axis; if the image is at infinity we take $t' = H_1$, H_1 being the distance from the axis of the corresponding ray in the image space. In either of these cases, the sine condition demands that

$$\frac{t}{t'} = \text{constant}. \tag{16}$$

Our problem may now be formulated as follows: Given (13) and (14), to find two surfaces S and S' which ensure that the pencil (U, H) goes over into the pencil (U', H') by successive refractions at the two surfaces; and, moreover, the corresponding rays in the two pencils must satisfy the relation (16).

Let n be the refractive index of the space which precedes S, n' the refractive index of the space which follows S', and $n^\star$ that of the space between them. Further let $\mathbf{s}$ be the unit vector along the ray incident at the point $T(Z, Y)$ and $\mathbf{s}^\star$ the unit vector along the refracted ray (see Fig. 4.42).

According to the law of refraction (cf. § 3.2.2) the vector $\mathbf{N} = n\mathbf{s} - n^\star\mathbf{s}^\star$ must be in the direction of the surface normal at T. Hence if $\boldsymbol{\tau}$ is the unit tangent at T of the meridional section of the surface,

$$(n\mathbf{s} - n^\star\mathbf{s}^\star) \cdot \boldsymbol{\tau} = 0. \tag{17}$$

Now the components of the vectors $\mathbf{s}$, $\mathbf{s}^\star$ and $\boldsymbol{\tau}$ are:

$$\left.\begin{array}{llll} \mathbf{s}: & 0, & -\sin U, & \cos U, \\ \mathbf{s}^\star: & 0, & -\sin U^\star, & \cos U^\star, \\ \boldsymbol{\tau}: & 0, & \dfrac{\dot{Y}}{\sqrt{\dot{Y}^2 + \dot{Z}^2}}, & \dfrac{\dot{Z}}{\sqrt{\dot{Y}^2 + \dot{Z}^2}} \end{array}\right\} \tag{18}$$

here $U^\star$ is the angle which the refracted ray TT' makes with the axis of the system, and the dot denotes differentiation with respect to the parameter t. Equation (17) becomes

$$n(\dot{Z} \cos U - \dot{Y} \sin U) = n^\star(\dot{Z} \cos U^\star - \dot{Y} \sin U^\star). \tag{19}$$

Now if D is the distance from T to T', D_y and D_z the projections of D on to the Y and Z-axes, and d the axial distance from O to O',

$$\cos U^\star = \frac{D_z}{D}, \qquad \sin U^\star = \frac{D_y}{D}, \tag{20}$$

with

$$D_y = Y - Y', \qquad D_z = d + Z' - Z, \qquad D = \sqrt{D_y^2 + D_z^2}. \tag{21}$$

Also, from the figure,

$$Y = H - Z \tan U, \tag{22}$$

$$Y' = H' - Z' \tan U'. \tag{23}$$

---

* If instead of the sine condition we wished (in addition to axial stigmatism) to satisfy the Herschel condition, we would choose

$$t = \sin U_0/2, \qquad t' = \sin U_1/2.$$

On substituting in (19) for $\cos U^\star$ and $\sin U^\star$ from (20), and for $\dot{Y}$ from (22), we find that

$$\frac{dZ}{dt} = \left(\frac{nD\cos U - n^\star D_z}{nD\sin U - n^\star D_y} + \tan U\right)^{-1}\left(\frac{dH}{dt} - Z\frac{d}{dt}(\tan U)\right). \tag{24}$$

Similarly

$$\frac{dZ'}{dt'} = \left(\frac{n'D\cos U' - n^\star D_z}{n'D\sin U' - n^\star D_y} + \tan U'\right)^{-1}\left(\frac{dH'}{dt'} - Z'\frac{d}{dt'}(\tan U')\right). \tag{25}$$

Equations (21)–(25), subject to the relation (16) and the boundary conditions $Z = Z' = 0$ when $t = t' = 0$, enable a complete computation of the two correcting surfaces to be carried out. For, using (21), (22) and (23) we may eliminate Y and Y' from (24) and (25); and, using (16), we then obtain two first-order simultaneous differential equations for Z and Z' of the type

$$\frac{dZ}{dt} = f(Z, Z', t), \qquad \frac{dZ'}{dt} = g(Z, Z', t). \tag{26}$$

These may be integrated by standard methods.* Since, however, it is necessary to determine not only Z and Z' but also Y and Y' for a selected range of the parameter t, it is preferable not to eliminate Y and Y' but rather to solve for the unknown quantities step by step.

* For example, by Adams' method (cf. E. T. Whittaker and G. Robinson, *The Calculus of Observations*, Glasgow, Blackie & Son, 4th ed., 1946, 363), or by the method of Runge and Kutta (cf. C. Runge and H. König, *Numerisches Rechnen*, Berlin, Springer, 1924).

CHAPTER V

GEOMETRICAL THEORY OF ABERRATIONS

IN § 4.9 it was mentioned that within the domain of geometrical optics the departure of the path of light from the predictions of the Gaussian theory may be studied either with the help of ray tracing or by means of algebraic analysis. In the latter treatment, which forms the subject matter of this chapter, terms which involve off-axis distances in powers higher than the second in the expansion of the characteristic functions are retained. These terms represent *geometrical aberrations.*

The discovery of photography in 1839 by DAGUERRE (1789–1851) was chiefly responsible for early attempts to extend the Gaussian theory. Practical optics, which until then was mainly concerned with the construction of telescope objectives, was confronted with the new task of producing objectives with large apertures and large fields. J. PETZVAL, a Hungarian mathematician, attacked with considerable success the related problem of supplementing the Gaussian formulae by terms involving higher powers of the angles of inclination of rays with the axis. Unfortunately, PETZVAL's extensive manuscript on the subject was destroyed by thieves; what is known about this work comes chiefly from semi-popular reports.* PETZVAL demonstrated the practical value of his calculations by constructing in about 1840 his well-known portrait lens (shown in Fig. 6.3(b)) which proved greatly superior to any then in existence. The earliest systematic treatment of geometrical aberrations which was published in full is due to SEIDEL,† who took into account all the terms of the third order in a general centred system of spherical surfaces. Since then, his analysis has been extended and simplified by many writers.

As the wave-fronts are the orthogonal trajectories of the pencils of rays it follows that the lack of homocentricity of the image-forming pencil is accompanied by departures of the associated wave-fronts from the spherical form. The knowledge of the shape of the wave-fronts is of particular importance in more refined treatments of aberrations, based on diffraction theory (see Chapter IX). For this reason, and also because of the intimate connection between the non-homocentricities of the ray pencils and the asphericities of the associated wave-fronts, we shall consider these two defects side by side. Our analysis will in part be based on important investigations of SCHWARZSCHILD‡ adapted somewhat to this purpose.

5.1 WAVE AND RAY ABERRATIONS; THE ABERRATION FUNCTION

Consider a rotationally symmetrical optical system. Let P_0', P_1' and P_1 be the points in which a ray from an object point P_0 intersects the plane of the entrance pupil, the

* J. PETZVAL, *Bericht über die Ergebnisse einiger dioptrischer Untersuchungen* (Pesth, 1843); also "Bericht über optische Untersuchungen", *Ber. Kais. Akad. Wien, Math. naturwiss. Kl.*, **24** (1857), 50, 92, 129.

† L. SEIDEL, *Astr. Nachr.*, **43** (1856), No. 1027, 289, No. 1028, 305, No. 1029, 321.

‡ K. SCHWARZSCHILD, *Abh. Königl. Ges. Wis. Göttingen. Math-phys. Kl.*, **4** (1905–1906), Nos. 1, 2, 3. Reprinted in *Astr. Mitt. Königl. Sternwarte Göttingen* (1905). An extension of Schwarzschild's analysis to systems without rotational symmetry was made by G. D. RABINOVICH, *Akad. Nauk. SSSR, Zh. Eksp. Teor. Fiz.*, **16** (1946), 161.

exit pupil, and the Gaussian image plane respectively. If P_1^* is the Gaussian image of P_0, the vector $\boldsymbol{\delta}_1 = \overrightarrow{P_1^* P_1}$ will be called *the aberration of the ray*, or simply *the ray aberration* (see Fig. 5.1).

Let W be the wave-front through the centre O_1' of the exit pupil, associated with the image-forming pencil which reaches the image space from P_0. In the absence of aberrations, W coincides with a sphere S which is centred on the Gaussian image point P^*

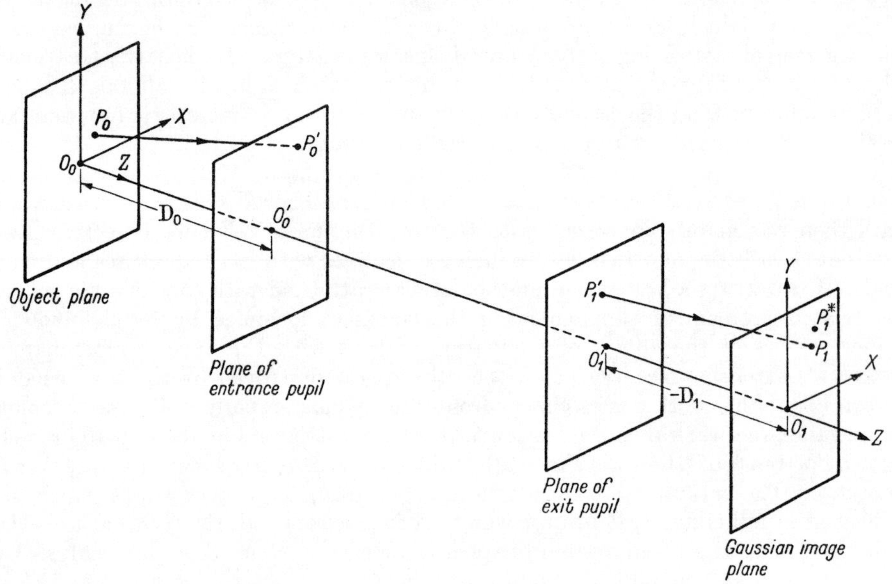

Fig. 5.1. The object plane, the image plane and the pupil planes.

and which passes through O_1'. S will be called the *Gaussian reference sphere* (see Fig. 5.2).

Let Q and $\bar{Q}$ be the points of intersection of the ray $P_1'P_1$ with the Gaussian reference sphere and with the wave-front W respectively. The optical path length $\Phi = [\bar{Q}Q]$ may be called *the aberration of the wave element at* Q, or simply the *wave aberration*, and will be regarded as positive if $\bar{Q}$ and P_1 are on the opposite sides of Q. In ordinary instruments, the wave aberrations may be as much as 40 or 50 wavelengths, but in instruments used for more precise work (such as astronomical telescopes or microscopes) they must be reduced to a much smaller value, only a fraction of a wavelength.

One can easily derive expressions for the wave aberration in terms of HAMILTON's point characteristic function of the system. If brackets [. . .] denote as before the optical path length, we have

$$\begin{aligned}\Phi &= [\bar{Q}Q]\\ &= [P_0Q] - [P_0\bar{Q}]\\ &= [P_0Q] - [P_0O_1'].\end{aligned} \tag{1}$$

Here use was made of the fact that $\bar{Q}$ and O_1' are on the same wave-front, so that $[P_0\bar{Q}] = [P_0O_1']$.

Let us introduce two sets of mutually parallel Cartesian rectangular axes, with origins at the axial points O_0 and O_1 of the object and image planes, and with the Z-directions along the axis of the system. Points in the object space will be referred to the axes at O_0, and points in the image space to the axes at O_1. The Z coordinates of the pupil planes will be denoted by D_0 and D_1 (D_1 is negative in Fig. 5.1).

In terms of the point characteristic V, the wave aberration is, according to (1), given by

$$\Phi = V(X_0, Y_0, 0; X, Y, Z) - V(X_0, Y_0, 0; 0, 0, \mathrm{D}_1), \tag{2}$$

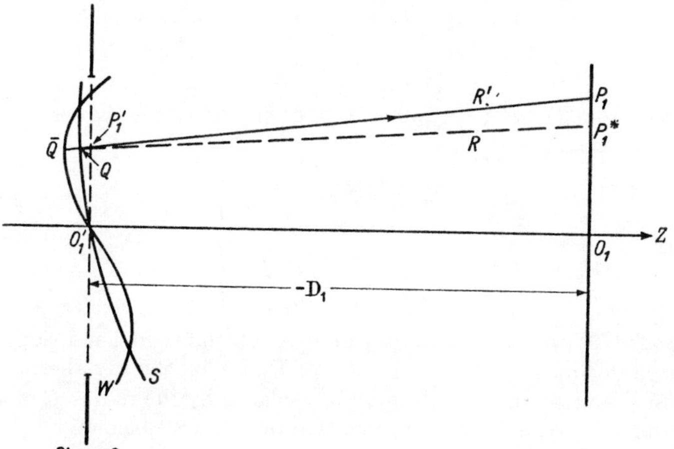

Fig. 5.2. Wave aberration and ray aberration.
(The ray $\bar{Q}QP_1$ does not necessarily lie in the meridional plane $O_1'O_1P_1^\star$.)

(X_0, Y_0) being the coordinates of P_0 and (X, Y, Z) those of Q. Now the coordinates (X, Y, Z) are not independent; they are connected by a relation which expresses the fact that Q lies on the Gaussian reference sphere:

$$(X - X_1^\star)^2 + (Y - Y_1^\star)^2 + Z^2 = R^2. \tag{3}$$

Here

$$X_1^\star = MX_0, \qquad Y_1^\star = MY_0, \tag{4}$$

are the coordinates of the Gaussian image point $P_1^\star$, M is the Gaussian lateral magnification between the object and image planes, and R is the radius of the Gaussian reference sphere

$$R = (X_1^{\star 2} + Y_1^{\star 2} + \mathrm{D}_1{}^2)^{1/2}. \tag{5}$$

Using (3), Z may be eliminated from (2), so that Φ may be regarded as a function of X_0, Y_0, X and Y only, i.e.*

$$\Phi = \Phi(X_0, Y_0; X, Y).$$

In terms of the aberration function $\Phi(X_0, Y_0; X, Y)$, simple expressions for the ray aberrations may be obtained. We have, from (2),

$$\frac{\partial \Phi}{\partial X} = \frac{\partial V}{\partial X} + \frac{\partial V}{\partial Z}\frac{\partial Z}{\partial X}. \tag{6}$$

* A more general aberration function, suitable for discussing the imaging of extended objects, was introduced by E. Wolf, *J. Opt. Soc. Amer.*, **42** (1952), 547.

If α_1, β_1 and γ_1 are the angles which the ray QP_1 makes with the axes, and (X, Y, Z) and $(X_1, Y_1, 0)$ are the coordinates of the points Q and P_1 respectively, we have, according to § 4.1 (7), and from Fig. 5.2,

$$\frac{\partial V}{\partial X} = n_1 \cos \alpha_1 = n_1 \frac{X_1 - X}{R'}, \qquad \frac{\partial V}{\partial Z} = n_1 \cos \gamma_1 = - n_1 \frac{Z}{R'}, \tag{7}$$

where

$$R' = \{(X_1 - X)^2 + (Y_1 - Y)^2 + Z^2\}^{1/2} \tag{8}$$

is the distance from Q to P_1 and n_1 is the refractive index of the image space. Further, from (3),

$$\frac{\partial Z}{\partial X} = - \frac{X - X_1^{\star}}{Z}. \tag{9}$$

Substitution from (7) and (9) into (6) gives the following expressions for the ray aberration components:

$$\left.\begin{aligned} X_1 - X_1^{\star} &= \frac{R'}{n_1} \frac{\partial \Phi}{\partial X}, \\ &\text{and similarly} \\ Y_1 - Y_1^{\star} &= \frac{R'}{n_1} \frac{\partial \Phi}{\partial Y}. \end{aligned}\right\} \tag{10}$$

The relations (10) are exact, but they involve, on the right-hand side, the distance R', which itself depends on the coordinates of P_1, i.e. on the ray aberrations*. However, for most practical purposes R' may be replaced by the radius R of the Gaussian reference sphere or by another approximation (cf. eq. (15) below).

It is easily seen that, on account of symmetry, Φ depends on the four variables only through the three combinations $X_0^2 + Y_0^2$, $X^2 + Y^2$ and $X_0 X + Y_0 Y$. For if polar coordinates are introduced in the XY-planes, i.e. if we set

$$\left.\begin{aligned} X_0 &= r_0 \cos \theta_0, \qquad & X &= r \cos \theta, \\ Y_0 &= r_0 \sin \theta_0, \qquad & Y &= r \sin \theta, \end{aligned}\right\} \tag{11}$$

Φ becomes a function of r_0, r, θ_0 and θ; or, what amounts to the same thing, a function of r_0, r, $\theta_0 - \theta$ and θ. Suppose now that the X- and Y-axes at O_0 and O_1 are rotated through the same angle and in the same sense about the axis of the system. This leaves r_0, r and $\theta_0 - \theta$ unchanged, whilst θ increases by the angle of rotation. Since Φ is invariant with respect to such rotations, it must be independent of the last variable, i.e. it depends on r_0, r and $\theta_0 - \theta$ only. Hence the aberration function Φ may be expressed as a function of the three scalar products

$$\mathbf{r}_0^2 = X_0^2 + Y_0^2, \quad \mathbf{r}^2 = X^2 + Y^2 \quad \text{and} \quad \mathbf{r}_0 \cdot \mathbf{r} = X_0 X + Y_0 Y, \tag{12}$$

of the two vectors $\mathbf{r}_0(X_0, Y_0)$ and $\mathbf{r}(X, Y)$.

From this result it follows that, if Φ is expanded into a power series with respect to the four coordinates, the expansion will contain only terms of even degree. There will be no terms of zero degree, since $\Phi(0, 0; 0, 0) = 0$. Moreover, there will be no terms of the second degree, except possibly for a term proportional to $(X_0^2 + Y_0^2)$; for according to (10) such terms would give rise to ray aberrations depending linearly on the coordinates, and this contradicts the fact that $P_1^{\star}$ is the Gaussian image of P_0. Hence the expansion is of the form

$$\Phi = c(X_0^2 + Y_0^2) + \Phi^{(4)} + \Phi^{(6)} + \ldots, \tag{13}$$

---

* A somewhat different pair of exact equations which connect the ray aberrations and the wave aberrations was derived by J. L. Rayces, *Optica Acta*, **11** (1964), 85.

where c is a constant and $\Phi^{(2k)}$ is a polynomial of degree $2k$ in the coordinates, and contains these coordinates in powers of the three scalar invariants (12) only. A term of a particular degree $2k$ is said to represent a *wave aberration of order* $2k$. The aberrations of the lowest order ($2k = 4$) are usually called *primary* or *Seidel aberrations** and will be studied in detail in § 5.3.

To show the order of magnitude of certain expressions and the degree of approximation involved in some of our calculations, it is convenient to introduce a parameter μ. This may be any quantity of the first order, e.g. the angular aperture of the system. Then all the rays passing through the system may be assumed to make angles $O(\mu)$ with the optical axis, where the symbol $O(\mu)$ means "not exceeding a moderate multiple" of μ.

Consider the error involved on replacing R' in the basic relations (10) by quantities independent of X_1 and Y_1. From (3) and (5),

$$Z^2 = \mathrm{D}_1{}^2 - (X^2 + Y^2) + 2(XX_1^* + YY_1^*) \tag{14}$$

and (8) then gives

$$\begin{aligned} R' &= -\mathrm{D}_1\left[1 + \frac{X_1{}^2 + Y_1{}^2 - 2X(X_1 - X_1^*) - 2Y(Y_1 - Y_1^*)}{\mathrm{D}_1{}^2}\right]^{1/2} \\ &= -\mathrm{D}_1 - \frac{X_1^{*2} + Y_1^{*2}}{2\mathrm{D}_1} + O(\mathrm{D}_1\mu^4). \end{aligned} \tag{15}$$

The relations (10) for the ray aberration components then become

$$X_1 - X_1^* = -\frac{1}{n_1}\left(\mathrm{D}_1 + \frac{X_1^{*2} + Y_1^{*2}}{2\mathrm{D}_1}\right)\frac{\partial\Phi}{\partial X} + O(\mathrm{D}_1\mu^7) \tag{16a}$$

$$= -\frac{\mathrm{D}_1}{n_1}\frac{\partial\Phi}{\partial X} + O(\mathrm{D}_1\mu^5) \tag{16b}$$

$$Y_1 - Y_1^* = -\frac{1}{n_1}\left(\mathrm{D}_1 + \frac{X_1^{*2} + Y_1^{*2}}{2\mathrm{D}_1}\right)\frac{\partial\Phi}{\partial Y} + O(\mathrm{D}_1\mu^7) \tag{17a}$$

$$= -\frac{\mathrm{D}_1}{n_1}\frac{\partial\Phi}{\partial Y} + O(\mathrm{D}_1\mu^5). \tag{17b}$$

5.2 THE PERTURBATION EIKONAL OF SCHWARZSCHILD

In his researches on geometrical aberrations, SCHWARZSCHILD employed a method similar to that used in the calculations of orbital elements in celestial mechanics. In such calculations variables are introduced which remain constant in the unperturbed motion, and the small changes which these quantities undergo in the actual motion are then determined with the help of a perturbation function. By analogy with this procedure, SCHWARZSCHILD introduced, in the papers already referred to, certain variables which, within the accuracy of Gaussian optics, have constant values along each ray passing through the optical system. With the help of a certain perturbation function which he introduced and called the *Seidel eikonal*, he then studied the changes which these variables undergo when the fourth-order terms in the expansion

* Since the *ray aberrations* associated with wave aberrations of this order are of the third degree in the coordinates, they are also sometimes called the *third-order aberrations*.

of the characteristic function are taken into account.* SCHWARZSCHILD called these special variables the *Seidel variables*, as they are related to those employed previously by SEIDEL.

Within the accuracy of the SEIDEL theory, the aberration function Φ defined in the preceding section is very closely related to the perturbation eikonal of SCHWARZSCHILD; and by following closely SCHWARZSCHILD's procedure one can derive the expressions for the fourth coefficients in the expansion of the aberration function for any centred system. This will be done in detail in § 5.5. Here the Seidel variables will be defined and the connection between our aberration function and the perturbation function of SCHWARZSCHILD will be examined.

Let us introduce new units of length l_0 in the object plane and l_1 in the Gaussian image plane, such that

$$\frac{l_1}{l_0} = M \tag{1}$$

is the lateral Gaussian magnification between the two planes. Points in the object plane will be specified by the coordinates x_0, y_0, and points in the image space by the coordinates x_1, y_1, such that

$$\left.\begin{aligned} x_0 &= \mathrm{C}\frac{X_0}{l_0}, \qquad & x_1 &= \mathrm{C}\frac{X_1}{l_1} \\ y_0 &= \mathrm{C}\frac{Y_0}{l_0}, \qquad & y_1 &= \mathrm{C}\frac{Y_1}{l_1} \end{aligned}\right\} \tag{2}$$

where (X_0, Y_0), (X_1, Y_1) are the ordinary coordinates of P_0 and P_1 (see Fig. 5.1), and C is a constant to be specified later. Within the accuracy of Gaussian optics, $x_1 = x_0$ and $y_1 = y_0$.

The coordinates (X_0', Y_0') of the points where the ray from (X_0, Y_0) meets the entrance pupil are connected with the ray components by the relations

$$\frac{X_0' - X_0}{\mathrm{D}_0} = \frac{p_0}{\sqrt{n_0^2 - p_0^2 - q_0^2}}, \qquad \frac{Y_0' - Y_0}{\mathrm{D}_0} = \frac{q_0}{\sqrt{n_0^2 - p_0^2 - q_0^2}}, \tag{3}$$

the expressions for the intersection points of the ray with the exit pupil being strictly similar. Within the accuracy of Gaussian optics, the square roots in the denominators may be replaced by n_0 and n_1 respectively, and the following linear relations between the coordinates are then obtained:

$$\left.\begin{aligned} \frac{X_0' - X_0}{\mathrm{D}_0} &= \frac{p_0}{n_0}, \qquad & \frac{X_1' - X_1}{\mathrm{D}_1} &= \frac{p_1}{n_1}, \\ \frac{Y_0' - Y_0}{\mathrm{D}_0} &= \frac{q_0}{n_0}, \qquad & \frac{Y_1' - Y_1}{\mathrm{D}_1} &= \frac{q_1}{n_1}. \end{aligned}\right\} \tag{4}$$

---

* SCHWARZSCHILD also considered the general types of fifth-order ray aberrations. Expressions for the associated sixth-order coefficients in the expansion of the characteristic function (SCHWARZSCHILD's perturbation eikonal) were first given by his pupil A. KOHLSCHÜTTER in his Dissertation (Göttingen, 1908).

Fifth-order aberrations were also investigated by M. HERZBERGER, *J. Opt. Soc. Amer.*, **29** (1939), 395, and also his *Modern Geometrical Optics* (New York, Interscience Publishers, 1958), H. A. BUCHDAHL, *Optical Aberration Coefficients* (Oxford University Press, 1954) and by other workers. A review of higher order aberration theory was published by J. FOCKE in *Progress in Optics*, Vol. 4, ed. E. WOLF (Amsterdam, North Holland Publishing Company and New York, J. Wiley and Sons, 1965), p. 1.

Next we introduce new units of length λ_0 and λ_1 in the plane of the entrance and the exit pupil such that

$$\frac{\lambda_1}{\lambda_0} = M' \tag{5}$$

is the lateral magnification between the two planes. In place of X'_0, Y'_0, X'_1, Y'_1, the following variables will be used:

$$\left.\begin{aligned} \xi_0 &= \frac{X'_0}{\lambda_0} = \frac{X_0}{\lambda_0} + \frac{\mathrm{D}_0 p_0}{\lambda_0 n_0}, & \xi_1 &= \frac{X'_1}{\lambda_1} = \frac{X_1}{\lambda_1} + \frac{\mathrm{D}_1 p_1}{\lambda_1 n_1}, \\ \eta_0 &= \frac{Y'_0}{\lambda_0} = \frac{Y_0}{\lambda_0} + \frac{\mathrm{D}_0 q_0}{\lambda_0 n_0}, & \eta_1 &= \frac{Y'_1}{\lambda_1} = \frac{Y_1}{\lambda_1} + \frac{\mathrm{D}_1 q_1}{\lambda_1 n_1}. \end{aligned}\right\} \tag{6}$$

Within the accuracy of Gaussian optics, $\xi_1 = \xi_0$, $\eta_1 = \eta_0$.

In order to simplify later calculations, it will be convenient to choose C as*

$$\mathrm{C} = \frac{n_0 l_0 \lambda_0}{\mathrm{D}_0} = \frac{n_1 l_1 \lambda_1}{\mathrm{D}_1}. \tag{7}$$

The equality of the two terms on the right-hand side follows from the SMITH–HELMHOLTZ formula (§ 4.4 (48)).

The quantities defined by (2) and (6) are the *Seidel variables*. The inverse relations expressing the old variables in terms of the Seidel variables will also be needed. We have, on solving (2) and (6):

$$\left.\begin{aligned} X_0 &= \frac{\mathrm{D}_0}{n_0 \lambda_0} x_0, & X_1 &= \frac{\mathrm{D}_1}{n_1 \lambda_1} x_1, \\ Y_0 &= \frac{\mathrm{D}_0}{n_0 \lambda_0} y_0, & Y_1 &= \frac{\mathrm{D}_1}{n_1 \lambda_1} y_1, \end{aligned}\right\} \tag{8}$$

$$\left.\begin{aligned} p_0 &= \frac{n_0 \lambda_0}{\mathrm{D}_0} \xi_0 - \frac{1}{\lambda_0} x_0, & p_1 &= \frac{n_1 \lambda_1}{\mathrm{D}_1} \xi_1 - \frac{1}{\lambda_1} x_1, \\ q_0 &= \frac{n_0 \lambda_0}{\mathrm{D}_0} \eta_0 - \frac{1}{\lambda_0} y_0, & q_1 &= \frac{n_1 \lambda_1}{\mathrm{D}_1} \eta_1 - \frac{1}{\lambda_1} y_1. \end{aligned}\right\} \tag{9}$$

Next the aberration function will be expressed in terms of the Seidel variables. We note first of all that the arguments X and Y may be replaced by X'_1 and Y'_1 in Φ without changing the error term in equations (16b) and (17b) of § 5.1. Let us denote by ϕ the aberration function, when regarded as a function of the Seidel variables,

$$\Phi(X_0, Y_0; X'_1, Y'_1) = \phi(x_0, y_0; \xi_1, \eta_1). \tag{10}$$

---

* Another choice of C, which is often useful in the study of images on the basis of diffraction theory, is

$$C = k n_0 l_0 \sin \theta_0 = -k n_1 l_1 \sin \theta_1.$$

Here k is the vacuum wave number of the light and $2\theta_0$ and $2\theta_1$ are the angular apertures of the system, i.e. the angles which the entrance pupil and the exit pupil subtend at the axial object and image points respectively. With this choice of C,

$$\begin{aligned} x_0 &= k n_0 X_0 \sin \theta_0, & x_1 &= -k n_1 X_1 \sin \theta_1. \\ y_0 &= k n_0 Y_0 \sin \theta_0, & y_1 &= -k n_1 Y_1 \sin \theta_1. \end{aligned}$$

Then

$$\frac{\partial \Phi}{\partial X_1'} = \frac{\partial \phi}{\partial \xi_1}\frac{\partial \xi_1}{\partial X_1'} = \frac{1}{\lambda_1}\frac{\partial \phi}{\partial \xi_1}, \tag{11a}$$

and from eqs. (2), (1) and (7) and § 5.1 (4)

$$X_1 - X_1^{\star} = \frac{D_1}{n_1\lambda_1}(x_1 - x_0). \tag{11b}$$

With the help of eqs. (11) the formulae (16b) and (17b) of § 5.1 give

$$\left.\begin{aligned} x_1 - x_0 &= -\frac{\partial \phi}{\partial \xi_1} + O(D_1\mu^5),\\ y_1 - y_0 &= -\frac{\partial \phi}{\partial \eta_1} + O(D_1\mu^5). \end{aligned}\right\} \tag{12}$$

It was mentioned earlier that within the accuracy of the Seidel theory, ϕ is closely related to a perturbation function introduced by SCHWARZSCHILD and called by him the *Seidel eikonal.* This perturbation function is defined by the expression

$$\psi = T + \frac{D_0}{2n_0\lambda_0^2}(x_0^2 + y_0^2) - \frac{D_1}{2n_1\lambda_1^2}(x_1^2 + y_1^2) + x_0(\xi_1 - \xi_0) + y_0(\eta_1 - \eta_0), \tag{13}$$

where $T = T(p_0, q_0; p_1, q_1)$ is the angle characteristic, referred to origins at O_0 and O_1. Consider now the effect of small variations in the coordinates. We have, according to eq. (27) of § 4.1, with $Z_0 = Z_1 = 0$, that the corresponding change in T is given by

$$\delta T = X_0 dp_0 + Y_0 dq_0 - X_1 dp_1 - Y_1 dq_1, \tag{14}$$

or, in terms of the Seidel variables,

$$\begin{aligned} \delta T = x_0\left(d\xi_0 - \frac{D_0}{n_0\lambda_0^2}dx_0\right) + y_0\left(d\eta_0 - \frac{D_0}{n_0\lambda_0^2}dy_0\right)\\ - x_1\left(d\xi_1 - \frac{D_1}{n_1\lambda_1^2}dx_1\right) - y_1\left(d\eta_1 - \frac{D_1}{n_1\lambda_1^2}dy_1\right). \end{aligned} \tag{15}$$

Using (15), we deduce from (13) that a small variation in the variables leads to a change in ψ given by

$$d\psi = (\xi_1 - \xi_0)dx_0 + (\eta_1 - \eta_0)dy_0 + (x_0 - x_1)d\xi_1 + (y_0 - y_1)d\eta_1, \tag{16}$$

so that ψ may be expressed as function of x_0, y_0, ξ_1, η_1 and we then have rigorously

$$\left.\begin{aligned} \xi_1 - \xi_0 &= \frac{\partial \psi}{\partial x_0}, & x_1 - x_0 &= -\frac{\partial \psi}{\partial \xi_1},\\ \eta_1 - \eta_0 &= \frac{\partial \psi}{\partial y_0}, & y_1 - y_0 &= -\frac{\partial \psi}{\partial \eta_1}. \end{aligned}\right\} \tag{17}$$

Hence from the knowledge of ψ the ray aberrations can be determined, both in the image plane and in the plane of the exit pupil by simple differentiation.

Comparison of (17) and (12) shows that within the accuracy of the Seidel theory $\phi - \psi$ must be independent of ξ_1 and η_1, i.e.

$$\phi(x_0, y_0; \xi_1, \eta_1) = \psi(x_0, y_0; \xi_1, \eta_1) + \chi(x_0, y_0) + O(D_1\mu^6), \tag{18}$$

where $\chi(x_0, y_0)$ is some function of x_0 and y_0. Now from the definition of ϕ, $\phi(x_0, y_0; 0, 0) = 0$. Hence $\chi(x_0, y_0) = -\psi(x_0, y_0; 0, 0)$ and consequently

$$\phi(x_0, y_0; \xi_1, \eta_1) = \psi(x_0, y_0; \xi_1, \eta_1) - \psi(x_0, y_0; 0, 0) + O(D_1\mu^6). \tag{19}$$

Within the range of validity of the Seidel theory, the error terms in (12) may be neglected. If, however, terms of order higher than fourth are taken into account, the expressions for the ray aberration components in terms of the aberration function ϕ are more complicated. On the other hand, the simple relations (17) for the ray aberration components in terms of the perturbation eikonal are exact; the perturbation eikonal appears, however, to have no simple physical meaning.

The determination of terms of order higher than fourth is very laborious in all but the simplest cases. For this reason, algebraic calculations are usually restricted to the domain of the Seidel theory, supplemented where necessary by ray tracing.

5.3 THE PRIMARY (SEIDEL) ABERRATIONS

By means of considerations strictly similar to those of § 5.1 relating to the aberration function, it follows on account of symmetry that the power series expansion of the perturbation eikonal of Schwarzschild is of the form

$$\psi = \psi^{(0)} + \psi^{(4)} + \psi^{(6)} + \psi^{(8)} + \dots \tag{1}$$

where $\psi^{(2k)}$ is a polynomial of degree $2k$ in the four variables; and moreover, that the four variables enter only in the three combinations

$$\mathsf{r}^2 = x_0^{\,2} + y_0^{\,2}, \qquad \rho^2 = \xi_1^{\,2} + \eta_1^{\,2}, \qquad \kappa^2 = x_0\xi_1 + y_0\eta_1. \tag{2}$$

In (1) the term of the second degree is absent, for the presence of such a term would, according to § 5.2 (17), contradict the fact that within the accuracy of Gaussian optics, $x_1 = x_0$, $y_1 = y_0$, $\xi_1 = \xi_0$, and $\eta_1 = \eta_0$.

Since the variables enter only in the combinations (2), it follows that $\psi^{(4)}$ must be of the form

$$\psi^{(4)} = -\tfrac{1}{4}A\mathsf{r}^4 - \tfrac{1}{4}B\rho^4 - C\kappa^4 - \tfrac{1}{2}D\mathsf{r}^2\rho^2 + E\mathsf{r}^2\kappa^2 + F\rho^2\kappa^2, \tag{3}$$

where A, B, . . . are constants. The signs and the numerical factors in (3) have been chosen in agreement with the usual practice; they lead to convenient expressions for the ray aberrations. The evaluation of the aberration constants for any centred system will be discussed in § 5.5.

Clearly the power series expansion of ϕ is of the same form as (1), but it contains no term of zero order ($\phi^{(0)} = 0$), and the leading term $\phi^{(4)}$ differs from $\psi^{(4)}$ by the absence of the term $-\tfrac{1}{4}A\mathsf{r}^4$, as is immediately obvious from eq. (19) of § 5.2. Hence the general expression for the wave aberration of the lowest order (fourth) is

$$\phi^{(4)} = -\tfrac{1}{4}B\rho^4 - C\kappa^4 - \tfrac{1}{2}D\mathsf{r}^2\rho^2 + E\mathsf{r}^2\kappa^2 + F\rho^2\kappa^2 \tag{4}$$

B, C, . . . being the same coefficients as in (3).

Substitution from (4) into eq. (12) of § 5.2 gives the general expression for the ray aberration components of the lowest (third) order:

$$\left.\begin{aligned} \Delta^{(3)}x &= x_1 - x_0 = \frac{n_1\lambda_1}{\mathrm{D}_1}(X_1 - X_1^\star) \\ &= x_0(2C\kappa^2 - E\mathsf{r}^2 - F\rho^2) + \xi_1(B\rho^2 + D\mathsf{r}^2 - 2F\kappa^2), \\ \Delta^{(3)}y &= y_1 - y_0 = \frac{n_1\lambda_1}{\mathrm{D}_1}(Y_1 - Y_1^\star) \\ &= y_0(2C\kappa^2 - E\mathsf{r}^2 - F\rho^2) + \eta_1(B\rho^2 + D\mathsf{r}^2 - 2F\kappa^2). \end{aligned}\right\} \tag{5}$$

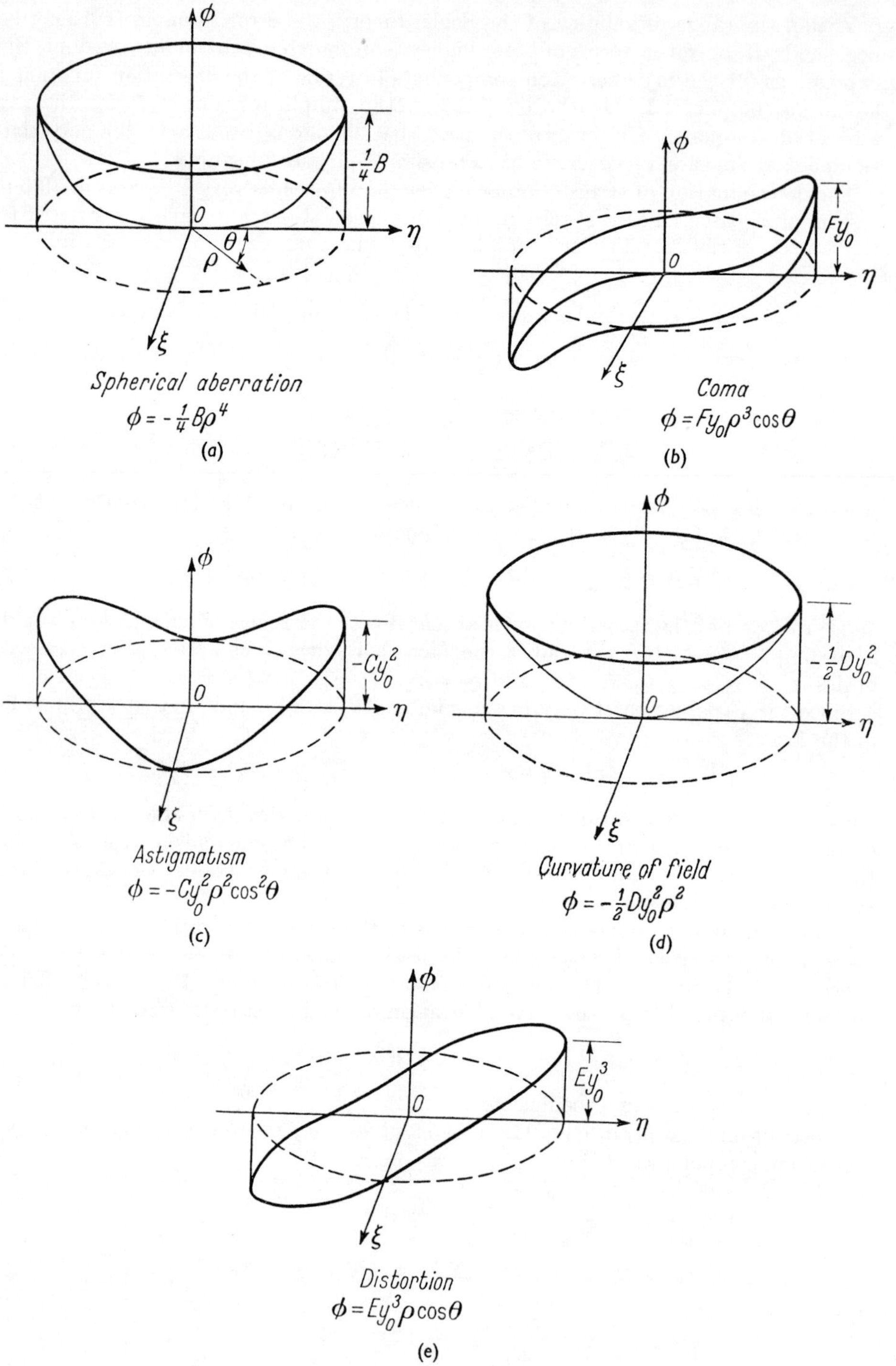

Fig. 5.3. The primary wave aberrations.

The coefficient A does not enter (4) and (5), so that there are altogether five types of aberrations of the lowest order characterized by the five coefficients B, C, D, E and F. As already mentioned on p. 207 these aberrations are known as *primary* or *Seidel aberrations*.

To discuss the effect of the Seidel aberrations it will be convenient to choose the axes so that the yz-plane passes through the object point; then $x_0 = 0$. If further polar coordinates are introduced, so that

$$\xi_1 = \rho \sin \theta, \qquad \eta_1 = \rho \cos \theta, \tag{6}$$

(4) becomes

$$\phi^{(4)} = -\tfrac{1}{4}B\rho^4 - Cy_0^2\rho^2 \cos^2 \theta - \tfrac{1}{2}Dy_0^2\rho^2 + Ey_0^3\rho \cos \theta + Fy_0\rho^3 \cos \theta, \tag{7}$$

and (5) takes the form

$$\left.\begin{aligned} \Delta^{(3)}x &= B\rho^3 \sin \theta - 2Fy_0\rho^2 \sin \theta \cos \theta + Dy_0^2\rho \sin \theta, \\ \Delta^{(3)}y &= B\rho^3 \cos \theta - Fy_0\rho^2(1 + 2\cos^2 \theta) + (2C + D)y_0^2\rho \cos \theta - Ey_0^3. \end{aligned}\right\} \tag{8}$$

In the special case when all the coefficients in (7) have zero values, the wave-front in the exit pupil coincides (within the present degree of accuracy) with the Gaussian reference sphere (cf. Fig. 5.2). In general, the coefficients will have finite values. Each term then represents a particular type of departure of the wave-front from the ideal spherical form; the five different types are illustrated in Fig. 5.3.

The significance of the ray aberrations associated with a given object point may be illustrated graphically by means of so-called *aberration* (or *characteristic*) *curves.* These are the curves traced out in the image plane by the intersection points of all the rays emerging from a fixed zone ρ = constant of the exit pupil. The area covered by the aberration curves which correspond to all the possible values of ρ then represent the imperfect image.

We shall consider in turn each of the Seidel aberrations.*

I. *Spherical aberration* ($B \neq 0$)

When all the coefficients except B are zero, (8) reduces to

$$\left.\begin{aligned} \Delta^{(3)}x &= B\rho^3 \sin \theta, \\ \Delta^{(3)}y &= B\rho^3 \cos \theta. \end{aligned}\right\} \tag{9}$$

The aberration curves are therefore concentric circles whose centres are at the Gaussian image point and whose radii increase with the third power of the zonal radius ρ, but are independent of the position (y_0) of the object in the field of view. This defect of the image is known as the *spherical aberration.*

The spherical aberration, being independent of y_0, affects the off-axis as well as the axial image. The rays from an axial object point which make an appreciable angle with the axis will intersect the axis in points which lie in front of or behind the Gaussian focus (Fig. 5.4). The point at which the rays from the edge of the aperture intersect the axis is called the *marginal focus.* If a screen is placed in the image region at right angles to the axis, there is a position for which the circular image spot appearing on the screen is a minimum; this minimal "image" is called the *circle of least confusion.*

---

* Aberration curves associated with aberrations of higher orders are discussed by G. C. STEWARD, *Trans. Camb. Phil. Soc.*, **23** (1926), 235 and by N. CHAKO, *Trans. Chalmers University of Technology* Gothenburg), Nr. 191 (1957).

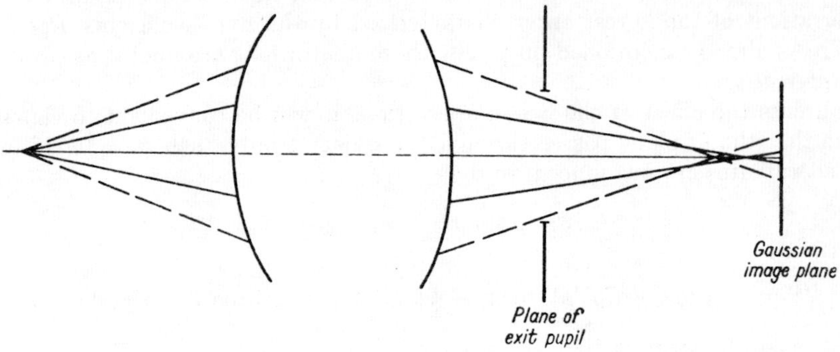

Fig. 5.4. Spherical aberration.

II. *Coma* ($F \neq 0$)

The aberration characterized by the coefficient F is known as coma. According to (8) the ray aberration components are in this case

$$\left.\begin{aligned}\Delta^{(3)}x &= -\,2F\rho^2 y_0 \sin\theta\cos\theta = -\,F y_0\rho^2 \sin 2\theta,\\ \Delta^{(3)}y &= -\,F\rho^2 y_0(1 + 2\cos^2\theta) = -\,F y_0\rho^2(2 + \cos 2\theta).\end{aligned}\right\} \qquad (10)$$

It is seen that, if y_0 is fixed and the zonal radius ρ is kept constant, the point P_1 in the image plane describes a circle twice over, as θ runs through the range $0 \leqslant \theta < 2\pi$. The circle is of radius $|F y_0 \rho^2|$ and its centre is at a distance $-\,2F\rho^2 y_0$ from the Gaussian focus, in the y-direction. The circle therefore touches the two straight lines which pass through the Gaussian image and which are inclined to the y-axis at 30°. As ρ takes on all possible values, the circles cover a region bounded by segments of the two straight lines and by an arc of the largest aberration circle (see Fig. 5.5). The overall size of this pattern increases linearly with the off-axis distance of the object point. Now in § 4.5 it was shown that, if the Abbe sine condition is satisfied, an element of the object plane in the immediate neighbourhood of the axis will be imaged sharply by the system. In such a case the expansion of the aberration function cannot therefore contain terms which depend linearly on y_0. Hence if the sine condition is satisfied, the primary coma, in particular, will be absent.

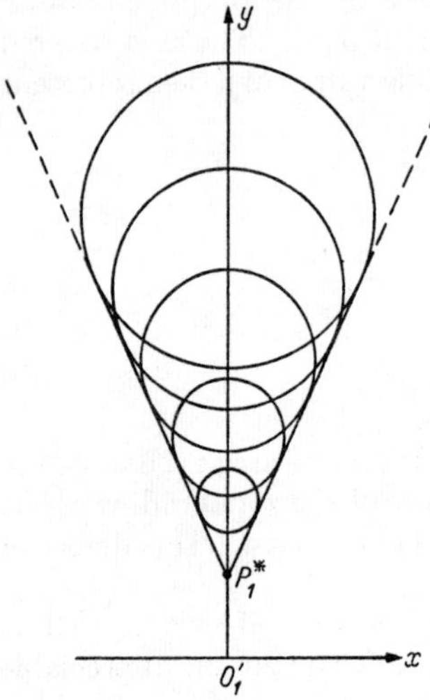

Fig. 5.5. Coma.

III. *Astigmatism* ($C \neq 0$) *and curvature of field* ($D \neq 0$)

The effects of the aberrations which are characterized by the coefficients C and D can best be studied together. We have from (8), if all the other coefficients have zero values,

$$\left.\begin{aligned}\Delta^{(3)}x &= D\rho y_0^{\,2}\sin\theta,\\ \Delta^{(3)}y &= (2C + D)\rho y_0^{\,2}\cos\theta.\end{aligned}\right\} \qquad (11)$$

To see the significance of these aberrations, assume to begin with that the image-forming pencil is very narrow. According to § 4.6, the rays of such a pencil intersect two short lines, of which one (the tangential focal line) is at right angles to the meridional plane, and the other (the sagittal focal line) lies in this plane. Consider now light from all the points in a finite region of the object plane. The focal lines in the image space give rise to two surfaces, the *tangential* and the *sagittal focal surface.* To a first approximation these surfaces may be considered to be spheres. Let R_t and R_s be their radii, regarded as positive if the corresponding centre of curvature lies on

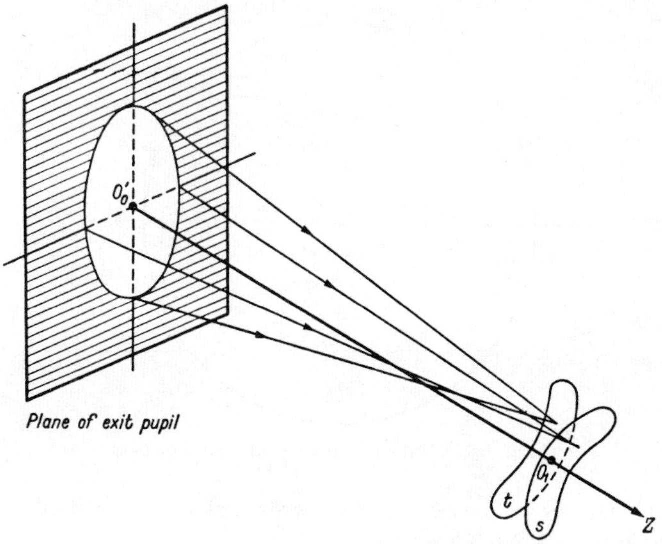

Fig. 5.6. The tangential and sagittal focal surfaces.

the side of the image plane from which the light is propagated ($R_t > 0$, $R_s < 0$ in Fig. 5.6).

The radii of curvature may be expressed in terms of the coefficients C and D. To show this it will be convenient to calculate the ray aberrations in the presence of curvature, using first the ordinary rather than the Seidel coordinates. We have (see Fig. 5.7)

$$\frac{\Delta^{(3)} Y_1}{Y_1'} = \frac{u}{\mathrm{D}_1 + u} \tag{12}$$

where u is the small distance between the sagittal focal line and the image plane. If v denotes the distance of this focal line from the axis,

$$R_t^2 = v^2 + (R_t - u)^2$$

i.e.

$$v^2 = 2R_t u - u^2.$$

If u is regarded as a small quantity of the first order, v may be replaced by Y_1, and u^2 may be neglected in the last equation, so that

$$u \sim \frac{Y_1^2}{2R_t}, \tag{13}$$

and (12) gives, if we also neglect u in comparison with D_1,

$$\Delta^{(3)} Y_1 = \frac{Y_1^2}{2R_t} \frac{Y_1'}{D_1}. \tag{14}$$

Similarly

$$\Delta^{(3)} X_1 = \frac{Y_1^2}{2R_s} \frac{X_1'}{D_1}. \tag{15}$$

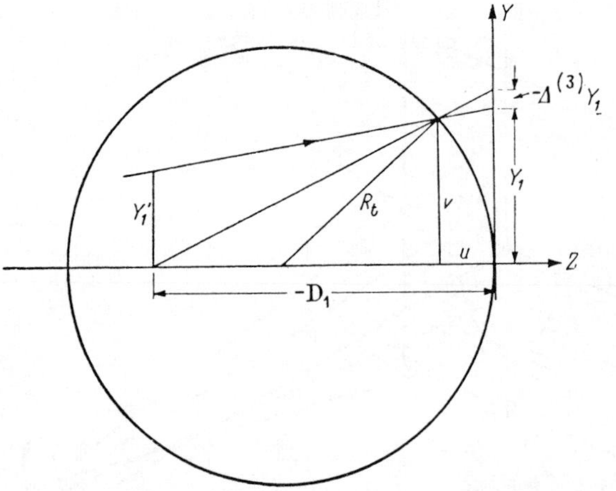

Fig. 5.7. Astigmatism and curvature of field.

Next these relations will be expressed in terms of the Seidel variables. Substitution from eqs. (6) and (8) of § 5.2 leads to

$$\Delta^{(3)} y = \frac{n_1 \lambda_1}{D_1} \Delta^{(3)} Y_1 = \frac{n_1 \lambda_1}{D_1} \frac{Y_1^2}{2R_t} \frac{Y_1'}{D_1} = \frac{n_1 \lambda_1}{2D_1^2 R_t} \frac{D_1^2}{n_1^2 \lambda_1^2} y_1^2 \lambda_1 \eta_1$$

i.e.

$$\Delta^{(3)} y = \frac{y_1^2 \eta_1}{2n_1 R_t}, \tag{16}$$

and similarly

$$\Delta^{(3)} x = \frac{y_1^2 \xi_1}{2n_1 R_s}. \tag{17}$$

In (16) and (17) y_1 may be replaced by y_0, and we finally obtain, on comparison with (11) and on using (6),

$$\frac{1}{R_t} = 2n_1(2C + D), \qquad \frac{1}{R_s} = 2n_1 D. \tag{18}$$

The quantity $2C + D$ is usually called the *tangential field curvature* and D the *sagittal field curvature*, and the quantity

$$\frac{1}{R} = \frac{1}{2}\left(\frac{1}{R_t} + \frac{1}{R_s}\right) = 2n_1(C + D), \tag{19}$$

which is proportional to their arithmetic mean is called, simply, *the field curvature.*

From (13) and (18) it is seen that, at height Y_1 from the axis, the separation between the two focal surfaces (i.e. the astigmatic distance of the image forming pencil) is

$$\frac{Y_1^2}{2}\left(\frac{1}{R_t}-\frac{1}{R_s}\right) = 2n_1CY_1^2. \tag{20}$$

The semi-difference

$$\frac{1}{2}\left(\frac{1}{R_t}-\frac{1}{R_s}\right) = 2n_1C \tag{21}$$

is accordingly called *astigmatism*. In the absence of astigmatism ($C = 0$), $R_t = R_s = R$. It will be shown in § 5.5.3 that the radius R of the common focal surface may then be calculated from a simple formula which only involves the radii of curvature of the individual surfaces of the system and the refractive indices of all the medii.

IV. *Distortion* ($E \neq 0$)

If the E coefficient alone differs from zero, we have according to (8)

$$\left.\begin{aligned} \Delta^{(3)}x &= 0, \\ \Delta^{(3)}y &= -Ey_0^3. \end{aligned}\right\} \tag{22}$$

Since these expressions are independent of ρ and θ, the imaging will be stigmatic and independent of the radius of the exit pupil; the off-axis distance of the image will, however, not be proportional to that of the object. This aberration is therefore called *distortion.*

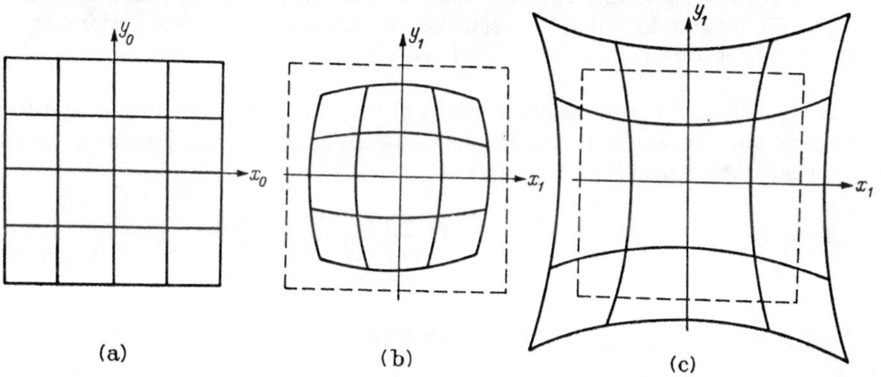

Fig. 5.8.

(a) Object

(b) Image in the presence of barrel distortion ($E > 0$)

(c) Image in the presence of pincushion distortion ($E < 0$)

If distortion is present, the image of any straight line in the object plane which meets the axis will itself be a straight line, but the image of any other straight line will be curved. This effect is shown in Fig. 5.8 (a), where the object has the form of a mesh formed by equidistant lines parallel to the x- and y-axes. Fig. 5.8 (b) illustrates the so-called *barrel distortion* ($E > 0$), and Fig. 5.8 (c) illustrates *pincushion distortion* ($E < 0$).

It has been seen that, of the five Seidel aberrations, three (namely spherical aberration, coma and astigmatism) are responsible for lack of sharpness of the image.

The other two (namely curvature of field and distortion) are related to the position and the form of the image. In general it is impossible to design a system which is free from all the primary as well as the higher order aberrations, and a suitable compromise as to their relative magnitudes has, therefore, to be made. In some cases the effects of the Seidel aberrations are reduced by balancing them against aberrations of higher orders. In others one has to eliminate certain aberrations completely even at the price of introducing aberrations of other types. For example, because of the asymmetric appearance of an image in the presence of coma, this aberration must always be suppressed in telescopes, as it would make precise positional measurements impracticable. On the other hand a certain amount of curvature of the field and of distortion is then relatively harmless as these can be eliminated by calculation.

So far we have studied the aberration effects on the basis of geometrical optics only. If, however, the aberrations are very small (wave aberrations of the order of a wavelength or less), diffraction plays an important part. The geometrical theory must then be supplemented by more refined considerations. This will be done in Chapter IX.

5.4 ADDITION THEOREM FOR THE PRIMARY ABERRATIONS

Having considered the significance of the primary aberrations, we must turn our attention to the more difficult task of calculating the primary aberration coefficients for a general centred system. As was seen, this is equivalent to determining the fourth-order terms in the power series expansion of the perturbation eikonal of SCHWARZSCHILD. In order not to interrupt the main calculations, it will be convenient to consider first the manner in which the perturbation eikonal of an optical system depends on the perturbation eikonals associated with the imaging by each surface of the system.

Consider a centred system which consists of two surfaces of revolution, and let O_0 be the axial object point and O_1 and O_2 its Gaussian images by the first surface and by both the surfaces respectively. Further let

$$T_1 = T_1(p_0, q_0; p_1, q_1) \tag{1}$$

be the angle characteristic for refraction at the first surface and,

$$T_2 = T_2(p_1, q_1; p_2, q_2) \tag{2}$$

the angle characteristic for refraction at the second surface, the former being referred to axes at O_0 and O_1 and the latter to parallel axes at O_1 and O_2; the Z-axes are taken along the axis of the system. Since the media are assumed to be homogeneous, the angle characteristic represents the optical path between the feet of the perpendiculars dropped on to the initial and final portions of the ray from the two origins (see Fig. 4.3). Hence the angle characteristic T of the system referred to axes at O_0 and O_2 is obtained from T_1 and T_2 by addition:

$$T = T_1 + T_2. \tag{3}$$

In this expression, the variables p_1, q_1 referring to the ray in the intermediate space must be eliminated with the help of formulae describing the imaging by each surface. It will be convenient to carry out the elimination explicitly in the corresponding relation for the perturbation eikonal.

According to § 5.2 (13), the perturbation eikonals ψ_1 and ψ_2 are given by

$$\psi_1 = T_1 + \frac{\mathrm{D}_0}{2n_0\lambda_0^2}(x_0^2 + y_0^2) - \frac{\mathrm{D}_1}{2n_1\lambda_1^2}(x_1^2 + y_1^2) + x_0(\xi_1 - \xi_0) + y_0(\eta_1 - \eta_0),$$

and

$$\psi_2 = T_2 + \frac{\mathrm{D}_1}{2n_1\lambda_1^2}(x_1^2 + y_1^2) - \frac{\mathrm{D}_2}{2n_2\lambda_2^2}(x_2^2 + y_2^2) + x_1(\xi_2 - \xi_1) + y_1(\eta_2 - \eta_1);$$

and the perturbation eikonal ψ of the combination is

$$\psi = T + \frac{\mathrm{D}_0}{2n_0\lambda_0^2}(x_0^2 + y_0^2) - \frac{\mathrm{D}_2}{2n_2\lambda_2^2}(x_2^2 + y_2^2) + x_0(\xi_2 - \xi_0) + y_0(\eta_2 - \eta_0).$$

Hence, using (3),

$$\psi = \psi_1 + \psi_2 + (x_0 - x_1)(\xi_2 - \xi_1) + (y_0 - y_1)(\eta_2 - \eta_1),$$

or, using eq. (17) of § 5.2

$$\psi = \psi_1 + \psi_2 + \frac{\partial \psi_1}{\partial \xi_1}\frac{\partial \psi_2}{\partial x_1} + \frac{\partial \psi_1}{\partial \eta_1}\frac{\partial \psi_2}{\partial y_1}. \tag{4}$$

If in (4) the power series expansions for ψ_1 and ψ_2 are substituted, it follows (since according to § 5.3 (1) there are no terms of second order), that

$$\psi = \psi_1^{(0)} + \psi_2^{(0)} + \psi_1^{(4)} + \psi_2^{(4)} + \dots, \tag{5}$$

the terms not shown being of sixth and higher orders. Now in $\psi_1^{(4)}$ and $\psi_2^{(4)}$ we may obviously replace the arguments by their Gaussian values. Hence the variables relating to the intermediate space may be eliminated by means of the following rule: In $\psi_1^{(4)}(x_0, y_0;\ \xi_1, \eta_1)$ replace ξ_1, η_1 by ξ_2, η_2 and in $\psi_2^{(4)}(x_1, y_1;\ \xi_2, \eta_2)$ replace x_1, y_1 by x_0, y_0, and add the resulting expressions. This gives $\psi^{(4)}$ as function of the four variables x_0, y_0, and ξ_2, η_2 as required. This elimination will now be carried out more explicitly.

According to § 5.3 (3), $\psi_1^{(4)}$ and $\psi_2^{(4)}$ are of the form

$$\left.\begin{aligned} \psi_1^{(4)} &= -\tfrac{1}{4}A_1\mathrm{r}_0^4 - \tfrac{1}{4}B_1\rho_1^4 - C_1\kappa_{01}^4 - \tfrac{1}{2}D_1\mathrm{r}_0^2\rho_1^2 + E_1\mathrm{r}_0^2\kappa_{01}^2 + F_1\rho_1^2\kappa_{01}^2, \\ \text{with}\quad & \\ \mathrm{r}_0^2 &= x_0^2 + y_0^2, \qquad \rho_1^2 = \xi_1^2 + \eta_1^2, \qquad \kappa_{01}^2 = x_0\xi_1 + y_0\eta_1; \end{aligned}\right\} \tag{6}$$

$$\left.\begin{aligned} \psi_2^{(4)} &= -\tfrac{1}{4}A_2\mathrm{r}_1^4 - \tfrac{1}{4}B_2\rho_2^4 - C_2\kappa_{12}^4 - \tfrac{1}{2}D_2\mathrm{r}_1^2\rho_2^2 + E\mathrm{r}_1^2\kappa_{12}^2 + F_2\rho_2^2\kappa_{12}^2, \\ \text{with}\quad & \\ \mathrm{r}_1^2 &= x_1^2 + y_1^2, \qquad \rho_2^2 = \xi_2^2 + \eta_2^2, \qquad \kappa_{12}^2 = x_1\xi_2 + y_1\eta_2; \end{aligned}\right\} \tag{7}$$

Replacing ξ_1, η_1 by ξ_2, η_2 and x_1, y_1 by x_0, y_0, i.e. ρ_1 by ρ_2, r_1 by r_0, κ_{12}^2 and κ_{01}^2 by

$$\kappa_{02}^2 = x_0\xi_2 + y_0\eta_2, \tag{8}$$

we obtain on addition

$$\begin{aligned} \psi^{(4)} = &-\tfrac{1}{4}(A_1 + A_2)\mathrm{r}_0^4 - \tfrac{1}{4}(B_1 + B_2)\rho_2^4 - (C_1 + C_2)\kappa_{02}^4 - \tfrac{1}{2}(D_1 + D_2)\mathrm{r}_0^2\rho_2^2 \\ &+ (E_1 + E_2)\mathrm{r}_0^2\kappa_{02}^2 + (F_1 + F_2)\rho_2^2\kappa_{02}^2. \end{aligned} \tag{9}$$

A similar result obviously holds for a centred system consisting of any number of surfaces. We have thus proved the following theorem:

Each primary aberration coefficient of a centred system is the sum of the corresponding aberration coefficients associated with the individual surfaces of the system.

It is at this point of our analysis that the advantage of the Seidel variables is particularly evident; for this simple and important result depends basically upon the use of the Seidel variables and has no analogy when ordinary variables are used.

5.5 THE PRIMARY ABERRATION COEFFICIENTS OF A GENERAL CENTRED LENS SYSTEM

The theorem established in the preceding section reduces the problem of determining the primary aberration coefficients of a general centred system to that of calculating the corresponding coefficients for each of its surfaces. This calculation will now be carried out.

5.5.1 The Seidel formulae in terms of two paraxial rays

It may be recalled that the Seidel aberration coefficients may (apart from simple numerical factors) be identified with the coefficients of the fourth-order terms in the power series expansion of the perturbation eikonal ψ of SCHWARZSCHILD. According to § 5.2 (13), this function is obtained by adding to the angle characteristic T certain quadratic terms, and by expressing the resulting expression in terms of the Seidel variables. Since the relations between the Seidel variables and the ray components are linear, the order of the terms does not change by transition from the one set of variables to the other. Hence

$$\psi^{(4)}(x_0, y_0; \xi_1, \eta_1) = T^{(4)}(p_0, q_0; p_1, q_1). \tag{1}$$

The expansion of the angle characteristic up to the fourth order for a refracting surface of revolution was derived in § 4.1. The fourth-order contribution may, according to § 4.1 (42), be written in the form*

$$T^{(4)}(p_0, q_0; p_1, q_1) = -\frac{r}{4(n_1-n_0)^2}[(p_0-p_1)^2+(q_0-q_1)^2]\left[\frac{p_0^2+q_0^2}{n_0}-\frac{p_1^2+q_1^2}{n_1}\right]$$
$$-\frac{(1+b)r}{8(n_1-n_0)^3}[(p_0-p_1)^2+(q_0-q_1)^2]^2+\frac{a_0}{8n_0^3}(p_0^2+q_0^2)^2-\frac{a_1}{8n_1^3}(p_1^2+q_1^2)^2. \tag{2}$$

It will be convenient to choose the axial points $z = a_0$, $z = a_1$ as the axial object point and its Gaussian image, and to set (see Fig. 5.9)

$$s = a_0, \qquad s' = a_1, \qquad t = a_0 + \mathrm{D}_0, \qquad t' = a_1 + \mathrm{D}_1. \tag{3}$$

The associated Abbe invariants (§ 4.4 (7)) will be denoted by K and L respectively:

$$n_0\left(\frac{1}{r}-\frac{1}{s}\right) = n_1\left(\frac{1}{r}-\frac{1}{s'}\right) = K, \tag{4}$$

$$n_0\left(\frac{1}{r}-\frac{1}{t}\right) = n_1\left(\frac{1}{r}-\frac{1}{t'}\right) = L. \tag{5}$$

---

* As in § 4.1, r denotes the radius of the surface. This symbol should not be confused with the symbol r which denotes the rotational invariant $\sqrt{x_0^2+y_0^2}$ introduced in § 5.3 (2).

Before substituting into (1) the expressions for the ray components in terms of the Seidel variables, it will be useful to re-write (1) in a slightly different form. Because of (4) we have

$$\frac{r}{8(n_1-n_0)^3}[(p_0-p_1)^2+(q_0-q_1)^2]^2 = \frac{r^2}{8(n_1-n_0)^4}\left[\frac{n_1}{s'}-\frac{n_0}{s}\right][(p_0-p_1)^2+(q_0-q_1)^2]^2.$$

Using this relation, (2) may be written as

$$\begin{aligned} T^{(4)} = {} & \frac{1}{8n_0 s}\left\{\frac{n_0 r}{(n_1-n_0)^2}[(p_0-p_1)^2+(q_0-q_1)^2] - \frac{s}{n_0}(p_0^2+q_0^2)\right\}^2 \\ & - \frac{1}{8n_1 s'}\left\{\frac{n_1 r}{(n_1-n_0)^2}[(p_0-p_1)^2+(q_0-q_1)^2] - \frac{s'}{n_1}(p_1^2+q_1^2)\right\}^2 \\ & - \frac{br}{8(n_1-n_0)^3}\{(p_0-p_1)^2+(q_0-q_1)^2\}^2. \end{aligned} \tag{6}$$

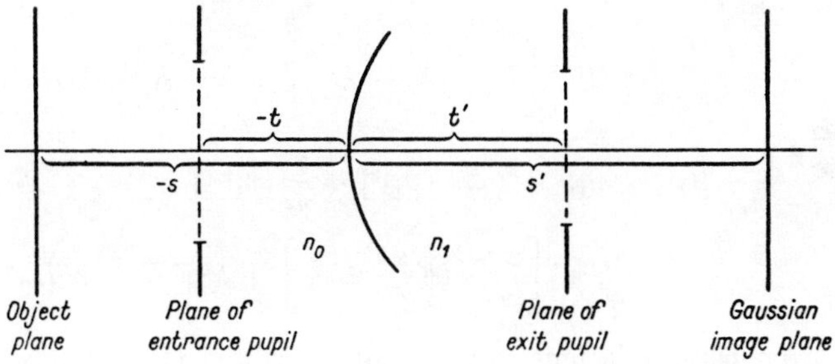

Fig. 5.9. Notation used in the calculation of the primary aberration coefficients.

In (6), the arguments may be replaced by their Gaussian approximations; in particular, the Seidel variables referring to points on the incident and the refracted ray may be interchanged. In order to obtain $\psi^{(4)}$ as a function of x_0, y_0, ξ_1 and η_1, we may then use in place of § 5.2 (9) the relations

$$\left.\begin{aligned} p_0 &= \frac{n_0\lambda_0}{D_0}\xi_1 - \frac{1}{\lambda_0}x_0, & p_1 &= \frac{n_1\lambda_1}{D_1}\xi_1 - \frac{1}{\lambda_1}x_0, \\ q_0 &= \frac{n_0\lambda_0}{D_0}\eta_1 - \frac{1}{\lambda_0}y_0, & q_1 &= \frac{n_1\lambda_1}{D_1}\eta_1 - \frac{1}{\lambda_1}y_0. \end{aligned}\right\} \tag{7}$$

It will be useful to make one further modification. The Gaussian lateral magnification between the object and the image plane (l_1/l_0) and between the planes of the entrance and the exit pupil (λ_1/λ_0) may be obtained from § 4.4 (14) and § 4.4 (10), or more simply by noting that imaging by a spherical surface is a projection from the centre of the sphere. Hence, if (4) is also used,

$$\frac{l_1}{l_0} = \frac{r-s'}{r-s} = \frac{n_0 s'}{n_1 s}, \qquad \frac{\lambda_1}{\lambda_0} = \frac{r-t'}{r-t} = \frac{n_0 t'}{n_1 t}. \tag{8}$$

Introducing the abbreviations

$$h = \frac{\lambda_0 s}{D_0} = \frac{\lambda_1 s'}{D_1}, \qquad H = \frac{t}{\lambda_0 n_0} = \frac{t'}{\lambda_1 n_1}, \tag{9}$$

where (8) and § 5.2 (7) was used, (7) becomes

$$\left.\begin{aligned} p_0 &= n_0\left(\frac{h}{s}\,\xi_1 - \frac{H}{t}\,x_0\right), & p_1 &= n_1\left(\frac{h}{s'}\,\xi_1 - \frac{H}{t'}\,x_0\right),\\ q_0 &= n_0\left(\frac{h}{s}\,\eta_1 - \frac{H}{t}\,y_0\right), & q_1 &= n_1\left(\frac{h}{s'}\,\eta_1 - \frac{H}{t'}\,y_0\right). \end{aligned}\right\} \tag{10}$$

If as before, r^2, ρ^2 and κ^2 denote the three rotational invariants

$$\mathsf{r}^2 = x_0^2 + y_0^2, \qquad \rho^2 = \xi_1^2 + \eta_1^2, \qquad \kappa^2 = x_0\xi_1 + y_0\eta_1, \tag{11}$$

the terms in the curly brackets of (6) become

$$(p_0 - p_1)^2 + (q_0 - q_1)^2 = \left(\frac{n_1 - n_0}{\mathsf{r}}\right)^2 [H^2\mathsf{r}^2 + h^2\rho^2 - 2hH\kappa^2],$$

$$\frac{n_0 r}{(n_1 - n_0)^2}[(p_0 - p_1)^2 + (q_0 - q_1)^2] - \frac{s}{n_0}(p_0^2 + q_0^2)$$
$$= H^2\mathsf{r}^2\left[L - (K - L)\frac{s}{t}\right] + h^2\rho^2 K - 2hH\kappa^2 L,$$

$$\frac{n_1 r}{(n_1 - n_0)^2}[(p_0 - p_1)^2 + (q_0 - q_1)^2] - \frac{s'}{n_1}(p_1^2 + q_1^2)$$
$$= H^2\mathsf{r}^2\left[L - (K - L)\frac{s'}{t'}\right] + h^2\rho^2 K - 2hH\kappa^2 L.$$

Substituting from these relations into (6) and recalling (1), we finally obtain the required expression for $\psi^{(4)}$:

$$\begin{aligned} \psi^{(4)} = \frac{1}{8}\,\mathsf{r}^4H^4 &\left\{\frac{b}{r^3}(n_0 - n_1) + L^2\left(\frac{1}{n_0 s} - \frac{1}{n_1 s'}\right) - 2L(K - L)\left(\frac{1}{n_0 t} - \frac{1}{n_1 t'}\right)\right.\\ &\left. + (K - L)^2\left(\frac{s}{n_0 t^2} - \frac{s'}{n_1 t'^2}\right)\right\}\\ &+ \frac{1}{8}\rho^4 h^4\left\{\frac{b}{r^3}(n_0 - n_1) + K^2\left(\frac{1}{n_0 s} - \frac{1}{n_1 s'}\right)\right\}\\ &+ \frac{1}{2}\kappa^4 H^2 h^2\left\{\frac{b}{r^3}(n_0 - n_1) + L^2\left(\frac{1}{n_0 s} - \frac{1}{n_1 s'}\right)\right\}\\ &+ \frac{1}{4}\mathsf{r}^2\rho^2 H^2 h^2\left\{\frac{b}{r^3}(n_0 - n_1) + KL\left(\frac{1}{n_0 s} - \frac{1}{n_1 s'}\right) - K(K - L)\left(\frac{1}{n_0 t} - \frac{1}{n_1 t'}\right)\right\}\\ &- \frac{1}{2}\mathsf{r}^2\kappa^2 H^3 h\left\{\frac{b}{r^3}(n_0 - n_1) + L^2\left(\frac{1}{n_0 s} - \frac{1}{n_1 s'}\right) - L(K - L)\left(\frac{1}{n_0 t} - \frac{1}{n_1 t'}\right)\right\}\\ &- \frac{1}{2}\rho^2\kappa^2 H h^3\left\{\frac{b}{r^3}(n_0 - n_1) + KL\left(\frac{1}{n_0 s} - \frac{1}{n_1 s'}\right)\right\}. \end{aligned} \tag{12}$$

This formula gives, on comparison with the general expression § 5.3 (3), the fourth-order coefficients A, B, . . . F of the perturbation eikonal of a refracting surface of revolution.

Generalization to a centred system consisting of any number of refracting surfaces is now straightforward. Let us denote by the suffix i quantities referring to the ith surface, and let n_i be the refractive index of the medium which follows the ith surface. Then, using the addition theorem of § 5.4 it follows from (12) on comparison with § 5.3 (3) that

$$B = \frac{1}{2}\sum_i h_i^4 \left\{\frac{b_i}{r_i^3}(n_i - n_{i-1}) + K_i^2\left(\frac{1}{n_i s_i'} - \frac{1}{n_{i-1}s_i}\right)\right\},$$

$$C = \frac{1}{2}\sum_i H_i^2 h_i^2 \left\{\frac{b_i}{r_i^3}(n_i - n_{i-1}) + L_i^2\left(\frac{1}{n_i s_i'} - \frac{1}{n_{i-1}s_i}\right)\right\},$$

$$D = \frac{1}{2}\sum_i H_i^2 h_i^2 \left\{\frac{b_i}{r_i^3}(n_i - n_{i-1}) + K_i L_i\left(\frac{1}{n_i s_i'} - \frac{1}{n_{i-1}s_i}\right) - K_i(K_i - L_i)\left(\frac{1}{n_i t_i'} - \frac{1}{n_{i-1}t_i}\right)\right\},$$

$$E = \frac{1}{2}\sum_i H_i^3 h_i \left\{\frac{b_i}{r_i^3}(n_i - n_{i-1}) + L_i^2\left(\frac{1}{n_i s_i'} - \frac{1}{n_{i-1}s_i}\right) - L_i(K_i - L_i)\left(\frac{1}{n_i t_i'} - \frac{1}{n_{i-1}t_i}\right)\right\},$$

$$F = \frac{1}{2}\sum_i H_i h_i^3 \left\{\frac{b_i}{r_i^3}(n_i - n_{i-1}) + K_i L_i\left(\frac{1}{n_i s_i'} - \frac{1}{n_{i-1}s_i}\right)\right\}. \tag{13}$$

These are the *Seidel formulae* for the primary aberration coefficients of a general centred system of refracting surfaces.*

Equations (13) express the primary aberration coefficients in terms of data specifying the passage of two paraxial rays through the system, namely a ray from the axial object point and a ray from the centre of the entrance pupil. It will be useful to summarize the relevant Gaussian formulae. Let d_i be the distance between the poles of the ith and the $(i + 1)$th surface. Since the Gaussian image formed by the first i surfaces of the system is the object for the $(i + 1)$th surface, we have the transfer formulae

$$s_{i+1} = s_i' - d_i, \qquad t_{i+1} = t_i' - d_i. \tag{14}$$

Given the distances s_1 and t_1 of the object plane and the plane of the entrance pupil from the pole of the first surface, the distances $s_1', t_1', s_2, t_2, s_2', t_2', \ldots$ and the corresponding values $K_1, L_1, K_2, L_2, \ldots$ may then be calculated successively from the Abbe relations

$$\left.\begin{aligned} n_{i-1}\left(\frac{1}{r_i} - \frac{1}{s_i}\right) &= n_i\left(\frac{1}{r_i} - \frac{1}{s_i'}\right) = K_i, \\ n_{i-1}\left(\frac{1}{r_i} - \frac{1}{t_i}\right) &= n_i\left(\frac{1}{r_i} - \frac{1}{t_i'}\right) = L_i, \end{aligned}\right\} \tag{15}$$

and from (14). We also have to determine the quantities h_i and H_i. If for simplicity the arbitrary length λ_0 in the plane of the entrance pupil is taken equal to *unity*, and

---

* SEIDEL actually considered centred systems consisting of spherical surfaces only. The effects of asphericities (characterized by the constants b_i) was taken into account by later writers (cf. M. VON ROHR, *Geometrical Investigation of Formation of Images in Optical Systems*, translated from German by R. KANTHACK (London, H.M. Stationery Office, 1920), 344.

the relations $D_i = t_i' - s_i' = t_{i+1} - s_{i+1}$ are used, it is seen from (9) that h_i and H_i may be calculated in succession from the relations*

$$\left.\begin{aligned} H_1 &= \frac{t_1}{n_0}, & h_1 &= \frac{s_1}{t_1 - s_1}, \\ \frac{H_{i+1}}{H_i} &= \frac{t_{i+1}}{t_i'}, & \frac{h_{i+1}}{h_i} &= \frac{s_{i+1}}{s_i'}. \end{aligned}\right\} \tag{16}$$

From (9) and from the Abbe relations (4) and (5) we obtain the following relation, which may be used as check on the calculations and which will be needed later:

$$h_i H_i = \frac{s_i t_i}{n_{i-1}(t_i - s_i)} = \frac{s_i' t_i'}{n_i(t_i' - s_i')} = \frac{1}{L_i - K_i}. \tag{17}$$

5.5.2 The Seidel formulae in terms of one paraxial ray

It is often desirable to express the primary aberration coefficients in a form which shows as clearly as possible the dependence of the coefficients on parameters which specify the optical system, and for this purpose (13) are not the most convenient set of formulae. To evaluate them, two rays must be traced through the system in accordance with the laws of Gaussian optics, namely a ray from the axial object point and a ray from the centre of the entrance pupil. It was, however, shown by SEIDEL that we may eliminate the data relating to the second ray and thus express the primary aberration coefficients by means of quantities which relate to the first ray only. Naturally one quantity depending on the position of the entrance pupil remains in the formulae, since alteration of the position of the stop will obviously influence the aberrations.

To eliminate the data referring to the ray from the centre of the entrance pupil, it is necessary first of all to express the heights H_i in terms of h_i. From (16), (14) and (17)

$$\begin{aligned} \frac{H_{i+1}}{h_{i+1}} - \frac{H_i}{h_i} &= \frac{1}{h_i h_{i+1}} H_i h_i \left(\frac{t_{i+1}}{t_i'} - \frac{s_{i+1}}{s_i'}\right) \\ &= \frac{1}{h_i h_{i+1}} H_i h_i d_i \left(\frac{1}{s_i'} - \frac{1}{t_i'}\right) \\ &= \frac{d_i}{n_i h_i h_{i+1}}. \end{aligned} \tag{18}$$

If quantities k_i are defined by

$$H_i = k_i h_i, \tag{19}$$

it follows from (18) that

$$k_{i+1} = k_1 + \sum_{j=1}^{i} \frac{d_j}{n_j h_j h_{j+1}}, \tag{20}$$

where, according to (16),

$$k_1 = \frac{H_1}{h_1} = \frac{t_1(t_1 - s_1)}{n_0 s_1}. \tag{21}$$

---

* The relation $h_{i+1}/h_i = s_{i+1}/s_i'$ implies that h_i is proportional to the height from the axis, at which a paraxial ray from the axial object point intersects the ith surface. The relation $H_{i+1}/H_i = t_{i+1}/t_i'$ gives a similar interpretation to H_i, in terms of a paraxial ray from the centre of the entrance pupil.

The other quantities in (13) which relate to the paraxial ray from the centre of the entrance pupil may also easily be expressed in terms of quantities referring to the other ray. From (17) and (19)

$$L_i = K_i + \frac{1}{k_i h_i^2}. \tag{22}$$

Further from (15) and (22)

$$\begin{aligned}
\frac{1}{n_i t_i'} - \frac{1}{n_{i-1} t_i} &= \frac{1}{n_i r_i} - \frac{1}{n_{i-1} r_i} - \left(\frac{1}{n_i^2} - \frac{1}{n_{i-1}^2}\right) L_i \\
&= \frac{1}{n_i}\left(\frac{K_i}{n_i} + \frac{1}{s_i'}\right) - \frac{1}{n_{i-1}}\left(\frac{K_i}{n_{i-1}} + \frac{1}{s_i}\right) - \left(\frac{1}{n_i^2} - \frac{1}{n_{i-1}^2}\right)\left(K_i + \frac{1}{k_i h_i^2}\right) \\
&= \frac{1}{n_i s_i'} - \frac{1}{n_{i-1} s_i} - \frac{1}{k_i h_i^2}\left(\frac{1}{n_i^2} - \frac{1}{n_{i-1}^2}\right).
\end{aligned} \tag{23}$$

Substitution from (19), (22), and (23) into (13) finally gives

$$\left.\begin{aligned}
B &= \frac{1}{2}\sum_i h_i^4 \frac{b_i}{r_i^3}(n_i - n_{i-1}) + h_i^4 K_i^2 \left(\frac{1}{n_i s_i'} - \frac{1}{n_{i-1} s_i}\right), \\
C &= \frac{1}{2}\sum_i h_i^4 k_i^2 \frac{b_i}{r_i^3}(n_i - n_{i-1}) + (1 + h_i^2 k_i K_i)^2 \left(\frac{1}{n_i s_i'} - \frac{1}{n_{i-1} s_i}\right), \\
D &= \frac{1}{2}\sum_i h_i^4 k_i^2 \frac{b_i}{r_i^3}(n_i - n_{i-1}) + h_i^2 k_i K_i (2 + h_i^2 k_i K_i) \left(\frac{1}{n_i s_i'} - \frac{1}{n_{i-1} s_i}\right) \\
&\qquad - K_i \left(\frac{1}{n_i^2} - \frac{1}{n_{i-1}^2}\right), \\
E &= \frac{1}{2}\sum_i h_i^4 k_i^3 \frac{b_i}{r_i^3}(n_i - n_{i-1}) + k_i(1 + h_i^2 k_i K_i)(2 + h_i^2 k_i K_i) \left(\frac{1}{n_i s_i'} - \frac{1}{n_{i-1} s_i}\right) \\
&\qquad - \frac{1 + h_i^2 k_i K_i}{h_i^2}\left(\frac{1}{n_i^2} - \frac{1}{n_{i-1}^2}\right), \\
F &= \frac{1}{2}\sum_i h_i^4 k_i \frac{b_i}{r_i^3}(n_i - n_{i-1}) + h_i^2 K_i (1 + k_i h_i^2 K_i) \left(\frac{1}{n_i s_i'} - \frac{1}{n_{i-1} s_i}\right).
\end{aligned}\right\} \tag{24}$$

This is the required form of the Seidel formulae. The position of the entrance pupil enters here only through the factor k_1 which is related to the distance t_1 of the plane of the entrance pupil from the pole of the first surface by (21). The quantities s_i, s_i', K_i and h_i are again computed from the appropriate relations in (14)–(16), whilst the k_i's are determined from (20).

5.5.3 Petzval's theorem

From the expressions for the coefficients of astigmatism and curvature we can derive an interesting relation due to PETZVAL. We have

$$C - D = \frac{1}{2}\sum_i \left\{[(1 + h_i^2k_iK_i)^2 - 2h_i^2k_iK_i - h_i^4k_i^2K_i^2]\left[\frac{1}{n_is_i'} - \frac{1}{n_{i-1}s_i}\right] + K_i\left(\frac{1}{n_i^2} - \frac{1}{n_{i-1}^2}\right)\right\}$$

$$= \frac{1}{2}\sum_i \left\{\frac{1}{n_is_i'} - \frac{1}{n_{i-1}s_i} + K_i\left(\frac{1}{n_i^2} - \frac{1}{n_{i-1}^2}\right)\right\}$$

$$= \frac{1}{2}\sum_i \frac{1}{r_i}\left(\frac{1}{n_i} - \frac{1}{n_{i-1}}\right), \tag{25}$$

where (15) was used.

According to § 5.3, C and D determine the sagittal and the tangential field curvature. If n_α denotes the refractive index of the last medium, it follows from eqs. (18) and (21) of § 5.3 that

$$C = \frac{1}{4n_\alpha}\left(\frac{1}{R_t} - \frac{1}{R_s}\right), \qquad D = \frac{1}{2n_\alpha}\frac{1}{R_s}. \tag{26}$$

Hence (25) may be written as

$$\frac{1}{R_t} - \frac{3}{R_s} = 2n_\alpha \sum \frac{1}{r_i}\left(\frac{1}{n_i} - \frac{1}{n_{i-1}}\right). \tag{27}$$

We thus obtain a relation between the curvatures of the two focal surfaces which contains only the radii of the refracting surfaces of the system and the corresponding refractive indices. If the system is free of spherical aberration, coma, and astigmatism, then a sharp image is formed on a surface of radius $R_s = R_t = R$; and the radius of this surface is according to (27) given by

$$\frac{1}{R} = -n_\alpha \sum_i \frac{1}{r_i}\left(\frac{1}{n_i} - \frac{1}{n_{i-1}}\right). \tag{28}$$

This result is known as *Petzval's theorem.*

The condition

$$\sum_i \frac{1}{r_i}\left(\frac{1}{n_i} - \frac{1}{n_{i-1}}\right) = 0 \tag{29}$$

is known as *Petzval's condition.* It is a necessary condition for *flatness of the field.* It should, however, be remembered that this condition belongs to the domain of the Seidel theory; it loses its significance outside this domain.

Whether or not aberrations are present, the spherical surface which touches the two focal surfaces at their common axial point and whose radius R is given by (28) is called the *Petzval surface.* According to (27) and (28) the radii of curvature of the sagittal focal surface, the tangential focal surface, and the Petzval surface are related by

$$\frac{3}{R_s} - \frac{1}{R_t} = \frac{2}{R}. \tag{30}$$

5.6 EXAMPLE: THE PRIMARY ABERRATIONS OF A THIN LENS

The Seidel formulae will now be used to find the primary aberration coefficients of a thin lens of refractive index n, situated in air (vacuum). In this case

$$n_0 = n_2 = 1, \qquad n_1 = n. \tag{1}$$

Since the thickness d of the lens is assumed to be negligible, we have according to § 5.5 (14)

$$s_2 = s_1'; \tag{2}$$

and § 5.5 (15) gives

$$\left.\begin{aligned} \frac{1}{r_1} - \frac{1}{s_1} &= n\left(\frac{1}{r_1} - \frac{1}{s_1'}\right) = K_1, \\ n\left(\frac{1}{r_2} - \frac{1}{s_2}\right) &= \left(\frac{1}{r_2} - \frac{1}{s_2'}\right) = K_2. \end{aligned}\right\} \tag{3}$$

Further from § 5.5 (16) and § 5.5 (20), using (2)

$$\left.\begin{aligned} h_1 = h_2 = h, \qquad k_1 = k_2 = k, \\ h = \frac{s_1}{t_1 - s_1}, \qquad k = \frac{t_1(t_1 - s_1)}{s_1}. \end{aligned}\right\} \tag{4}$$

where

It will be convenient to characterize the lens by a number of simple parameters. Let $\mathscr{P}$ be the power of the lens (cf. eqs. (35) and (36) of § 4.4)

$$\mathscr{P} = \frac{1}{f} = (n-1)\left(\frac{1}{r_1} - \frac{1}{r_2}\right), \tag{5}$$

and let

$$\sigma = (n-1)\left(\frac{1}{r_1} + \frac{1}{r_2}\right). \tag{6}$$

From (3),

$$(n-1)\frac{1}{r_1} = \frac{n}{s_1'} - \frac{1}{s_1}, \qquad (n-1)\frac{1}{r_2} = \frac{n}{s_2} - \frac{1}{s_2'}, \tag{7}$$

so that, because of (2),

$$\frac{1}{s_2'} - \frac{1}{s_1} = \mathscr{P}. \tag{8}$$

This relation will be written in the form

$$-\frac{1}{s_1} - \frac{\mathscr{P}}{2} = -\frac{1}{s_2'} + \frac{\mathscr{P}}{2} = \mathscr{K}; \tag{9}$$

$\mathscr{K}$ plays a similar role for the lens as K does for a single surface. Accordingly, $\mathscr{K}$ will be referred to as the *Abbe invariant of the lens.*

The deformation coefficients b_1 and b_2 of the two surfaces of the lens will be seen later to enter the formulae only through the quantity

$$\beta = (n-1)\left(\frac{b_1}{r_1^3} - \frac{b_2}{r_2^3}\right); \tag{10}$$

this quantity may be called the *deformation coefficient of the lens.*

We now express the various quantities which enter the Seidel formulae § 5.5 (24) in terms of the parameters $\mathscr{P}$, σ, β and $\mathscr{K}$. The first three specify the lens and the last specifies the position of the object. First, we obtain from (3)

$$K_1 = \mathscr{K} + \frac{\sigma + n\mathscr{P}}{2(n-1)}, \qquad K_2 = \mathscr{K} + \frac{\sigma - n\mathscr{P}}{2(n-1)}. \tag{11}$$

Further

$$\left.\begin{aligned} \frac{1}{ns_1'} - \frac{1}{s_1} &= \frac{n^2-1}{n^2}\mathscr{K} + \frac{\sigma}{2n^2} + \frac{\mathscr{P}}{2}, \\ \frac{1}{s_2'} - \frac{1}{ns_2} &= -\frac{n^2-1}{n^2}\mathscr{K} - \frac{\sigma}{2n^2} + \frac{\mathscr{P}}{2}. \end{aligned}\right\} \tag{12}$$

Substitution into the Seidel formulae § 5.5 (24) leads to the following expressions for the primary aberration coefficients of a thin lens:

$$\left.\begin{aligned} B &= h^4U, \\ F &= h^4kU + h^2V, \\ C &= h^4k^2U + 2h^2kV + \tfrac{1}{2}\mathscr{P}, \\ D &= h^4k^2U + 2h^2kV + \frac{n+1}{2n}\mathscr{P}, \\ E &= h^4k^3U + 3h^2k^2V + k\frac{3n+1}{2n}\mathscr{P}, \end{aligned}\right\} \tag{13}$$

where

$$\left.\begin{aligned} U &= \frac{1}{2}\beta + \frac{n^2}{8(n-1)^2}\mathscr{P}^3 - \frac{n}{2(n+2)}\mathscr{K}^2\mathscr{P} \\ &\qquad + \frac{1}{2n(n+2)}\mathscr{P}\left[\frac{n+2}{2(n-1)}\sigma + 2(n+1)\mathscr{K}\right]^2, \\ V &= \frac{1}{2n}\mathscr{P}\left[\frac{n+1}{2(n-1)}\sigma + (2n+1)\mathscr{K}\right]. \end{aligned}\right\} \tag{14}$$

Only the case when the *entrance pupil coincides with the lens* ($t_1 = 0$) will be considered here. According to (4) we then have

$$h = -1, \qquad k = 0,$$

and the formulae (13) become

$$\left.\begin{aligned} B &= U, \\ F &= V, \\ \frac{1}{2}(C+D) &= \frac{2n+1}{4n}\mathscr{P}, \\ \frac{1}{2}(C-D) &= -\frac{1}{4n}\mathscr{P}, \\ E &= 0. \end{aligned}\right\} \tag{15}$$

The quantities U and V defined by (14) are seen to represent the spherical aberration and coma of a lens working with the stop in the plane of the lens. Such a lens is free from distortion ($E = 0$) but astigmatism and field curvature are always present ($C \neq 0$, $D \neq 0$).

Let us examine whether, for a given lens which is working with the stop in the plane of the lens, there exists a pair of aplanatic points ($B = F = 0$). For coma to be absent we must have, according to (15), $V = 0$,

i.e.
$$\frac{n+1}{2(n-1)}\sigma + (2n+1)\mathscr{K} = 0. \tag{16}$$

Eq. (9) then gives the object distance s_1:

$$\frac{1}{s_1} = -\frac{1}{2}\mathscr{P} + \frac{n+1}{2(n-1)(2n+1)}\sigma. \tag{17}$$

With this choice of s_1, the spherical aberration of the lens is fully determined. Hence in general there is no aplanatic point pair for a given lens working with the stop at the lens. However, if only $\mathscr{P}$ and σ are given, one may eliminate the spherical aberration by an appropriate choice of the deformation coefficient β.

Fig. 5.10. Coma-free thin lens (object at infinity).

Conversely, if the object distance s_1 and the focal length $f = 1/\mathscr{P}$ are given one may first choose the radii r_1 and r_2 in such a way that coma vanishes. According to (17), (5) and (6) we have in this case,

$$\left.\begin{aligned}\frac{1}{r_1} &= \frac{2n+1}{n+1}\frac{1}{s_1} + \frac{n^2}{n^2-1}\mathscr{P},\\ \frac{1}{r_2} &= \frac{2n+1}{n+1}\frac{1}{s_1} + \frac{n^2-n-1}{n^2-1}\mathscr{P}.\end{aligned}\right\} \tag{18}$$

This gives, for example, for an object at infinity ($s_1 = \infty$) and with $n = 1{\cdot}5$,

$$r_1 = \frac{5}{9}\frac{1}{\mathscr{P}} = \frac{5}{9}f, \qquad r_2 = -5\frac{1}{\mathscr{P}} = -5f. \tag{19}$$

Such a coma-free lens is shown in Fig. 5.10, where C_1 and C_2 are the centres of curvature of the two surfaces of the lens.

Next, the spherical aberration may be eliminated by a suitable choice of β. When the object is at infinity, we have, according to (9), $\mathscr{K} = -\mathscr{P}/2$, and hence from (16) $\sigma = (2n+1)(n-1)\mathscr{P}/(n+1)$; the condition $U = 0$ for the absence of spherical aberration then gives the following expression for β:

$$-\frac{1}{2}\beta = \mathscr{P}^3\left\{\frac{n^2}{8(n-1)^2} - \frac{n}{8(n+2)} + \frac{1}{2n(n+2)}\left[\frac{(n+2)(2n+1)}{2(n+1)} - (n+1)\right]^2\right\},$$

whence

$$\beta = -\frac{n^3}{(n^2-1)^2}\mathscr{P}^3. \tag{20}$$

For $n = 1{\cdot}5$ this gives

$$\beta = -\tfrac{54}{25}\mathscr{P}^3, \tag{21}$$

so that by (19) and (10)

$$729b_1 + b_2 = -540. \tag{22}$$

This may be satisfied, for example, by taking $b_1 = b_2 = -0{\cdot}74$.

Instead of eliminating coma, one may first attempt to reduce the spherical aberration as much as possible. It is then found that, if the focal length is positive, the spherical aberration may be removed completely only if the object lies in a certain limited range; for $n = 1{\cdot}5$ this range extends from $0{\cdot}36f$ to $0{\cdot}44f$ behind the lens. For any other position of the object a certain amount of spherical aberration will always be present.

In order to affect the curvature of field, the lens must be used with a stop placed in an appropriate position, which may be determined from the general formulae (13). For a lens with a positive focal lens and with $n = 1{\cdot}5$, the curvature may be eliminated only if the object lies within a range extending from one focal length in front of the lens to half of the focal length behind it.

For a system which consists of a number of lenses, the calculations become more complicated. The removal of the chromatic aberration is then of primary importance. The primary aberrations of an achromatic telescope objective which consists of two cemented thin lenses may still be investigated with relative ease, and it is found that all the primary aberrations except astigmatism and curvature of field may be eliminated; it may therefore be used over a narrow field only (at most up to about 3° for an $f/10$ system).

5.7 THE CHROMATIC ABERRATION OF A GENERAL CENTRED LENS SYSTEM

The chromatic aberration has already been briefly discussed in § 4.7. Here explicit expressions will be derived for the chromatic change in the position of the image, and in the magnification, using the general formalism of the preceding sections.* The system will be assumed to be perfectly corrected for monochromatic light. This is a permissible idealization if effects of the leading order only are being considered; for changes in the monochromatic aberrations associated with a small change in the wavelength may be assumed to be small in comparison with the magnitudes of the aberrations themselves; in general these changes will therefore be of the same order of magnitude as the terms neglected in the Seidel theory.

According to § 5.5 (15) and § 5.5 (14),

$$n_{i-1}\left(\frac{1}{r_i} - \frac{1}{s_i}\right) = n_i\left(\frac{1}{r_i} - \frac{1}{s'_i}\right) = K_i, \tag{1}$$

$$s_{i+1} = s'_i - d_i. \tag{2}$$

If δ denotes the change in a quantity associated with a small change $\delta\lambda$ in the wavelength, we have from (1) and (2),

$$n_i\frac{\delta s'_i}{s'^2_i} - n_{i-1}\frac{\delta s_i}{s_i^2} = -\left(\frac{\delta n_i}{n_i} - \frac{\delta n_{i-1}}{n_{i-1}}\right)K_i, \tag{3}$$

$$\delta s_{i+1} = \delta s'_i. \tag{4}$$

Multiplying (3) by h_i^2 and using the relation

$$\frac{h_{i+1}}{h_i} = \frac{s_{i+1}}{s'_i}, \tag{5}$$

---

* We again follow substantially the analysis of K. SCHWARZSCHILD, *Abh. Königl. Ges. Wiss. Göttingen, Math.-phys. Kl.*, **4** (1905–1906), No. 3. Reprinted in *Astr. Mitt. Königl. Sternwarte Göttingen* (1905).

we find that

$$n_i \left(\frac{h_{i+1}}{s_{i+1}}\right)^2 \delta s_{i+1} - n_{i-1}\left(\frac{h_i}{s_i}\right)^2 \delta s_i = -\, h_i^2 K_i \left(\frac{\delta n_i}{n_i} - \frac{\delta n_{i-1}}{n_{i-1}}\right). \tag{6}$$

Let α denote the number of surfaces in the system. Then by adding all the equations of the form (6) it follows that

$$n_\alpha \left(\frac{h_{\alpha+1}}{s_{\alpha+1}}\right)^2 \delta s_{\alpha+1} = n_\alpha \left(\frac{h_\alpha}{s'_\alpha}\right)^2 \delta s'_\alpha = n_0 \left(\frac{h_1}{s_1}\right)^2 \delta s_1 - \sum_{i=1}^{\alpha} h_i^2 K_i \left(\frac{\delta n_i}{n_i} - \frac{\delta n_{i-1}}{n_{i-1}}\right).$$

Since the position of the object is independent of the wavelength, $\delta s_1 = 0$, and we obtain the following expression for the *chromatic change* $\delta s'_\alpha$ *in the position of the image:*

$$\delta s'_\alpha = -\frac{1}{n_\alpha}\left(\frac{s'_\alpha}{h_\alpha}\right)^2 \sum_{i=1}^{\alpha} h_i^2 K_i \left(\frac{\delta n_i}{n_i} - \frac{\delta n_{i-1}}{n_{i-1}}\right). \tag{7}$$

Next consider the chromatic change in the magnification M of the system. We have

$$M = \frac{l_1}{l_0}\frac{l_2}{l_1}\frac{l_3}{l_2}\cdots\frac{l_\alpha}{l_{\alpha-1}}. \tag{8}$$

Now by § 5.5 (8),

$$\frac{l_i}{l_{i-1}} = \frac{n_{i-1}}{n_i}\frac{s'_i}{s_i},$$

so that

$$M = \frac{n_0}{n_\alpha}\left(\frac{s'_1}{s_1}\frac{s'_2}{s_2}\frac{s'_3}{s_3}\cdots\frac{s'_\alpha}{s_\alpha}\right). \tag{9}$$

By logarithmic differentiation,

$$\begin{aligned}\frac{\delta M}{M} &= \frac{\delta n_0}{n_0} - \frac{\delta n_\alpha}{n_\alpha} + \sum_{i=1}^{\alpha}\left(\frac{\delta s'_i}{s'_i} - \frac{\delta s_i}{s_i}\right)\\ &= \left(\frac{\delta n_0}{n_0} - \frac{\delta n_\alpha}{n_\alpha}\right) + \frac{\delta s'_\alpha}{s'_\alpha} - \frac{\delta s_1}{s_1} - \sum_{i=2}^{\alpha}\delta s_i\left(\frac{1}{s_i} - \frac{1}{s'_{i-1}}\right).\end{aligned} \tag{10}$$

The sum entering this expression may be rewritten as follows

$$\sum_{i=2}^{\alpha}\delta s_i\left(\frac{1}{s_i} - \frac{1}{s'_{i-1}}\right) = \sum_{i=2}^{\alpha}\frac{d_{i-1}\delta s_i}{s_i s'_{i-1}} = \sum_{i=2}^{\alpha}\frac{d_{i-1}}{s_i^2}\frac{h_i}{h_{i-1}}\delta s_i.$$

But by § 5.5 (18),

$$\frac{d_{i-1}}{h_{i-1}} = (k_i - k_{i-1})h_i n_{i-1},$$

so that

$$\sum_{i=2}^{\alpha}\delta s_i\left(\frac{1}{s_i} - \frac{1}{s'_{i-1}}\right) = \sum_{i=2}^{\alpha} h_i^2 n_{i-1}\left(\frac{k_i - k_{i-1}}{s_i^2}\right)\delta s_i,$$

or, re-arranging according to the k_i's,

$$\begin{aligned}\sum_{i=2}^{\alpha}\delta s_i\left(\frac{1}{s_i} - \frac{1}{s'_{i-1}}\right) = &-n_0\left(\frac{h_1}{s_1}\right)^2 k_1\delta s_1 + n_\alpha\left(\frac{h_{\alpha+1}}{s_{\alpha+1}}\right)^2 k_\alpha \delta s_{\alpha+1}\\ &- \sum_{i=1}^{\alpha} k_i\left[n_i\left(\frac{h_{i+1}}{s_{i+1}}\right)^2\delta s_{i+1} - n_{i-1}\left(\frac{h_i}{s_i}\right)^2\delta s_i\right].\end{aligned}$$

Rewriting the second term on the right with the help of (4) and (5), and the last term with the help of (6), we obtain

$$\sum_{i=2}^{\alpha} \delta s_i \left(\frac{1}{s_i} - \frac{1}{s'_{i-1}}\right) = -n_0 \left(\frac{h_1}{s_1}\right)^2 k_1 \delta s_1 + n_\alpha \left(\frac{h_\alpha}{s'_\alpha}\right)^2 k_\alpha \delta s'_\alpha + \sum_{i=1}^{\alpha} h_i^2 k_i K_i \left(\frac{\delta n_i}{n_i} - \frac{\delta n_{i-1}}{n_{i-1}}\right).$$

On substituting this expression into (10) and on using the fact that $\delta s_1 = 0$, we obtain

$$\frac{\delta M}{M} = \left(\frac{\delta n_0}{n_0} - \frac{\delta n_\alpha}{n_\alpha}\right) + \frac{\delta s'_\alpha}{s'_\alpha}\left(1 - \frac{n_\alpha}{s'_\alpha} h_\alpha^2 k_\alpha\right) - \sum_{i=1}^{\alpha} h_i^2 k_i K_i \left(\frac{\delta n_i}{n_i} - \frac{\delta n_{i-1}}{n_{i-1}}\right).$$

But, by § 5.5 (22) and § 5.5 (15),

$$\frac{\delta s'_\alpha}{s'_\alpha}\left[1 - \frac{n_\alpha}{s'_\alpha} h_\alpha^2 k_\alpha\right] = \frac{\delta s'_\alpha}{s'_\alpha}\left[1 - \frac{n_\alpha}{s'_\alpha(L_\alpha - K_\alpha)}\right]$$

$$= \frac{\delta s'_\alpha}{s'_\alpha - t'_\alpha},$$

so that we finally obtain the following expression for the *chromatic change δM in the magnification*

$$\frac{\delta M}{M} = \left(\frac{\delta n_0}{n_0} - \frac{\delta n_\alpha}{n_\alpha}\right) + \frac{\delta s'_\alpha}{s'_\alpha - t'_\alpha} - \sum_{i=1}^{\alpha} h_i^2 k_i K_i \left(\frac{\delta n_i}{n_i} - \frac{\delta n_{i-1}}{n_{i-1}}\right). \tag{11}$$

If the refractive indices of the object and the image space are equal, as is usually the case, the first two terms on the right-hand side cancel out. If the system is achromatized with respect to the position of the image, $\delta s'_\alpha$ (given by (7)) also vanishes, and we then obtain the following two conditions for the achromatization of the position of the image and of the magnification

$$\sum_{i=1}^{\alpha} h_i^2 K_i \left(\frac{\delta n_i}{n_i} - \frac{\delta n_{i-1}}{n_{i-1}}\right) = 0, \tag{12}$$

and

$$\sum_{i=1}^{\alpha} h_i^2 k_i K_i \left(\frac{\delta n_i}{n_i} - \frac{\delta n_{i-1}}{n_{i-1}}\right) = 0. \tag{13}$$

For a *thin lens in air*, we obtain, on using the appropriate formulae § 5.6, the following expressions for the two sums:

$$\Sigma h_i^2 K_i \left(\frac{\delta n_i}{n_i} - \frac{\delta n_{i-1}}{n_{i-1}}\right) = h^2 \mathscr{P} \frac{\delta n}{n - 1}, \tag{14}$$

and

$$\Sigma h_i^2 k_i K_i \left(\frac{\delta n_i}{n_i} - \frac{\delta n_{i-1}}{n_{i-1}}\right) = h^2 k \mathscr{P} \frac{\delta n}{n - 1}, \tag{15}$$

where h, k, and the power $\mathscr{P}$ of the lens are given by § 5.6 (4) and § 5.6 (5). The corresponding expressions for the chromatic change in the position of the image and in the magnification are then obtained by substituting from (14) into (7) and from (15) into (11) respectively.

CHAPTER VI

IMAGE-FORMING INSTRUMENTS

THE three preceding chapters give an account of the geometrical theory of optical imaging, using for the main part the predictions of Gaussian optics and of the Seidel theory. An outstanding instance of the invaluable service rendered by this branch of optics lies in its ability to present the working principles of optical instruments in an easily visualized form. Although the quality of optical systems cannot be estimated by means of Gaussian theory alone, the purpose served by the separate optical elements can be indicated in this way, so that a simple, though somewhat approximate, picture of the action of the system can often be obtained without entering into the full intricacy of the techniques of optical design.

The development of optical instruments in the past has proceeded just as fast as technical difficulties have been overcome. It is hardly possible to give a step-by-step account of the design of optical systems, for two reasons. Firstly, the limitations of a given arrangement are not indicated by the predictions of the simple theory; in particular cases this needs to be supplemented by a fuller analysis often involving tedious calculations. Secondly, difficulties of a practical nature may prevent an otherwise praiseworthy arrangement from being used. It is not intended in this account to discuss the theoretical and practical limitations in individual cases; only the basic principles underlying the arrangement of some of the more important optical instruments will be given, in order to provide a framework for some of the later chapters which deal with the more detailed theories of optical image formation.*

6.1 THE EYE

Perhaps the simplest of optical instruments is that consisting of a single convergent lens forming a real image of an object upon a light-sensitive surface. Examples of such an optical system are found in the photographic camera and in the eye (Fig. 6.1). The back inner surface of the eye,† called the *retina*, consists of a layer of light-sensitive cells. Inasmuch as the eye forms an integral part of many optical systems, an understanding of its characteristics is an essential part of instrumental optics. The present section will include a description of a few of its features.

Refraction in the eye takes place mostly at the first surface or *cornea* of the eye, but also at the surfaces of the *lens* of the eye situated behind a contractible *iris*.

---

* For fuller accounts of optical instruments see, for example, *Dictionary of Applied Physics*, ed. R. GLAZEBROOK (London, Macmillan, 1923), Vol. IV; L. C. MARTIN, *An Introduction to Applied Optics*, Vol. II (London, Pitman, 1950); G. A. BOUTRY, *Optique Instrumentale* (Paris, Masson, 1946). References to books and articles dealing with particular types of instruments will be found in D. A. JACOBS, *Fundamentals of Optical Engineering* (New York, McGraw-Hill, 1943).

† For a detailed account of physiological optics the reader is referred to the standard work of H. HELMHOLTZ, *Handbuch der Physiologischen Optik* (Hamburg and Leipzig); L. Voss, 3 Aufl., I (1909), II (1911), III (1910). English translation edited by J. P. C. SOUTHALL (published by the Optical Society of America, 1924); and to J. P. C. SOUTHALL, *An Introduction to Physiological Optics* (Oxford University Press, 1937).

The latter gives an aperture-diameter which may be varied by unconscious muscular action, according to the total flux of light entering the eye, between approximately 1·5 mm and 6 mm. The lens is doubly convex, and its density, and hence its refractive index, is not uniform but decreases towards the outer zones. This serves partly to correct for spherical aberration (see § 5.3). The average value of the refractive index of the lens is 1·44, while that of the *vitreous humour* between the lens and the retina is approximately 1·34, near to that of water.

The radius of curvature of the lens surfaces may be altered by muscular contraction, to serve the purpose of focusing. This power of adjustment is called *accommodation*, and the least distance of distinct vision of a normal or *emmetropic* eye is around 25 cm. The accommodation of the eye varies greatly from one person to another and, in each person, with age.* The correction of excessive departures from normality by the introduction of supplementary eye-lenses forms the subject of *ophthalmics*. There are three principal abnormalities which occur in the eye:

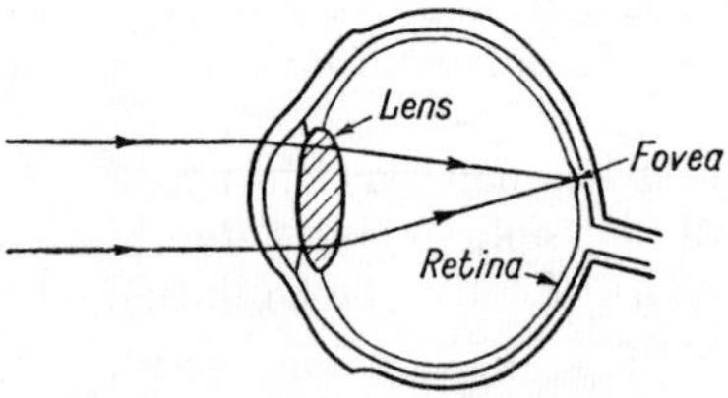

Fig. 6.1. The human eye.

(a) *Myopia* or short sight, in which the rays from an infinitely distant object-point reach a focus in front of the retina. This is corrected by means of a divergent lens placed in front of the eye.

(b) *Hypermetropia* or long sight, in which the true focus of an infinitely distant object lies behind the retina. Correction is obtained by means of a convergent lens.

(c) *Astigmatism,* in which the power of the eye differs in different planes containing its optical axis. This defect is corrected by means of a cylindrical or toroidal surface on the lens of the eye-glass.

The retina contains two principal types of light-sensitive cells, the *rods* and the *cones.* The latter, according to the usual theory of colour-vision, may be subdivided into the three types of colour-sensitivity. They predominate and are of greatest concentration in the *fovea*, the central part of the retina. The field of most distinct vision defined by this region and determining the *fixation*, or direction of "seeing", is slightly less than one degree in angular extent. The diameter of the cones in the fovea varies between 0·0015 mm and 0·005 mm. The rods, which are not so highly selective in colour-range, predominate in the outer regions and are the most responsive to low-intensity sources. This is the cause of the phenomenon of "averted vision" in scanning, for instance, the night sky.

There is an ultimate limit to the clarity of detail which one can attain with any given optical system. In the last analysis this limit arises, as will be seen in § 8.6.2, from the wave nature of light. The practical limit down to which two object points separated by a small distance may be distinguished by the eye, its *visual acuity,* is clearly a quantity of the greatest interest to the optical designer, for upon it depend the tolerances, both optical and mechanical, to which a visual instrument ought to be made. For a normal eye, this limiting angular separation is about one minute of arc in angular measure, corresponding to separation of about 0·0045 mm on the retina, if a focal length of 15 mm is assumed. When the aperture of the eye is increased beyond 5 mm, the effect of chromatic aberration and spherical aberration will decrease the resolving power. In designing a visual instrument one therefore usually assumes that

* At 10 years the power of the eye can cover 14 dioptres, at 25 years 10 dioptres, and at 50 years 2·5 dioptres.

a pencil not more than 4–5 mm in diameter is to enter the eye. Visual instruments are seldom, if ever, designed with a view to compensating the defects of the eye, because the amounts present in the eye vary from person to person.

6.2 THE CAMERA

In the *photographic camera* (Fig. 6.2), a real inverted image of an object is formed by a lens or combination of lenses upon the surface of a photographic film or plate. The object may be stationary or in motion relative to the camera. In the latter case short exposures are necessary, and the aperture must therefore be as large as possible, in order to collect sufficient light. It follows from § 4.8 (24) that the light reaching the image per unit area (the illumination E) from an extended object is proportional to the ratio d^2/f^2, d being the diameter of the entrance pupil and f the focal length, i.e.

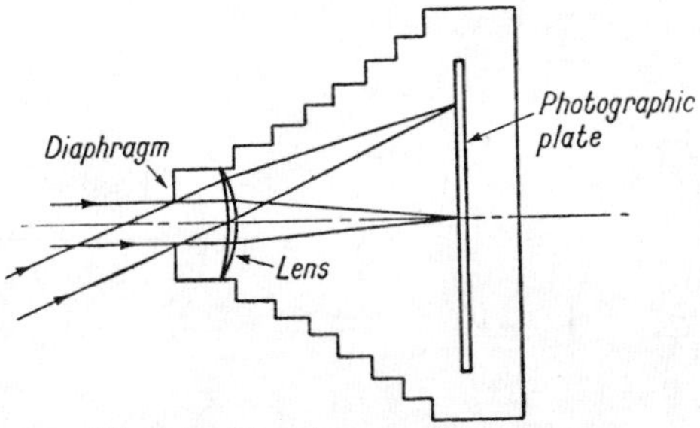

Fig. 6.2. The photographic camera.

it is inversely proportional to the square of the F number (nominal focal ratio). This may usually be varied by means of a variable *diaphragm.*

In the absence of aberrations, the image of a distant object which subtends an angle $\delta\theta$ at the first nodal point is of linear dimensions $f\delta\theta$. Hence, in order to produce a large image, the focal length must be large. Thus the two main characteristics of a camera are the focal length of its optical system and the range of the focal ratios at which it can operate. Another possible requirement of a camera is that it should cover a large angular field. Cameras are often designed to cover a field between ten and ninety degrees, using a focal length in the range from one inch to six feet. Greater focal lengths (6–60 ft) are needed in astronomical photography, where fields of much smaller angular extent ($\sim 1°$) containing effective point sources must often be reproduced on a large scale.

The simplest of small cameras use a single convergent lens, usually of meniscus form (concavo-convex) with a stop somewhat displaced from the lens, as shown in Fig. 6.2. No correction is made for chromatic effects and therefore such lenses were more satisfactory when photographic emulsions were generally sensitive to only a small region at the blue end of the spectrum (4000–4800 Å). Such lenses are often used in cheaper types of camera at a focal ratio $f/11$ over a 45° field.

The "*landscape lens*", as this meniscus form was once called, originated about 1812 and was followed, soon after the introduction of photography, by the achromatic doublet of Chevalier (Fig. 6.3(a)). The aim of the achromatic combination was to make the focus of the photographically active light coincident with that of the visual light (brightest around $\lambda = 5800$ Å); good images could not be obtained at focal ratios lower than $f/16$. The need for a much "faster" lens (i.e. one with a lower focal ratio) for portrait photography resulted in the widespread use of the *Petzval portrait lens* invented about 1840 (Fig. 6.3(b)) and comprising two separated doublets. This arrangement gave a field-surface with considerable curvature but good definition was obtainable over a field of a few degrees at focal ratios down to $f/4$. A development of the Petzval lens by Dallmeyer has been used extensively in ciné-projectors.

It has been known since about 1841 that a symmetrical lens would automatically be free from distortion. This is strictly true only for the case of unit magnification, but is substantially true for other working distances. This principle has been the basis of

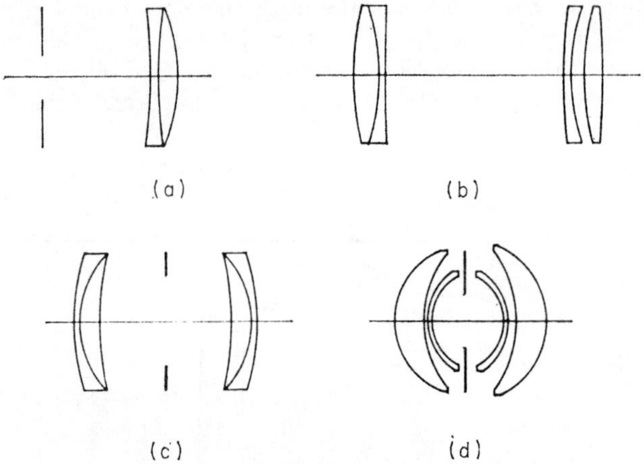

Fig. 6.3.

(a) The Chevalier lens.
(b) The Petzval lens.
(c) The Rapid-Rectilinear lens.
(d) The Topogon lens.

design of several long lines of successful photographic lenses. The *Rapid-Rectilinear* of Dallmeyer (Fig. 6.3(c)) dates from 1866. It was used in many cameras and eventually at focal ratios down to $f/8$ (as late as 1920), over fields of about 45°. Many firms manufactured such lenses and they became known under the general term of "*aplanat*". The same principle of symmetry, leading to freedom from distortion, has been continuously followed in the design of lenses intended to cover very wide fields. The original lens of this kind, the *Goerz Hypergon* of 1900, could, at a focal ratio around 25, cover a flat field up to 150°, although it has more often been sufficient to cover 90° at $f/16$. The four-element *Topogon* (Fig. 6.3(d)) of 1933 which will cover a field of 90° at $f/6{\cdot}3$, represents the basic design for modern wide-angle lenses.

During most of the nineteenth century, lens-designers were restricted to the use of two main types of glass; boro-silicate crown, and dense flint. Around 1890 the Jena Company introduced a radically new type of glass—the barium crown—having a high refractive index but low dispersive power. Incorporating this additional type of glass into lens systems permitted higher degrees of correction to be obtained. Rudolph of Jena, whose name is associated with the origin of many types of photographic lenses, coined the word "*anastigmatic*" to describe a lens for which the astigmatism at one off-axis angle could be reduced to zero. Using the older types of glass only, Rudolph had produced an anastigmat which later became known as the *Zeiss Protar* (Fig. 6.4(a)), but using three types of glass he was able to make an anastigmatic triplet. Two such lenses used symmetrically constituted a highly corrected combination. Such a lens was produced by ZEISS under the name *Triple Protar* and by GOERZ as the *Dagor* or *double anastigmat* (Fig. 6.4(b)). A typical lens of this type might work over a 50° field at *f*/4·5. RUDOLPH found further that if each half were formed of four cemented

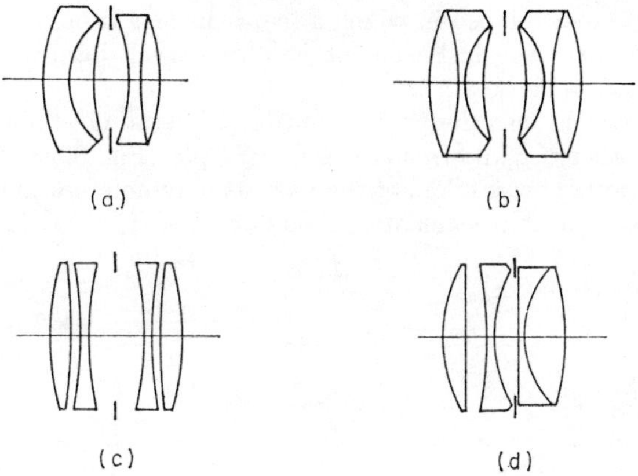

Fig. 6.4.

(a) The Protar lens. (b) The Dagor lens.
(c) The Celor lens. (d) The Tessar lens.

components the system could be made "convertible", i.e. the halves, having different focal lengths, could be used separately or in combination. This was produced as the *Double Protar* of 1894.

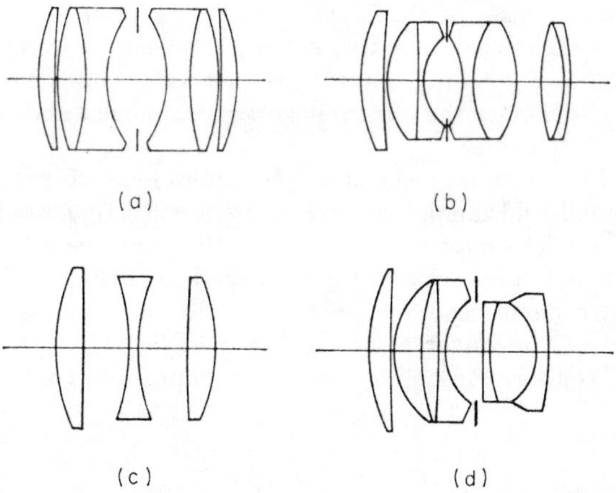

Fig. 6.5.

(a) The Planar lens. (b) The Zeiss Biotar lens.
(c) The Cooke Triplet. (d) The Sonnar lens.

A lens consisting of four separate elements was developed by GOERZ in 1898 and called the *Celor* (Fig. 6.4(c)). This became the basis of design of several important lenses at focal ratios down to *f*/4·5, such as the *Unifocal* of STEINHEIL and the *Cooke Aviar lens* for aerial photography. RUDOLPH was working along similar lines and his *Tessar* of 1902 (Fig. 6.4(d)) is still well known as a relatively inexpensive lens giving excellent images over fields up to 60° and at focal ratios as low as $f/3{\cdot}5$. It has been constructed with a great variety of focal lengths.

An earlier symmetrical design by RUDOLPH, the *Planar* of 1895, incorporated two meniscus-type lenses as the first and last elements (Fig. 6.5(a)). Since 1920 many very fast lenses have been developed in which a meniscus lens is used as the first element, although in most cases the last element is a cemented doublet as for example in MERTÉ's *Zeiss Biotar* (Fig. 6.5(b)).

A separate stream in lens design was initiated by the development of the three-lens anastigmat, otherwise known as the *Cooke Triplet* (Fig. 6.5(*c*)), by H. D. TAYLOR in 1895. Full advantage was taken of the various parameters which can be altered in a system of three separated elements. This "anastigmatic" design largely replaced

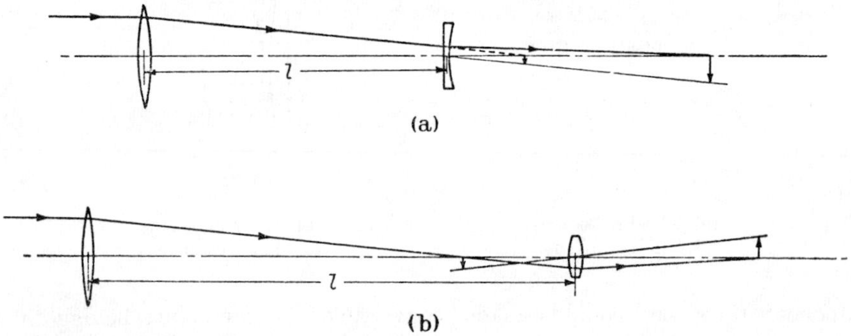

Fig. 6.6.

(a) The telephoto lens. (b) The photoheliograph.

the older "aplanatic" lenses of the rapid-rectilinear type in low- to medium-priced cameras. Types of high-speed lens whose design developed from the Cooke Triplet (e.g. the *Sonnar*, $f/1{\cdot}5$, Fig. 6.5(d)), have been used widely in ciné-cameras and miniature cameras. In the Sonnar the central negative component and the second positive component are both triplets.

In cameras used for certain purposes which require long effective focal lengths and correspondingly small fields an optical arrangement consisting essentially of an objective and an enlarging lens may be used. In the *telephoto lens*, the enlarging lens is a divergent element placed anterior to the primary image (Fig. 6.6(a)). In the *photoheliograph*, an instrument used for photographing the sun on a large scale, the enlarging lens is usually a convergent system (Fig. 6.6(b)). In these systems the effective focal length f is approximately given by the expression § 4.7 (3):

$$\frac{1}{f} = \frac{1}{f_1} + \frac{1}{f_2} - \frac{l}{f_1 f_2},$$

where f_1 and f_2 are respectively the focal lengths of the objective and the enlarging lens, and l is their separation. The effective focal length may be greater than the overall length from the objective to the focal plane by a factor up to fifteen.*

* Fuller discussions of photographic lenses will be found in W. MERTÉ, R. RICHTER and M. VON ROHR, *Das Photographische Objectiv* (Vienna, J. Springer, 1932). More recent designs are described in articles by W. TAYLOR and H. W. LEE, *Proc. Phys. Soc.*, **47** (1935), 502; H. W. LEE, *Rep. Progr. Phys.* (London, Physical Society), **7** (1940), 130; R. KINGSLAKE, *Proceedings of the London Conference on Optical Instruments* (London, Chapman & Hall, 1951), 1 and C. G. WYNNE, *Rep. Progr. Phys.* (London, Physical Society), **19** (1956), 298.

Finally it is to be noted that the F number of a photographic objective is an approximate measure of the "speed of action" of the system only where the object is an extended one. With a point-source object (as is effectively the case in astronomical photography), the light reaching the image plane would, under ideal conditions, be concentrated in a vanishingly small area, so that the square of the aperture diameter, rather than F number, would characterize its light-gathering power. In reality the situation is more complicated, as several factors contribute to a spreading of the light over a finite (though often a very small) area. Chief amongst these are diffraction (see Chapter VIII), the granularity of the photographic emulsion, and the unsteadiness of the atmosphere.

6.3 THE REFRACTING TELESCOPE*

The *telescope* is an optical system by means of which an enlarged image of a distant object may be viewed. The principle of the *astronomical refracting telescope* is shown

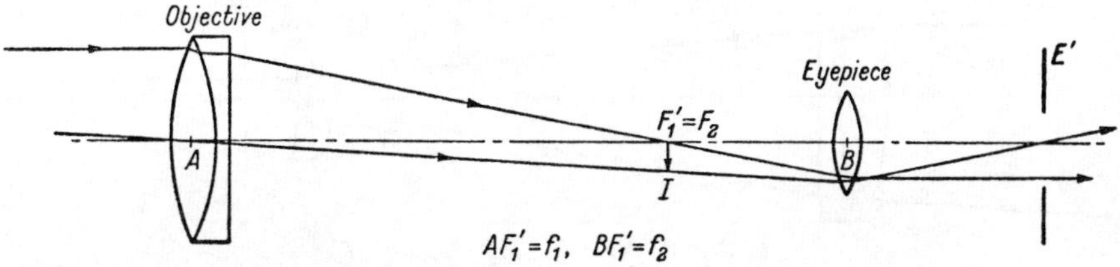

Fig. 6.7. The astronomical refracting telescope.

in Fig. 6.7. It consists of two convergent lenses the first of which, the *objective*, is usually an achromatic doublet forming a real inverted image I, which is examined using the second lens, the *eyepiece*. In the normal state of adjustment the second focal plane of the objective coincides with the first focal plane of the eyepiece, so that an incident pencil of parallel rays emerges as a parallel pencil. The image may be erected by the use of an auxiliary lens.

The entrance pupil of the system coincides with the objective; its image by the remainder of the system, the exit pupil, is denoted by E' in Fig. 6.7. The eye should be so placed that its entrance pupil coincides with E'; then all the light entering the objective at different off-axis angles will reach the eye.

The *magnifying power* of an instrument used for examination of objects at infinity is defined as the angular magnification at the pupils. By the SMITH–HELMHOLTZ formula, § 4.4 (49), this is the reciprocal of the linear magnification at the pupil planes, so that the magnifying power is equal to the ratio of the radius of the entrance pupil to that of the exit pupil. An expression for the magnifying power is immediately obtained from § 4.3 (31) where, because the angles are assumed to be small, γ and γ' may be written in place of $\tan\gamma$ and $\tan\gamma'$. This gives

$$\frac{\gamma'}{\gamma} = \frac{f_1}{f_2'}.$$

Now the eye may be accommodated to distances other than infinity and the telescope may easily be adjusted to form a virtual image at, say, a distance D from

* For a fuller discussion of telescopes see, for example, A. DANJON and A. COUDER, "Lunettes et Télescopes" (Paris, Revue d'Optique, 1935).

the eye.* D should, however, not be decreased below the value of about 25 cm which, as already mentioned, is the closest distance at which an object can be viewed distinctly without eye-strain with a normal eye.

The earliest telescope of which there exists definite knowledge is that of GALILEO, constructed in 1609. In this system (Fig. 6.8) the objective was a convergent lens, but the eyepiece was a divergent lens, placed anterior to the primary image in such a way that the focal points of the two lenses coincided beyond the eyepiece. An erect image was then obtained at infinity, there being no intermediate image. In this type of telescope, known as the *Galilean telescope*, the aperture stop is not in the plane of the objective, for the image of the objective formed by the eyepiece lies between the two lenses and is therefore not accessible to the eye. The eye is placed close behind the eyepiece, so that the pupil of the eye becomes the aperture stop and the exit pupil. The principal ray for points at the limit of the field will pass very near to

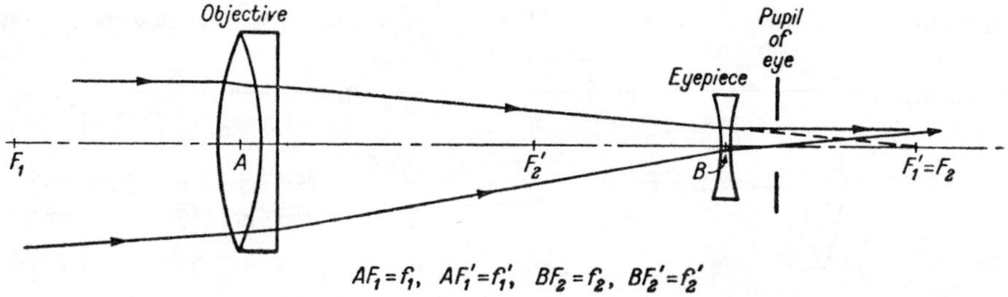

Fig. 6.8. The Galilean telescope.

the edge of the objective, so that the objective serves as a field stop rather than as an aperture stop.

A disadvantage of the Galilean telescope is that it is limited to small fields and that it does not possess a real image where cross-wires, or a graticule, may be placed. On the other hand, its small overall length and the fact that it gives an erect image make the system very suitable for use in opera glasses, usually with a magnification of two to three times.

In the *binocular*, which uses the optical arrangement similar to that of an astronomical telescope, an erect image is obtained by means of four reflections, as shown in Fig. 6.9. These reflections take place at an incidence angle of 45° at the glass–air surfaces of so-called *Porro prisms*; this angle being, of course, beyond the critical angle. A reflection of this kind forms a convenient and efficient device often used in small optical instruments. An alternative form of binocular uses the *König erecting prism*. This incorporates the device of the "roof" where two adjacent reflecting surfaces, at right angles to each other within a second of arc, are placed athwart the optical beam, as shown in Fig. 6.10. The larger spacing in the more usual type of binocular of the two objectives is of advantage, however, since it serves to enhance the stereoscopic effect.†

---

* The magnification is then changed to the value:

$$\frac{\gamma'}{\gamma} = \frac{f_1}{f_2'} + \frac{f_1}{D}.$$

† For a fuller account of binoculars see M. VON ROHR, *Die Binokularen Instrumente* (Berlin, J. Springer, 1920).

Terrestrial telescopes are generally similar to the astronomical telescope; but an erect image must be obtained, either by the help of erecting prisms, as in the binocular, or with the use of additional lenses. A typical system is shown in Fig. 6.11.

The telescope objective is usually an achromatic combination of two lenses in which the radii of curvature are such as to give, with spherical surfaces, little spherical

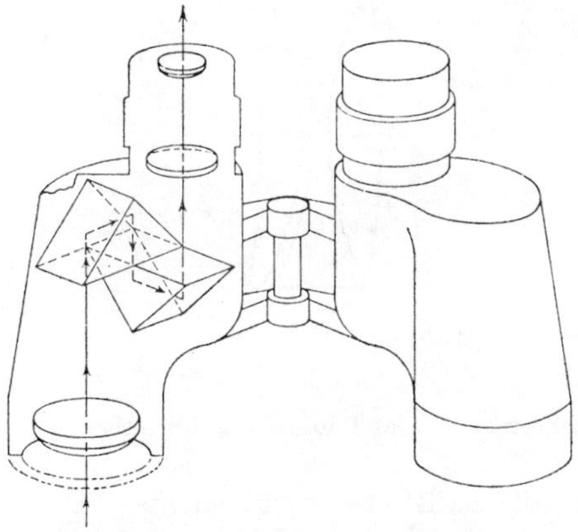

Fig. 6.9. The binocular telescope.

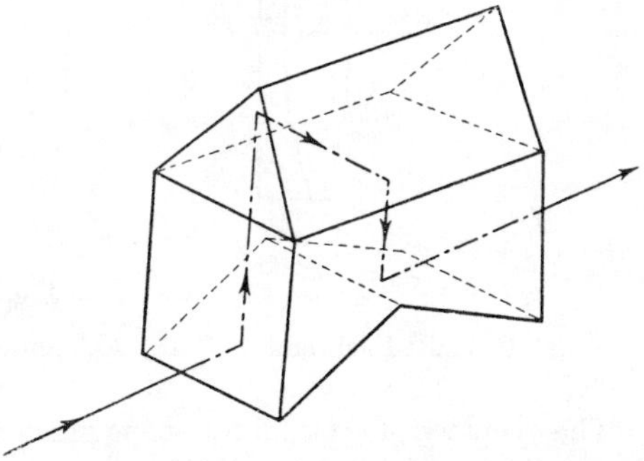

Fig. 6.10. The König erecting prism.

aberration. If a cemented doublet is to be used, a design in which the first (crown) component is equi-convex and the second (flint) component is plano-concave is often a suitable solution (Fig. 6.12(a)). If freedom from coma is desired as well as freedom from spherical aberration, as in a high-quality astronomical telescope, it is necessary to utilize the extra flexibility in design of two non-cemented components, and to choose the radii to give an aplanatic design. The design may then be approximately that

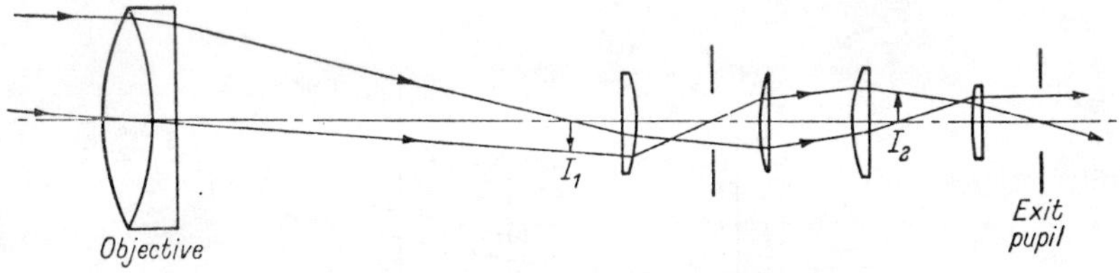

Fig. 6.11. The terrestrial telescope.

shown in Fig. 6.12(b). This includes a concavo-convex flint component. The steep concave curve of the first surface of this component can be reduced if a coma-free design with aspherical figuring on one surface to give axial stigmatism is adopted.

An objective with the crown lens in front is the usual form, originally known as the *Fraunhofer objective*. The *Steinheil*, or "*flint-in-front*", *objective* is less used because of the greater susceptibility of the flint glass (less stable than crown glass) to atmospheric attack.

The *photo-visual objective*, shown in Fig. 6.12(c), is the third type of objective commonly used in refracting telescopes. The use of three different glasses permits a considerable reduction of chromatic difference of focus over a given spectral range to be obtained. This objective is almost equally good for "photographic" and "visual" wavelengths, and at one time was popular for astronomical use at apertures up to about 12 in. The steep curves of the centre component, making centering difficult, and the early use of unstable varieties of flint glass, have detracted, however, from the value of these lenses.

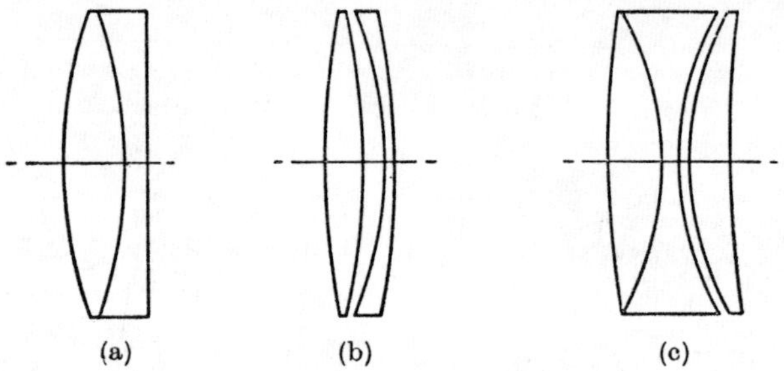

Fig. 6.12.

(a) Cemented achromat. (b) Aplanatic achromat. (c) Photo-visual objective.

The eyepiece of a visual telescope must, broadly, satisfy two requirements. Firstly, it must have a focal length which will give the required degree of magnification. Secondly, it must have sufficient aperture to collect the light from an extended object. These requirements can be achieved by means of a single lens, as shown in Fig. 6.13(a), but more conveniently two smaller lenses can be used as in Fig. 6.13(b), which shows the optical arrangement of the *Ramsden eyepiece*.

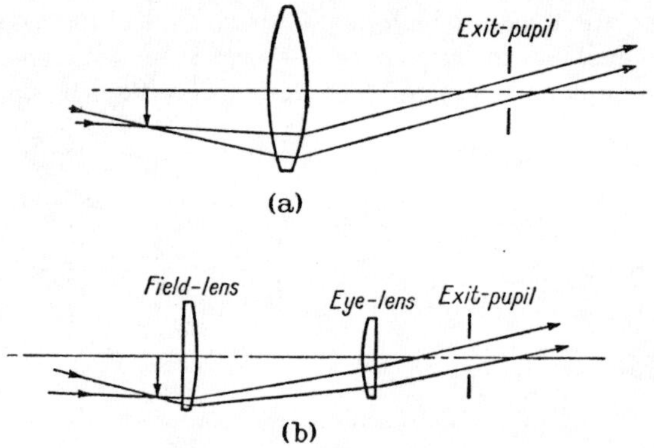

Fig. 6.13.

(a) Single-lens eyepiece. (b) The Ramsden eyepiece.

The eyepiece should form an image of the aperture stop at a position convenient for placing the eye. The distance of this image behind the last surface of the eyepiece is called the *eye-relief*.

The first component of a two-component eyepiece is called the *field-lens* because it is often very near to the primary image. It is separated from this image partly to prevent any dust or imperfection on the lens being visible, and partly to render the plane of the image accessible to cross-wires, or a graticule. The second component is called the *eye-lens*.

The most common eyepiece in use is the Ramsden eyepiece, and the other chief type is the *Huygenian eyepiece,* shown in Fig. 6.14(a), which has the field lens anterior to the primary image. The disadvantages of the Huygenian eyepiece are firstly the rather short eye-relief obtainable, and secondly the fact that it cannot be used with

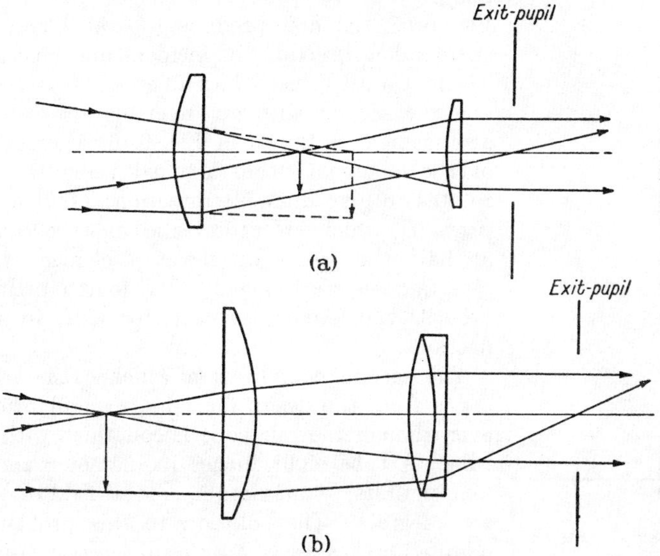

Fig. 6.14.

(a) The Huygenian eyepiece. (b) The Kellner eyepiece.

cross-wires external to itself. It cannot give such a high correction for spherical aberration as the Ramsden eyepiece, but is free from lateral chromatic aberration and off-axis coma.

The use of two lenses (see Fig. 6.14) instead of one aids the removal of harmful chromatic aberrations, and correction in the Ramsden eyepiece can be greatly

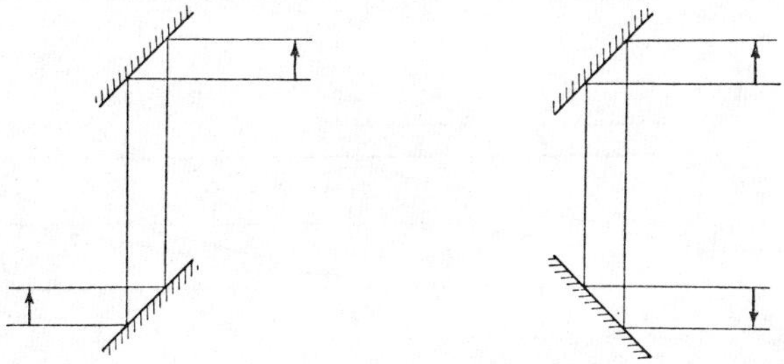

Fig. 6.15. Periscopic mirrors.

improved by using a doublet as the eye-lens. More complicated are the "orthoscopic" and "symmetrical" eyepieces, each using four components but generally similar to the Ramsden eyepiece. A higher degree of correction for aberrations is obtained, especially with regard to distortion, and longer eye-reliefs are available for a given power.

The principle of the *periscope*, i.e. an instrument used for viewing objects which are situated so that direct viewing is precluded by an obstacle, is the same as that of the telescope, but two requirements make the designing of such an instrument more complicated. In the first place, the periscope must operate with a bent ray-path and give an erect non-reversed image. An even number of reflections, or none, are necessary to give a non-reversed image, and two simple arrangements are shown in Fig. 6.15, the first producing parallel and the second antiparallel directions of incident and emergent beams. The first must be used with an erecting telescopic system and the second with an inverting telescope. Another arrangement is shown in Fig. 6.16, where the direction of the emergent beam is fixed while the first prism rotates to give different directions of view. The *Dove prism D*, which incorporates a single reflection, rotates at half the speed of the top element in order to maintain an erect image. The fourth prism is of the "roof" type (Amici) so as to provide, in all, four reflections.

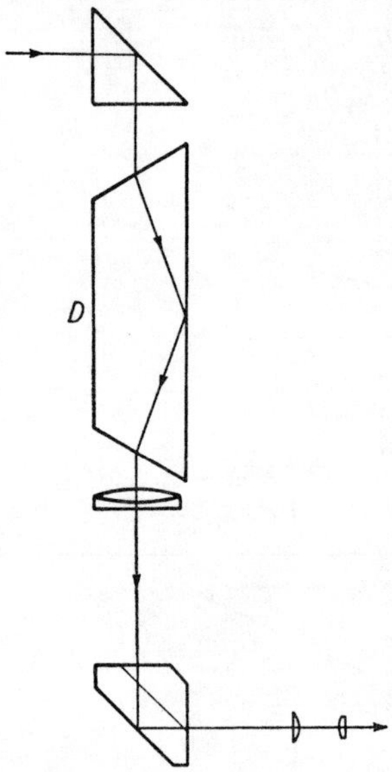

Fig. 6.16. Periscope with Dove prism.

The second complication which exists in periscopes arises from the desire for a large angular field of view, even though the telescope is contained within a narrow tube. A tube eight inches in diameter and forty feet long permits, at maximum, a total field of view of only two degrees. The solution to this problem which is adopted makes use of a series of lenses passing the light down the tube. Different spacings of the lenses are possible. In Fig. 6.17 three arrangements are shown of a system of unit magnification which employs three lenses.

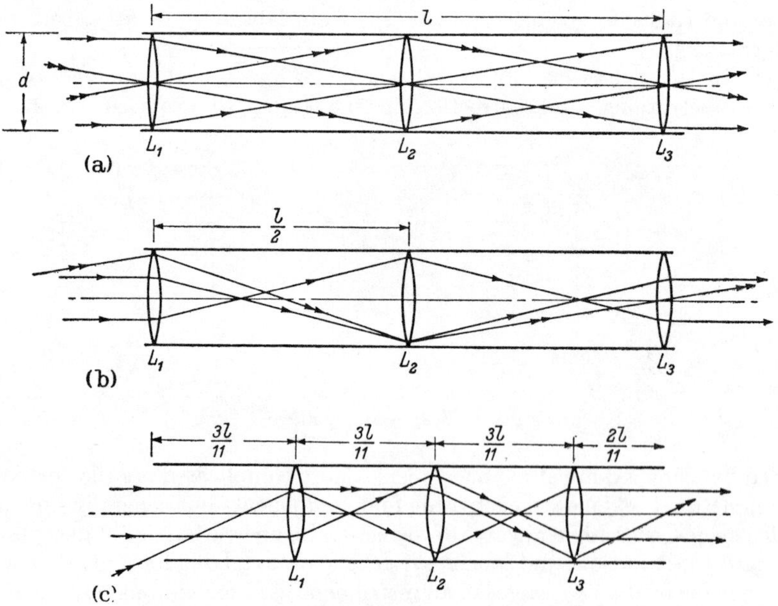

Fig. 6.17. Three-lens systems; unit magnification.

TABLE XIV

	Focal lengths of lenses			Image	Aperture on-axis	Field (radians)	Vignetting at limit of field
	L_1	L_2	L_3				
(a)	$l/2$	$l/4$	$l/2$	Inverted	d	$2d/l$	0
(b)	$l/6$	$l/6$	$l/6$	Erect	$d/2$	$2d/l$	100%
(c)	$l/11$	$l/11$	$l/11$	Erect	$d/2$	$11d/2l$	$\sim$ 86%

The range of fields of view of these three systems and their effective apertures are shown in Table XIV. It will be clear that, with a larger number of lenses, a wider range of possibilities exists.*

6.4 THE REFLECTING TELESCOPE

In the *astronomical reflecting telescope*, the light from a celestial objective is received on a concave primary mirror. The mirror serves the same purpose as the objective in a refracting telescope, namely to form a real image of the object in its focal plane. This image is either received directly on a photographic plate, or is examined visually by an eyepiece. This type of arrangement is the most common form of astronomical telescope used to-day. As in the case of the refracting telescope, the magnifying power is equal to the ratio of the focal length of the objective to that of the eyepiece.

The first telescope of this type was made in 1668 by NEWTON. In the arrangement which he used a small plane mirror was placed in the path of the rays reflected by the primary mirror, so as to divert the rays to one side of the tube, where the image could be conveniently examined (Fig. 6.18). This arrangement is known as the *Newtonian telescope*.

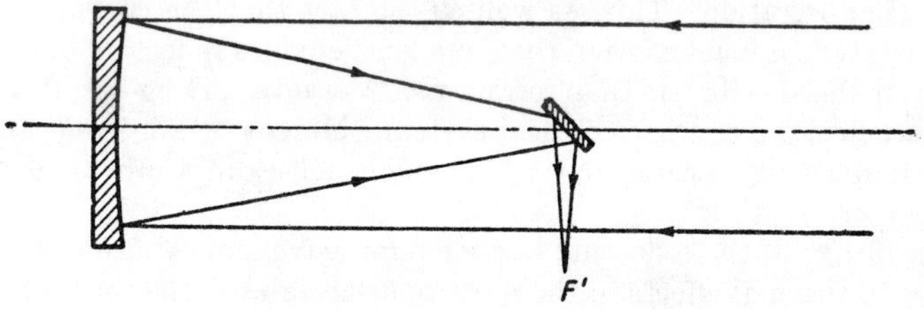

Fig. 6.18. The Newtonian telescope.

In order to have a stigmatic axial image, the figure of the mirror must be a paraboloid. The fact that off-axis images formed by a parabolic mirror suffer strongly from coma, together with the circumstance that large relative apertures are employed, makes the usable field very small. In the case of a 36 in. $f/6$ mirror the total field which can be covered is of the order of 20 minutes of arc, while in the case of the largest telescope in existence to-day, the Hale telescope, which has a primary mirror of diameter 200 in., the field is about 45 minutes in angular diameter.†

* Further details will be found in an article by T. SMITH in the *Dictionary of Applied Physics*, Ed. R. GLAZEBROOK (London, Macmillan, 1923) Vol. IV, p. 350.

† F. E. ROSS, *Publ. Astr. Soc. Pac.*, **46** (1934), 342. For a general account of the 200 in. telescope see articles by J. A. ANDERSON, *Publ. Astr. Soc. Pac.*, **60** (1948), 221; and B. RULE *idem*, p. 225.

Other types of reflecting telescope employ two principal mirrors, a primary concave mirror and a secondary mirror. In the *Cassegrain telescope* (Fig. 6.19) the secondary mirror is convex. In a large astronomical telescope, such an arrangement is often an alternative to the Newtonian arrangement. Different mirrors can be utilized to provide a variety of focal lengths. If the figure of the primary mirror is a paraboloid, the secondary mirror must have a hyperbolic figure. Other figures for the two mirrors can be chosen which give a stigmatic axial image and different amounts of off-axis coma. In § 4.10 we mentioned a two-mirror system proposed by SCHWARZSCHILD, which is completely aplanatic; the figures of the two mirrors are, however, more complicated. The Cassegrain system is often employed in terrestrial telescopes used for viewing landscapes as it provides a compact and inexpensive instrument with a long focal length, and a high degree of enlargement can be obtained without the need for a short eye-relief.

The *Gregorian telescope*, less common than the Cassegrain, employs a concave secondary mirror placed beyond the prime focus of the primary mirror.

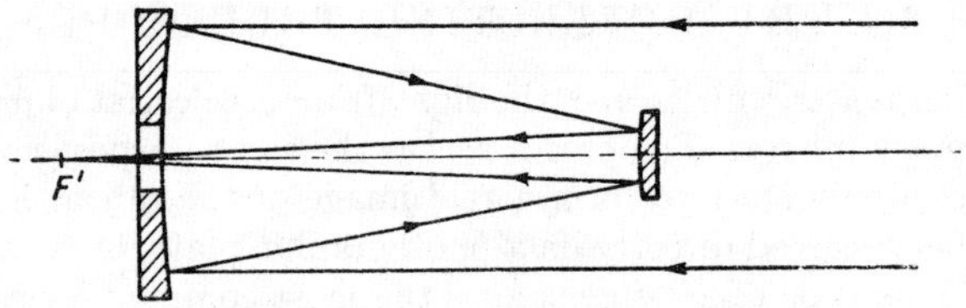

Fig. 6.19. The Cassegrain telescope.

In the largest reflecting telescopes, very great focal lengths, greater than that of the largest refractor, may be obtained by the use of a secondary mirror. In the *Coudé arrangement* the beam is reflected by a third (plane) mirror down the polar axis (about which the telescope rotates), so that it can be made to feed a fixed spectrograph.

One of the principal advantages of the reflecting telescope is its complete freedom from chromatic aberration. This, as well as the fact that the curvature required in the mirror-surface is much less than that in a lens, enables reflectors to be built which have a much smaller focal ratio than refractors. As well as giving brighter images, this results in a much more compact optical system. Moreover, a mirror can be built in larger sizes than a lens, because optical inhomogeneities in a mirror-block are of no significance.

The disadvantage of the reflecting telescope for astronomical use lies principally in its sensitivity to thermal effects in the mirror-surface and in the telescope-tube, in its inability to cover more than a small angular field of view, and in the mechanical difficulties of making a sufficiently rigid mounting. It is this last aspect, in fact, which produced the most difficult of the problems which confronted the builders of the Hale telescope.

The successful use of the reflecting telescope with photographic plates has produced the desire for covering larger fields, and one of the methods available for increasing the usable field consists in employing field lenses just in front of the photographic plate. These lenses are designed to introduce off-axis coma such as to compensate that of the main mirror. Fields of the order of $1\frac{1}{2}°$ may be obtained with a 36 in. telescope.*

Much more successful in an attempt to photograph wide angular fields with large

* See F. E. Ross, *Astrophys. J.*, **81** (1935), 156.

reflecting systems has been the optical arrangement already referred to in § 4.10, invented by B. SCHMIDT around 1930.* While a parabolic mirror gives perfect imaging for the axial rays and badly comatic images a short distance off-axis, a

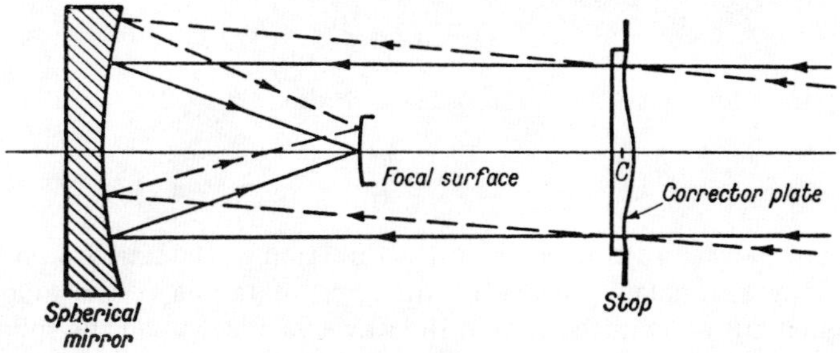

Fig. 6.20. The Schmidt camera.
(The figuring of the corrector plate is greatly exaggerated.)

spherical mirror with an aperture-stop at its centre of curvature C would give uniform images over a wide spherical field-surface concentric with itself, each image suffering from a large amount of spherical aberration. SCHMIDT introduced into the aperture stop a thin, nearly plane-parallel plate, called the *corrector plate*, one face of which was plane, whilst the other was figured (Figs. 6.20, 6.21). The purpose of this plate was

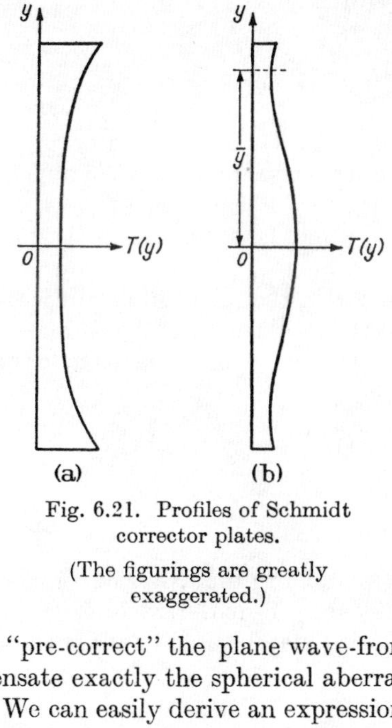

Fig. 6.21. Profiles of Schmidt corrector plates.
(The figurings are greatly exaggerated.)

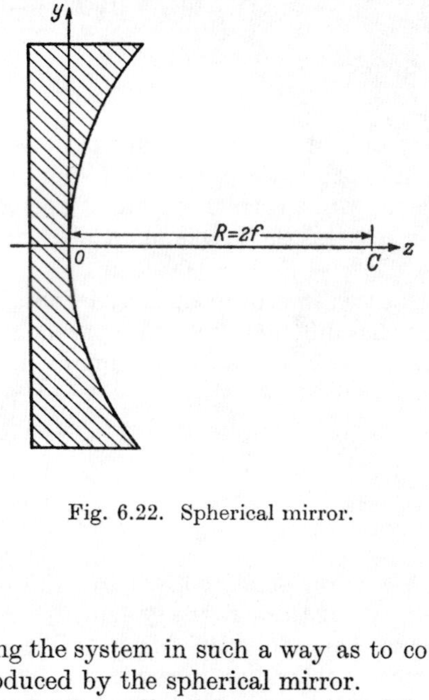

Fig. 6.22. Spherical mirror.

to "pre-correct" the plane wave-fronts entering the system in such a way as to compensate exactly the spherical aberration introduced by the spherical mirror.

We can easily derive an expression for the aspheric profile of the corrector plate of the *Schmidt camera*. For this purpose let us compare the expressions for a spherical

* B. SCHMIDT, *Central-Zeitung f. Optik u. Mechanik*, **52** (1931), Heft 2, and *Mitt. Hamburg. Sternw. Bergedorf*, **7** (1932), No. 36, 15. See also R. SCHORR, *Zeitschr. f. Instrumkde*, **56** (1936), 336, *Mitt. Hamburg. Sternw. Bergedorf*, **7** (1936), Nr. 42, 175, and *Astr. Nachr.*, **258** (1936), 45.

and a paraboloidal mirror. The equation of a spherical mirror of radius $R = 2f$ is given by (see Fig. 6.22)

$$(z^{(s)} - 2f)^2 + y^2 = (2f)^2,$$

i.e.

$$z^{(s)} = \frac{y^2}{4f} + \frac{y^4}{64f^3} + \ldots,$$

whilst for a paraboloid with the same paraxial radius ($R = 2f$),

$$z^{(p)} = \frac{y^2}{4f}.$$

Now a pencil of parallel rays incident in the direction of the axis would be rendered stigmatic by the paraboloid; hence at the zone of radius y, the amount of pre-correction which has to be introduced in the waves incident upon the spherical mirror is given by

$$2(z^{(s)} - z^{(p)}) = \frac{y^4}{32f^3}$$

approximately, terms involving sixth and higher powers of y being neglected.* After passage through the plate, the rays are still nearly parallel, so that it makes no appreciable difference whether y is measured at the mirror or at the plate. Hence if n is the refractive index of the plate, its thickness $T(y)$ for a zone of radius y must exceed its axial thickness $T(0)$ by an amount given by

$$(n - 1)[T(y) - T(0)] = 2(z^{(s)} - z^{(p)})$$

i.e.

$$T(y) - T(0) = \frac{y^4}{32(n - 1)f^3}. \tag{1}$$

The profile of such a corrector plate is shown in Fig. 6.21(a).

The asphericities required are very small. Consider for example an $f/3{\cdot}5$ Schmidt camera, with a corrector plate of aperture diameter $2y_0 = 40$ cm. Then $f = 140$ cm. With $n = 1{\cdot}5$ the maximum asphericity $[T(y) - T(0)]_{\max}$ is, according to (1), equal to $y_0^4/32(n - 1)f^3 = 0{\cdot}0036$ cm.

A further improvement may be obtained by comparing the spherical mirror with a somewhat different "reference paraboloid". If f' is the focal length of the paraboloid, then $z^{(p)} = y^2/4f'$, and one obtains in place of (1) the following expression for the plate profile:

$$T(y) - T(0) = \frac{y^4 - Ay^2}{32(n - 1)f^3}, \tag{2}$$

where

$$A = 16f^3\left(\frac{1}{f'} - \frac{1}{f}\right).$$

With a corrector plate given by (2), rays incident parallel to the axis will converge, after passing through the system, to a point at a distance f' from the axial point of the mirror. This additional degree of freedom (choice of f') may be used to minimize the chromatic aberration introduced by the plate. Since

$$\frac{dT}{dy} = 0 \qquad \text{when} \qquad y(2y^2 - A) = 0,$$

* This may be shown to be a permissible approximation if the focal ratio of the Schmidt camera is not lower than about $f/3$.

it follows that a ray parallel to the axis and incident at height $\bar{y} = \sqrt{A/2}$ is not deflected by the plate. The zone of radius $\bar{y} = \sqrt{A/2}$ is known as the *neutral zone*. In order to minimize the chromatic aberration, the neutral zone must be fairly close to the edge of the plate,* as shown in Fig. 6.21(b).

It is evident that the corrector plate does not impart to rays which pass obliquely through the plate precisely the same deviations as to those passing normally. In consequence, the system will not be free from off-axis aberrations. Judged by ordinary standards, the images formed on the spherical receiving surface concentric with the

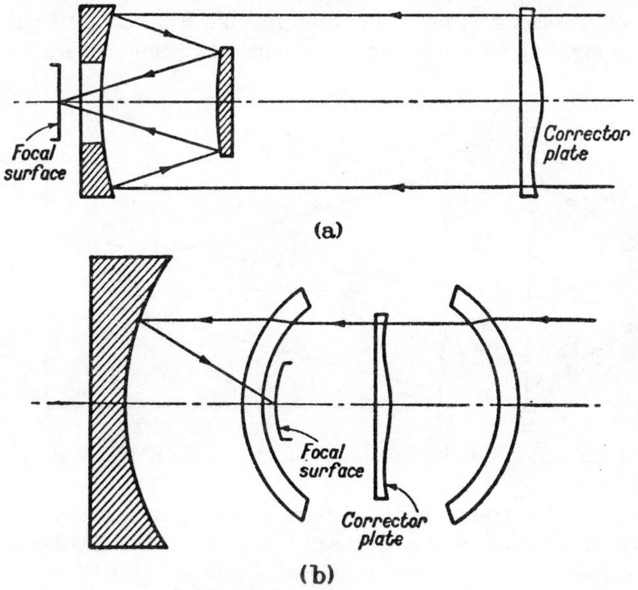

Fig. 6.23.

(a) The Schmidt–Cassegrain camera.
(b) The Baker Super-Schmidt camera.

mirror are, however, of excellent quality;† this is so even in a Schmidt camera of low focal ratio, working over a field many times larger than is possible with systems of a more conventional design.

More complex systems using the principle of the Schmidt camera, have been considered‡ and in some cases built. Most important amongst these are the *Schmidt–Cassegrain camera* which employ two (spherical or aspheric) mirrors together with a

---

* More precisely, the diameter of the colour confusion circle may be shown to be least when the neutral zone is at about 0·87 of the full radius y_0 of the plate. This corresponds to a choice $A = \frac{3}{4}y_0^2$. For a discussion of this point see B. Strömgren, *Vierteljahrsschrift der Astr. Ges.*, **70** (1935), 82, and E. H. Linfoot, *Mon. Not. Roy. Astr. Soc.*, **109** (1949), 279. [Also his *Recent Advances in Optics* (Oxford, Clarendon Press, 1955) 182 and 192.]

† This is partly due to the fact that field curvature may be shown to be the only primary aberration of the system. (B. Strömgren, *loc. cit.*) The fifth-order aberrations of the Schmidt camera were investigated by C. Carathéodory, *Elementare Theorie des Spiegelteleskops von B. Schmidt* (Leipzig, Teubner, 1940), and E. H. Linfoot, *loc. cit.*

‡ For a review of such systems see H. Slevogt, *Zeitschr. f. Instrumkde*, **62** (1942), 312; H. Köhler, *Astr. Nachr.*, **278** (1949), 1; E. H. Linfoot, *Mon. Not. Roy. Astr. Soc.*, **108** (1948), 81. [Also his *Recent Advances in Optics* (Oxford, Clarendon Press, 1955) Chapter IV.] For a fuller description of the Baker Super-Schmidt camera see F. L. Whipple, *Sky and Telescope*, **8** (1949), 90.

corrector plate (Fig. 6.23(a)) and the *Baker Super-Schmidt camera,* consisting of a spherical mirror, two menisci, and a corrector plate (Fig. 6.23(b)). The former possesses the advantage that it operates with a flat field surface, which can be made more accessible than in the Schmidt camera; also, the overall length of this system is shorter than that of the Schmidt camera with the same focal length. The Baker Super-Schmidt camera can operate with extremely low focal ratios.

6.5 INSTRUMENTS OF ILLUMINATION

The function of collecting as much light as possible from an artificial source and of deflecting it through the entrance pupil of an optical system, covering specified angles

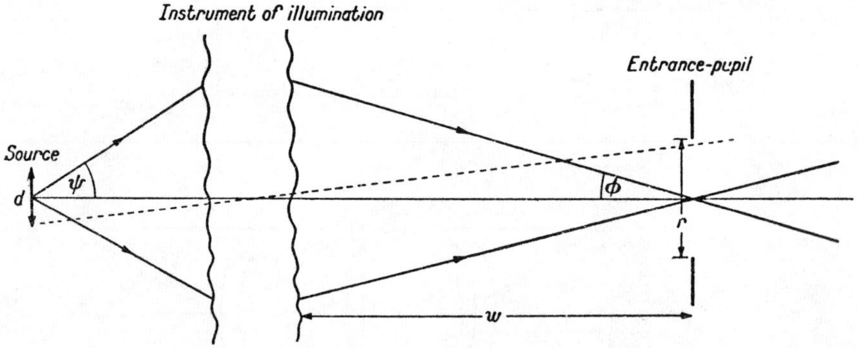

Fig. 6.24. Notation relating to an instrument of illumination.

of acceptance at that entrance pupil, is carried out by an *instrument of illumination.* Light can only be obtained from a source of finite size and so the design of the instrument depends upon the nature of the source as well as upon the specification of the illumination required.

In Fig. 6.24, the entrance pupil of diameter r is shown, of an optical system requiring illumination over directions making angles with the axis up to a value ϕ. Let d be the diameter of the source, supposed circular. For greatest efficiency the illuminating system must work with magnification r/d, while its aperture must subtend an angle 2ϕ from the entrance pupil. The working distance w must be greater than a certain prescribed value. It is clear from a consideration of Fig. 6.24 that for lower values of d the *angle of collection*, 2ψ, will be greater, hence less light will be wasted. Also, large angles of illumination (large ϕ) will often make possible smaller values of w/r, so that larger sources can then be used with reasonable efficiency. Thus it is clear that in any particular case there is a restriction on the largest source which will give reasonable efficiency and that this restriction is less stringent when large angles of illumination are required. In a photographic enlarger ($\phi \sim 20°$), large sources (frosted bulbs or cold-cathode tubes) often provide reasonably efficient illumination. In a cinematograph projector ($\phi \sim 5°$) requiring maximum brightness, an electric filament is suitable. With a searchlight, where a strong beam over only one or two degrees is required, especially small bright sources are the only possible ones, unless the size of the instrument is to be unduly great.

In many instruments, the illumination can be provided by means of a suitable source and a condenser lens. This lens must be sufficiently free from aberrations to be able to utilize the smallness of the source. Every condenser may be said to possess a "focal sphere", which represents the minimum source size with which it can efficiently be used. A design which is usually sufficiently good to give satisfactory results is that of the double plano-convex condenser (Fig. 6.25(a)), though sometimes a triple condenser is used (Fig. 6.25(b)). In other cases a single lens may be adequate.

The illumination required by microscopes involves more complex consideration and will be briefly discussed later (§ 8.6.3 and § 10.5.2).

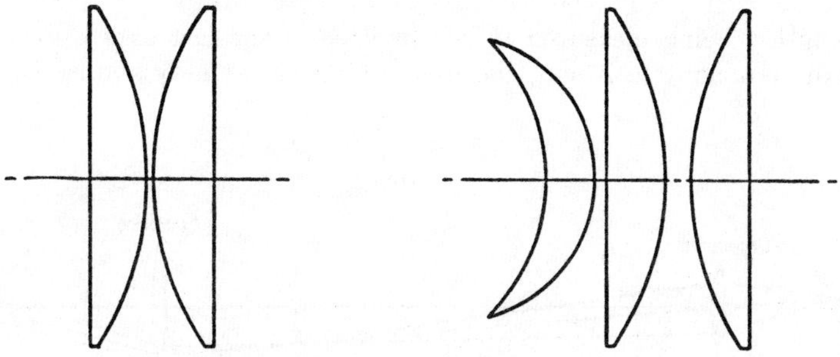

Fig. 6.25. Condenser lenses.

6.6 THE MICROSCOPE

The apparent size of an object is determined by the size of its retinal image. If the eye is unaided, this apparent size depends on the angle which the object subtends at the eye. For a normal eye, the least distance of distinct vision is, as mentioned in § 6.1, about 25 cm. This is the most favourable distance at which to examine the detail of any object. If a convergent lens is placed before the eye, the object may be brought much closer; for the lens will form an enlarged virtual image at a distance greater

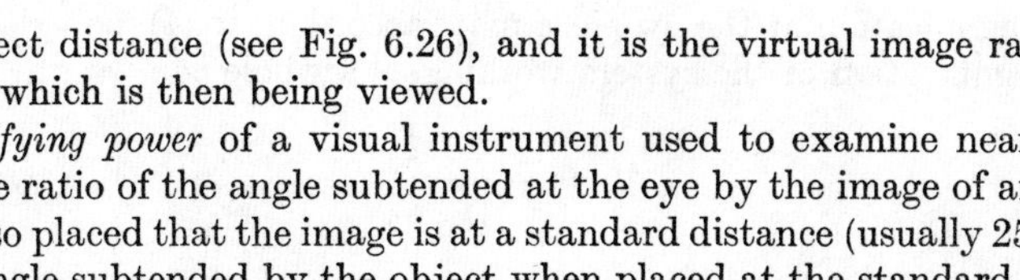

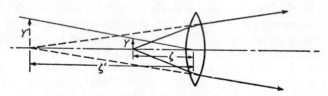

Fig. 6.26. The magnifier.

than the object distance (see Fig. 6.26), and it is the virtual image rather than the object itself, which is then being viewed.

The *magnifying power* of a visual instrument used to examine nearby objects is defined as the ratio of the angle subtended at the eye by the image of an object when the object is so placed that the image is at a standard distance (usually 25 cm) from the eye, to the angle subtended by the object when placed at the standard distance from the eye and viewed directly. This is equivalent to defining the magnifying power M as the ordinary linear magnification when the image is situated at the standard distance in front of the exit pupil of the instrument. For a simple lens, we have (see Fig. 6.26)

$$M = \frac{Y'}{Y} = \frac{\zeta'}{\zeta},$$

where $\zeta' = 25$ cm is the standard distance. By § 4.4 (31),

$$\frac{1}{\zeta} - \frac{1}{\zeta'} = \frac{1}{|f|}, \tag{2}$$

so that

$$M = 1 + \frac{\zeta'}{|f|} = 1 + \frac{25}{|f|}, \tag{3}$$

the focal length f being measured in centimetres in the last expression. Since f is usually small compared to 25 cm, the magnifying power may simply be written as

$$M \sim \frac{25}{|f|}. \tag{4}$$

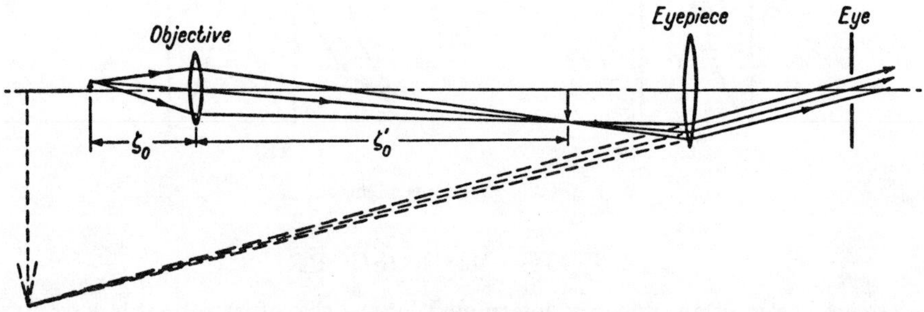

Fig. 6.27. Illustrating the principle of the microscope.

Various forms of magnifiers are used; the simple double-convex lens, or an achromatic doublet, is the most usual in pocket magnifiers or watchmaker's eye-glasses. A separated pair of plano-convex lenses similar to the Ramsden eyepiece (Fig. 6.13(b)) is often used. It is evident that the distance between the object and the eye will become inconveniently small if larger apparent fields of view are required with large magnification. For this reason the optical system of the *microscope* (invented by GALILEO about 1610), employing an objective of short focal length and a magnifying eyepiece, was evolved, the magnification being achieved in two stages (Fig. 6.27).

The *objective* of the microscope forms an enlarged image of the object in a position suitable for viewing through the *eyepiece*. The magnification at the objective is given by $-\zeta_0'/\zeta_0$ where ζ_0 and ζ_0' are the working distances of the objective, satisfying a relation of the form (2). If f_1 is the focal length of the eyepiece, measured in centimetres, the magnification at the eyepiece may, according to (4), be written as $25/|f_1|$. Hence the magnification of the system working as a whole is

$$M = -\frac{25}{|f_1|}\frac{\zeta_0'}{\zeta_0},$$

the negative value indicating that the image is inverted. The magnifications of the objective and eyepiece are usually stated separately in microscope specifications.

The other quantity of interest in the general specification of a microscope objective is its *numerical aperture* (N.A.) already defined in § 4.8.2 as the product $n \sin \theta$, where θ is the angular semi-aperture on the object side (i.e. the semi-angle of the cone of rays from the axial object point which is received by the objective) and n is the refractive index of the medium of the object space. This quantity is a measure not only of the light-gathering power of the objective, but also, as will be shown in § 8.6.3, of its resolving power, i.e. of the ultimate limit to the clarity of detail which can be obtained with it. Here we shall describe how the highest numerical apertures are attained.

Microscope objectives must, in general, be highly corrected for spherical aberration and coma and for chromatic aberration, since they are to receive rays over as large an aperture as possible. In objectives of low magnification ($\sim 10\times$) it is possible to employ two separated cemented achromats (Fig. 6.29(a)), the combination being corrected for spherical aberration and coma. Such lens combinations are, however, not suitable for objectives of high magnification ($\sim 50\times$ or more). One then uses different lens systems, which utilize the existence of aplanatic points of a spherical surface (cf. § 4.2.3) in the following way:

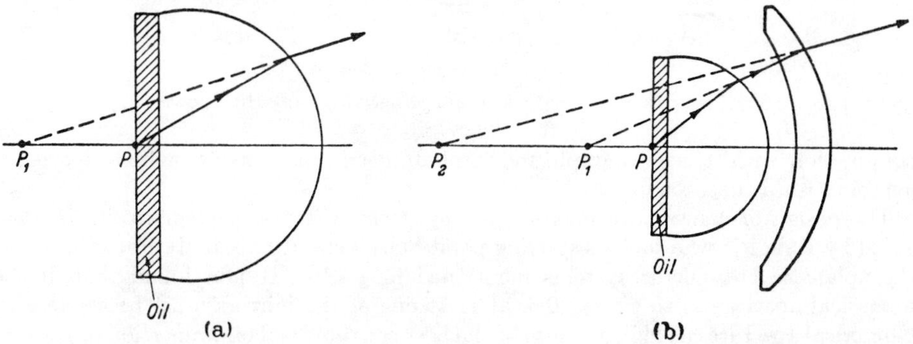

Fig. 6.28. Reduction of the angular divergence of rays in microscopes.

The object is placed at a point P close to the plano-convex lens (Fig. 6.28(a)) and the space between the object and the lens is filled with oil, the refractive index n of which is nearly equal to that of the lens. If C is the centre of curvature of the curved portion of the lens, and r the radius of curvature, then a point at distance r_1/n_1 from C will give rise to a virtual image P_1 at distance n_1r_1 from C, the imaging between the two points being aplanatic. In this way a higher numerical aperture may be employed, the angular divergence of the rays from P being reduced within the system, without the introduction of monochromatic aberrations. Chromatic aberration is, however, introduced and this must be compensated by the rest of the system.

The angular divergence of the rays may be reduced still further by the use of a convergent meniscus* (Fig. 6.28(b)). The front surface of the meniscus has its centre of curvature at P_1, and the radius of its rear surface is such that P is an aplanatic point with respect to it. The rays refracted from this surface form a virtual image at the other aplanatic point P_2. Adding further menisci, it is possible to produce successive virtual images P_3, P_4, . . . lying further and further away from P, thus reducing the angular divergence of the rays more and more. However, more than two lenses are seldom employed as otherwise the chromatic aberration cannot be adequately compensated.

The *oil-immersion objectives* just described admit a wider cone of rays than a *dry objective* of the same diameter would. For in a dry objective, the cone of rays emerging from the cover glass which protects the object pass into air, and consequently the rays are bent outwards by the refraction, the angular divergence being thus increased;

* This method is due to G. B. Amici, *Ann. de chim. et phys.* (3), **12** (1844), 117.

in an oil-immersion objective, on the other hand, they are not refracted on emergence from the cover glass. It is only with oil-immersion objectives that the highest numerical apertures ($\sim 1{\cdot}4$) can be attained.

The demand has also arisen for microscope objectives which can operate with light which includes the near ultra-violet wavelengths ($\lambda \sim 2500$ Å) and wavelengths of the visible and the near infra-red regions. For example, if a microscope is to be used in

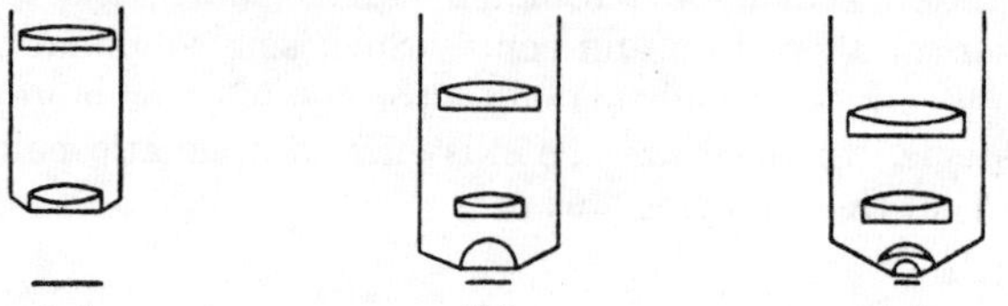

Fig. 6.29. Microscope objectives.
(a) Low power. (b) Medium power. (c) High power.

conjunction with a spectrograph and a photometric instrument, a high degree of achromatism is necessary.

The design of achromatic microscope objectives was first systematically investigated by BURCH,* who took as starting point SCHWARZSCHILD's analytical solution of the aplanatic two-mirror systems mentioned in § 4.10. BURCH found that, if the numerical aperture is to exceed 0·5, at least one of the mirrors must be aspherical. Numerical apertures as high as those of the best conventional oil-immersion objectives cannot however be realized in a reflecting system. A typical Burch achromatic

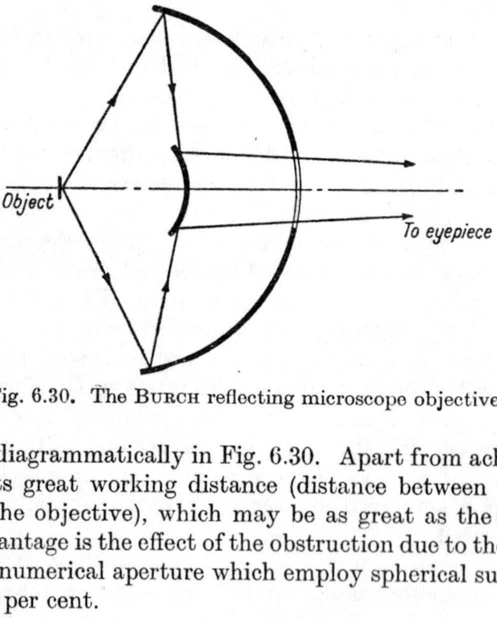

Fig. 6.30. The BURCH reflecting microscope objective.

objective is shown diagrammatically in Fig. 6.30. Apart from achromatism, its chief advantage lies in its great working distance (distance between the object and the nearest surface of the objective), which may be as great as the focal length of the objective. A disadvantage is the effect of the obstruction due to the secondary mirror; in systems of lower numerical aperture which employ spherical surfaces, the obstruction may exceed 45 per cent.

Microscope objectives which combine reflection and refraction have also been

* C. R. BURCH, *Proc. Phys. Soc.*, **59** (1947), 41; *ibid.*, **59** (1947), 47. The first paper contains also a review of earlier researches on achromatic microscope objectives.

considered. Thus BOUWERS* has constructed a system in which the reflections take place at concentric spherical silvered surfaces within glass, the silvering on the second surface being so distributed that the effective obstruction is reduced to 20 per cent. A two-mirror objective with low obstruction ratio was also described by GREY.† It uses quartz and fluorite components of considerable power which, whilst maintaining adequate colour correction throughout the ultra-violet and visible wavelengths, effectively replace the figuring which would otherwise be necessary on the mirrors.

The objects examined under a microscope are usually non-luminous and have therefore to be illuminated. In most cases the object has the form of a thin slice of transparent material. It is then illuminated from behind, or *transilluminated.* In systems of low numerical aperture (N.A. up to 0·25), diffuse sky-light reflected at oblique incidence from a concave mirror will suffice, but otherwise illumination from an artificial source is necessary. To obtain sufficient concentration of light, an auxiliary lens system, a condenser, is used. There are various methods of illumination, two of which will be described in Chapter X.

The present section has included only the basic principles of a microscope. An adequate discussion of the image formation in a microscope needs more refined methods (see § 8.6.3, § 9.5 and § 10.5).

---

* A. BOUWERS, *Achievements in Optics* (Amsterdam, Elsevier, 1946).

† D. S. GREY, *Proceedings of the London Conference on Optical Instruments* (London, Chapman & Hall, 1951), 65.

CHAPTER VII

ELEMENTS OF THE THEORY OF INTERFERENCE AND INTERFEROMETERS

7.1 INTRODUCTION

In Chapter III a geometrical model of the propagation of light was derived from the basic equations of Electromagnetic Theory, and it was shown that, with certain approximations, variations of intensity in a beam of light can be described in terms of changes in the cross-sectional area of a tube of rays. When two or more light beams are superposed, the distribution of intensity can no longer in general be described in such a simple manner. Thus if light from a source is divided by suitable apparatus into two beams which are then superposed, the intensity in the region of superposition is found to vary from point to point between maxima which exceed the sum of the intensities in the beams, and minima which may be zero. This phenomenon is called *interference*. We shall see shortly that the superposition of beams of strictly monochromatic light always gives rise to interference. However, light produced by a real physical source is never strictly monochromatic but, as we learn from atomistic theory, the amplitude and phase undergo irregular fluctuations much too rapid for the eye or an ordinary physical detector to follow. If the two beams originate in the same source, the fluctuations in the two beams are in general correlated, and the beams are said to be completely or partially *coherent* depending on whether the correlation is complete or partial. In beams from different sources, the fluctuations are completely uncorrelated, and the beams are said to be mutually *incoherent*. When such beams from different sources are superposed, no interference is observed under ordinary experimental conditions, the total intensity being everywhere the sum of the intensities of the individual beams. We shall see later (Chapter X) that the "degree of correlation" that exists between the fluctuations in two light beams determines, and conversely is revealed by, the "distinctness" of the interference effects to which the beams give rise on superposition.

There are two general methods of obtaining beams from a single beam of light, and these provide a basis for classifying the arrangements used to produce interference. In one the beam is divided by passage through apertures placed side by side. This method, which is called *division of wave-front*, is useful only with sufficiently small sources. Alternatively the beam is divided at one or more partially reflecting surfaces, at each of which part of the light is reflected and part transmitted. This method is called *division of amplitude*; it can be used with extended sources, and so the effects may be of greater intensity than with division of wave-front. In either case, it is convenient to consider separately the effects which result from the superposition of two beams (*two-beam interference*), and those which result from the superposition of more than two beams (*multiple-beam interference*).

Historically, interference phenomena have been the means of establishing the wave nature of light (see Historical Introduction) and to-day they have important practical uses, for example in spectroscopy and metrology. In this chapter we shall be concerned mainly with the idealized case of interference between beams from perfectly monochromatic sources. The elementary monochromatic theory is adequate to describe the action of the apparatus used in the great majority of interference experiments. Where necessary, we shall, with the help of FOURIER's theorem, take explicit account of the fact that real sources are not monochromatic; but this fact is always implied by our treatment of extended primary sources, which we consider to be made up of a large number of mutually incoherent point sources. Throughout the chapter we assume, whenever possible, that the individual beams obey the laws of geometrical optics, neglecting diffraction effects which, as briefly explained in § 3.1.4, arise in the neighbourhood of focal points and shadow boundaries. These effects will be considered in detail in Chapter VIII for the case of monochromatic light. The general case of interference and diffraction with polychromatic partially coherent light will be considered in Chapter X.

7.2 INTERFERENCE OF TWO MONOCHROMATIC WAVES

The intensity I of light has been defined as the time average of the amount of energy which crosses in unit time a unit area perpendicular to the direction of the energy flow. For a plane wave, according to § 1.4 (8), (9),

$$I = v\langle w \rangle = \frac{c}{4\pi}\sqrt{\frac{\varepsilon}{\mu}}\langle \mathbf{E}^2 \rangle = \frac{c}{4\pi}\sqrt{\frac{\mu}{\varepsilon}}\langle \mathbf{H}^2 \rangle, \tag{1}$$

and we have seen in Chapter III that these relations hold, at least as an approximation, for waves of more general type. Since we shall be comparing intensities in the same medium we may take the quantity $\langle \mathbf{E}^2 \rangle$ as measure of intensity. We shall be concerned mainly with monochromatic fields, and represent the electric vector $\mathbf{E}$ in the form

$$\mathbf{E}(\mathbf{r},t) = \mathscr{R}\{\mathbf{A}(\mathbf{r})e^{-i\omega t}\} = \tfrac{1}{2}[\mathbf{A}(\mathbf{r})e^{-i\omega t} + \mathbf{A}^{\star}(\mathbf{r})e^{i\omega t}]. \tag{2}$$

Here $\mathbf{A}$ is a complex vector with Cartesian rectangular components

$$A_x = a_1(\mathbf{r})e^{ig_1(\mathbf{r})}, \qquad A_y = a_2(\mathbf{r})e^{ig_2(\mathbf{r})}, \qquad A_z = a_3(\mathbf{r})e^{ig_3(\mathbf{r})}, \tag{3}$$

where a_j and g_j $(j = 1, 2, 3)$ are real functions. For a homogeneous plane wave the amplitudes a_j are constant, whilst the phase functions g_j are of the form $g_j(\mathbf{r}) = \mathbf{k} \cdot \mathbf{r} - \delta_j$, where $\mathbf{k}$ is the propagation vector and the δ_j's are phase constants which specify the state of polarization.

From (2),

$$\mathbf{E}^2 = \tfrac{1}{4}(\mathbf{A}^2 e^{-2i\omega t} + \mathbf{A}^{\star 2} e^{2i\omega t} + 2\mathbf{A} \cdot \mathbf{A}^{\star}), \tag{4}$$

whence, taking the time average over an interval large compared with the period $T = 2\pi/\omega$,

$$\langle \mathbf{E}^2 \rangle = \tfrac{1}{2}\mathbf{A} \cdot \mathbf{A}^{\star} = \tfrac{1}{2}(|A_x|^2 + |A_y|^2 + |A_z|^2) = \tfrac{1}{2}(a_1^2 + a_2^2 + a_3^2). \tag{5}$$

Suppose now that two monochromatic waves $\mathbf{E}_1$ and $\mathbf{E}_2$ are superposed at some point P. The total electric field at P is

$$\mathbf{E} = \mathbf{E}_1 + \mathbf{E}_2, \tag{6}$$

so that

$$\mathbf{E}^2 = \mathbf{E}_1^2 + \mathbf{E}_2^2 + 2\mathbf{E}_1 \cdot \mathbf{E}_2. \tag{7}$$

Hence the total intensity at P is

$$I = I_1 + I_2 + J_{12}, \tag{8}$$

where

$$I_1 = \langle \mathbf{E}_1^2 \rangle, \qquad I_2 = \langle \mathbf{E}_2^2 \rangle \tag{9a}$$

are the intensities of the two waves, and

$$J_{12} = 2\langle \mathbf{E}_1 \cdot \mathbf{E}_2 \rangle \tag{9b}$$

is the *interference term*. Let $\mathbf{A}$ and $\mathbf{B}$ be the complex amplitudes of the two waves, where

$$A_x = a_1 e^{ig_1}, \ldots, \qquad B_x = b_1 e^{ih_1}, \ldots \tag{10}$$

The (real) phases g_j and h_j of the two waves will in general be different, since the waves will have travelled to P by different paths; but if the experimental conditions are such that the same *phase difference* δ is introduced between the corresponding components, we have

$$g_1 - h_1 = g_2 - h_2 = g_3 - h_3 = \delta = \frac{2\pi}{\lambda_0} \Delta\mathscr{S}, \tag{11}$$

where $\Delta\mathscr{S}$ is the difference between the optical paths for the two waves from their common source to P, and λ_0 is the wavelength in vacuum. In terms of $\mathbf{A}$ and $\mathbf{B}$,

$$\begin{aligned} \mathbf{E}_1 \cdot \mathbf{E}_2 &= \tfrac{1}{4}(\mathbf{A}e^{-i\omega t} + \mathbf{A}^\star e^{i\omega t}) \cdot (\mathbf{B}e^{-i\omega t} + \mathbf{B}^\star e^{i\omega t}) \\ &= \tfrac{1}{4}(\mathbf{A} \cdot \mathbf{B}e^{-2i\omega t} + \mathbf{A}^\star \cdot \mathbf{B}^\star e^{2i\omega t} + \mathbf{A} \cdot \mathbf{B}^\star + \mathbf{A}^\star \cdot \mathbf{B}), \end{aligned} \tag{12}$$

so that

$$\begin{aligned} J_{12} = 2\langle \mathbf{E}_1 \cdot \mathbf{E}_2 \rangle &= \tfrac{1}{2}(\mathbf{A} \cdot \mathbf{B}^\star + \mathbf{A}^\star \cdot \mathbf{B}) \\ &= a_1 b_1 \cos(g_1 - h_1) + a_2 b_2 \cos(g_2 - h_2) + a_3 b_3 \cos(g_3 - h_3) \\ &= (a_1 b_1 + a_2 b_2 + a_3 b_3) \cos\delta. \end{aligned} \tag{13}$$

This expression shows the dependence of the interference term on the amplitude components and on the phase difference of the two waves.

In the derivation of (13) we have made no use of electromagnetic theory, and in particular, no use of the fact that the vibrations are transverse. Now in the Historical Introduction it was mentioned that Fresnel and Arago found that two light beams polarized at right angles to each other do not interfere, from which they concluded that light vibrations must be transverse. This conclusion is easily verified from (13). Suppose the two waves are propagated in the z-direction, and let the electric vector of the first wave be in the xz-plane and that of the second wave in the yz-plane. Then

$$a_2 = 0, \qquad b_1 = 0,$$

and from (13) the interference term is

$$J_{12} = a_3 b_3 \cos\delta.$$

Since the observations of Fresnel and Arago showed that no interference takes place under these circumstances, we must conclude that $a_3 = b_3 = 0$, i.e. that the electric vectors of the two waves must be perpendicular to the z-direction. Hence light waves must be transverse, in agreement with our earlier deduction from electromagnetic theory.

Let us consider the distribution of intensity resulting from the superposition of two waves which are propagated in the z-direction, and are linearly polarized with their E vectors in the x-direction. Then

$$a_2 = a_3 = b_2 = b_3 = 0,$$

so that using (5), (9a) and (13),

$$\left.\begin{aligned} I_1 = \tfrac{1}{2}a_1^2, \qquad I_2 = \tfrac{1}{2}b_1^2, \\ \text{and} \qquad J_{12} = a_1 b_1 \cos\delta = 2\sqrt{I_1 I_2}\cos\delta. \end{aligned}\right\} \tag{14}$$

The total intensity is then given by (8) as

$$I = I_1 + I_2 + 2\sqrt{I_1 I_2}\cos\delta. \tag{15}$$

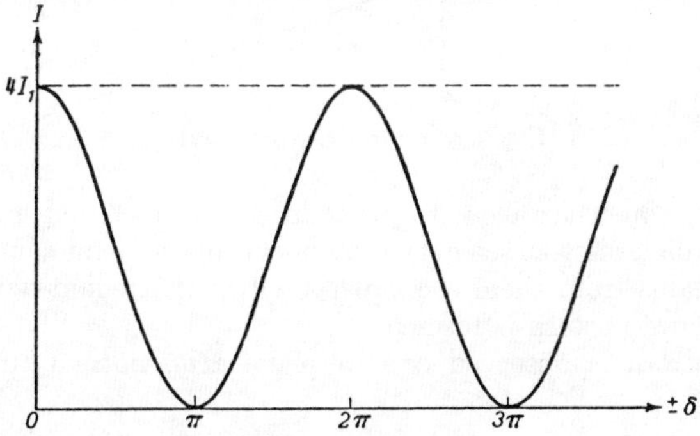

Fig. 7.1. Interference of two beams of equal intensity; variation of intensity with phase difference.

Evidently there will be maxima of intensity

$$\left.\begin{aligned} I_{\max} = I_1 + I_2 + 2\sqrt{I_1 I_2} \\ \text{when} \qquad |\delta| = 0,\ 2\pi,\ 4\pi,\ \ldots, \end{aligned}\right\} \tag{16a}$$

and minima of intensity

$$\left.\begin{aligned} I_{\min} = I_1 + I_2 - 2\sqrt{I_1 I_2} \\ \text{when} \qquad |\delta| = \pi,\ 3\pi,\ \ldots \end{aligned}\right\} \tag{16b}$$

In the special case when $I_1 = I_2$, (15) reduces to

$$I = 2I_1(1 + \cos\delta) = 4I_1 \cos^2\frac{\delta}{2}, \tag{17}$$

and the intensity varies between a maximum value $I_{\max} = 4I_1$, and a minimum value $I_{\min} = 0$ (Fig. 7.1).

The same formulae hold also for natural unpolarized light since, as will be seen later (§ 10.8.2), a beam of natural light may be represented as a superposition of two incoherent beams linearly polarized at right angles to each other (say in the x and y-directions). The interference between the x-components and y-components may then be considered separately, and the total intensity is obtained by addition of the separate intensities. Since δ has the same value in each case, the above formulae are again obtained.

7.3. TWO-BEAM INTERFERENCE: DIVISION OF WAVE-FRONT

7.3.1 Young's experiment

The earliest experimental arrangement for demonstrating the interference of light is due to YOUNG. Light from a monochromatic point source S falls on two

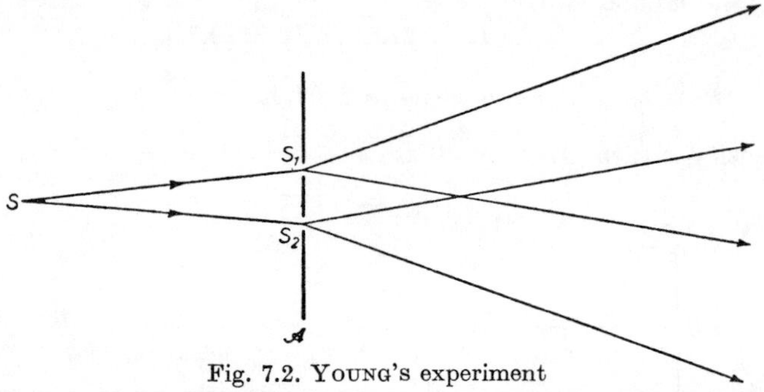

Fig. 7.2. YOUNG's experiment

pinholes S_1 and S_2 which are close together in a screen $\mathscr{A}$ and equidistant from S (Fig. 7.2). The pinholes act as secondary monochromatic point sources* which are in phase, and the beams from them are superposed in the region beyond $\mathscr{A}$. In this region an interference pattern is formed.

Suppose the pattern is observed over a plane xOy, normal to a perpendicular

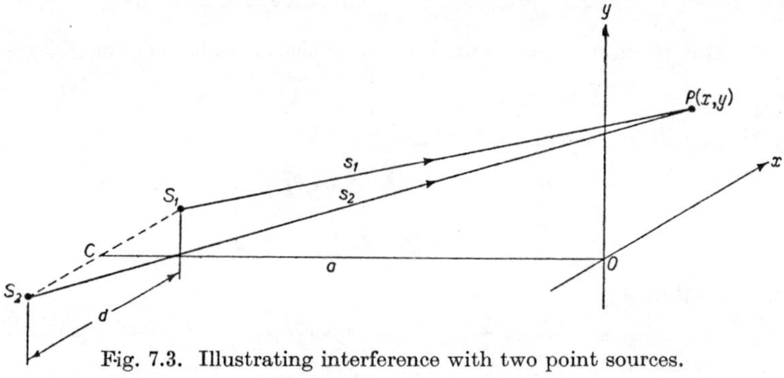

Fig. 7.3. Illustrating interference with two point sources.

bisector CO of S_1S_2, and with the x-axis parallel to S_1S_2 (Fig. 7.3). Let d be the separation of the pinholes, and a the distance between the line joining the pinholes and the plane of observation. For a point $P(x, y)$ in the plane of observation,

$$s_1 = S_1P = \sqrt{a^2 + y^2 + \left(x - \frac{d}{2}\right)^2}, \tag{1a}$$

$$s_2 = S_2P = \sqrt{a^2 + y^2 + \left(x + \frac{d}{2}\right)^2}, \tag{1b}$$

so that

$$s_2^2 - s_1^2 = 2xd. \tag{2}$$

---

* These secondary sources have directional properties which will be discussed in detail in the theory of diffraction (Chapter VIII).

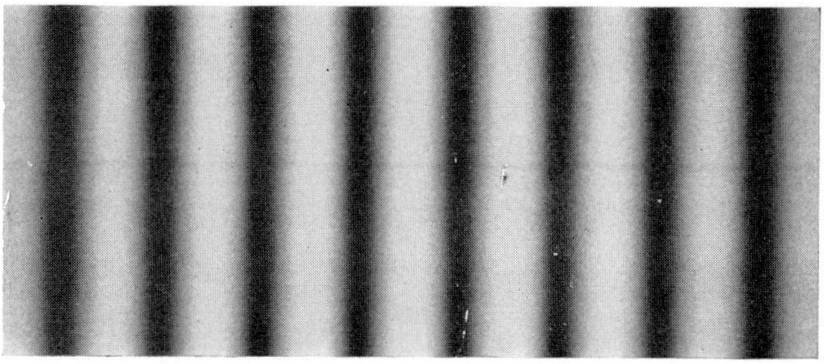

Fig. 7.4. YOUNG's fringes.

The difference of geometrical path for light reaching P from S_2 and S_1 may therefore be expressed in the form

$$\Delta s = s_2 - s_1 = \frac{2xd}{s_2 + s_1}. \tag{3}$$

In practice, because of the short wavelength of visible light, the pattern can be observed conveniently only if d is much smaller than a. Then provided x and y are also small compared with a,

$$s_2 + s_1 \sim 2a, \tag{4}$$

so that apart from terms of the second and higher orders in d/a, x/a, y/a,

$$\Delta s = \frac{xd}{a}. \tag{5}$$

If n is the refractive index of the medium (assumed homogeneous) in which the experiment is made, the difference of optical path from S_2 and S_1 to P is therefore

$$\Delta\mathscr{S} = n\Delta s = \frac{nxd}{a}, \tag{6}$$

and the corresponding phase difference is

$$\delta = \frac{2\pi}{\lambda_0}\frac{nxd}{a}. \tag{7}$$

Since the angle S_1PS_2 is very small, we can consider the waves from S_1 and S_2 to be propagated in the same direction at P, so that the intensity can be calculated from § 7.2 (15); according to (7) and § 7.2 (16) there are maxima of intensity when

$$x = \frac{ma\lambda_0}{nd}, \qquad |m| = 0, 1, 2, \ldots, \tag{8a}$$

and minima of intensity when

$$x = \frac{ma\lambda_0}{nd}, \qquad |m| = \tfrac{1}{2}, \tfrac{3}{2}, \tfrac{5}{2}, \ldots \tag{8b}$$

The interference pattern in the immediate vicinity of O thus consists of bright and dark bands called *interference fringes* (Fig. 7.4), equidistant and running at right angles to the line S_1S_2 joining the two sources. The separation of adjacent bright fringes is $a\lambda_0/nd$. At any point of the interference pattern the number m defined by

$$m = \frac{\delta}{2\pi} = \frac{\Delta\mathscr{S}}{\lambda_0} \tag{9}$$

is called the *order of interference* at the point: thus the bright fringes correspond to integral orders.

7.3.2 Fresnel's mirrors and similar arrangements

Young's experiment is of historical importance as a crucial step in establishing the wave theory of light. It also provides a method, though one of low precision, of measuring the wavelength of monochromatic light with extremely simple apparatus; it is necessary to measure only d, a, and the fringe spacing, which in air ($n \sim 1$) is $a\lambda_0/d$. However, the light from the primary source S does not reach the region of superposition by uninterrupted rectilinear paths of the kind to which geometrical optics applies, and in order to show that this circumstance is not essential to the

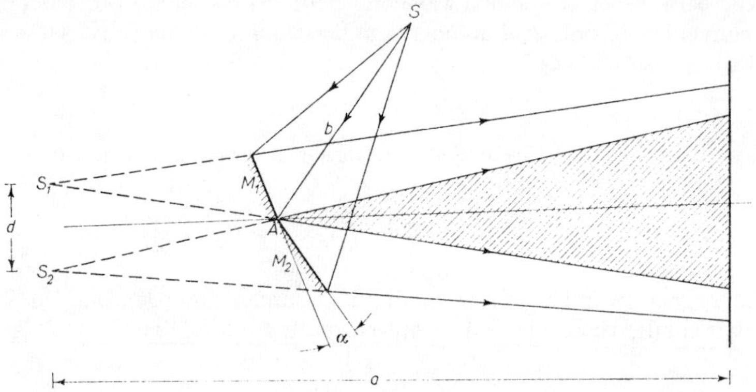

Fig. 7.5. FRESNEL's mirrors.

production of interference effects a number of alternative ways of producing two coherent sources were subsequently devised.

One example is the arrangement known as *Fresnel's mirrors* (Fig. 7.5). Light from a point source S is incident on two plane mirrors M_1 and M_2, mutually inclined at a small angle, and reflection at the mirrors gives rise to two virtual images S_1, S_2 of S

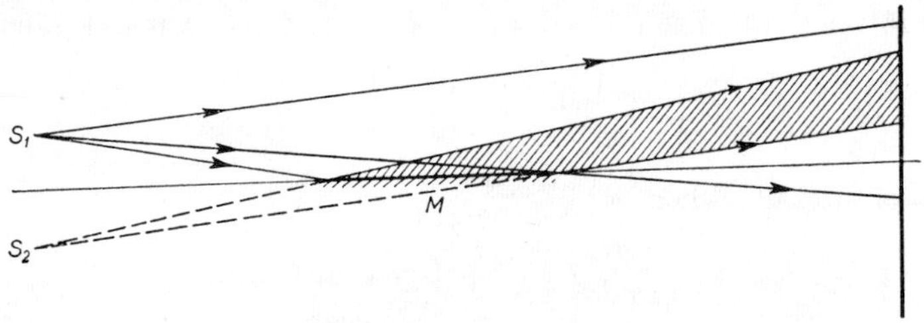

Fig. 7.6. LLOYD's mirror.

which act as coherent sources. The plane $S\,S_1S_2$ is evidently normal to the line of intersection of the mirrors, which it meets in A. If $SA = b$, then

$$S_1A = S_2A = b,$$

so that the perpendicular bisector of S_1S_2 also passes through A. The separation of S_1 and S_2 is

$$d = 2b \sin \alpha, \qquad (10)$$

where α is the angle between the mirrors.

An even simpler arrangement is *Lloyd's mirror* (Fig. 7.6). A point source S_1 is placed some distance away from a plane mirror M and close to the plane of the mirror surface, so that light is reflected at nearly grazing incidence. The coherent sources are the primary source S_1 and its virtual image S_2 in the mirror. The perpendicular bisector of S_1S_2 then lies in the plane of the mirror surface.

We may mention two other similar devices. *Fresnel's bi-prism* (Fig. 7.7) is formed by two equal prisms of small refracting angle placed together base to base with their refracting edges parallel. A pencil of light from a point source S is divided by refraction into two overlapping pencils. These refracted pencils are not strictly stigmatic,

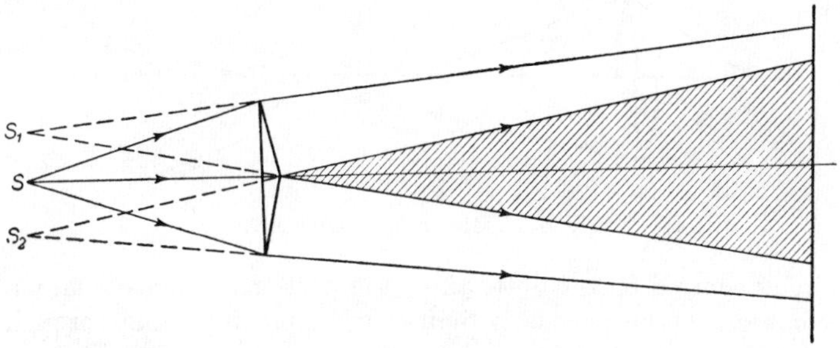

Fig. 7.7. FRESNEL's bi-prism.

but because of the smallness of the refracting angle and of the angular aperture of the pencils used we may neglect this aberration and suppose the prisms to form two virtual images S_1, S_2 of S. *Billet's split lens* (Fig. 7.8) consists of a convex lens cut diametrically into two, the halves being separated by a small distance at right angles to the optic axis. Two real images S_1, S_2 are then produced from a single point source S.

With all these arrangements which use a primary point source, interference fringes are visible in monochromatic light over any plane in the region common to the diverging pencils from the sources S_1 and S_2 (shown hatched in Figs. 7.5 to 7.8).

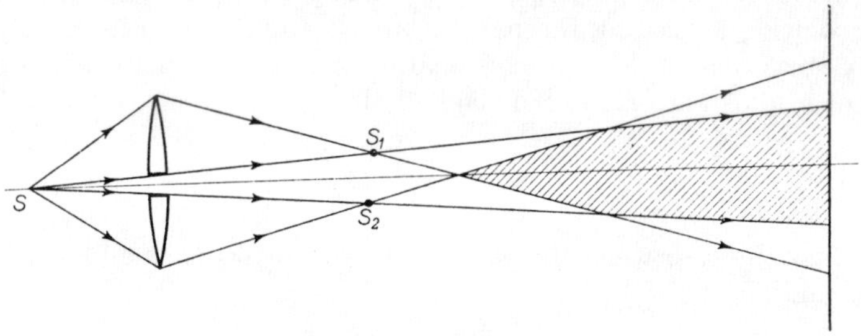

Fig. 7.8. BILLET's split lens.

Such fringes are said to be *non-localized*. Denoting as before the distances S_1P and S_2P by s_1 and s_2, the locus of points P for which the phase difference between waves from S_2 and S_1 is constant is the surface defined by

$$s_2 - s_1 = \text{constant}. \tag{11}$$

Hence the maxima and minima of the resultant intensity form a family of hyperboloids of revolution about S_1S_2 as axis, and having S_1 and S_2 as common foci. The fringes in a plane normal to the perpendicular bisector CO of S_1S_2 are sections of these hyperboloids, and are themselves hyperbolae; but near O they approximate, as we have seen in § 7.3.1, to equidistant straight lines running at right angles to S_1S_2. In a plane normal to S_1S_2 the sections of the hyperboloids are concentric circles, but such fringes cannot be observed with the arrangements described above. They can however be seen in *Meslin's experiment*, in which the halves of BILLET's split lens are separated along the optic axis instead of transversely (Fig. 7.9). The sources S_1 and S_2 are then at different points of the optic axis, and the corresponding pencils overlap in a region between them. Here fringes can be observed which in planes normal to S_1S_2 are

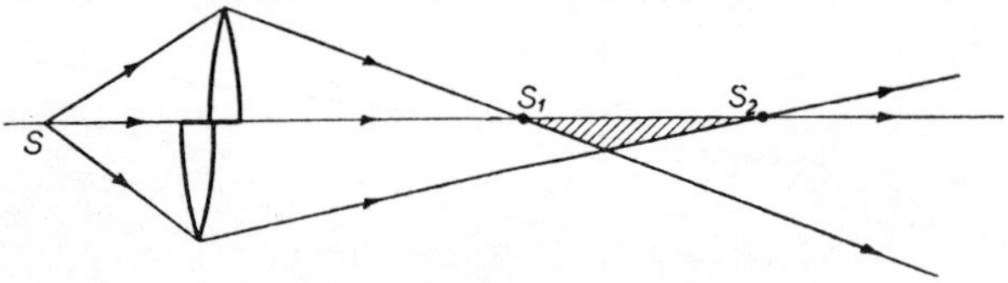

Fig. 7.9. MESLIN's experiment.

concentric circles centred on the optic axis; the portions seen are semi-circular since the region common to the pencils is limited by a plane passing through the optic axis.

7.3.3 Fringes with quasi-monochromatic and white light

So far we have assumed that the primary point source is monochromatic. We now remove this restriction and suppose that a point source S of polychromatic light is used, for example with FRESNEL's mirrors (Fig. 7.5). As we shall see later (§ 7.5.8), such light can be represented by a combination of mutually incoherent monochromatic components extending over a range of frequencies. Each component produces an interference pattern as described above, and the total intensity is everywhere the sum of the intensities in these monochromatic patterns. Suppose that the components cover a wavelength range $\Delta\lambda_0$ around a mean wavelength $\bar{\lambda}_0$. The central maxima of all the monochromatic patterns, corresponding to equality of the paths from S_1 and S_2, coincide in passing through O, but elsewhere the patterns are mutually displaced because their scale is proportional to wavelength; the maxima of order m are spread over a distance Δx in the plane of observation where, by (8),

$$\Delta x = \frac{|m|a}{nd}\,\Delta\lambda_0. \tag{12}$$

Consider first the case when the wavelength range $\Delta\lambda_0$ is small compared to the mean wavelength $\bar{\lambda}_0$, i.e.

$$\frac{\Delta\lambda_0}{\bar{\lambda}_0} \ll 1. \tag{13}$$

We shall call light which satisfies this condition *quasi-monochromatic light*.*

If over the field of observation

$$|m| \ll \frac{\bar{\lambda}_0}{\Delta\lambda_0}, \tag{14}$$

---

* Since $\lambda_0 = 2\pi c/\omega_0$, (13) may also be expressed in the form

$$\frac{\Delta\omega_0}{\bar{\omega}_0} \ll 1,$$

where $\Delta\omega_0$ is the effective frequency range and $\bar{\omega}_0$ the mean frequency.

or by (9), if

$$|\Delta\mathscr{S}| \ll \frac{\bar{\lambda}_0^2}{\Delta\lambda_0}, \tag{15}$$

we can neglect Δx compared with the mean separation $a\lambda_0/nd$ of adjacent maxima, and take the component patterns to be in coincidence. There are then fringes in the plane of observations which have the same appearance as those given by a strictly monochromatic source of wavelength $\bar{\lambda}_0$.

If the light is quasi-monochromatic but (15) is not satisfied, the fringes are less distinct than with monochromatic light, the total intensity depending on the distribution of intensity amongst the monochromatic components.

If the light is not quasi-monochromatic, i.e. if (13) is not satisfied, what is observed depends also on the spectral response of the radiation detector used. A case of practical importance is when the light is white and the observation is visual, so that the effective wavelength range extends from about 4000 Å to about 7000 Å and $\Delta\lambda_0/\bar{\lambda}_0$ is of the order of $\frac{1}{2}$. There is then a central white fringe in the position of the monochromatic fringe of order zero, with a few coloured maxima and minima on either side, and farther away what appears to the eye to be uniform white illumination. This light at some distance from the centre of the pattern is not normal white light. Thus at distance x from the central fringe there are according to (8) maxima of intensity for

$$\lambda_0 = \frac{nd}{a}\frac{x}{m}, \qquad |m| = 1, 2, 3, \ldots, \tag{16a}$$

and minima of intensity for

$$\lambda_0 = \frac{nd}{a}\frac{x}{m}, \qquad |m| = \tfrac{1}{2}, \tfrac{3}{2}, \tfrac{5}{2} \ldots \tag{16b}$$

Hence if the light enters a spectroscope with its slit parallel to the direction of monochromatic fringes, so that x is constant for the light admitted, the spectrum is crossed by light and dark bands parallel to the slit—an example of a so-called *channelled spectrum*. The bands are equally spaced in spectroscopic wave-number $1/\lambda_0$, the separation of adjacent bright bands being $a/nd|x|$.

As will appear later, the white light fringe pattern is useful in interferometry because in particular cases it enables the monochromatic fringe corresponding to zero path difference to be recognized. We have referred to the central fringe obtained with FRESNEL's mirrors as an intensity maximum, and this is the case also with FRESNEL's bi-prism, BILLET's split lens, and in YOUNG's experiment. With LLOYD's mirror, however, the fringe which lies in the plane of the mirror surface is an intensity minimum, and with a white light source it appears black. This is because the phase of the wave reflected from the mirror is changed by π on reflection (see § 1.5.2), irrespective of its wavelength. In MESLIN's experiment also the centre of the pattern is an intensity minimum, in this case because, as we shall see later (§ 8.8.4), the wave suffers a phase change of π on its passage through focus.

7.3.4 Use of slit sources; visibility of fringes

The foregoing discussion has been in terms of a primary point source, but since practical sources have finite extension we must examine how such extension affects the fringes. The description of a true physical source involves atomistic theory, which is outside the scope of this book, but for our purposes it is adequate to idealize the situation by regarding the source to be made up of a large number of point sources which are mutually incoherent. The intensity at any point in the wave-field is then the sum of the intensities from the individual point sources.

With the arrangements we have described (MESLIN's experiment excepted) the fringes are perpendicular to the plane which contains the primary point source S and the derived sources S_1 and S_2, and it follows that if S is displaced perpendicular to this plane the fringes are merely displaced parallel to their length. Hence a line source—or in practice a sufficiently narrow slit source—in this direction may be used without impairing the distinctness of the fringes, at least so far as their curvature is negligible. Similarly the pinholes of YOUNG's experiment may be replaced by narrow

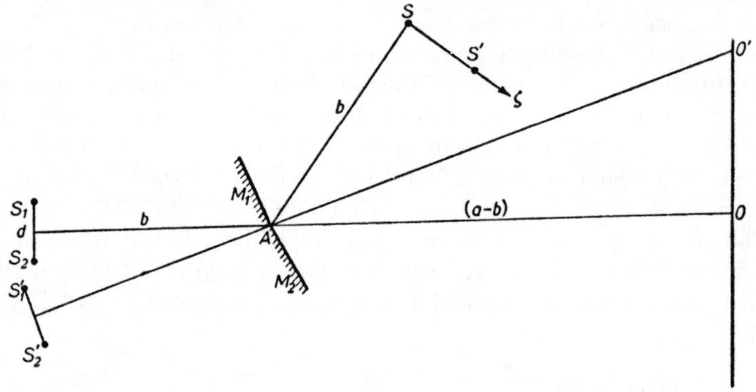

Fig. 7.10. Illustrating FRESNEL's mirrors with slit source.

slits parallel to the slit source. In this way the intensity of the patterns can be increased, though there is an additional experimental difficulty in securing correct orientation of the source.

To obtain even more light in the patterns it is necessary to increase the width of the source slit, but then the fringes become less distinct. Consider for example the case of FRESNEL's mirrors. If the source S is displaced to S' at right angles to SA and in the plane SS_1S_2 (Fig. 7.10), the secondary sources S_1, S_2 are displaced to S_1', S_2', and apart from terms of second order their separation d—and therefore the fringe spacing —is unchanged; but the central fringe is displaced from O to O', since it lies on the perpendicular bisector of $S_1'S_2'$ which passes through A. If $SS' = \zeta$ (taken positive in Fig. 7.10) then $S_1S_1' = S_2S_2' = \zeta$, so that

$$OO' = \frac{(a-b)\zeta}{b}. \tag{17}$$

Hence if at P the optical path difference for light from the source S is $\Delta\mathscr{S}$, the optical path difference for light from the source S' is, by (6),

$$\Delta\mathscr{S}' = \Delta\mathscr{S} - \frac{nd}{a}OO' = \Delta\mathscr{S} - D\zeta, \tag{18}$$

where

$$D = \frac{(a-b)nd}{ab}. \tag{19}$$

The corresponding phase difference is

$$\delta(\zeta, \Delta\mathscr{S}) = \frac{2\pi}{\lambda_0}(\Delta\mathscr{S} - D\zeta). \tag{20}$$

Suppose now that the source is a slit of width e centred on S. We assume that the number of point sources which constitute the slit source is so large that we may treat it as effectively continuous, and imagine it divided into elementary strips perpendicular to the plane SS_1S_2. If $i_1 d\zeta$ is the intensity at P from an elementary strip when the light reaches P from one mirror only, the intensity at P from the elementary strip through S' is, by § 7.2 (17),

$$i(\zeta, \Delta\mathscr{S})d\zeta = 2i_1(1 + \cos\delta)d\zeta,$$

and the total intensity at P is given by

$$I(e, \Delta\mathscr{S}) = 2i_1 \int_{-e/2}^{e/2} (1 + \cos\delta)d\zeta. \tag{21}$$

On substituting from (20) into (21) and evaluating the integral we obtain

$$I(e, \Delta\mathscr{S}) = 2I_1\left\{1 + \left(\frac{\sin\dfrac{\pi De}{\lambda_0}}{\dfrac{\pi De}{\lambda_0}}\right)\cos\left(\frac{2\pi}{\lambda_0}\Delta\mathscr{S}\right)\right\}, \tag{22}$$

where $I_1 = i_1 e$.

Following MICHELSON we take as measure of the distinctness of the fringes at P their *visibility* $\mathscr{V}$ defined by

$$\mathscr{V} = \frac{I_{\max} - I_{\min}}{I_{\max} + I_{\min}}, \tag{23}$$

where $I_{\max}$ and $I_{\min}$ are the maximum and minimum intensities in the immediate neighbourhood of P. Evidently $\mathscr{V}$ has a maximum value of unity when $I_{\min} = 0$, as is the case for fringes from two equal monochromatic point sources, and declines to zero when $I_{\max} = I_{\min}$ and the fringes disappear. In the present case

$$I_{\max} = 2I_1\left(1 + \frac{\left|\sin\dfrac{\pi De}{\lambda_0}\right|}{\dfrac{\pi De}{\lambda_0}}\right), \qquad I_{\min} = 2I_1\left(1 - \frac{\left|\sin\dfrac{\pi De}{\lambda_0}\right|}{\dfrac{\pi De}{\lambda_0}}\right), \tag{24}$$

so that

$$\mathscr{V} = \frac{\left|\sin\dfrac{\pi De}{\lambda_0}\right|}{\dfrac{\pi De}{\lambda_0}}. \tag{25}$$

The visibility $\mathscr{V}$ given by (25) is shown as a function of the source width e in Fig. 7.11. We see that the visibility is greater than about 0·9 when e does not exceed $\lambda_0/4D$, or from (20), when the range of phase difference at P corresponding to the elements of the source does not exceed $\pi/2$. If we somewhat arbitrarily take this condition to define the maximum slit width which gives rise to satisfactory fringes, we have from (19) $e \leqslant \lambda_0 ab/4(a - b)nd$, or, using (10),

$$e \leqslant \frac{\lambda_0 a}{8(a - b)n \sin\alpha}, \tag{26}$$

where, as before, α is the angle between the mirrors. Evidently the tolerable value of e increases as $(a - b)$ decreases, i.e. as the plane of observation approaches the mirror junction. For example, with typical values $\alpha = 2$ minutes of arc, $a = 120$ cm, $b = 100$ cm, $\lambda_0 = 5500$ Å, $n = 1$, we have $e \leqslant 0{\cdot}7$ mm; the spacing of the bright fringes is according to (8) and (10) given by the formula $a\lambda_0/2bn \sin \alpha$; in the present case it is equal to 0·57 mm.

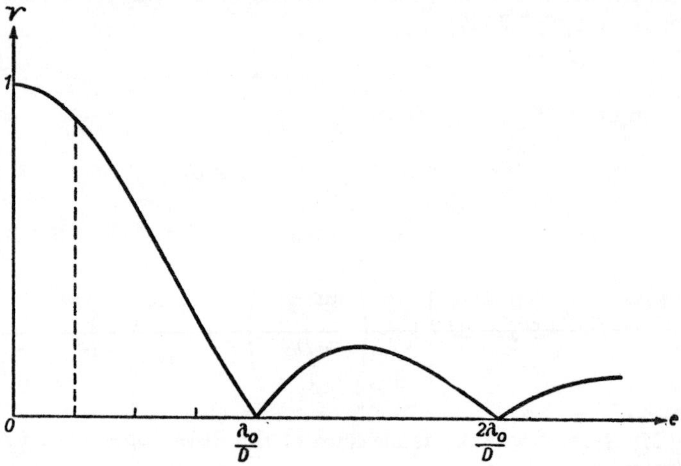

Fig. 7.11. Variation of fringe visibility with width of source slit. (FRESNEL's mirrors.)

Similar considerations apply to the source width with FRESNEL's bi-prism, BILLET's split lens, and in YOUNG's experiment. With LLOYD's mirror the situation is different since displacement of the source S normal to the mirror plane results in displacement of its image in the opposite direction. With a slit source of finite width the elementary patterns therefore have minima which coincide in the plane of the mirror but are of different spacing, so that the visibility of the fringes decreases with increasing distance from the mirror plane.

7.3.5 Application to the measurement of optical path difference: the Rayleigh interferometer

We will anticipate a result from the theory of diffraction by saying that the light from the secondary sources S_1 and S_2 in YOUNG's experiment has greatest intensity

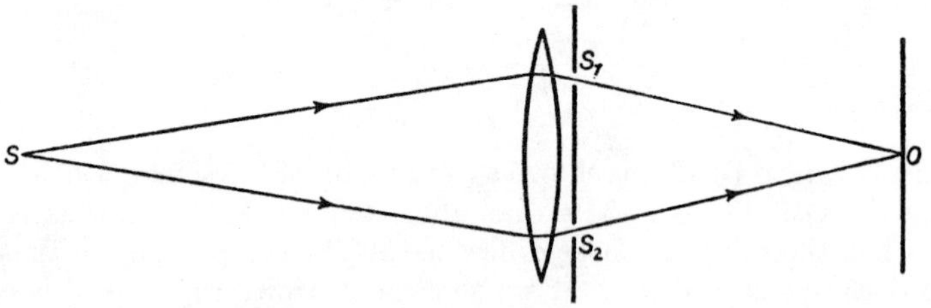

Fig. 7.12. Illustrating the use of a lens with YOUNG's arrangement.

in the direction of the geometrical rays from the primary source S. In YOUNG's experiment these rays diverge beyond S_1 and S_2, but if a lens is introduced in front of the apertures (Fig. 7.12) they may be brought together at O, conjugate to S with respect to the lens. In this way the intensity in the interference pattern near O is increased, and the fringes can be observed with the apertures S_1 and S_2 farther apart.

The separation of adjacent bright fringes is still $a\lambda_0/nd$, and by the principle of equal optical path, if the lens is stigmatic for S the fringe of order zero is at O. If the lens is not stigmatic for S the fringe of order zero will be displaced from O by an amount which depends on the difference of optical path from S to O through the two apertures; for a difference of optical path $\Delta\mathscr{S}$, the displacement will be Δm times the separation of adjacent bright fringes where

$$\Delta m = \frac{\Delta\mathscr{S}}{\lambda_0}. \tag{27}$$

Evidently the arrangement may be used as a quantitative test of the performance of the lens, as was done by MICHELSON.* If one aperture is fixed over the centre of the

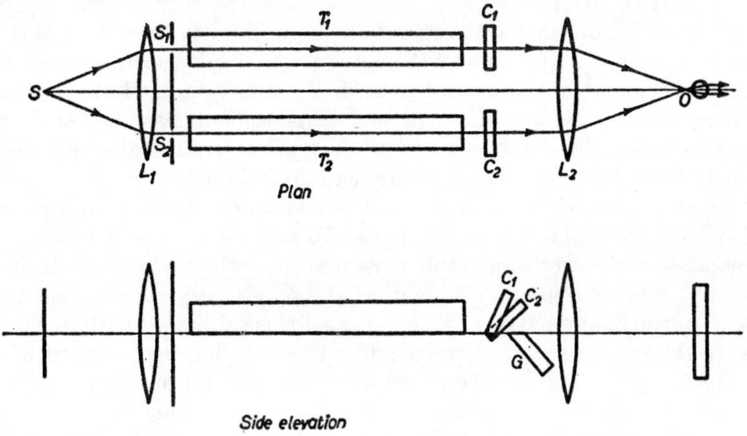

Fig. 7.13. The RAYLEIGH interferometer.

lens, measurements of Δm for different radial positions of the other aperture give the departure from sphericity of the wave-front from S after passage through the lens (the wave aberration).

Similarly, if a plate of transparent material of thickness l and refractive index n' is introduced into the path of the light from S_2, the optical path $[SS_2O]$ is increased by $(n' - n)l$ and there will be a change Δm in the order of interference at O given by

$$\Delta m = \frac{(n' - n)l}{\lambda_0}. \tag{28}$$

From measurement of Δm, l and λ_0, the difference $(n' - n)$ between the refractive index of the plate and that of the surrounding medium can be determined, and used in this way the arrangement forms the basis of the *Rayleigh interferometer*,† which is used for the precision determination of the refractive indices of gases. A modern form of this instrument is shown in plan and side elevation in Fig. 7.13. Light from a slit source S is collimated by the lens L_1 before falling on slit apertures S_1, S_2 which are parallel to S; the geometrical rays from S_1 and S_2 are then parallel in passing through separate gas chambers T_1 and T_2 before being re-combined by the lens L_2 in its focal

---

* A. A. MICHELSON, *Astrophys. J.*, **47** (1918), 283.

† Lord RAYLEIGH, *Proc. Roy. Soc.*, **59** (1896), 198.
F. HABER and F. LÖWE, *Zeits f. angew Chem.*, **23** (1910), 1393.

plane, where interference fringes are formed, parallel to the slits.* The presence of the gas chambers requires S_1 and S_2 to be widely separated, so that the fringes are closely spaced and a high magnification is needed to observe them; further, the restriction on the width of the source slit S is correspondingly severe, so that the amount of light in the pattern is small. Since magnification is needed only in the direction at right angles to the fringes it is best obtained by a cylindrical eyepiece in the form of a thin glass rod parallel to the fringes. The pattern viewed in this way appears much brighter than if a spherical eyepiece were used, but the use of a cylindrical eyepiece has a further important advantage. It allows a second fixed system of fringes, with the same spacing as the main system but formed by light from S_1 and S_2 which passes beneath the gas chambers, to be used as a fiducial mark. By means of the glass plate G this fiducial system is displaced vertically so that its upper edge meets the lower edge of the main system in a sharp dividing line which is the edge of G as viewed through L_2. In this way the detection of displacement of the main fringe system due to changes in the optical path in T_1 and T_2 is made to depend on the vernier acuity of the eye, which is high, and displacements as small as $\frac{1}{40}$ order can be detected. Accidental displacements of the optical system also become less important in so far as both fringe systems are equally affected.

In the technical use of the instrument it is convenient to compensate the optical path difference, rather than to count fringes. To achieve this the light emerging from the gas chambers is passed through thin glass plates, of which one, C_1, is fixed while the other, C_2, can be rotated about a horizontal axis to give reproducible changes in the optical path of the light from S_2. The compensator is calibrated with monochromatic light to give the mean rotation corresponding to one order displacement of the main system. The fringe systems are then used as a null indicator of equality of the optical paths $[SS_1O]$ and $[SS_2O]$. In normal use the gas chambers are evacuated, and with white light the central fringes of the main and fiducial systems are brought into approximate coincidence by means of the compensator. An exact setting for coincidence of the zero orders is then made with monochromatic light. Next the gas under investigation is admitted to one chamber and again, first with white light and then with monochromatic light, the compensator is adjusted to bring the zero orders into coincidence.† The difference between the two settings of the compensator in terms of

* The exact distribution of intensity in the focal plane of L_2 can be found by the theory of Fraunhofer diffraction (§ 8.5).

† The optical paths from S_1 and S_2 to the main pattern involve media of differing dispersion, so that, unlike in the simple case considered in § 7.3.3, the zero orders for light of different wavelengths do not in general coincide, and with white light there is no perfectly white fringe. The fringe showing least colour is that for which $\partial m/\partial\lambda_0 = 0$ at some mean wavelength $\lambda = \bar{\lambda}_0$ in the visible spectrum which depends on the wavelength response of the eye, and this is called the *achromatic fringe* by analogy with the term achromatic as applied to a lens. If the compensator introduces an optical path difference $\Delta\mathscr{S}$, the order of interference at O is

$$m = [(n' - 1)l + \Delta\mathscr{S}]\frac{1}{\lambda_0}$$

so that

$$\frac{\partial m}{\partial\lambda_0} = -\left(n' - 1 - \lambda_0\frac{\partial n'}{\partial\lambda_0}\right)\frac{l}{\lambda_0^2} - \left(\Delta\mathscr{S} - \lambda_0\frac{\partial}{\partial\lambda_0}(\Delta\mathscr{S})\right)\frac{1}{\lambda_0^2}$$

Hence the achromatic fringe is at O when

$$\left(\Delta\mathscr{S} - \lambda_0\frac{\partial}{\partial\lambda_0}(\Delta\mathscr{S})\right)_{\bar{\lambda}_0} = -l\left(n' - 1 - \lambda_0\frac{\partial n'}{\partial\lambda_0}\right)_{\bar{\lambda}_0},$$

[*Footnote continued on page* 271

its calibration gives the order displacement Δm of the main system due to the introduction of the gas, and its refractive index n' is obtained from (28):

$$(n' - 1) = \frac{\lambda_0}{l} \Delta m, \tag{29}$$

where l is the length of the gas chamber. With typical values $l = 100$ cm, $\lambda_0 = 5500$Å and settings correct to $\frac{1}{40}$ order, the detectable change of $(n' - 1)$ is about 10^{-8}.*

Higher sensitivity can in principle be obtained by increasing l, but this is limited in practice by difficulties of temperature control. For the same reason, versions of the instrument intended for the measurement of differences of refractive index of liquids employ only short chambers. Further, the path difference that can be compensated is limited, so that if the difference of refractive index in the two chambers is large the length of the chambers must be proportionately reduced.

7.3.6 Application to the measurement of angular dimensions of sources: the Michelson stellar interferometer

We have already seen (§ 7.3.4) that in YOUNG's experiment the distinctness of the fringes is affected by extension of the source in the direction joining the apertures S_1, S_2. This effect is the basis of methods of measuring the angular dimensions of small sources.

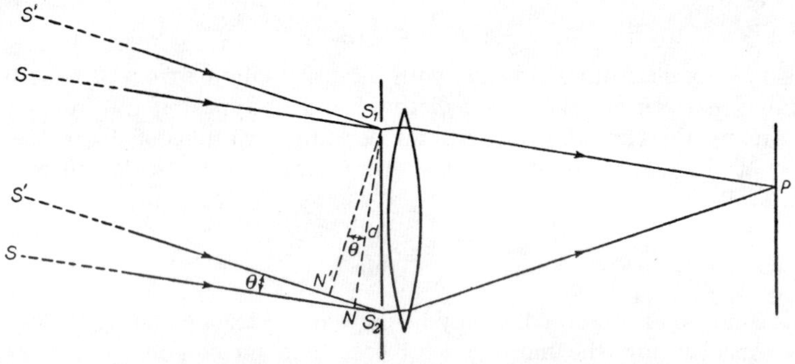

Fig. 7.14. Illustrating a telescope objective diaphragmed by two apertures and illuminated by two distant point sources.

Suppose a telescope objective, diaphragmed by two equal small apertures S_1, S_2 of separation d, is used to view two distant quasi-monochromatic point sources S and S', of effective wavelength λ_0, separated by angle θ in the direction joining the apertures (Fig. 7.14). S and S' each give an interference pattern in the focal plane with the same fringe spacing, and if S and S' are incoherent sources, the combined pattern

and with this setting of the compensator the zero order of the monochromatic pattern is not in general at O since this requires

$$(\Delta \mathscr{S})_{\bar{\lambda}_0} = -l(n' - 1)_{\bar{\lambda}_0}.$$

The effect may be large enough to make identification of the monochromatic fringe of zero order uncertain, in which case a preliminary measurement must be made at low pressure or with short chambers.

We note further that the achromatic fringe is recognizable only if, at the point of the pattern where $(\partial m/\partial \lambda_0)_{\bar{\lambda}_0} = 0$, the range of m for wavelengths in the visible spectrum is sufficiently small. For this reason, in interferometers which are required to show fringes in white light, the interfering waves are arranged to have as far as possible equal paths in media of identical dispersion.

* The value of $(n' - 1)$ for common gases is of the order of 10^{-4} (cf. Table I, p. 13).

is formed by summing at each point the intensities in these two patterns. If N is the foot of the perpendicular from S_1 to SS_2, S_1N lies in a plane wave-front from S so that the optical path difference for light from S at a point P in the focal plane is

$$\begin{aligned}\Delta\mathscr{S} &= [SS_2P] - [SS_1P] \\ &= [NS_2] + [S_2P] - [S_1P]. \end{aligned} \tag{30}$$

Similarly if N' is the foot of the perpendicular from S_1 to $S'S_2$, the optical path difference at P for light from S' is

$$\Delta\mathscr{S}' = [N'S_2] + [S_2P] - [S_1P]. \tag{31}$$

Hence the patterns from S' and S are mutually displaced through Δm orders, where

$$\Delta m = \frac{|\Delta\mathscr{S}' - \Delta\mathscr{S}|}{\lambda_0} = \frac{|[N'S_2] - [NS_2]|}{\lambda_0} \sim \frac{\theta d}{\lambda_0} \tag{32}$$

for $n \sim 1$ and θ small. When

$$\Delta m = 0,\ 1,\ 2,\ \ldots, \tag{33a}$$

the intensity maxima from S and S' coincide and the fringes in the combined pattern are most distinct. On the other hand when

$$\Delta m = \tfrac{1}{2},\ \tfrac{3}{2},\ \tfrac{5}{2},\ \ldots, \tag{33b}$$

the intensity maxima from S coincide with intensity minima from S'; the fringes in the combined pattern are then least distinct, and disappear if the intensities from S and S' are equal. Thus the fringes show periodic variations of distinctness as the separation of the apertures is increased from zero. In particular there is a first minimum of distinctness when

$$d = \frac{\lambda_0}{2\theta}, \tag{34}$$

and if this condition is observed, θ may be determined from d and λ_0.

Let us consider now the more general case of a quasi-monochromatic primary source extended about an origin S. As before we may take such a source to be made up of mutually incoherent point sources. We may imagine it divided into elementary strips at right angles to the direction joining the apertures S_1 and S_2, and find the intensity at a point P in the focal plane as the sum of the intensity contributions from these elements. Suppose that $\Delta\mathscr{S}$ is the optical path difference at P for light from the element through the origin S, so that for given d the position of P is defined by $\Delta\mathscr{S}$. The optical path difference for light from the element at angle α from S is, by (32), $\Delta\mathscr{S} + \alpha d$, and the corresponding phase difference is

$$\delta(\alpha, \Delta\mathscr{S}) = \frac{2\pi}{\lambda_0}(\Delta\mathscr{S} + \alpha d). \tag{35}$$

We again assume that the number of the point sources that make up the extended source is so large that we may treat the extended source as effectively continuous. The total intensity at P is then by § 7.2 (15)

$$I(d, \Delta\mathscr{S}) = \int i_1 d\alpha + \int i_2 d\alpha + 2\int \sqrt{i_1 i_2}\cos\delta\, d\alpha, \tag{36}$$

where $i_1(\alpha, \Delta\mathscr{S})d\alpha$ and $i_2(\alpha, \Delta\mathscr{S})d\alpha$ are the intensities at P due to an elementary strip when the light reaches P through one of the apertures only, and the integration is taken over the range of values of α subtended by all the source elements.

We have already mentioned in § 7.3.5 that, because of diffraction, the secondary sources formed by the apertures S_1 and S_2 have directional properties, but if we can neglect these* over the range of values of α involved in (36), we may write

$$i_1(\alpha, \Delta\mathscr{S}) = i_2(\alpha, \Delta\mathscr{S}) = i(\alpha)f(\Delta\mathscr{S}), \tag{37}$$

where $i(\alpha)$ is proportional to the intensity of the corresponding strip of the source, and $f(\Delta\mathscr{S})$ characterizes the directional properties of S_1 (or S_2). From (36), using (37) and (35) we then have

$$\begin{aligned} I(d, \Delta\mathscr{S}) &= 2f(\Delta\mathscr{S})\int i(\alpha)\{1 + \cos\delta\}\,d\alpha \\ &= f(\Delta\mathscr{S})\left\{P + C(d)\cos\left(\frac{2\pi}{\lambda_0}\Delta\mathscr{S}\right) - S(d)\sin\left(\frac{2\pi}{\lambda_0}\Delta\mathscr{S}\right)\right\}, \end{aligned} \tag{38}$$

where

$$\left.\begin{aligned} P &= 2\int i(\alpha)d\alpha, \\ C(d) &= 2\int i(\alpha)\cos\left(\frac{2\pi}{\lambda_0}\alpha d\right)d\alpha, \\ S(d) &= 2\int i(\alpha)\sin\left(\frac{2\pi}{\lambda_0}\alpha d\right)d\alpha. \end{aligned}\right\} \tag{39}$$

If we further assume that the variation of $f(\Delta\mathscr{S})$ is slow compared with that of $\cos(2\pi\Delta\mathscr{S}/\lambda_0)$ and $\sin(2\pi\Delta\mathscr{S}/\lambda_0)$*, the positions of the maxima and minima of I are given by

$$\frac{\partial I}{\partial(\Delta\mathscr{S})} = 0 = -\frac{2\pi}{\lambda_0}\left\{C\sin\left(\frac{2\pi}{\lambda_0}\Delta\mathscr{S}\right) + S\cos\left(\frac{2\pi}{\lambda_0}\Delta\mathscr{S}\right)\right\},$$

i.e. by

$$\tan\left(\frac{2\pi}{\lambda_0}\Delta\mathscr{S}\right) = -\frac{S}{C}. \tag{40}$$

From (38) and (40) the extreme values of I are therefore

$$I_{\text{ext}} = f(\Delta\mathscr{S})[P \pm \sqrt{C^2 + S^2}] \tag{41}$$

and hence the visibility of the fringes, defined by (23), is

$$\mathscr{V}(d) = \frac{\sqrt{C^2 + S^2}}{P}. \tag{42}$$

The functions $\mathscr{V}(d)$ for a number of different functions $i(\alpha)$ are shown in Fig. 7.15. Fig. 7.15(a) corresponds to the case of two point sources, which we treated simply at the beginning of this section. In Fig. 7.15(b) the source is of uniform intensity and rectangular, with sides parallel to the line joining the apertures S_1 and S_2; evidently $\mathscr{V}(d)$ in this case is the same as the visibility function of Fig. 7.11. In Fig. 7.15(c) the source is a circular disc with radial symmetry; the left-hand curve corresponds to uniform intensity, and the others to different degrees of darkening from centre to edge.

* This approximation is justified if the apertures are sufficiently narrow, as will be clear from § 8.5.

Conversely, if the positions of the fringes and their visibility are measured as functions of the aperture separation d, the functions C and S can be determined from (40) and (42), apart from the constant factor of proportionality P and the sign; the latter may usually be fixed from considerations of physical plausibility. The intensity distribution $i(\alpha)$ of the source can then be obtained from (39) by the FOURIER

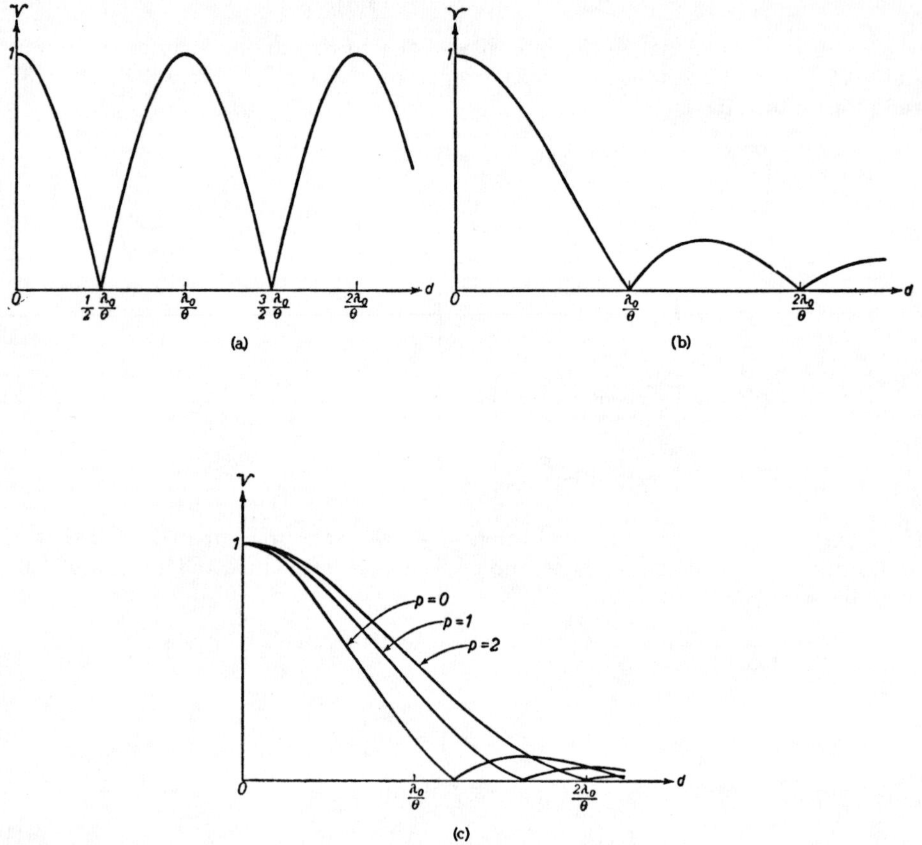

Fig. 7.15. Variation of fringe visibility with aperture separation (arrangement of Fig. 7.14).

(a) Two equal point sources with angular separation θ in the direction of the line joining the apertures.
(b) Uniform rectangular source with sides parallel to the line joining the apertures and of angular width θ.
(c) Circular disc source of angular diameter $\theta = 2\beta_0$, with intensity distribution $I(\beta) \propto (\beta_0^2 - \beta^2)^p$, where β is the angular radius from the centre. [After A. A. MICHELSON and F. G. PEASE, *Astrophys. J.*, **53** (1921), 256.]

inversion theorem. Measurements of this kind, though in principle possible, are difficult. However, if the source is known in advance to have one of the forms to which Fig. 7.15 refers, its angular dimensions can be determined simply by observing the smallest value of d for which the visibility of the fringes is minimum. This condition occurs when

$$d = \frac{A\lambda_0}{\theta}, \tag{43}$$

where $A = 0{\cdot}5$ for two point sources of angular separation θ, as we have seen in (34); $A = 1{\cdot}22$ for a uniform circular disc source of angular diameter θ; and $A > 1{\cdot}22$ for a circular disc source darker at the edges than at the centre.

This method was suggested first by FIZEAU* and later by MICHELSON† as a means of determining the angular dimensions of astronomical objects which are too small or distant to be measured with an undiaphragmed telescope (cf. § 8.6.2). Such objects emit white light, and because of intensity considerations the observations have to be made with white light fringes.‡ It is therefore necessary to assume for λ_0 in (43) an effective wavelength which depends on the wavelength distribution of intensity of the light and the colour response of the eye. With this limitation the method was successfully used to measure the angular diameters of planetary satellites § and the angular

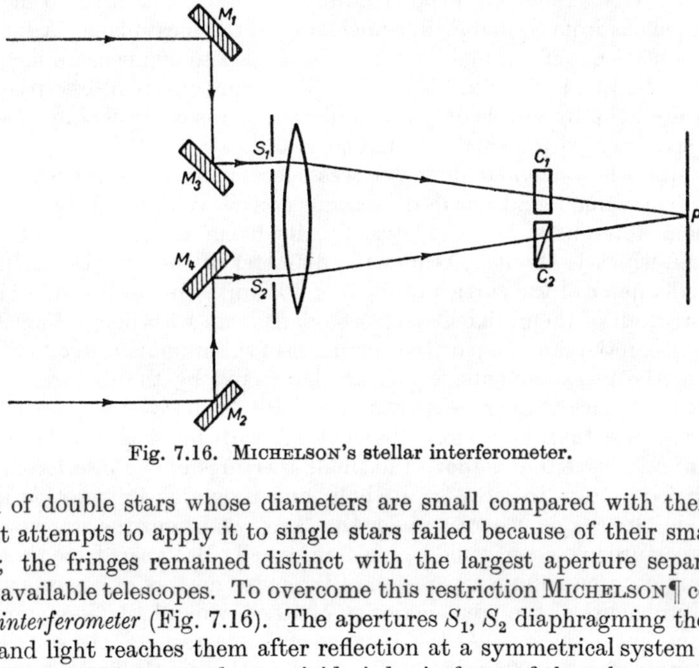

Fig. 7.16. MICHELSON's stellar interferometer.

separation of double stars whose diameters are small compared with their distance apart ||, but attempts to apply it to single stars failed because of their small angular diameters; the fringes remained distinct with the largest aperture separation permitted by available telescopes. To overcome this restriction MICHELSON¶ constructed his *stellar interferometer* (Fig. 7.16). The apertures S_1, S_2 diaphragming the telescope are fixed, and light reaches them after reflection at a symmetrical system of mirrors M_1, M_2, M_3, and M_4, mounted on a rigid girder in front of the telescope. The inner mirrors M_3 and M_4 are fixed, but the outer mirrors M_1 and M_2 can be separated symmetrically in the direction joining S_1 and S_2. If the optical paths $[M_1M_3S_1]$ and $[M_2M_4S_2]$ are maintained equal, the optical path difference for light from a distant point source is the same at S_1 and S_2 as at M_1 and M_2, so that the outer mirrors play the part of the movable apertures in the FIZEAU method. The smallest angular diameter that can be measured with the arrangement is thus determined not by the

* H. FIZEAU, *C. R. Acad. Sci. Paris*, **66** (1868), 934.

† A. A. MICHELSON, *Phil. Mag.* (5), **30** (1890), 1.

‡ The general problem of fringes produced by extended sources of finite wavelength range is elegantly treated by the theory of partial coherence (Chapter X). In § 10.4, p. 512, the action of MICHELSON's stellar interferometer is briefly discussed from the standpoint of this theory.

§ A. A. MICHELSON, *Nature, Lond.*, **45** (1891), 160.

|| J. A. ANDERSON, *Astrophys. J.*, **51** (1920), 263.

¶ A. A. MICHELSON, *Astrophys. J.*, **51** (1920), 257; A. A. MICHELSON and F. G. PEASE, *Astrophys. J.*, **53** (1921), 249.

diameter of the telescope objective, but by the maximum separation of the outer mirrors. There is the further advantage that the fringe spacing, which depends on the separation of S_1 and S_2, remains constant as the separation of the movable apertures is varied. The interferometer was mounted on the large reflecting telescope (diameter 100 in.), of the Mount Wilson Observatory, which was used simply because of its mechanical strength. The apertures S_1 and S_2 were 114 cm apart, giving a fringe spacing of about 0·02 mm in the focal plane.

The maximum separation of the outer mirrors was 6·1 m, so that the smallest measurable angular diameter (with $\lambda_0 = 5500$ Å) was about 0·02 seconds of arc. Because of inevitable mechanical imperfections, two auxiliary devices were necessary to ensure correct adjustment in all positions of the outer mirrors. A plane parallel glass plate C_1, which could be inclined in any direction, was used to maintain the geometrical pencils from S_1 and S_2 in coincidence in the focal plane. A further plane parallel glass plate C_2, of variable thickness, was used to compensate inequalities of the optical paths $[M_1M_3S_1]$ and $[M_2M_4S_2]$. This compensation, essential since the fringes in white light are visible only near order zero, was controlled by observing the channelled spectrum with a small spectroscope.

The first star whose angular diameter was successfully measured was Betelgeuse (α Orionis). The value found was 0·047 seconds of arc. With the distance of this star from the sun, determined trigonometrically, its linear diameter is then found as $4{\cdot}1 \times 10^8$ km, which is about 300 times the diameter of the sun ($1{\cdot}4 \times 10^6$ km) and exceeds the diameter of the earth's orbit (3×10^8 km). Only a few other stars have been measured, all of them, like Betelgeuse, giant stars with linear diameters many times greater than the sun. In part the smallness of this number is due to the inherent difficulties of the measurements, which are hampered by the disturbing effects of atmospheric turbulence, though MICHELSON and PEASE discovered these effects to be of less consequence than in normal observations with telescopes of large aperture. Variations of refractive index above the small apertures of the interferometer cause the interference pattern to move as a whole, and providing this motion is slow the fringes remain observable, whereas under the same conditions the star image formed by the full aperture telescope would be much impaired. But apart from these observational difficulties, a maximum separation of the outer mirrors of 6 m is insufficient to permit measurements of the great majority of stars, which have diameters not very different from that of the sun. At the distance of the nearest star the sun's disc would subtend an angle of only 0·007 seconds of arc, and to observe the first disappearance of the fringes a mirror separation of about 20 m would be necessary. The construction of such a large interferometer would be a difficult undertaking because of the requirement of rigid mechanical connection between the collecting mirrors and the eyepiece.

An instrument analogous to the MICHELSON stellar interferometer has been used in radio-astronomy to determine the angular size of celestial radio sources.* It consists of two separated aerials supplying signals to a common detector system. In this case also it is technically difficult to increase the separation of the aerials without introducing inconstant phase differences into the paths between aerials and detector. To overcome this difficulty HANBURY BROWN and TWISS† have devised another form of radio interferometer, in which the signals at the aerials are detected separately and the angular size of the source is obtained from measurements of the correlation of the

* See, for example, J. L. PAWSEY and R. N. BRACEWELL, *Radio Astronomy* (Oxford, Clarendon Press, 1955), Chapter II; or R. N. BRACEWELL, "Radio Astronomy Techniques" in *Encyclopedia of Physics*, ed. S. FLÜGGE (Berlin, Springer), **54** (1959), Chapter V.

† R. HANBURY BROWN and R. Q. TWISS, *Phil. Mag.* (7), **45** (1954), 663.

intensity fluctuations of the signals as a function of aerial separation. They have also shown* that an equivalent arrangement may be used with visible light. The light from a star, collected by two concave mirrors, is focused on to two photoelectric cells, and the correlation of fluctuations in their photo-currents is measured as the mirror separation is varied. With this arrangement (called *intensity interferometer*) large mirror separations present no difficulty, so that measurements of stars of much smaller diameter than hitherto become possible. The principle of the method is best understood from the standpoint of the theory of partial coherence and is briefly discussed in § 10.4, p. 512.

7.4. STANDING WAVES

With the arrangements we have considered so far, the two waves which interfere are propagated in approximately the same direction at the point of observation. We now consider the interference effects of two waves propagated in different directions, taking as example the interference of the incident and reflected waves when a plane monochromatic wave falls on a highly reflecting plane surface.

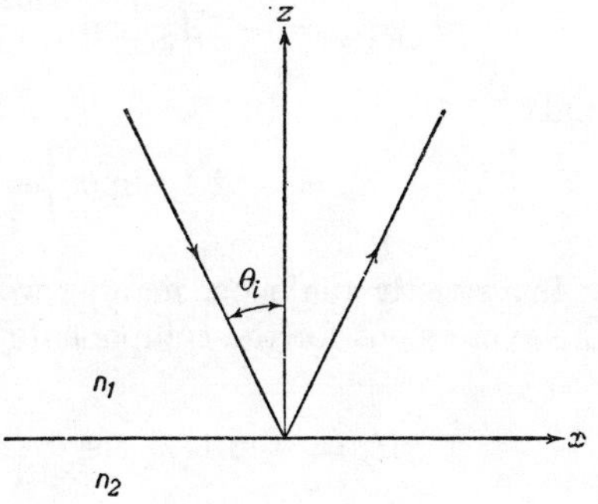

Fig. 7.17. Reflection of a plane wave at a plane surface.

Suppose the surface is the plane $z = 0$, with the positive direction of z pointing into the medium from which the wave is incident, and let n_1 and n_2 be the refractive indices of the two media (Fig. 7.17).

Let θ_i be the angle of incidence, and let the xz-plane be the plane of incidence. If $\mathbf{E}^{(i)}$ is the electric vector of the incident wave, and if as in § 1.5.2 we denote by $A_{\parallel}$ and $A_{\perp}$ the amplitudes of the components of $\mathbf{E}^{(i)}$ parallel and perpendicular respectively to the plane of incidence, the Cartesian components of $\mathbf{E}^{(i)}$ are given by § 1.5 (11), with z replaced by $-z$, i.e.

$$E_x^{(i)} = -A_{\parallel}\cos\theta_i e^{-i\tau_i}, \quad E_y^{(i)} = A_{\perp}e^{-i\tau_i}, \quad E_z^{(i)} = -A_{\parallel}\sin\theta_i e^{-i\tau_i}, \tag{1}$$

(as usual the real parts being understood), where

$$\tau_i = \omega\left(t - \frac{x\sin\theta_i - z\cos\theta_i}{v_1}\right), \tag{2}$$

and v_1 is the velocity of propagation in the first medium. The Cartesian components of the electric vector $\mathbf{E}^{(r)}$ of the reflected wave are given by similar expressions (eqn. (16), § 1.5), which, if the law of reflection § 1.5 (7) is used, are

$$E_x^{(r)} = R_{\parallel}\cos\theta_i e^{-i\tau_r}, \quad E_y^{(r)} = R_{\perp}e^{-i\tau_r}, \quad E_z^{(r)} = -R_{\parallel}\sin\theta_i e^{-i\tau_r}, \tag{3}$$

where

$$\tau_r = \omega\left(t - \frac{x\sin\theta_i + z\cos\theta_i}{v_1}\right). \tag{4}$$

The amplitudes $R_{\parallel}$, $R_{\perp}$ of the reflected wave are related to the amplitudes $A_{\parallel}$, $A_{\perp}$ of the incident wave by the Fresnel formulae (21) of § 1.5, viz.

$$R_{\parallel} = \frac{n_2\cos\theta_i - n_1\cos\theta_t}{n_2\cos\theta_i + n_1\cos\theta_t}A_{\parallel}, \qquad R_{\perp} = \frac{n_1\cos\theta_i - n_2\cos\theta_t}{n_1\cos\theta_i + n_2\cos\theta_t}A_{\perp}. \tag{5}$$

* R. Hanbury Brown and R. Q. Twiss, *Nature*, **178** (1956), 1046; *Proc. Roy. Soc.* A**248** (1958), 199, 222.

To simplify the discussion we suppose n_2/n_1 to be so large that the reflectivity may be taken as unity; (5) then gives, in the limit as $n_2/n_1 \to \infty$,*

$$R_{\parallel} = A_{\parallel}, \qquad R_{\perp} = -A_{\perp}. \tag{6}$$

By substituting from (6) into (3), and adding the expressions for the incident and reflected fields, we obtain the total field. Thus the x-component of the electric vector of the total field is

$$E_x = E_x^{(i)} + E_x^{(r)}$$
$$= 2A_{\parallel} \cos\theta_i \left\{\sin\left(\frac{\omega z \cos\theta_i}{v_1}\right)\right\} e^{-i\left[\omega\left(t - \frac{x \sin\theta_i}{v_1}\right) - \frac{\pi}{2}\right]}. \tag{7a}$$

Similarly,

$$E_y = -2A_{\perp} \left\{\sin\left(\frac{\omega z \cos\theta_i}{v_1}\right)\right\} e^{-i\left[\omega\left(t - \frac{x \sin\theta_i}{v_1}\right) - \frac{\pi}{2}\right]}, \tag{7b}$$

and

$$E_z = -2A_{\parallel} \sin\theta_i \left\{\cos\left(\frac{\omega z \cos\theta_i}{v_1}\right)\right\} e^{-i\left[\omega\left(t - \frac{x \sin\theta_i}{v_1}\right)\right]}. \tag{7c}$$

In a strictly analogous manner we obtain from eqs. (13) and (16) of § 1.5 the following expressions for the components of the magnetic vector of the total field:

$$H_x = -2A_{\perp} n_1 \cos\theta_i \left\{\cos\left(\frac{\omega z \cos\theta_i}{v_1}\right)\right\} e^{-i\left[\omega\left(t - \frac{x \sin\theta_i}{v_1}\right)\right]}, \tag{8a}$$

$$H_y = -2A_{\parallel} n_1 \left\{\cos\left(\frac{\omega z \cos\theta_i}{v_1}\right)\right\} e^{-i\left[\omega\left(t - \frac{x \sin\theta_i}{v_1}\right)\right]}, \tag{8b}$$

$$H_z = 2A_{\perp} n_1 \left\{\sin\theta_i \sin\left(\frac{\omega z \cos\theta_i}{v_1}\right)\right\} e^{-i\left[\omega\left(t - \frac{x \sin\theta_i}{v_1}\right) - \frac{\pi}{2}\right]}, \tag{8c}$$

where the Maxwell relation $n_1 = \sqrt{\varepsilon_1}$ has been used. Each of the expressions (7) and (8) represents a wave propagated in the x-direction with velocity $v_1/\sin\theta_i$. The amplitude of the wave is not constant, but varies periodically in the z-direction, with period $2\pi v_1/\omega \cos\theta_i = \lambda_0/n_1 \cos\theta_i$, where λ_0 is the wavelength in vacuum.

The case of normal incidence ($\theta_i = 0$) is of particular interest. If A_x, A_y are the amplitude components of the electric vector, we may write, from (1), $A_{\parallel} = -A_x$, $A_{\perp} = A_y$; and (7) and (8) give

$$\left.\begin{aligned} E_x &= -2A_x \left\{\sin\left(\frac{\omega z}{v_1}\right)\right\} e^{-i\left(\omega t - \frac{\pi}{2}\right)}, \\ E_y &= -2A_y \left\{\sin\left(\frac{\omega z}{v_1}\right)\right\} e^{-i\left(\omega t - \frac{\pi}{2}\right)}, \\ E_z &= 0, \end{aligned}\right\} \tag{9}$$

* We exclude grazing incidence ($\cos\theta_i \to 0$), for which the reflectivity approaches unity even when n_2/n_1 is not large; in this case (5) gives $R_{\parallel} = -A_{\parallel}$, $R_{\perp} = -A_{\perp}$. This corresponds to the case of LLOYD's mirror (§ 7.3.2).

$$\left.\begin{aligned} H_x &= -\,2A_y n_1 \left\{\cos\left(\frac{\omega z}{v_1}\right)\right\} e^{-i\omega t}, \\ H_y &= 2A_x n_1 \left\{\cos\left(\frac{\omega z}{v_1}\right)\right\} e^{-i\omega t}, \\ H_z &= 0. \end{aligned}\right\} \tag{10}$$

We see that at each instant of time the phase is constant throughout the first medium. There is no finite velocity of propagation, and we speak of a *standing wave*. The amplitudes of the electric and magnetic vectors are periodic functions of z; the planes

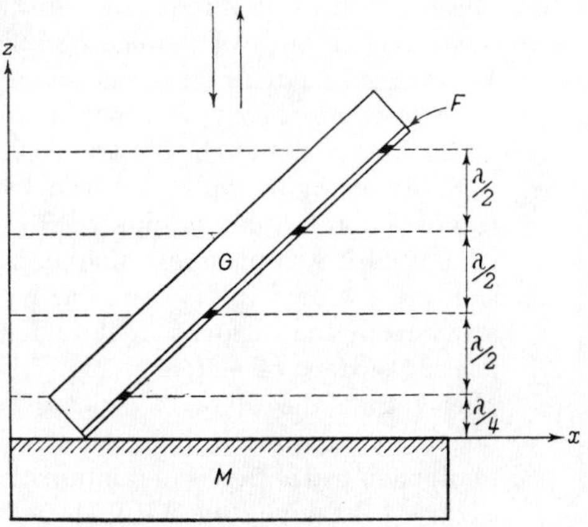

Fig. 7.18. WIENER's experiment on standing waves. (The inclination of the plate to the mirror is greatly exaggerated.)

of zero amplitude are called *nodes*, and the planes where the amplitude has extreme values are called *anti-nodes*. From (9), the nodes of the electric field are given by

$$z = \frac{m\pi v_1}{\omega} = \frac{m\lambda_0}{2n_1}, \qquad m = 0, 1, 2, \ldots, \tag{11a}$$

and the anti-nodes by

$$z = \frac{m\lambda_0}{2n_1}, \qquad m = \tfrac{1}{2}, \tfrac{3}{2}, \tfrac{5}{2}, \ldots \tag{11b}$$

From (10), the nodes of the magnetic field coincide with the anti-nodes of the electric field, and vice versa. In particular, the reflecting surface is a node of the electric field and an anti-node of the magnetic field.

The existence of standing light waves was first demonstrated experimentally by WIENER.* His arrangement is illustrated in Fig. 7.18. A plane mirror M, silvered on the front surface, was illuminated normally by a parallel beam of quasi-monochromatic light. A film F of transparent photographic emulsion, coated on the plane surface of a glass plate G and less than 1/20 wavelength thick, was placed in front of M and inclined to it at a small angle. On development, the emulsion was found to be blackened in equidistant parallel bands with transparent regions between. The

* O. WIENER, *Ann. d. Physik*, **40** (1890), 203.

maxima of blackening corresponded to the intersection of F with the anti-nodal planes of either the electric or the magnetic field. From further experiments, in which the emulsion coated plate was pressed in contact with a convex spherical reflecting surface, WIENER concluded that there was no blackening at the surface of the mirror, which we have seen to be an anti-node of the magnetic field.* The blackened regions therefore correspond to anti-nodes of the electric field, i.e. *the photochemical action is directly related to the electric and not to the magnetic vector.* This conclusion is, of course, to be expected from electron theory. The photographic process is an ionization process, in which an electron is removed from an atomic bond of silver halide, and the electromagnetic force on a charged particle at rest is proportional to the electric vector (cf. § 1.1(34)).

Similar experiments have been made using fluorescent films,† and photo-emissive films,‡ as detectors of standing waves in place of the photographic emulsion used by WIENER. In both cases, and again as is to be expected from electron theory, the maximum response was found at the anti-nodes of the electric field.

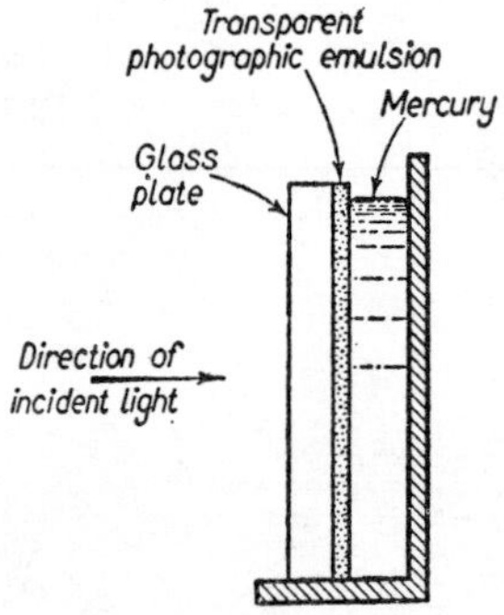

Fig. 7.19. LIPPMANN'S arrangement for colour photography.

Standing light waves are the basis of a method of colour photography originated by LIPPMANN §. Plates coated with a transparent fine-grain photographic emulsion are exposed in the camera, with the emulsion side away from the incident light and in contact with a reflecting surface of mercury (Fig. 7.19). Suppose for simplicity that the plate is exposed to normally incident quasi-monochromatic light of wavelength λ_0. Since the photochemical action is maximum at the anti-nodes of the electric field, given by (11b), the silver in the developed plate forms a system of equidistant layers, parallel to the surface of the emulsion and with optical separation $\lambda_0/2$. If the plate is now illuminated normally with white light, these silver layers act as partially reflecting surfaces, so that the reflected light consists of a series of beams with optical path differences which are integral multiples of λ_0. We shall consider later (§ 7.6) the interference of such a series of beams. The analysis shows that there is a maximum of resultant intensity for wavelength λ_0, which is quite sharp if the number of beams is large. The LIPPMANN plate thus acts as a selective reflector for light of the wavelength used to prepare it.

WIENER|| also used his arrangement to examine the interference effects when the angle of incidence is 45° and the incident light is linearly polarized. He found that with the direction of electric vibrations in the incident light perpendicular to the plane of incidence, the emulsion was blackened in a system of equidistant parallel bands; but that with the direction of electric vibrations in the incident light in the plane of incidence, the blackening was uniform. This result again confirms that the photochemical effect is directly related to the electric and not to the magnetic field. For

* We here assume that equation (6), which was derived for dielectric media, is valid for a silvered reflecting surface. The theory of metallic reflection, which is discussed in Chapter XIII, shows that this is justified under the conditions of WIENER's experiment.

† P. DRUDE and W. NERNST, *Wiedem. Ann.*, **45** (1892), 460.

‡ H. E. IVES and T. C. FRY, *J. Opt. Soc. Amer.*, **23** (1933), 73.

§ G. LIPPMANN, *C. R. Acad. Sci. Paris*, **112** (1891), 274.

|| O. WIENER, *loc. cit.*

when the direction of electric vibrations is perpendicular to the plane of incidence, $A_{\|} = 0$, and with $\theta_i = 45°$, (7) gives

$$\left.\begin{aligned} E_x &= 0, \\ E_y &= -2A_{\perp}\left\{\sin\left(\frac{\omega z}{\sqrt{2}\, v_1}\right)\right\} e^{-i\left[\omega\left(t-\frac{x}{\sqrt{2}v_1}\right)-\frac{\pi}{2}\right]}, \\ E_z &= 0; \end{aligned}\right\} \tag{12}$$

so that the amplitude of the electric vector and also the time-averaged electric energy density vary periodically in the z-direction. When the direction of vibration is in the plane of incidence $A_{\perp} = 0$, and from (7),

$$\left.\begin{aligned} E_x &= \sqrt{2}\, A_{\|} \sin\left(\frac{\omega z}{\sqrt{2}\, v_1}\right) e^{-i\left[\omega\left(t-\frac{x}{\sqrt{2}v_1}\right)-\frac{\pi}{2}\right]}, \\ E_y &= 0, \\ E_z &= -\sqrt{2}\, A_{\|} \cos\left(\frac{\omega z}{\sqrt{2}\, v_1}\right) e^{-i\left[\omega\left(t-\frac{x}{\sqrt{2}v_1}\right)\right]}. \end{aligned}\right\} \tag{13}$$

From (13) and § 1.4 (54) it follows that the time-averaged energy density is in this case equal to

$$\langle w_e \rangle = \frac{n_1^2}{16\pi} \mathbf{E} \cdot \mathbf{E}^{\star} = \frac{n_1^2}{16\pi} (E_x E_x^{\star} + E_z E_z^{\star}) = \frac{n_1^2}{8\pi} A_{\|} A_{\|}^{\star}, \tag{14}$$

and so on time average the electric energy density is independent of z. On the other hand it follows from (8) that for the magnetic field the situation is reversed; the time averaged magnetic energy density varies periodically with z when the direction of magnetic vibrations in the incident light is in the plane of incidence, but is on time average independent of z when the direction of magnetic vibrations in the incident light is perpendicular to the plane of incidence.

7.5. TWO-BEAM INTERFERENCE: DIVISION OF AMPLITUDE

7.5.1 Fringes with a plane parallel plate

Suppose a plane parallel plate of transparent material is illuminated by a point source S of quasi-monochromatic light (Fig. 7.20). Whatever its position, a point P on the same side of the plate as S is reached by two rays—one reflected at the upper surface and the other at the lower surface of the plate—so that there is a non-localized interference pattern on the same side of the plate as S. From considerations of symmetry, the fringes in planes parallel to the plate are circular about the normal SN to the plate as axis, so that at any position of P they run perpendicular to the plane SNP. We may expect from the discussion of § 7.3.4 that the visibility of these fringes will be reduced if the source is extended parallel to the plane SNP, but to this there is an important exception when P is at infinity, i.e. when the fringes are observed with a relaxed eye, or in the focal plane of a telescope objective. Under these conditions the two rays from S to P, namely $SADP$ and $SABCEP$ (Fig. 7.21), derive from the same incident ray and are parallel after leaving the plate. The difference of the optical paths along them is

$$\Delta\mathscr{S} = n'(AB + BC) - nAN, \tag{1}$$

where n', n are the refractive indices of the plate and the surrounding medium, and N is the foot of the perpendicular from C to AD. If h is the thickness of the plate, and θ, θ' are the angles of incidence and refraction at the upper surface, we have

$$AB = BC = \frac{h}{\cos\theta'}, \tag{2}$$

$$AN = AC \sin\theta = 2h \tan\theta' \sin\theta, \tag{3}$$

$$n' \sin\theta' = n \sin\theta. \tag{4}$$

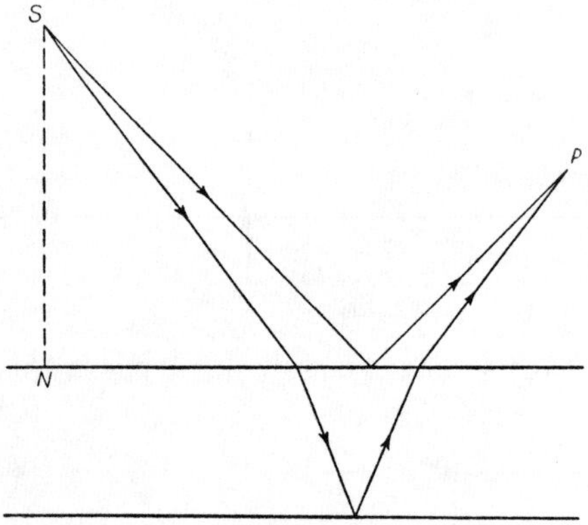

Fig. 7.20. Plane parallel plate with point source.

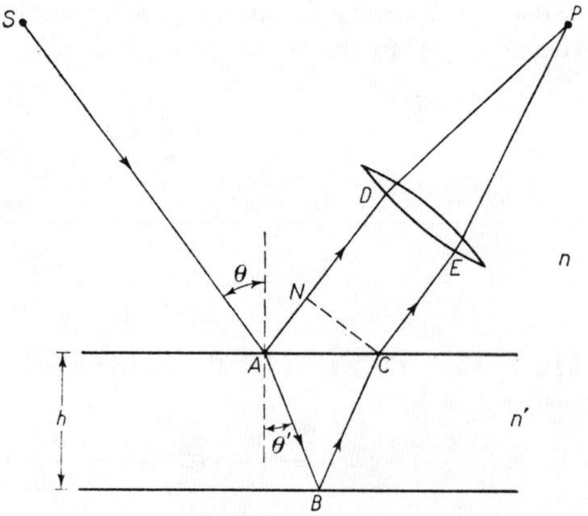

Fig. 7.21. Plane parallel plate: illustrating formation of fringes localized at infinity.

Hence from (1), (2), (3), and (4),

$$\Delta\mathscr{S} = 2n'h \cos\theta', \tag{5}$$

and the corresponding phase difference is

$$\delta = \frac{4\pi}{\lambda_0} n'h \cos\theta'. \tag{6}$$

We must also take into account the phase change of π which, according to the Fresnel formulae of § 1.5 (21a), occurs on reflection at either the upper or lower surface. The total phase difference at P is therefore

$$\delta = \frac{4\pi}{\lambda_0} n'h \cos\theta' \pm \pi \tag{7a}$$

$$= \frac{4\pi h}{\lambda_0} \sqrt{n'^2 - n^2 \sin^2\theta} \pm \pi. \tag{7b}$$

Now θ is determined only by the position of P in the focal plane of the telescope, so that δ is independent of the position of S. It follows that the fringes are just as distinct with an extended source as with a point source. Since this is true only for one particular plane of observation, the fringes are said to be *localized*—in this case—localized at infinity.

The intensity in the pattern varies according to the relation § 7.2 (15); from (7) and § 7.2 (16) there are bright fringes when

$$2n'h \cos \theta' \pm \frac{\lambda_0}{2} = m\lambda_0, \qquad m = 0, 1, 2, \ldots, \tag{8a}$$

and dark fringes when

$$2n'h \cos \theta' \pm \frac{\lambda_0}{2} = m\lambda_0, \qquad m = \tfrac{1}{2}, \tfrac{3}{2}, \tfrac{5}{2}, \ldots \tag{8b}$$

A given fringe is thus characterized by a constant value of θ' (and therefore θ), and so is formed by light incident on the plate at a particular angle. For this reason the fringes are often called *fringes of equal inclination.* When the axis of the telescope objective is normal to the plate, the fringes are concentric circles about the focal point for normally reflected light ($\theta = \theta' = 0$). The order of interference is highest at the centre of the pattern, where it has the value m_0 given by

$$2n'h \pm \frac{\lambda_0}{2} = m_0\lambda_0; \tag{9}$$

m_0 is not necessarily an integer, and we may write

$$m_0 = m_1 + e, \tag{10}$$

where m_1 is the integral order of the innermost bright fringe, and e, which is less than unity, is called the *fractional order at the centre.* For the pth bright fringe from the centre, of angular radius θ_p, the order of interference is m_p, where by (8a),

$$\begin{aligned} 2n'h \cos \theta'_p \pm \frac{\lambda_0}{2} &= m_p\lambda_0 \\ &= [m_1 - (p-1)]\lambda_0; \end{aligned} \tag{11}$$

and from (9), (10), and (11),

$$2n'h\,(1 - \cos \theta'_p) = (p - 1 + e)\lambda_0. \tag{12}$$

Now if θ'_p is small, we have from the law of refraction $n' \sim n\theta_p/\theta'_p$, and $1 - \cos \theta'_p \sim \theta'^2_p/2 \sim n^2\theta_p{}^2/2n'^2$; hence from (12),

$$\theta_p \sim \frac{1}{n}\sqrt{\frac{n'\lambda_0}{h}}\,\sqrt{p - 1 + e}. \tag{13}$$

The angular scale of the pattern is thus proportional to $\sqrt{1/h}$, and if $e = 0$, so that there is an intensity maximum at the centre, the radii of the bright fringes are proportional to the square roots of the positive integers.

Similar fringes localized at infinity may be obtained with a parallel plate of air, bounded by the inner plane surfaces of two transparent plates (Fig. 7.22). With this arrangement the fringes can be observed as the thickness h is varied continuously by separating the plates. As h increases, the fringes expand from the centre of the pattern, and a new fringe appears there each time h increases by $\lambda_0/2$ (taking $n' = 1$ for air). To avoid disturbing effects of the reflections at the outer surfaces, the plates are made slightly wedge-shaped. The difference of optical path from S to P for light reflected at the outer surfaces then varies with the position of S, so that with an extended source this light gives, on average, uniform intensity over the focal plane. The fringes formed by light reflected from the inner parallel surfaces are superposed on this background.

Unless the plate is very thin, the fringes correspond to high orders of interference and so are not visible in white light. For example, if $h = 1$ cm and $n' = 1{\cdot}5$, the order of interference at the centre is about 75,000 for $\lambda_0 = 4000$ Å and about 43,000 for $\lambda_0 = 7000$ Å; hence in the visible spectrum there are about 32,000 wavelengths giving intensity maxima at the centre. It is clear from (5) and § 7.3 (15) that if the fringes are to be distinct, the departure from strict monochromatism of the source becomes more severely restricted as the optical thickness of the plate is increased.

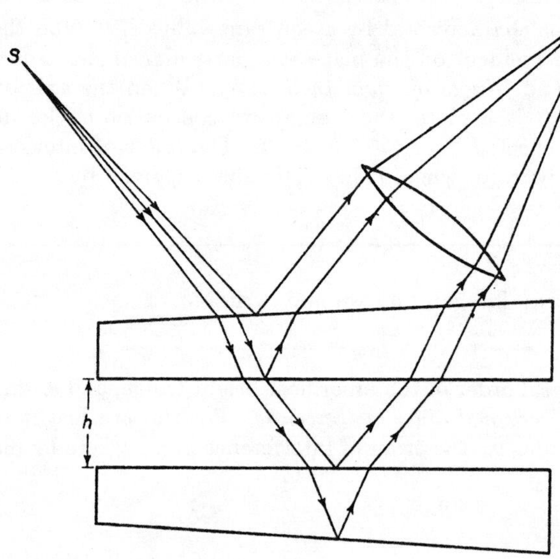

Fig. 7.22. Illustrating formation of fringes localized at infinity with plane parallel plate of air.

As we shall see later (§ 7.5.8), there is an upper limit to the optical thickness of a plate that will yield fringes with available sources.

So far we have assumed that the optical thickness of the plate is everywhere the same. In practice this assumption may be justified by using a suitable diaphragm to limit the illuminated area of the plate. From (8), a change of optical thickness $\Delta(n'h)$ results in a displacement of the pattern through Δm orders, where

$$\Delta m = \frac{2 \cos \theta'}{\lambda_0} \Delta(n'h), \tag{14}$$

and at the centre, where $\theta' = 0$,

$$\Delta m = \frac{2}{\lambda_0} \Delta(n'h). \tag{15}$$

Thus if the plate is moved relative to the diaphragm, variations of $n'h$ may be determined from changes of the order of interference at the centre. The method has been used in optical workshops for testing plates which are required to be of uniform optical thickness.*

* F. Twyman, *Prism and Lens Making* (London, Hilger and Watts, Ltd., 2nd edition, 1952), p. 388.

We have considered so far only the light reflected from the plate, but evidently similar considerations apply also to the transmitted light. In this case (Fig. 7.23) two rays from S reach the focal plane of the telescope at P, one directly transmitted and the other after two internal reflections. The difference of optical path along them is found, in a way similar to the derivation of (5), to be

$$\Delta\mathscr{S} = 2n'h\cos\theta', \tag{16}$$

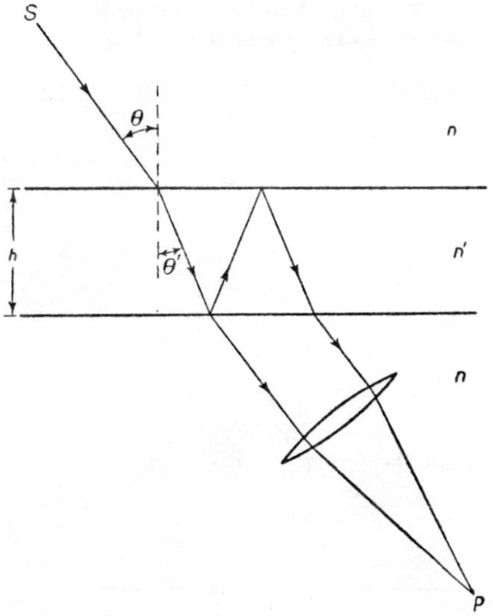

Fig. 7.23. Plane parallel plate: illustrating the formation of fringes localizod at infinity in transmitted light.

so that the corresponding phase difference is

$$\delta = \frac{4\pi}{\lambda_0} n'h\cos\theta'. \tag{17}$$

There is no additional phase difference from phase changes on reflection, since the internal reflections take place at the two surfaces under identical conditions. With an extended source there is again an interference pattern localized at infinity, and comparison of (17) and (7a) shows that this pattern in transmitted light and the pattern in reflected light are complementary, in the sense that the bright fringes of the one and the dark fringes of the other are in the same angular positions relative to the plate normal. However, if the reflectivity of the plate surfaces is low (as for example with glass-air boundaries, for which the reflectivity at normal incidence is about 0·04), the two beams which form the transmitted pattern are of very different intensity; by § 7.2 (16), the difference of intensity between maxima and minima is small, and the fringes are of low visibility.

The foregoing discussion is only approximate because we have ignored the effect of multiple internal reflections in the plate. In reality a series of beams reaches P from S, and not just two as we have supposed. So long as the reflectivity of the plate surfaces is low, the approximation is good because the beams after the first two carry negligible energy. From the more exact discussion given later (§ 7.6) we shall see that the positions of the maxima and minima are given correctly by (8); but if the reflectivity of the plate surfaces is high, the multiple reflections greatly modify the distribution of intensity in the fringes.

7.5.2 Fringes with thin films; the Fizeau interferometer

Suppose a transparent film with plane reflecting surfaces, not necessarily parallel, is illuminated by a point source S of quasi-monochromatic light. Two rays* from S, namely SAP and $SBCDP$ (Fig. 7.24), reach any point P on the same side of the film as S, so that there is a non-localized interference pattern in this region. The difference between these two optical paths from S to P is

$$\Delta\mathscr{S} = n(SB + DP - SA - AP) + n'(BC + CD), \tag{18}$$

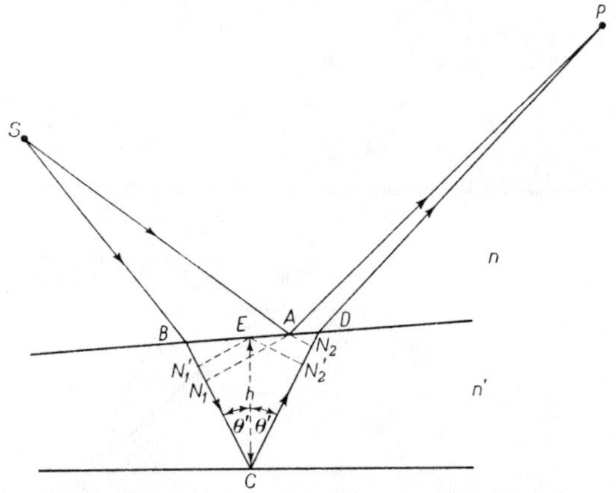

Fig. 7.24. Thin film with point source.

where n', n are respectively the refractive indices of the film and of the surrounding medium. The exact value of $\Delta\mathscr{S}$ may be difficult to calculate, but if the film is sufficiently thin, B, A, and D are close together on the upper surface, so that

$$nSA \sim nSB + n'BN_1, \tag{19a}$$

and

$$nAP \sim nDP + n'N_2D, \tag{19b}$$

where AN_1, AN_2 are respectively perpendicular to BC, CD. From (18) and (19),

$$\Delta\mathscr{S} \sim n'(N_1C + CN_2). \tag{20}$$

Further, if the angle between the surfaces of the film is sufficiently small,

$$N_1C + CN_2 \sim N_1'C + CN_2', \tag{21}$$

where N_1', N_2' are respectively the feet of perpendiculars from E to BC, CD; and E is the intersection with the upper surface of the normal to the lower surface at C. Now

$$N_1'C = CN_2' = h\cos\theta', \tag{22}$$

where $h = CE$ is the thickness of the film at C, measured normal to the lower surface, and θ' is the angle of reflection in the film. Hence for a thin film of small angle we may write, from (20), (21), and (22)

$$\Delta\mathscr{S} = 2n'h\cos\theta', \tag{23}$$

* We again neglect multiple reflections; their effect is considered later (§ 7.6).

and the corresponding phase difference at P is

$$\delta = \frac{4\pi}{\lambda_0} n'h \cos \theta'. \tag{24}$$

In general, for a given P, both h and θ' vary with the position of S, and a small extension of the source makes the range of δ at P so large that the fringes disappear. There is, however, a special case when P is in the film, as when observations are made with a microscope focused on the film, or with the eye accommodated for it. Under these circumstances h is practically the same for all pairs of rays from an extended source reaching P', conjugate to P (Fig. 7.25), and differences of δ at P' are due mainly to differences of $\cos \theta'$. If the range of values of $\cos \theta'$ is sufficiently small, the range of δ at P' may be much less than 2π, even with a source of appreciable extension, and distinct fringes are then visible, apparently localized in the film. In

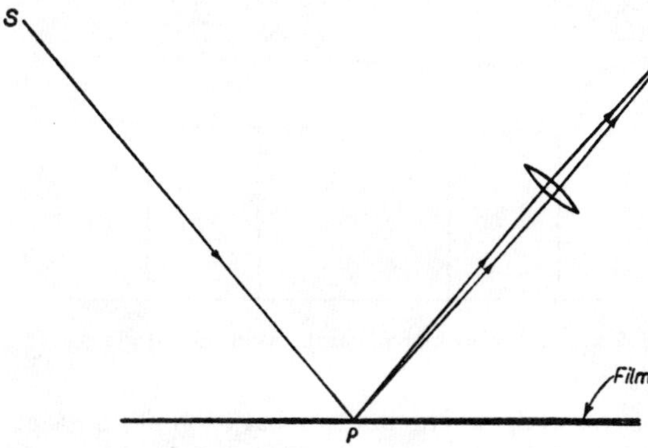

Fig. 7.25. Thin film: illustrating formation of fringes localized in the film.

practice, the condition of a small range of $\cos \theta'$ can be satisfied by observing near normal incidence, and also by restricting the entrance pupil, though the pupil of the unaided eye may itself be sufficiently small. Taking into account the phase change of π on reflection at one of the surfaces of the film, there are, by (24) and § 7.2 (16) maxima of intensity at P' (and so apparently at P) when

$$2n'h \overline{\cos \theta'} \pm \frac{\lambda_0}{2} = m\lambda_0, \qquad m = 0, 1, 2, \ldots, \tag{25a}$$

and minima of intensity when

$$2n'h \overline{\cos \theta'} \pm \frac{\lambda_0}{2} = m\lambda_0, \qquad m = \tfrac{1}{2}, \tfrac{3}{2}, \tfrac{5}{2}, \ldots, \tag{25b}$$

where $\overline{\cos \theta'}$ is a mean value of $\cos \theta'$ for the points of the source which contribute light to P'. The quantity $n'h$ which appears in these relations is the optical thickness of the film at P; as far as our approximations are justified, the state of interference at P is unaffected by the film thickness elsewhere. It follows that (25) holds even if the bounding surfaces of the thin film are not plane, so long as the angle between them remains small. Then if $\overline{\cos \theta'}$ is effectively constant, the fringes are loci of points in the film at which the optical thickness is constant, and for this reason they are often called *fringes of equal thickness* (Fig. 7.26).

Such fringes may be observed in a thin air film between two reflecting surfaces of two transparent plates. Near normal incidence, the condition (25) for a dark fringe then becomes, with $\overline{\cos\theta'} = 1$, and the wavelength $\lambda = \lambda_0/n$ in air,

$$h = \frac{m\lambda}{2}, \qquad m = 0, 1, 2, \ldots \tag{26}$$

The fringes are thus contours of the film at thickness intervals $\lambda/2$. If the film is of constant thickness the intensity over it is uniform; this effect is commonly used to test the figure of an optical surface by observing the film between the surface and a reference surface (proof plate) of equal and opposite curvature. When the air film is a wedge formed between plane surfaces, the fringes are equidistant and

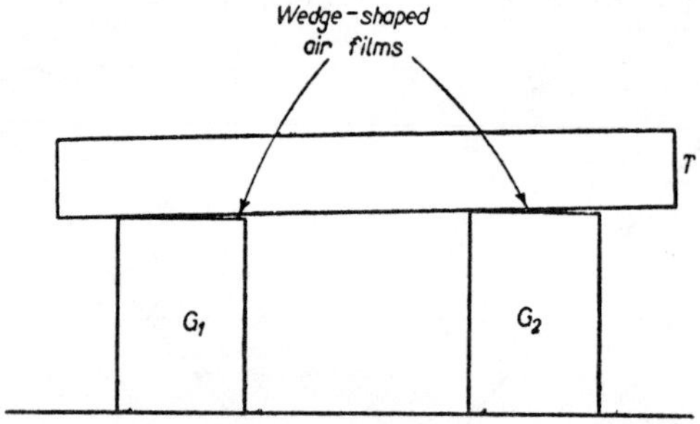

Fig. 7.27. Interferometric comparison of end gauges.

parallel to the apex of the wedge. The linear separation of adjacent bright fringes is $\lambda/2\alpha$, where α is the wedge angle; for example with $\alpha = 1$ minute of arc, $\lambda = 5500$ Å, the separation is about 1·9 mm, showing that the wedge angle must be very small if the fringes are to be reasonably spaced. Wedge fringes are used in testing end gauges which serve as standards of length in mechanical workshops. The gauge G_1 (Fig. 7.27) is a steel block with two opposite surfaces, which define its length, polished plane and parallel. One of these surfaces, and one surface of a reference gauge G_2 of the same nominal length, is wrung in contact with a plane steel surface, and the plane surface of a transparent plate T is allowed to rest on the upper surfaces of the gauges. In general there are wedge-shaped air films between the plate and the gauges, and fringes can be observed in them with monochromatic light. The difference in the lengths of the gauges can be found from their distance apart and the fringe spacing.*

The fringes called *Newton's rings* (Fig. 7.28), which are of historical interest in connection with NEWTON's views on the nature of light, are another example of fringes of equal thickness. They are observed in the air film between the convex spherical surface of a lens and a plane glass surface in contact (Fig. 7.29). The fringes are circles about the point of contact C. If R is the radius of curvature OC of the convex surface, the thickness of the film at distance r from C is

$$h = R - \sqrt{R^2 - r^2} \sim \frac{r^2}{2R}, \tag{27}$$

* The phase change on reflection at the metallic surface is not strictly π, as assumed in (25), but this does not affect the fringe spacing.

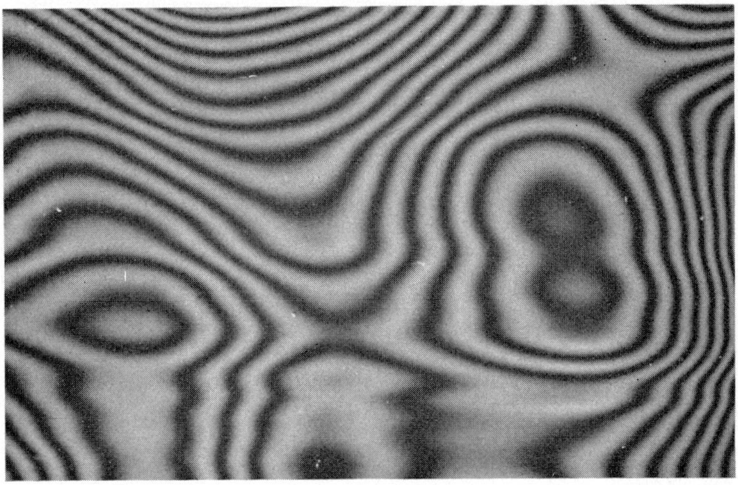

Fig. 7.26. Fringes of equal thickness given by a thin sheet of glass.

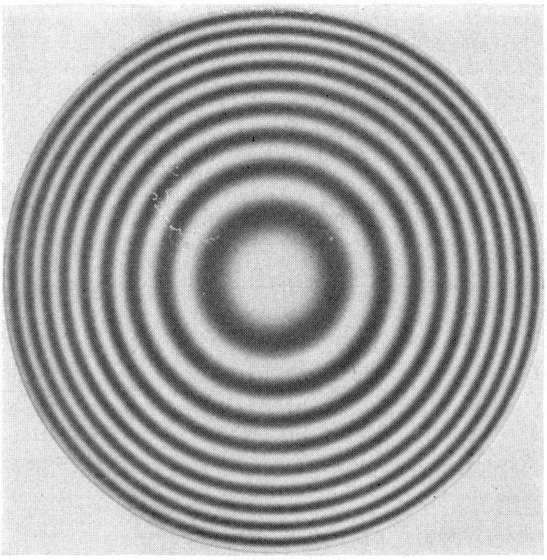

Fig. 7.28. NEWTON's rings.

if we neglect terms of the fourth order. With normal incidence, the condition for a dark fringe is therefore, by (26) and (27),

$$r = \sqrt{mR\lambda}, \qquad m = 0, 1, 2, \ldots, \tag{28}$$

so that the radii of the dark fringes are proportional to the square roots of the positive integers. If the lens and plate are separated, the points of the film with given h move inwards and the fringes collapse towards the centre, where one disappears each time

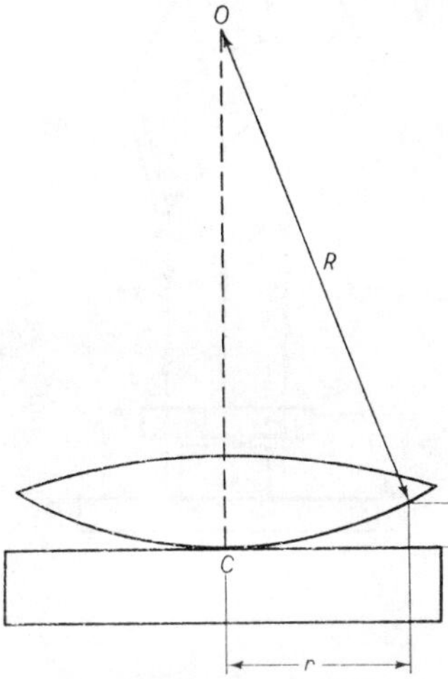

Fig. 7.29. Illustrating the formation of NEWTON's rings.

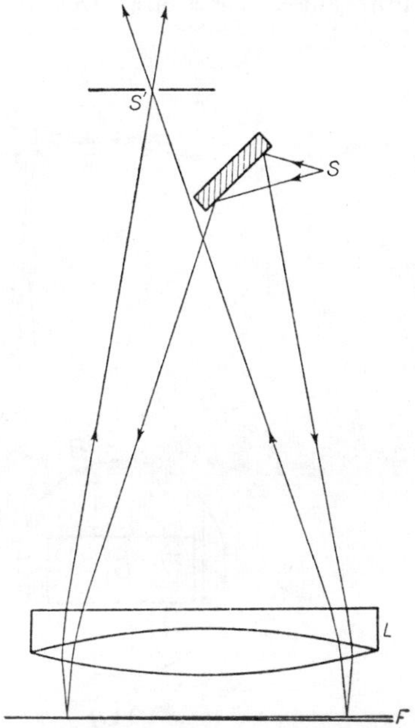

Fig. 7.30. The FIZEAU interferometer.

the separation increases by $\lambda/2$. It is interesting to note that, like YOUNG's experiment, the arrangement provides a means of determining the approximate wavelength of light with very simple apparatus.

If the film thickness is only a few half-wavelengths, the orders of interference in the monochromatic pattern are very low, and fringes are visible with a white light source. The reflected colours exhibited by soap bubbles, and by oil films on water, are examples of such fringes. With the arrangement for observing NEWTON's rings (Fig. 7.29), the phase difference at the centre is π for all wavelengths when the lens and plate are in contact, so that with white light there is a black central spot. Away from this the patterns from the different monochromatic components of the source become increasingly out of step; to visual observation there are coloured rings immediately surrounding the centre, in a characteristic sequence known as *Newton's colours*, and farther out is what appears to the eye to be uniform white illumination (cf. § 7.3.3). Similarly, with a wedge-shaped air film, there is in white light a black fringe defining the apex of the wedge.

We have been concerned so far with light reflected from the film, but an interference pattern localized in the film is visible also in transmitted light. As in the case of a plane parallel plate, the patterns with reflected and transmitted light are complementary; the bright fringes of one appear at the same points of the film as the dark fringes of the other. With surfaces of low reflectivity, the visibility of the fringes in the transmitted pattern is low because of the disparity of intensities of the interfering beams.

We have seen that if the fringes are to be distinct the range of values of cos θ' corresponding to each point of the film must be restricted, and that the fringes follow lines of equal optical thickness only if $\overline{\cos\theta'} \sim 1$. These conditions are satisfied simultaneously over a large area of the film in the *Fizeau interferometer** (Fig. 7.30).

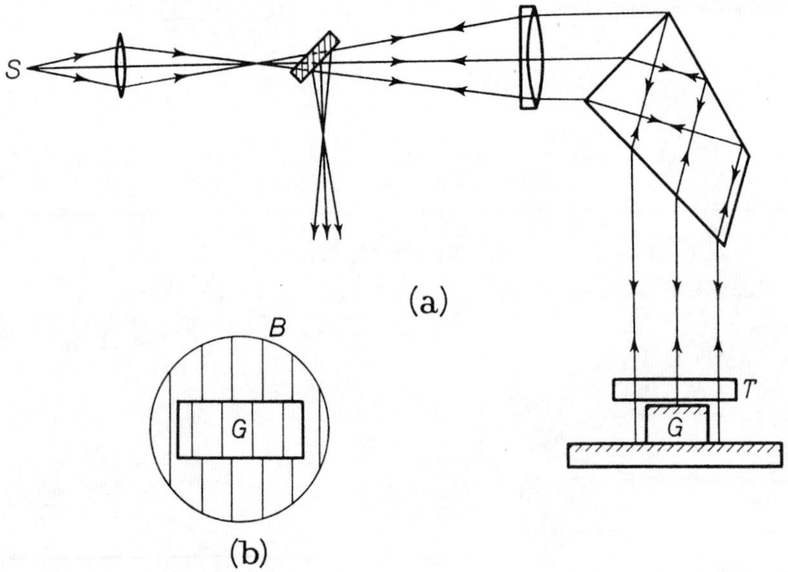

Fig. 7.31. N.P.L. gauge-measuring interferometer: (a) arrangement; (b) field of view.

Light from a quasi-monochromatic source S, after reflection at a small mirror, is collimated by the lens L and falls on the film F at nearly normal incidence. The light reflected from the surfaces of the film returns through L and converges to an aperture S' in the focal plane of L. To an eye placed immediately behind S' and accommodated for the film, fringes are visible over the whole area illuminated by L, following lines of equal optical thickness. These fringes are often called *Fizeau fringes*. As we shall see later (§ 7.5.3), FIZEAU fringes may be obtained with films which are not thin providing the source is sufficiently small, and the interferometer is used in optical workshops to test the uniformity of optical thickness of plane parallel transparent plates. It is also used in the National Physical Laboratory (N.P.L.) in Great Britain to measure the lengths of end gauges.† The arrangement for this is shown in Fig. 7.31(a). One surface of the gauge G is wrung in contact with the plane surface of a polished steel plate B so that the upper surfaces of G and B are parallel, and a transparent plate T, whose lower surface is plane, is mounted above them. In general there are wedge-shaped air films between T and G, and between T and B, in each of which an interference pattern can be observed. The fringes are parallel straight

* H. FIZEAU, *Ann. Chim. Phys.*, (3), **66** (1862), 429.
† F. H. ROLT, *Engineering*, **144** (1937), 162.

lines with the same spacing in each pattern, and by adjusting the inclination of T the fringes are arranged to run at right angles to one edge of the gauge; the appearance is then as in Fig. 7.31(b). If at a point, on the edge of the gauge the air films over G and B are of thickness h_1 and h_2 respectively, and if the corresponding orders of interference are m_1, m_2, the length d of the gauge is given by (26) as

$$d = h_2 - h_1 = (m_2 - m_1)\frac{\lambda}{2} = \Delta m \frac{\lambda}{2}, \tag{29}$$

where Δm is the order displacement of the two patterns. Thus we may write

$$d = (x + e)\frac{\lambda}{2}, \tag{30}$$

where x is an unknown integer, and $e < 1$ is the fractional order displacement, which can be measured by interpolation. The value of x is found by the *method of excess fractions*, which was first used in interferometry by BENOIT.* The fractional order displacement of the patterns is measured with light of four known wavelengths, which are conveniently isolated in turn from a suitable source by rotation of the constant deviation prism D. There are then relations of the form

$$d = (x_1 + e_1)\frac{\lambda_1}{2} = (x_2 + e_2)\frac{\lambda_2}{2} = (x_3 + e_3)\frac{\lambda_3}{2} = (x_4 + e_4)\frac{\lambda_4}{2}, \tag{31}$$

where x_1, x_2, x_3, and x_4 are unknown integers; and these relations define a set of tetrads of possible values of x_1, x_2, x_3, and x_4. An approximate measurement of d with a micrometer is sufficient to decide which of these tetrads is correct, and hence a precise value of d is obtained. With proper precautions, measurements of gauges up to about 10 cm in length can be made; if the fractional order is correct to 0·1, the accuracy is $\pm$ 2·5 $\times$ 10^{-6} cm for $\lambda = 5000$ Å.

It is evident that there is a limiting case of the FIZEAU arrangement when the source S (Fig. 7.30) is reduced to a point, so that θ' has a unique value at each point of the film. Under these circumstances, however, at least if the film surfaces are plane, the fringes must be non-localized, i.e. the fringes over any plane common to the reflected beams are just as distinct as those in the film. We are therefore led to examine more closely the concept of fringe localization, and its relation to the extension of the source.

7.5.3 Localization of fringes†

The arrangements for producing interference effects that we have considered so far may be represented generally as devices by which light from a source is made to reach points in a region of space by two different paths. Let P be a point in this region, and suppose first that the light originates in a quasi-monochromatic point source S, of wavelength λ_0. If SA_1B_1P, SA_2B_2P are the two rays from S to P (Fig. 7.32), the phase difference at P is

$$\delta_0 = \frac{2\pi}{\lambda_0}\{[SA_2B_2P] - [SA_1B_1P]\}. \tag{32}$$

The value of δ_0 depends on the position of P but it is uniquely determined for all P, so that interference fringes, following the loci of points for which δ_0 is constant,

* J. R. BENOIT, *Journ. de Phys.* (3), **7** (1898), 57.

† For a more extensive discussion of the localization of fringes see J. MACÉ DE LÉPINAY and C. FABRY, *Journ. de Phys.* (2), **10** (1891), 5.

are formed over any plane in the region common to the two paths from S. We say that the fringes are *non-localized*; their visibility depends only on the relative intensities of the two waves. Such fringes are always obtained with a point source.

Suppose now that the source is a quasi-monochromatic primary source extended about S. As in § 7.3.4 we assume that such a source is made up of incoherent point sources, each of which gives rise to a non-localized interference pattern; the total intensity is then, at each point, the sum of the intensities in these elementary patterns. If the phase difference at P is not the same for all points of the extended source, the elementary patterns are mutually displaced in the vicinity of P, and the visibility of the fringes at P is less than with a point source. In general, as we shall see later,

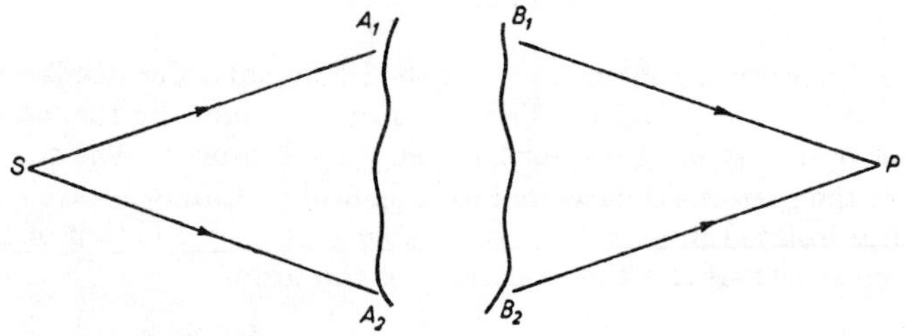

Fig. 7.32. Illustrating interference of two beams from a point source.

the mutual displacement—and hence the reduction of visibility—increases with the extension of the source, but at a rate which depends on the position of P. Thus as the source is extended about S, the visibility at some positions of P may remain at or near its value with a point source, when elsewhere it has fallen effectively to zero. We then say that the fringes are *localized*; such fringes are characteristic of an extended source.

The reduction of visibility associated with a given source extension is in general difficult to calculate, since it depends on the relative intensities of the elementary patterns as well as their mutual displacement. However, we can easily recognize two extreme cases. A mutual displacement of the elementary patterns which is small compared with one order will obviously have little effect on the fringes; and in the special circumstances to which Fig. 7.11 refers we have seen that the loss of visibility remains inappreciable, at least for visual observations, so long as

$$\delta_{\max} - \delta_{\min} \lesssim \frac{\pi}{2}, \tag{33}$$

where $\delta_{\max}$, $\delta_{\min}$ are the maximum and minimum phase differences at P for points of the extended source. This criterion is valid also for the source distributions of Fig. 7.15 (b), (c), and we may assume it to be generally satisfactory. On the other hand, when

$$\delta_{\max} - \delta_{\min} \gg \pi, \tag{34}$$

the elementary patterns are mutually displaced through many orders, and the visibility of the fringes will be very small.

Let us now examine how δ for a point S' of the extended source depends on the position of S' relative to S. The points S and P are "conjugate" in the sense that a point source of wavelength λ_0 at P would give rise to two waves at S with phase difference δ_0. If W_1, W_2 are the wave-fronts through S for these two waves (Fig. 7.33), W_1, W_2 are normal to SA_1, SA_2 respectively; their curvatures depend on the positions of S and P, and on the optical properties of the particular arrangement. Let the normals from S' to W_1, W_2 meet them in N_1, N_2 respectively. The optical path from N_1 to P is equal to $[SA_1B_1P]$, and similarly the optical path from N_2 to P is equal to $[SA_2B_2P]$. The phase difference at P corresponding to S' is therefore

$$\delta = \frac{2\pi}{\lambda_0}\{[S'N_2] + [SA_2B_2P] - [S'N_1] - [SA_1B_1P]\}, \tag{35}$$

or by (32),

$$\delta - \delta_0 = \frac{2\pi n}{\lambda_0}(s_2 - s_1), \tag{36}$$

where $s_1 = S'N_1$, $s_2 = S'N_2$, and n is the refractive index of the medium surrounding the source.

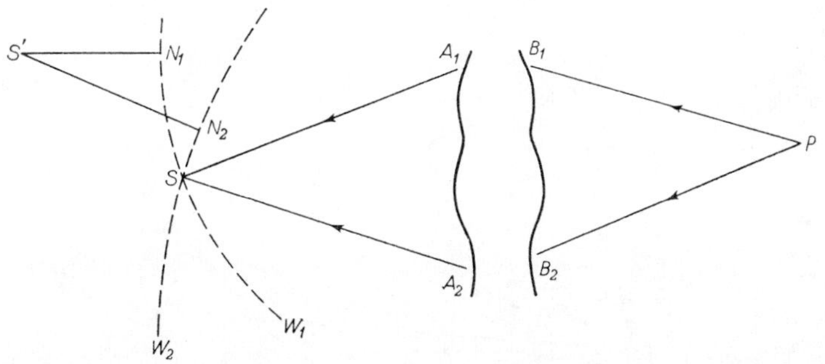

Fig. 7.33. Illustrating interference of two beams from an extended source.

Let S be the origin of rectangular coordinate axes SX, SY, SZ, with SZ and SX respectively internal and external bisectors of the angle A_1SA_2 (Fig. 7.34). If W_1, W_2 have radii of curvature R_1, R_2 in the planes $S'SA_1$, $S'SA_2$ respectively, the corresponding centres of curvature C_1, C_2 are at $\{R_1 \sin(\beta/2),\ 0,\ R_1 \cos(\beta/2)),$ $(-R_2 \sin(\beta/2),\ 0,\ R_2 \cos(\beta/2)\}$ respectively, where β denotes the angle A_1SA_2. If (x, y, z) are the coordinates of S', then

$$C_1S' = s_1 + R_1 = \sqrt{\left(x - R_1 \sin\frac{\beta}{2}\right)^2 + y^2 + \left(z - R_1 \cos\frac{\beta}{2}\right)^2}$$

$$= R_1\sqrt{1 - \frac{2}{R_1}\left(x \sin\frac{\beta}{2} + z\cos\frac{\beta}{2}\right) + \frac{(x^2 + y^2 + z^2)}{R_1{}^2}}. \tag{37}$$

When the source has linear dimensions small compared with R_1 we may neglect powers of x/R_1, y/R_1, z/R_1 higher than the second, so that

$$s_1 \sim -x\sin\frac{\beta}{2} - z\cos\frac{\beta}{2} + \frac{1}{2R_1}\left\{x^2 + y^2 + z^2 - \left(x\sin\frac{\beta}{2} + z\cos\frac{\beta}{2}\right)^2\right\}. \tag{38}$$

Similarly,

$$s_2 \sim x\sin\frac{\beta}{2} - z\cos\frac{\beta}{2} + \frac{1}{2R_2}\left\{x^2 + y^2 + z^2 - \left(x\sin\frac{\beta}{2} - z\cos\frac{\beta}{2}\right)^2\right\}. \tag{39}$$

In most cases of practical interest β is small and terms in the expansions of (38) and (39) which involve β^2 and $\beta z/R$ may be neglected. From (36), (38), and (39) we then obtain

$$\delta - \delta_0 \sim \frac{2\pi n}{\lambda_0}\left\{\beta x + \tfrac{1}{2}\left(\frac{1}{R_2} - \frac{1}{R_1}\right)(x^2 + y^2)\right\}. \tag{40}$$

When the term of (40) independent of β can be neglected,

$$\delta - \delta_0 \sim \frac{2\pi n}{\lambda_0}\, \beta x. \tag{41}$$

With this approximation, $\delta = \delta_0$ when $x = 0$, so that the phase difference (and hence reduction of the visibility) is negligible when the source is extended in the plane YSZ. By (33) and (41), the reduction of visibility remains inappreciable if the source

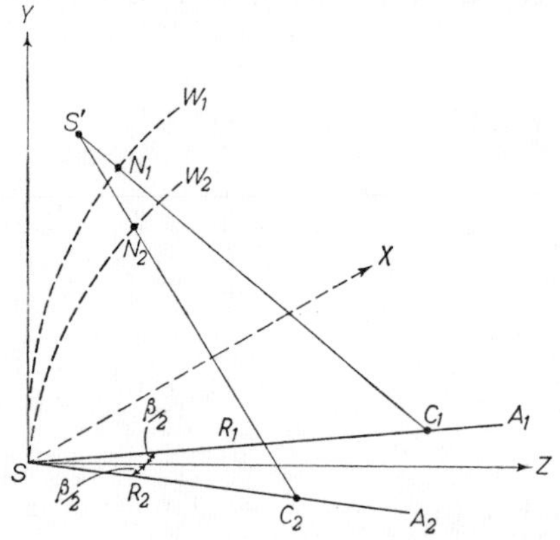

Fig. 7.34. Illustrating the discussion of fringe localization with a two-beam interference arrangement.

extension normal to this plane does not exceed $\lambda_0/4n\beta$. Now the orientations of SA_1 and SA_2, and hence of the coordinate axes we have chosen, depend in general on the position of P. If for all P the planes YSZ have a common line intersection, a line source (or in practice a slit source of width less than $\lambda_0/4n\beta$) along this intersection gives fringes which have everywhere the same visibility as with a point source; the fringes remain non-localized. This is so for the arrangements of § 7.3.2 (except Meslin's experiment). With Fresnel's mirrors, for example, whatever the position of P, the plane YSZ contains the line through S parallel to the mirror junction, and a slit source oriented along this line gives non-localized fringes, as we have seen in § 7.3.4. It is easy to verify that the expression of § 7.3. (26) for the slit width is equivalent to $e \leqslant \lambda_0/4n\beta$. More generally, however, the planes YSZ for all P have no common intersection. A line source may lie in the plane YSZ for points P in a restricted region; but elsewhere the source extends outside this plane to distances large compared with $\lambda_0/2n\beta$, and according to (34) and (41) this means that the visibility is very small. Under these circumstances the fringes given by a slit source are localized, and the region of localization depends on the orientation of the slit.

The value of β depends on the position of P, and by (40), if we neglect the dependence of R_1 and R_2 on β, $\delta - \delta_0$ for given x decreases as β decreases. It follows that, as the source extension normal to the plane YSZ is increased, the fringes become localized in regions corresponding to sufficiently small values of β. In particular, $\beta = 0$ when SA_1 and SA_2 coincide; the region of localization includes those points P—if they exist—which lie at the intersection of two rays derived from a single incident ray from S. In the vicinity of these points, β is negligible, and we have from (40),

$$\delta - \delta_0 \sim \frac{\pi n}{\lambda_0}\left(\frac{1}{R_2} - \frac{1}{R_1}\right)(x^2 + y^2). \tag{42}$$

If W_1, W_2 are spherical, R_1, R_2 are independent of x and y, and the reduction of visibility is independent of the direction in which the source is extended in the plane YSZ. For a circular source in this plane, the reduction of visability is, according to (33) and (42), inappreciable providing the radius of the source does not exceed about $\sqrt{\lambda_0 R_1 R_2/2n|R_1 - R_2|}$.

In the case of a plane parallel plate observed with a telescope (§ 7.5.1), $\beta = 0$ when P is in the focal plane of the telescope objective. R_1 and R_2 are then both infinite for

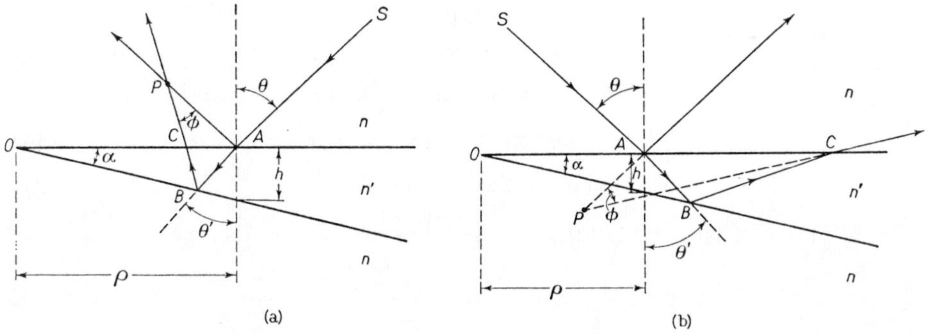

Fig. 7.35. Rays reflected and refracted at the surfaces of a plane wedge.

all positions of S, so that by (42), $\delta = \delta_0$ for all x and y. Thus the fringes in the focal plane of the objective suffer no reduction of visibility whatever the position and extension of the source. In practice, the aperture of the telescope always sets an upper limit to the effective extension of the source. Hence if the fringes are observed outside the focal plane of the objective, as when the telescope is imperfectly adjusted, the reduction of visibility remains inappreciable so long as this effective extension of the source does not exceed the value allowed by (33) and (40).

As a further example*, let us consider the fringes in reflected light from a transparent wedge of refractive index n', bounded by plane surfaces inclined at a small angle α, and situated in a medium of refractive index n. Suppose S is a quasi-monochromatic point source, and consider a ray in a principal section of the wedge (section at right angles to the apex), which meets the front surface of the wedge in A (Fig. 7.35). This incident ray gives rise to a reflected ray AP, and a refracted ray AB which, after reflection at B on the rear surface and refraction at C on the front surface, emerges along CP to meet the first ray in P. When the source is extended about S, the fringes near the plane SAP become localized in the vicinity of P.

Let ρ be the distance of A from the apex O of the wedge; let θ, θ' be respectively the angles of incidence and refraction at A; and let ϕ be the angle APC. When B and O are on the same side of the normal from A to the rear surface of the wedge,

* For fuller treatments of the fringes given by a plane wedge see J. Macé de Lépinay, *Journ. de Phys.* (2), **9** (1890), 121; W. Feussner and L. Janicki, *Handbuch der Physikalischen Optik*, *I* (ed. E. Gehrcke, Leipzig, Barth, 1926), 396; G. F. C. Searle, *Phil. Mag.* (7), **37** (1946), 361.

P is real (Fig. 7.35 (a)); the angle of reflection at B is $(\theta' - \alpha)$, the angle of emergence at C is $(\theta - \phi)$, and we have by elementary geometry,

$$AB = \frac{\rho \sin \alpha}{\cos (\theta' - \alpha)}, \tag{43}$$

$$\frac{BC}{AB} = \frac{\cos \theta'}{\cos (\theta' - 2\alpha)}, \tag{44}$$

$$\frac{AC}{AB} = \frac{\sin 2(\theta' - \alpha)}{\cos (\theta' - 2\alpha)}, \tag{45}$$

$$\frac{AP}{AC} = \frac{\cos (\theta - \phi)}{\sin \phi}, \tag{46}$$

$$\frac{CP}{AC} = \frac{\cos \theta}{\sin \phi}. \tag{47}$$

We also have, by SNELL's law, for refraction at A and C respectively.

$$n' \sin \theta' = n \sin \theta. \tag{48}$$

$$n' \sin (\theta' - 2\alpha) = n \sin (\theta - \phi). \tag{49}$$

From (48) and (49), by subtraction and the use of the identity

$$\sin a - \sin b = 2 \cos \frac{a+b}{2} \sin \frac{a-b}{2},$$

we have

$$\sin \frac{\phi}{2} = \frac{n' \sin \alpha \cos (\theta' - \alpha)}{n \cos \left(\theta - \dfrac{\phi}{2}\right)}; \tag{50}$$

and providing θ is not too near $\pi/2$, we may write, for α small,

$$\phi \sim \frac{2n'\alpha \cos \theta'}{n \cos \theta}. \tag{51}$$

From (43), (45), and (46),

$$AP = \frac{2\rho \sin \alpha \sin (\theta' - \alpha) \cos (\theta - \phi)}{\cos (\theta' - 2\alpha) \sin \phi}. \tag{52}$$

For α small we may expand (52) in powers of α, using (51). If we then retain only the leading term, we obtain, if we use also (48),

$$AP \sim \frac{\rho n^2 \sin \theta \cos^2 \theta}{n'^2 - n^2 \sin^2 \theta}. \tag{53}$$

The quantity neglected in (53) is of the order of $\rho\alpha$, i.e. it is of the order of the thickness of the wedge A. When S and O are on the same side of the normal from A to the rear surface of the wedge (Fig. 7.35 (b)), P is virtual, but the approximate formula (53) holds.

If the wedge is an air film between glass plates, and if we ignore the effects of refraction in the plate on the front side, we may put $n = n' = 1$ in (53) and find $AP \sim \rho \sin \theta$, so that the angle $OPA \sim \pi/2$. The locus of P is a circle on OS_1 as diameter, where S_1 is the reflected image of S in the front surface of the wedge (Fig. 7.36).

If S is so far from the wedge that all the incident rays may be taken to be parallel* to SA, θ and hence θ' and ϕ are independent of ρ, and we see from (52) that AP is

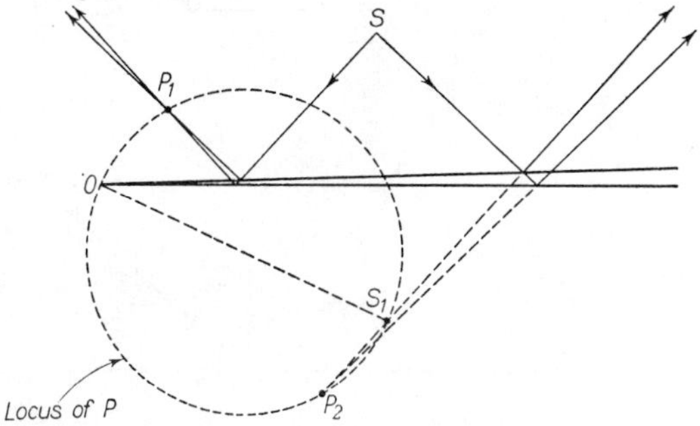

Fig. 7.36. Localization of fringes in the principal section of an air wedge (neglecting effects of refraction at the bounding surfaces).

proportional to ρ. The locus of P is now a plane passing through the vertex of the wedge (Fig. 7.37); the angle γ between this plane of localization and the plane normal to the reflected beam is found, using (53), and the relation $AN = \rho \sin \theta$, to be

$$\tan \gamma = \frac{AN - AP}{ON} \sim \left(\frac{n'^2 - n^2}{n'^2 - n^2 \sin^2 \theta} \right) \tan \theta. \tag{54}$$

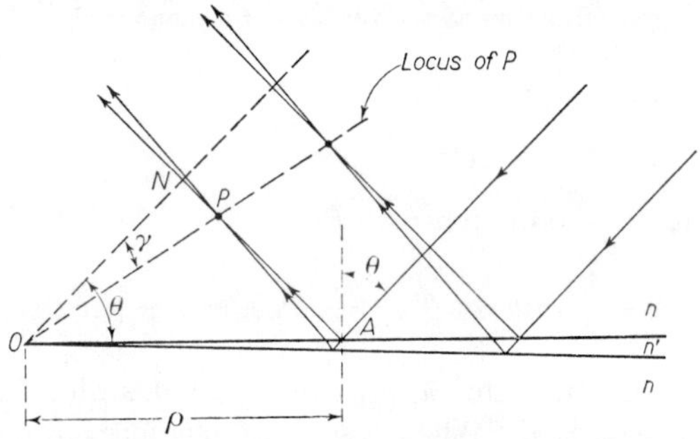

Fig. 7.37. Localization of wedge fringes when the source is at infinity.

Evidently when $\theta = 0$, $\gamma \sim 0$, so that for normally incident light the plane of localization effectively coincides with the front surface of the wedge.

Let us now consider the difference of optical path $\Delta \mathscr{S}$ at P. From Fig. 7.35 (a), when P is real,

$$\Delta \mathscr{S} = n'(AB + BC) + n(CP - AP). \tag{55}$$

By (43) and (44), for α small and θ' not too near $\pi/2$,

$$AB + BC = \frac{2\rho \sin \alpha \cos \alpha}{\cos (\theta' - 2\alpha)} \sim \frac{2h}{\cos \theta'}, \tag{56}$$

where

$$h = \rho \tan \alpha \tag{57}$$

* This condition is satisfied exactly in the Fizeau arrangement (§ 7.5.2).

is the thickness of the wedge at A, and in the approximation we have neglected terms involving powers of α. Similarly, by (43), (45), (46), and (47),

$$CP - AP = \frac{-2\rho \sin\alpha \sin\left(\theta - \dfrac{\phi}{2}\right) \sin(\theta' - \alpha)}{\cos\dfrac{\phi}{2}\cos(\theta' - 2\alpha)} \sim -\frac{2h \sin\theta \sin\theta'}{\cos\theta'}. \tag{58}$$

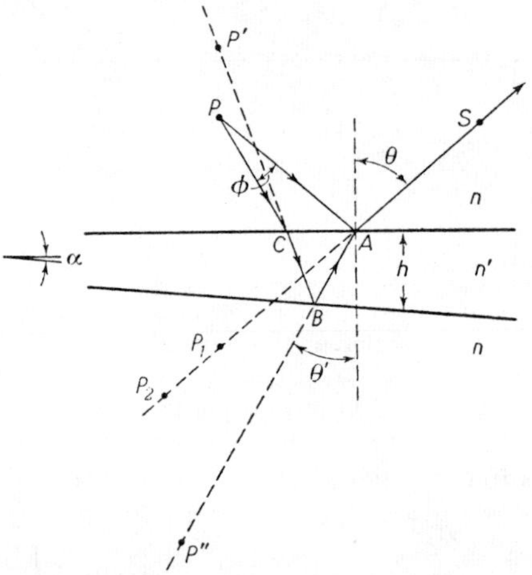

Fig. 7.38. Centres of curvature of wave-fronts from a point source after reflection and refraction at the surfaces of a plane wedge.

Hence by (55), (56), (58), and (48),

$$\Delta\mathscr{S} \sim 2n'h\cos\theta', \tag{59}$$

and the corresponding phase difference at P is

$$\delta \sim \frac{4\pi}{\lambda_0} n'h\cos\theta' = \frac{4\pi h}{\lambda_0}\sqrt{n'^2 - n^2\sin^2\theta}. \tag{60}$$

In a similar way we find that this approximation holds also for the conditions of Fig. 7.35 (b), when P is virtual. When the phase change π which occurs on reflection at one of the wedge surfaces is taken into account, there are, by (60) and § 7.2 (16), maxima of intensity at P when

$$2n'h\cos\theta' \pm \frac{\lambda_0}{2} = m\lambda_0, \qquad m = 0, 1, 2, \ldots, \tag{61a}$$

and minima of intensity when

$$2n'h\cos\theta' \pm \frac{\lambda_0}{2} = m\lambda_0, \qquad m = \tfrac{1}{2}, \tfrac{3}{2}, \tfrac{5}{2}, \ldots \tag{61b}$$

Evidently, for θ' constant, the fringes in the plane of localization are equidistant and parallel to the apex of the wedge.

Finally, let us see how far the source may be extended about S without appreciably reducing the visibility of the fringes at P. For this we imagine a point source placed at P, and determine the radii of curvature R_1, R_2 of the wave-fronts W_1, W_2 for light reaching S after reflection at the front and rear surfaces of the wedge (Fig. 7.38).

W_1 is spherical; its centre of curvature is at P_1 on AS, where P_1 is the reflected image of P in the front surface of the wedge. Thus

$$R_1 = SP_1 = SA + AP. \tag{62}$$

W_2 is in general not spherical. Consider first its radius of curvature in the principal section of the wedge. After refraction at C, the centre of curvature P' is situated on BC at a position given by formula (21) of § 4.6. With the notation of § 4.6 we have $n_0 = n$, $n_1 = n'$, $r_y = \infty$, $\theta_0 = \theta - \phi$, $\theta_1 = \theta' - 2\alpha$, $d_0^{(t)} = CP$, $d_1^{(t)} = CP'$, and § 4.6 (21) gives

$$\frac{CP'}{CP} = \frac{n' \cos^2 (\theta' - 2\alpha)}{n \cos^2 (\theta - \phi)}. \tag{63}$$

After reflection at B, the centre of curvature is at P'' on AB, where P'' is the reflected image of P' in the rear surface of the wedge; and after refraction at A, the centre of curvature is at P_2 on AS, where according to § 4.6 (21) (with $n_0 = n'$, $n_1 = n$, $r_y = \infty$, $\theta_0 = \theta'$, $\theta_1 = \theta$, $d_0^{(t)} = AP''$, $d_1^{(t)} = A_2P$),

$$\frac{AP_2}{AP''} = \frac{n \cos^2 \theta}{n' \cos^2 \theta'}. \tag{64}$$

The radius of curvature R_2 of the section of W_2 in the principal section of the wedge is therefore

$$R_2 = SP_2 = SA + AP_2, \tag{65}$$

where according to (64), and the relation $AP'' = AB + BC + CP'$,

$$AP_2 = \frac{n \cos^2 \theta}{n' \cos^2 \theta'} (AB + BC + CP'); \tag{66}$$

or, using (63),

$$AP_2 = \frac{n \cos^2 \theta}{n' \cos^2 \theta'} (AB + BC) + \frac{\cos^2 \theta \cos^2 (\theta' - 2\alpha)}{\cos^2 \theta' \cos^2 (\theta - \phi)} CP. \tag{67}$$

Now for small α this reduces to

$$\begin{aligned} AP_2 &\sim \frac{n \cos^2 \theta}{n' \cos^2 \theta'} (AB + BC) + CP \\ &\sim \frac{n \cos^2 \theta}{n' \cos^2 \theta'} \cdot \frac{2h}{\cos \theta'} + AP - \frac{2h \sin \theta \sin \theta'}{\cos \theta'}, \end{aligned} \tag{68}$$

where the relations (56) and (58) have been used. From (62), (65), and (68) we finally obtain

$$R_2 - R_1 \sim \frac{2h}{\cos \theta'} \left(\frac{n \cos^2 \theta}{n' \cos^2 \theta'} - \sin \theta \sin \theta' \right). \tag{69}$$

In a similar way we may derive the corresponding expression relating to the section of W_2 at right angles to the principal section of the wedge. In place of § 4.6 (21) we must now use § 4.6 (22); this gives, instead of (63) and (64),

$$\frac{CP'}{CP} = \frac{n'}{n}, \qquad \frac{AP_2}{AP''} = \frac{n}{n'}, \tag{70}$$

and leads to the following formula for the difference in the radii of curvature of the sections of W_1 and W_2 at right angles to the principal section:

$$R_2 - R_1 \sim \frac{2h}{\cos\theta'}\left(\frac{n}{n'} - \sin\theta\sin\theta'\right). \tag{71}$$

As we have noted earlier, (33) and (42) imply that the reduction of visibility at P is inappreciable if the radial extension of the source about SA does not exceed $\sqrt{\lambda_0 R_1 R_2/2n|R_1 - R_2|}$. Now when θ is not too close to $\pi/2$ we see from (69) and (71) that $R_2 - R_1$ is of the order of h; then if R_1 is sufficiently large compared with h, as with the Fizeau arrangement, we may take $R_2/R_1 \sim 1$. With this approximation the tolerable *angular* radius of the source, as viewed from P_1, is

$$\varepsilon \sim \frac{1}{R_1}\sqrt{\frac{\lambda_0 R_1 R_2}{2n|R_1 - R_2|}} \sim \sqrt{\frac{\lambda_0}{2n|R_1 - R_2|}}. \tag{72}$$

In particular, near normal incidence, (69) and (71) give $R_2 - R_1 \sim 2hn/n'$, and (72) becomes

$$\varepsilon \sim \frac{1}{2n}\sqrt{\frac{\lambda_0 n'}{h}}. \tag{73}$$

With typical values for a thin air film, $h = 0{\cdot}01$ cm, $n' = n \sim 1$, $\lambda_0 = 5500$ Å, (73) gives $\varepsilon \sim 2°$. Evidently the tolerable source extension is proportional to $\sqrt{1/h}$; for example, in the above case, but with $h = 1$ cm, ε according to (73) is about 12 minutes of arc. Thus, as we have mentioned in discussing the Fizeau interferometer, fringes of high visibility may be obtained when the wedge is not thin, providing the source is sufficiently small.

7.5.4 The Michelson interferometer

In the arrangements of § 7.5.1 and § 7.5.2 the two beams are superposed except in the region between the bounding surfaces of the plate or film. For some purposes this is inconvenient; it may be avoided by using an auxiliary semi-reflecting surface to produce the beams, which may then be clearly separated before re-combination. Such an arrangement is the basis of the *Michelson interferometer*.*

The simplest form of the instrument is shown in Fig. 7.39. Light from an extended source S is divided at the semi-reflecting surface $\mathscr{A}$ of a plane parallel glass plate D into two beams at right angles. These are reflected at plane mirrors M_1, M_2, and return to D, where they are re-combined to enter the observing telescope T. M_2 is fixed, while M_1 is mounted on a carriage and can be moved towards or away from D by means of a micrometer screw. The beam reflected from M_1 traverses the dispersive material of D three times before reaching T, compared with a single passage for the beam reflected from M_2. To remove this asymmetry, which would otherwise prevent the use of white light fringes, a compensating plate C, of material and thickness identical to D and parallel to it, is introduced between D and M_2.

Suppose M_2' is the image of M_2 in the beam divider. The optical path between S and the point P along a ray SI_1JI_2P, transmitted at $\mathscr{A}$ and reflected at M_2, is equal to the optical path between S and P along the ray SI_1KI_2P, reflected at $\mathscr{A}$ and reflected at the virtual surface M_2'. The interference pattern observed with the telescope may therefore be considered to arise from an air film bounded by the real reflecting surface M_1 and the virtual reflecting surface M_2', providing we associate

* A. A. Michelson, *Amer. J. Sci.* (3), **22** (1881), 120; *Phil. Mag.* (5), **13** (1882), 236.

with the latter a phase change ϕ equal to the difference between the phase changes for external and internal reflection at $\mathscr{A}$. The value of ϕ depends on the nature of the semi-reflector $\mathscr{A}$.

When M_1 and M_2' are parallel, the fringes given by a quasi-monochromatic source are circular and localized at infinity. They differ from the fringes of equal inclination considered in § 7.5.1 only in that here there are no multiple reflections, so that the intensity distribution is strictly in accordance with § 7.2 (15). If M_1 is moved so that it approaches M_2', the fringes contract towards the centre, but the angular scale of

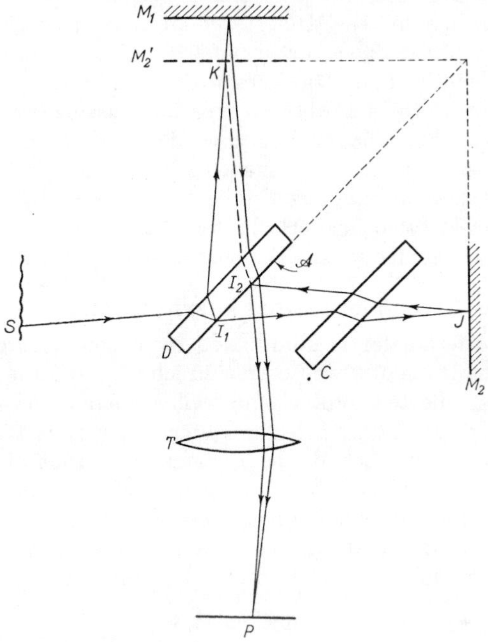

Fig. 7.39. The MICHELSON interferometer.

the pattern increases until, when M_1 coincides with M_2', the illumination over the field of view is uniform at a level which depends on ϕ. The mirrors M_1 and M_2 are then said to be in *optical contact*. When M_1 and M_2' are close together but mutually inclined to form a wedge of small angle, there are fringes localized at or near the surface of this wedge. If the separation of M_1 and M_2' is sufficiently small, these fringes are fringes of equal thickness and so are equidistant straight lines parallel to the apex of the wedge. As the separation increases, however, the range of incidence angle corresponding to each point of the field of view, and the variation of mean incidence angle over the field of view, cease to be negligible; the visibility of the fringes decreases, and they become curved with convex side towards the wedge apex.

Whether M_1 and M_2' are parallel or inclined, a change $\Delta m \,.\, \lambda_0$ of the optical path in either arm of the instrument results in a displacement of the pattern through Δm orders. Displacements can be estimated visually to about 1/20th order, but in certain circumstances displacements as small as 1/1000 order can be detected* by a method due to KENNEDY.†

* K. K. ILLINGWORTH, *Phys. Rev.* (2), **30** (1927), 692.
† R. J. KENNEDY, *Proc. Nat. Acad. Sci.*, **12** (1926), 621.

When the separation of M_1 and M_2' is only a few wavelengths, fringes are visible with white light. They are used to recognize a reference fringe in the monochromatic pattern. If $|\phi| = \pi$, the central fringe of the white light pattern is black and defines the intersection of M_1 and M_2', so that it is in the same position as the monochromatic fringe with $|m| = \frac{1}{2}$; otherwise the achromatic fringe does not in general coincide with a bright or dark fringe of the monochromatic pattern, but this presents no difficulty if the transfer between white light and monochromatic patterns is made consistently.

Except in modified forms using collimated illumination (§ 7.5.5), the Michelson interferometer is now obsolete, but it is famous because of its use by MICHELSON in three important experiments: the MICHELSON–MORLEY ether-drift experiment;* the first systematic study of the fine structure of spectral lines; and the first direct comparison of the wavelength of spectral lines with the standard metre.† In this book we are not concerned with the first of these experiments, since we confine our attention to the optics of stationary media; nor shall we describe the third, since more precise measurements have subsequently been made by other methods (§ 7.7). MICHELSON's method of analysing spectral lines has also been superseded by more direct methods, but because of its considerable theoretical interest and because of its connection with the theory of partial coherence we shall discuss it in detail later (§ 7.5.8).

7.5.5 The Twyman–Green and related interferometers

If a MICHELSON interferometer is illuminated by a point source S of quasi-monochromatic light at the focus of a well-corrected lens L_1, and the light emerging from the interferometer is collected by a second well-corrected lens L_2, the arrangement becomes equivalent to the FIZEAU interferometer, but with the beams having clearly separated paths (Fig. 7.40). Let W_1 be a plane wave-front in the beam returning from M_1, W_2 the corresponding plane wave-front in the beam returning from M_2; and let W_1' be the virtual plane wave-front returning from M_2 which would emerge from the beam divider coincident and co-phasal with W_1. The difference of optical path between the emergent rays which intersect virtually at a point P on W_2 is then

$$\Delta\mathscr{S} = nh, \tag{74}$$

where $h = PN$ is the normal distance from W_1' to P, and n is the refractive index of the medium between W_1' and W_2. The corresponding phase difference is

$$\delta = \frac{2\pi}{\lambda_0} nh; \tag{75}$$

and by (75) and § 7.2 (16), an eye placed in the focal plane of L_2 and focused on W_2 (with the aid of an auxiliary lens if necessary) will see at P a bright fringe if

$$nh = m\lambda_0, \qquad |m| = 0, 1, 2, \ldots, \tag{76a}$$

and a dark fringe if

$$nh = m\lambda_0, \qquad |m| = \tfrac{1}{2}, \tfrac{3}{2}, \tfrac{5}{2}, \ldots \tag{76b}$$

Thus there are in general straight line fringes, parallel to the apex of the wedge formed by W_1' and W_2; if the latter are made parallel by suitably adjusting the orientation of M_1, the field of view is uniformly illuminated. If the source were a point, the fringes would be non-localized, but in practice, because of intensity considerations,

---

* A. A. MICHELSON and E. W. MORLEY, *Phil. Mag.* (5), **24** (1887), 449.

† A. A. MICHELSON and J. R. BENOIT, *Trav. et Mem. Int. Bur. Poids et Mes.*, **11** (1895), 1.

the extension of the source is not negligible. Since the paths of the emergent rays correspond to reflection from a wedge formed by M_2 and M_1', where M_1' is the virtual image of M_1 in the beam divider, the fringes with an extended source are virtually localized in the vicinity of this wedge, as in the case of FIZEAU fringes (p. 297); and the tolerable source extension is greatest when M_1' and M_2 coincide. On the other hand, because of the departure from monochromatism of the source, the fringes are visible only if the optical paths in the two arms are sufficiently near to equality. It is important to note that the conditions of equality of optical paths, and coincidence

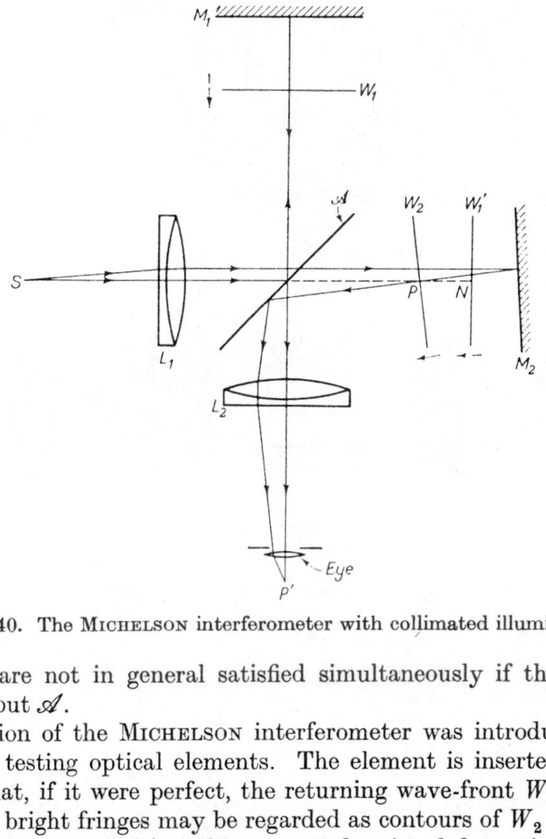

Fig. 7.40. The MICHELSON interferometer with collimated illumination.

of M_1' and M_2, are not in general satisfied simultaneously if the arrangement is asymmetrical about $\mathscr{A}$.

This modification of the MICHELSON interferometer was introduced by TWYMAN and GREEN* for testing optical elements. The element is inserted in the arm M_2 in such a way that, if it were perfect, the returning wave-front W_2 would be plane. Since by (76) the bright fringes may be regarded as contours of W_2 defined by planes parallel to W_1' at intervals of λ_0 (taking $n = 1$ for air), deformation of W_2 resulting from the double passage of light through the element may be measured. The sign of the deformation can be determined from the direction of motion of the fringes when the distance of M_1 from the beam divider is increased. The arrangement† for testing a prism at minimum deviation is shown in Fig. 7.41 (a). The fringes are observed coincident with one of the prism surfaces, and may be traced there by the observer as a guide to subsequent local polishing. In this way internal variations in refractive index of the prism material can be compensated. Fig. 7.41 (b) shows the arrangement used for testing a camera lens.‡ M_2 is a spherical convex mirror with its centre

* F. TWYMAN and A. GREEN, British Patent No. 103832 (1916).
† F. TWYMAN, *Phil. Mag.* (6), **35** (1918), 49.
‡ F. TWYMAN, *Phil. Mag.* (6), **42** (1921), 777.

of curvature at the focus of the lens C to be tested. The latter can be rotated about a line perpendicular to its axis to allow tests at different obliquities, and a mechanical linkage ensures that the centre of curvature of M_2 remains in the focal plane as the lens is rotated. There is some uncertainty in relating defects shown by the fringes

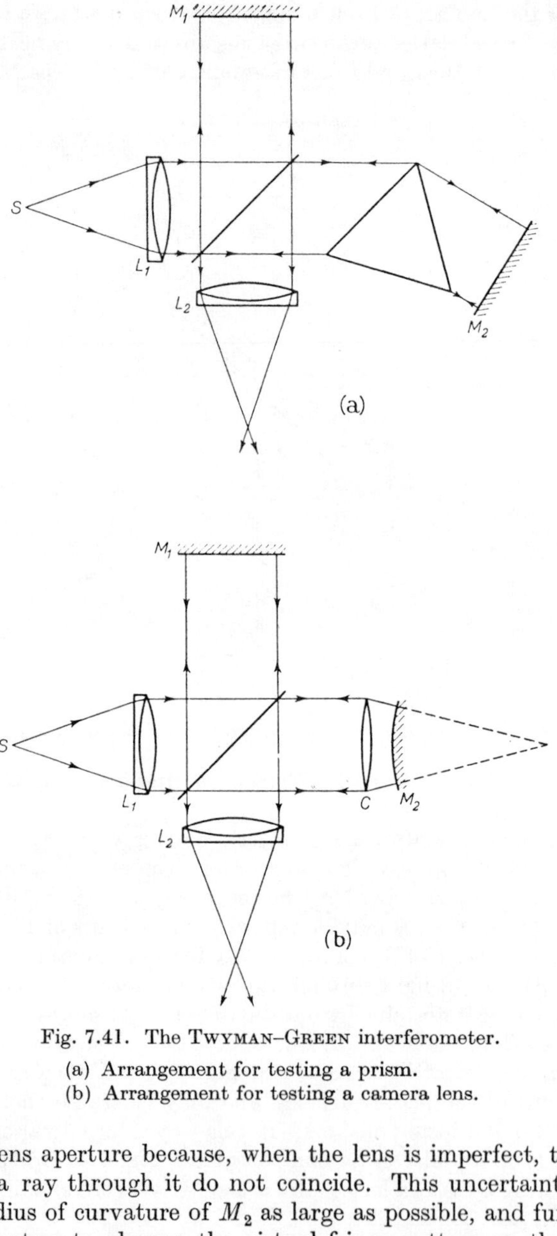

Fig. 7.41. The TWYMAN–GREEN interferometer.

(a) Arrangement for testing a prism.
(b) Arrangement for testing a camera lens.

to points of the lens aperture because, when the lens is imperfect, the outward and return paths of a ray through it do not coincide. This uncertainty is diminished by having the radius of curvature of M_2 as large as possible, and further by using a suitable optical system to observe the virtual fringe pattern on the surface of M_2. Photographs of such fringe patterns, together with the corresponding computed patterns, are shown in Fig. 7.42.

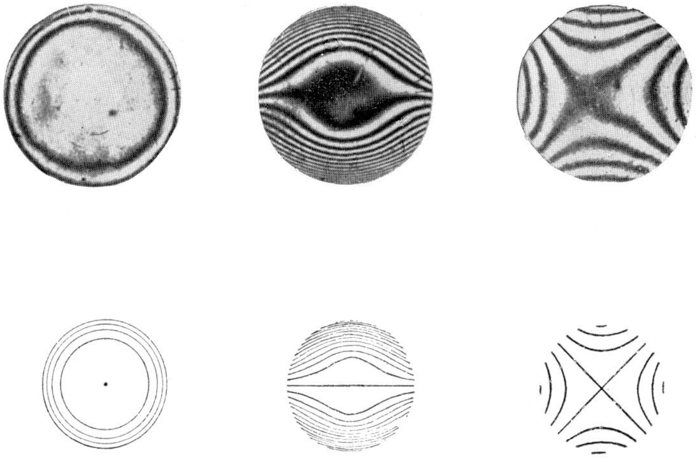

(*a*) *Spherical aberration*
$\Phi_{max} = 0{\cdot}4\lambda_0$
Gaussian focal plane

(*b*) *Coma*
$\Phi_{max} = 2\lambda_0$
Gaussian focal plane

(*c*) *Astigmatism*
$\Phi_{max} = 2{\cdot}1\lambda_0$
Central plane

Fig. 7.42. Twyman–Green interference patterns from lenses showing primary aberrations. Φ_{max} is the maximum wave aberration in the exit pupil. The patterns above are observed, those below are calculated.

(After R. Kingslake, *Trans. Opt. Soc., London*, **27** (1927), 94.)

The TWYMAN–GREEN arrangement is applied to the measurement of the length of end gauges in the *Kösters interferometer** (Fig. 7.43 (a)). The gauge G is wrung to the mirror M_2; and the mirror M_1 is positioned so that the equivalent virtual reflecting surface M_1' is about midway between M_2 and the upper surface of G, and inclined to them at a suitable small angle. With a quasi-monochromatic source, parallel equidistant fringes are then visible over M_2 and G, running at right angles to one edge of G (Fig. 7.43 (b)). The length of G is found by the method of excess

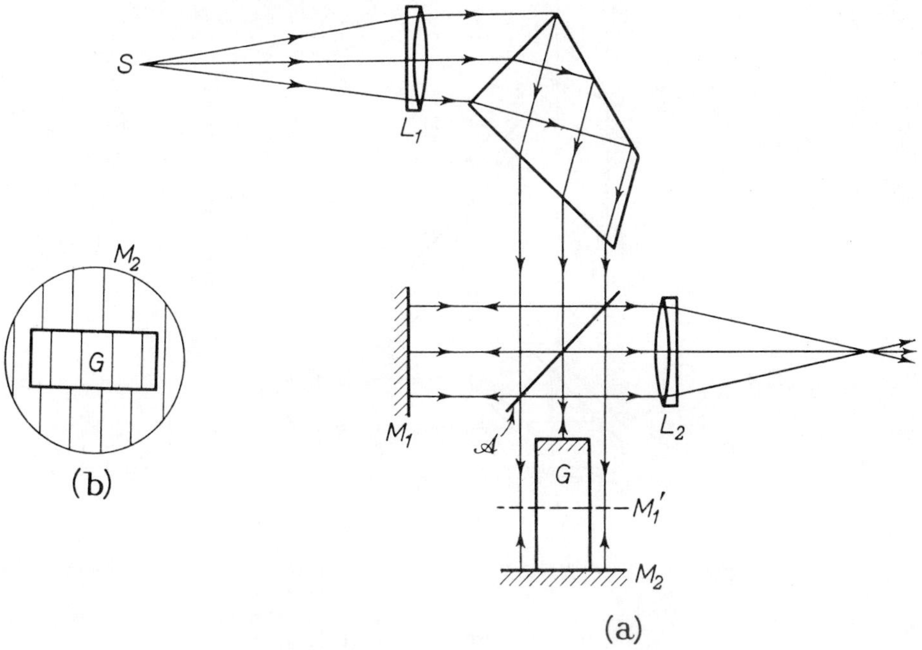

Fig. 7.43. The KÖSTERS interferometer: (a) arrangement; (b) field of view.

fractions, as with the N.P.L. gauge measuring interferometer already described (§ 7.5.2). There is evidently a close resemblance between these two instruments.†

We may mention also a similar interferometer—the *Dowell end-gauge interferometer*,‡ (Fig. 7.44 (a))—which enables the lengths of two end gauges to be compared without need for wringing contacts. Suppose A_1', A_2' are the virtual images in M_1 of the end surfaces A_1, A_2 of the gauges G_1, G_2; and B_1', B_2' are the virtual images in M_2 and the beam divider of the end surfaces B_1, B_2. It is arranged that the plane virtual reflecting surfaces A_1', A_2', B_1', B_2' appear to overlap (Fig. 7.44 (b)), and the interference effects may be considered to arise from the wedge-shaped air films between them. By suitably

* W. KÖSTERS, *Handbuch der Physikalischen Optik*, Vol. I (Ed. by E. Gehrcke, Leipzig, *Barth*, 1927), p. 484.

† With both arrangements there is an upper limit to the separation of reflecting surfaces for which fringes of adequate visibility can be observed; this limit depends on the angular size of the source (§ 7.5.3), and its departure from monochromatism (§ 7.5.8). Since the KÖSTERS arrangement allows the auxiliary reflecting surface M_1' to be positioned midway between the ends of the gauge, it can be used with a given source to measure gauges about twice as long as the maximum with the N.P.L. instrument.

‡ J. H. DOWELL, British Patent No. 555672, (1942).

adjusting G_1, a pattern (i) of horizontal fringes is obtained with monochromatic light over the region common to A_1' and B_1', and the zero order fringe, which can be recognized in white light, is brought to the centre of the field. G_2 is next adjusted until A_2' is co-planar with A_1'. Then since B_2' is parallel to B_1', there are similar patterns (ii), (iii) of horizontal fringes over the regions common to B_2' and A_1', A_2', with zero orders in coincidence. In this condition, the distance d between B_2' and B_1', which is the difference in

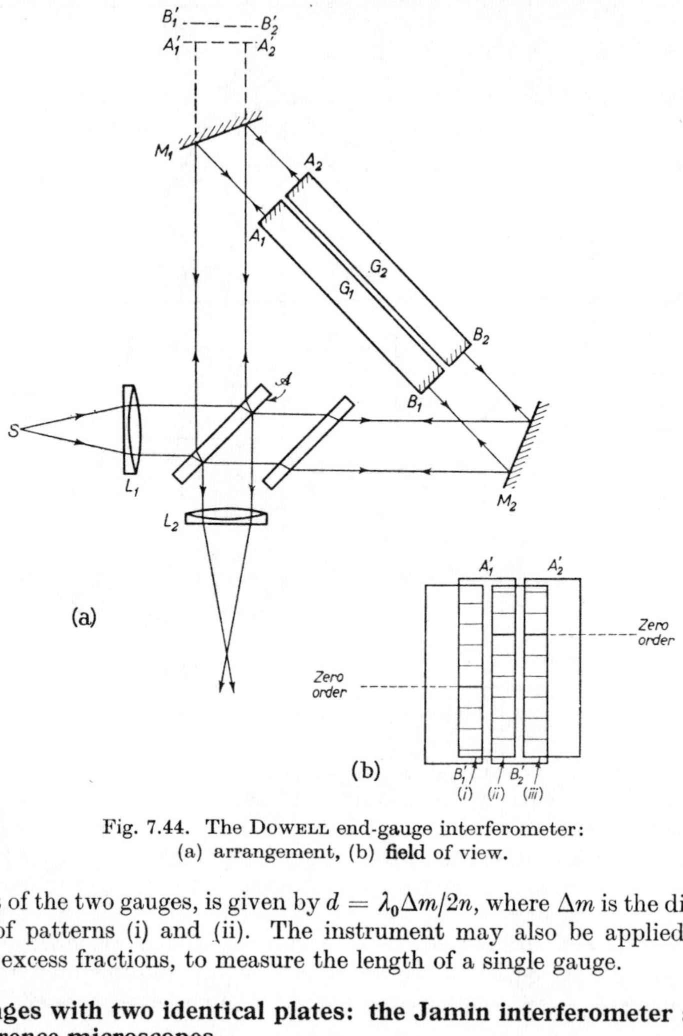

Fig. 7.44. The DOWELL end-gauge interferometer: (a) arrangement, (b) field of view.

the lengths of the two gauges, is given by $d = \lambda_0 \Delta m / 2n$, where Δm is the displacement in orders of patterns (i) and (ii). The instrument may also be applied, using the method of excess fractions, to measure the length of a single gauge.

7.5.6 Fringes with two identical plates: the Jamin interferometer and interference microscopes

Suppose that light from a quasi-monochromatic point source S is incident on two transparent plane parallel plates, each of thickness h and refractive index n', placed one behind the other and mutually inclined at a small angle α (Fig. 7.45). If we neglect rays which have suffered more than two reflections at the plate surfaces, there are two groups of parallel transmitted rays derived from an incident ray SA. The first group consists of the directly transmitted ray, and the rays reflected at both surfaces of either plate. The second group—$SABDH$, $SABCFI$,

SABDEJ and *SABCFGK*—consists of rays reflected at one surface of each plate; these rays traverse the gap between the plates three times, and are inclined at angle 2α to the rays of the first group.

The rays of the second group may be re-combined at point P in the focal plane of a lens L, and if we denote the optical paths between S and P along them by $\mathscr{S}_1$, $\mathscr{S}_2$, $\mathscr{S}_3$, and $\mathscr{S}_4$ respectively, we have, using § 7.5 (5),

$$\mathscr{S}_2 - \mathscr{S}_1 = \Delta\mathscr{S}_{21} = 2n'h \cos\theta_1', \tag{77a}$$

$$\mathscr{S}_3 - \mathscr{S}_2 = \Delta\mathscr{S}_{32} = 2n'h (\cos\theta_2' - \cos\theta_1'), \tag{77b}$$

$$\mathscr{S}_4 - \mathscr{S}_3 = \Delta\mathscr{S}_{43} = 2n'h \cos\theta_1', \tag{77c}$$

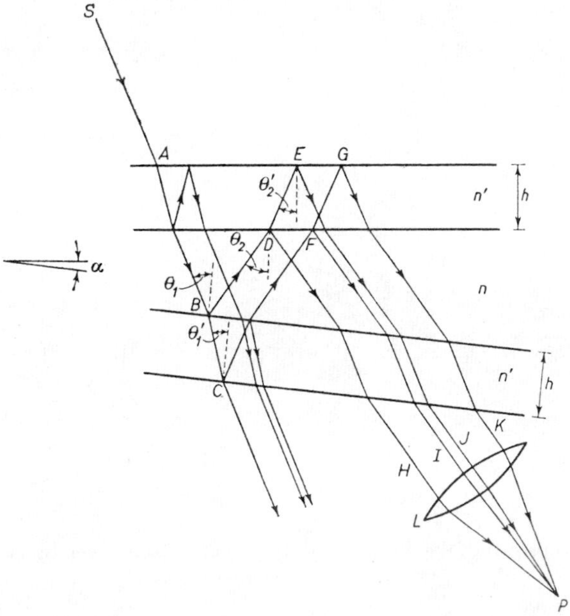

Fig. 7.45. Two identical plane parallel plates: illustrating formation of fringes localized at infinity.

where θ_1' and θ_2' are respectively the angles of refraction in the second plate at B and the first plate at D. Since, for given P, the angles θ_1' and θ_2' depend only on the orientation of the plates, the optical path differences $\Delta\mathscr{S}$ are independent of the position of S, and an interference pattern is formed in the focal plane of L with an extended source. Providing the range of angles of incidence is not too large, this pattern is not overlapped by light from rays of the first group. Further, since $\cos\theta_1' \sim \cos\theta_2'$, $\Delta\mathscr{S}_{21}$ and $\Delta\mathscr{S}_{43}$ are large compared with $\Delta\mathscr{S}_{32}$ when h is sufficiently large, so that we may employ a source for which the condition § 7.3 (15) for distinct fringes holds with respect to $\Delta\mathscr{S}_{32}$ but not with respect to $\Delta\mathscr{S}_{21}$ and $\Delta\mathscr{S}_{43}$. In these circumstances we may consider the fringe pattern to be associated only with rays such as *SABCFI* and *SABDEJ*; the remainder of the light produces an effectively uniform background which merely reduces the visibility of the fringes.

From (77b), the phase difference at P corresponding to $\Delta\mathscr{S}_{32}$ is

$$\delta = \frac{4\pi}{\lambda_0} n'h (\cos\theta_2' - \cos\theta_1'); \tag{78}$$

i.e. using the law of refraction,

$$\delta = \frac{4\pi h}{\lambda_0}\left(\sqrt{n'^2 - n^2 \sin^2\theta_2} - \sqrt{n'^2 - n^2 \sin^2\theta_1}\right), \tag{79}$$

where θ_1 and θ_2 are the angles of incidence at B and D respectively, and n is the refractive index of the medium surrounding the plates. To find the form of the fringes, consider rectangular coordinate axes OX, OY, OZ, with origin O at the second nodal point of the lens L, and OZ parallel to the apex of the wedge between the plates (Fig. 7.46). Let N_1, in the plane XOY, be the focal point for light reflected normally at B, $(\theta_1 = 0)$; let N_2, also in the plane XOY, be the focal point for light reflected

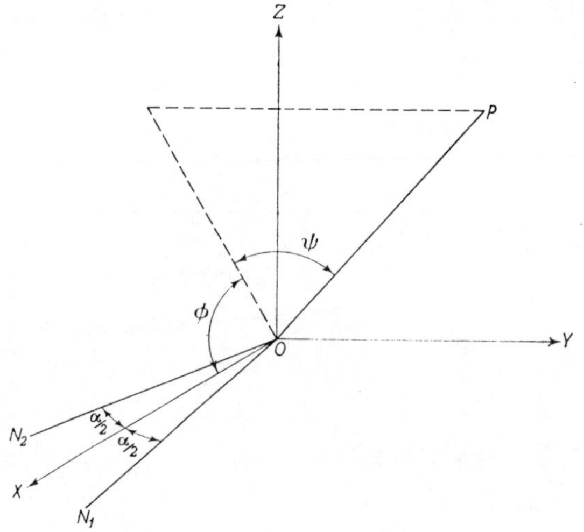

Fig. 7.46. Illustrating the discussion of fringes at infinity given by two mutually inclined plane parallel plates.

normally at D $(\theta_2 = 0)$; and let OX be the internal bisector of ON_1 and ON_2. Then if the wedge opens out in the direction OY, the direction cosines of ON_1 and ON_2 are respectively $\{\cos(\alpha/2), \sin(\alpha/2), 0\}$, $\{\cos(\alpha/2), -\sin(\alpha/2), 0\}$. Let OP make angle ψ with the plane XOZ, and let the projection of OP on this plane make angle ϕ with OX. The direction cosines of OP are $(\cos\psi\,\cos\phi,\ \sin\psi,\ \cos\psi\,\sin\phi)$, and since the angles $PON_1 = \theta_1$, $PON_2 = \theta_2$, we have

$$\cos\theta_1 = \cos\frac{\alpha}{2}\cos\psi\cos\phi + \sin\frac{\alpha}{2}\sin\psi, \tag{80a}$$

$$\cos\theta_2 = \cos\frac{\alpha}{2}\cos\psi\cos\phi - \sin\frac{\alpha}{2}\sin\psi. \tag{80b}$$

For α small we may neglect terms involving the second and higher powers of α in the expansion of $\cos(\alpha/2)$ and $\sin(\alpha/2)$, so that from (80),

$$\cos^2\theta_1 = \cos^2\psi\cos^2\phi + \alpha\sin\psi\cos\psi\cos\phi, \tag{81a}$$

$$\cos^2\theta_2 = \cos^2\psi\cos^2\phi - \alpha\sin\psi\cos\psi\cos\phi. \tag{81b}$$

Then from (79) and (81),

$$\delta = \frac{4\pi h}{\lambda_0}\{\sqrt{n'^2 - n^2(1 - \cos^2\psi\cos^2\phi + \alpha\sin\psi\cos\psi\cos\phi)}$$
$$- \sqrt{n'^2 - n^2(1 - \cos^2\psi\cos^2\phi - \alpha\sin\psi\cos\psi\cos\phi)}\}, \tag{82}$$

and after expansion in powers of α, and again neglecting terms in powers of α higher than the first, we obtain

$$\delta = -\frac{4\pi h}{\lambda_0}\frac{n^2\alpha\sin\psi\cos\psi\cos\phi}{\sqrt{n'^2 - n^2(1 - \cos^2\psi\cos^2\phi)}}. \tag{83}$$

Thus according to (83) and § 7.2 (16), there is a bright fringe at P when

$$\frac{\sin\psi\cos\psi\cos\phi}{\sqrt{n'^2 - n^2(1 - \cos^2\psi\cos^2\phi)}} = \frac{m\lambda_0}{2n^2h\alpha}, \qquad |m| = 0, 1, 2, \ldots, \tag{84a}$$

and a dark fringe when

$$\frac{\sin\psi\cos\psi\cos\phi}{\sqrt{n'^2 - n^2(1 - \cos^2\psi\cos^2\phi)}} = \frac{m\lambda_0}{2n^2h\alpha}, \qquad |m| = \tfrac{1}{2}, \tfrac{3}{2}, \tfrac{5}{2}, \ldots \tag{84b}$$

When P is close to the plane XOZ, so that ψ is small, the order of interference m is low, and in this region fringes can be observed in white light. The central white fringe ($m = 0$) lies in the plane XOZ and is formed by light for which $\theta_1 = \theta_2$.

In the special case of observation near normal incidence, ψ and ϕ are both small, and neglecting terms of the second and higher powers in ψ and ϕ, (84a) reduces to

$$\psi = \frac{mn'\lambda_0}{2n^2h\alpha}, \qquad |m| = 0, 1, 2, \ldots \tag{85}$$

The fringes in quasi-monochromatic light are thus equidistant straight lines, parallel to the apex of the wedge between the plates. The angular separation of adjacent bright fringes is proportional to the refractive index of the plates, and inversely proportional to their thickness and the angle between them. Such fringes were first observed by Brewster, and are called *Brewster's fringes*.

Similar fringes are used in the *Jamin interferometer*,* which was at one time widely used for measurements of the refractive index of gases, though it is now superseded by the Rayleigh interferometer (§ 7.3.5). The instrument consists essentially of two equally thick plane parallel glass plates of the same refractive index, opaquely silvered on the surfaces M_1, M_2, and arranged as shown in Fig. 7.47. A beam of light from an extended source, incident on one plate at about 45°, gives rise to two beams, one reflected from the front surface of the first plate and the rear surface of the second, the other reflected from the rear surface of the first plate and the front surface of the second; and the two are recombined to give an interference pattern in the focal plane of the telescope T. The thickness of the plates is such that, when suitably diaphragmed, the beams are clearly separated between the plates, and here are inserted gas chambers G_1, G_2 and a compensator C_1, C_2 similar to that described in § 7.3.5. In use, the plates are inclined so that they make a wedge of small angle with apex parallel to the plane of Fig. 7.47, which we take to be horizontal. The plate surfaces are approximately vertical, so that, with the notation of Fig. 7.46, the plane XOZ is

* J. Jamin, *C. R. Acad. Sci. Paris*, **42** (1856), 482.

approximately horizontal; the region of observation then corresponds to ψ small, and from (84a) the bright fringes are given by

$$\frac{\cos\phi}{\sqrt{n'^2 - n^2\sin^2\phi}}\,\psi = \frac{m\lambda_0}{2n^2h\alpha}, \qquad |m| = 0, 1, 2, \ldots \tag{86}$$

Now $\phi \sim 45°$, and over a small angular field the variation of the term in ϕ is negligible. The fringes therefore follow loci of constant ψ, i.e. they are horizontal, and are equidistant. They are of low order, and by suitable adjustment the zero order fringe

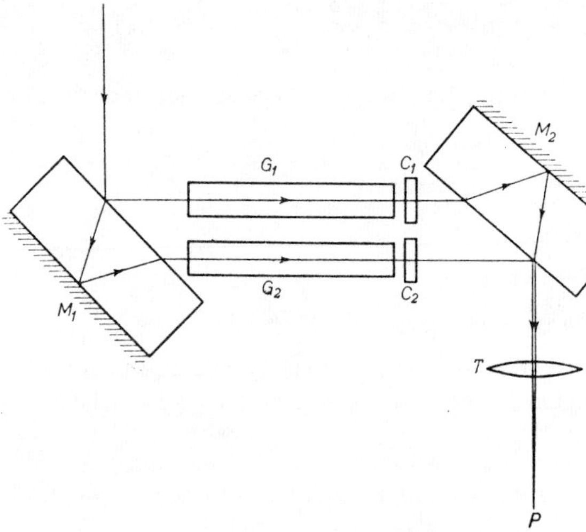

Fig. 7.47. The JAMIN interferometer.

($\psi = 0$), recognizable in white light, can be brought to the centre of the field of view.*

The interferometer is used for measurements in a manner similar to that already described for the RAYLEIGH interferometer, except that there is no second system of fringes to serve as a fiducial mark, and settings are made on a cross wire in the telescope. In consequence the instrument is more sensitive to disturbances of the optical system than the RAYLEIGH instrument, and the precision of measurement is lower.

A variant of the JAMIN interferometer, using plates which are slightly wedge-shaped instead of plane parallel, was developed by SIRKS,† and later by PRINGSHEIM,‡ for the measurement of the refractive index of small objects. The plates are set with the wedge apexes anti-parallel and the inner unsilvered surfaces approximately

---

* If the plates are inclined so that they form a wedge with apex vertical, the plane XOY is horizontal. The region of observation then corresponds to ϕ small, and the intensity maxima are given by (84a) as

$$\frac{\cos\psi\sin\psi}{\sqrt{n'^2 - n^2\sin^2\psi}} = \frac{m\lambda_0}{2n^2h\alpha}, \qquad |m| = 0, 1, 2, \ldots$$

The fringes follow loci of constant ψ, which now means that they are vertical, but with $\psi \sim 45°$ the order of interference is not near zero. White light fringes cannot be obtained with this orientation of the plates.

† J. A. SIRKS, *Hd. Ned. Nat. en Geneesk. Congr., Groningen* (1893), p. 92.

‡ E. PRINGSHEIM, *Verh. Phys. Ges. Berlin*, **17** (1898), 152.

parallel (Fig. 7.48), and are illuminated with collimated light. Corresponding to an incident ray SA in a principal section of the wedges, the two rays $SABCG$, $SADEF$ which leave the second plate intersect virtually at a point P behind the second plate; and with a quasi-monochromatic source of not too large extension there are fringes apparently localized in the vicinity of P, which can be observed with the microscope M. The fringes at P run at right angles to the plane defined by the two emergent rays, i.e. parallel to the wedge apexes. The object O to be examined is placed between the plates in the path of the ray CG. The image P' of P in the front surface of the second plate also lies on CG, at a position which depends on the inclination of the plates. According to PRINGSHEIM (*loc. cit.*), P' is approximately midway between

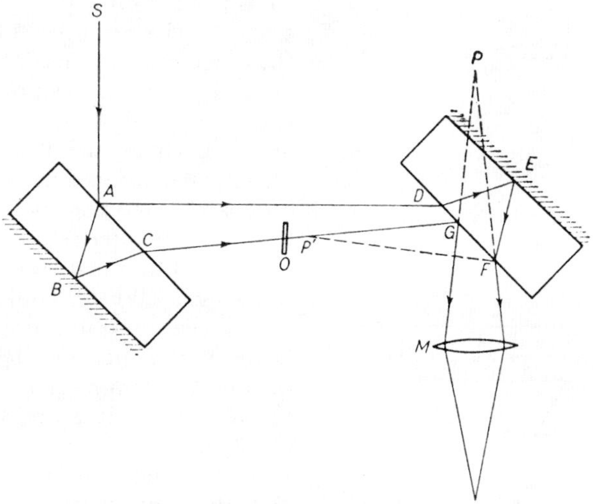

Fig. 7.48. The SIRKS–PRINGSHEIM interferometer.

the plates when the inner plate surfaces are parallel, but a small rotation of either plate about an axis parallel to the wedge apexes results in an appreciable displacement of P'. By means of such an adjustment, P' can be made to lie in O, and the object and fringe system then appear superposed in the field of the microscope. The change in the order of interference at P' which results from the introduction of the object can be determined by means of a compensator, and hence the refractive index of the object at P' may be found if its thickness is known.

More recently, a similar combination of microscope and interferometer has been developed by DYSON* (Fig. 7.49). The object to be examined, mounted on a glass slide, is placed at O between two identical glass plates G_1, G_2 which are plane wedges of small angle with their apexes anti-parallel. The upper surface of the lower plate G_1 is coated with a partially transparent silver film, and its lower surface has a small opaquely silvered central spot C somewhat larger than the field of view of the microscope. The upper plate G_2 is coated on each side with a partially transparent silver film. Spaces between the object slide and the glass plates are filled with medium of the same refractive index as the glass. The arrangement is illuminated by light from the microscope condenser converging to an image of the source in the plane of O. Part of this light—the object beam—passes through O and emerges from G_2 after reflection at the upper and lower surfaces. Another part—the reference beam—is reflected at

* J. DYSON, *Proc. Roy. Soc.*, A, **204** (1950), 170.

the upper surface of G_1 and converges to C; from here it is reflected, and after passing outside O, which lies in the shadow of C, it is directly transmitted through G_2. A glass block, with spherical upper surface R opaquely silvered except for a small clear aperture A on the axis, is cemented to G_2 and is so designed that, after reflection at R and the upper surface of G_2, the object and reference beams converge to A. Near A the reference beam then forms a real image σ_1 of the source, and the object beam forms a real image σ_2 of the source with a real image Π of the object plane superposed. The latter is observed by means of a conventional microscope.*

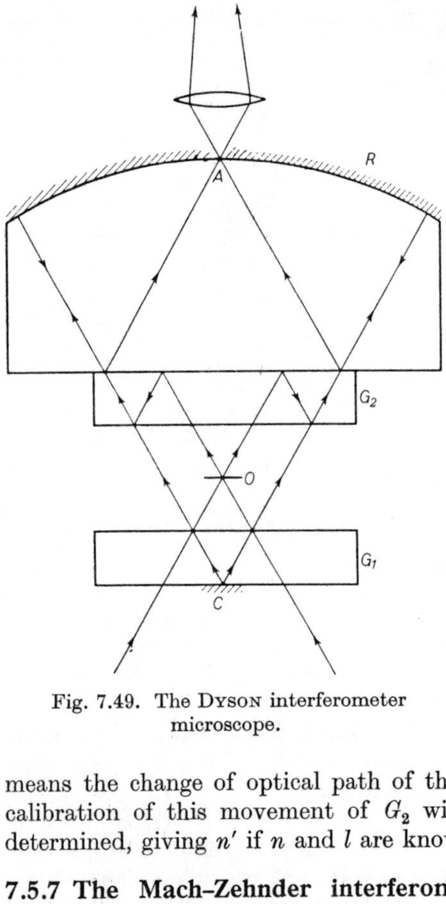

Fig. 7.49. The DYSON interferometer microscope.

Corresponding points of σ_1 and σ_2 are images of the same point of the source, and so act as mutually coherent secondary sources. The arrangement is adjusted so that such points coincide. In the absence of an object, the optical paths of the object and reference beams to any point P of Π are then equal, and if we neglect small phase differences due to differing conditions of reflection at silvered surfaces, the order of interference is zero over all Π—a condition that can be recognized in white light. When an object is introduced at O, the optical path of the object beam to P is increased by approximately $(n' - n)l$, where n' and l are respectively the refractive index and thickness of the object at the point P' conjugate to P, and n is the refractive index of the surrounding medium. Variations of n' and l over the object thus give rise to variations of intensity over Π in quasi-monochromatic light, and to variations of colour in white light. If the plate G_2 is moved in a direction normal to the wedge vertex and the optical axis, the optical path of the reference beam is altered, and by this means the change of optical path of the object beam can be compensated. From calibration of this movement of G_2 with monochromatic light, $(n' - n)l$ can be determined, giving n' if n and l are known.†

7.5.7 The Mach–Zehnder interferometer; the Bates wave-front shearing interferometer

In the JAMIN interferometer (§ 7.5.6), the front surfaces of the two plates act as beam-dividers and the rear surfaces as plane mirrors, but these elements cannot be

* For beams other than those mentioned, A is completely covered by the shadow of C, so that light from these beams does not enter the microscope.

† In practice, the adjustment for coincidence of σ_1 and σ_2 is never perfect. Further, the introduction of a refracting object, and the use of G_1 as a compensator, each results in a mutual displacement of σ_1 and σ_2 parallel to the optical axis. A satisfactory account of the interference effects in the region of σ_1 and σ_2 when corresponding points of these images do not coincide is beyond the scope of the elementary treatment of this chapter; it may be treated elegantly by the theory of partial coherence (Chapter X).

adjusted independently, and the separation of the two beams is limited by the thickness of the plates. A much more versatile instrument, in which the beams may be widely separated, is obtained when the beam dividers and mirrors are separate elements. This is the basis of the *Mach–Zehnder interferometer*,* which is used to measure variations of refractive index, and hence of density, in compressible gas flows.

The arrangement is shown in Fig. 7.50. Light from a source S in the focal plane of a well-corrected lens L_1 is divided at the semi-reflecting surface $\mathscr{A}_1$ of a plane parallel glass plate D_1 into two beams, which, after reflection at plane mirrors M_1, M_2, are recombined at the semi-reflecting surface $\mathscr{A}_2$ of a second identical plane parallel plate D_2, and emerge to a well-corrected collecting lens L_2. The four reflecting surfaces are usually arranged to be approximately parallel, with their centres at the corners of a parallelogram. Suppose the source is a point source of quasi-monochromatic light. Let W_1 be a plane wave-front in the beam between M_1 and D_2, W_2 the corresponding plane wave-front in the beam between M_2 and D_2, and W_1' the virtual plane wave-front between M_2 and D_2 which would emerge from D_2 coincident and co-phasal with W_1. At a point P on W_2, the virtual phase difference between the emergent beams is then

$$\delta = \frac{2\pi}{\lambda_0} nh, \qquad (87)$$

where $h = PN$ is the normal distance from P to W_1', and n is the refractive index of the medium between W_2 and W_1'. At the point P' in the emergent beams, conjugate to P, there will by § 7.2 (16) be a bright fringe if

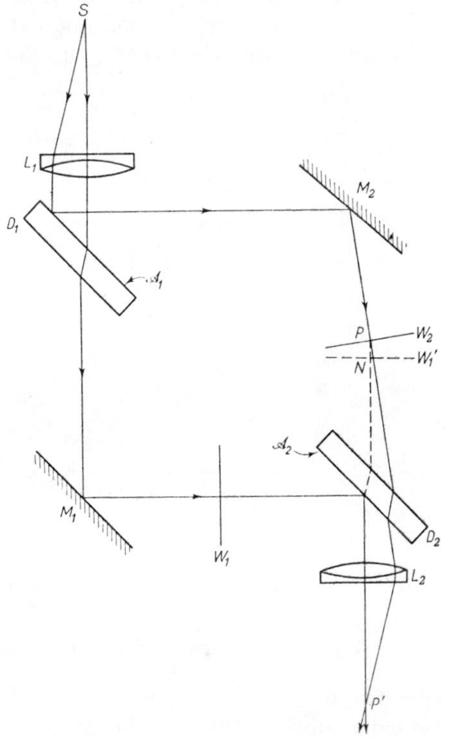

Fig. 7.50. The MACH–ZEHNDER interferometer.

$$nh = m\lambda_0, \qquad |m| = 0, 1, 2, \ldots, \qquad (88a)$$

and a dark fringe if

$$nh = m\lambda_0, \qquad |m| = \tfrac{1}{2}, \tfrac{3}{2}, \tfrac{5}{2}, \ldots \qquad (88b)$$

When W_1' and W_2 are parallel, the intensity is the same for all P, and under these circumstances an extended source would give fringes at infinity (i.e. in the focal plane of L_2) similar to those of the JAMIN interferometer. In general, however, W_1' and W_2 are mutually inclined, and the fringes are straight lines parallel to their intersection. It is these wedge fringes that are normally used in the examination of gas flows, and because of intensity considerations it is desirable to form them with the largest source extension possible without loss of visibility. As we have seen in § 7.5.3, the fringes then become localized in the region where intersecting rays have the smallest angular separation on leaving S. The position of this region of localization can be varied by varying the combination of rotations of the elements used to produce

* L. ZEHNDER, *Zeitschr. f. Instrkde*, **11** (1891), 275. L. MACH, *Zeitschr. f. Instrkde*, **12** (1892), 89.

the mutual inclination of W_1' and W_2. For example, if the reflecting surfaces are initially parallel, and if for simplicity we consider the case of rotations about axes perpendicular to the plane of centres, the virtual region of localization is near M_2 when M_2 is rotated (Fig. 7.51(a)), but lies between M_2 and D_2 when both M_2 and D_2 are rotated (Fig. 7.51(b)). This property distinguishes the wedge fringes of the MACH–ZEHNDER interferometer from those given by the MICHELSON interferometer with collimated light (§ 7.5.5), which are virtually localized in the vicinity of the mirrors.

In the technical use of the instrument, the region C_1 where the gas flow is to be examined—commonly the working section of a wind tunnel or shock-wave tube—and a compensating chamber C_2, are in opposite arms of the interferometer, which is

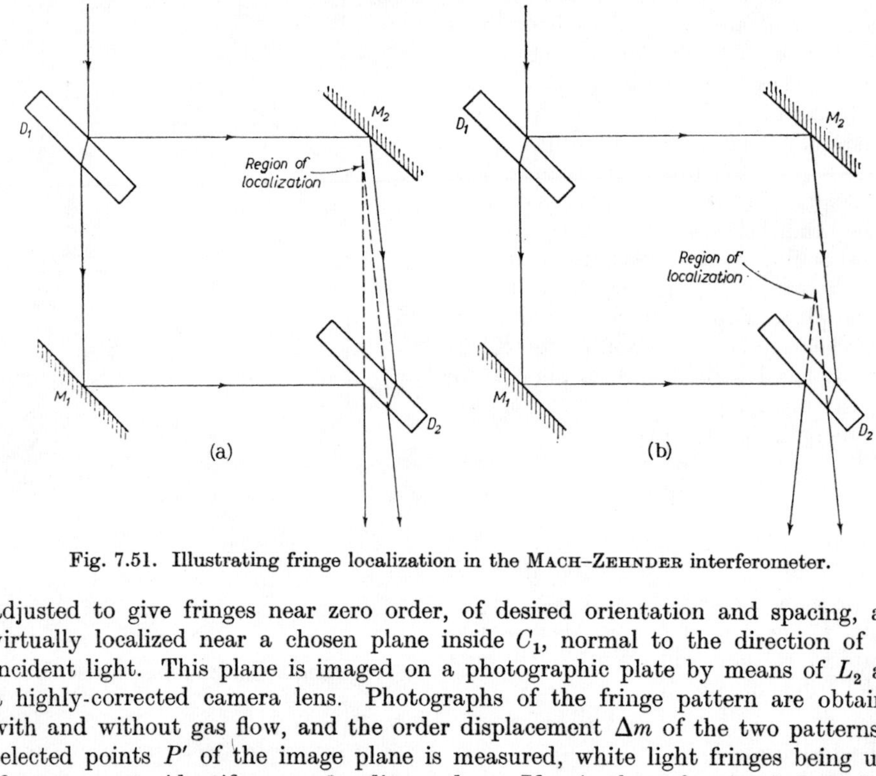

Fig. 7.51. Illustrating fringe localization in the MACH–ZEHNDER interferometer.

adjusted to give fringes near zero order, of desired orientation and spacing, and virtually localized near a chosen plane inside C_1, normal to the direction of the incident light. This plane is imaged on a photographic plate by means of L_2 and a highly-corrected camera lens. Photographs of the fringe pattern are obtained with and without gas flow, and the order displacement Δm of the two patterns at selected points P' of the image plane is measured, white light fringes being used if necessary to identify corresponding orders. If n is the refractive index of the undisturbed gas in C_1, and n' the refractive index under flow conditions, we have

$$\Delta m = \frac{1}{\lambda_0}\int (n' - n)ds, \tag{89}$$

where the integration is taken along the path of the ray passing through C_1 and reaching P'. Let Ox, Oy, Oz be rectangular coordinate axes in C_1, with origin O in the chosen plane in C_1 and OZ in the direction of the incident light; and let P, coordinates $(x, y, 0)$, be the point conjugate to P' when there is no flow in C_1. Then if refractive deviations of rays caused by the flow are negligible, (89) may be written

$$\Delta m(x, y) = \frac{1}{\lambda_0}\int_0^s \{n'(x, y, z) - n\}dz, \tag{90}$$

where s is the length of C_1; and when the flow satisfies certain symmetry conditions* (90) may be solved for $(n' - n)$ in terms of the measured Δm. The density change $\Delta\rho$ due to the flow may then be determined since $\Delta\rho$ is proportional to $(n' - n)$.

In a modified form due to BATES,† the MACH–ZEHNDER arrangement may also be used to measure the asphericity of convergent wave-fronts without the need for a substantially error-free reference wave-front, such as is required for the TWYMAN–GREEN method (p. 303). The arrangement, which is particularly advantageous for testing the performance of large aperture systems, is shown in Fig. 7.52 (a). The convergent

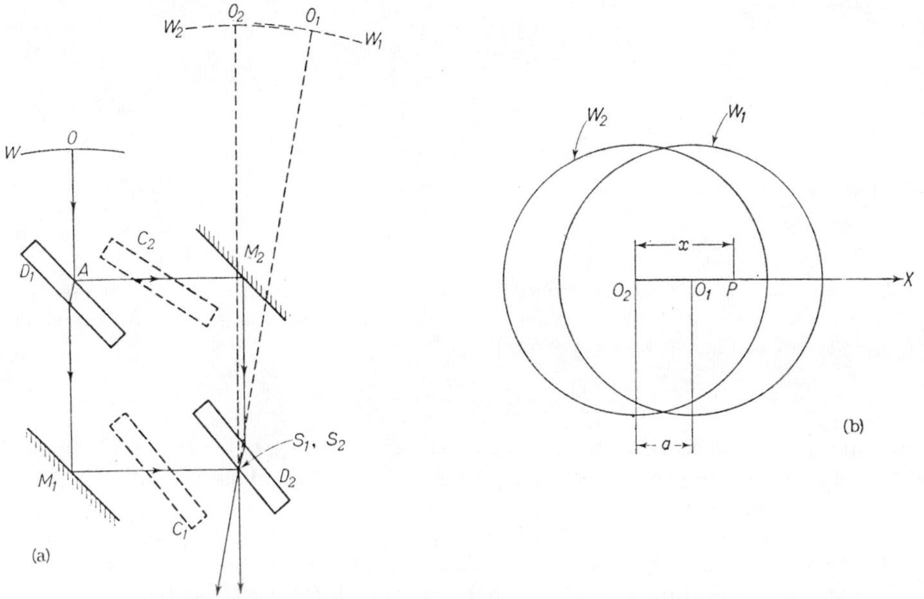

Fig. 7.52. The Bates wave-front shearing interferometer: (a) arrangement, (b) field of view with sheared wave-fronts.

beam to be tested, with principal axis OA which we take to be horizontal, is divided at D_1 into two beams converging to images S_1, S_2 of a sufficiently small quasi-monochromatic source. Initially the four reflecting surfaces are vertical and parallel, and positioned so that S_1 and S_2 coincide on the semi-reflecting surface of D_2. Corresponding to an incident wave-front W, the virtual emergent wave-fronts W_1 and W_2, with principal axes O_1S_1 and O_2S_2, are then exactly superposed, and an eye placed behind D_2 will see the field of view uniformly illuminated. D_1 and M_1 are now rotated as a whole about an axis parallel to OA, so that O_1S_1 and O_2S_2 are given a small vertical separation. This is substantially equivalent to tilting W_2 relative to W_1 about a horizontal axis, and the field is crossed by equidistant horizontal fringes, which are visible in white light.

Suppose now that D_2 is rotated about a vertical axis through S_1 and S_2; O_1S_1 then rotates about S_1 in a horizontal plane, i.e. W_1 is sheared relative to W_2. When W_1 and W_2 are perfectly spherical, the fringe pattern observed in the region where they

* See, for example, R. LADENBURG and D. BERSHADER, *Interferometry in High Speed Aerodynamics and Jet Propulsion*, Vol. IX, Physical Measurements in Gas Dynamics and Combustion (London, Oxford University Press, 1955), Article A.3.

† W. J. BATES, *Proc. Phys. Soc.*, **59** (1947), 940.

overlap is unaffected by the shear, but otherwise the fringes are displaced by an amount which depends on the asphericity of W. Thus if O_2X is a coordinate axis in the shear direction, with origin O_2, the order displacement $\Delta m(x; a)$ at the point P, coordinate x, is

$$\Delta m(x; a) = \frac{1}{\lambda_0} \{\mathscr{S}(x) - \mathscr{S}(x - a)\}, \tag{91}$$

where $\mathscr{S}$ is the optical path between W_2 and a sphere with centre S_2 and radius O_2S_2, and a is the shear distance (Fig. 7.52 (b)). From (91), since $\mathscr{S}$ is zero at O_2 $(x = 0)$,

$$\Delta m(a; a) = \frac{1}{\lambda_0} \mathscr{S}(a); \tag{92a}$$

and similarly

$$\Delta m(2a; a) = \frac{1}{\lambda_0} \{\mathscr{S}(2a) - \mathscr{S}(a)\},$$

so that

$$\Delta m(a; a) + \Delta m(2a; a) = \frac{1}{\lambda_0} \mathscr{S}(2a). \tag{92b}$$

In a similar way we may find expressions $\mathscr{S}(3a)$, $\mathscr{S}(4a)$ . . . etc., and we see that $\mathscr{S}(x)$ can be determined at intervals of a from measurements of Δm. Alternatively, when a is not too large, we have from (91)

$$\Delta m(x; a) \sim \frac{a}{\lambda_0} \frac{d\mathscr{S}(x)}{dx}, \tag{93}$$

showing that Δm is proportional to the angular aberration $d\mathscr{S}/dx$ of the ray leaving P. Evidently when there is no symmetry of revolution the whole wave-front may in principle be examined by varying the shear direction.

When the wave-fronts are sheared, the emergent rays which intersect virtually at P traverse the beam dividers at different angles, and if the fringe displacement on shearing is to depend only on the asphericity, this difference must be compensated. For this purpose, two compensating plates, identical to the plates used for the beam dividers, are introduced into the interferometer arms. One of them, C_1, is fixed parallel to D_2 and rotates with it during shearing; the other, C_2, is connected to D_2 by a mechanical linkage, which causes it to rotate at twice the rate of D_2 in the opposite sense. More recently DREW* has developed a simpler form of the interferometer in which compensating plates are unnecessary.

7.5.8 The coherence length; the application of two-beam interference to the study of the fine structure of spectral lines

When a gas, for example cadmium vapour, is excited under suitable conditions by an electrical discharge, it emits light whose spectrum consists of sharp bright lines separated by dark regions—a so-called *emission line spectrum*. If the light of one of these lines is isolated and used to illuminate, for example, a MICHELSON interferometer adjusted to give circular fringes, it is found that the fringes are distinct if the optical paths of the two interfering beams are nearly equal; but that as the difference of optical path is increased, the visibility of the fringes decreases (in general not monotonically), and they eventually disappear.

---

* R. L. DREW, *Proc. Phys. Soc.* B, **64** (1951), 1005.

Applications of shearing interferometry are discussed in an article by O. BRYNGDAHL in *Progress in Optics*, Vol. 4, ed. E. WOLF (Amsterdam, North Holland Publishing Company and New York, J. Wiley and Sons, 1965), p. 37.

We can account for this disappearance of the fringes by supposing that the light of the spectral line is not strictly monochromatic, but is made up of wave trains of finite length, of which *a large number pass at random time intervals during the time required to make an observation.* Let us assume for the moment that all these wave trains are identical. Each one entering the interferometer is divided into two trains of equal length; and when the difference of optical path in the arms of the interferometer is greater than this length, one of these two wave trains has passed the point of observation P before the other arrives. There is then no interference at P due to pairs of wave trains derived from the *same* incident wave train; the wave trains superposed at P at any instant are derived from *different* incident wave trains, and because these arrive at random, and in rapid succession, their contributions to the interference term average to zero over the relatively long time required to make an observation.

We can put this explanation into another form, which is mathematically more convenient for describing the variation of fringe visibility with path difference, by using the method of FOURIER integral analysis. Let $F(t)$ be the light disturbance at a point at time t, due to a single wave train. We assume F to be zero for $|t| \geqslant t_0$, and express it as a FOURIER integral

$$F(t) = \int_{-\infty}^{\infty} f(\nu) e^{-2\pi i\nu t}\, d\nu, \tag{94}$$

where by the FOURIER inversion theorem,

$$f(\nu) = \int_{-\infty}^{\infty} F(t) e^{2\pi i\nu t}\, dt. \tag{95}$$

If N such wave trains pass the point during the time required to make an observation, the total light disturbance involved in the observation may be written

$$V(t) = \sum_{n=1}^{N} F(t - t_n), \tag{96}$$

where the t_n's denote the times of arrival of the wave trains. The light intensity averaged over the time interval $2T$ needed to make an observation is

$$I = \frac{1}{2T} \int_{-T}^{T} |V(t)|^2 dt \sim \frac{1}{2T} \int_{-\infty}^{+\infty} |V(t)|^2 dt \tag{97}$$

if T is large compared to the half-duration t_0 of each wave train. Now from (94) and (96),

$$V(t) = \int_{-\infty}^{+\infty} v(\nu) e^{-2\pi i\nu t}\, d\nu, \tag{98}$$

where

$$v(\nu) = f(\nu) \sum_{n=1}^{N} e^{2\pi i\nu t_n}; \tag{99}$$

so that by PARSEVAL's theorem,

$$\int_{-\infty}^{+\infty} |V(t)|^2 dt = \int_{-\infty}^{+\infty} |v(\nu)|^2 d\nu = \int_{-\infty}^{+\infty} |f(\nu)|^2 \sum_{n=1}^{N} \sum_{m=1}^{N} e^{2\pi i\nu(t_n - t_m)}\, d\nu. \tag{100}$$

We have

$$\sum_{n=1}^{N}\sum_{m=1}^{N} e^{2\pi i\nu(t_n-t_m)} = N + \sum_{n\neq m} e^{2\pi i\nu(t_n-t_m)}$$

$$= N + 2\sum_{n<m}\cos 2\pi\nu(t_n - t_m). \tag{101}$$

Now since the t_n's are distributed at random, there is an equal likelihood for each cosine term to be positive or negative. Hence the average value of the double sum in (100) in a large number of similar experiments is N, and it follows from (97) and (100) that the mean intensity is given by

$$I = \frac{N}{2T}\int_{-\infty}^{+\infty} |f(\nu)|^2 d\nu, \tag{102}$$

i.e. it is proportional to the integral of the intensities $i(\nu) = |f(\nu)|^2$ (incoherent superposition) of the monochromatic components of which a single wave train is made up.* With the interferometer, each monochromatic component produces an interference pattern as described in § 7.5.4, and as the path difference is increased from zero, these component patterns show increasing mutual displacement because of the difference of wavelength. The visibility of the fringes therefore decreases, and they disappear altogether when the optical path difference is sufficiently large.

These two ways of interpreting the absence of fringes at sufficiently large path differences—in terms of either a random succession of finite wave trains, or a superposition of monochromatic components distributed over a range of frequency—are for most practical purposes equivalent; and from the discussion above we must expect that the longer the wave trains, the narrower the frequency range over which the Fourier components have appreciable intensity. We may illustrate this relationship by a simple example. Suppose the wave trains are all of duration Δt, during which $F(t)$ is simply periodic with frequency ν_0, i.e.

$$\left.\begin{aligned} F(t) &= f_0 e^{-2\pi i\nu_0 t} \quad \text{when} \quad |t| \leqslant \frac{\Delta t}{2}, \\ &= 0 \qquad\qquad\quad \text{when} \quad |t| > \frac{\Delta t}{2}, \end{aligned}\right\} \tag{103}$$

where f_0 is constant. Then from (95) and (103),

$$f(\nu) = f_0\int_{-\frac{\Delta t}{2}}^{\frac{\Delta t}{2}} e^{2\pi i(\nu-\nu_0)t}\, dt$$

$$= f_0 \Delta t\left[\frac{\sin\{\pi(\nu-\nu_0)\Delta t\}}{\pi(\nu-\nu_0)\Delta t}\right]. \tag{104}$$

The function $[\sin\{\pi(\nu-\nu_0)\Delta t\}/\pi(\nu-\nu_0)\Delta t]^2$, which governs the intensity distribution of the Fourier components of (103), is shown in Fig. 7.53. The frequency interval

* A rigorous formulation of the result is given by a theorem due to N. Campbell, well known in the analysis of random noise, especially in connection with the shot effect (fluctuations in intensity of a stream of electrons in vacuum tubes). Cf. S. O. Rice, *Bell Tech. J.*, **23** (1944), 282. [Reprinted in *Selected Papers on Noise and Stochastic Processes* (ed. N. Wax, New York, Dover Publications, 1954), 133.]

$\nu_0 - \Delta\nu/2 \leqslant \nu \leqslant \nu_0 + \Delta\nu/2$ over which the intensity may be said to be appreciable is somewhat arbitrary, but since the first zero (which occurs when the argument of the sine term is equal to π) corresponds to $\nu - \nu_0 = \pm 1/\Delta t$, it is clear that

$$\Delta\nu \sim \frac{1}{\Delta t}. \tag{105}$$

Thus the effective frequency range of the Fourier spectrum is of the order of the reciprocal of the duration of a single wave train.

This example, in which the wave trains are all identical and of simple form, is only an idealization of the light from real sources. According to atomic theory, the loss of energy by atoms during emission results in damping of the wave trains. Further, the

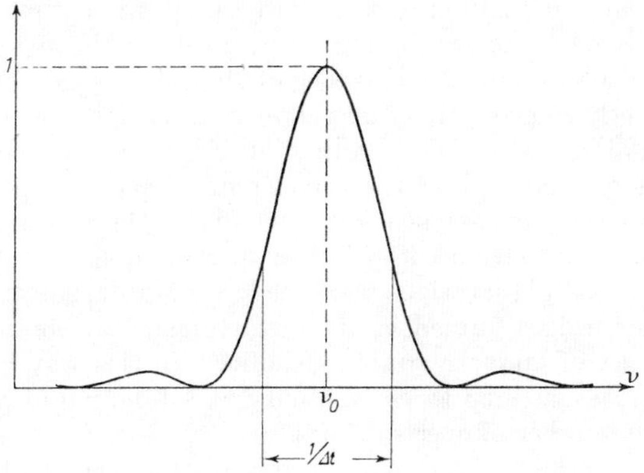

Fig. 7.53. The function $\left[\frac{\sin\{\pi(\nu-\nu_0)\Delta t\}}{\pi(\nu-\nu_0)\Delta t}\right]^2$.

atoms are in random thermal motion relative to the observer, so that observed spectra are modified by the DOPPLER effect. Again, emitting atoms are disturbed by their neighbours, with the result that the wave trains are irregularly modified. For these reasons we cannot expect with real light to give simple meanings to the terms "duration of the wave trains" and "frequency range of the FOURIER spectrum". However, for any light disturbance $V(t)$, and its FOURIER inverse $v(\nu)$, it is possible to define quantities Δt and $\Delta\nu$ which may respectively be regarded as the average duration of the wave trains of which V is composed, and the effective frequency range of the FOURIER spectrum; and it can be shown that these averages always satisfy the reciprocity relation

$$\Delta t \Delta\nu \geqslant \frac{1}{4\pi}. \tag{106}$$

This inequality, which in a sense is analogous to the HEISENBERG uncertainty relation in quantum mechanics, is derived and discussed in § 10.7.3. Here we remark only that in most cases of practical interest the inequality in (106) may be replaced by the order of magnitude sign.

The time Δt of (106) is known as the *coherence time* of the light; and if $\overline{\lambda}_0$ is the mean wavelength, the length Δl defined by

$$\Delta l = c\Delta t \sim \frac{c}{\Delta\nu} = \frac{(\overline{\lambda}_0)^2}{\Delta\lambda_0}, \tag{107}$$

is known as the *coherence length*. By comparison of (107) and § 7.3 (15) we see that our earlier restriction on the path difference between quasi-monochromatic beams implies that the path difference must be small compared to the coherence length of the light. When the difference of optical path is of the order of, or much greater than, the coherence length, interference effects are no longer appreciable.

It is clear from the foregoing that observations of the variation of fringe visibility with optical path difference in a suitable interference experiment must yield information about the spectral intensity distribution of the light used. The first observations of this kind were made by FIZEAU.* Using his interferometer (§ 7.5.2) illuminated by the yellow light of a sodium flame, he obtained NEWTON's rings and observed their appearance as the separation of the lens and plate was increased. He found that while the rings were distinct when the lens and plate were in contact, they were almost invisible near the position of the 490th ring, but had regained almost their original distinctness near the 980th ring; and he was able to follow such periodic variations of visibility through 52 cycles each of about 980 rings. From these observations FIZEAU correctly inferred that the yellow sodium light has two components of approximately equal intensity. The maxima of fringe visibility occur when the path difference is an integral multiple of the wavelength of each component, so that their wavelengths are in the approximate ratio 981/980. He was able to confirm this conclusion by direct observation with a prism spectroscope.

Later, more elaborate observations were made by MICHELSON,† who measured the visibility of circular fringes formed in his interferometer by comparing them with a set of circular fringes of known variable visibility. In this way he was able to construct *visibility curves*, showing fringe visibility as a function of path difference, for the light of a large number of spectral lines.

Let us examine how the visibility curve is related to the spectral intensity distribution. For simplicity we assume that the intensities of the two interfering beams are equal. For optical path difference $\Delta\mathscr{S}$, the phase difference is

$$\delta(k_0, \Delta\mathscr{S}) = k_0\Delta\mathscr{S}, \tag{108}$$

where $k_0 = 2\pi/\lambda_0$ is the wave number; so that by § 7.2 (17), the intensity due to the components in the elementary wave number range dk_0 is

$$i(k_0, \Delta\mathscr{S})dk_0 = 2i_1(k_0)\{1 + \cos (k_0\Delta\mathscr{S})\}dk_0, \tag{109}$$

where $i_1(k_0)$ represents the spectral distribution of intensity of either beam. Since, as already shown, the different spectral components add incoherently, the total intensity in the interference pattern is

$$I(\Delta\mathscr{S}) = 2\int i_1(k_0)\{1 + \cos (k_0\Delta\mathscr{S})\}dk_0. \tag{110}$$

For the light of a spectral line, $i_1(k_0)$ will be negligible except in a small range of k_0 about some mean wave number $\bar{k}_0$. If then we put

$$\left.\begin{aligned} x &= k_0 - \bar{k}_0, \\ j(x) &= i_1(\bar{k}_0 + x), \end{aligned}\right\} \tag{111}$$

* H. FIZEAU, *Ann. Chim. Phys.* (3), **66** (1862), 429.

† A. A. MICHELSON, *Phil. Mag.* (5), **31** (1891), 338; *ibid.*, **34** (1892), 280.

we may write in place of (110)

$$I = 2\int j(x)\{1 + \cos[(\bar{k}_0 + x)\Delta\mathscr{S}]\}dx. \tag{112}$$

From this point the analysis is similar to that of p. 272. Thus (112) may be written (cf. § 7.3 (38))

$$I(\Delta\mathscr{S}) = P + C(\Delta\mathscr{S})\cos(\bar{k}_0\Delta\mathscr{S}) - S(\Delta\mathscr{S})\sin(\bar{k}_0\Delta\mathscr{S}), \tag{113}$$

where

$$\left.\begin{aligned} P &= 2\int j(x)dx, \\ C(\Delta\mathscr{S}) &= 2\int j(x)\cos(x\Delta\mathscr{S})dx, \\ S(\Delta\mathscr{S}) &= 2\int j(x)\sin(x\Delta\mathscr{S})dx. \end{aligned}\right\} \tag{114}$$

Since, for a spectral line, $j(x)$ has non-zero values only for $|x| \ll \bar{k}_0$, the variation of C and S is negligible compared with that of $\cos(\bar{k}_0\Delta\mathscr{S})$ and $\sin(\bar{k}_0\Delta\mathscr{S})$, and it follows that the positions of the extrema of I are to a good approximation given by

$$\frac{dI}{d(\Delta\mathscr{S})} = -\bar{k}_0[C\sin(\bar{k}_0\Delta\mathscr{S}) + S\cos(\bar{k}_0\Delta\mathscr{S})] = 0,$$

i.e. by

$$\tan(\bar{k}_0\Delta\mathscr{S}) = -\frac{S}{C}. \tag{115}$$

From (113) and (115), the extreme values of I are therefore

$$I_{\text{ext}} = P \pm \sqrt{C^2 + S^2}. \tag{116}$$

Hence the visibility curve is given by

$$\mathscr{V}(\Delta\mathscr{S}) = \frac{I_{\max} - I_{\min}}{I_{\max} + I_{\min}} = \frac{\sqrt{C^2 + S^2}}{P}. \tag{117}$$

We note that (113) may also be written as

$$I = P\left[1 + \frac{\sqrt{C^2 + S^2}}{P}\cos(\phi + \bar{k}_0\Delta\mathscr{S})\right], \tag{118}$$

where $\tan\phi = S/C$, so that the visibility curve is the envelope of the normalized intensity curve I/P.

The calculated visibility curves for a number of assumed spectral distributions of intensity are shown in Fig. 7.54. But in practice we have the converse problem of determining the spectral distribution from the observed visibility curve. If $j(x)$ is symmetrical, $S = 0$ and (117) reduces to

$$\mathscr{V} = \frac{|C|}{P}. \tag{119}$$

In this case the visibility curve determines C (apart from the constant factor of proportionality P, and the sign, which may usually be fixed from considerations of physical plausibility); and $j(x)$ can be obtained from (114) by FOURIER inversion. In general, however, the visibility curve determines only $\sqrt{C^2 + S^2}$, and alone it is in general insufficient to determine $j(x)$ since this requires both C and S to be known separately. It was pointed out by RAYLEIGH* that both C and S can be determined if measurements are made not only of the visibility of the fringes but also of their

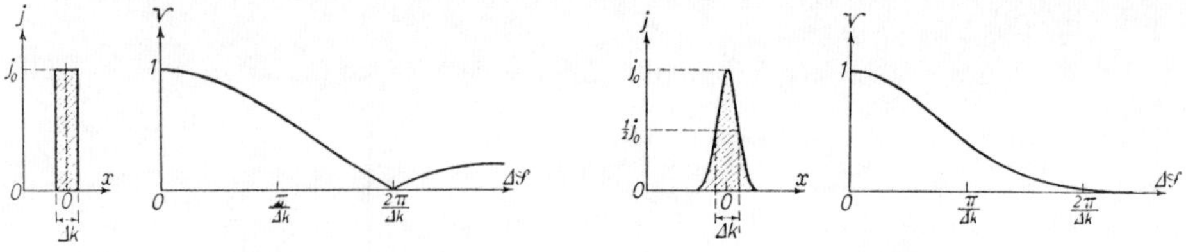

(a)

$$j = j_0 \quad \text{when} \quad |x| \leqslant \tfrac{1}{2}\Delta k,$$
$$j = 0 \quad \text{when} \quad |x| > \tfrac{1}{2}\Delta k,$$
$$\mathscr{V} = \frac{|\sin(\frac{1}{2}\Delta k \Delta \mathscr{S})|}{|\frac{1}{2}\Delta k \Delta \mathscr{S}|}.$$

(b)

$$j = j_0 e^{-\alpha^2 x^2},$$
$$\mathscr{V} \sim e^{-\left(\frac{\Delta \mathscr{S}}{2\alpha}\right)^2}.$$

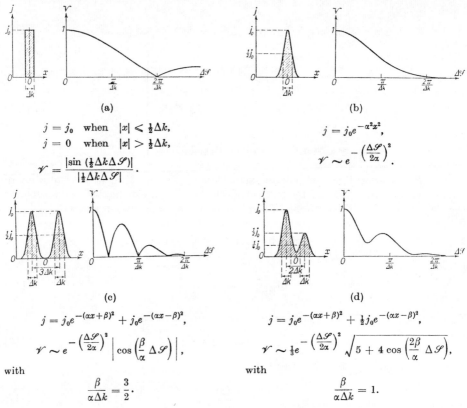

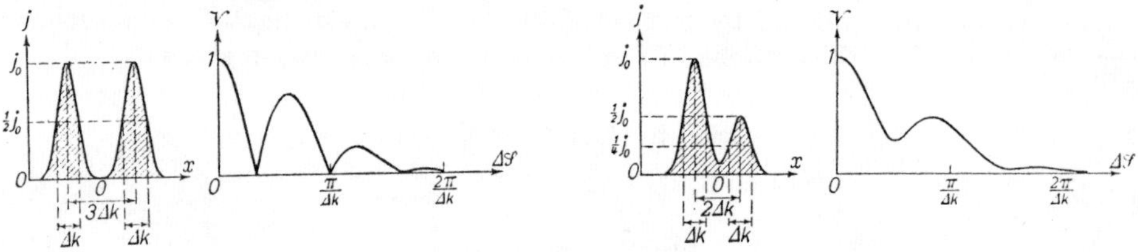

(c)

$$j = j_0 e^{-(\alpha x+\beta)^2} + j_0 e^{-(\alpha x-\beta)^2},$$
$$\mathscr{V} \sim e^{-\left(\frac{\Delta \mathscr{S}}{2\alpha}\right)^2} \left|\cos\left(\frac{\beta}{\alpha}\Delta \mathscr{S}\right)\right|,$$

with

$$\frac{\beta}{\alpha \Delta k} = \frac{3}{2}.$$

(d)

$$j = j_0 e^{-(\alpha x+\beta)^2} + \tfrac{1}{2} j_0 e^{-(\alpha x-\beta)^2},$$
$$\mathscr{V} \sim \tfrac{1}{3} e^{-\left(\frac{\Delta \mathscr{S}}{2\alpha}\right)^2} \sqrt{5 + 4\cos\left(\frac{2\beta}{\alpha}\Delta \mathscr{S}\right)},$$

with

$$\frac{\beta}{\alpha \Delta k} = 1.$$

Fig. 7.54. Visibility curves corresponding to various spectral distributions. (The analytical expressions for $\mathscr{V}$ were derived on the assumption that the light is quasi-monochromatic.)
In figures (b), (c), (d), $\Delta k = 2\sqrt{\log_e 2}/\alpha \sim 1{\cdot}66/\alpha$.

position, since the latter would by (115) give the ratio C/S; but such measurements are very difficult to make.†

In spite of this limitation, MICHELSON was able to deduce structures for simple spectral lines which have been broadly confirmed by later work. In particular, he found the red line ($\lambda = 6438$ Å) of cadmium to be the most nearly monochromatic of all those he examined: its visibility curve (Fig. 7.55) corresponded to a symmetrical spectral distribution of Gaussian form with a width of only 0·013 Å at half peak intensity, and he was able to observe fringes with a path difference exceeding 500,000 wavelengths ($\sim$ 30 cm).

* LORD RAYLEIGH, *Phil. Mag.* (5). **34** (1892), 407; see also E. WOLF, *Proc. Phys. Soc.*, **80** (1962), 1269.

† A number of such measurements was reported by A. PÉRARD, *Rev. d'Optique*, **7** (1928), 1; *Réunions l'Institut d'Optique* (Paris: Revue d'Optique), 6 éme année, (1935), 10,

This method of analysing spectra by means of two-beam interference, which is of historical importance as the first use of interference in spectroscopy, was superseded by methods involving multiple-beam interference (§ 7.6). More recently, however, interest in the use of the two-beam method has been revived partly because for technical reasons it has advantages in the infra-red region of the spectrum.*

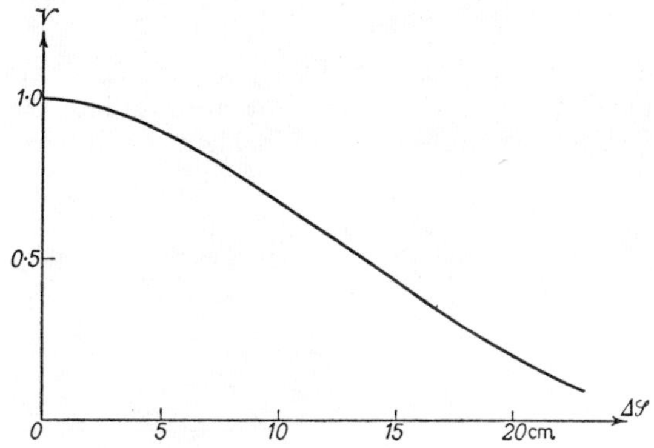

Fig. 7.55. Visibility curve for the red cadmium line.
(After A. A. Michelson, *Phil. Mag.* (5), **4** (1892), 291.)

7.6 MULTIPLE-BEAM INTERFERENCE

When a beam of light is incident on a transparent plate, there are multiple reflections at the plate surfaces, with the result that a series of beams of diminishing amplitude emerges on each side of the plate. In discussing the interference effects produced by such plates (§ 7.5.1, § 7.5.2, § 7.5.6) we neglected the contribution to the resultant intensity made by those beams which suffer more than two reflections—an approximation which is justified when the reflectivity of the surfaces is low. We will now take into account all the reflected beams: we shall see that if the reflectivity of the surfaces is high, the intensity distribution in the fringe patterns is modified in a way which has important practical uses.

7.6.1 Multiple-beam fringes with a plane parallel plate

Consider a plane parallel transparent plate of refractive index n', surrounded by a medium of refractive index n, and suppose that a plane wave of monochromatic light is incident upon the plate at angle θ. Let the ray SB_1 (Fig. 7.56) represent the direction of propagation of the incident wave. At the first surface this wave is divided into two plane waves, one reflected in the direction B_1C_1 and the other transmitted into the plate in the direction B_1D_1. The latter wave is incident on the second surface at angle θ' and is there divided into two plane waves, one transmitted in the direction D_1E_1, the other reflected back into the plate in the direction D_1B_2; and the process of division of the wave remaining inside the plate continues as indicated in the figure.

Let $A^{(i)}$ be the amplitude of the electric vector of the incident wave, which we assume to be linearly polarized, with the electric vector either parallel or perpendicular to the plane of incidence. As in § 1.5.2 we take $A^{(i)}$ to be complex, with its phase equal to the constant part of the phase of the wave function. For each member of

* See, for example P. Fellgett, *Journ. de Phys.*, **19** (1958), 187, 237. See also the review article on Fourier spectroscopy by G. A. Vanasse and H. Sakai in *Progress in Optics*, Vol. 6, ed. E. Wolf (Amsterdam, North Holland Publishing Company and New York, J. Wiley and Sons, 1967), p. 259.

either the reflected or the transmitted set of waves, the variable part of the phase of the wave function differs from that of the preceding member by an amount which corresponds to a double traversal of the plate. According to § 7.5 (6) this phase difference is

$$\delta = \frac{4\pi}{\lambda_0} n'h \cos \theta', \tag{1}$$

where h is the thickness of the plate and λ_0 is the wavelength in vacuum. For a wave travelling from the surrounding medium into the plate, let r be the reflection coefficient (ratio of reflected and incident amplitudes), and t the transmission coefficient

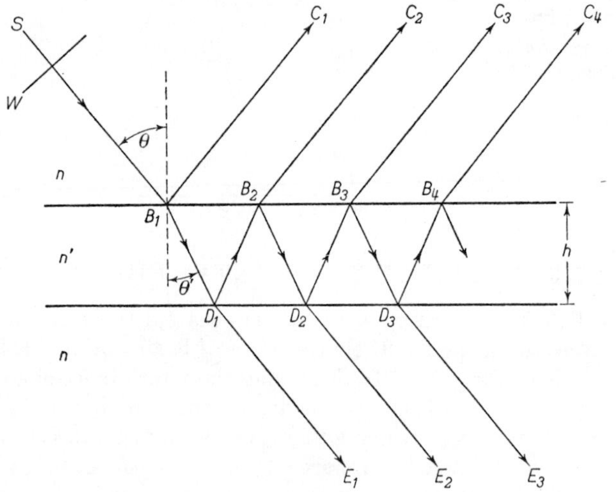

Fig. 7.56. Reflection of a plane wave in a plane parallel plate.

(ratio of transmitted and incident amplitudes); and let r', t' be the corresponding coefficients for a wave travelling from the plate to the surrounding medium. The complex amplitudes of the waves reflected from the plate are then

$$rA^{(i)}, \quad tt'r'A^{(i)}e^{i\delta}, \quad tt'r'^3A^{(i)}e^{2i\delta}, \ldots \ldots tt'r'^{(2p-3)}A^{(i)}e^{i(p-1)\delta}, \ldots$$

Similarly, the complex amplitudes of the waves transmitted through the plate are, apart from an unimportant constant phase factor,

$$tt'A^{(i)}, \quad tt'r'^2A^{(i)}e^{i\delta}, \quad tt'r'^4A^{(i)}e^{2i\delta}, \ldots \ldots tt'r'^{2(p-1)}A^{(i)}e^{i(p-1)\delta}, \ldots$$

The quantities r, r', t, t' are given in terms of n, n', θ, θ' by the Fresnel formulae of § 1.5.2. For our present purposes we do not need these explicit expressions, but only relations between them. Thus for either polarized component, we see from relations (20a) and (35) of § 1.5 that

$$tt' = \mathscr{T}; \tag{2}$$

similarly using equation (21(a)) of § 1.5 we have

$$r = -r', \tag{3}$$

whence by § 1.5 (33),

$$r^2 = r'^2 = \mathscr{R}, \tag{4}$$

where $\mathscr{R}$ and $\mathscr{T}$, respectively the reflectivity and transmissivity of the plate surfaces, are related by

$$\mathscr{R} + \mathscr{T} = 1. \tag{5}$$

If the first p reflected waves are superposed, the amplitude $A^{(r)}(p)$ of the electric vector of the reflected light is given by the expression

$$\begin{aligned} A^{(r)}(p) &= \{r + tt'r'e^{i\delta}(1 + r'^2e^{i\delta} + \dots + r'^{2(p-2)}e^{i(p-2)\delta})\}A^{(i)} \\ &= \left\{r + \left(\frac{1 - r'^{2(p-1)}e^{i(p-1)\delta}}{1 - r'^2e^{i\delta}}\right) tt'r'e^{i\delta}\right\} A^{(i)}. \end{aligned} \tag{6}$$

If the plate is sufficiently long, the number of reflected waves is large; and in the limit as $p \to \infty$, we have from (6), if (3) is also used,

$$A^{(r)} \equiv A^{(r)}(\infty) = -\frac{r'\{1 - (r'^2 + tt')e^{i\delta}\}}{1 - r'^2e^{i\delta}} A^{(i)}. \tag{7}$$

From (2), (4) and (5) we hence obtain

$$A^{(r)} = \frac{(1 - e^{i\delta})\sqrt{\mathscr{R}}}{1 - \mathscr{R}e^{i\delta}} A^{(i)}, \tag{8}$$

so that the intensity $I^{(r)} = A^{(r)}A^{(r)\star}$ of the reflected light is

$$I^{(r)} = \frac{(2 - 2\cos\delta)\mathscr{R}}{1 + \mathscr{R}^2 - 2\mathscr{R}\cos\delta} I^{(i)} = \frac{4\mathscr{R}\sin^2\dfrac{\delta}{2}}{(1 - \mathscr{R})^2 + 4\mathscr{R}\sin^2\dfrac{\delta}{2}} I^{(i)}, \tag{9}$$

where $I^{(i)} = A^{(i)}A^{(i)\star}$ is the intensity of the incident light.

In a similar way we obtain the following expression for the amplitude $A^{(t)}$ of the transmitted light:

$$\begin{aligned} A^{(t)}(p) &= tt'(1 + r'^2e^{i\delta} + \dots + r'^{2(p-1)}e^{i(p-1)\delta})A^{(i)} \\ &= \left(\frac{1 - r'^{2p}e^{ip\delta}}{1 - r'^2e^{i\delta}}\right) tt'A^{(i)}. \end{aligned} \tag{10}$$

In the limit as $p \to \infty$, (10) reduces to

$$A^{(t)} \equiv A^{(t)}(\infty) = \frac{tt'}{1 - r'^2e^{i\delta}} A^{(i)}. \tag{11}$$

Hence, using (2) and (4), we have

$$A^{(t)} = \frac{\mathscr{T}}{1 - \mathscr{R}e^{i\delta}} A^{(i)}, \tag{12}$$

and the corresponding intensity $I^{(t)} = A^{(t)}A^{(t)\star}$ of the transmitted light is

$$I^{(t)} = \frac{\mathscr{T}^2}{1 + \mathscr{R}^2 - 2\mathscr{R}\cos\delta} I^{(i)} = \frac{\mathscr{T}^2}{(1 - \mathscr{R})^2 + 4\mathscr{R}\sin^2\dfrac{\delta}{2}} I^{(i)}. \tag{13}$$

The formulae (9) and (13), which are known as *Airy's formulae*, are in agreement with results already derived from the general theory of propagation in stratified media (§ 1.6); for if in equations (60) of § 1.6 we put $r_{12} = r$, $r_{23} = r'$, $t_{12} = t$, $t_{23} = t'$, $2\beta = \delta$, and use the relations (2) and (3) above, the expressions (9) and (13) follow.*

Suppose now that plane waves of equal intensity are incident over a range of angles, and the transmitted light is collected by a lens L (Fig. 7.57). At a point P in the focal plane of L, the intensity is in the ratio $I^{(t)}/I^{(i)}$ to the intensity at P when the plate is

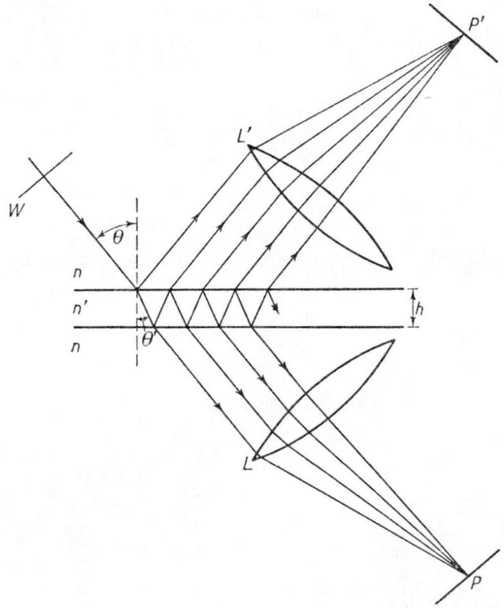

Fig. 7.57. Illustrating formation of multiple beam fringes of equal inclination with a plane parallel plate.

removed; hence according to (13) there will, when the plate is present, be maxima of intensity at P when the order of interference m, defined by

$$m = \frac{\delta}{2\pi} = \frac{2n'h \cos \theta'}{\lambda_0}, \tag{14}$$

has integral values 1, 2, . . ., and minima when it has half-integral values $\frac{1}{2}, \frac{3}{2}, \frac{5}{2}, \ldots$ Evidently there are fringes of equal inclination in the focal plane of L, following loci of constant θ' (and therefore θ). Similarly, if the light reflected from the plate is collected by a lens L', there are fringes of equal inclination in the focal plane of L'; and from (9) we find that the intensity maxima of this reflected pattern correspond to half-integral values $\frac{1}{2}, \frac{3}{2}, \frac{5}{2}, \ldots$ of the order of interference m, and minima to integral values 1, 2, . . . Thus the fringes of both patterns are in the positions given by the approximate treatment of § 7.5.1, in which we took into account only the first two beams from each side of the plate.

---

* When this comparison is made it must be remembered that, unlike here, the symbols $\mathscr{R}$ and $\mathscr{T}$ denote in § 1.6 the reflectivity and transmissivity of the whole plate.

The intensity distributions of the reflected and transmitted patterns are given by (9) and (13), which, using (5), we may write as

$$\frac{I^{(r)}}{I^{(i)}} = \frac{F\sin^2\dfrac{\delta}{2}}{1 + F\sin^2\dfrac{\delta}{2}}, \tag{15a}$$

$$\frac{I^{(t)}}{I^{(i)}} = \frac{1}{1 + F\sin^2\dfrac{\delta}{2}}, \tag{15b}$$

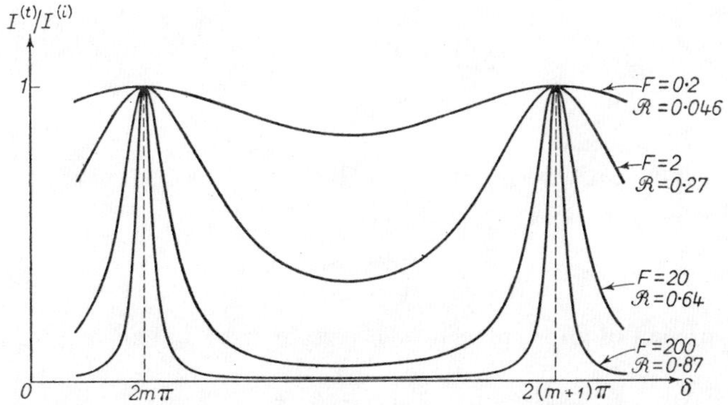

Fig. 7.58. Multiple beam fringes of equal inclination in transmitted light: ratio $I^{(t)}/I^{(i)}$ of transmitted and incident intensities as a function of phase difference δ (m is an integer).

where the parameter F is defined by the formula

$$F = \frac{4\mathscr{R}}{(1-\mathscr{R})^2}. \tag{16}$$

Evidently the two patterns are complementary, in the sense that

$$\frac{I^{(r)}}{I^{(i)}} + \frac{I^{(t)}}{I^{(i)}} = 1. \tag{17}$$

The behaviour of $I^{(t)}/I^{(i)}$ as a function of the phase difference δ is shown for various values of F in Fig. 7.58. When $\mathscr{R}$ is small compared with unity, F also is small compared with unity, so that we may expand $1/(1 + F\sin^2\delta/2)$ in (15), and retain terms up to only the first power in F. This gives

$$\frac{I^{(r)}}{I^{(i)}} \sim F\sin^2\frac{\delta}{2} = \frac{F}{2}(1-\cos\delta), \tag{18a}$$

$$\frac{I^{(t)}}{I^{(i)}} \sim 1 - F\sin^2\frac{\delta}{2} = 1 - \frac{F}{2}(1-\cos\delta), \tag{18b}$$

i.e. the intensity variations are of the form of § 7.2 (15), characteristic of two interfering beams. If $\mathscr{R}$ is increased, the intensity of the minima of the transmitted pattern falls, and the maxima become sharper until, when $\mathscr{R}$ approaches unity so that F is large, the intensity of the transmitted light is very small except in the immediate neighbourhood of those maxima. The pattern in transmitted light then consists of narrow bright fringes on an almost completely dark background. Similarly the pattern in reflected light becomes one of narrow dark fringes on an otherwise nearly uniform bright background. The sharpness of the fringes is conveniently measured by their *half-intensity width*, or *half-width*, which, in the case of the pattern in transmitted light, is the width between the points on either side of a maximum where the intensity has fallen to half its maximum value. The ratio of the separation of adjacent fringes and the half-width we shall call the *finesse* $\mathscr{F}$ of the fringes. For the fringe of integral order m, the points where the intensity is half its maximum value are at

$$\delta = 2m\pi \pm \frac{\varepsilon}{2} \tag{19}$$

where by (15b)

$$\frac{1}{1 + F\sin^2\dfrac{\varepsilon}{4}} = \tfrac{1}{2}; \tag{20}$$

and when F is sufficiently large, ε is so small that we may put $\sin(\varepsilon/4) = \varepsilon/4$ in (20) and obtain the half-width as

$$\varepsilon = \frac{4}{\sqrt{F}}. \tag{21}$$

Since the separation of adjacent fringes corresponds to a change 2π of δ, the finesse is then

$$\mathscr{F} = \frac{2\pi}{\varepsilon} = \frac{\pi\sqrt{F}}{2}. \tag{22}$$

So far we have assumed the light to be strictly monochromatic. With quasi-monochromatic light, the intensity distribution is the sum of intensity distributions of the form of (15) due to each monochromatic component; and if these components cover a wavelength range $\Delta\lambda_0$ about mean wavelength $\bar{\lambda}_0$, the maxima of order m spread over a distance which corresponds to $|\Delta\delta|$ in the pattern of wavelength $\bar{\lambda}_0$, where from (14), if we neglect dependence of $n'h$ on wavelength, $|\Delta\delta| = 2\pi m\Delta\lambda_0/\bar{\lambda}_0$. We can take these component patterns to be in coincidence, and the resultant intensity distribution to be the same as that from a strictly monochromatic source of wavelength $\bar{\lambda}_0$, providing $|\Delta\delta|$ is negligible compared with the half-width ε of a monochromatic fringe. From (22) this condition is $m\mathscr{F} \ll \bar{\lambda}_0/\Delta\lambda_0$, i.e.

$$\mathscr{F}\Delta\mathscr{S} \ll \frac{\bar{\lambda}_0^{\,2}}{\Delta\lambda_0}, \tag{23}$$

where

$$\Delta\mathscr{S} = \frac{\bar{\lambda}_0}{2\pi}\,\delta = m\bar{\lambda}_0 \tag{24}$$

is the difference of optical path of successive interfering beams; (23) may be compared with the analogous inequality § 7.3 (15) relating to two-beam interference. The quantity on the right of (23) will be recognized as the coherence length of the light (cf. § 7.5 (107)).

From the foregoing it is evident that as the reflectivity of the surfaces—and hence $\mathscr{F}$—is increased, the intensity distributions become more favourable for the measurement of fringe positions, and fringes due to different monochromatic components may become clearly separated in the transmitted pattern. It is for these reasons that multiple-beam interferometry is of practical importance. At normal incidence the reflectivity of a surface between available dielectrics is low at optical wavelengths: for example, for an interface between air ($n \sim 1$) and glass ($n' \sim 1{\cdot}5$), § 1.5 (37) gives $\mathscr{R} \sim 0{\cdot}04$. Increased reflectivity can be obtained at oblique incidence, and as we have seen in § 1.5, $\mathscr{R} \to 1$ when the direction of propagation in the optically denser medium approaches the critical angle. Reflectivity approaching unity may also be obtained near normal incidence by coating the dielectric surface with a multi-layer system of suitable dielectric films (as shown in § 1.6) or with a partially transparent film of metal. The theory of metal films on dielectric substrates is given in Chapter XIII, but here we may anticipate it by saying that such films absorb light, that the phase changes on reflection are no longer necessarily zero or π, and that the reflectivities and phase changes on the two sides of the film are different when the bounding dielectric media are of different refractive index. In consequence the analysis above is not valid for a plate coated with metal films. However, when the films on the two surfaces of the plate are identical, equations (12) and (13) hold providing we interpret $\mathscr{R}$ as the reflectivity for internal reflection, and replace δ as defined by (1) by

$$\delta = \frac{4\pi}{\lambda_0} n'h \cos \theta' + 2\phi, \tag{25}$$

where ϕ is the phase change on internal reflection. If then we put

$$\mathscr{R} + \mathscr{T} + \mathscr{A} = 1, \tag{26}$$

where $\mathscr{A}$ is the fraction of light absorbed by the metal, we have from (13), using (16)

$$\frac{I^{(t)}}{I^{(i)}} = \left(1 - \frac{\mathscr{A}}{1 - \mathscr{R}}\right)^2 \frac{1}{1 + F \sin^2 \dfrac{\delta}{2}}. \tag{27}$$

Comparing (27) and (15b) we see that, for given F, the effect of absorption is to diminish the intensity of the transmitted pattern by a factor $[(1 - \mathscr{A}/(1 - \mathscr{R})]^{-2}$.* Near normal incidence, the effect of the phase change ϕ at reflection is equivalent to an increase $\phi\lambda_0/2\pi$ in the optical thickness of the plate: at oblique incidence there is the further complication that the phase change depends on the state of polarization.

7.6.2 The Fabry–Perot interferometer

The multiple beam interference fringes from a plane parallel plate illuminated near normal incidence are used in the *Fabry–Perot interferometer*.† This instrument consists essentially of two glass or quartz plates P_1, P_2 (Fig. 7.59) with plane surfaces. The inner surfaces are coated with partially transparent films of high reflectivity, and are parallel, so that they enclose a plane parallel plate of air. The plates themselves are made slightly prismatic, in order to avoid disturbing effects due to reflections at the outer uncoated surfaces. In the original form of the instrument, one plate

---

* With multilayer dielectric reflecting films also, the intensity distribution of the transmitted pattern is found to be of the form of (27); in this case the reduction of intensity is due principally to scattering: see P. Giacomo, *Rev. d'Opt.*, **35** (1956), 442.

† C. Fabry and A. Perot, *Ann. Chim. Phys.* (7), **16** (1899), 115.

was fixed and the other was mounted on a screw-controlled carriage to allow continuous variation of the plate separation, but because of difficulties of mechanical construction this arrangement is obsolete. Instead the plates are separated by a fixed spacer D, which is commonly a hollow cylinder of invar or silica with three projecting studs at each end, and the plates are kept in place by the pressure of springs. The spacer is optically worked so that the planes defined by the studs are as nearly parallel as possible, and fine adjustments can be made by varying the spring pressure. This form of the interferometer, with fixed plate separation, is sometimes called a *Fabry–Perot etalon.*

As we have seen in § 7.6.1, light from an extended quasi-monochromatic source S which satisfies (23) forms narrow bright fringes of equal inclination in the focal plane of the lens L. From (25) the order of interference is given by

$$m = \frac{\delta}{2\pi} = \frac{2n'h\cos\theta'}{\lambda_0} + \frac{\phi}{\pi}, \tag{28}$$

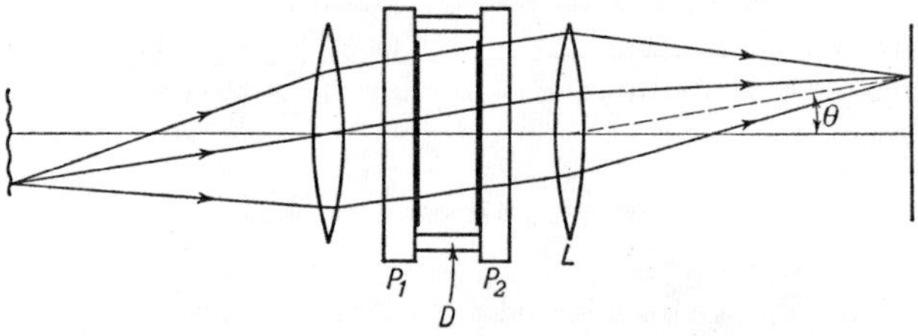

Fig. 7.59. The FABRY–PEROT interferometer.

where n' is the refractive index of the air between the plates, h is the separation of the reflecting surfaces, θ' is the angle of reflection, and ϕ is the phase change. The axis of the lens is usually normal to the plates, and the bright fringes, corresponding to integral values of m, are then circles with common centre at the focal point for normally transmitted light (Fig. 7.60). At this point, m has its maximum value m_0 given by

$$m_0 = \frac{2n'h}{\lambda_0} + \frac{\phi}{\pi}. \tag{29}$$

In general m_0 is not an integer, and we may write

$$m_0 = m_1 + e, \tag{30}$$

where m_1 is the integral order of the innermost bright fringe, and e, which is less than unity, is the fractional order at the centre. From (28), (29) and (30), in an exactly similar way to the derivation of equation (13) of § 7.5, we obtain the angular radius θ_p of the pth bright fringe from the centre, when θ_p is not too large, as

$$\theta_p = \frac{1}{n}\sqrt{\frac{n'\lambda_0}{h}}\sqrt{p-1+e}, \tag{31}$$

where n is the refractive index of the air outside the plates. The diameter D_p of this fringe is therefore given by

$$D_p^2 = (2f\theta_p)^2 = \frac{4n'\lambda_0 f^2}{n^2 h}(p-1+e), \tag{32}$$

where f is the focal length of the lens L.

Fig. 7.60. FABRY–PEROT fringes.

The characteristics of the FABRY–PEROT interferometer which are of importance in practice are the finesse $\mathscr{F}$, already defined as the ratio of fringe separation and half-width; the *peak-transmission*

$$\tau = \left(\frac{I^{(t)}}{I^{(i)}}\right)_{\max}; \tag{33}$$

and the *contrast factor*

$$\mathscr{C} = \left(\frac{I^{(t)}}{I^{(i)}}\right)_{\max} \Big/ \left(\frac{I^{(t)}}{I^{(i)}}\right)_{\min}, \tag{34}$$

where $I^{(t)}$ is the intensity in the fringe pattern, and $I^{(i)}$ is the corresponding intensity when the interferometer is removed. If we assume for the moment that the inner surfaces of the plates are plane and parallel, and neglect the reflections at the outer surfaces, $I^{(t)}/I^{(i)}$ is given by (27). Comparing (27) with (15b), we see that the finesse is given by (22), and using (16) we have

$$\mathscr{F} = \frac{\pi\sqrt{\mathscr{R}}}{1-\mathscr{R}}. \tag{35}$$

From (27) the peak-transmission is

$$\tau = \left(1 - \frac{\mathscr{A}}{1-\mathscr{R}}\right)^2; \tag{36}$$

and also from (22) and (16), the contrast factor is

$$\mathscr{C} = 1 + F = \left(\frac{1+\mathscr{R}}{1-\mathscr{R}}\right)^2 = 1 + \frac{4\mathscr{F}^2}{\pi^2}. \tag{37}$$

As we have mentioned in § 7.6.1, the plate coatings may be either metal films, usually of silver or aluminium, or dielectric films, formed of alternate layers each $\lambda_0/4$ in optical thickness, of high and low refractive index media—for example, zinc sulphide and cryolite.* Both kinds of film are prepared by thermal evaporation in vacuum. At a given wavelength, $\mathscr{R}$ in general increases with increasing film thickness, in the case of metals, and with increasing number of layers, in the case of dielectrics (cf. Table III, § 1.6). However, it is found with both types of coating that at the high values of $\mathscr{R}$ which are of practical interest, an increase of $\mathscr{R}$ is accompanied by a increase of $\mathscr{A}/(1-\mathscr{R})$, i.e. from (35) and (36), τ decreases with increasing $\mathscr{F}$. Thus high values of both peak-transmission and finesse (or contrast factor) are incompatible requirements, and in practice some compromise must be made between them. Figure 7.61 shows values of τ and $\mathscr{F}$ derived from measurements on typical films. We see that, except in the extreme red, the highest values of finesse for given peak-transmission are obtained with dielectric films. We must note, however, that the reflection coefficient of a multilayer dielectric film is high only in a limited wavelength region about the wavelength λ_0 for which the layers have optical thickness $\lambda_0/4$, so that dielectric coatings are unsuited to applications where a single interferometer is to be used with light covering a wide range of wavelengths. At wavelengths in the visible spectrum, silver gives sharper fringes for given peak-transmission than aluminium, but below about 4000 Å the converse is true, and aluminium coatings have been used for work in the ultra-violet as far as 2000 Å.

The discussion above refers to an ideal interferometer in which the reflecting

* Combinations of metal films and dielectric films have also been used: see C. DUFOUR, *Ann. de Physique* (12), **6** (1951), 5.

surfaces are perfectly plane and parallel.* In practice the surfaces of interferometer plates cannot be worked perfectly plane, and in consequence the plate separation h always varies over the aperture. The effect of such variations has been considered by DUFOUR and PICCA† and by CHABBAL.‡ They have shown that the finesse and peak-transmission are always less than the values given by (35) and (36), and that

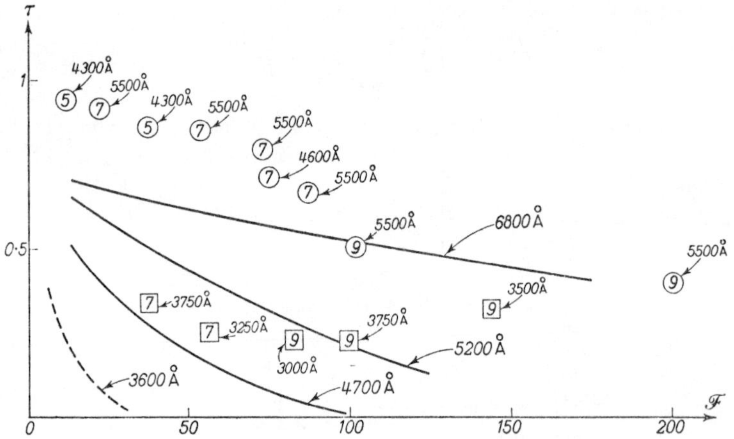

Fig. 7.61. Peak-transmission τ and finesse $\mathscr{F}$ with various reflecting coatings, assuming perfectly plane and parallel surfaces.

— *Fresh silver films*

(From H. KUHN and B. A. WILSON, *Proc. Phys. Soc.* B, **63** (1950), 745.)

--- *Aluminium films*

(From J. C. BURRIDGE, H. KUHN and A. PERY, *Proc. Phys. Soc.* B, **66** (1953), 963.)

○ *Zinc sulphide–cryolite films*

(From P. GIACOMO, *Rev. d'Opt.*, **35** (1956), 317.
J. RING and W. L. WILCOCK, *Nature*, **171** (1953), 648; *ibid.* **173** (1954), 994.)

□ *Lead chloride–magnesium fluoride films*

(From S. PENSELIN and A. STEUDEL, *Z. Phys.*, **142** (1955), 21.

The number of layers and the optimum wavelength for the dielectric films are indicated.

as $\mathscr{R} \to 1$, the finesse approaches a limit $\mathscr{F}_d$ which depends only on the defects of the plates, i.e. with given plates there is an upper limit of fringe sharpness which cannot be exceeded whatever the reflectivity of the coatings. The value of $\mathscr{F}_d$ depends on the form and magnitude of the departure from plane parallelism. For the particular case where the defect is a slight spherical curvature of the plates, such that $n'h$ changes by λ_0/q between centre and edge of the interferometer aperture, $\mathscr{F}_d = q/2$. As an example we may take the case of an interferometer with $q = 100$ (i.e. $\mathscr{F}_d = 50$), coated with fresh silver films and used at $\lambda \sim 5200$ Å. The finesse and peak-transmission, compared with values taken from Fig. 7.61, are shown in

* A modified version of the FABRY–PEROT interferometer in which the plane mirrors are replaced by spherical ones of equal radii of curvature and with coincident foci was described by P. CONNES, *Rev. d'Opt.*, **35** (1956), 37; *Journ. de Phys.*, **19** (1958), 262. Interferometers of this type are used, for example, as resonators for optical masers. (Cf. A. G. FOX and T. LI, *Bell Tech. J.*, **40** (1961), 453, and G. D. BOYD and J. P. GORDON, *ibid*, **40** (1961), 489.)

† C. DUFOUR and R. PICCA, *Rev. d'Opt.*, **24** (1945), 19.

‡ R. CHABBAL, *J. Rech. Cent. Nat. Rech. Sci.*, Labs. Bellevue (Paris), No. 24 (1953), 138.

Table XV. We see that when $\mathscr{F} \gtrsim \mathscr{F}_d$, increase of reflectivity results in a reduction of peak-transmission with little compensating increase of the finesse. The example also illustrates the high degree of flatness of the plates that would be required to take full advantage of attainable reflectivities.

TABLE XV

Fabry–Perot interferometer with fresh silver films, used at wavelength 5200 Å. *Finesse* $\mathscr{F}$ *and peak-transmission* τ *for* (*a*) *plane plates and* (*b*) *spherically curved plates such that nh differs by* 52 Å *between centre and edge.*

The values for (*a*) are from Fig. 7.61. The values for (*b*) were obtained from theoretical curves given by R. CHABBAL, *J. Rech. Cent. Nat. Rech. Sci.*, Labs. Bellevue (Paris), No. 24 (1953), 138.

(*a*) Plane plates		(*b*) Spherically curved plates ($\mathscr{F}_d = 50$)	
$\mathscr{F}$	τ	$\mathscr{F}$	τ
25	0·59	22	0·55
50	0·44	36	0·34
75	0·30	42	0·20
100	0·20	45	0·11
125	0·13	46	0·06

7.6.3 The application of the Fabry–Perot interferometer to the study of the fine structure of spectral lines

When a FABRY–PEROT interferometer is illuminated by quasi-monochromatic light which does not satisfy the condition (23), the form of the intensity distribution of the transmitted light differs from that given by (27), and yields some information about the spectral distribution of the light used. In particular, suppose the light has two monochromatic components. If we imagine that their wavelength difference is gradually increased, and providing they do not differ too greatly in intensity, their presence will eventually be evident from the presence of two mutually displaced sets of maxima in the interference pattern. The components are then said to be *resolved* by the interferometer. In this way FABRY and PEROT* were able to observe directly the fine structure of spectral lines which MICHELSON could only infer (§ 7.5.8), and the FABRY–PEROT interferometer has since played a dominant role in this branch of spectroscopy.

In order to compare the powers of different instruments to resolve spectral structure, it is convenient to consider the case when the two components are of equal intensity, and to fix somewhat arbitrarily a displacement of maxima at which the components may be said to be "just resolved." If $\lambda_0 \pm \frac{1}{2}\Delta\lambda_0$ are the wavelengths of the two components, the quantity $\lambda_0/\Delta\lambda_0$ is called the *resolving power* of the instrument. Such a criterion of resolution was first introduced by Lord RAYLEIGH,† in connection with prism and grating spectroscopes, where the intensity distribution for monochromatic light is of the form $I(\delta) = \{\sin(\delta/2)/(\delta/2)\}^2 I_{max}$. RAYLEIGH proposed that, in this case, two components of equal intensity should be considered to be just resolved when the principal intensity maximum of one coincides with the first intensity minimum of the other (Fig. 7.62); in the combined distribution, the ratio of the intensity at the mid-point to that at the maxima is then $8/\pi^2 = 0{\cdot}811$.

Let us adopt this saddle-to-peak intensity ratio as a criterion of resolution in the

* C. FABRY and A. PEROT, *Ann. Chim. Phys.* (7), **16** (1899), 115.

† LORD RAYLEIGH, *Phil. Mag.* (5), **8** (1879), 261.

case of the FABRY–PEROT interferometer.* If the reflectivity of the plates is not so high that their imperfections are important, the intensity distribution $I(\delta)$ due to a single monochromatic component may be written, from (27), in the form

$$I(\delta) = \frac{I_0}{1 + F \sin^2 \dfrac{\delta}{2}}; \tag{38}$$

and the total intensity $I_{tot}(\delta, \varepsilon)$ resulting from the superposition of two such components, whose relative displacement corresponds to a change of δ by an amount ε, is

$$\begin{aligned} I_{tot}(\delta, \varepsilon) &= I(\delta + \tfrac{1}{2}\varepsilon) + I(\delta - \tfrac{1}{2}\varepsilon) \\ &= \frac{I_0}{1 + F \sin^2 \dfrac{(\delta + \frac{1}{2}\varepsilon)}{2}} + \frac{I_0}{1 + F \sin^2 \dfrac{(\delta - \frac{1}{2}\varepsilon)}{2}}. \end{aligned} \tag{39}$$

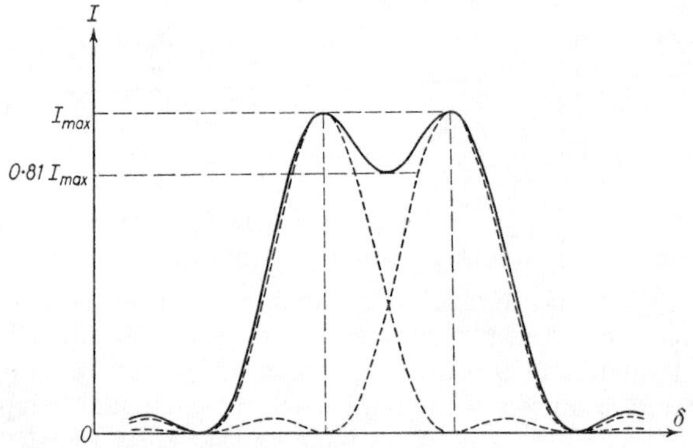

Fig. 7.62. Illustrating two monochromatic spectral components just resolved according to the RAYLEIGH criterion.

The total intensity at the point mid-way between the intensity maxima of the two components is $I_{tot}(2m\pi, \varepsilon)$ where m is an integer; and if, when the components are just resolved, we take the maxima of total intensity to coincide with the intensity maxima of the components, the total intensity at the maxima is $I_{tot}(2m\pi \pm \frac{1}{2}\,\varepsilon, \varepsilon)$. Hence, with the criterion we have chosen, the two lines are just resolved when ε is such that

$$\frac{2I_0}{1 + F \sin^2 \dfrac{\varepsilon}{4}} = 0{\cdot}81 \left\{ I_0 + \frac{I_0}{1 + F \sin^2 \dfrac{\varepsilon}{2}} \right\}. \tag{40}$$

If the finesse of the fringes is high, ε in (40) is small compared with $\pi/2$ and we may put $\sin \varepsilon = \varepsilon$; (40) then reduces to

$$F^2 \varepsilon^4 - 15{\cdot}5 F \varepsilon^2 - 30 = 0,$$

which gives

$$\varepsilon = \frac{4{\cdot}15}{\sqrt{F}} = \frac{2{\cdot}07\pi}{\mathscr{F}} \tag{41}$$

because of (22).

* No special physical significance is to be attached to the RAYLEIGH criterion, and from time to time other criteria of resolution have been proposed. For purposes of comparison of different instruments the particular choice of criterion is of little importance.

Now from (28), if we take n' to be independent of wavelength, and if h is so large that we can neglect ϕ compared with δ, we have

$$|\Delta\delta| = \frac{4\pi n' h \cos\theta'}{\lambda_0^2}\,\Delta\lambda_0 = 2\pi m\,\frac{\Delta\lambda_0}{\lambda_0}. \tag{42}$$

At the limit of resolution, $|\Delta\delta|$ is equal to the value ε given by (41), so that the resolving power of the interferometer is

$$\frac{\lambda_0}{\Delta\lambda_0} = 0{\cdot}97 m\mathscr{F}. \tag{43}$$

By analogy with the expression (14) of § 8.6 for the resolving power of a diffraction grating, where one is concerned with a *finite* number of interfering beams of *equal* intensity, the factor $0{\cdot}97\mathscr{F}$ is sometimes called the *effective number of beams*. Near normal incidence, $m \sim 2n'h/\lambda_0$ and we may take the resolving power to be

$$\frac{\lambda_0}{\Delta\lambda_0} \sim \frac{2\mathscr{F} n' h}{\lambda_0}. \tag{44}$$

Thus the resolving power of the interferometer is proportional to the finesse and to the optical separation of the plates. As an example we may take $\mathscr{F} = 30$ ($\mathscr{R} \sim 0{\cdot}9$), a value which can easily be attained in the visible spectrum; with $n'h = 4$ mm, the resolving power for $\lambda_0 = 5000$ Å is then about 5×10^5, which is of the order of the resolving power attainable with the largest line gratings. For spectroscopic purposes, it is also convenient to use the interval $\Delta\kappa_0$ of spectroscopic wave number ($\kappa_0 = 1/\lambda_0$) corresponding to the least resolvable wavelength difference $\Delta\lambda_0$:

$$\Delta\kappa_0 = \frac{\Delta\lambda_0}{\lambda_0^2} \sim \frac{1}{2\mathscr{F} n' h}. \tag{45}$$

The quantity $\Delta\kappa_0$ of (45) is sometimes called the *resolving limit* of the interferometer. For the above example, the resolving limit is about $0{\cdot}04$ cm^{-1}.

If the wavelength separation of two components is sufficiently large, the displacement between the two patterns is greater than the distance between adjacent maxima of either, and there is said to be "overlapping" of orders. The wavelength difference $(\Delta\lambda_0)_{S.R.}$ corresponding to a displacement of one order ($|\Delta\delta| = 2\pi$) is called the *spectral range* of the interferometer; near normal incidence, from (42),

$$(\Delta\lambda_0)_{S.R.} = \frac{\lambda_0}{m} \sim \frac{\lambda_0^2}{2n'h}. \tag{46}$$

In terms of spectroscopic wave-number, the spectral range is

$$(\Delta\kappa_0)_{S.R.} = \frac{(\Delta\lambda_0)_{S.R.}}{\lambda_0^2} \sim \frac{1}{2n'h}. \tag{47}$$

It is seen that the spectral range is inversely proportional to the plate separation, so that an increase of resolving power obtained by increasing the plate separation is accompanied by a proportionate reduction of the spectral range. Comparing (44) and (46), we see also that the spectral range $(\Delta\lambda_0)_{S.R.}$ is approximately $\mathscr{F}$ times the least resolvable wavelength difference. With the values of $\mathscr{F}$ that can be reached in practice, the spectral range is very small when the resolving power is high. For the typical case considered above ($\mathscr{F} = 30, n'h = 4$ mm, $\lambda_0 = 5000$ Å), $(\Delta\lambda_0)_{S.R.} \sim 0{\cdot}3$ Å. In § 7.6.8 we shall see that a greater spectral range can be obtained by using two FABRY–PEROT interferometers in series. Nevertheless, when complex spectra are to be examined, it is necessary to separate the patterns of all but the closest lines by means of auxiliary apparatus, as we shall describe later.

When two close components are resolved, the interval of spectroscopic wave-number between them is easily determined from measurements of the ring diameters and the plate separation. From (29) and (30), for spectroscopic wave numbers κ_0, κ_0', we may write

$$m_1 + e_1 = 2n'h\kappa_0 + \frac{\phi}{\pi}, \qquad m_1' + e_1' = 2n'h\kappa_0' + \frac{\phi}{\pi}, \tag{48}$$

where m_1, m_1' are the integral orders of the first bright rings, e_1, e_1' are the fractional orders at the centre, and we have assumed ϕ to be constant over the interval from κ_0 to κ_0'. Hence, by subtraction

$$\kappa_0' - \kappa_0 = \frac{(m_1' - m_1) + (e_1' - e_1)}{2n'h}. \tag{49}$$

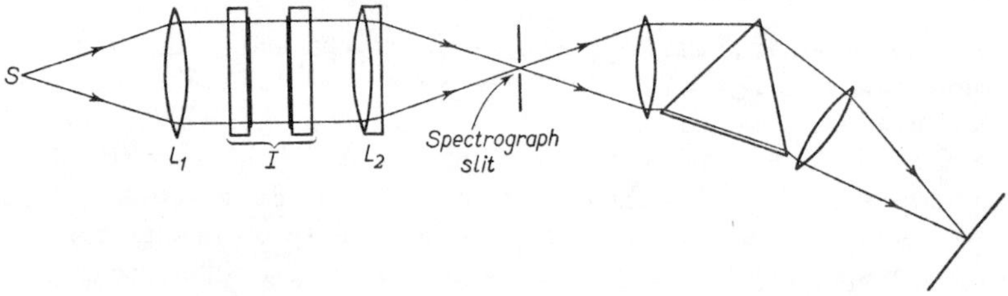

Fig. 7.63. A FABRY–PEROT interferometer crossed with a prism spectrograph.

For each line, the fractional orders may be obtained from measurements of the diameters of two bright rings. Thus from (32) the diameters D_p, D_q of the pth and qth rings are related by

$$\frac{D_p^2}{D_q^2} = \frac{p - 1 + e}{q - 1 + e},$$

whence

$$e = \frac{(q-1)D_p^2 - (p-1)D_q^2}{D_q^2 - D_p^2}. \tag{50}$$

If measurements are made on more than two rings, a mean value of e can be found by the least squares method. The value of the integer $(m_1' - m_1)$, which is the number of orders of overlap between the two patterns, is decided from observations of the change in the patterns when the plate separation is reduced. Usually the accuracy of the measured order separation $(m_1' - m_1) + (e_1' - e_1)$ does not exceed 1 part in 10^3; since plate separations less than about 1 mm are rarely used, h is then obtained accurately enough by means of a micrometer, and n' for air can be taken as unity.

We have already mentioned that because of the small spectral range of the interferometer, the patterns from all but very close lines must be separated. When the observations are photographic, this separation is usually achieved by "crossing" the interferometer with a stigmatic prism or grating spectrograph. One arrangement is shown in Fig. 7.63. The interferometer I is illuminated with light from the source S in the focal plane of the lens L_1, and the interference pattern is projected on to the plane of the spectrograph slit by means of the well-corrected lens L_2.* The orientation of the interferometer is adjusted so that the centre of the pattern coincides with the centre of the slit. The slit then selects a diametrical section from the ring system of each line present, and the action of the spectrograph is to separate these sections. When the spectrum to be examined consists of emission lines, the slit may usually be relatively wide. The appearance in the focal plane of the spectrograph is then as in Fig. 7.64, the images of the slit formed in the light of each line being crossed by short arcs of the bright fringes. A similar appearance is obtained with absorption lines, provided the continuous background associated with each line extends over a wavelength range less than the spectral range of the interferometer. The absorption lines appear as dark lines across the broader bright fringes from the background (Fig. 7.65).

When the continuous background extends over a wide wavelength range, as for example in the solar spectrum, the situation is more complicated.† To understand it, let us suppose for the moment that the width of the spectrograph slit is negligibly small, and that the resolving power of the spectrograph is then infinite. Let Ox, Oy be rectangular coordinate axes in the focal plane of the spectrograph, with origin O on the line defined by light passing through the centre of the slit, and Oy parallel to the slit. Along a line $x =$ constant, the wavelength λ_0 is constant, and the maxima of intensity are in positions determined by (28), which we may write

$$\left(m - \frac{\phi}{\pi}\right) \lambda_0 = 2n'h \cos \theta' = 2n'h \left(1 - \frac{\theta'^2}{2}\right) = 2h \left(1 - \frac{\theta^2}{2}\right), \qquad m = 1, 2, \ldots \quad (51)$$

where we have assumed θ' to be small, and put $n' = n = 1$ for air. The angle θ is the angle of emergence from the interferometer; for θ small, it is related to the coordinate y by $y = Mf\theta$, where f is the focal length of the lens L_2 used to project the fringes on the slit, and M is the magnification introduced by the spectrograph. Further, λ_0 is related to the coordinate x, say $F(x) = \lambda_0$, where the function F is characteristic of the dispersion of the spectrograph. Using these relations to eliminate λ_0 and θ from (51), we obtain the loci of intensity maxima when the source emits a continuous spectrum:

$$\left(m - \frac{\phi}{\pi}\right) F(x) = 2h \left(1 - \frac{y^2}{2M^2f^2}\right), \qquad m = 1, 2, \ldots \quad (52)$$

Between these maxima the relative intensity distribution is characteristic of the interferometer; the spectrum in the focal plane of the spectrograph is thus channelled with narrow bright fringes separated by broader dark intervals. The channels are symmetrical about the line $y = 0$, convex on the side towards longer wavelengths, and for $y =$ constant their wavelength separation is evidently equal to the spectral range of the interferometer. In the special case when $F(x)$ is a linear function of x, the channels are parabolic.

When absorption lines are present, the bright channels are interrupted by dark spots where they intersect the absorption lines, i.e., these dark spots have exactly the same positions as bright maxima would have if they were formed by emission lines of the same wavelength as the absorption lines. If the slit is opened symmetrically,

* Other arrangements of interferometer and spectrograph have been used. The choice of a suitable arrangement for a particular purpose depends on such considerations as compactness and expense, the size, intensity and uniformity of the source, and the importance of "ghost" patterns which arise from unwanted reflections at surfaces in the system. [See S. Tolansky, *High Resolution Spectroscopy*, London, Methuen, 1947, Chapt. 9.]

† It was first discussed by C. Fabry and H. Buisson, *Journ. de Phys.* (4), **9** (1910), 197.

the dark spots extend in the x-direction; at the same time, the bright channels similarly broaden in the x-direction, until eventually they touch, and coalesce. In this condition, there is what appears to be a continuous spectrum containing dark "fringe" patterns from the absorption lines (Fig. 7.66), which may be measured in the same way as bright patterns from emission lines. In practice, because of diffraction, the resolving power of the spectrograph is not infinite and in consequence the light of any given wavelength is initially spread over a finite distance in the x-direction. The effect of this is similar to that of opening the slit, i.e. the bright channels are broadened in the x-direction, and they will overlap unless the spectrograph is able to resolve clearly wavelengths whose separation is that of adjacent bright channels. We have seen that this separation is equal to the spectral range of the interferometer, which decreases with increasing plate separation. The finite resolving power of the auxiliary spectrograph thus sets an upper limit to the plate separation that can be employed.

The Fabry–Perot interferometer may also be used for spectroscopy in conjunction with photoelectric detectors.* Light of the spectral line to be examined, after isolation by means of a preliminary monochromator, is fed to the interferometer, and the interference pattern is projected on to an annular aperture concentric with the rings, which allows the light in a small fraction of an order to reach the photocell. The pattern is explored by varying the optical separation of the interferometer plates, which causes the rings to expand or contract over the aperture. The arrangement has considerable technical importance because, as Jacquinot† has shown, the Fabry–Perot interferometer can pass much greater light flux for given resolving power than conventional prism and grating monochromators.

7.6.4 The application of the Fabry–Perot interferometer to the comparison of wavelengths

Measurements of wavelengths in spectra produced by prism and grating spectrographs are carried out by interpolation, the wavelength of any line being expressed in terms of the known wavelengths of neighbouring standard lines. Since with large spectrographs such comparisons can be made with an accuracy of about 1 part in 10^6, it is necessary to know the relative wavelengths of the standard lines with an accuracy at least equal to this, and they must be sufficiently numerous to allow convenient interpolation. Further, since wavelengths are actual lengths, it is desirable to refer them to the metre. The establishment of a satisfactory system of wavelengths for spectroscopic purposes thus involves (*a*) the comparison of the wavelength of a chosen primary standard line with the material standard which represents the metre, and (*b*) the comparison of the wavelength of the primary standard with the wavelengths of other lines which are to serve as secondary standards throughout the spectrum. The primary standard, chosen in 1907,‡ is the red line (6438 Å) of cadmium, excited under prescribed conditions.§ The measurement of the wavelength of this line in terms of the metre is a metrological process requiring special methods, which we shall describe in § 7.7. The value found is $6438{\cdot}4696 \times 10^{-10}$ metre, and is taken to be 6438·4696 Å. This is the definition of the Ångström; it does not differ from 10^{-10} metre by more than 1 part in 3×10^6. With the primary standard of wave-

* P. Jacquinot and C. Dufour, *J. Rech. Cent. Nat. Rech. Sci.*, Labs. Bellevue (Paris), No. 6 (1948), 91.

† P. Jacquinot, *J. Opt. Soc. Amer.*, **44** (1954), 761.

‡ *Trans. Int. Union for Co-operation in Solar Research*, **2** (Manchester Univ. Press, 1908), p. 17.

§ *Proc. Verb. Com. Int. Poids et Mes.* (2), **17** (1935), 91.

length so defined, the comparison of other wavelengths with it is a purely optical process which has been one of the important applications of the FABRY–PEROT interferometer.

The interferometer is illuminated (preferably simultaneously) with light of the primary standard and unknown lines, and the patterns are separated and photographed by means of a suitable arrangement. Let λ_s be the wavelength of the standard line in the air of refractive index n' between the interferometer plates, and similarly let $\lambda_1, \lambda_2, \ldots$ be the wavelengths of the unknown lines. Then from (29) and (30), if we neglect the variation of $\phi\lambda$ with wavelength, we have

$$(m_{1s} + e_s)\lambda_s = (m_{11} + e_1)\lambda_1 = (m_{12} + e_2)\lambda_2 = \ldots\ldots = 2H, \tag{53a}$$

$$2H = 2h + \frac{\phi\lambda}{\pi}, \tag{53b}$$

where $m_{1s}, m_{11}, m_{12}, \ldots$ are the integral orders of the first bright rings, and e_s, e_1, $e_2, \ldots$ are the fractional orders at the centre. For each line, the fractional orders may be obtained from measurements of the ring diameters, as we have already described (p. 337). The integers $m_{1s}, m_{11}, m_{12}, \ldots$ are found by the method of excess fractions, to which we referred briefly on p. 291. For this, the starting point is a measurement of the separation h of the plates by means of a micrometer. The integer m'_{1s} nearest $2h/\lambda_s$ is then an approximate value of m_{1s}, and we may write

$$m_{1s} = m'_{1s} + x, \tag{54}$$

where x is an unknown integer. If the uncertainty of the measurement of h is Δh, then

$$|x| \lesssim \frac{2\Delta h}{\lambda_s}. \tag{55}$$

For example, with $\Delta h \sim 0{\cdot}01$ mm and $\lambda_s = 6438$ Å, $|x| \lesssim 30$. Next, with known approximate values $\lambda'_1, \lambda'_2, \ldots$ of $\lambda_1, \lambda_2, \ldots$, the approximate orders of interference for the unknown lines corresponding to order $(m'_{1s} + e_s)$ of λ_s are calculated from (53a) as

$$\left.\begin{aligned} m'_{11} + e'_1 &= (m'_{1s} + e_s)\frac{\lambda_s}{\lambda'_1}, \\ m'_{12} + e'_2 &= (m'_{1s} + e_s)\frac{\lambda_s}{\lambda'_2}, \quad \ldots \end{aligned}\right\} \tag{56}$$

If we put $\lambda_1 = \lambda'_1 + \Delta\lambda'_1$, we have, from (53a) and (54),

$$\begin{aligned} (m_{11} + e_1)(\lambda'_1 + \Delta\lambda'_1) &= (m_{1s} + e_s)\lambda_s \\ &= (m'_{1s} + x + e_s)\lambda_s, \end{aligned} \tag{57}$$

whence, using (56)

$$m_{11} + e_1 = m'_{11} + x + \left[e'_1 + \left(\frac{\lambda_s - \lambda'_1}{\lambda'_1}\right)x\right] - (m_{11} + e_1)\frac{\Delta\lambda'_1}{\lambda'_1}. \tag{58}$$

The fractional part contributed by the expression in square brackets on the R.H.S. of (58) is now calculated for the possible values of x allowed by (55). For the correct x, this calculated fraction must agree with the measured fraction e_1 on the L.H.S. of (58) within the uncertainty $(m_{11} + e_1)\Delta\lambda'_1/\lambda'_1$, which is small compared with unity if the plate separation is sufficiently small; for example, with approximate wavelengths obtained from grating measurements, $\Delta\lambda'_1/\lambda'_1 \sim 10^{-5}$, so that $(m_{11} + e_1)\Delta\lambda'_1/\lambda'_1 \sim 0{\cdot}1$ for $m_{11} \sim 10^4$, which corresponds to $h \sim m_{11}\lambda_1/2 \sim 3$ mm for wavelengths in the visible spectrum. The comparison of calculated and measured fractions then shows some values of x to be inadmissible, and similar calculations with approximate wavelengths λ_2', λ_3', . . . enable x to be determined unambiguously. Usually three lines suffice, if their wavelengths are suitably spaced, with perhaps a fourth to serve as a check.

With x known, m_{1s}, m_{11}, m_{12}, ... are given by (54) and (58), and from (53a) we have

$$\lambda_1 = \frac{m_{1s} + e_s}{m_{11} + e_1}\lambda_s, \qquad \lambda_2 = \frac{m_{1s} + e_s}{m_{12} + e_2}\lambda_s, \ldots. \tag{59}$$

Since the integral orders are not in doubt, the uncertainties $\Delta\lambda_1$, $\Delta\lambda_2$, . . . of these wavelengths derive only from the uncertainties of the measured fractions e_s, e_1, e_2, ... If the latter are within $\pm$ 0·01, a precision which is not difficult to attain with high reflectivity coatings, and with m_{1s}, m_{11}, m_{12}, . . . of the order of 10^4 as in the example above, $\Delta\lambda_1/\lambda_1$, $\Delta\lambda_2/\lambda_2$, . . . are of the order of 10^{-6}, i.e. the measurements have increased the accuracy with which λ_1, λ_2, . . . are known by a factor of the order of ten. The improved wavelengths may now be used in a repetition of the experiment with larger plate separation, which in turn yields still more accurate wavelength values; and the process may be continued as far as is justified by the sharpness of the lines to be measured.

We have assumed above that, apart from the primary standard, only relatively inaccurate values of wavelength are available.* Once a number of secondary standards have been measured with an accuracy approaching that with which the primary standard is defined, the determination of further wavelengths is simplified. The standard wavelengths are used with the method of excess fractions† to determine $2H$, which need only be small enough to allow m_{11}, m_{12}, . . . to be determined unambiguously from the approximate wavelengths λ_1', λ_2', ... Thus from (53a) and (57),

$$m_{11} + e_1 = \frac{2H}{\lambda_1'} - (m_{11} + e_1)\frac{\Delta\lambda_1'}{\lambda_1'}, \tag{60}$$

and m_{11} is obtained unambiguously if $(m_{11} + e_1)\Delta\lambda_1'/\lambda_1' \sim 0{\cdot}3$ say. If as before $\Delta\lambda_1'/\lambda_1' \sim 10^{-5}$, this implies $m_{11} \sim 3 \times 10^4$, which corresponds to $H \sim 10$ mm for wavelengths in the visible spectrum; and with the measured fraction e_1 within $\pm$ 0·01, the uncertainty $\Delta\lambda_1/\lambda_1$ of the wavelength $\lambda_1 = 2H/(m_{11} + e_1)$ is about 3×10^{-7}, i.e. the single application of the interferometer increases the accuracy by a factor of about thirty.

In practice, the assumption that $\phi\lambda$ is independent of wavelength may not be justified if the lines to be compared are widely separated in the spectrum. Systematic errors from this cause can be eliminated if measurements are made with two values

* For their earliest comparisons of wavelengths with that of the red cadmium line, FABRY and PEROT (*Ann. Chim. Phys.*, **16** (1899), 289) employed the sliding plate form of their interferometer. They made use of accurate values of the wavelengths of the green (5086 Å) and blue (4800 Å) cadmium lines obtained by two-beam interferometry (A. A. MICHELSON and J. R. BENOIT, *Trav. et Mem. Bur. Int. Poids et Mes.*, **11** (1895), 1), and determined integral orders by the so-called "*method of coincidences*". In principle this is the same as the method of excess fractions we have described, but it requires visual observation and is laborious.

† If the standard wavelengths are sufficiently close together, it is possible to calculate quickly an approximate value of x, and so reduce the amount of computation (see, for example, C. V. JACKSON, *Phil. Trans. Roy. Soc.* A., **236** (1936), 1). We may mention also that G. R. HARRISON (*J. Opt. Soc. Amer.*, **36** (1946), 644) has described a machine by means of which wavelengths are determined directly from interferometer patterns, the necessary computations being carried out automatically.

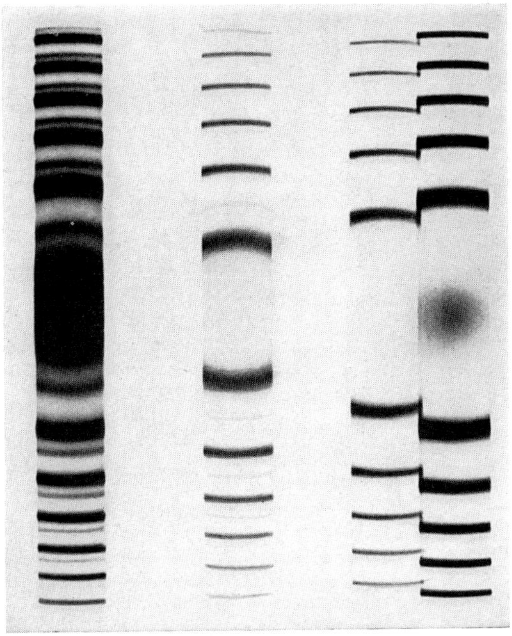

Fig. 7.64. Fabry–Perot fringes from lines in the emission spectrum of helium (photographic negative).

(After K. W. Meissner, *J. Opt. Soc. Amer.*, **31** (1941), 416.)

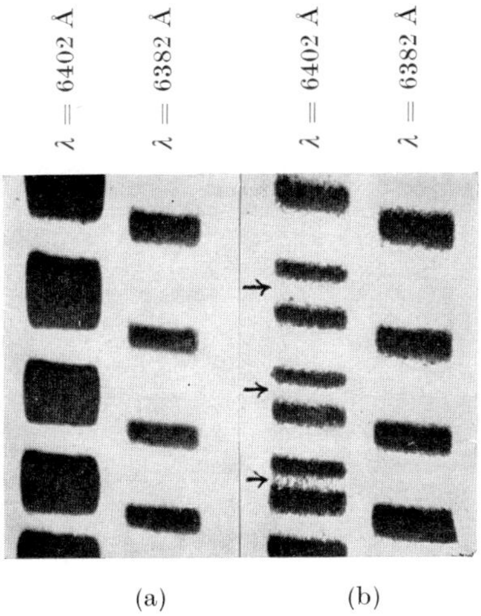

Fig. 7.65. Fabry–Perot fringes from two lines in the spectrum of neon (photographic negative): (a) emission only; (b) central absorption of the line of wavelength 6402 Å, indicated by arrows.

(After K. W. Meissner, *J. Opt. Soc. Amer.*, **32** (1942), 191.)

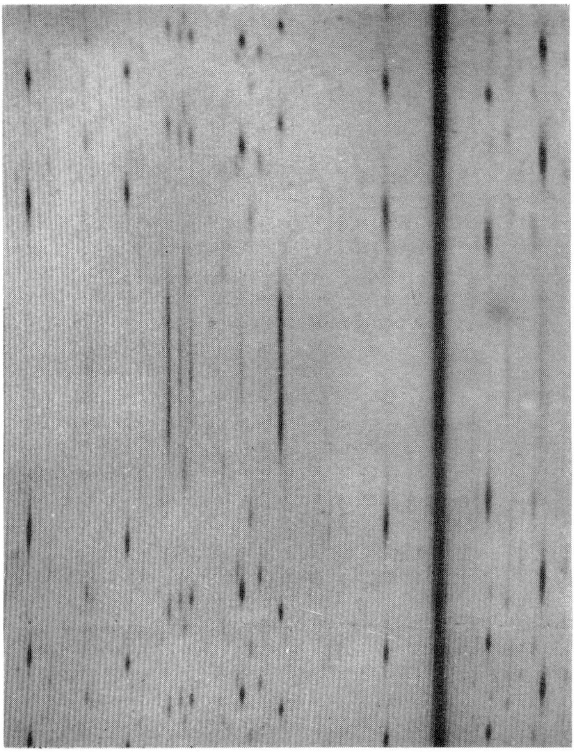

Fig. 7.66. FABRY–PEROT fringes from absorption lines in the solar spectrum. (From H. D. BABCOCK, *Astrophys. J.*, **65** (1927), 140.)

$h_{\rm I}$, $h_{\rm II}$ of h. For then, from (29) and (30), with the same notation as in (53), we may write

$$\left\{(m_{1s} + e_s)_{\rm I} - \frac{\phi_s}{\pi}\right\}\lambda_s = \left\{(m_{11} + e_1)_{\rm I} - \frac{\phi_1}{\pi}\right\}\lambda_1 = \ldots = 2h_{\rm I},$$

$$\left\{(m_{1s} + e_s)_{\rm II} - \frac{\phi_s}{\pi}\right\}\lambda_s = \left\{(m_{11} + e_1)_{\rm II} - \frac{\phi_1}{\pi}\right\}\lambda_1 = \ldots = 2h_{\rm II},$$

whence

$$\lambda_1 = \frac{(m_{1s} + e_s)_{\rm II} - (m_{1s} + e_s)_{\rm I}}{(m_{11} + e_1)_{\rm II} - (m_{11} + e_1)_{\rm I}}\,\lambda_s. \tag{61}$$

The wavelengths obtained after correction for phase change must be reduced to their values under the standard conditions of dry air at 15°C and 760 mm Hg pressure for which the primary standard is defined. For this reduction it is necessary to know the dispersion of air. If the wavelengths are to be further reduced to values in vacuum, the refractive index of air must also be known. Values of the refractive index and dispersion of air sufficiently accurate for this purpose have been obtained by multiple-beam methods (§ 7.6.8).

7.6.5 The Lummer–Gehrcke interferometer

We have already mentioned in § 7.6.1 that the reflection coefficient of a dielectric interface approaches unity when the angle of incidence in the denser medium nears the critical angle. This effect is used in the multiple beam interferometer devised by LUMMER and GEHRCKE,* which consists essentially of a long plane parallel plate of glass or crystalline quartz (Fig. 7.67). A beam of light from a source on the long axis of the plate is admitted through the prism P fixed to one end of the plate, and meets the plate surfaces at an angle slightly smaller than the critical angle. A series of beams then leaves each side of the plate near grazing emergence, and these are collected by the lens L to form an interference pattern in its focal plane.

For monochromatic light of wavelength λ_0, the phase difference δ between successive beams is given by (1), viz.

$$\delta = \frac{4\pi}{\lambda_0} n'h\cos\theta' = \frac{4\pi h}{\lambda_0}\sqrt{n'^2 - n^2\sin^2\theta}, \tag{62}$$

where h and n' are the thickness and refractive index of the plate, n is the refractive index of the surrounding air, θ' is the angle of reflection in the plate and θ is the angle of emergence. The corresponding order of interference m is

$$m = \frac{\delta}{2\pi} = \frac{2h}{\lambda_0}\sqrt{n'^2 - n^2\sin^2\theta}. \tag{63}$$

The fringes, corresponding to θ constant, form a family of hyperbolae, and near the centre O of the pattern they approximate to straight lines parallel to the plate surfaces. From (63), the order of interference m_0 at the centre ($\theta = \pi/2$) is

$$m_0 = \frac{2h}{\lambda_0}\sqrt{n'^2 - n^2}, \tag{64}$$

---

* O. LUMMER, *Verh. Deutsch. Phys. Ges.*, **3** (1901), 85.
O. LUMMER and E. GEHRCKE, *Ann. d. Physik*, (4), **10** (1903), 457.

and,

$$m - m_0 = \frac{2h}{\lambda_0}\left(\sqrt{n'^2 - n^2 \sin^2 \theta} - \sqrt{n'^2 - n^2}\right). \tag{65}$$

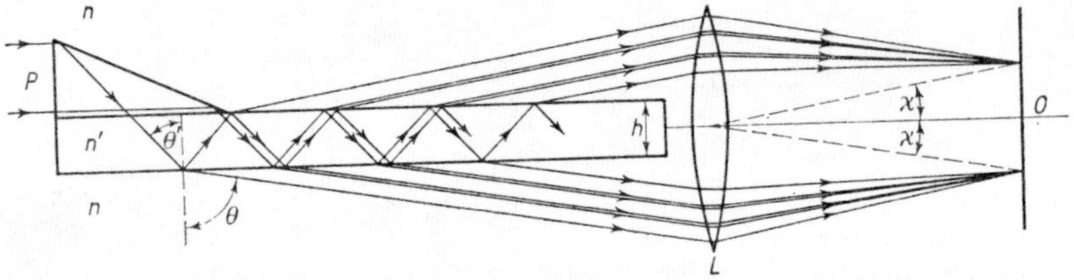

Fig. 7.67. The LUMMER–GEHRCKE interferometer.

If we put $\chi = (\pi/2) - \theta$, χ is small near the centre, so that we may take $\sin^2 \theta = 1 - \chi^2$; then after expanding (65), and neglecting powers of χ higher than the second, we obtain

$$m - m_0 = \frac{n^2}{\sqrt{n'^2 - n^2}} \frac{h\chi^2}{\lambda_0}. \tag{66}$$

The angle χ_q corresponding to the qth bright fringe from the centre, of integral order m_q, is given by

$$\chi_q = \frac{\sqrt[4]{n'^2 - n^2}}{n} \sqrt{\frac{\lambda_0}{h}} \sqrt{m_q - m_0} = \frac{\sqrt[4]{n'^2 - n^2}}{n} \sqrt{\frac{\lambda_0}{h}} \sqrt{q - e}, \tag{67}$$

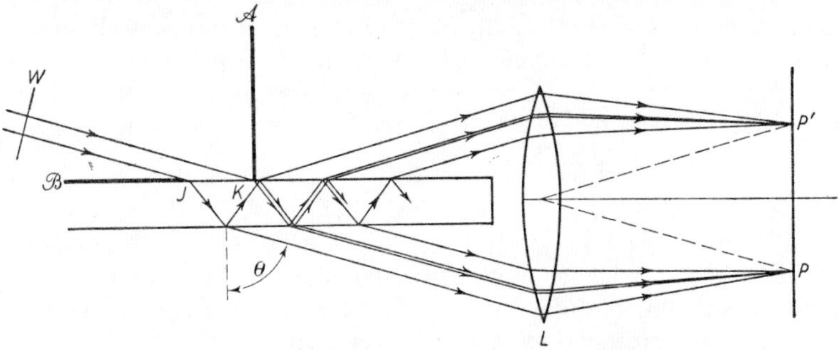

Fig. 7.68. Illustrating multiple reflections in a plane parallel plate when the externally reflected beam is suppressed.

where e is the fractional order at the centre. The angular scale of the pattern is thus proportional to $\sqrt{\lambda_0/h}$, and if $e = 0$ the distances of the bright fringes from the axis are proportional to the square roots of the positive integers.

The intensity distribution in the pattern differs from that considered in § 7.6.1 because there is now no contribution from light externally reflected at the plate surfaces. To simplify discussion of this point it is convenient to consider the arrangement of Fig. 7.68, in which light is refracted into the plate through an aperture JK bounded by the diaphragm $\mathscr{A}\mathscr{B}$. It is evident that, as far as the action of the plate is concerned, the prism of Fig. 7.67 behaves as such an aperture, but with negligible reflection coefficient.* Suppose P and P' are the focal points of L for light emerging from the lower and upper surfaces of the plate at angle θ, and let A be the complex amplitude of the light at P when the plate is removed. The complex amplitudes at P of the beams of the lower set are (cf. § 7.6.1), apart from an unimportant constant phase factor,

$$tt'A, \quad tt'r'^2Ae^{i\delta}, \ldots\ldots \quad tt'r'^{2(p-1)}Ae^{i(p-1)\delta},$$

and the amplitude at P resulting from the superposition of the first p beams is therefore

$$A^{(t)}(p) = (1 + r'^2e^{i\delta} + \ldots\ldots + r'^{2(p-1)}e^{i(p-1)\delta})tt'A$$

$$= \frac{1 - \mathscr{R}^p e^{ip\delta}}{1 - \mathscr{R}e^{i\delta}} \mathscr{T} A. \tag{68}$$

If we take the limit as the number of beams $p \to \infty$, we obtain

$$A^{(t)} \equiv A^{(t)}(\infty) = \frac{\mathscr{T}}{1 - \mathscr{R}e^{i\delta}} A, \tag{69}$$

and the intensity $I^{(t)} = A^{(t)}A^{(t)\star}$ at P is then

$$I^{(t)} \equiv I^{(t)}(\infty) = \frac{\mathscr{T}^2}{1 + \mathscr{R}^2 - 2\mathscr{R}\cos\delta} I = \frac{1}{1 + F\sin^2\dfrac{\delta}{2}} I, \tag{70}$$

where $I = AA^\star$ is the intensity at P when the plate is removed. Similarly, the amplitudes at P' of the beams of the upper set are

$$tt'r'A, \quad tt'r'^3e^{i\delta}A, \quad \ldots\ldots\ldots tt'(r')^{2p-1}e^{i(p-1)\delta}A \ldots\ldots\ldots,$$

so that the amplitude at P' resulting from the superposition of the first p beams is

$$A^{(r)}(p) = (1 + r'^2e^{i\delta} + \ldots + r'^{2(p-1)}e^{i(p-1)\delta})tt'r'A$$

$$= \frac{1 - \mathscr{R}^p e^{ip\delta}}{1 - \mathscr{R}e^{i\delta}} \sqrt{\mathscr{R}}\mathscr{T} A; \tag{71}$$

and in the limit as $p \to \infty$ the intensity at P' is

$$I^{(r)} \equiv I^{(r)}(\infty) = \frac{\mathscr{R}\mathscr{T}^2}{1 + \mathscr{R}^2 - 2\mathscr{R}\cos\delta} I = \mathscr{R}I^{(t)}. \tag{72}$$

From (70) and (72) it is evident that the maxima of the patterns above and below the plate are at the same angular positions relative to the plate normal.

Successive beams emerging from one side of the plate are evidently separated by a distance $2h\tan\theta'$ along the plate, so that for a plate of length l, the number of beams p is given by

$$p \sim \frac{l}{2h}\cot\theta' = \frac{l}{2h}\sqrt{\frac{n'^2}{n^2\sin^2\theta} - 1}; \tag{73}$$

---

* A similar arrangement is realized in practice if light is admitted to a FABRY–PEROT interferometer through a clear aperture in one of the reflecting coatings; its properties have been discussed by C. DUFOUR in *Rev. d'Opt.*, **24** (1945), 11.

or, for $\theta \sim \pi/2$, and with $n = 1$ for air,

$$p \sim \frac{l}{2h}\sqrt{n'^2 - 1}. \tag{74}$$

As θ' approaches the critical angle, $\mathscr{R} \to 1$ and $\mathscr{R}^p$ ceases to be negligible, and the use of the limiting expressions (70), (72) with $p \to \infty$ is then not justified. In these

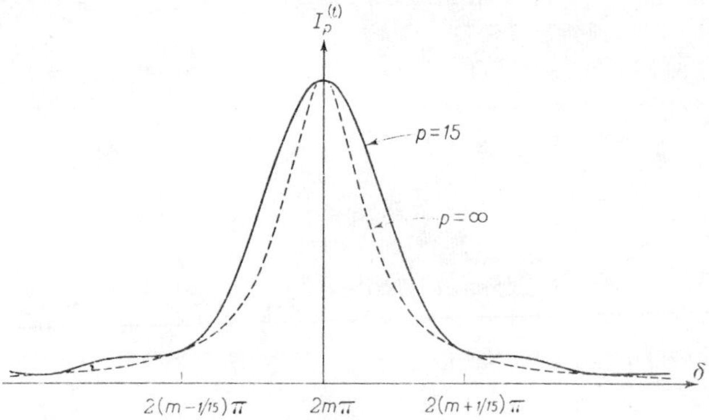

Fig. 7.69. Fringes with a LUMMER–GEHRCKE interferometer: intensity as a function of phase difference ($\mathscr{R} = 0{\cdot}87$).

circumstances the intensity distribution in the upper pattern is obtained from (68) as

$$I^{(t)}(p) = \frac{1 + \mathscr{R}^{2p} - 2\mathscr{R}^p \cos p\delta}{1 + \mathscr{R}^2 - 2\mathscr{R}\cos\delta}\mathscr{T}^2 I$$

$$= \frac{(1 - \mathscr{R}^p)^2 + 4\mathscr{R}^p \sin^2 \dfrac{p\delta}{2}}{(1 - \mathscr{R})^2 + 4\mathscr{R}\sin^2\dfrac{\delta}{2}}\mathscr{T}^2 I$$

$$= \frac{1 + G_p \sin^2 \dfrac{p\delta}{2}}{1 + F\sin^2\dfrac{\delta}{2}}(1 - \mathscr{R}^p)^2 I, \tag{75}$$

where

$$G_p = \frac{4\mathscr{R}^p}{(1 - \mathscr{R}^p)^2}, \tag{76}$$

and F is given by (16). Thus the intensity distribution is governed by the function $(1 + G_p \sin^2 p\delta/2)/(1 + F\sin^2\delta/2)$, which differs from the distribution function for $p \to \infty$ by the addition of the term $(G_p \sin^2 p\delta/2)/(1 + F\sin^2\delta/2)$. This additional term does not affect the positions of the absolute maxima, which remain at $\delta/2 = m\pi$, where m is an integer; but since $(G_p \sin^2 p\delta/2)/(1 + F\sin^2\delta/2)$ is zero for $\delta/2 = (m + q/p)\pi$, $q = 0, 1, 2, \ldots, p$, and is otherwise positive, these absolute maxima are broader than those for $p \to \infty$, and there are secondary maxima between them (Fig. 7.69).

The LUMMER–GEHRCKE interferometer has been employed solely for the examination of the fine structure of spectral lines, for which purpose it is used, like the FABRY–PEROT interferometer, in combination with an auxiliary dispersing instrument. The spectral range of the interferometer may be obtained from (63), which, taking $n = 1$, we may write as

$$m^2\lambda_0^2 = 4h^2(n'^2 - \sin^2\theta). \tag{77}$$

Hence, at a given angular position, the change Δm in order corresponding to a change $\Delta\lambda_0$ in wavelength is, to first order of small quantities,

$$\Delta m = \frac{4h^2 n' \dfrac{dn'}{d\lambda_0} - m^2\lambda_0}{m\lambda_0^2}\Delta\lambda_0. \tag{78}$$

The spectral range which corresponds to a change of unity in m, is therefore

$$(\Delta\lambda_0)_{S.R.} = \frac{m\lambda_0^2}{\left|m^2\lambda_0 - 4h^2n'\dfrac{dn'}{d\lambda_0}\right|}; \tag{79}$$

or, if we substitute for m from (77), and put $\sin\theta \sim 1$ near grazing emergence,

$$(\Delta\lambda_0)_{S.R.} \sim \frac{\lambda_0^2}{2h}\frac{\sqrt{n'^2-1}}{\left|n'^2 - n'\lambda_0\dfrac{dn'}{d\lambda_0} - 1\right|}. \tag{80}$$

We see that $(\Delta\lambda_0)_{S.R.}$ is inversely proportional to the plate thickness. For quartz the factor $\sqrt{n'^2-1}\Big/\left(n'^2 - n'\lambda_0\dfrac{dn'}{d\lambda_0} - 1\right)$ varies from about 0·6 at $\lambda_0 = 2000$ Å to about 0·8 at $\lambda_0 = 6000$ Å.

If we assume $\mathscr{R}$ constant, the points in the fringe pattern where $I^{(t)}(p)$ has fallen to half its maximum values correspond to $\delta = 2m\pi \pm \varepsilon/2$ (m integral), where from (75)

$$\frac{1 + G_p\sin^2\left(\dfrac{p\varepsilon}{4}\right)}{1 + F\sin^2\left(\dfrac{\varepsilon}{4}\right)} = \tfrac{1}{2},$$

i.e.

$$\sin^2\left(\frac{p\varepsilon}{4}\right) - \frac{F}{2G_p}\sin^2\left(\frac{\varepsilon}{4}\right) + \frac{1}{2G_p} = 0. \tag{81}$$

Evidently the finesse $\mathscr{F} = 2\pi/\varepsilon$ depends on both $\mathscr{R}$ and p, and it is shown as a function of $\mathscr{R}$ for several values of p in Fig. 7.70. By (74), p is determined essentially by the ratio l/h. $\mathscr{R}$ is of course a function of θ', and therefore of χ and for given χ, $\mathscr{R}_\perp$ is greater than $\mathscr{R}_\parallel$ (for $n' > n$); it is therefore advantageous to select the polarized component vibrating perpendicular to the plane of incidence. The limiting case $p \to \infty$ corresponds to a plate of infinite length, for which the intensity distribution is similar to that for a FABRY–PEROT interferometer (cf. (70) and (27)); in this case, the finesse is given by (35). When p is finite, Fig. 7.70 shows that $\mathscr{F}$ has the value corresponding to $p = \infty$ so long as $\mathscr{R}$ is not too large. For such values of $\mathscr{R}$, $\mathscr{R}^p$ (and hence G_p) is negligible, and substantially all the light which enters the plate emerges to form the fringes. For larger values of $\mathscr{R}$ the value of $\mathscr{F}$ is smaller than that corresponding to $p = \infty$, and as $\mathscr{R}$ approaches unity, $\mathscr{F}$ approaches a maximum value

$$\mathscr{F}_l = \frac{2\pi}{\varepsilon_l}, \tag{82}$$

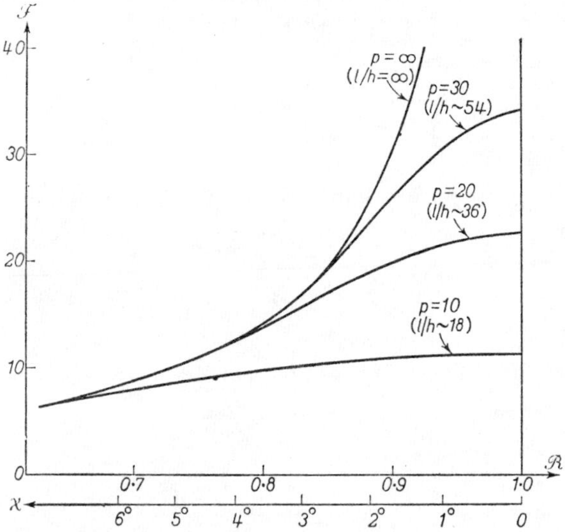

Fig. 7.70. Variation of finesse $\mathscr{F}$ with the reflectivity $\mathscr{R}$ and the number p of reflections for a LUMMER–GEHRCKE interferometer. (For comparison, values of the angle χ from the centre of the pattern, and the ratio of plate length to plate thickness l/h are shown for a quartz plate.)

where the limiting value ε_l can be obtained with the help of (81). Thus setting $\mathscr{R} = 1 - \mathscr{T}$, we have, for $\mathscr{T}$ small, $1 - \mathscr{R}^p \sim p\mathscr{T}$, so that $F/G_p = (1 - \mathscr{R}^p)^2/\mathscr{R}^{p-1}(1 - \mathscr{R})^2 \sim p^2/(1 - \mathscr{T})^{p-1} \to p^2$ as $\mathscr{T} \to 0$; and since $1/G_p \to 0$ as $\mathscr{T} \to 0$, (81) gives

$$\sin^2\left(\frac{p\varepsilon_l}{4}\right) - \frac{p^2}{2}\sin^2\left(\frac{\varepsilon_l}{4}\right) = 0. \tag{83}$$

Hence for ε_l small,

$$\frac{\sin\dfrac{p\varepsilon_l}{4}}{\dfrac{p\varepsilon_l}{4}} = \frac{1}{\sqrt{2}}, \qquad \frac{p\varepsilon_l}{4} = 0{\cdot}45\pi, \qquad \mathscr{F}_l = 1{\cdot}1p. \tag{84}$$

However, (75) shows that as $\mathscr{R} \to 1$ the intensity $I^{(t)}(p)$ in the pattern tends to zero; an increasing proportion of the light which enters the plate remains inside after the last reflection and is lost at the far end of the plate. If, as is usual, any appreciable loss of light is unacceptable, the interferometer must be used in the region where $\mathscr{F}$ has the value corresponding to $p = \infty$, and it can be seen from Fig. 7.70 that the highest value of $\mathscr{F}$ is then approximately $2\mathscr{F}_l/3 \sim 0{\cdot}7p$. In this region the form of the relative intensity distribution is similar to that given by the Fabry–Perot interferometer, so that, as we have seen on p. 335 we may take the smallest resolvable wavelength difference to be approximately $1/\mathscr{F}$ times the spectral range $(\Delta\lambda_0)_{S.R.}$. Using (74) and (80) we thus obtain the resolving power corresponding to $\mathscr{F} \sim 0{\cdot}7p$ as

$$\frac{\lambda_0}{\Delta\lambda_0} \sim \frac{0{\cdot}7p\lambda_0}{(\Delta\lambda_0)_{S.R.}} \sim 0{\cdot}7\frac{l}{\lambda_0}\left|n'^2 - n'\lambda_0\frac{dn'}{d\lambda_0} - 1\right|. \quad (85)$$

The resolving power given by (85) depends only on the length of the plate, and not at all on its thickness, but we must remember that this resolving power is reached at a value of θ which approaches $\pi/2$ as l increases. In practice, the length l of the plate is limited by technical difficulties of manufacture.

When made of crystalline quartz the Lummer–Gehrcke interferometer is transparent down to about 2000 Å, and for long it was the best available interferometer for high resolution spectroscopy in the ultra-violet region. However, the development of ultra-violet reflecting coatings suitable for use in the Fabry–Perot interferometer, and the successful construction of reflecting echelon gratings, have removed this advantage, and the Lummer–Gehrcke interferometer is now only rarely employed in research.

7.6.6 Interference filters

Suppose a parallel beam of white light is incident normally on a plane parallel plate having surfaces of high reflectivity. We see from (27) that there are maxima of the intensity of the transmitted light when the phase difference δ is an integral multiple of 2π, or, by (25), when $\lambda_0 = \lambda_0^{(m)}$ where

$$\lambda_0^{(m)} = \frac{2n'h}{(m - \phi/\pi)}, \qquad m = 1, 2, \ldots; \quad (86)$$

and that at wavelengths on either side of these maxima the intensity of the transmitted light falls sharply to low values. The plate thus acts as a wavelength filter, with multiple transmission bands associated with integral values of the order m. In particular, if the optical thickness of the plate is only a few half-wavelengths of visible light, the transmission bands in the visible region are of low order and are widely separated in wavelength, so that it is usually possible to suppress the light in all but one of them, either by means of auxiliary absorbing filters, or by virtue of selective wavelength response of the detector used to make observations.

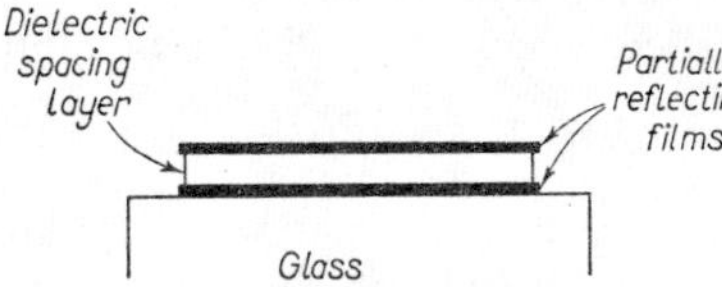

Fig. 7.71. Fabry–Perot type interference filter.

A filter of this kind may be prepared on the plane surface of a glass plate by depositing two reflecting films, separated by a spacing film of dielectric material (Fig. 7.71). If for normal incidence the filter is to have a transmission band of order m at wavelength $\lambda_0^{(m)}$, the optical thickness $n'h$ of the spacing film must by (86) be $(m - \phi/\pi)\lambda_0^{(m)}/2$. With metal reflecting films, ϕ depends on λ_0 and the film thickness (and hence on the reflectivity $\mathscr{R}$). With multilayer dielectric reflecting films, $\mathscr{R}(\lambda_0^{(m)})$ is maximum when the optical thickness of each layer is $\lambda_0^{(m)}/4$, and $\phi(\lambda_0^{(m)})$ is then zero (cf. § 1.6.); hence in this case the optical thickness of the spacer required is $m\lambda_0^{(m)}/2$.

Important characteristics of the filter are the *peak-transmission* τ, defined as for the Fabry–Perot interferometer (eq.(33)); and the wavelength *half-width* $(\Delta\lambda_0)_{H.W.}$, which is defined as the interval between wavelengths in the transmission band at which

$I^{(t)}/I^{(i)}$ has fallen to half its maximum value. If the optical separation of the reflecting surfaces is constant over the aperture of the filter, τ is given by (36); and by (21), $(\Delta\lambda_0)_{H.W.}$ corresponds to a change of δ by an amount $4/\sqrt{F}$, providing we can neglect variations of $\mathscr{R}$ and $\mathscr{T}$ with wavelength over this wavelength interval. Now from (25), for normal incidence, and if we neglect dispersion in the spacer, a change $\Delta\delta$ in δ corresponds to a change $\Delta\lambda_0$ in wavelength given to the first order of small quantities by

$$\Delta\delta = \left(-\frac{4\pi}{\lambda_0^2}n'h + 2\frac{d\phi}{d\lambda_0}\right)\Delta\lambda_0; \tag{87}$$

hence near the transmission band of order m, we have, using (86),

$$\Delta\delta = \frac{2\pi}{\lambda_0^{(m)}}\left\{-m + \frac{1}{\pi}\left[\frac{d}{d\lambda_0}(\phi\lambda_0)\right]_{\lambda_0^{(m)}}\right\}\Delta\lambda_0. \tag{88}$$

The wavelength half-width of the filter, which as we have just seen corresponds to a change in δ of $4/\sqrt{F}$, is therefore

$$(\Delta\lambda_0)_{H.W.} = \frac{2\lambda_0^{(m)}}{\pi\sqrt{F}\left|m - \frac{1}{\pi}\left[\frac{d}{d\lambda_0}(\phi\lambda_0)\right]_{\lambda_0^{(m)}}\right|} = \frac{\lambda_0^{(m)}}{\mathscr{F}\left|m - \frac{1}{\pi}\left(\frac{d}{d\lambda_0}(\phi\lambda_0)\right)_{\lambda_0^{(m)}}\right|}, \tag{89}$$

where the relation (22) has been used. When the reflecting films are of metal, $d\phi/d\lambda_0$ is so small that $d(\phi\lambda_0)/d\lambda_0 \sim \phi$. The phase change effects when the reflecting films involve dielectric multilayers have been discussed by DUFOUR,* who has shown that in the visible spectrum, with zinc sulphide and cryolite as the dielectric materials, $\frac{d}{d\lambda_0}(\phi\lambda_0)/\pi$ lies between about $-1{\cdot}0$ and $-1{\cdot}5$†.

We have seen that with available reflecting films τ decreases with increasing $\mathscr{F}$, so that for filters of given order prepared with given materials, smaller half-widths are associated with lower peak transmissions. In practice, as with the FABRY–PEROT interferometer, an upper limit to the useful reflectivity is set by variations of the optical separation of the reflecting surfaces over the aperture. In this connection, however, departures from planeness of the glass surface on which the filter is prepared are not important, since the evaporated films follow the contours of the surface on which they are deposited. With a suitable arrangement for preparing the filter, the variation of optical separation of the reflecting surfaces can be made much smaller than is possible for the two independently worked surfaces of the FABRY–PEROT interferometer, and correspondingly higher reflectivities may be employed. Typical performance data for filters with metal and metal–dielectric reflecting films are shown in Table XVI, and for all dielectric filters in Table XVII. It is seen that the highest peak-transmissions for given half-width are obtained with reflecting films which involve dielectric layers. We may note that, because of the wavelength dependence of the reflecting properties of such films, filters in which they are used have secondary transmission bands in addition to those given by (86).

Since filters are usually required to transmit a specified wavelength, the optical thickness of the spacer must be accurately controlled. From (86), a change $\Delta(n'h)$ in $n'h$ corresponds to a displacement $\Delta\lambda_0^{(m)}$ of the transmission band of order m given

---

* C. DUFOUR, *Rev. d'Opt.*, **31** (1952), p. 1.

† In discussing the resolving power of the FABRY–PEROT interferometer, we neglected wavelength dependence of ϕ (cf. (42) and (88)). This is evidently justified when m is sufficiently large.

TABLE XVI

Performance of interference filters with metal and metal–dielectric reflecting films. M denotes a metal film (silver), L denotes a quarter wave layer of low refractive index dielectric (magnesium fluoride), and H denotes a quarter wave layer of high refractive index dielectric (zinc sulphide).

From A. F. TURNER, *J. Phys.*, **11** (1950), 457.

Type of filter	Wavelength $\lambda_0^{(m)}$ of maximum transmission	Peak transmission τ	Half-width $(\Delta\lambda_0)_{H.W.}$
M—2L—M . . .	5310 Å	0·30	130 Å
M—4L—M . . .	5350 Å	0·26	70 Å
MLH—2L—HLM . .	5470 Å	0·43	48 Å
MLHLH—2L—HLHLM .	6050 Å	0·38	20 Å

TABLE XVII

Performance of zinc sulphide–cryolite interference filters. L denotes a quarter wave layer of cryolite, H denotes a quarter wave layer of zinc sulphide.

From P. H. LISSBERGER and J. RING, *Optica Acta*, **2** (1955), 45.

Type of filter	Wavelength $\lambda_0^{(m)}$ of maximum transmission	Peak transmission τ	Half-width $(\Delta\lambda_0)_{H.W.}$
HLH—2L—HLH	5185 Å	0·90	380 Å
HLHLH—2L—HLHLH . . .	4750 Å	0·85	110 Å
HLHLHL—2H—LHLHLH . . .	6565 Å	0·90	65 Å
HLHLHLH—2L—HLHLHLH . .	5200 Å	0·70	40 Å
HLHLHLHL—2H—LHLHLHLH .	6500 Å	0·80	35 Å
HLHLHLHLHL—2H—LHLHLHLHLH	6600 Å	0·50	20 Å

by $\Delta\lambda_0^{(m)} = \lambda_0^{(m)}\Delta(n'h)/n'h$. Thus with $\lambda_0^{(m)} = 5000$ Å, for example, an error of 1 per cent in the optical thickness of the spacer results in an error of 50 Å in the position of the pass-band, which is greater than the half-width of narrow band filters. Suitable methods for monitoring the thickness of the spacer during deposition have been described by GREENLAND and BILLINGTON,* GIACOMO and JACQUINOT† and LISSBERGER and RING‡. We see also from (25) that the pass-band may be adjusted to shorter wavelengths by tilting the filter so that the incidence is no longer normal, but the performance deteriorates as the tilt increases. In particular, when the reflecting films are of metal, the phase changes $\phi_{\parallel}$ and $\phi_{\perp}$ for light polarized with its electric vector parallel and perpendicular to the plane of incidence become unequal, and the pass-bands for the two polarized components are at different wavelengths.

* K. M. GREENLAND and C. BILLINGTON, *Journ. de Phys.*, **11** (1950), 418.

† P. GIACOMO and P. JACQUINOT, *J. Phys.*, **13** (1952), 59A.

‡ P. H. LISSBERGER and J. RING, *Optica Acta* **2** (1955), 42.

The wavelength displacement of the pass-band for given tilt depends on θ', the angle of reflection in the spacer, and so is lower the higher the refractive index of the spacer. For this reason it is advantageous to make the spacer of high refractive index material if the filter is to be used in a convergent beam.

In another form of interference filter—the so-called *frustrated total reflection filter*, first described by LEURGANS and TURNER*—each reflecting film is a thin layer of low refractive index material between media of higher refractive index. We have seen in § 1.5.4 that if such a layer is sufficiently thin, and is illuminated at an angle greater than the critical angle, the reflection is not total; some light passes through

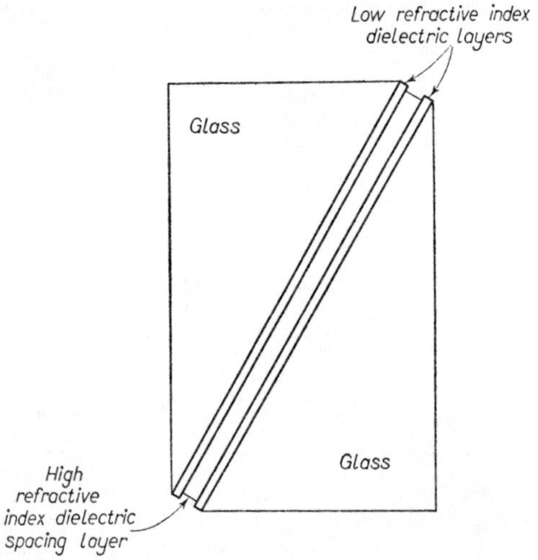

Fig. 7.72. A frustrated total reflection interference filter.

the layer, which thus acts as a non-absorbing reflector, and any desired reflectivity may be obtained by adjusting the layer thickness. The construction of the filter is shown in Fig. 7 72. The hypotenuse face of a prism of dense glass is coated with a

TABLE XVIII

Performance data of some frustrated total reflection filters, with reflecting layers of magnesium fluoride on dense flint glass, and spacing layers of zinc sulphide.

From A. R. TURNER, *J. Phys.*, **11** (1950), 458.

Order of interference	1		2		1	
Polarization	$\parallel$	$\perp$	$\parallel$	$\perp$	$\parallel$	$\perp$
Wavelength of maximum transmission $\lambda_0^{(m)}$	4600 Å	5300 Å	4630 Å	5080 Å	—	5110 Å
Peak-transmission τ	$>$ 0·93	$>$ 0·93	0·90	0·90	—	0·12
Half-width $\lvert\Delta\lambda_0\rvert_{H.W.}$	66 Å	120 Å	50 Å	49 Å	—	30 Å

* P. LEURGANS and A. F. TURNER, *J. Opt. Soc. Amer.*, **37** (1947), 983.

layer of low refractive index, followed by a spacing layer of high refractive index and a further layer of low refractive index; and a second similar prism is attached to the first by means of a cement of the same refractive index as the glass. The prism angle and the refractive indices are chosen so that light incident normally on the prism base meets the low index layers at an angle greater than the critical angle. The wavelengths of the transmission bands depend on the optical thickness of the spacer, the angle of reflection in it, and the phase change on reflection; the reflectivity (and hence the half-width for given order), depends on the thickness of the low refractive index layers. The phase change depends on the state of polarization, and so the transmission bands of given order for the components vibrating parallel and perpendicular to the plane of incidence are at different wavelengths. Performance data for typical filters prepared for use in the visible spectrum are given in Table XVIII. Filters of this type have also been made for use with centimetre waves.*

7.6.7 Multiple-beam fringes with thin films

In § 7.6.1 we have seen how the fringes of equal inclination from a plane parallel plate become much narrower when the reflectivity of the surfaces is increased. Increased reflectivity has a similar effect on the intensity distribution of the FIZEAU fringes of equal thickness (§ 7.5.2) given by a thin film, with the result that finer details of the thickness contours of the film are revealed.

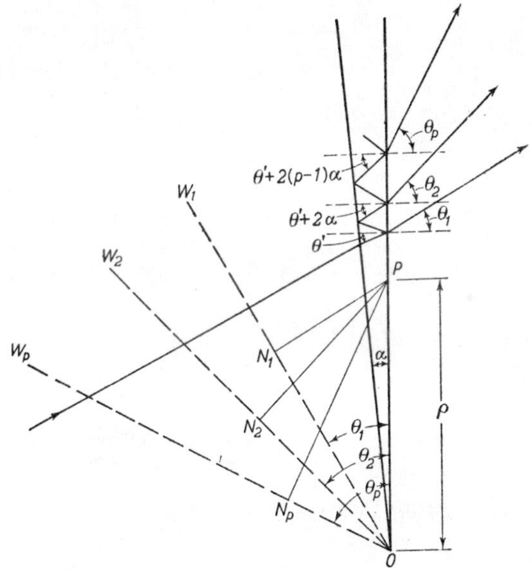

Fig. 7.73. Illustrating multiple reflections in a wedge.

For simplicity we consider first the case of a film in the form of a wedge with plane surfaces inclined at a small angle α, illuminated with a plane wave of monochromatic light propagated at right angles to the wedge apex O.† (Fig. 7.73.) Because of multiple reflections at the surfaces, the transmitted light consists of a set of plane waves propagated in different directions. If the wave transmitted through the first surface approaches the second surface from the side of the normal nearer the wedge apex

* B. H. BILLINGS, *J. Opt. Soc. Amer.*, **39** (1949), 634.

† The analysis given here is essentially due to J. BROSSEL, *Proc. Phys. Soc.*, **59** (1947), 224.

and is incident at angle θ', the pth wave of the transmitted set emerges at angle θ_p where, from the laws of refraction and reflection,

$$n \sin \theta_p = n' \sin [\theta' + 2(p-1)\alpha], \tag{90}$$

n' being the refractive index of the film, and n that of the surrounding medium. The virtual transmitted wave-fronts $W_1, W_2, \ldots W_p, \ldots$, which contain the wedge apex would be co-phasal if there were no phase change at reflection. At a point P on the second surface, at distance ρ from O, the difference of optical path of the pth wave and the wave which is directly transmitted ($p = 1$) is therefore

$$\Delta \mathscr{S}_p = n(PN_p - PN_1) = n\rho (\sin \theta_p - \sin \theta_1), \tag{91}$$

where N_p, N_1 are the feet of the perpendiculars from P to W_p, W_1 respectively. If ϕ is the phase change arising from a single reflection at either surface of the wedge, the total phase difference δ_p between the pth wave and the directly transmitted wave is, if (90) and (91) are used,

$$\begin{aligned}\delta_p &= \frac{2\pi}{\lambda_0} \Delta \mathscr{S}_p + 2(p-1)\phi \\ &= \frac{2\pi}{\lambda_0} n'\rho\{\sin [\theta' + 2(p-1)\alpha] - \sin \theta'\} + 2(p-1)\phi \\ &= \frac{4\pi}{\lambda_0} n'h \cos \theta' \frac{\sin (p-1)\alpha}{\tan \alpha} \{\cos (p-1)\alpha - \tan \theta' \sin (p-1)\alpha\} + 2(p-1)\phi,\end{aligned} \tag{92}$$

where

$$h = \rho \tan \alpha$$

is the thickness of the film at P. If the wave transmitted through the first surface meets the second surface from the side of the normal away from the wedge apex, the wave-front W_1 is real. In this case δ_p is given by (92), but with the sign of the term in $\tan \theta'$ changed.

With notation similar to that of § 7.6.1, the amplitudes of the transmitted waves are

$$tt'A^{(i)}e^{i\delta_1}, \qquad tt'r'^2A^{(i)}e^{i\delta_2}, \ldots \ldots tt'r'^{2(p-1)}A^{(i)}e^{i\delta_p}, \ldots,$$

and the amplitude at P resulting from the superposition of the infinity of such waves is

$$A^{(t)} = A^{(i)}tt' \sum_{p=1}^{\infty} r'^{2(p-1)}e^{i\delta_p} = A^{(i)}\mathscr{T} \sum_{p=1}^{\infty} \mathscr{R}^{p-1}e^{i\delta_p}. \tag{93}$$

The corresponding intensity is

$$I^{(t)} = I^{(i)}\mathscr{T}^2 \left| \sum_{p=1}^{\infty} \mathscr{R}^{p-1}e^{i\delta_p} \right|^2. \tag{94}$$

Let us examine the consequences of retaining only a finite number of terms in the series in (93) and (94). $|A^{(t)}|$ cannot be greater than $|A^{(i)}|\mathscr{T}/(1-\mathscr{R})$, and if only the first p terms are taken into account, the error $|\Delta A^{(t)}|$ in $A^{(t)}$ cannot exceed $|A^{(i)}|\mathscr{T}\mathscr{R}^p/(1-\mathscr{R})$. The corresponding relative error $|\Delta I^{(t)}|/I^{(t)}$ satisfies the relation

$$\frac{|\Delta I^{(t)}|}{I^{(t)}} \leqslant 2\left|\frac{\Delta A^{(t)}}{A^{(t)}}\right| \leqslant 2\left|\frac{A^{(i)}}{A^{(t)}}\right| \frac{\mathscr{T}\mathscr{R}^p}{(1-\mathscr{R})}. \tag{95}$$

We shall see later that in the conditions of most practical interest, $I^{(t)}$ is given nearly enough by the AIRY formula (13), according to which the minimum intensity corresponds to $|A^{(t)}| = |A^{(i)}|\mathscr{T}/(1+\mathscr{R})$. Thus if for this particular value of $|A^{(t)}|$ we wish $|\Delta I^{(t)}|/I^{(t)}$ to be not greater than 0·01 say, which means that the relative error of intensities in the neighbourhood of maxima of $I^{(t)}$ will be much less than 0·01, we find from (95) that the number p of terms which have to be retained in the series is given by the relation

$$\mathscr{R}^p \sim \frac{1}{200}\frac{1-\mathscr{R}}{1+\mathscr{R}}. \tag{96}$$

With $\mathscr{R} = 0{\cdot}93$, for example, $p \sim 120$.

With only a finite number of terms in the summation of (94), and providing α is sufficiently small, we may expand (92) in powers of α and retain terms only up to the second power, whence

$$\delta_p = (p-1)\left\{\frac{4\pi}{\lambda_0} n'h \cos\theta' + 2\phi\right\} - (p-1)^2\alpha\frac{4\pi}{\lambda_0} n'h \sin\theta' - \frac{(p-1)(2p^2-4p+3)\alpha^2}{3}\cdot\frac{4\pi n'h\cos\theta'}{\lambda_0}. \tag{97}$$

In particular, for normal incidence ($\theta' = 0$),

$$\delta_p = (p-1)\left(\frac{4\pi n'h}{\lambda_0} + 2\phi\right) - \frac{(p-1)(2p^2-4p+3)\alpha^2}{3}\cdot\frac{4\pi n'h}{\lambda_0}. \tag{98}$$

When the term of (98) in α^2 is small compared with π, we may neglect it and take the phase difference $\delta_p - \delta_{p-1}$ of successive waves to be constant and equal to $\delta = 4\pi n'h/\lambda_0 + 2\phi$. $I^{(t)}$ is then given by the AIRY formula (13), so that there are fringes in the plane of the wedge with an intensity distribution similar to that of the fringes at infinity from a plane parallel plate (Fig. 7.58). The positions of the intensity maxima, which correspond to $\delta = 2m\pi$, are given by

$$2n'h = \left(m - \frac{\phi}{\pi}\right)\lambda_0, \qquad m = 1, 2, \ldots \tag{99}$$

Thus the fringes are parallel to the wedge apex; they are spaced at intervals $\lambda_0/2n'\alpha$. If the reflectivity is high ($\mathscr{R} \gtrsim 0{\cdot}9$), so that p is large and we may replace the term $(p-1)(2p^2-4p+3)$ by $2p^3$, the condition that the term of (98) in α^2 should be small compared with π requires

$$n'h \ll \frac{3\lambda_0}{8p^3\alpha^2}. \tag{100}$$

With values of α corresponding to a distance between adjacent fringes of the order of 1 cm or less, this condition represents a severe restriction of $n'h$, and the fringes have the AIRY intensity distribution only in the region of the wedge near the apex where the optical thickness is a few wavelengths. For example, with $\alpha \sim 2{\cdot}5 \times 10^{-4}$ (which with $n' \sim 1$, $\lambda_0 = 5500$ Å corresponds to about 1 fringe per mm), $p \sim 50$ ($\mathscr{R} \sim 0{\cdot}9$), (100) gives $n'h \ll 50\lambda_0$. Further from the wedge apex, the term of (98) in α^2 becomes increasingly significant, and numerical calculations show (Fig. 7.74) that $I^{(t)}$ begins to be affected in three ways: the intensities of the maxima become less, and their half-widths greater, than the values given by the AIRY formula; the maxima are displaced from the positions given by (99), away from the wedge apex; and the fringes become asymmetrical, with secondary maxima on the side away from the wedge apex.

According to (99) the fringes follow lines of equal optical thickness when the surfaces of the film are plane. However, the important practical application of the fringes is to the examination of films whose surfaces are not plane, and we must now consider how closely the fringes follow thickness contours in such circumstances. For the fringes to contour an irregular film exactly, the intensity at each point P must be

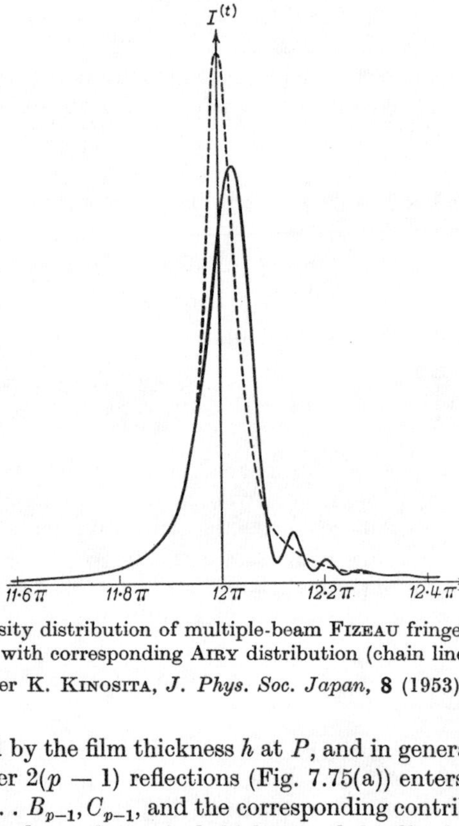

Fig. 7.74. Intensity distribution of multiple-beam FIZEAU fringe (full line), compared with corresponding AIRY distribution (chain line).
(After K. KINOSITA, *J. Phys. Soc. Japan*, **8** (1953), 219.)

uniquely determined by the film thickness h at P, and in general this is not so. A ray which reaches P after $2(p-1)$ reflections (Fig. 7.75(a)) enters the film at A_1 and is reflected at $B_1, C_1, \ldots B_{p-1}, C_{p-1}$, and the corresponding contribution to the resultant amplitude at P depends on the optical thickness of the film between A_1 and P. The fringes may thus be displaced from the true contours of optical thickness by an amount which depends on the shape of the film in the vicinity of P. The effect is evidently more important the higher the reflectivity, but for given reflectivity it is smallest where the film thickness is smallest, and when the incidence is near normal. We may estimate the extent of the region of the film which influences the intensity at P if we assume that over it the surfaces are plane (Fig. 7.75(b)). Let $A_1', A_1'', \ldots A_{p-1}', A_{p-1}''$ be the images of A_1 corresponding to successive reflections at the second and first surfaces of the film. These images, which are in the principal section of the wedge through A_1, lie on a circle with centre O at the wedge apex and radius $OA_1 = \rho'$; and the angle $A_1OA_{p-1}'' = 2(p-1)\alpha$. The image A_{p-1}'' is associated with the portion $C_{p-1}P$ of the ray so that $C_{p-1}P$ produced must pass through A_{p-1}''. If we take the ray at B_1 to be in a principal section of the wedge, and to be incident at angle θ' from the side of the normal nearer O, the angle of incidence of the ray when it reaches P is $\theta' + 2(p-1)\alpha$. In the triangle $A''_{p-1}PO$ we then have $OA''_{p-1} = \rho'$, and the angles

$$A''_pPO = \pi/2 - [\theta' + 2(p-1)\alpha], \qquad PA''_pO = \pi/2 + (\theta' - \alpha);$$

so that putting $OP = \rho$, we have $\rho'/\cos[\theta' + 2(p-1)\alpha] = \rho/\cos(\theta' - \alpha)$. Hence

$$\Delta\rho = \rho - \rho' = \frac{h}{\tan\alpha}\left\{1 - \frac{\cos[\theta' + 2(p-1)\alpha]}{\cos(\theta' - \alpha)}\right\}, \tag{101}$$

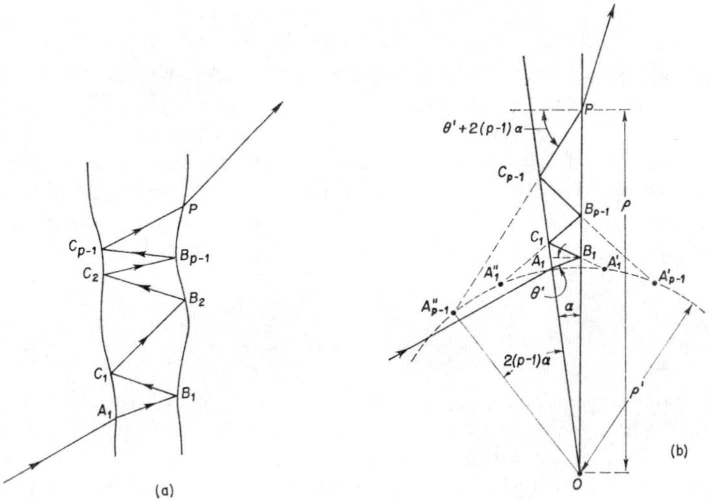

Fig. 7.75. Illustrating the displacement of a multiple reflected ray along a wedge-shaped film. (a) Irregular surfaces; (b) Plane surfaces.

and for $\theta' = 0$, and neglecting powers of α higher than the second, this reduces to

$$\Delta\rho = \frac{4p^2 - 8p + 3}{2} h\alpha. \tag{102}$$

Thus if there are p significant beams we may expect the intensity at P to be influenced by the film thickness over a distance of the order of $2p^2h\alpha$ about P. Even with high reflectivities this distance is small when h is only a few wavelengths of visible light. For example, with $p \sim 50$, $h = 2{\cdot}5 \times 10^{-4}$ cm (about five wavelengths), and $\alpha = 2{\cdot}5 \times 10^{-4}$, it amounts to about 3×10^{-3} mm. Providing small irregularities of the fringes are interpreted with caution, the fringes delineate optical thickness contours of irregular films when the film thickness is sufficiently small.

We have assumed so far that the source is a point, but in practice, because of intensity considerations, it is desirable to use an extended source of maximum possible size. According to (97), if we can neglect terms in α and α^2, the intensity maxima for non-normal incidence occur when

$$2n'h\cos\theta' = \left(m - \frac{\phi}{\pi}\right)\lambda_0, \qquad m = 1, 2, \ldots, \tag{103}$$

and since this relation rests on neglect of the wedge angle it is valid for all planes of incidence. Hence if the source is extended radially, so that the angles of incidence cover the range $0 \leqslant \theta' \leqslant \varepsilon'$, the maxima of given order spread over a distance corresponding to Δm orders, where from (103), for ε' small,

$$\Delta m = \frac{2}{\lambda_0} n'h|\Delta(\cos\theta')| = \frac{n'h\varepsilon'^2}{\lambda_0}. \tag{104}$$

The importance of this fringe broadening due to the source extension depends on the finesse $\mathscr{F}$ of the fringes formed by a single element of the source: an increase of ε' will result mainly in increased intensity of the maxima when $n'h\varepsilon'^2/\lambda_0 \ll 1/\mathscr{F}$, but mainly in increased fringe width when $n'h\varepsilon'^2/\lambda_0 \gg 1/\mathscr{F}$. For a closer examination of the effect we may take advantage of its formal similarity to the broadening of Fabry–Perot fringes which results from spherical curvature of the interferometer plates (pp. 332–333). Appropriate modification of the results of Dufour and Picca* shows that there is little increase of intensity of the maxima once $n'h\varepsilon'^2/\lambda_0$ exceeds $1/\mathscr{F}$, and a reasonable criterion is that it should have half this value, the fringes then being about 10 per cent broader than with a point source. Taking into account refraction at the first surface of the film, we may therefore take the tolerable angular radius ε of the source to be

$$\varepsilon \sim \frac{n'\varepsilon'}{n} \sim \frac{1}{n}\sqrt{\frac{\lambda_0 n'}{2h\mathscr{F}}}, \tag{105}$$

which may be compared with the analogous relation § 7.5 (73) relating to two-beam interference. When h is only a few wavelengths, ε is not very small; for example, with $h \sim 5\lambda_0$, $\mathscr{F} = 30$ ($\mathscr{R} \sim 0{\cdot}9$), $n' \sim n \sim 1$, (105) gives $\varepsilon \sim 3°$. Thus the incident light need not be strictly collimated, and the large sources permitted are advantageous for observation of the fringes under high magnification.

When the light is quasi-monochromatic, with wavelength components covering a wavelength range $\Delta\lambda_0$ about mean wavelength $\overline{\lambda}_0$, the maxima of given order spread over a distance which corresponds to Δm orders, where from (99), if we neglect wavelength dependence of n' and ϕ, $\Delta m = 2n'h\Delta\lambda_0/\overline{\lambda}_0^2$. The fringes are not significantly broadened by a departure from strict monochromatism if this quantity is small compared with $1/\mathscr{F}$, i.e. if

$$\Delta\lambda_0 \ll \frac{(\overline{\lambda}_0)^2}{2n'h\mathscr{F}}. \tag{106}$$

Again, when $n'h$ is only a few wavelengths, this is not a severe condition; for example, with $n'h \sim 5\lambda_0$, $\mathscr{F} = 30$, $\lambda_0 = 5500$ Å, (106) gives $\Delta\lambda_0 \ll 18$ Å. In consequence the source is not restricted to the type familiar in interferometry with large path differences, but instead high pressure arc discharges may be employed, which are very bright and are further advantageous for observations with high magnification.

We have been concerned above with light transmitted through the film, but a multiple-beam interference pattern may also be observed with reflected light. When there is no absorption at the reflecting surfaces, this pattern is complementary to the transmitted pattern in the sense that the sum of the intensities in the two is at each point equal to the incident intensity. The effect of absorption, which is always

---

* C. Dufour and R. Picca, *Rev. d'Opt.*, **24** (1945), 19.

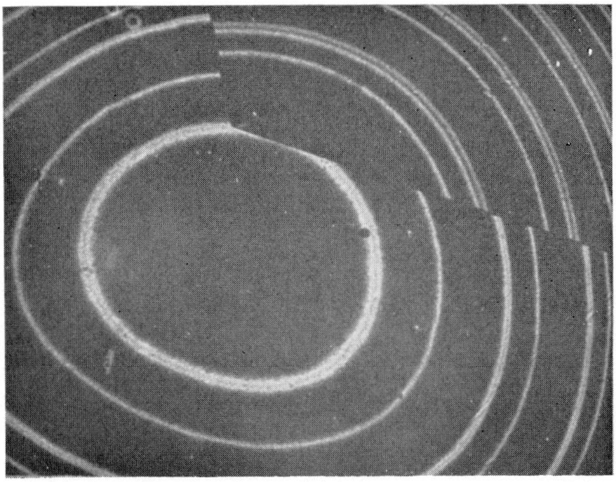

Fig. 7.76. Multiple-beam FIZEAU fringes over a cleavage surface of mica (transmitted light). The fringes are given by the green (5461 Å) and yellow (5770 Å, 5790 Å) lines of mercury.

(After W. L. WILCOCK.)

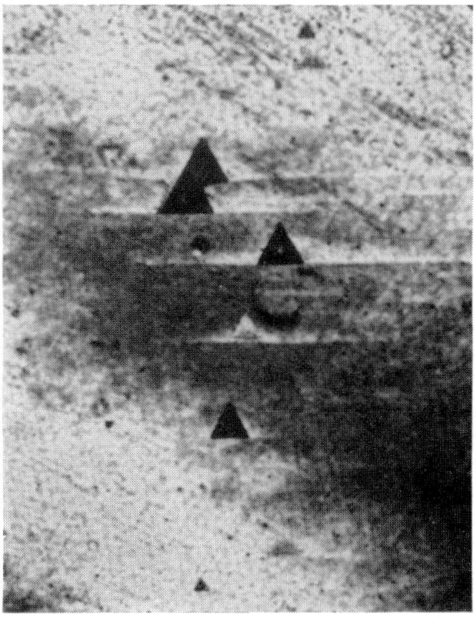

Fig. 7.77. Multiple-beam FIZEAU fringe over a crystal surface of diamond.

(After W. L. WILCOCK.)

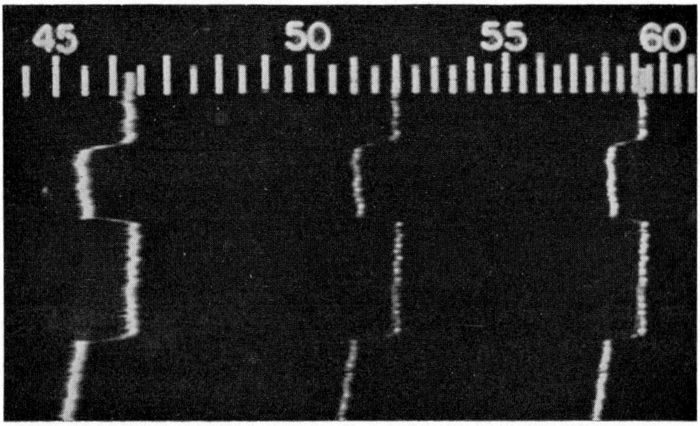

Fig. 7.79. Fringes of equal chromatic order given by a section of a diamond crystal surface. The scale is of wavelength in hundreds of Ångstroms.

(After S. Tolansky and W. L. Wilcock, *Proc. Roy. Soc.*, A, **191** (1947), 192.)

associated with practical reflecting coatings, has been considered by HOLDEN.* He has shown that, with silver coatings of high reflectivity and low absorption, the pattern in reflected light is of dark fringes on a nearly uniform background; that these fringes are narrower than the corresponding bright fringes in transmitted light; but that the intensity at the minima of the reflected pattern is critically dependent on the absorption in the coating on the entrant surface of the film, and if this absorption is too high the fringes are invisible.

Multiple-beam FIZEAU fringes are used in optical workshops to test the figure of high grade optical flats such as are required for the FABRY–PEROT interferometer, and they have been employed extensively by TOLANSKY and his co-workers to study the topography of crystal and metal surfaces which are nearly plane.† For this, the surface to be examined and the surface of an optical flat—each with a reflecting coating of silver—are held as close together as possible, and fringes in the air film between them are observed with a microscope of sufficient angular aperture to collect all the significant beams. If $\lambda = \lambda_0/n'$ is the wavelength in air, the fringes are contours of the unknown surface defined by planes parallel to the optical flat at intervals of $\lambda/2$. When the mutual inclination of the surfaces is sufficient to give a number of fringes in the field of view (see, for example, Fig. 7.76), the topography of the unknown surface can be quantitatively evaluated by measuring the deviation of the fringes from straight lines. In crossing a surface feature of height Δh, a fringe suffers a lateral displacement corresponding to Δm orders, where

$$\Delta h = \frac{\lambda}{2}\,\Delta m; \tag{107}$$

so that Δh can be determined if Δm is known, and Δm is, at least approximately, the ratio of the fringe displacement and the distance between adjacent fringes. When the reflectivity is high, surface features of very small height can be measured in this way. A displacement equal to the fringe half-width, which is easily measurable, corresponds to $\Delta h = \lambda/2\mathscr{F}$, and with $\lambda = 5500$ Å, $\mathscr{F} = 40$, this is only about 70 Å.‡ Whether a surface feature is an elevation or a depression can often be decided by observing the direction of motion of the fringes when the separation of the surfaces is altered, but in general it is necessary to use more than one wavelength. With fringes in transmitted light the several wavelengths can be used simultaneously, and the yellow lines of mercury $\lambda = 5770$ Å, 5790 Å are convenient. The fringes then occur in close pairs whose separation increases with increasing order of interference; the direction of the local wedge angle is thus immediately visible, and if there is a discontinuity in the fringes, as when they cross a crystal cleavage step, it is possible to decide which are corresponding orders on opposite sides.

If the unknown surface is nearly plane, its inclination to the optical flat may be reduced until the whole field of view is occupied by only one fringe (see, for example, Fig. 7.77). Under these conditions extremely small variations of height can lead to detectable variations of intensity. Let I, $I + \Delta I$ be the intensities over adjacent areas with a small difference of height Δh, which corresponds to a change $\Delta\delta$ in δ.

---

* J. HOLDEN, *Proc. Phys. Soc.* B., **62** (1949), 405.

† S. TOLANSKY, *Multiple-beam Interferometry of Surfaces and Films* (Oxford, Clarendon Press, 1948).

‡ Although the silver coating is about 500 Å thick, it has been established that it does not alter the surface topography to an extent which is detectable by the technique.

The contrast, defined by $|\Delta I|/I$, is maximum when δ is adjusted to make $|dI/d\delta|/I$ maximum. Considering the pattern in transmitted light, we have from (27)

$$\frac{I^{(t)}}{I^{(t)}_{\max}} = \frac{1}{1 + F \sin^2 \dfrac{\delta}{2}}, \tag{108}$$

whence

$$\frac{d}{d\delta}\left(\frac{I^{(t)}}{I^{(t)}_{\max}}\right) = \frac{-F \sin \delta}{2\left(1 + F \sin^2 \dfrac{\delta}{2}\right)^2}. \tag{109}$$

Hence

$$\frac{1}{I^{(t)}}\left|\frac{dI^{(t)}}{d\delta}\right| = \frac{F \sin \delta}{2\left(1 + F \sin^2 \dfrac{\delta}{2}\right)}, \tag{110}$$

so that for maximum contrast,

$$2\left(1 + F \sin^2 \frac{\delta}{2}\right) \cos \delta - F \sin^2 \delta = 0. \tag{111}$$

When the reflectivity is high, F is large and the values of δ which satisfy (111) are close to integral multiples of 2π, so that we may write $\delta = 2m\pi \pm \varepsilon$, where m is an integer and ε is small compared with $\pi/2$. We then obtain from (111), neglecting powers of ε higher than the second

$$\varepsilon = \frac{2}{\sqrt{F + 2}} \sim \frac{2}{\sqrt{F}}, \tag{112}$$

and (110) then gives

$$\frac{1}{I^{(t)}}\left|\frac{dI^{(t)}}{d\delta}\right|_{\max} \sim \frac{F\varepsilon}{2\left(1 + \dfrac{F\varepsilon^2}{4}\right)} \sim \frac{\sqrt{F}}{2}. \tag{113}$$

Hence to first order the maximum contrast is given by

$$\left|\frac{\Delta I^{(t)}}{I^{(t)}}\right|_{\max} = \frac{1}{I^{(t)}}\left|\frac{dI^{(t)}}{d\delta}\right|_{\max} |\Delta\delta| \sim \frac{\sqrt{F}}{2} |\Delta\delta|. \tag{114}$$

Since, from (99), $\Delta\delta$ and Δh are related by $|\Delta\delta| = 4\pi|\Delta h|/\lambda$, and if the relation (22) is used, the maximum contrast may be expressed in the form

$$\left|\frac{\Delta I^{(t)}}{I^{(t)}}\right|_{\max} \sim \frac{4\mathscr{F}}{\lambda} |\Delta h|. \tag{115}$$

With $\lambda = 5500$ Å, $\mathscr{F} = 40$, (115) gives $|\Delta I^{(t)}/I^{(t)}|_{\max} \sim 0{\cdot}1$ when Δh is only $\sim 3{\cdot}5$ Å, which is of the order of molecular dimensions.

We have considered so far the effects with quasi-monochromatic light. Suppose now that the film is illuminated at normal incidence with white light, and that by means of an achromatic lens L the transmitted light is used to form an image of the film in the plane of the slit of a spectrograph (Fig. 7.78). For each wavelength component, the phase relationship of the beams reaching a point P' of the slit is the same as at the point P of the film conjugate to P'. Hence if the film is sufficiently thin to satisfy condition (100), there are according to (99) maxima of intensity at P for those wavelengths $\lambda_0^{(m)}$ which satisfy the relation

$$\lambda_0^{(m)} = \frac{2n'h}{m - \dfrac{\phi}{\pi}}, \qquad m = 1, 2, \ldots, \tag{116}$$

where $n'h$ is the optical thickness of the film at P; and the intensity at wavelengths between these maxima is given by the AIRY formula. Thus when the reflectivity is high, the spectrum in the focal plane of the spectrograph is channelled with narrow bright fringes separated by much broader dark intervals, the wavelength separation of adjacent fringes being greater the thinner the film.

In the special case when $n'h$ is constant, $\lambda_0^{(m)}$ is constant and the fringes are straight lines parallel to the slit. These fringes are sometimes called *Edser–Butler fringes.* If ϕ is independent of wavelength they are spaced at equal intervals $\Delta\kappa_0 = 1/2n'h$

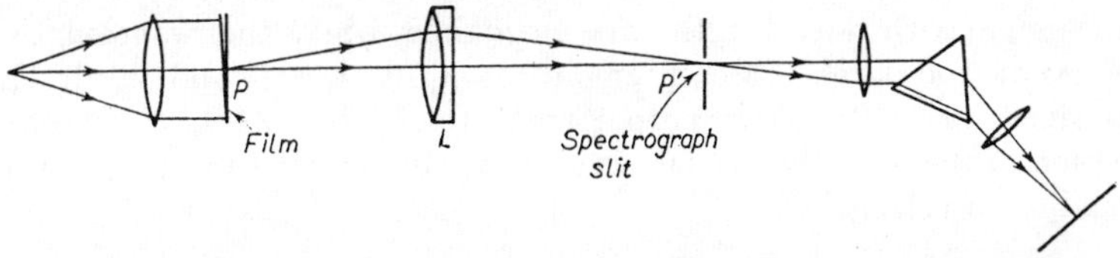

Fig. 7.78. Arrangement for the observation of white light fringes from a thin film.

of spectroscopic wave-number $\kappa_0 = 1/\lambda_0$, and because of this property they are useful for the calibration of spectrographs, particularly in the infra-red where standard wavelengths are not numerous. More generally, $\lambda_0^{(m)}$ varies along a fringe in a way which depends on the variation of optical thickness along the section of the film conjugate to the slit. In particular, if the film is an air film formed between an unknown surface and an optical flat, the fringes provide a powerful means of measuring the profile of a chosen section of the unknown surface. Such fringes, first described by TOLANSKY,* are sometimes called *fringes of equal chromatic order* (Fig. 7.79, f. p. 357).

Suppose the thicknesses of the film at two points P_1, P_2 are h_1, h_2, and let the maximum of order m at the conjugate points P_1', P_2' on the slit be at wavelengths in air $\lambda_1^{(m)}$, $\lambda_2^{(m)}$ respectively. Then using (116) we obtain, since $\lambda^{(m)} = \lambda_0^{(m)}/n'$,

$$\Delta h = h_2 - h_1 = \left(m - \frac{\phi_1}{\pi}\right) \frac{\lambda_2^{(m)} - \lambda_1^{(m)}}{2} + \frac{a}{2}\lambda_2^{(m)}, \tag{117a}$$

where

$$a = \frac{\phi_1 - \phi_2}{\pi}, \tag{117b}$$

and ϕ_1, ϕ_2 are the phase changes at reflection for wavelengths $\lambda_1^{(m)}$, $\lambda_2^{(m)}$. The quantity $m - \phi_1/\pi$ may be determined by measuring the wavelength separation of adjacent fringes. Thus if, for the point P_1' of the slit, the maximum of order $m + 1$ is at wavelength $\lambda_1^{(m+1)}$ we have also from (116)

$$\left(m - \frac{\phi_1}{\pi}\right) \lambda_1^{(m)} = \left(m + 1 - \frac{\phi_1'}{\pi}\right) \lambda_1^{(m+1)},$$

* S. TOLANSKY, *Phil. Mag.* (7), **36** (1945), 225.

i.e.

$$m - \frac{\phi_1}{\pi} = (1 + b)\frac{\lambda_1{}^{(m+1)}}{\lambda_1{}^{(m)} - \lambda_1{}^{(m+1)}}, \tag{118a}$$

where

$$b = \frac{\phi_1 - \phi_1'}{\pi}, \tag{118b}$$

and ϕ_1' is the phase change at reflection for wavelength $\lambda_1{}^{(m+1)}$. If the fringes are discontinuous, the wavelength separation of adjacent fringes on each side of the discontinuity must be measured in order to identify corresponding orders. From (117) and (118),

$$\Delta h = (1 + b)\frac{\lambda_1{}^{(m+1)}}{\lambda_1{}^{(m)} - \lambda_1{}^{(m+1)}}\frac{\lambda_2{}^{(m)} - \lambda_1{}^{(m)}}{2} + \frac{a}{2}\lambda_2{}^{(m)}. \tag{119}$$

With reflecting coatings of silver, the quantities a and b, which arise from variation of phase change with wavelength, are negligible when the wavelength interval involved is not too large. Measurement of $\lambda_1{}^{(m)}$, $\lambda_1{}^{(m+1)}$ and $\lambda_2{}^{(m)}$) then gives Δh; and since by (119) Δh is proportional to $\lambda_2{}^{(m)} - \lambda_1{}^{(m)}$, the profile of the chosen section of the unknown surface is obtained by plotting a single fringe on a scale proportional to wavelength.

7.6.8 Multiple-beam fringes with two plane parallel plates

(a) *Fringes with monochromatic and quasi-monochromatic light*

Consider two plane parallel plates, with surfaces of high reflectivity, placed one behind the other and illuminated with plane waves of monochromatic light (Fig. 7.80). If we use the notation of § 7.6.1, and distinguish the two plates by subscripts 1 and 2, the intensity of the light transmitted through the first plate is, according to equations (27) and (26),

$$I_1{}^{(t)} = \frac{\mathscr{T}_1^2}{(1 - \mathscr{R}_1)^2}\frac{1}{1 + F_1 \sin^2 \dfrac{\delta_1}{2}} I^{(i)}, \tag{120}$$

where $I^{(i)}$ is the intensity of the incident light. Similarly, the intensity of the light transmitted through the second plate, corresponding to incident intensity $I_1{}^{(t)}$, is

$$I_2{}^{(t)} = \frac{\mathscr{T}_2^2}{(1 - \mathscr{R}_2)^2}\frac{1}{1 + F_2 \sin^2 \dfrac{\delta_2}{2}} I_1{}^{(t)}. \tag{121}$$

We will suppose that light reflected back from the second plate, and subsequently reflected forward from the first plate, is excluded; this implies that the distance between the plates is sufficiently large compared with their apertures, and if the plates are mutually parallel, that the incidence is not too close to normal. The intensity $I^{(t)}$ of the total transmitted light is then equal to $I_2{}^{(t)}$, and from (120) and (121)

$$\frac{I^{(t)}}{I^{(i)}} = \frac{\mathscr{T}_1^2}{(1 - \mathscr{R}_1)^2}\frac{\mathscr{T}_2^2}{(1 - \mathscr{R}_2)^2}\frac{1}{\left(1 + F_1 \sin^2 \dfrac{\delta_1}{2}\right)}\frac{1}{\left(1 + F_2 \sin^2 \dfrac{\delta_2}{2}\right)}. \tag{122}$$

There are thus fringes in the focal plane of the lens L, with a relative intensity distribution which is the product of the relative intensity distributions of the fringes given by each plate alone.

The phase differences δ_1 and δ_2 are, according to (25),

$$\delta_1 = \frac{4\pi}{\lambda_0} n_1' h_1 \cos\theta_1' + 2\phi_1, \qquad \delta_2 = \frac{4\pi}{\lambda_0} n_2' h_2 \cos\theta_2' + 2\phi_2. \tag{123}$$

Let us consider the case when the plates are mutually parallel, so that the ring systems which would be given by each plate alone are concentric. When the angle of incidence

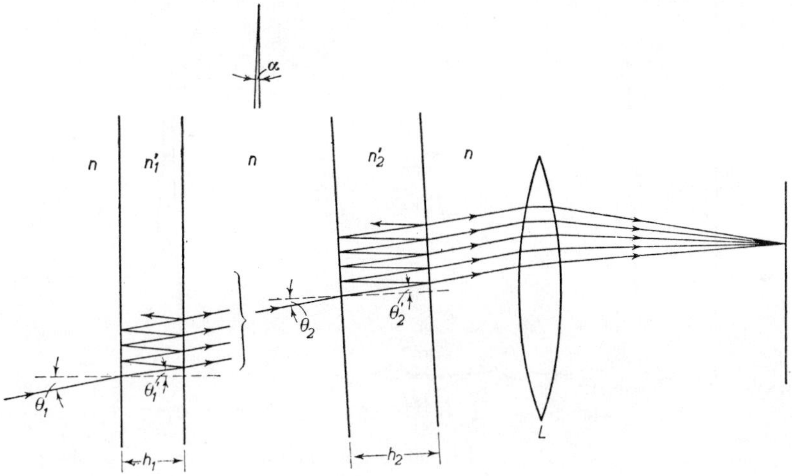

Fig. 7.80. Illustrating formation of fringes at infinity after multiple reflections in two plane parallel plates.

$\theta_1 = \theta_2 = \theta$ is small, $\cos\theta_1' \sim 1 - \theta_1'^2/2$, $\cos\theta_2' \sim 1 - \theta_2'^2/2$, and if n is the refractive index of the surrounding medium, we have from the law of refraction $\theta_1' \sim n\theta/n_1'$, $\theta_2' \sim n\theta/n_2'$. Hence from (123), to the second order in θ,

$$\delta_1 = \frac{4\pi}{\lambda_0} h_1 \left(n_1' - \frac{n^2\theta^2}{2n_1'}\right) + 2\phi_1, \qquad \delta_2 = \frac{4\pi}{\lambda_0} h_2 \left(n_2' - \frac{n^2\theta^2}{2n_2'}\right) + 2\phi_2. \tag{124}$$

From (124) it follows that for a change of θ the corresponding changes of the δ's are in the ratio

$$\frac{\Delta\delta_2}{\Delta\delta_1} = \frac{n_1' h_2}{n_2' h_1}, \tag{125}$$

where ϕ has been taken to be independent of θ. If the refractive indices and thicknesses of the plates are such that in some direction $\theta = \theta_0$, the order of interference is integral for each plate, i.e. if

$$\theta = \theta_0, \qquad \delta_1 = 2m_1\pi, \qquad \delta_2 = 2m_2\pi, \qquad (m_1 \text{ and } m_2 \text{ integers}), \tag{126}$$

the terms of (122) in F_1 and F_2 both have their maximum value of unity, and there is an absolute maximum of the intensity of the transmitted light, for $\theta = \theta_0$. If also

$$\frac{n_1' h_2}{n_2' h_1} = a, \tag{127}$$

where a is a positive integer, we see from (125) that the integral orders $(m_2 - a)$, $(m_2 - 2a)$, . . . of the second plate correspond to the same values of θ as the integral orders $(m_1 - 1)$, $(m_1 - 2)$, . . . of the first plate, so that there are absolute maxima of $I^{(t)}/I^{(i)}$ in these directions also.*

The form of the intensity curve of the transmitted light is shown in Fig. 7.81. Between successive principal maxima there are $(a - 1)$ secondary maxima, corresponding to integral orders $(m_2 - 1)$, $(m_2 - 2)$, . . . $(m_2 - a + 1)$, ... of the second plate; but if F_1 is large, and a is not too large, the corresponding values of the term in

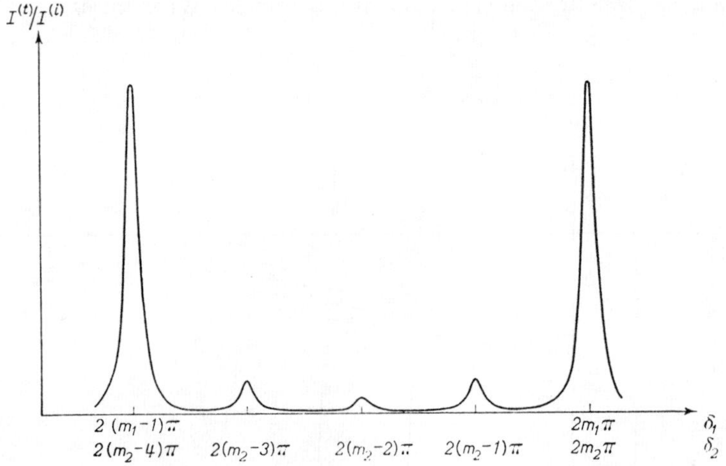

Fig. 7.81. Multiple-beam fringes of equal inclination from two plane parallel plates: ratio of transmitted and incident intensities as a function of phase differences δ_1 and δ_2. ($\mathscr{R}_1 = 0{\cdot}64$, $\mathscr{R}_2 = 0{\cdot}64$, $a = 4$).

F_1 are small compared with unity, so that these secondary maxima are much weaker than the principal maxima.

Fringes of this kind from two FABRY-PEROT interferometers have been used to examine the fine structure of spectral lines. The spacers are worked so that h_2/h_1 is approximately equal to the desired integer a, and fine adjustments are conveniently

---

* In practice only a finite number p of coincidences is required, and then $n_1'h_2/n_2'h_1$ is sufficiently close to an integer a if the mutual displacement of the maxima of the terms in F_1, F_2 at $\delta_1 = 2(m_1 - p)\pi$, $\delta_2 = 2(m_2 - pa)\pi$ is small compared with their half-widths. Using (125) and equation (21), these half-widths, in terms of δ_2, are $4n_1'h_2/n_2'h_1\sqrt{F_1}$, $4/\sqrt{F_2}$. Hence the required condition is

$$\left|\frac{n_1'h_2}{n_2'h_1}2p\pi - 2pa\pi\right| \ll \frac{n_1'h_2}{n_2'h_1}\frac{4}{\sqrt{F_1}} + \frac{4}{\sqrt{F_2}},$$

which we may write, for $(n_1'h_2/n_2'h_1) \sim a$,

$$\left|\frac{n_1'h_2}{n_2'h_1} - a\right| \ll \frac{1}{p}\left(\frac{a}{\mathscr{F}_1} + \frac{1}{\mathscr{F}_2}\right), \tag{126a}$$

where $\mathscr{F}_1$, $\mathscr{F}_2$ are the values of finesse for the fringes given by each plate alone. So long as p is not too large and the order of interference m_1 for the first plate is not too small, relations (126) and (126a) may in practice be satisfied simultaneously. Thus with $\mathscr{F}_1 = \mathscr{F}_2 = 30$ and $a = 1$, (126a) requires $|(n_1'h_2/n_2'h_1) - a| \ll 1/15p$; whilst (126) can always be satisfied by a change of the order of $\lambda_0/4$ in $n_2'h_2$, and the corresponding change of $n_1'h_2/n_2'h_1$ is about $n_1'\lambda_0/4n_2'^2h_1 \sim n_1^2/2n_2'^2m_1$, since $m_1 \sim 2n_1'h_1/\lambda_0$.

made by enclosing one of the interferometers in an air-tight container and varying the pressure—and hence the refractive index—of the air inside.*

When a is greater than unity, the combination has the advantage of greater spectral range for given resolving power than can be obtained with a single interferometer.† The change of wavelength required to displace the pattern by an amount equal to the separation of adjacent principal maxima corresponds to a change of unity in the order of interference in the first plate, and a change of a in the order of interference in the second plate. The spectral range without overlap of principal maxima is thus equal to the spectral range of the thinner interferometer, and is a times the spectral range of the thicker interferometer (cf. (46)). The half-width of the principal maxima corresponds to $\delta_1 = 2m_1\pi \pm \varepsilon_1/2$ or $\delta_2 = 2m_2\pi \pm \varepsilon_2/2$, ($m_1$, m_2 integers), where from (122),

$$\left(1 + F_1 \sin^2 \frac{\varepsilon_1}{4}\right)\left(1 + F_2 \sin^2 \frac{\varepsilon_2}{4}\right) = 2,$$

or by (125) and (127),

$$\left(1 + F_1 \sin^2 \frac{\varepsilon_2}{4a}\right)\left(1 + F_2 \sin^2 \frac{\varepsilon_2}{4}\right) = 2. \tag{128}$$

For F_2 sufficiently large, ε_2 is much less than $\pi/2$ and we may take $\sin \varepsilon_2 = \varepsilon_2$. Equation (128) then reduces to

$$\left(\frac{4}{\varepsilon_2}\right)^4 - \left(F_2 + \frac{F_1}{a^2}\right)\left(\frac{4}{\varepsilon_2}\right)^2 - \frac{F_1 F_2}{a^2} = 0, \tag{129}$$

whence

$$\varepsilon_2 = \frac{4\sqrt{2}}{\sqrt{\left(F_2 + \dfrac{F_1}{a^2}\right) + \sqrt{F_2^{\,2} + \dfrac{6F_1F_2}{a^2} + \dfrac{F_1^{\,2}}{a^4}}}}. \tag{130}$$

When the thinner interferometer is removed, $F_1 = 0$ and (130) reduces to $\varepsilon_2 = 4/\sqrt{F_2}$, in agreement with (21). Otherwise ε_2 is less than $4/\sqrt{F_2}$, so that the half-width of the principal maxima is less than the half-width of the fringes from the thicker interferometer. In practice a is made equal to 3 or more, and we see from (130) that an increase of F_2 results in a much greater reduction of ε_2 than the same increase of F_1. We must also remember that the peak-transmission of the combination is equal to the product $\tau_1\tau_2$ of the peak-transmissions of the two interferometers, as may be seen from (122) and (33); and we have seen in § 7.6.2 that, with available reflecting coatings, increases of $\mathscr{R}_1$ and $\mathscr{R}_2$ (and hence of F_1 and F_2) are accompanied by decreases of τ_1 and τ_2. For these reasons it is advantageous to make F_1 no larger than is necessary to suppress the secondary maxima to a satisfactory degree, and to make F_2 as large as intensity considerations permit. In these circumstances, F_1/a^2 is so small compared with F_2 that $\varepsilon_2 \sim 4/\sqrt{F_2}$ and the resolving power of the combination is nearly enough that of the thicker interferometer. The presence of secondary maxima is inconvenient if the source to be examined has spectral components of widely different intensities, since the principal maxima from a weak component

* It is also possible, though less satisfactory, to make fine adjustments by tilting one interferometer relative to the other.

† The arrangement was first described by W. V. Houston, *Phys. Rev.* (2), **29** (1927), 478, and by E. Gehrcke and E. Lau, *Zeits. f. tech. Physik*, **8** (1927), 157.

may be confused with the secondary maxima from a strong component. Such doubts may be removed by making observations with different thickness combinations.

When $a = 1$ there are no secondary maxima, and from (122) and (34) the contrast factor of the combination is equal to the product $\mathscr{C}_1\mathscr{C}_2$ of the contrast factors of the two interferometers. For given peak-transmission it is possible to obtain a much higher contrast factor than with a single interferometer, and the arrangement is of value for observations of weak satellites of spectral lines.*

(b) *Fringes of superposition*

Fringes of great practical importance may be observed when two plane parallel plates, mutually inclined at angle α, are illuminated near normal incidence with light which is so far from monochromatic that no fringes can be observed with either plate alone. To understand this,† consider first the transmission of a monochromatic wave of wave number $k_0 = 2\pi/\lambda_0$ through one plate. Let $A^{(i)}(k_0)$ and $A^{(t)}(k_0)$ be the complex amplitude of the incident and of the transmitted wave respectively. Taking all the reflections into account, we have, according to (10), when (2) and (4) are also used,

$$A^{(t)}(k_0) = A^{(i)}(k_0)\mathscr{T} \sum_{p=0}^{\infty} \mathscr{R}^p e^{ip\delta}, \tag{131}$$

so that the intensity $I^{(t)} = A^{(t)}A^{(t)\star}$ of the transmitted light is

$$I^{(t)}(k_0) = I^{(i)}(k_0)\mathscr{T}^2 \sum_{p=0}^{\infty} \sum_{p'=0}^{\infty} \mathscr{R}^p\mathscr{R}^{p'} e^{i(p-p')\delta}, \tag{132}$$

where $I^{(i)} = A^{(i)}A^{(i)\star}$ is the intensity of the incident light. If we set $|p - p'| = q$, (132) may be re-written in the form

$$\begin{aligned} I^{(t)}(k_0) &= I^{(i)}(k_0)\mathscr{T}^2 \sum_{p=0}^{\infty} \mathscr{R}^{2p} \left\{1 + \sum_{q=1}^{\infty} \mathscr{R}^q(e^{iq\delta} + e^{-iq\delta})\right\} \\ &= I^{(i)}(k_0) \frac{\mathscr{T}^2}{1 - \mathscr{R}^2} \left(1 + 2 \sum_{q=1}^{\infty} \mathscr{R}^q \cos q\delta\right). \end{aligned} \tag{133}$$

With two plates in series, if we neglect the light reflected backwards and forwards between them, the intensity of the monochromatic light transmitted by both plates therefore is

$$I^{(t)}(k_0) = I^{(i)}(k_0) \frac{\mathscr{T}_1^2\mathscr{T}_2^2}{(1 - \mathscr{R}_1^2)(1 - \mathscr{R}_2^2)} \left(1 + 2 \sum_{r=1}^{\infty} \mathscr{R}_1^r \cos r\delta_1\right)\left(1 + 2 \sum_{s=1}^{\infty} \mathscr{R}_2^s \cos s\delta_2\right), \tag{134}$$

where the suffixes 1 and 2 refer to the first and the second plate respectively. The phase differences δ_1 and δ_2 are given by the expressions

$$\left.\begin{aligned} \delta_1 &= k_0\Delta\mathscr{S}_1 + 2\phi_1 = k_0 2n_1'h_1 \cos\theta_1' + 2\phi_1, \\ \delta_2 &= k_0\Delta\mathscr{S}_2 + 2\phi_2 = k_0 2n_2'h_2 \cos\theta_2' + 2\phi_2, \end{aligned}\right\} \tag{135}$$

where the various symbols have the same meaning as before.

---

* L. C. Bradley and H. Kuhn, *Nature, London*, **162** (1948), 412.
C. Dufour, *Ann. de Physique, Paris* (12), **6** (1951), 5.

† The analysis given here follows essentially that of J. R. Benoit, C. Fabry and A. Perot, *Trav. et Mem. Bur. Int. Poids et Mes.*, **15** (1913), 1.

Now if the light is not monochromatic we may regard it as superposition of monochromatic components of different frequencies. According to § 7.5.8 the different components add incoherently, so that the total intensity is given by the sum (integral) of the intensities of the individual components. Thus from (134) we obtain the following expression for the total intensity of the light transmitted by both plates:

$$I^{(t)} = \int \frac{\mathcal{T}_1^2 \mathcal{T}_2^2}{(1-\mathcal{R}_1^2)(1-\mathcal{R}_2^2)} i^{(i)}(k_0)\left(1+2\sum_{r=1}^{\infty}\mathcal{R}_1^r \cos r\delta_1\right)\left(1+2\sum_{s=1}^{\infty}\mathcal{R}_2^s \cos s\delta_2\right)dk_0, \tag{136}$$

where $i^{(i)}(k_0)$ represents the spectral intensity distribution of the incident light. The quantities $\mathcal{R}$ and $\mathcal{T}$ are in general functions of k_0, but we will assume that their variation with k_0 is negligible in the range over which $i^{(i)}(k_0)$ is appreciable. The first factors in (136) may then be taken outside the integral, and we may re-write the resulting expression in the form

$$I^{(t)} = \frac{\mathcal{T}_1^2 \mathcal{T}_2^2}{(1-\mathcal{R}_1^2)(1-\mathcal{R}_2^2)} \int i^{(i)}(k_0)\{1+2[\sigma_1+\sigma_2+\sigma_{12}^{+}+\sigma_{12}^{-})\}dk_0, \tag{137a}$$

where

$$\left.\begin{aligned}
\sigma_1 &= \sum_{r=1}^{\infty}\mathcal{R}_1^r \cos r\delta_1,\\
\sigma_2 &= \sum_{s=1}^{\infty}\mathcal{R}_2^s \cos s\delta_2,\\
\sigma_{12}^{+} &= \sum_{r=1}^{\infty}\sum_{s=1}^{\infty}\mathcal{R}_1^r\mathcal{R}_2^s \cos(r\delta_1+s\delta_2),\\
\sigma_{12}^{-} &= \sum_{r=1}^{\infty}\sum_{s=1}^{\infty}\mathcal{R}_1^r\mathcal{R}_2^s \cos(r\delta_1-s\delta_2).
\end{aligned}\right\} \tag{137b}$$

If the monochromatic components cover a wavelength range $\Delta\lambda_0$ about a mean wavelength $\bar{\lambda}_0$, the corresponding range of k_0 is $2\pi\Delta\lambda_0/\bar{\lambda}_0^2$ and when $\Delta\mathcal{S}_1$ and $\Delta\mathcal{S}_2$ are sufficiently large compared with the coherence length $\bar{\lambda}_0^2/\Delta\lambda$, the ranges of δ_1 and δ_2 are large compared with 2π*. In these circumstances σ_1, σ_2 and σ_{12}^{+} will vary rapidly over the domain of integration and each will change sign many times. In consequence these terms will not contribute appreciably to $I^{(t)}$, and (137) reduces to†

$$I^{(t)} \sim \frac{\mathcal{T}_1^2 \mathcal{T}_2^2}{(1-\mathcal{R}_1^2)(1-\mathcal{R}_2^2)} \int i^{(i)}(k_0)\left(1+2\sum_{r=1}^{\infty}\sum_{s=1}^{\infty}\mathcal{R}_1^r\mathcal{R}_2^s \cos(r\delta_1-s\delta_2)\right)dk_0. \tag{138}$$

In general the quantities $r\delta_1 - s\delta_2$ will also be at least of the order of δ_1 or δ_2, so that the integrals of the cosine terms in (138) will also be negligible, and $I^{(t)}$ is then effectively independent of δ_1 and δ_2. There is, however, an exception when

$$a\left(\Delta\mathcal{S}_1+\frac{2\phi_1}{k_0}\right) - b\left(\Delta\mathcal{S}_2+\frac{2\phi_2}{k_0}\right) = \varepsilon, \tag{139}$$

---

* As is evident on integrating (133) over all spectral components, this implies that with either plate alone the transmitted intensity is effectively independent of δ, i.e. there are no fringes.

† A rigorous justification of the transition from (137a) to (142) would require more refined considerations than those given here.

where a and b are small integers without a common factor, and $|\varepsilon|$ is not large compared with $\bar{\lambda}_0^2/\Delta\lambda_0$. In this case, for

$$r = qa, \qquad s = qb, \qquad (q = 1, 2, 3, \ldots), \tag{140}$$

we have

$$r\delta_1 - s\delta_2 = qk_0\varepsilon, \tag{141}$$

and the range of values of $|r\delta_1 - s\delta_2|$ is $|q\varepsilon\Delta k_0|$, which is not large compared with $2\pi q$; for these values of r and s the integrals of the cosine terms in (138) are not necessarily negligible, and we obtain

$$I^{(t)}(\varepsilon) \sim \frac{\mathscr{T}_1^2\mathscr{T}_2^2}{(1-\mathscr{R}_1^2)(1-\mathscr{R}_2^2)} \int i^{(i)}(k_0)\left(1 + 2\sum_{q=1}^{\infty}(\mathscr{R}_1^a\mathscr{R}_2^b)^q \cos qk_0\varepsilon\right) dk_0. \tag{142}$$

The series under the integral sum of (142) is identical with the series entering the expression (133) for the intensity of a monochromatic wave transmitted by a single plate, and in § 7.6.1 it has been evaluated in a closed form [eq. (13)]. Hence the sum of the series may immediately be written down by comparison,

$$1 + 2\sum_{q=1}^{\infty}(\mathscr{R}_1^a\mathscr{R}_2^b)^q \cos qk_0\varepsilon = \frac{1-(\mathscr{R}_1^a\mathscr{R}_2^b)^2}{(1-\mathscr{R}_1^a\mathscr{R}_2^b)^2 + 4\mathscr{R}_1^a\mathscr{R}_2^b \sin^2 \dfrac{k_0\varepsilon}{2}}. \tag{143}$$

From (142) and (143),

$$I^{(t)}(\varepsilon) = \frac{\mathscr{T}_1^2\mathscr{T}_2^2\{1-(\mathscr{R}_1^a\mathscr{R}_2^b)^2\}}{(1-\mathscr{R}_1^2)(1-\mathscr{R}_2^2)(1-\mathscr{R}_1^a\mathscr{R}_2^b)^2} \int \frac{i^{(i)}(k_0)}{1 + B\sin^2\left(\dfrac{k_0\varepsilon}{2}\right)} dk_0, \tag{144}$$

where

$$B = \frac{4\mathscr{R}_1^a\mathscr{R}_2^b}{(1-\mathscr{R}_1^a\mathscr{R}_2^b)^2}. \tag{145}$$

By comparison with (15b) we see that the intensity distribution (144) is equivalent to a superposition of monochromatic intensity distributions, each of the form shown in Fig. 7.58. The intensity maxima of these distributions occur when $k_0\varepsilon$ is an integral multiple of 2π, i.e. when

$$\varepsilon = m\lambda_0, \qquad |m| = 0, 1, 2, \ldots \tag{146}$$

If we take n_1', n_2', ϕ_1/k_0 and ϕ_2/k_0 to be constant over the spectral range involved in (144), the zero order maxima ($\varepsilon = 0$) of the monochromatic intensity distributions coincide in a central fringe. By (139) and (135), the position of this fringe in the focal plane of the lens L (Fig. 7.80) is defined by

$$\frac{n_2'h_2\cos\theta_2' + c_2}{n_1'h_1\cos\theta_1' + c_1} = \frac{a}{b}, \tag{147}$$

where c_1, c_2 are the values of ϕ_1/k_0, ϕ_2/k_0 respectively. On either side of this fringe, as $|\varepsilon|$ increases, the component patterns are mutually displaced because the scale of each is proportional to wavelength, and the fringes become less distinct. With white light there is a central white fringe in the position given by (147), with coloured maxima and minima on either side, and further away what appears to the eye to be uniform illumination. These fringes are called *fringes of superposition*. They are the multiple-beam form of BREWSTER's fringes (§ 7.5.6), and like them are straight lines, parallel to the apex of the wedge formed between the plates, and with spacing inversely proportional to the wedge angle α.

Fringes of superposition can be used, in a way suggested by FABRY and BUISSON,* to determine the difference of optical thickness of two FABRY–PEROT etalons whose optical thicknesses are very nearly in integral ratio a. For this purpose, one etalon is fixed and the other is tilted until the central white fringe passes through O, the focal point of the lens L for light transmitted normally through the fixed etalon (Fig. 7.80). If the first etalon is fixed we then have $\theta_1 = \theta_1' = 0$, and $\theta_2 = \alpha$, where α is the angle between the etalons. For α small we have, from the law of refraction, $\theta_2' \sim n\alpha/n_2'$, where n is the refractive index of the air surrounding the etalons; $\cos\theta_2' \sim 1 - \theta_2'^2/2 \sim 1 - n^2\alpha^2/2n_2'^2$; and from (147), putting $b = 1$, we have, to the second power of α,

$$(n_2'h_2 + c_2) - a\,(n_1'h_1 + c_1) = \frac{n^2h_2}{2n_2'}\,\alpha^2. \tag{148a}$$

Alternatively, if the second etalon is fixed the white fringe is at O when

$$a(n_1'h_1 + c_1) - (n_2'h_2 + c_2) = \frac{n^2ah_1}{2n_1'}\,\alpha^2. \tag{148b}$$

Even when the etalons are evacuated, as is sometimes the case, we may take $n_1' = n_2' = n$ if the difference $|a(n_1'h_1 + c_1) - (n_2'h_2 + c_2)|$ is sufficiently small, and measurement of α gives this difference in terms of the optical thickness of one of the etalons. Measurements with values of a as high as ten have been made in this way.

A similar arrangement, but with etalons of approximately equal thickness, has been used to measure the refractive index and dispersion of air.† The etalons are first evacuated, and one is tilted to give fringes of convenient spacing with nearly monochromatic light of mean wavelength $\bar{\lambda}_0$. Air is then slowly admitted to the fixed etalon, of thickness h, causing a change $2(n-1)h$ of ε, where n is the refractive index of the air. The corresponding number Δm of fringes passing O is counted, and from (146),

$$\Delta m = \frac{\Delta\varepsilon}{\bar{\lambda}_0} = \frac{2(n-1)h}{\bar{\lambda}_0}, \tag{149}$$

giving $(n-1)$ if h and $\bar{\lambda}_0$ are known.

7.7 THE COMPARISON OF WAVELENGTHS WITH THE STANDARD METRE

The standard of length is the distance between two lines engraved on a bar of platinum–iridium alloy at a temperature of 0°C. This bar, which is kept in France, is called the *International Prototype Metre*, and, as we have already mentioned in § 7.6.4, the relation of optical wavelengths to actual lengths rests on a comparison of the wavelength of the red line (6438 Å) of cadmium with this standard. The comparison was first made in 1892 by MICHELSON and BENOIT‡ using a form of the MICHELSON interferometer, but in 1905 the measurement was repeated with increased accuracy by BENOIT, FABRY and PEROT.§ They used five FABRY–PEROT etalons of

* C. FABRY and H. BUISSON, *Journ. de Phys.* (5), **9** (1919), 189.
† H. BARRELL and J. E. SEARS, *Phil. Trans. Roy. Soc.*, A., **238** (1939), 1.
‡ A. A. MICHELSON and J. R. BENOIT, *Trav. et Mem. Bur. Int. Poids et Mes.*, **11** (1895), 1.
§ J. R. BENOIT, C. FABRY and A. PEROT, *Trav. et Mem. Bur. Int. Poids et Mes.*, **15** (1913), 1.

lengths approximately 6·25, 12·5, 25, 50 and 100 cm. The number of wavelengths of the cadmium red line in the length of the shortest etalon was determined by the method of excess fractions, which requires a knowledge of sufficiently accurate wavelength ratios (§ 7.6.4) but not actual wavelengths. The length of each etalon was then compared with that of the next longer by means of white light fringes. For this, the two etalons were set parallel and illuminated by white light, and the transmitted light was passed through a thin wedge of air formed between two half-silvered plane surfaces. Under these circumstances, fringes similar to those described in § 7.6.8(*b*) are localized in the wedge, the central fringe being along the line where the thickness of the wedge is equal to the difference between the length of the longer etalon and twice the length of the shorter etalon. This difference was obtained in wavelengths of the red cadmium line from a previous calibration of the wedge thickness, and so after four inter-comparisons the number of wavelengths in the length of the 100 cm etalon was determined. Finally, the difference between the length of this etalon and a copy of the prototype metre was measured. This part of the experiment involved making settings with high-quality travelling microscopes on the terminal graduation lines on the metre and on similar graduation lines engraved on the edges of the end plates of the etalon, the distance between these latter lines and the reflecting surfaces of the etalon being determined by a subsidiary experiment. The final result obtained for the wavelength of the red cadmium line in dry air at 15°C and 760 mm Hg pressure was $6438{\cdot}4696 \times 10^{-10}$ m, with a probable error for the interferometric measurements of about 1 part in 10^7.

Further re-determinations* of the relationship between the wavelength of the red cadmium line and the metre were made in a number of standardizing laboratories with methods the same as, or similar in principle to, that of Benoit, Fabry and Perot. The experiment of Sears and Barrell† is of interest because they were able to make direct measurements of the wavelength in vacuum. They used only three Fabry–Perot etalons, the longest being slightly over a metre in length, and the others approximately one-third and one-ninth of this. The spacers were invar cylinders with optically flat chromium-plated ends to which the etalon plates were wrung, and the joint so formed was airtight so that the etalons could be evacuated. The number of wavelengths in the length of the shortest etalon was determined by the method of excess fractions, and the inter-comparison of the etalons was made with the fringes of superposition described in § 7.6.8. The longest etalon was large enough to accommodate a steel end gauge, nominally a metre long, and the distances in wavelengths between the polished ends of the gauge and the reflecting surfaces of the etalon were determined from observations of fringes at infinity with reflected light. In this way the length of the end gauge was obtained in terms of the wavelength of the red cadmium line with an accuracy of about 2 parts in 10^8. Finally, the end gauge was compared with a copy of the metre by well-established procedure for comparison of an end standard with a line standard.‡

When the results of all determinations are reduced to the wavelength in standard air (air at 15°C and 760 mm Hg pressure, containing no water vapour and 0·03 per cent of carbon dioxide), the mean value is found to be $6438{\cdot}4696 \times 10^{-10}$ m, which by chance is the value used for the definition of the Ångström. The greatest departure of any one determination from the mean is 1 part in about 3×10^6, or about

* For references see H. Barrell, *Proc. Roy. Soc.*, A, **186** (1946), 164.

† J. E. Sears and H. Barrell, *Phil. Trans. Roy. Soc.*, A, **231** (1932), 75; **233** (1934), 143.

‡ For details see J. E. Sears, *Proc. Roy. Soc.*, A, **186** (1946), 152.

3×10^{-4} mm in a metre. This is much less than the width of the graduation lines on the Prototype Metre and on the copies of it which have been used in the experiments, and there is no doubt that it represents the uncertainty in the comparisons made with travelling microscopes. The interferometric measurements can be made with much higher precision, so that it is natural to consider the replacement of the material standard by a definition of the metre in terms of the wavelength of some spectral line.

The International Committee of Weights and Measures accepted a proposal to this effect in 1954,* with the proviso that, in order to preserve as far as possible continuity in the standard of length, the re-definition should be consistent with the value $6438{\cdot}4696 \times 10^{-10}$ m for the wavelength of the red cadmium line. In 1958 the Committee decided† that the best available choice of spectral line is the line of approximate wavelength 6056 Å corresponding to the transition between levels $2p_{10}$ and $5d_5$ of the krypton atom of mass number 86; and on the basis of measurements in five different laboratories it proposed that the metre should be defined as exactly 1,650,763·73 wavelengths in vacuum of this radiation. This definition was approved unanimously in 1960 by the 11th General Conference of Weights and Measures.‡ The primary standards of wavelength and the primary standard of length are therefore now identical, and the Ångström is exactly 10^{-10} m.

---

* *Proc. Verb. Com. Int. Poids et Mes.*, **24** (1954) 2.

† *Proc. Verb. Com. Int. Poids et Mes.* (2), **26–B** (1958) M.30.

‡ *C.R.* 11me *Conf. Gén. Poids et Mes.* (Paris, Gauthier-Villars, 1960) pp. 51, 85.

CHAPTER VIII

ELEMENTS OF THE THEORY OF DIFFRACTION

8.1. INTRODUCTION

In carrying out the transition from the general electromagnetic field to the optical field, which is characterized by very high frequencies (short wavelengths), we found that in certain regions the simple geometrical model of energy propagation was inadequate. In particular, we saw that deviations from this model must be expected in the immediate neighbourhood of the boundaries of shadows and in regions where a large number of rays meet. These deviations are manifested by the appearance of dark and bright bands, the diffraction fringes. Diffraction theory is mainly concerned with the field in these special regions; such regions are of great practical interest as they include the part of the image space in which the optical image is situated (region of focus).

The first reference to diffraction phenomena appears in the work of Leonardo da Vinci (1452–1519). Such phenomena were, however, first accurately described by Grimaldi in a book, published in 1665, two years after his death. The corpuscular theory, which, at the time, was widely believed to describe correctly the propagation of light, could not explain diffraction. Huygens, the first proponent of the wave theory, seems to have been unaware of Grimaldi's discoveries; otherwise he would have undoubtedly quoted them in support of his views. The possibility of explaining diffraction effects on the basis of a wave theory was not noticed until about 1818. In that year there appeared the celebrated memoir of Fresnel (cf. Historical Introduction) in which he showed that diffraction can be explained by the application of Huygens' construction (cf. § 3.3.3) together with the principle of interference. Fresnel's analysis was later put on a sound mathematical basis by Kirchhoff (1882), and the subject has since then been extensively discussed by many writers.*

Diffraction problems are amongst the most difficult ones encountered in optics. Solutions which, in some sense, can be regarded as rigorous are very rare in diffraction theory. The first such solution was given as late as 1896 by A. Sommerfeld when, in an important paper, he discussed the diffraction of a plane wave by a perfectly conducting semi-infinite plane screen. Since then rigorous solutions of a small number of other diffraction problems (mainly two-dimensional) have also been found (cf. Chapter XI), but, because of mathematical difficulties, approximate methods must be used in most cases of practical interest. Of these the theory of Huygens and Fresnel is by far the most powerful and is adequate for the treatment of the majority of problems encountered in instrumental optics. This theory and some of its applications form the main subject matter of this chapter.

8.2. THE HUYGENS–FRESNEL PRINCIPLE

According to Huygens' construction (§ 3.3.3), every point of a wave-front may be considered as a centre of a secondary disturbance which gives rise to spherical wavelets,

* For a fuller historical account of the development of the subject see C. F. Meyer, *The Diffraction of Light, X-rays, and Material Particles* (Chicago, The University Press, 1934).

and the wave-front at any later instant may be regarded as the envelope of these wavelets. FRESNEL was able to account for diffraction by supplementing HUYGENS' construction with the postulate that the secondary wavelets mutually interfere. This combination of HUYGENS' construction with the principle of interference is called the *Huygens–Fresnel Principle.* Before applying it to the study of diffraction effects we shall verify that (with certain simple additional assumptions) the principle correctly describes the propagation of light in free space.

Let S (Fig. 8.1) be the instantaneous position of a spherical monochromatic wave-front of radius r_0 which proceeds from a point source P_0, and let P be a point at

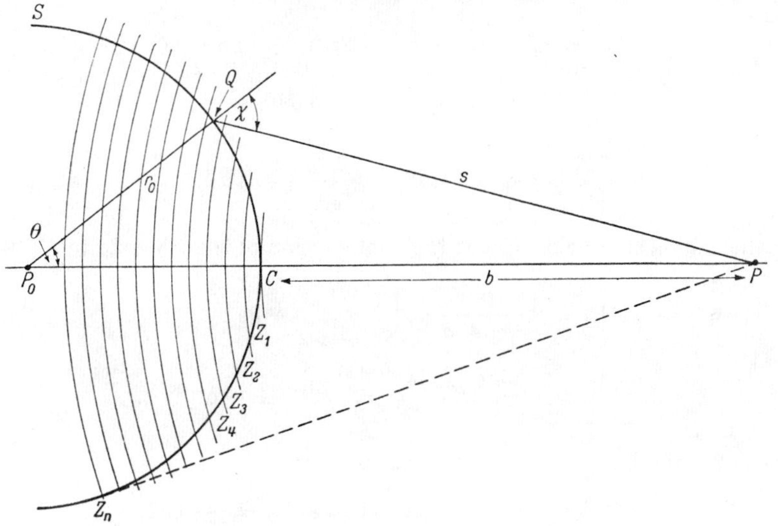

Fig. 8.1. FRESNEL's zone construction.

which the light disturbance is to be determined. The time periodic factor $e^{-i\omega t}$ being omitted, the disturbance at a point Q on the wave-front may be represented by Ae^{ikr_0}/r_0, where A is the amplitude at unit distance from the source. In accordance with the HUYGENS–FRESNEL principle we regard each element of the wave-front as the centre of a secondary disturbance which is propagated in the form of spherical wavelets, and obtain for the contribution $dU(P)$ due to the element dS at Q the expression

$$dU(P) = K(\chi)\frac{Ae^{ikr_0}}{r_0}\frac{e^{iks}}{s}\,dS,$$

where $s = QP$ and $K(\chi)$ is an *inclination factor* which describes the variation with direction of the amplitude of the secondary waves, χ being the angle (often called the *angle of diffraction*) between the normal at Q and the direction QP. Following FRESNEL we assume that K is maximum in the original direction of propagation, i.e. for $\chi = 0$, and that it rapidly decreases with increasing χ, being zero when QP is tangential to the wave-front, i.e. when $\chi = \pi/2$; and finally, that only that part S' of the primary wave contributes to the effect at P, which is not obstructed by obstacles which may be situated between P_0 and P. Hence the total disturbance at P is given by

$$U(P) = \frac{Ae^{ikr_0}}{r_0}\iint_S \frac{e^{iks}}{s}\,K(\chi)dS. \qquad (1)$$

To evaluate (1) we shall use the so-called *zone construction* of FRESNEL. With centre at P, we construct spheres of radii

$$b, \quad b+\frac{\lambda}{2}, \quad b+\frac{2\lambda}{2}, \quad b+\frac{3\lambda}{2}, \ldots \ldots, b+\frac{j\lambda}{2}, \ldots .$$

where $b = CP$, C being the point of intersection of P_0P with the wave-front S (see Fig. 8.1). The spheres divide S into a number of zones $Z_1, Z_2, Z_3 \ldots Z_j \ldots$.

We assume that both r_0 and b are large compared to the wavelength; then K may be assumed to have the same value, K_j, for points on one and the same zone. From the figure

$$s^2 = r_0^2 + (r_0+b)^2 - 2r_0(r_0+b)\cos\theta,$$

so that

$$s ds = r_0(r_0+b)\sin\theta d\theta, \tag{2}$$

and therefore

$$dS = r_0^2 \sin\theta d\theta d\phi = \frac{r_0}{r_0+b}\, s\, ds\, d\phi,$$

ϕ being the azimuthal angle. Hence the contribution of the jth zone to $U(P)$ is

$$U_j(P) = 2\pi \frac{Ae^{ikr_0}}{r_0+b} K_j \int_{b+(j-1)\lambda/2}^{b+j\lambda/2} e^{iks}\, ds$$

$$= -\frac{2\pi i}{k} K_j \frac{Ae^{ik(r_0+b)}}{r_0+b} e^{ikj\lambda/2}(1-e^{-ik\lambda/2}).$$

Since $k\lambda = 2\pi$, the last two factors reduce to

$$e^{ikj\lambda/2}(1-e^{-ik\lambda/2}) = e^{i\pi j}(1-e^{-i\pi}) = (-1)^j 2,$$

so that

$$U_j(P) = 2i\lambda(-1)^{j+1} K_j \frac{Ae^{ik(r_0+b)}}{r_0+b} \tag{3}$$

We note that the contributions of the successive zones are alternately positive and negative. The total effect at P is obtained by summing all the contributions:

$$U(P) = 2i\lambda \frac{Ae^{ik(r_0+b)}}{r_0+b} \sum_{j=1}^{n} (-1)^{j+1} K_j. \tag{4}$$

The series

$$\Sigma = \sum_{j=1}^{n} (-1)^{j+1} K_j = K_1 - K_2 + K_3 - \ldots + (-1)^{n+1} K_n \tag{5}$$

can now be approximately summed by a method due to SCHUSTER*.

First we write (5) in the form

$$\Sigma = \frac{K_1}{2} + \left(\frac{K_1}{2} - K_2 + \frac{K_3}{2}\right) + \left(\frac{K_3}{2} - K_4 + \frac{K_5}{2}\right) + \ldots \ldots \tag{6}$$

the last term being $\frac{1}{2}K_n$ or $\frac{1}{2}K_{n-1} - K_n$ according to n being odd or even. Let us assume for the moment that the law which specifies the directional variation is such

* A. SCHUSTER, *Phil. Mag.* (5), **31** (1891), p. 77.

that K_j is *greater* than the arithmetic mean of its two neighbours K_{j-1} and K_{j+1}. Then each of the bracketed terms in (6) is negative and it follows that

$$\left.\begin{aligned} \Sigma &< \frac{K_1}{2} + \frac{K_n}{2} \qquad \text{when } n \text{ is odd} \\ \text{and} \qquad & \\ \Sigma &< \frac{K_1}{2} + \frac{K_{n-1}}{2} - K_n \qquad \text{when } n \text{ is even.} \end{aligned}\right\} \tag{7}$$

We can also write (5) in the form

$$\Sigma = K_1 - \frac{K_2}{2} - \left(\frac{K_2}{2} - K_3 + \frac{K_4}{2}\right) - \left(\frac{K_4}{2} - K_5 + \frac{K_6}{2}\right) - \cdots\cdots \tag{8}$$

the last term now being $-\frac{1}{2}K_{n-1} + K_n$ when n is odd and $-\frac{1}{2}K_n$ when n is even. Hence

$$\left.\begin{aligned} \Sigma &> K_1 - \frac{K_2}{2} - \frac{K_{n-1}}{2} + K_n \qquad (n \text{ odd}) \\ \text{and} \qquad & \\ \Sigma &> K_1 - \frac{K_2}{2} - \frac{K_n}{2} \qquad (n \text{ even}). \end{aligned}\right\} \tag{9}$$

Now each K_j differs only slightly from its neighbouring values K_{j-1} and K_{j+1} so that the right-hand sides of the corresponding relations in (7) and (9) are practically equal, and, therefore, approximately,

$$\left.\begin{aligned} \Sigma &= \frac{K_1}{2} + \frac{K_n}{2} \qquad (n \text{ odd}) \\ \text{and} \qquad & \\ \Sigma &= \frac{K_1}{2} - \frac{K_n}{2} \qquad (n \text{ even}). \end{aligned}\right\} \tag{10}$$

It may easily be verified that (10) remains valid when each K_j is *smaller* than the arithmetic mean of its two neighbours, each of the bracketed terms in (6) and (8) then being positive. Moreover, (10) may be expected to remain valid even when only some of the bracketed terms are negative whilst the others are positive, for the series may then be divided into two parts according to the signs of the bracketed terms and a similar argument may be applied to each part. We may, therefore, conclude that the sum of the series is given by (10) unless the bracketed terms in (6) and (8) change sign so frequently that the error terms add up to an appreciable amount. If we exclude the later case it follows from (10) and (4) that

$$U(P) = i\lambda(K_1 \pm K_n)\frac{Ae^{ik(r_0+b)}}{r_0 + b}, \tag{11}$$

the upper or lower sign being taken according as n is odd or even. Using (3), equation (11) may be also written in the form

$$U(P) = \tfrac{1}{2}[U_1(P) + U_n(P)]. \tag{12}$$

For the last zone (Z_n) that can be seen from P, QP is a tangent to the wave, i.e. $\chi = \pi/2$, and for this value of χ, as already mentioned, K was assumed to be zero Hence $K_n = 0$ and (11) reduces to

$$U(P) = i\lambda K_1 \frac{Ae^{ik(r_0+b)}}{r_0 + b} = \tfrac{1}{2}U_1(P), \tag{13}$$

showing that *the total disturbance at P is equal to half of the disturbance due to the first zone.*

Equation (13) is in agreement with the expression for the effect of the spherical wave if

$$i\lambda K_1 = 1,$$

i.e. if

$$K_1 = -\frac{i}{\lambda} = \frac{e^{-i\pi/2}}{\lambda}. \tag{14}$$

The factor $e^{-i\pi/2}$ may be accounted for by assuming that the secondary waves oscillate a quarter of a period out of phase with the primary wave; the other factor can be explained by assuming that the amplitudes of the secondary vibrations are to the amplitudes of the primary vibrations in the ratio $1:\lambda$. We can therefore conclude that, with these assumptions about the amplitude and phase of the secondary waves, the HUYGENS–FRESNEL principle leads to the correct expression for the propagation of a spherical wave in free space. The additional assumptions must, however, be regarded as purely a convenient way of interpreting the mathematical expressions and as being devoid of any physical significance; the real justification of the factor (14) will become evident later (§ 8.3).

Still following FRESNEL, let us consider the effect at P when some of the zones are obstructed by a plane screen with a circular opening, perpendicular to P_0P and with its centre on this line. The total disturbance at P must now be regarded as due to wavelets from only those zones that are not obstructed by the screen. When the screen covers all but half of the first zone, (3) gives, on setting $j = 1$, and multiplying by $\frac{1}{2}$,

$$U(P) = i\lambda K_1 \frac{Ae^{ik(r_0+b)}}{r_0 + b} = \frac{Ae^{ik(r_0+b)}}{r_0 + b}; \tag{15}$$

hence the disturbance at P is now the same as would be obtained if no screen were present. When all the zones are covered except the first, (3) gives

$$U(P) = 2i\lambda K_1 \frac{Ae^{ik(r_0+b)}}{r_0 + b} = 2\frac{Ae^{ik(r_0+b)}}{r_0 + b}, \tag{16}$$

so that the intensity $I(P) = |U(P)|^2$ is four times larger than if the screen were absent. When the opening is increased still further the intensity will decrease, since the first two terms in (4) have different signs. Moreover, since K_1 and K_2 are nearly equal, it follows that there will be almost complete darkness at P when the opening is approximately equal to the first two zones. Thus, when the size of the opening is varied, there is a periodic fluctuation in the intensity at P. Similar results are obtained when the size of the opening and the source are fixed, but the position of the point P of observation is varied along the axis; for then, as P gradually approaches the screen, an increasingly larger number of zones is required to fill the opening completely.

All these results were found to be in good agreement with experiment. One prediction of FRESNEL's theory made a strong impression on his contemporaries, and was, in fact, one of the decisive factors which temporarily ended the long battle between the corpuscular and the wave theory of light in favour of the latter. It concerns the effect which arises when the first zone is obstructed by a small circular disc placed at right angles to P_0P. According to (5) the complex amplitude at P is then given by

$$U(P) = 2i\lambda \frac{Ae^{ik(r_0+b)}}{r_0 + b}[- K_2 + K_3 - K_4 + \dots], \tag{17}$$

and, by a similar argument as before, the sum of the series in the brackets is $- K_2/2$. Since K_2 is assumed to differ only slightly from $K_1 = 1/i\lambda$, it follows that there is light in the geometrical shadow of the disc, and, moreover, that the intensity there is the same as if no disc were present.*

8.3. KIRCHHOFF'S DIFFRACTION THEORY

8.3.1 The integral theorem of Kirchhoff

The basic idea of the HUYGENS–FRESNEL theory is that the light disturbance at a point P arises from the superposition of secondary waves that proceed from a surface situated between this point and the light source. This idea was put on a sounder mathematical basis by KIRCHHOFF†, who showed that the HUYGENS-FRESNEL principle may be regarded as an approximate form of a certain integral theorem‡ which expresses the solution of the homogeneous wave equation, at an arbitrary point in the field, in terms of the values of the solution and its first derivatives at all points on an arbitrary closed surface surrounding P.

We consider first a strictly monochromatic scalar wave

$$V(x, y, z, t) = U(x, y, z)e^{-i\omega t}. \tag{1}$$

In vacuum the space-dependent part then satisfies the time-independent wave equation

$$(\nabla^2 + k^2)U = 0, \tag{2}$$

where $k = \omega/c$. Equation (2) is also known as the HELMHOLTZ equation.

---

* That a bright spot should appear at the centre of the shadow of a small disc was deduced from FRESNEL's theory by S. D. POISSON in 1818. POISSON, who was a member of the committee of the French Academy which reviewed FRESNEL's prize memoir, appears to have considered this conclusion contrary to experiment and so refuting FRESNEL's theory. However, ARAGO, another member of the committee, performed the experiment and found that the surprising prediction was correct. A similar observation had been made a century earlier by MARALDI but had been forgotten.

† G. KIRCHHOFF, *Berl. Ber.* (1882), 641; *Ann. d. Physik.* (2) **18** (1883), 663; *Ges. Abh. Nachtr.*, 22.

KIRCHHOFF's theory applies to the diffraction of scalar waves. As will be shown in § 8.4 a scalar theory is usually quite adequate for the treatment of the majority of problems of instrumental optics.

Vectorial generalizations of the Huygens–Fresnel principle have been proposed by many authors. The first satisfactory generalization is due to F. KOTTLER, *Ann. d. Physik*, **71** (1923), 457; **72** (1923), 320. [Cf. B. B. BAKER and E. T. COPSON, *The Mathematical Theory of Huygens' Principle*, (Oxford, Clarendon Press, 2nd edition, 1950), p. 114].

‡ For monochromatic waves this theorem was derived earlier in acoustics by H. VON HELMHOLTZ, *J. f. Math.*, **57** (1859), 7.

Let v be a volume bounded by a closed surface S, and let P be any point within it; we assume that U possesses continuous first- and second-order partial derivatives within and on this surface. If U' is any other function which satisfies the same continuity requirements as U, we have by GREEN's theorem

$$\iiint_v (U\nabla^2 U' - U'\nabla^2 U)dv = -\iint_S \left(U\frac{\partial U'}{\partial n} - U'\frac{\partial U}{\partial n}\right) dS, \tag{3}$$

where $\partial/\partial n$ denotes differentiation along the *inward** normal to S. In particular, if U' also satisfies the time-independent wave equation, i.e. if

$$(\nabla^2 + k^2)U' = 0, \tag{4}$$

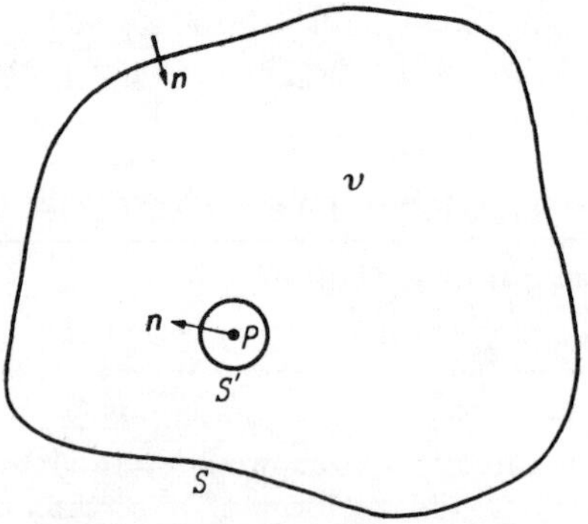

Fig. 8.2. Derivation of the HELMHOLTZ–KIRCHHOFF integral theorem: region of integration.

then it follows at once from (2) and (4) that the integrand on the left of (3) vanishes at every point of v, and consequently

$$\iint_S \left(U\frac{\partial U'}{\partial n} - U'\frac{\partial U}{\partial n}\right) dS = 0. \tag{5}$$

Suppose we take $U'(x, y, z) = e^{iks}/s$, where s denotes the distance from P to the point (x, y, z). This function has a singularity for $s = 0$, and since U' was assumed to be continuous and differentiable, P must be excluded from the domain of integration. We shall therefore surround P by a small sphere of radius ε and extend the integration throughout the volume between S and the surface S' of this sphere (Fig. 8.2). In place of (5), we then have

$$\iint_S + \iint_{S'} \left\{U\frac{\partial}{\partial n}\left(\frac{e^{iks}}{s}\right) - \frac{e^{iks}}{s}\frac{\partial U}{\partial n}\right\} dS = 0,$$

whence

$$\iint_S \left\{U\frac{\partial}{\partial n}\left(\frac{e^{iks}}{s}\right) - \frac{e^{iks}}{s}\frac{\partial U}{\partial n}\right\} dS = -\iint_{S'} \left\{U\frac{e^{iks}}{s}\left(ik - \frac{1}{s}\right) - \frac{e^{iks}}{s}\frac{\partial U}{\partial n}\right\} dS'$$

$$= -\iint_{\Omega} \left\{U\frac{e^{ik\varepsilon}}{\varepsilon}\left(ik - \frac{1}{\varepsilon}\right) - \frac{e^{ik\varepsilon}}{\varepsilon}\frac{\partial U}{\partial s}\right\} \varepsilon^2 d\Omega, \tag{6}$$

where $d\Omega$ denotes an element of the solid angle. Since the integral over S is independent of ε, we may replace the integral on the right-hand side by its limiting value

* GREEN's theorem is usually expressed in terms of the outward normal, but the inward normal is more convenient in the present application.

as $\varepsilon \to 0$; the first and third terms in this integral give no contribution in the limit, and the total contribution of the second term is $4\pi U(P)$. Hence

$$U(P) = \frac{1}{4\pi} \iint_S \left\{ U \frac{\partial}{\partial n} \left(\frac{e^{iks}}{s} \right) - \frac{e^{iks}}{s} \frac{\partial U}{\partial n} \right\} dS. \tag{7}$$

This is one form of the *integral theorem of Helmholtz and Kirchhoff.**

We note, that as $k \to 0$, the time-independent wave equation (2) reduces to LAPLACE's equation $\nabla^2 U = 0$, and (7) then goes over into the well-known formula of potential theory

$$U(P) = \frac{1}{4\pi} \iint_S \left\{ U \frac{\partial}{\partial n} \left(\frac{1}{s} \right) - \frac{1}{s} \frac{\partial U}{\partial n} \right\} dS. \tag{8}$$

If P lies outside the surface S, but U is still assumed to be continuous and differentiable up to the second order within S, and if as before we take $U' = e^{iks}/s$, equation (3) remains valid throughout the whole volume within S. According to (5) the surface integral then has the value zero.

There is a complementary form of the HELMHOLTZ–KIRCHHOFF theorem for the case when U is continuous and differentiable up to the second order *outside* and on a closed surface S (sources inside). In this case, however, as in other problems of propagation in an infinite medium, the boundary values on S are no longer sufficient to specify the solution uniquely and additional assumptions must be made about the behaviour of the solution as $s \to \infty$. For a discussion of this case we must, however, refer elsewhere.†

So far we have considered strictly monochromatic waves. We now derive the general form of KIRCHHOFF's theorem which applies to waves that are not necessarily monochromatic.

Let $V(x, y, z, t)$ be a solution of the wave equation

$$\nabla^2 V = \frac{1}{c^2} \frac{\partial^2 V}{\partial t^2}, \tag{9}$$

and assume that V can be represented in the form of a FOURIER integral

$$V(x, y, z, t) = \frac{1}{\sqrt{2\pi}} \int_{-\infty}^{+\infty} U_\omega(x, y, z) e^{-i\omega t} \, d\omega. \tag{10}$$

Then, by the FOURIER inversion formula

$$U_\omega(x, y, z) = \frac{1}{\sqrt{2\pi}} \int_{-\infty}^{+\infty} V(x, y, z, t) e^{i\omega t} \, dt. \tag{11}$$

Since $V(x, y, z, t)$ is assumed to satisfy the wave equation (9), $U_\omega(x, y, z)$ will satisfy the time independent wave equation (2). If moreover V obeys the appropriate regularity

---

* This theorem expresses $U(P)$ in terms of the values of both U and $\partial U/\partial n$ on S. It may, however, be shown from the theory of GREEN's functions that the values of either U or $\partial U/\partial n$ on S are sufficient to specify U at every point P within S. (See for example F. POCKELS: *Über die Partielle Differentialgleichung* $(\nabla^2 + k^2)U = 0$ (Leipzig, Teubner, 1891.) However, only in the simplest cases, e.g. when S is a plane, is it possible to determine the appropriate GREEN's function [cf. A. SOMMERFELD, *Optics* (New York, Academic Press, 1954), p. 199].

† See for example B. B. BAKER and E. T. COPSON, *The Mathematical Theory of Huygens' Principle* (Oxford, Clarendon Press, 2nd ed., 1950), pp. 24–25.

conditions within and on a closed surface S, we may apply the KIRCHHOFF formula separately to each FOURIER component $U_\omega(x, y, z) = U_\omega(P)$:

$$U_\omega(P) = \frac{1}{4\pi}\iint_S \left\{U_\omega \frac{\partial}{\partial n}\left(\frac{e^{iks}}{s}\right) - \frac{e^{iks}}{s}\frac{\partial U_\omega}{\partial n}\right\} dS. \tag{12}$$

When we change the order of integration and set $k = \omega/c$, (10) becomes,

$$\begin{aligned} V(P, t) &= \frac{1}{4\pi}\iint_S dS \frac{1}{\sqrt{2\pi}}\int_{-\infty}^{+\infty}\left\{U_\omega \frac{\partial}{\partial n}\left(\frac{e^{-i\omega(t-s/c)}}{s}\right) - \frac{e^{-i\omega(t-s/c)}}{s}\frac{\partial U_\omega}{\partial n}\right\} d\omega \\ &= \frac{1}{4\pi}\iint_S dS \frac{1}{\sqrt{2\pi}}\int_{-\infty}^{+\infty}\left\{U_\omega \left\{\frac{\partial}{\partial n}\left(\frac{1}{s}\right) + \frac{i\omega}{sc}\frac{\partial s}{\partial n}\right\} e^{-i\omega(t-s/c)} - \frac{e^{-i\omega(t-s/c)}}{s}\frac{\partial U_\omega}{\partial n}\right\} d\omega \end{aligned}$$

or using (10),

$$V(P, t) = \frac{1}{4\pi}\iint_S \left\{[V]\frac{\partial}{\partial n}\left(\frac{1}{s}\right) - \frac{1}{cs}\frac{\partial s}{\partial n}\left[\frac{\partial V}{\partial t}\right] - \frac{1}{s}\left[\frac{\partial V}{\partial n}\right]\right\} dS. \tag{13}$$

The square brackets denote "retarded values", i.e. values of the functions taken at the time $t - s/c$. The formula (13) is the general form of *Kirchhoff's theorem.*

It can also be seen by analogy with the previous case, that the value of the integra in (13) is zero when P is outside S.

The last term in (13) represents the contribution of a distribution of sources of strength $-\dfrac{1}{4\pi}\dfrac{\partial V}{\partial n}$ per unit area, whilst the first two terms may be shown to represent a contribution of doublets of strength $V/4\pi$ per unit area, directed normally to the surface. Naturally these sources and doublets are fictitious, there being no deep physical significance behind such an interpretation.

8.3.2 Kirchhoff's diffraction theory

Whilst the integral theorem of KIRCHHOFF embodies the basic idea of the HUYGENS–FRESNEL principle, the laws governing the contributions from different elements of the surface are more complicated than FRESNEL assumed. KIRCHHOFF showed, however, that in many cases the theorem may be reduced to an approximate but much simpler form, which is essentially equivalent to the formulation of FRESNEL, but which in addition gives an explicit formula for the inclination factor that remained undetermined in FRESNEL's theory.

Consider a monochromatic wave, from a point source P_0, propagated through an opening in a plane opaque screen, and let P as before be the point at which the light disturbance is to be determined. We assume that the linear dimensions of the opening, although large compared to the wavelength, are small compared to the distances of both P_0 and P from the screen.

To find the disturbance at P we take KIRCHHOFF's integral over a surface S formed by (see Fig. 8.3(a)): (1) the opening $\mathscr{A}$, (2) a portion $\mathscr{B}$ of the non-illuminated side of the screen, and (3) a portion $\mathscr{C}$ of a large sphere of radius R, centred at P which, together with $\mathscr{A}$ and $\mathscr{B}$, forms a closed surface.

KIRCHHOFF's theorem, expressed by equation (7), then gives

$$U(P) = \frac{1}{4\pi}\left[\iint_{\mathscr{A}} + \iint_{\mathscr{B}} + \iint_{\mathscr{C}}\right]\left\{U\frac{\partial}{\partial n}\left(\frac{e^{iks}}{s}\right) - \left(\frac{e^{iks}}{s}\right)\frac{\partial U}{\partial n}\right\} dS, \tag{14}$$

where, as before, s is the distance of the element dS from P and $\partial/\partial n$ denotes differentiation along the inward normal to the surface of integration.

The difficulty is encountered that the values of U and $\partial U/\partial n$ on $\mathscr{A}$, $\mathscr{B}$, and $\mathscr{C}$ which should be substituted into (14) are never known exactly. However, it is reasonable to suppose that everywhere on $\mathscr{A}$, except in the immediate neighbourhood of the rim of the opening, U and $\partial U/\partial n$ will not appreciably differ from the values

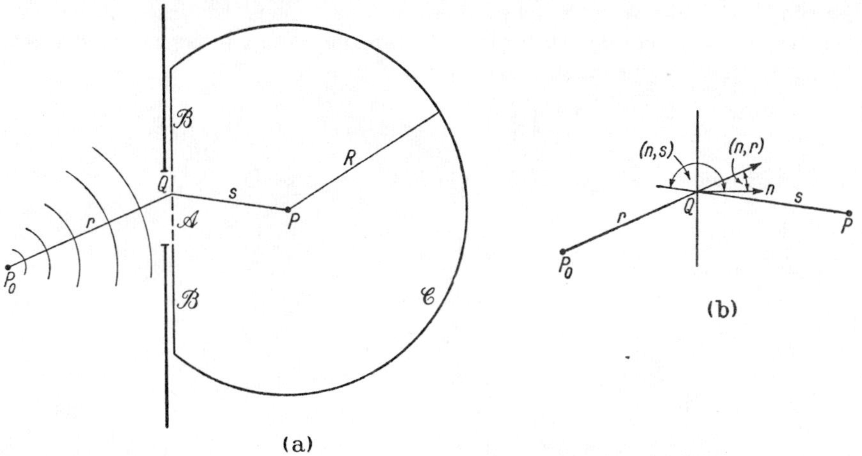

Fig. 8.3. Illustrating the derivation of the Fresnel–Kirchhoff diffraction formula.

obtained in the absence of the screen, and that on $\mathscr{B}$ these quantities will be approximately zero. Kirchhoff accordingly set

$$\left.\begin{aligned} \text{on } \mathscr{A}: \quad & U = U^{(i)}, \quad & \frac{\partial U}{\partial n} = \frac{\partial U^{(i)}}{\partial n}, \\ \text{on } \mathscr{B}: \quad & U = 0 & \frac{\partial U}{\partial n} = 0, \end{aligned}\right\} \tag{15}$$

where

$$U^{(i)} = \frac{Ae^{ikr}}{r}, \qquad \frac{\partial U^{(i)}}{\partial n} = \frac{Ae^{ikr}}{r}\left[ik - \frac{1}{r}\right]\cos(n, r) \tag{16}$$

are the values relating to the incident field (see Fig. 8.3 (b)) and A is a constant. The approximations (15) are called *Kirchhoff's boundary conditions* and are the basis of *Kirchhoff's diffraction theory*.

It remains to consider the contribution from the spherical portion $\mathscr{C}$. Now it is evident that by taking the radius R sufficiently large, the values of U and $\partial U/\partial n$ on $\mathscr{C}$ may be made arbitrarily small, which suggests that the contribution from $\mathscr{C}$ may be neglected. However, by letting R increase indefinitely, the area of $\mathscr{C}$ also increases beyond all limits, so that the condition $U \to 0$ and $\partial U/\partial n \to 0$ as $R \to \infty$ is not sufficient to make the integral vanish. A more precise assumption about the behaviour of the wave function at a large distance from the screen must therefore be made, a point which we have already touched upon on p. 377 in connection with the uniqueness of solutions in problems involving an infinite medium. For our purposes it is sufficient to make the physically obvious assumption that the radiation field does not exist at all times but that it is produced by a source that begins to radiate at some particular instant of time $t = t_0$.* (This, of course, implies, that we now depart from strict monochromacy, since a perfectly monochromatic field would exist for all

* This assumption is not essential but shortens the discussion. For a more formal argument see M. Born, *Optik* (Berlin, Springer, 1933), p. 149.

times.) Then at any time $t > t_0$, the field fills a region of space the outer boundary of which is at distance not greater than $c(t - t_0)$ from P_0, c being the velocity of light. Hence if the radius R is chosen so large that at the time when the disturbance at P is considered no contributions from $\mathscr{C}$ could have reached P because at the appropriate earlier time the field has not reached these distant regions, the integral over $\mathscr{C}$ will vanish. Thus finally, on substituting into (14), and neglecting in the normal derivatives the terms $1/r$ and $1/s$ in comparison to k, we obtain

$$U(P) = -\frac{iA}{2\lambda} \iint_{\mathscr{A}} \frac{e^{ik(r+s)}}{rs} [\cos(n, r) - \cos(n, s)] dS, \tag{17}$$

which is known as the *Fresnel–Kirchhoff diffraction formula.*

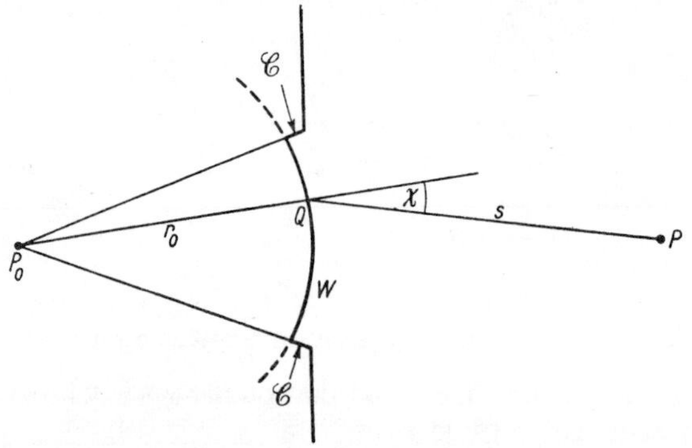

Fig. 8.4. Illustrating the diffraction formula (18).

It is evident that in place of $\mathscr{A}$ any other open surface, the rim of which coincides with the edge of the aperture, could have been chosen. In particular we may choose instead of $\mathscr{A}$ a portion W of an incident wave front which approximately fills the aperture, together with a portion $\mathscr{C}$ of a cone with vertex at P_0 and with generators through the rim of the aperture (Fig. 8.4). If the radius of curvature of the wave is sufficiently large, the contribution from $\mathscr{C}$ may obviously be neglected. Also, on W, $\cos(n, r_0) = 1$. If further we set $\chi = \pi - (r_0, s)$ we obtain, in place of (17),

$$U(P) = -\frac{i}{2\lambda} \frac{Ae^{ikr_0}}{r_0} \iint_W \frac{e^{iks}}{s} (1 + \cos\chi) dS, \tag{18}$$

where r_0 is the radius of the wave-front W. This result is in agreement with FRESNEL's formulation of HUYGENS' principle if, as the contribution from the element dW of the wave-front we take

$$-\frac{i}{2\lambda} \frac{Ae^{ikr_0}}{r_0} \frac{e^{iks}}{s} (1 + \cos\chi) dS. \tag{19}$$

Comparison of (18) with § 8.2 (1) gives for the inclination factor of FRESNEL's theory the expression*

$$K(\chi) = -\frac{i}{2\lambda} (1 + \cos\chi) \tag{20}$$

For the central zone $\chi = 0$, and (20) gives $K_1 = K(0) = -i/\lambda$ in agreement with § 8.2. (14). It is however not true, as FRESNEL assumed, that $K(\pi/2) = 0$.

* Expression (20) for the inclination factor was first derived by G. G. STOKES, *Trans. Camb. Phil. Soc.*, **9** (1849), 1; reprinted in his *Math. and Phys. Papers* (Cambridge University Press, **2** (1883), 243).

Returning now to the FRESNEL–KIRCHHOFF diffraction formula (17), we note that it is symmetrical with respect to the source and the point of observation. This implies that *a point source at P_0 will produce at P the same effect as a point source of equal intensity placed at P will produce at P_0*. This result is sometimes referred to as the *reciprocity theorem* (or the *reversion theorem*) *of Helmholtz*.

So far we have assumed that the light on its passage from the source to P does not encounter any other surface than the diffracting screen; the incident waves are then spherical. The analysis can be easily extended to cover more complicated cases, where the waves are no longer of such simple form. It is again found that, provided the radii of curvature at each point of the wave-front are large compared to the wave-length, and provided that the angles involved are sufficiently small, the results of KIRCHHOFF's theory are substantially equivalent to predictions based on the HUYGENS–FRESNEL principle.

From the preceding discussion we can also immediately draw a conclusion which concerns the distribution of light diffracted by complementary screens, i.e. by screens which are such that the openings in one correspond exactly to the opaque portions of the others and vice versa. Let $U_1(P)$ and $U_2(P)$ denote respectively the values of the complex displacement when the first or the second screen alone is placed between the source and the point P of observation, and let $U(P)$ be the value when no screen is present. Then, since U_1 and U_2 can be expressed as integrals over the openings, and since the openings in the two screens just add up to fill the whole plane,

$$U_1 + U_2 = U. \tag{21}$$

This result is known as *Babinet's principle.**

From BABINET's principle two conclusions follow at once: If $U_1 = 0$, then $U_2 = U$; hence at points at which the intensity is zero in the presence of one of the screens, the intensity in the presence of the other is the same as if no screen was present. Further if $U = 0$, then $U_1 = - U_2$; this implies that, at points where U is zero, the phases of U_1 and U_2 differ by π and the intensities $I_1 = |U_1|^2$, $I_2 = |U_2|^2$ are equal. If for example a point source is imaged by an error-free lens, the light distribution U in the image plane will be zero except in the immediate neighbourhood of the image O of the source. If then complementary screens are placed between the object and the image one has $I_1 = I_2$ except in the neighbourhood of O.

The consequences of the basic approximation (15) of KIRCHHOFF's theory have been subject to many critical discussions, which showed, for example, that KIRCHHOFF's solution does not reproduce the assumed values in the plane of the aperture†. However, more recently it was shown by WOLF and MARCHAND‡ that KIRCHHOFF's theory

---

* A. BABINET, *Compt. Rend.*, **4** (1837), 638. An analogous theorem of this type, which involves the electromagnetic field vectors rather than the single scalar U and which may be considered as rigorous formulation of BABINET's principle, is given in § 11.3.

† H. POINCARÉ, *Théorie mathématique de la lumière* (Paris, George Carré, II (1892)), pp. 187–8. See also B. B. BAKER and E. T. COPSON, *The Mathematical Theory of Huygens' Principle* (Oxford, Clarendon Press, 2nd ed., 1950), pp. 71–72 and G. TORALDO DI FRANCIA, *Atti Fond. Giorgio Ronchi*, **XI** (1956), § 6.

‡ E. WOLF and E. W. MARCHAND, *J. Opt. Soc. Amer.*, **56** (1966), 1712. Also, it has been shown by F. KOTTLER, *Ann. der Physik*, **70** (1923), 405 that KIRCHHOFF's theory may be regarded as providing a rigorous solution to a certain saltus problem (problem with prescribed discontinuities rather than prescribed boundary values). This interpretation is of particular interest in connection with the problem of diffraction at a black (completely absorbing) screen. [See also F. KOTTLER, *Progress in Optics*, Vol. 4, ed. E. WOLF (Amsterdam, North Holland Publishing Company and New York, J. Wiley and Sons, 1964), p. 281 and B. B. BAKER and E. T. COPSON, *loc. cit.*, p. 98.]

An article by C. J. BOUWKAMP, *Rep. Progr. Phys.* (London, Physical Society), **17** (1954), 35, contains references to numerous papers concerned with various modifications of KIRCHHOFF's theory.

may be interpreted in a mathematically consistent way, as providing an exact solution to a somewhat different boundary value problem than that specified by eqs. (15) and (16). It turns out that KIRCHHOFF's theory is entirely adequate for the treatment of the majority of problems encountered in instrumental optics. This is mainly due to the smallness of the optical wavelengths in comparison with the dimensions of the diffracting obstacles.* In other problems, such as those relating to the behaviour of the field in the immediate neighbourhood of screens and obstacles, more refined methods have to be used; they must then be considered as boundary-value problems of electromagnetic theory, with the sources as appropriate singularities of the wave functions. Only in a very limited number of cases have such solutions been found; some of them will be discussed in Chapter XI.

8.3.3 Fraunhofer and Fresnel diffraction

We now examine more closely the FRESNEL–KIRCHHOFF diffraction integral (17),

$$U(P) = -\frac{Ai}{2\lambda}\iint_{\mathscr{A}} \frac{e^{ik(r+s)}}{rs}[\cos(n, r) - \cos(n, s)]dS. \tag{22}$$

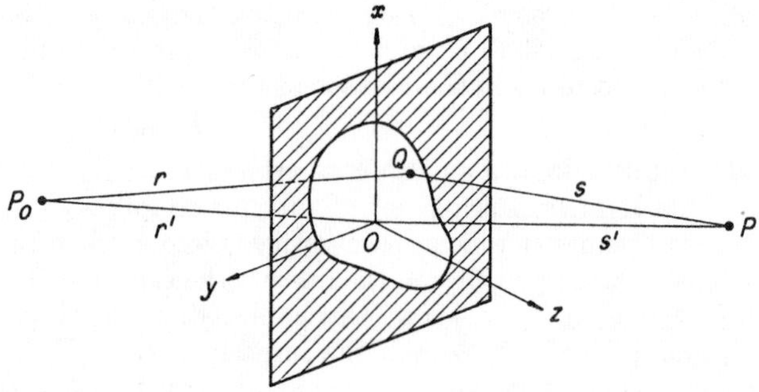

Fig. 8.5. Diffraction at an aperture in a plane screen.

As the element dS explores the domain of integration, $r + s$ will in general change by very many wavelengths, so that, the factor $e^{ik(r+s)}$ will oscillate rapidly. On the other hand, if the distances of the points P_0 and P from the screen are large compared to the linear dimensions of the aperture, the factor $[\cos(n, r) - \cos(n, s)]$ will not vary appreciably over the aperture. Further, we assume that if O is any point in the aperture, the angles which the lines P_0O and OP make with P_0P are not too large. We may then replace this factor by $2\cos\delta$,where δ is the angle between the line P_0P and the normal to the screen. Finally the factor $1/rs$ may be replaced by $1/r's'$, where r' and s' are the distance of P_0 and P from the origin and (22) then reduces to

$$U(P) \sim -\frac{Ai}{\lambda}\frac{\cos\delta}{r's'}\iint_{\mathscr{A}} e^{ik(r+s)}\,dS. \tag{23}$$

We take a Cartesian reference system with origin in the aperture and with the x— and y — axes in the plane of the aperture and choose the positive z direction to point into the half-space that contains the point P of observation (Fig. 8.5).

If (x_0, y_0, z_0) and (x, y, z) are the coordinates of P_0 and of P respectively, and (ξ, η) the coordinates of a point Q in the aperture, we have

$$\left.\begin{aligned} r^2 &= (x_0 - \xi)^2 + (y_0 - \eta)^2 + {z_0}^2, \\ s^2 &= (x - \xi)^2 + (y - \eta)^2 + z^2, \end{aligned}\right\} \tag{24}$$

* Cf. S. SILVER, *J. Opt. Soc. Amer.*, **52** (1962), 131.

$$\left.\begin{aligned} r'^2 &= x_0^2 + y_0^2 + z_0^2, \\ s'^2 &= x^2 + y^2 + z^2. \end{aligned}\right\} \tag{25}$$

Hence

$$\left.\begin{aligned} r^2 &= r'^2 - 2(x_0\xi + y_0\eta) + \xi^2 + \eta^2, \\ s^2 &= s'^2 - 2(x\xi + y\eta) + \xi^2 + \eta^2. \end{aligned}\right\} \tag{26}$$

Since we assumed that the linear dimensions of the aperture are small compared to both r' and s' we may expand r and s as power series in ξ/r', η/r', ξ/s' and η/s'. We then obtain

$$\left.\begin{aligned} r &\sim r' - \frac{x_0\xi + y_0\eta}{r'} + \frac{\xi^2 + \eta^2}{2r'} - \frac{(x_0\xi + y_0\eta)^2}{2r'^3} - \dots \\ s &\sim s' - \frac{x\xi + y\eta}{s'} + \frac{\xi^2 + \eta^2}{2s'} - \frac{(x\xi + y\eta)^2}{2s'^3} - \dots \end{aligned}\right\} \tag{27}$$

Substitution from (27) into (23) gives

$$U(P) = -\frac{i \cos \delta}{\lambda} \frac{A e^{ik(r'+s')}}{r's'} \iint_{\mathscr{A}} e^{ikf(\xi,\eta)}\, d\xi d\eta, \tag{28}$$

where

$$\begin{aligned} f(\xi, \eta) = {} & -\frac{x_0\xi + y_0\eta}{r'} - \frac{x\xi + y\eta}{s'} + \frac{\xi^2 + \eta^2}{2r'} + \frac{\xi^2 + \eta^2}{2s'} \\ & - \frac{(x_0\xi + y_0\eta)^2}{2r'^3} - \frac{(x\xi + y\eta)^2}{2s'^3} \dots \end{aligned} \tag{29}$$

If we denote by (l_0, m_0) and (l, m) the first two direction cosines

$$\left.\begin{aligned} l_0 &= -\frac{x_0}{r'}, & l &= \frac{x}{s'}, \\ m_0 &= -\frac{y_0}{r'}, & m &= \frac{y}{s'}, \end{aligned}\right\} \tag{30}$$

(29) may be written in the form

$$\begin{aligned} f(\xi, \eta) = {} & (l_0 - l)\xi + (m_0 - m)\eta + \tfrac{1}{2}\left\{\left(\frac{1}{r'} + \frac{1}{s'}\right)(\xi^2 + \eta^2) \right. \\ & \left. - \frac{(l_0\xi + m_0\eta)^2}{r'} - \frac{(l\xi + m\eta)^2}{s'}\right\} \dots \end{aligned} \tag{31}$$

We have reduced the problem of determining the light disturbance at P to the evaluation of the integral (28). Naturally the evaluation is simpler to carry out when the quadratic and higher order terms in ξ and η may be neglected in f. In this case one speaks of *Fraunhofer diffraction*; when the quadratic terms cannot be neglected, one speaks of *Fresnel diffraction*. Fortunately the simpler case of FRAUNHOFER diffraction is of much greater importance in optics.

Strictly speaking, the second and higher order terms disappear only in the limiting case $r' \to \infty$, $s' \to \infty$, i.e. when both the source and the point of observation are at infinity; (the factor A outside the integral must then be assumed to tend to infinity like $r's'$). It is, however, evident that the second order terms do not appreciably contribute to the integral if

$$\tfrac{1}{2}k\left|\left(\frac{1}{r'}+\frac{1}{s'}\right)(\xi^2+\eta^2)-\frac{(l_0\xi+m_0\eta)^2}{r'}-\frac{(l\xi+m\eta)^2}{s'}\right|\ll 2\pi. \tag{32}$$

We can immediately recognize certain conditions under which (32) will be satisfied. If we make use of inequalities of the form $(l_0\xi+m_0\eta)^2 \leqslant (l_0^2+m_0^2)(\xi^2+\eta^2)$ and remember that l_0^2, m_0^2, l^2 and m^2 cannot exceed unity, we find that (32) will be satisfied if

$$|r'| \gg \frac{(\xi^2+\eta^2)_{\max}}{\lambda} \quad \text{and} \quad |s'| \gg \frac{(\xi^2+\eta^2)_{\max}}{\lambda}, \tag{33}$$

or if

$$\frac{1}{r'}+\frac{1}{s'}=0 \quad \text{and} \quad l_0^2, m_0^2, l^2, m^2 \ll \frac{|r'|\lambda}{(\xi^2+\eta^2)_{\max}}. \tag{34}$$

The conditions (33) give an estimate of the distances r' and s' for which the FRAUNHOFER representation may be used. Conditions (34) imply that FRAUNHOFER diffraction also occurs when the point of observation is situated in a plane parallel to that of the aperture, provided that both the point of observation and the source are sufficiently close to the z axis. Here two cases may be distinguished: When r' is negative, the wave-fronts incident upon the aperture are concave to the direction of the propagation, i.e. P_0 is a centre of convergence and not of divergence of the incident wave. This case is of great practical importance, as it arises in the image space of a well-corrected centred system that images a point source which is not far from the axis. A FRAUNHOFER pattern is then formed in the Gaussian image plane and may be considered as arising from the diffraction of the image-forming wave on the exit pupil. When r' is positive, the wave-fronts are convex to the direction of propagation. The diffraction phenomena are virtual, being apparently formed on a screen through the source P_0. This case arises, for example, when an aperture is held in front of the eye, or the object glass of a telescope adjusted for distant vision of the light source.

To understand in more physical terms why FRAUNHOFER phenomena are observed in the focal plane of a well-corrected lens, let us compare first the two situations illustrated in Fig. 8.6. In (*a*) a pencil of rays from an infinitely distant point is incident upon the aperture in the direction specified by the direction cosines l_0, m_0, n_0. The effect observed at a very distant point P in the direction l, m, n may be regarded as arising from the superposition of plane waves originating at each point of the aperture and propagated in this direction. These waves (which have no existence in the domain of geometrical optics) may be called *diffracted waves* and the corresponding wave-normals the *diffracted rays*.

If now a well-corrected lens is placed behind the screen (Fig. 8.6 (b)) all the light diffracted in the (l, m, n) direction will be brought to a focus P' in the focal plane of the lens. Since the optical path from a wave-front of the diffracted pencil to P' is the same for all the rays, one obtains substantially the same interference effects as in the first case; it being assumed, of course, that the lens is so large that it introduces no additional diffraction. More generally still, the restriction that the wave incident upon the aperture is plane may also be removed, provided that the path lengths from the source to P' are substantially the same for all the rays.

In the case of FRAUNHOFER diffraction, the four quantities l_0, m_0, l, m enter (31) only in the combinations

$$p = l - l_0, \qquad q = m - m_0. \tag{35}$$

Hence, within the range of validity of the above approximation, the effect is unchanged when the aperture is displaced in its own plane.

We shall write the integral governing FRAUNHOFER diffraction in the form

$$U(P) = C \iint_{\mathscr{A}} e^{-ik(p\xi + q\eta)}\, d\xi d\eta, \tag{36}$$

C being the constant appearing in front of the integral (28). C is defined in terms of quantities depending on the position of the source and of the point of observation, but in practice it is often more convenient to express it in terms of other quantities. Let

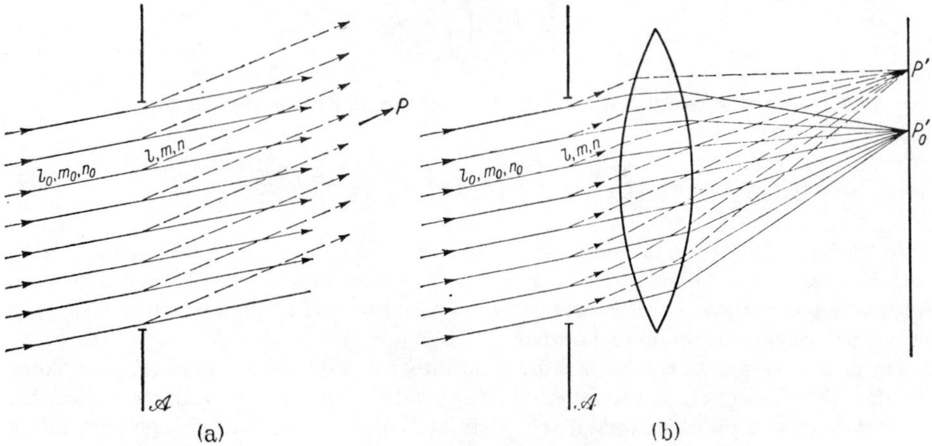

Fig. 8.6. Comparison of two cases of FRAUNHOFER diffraction.

E be the total energy incident upon the aperture. By the law of conservation of energy the total energy that reaches the plane of observation must also be equal to E, so that we have the normalizing condition

$$\iint |U(p, q)|^2 dp dq = E, \tag{37}$$

the integration extending over all possible values of p and q. Equation (36) may be re-written as a FOURIER integral

$$U(p, q) = \iint G(\xi, \eta) e^{-\frac{2\pi i}{\lambda}(p\xi + q\eta)}\, d\xi d\eta, \tag{38}$$

where the *pupil function** G is given by

$$\begin{aligned} G(\xi, \eta) &= \text{constant } (C) \text{ at points in the opening} \\ &= 0 \qquad\qquad\quad \text{at points outside opening} \end{aligned} \tag{39}$$

and the integral extends over the whole (ξ, η) plane.

By PARSEVAL's theorem for FOURIER transforms†

$$\iint |G(\xi, \eta)|^2 d\xi d\eta = \left(\frac{1}{\lambda}\right)^2 \iint |U(p, q)|^2 dp dq, \tag{40}$$

* More general pupil functions will be considered in § 8.6 and § 9.5.

† Cf. I. N. SNEDDON, *Fourier Transforms* (New York, McGraw-Hill, 1951), pp. 25 and 44.

or, substituting from (37) and (39) and denoting by D the area of the opening,

$$\frac{1}{\lambda^2} E = |C|^2 D \tag{41}$$

whence*

$$C = \frac{1}{\lambda}\sqrt{\frac{E}{D}}. \tag{42}$$

The basic integral for FRAUNHOFER diffraction then takes the form

$$U(p, q) = \frac{1}{\lambda}\sqrt{\frac{E}{D}} \iint_{\mathscr{A}} e^{-ik(p\xi+q\eta)}\, d\xi d\eta. \tag{43}$$

We note that the intensity $I_0 = |U(0, 0)|^2$ at the centre of the pattern $p = q = 0$ is given by

$$I_0 = \left(\frac{1}{\lambda}\right)^2 \frac{E}{D}\left(\iint_{\mathscr{A}} d\xi d\eta\right)^2 = \frac{ED}{\lambda^2} = C^2 D^2. \tag{44}$$

In deriving (43) we have disregarded the fact that (36) was obtained subject to certain restrictions on the range of p and q. The errors introduced by extending the integration in (40) over all p and q values is, however, negligible, since $U(p, q)$ is very small except in the neighbourhood of $p = q = 0$.

Let us now return to the basic diffraction integral (28). As the point (ξ, η) explores the domain of integration, the function $f(\xi, \eta)$ will change by very many wavelengths, so that both the real and imaginary parts of the integrand will change sign many times. In consequence the contributions from the various elements will in general virtually cancel each other out (destructive interference). The situation is, however, different for an element which surrounds a point (called *critical point* or *pole*) where $f(\xi, \eta)$ is stationary. Here the integrand varies much more slowly and may be expected to give a significant contribution. Hence, when the wavelength is sufficiently small, the value of the integral is determined substantially by the behaviour of f in the neighbourhood of points where f is stationary. This is the principle of the *method of stationary phase* for determining the asymptotic behaviour of a certain class of integrals, and is discussed more fully in Appendix III. Here we only note the bearing of this result on the classification of diffraction phenomena:

On comparing (22) and (28) we see that $r + s = r' + s' + f$, so that (see Fig. 8.5)

$$f = P_0Q + QP + \text{constant}. \tag{45}$$

Obviously f considered as function of Q will be stationary when Q is collinear with P_0 and P. Hence the main contribution to the disturbance at P comes from the immediate neighbourhood of the point $\bar{Q}$ where the line joining the source to the point of observation intersects the plane of the aperture. Now in the special case of FRAUNHOFER diffraction, P_0 and P are effectively at infinity, so that there is no preferential point $\bar{Q}$. In this case the behaviour of the diffraction integral must, therefore, be expected to be somewhat exceptional.

In § 8.5–§ 8.8 we shall study the most important cases of FRAUNHOFER and FRESNEL diffraction. But first we must justify the use of the single scalar wave function U in calculations of the light intensity.

---

* We omit here a constant phase factor as it contributes nothing to the intensity $I = |U|^2$.

8.4. TRANSITION TO A SCALAR THEORY*

The only property of the U function that we used in the derivation of the KIRCHHOFF integral theorem was that it satisfies the homogeneous scalar wave equation. It therefore follows that this theorem and the conclusion of the preceding section apply to each of the Cartesian components of the field vectors, the vector potential, the HERTZ vectors, etc., in regions where there are no currents and charges. To obtain a complete description of the field, the theorem must be applied separately to each of the Cartesian components. Fortunately it turns out that in the majority of problems encountered in optics an approximate description in terms of a single complex scalar wave function is adequate.

A complete description of an electromagnetic field requires the specification of the magnitude of the field vectors as well as their direction (polarization), both as functions of position and time. However, because of the very high frequencies of optical fields (of the order of 10^{14}/sec.), one cannot measure the instantaneous values of any of these quantities, but only certain time averages over intervals that are large compared to the optical periods. Moreover, one usually deals with natural light, so that there is no preferential polarization direction of the observable (macroscopic) field. The quantity which is then of primary interest is the *intensity* I defined in § 1.1.4 as the time average of the energy which crosses a unit area containing the electric and magnetic vector in unit time

$$I = \frac{c}{4\pi} |\langle \boldsymbol{E} \wedge \boldsymbol{H} \rangle|.$$

We shall show that the electromagnetic field which is associated with the passage of natural light through an optical instrument of moderate aperture and of conventional design is such that the intensity may approximately be represented in terms of a single complex scalar wave function by means of the formula†

$$I = |U|^2,$$

and that the function U may be calculated from the knowledge of the eikonal function of the system.

8.4.1 The image field due to a monochromatic oscillator

We consider a symmetrical optical system with a point source at P_0 (Fig. 8.7) emitting natural, quasi-monochromatic light of frequency ω_0. We assume that the inclination to the axis of the rays which pass through the system is not large, say not more than 10° or 15°. At P_0 we choose a set of Cartesian axes (x_1, x_2, x_3) with the x_3 direction along the principal ray. The source may be regarded as a dipole of moment $\boldsymbol{Q}(t)$ which varies both in magnitude and direction with time t. The components of $\boldsymbol{Q}(t)$ in the three directions will be written in the form of FOURIER integrals,

$$Q_j(t) = \frac{1}{\sqrt{2\pi}} \int_{-\infty}^{+\infty} q_j(\omega) e^{-i\omega t} \, d\omega \qquad (j = 1, 2, 3). \tag{1}$$

---

* We follow here substantially the analysis of O. THEIMER, G. D. WASSERMANN and E. WOLF, *Proc. Roy. Soc.*, A., **212** (1952), 426.

† More generally it was shown by E. WOLF, *Proc. Phys. Soc.*, **74** (1959), 269, that both the (time-averaged) energy density and energy flow in an unpolarized quasi-monochromatic field may always be derived from one complex time-harmonic scalar wave function.

Since $Q_j(t)$ is real it follows that the complex quantities $q_j(\omega)$ satisfy the relations

$$q_j(-\omega) = q_j^\star(\omega) \tag{2}$$

where the asterisk denotes the complex conjugate. Consequently (1) may be written as

$$Q_j(t) = \mathscr{R}\left\{\sqrt{\frac{2}{\pi}}\int_0^\infty q_j(\omega)e^{-i\omega t}\,d\omega\right\} \qquad (j = 1, 2, 3), \tag{3}$$

$\mathscr{R}$ denoting the real part. Each FOURIER component of (3) represents a monochromatic Hertzian oscillator with its axis along the x_j direction.

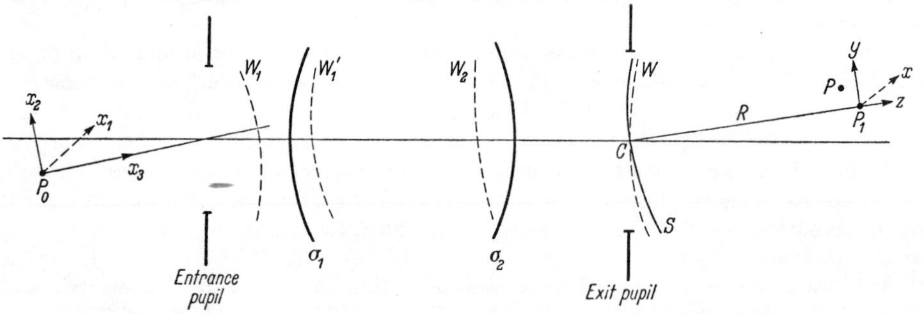

Fig. 8.7. Propagation of an electromagnetic wave through an optical system.

Let $|q_j(\omega)|$ and $\delta_j(\omega)$ be the amplitude and the phase of $q_j(\omega)$,

$$q_j(\omega) = |q_j(\omega)|e^{i\delta_j(\omega)}. \tag{4}$$

Since the source is assumed to emit quasi-monochromatic light, the modulus $|q_j(\omega)|$ will, for each j, differ appreciably from zero only within a narrow interval $(\omega_0 - \frac{1}{2}\Delta\omega, \omega_0 + \frac{1}{2}\Delta\omega)$. The assumption that the light is natural implies that $\delta_j(\omega)$ are rapidly and irregularly varying functions over the frequency range.*

Since the field may be regarded as a superposition of strictly monochromatic fields, it will be convenient to examine first the contributions from a single monochromatic Hertzian oscillator at P_0. As the field of such an oscillator is weak in the neighbourhood of its axis, and as we assume that the angles which the diameters of the entrance pupil subtend at P_0 are small, it follows that only the components $Q_1(t)$ and $Q_2(t)$ of $\boldsymbol{Q}(t)$ will substantially contribute to the field. We shall therefore take as our typical oscillator one which has its axis in the x_1x_2-plane.

Let

$$\mathscr{R}\{q(\omega)\mathbf{p}_0(\omega)e^{-i\omega t}\} \tag{5}$$

be the moment of this typical dipole, $\mathbf{p}_0(\omega)$ being a unit vector in the direction of its axis. Such a dipole will produce at a point T in vacuum, whose distance from P_0 is large compared to the wavelength $\lambda = (2\pi c/\omega)$, a field given by [see § 2.2 (64)]:

$$\left.\begin{aligned} \boldsymbol{E}_\omega &= \mathscr{R}\left\{\frac{\omega^2}{c^2 r}|q(\omega)|\boldsymbol{r}_0 \wedge (\mathbf{p}_0(\omega) \wedge \boldsymbol{r}_0)e^{i[\delta(\omega)-\omega(t-r/c)]}\right\}, \\ \boldsymbol{H}_\omega &= \mathscr{R}\left\{\frac{\omega^2}{c^2 r}|q(\omega)|\boldsymbol{r}_0 \wedge \mathbf{p}_0(\omega)e^{i[\delta(\omega)-\omega(t-r/c)]}\right\}, \end{aligned}\right\} \tag{6}$$

where $\boldsymbol{r}_0$ denotes the unit radial vector.

* For a detailed discussion of this point see M. PLANCK, *Ann. d. Physik*, (4), **1** (1900), 61.

Let W_1 be a typical geometrical wave-front in the object space at a distance from P_0 which is large compared with the wavelength. Since we assume that the angles which the rays make with the axis of the system are small, it follows immediately from (6) that at any particular instant of time the vectors $\boldsymbol{E}_\omega$ and $\boldsymbol{H}_\omega$ do not vary appreciably in magnitude and direction over W_1.

The effect of the first surface* σ_1 on the incident field is twofold. First, the amplitudes of the field vectors are diminished on account of reflection losses; secondly, the directions of vibrations are changed. FRESNEL's formulae show that both these effects depend mainly on the magnitudes of the angle of incidence. If this angle is small (say 10° or so), reflection losses are also small (approx. 5 per cent) and the rotations of the planes of vibration do not exceed a few degrees (cf. § 1.5). Moreover, these effects are practically uniform over σ_1. Since the time-independent parts of $\boldsymbol{E}_\omega$ and $\boldsymbol{H}_\omega$ do not vary appreciably with position over the wave-front W_1, they will also not vary appreciably over the refracted wave-front W_1' which follows the surface σ_1 (see Fig. 8.7). The same applies to the behaviour of the two fields over any other wave-front in the space between σ_1 and the second surface σ_2. For, as we showed in § 3.1.3, in a homogeneous medium the direction of vibration along each ray remains constant, and also, since the wave-fronts are nearly spherical (centred on the Gaussian image of P_0 by the first surface), the amplitudes will be diminished almost in the ratio of their paraxial radii of curvature.

Repeating these arguments we finally arrive at a wave-front W which passes through the centre C of the exit pupil and find again that the time-independent parts of $\boldsymbol{E}_\omega$ and $\boldsymbol{H}_\omega$ do not vary appreciably over this wave-front. This result makes it immediately possible to write down an approximate mathematical representation for the field vectors in the region of the image.

We take rectangular Cartesian axes (x, y, z) with origin at the Gaussian image P_1 of P_0, with the z direction along CP_1. The field at all points in the region of the aperture except those in the immediate neighbourhood of the edge of the aperture can be approximately expressed in the form (cf. Chapter III)

$$\left.\begin{aligned} \boldsymbol{E}_\omega(x, y, z, t) &= \mathscr{R}\left\{\frac{\omega^2}{c^2}\,\boldsymbol{e}_\omega(x, y, z)e^{i\left\{\delta(\omega)-\omega\left[t-\frac{1}{c}\mathscr{S}_\omega(x,y,z)\right]\right\}}\right\}, \\ \boldsymbol{H}_\omega(x, y, z, t) &= \mathscr{R}\left\{\frac{\omega^2}{c^2}\,\boldsymbol{h}_\omega(x, y, z)e^{i\left\{\delta(\omega)-\omega\left[t-\frac{1}{c}\mathscr{S}_\omega(x,y,z)\right]\right\}}\right\}, \end{aligned}\right\} \tag{7}$$

which may be regarded as generalization of (6). Here $\mathscr{S}_\omega(x, y, z)$ is the optical length from the object point to the point (x, y, z), and $\boldsymbol{e}_\omega(x, y, z)$ and $\boldsymbol{h}_\omega(x, y, z)$ are mutually orthogonal real vectors.† In a homogeneous non-magnetic medium of refractive index n, these vectors satisfy the relation (see eq. (19) and (20) of § 3.1)

$$|\boldsymbol{h}_\omega| = n|\boldsymbol{e}_\omega|. \tag{8}$$

We take a reference sphere S, centred on P_1, which passes through the centre C of the exit pupil, and denote by R its radius CP_1. In practice the distance between

---

* We assume here that σ_1 is a refracting surface. If σ_1 is a mirror, no essential modifications of our argument are necessary, as is seen by inspection of FRESNEL's formulae.

† That $\boldsymbol{e}_\omega$ and $\boldsymbol{h}_\omega$ are real follows from the fact that the corresponding vectors in (6) are real (linear polarization) and that the state of polarization remains linear on each refraction [cf. p. 40]. Moreover, between any two consecutive surfaces, the state of polarization is constant along each ray, as shown in § 3.1.3.

S and W will nowhere exceed a few dozen wavelengths. Consequently on S just as on W the amplitude vectors $\boldsymbol{e}_\omega$ and $\boldsymbol{h}_\omega$ will be practically constant in magnitude and direction.

Let $P(X, Y, Z)$ be a point in the region of the image where the intensity is to be determined. If the angles which the diameters of the exit pupil subtend at P are small, we may apply KIRCHHOFF's formula with the same approximation as in the previous section, and we find on integrating the expressions (7) over that part S' of S which approximately fills the exit pupil, if in addition we also neglect the variation of the inclination factor over S', that

$$\left.\begin{aligned} \boldsymbol{E}_\omega(X, Y, Z, t) &= \mathscr{R}\, \frac{\omega^3}{2\pi i c^3}\, e^{i[\delta(\omega)-\omega t]} \iint_{S'} \frac{1}{s}\, \boldsymbol{e}_\omega(x', y', z') e^{i\frac{\omega}{c}[\mathscr{S}_\omega(x',y',z')+s]}\, dS, \\ \boldsymbol{H}_\omega(X, Y, Z, t) &= \mathscr{R}\, \frac{\omega^3}{2\pi i c^3}\, e^{i[\delta(\omega)-\omega t]} \iint_{S'} \frac{1}{s}\, \boldsymbol{h}_\omega(x', y', z') e^{i\frac{\omega}{c}[\mathscr{S}_\omega(x',y',z')+s]}\, dS, \end{aligned}\right\} \tag{9}$$

where s is the distance from a typical point (x', y', z') on the reference sphere to P.

Since the vectors $\boldsymbol{e}_\omega(x', y', z')$ and $\boldsymbol{h}_\omega(x', y', z')$ do not vary appreciably over the surface of integration we may replace them by the values $\boldsymbol{e}_\omega(0, 0, -R)$ and $\boldsymbol{h}_\omega(0, 0, -R)$ which they take at the centre C of the exit pupil. Now these vectors are orthogonal and satisfy (8), so that we may set, if in addition we take $n = 1$,

$$\left.\begin{aligned} \boldsymbol{e}_\omega(0, 0, -R) &= a(\omega)\boldsymbol{\alpha}(\omega), \\ \boldsymbol{h}_\omega(0, 0, -R) &= a(\omega)\boldsymbol{\beta}(\omega), \end{aligned}\right\} \tag{10}$$

where $\boldsymbol{\alpha}(\omega)$ and $\boldsymbol{\beta}(\omega)$ are orthogonal unit vectors in the plane perpendicular to the z direction. The relations (9) then become

$$\left.\begin{aligned} \boldsymbol{E}_\omega(X, Y, Z, t) &= \mathscr{R}\left\{\frac{\omega^2}{c^2}\, U_\omega(X, Y, Z) a(\omega)\boldsymbol{\alpha}(\omega) e^{i[\delta(\omega)-\omega t]}\right\}, \\ \boldsymbol{H}_\omega(X, Y, Z, t) &= \mathscr{R}\left\{\frac{\omega^2}{c^2}\, U_\omega(X, Y, Z) a(\omega)\boldsymbol{\beta}(\omega) e^{i[\delta(\omega)-\omega t]}\right\}, \end{aligned}\right\} \tag{11}$$

where U_ω is the scalar wave function

$$U_\omega(X, Y, Z) = \frac{\omega}{2\pi i c} \iint_{S'} \frac{1}{s}\, e^{i\frac{\omega}{c}[\mathscr{S}_\omega(x',y',z')+s]}\, dS. \tag{12}$$

From (11) we can immediately deduce by calculating the POYNTING vector $\boldsymbol{S}_\omega = c[\boldsymbol{E}_\omega \wedge \boldsymbol{H}_\omega]/4\pi$ and taking the time average, that the intensity at the point $P(X, Y, Z)$ due to the single dipole (represented by (5)) at P_0 is proportional to the square of the modulus of the scalar wave function $U_\omega(X, Y, Z)$. However, to justify the use of a single scalar wave function in calculating the intensity we must carry out the time averaging not for the monochromatic component but for the total field.

8.4.2 The total image field

We saw that the contributions of each frequency component to the total field may be regarded as arising essentially from two dipoles at P_0 with their axes along the x_1 and x_2 directions. Hence it follows from (1) and (11), if we also define contributions from negative frequencies by relations of the form (2), that the total field in the image region may be expressed approximately in the form

$$\left.\begin{aligned} E(X,Y,Z,t) &= \frac{1}{\sqrt{2\pi}}\int_{-\infty}^{\infty}\frac{\omega^2}{c^2}U_\omega(X,Y,Z)[a_1(\omega)\boldsymbol{\alpha}_1(\omega)e^{i\delta_1(\omega)} \\ &\qquad + a_2(\omega)\boldsymbol{\alpha}_2(\omega)e^{i\delta_2(\omega)}]e^{-i\omega t}\,d\omega, \\ H(X,Y,Z,t) &= \frac{1}{\sqrt{2\pi}}\int_{-\infty}^{\infty}\frac{\omega^2}{c^2}U_\omega(X,Y,Z)[a_1(\omega)\boldsymbol{\beta}_1(\omega)e^{i\delta_1(\omega)} \\ &\qquad + a_2(\omega)\boldsymbol{\beta}_2(\omega)e^{i\delta_2(\omega)}]e^{-i\omega t}\,d\omega. \end{aligned}\right\} \tag{13}$$

Here suffixes 1 and 2 refer to the contributions from oscillators which have their axes along the x_1 and x_2 directions.

In order to determine the intensity in the image region it will be convenient to write down separate expressions for each of the Cartesian components of $\boldsymbol{E}$ and $\boldsymbol{H}$. Let $\theta_1(\omega)$ and $\theta_2(\omega)$ denote the angles which the unit vectors $\boldsymbol{\alpha}_1(\omega)$ and $\boldsymbol{\alpha}_2(\omega)$ make with the x direction in the image space. Since $\boldsymbol{\alpha}_1(\omega)$ and $\boldsymbol{\beta}_1(\omega)$ and $\boldsymbol{\alpha}_2(\omega)$ and $\boldsymbol{\beta}_2(\omega)$ are real, mutually orthogonal vectors which lie in a plane perpendicular to the z direction, it follows from (13) that the components of $\boldsymbol{E}$ and $\boldsymbol{H}$ are approximately given by*

$$\left.\begin{aligned} E_x(X,Y,Z,t) = H_y(X,Y,Z,t) &= \frac{1}{\sqrt{2\pi}}\int_{-\infty}^{+\infty}U_\omega(X,Y,Z)f(\omega)e^{-i\omega t}\,d\omega, \\ E_y(X,Y,Z,t) = -H_x(X,Y,Z,t) &= \frac{1}{\sqrt{2\pi}}\int_{-\infty}^{+\infty}U_\omega(X,Y,Z)g(\omega)e^{-i\omega t}\,d\omega, \\ E_z(X,Y,Z,t) = H_z(X,Y,Z,t) &= 0 \end{aligned}\right\} \tag{14}$$

where

$$\left.\begin{aligned} f(\omega) &= \frac{\omega^2}{c^2}[a_1(\omega)\cos\theta_1(\omega)e^{i\delta_1(\omega)} + a_2(\omega)\cos\theta_2(\omega)e^{i\delta_2(\omega)}], \\ g(\omega) &= \frac{\omega^2}{c^2}[a_1(\omega)\sin\theta_1(\omega)e^{i\delta_1(\omega)} + a_2(\omega)\sin\theta_2(\omega)e^{i\delta_2(\omega)}]. \end{aligned}\right\} \tag{15}$$

It follows from (14) that the magnitude of the Poynting vector $\boldsymbol{S} = c[\boldsymbol{E} \wedge \boldsymbol{H}]/4\pi$ can be expressed approximately in the form

$$|\boldsymbol{S}| = \frac{c}{4\pi}[E_x^{\,2} + E_y^{\,2}] = \frac{c}{4\pi}[H_x^{\,2} + H_y^{\,2}]. \tag{16}$$

We must now determine the time average of this quantity.†

For reasons of convergence we assume that the radiation field exists only between the instants $t = -T$ and $t = T$, where $T \gg 2\pi/\omega_0$; it is easy to pass to the limit $T \to \infty$ subsequently. It follows from (14), by the Fourier inversion theorem, that

$$U_\omega(X,Y,Z)f(\omega) = \frac{1}{\sqrt{2\pi}}\int_{-T}^{T}E_x(X,Y,Z,t)e^{i\omega t}\,dt, \tag{17}$$

---

* It would be incorrect to conclude from (14) that the direction of the energy flow in the image region is necessarily everywhere parallel to z. For the relative errors in (14) may substantially affect the calculations of direction in regions where the intensity is small, e.g. in the neighbourhood of the dark rings in the Airy pattern.

† Our method of evaluating the average is due to M. Born and P. Jordan, *Z. f. Phys.*, **33** (1925), 479.

with similar expressions involving E_y, H_x, and H_y. Now we have by (14)

$$\langle E_x^2 \rangle = \frac{1}{2T}\int_{-T}^{T} E_x^2 dt = \frac{1}{2T}\int_{-T}^{T} E_x dt \frac{1}{\sqrt{2\pi}}\int_{-\infty}^{+\infty} U_\omega f(\omega) e^{-i\omega t}\, d\omega, \tag{18}$$

or, inverting the order of integration,

$$\begin{aligned}\langle E_x^2 \rangle &= \frac{1}{2T}\int_{-\infty}^{+\infty} U_\omega f(\omega) d\omega \frac{1}{\sqrt{2\pi}}\int_{-T}^{T} E_x e^{-i\omega t}\, dt \\ &= \frac{1}{2T}\int_{-\infty}^{+\infty} U_\omega f_\omega U_\omega^\star f_\omega^\star\, d\omega \qquad \text{(by 17)} \\ &= \frac{1}{T}\int_0^\infty |U_\omega|^2 |f(\omega)|^2\, d\omega, \end{aligned} \tag{19}$$

since $U_{-\omega} f(-\omega) = U_\omega^\star f^\star(\omega)$. Similarly

$$\langle E_y^2 \rangle = \frac{1}{T}\int_0^\infty |U_\omega|^2 |g(\omega)|^2\, d\omega. \tag{20}$$

Hence, the intensity $I(X, Y, Z)$, defined as the time average of the magnitude of the POYNTING vector, is, according to (16), (19), and (20)

$$I(X, Y, Z) = \frac{c}{4\pi T}\int_0^\infty |U_\omega(X, Y, Z)|^2 [|f(\omega)|^2 + |g(\omega)|^2]\, d\omega. \tag{21}$$

Now if $|\Delta\omega|$ is sufficiently small $|U_\omega|$ will be practically independent of ω over the effective frequency range, so that $|U_\omega|$ may then be replaced by $|U_{\omega_0}|$ and taken outside the integral. The remaining term

$$\frac{c}{4\pi T}\int_0^\infty \{|f(\omega)|^2 + |g(\omega)|^2\}\, d\omega, \tag{22}$$

which is independent of X, Y, and Z must also be independent of T (implicitly contained in f and g on account of (17)), if a stationary phenomenon is observed. Hence (22) must be a constant (C say) and the intensity may therefore be finally written in the form

$$I(X, Y, Z) = C|U_{\omega_0}(X, Y, Z)|^2. \tag{23}$$

The constant C depends in a complicated manner on the source and on the optical instrument; however one is usually only interested in the relative distribution of the intensity and not in its absolute value. The intensity may then simply be measured by the quantity $|U_{\omega_0}|^2$. Thus the complex scalar function (12) is adequate for calculating the intensity distribution in the image formed with a source of natural light by an optical system of moderate numerical aperture.

8.5. FRAUNHOFER DIFFRACTION AT APERTURES OF VARIOUS FORMS

We shall now investigate the FRAUNHOFER diffraction pattern for apertures of various forms.

8.5.1 The rectangular aperture and the slit

Consider first a rectangular aperture of sides $2a$ and $2b$. With origin O at the centre of the rectangle and with $O\xi$ and $O\eta$ axes parallel to the sides (Fig. 8.8), the FRAUNHOFER diffraction integral § 8.3 (36) becomes

$$U(P) = C\int_{-a}^{a}\int_{-b}^{b} e^{-ik(p\xi+q\eta)}\, d\xi d\eta = C\int_{-a}^{a} e^{-ikp\xi}\, d\xi \int_{-b}^{b} e^{-ikq\eta}\, d\eta.$$

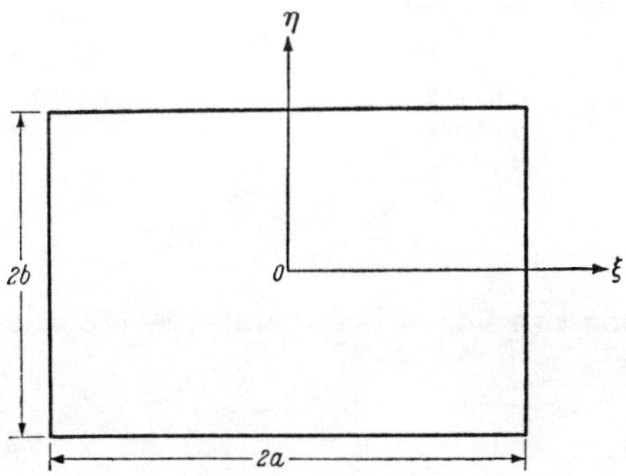

Fig. 8.8. Rectangular aperture.

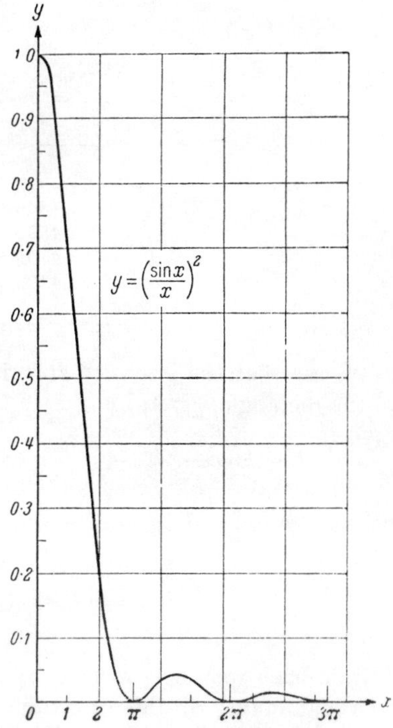

Fig. 8.9. FRAUNHOFER diffraction at a rectangular aperture. The function

$$y = \left(\frac{\sin x}{x}\right)^2.$$

Now

$$\int_{-a}^{a} e^{-ikp\xi}\, d\xi = -\frac{1}{ikp}\left[e^{-ikpa} - e^{ikpa}\right] = 2\,\frac{\sin kpa}{kp},$$

with a similar expression for the other integral. Hence the intensity is given by

$$I(P) = |U(P)|^2 = \left(\frac{\sin kpa}{kpa}\right)^2\left(\frac{\sin kqb}{kqb}\right)^2 I_0, \qquad (1)$$

where by § 8.3 (44) $I_0 = C^2D^2 = ED/\lambda^2$ is the intensity at the centre of the pattern, E being the total energy incident upon the aperture and $D = 4ab$ the area of the rectangle.

The function $y = (\sin x/x)^2$ is displayed in Fig. 8.9. It has a principal maximum $y = 1$ at $x = 0$ and zero minima at $x = \pm\pi, \pm 2\pi, \pm 3\pi, \ldots$ The minima separate the secondary maxima whose positions are given by the roots of the equation $\tan x - x = 0$ (see Table XIX). The roots asymptotically approach the values $x = (2m + 1)\pi/2$, m being an integer.

TABLE XIX

The first five maxima of the function

$$y = \left(\frac{\sin x}{x}\right)^2.$$

x	$y = \left(\frac{\sin x}{x}\right)^2$
0	1
$1{\cdot}430\pi = 4{\cdot}493$	0·04718
$2{\cdot}459\pi = 7{\cdot}725$	0·01694
$3{\cdot}470\pi = 10{\cdot}90$	0·00834
$4{\cdot}479\pi = 14{\cdot}07$	0·00503

We see that the intensity $I(P)$ is zero along two sets of lines parallel to the sides of the rectangle, given by

$$kpa = \pm u\pi, \qquad kqb = \pm v\pi, \qquad (u, v = 1, 2, 3, \ldots) \tag{2}$$

or, since $p = l - l_0$, $q = m - m_0$, $k = 2\pi/\lambda$,

$$l - l_0 = \pm \frac{u\lambda}{2a}, \qquad m - m_0 = \pm \frac{v\lambda}{2b}. \tag{3}$$

Within each rectangle formed by pairs of consecutive dark lines the intensity rises to a maximum; all these maxima are, however, only a small fraction of the central maximum, and decrease rapidly with increasing distance from the centre (Fig. 8.10). The larger the opening, the smaller is the effective size of the diffraction pattern.

From the elementary diffraction pattern formed by coherent light from a point source, the diffraction pattern due to light from an extended source may be found by integration. If the source is coherent, it is the complex amplitude, and if it is incoherent, it is the intensity that must be integrated. The pattern due to a partially coherent source may also be determined from this elementary solution by a process of integration, taking into account the correlation which exists between the light from the different elements of the source (cf. Chapter X). A case of particular importance is that of a very long incoherent line source (e.g. a luminous wire), the light from which is diffracted by a narrow slit, parallel to the source. For simplicity of calculations we assume that the luminous wire as well as the slit are effectively infinitely long, and take the y-axis in the direction of the source. Since $q = m - m_0$ where m_0 specifies the position of a point source, it follows that the intensity I' due to the line source is obtained by integrating (1) with respect to q:

$$I' = \int_{-\infty}^{+\infty} I(P)dq = \frac{1}{kb}\left(\frac{\sin kpa}{kpa}\right)^2 I_0 \int_{-\infty}^{+\infty}\left(\frac{\sin t}{t}\right)^2 dt.$$

Now*

$$\int_{-\infty}^{+\infty}\left(\frac{\sin t}{t}\right)^2 dt = \pi,$$

so that

$$I' = \left(\frac{\sin kpa}{kpa}\right)^2 I'_0, \tag{4}$$

where

$$I'_0 = \frac{\lambda}{2b} I_0 = \frac{2aE}{\lambda}. \tag{5}$$

The pattern is again characterized by the function $(\sin x/x)^2$, and consists of a succession of bright and dark fringes parallel to the line source and the slit. The constant I'_0 is the intensity at the central position $p = 0$.

8.5.2 The circular aperture

In a similar way we may investigate FRAUNHOFER diffraction at a circular aperture. It is now appropriate to use polar instead of rectangular coordinates. Let (ρ, θ) be the polar coordinates of a typical point in the aperture:

$$\rho \cos \theta = \xi, \qquad \rho \sin \theta = \eta; \tag{6}$$

and let (w, ψ) be the coordinates of a point P in the diffraction pattern, referred to the geometrical image of the source:

$$w \cos \psi = p, \qquad w \sin \psi = q. \tag{7}$$

From the definition of p and q it follows that $w = \sqrt{p^2 + q^2}$ is the sine of the angle which the direction (p, q) makes with the central direction $p = q = 0$. The diffraction integral § 8.3 (36) now becomes, if a is the radius of the circular aperture,

$$U(P) = C \int_0^a \int_0^{2\pi} e^{-ik\rho w \cos(\theta-\psi)} \rho d\rho d\theta. \tag{8}$$

Now we have the well-known integral representation of the BESSEL functions† $J_n(z)$:

$$\frac{i^{-n}}{2\pi} \int_0^{2\pi} e^{ix\cos\alpha} e^{in\alpha} d\alpha = J_n(x). \tag{9}$$

Equation (8) therefore reduces to

$$U(P) = 2\pi C \int_0^a J_0(k\rho w)\rho d\rho. \tag{10}$$

Also, there is the well-known recurrence relation‡

$$\frac{d}{dx}\{x^{n+1} J_{n+1}(x)\} = x^{n+1} J_n(x), \tag{11}$$

---

* See, for example, W. GRÖBNER and N. HOFREITER, *Integraltafel*, Vol. II (Wien, Springer, 1950), p. 333.

† See, for example, E. JAHNKE and F. EMDE, *Tables of Functions with Formulae and Curves* (Leipzig, Teubner, 1933; reprinted by Dover Publications, New York, 4th ed., 1945), p. 149; or G. N. WATSON, *A Treatise on the Theory of Bessel Functions* (Cambridge University Press, 1922), p. 20, equation 5 (with an obvious substitution).

‡ See for example E. JAHNKE and F. EMDE, *loc. cit.*, p. 145 or E. T. WHITTAKER and G. N. WATSON, *A Course of Modern Analysis* (Cambridge University Press, 4th ed., 1952), pp. 360–361.

giving, for $n = 0$, on integration

$$\int_0^x x' J_0(x') dx' = x J_1(x). \tag{12}$$

From (10) and (12) it follows that

$$U(P) = CD \left[\frac{2J_1(kaw)}{kaw}\right], \tag{13}$$

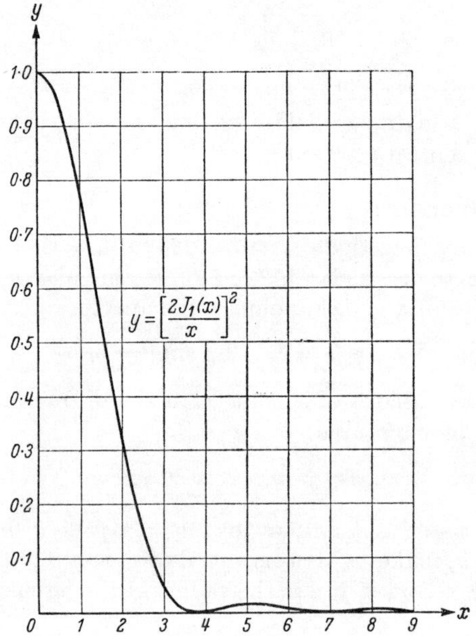

Fig. 8.11. FRAUNHOFER diffraction at a circular aperture. The function

$$y = \left(\frac{2J_1(x)}{x}\right)^2.$$

where $D = \pi a^2$. Hence the intensity is given by

$$I(P) = |U(P)|^2 = \left[\frac{2J_1(kaw)}{kaw}\right]^2 I_0, \tag{14}$$

where by § 8.3 (44) $I_0 = C^2 D^2 = ED/\lambda^2$. This is a celebrated formula first derived in a somewhat different form by AIRY.*

The intensity distriburion in the neighbourhood of the geometrical image is characterized by the function $y = (2J_1(x)/x)^2$ shown in Fig. 8.11. It has its principal maximum $y = 1$ at $x = 0$, and with increasing x it oscillates with gradually diminishing amplitude, in a similar way to the function $(\sin x/x)^2$ which we discussed in § 8.5.1. The intensity is zero (minimum) for values of x given by $J_1(x) = 0$. The minima are

* G. B. AIRY, *Trans. Camb. Phil. Soc.*, **5** (1835), 283. Almost at the same time as AIRY, SCHWERD obtained an approximate solution by replacing the circle by a regular polygon with 180 sides.

Vectorial treatments of diffraction of a convergent spherical wave at a circular aperture, which take into account polarization properties of the field were published by W. IGNATOWSKI, *Trans. Opt. Inst. Petrograd.*, **1** (1919) IV; V. A. FOCK, *ibid.*, **3** (1924), 24; H. H. HOPKINS, *Proc. Phys. Soc.* **55** (1943), 116; R. BURTIN, *Optica Acta*, **3** (1956), 104; B. RICHARDS and E. WOLF, *Proc. Roy. Soc.*, A, **253** (1959), 358; A. BOIVIN and E. WOLF, *Phys. Rev.*, **138** (1965), B 1561; A. BOIVIN, J. DOW and E. WOLF, *J. Opt. Soc. Amer.*, **57** (1967), 1171,

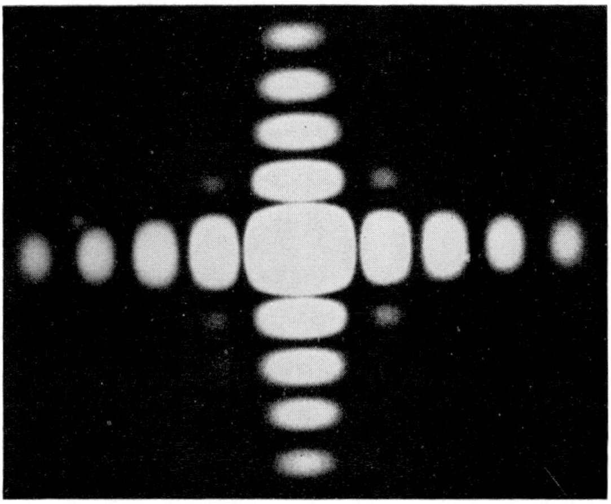

Fig. 8.10. FRAUNHOFER diffraction pattern of a rectangular aperture 8 mm × 7 mm, magnification 50×, mercury yellow light $\lambda = 5790$ Å. To show the existence of the weak secondary maxima the central portion was overexposed.

(After H. LIPSON, C. A. TAYLOR, and B. J. THOMPSON.)

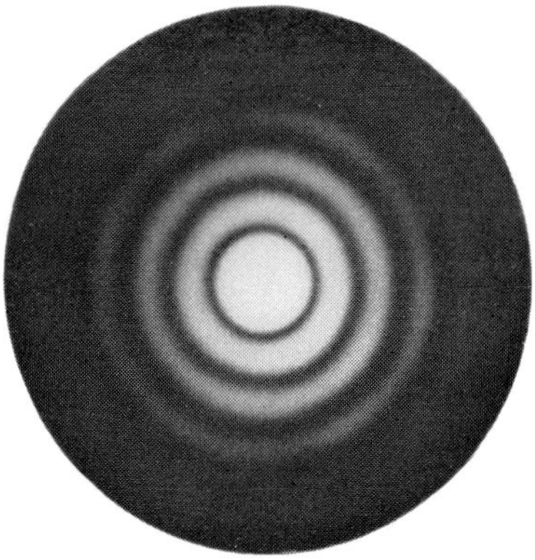

Fig. 8.12. FRAUNHOFER diffraction pattern of a circular aperture (the AIRY pattern) 6 mm in diameter, magnification 50×, mercury yellow light $\lambda = 5790$ Å. To show the existence of the weak subsidiary maxima, the central portion was overexposed.

(After H. LIPSON, C. A. TAYLOR, and B. J. THOMPSON.)

TABLE XX

The first few maxima and minima of the function

$$y = \left(\frac{2J_1(x)}{x}\right)^2.$$

x	$\left(\frac{2J_1(x)}{x}\right)^2$	
0	1	Max.
$1{\cdot}220\pi = 3{\cdot}833$	0	Min.
$1{\cdot}635\pi = 5{\cdot}136$	0·0175	Max.
$2{\cdot}233\pi = 7{\cdot}016$	0	Min.
$2{\cdot}679\pi = 8{\cdot}417$	0·0042	Max.
$3{\cdot}238\pi = 10{\cdot}174$	0	Min.
$3{\cdot}699\pi = 11{\cdot}620$	0·0016	Max.

no longer strictly equidistant [see Table XX]. The positions of the secondary maxima are given by the values of x that satisfy the equation

$$\frac{d}{dx}[J_1(x)/x] = 0,$$

or using the formula* (analogous to (11))

$$\frac{d}{dx}\left[x^{-n} J_n(x)\right] = -x^{-n} J_{n+1}(x), \tag{15}$$

by the roots of the equations $J_2(x) = 0$. With increasing x the separation between two successive minima or two successive maxima approaches the value π, as in the previous case.

The results show that the pattern consists of a bright disc, centred on the geometrical image $p = q = 0$ of the source, surrounded by concentric bright and dark rings (see Figs. 8.11 and 8.12). The intensity of the bright rings decreases rapidly with their radius and normally only the first one or two rings being bright enough to be visible to the naked eye. From Table XX it follows, since $x = 2\pi a w/\lambda$, that the radii of the dark rings are

$$w = \sqrt{p^2 + q^2} = 0{\cdot}610\,\frac{\lambda}{a}, \qquad 1{\cdot}116\,\frac{\lambda}{a}, \qquad 1{\cdot}619\,\frac{\lambda}{a}, \ldots \tag{16}$$

The separation between two neighbouring rings approaches asymptotically the value $\lambda/2a$. The effective size of the diffraction pattern is again seen to be inversely proportional to the linear dimensions of the aperture.

It is also of interest to examine what fraction of the total incident energy is contained within the central core of the diffraction pattern. Denoting by $L(w_0)$ the fraction of the total energy contained within a circle of radius w_0 in the image plane, centred on the geometrical image, we have

* See for example E. JAHNKE and F. EMDE, *loc. cit.*, p. 145, or E. T. WHITTAKER and G. N. WATSON, *loc. cit.*, p. 361.

$$L(w_0) = \frac{1}{E}\int^{w_0}\int^{2\pi} I(w)wdwd\psi$$

$$= \frac{D}{\lambda^2}\int_0^{w_0}\int_0^{2\pi}\left[\frac{2J_1(kaw)}{kaw}\right]^2 wdwd\psi$$

$$= 2\int_0^{kaw_0}\frac{J_1^2(x)}{x}\,dx. \tag{17}$$

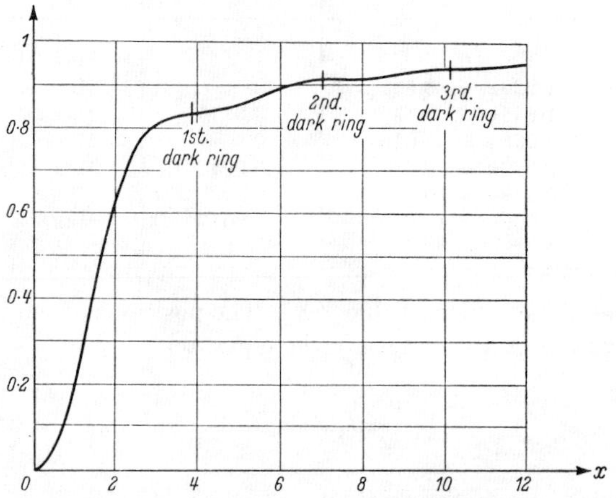

Fig. 8.13. The function $1 - J_0^2(x) - J_1^2(x)$ representing the fraction of the total energy contained within circles of prescribed radii in the FRAUNHOFER diffraction pattern of a circular aperture.

Now from (11) for $n = 0$, we have, on multiplying by $J_1(x)$ and using (15) with $n = 0$,

$$\frac{J_1^2(x)}{x} = J_0(x)J_1(x) - \frac{dJ_1(x)}{dx}J_1(x)$$

$$= -\frac{1}{2}\frac{d}{dx}[J_0^2(x) + J_1^2(x)].$$

The expression (17) now becomes, remembering that $J_0(0) = 1$, $J_1(0) = 0$,

$$L(w_0) = 1 - J_0^2(kaw_0) - J_1^2(kaw_0), \tag{18}$$

a formula due to RAYLEIGH.* This function is shown in Fig. 8.13. For the dark rings $J_1(kaw_0) = 0$, so that the fraction of the total energy outside any dark ring is simply $J_0^2(kaw_0)$. For the first, second, and third dark ring, $J_0^2(kaw_0)$ is equal to 0·162, 0·090, and 0·062 respectively. Thus more than 90 per cent of the light is contained within the circle bounded by the second dark ring.

8.5.3 Other forms of aperture

FRAUNHOFER diffraction at apertures of other forms may be studied in a similar manner, the calculations being particularly simple when curvilinear coordinates can be chosen so that one of the coordinate lines coincides with the boundary of the aperture.

* Lord RAYLEIGH, *Phil. Mag.* (5), **11** (1881), 214. Also his *Scientific Papers*, **1**, 513.

We cannot discuss other cases in detail here,* but we shall derive a useful theorem concerning the modification of the pattern when the aperture is uniformly extended (or contracted) in one direction, and also consider FRAUNHOFER diffraction at a screen containing a large number of openings of the same size and shape.

Let $\mathscr{A}_1$ and $\mathscr{A}_2$ be two apertures such that the extension of $\mathscr{A}_2$ in a particular direction ($O\xi$) is μ times that of $\mathscr{A}_1$. For FRAUNHOFER diffraction at $\mathscr{A}_1$, we have

$$U_1(p, q) = C\iint_{\mathscr{A}_1} e^{-ik(p\xi+q\eta)}\, d\xi d\eta. \tag{19}$$

Similarly for FRAUNHOFER diffraction at $\mathscr{A}_2$,

$$U_2(p, q) = C\iint_{\mathscr{A}_2} e^{-ik(p\xi+q\eta)}\, d\xi d\eta. \tag{20}$$

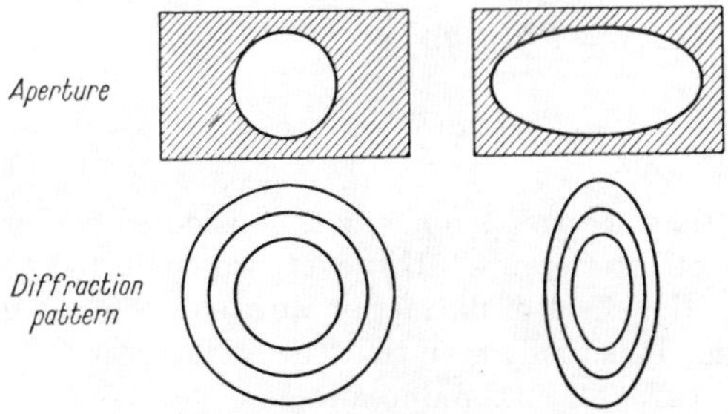

Fig. 8.14. Comparison of FRAUNHOFER diffraction at a circular and an elliptical aperture.

If in (20) we change the variables of integration from (ξ, η) to (ξ', η') where

$$\xi' = \frac{1}{\mu}\xi, \qquad \eta' = \eta, \tag{21}$$

we obtain

$$U_2(p, q) = \mu C\iint_{\mathscr{A}_1} e^{-ik(\mu p\xi'+q\eta')}\, d\xi' d\eta' = \mu U_1(\mu p, q). \tag{22}$$

This shows that *when the aperture is uniformly extended in the ratio* $\mu:1$ *in a particular direction, the Fraunhofer pattern contracts in the same direction in the ratio* $1:\mu$; *and the intensity in the new pattern is* μ^2 *times the intensity at the corresponding point of the original pattern.* Using this result we may, for example, immediately determine the FRAUNHOFER pattern of an aperture which has the form of an ellipse or a parallelogram from that of a circle or rectangle respectively. Fig. 8.14 illustrates the case of an elliptical aperture.

We now consider the important case of a screen that contains a large number of identical and similarly oriented apertures. (According to BABINET's principle the results will also apply to the complementary distribution of obstacles.) Let O_1, O_2,

---

* FRAUNHOFER diffraction at an annular aperture is briefly considered in connection with resolving power in § 8.6.2.

Photographs of FRAUNHOFER diffraction patterns for apertures of various forms can be found in a paper by J. SCHEINER and S. HIRAYAMA, *Abh. d. Königl. Akad. Wissensch.*, Berlin (1894), Anhang I. Photographs of FRESNEL patterns were published by Y. V. KATHAVATE, *Proc. Ind. Acad. Sci.*, **21** (1945), 177–210.

. . . O_N be a set of similarly situated points, one in each aperture, and let the coordinates of those points referred to a fixed set of axes in the plane of the apertures be $(\xi_1, \eta_1), (\xi_2, \eta_2), \ldots (\xi_N, \eta_N)$. The light distribution in the FRAUNHOFER diffraction pattern is then given by

$$U(p, q) = C \sum_n \iint_{\mathscr{A}} e^{-ik[(\xi_n+\xi')p+(\eta_n+\eta')q]} \, d\xi' d\eta'$$

$$= C \sum_n e^{-ik[p\xi_n+q\eta_n]} \iint_{\mathscr{A}} e^{-ik(p\xi'+q\eta')} \, d\xi' d\eta', \tag{23}$$

where the integration extends over any one opening $\mathscr{A}$ of the set. The integral expresses the effect of a single aperture, whilst the sum represents the superposition of the coherent diffraction patterns. If $I^{(0)}(p, q)$ is the intensity distribution arising from a single aperture, then, according to (23), the total intensity is given by

$$I(p, q) = I^{(0)}(p, q) \left| \sum_n e^{-ik(p\xi_n+q\eta_n)} \right|^2$$

$$= I^{(0)}(p, q) \sum_n \sum_m e^{-ik[p(\xi_n-\xi_m)+q(\eta_n-\eta_m)]}. \tag{24}$$

The simplest case, that of two openings, was considered earlier in § 7.2, in connection with the theory of interference. However, we neglected there the dependence of $I^{(0)}$ on p and q (i.e. the effect of diffraction at each opening) and only studied the effect of superposition. It is easily seen that the earlier result (§ 7.2 (17)) is in agreement with (24). For if $N = 2$, (24) reduces to

$$I = I^{(0)}\{2 + e^{-ik[p(\xi_1-\xi_2)+q(\eta_1-\eta_2)]} + e^{-ik[p(\xi_2-\xi_1)+q(\eta_2-\eta_1)]}\}$$
$$= 4I^{(0)} \cos^2 \tfrac{1}{2}\delta,$$

with

$$\delta = k[p(\xi_2 - \xi_1) + q(\eta_2 - \eta_1)].$$

Let us now consider the effect of a large number of apertures. We shall see that quite different results are obtained, depending on whether the apertures are distributed regularly or irregularly over the screen.

When the apertures are distributed irregularly over the screen, terms with different values of m and n in the double sum will fluctuate rapidly between $+1$ and -1 as m and n take on different values, and in consequence the sum of such terms will have zero mean value. Each remaining term ($m = n$) has the value unity. Hence it follows that except for local fluctuations* the total intensity is N times the intensity of the light diffracted by a single aperture:

$$I(p, q) \sim NI^{(0)}(p, q). \tag{25}$$

Diffraction effects of this type or, more often still, complementary effects (in the sense of BABINET's principle) may be easily observed, for example when a glass plate, dusted with lycopodium powder or covered with other particles of equal size and shape, is held in the path of light from a distant source. A piece of tin-foil pierced indiscriminately by a pin will also act as a diffracting screen of the type just considered.

---

* Fluctuations of a somewhat different type arise when the apertures are not of the same form, but are distributed regularly or according to some statistical law (cf. M. v. LAUE, *Berl. Ber.*, (1914), 1144). Similar effects are observed in connection with diffraction of X-rays by liquids (cf. J. A. PRINS, *Naturwiss*, **19** (1931), 435).

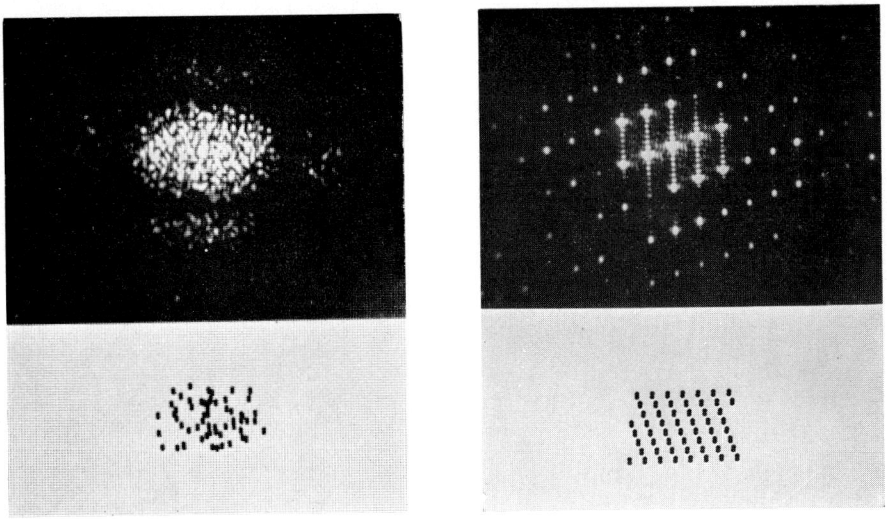

(a) (b)

Fig. 8.15. FRAUNHOFER diffraction pattern from an irregular distribution (a), and a regular distribution (b), of 56 identical and similarly situated apertures in a plane screen. The form and distribution of the apertures is shown in the lower portions of the figures. Light: Mercury yellow, $\lambda = 5790$ Å.

(After H. LIPSON, C. A. TAYLOR, and B. J. THOMPSON.)

The results are quite different when the openings are distributed regularly, for the terms with $m \neq n$ may now give appreciable contributions for certain values of p and q. For example, if the points O_n are so situated that for certain values of p and q the phases of all the terms for which $m \neq n$ are exact multiple of 2π their sum will be equal to $N(N - 1)$ and so for large N will be of the order of N^2. This enormous increase in intensity for particular directions, clearly illustrated in Fig. 8.15, is, as we shall see in the next section, of great importance in practice.

8.6. FRAUNHOFER DIFFRACTION IN OPTICAL INSTRUMENTS

8.6.1 Diffraction gratings

(a) *The principle of the diffraction grating*

A diffraction grating may be defined as any arrangement which imposes on an incident wave a periodic variation of amplitude or phase, or both. We may characterize any particular grating by its *transmission function,* defined as follows:

Let a transparent or semi-transparent object (not necessarily periodic) cover a portion of a fictitious reference plane $\xi\eta$, and let it be illuminated by a plane monochromatic wave incident in a direction specified by the direction cosines l_0, m_0. Fig. 8.16 illustrates the arrangement, the η-axis being perpendicular to the plane of the

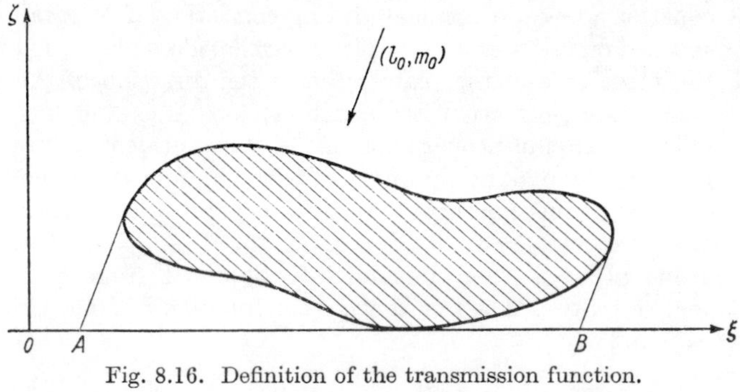

Fig. 8.16. Definition of the transmission function.

drawing. If no object were present, the disturbance in the $\xi\eta$-plane would be represented by the function $V_0(\xi, \eta) = A \exp\{ik(l_0\xi + m_0\eta)\}$, the factor $\exp(-i\omega t)$ being, as usual, omitted. Because of the presence of the object the disturbance will be modified and may be represented by some other function, which we denote by $V(\xi, \eta)$. The *transmission function* of the object is then defined as

$$F(\xi, \eta) = \frac{V(\xi, \eta)}{V_0(\xi, \eta)}. \tag{1}$$

In general F depends, of course, not only on ξ and η but also on the direction (l_0, m_0) of illumination. The transmission function is in general complex, since both the amplitude and the phase of the light may be altered on passing through the object. In the special case when the object alters the amplitude but not the phase of the incident wave (i.e. if $\arg F \equiv 0$), we speak of an *amplitude object*; if it alters the phase but not the amplitude (i.e. $|F| = 1$) we speak of a *phase object.*

If we are concerned with reflected light rather than with light that is transmitted by an object, it is more appropriate to speak of a *reflection function,* defined in a similar way, the only difference being that the reference plane is on the same side of the object as the incident light.

The ratio $|V/V_0|$ is practically unity for points outside the geometrical shadow (whose boundary is represented by points A and B in Fig. 8.16) cast by the object. If the portion outside the shadow region is covered by an opaque screen, the arrangements act as a diffracting aperture $\mathscr{A}$ with a non-uniform pupil function (cf. § 8.3 (39)). If the linear dimensions of $\mathscr{A}$ are large compared to the wavelength and if F remains sensibly constant over regions whose dimensions are of the same order as the wavelength, the diffraction formula § 8.3 (23) remains valid under the same conditions as before, provided that the integrand of the diffraction integral is multiplied by F.

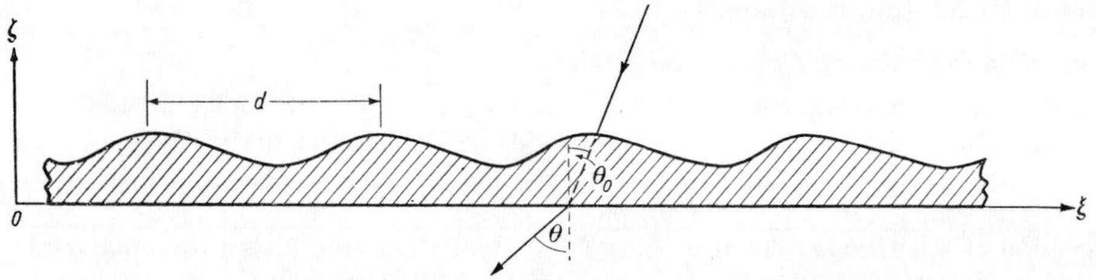

Fig. 8.17. Illustrating the theory of the diffraction grating.

Let us now consider a one-dimensional grating consisting of N parallel grooves of arbitrary profile, ruled on one surface of a plane-parallel glass plate. Let the $\xi\eta$-plane coincide with the plane face of the plate, η being the direction of the grooves and let d be the period in the ξ-direction (see Fig. 8.17).

Assume that the direction of propagation of the wave incident upon the grating is in the plane of the figure, making an angle θ_0 with $O\zeta$, and let θ denote the angle which $O\zeta$ makes with the line joining a very distant point of observation P with the grating. As before we set $l_0 = \sin\theta_0$, $l = \sin\theta$, $p = l - l_0 = \sin\theta - \sin\theta_0$. The complex amplitude at P is then immediately obtained from § 8.5 (23), where the integrand must be multiplied by the transmission function F of one periodic element. We may set $q = 0$ and

$$\xi_n = nd, \qquad \eta_n = 0, \qquad (n = 0, 1, \ldots N-1). \tag{2}$$

We then obtain

$$U(p) = U^{(0)}(p) \sum_{n=0}^{N-1} e^{-ikndp} = U^{(0)}(p) \frac{1 - e^{-iNkdp}}{1 - e^{-ikdp}}, \tag{3}$$

where*

$$U^{(0)}(p) = C \int_{\mathscr{A}} F(\xi) e^{-ikp\xi}\, d\xi. \tag{4}$$

Hence

$$I(p) = |U(p)|^2 = \frac{(1 - e^{-iNkdp})}{(1 - e^{-ikdp})} \cdot \frac{(1 - e^{iNkdp})}{(1 - e^{ikdp})} |U^{(0)}(p)|^2$$

$$= \frac{1 - \cos Nkdp}{1 - \cos kdp} I^{(0)}(p), \tag{5}$$

* Since F depends on l_0, the quantities $U^{(0)}$ and $I^{(0)}$ now depend both on l and l_0 and not on the difference $l - l_0$ only. As we are only interested in effects for a fixed direction of incidence, we may regard l_0 as a constant and retain the previous notation.

where $I^{(0)}(p) = |U^{(0)}(p)|^2$. If we introduce the function

$$H(N, x) = \left(\frac{\sin Nx}{\sin x}\right)^2, \tag{6}$$

the formula (5) for the intensity may be written as

$$I(p) = H\left(N, \frac{kdp}{2}\right) I^{(0)}(p). \tag{5a}$$

Before discussing the implications of this basic formula we note that according to (3) the light distribution is the same as that due to a set of coherent secondary sources each characterized by the same amplitude function $|U^{(0)}(p)|$ and with phases that

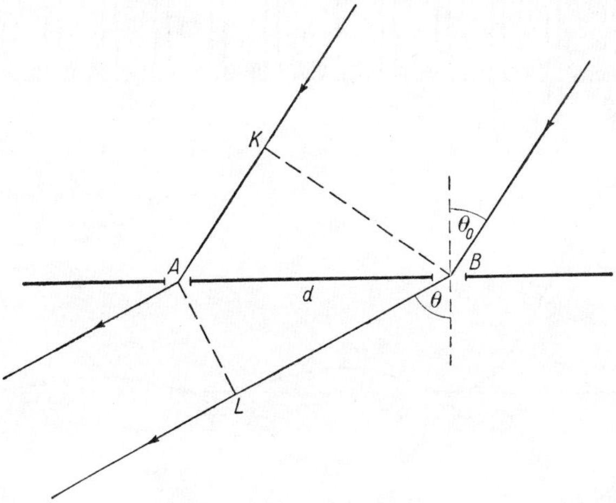

Fig. 8.18. Illustrating the theory of the diffraction grating.

differ from each other by integral multiples of kdp. To see the significance of this phase difference consider two corresponding points A and B in neighbouring grooves of the grating (Fig. 8.18). Since the effect of the grating is to impress a periodic variation on to the incident wave, it follows that the path difference between the light arriving at A and at B is the same as in the absence of the grating, i.e. it is equal to $AK = d \sin \theta_0$, K denoting the foot of the perpendicular from B on to the ray incident at A. Further, the light path from B in the direction θ exceeds the light path from A by $BL = d \sin \theta$, L being the foot of the perpendicular from A on to the ray diffracted at B in the direction θ. Hence the total path difference between light arriving at the distant point of observation from corresponding points in two neighbouring grooves is

$$BL - AK = d(\sin \theta - \sin \theta_0) = dp, \tag{7}$$

and the corresponding phase difference is $2\pi dp/\lambda = kdp$.

The formula (5a) expresses $I(p)$ as the product of two functions: one of them, $I^{(0)}$, represents the effect of a single period of the grating; the other, H, represents the effect of interference of light from different periods. The function $H(N, x)$ has maxima, each of height N^2, at all points where the denominator $\sin^2 x$ vanishes, i.e. where x is zero or an integral multiple of π. Hence $H(N, kdp/2)$ has maxima of height N^2 when

$$p \equiv \sin \theta - \sin \theta_0 = \frac{m\lambda}{d}. \qquad (m = 0, \pm 1, \pm 2, \ldots). \tag{8}$$

The integer m represents, according to (7), the path difference in wavelengths between light diffracted in the direction of the maximum, from corresponding points in two neighbouring grooves. In agreement with our earlier definition (§ 7.3.1), we call m the *order of interference*. Between these principal maxima there are weak secondary maxima (see Fig. 8.19 (a)), the first secondary maximum being only a few per cent of

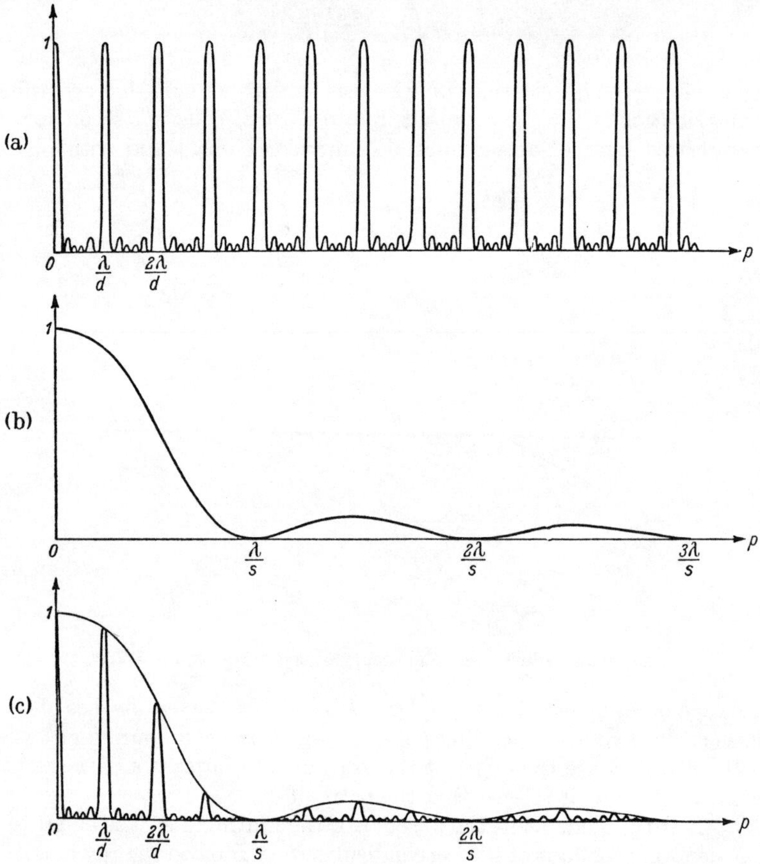

Fig. 8.19.

(a) The normalized interference function

$$\frac{1}{N^2} H(N, kdp/2) = \left[\frac{\sin (Nkdp/2)}{N \sin (kdp/2)}\right]^2.$$

(b) The normalized intensity function of a slit

$$I^{(0)}(p) = \left[\frac{\sin ksp/2}{ksp/2}\right]^2.$$

(c) The normalized intensity function of a grating consisting of N similar equidistant parallel slits

$$\frac{1}{N^2} I(p) = \left[\frac{\sin (Nkdp/2)}{N \sin (kdp/2)}\right]^2 \left[\frac{\sin ksp/2}{ksp/2}\right]^2.$$

Only the range $p \geqslant 0$ is shown, all the curves being symmetrical about the vertical axis $p = 0$.

the principal maximum when N is large. The maxima are separated by points of zero intensity at $x = kdp/2 = \pm n\pi/N$, i.e. in directions given by

$$p \equiv \sin\theta - \sin\theta_0 = \frac{n\lambda}{Nd}, \qquad (n = \pm 1, \pm 2, \ldots), \tag{9}$$

the case where n/N is an integer being excluded.

The function $I^{(0)}(p)$ depends on the form of the grooves. Suppose that it has a principal maximum for some direction $p = p'$ and that on both sides of the maximum it falls off slowly in comparison with H. Then $I(p)$ will have the general form of the interference function H, but will be "modulated" by $I^{(0)}$. Thus $I(p)$ will still have fairly sharp maxima near the directions $p = m\lambda/d$. Since these directions (except for $m = 0$) depend on the wavelength, we see that the grating will decompose a beam of non-monochromatic light into *spectral orders*.

To illustrate these remarks let us consider a grating consisting of a succession of long equidistant slits (Fig. 8.20), each of width s and length L, in an opaque screen.

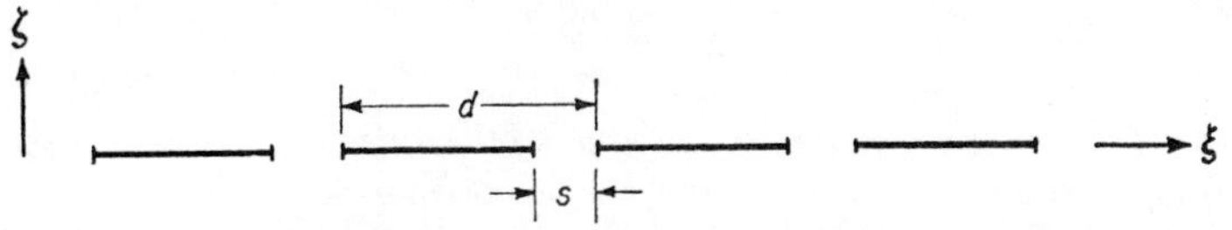

Fig. 8.20. Profile of a simple line grating.

If the grating is illuminated from a very distant line source parallel to the slits, the intensity $I^{(0)}$ is given by the expression § 8.5 (4) [with $2a = s$, $2b = L$] and we obtain

$$I(p) = \frac{sE}{\lambda}\left(\frac{\sin\dfrac{Nkdp}{2}}{\sin\dfrac{kdp}{2}}\right)^2\left(\frac{\sin\dfrac{ksp}{2}}{\dfrac{ksp}{2}}\right)^2. \tag{10}$$

Curves representing the two factors in (10) and their product are shown in Fig. 8.19. The last factor in (10), which represents the effect of a single slit, has a principal maximum at $p = 0$ and minima given by $ksp/2 = n\pi$, i.e. at

$$p = \frac{n\lambda}{s}, \qquad (n = \pm 1, \pm 2, \ldots) \tag{11}$$

separated by weak secondary maxima. We see that if $\lambda/s \gg \lambda/d$, i.e. if the width of each slit is small compared to d, the intensity $I(p)$ has in addition to a principal maximum at $p = 0$ a series of sharp, but progressively decreasing, maxima on either side of it, near directions given by (8).

Returning to the general case, it is evident that if the width of each groove is very small, of the order of a wavelength (as is often the case in practice) the formula (4), derived on the basis of KIRCHHOFF's approximation, can evidently no longer be expected to hold. In such cases more refined considerations must be made to determine the detailed distribution of the intensity. We may, however, expect that the main qualitative features indicated by our elementary theory, namely the existence of sharp maxima whose positions are substantially determined by the interference function H, remain even when the grooves are very narrow, provided, of course, that the intensity function of a single period varies slowly in an interval of the order $\Delta p = \lambda/d$.

Let us now consider the resolution that may be attained with a grating. The separation between a primary maximum of order m and a neighbouring minimum is, according to (9), given by

$$\Delta p = \frac{\lambda}{Nd}. \tag{12}$$

If the wavelength is changed by an amount $\Delta\lambda$, the mth order maximum is, according to (8), displaced by an amount

$$\Delta' p = \frac{|m|}{d}\,\Delta\lambda. \tag{13}$$

Assuming that the lines of wavelength $\lambda \pm \frac{1}{2}\Delta\lambda$ will just be resolved when the maximum of the one wavelength coincides with the first minimum of the other (cf. p. 334) we have on the limit of resolution in the mth order, $\Delta p = \Delta' p$, i.e.

$$\frac{\lambda}{\Delta\lambda} = |m|N. \tag{14}$$

Thus, *the resolving power is equal to the product of the order number m and the number N of the grooves.* For the mth order we have, according to (8), that $d(\sin\theta - \sin\theta_0) = m\lambda$, so that we may also express the resolving power in the form

$$\frac{\lambda}{\Delta\lambda} = \frac{Nd|\sin\theta - \sin\theta_0|}{\lambda}. \tag{14a}$$

Because of (7) this implies that *the resolving power is equal to the number of wavelengths in the path difference between rays that are diffracted in the direction θ from the two extreme ends (separated by distance Nd) of the grating.* It is to be noted that since $|\sin\theta - \sin\theta_0|$ cannot exceed 2, the resolving power that can be attained with a grating of overall width w can never exceed the value $2w/\lambda$.

Let us illustrate the formula (14) by determining the number of grooves that a grating must have, in order to separate two lines which are a tenth of an Ångström unit apart, near the centre of the visible region of the spectrum. In this case $\lambda \sim 5500$ Å, $\Delta\lambda = 10^{-1}$ Å, and if we observe in the second order ($m = 2$), we must have, according to (14) $N \gtrsim 5{\cdot}5\,.\,10^3/2.10^{-1} = 27{,}500$, i.e. the grating must have at least 27,500 grooves.

For comparison let us consider the resolving power of a prism, in the position of minimum deviation, with a line source that is parallel to the edge A of the prism (slit of the spectrograph). A pencil of parallel rays will be incident upon the prism and will be diffracted at a rectangle of width $l_1 = l_2$ (see Fig. 4.28). According to § 8.5 (2), the first minimum of the intensity is at an angular distance (assumed to be small)

$$p = \frac{\lambda}{l_1} \tag{15}$$

from the geometrical image of the slit. The change in the angular dispersion corresponding to a change of wavelength by amount $\Delta\lambda$, is, according to § 4.7 (36),

$$\Delta\varepsilon = \frac{t}{l_1}\frac{dn}{d\lambda}\,\Delta\lambda, \tag{16}$$

where t is the greatest thickness of the glass through which one of the extreme rays has passed, and n is the refractive index of the material of the prism. Since at the limit of resolution $p \sim \Delta\varepsilon$ the resolving power is given by

$$\frac{\lambda}{\Delta\lambda} = t\left|\frac{dn}{d\lambda}\right|. \tag{17}$$

Equation (17) shows, that, *with given glass, the resolving power of a prism depends only on the greatest thickness of the glass traversed by the rays; in particular the resolving power is independent of the angle of the prism.* As an example, suppose that the length of the base of the prism is equal to 5 cm and that it is made of heavy flint glass, for which $dn/d\lambda \sim 1000\ \text{cm}^{-1}$ at wavelength $\lambda = 5500$ Å. If the prism is completely filled with light, then according to (17) it will resolve lines near the centre of the visible region which are not less than $\Delta\lambda$ apart, where $\Delta\lambda \sim 5{\cdot}5\ .\ 10^{-5}\ \text{cm}/5.10^3 = 1{\cdot}1$ Å. Thus a prism of this considerable size has a resolving power 10 times smaller than the grating of 27,500 grooves discussed before.

We have so far considered one-dimensional gratings only, but the analysis may easily be extended to two- and three-dimensional periodic arrangements of diffracting bodies. Two-dimensional gratings (called cross-gratings) find no practical applications, though their effects can often be observed, for example when looking at a bright source through a finely woven material (e.g. a handkerchief). The theory of three-dimensional gratings is, on the other hand, of great importance, such gratings being formed by a regular arrangement of atoms in a crystal. The lattice distances (distances between neighbouring atoms) are of the order of an Ångström unit (10^{-8} cm), this being also the order of magnitude of the wavelengths of X-rays. Hence, by sending a beam of X-rays through a crystal, diffraction patterns are produced, and from their analysis information about the structure of the crystal may be deduced.* Diffraction of X-rays by crystals was predicted by VON LAUE† in 1912 and first observed by FRIEDRICH and KNIPPING.

Another example of a grating-like structure is presented by ultrasonic waves in liquids. These are elastic waves produced by a piezo-electric oscillator, differing from ordinary sound waves only in having a frequency well above the upper limit of audibility. Such waves give rise to rarefactions and condensations in the liquid which then act on the incident light like a grating. The theory of this phenomenon is discussed in Chapter XII. For the rest of this section we shall restrict our attention to one-dimensional gratings as used in spectroscopic work.

(b) *Types of grating*‡

The principle of the diffraction grating was discovered by RITTENHOUSE in 1785,§ but this discovery attracted practically no attention. The principle was rediscovered

---

* See for example M. VON LAUE, *Röntgenstrahl-Interferenzen*, Leipzig, Akademische Verlagsgesellschaft 2 Aufl. (1948); or *The Crystalline State*, ed. Sir L. BRAGG (London, Bell and Sons), Vol. I, W. L. BRAGG, *A General Survey* (1933); Vol. II, R. W. JAMES, *The Optical Principles of the Diffraction of X-rays* (1948); Vol. III, H. LIPSON and W. COCHRAN, *The Determination of Crystal Structures* (1953).

† W. FRIEDRICH, P. KNIPPING and M. VON LAUE, *Münchener Sitzungsber.*, 1912, p. 303; *Ann. d. Physik*, (4), **61** (1913), 971.

‡ A fuller account of methods of production of gratings and their development, see G. R. HARRISON, *J. Opt. Soc. Amer.*, **39** (1949), 413.

§ D. RITTENHOUSE, *Trans. Amer. Phil. Soc.*, **2** (1786), 201. See also article by T. D. COPE in *Journ. Franklin Inst.*, **214** (1932), 99.

by FRAUNHOFER* in 1819. FRAUNHOFER's first gratings were made by winding very fine wire round two parallel screws. Because of the relative ease with which wire gratings may be constructed these are occasionally used even to-day, particularly in the long-wavelength (infra-red) range. Later FRAUNHOFER made gratings with the help of a machine, by ruling through gold films deposited on a glass plate; also, using a diamond as a ruling point, he ruled the grooves directly on to the surface of glass.

Great advances in the technique of production of gratings were made by ROWLAND† who constructed several excellent ruling machines and also invented the so-called *concave grating* (discussed on pp. 412–414). ROWLAND's machine was able to rule gratings with grooves over 4 in. long over a length of 6 in., and his first machine ruled about 14,000 grooves per inch, giving a resolving power in excess of 150,000. Later MICHELSON ruled gratings considerably wider than 6 in., with a resolving power approaching 400,000.

Most of the early gratings were ruled on speculum metal and glass, but the more recent practice is to rule the grooves on evaporated layers of aluminium. Since aluminium is a soft metal it causes less wear on the ruling point (diamond) and it also reflects better in the ultra-violet.

A perfect grating would have all the grooves strictly parallel and of identical form, but in practice errors will naturally occur. Quite irregular errors lead to a blurring of the spectrum and are not so serious as systematic errors, such as periodic errors of spacing. These errors give rise to spurious lines in the spectrum, known as *ghosts*. Often they can be distinguished from true lines only with difficulty.

High resolving power is not always the only important requirement in spectroscopic applications. When little energy is available, as for example in the study of spectra of faint stars or nebulae, or for work in the infra-red region of the spectrum, it is essential that as much light as possible should be diffracted into one particular order. Moreover, for precise wavelength measurements, a grating that gives high dispersion must be used. According to (8), the angular dispersion (with a fixed angle of incidence) is given by

$$\frac{d\theta}{d\lambda} = \frac{1}{\cos\theta}\frac{m}{d}, \tag{18}$$

so that to obtain high dispersion the spacing d should be small or the observations must be made in high orders (m large). If, however, the grating is formed by a succession of opaque and transparent (or reflecting) strips, only a small fraction of the incident light is thrown into any one order. This drawback is overcome in modern practice by ruling the grooves to controlled shape. With a grating which consists of grooves of the form shown in Fig. 8.21, most of the light may be directed into one or two orders on one side of the central image. Gratings of this type, with fairly coarse grooves, are called *echelette gratings*, because they may be regarded as being intermediate between the older types of grating and the so-called echelon gratings which will be described later. Echelette gratings were first ruled by WOOD‡ on copper plates, using the natural edge of a selected carborundum crystal as a ruling point. Later they were ruled with diamond edges ground to the desired shape. They

* J. FRAUNHOFER, *Denkschr. Akad. Wiss. München*, **8** (1821–1822), 1. *Ann. d. Physik.*, **74** (1823), 337. Reprinted in his collected works (Munich, 1888), 51, 117.

† H. A. ROWLAND, *Phil. Mag.* (5), **13** (1882), 469. *Nature*, **26** (1882), 211. *Phil. Mag.*, **16** (1883), 297.

‡ R. W. WOOD, *Phil. Mag.*, **20** (1910), 770; *Ibid.* **23** (1912), 310; A. TROWBRIDGE and R. W. WOOD, *Ibid.*, **20** (1910), 886, 898.

had 2000–3000 grooves per inch, and when used with visible light they sent the greater part of the light into a group of spectra near the 15th or 30th order. Echelette gratings have considerable value in infra-red spectroscopy.

More recently methods have been developed for controlling the groove shape for gratings with much smaller groove spacing.* These *blazed gratings*, as they are called, have grooves of similar form as the echelettes, but form the most intense spectra in much lower orders (usually the first or second).

It appears that the resolving power of gratings of the type described is limited by practical considerations of manufacture to about 400,000. For some applications (e.g. for the study of ZEEMAN effects and the hyperfine and isotope structure patterns), a resolving power that exceeds this value is required. For the attainment of such a high resolving power, HARRISON† proposed the so-called *echelle grating*, which has wide, shallow grooves and is designed for use at an angle of incidence greater than 45°, the direction of incidence being normal to the narrow side of the step. These gratings

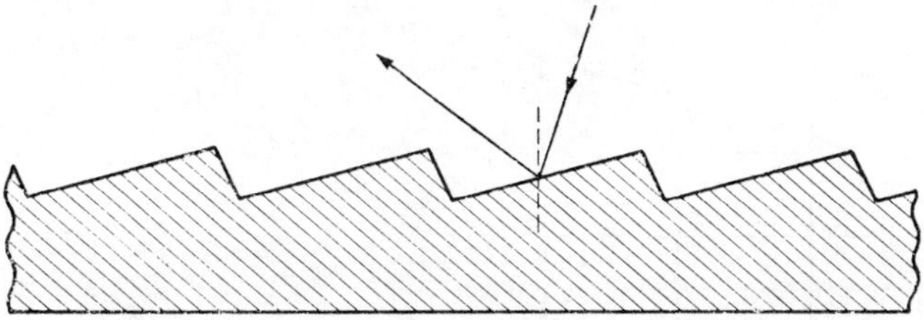

Fig. 8.21. Reflection grating with controlled groove form.

operate with relatively high orders ($m \sim 1000$). A 10 in. echelle with 100 grooves per inch, designed for observation in the 1000th order, has a resolving power of 1,000,000‡.

Because a grating of good quality is very difficult to produce, *replicas* of original rulings are often used.§ These are obtained by moulding from an original ruled master grating.

Finally we must mention a "grating" of an entirely different construction, the *echelon* invented by MICHELSON.‖ It consists of a series of strictly similar plane-parallel glass plates arranged in the form of a flight of steps (hence the name), as shown in Fig. 8.22. Each step retards the beam of light which passes through it by the same amount with respect to its neighbour. Because the breadth of each step is large compared to the wavelength, the effect of diffraction is confined to small angles, so that most of the light is concentrated in one or two spectra near the direction $\theta = 0$, and these correspond to very high orders, since the retardation introduced between successive beams is very many wavelengths.

The resolving power of the echelon depends not only on the path difference between the rays from the extreme ends of the grating but also (though to a much lesser extent) on the dispersion of the glass. If n is the refractive index, t the thickness of

* R. W. WOOD, *Nature*, **140** (1937), 723; *J. Opt. Soc. Amer.*, **34** (1944), 509; H. BABCOCK, **34** (1944), 1.

† G. R. HARRISON, *J. Opt. Soc. Amer.*, **39** (1949), 522.

‡ For a review of the theory and production of high resolution gratings, see G. W. STROKE, *Progress in Optics*, Vol. 2, ed. E. WOLF (Amsterdam, North Holland Publishing Company and New York, J. Wiley and Sons, 1963), 1.

§ First replicas were made by T. THORP, *British Patent No.* 11,460 (1899); and later by R. J. WALLACE, *Astrophys. J.*, **22** (1905), 123; *Ibid.*, **23** (1906), 96. Improved methods have been described by T. MERTON, *Proc. Roy. Soc. A*, **201** (1950), 187.

‖ A. A. MICHELSON, *Astrophys. J.*, **8** (1898), 37; *Proc. Amer. Acad. Arts Sci.*, **35** (1899), 109.

each step, and d its breadth (see Fig. 8.22), the path difference between rays diffracted from neighbouring steps is evidently $pd + (n - 1)t$, it being assumed that p is small. Hence the position of the principal maxima are given by

$$pd + (n-1)t = m\lambda, \qquad (m = 0, 1, 2, \ldots). \tag{19}$$

If the wavelength is changed by an amount $\Delta\lambda$, the mth order maximum is displaced by

$$\Delta' p = \left| m - t\frac{dn}{d\lambda} \right| \frac{\Delta\lambda}{d}. \tag{20}$$

The separation Δp between a principal maximum of order m and a neighbouring minimum is again given by (12), so that the condition $\Delta p = \Delta' p$ for the limit of resolution gives

$$\frac{\lambda}{\Delta\lambda} = N\left| m - \frac{dn}{d\lambda} t \right|. \tag{21}$$

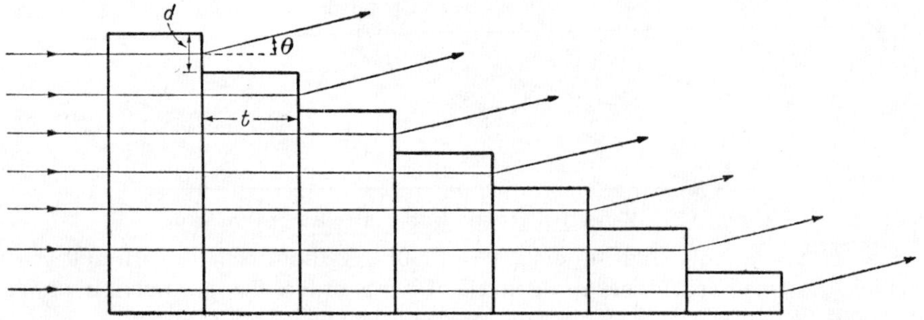

Fig. 8.22. MICHELSON's echelon.

Here we may substitute for m the value $(n - 1)t/\lambda$ obtained from (19) by neglecting the term pd, for the p values for which the intensity is appreciable are of the order of λ/d, i.e. pd is of the order of a wavelength, whilst $(n - 1)t$ is of the order of many thousand wavelengths. We thus obtain the following expression for the resolving power of the echelon:

$$\frac{\lambda}{\Delta\lambda} \sim N \left| \frac{n-1}{\lambda} - \frac{dn}{d\lambda} \right| t. \tag{22}$$

The ratio $\dfrac{dn}{d\lambda} \Big/ \dfrac{n-1}{\lambda}$ is small. For flint glass near the centre of the visible region it has a value near $-0{\cdot}05$ to $-0{\cdot}1$. Hence, under these circumstances, the resolving power of an echelon exceeds, by about 5 to 10 per cent, the resolving power of a line grating with N grooves, when observation is made in the order $m = (n - 1)t/\lambda$. One of MICHELSON's echelons consisted of twenty plates, each having a thickness $t \doteq 18$ mm, and the breadth d of each step was about 1 mm. Taking $n = 1{\cdot}5$, the retardation between two successive beams measured in wavelengths of green light $\lambda = 5.10^{-5}$ cm was $m \sim 0{\cdot}5 \,.\, 1{\cdot}8/5.10^{-5} \sim 20{,}000$. Assuming $\dfrac{dn}{d\lambda} \Big/ \dfrac{n-1}{\lambda} = -0{\cdot}1$, this gives a resolving power of about 20 $(20{,}000 + 0{\cdot}1 \times 20{,}000) = 440{,}000$.

More important is the *reflection echelon*. Here each step is made highly reflecting by means of metallic coating, and the spectra formed by reflected light are observed. With a reflection echelon the resolving power is three to four times as large than with a transmission echelon of corresponding dimensions, since each step introduces a retardation between successive beams of amount $2t$ instead of $(n - 1)t \sim t/2$. Like the echelle grating the reflection echelon is capable of giving resolving power of over one million. Another advantage of the reflection echelon over the transmission echelon is that it may be used in the ultra-violet region of the spectrum, where glass absorbs. Although MICHELSON realized that advantages would be gained by using the instrument with reflected rather than transmitted light, technical difficulties prevented the production of a satisfactory reflection echelon for nearly thirty years until they were overcome by WILLIAMS.* Because of difficulties in assembling a large number of plates of equal thickness within the narrow permissible tolerance, the number of steps is limited in practice to about forty.

Finally, a few remarks must be made about overlapping of orders. Restricting ourselves to the visible region, i.e. considering wavelengths in the range $\lambda_1 = 0{\cdot}4\mu$ to

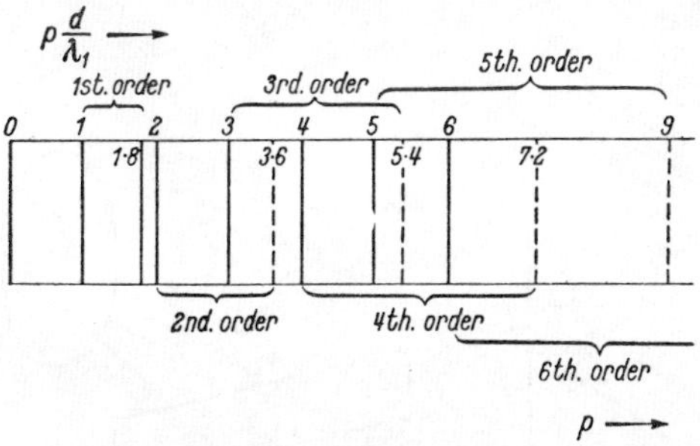

Fig. 8.23. The overlapping of grating spectra.

$\lambda_2 = 0{\cdot}75\mu$, we see that the first-order spectrum does not quite reach to the spectrum of the second order: for the first-order spectrum covers the range from $p = \lambda_1/d$ to $p = \lambda_2/d = 0{\cdot}75\lambda_1/0{\cdot}4d = 1{\cdot}8\lambda_1/d$, whilst the second order begins at $p = 2\lambda_1/d$. On the other hand the spectrum of the second order extends across a part of the third-order spectrum, namely from $p = 2\lambda_1/d$ to $p = 2\lambda_2/d$, whilst the third order begins already at $2 \times 1{\cdot}8\lambda_1/d$. As the order increases the successive spectra overlap more and more (see Fig. 8.23). If the lines of wavelength λ and $\lambda + \delta\lambda$ coincide in two successive orders $(m + 1)$th and mth, then

$$(m + 1)\lambda = m(\lambda + \delta\lambda),$$

i.e.

$$\frac{\delta\lambda}{\lambda} = \frac{1}{m}. \tag{23}$$

Thus the *"free spectral range" is inversely proportional to the order.*

The overlapping of orders was formerly used to compare wavelengths, in the so-called method of coincidences (cf. p. 340); this method has been superseded by simple interpolation between standard wavelengths determined interferometrically.

In conclusion let us summarize the main distinguishing features of the different types of gratings. We recall that, according to (14), high resolving power may be attained with either a large number of periods and relatively low orders, or with a moderate number of periods and large orders. Ordinary ruled gratings represent the low-order extreme ($m \sim 1$ to 5), whilst the echelons represent the extreme of high orders ($m \sim 20{,}000$). In between are the echellettes ($m \sim 15$ to 30) and the echelles

* W. E. WILLIAMS, *British Patent*, 312534 (1926); *Proc. Opt. Conv.*, **2** (1926), 982; *Proc. Phys. Soc.*, **45** (1933), 699.

($m \sim 1000$). For particular applications one must bear in mind that the angular dispersion is directly proportional to the spectral order and inversely proportional to the period whilst the free spectral range is inversely proportional to the order.

(c) *Grating spectrographs*

In a grating spectrograph coloured images of a slit source are produced in the various orders into which the grating separates the incident light. A simple arrangement is shown in Fig. 8.24. Collimated light from a slit source S in the focal plane of a lens L is incident on a reflection grating G, and the images of S formed by the diffracted rays are observed in the focal plane F of a telescope T. A modification of this arrangement,

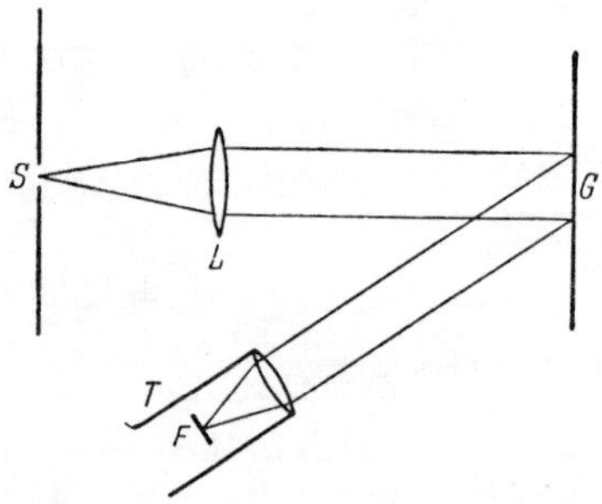

Fig. 8.24. A grating spectrograph.

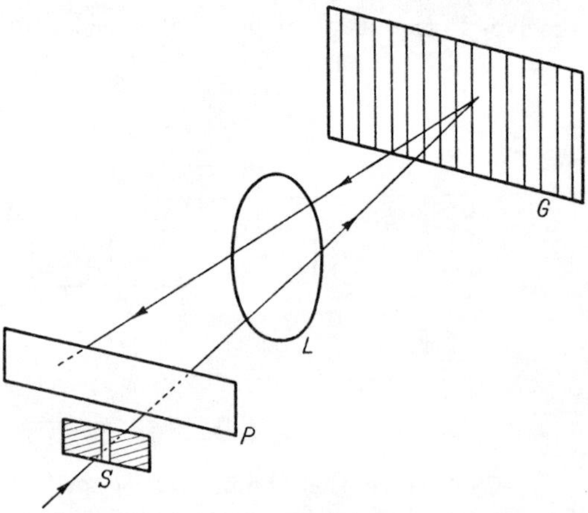

Fig. 8.25. A grating spectrograph: Littrow's mounting.

known as *Littrow's mounting*, which has the advantage of compactness, is shown in Fig. 8.25. This is an autocollimation device, which needs only one lens. The slit is just below the plate P and the lens is near the grating, which can be turned through a prescribed angle with respect to the direction of the incident beam.

In order to avoid losses of light which necessarily arise when the diffracted rays are focused by means of lenses, Rowland introduced the *concave grating*. Here the grooves are ruled on a concave highly reflecting metal surface, i.e. on a concave mirror, in such a way that their projections on a chord of the mirror surface are equidistant. A simple geometrical theorem indicates the possible positions of the slit and the plane of observation relative to the grating:

Let Q be the midpoint of the surface of the grating and C its centre of curvature, and describe a circle K with centre at the midpoint O of QC and with radius $r = OQ = OC$ (Fig. 8.26). We shall prove that light from any point S of the circle K will be approximately reflected to a point P and diffracted to points P', P'', . . . on the circle, each of these points being a focus for diffracted rays of a particular order. To show this, construct the reflected ray QP corresponding to the incident ray SQ. If $\alpha = SQC$ is the angle of incidence, the angle CQP of reflection is also equal to α and, moreover, the arc SC is equal to the arc CP. Consider now another ray from S incident upon the grating at a different point R. If the diameter of the circle is sufficiently large (in practice it is usually several feet), then no appreciable error is introduced by assuming R to lie on the circle K. Hence, since C is the centre of curvature of the grating, the angle of incidence SRC and consequently also the angle of reflection are again equal to α. Moreover, since the arc CP is equal to the arc SC it follows that the ray reflected at R again passes through P.

Similar considerations apply to the diffracted rays. Let β be the angle which a ray diffracted at Q makes with QP. The corresponding ray of the same order, diffracted at

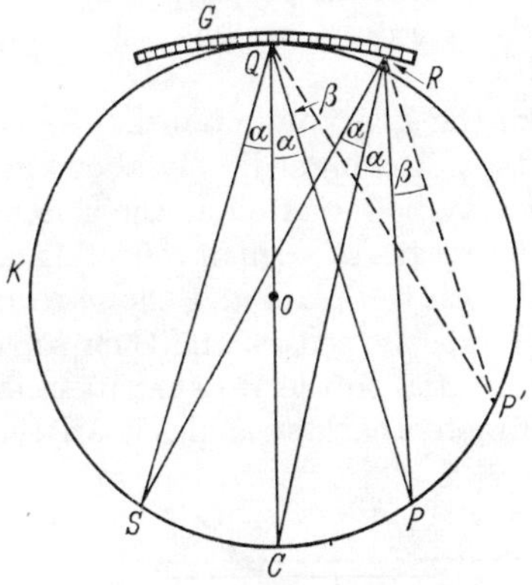

Fig. 8.26. Focusing with a concave grating (ROWLAND's circle).

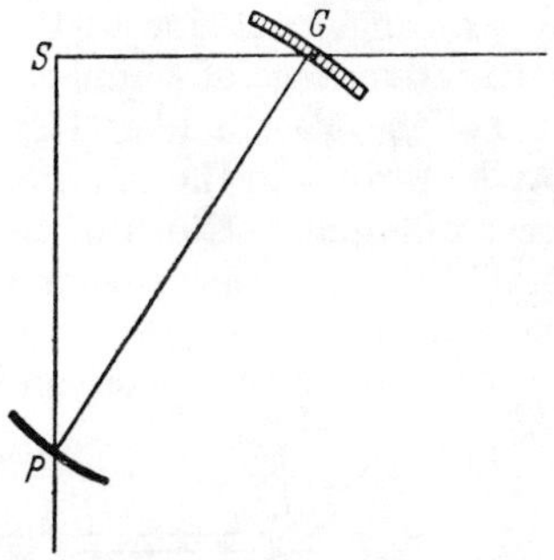

Fig. 8.27. ROWLAND's mounting for a concave grating.

R, will make the same angle (β) with RP. Hence the ray diffracted at Q makes the same angle with SQ as the ray diffracted at R makes with SR, namely $2\alpha + \beta$. The two diffracted rays, therefore, meet in a point P' of the circle K. Thus, *to obtain sharp lines, the slit, the grating, and the plane of observation (photographic plate) should be*

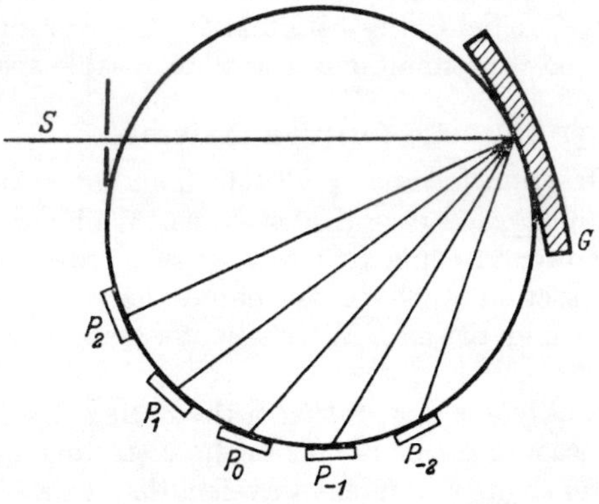

Fig. 8.28. PASCHEN's mounting for a concave grating.

situated on a circle, whose diameter is equal to the radius of curvature of the concave grating.

There are several mountings based on this principle. ROWLAND himself used the arrangement shown in Fig. 8.27. Here the grating G and the plate holder P are fixed to opposite ends of a movable girder, whose length is equal to the radius of curvature of the grating. The two ends of the girder are free to move along fixed tracks which are at right angles to each other. The slit S is mounted immediately above their intersection in such a way that light falling normally on the slit proceeds along SG.

The slit is thus situated on a ROWLAND circle with diameter PG and the order of the spectrum appearing on the plate depends on the position of the girder.

A different arrangement, shown in Fig. 8.28, avoids the use of mobile parts. Here a circular steel rail to which the slit S and the grating G are permanently attached plays the part of ROWLAND's circle. Round the rail a series of plate holders P_0, P_1, P_{-1}, . . . is set up, so that spectra of several orders can be photographed simultaneously. This arrangement, called *Paschen's mounting*, also has the advantage of great stability.

Another arrangement due to EAGLE has, like the LITTROW mounting for a plane grating, the advantage of compactness. Here the slit is immediately above or below the centre of the plate holder (Fig. 8.29), or it may be mounted at the side and the light may be thrown in the required direction by means of a small reflecting surface. To observe different portions of the spectrum, both the plate and the grating have to be rotated through the same amounts and in opposite senses, and their separation has to be changed, so that they are always tangential to the ROWLAND circle. One observes that part of the spectrum which is diffracted back at angles nearly equal to

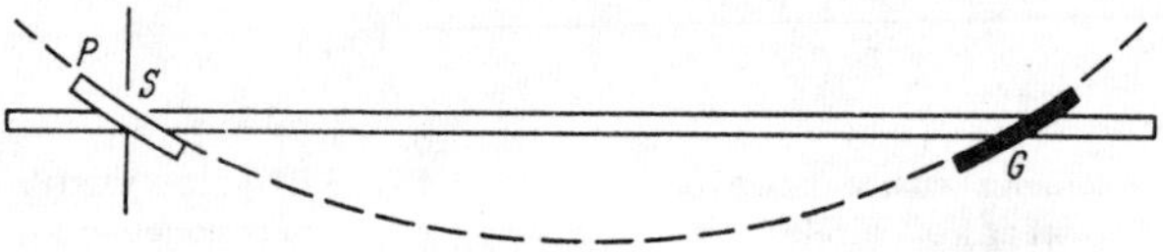

Fig. 8.29. EAGLE's mounting for a concave grating.

the angle of incidence. For this mounting to be strictly autocollimating, the slit S should be at the centre of the plate.

Spectral lines obtained with a concave grating show the same aberrations as images obtained with a concave mirror, chiefly astigmatism. If, however, the concave grating is used in parallel light, the astigmatism may be made to vanish on the grating normal and to be very small over the whole usable spectrum.*

8.6.2 Resolving power of image-forming systems

The FRAUNHOFER diffraction formula § 8.3 (36) finds important applications in the calculation of the *resolving power* of optical systems. We have already introduced the concept of resolving power in connexion with interference spectroscopes in § 7.6.3, and in the preceding section we have estimated the resolving power that can be attained with gratings and prisms. We shall now extend this concept to image-forming systems.

In a spectral apparatus (e.g. a line grating or the FABRY–PEROT interferometer), the resolving power is a measure of the ability of the instrument to separate two neighbouring spectral lines of slightly different wavelengths. In an image-forming system, it is a measure of the ability to separate images of two neighbouring object points. In the absence of aberrations each point object would, according to geometrical optics, give rise to a sharp point image. Because of diffraction the actual image will nevertheless always be a finite patch of light. And if two such image patches (diffraction patterns) overlap, it will be more and more difficult to detect the presence of two objects, the closer the central intensity maxima are to each other. The limit down to

* Such an astigmatic-free mounting was described by F. L. O. WADSWORTH, *Astrophys. J.*, **3** (1896), 54. For a discussion of the aberration theory of gratings and grating mountings see an article by W. T. WELFORD in *Progress in Optics*, Vol. 4, ed. E. WOLF (Amsterdam, North Holland Publishing Company and New York, J. Wiley and Sons, 1965), p. 241.

which the eye can detect the two objects is, of course, to some extent a matter of practical experience. With a photographic plate the contrast may be enhanced and so the limit of resolution decreased by suitable development. Nevertheless it is desirable to have some simple criterion which permits a rough comparison of the relative efficiency of different systems, and for this purpose RAYLEIGH's criterion discussed in § 7.6.3 may again be employed. According to this criterion two images are regarded as just resolved when the principal maximum of one coincides with the first minimum of the other. For a spectral apparatus, where the limit of resolution is a certain wavelength difference $\Delta\lambda$, the resolving power was defined as the quantity $\lambda/\Delta\lambda$. For an image-forming system the limit of resolution is some distance δx or angle $\delta\theta$ and the resolving power is defined as the reciprocal (i.e. $1/\delta x$ or $1/\delta\theta$) of this quantity.

Let us consider first the limit of resolution of a telescope. For a distant object, the edge of the entrance pupil coincides with the circular boundary of the objective, and acts as the diffracting aperture. If a is the radius of the objective aperture then, according to § 8.5 (16), the position of the first minimum of intensity referred to the central maximum is given by*

$$w = 0{\cdot}61\,\frac{\lambda}{a}. \tag{24}$$

Now $w = \sqrt{p^2 + q^2}$ represents the sine of the angle ϕ which the direction (p, q) makes with the central direction $p = q = 0$. This angle is usually so small that its sine may be replaced by the angle itself, and it then follows that (on the basis of RAYLEIGH's criterion) *the angular separation of two stars that can just be resolved is* $0{\cdot}61\lambda/a$.

With a given objective, the angular size of the image as seen by the eye depends on the magnification of the eyepiece. It is impossible, however, to bring out detail not present in the primary image by increasing the power of the eyepiece, for each element of the primary image is a small diffraction pattern, and the actual image, as seen by the eyepiece, is only the ensemble of the magnified images of these patterns.

The largest telescope in existence (at Mount Palomar) has a diameter $2a \sim 5$ m. Neglecting for the moment the effect of the central obstruction in the telescope, the theoretical limit of resolution for light near the centre of the visible range ($\lambda \sim 5{\cdot}6 \,.\, 10^{-5}$ cm) is seen to be

$$\phi \sim 0{\cdot}61\,\frac{5{\cdot}6 \,.\, 10^{-5}\ \text{cm}}{2{\cdot}5 \,.\, 10^{2}\ \text{cm}} \sim 1{\cdot}4 \,.\, 10^{-7}$$

or, in seconds of arc,

$$\phi \sim 0{\cdot}028''.$$

In § 6.1 we quoted the value of 1 min. of arc for the limit of resolution of the eye. We can now give a more precise estimate. Since the diameter of the pupil of the eye varies from about 1·5 mm to about 6 mm (depending on the intensity of the light), it follows that the limit of resolution lies in the range (again taking $\lambda = 5{\cdot}6 \,.\, 10^{-5}$ cm)

$$0{\cdot}61\,\frac{5{\cdot}6 \,.\, 10^{-5}}{0{\cdot}75 \,.\, 10^{-1}} > \phi > 0{\cdot}61\,\frac{5{\cdot}6 \,.\, 10^{-5}}{3 . 10^{-1}}$$

* According to § 7.6.3 the saddle-to-peak intensity ratio at the limit of resolution for diffraction at a slit aperture is $8/\pi^2 = 0{\cdot}811$. The corresponding value for the present case (circular aperture) is 0·735.

i.e.

$$4{\cdot}55\,.\,10^{-4} > \phi > 1{\cdot}14\,.\,10^{-4}$$

or, in minutes and seconds:

$$1'34'' > \phi > 0'24''.$$

So far we have assumed the aperture to be circular. Of considerable interest is also the case of an annular aperture, since in many telescopes, for example, the central portion of the circular aperture is obstructed by the presence of a secondary mirror. Suppose that the annular aperture is bounded by two concentric circles of radii a and εa, where ε is some positive number less than unity. The light distribution in the FRAUNHOFER pattern is then represented by an integral of the form § 8.5 (8), but with the ρ integration extending only over the domain $\varepsilon a \leqslant \rho \leqslant a$. In place of § 8.5 (13) we then obtain

$$U(P) = C\pi a^2 \left[\frac{2J_1(kaw)}{kaw}\right] - C\pi\varepsilon^2 a^2 \left[\frac{2J_1(k\varepsilon aw)}{k\varepsilon aw}\right], \tag{25}$$

so that the intensity is given by

$$I(P) = \frac{1}{(1-\varepsilon^2)^2}\left[\left(\frac{2J_1(kaw)}{kaw}\right) - \varepsilon^2\left(\frac{2J_1(k\varepsilon aw)}{k\varepsilon aw}\right)\right]^2 I_0, \tag{26}$$

where $I_0 = |C|^2\pi^2 a^4(1-\varepsilon^2)^2$ is the intensity at the centre $w = 0$ of the pattern. The position of the minima (zeros) of intensity are now given by the roots of the equation

$$J_1(kaw) - \varepsilon J_1(k\varepsilon aw) = 0, \qquad (w \neq 0), \tag{27}$$

whilst the maxima are given by the roots of

$$J_2(kaw) - \varepsilon^2 J_2(k\varepsilon aw) = 0. \tag{28}$$

In deriving (28), the relation § 8.5 (15) was used, as in the case of circular aperture. For the unobstructed aperture ($\varepsilon = 0$) the first root of (27) is, of course, the value given by (24), namely $w = 3{\cdot}83/ka = 0{\cdot}61\lambda/a$. As ε is increased, the first root of (27) decreases,* and with $\varepsilon = \frac{1}{2}$, for example, it is slightly less than $3{\cdot}15/ka = 0{\cdot}50\lambda/a$. As the principal maximum remains at $w = 0$ independently of ε, we see that on obstructing the central portion of the aperture the resolving power is increased. This improvement is, however, accompanied by a decrease in the brightness of the image. Also the secondary maxima become more pronounced so that the contrast is reduced. With $\varepsilon = \frac{1}{2}$, the first secondary maximum (at $w = 4{\cdot}8/ka$) is $0{\cdot}10$ of the principal maximum, as compared with the value $0{\cdot}018$ (at $w = 5{\cdot}14/ka$) for a circular aperture (see Fig. 8.30).

The obstruction of the central part of the circular aperture corresponds to the replacement of the pupil function (cf. § 8.3 (39)) $G(\xi, \eta) = C$ or 0 according as $0 \leqslant \rho \leqslant a$ or $\rho > a$ by $G(\xi, \eta) = C$ or 0 according as $\varepsilon a \leqslant \rho \leqslant a$ or $\rho < \varepsilon a, \rho > a$. Naturally it is possible to alter the pupil function in other ways. A general method for modifying a pupil function consists in depositing on one or more surfaces of the system a thin, partially transmitting film of suitable substance. The same effect may be achieved by means of a specially constructed "filter", for example, a hollow lens of appropriate form filled with an absorbing liquid. The problem arises of determining the form of the pupil function which, in some agreed sense, would give the best possible image.

* From (26) it follows by a simple calculation that as $\varepsilon \to 1$, $I/I_0 \to J_0^2(kaw)$. Since the first zero of the equation $J_0(x) = 0$ is at $x = 2{\cdot}40$, it follows that with increasing ε the radius of the first dark ring approaches the value given by $w = 2{\cdot}40/ka = 0{\cdot}38\lambda/a$.

This problem has been investigated by a number of workers.* Of particular interest is a result due to TORALDO DI FRANCIA,† that the pupil function may be so chosen as to make the radius of the first dark ring arbitrarily small and at the same time the dark zone surrounding the central ring arbitrarily large. However, by gradually decreasing the radius of the first dark ring, less and less light is utilized in the central

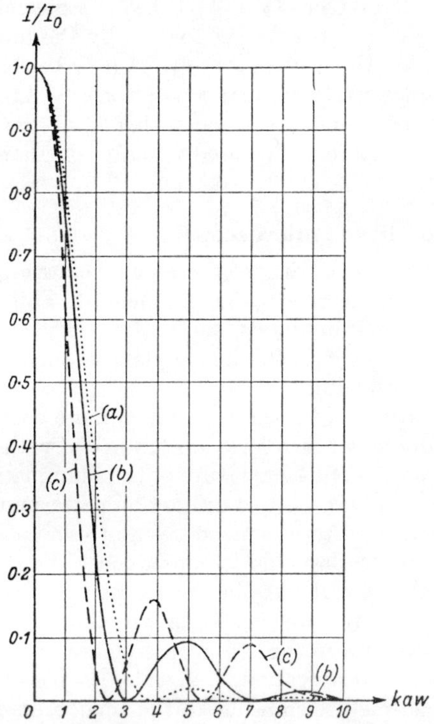

Fig. 8.30. Illustrating the effect of central obstruction on the resolution. Normalized intensity curves for FRAUNHOFER patterns of (a) Circular aperture, (b) Annular aperture with $\varepsilon = \frac{1}{2}$, and (c) Annular aperture with $\varepsilon \to 1$.

[After G. C. STEWARD, *The Symmetrical Optical System* (Cambridge University Press, 1928), p. 89.]

disc, so that the smallest practicable size of the disc and hence the resolving power is limited by the amount of light available.

A partial suppression of the secondary maxima by an appropriate modification of the pupil function is known as *apodization*.‡ In spectroscopic analysis it facilitates the detection of satellites of spectral lines, whilst in astronomical applications it facilitates the resolution of double stars of appreciably different apparent brightness.

---

* See, for example, R. STRAUBEL, *P. Zeeman Verh.* (Den Haag, Martinus Nijhoff, 1935), p. 302; R. K. LUNEBURG, *Mathematical Theory of Optics* (Mimeographed Lecture Notes, Brown University, Providence, R.I. (1944); printed version published by University of California Press, Berkeley and Los Angeles, 1964), § 50); H. OSTERBERG and J. E. WILKINS, *J. Opt. Soc. Amer.*, **39** (1949), 553; G. LANSRAUX, *Rev. d'Optique* **32** (1953), 475.

A brief review of investigations in this field is given in an article by E. WOLF, *Rep. Progr. Phys.* (London, Physical Society), **14** (1951), 109.

† G. TORALDO DI FRANCIA, *Suppl. Nuovo Cimento*, **9** (1952), 426.

‡ A thorough review of investigations on this subject was given by P. JACQUINOT and B. ROIZEN-DOSSIER in *Progress in Optics*, Vol. 3, ed. E. WOLF (Amsterdam, North Holland Publishing Company and New York, J. Wiley and Sons, 1964), p. 29.

The conventional theory of resolving power, as outlined in this section, is particularly appropriate to direct visual observations. With other methods of detection (e.g. photometric) the presence of two objects of much smaller angular separation than indicated by RAYLEIGH's criterion may often be revealed. In this connexion it is also of interest to compare the relative resolving efficiency of a telescope and a MICHELSON Stellar Interferometer (§ 7.3.6). If the presence of two stars is judged by means of the first vanishing of the fringes formed by the interferometer, and if d is the maximum distance by which its outer mirrors may be separated, then according to § 7.3 (34) double stars down to angular separation $\phi \sim \frac{1}{2}\lambda/d$ may be detected with this instrument. Comparison of this value with (24) shows that, to detect double stars of this separation with a telescope used visually, the diameter $2a$ of its objective would have to be about $2{\cdot}4d$.

8.6.3 Image formation in the microscope

In the elementary theory of resolving power which we have just outlined, light from the two object points was assumed to be incoherent. This assumption is justified when the two objects are self-luminous, e.g. with stars viewed by a telescope. The intensity observed at any point in the image plane is then equal to the sum of the intensities due to each of the object points.

In a microscope the situation is, as a rule, much more complicated. The object is usually non-luminous and must, therefore, be illuminated with the help of an auxiliary system. Owing to diffraction on the aperture of the illuminating system (condenser), each element of the source gives rise to a diffraction pattern in the object plane of the microscope. The diffraction patterns which have centres on points that are sufficiently close to each other partly overlap, and in consequence the light vibrations at neighbouring points of the object plane are in general partially correlated. Some of this light is transmitted through the object with or without a change of phase, whilst the rest is scattered, reflected, or absorbed. In consequence, it is in general impossible to obtain, by means of a single observation, or even by the use of one particular arrangement, a faithful enlarged picture showing all the small-scale structural variations of the object. Various methods of observation have, therefore, been developed, each suitable for the study of certain types of objects, or designed to bring out particular features.

We shall briefly outline the theory of image formation in a microscope, confining our attention first of all to the two extreme cases of completely incoherent and perfectly coherent illumination. Partially coherent illumination will be discussed in § 10.5.2.

(a) *Incoherent illumination*

We first consider a self-luminous object (e.g. an incandescent filament of an electric bulb). Let P be the axial point of the object and Q a neighbouring point in the object plane, at a distance Y from P, and let P' and Q' be the images of these points (Fig. 8.31). Further let θ and θ' be the angles which the marginal rays of the axial pencils make with the axis.

If a' is the radius of the region (assumed to be circular) in which the beam of light converging on P' intersects the back focal plane $\mathscr{F}'$ and if D' is the distance between the back focal plane and the image plane, then, since θ' is small,

$$\theta' = \frac{a'}{\mathrm{D}'}. \tag{29}$$

Further, if $w = \sqrt{p^2 + q^2}$ is the separation of Q' from P' measured in "diffraction units" [cf. § 8.3 (35) and § 8.5 (7)], i.e. the sine of the angle which the two points subtend at the centre of the diffracting aperture, then we have to a good approximation,

$$Y' = w\mathrm{D}'. \tag{30}$$

Let n and n' be the refractive indices, λ and λ' the wavelengths in the object and image spaces, and λ_0 the wavelength in vacuum. Then, since according to § 8.5 (16) the first minimum of the diffraction pattern of P is given by $w = 0{\cdot}61\lambda'/a'$, we have, at the limit of resolution

$$Y' = 0{\cdot}61\lambda'\frac{\mathrm{D}'}{a'} = 0{\cdot}61\,\frac{\lambda'}{\theta'} = 0{\cdot}61\,\frac{\lambda_0}{n'\theta'}. \tag{31}$$

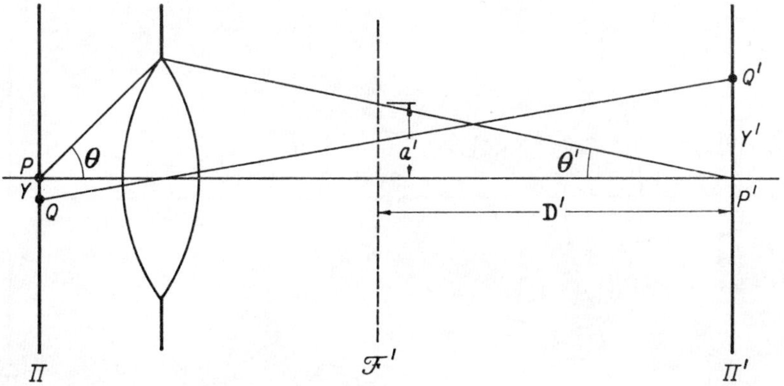

Fig. 8.31. Illustrating the theory of resolving power of the microscope.

A microscope must, of course, be so designed that it gives a sharp image not only of an axial point but also of neighbouring points of the object plane. According to § 4.5.1 the sine condition must therefore be satisfied, i.e.*

$$nY \sin \theta = -\, n'Y' \sin \theta'.$$

Since θ' is small we may replace $\sin \theta'$ by θ'. On substituting for Y' into (31), we finally obtain

$$|Y| \sim 0{\cdot}61\,\frac{\lambda_0}{n \sin \theta}. \tag{32}$$

This formula gives the distance between two object points which a microscope can just resolve when the illumination is *incoherent* and the aperture is circular.

The quantity $n \sin \theta$ which enters into (32) is the *numerical aperture*, [cf. § 4.8 (13)] and must be large if a high resolving power is to be achieved. Means for obtaining a large numerical aperture were discussed in § 6.6.

(b) *Coherent illumination—Abbe's theory*

We now consider the other extreme case, namely when the light emerging from the object may be treated as strictly coherent. This situation is approximately realized when a thin object of relatively simple structure is illuminated by light from a sufficiently small source via a condenser of low aperture (cf. § 10.5.2).

* The minus sign appears here because θ' corresponds to $-\gamma_1$ of § 4.5.

The first satisfactory theory of resolution with coherent illumination was formulated and also illustrated with beautiful experiments, by E. ABBE.* According to ABBE, the object acts as a diffraction grating, so that not only every element of the aperture of the objective, but also every element of the object must be taken into account in determining the complex disturbance at any particular point in the image plane. Expressed mathematically, the transition from the object to the image involves two integrations, one extending over the object plane, the other extending over the aperture. In ABBE's theory, diffraction by the object is first considered and the effect of the aperture is taken into account in the second stage. An alternative procedure, in which the order is reversed, is also permissible and leads naturally to the same result.†

To illustrate ABBE's theory we consider first the imaging of a grating-like object which is illuminated by a plane wave incident normally on to the object plane

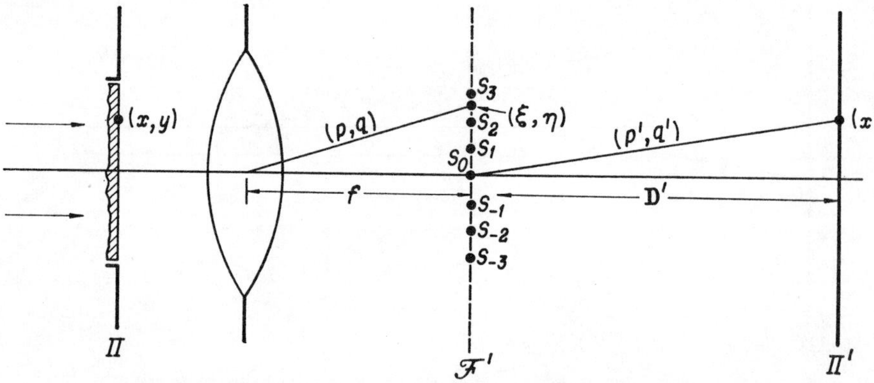

Fig. 8.32. Illustrating the ABBE theory of image formation in a microscope, with coherent illumination.

(KÖHLER's central illumination). The wave is diffracted by the object and gives rise to a FRAUNHOFER diffraction pattern of the grating (cf. § 8.6.1), in the back focal plane $\mathscr{F}'$ of the objective. In Fig. 8.32 the maxima (spectra of successive orders) of this pattern are denoted by . . . S_{-2}, S_{-1}, S_0, S_1, S_2, ... Every point in the focal plane may be considered to be a centre of a coherent secondary disturbance, whose strength is proportional to the amplitude at that point. The light waves that proceed from these secondary sources will then interfere with each other and will give rise to the image of the object in the image plane Π' of the objective. To obtain a faithful image it is necessary that all the spectra contribute to the formation of the image. Strictly this is never possible because of the finite aperture of the objective. We shall see later that the exclusion of some of the spectra may result in completely false detail appearing in the image. For practical purposes it is evidently sufficient that the aperture shall be large enough to admit all those spectra that carry an appreciable amount of energy.

* ERNST ABBE, *Archiv. f. Mikroskopische Anat.*, **9** (1873), 413. Also his *Gesammelte Abhandlungen*, **1** (Jena: G. FISCHER, 1904), 45 and O. LUMMER and F. REICHE, *Die Lehre von der Bildtstehung im Mikroskop von Ernst Abbe*, (Braunschweig, Vieweg, 1910). A good account of ABBE's theory was also given in A. B. PORTER, *Phil. Mag.* (6), **11** (1906), 154.

† A theory of image formation in the microscope which is equivalent to this alternative approach was formulated by Lord RAYLEIGH in *Phil. Mag.* (5), **42** (1896), 167; also his *Scientific Papers*, **4** (Cambridge University Press, 1903), p. 235.

Let us express these considerations in more precise terms without restricting ourselves to a grating-like object. If x, y are the coordinates of a typical point in the object plane and f is the distance of the focal plane $\mathscr{F}'$ from the lens objective, the disturbance at a point

$$\xi = pf, \qquad \eta = qf, \tag{33}$$

of the $\mathscr{F}'$ plane (see Fig. 8.32) is given by the FRAUNHOFER formula

$$U(\xi, \eta) = C_1 \int\int_{\mathscr{A}} F(x, y) e^{-ik\left[\frac{\xi}{f}x + \frac{\eta}{f}y\right]} dxdy, \tag{34}$$

where F is the transmission function of the object, C_1 is a constant, and the integration is taken over the area $\mathscr{A}$ of the object plane Π covered by the object.

Next consider the transition from the back focal plane $\mathscr{F}'$ to the image plane Π'. If, as before, D′ denotes the distance between $\mathscr{F}'$ and Π', and $V(x', y')$ is the disturbance at a typical point

$$x' = p'\mathrm{D}', \qquad y' = q'\mathrm{D}', \tag{35}$$

of the image plane, we have for FRAUNHOFER diffraction on the aperture $\mathscr{B}$ in $\mathscr{F}'$

$$V(x', y') = C_2 \int\int_{\mathscr{B}} U(\xi, \eta) e^{-ik\left[\frac{x'}{\mathrm{D}'}\xi + \frac{y'}{\mathrm{D}'}\eta\right]} d\xi d\eta, \tag{36}$$

it being assumed that $a'/D' \ll 1$ (see figure 8.31). Substitution from (34) into (36) gives

$$V(x', y') = C_1 C_2 \int\int_{\mathscr{A}} \int\int_{\mathscr{B}} F(x, y) e^{-i\frac{k}{f}\left[\left(x + \frac{f}{\mathrm{D}'}x'\right)\xi + \left(y + \frac{f}{\mathrm{D}'}y'\right)\eta\right]} dxdyd\xi d\eta. \tag{37}$$

Now if $F(x, y)$ is defined as zero for all points of the object plane that lie outside $\mathscr{A}$, the integration with respect to x and y may formally be extended from $-\infty$ to $+\infty$. Also, if the aperture $\mathscr{B}$ is so large that $|U(\xi, \eta)|$ is negligible for points of the $\mathscr{F}'$–plane that lie outside $\mathscr{B}$, the integrations with respect to ξ and η may likewise each be extended over the range from $-\infty$ to $+\infty$. Noting also that [cf. § 4.3 (10) where f' and Z' correspond to our f and $-$ D′ respectively]

$$\frac{f}{\mathrm{D}'} = -\frac{1}{M}, \tag{38}$$

where $M(< 0)$ is the magnification between Π and Π', we obtain by the application of the FOURIER integral theorem*

$$V(x', y') = CF\left(\frac{x'}{M}, \frac{y'}{M}\right) = CF(x, y), \tag{39}$$

where (x, y) is the object point whose image is at (x', y'), and

$$C = C_1 C_2 \lambda^2 f^2$$

is a constant. Hence to the accuracy here implied† the image is strictly similar to the object (but inverted), provided the aperture is large enough.

To show that completely false detail may appear in the image if some of the spectra that carry appreciable energy are excluded, we consider a one-dimensional grating-like object consisting of N equidistant congruent slits of width s, separated by

---

* See for example R. COURANT and D. HILBERT, *Methods of Mathematical Physics* (New York. Interscience Publishers, 1953), Vol. 1, p .79.

† As pointed out on p. 384, the FRAUNHOFER approximation used here is restricted to the case when the object points as well as the image points are sufficiently close to the axis.

opaque regions, with period d. For simplicity the aperture will be assumed to be rectangular with two of its sides parallel to the strips.

According to § 8.6 (3)

$$U(\xi) = C_1' \left(\frac{\sin \dfrac{k\xi s}{2f}}{\dfrac{k\xi s}{2f}} \right) \frac{1 - e^{-iNkd\xi/f}}{1 - e^{-ikd\xi/f}}, \tag{40}$$

where for $U^{(0)}$ there has been substituted the expression relating to diffraction on a rectangular aperture and C_1' is a constant (cf. § 8.5.1). If the rectangular aperture extends in the ξ direction throughout the range

$$-a \leqslant \xi \leqslant a,$$

the disturbance in the image plane is, by (36) and (40), given by (C' denoting a constant)

$$V(x') = C' \int_{-a}^{a} \frac{\sin \dfrac{ks\xi}{2f}}{\dfrac{ks\xi}{2f}} \frac{1 - e^{-iNkd\xi/f}}{1 - e^{-ikd\xi/f}} e^{-ikx'\xi/D'} \, d\xi. \tag{41}$$

The position of principal maxima of the integrand are given by the roots of the equation $1 - \exp[-ikd\xi/f] = 0$, i.e. by $\xi = mf\lambda/d$, where m is an integer. Between these principal maxima there are weak secondary maxima. If N is large, the principal maxima are very sharp and the secondary maxima negligible in comparison. To a good approximation we may then replace the integral by a sum of integrals, each extending from the midpoint Q_m of the interval between two successive principal maxima to the next midpoint Q_{m+1}. In each interval we may replace the argument by the central value $\xi = mf\lambda/d = 2\pi mf/kd$, and obtain for V the following expression:

$$V(x') \sim V_0 \sum_{-\overline{m}<m<\overline{m}} \frac{\sin \dfrac{m\pi s}{d}}{\dfrac{m\pi s}{d}} e^{\frac{2\pi i m x'}{d'}}. \tag{42}$$

Here

$$\overline{m} = \frac{ad}{\lambda f}, \qquad d' = Md = -\frac{D'}{f} d, \tag{43}$$

and V_0 is the integral

$$V_0 = C' \int_{Q_m}^{Q_{m+1}} \frac{1 - e^{-iNkd\xi/f}}{1 - e^{-ikd\xi/f}} \, d\xi, \tag{44}$$

which, apart from small correction terms in the high orders, is practically independent of m. The series (42) may be re-written in real form as

$$\frac{V(x')}{V_0} = 1 + 2 \sum_{1<m<\overline{m}} \frac{\sin \dfrac{m\pi s}{d}}{\dfrac{m\pi s}{d}} \cos \frac{2\pi m x'}{d'}. \tag{45}$$

Suppose first that the length a of the aperture is very large. The summation may then formally be extended over the whole infinite range ($\overline{m} = \infty$), and we can easily verify that the image is then strictly similar to the object. For this purpose we expand the transmission function F of the grating-like object (see Fig. 8.33),

$$\left.\begin{aligned} F(x) &= F_0 \quad \text{when} \quad 0 < |x| < s/2 \\ &= 0 \quad \text{when} \quad s/2 < |x| < d/2 \end{aligned}\right\} \tag{46}$$

into a FOURIER series

$$F(x) = c_0 + 2 \sum_{m=1}^{\infty} c_m \cos \frac{2\pi m x}{d}. \tag{47}$$

Then

$$c_0 = \frac{F_0 s}{d}, \qquad c_m = F_0 \frac{\sin \dfrac{\pi m s}{d}}{\pi m} \qquad (m = 1, 2, 3 \ldots). \tag{48}$$

We see that apart from a constant factor this series is the same as (45).

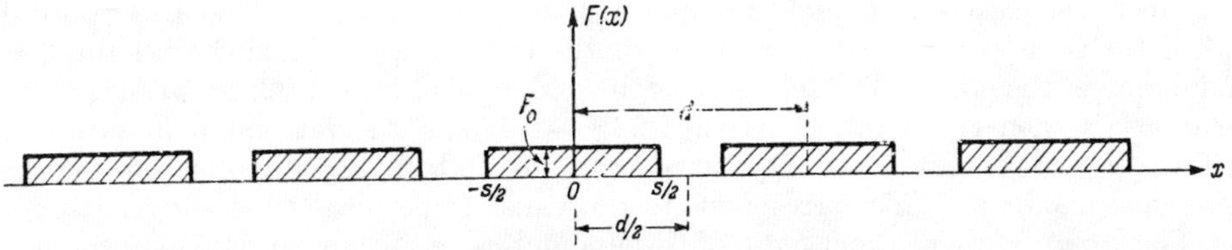

Fig. 8.33. A grating-like object.

Suppose now that the length a of the aperture is decreased. If a is so small that only the zero-order spectrum contributes to the image, i.e. if $\overline{m} = ad/\lambda f$ is only a fraction of unity, then according to (45) $V(x') =$ constant, so that the image plane is uniformly illuminated. (This result is, of course, not strictly true, as we have neglected certain error terms; in reality there is a weak drop in intensity towards the edge.)

If in addition to the zero-order spectrum the two spectra of the first order (S_1, S_{-1}) are also admitted by the aperture, i.e. if $\overline{m} = ad/\lambda f$ is slightly greater than unity, then we see from (45) that

$$\frac{V(x')}{V_0} = 1 + 2 \frac{\sin \dfrac{\pi s}{d}}{\dfrac{\pi s}{d}} \cos \frac{2\pi x'}{d'}. \tag{49}$$

The image has now the correct periodicity $x' = d'$, but a considerably flattened intensity distribution. By increasing the aperture more and more the image is seen to resemble the object more and more closely.

A completely false image is obtained when the lower orders are excluded. If for example all orders except the second are excluded, then

$$\frac{V(x')}{V_0} = 2 \frac{\sin \dfrac{2\pi s}{d}}{\dfrac{2\pi s}{d}} \cos \frac{4\pi x'}{d'}, \tag{50}$$

so that the image has the period $x' = d'/2$; the "image" shows twice the number of lines that are in fact present in the object.

Finally let us estimate the resolving power. Consider again the situation illustrated in Fig. 8.31, but assume now that the light from P and Q is coherent. Then the distribution in the image plane arises essentially from the coherent superposition of the two Airy diffraction patterns, one centred on P', the other on Q'. The complex amplitude at a point situated between P' and Q' at distance w_1 (measured in "diffraction units") from P' is given by

$$U(w_1) = \left[\frac{2J_1(kaw_1)}{kaw_1} + \frac{2J_1[ka(w - w_1)]}{ka(w - w_1)}\right] U_0, \tag{51}$$

w being the distance between P' and Q' and the other symbols having the same meaning as before. The intensity is, therefore, given by

$$I(w_1) = \left[\frac{2J_1(kaw_1)}{kaw_1} + \frac{2J_1(ka(w - w_1)}{ka(w - w_1)}\right]^2 I_0. \tag{52}$$

Now in the case of incoherent illumination, P' and Q' were considered as resolved when the principal intensity maximum of the one pattern coincided with the first minimum of the other. The intensity at the midpoint ($kaw \sim 1{\cdot}92$) between the two maxima is then equal to $2[2J_1(1{\cdot}92)/1{\cdot}92]^2 \sim 0{\cdot}735$ of the maximum intensity of either, i.e. the combined intensity curve has a dip of about 26·5 per cent between the principal maxima. (This corresponds to the value 19 per cent for a slit aperture—cf. Fig. 7.62.) If we consider a dip of this amount as again substantially determining the limit of resolution, the critical separation $w = 2w_1$ is obtained from the relation

$$\frac{I(w_1)}{I_0} = \left[2\,\frac{2J_1(kaw_1)}{kaw_1}\right]^2 = 0{\cdot}735. \tag{53}$$

The first root of this transcendental equation is $w_1 \sim 2{\cdot}41/ka$, so that the critical separation measured in ordinary units is

$$Y' = 2w_1\mathrm{D}' \sim \frac{2{\cdot}41}{\pi}\,\frac{\mathrm{D}'\lambda'}{a} = \frac{0{\cdot}77\lambda'}{\theta'} = \frac{0{\cdot}77\lambda_0}{n'\theta'}. \tag{54}$$

To relate Y' to the corresponding separation Y of the object points we use the sine condition (with the approximation $\sin\theta' \sim \theta'$), and finally obtain for the *limit of resolution with coherent illumination* the expression

$$|Y| = 0{\cdot}77\,\frac{\lambda_0}{n\sin\theta}. \tag{55}$$

Apart from a larger numerical factor (which in any case is somewhat arbitrary as it depends on the form of the object and aperture and on the sensitivity of the receptor), we obtain the same expression as in the case of incoherent illumination [eq. (32)]. Thus with light of a given wavelength the resolving power is again substantially determined by the numerical aperture of the objective.

(c) *Coherent illumination—Zernike's phase contrast method of observation**

We have defined a *phase object* as one which alters the phase but not the amplitude of the incident wave. An object of this type is of non-uniform optical thickness, but

* For fuller discussion of the phase contrast method see, for example, M. Françon, *Le contraste de phase en optique et en microscopie* (Paris, *Revue d'Optique*, 1950); and A. H. Bennett, H. Jupnik, H. Osterberg, and O. W. Richards, *Phase Microscopy* (New York, J. Wiley & Sons, 1952).

does not absorb any of the incident light. Such objects are frequently encountered in biology, crystallography, and other fields. It is evident from the preceding discussion that with ordinary methods of observation little information about phase objects can be obtained. For the complex amplitude function that specifies the disturbance in the image plane is then similar to the transmission function of the object* and, as the eye (or any other observing instrument) only distinguishes changes in intensity, one can only draw conclusions about the amplitude changes but not about the phase changes introduced by the object.

To obtain information about phase objects, special methods of observation must be used, for example, the so-called *central dark ground method of observation* where the central order is excluded by a stop, or the *Schlieren method,* where all the spectra on one side of the central order are excluded. The most powerful method, which has the advantage that it produces an intensity distribution which is directly proportional to the phase changes introduced by the object, is due to ZERNIKE† and was first described by him in 1935. It is known as *the phase contrast method.*

To explain the principle of the phase contrast method, consider first a transparent object in the form of a one-dimensional phase grating. The transmission function of such an object is by definition (see p. 401) of the form

$$F(x) = e^{i\phi(x)}, \tag{56}$$

where $\phi(x)$ is a real periodic function, whose period (d say) is equal to the period of the grating. We assume that the magnitude of ϕ is small compared to unity, so that we may write

$$F(x) \sim 1 + i\phi(x). \tag{57}$$

If we develop F into a FOURIER series

$$F(x) = \sum_{m=-\infty}^{\infty} c_m e^{\frac{2\pi imx}{d}}, \tag{58}$$

then, since F is of the form (56) and ϕ is real and numerically small compared to unity,

$$c_0 = 1, \qquad c_{-m} = -c_m^{\star} \qquad (m \neq 0). \tag{59}$$

The intensity of the mth order spectra is proportional to $|c_m|^2$.

In the phase contrast method of observation a thin plate of transparent material called the *phase plate* is placed in the back focal plane $\mathscr{F}'$ of the objective and by means of it the phase of the central order (S_0 in Fig. 8.32) is retarded or advanced with respect to the diffraction spectra ($S_1, S_{-1}, S_2, S_{-2}, \ldots$) by one-quarter of a period. This means that the complex amplitude distribution in the focal plane is altered from a distribution characterized by the coefficients c_m, to a distribution characterized by coefficients c'_m, where

$$c'_0 = c_0 e^{\pm i\pi/2} = \pm i, \qquad c'_m = c_m \qquad (m \neq 0), \tag{60}$$

---

* Strict similarity would actually be attained only if the objective had an infinite aperture. Because the aperture is always finite, some details of the phase structures can be seen. In some cases the visibility of such "images" is enhanced, at the expense of resolution, by a slight defocusing of the instrument (cf. H. H. HOPKINS, contribution in M. FRANÇON, *Le contraste de phase et le contraste par interférences* (Paris, *Revue d'Optique,* 1952), 142).

† F. ZERNIKE, *Z. Tech. Phys.*, **16** (1935), 454; *Phys. Z.*, **36** (1935), 848; *Physica*, **9** (1942), 686, 974.

the positive or negative sign being taken according as the phase of the central order is retarded or advanced. The resulting light distribution in the image plane will now no longer represent the phase grating (57), but rather a fictitious amplitude grating

$$G(x) = \pm\, i + i\phi(x). \tag{61}$$

Hence the intensity in the image plane will now be proportional to (neglecting ϕ^2 in comparison to unity)

$$I(x') = |G(x)|^2 = 1 \pm 2\phi(x), \tag{62}$$

where as before $x' = Mx$, M being the magnification. This relation shows that *with the phase contrast method of observation, phase changes introduced by the object are transformed into changes in intensity, the intensity at any point of the image plane being (apart from an additive constant) directly proportional to the phase change due to the corresponding element of the object.** When the phase of the central order is retarded with respect to the diffraction spectra (upper sign in (61)), regions of the object which have greater optical thickness will appear brighter than the mean illumination, and one then speaks of a *bright phase contrast*; when the phase of the central order is advanced, regions of greater spectral thickness will appear darker and one then speaks of a *dark phase contrast* (Figs. 8.34 and 8.35).

To obtain good resolution, the aperture of the illuminating system is often of annular rather than circular form (cf. § 8.6.2). In this case the annular region of $\mathscr{F}'$ through which the direct (undiffracted) light passes plays the role of the central order S_0 of Fig. 8.32, and it is this light which must then be retarded or advanced by a quarter period.

The phase-changing plate may be produced by evaporating a thin layer of a suitable dielectric substance on to a glass substrate. If n is the refractive index of the substance and d the thickness of the layer, then for a retardation of a quarter of a period one must have $d = \lambda/4(n-1)$. A retardation of the central order by this amount is, of course, equivalent to an advance of the diffracted spectra by three-quarters of a period, and vice versa. It is possible to increase the sensitivity of the method by using slightly absorbing instead of a dielectric coating. We shall return to this point later.

It remains to show that the phase contrast method is not restricted to phase objects of periodic structure. For this purpose we divide the integral (34) into two parts:

$$U(\xi, \eta) = U_0(\xi, \eta) + U_1(\xi, \eta) \tag{63}$$

where

$$\left.\begin{aligned} U_0 &= C_1 \iint_{\mathscr{A}} e^{-\frac{ik}{f}[\xi x + \eta y]}\, dx dy, \\ U_1 &= C_1 \iint_{\mathscr{A}} [F(x, y) - 1] e^{-\frac{ik}{f}[\xi x + \eta y]}\, dx dy. \end{aligned}\right\} \tag{64}$$

U_0 represents the light distribution that would be obtained in the plane $\mathscr{F}'$ if no object were present, whilst U_1 represents the effect of diffraction. Now the "direct light" U_0 (corresponding to the central order S_0 of Fig. 8.32), will be concentrated in only a small region $\mathscr{B}_0$ of the $\mathscr{F}'$-plane, around the axial point $\xi = \eta = 0$. On the

* The approximation implicit in the elementary theory of the phase contrast method are discussed by J. Picht, *Zeitschr. f. Instrkde*, **56** (1936), 481; *ibid.* **58** (1938), 1 and by F. D. Kahn, *Proc. Phys. Soc.*, B **68** (1955), 1073.

other hand a very small fraction of the diffracted light will, in general, reach this region, most of it being diffracted to other parts of this plane.*

Suppose that the region $\mathscr{B}_0$ through which the direct light passes is covered by a phase plate. The effect of the plate may be described by a transmission function

$$A = ae^{i\alpha}. \tag{65}$$

For a plate that only retards or advances the light which is incident upon it, $a = 1$; for a plate that also absorbs light $a < 1$. The light emerging from the aperture will be represented by

$$U'(\xi, \eta) = AU_0(\xi, \eta) + U_1(\xi, \eta) \tag{66}$$

so that, according to (36), the distribution of the complex amplitude in the image is given by

$$V(x', y') = V_0(x', y') + V_1(x', y'), \tag{67}$$

where

$$\left.\begin{aligned} V_0 &= AC_2 \iint_{\mathscr{B}} U_0(\xi, \eta) e^{-\frac{ik}{D'}[x'\xi + y'\eta]} d\xi d\eta, \\ V_1 &= C_2 \iint_{\mathscr{B}} U_1(\xi, \eta) e^{-\frac{ik}{D'}[x'\xi + y'\eta]} d\xi d\eta. \end{aligned}\right\} \tag{68}$$

Now the aperture $\mathscr{B}$ greatly exceeds in size the region $\mathscr{B}_0$, and since U_0 was seen to be practically zero outside $\mathscr{B}_0$, no appreciable error is introduced by extending the domain of integration in V_0 over the whole $\mathscr{F}'$-plane. Moreover, if $\mathscr{B}$ is assumed to be so large as to admit all the diffracted rays that carry any appreciable energy, the integral for V_1 may likewise be given infinite limits. Finally, if as before the transmission function $F(x, y)$ is defined as zero at points of the object plane outside the region covered by the object, the integrals (64) may also be taken with infinite limits. We then obtain, on substituting from (64) into (68), and using the FOURIER integral theorem and the relation (38),

$$\left.\begin{aligned} V_0(x', y') &= CA, \\ V_1(x', y') &= C\left[F\left(\frac{x'}{M}, \frac{y'}{M}\right) - 1\right] = C[F(x, y) - 1]. \end{aligned}\right\} \tag{69}$$

From (67) and (69) it follows that the intensity in the image plane is given by

$$I(x', y') = |V(x', y')|^2 = |C|^2 |A + F(x, y) - 1|^2. \tag{70}$$

With a phase object

$$F(x, y) = e^{i\phi(x,y)}, \tag{71}$$

and (70) reduces to†

$$I(x', y') = |C|^2[a^2 + 2\{1 - a\cos\alpha - \cos\phi(x, y) + a\cos(\alpha - \phi(x, y))\}]. \tag{72}$$

---

* This point was investigated in detail by J. PICHT, *Zeitschr. f. Instrkde*, **58** (1938), 1. See also F. ZERNIKE, *Mon. Not. Roy. Astr. Soc.*, **94** (1934), 382–383, where it is discussed in a somewhat different connexion.

† The special case when $a = 0$ corresponds to the dark-ground method of observation. According to (72), the intensity distribution is then given by

$$I(x', y') = 2C^2[1 - \cos\phi(x, y)].$$

Since ϕ was assumed to be small, (72) may be written as

$$I(x', y') = |C|^2[a^2 + 2a\phi(x, y) \sin \alpha], \tag{73}$$

and, if the phase difference introduced by the plate represents a retardation or advance by a quarter of a period, then $\alpha = \pm \pi/2$ and (73) reduces to

$$I(x', y') = |C|^2[a^2 \pm 2a\phi(x, y)]. \tag{74}$$

When the plate does not absorb any of the incident light ($a = 1$) we have again the expression (62). The intensity changes are then directly proportional to the phase variations of the object. With a plate that absorbs a fraction a^2 of the direct light the ratio of the second term to the first term in (73) has the value $\pm \phi/a$, so that the contrast of the image is enhanced. For example, by weakening the direct light to one-ninth of its original value, the sensitivity of the method is increased three times.

8.7. FRESNEL DIFFRACTION AT A STRAIGHT EDGE

8.7.1 The diffraction integral

Having considered various cases of FRAUNHOFER diffraction, we now turn our attention to the more general case of FRESNEL diffraction.

The basic diffraction integral § 8.3 (28) may be written in the form

$$U(P) = B(C + iS), \tag{1}$$

where

$$B = -A \frac{i}{\lambda} \cos \delta \frac{e^{ik(r'+s')}}{r's'}, \tag{2}$$

$$\left.\begin{aligned} C &= \iint_{\mathscr{A}} \cos \{kf(\xi, \eta)\} d\xi d\eta, \\ S &= \iint_{\mathscr{A}} \sin \{kf(\xi, \eta)\} d\xi d\eta. \end{aligned}\right\} \tag{3}$$

The intensity $I(P) = |U(P)|^2$ at the point P of observation is then given by

$$I(P) = |B|^2(C^2 + S^2). \tag{4}$$

We must now retain in the expansion § 8.3 (31) for $f(\xi, \eta)$ terms in ξ and η at least up to the second order.

As before we take the plane of the aperture $\mathscr{A}$ as the xy-plane. To simplify the calculations we choose as the x direction the projection of the line P_0P on to the plane of the aperture (Fig. 8.36). Thus with a source in a prescribed position our reference system will in general be different for different points of observation.

According to § 8.3 (30), we now have $l = l_0$, $m = m_0$, so that the linear terms in $f(\xi, \eta)$ disappear. The direction cosines of the rays P_0O and OP are

$$\left.\begin{aligned} l &= l_0 = \sin \delta, \\ m &= m_0 = 0, \\ n &= n_0 = \cos \delta, \end{aligned}\right\} \tag{5}$$

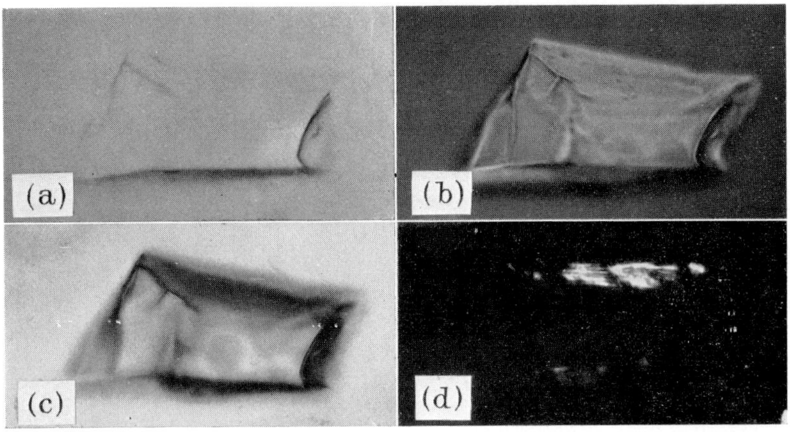

Fig. 8.34. Microscope images of glass fragments ($n = 1{\cdot}52$) mounted in clarite ($n = 1{\cdot}54$), 100 ×.
(a), Bright field image; (b) and (c), Phase contrast images; (d), Dark field image.
(After A. H. Bennett, H. Jupnik, H. Osterberg, and O. W. Richards, *Trans. Amer. Microscop. Soc.*, **65** (1946), 126.)

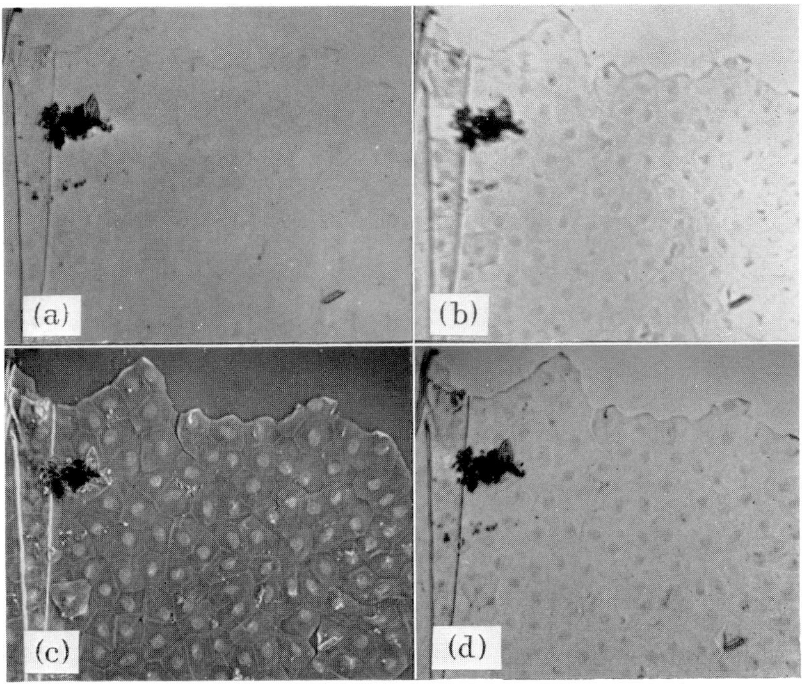

Fig. 8.35. Microscope images of epithelium from frog nictitating membrane, 100 ×.
(a), Bright field image at full aperture, N.A. = 0·25; (b), Bright field image with aperture half filled; (c), Phase-contrast image (bright contrast) at full aperture; (d), Phase-contrast image (dark contrast) at full aperture.
(After A. H. Bennett, H. Jupnik, H. Osterberg, and O. W. Richards. *Trans. Amer. Microscop. Soc.*, **65** (1946), 119.)

where, as before, δ denotes the angle between the line P_0P and the z-axis. The expression § 8.3 (31) for $f(\xi, \eta)$ reduces to

$$f(\xi, \eta) = \frac{1}{2}\left(\frac{1}{r'} + \frac{1}{s'}\right)(\xi^2 \cos^2 \delta + \eta^2) + \dots \tag{6}$$

If we neglect terms of third and higher order in ξ and η, the integrals (3) become

$$\left.\begin{aligned} C &= \iint_{\mathscr{A}} \cos\left\{\frac{\pi}{\lambda}\left(\frac{1}{r'} + \frac{1}{s'}\right)(\xi^2 \cos^2 \delta + \eta^2)\right\} d\xi d\eta, \\ S &= \iint_{\mathscr{A}} \sin\left\{\frac{\pi}{\lambda}\left(\frac{1}{r'} + \frac{1}{s'}\right)(\xi^2 \cos^2 \delta + \eta^2)\right\} d\xi d\eta. \end{aligned}\right\} \tag{7}$$

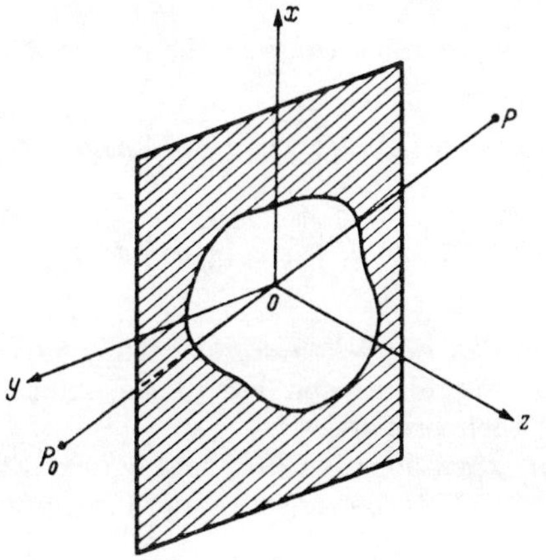

Fig. 8.36. FRESNEL diffraction at an opening in a plane opaque screen.

It is convenient to introduce new variables of integration u, v, defined by

$$\left.\begin{aligned} \frac{\pi}{\lambda}\left(\frac{1}{r'} + \frac{1}{s'}\right)\xi^2 \cos^2 \delta &= \frac{\pi}{2} u^2, \\ \frac{\pi}{\lambda}\left(\frac{1}{r'} + \frac{1}{s'}\right)\eta^2 &= \frac{\pi}{2} v^2. \end{aligned}\right\} \tag{8}$$

Then

$$d\xi d\eta = \frac{\lambda}{2} \frac{dudv}{\left(\frac{1}{r'} + \frac{1}{s'}\right)\cos \delta}$$

and our integrals become

$$\left.\begin{aligned} C &= b \iint_{\mathscr{A}'} \cos\left\{\frac{\pi}{2}(u^2 + v^2)\right\} dudv, \\ S &= b \iint_{\mathscr{A}'} \sin\left\{\frac{\pi}{2}(u^2 + v^2)\right\} dudv. \end{aligned}\right\} \tag{9}$$

where

$$b = \frac{\lambda}{2\left(\frac{1}{r'} + \frac{1}{s'}\right)\cos \delta}. \tag{10}$$

The integration now extends over the region $\mathscr{A}'$ of the (u, v) plane into which the region $\mathscr{A}$ of the aperture is transformed by means of (8).

8.7.2 Fresnel's integrals

If $\mathscr{A}'$ is a rectangle with sides parallel to the axes of u and v, the integrals are simplified still further by means of the identities

$$\left.\begin{aligned}\cos\left[\frac{\pi}{2}(u^2+v^2)\right] &= \cos\left(\frac{\pi}{2}u^2\right)\cos\left(\frac{\pi}{2}v^2\right) - \sin\left(\frac{\pi}{2}u^2\right)\sin\left(\frac{\pi}{2}v^2\right),\\ \sin\left[\frac{\pi}{2}(u^2+v^2)\right] &= \sin\left(\frac{\pi}{2}u^2\right)\cos\left(\frac{\pi}{2}v^2\right) + \cos\left(\frac{\pi}{2}u^2\right)\sin\left(\frac{\pi}{2}v^2\right).\end{aligned}\right\} \tag{11}$$

To evaluate (9) in this case we must consider the integrals

$$\left.\begin{aligned}\mathscr{C}(w) &= \int_0^w \cos\left(\frac{\pi}{2}\tau^2\right) d\tau,\\ \mathscr{S}(w) &= \int_0^w \sin\left(\frac{\pi}{2}\tau^2\right) d\tau.\end{aligned}\right\} \tag{12}$$

$\mathscr{C}(w)$ and $\mathscr{S}(w)$ are known as *Fresnel's integrals*. They are of importance in connexion with many diffraction problems and have been extensively studied. We must briefly consider some of their properties.*

First we derive series expressions for $\mathscr{C}(w)$ and $\mathscr{S}(w)$. Expanding the cosine and sine under the integral signs into power series and integrating term by term we find that

$$\left.\begin{aligned}\mathscr{C}(w) &= w\left[1 - \frac{1}{2!5}\left(\frac{\pi}{2}w^2\right)^2 + \frac{1}{4!9}\left(\frac{\pi}{2}w^2\right)^4 - \dots\right],\\ \mathscr{S}(w) &= w\left[\frac{1}{1!3}\left(\frac{\pi}{2}w^2\right) - \frac{1}{3!7}\left(\frac{\pi}{2}w^2\right)^3 + \frac{1}{5!11}\left(\frac{\pi}{2}w^2\right)^5 - \dots\right].\end{aligned}\right\} \tag{13}$$

The series (13) are convergent for all values of w but are suitable for computations only when w is small. When w is large the integrals may be evaluated from series in inverse powers of w. We re-write (12) as

$$\mathscr{C}(w) = \mathscr{C}(\infty) - \int_w^\infty \frac{d}{d\tau}\left(\sin\frac{\pi}{2}\tau^2\right)\frac{d\tau}{\pi\tau}. \tag{14}$$

Integration by parts gives

$$\mathscr{C}(w) = \mathscr{C}(\infty) + \frac{1}{\pi w}\sin\left(\frac{\pi}{2}w^2\right) + \int_w^\infty \frac{d}{d\tau}\left(\cos\frac{\pi}{2}\tau^2\right)\frac{d\tau}{\pi^2\tau^3}.$$

---

* Of the numerous tables of FRESNEL's integrals we may refer to the following: British Association Report (Oxford, 1926), 273–5. E. JAHNKE and F. EMDE, *Tables of Functions with Formulae and Curves* (Leipzig and Berlin, Teubner; reprinted by Dover Publications, New York, 4th ed., 1945), p. 35. G. N. WATSON, *A Treatise on the Theory of Bessel Functions* (Cambridge University Press, 2nd ed., 1944), p. 744. T. PEARCEY, *Tables of Fresnel Integrals to Six Decimal Places* (Cambridge University Press, 1956).

Integrating again by parts and continuing this process we obtain

$$\left.\begin{aligned}\mathscr{C}(w) &= \mathscr{C}(\infty) - \frac{1}{\pi w}\left[P(w)\cos\left(\frac{\pi}{2}w^2\right) - Q(w)\sin\left(\frac{\pi}{2}w^2\right)\right],\\ &\text{and similarly}\\ \mathscr{S}(w) &= \mathscr{S}(\infty) - \frac{1}{\pi w}\left[P(w)\sin\left(\frac{\pi}{2}w^2\right) + Q(w)\cos\left(\frac{\pi}{2}w^2\right)\right],\end{aligned}\right\} \tag{15}$$

where

$$\left.\begin{aligned}Q(w) &= 1 - \frac{1.3}{(\pi w^2)^2} + \frac{1.3.5.7}{(\pi w^2)^4} - \ldots\\ P(w) &= \frac{1}{\pi w^2} - \frac{1.3.5}{(\pi w^2)^3} + \frac{1.3.5.7.9}{(\pi w^2)^5}\ldots\end{aligned}\right\} \tag{16}$$

To evaluate the integrals $\mathscr{C}(\infty)$ and $\mathscr{S}(\infty)$, we combine them into a complex integral

$$\mathscr{C}(\infty) + i\mathscr{S}(\infty) = \int_0^\infty e^{i\frac{\pi}{2}\tau^2}\,d\tau \tag{17}$$

and introduce a new variable of integration

$$\zeta = \tau\sqrt{-\frac{i\pi}{2}} = \tau\frac{i-1}{2}\sqrt{\pi}, \qquad \tau = -\zeta\frac{i+1}{\sqrt{\pi}}.$$

The real path of integration $0 \leqslant \tau \leqslant \infty$ goes over into a path along a line through the origin and inclined at 45° to the real axis in the complex ζ plane. Now it is easy to see that if the integral is taken along a line parallel to the imaginary axis then, with increasing distance from the origin, the integral tends to zero. It then follows from CAUCHY's residue theorem that the integral taken along any oblique line through the origin is equal to the value of the integral taken along the real axis. Hence

$$\mathscr{C}(\infty) + i\mathscr{S}(\infty) = \frac{i+1}{\sqrt{\pi}}\int_0^\infty e^{-\zeta^2}\,d\zeta = \frac{i+1}{2}.$$

(The real integral with respect to ζ is the well known Gaussian error integral,* and has the value $\sqrt{\pi}/2$.) Thus

$$\left.\begin{aligned}\mathscr{C}(\infty) &= \int_0^\infty \cos\left(\frac{\pi}{2}\tau^2\right)d\tau = \tfrac{1}{2},\\ \mathscr{S}(\infty) &= \int_0^\infty \sin\left(\frac{\pi}{2}\tau^2\right)d\tau = \tfrac{1}{2}.\end{aligned}\right\} \tag{18}$$

The relations (15), together with (16) and (18), express FRESNEL's integrals in series of inverse powers of w. These are divergent (asymptotic) series, which provide a good approximation to the integrals when w is large by taking only a limited number of terms into account (cf. Appendix III).

The behaviour of FRESNEL's integrals may be illustrated by means of an elegant geometrical representation due to CORNU.† $\mathscr{C}$ and $\mathscr{S}$ are regarded as rectangular

* See, for example, R. COURANT, *Differential and Integral Calculus*, Vol. 1 (London and Glasgow. Blackie and Sons Ltd., 2nd ed., 1942), p. 496.

† A. CORNU, *Journ. de Phys.*, **3** (1874), 5, 44.

Cartesian coordinates of a point P. As w takes on all the possible values, the point P describes a curve. Since $\mathscr{C}(0) = \mathscr{S}(0) = 0$ the curve passes through the origin, and since

$$\mathscr{C}(-w) = -\mathscr{C}(w), \qquad \mathscr{S}(-w) = -\mathscr{S}(w), \tag{19}$$

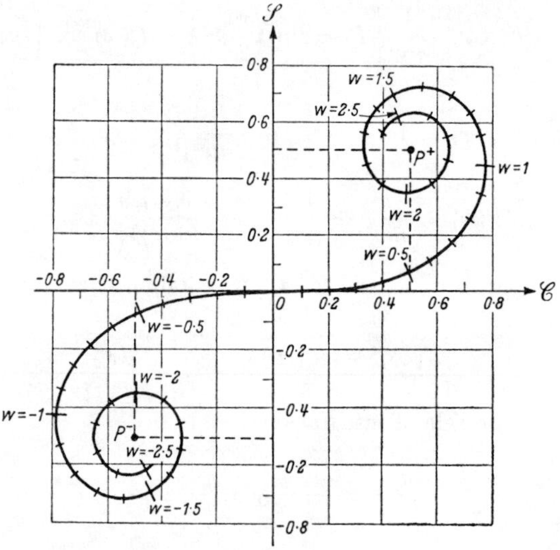

Fig. 8.37. The CORNU spiral.

it is antisymmetric with respect to both axes. If dl is an element of arc of the curve then

$$dl^2 = d\mathscr{C}^2 + d\mathscr{S}^2 = \left[\left(\frac{d\mathscr{C}}{dw}\right)^2 + \left(\frac{d\mathscr{S}}{dw}\right)^2\right](dw)^2$$

$$= \left[\cos^2\left(\frac{\pi}{2}w^2\right) + \sin^2\left(\frac{\pi}{2}w^2\right)\right](dw)^2,$$

i.e.

$$(dl)^2 = (dw)^2. \tag{20}$$

Hence if l is measured in the sense of increasing w, the parameter w represents the length of arc of the curve measured from the origin.

Let θ be the angle which the tangent to the curve makes with the $\mathscr{C}$ axis. Then

$$\tan\theta = \frac{d\mathscr{S}}{d\mathscr{C}} = \frac{\dfrac{d\mathscr{S}}{dw}}{\dfrac{d\mathscr{C}}{dw}} = \frac{\sin\left(\dfrac{\pi}{2}w^2\right)}{\cos\left(\dfrac{\pi}{2}w^2\right)} = \tan\left(\frac{\pi}{2}w^2\right)$$

i.e.

$$\theta = \frac{\pi}{2}w^2. \tag{21}$$

Thus θ increases monotonically with $|w|$. Since $\theta = 0$ when $w = 0$ the tangent touches the $\mathscr{C}$ axis at the origin. When $w^2 = 1$ then $\theta = \pi/2$, so that the tangent is then perpendicular to the $\mathscr{C}$ axis. When $w^2 = 2$, $\theta = \pi$ and the tangent is then parallel to the $\mathscr{C}$ axis again but is oriented in the negative direction. Since according to (18) and (19) $\mathscr{C}(\infty) = -\mathscr{C}(-\infty) = \frac{1}{2}$, $\mathscr{S}(\infty) = -\mathscr{S}(-\infty) = \frac{1}{2}$, the two branches of the curve approach the points P^+ and P^- with coordinates are $(\frac{1}{2}, \frac{1}{2})$ and $(-\frac{1}{2}, -\frac{1}{2})$ respectively. This curve is known as the *Cornu spiral* (see Fig. 8.37) and is useful in discussions of the general properties of FRESNEL diffraction patterns.

8.7.3 Fresnel diffraction at a straight edge

We now consider FRESNEL diffraction at a semi-infinite plane bounded by a sharp straight edge. This problem is of special interest with regard to the behaviour of the field near the boundaries of geometrical shadows. We restrict our attention to the case where the line P_0P and also its projection (our x axis) on to the half-plane is

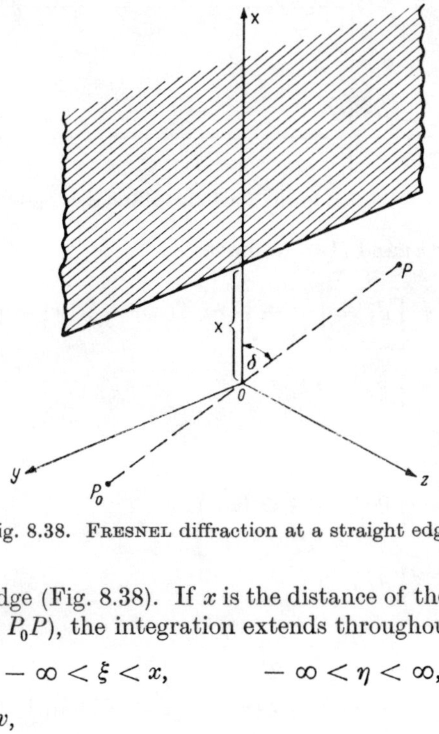

Fig. 8.38. FRESNEL diffraction at a straight edge.

perpendicular to the edge (Fig. 8.38). If x is the distance of the edge from the origin (which lies on the line P_0P), the integration extends throughout the region

$$-\infty < \xi < x, \qquad -\infty < \eta < \infty,$$

or, in terms of u and v,

$$-\infty < u < w, \qquad -\infty < v < \infty, \tag{22}$$

where

$$w = \sqrt{\frac{2}{\lambda}\left(\frac{1}{r'} + \frac{1}{s'}\right)}\, x \cos \delta. \tag{23}$$

The point of observation P lies in the illuminated region or in the geometrical shadow according as $x > 0$ or $x < 0$.

The diffraction integrals (9) become

$$\left.\begin{aligned} C &= b\int_{-\infty}^{w} du \int_{-\infty}^{+\infty} dv \left\{\cos\left(\frac{\pi}{2}u^2\right)\cos\left(\frac{\pi}{2}v^2\right) - \sin\left(\frac{\pi}{2}u^2\right)\sin\left(\frac{\pi}{2}v^2\right)\right\} \\ S &= b\int_{-\infty}^{w} du \int_{-\infty}^{\infty} dv \left\{\sin\left(\frac{\pi}{2}u^2\right)\cos\left(\frac{\pi}{2}v^2\right) + \cos\left(\frac{\pi}{2}u^2\right)\sin\left(\frac{\pi}{2}v^2\right)\right\}. \end{aligned}\right\} \tag{24}$$

We have violated here a condition used in the derivation of (9), namely that the linear dimensions of the domain of integration shall be small compared to the distances P_0O and OP. To justify the approximate validity of these formulae also in the present case, a more careful discussion of the error terms is necessary. We shall omit it here, since the diffraction by a half-plane will be treated again later by rigorous methods [§ 11.5].

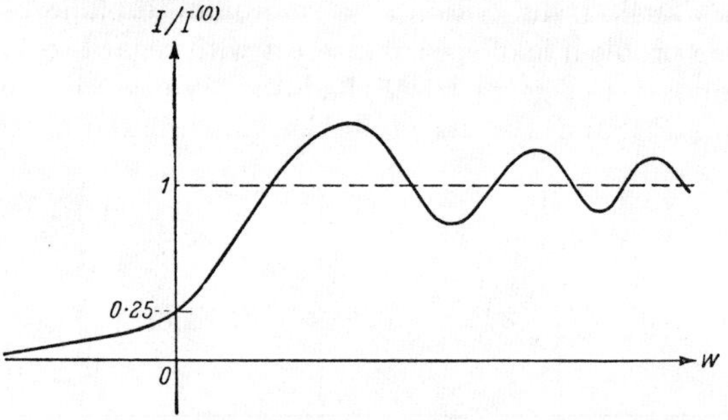

Fig. 8.39. Intensity distribution in the FRESNEL diffraction pattern of a straight edge.

From the relations (18) and (19), we have

$$\left.\begin{aligned}\int_{-\infty}^{w} \cos\left(\frac{\pi}{2}\tau^2\right) d\tau &= \int_{-\infty}^{0} + \int_{0}^{w} = \mathscr{C}(\infty) + \mathscr{C}(w) = \tfrac{1}{2} + \mathscr{C}(w),\\ \int_{-\infty}^{+\infty} \cos\left(\frac{\pi}{2}\tau^2\right) d\tau &= 1,\end{aligned}\right\} \tag{25}$$

and similarly

$$\left.\begin{aligned}\int_{-\infty}^{w} \sin\left(\frac{\pi}{2}\tau^2\right) d\tau &= \tfrac{1}{2} + \mathscr{S}(w),\\ \int_{-\infty}^{+\infty} \sin\left(\frac{\pi}{2}\tau^2\right) d\tau &= 1.\end{aligned}\right\} \tag{26}$$

Hence (24) becomes

$$\left.\begin{aligned}C &= b\{[\tfrac{1}{2} + \mathscr{C}(w)] - [\tfrac{1}{2} + \mathscr{S}(w)]\},\\ S &= b\{[\tfrac{1}{2} + \mathscr{C}(w)] + [\tfrac{1}{2} + \mathscr{S}(w)]\},\end{aligned}\right\} \tag{27}$$

and substitution into (4) gives finally an expression for the intensity:

$$I = \tfrac{1}{2}\{[\tfrac{1}{2} + \mathscr{C}(w)]^2 + [\tfrac{1}{2} + \mathscr{S}(w)]^2\}I^{(0)}, \tag{28}$$

where

$$I^{(0)} = 4|B|^2 b^2 = \frac{|A|^2}{(r' + s')^2}. \tag{29}$$

The behaviour of the intensity function (28) can be deduced from the CORNU spiral. The quantity $2I/I^{(0)}$ is seen to be equal to the square of the distance of the point w of the CORNU Spiral from the "asymptotic point" $P^-(-\tfrac{1}{2}, -\tfrac{1}{2})$. Thus if the point of observation is in the illuminated region ($w > 0$), $I/I^{(0)}$ oscillates with diminishing amplitudes as the distance from the edge increases and approaches asymptotically the value unity, as may be expected on the basis of geometrical optics. The maximum value of the intensity is not at the edge of the geometrical shadow, but some distance away from it, in the directly illuminated region (Fig. 8.39). On the edge of the shadow

($w = 0$), $I/I^{(0)} = \frac{1}{4}$. In the shadow region, $I/I^{(0)}$ decreases monotonically towards zero. These predictions are found to be in good agreement with experimental results.

8.8. THE THREE-DIMENSIONAL LIGHT DISTRIBUTION NEAR FOCUS

In § 8.3.3 it was shown that the light distribution in the focal plane of a well-corrected lens arises essentially from FRAUNHOFER diffraction on the aperture of the lens, and in § 8.5 the FRAUNHOFER diffraction patterns for apertures of various forms were studied in detail. To obtain a fuller knowledge of the structure of an optical image, we must study the light distribution not only in the geometrical focal plane but also in the neighbourhood of this plane. The knowledge of the three-dimensional (FRESNEL) distribution near focus is of particular importance in estimating the tolerance in the setting of the receiving plane in an image-forming system.

The properties of the out-of-focus monochromatic images of a point source by a circular aperture were first discussed in detail by E. LOMMEL in a classical memoir.* Starting from the HUYGENS–FRESNEL integral, LOMMEL succeeded in expressing the complex disturbance in terms of convergent series of BESSEL functions and also confirmed experimentally the phenomena predicted on the basis of these calculations. Almost at the same time as LOMMEL, H. STRUVE† published a similar though a less comprehensive analysis relating to the circular aperture. He did not work out the numerical consequences in such detail but gave useful approximations for the intensity near the edge of the geometrical shadow, where the series expansions are rather slowly convergent. Asymptotic approximations relating to points of observations at distances of many wavelengths from the focus were derived some years later by K. SCHWARZSCHILD.‡

The investigations of LOMMEL and STRUVE attracted relatively little attention, and in 1909 the problem was treated again by P. DEBYE,§ whose discussion established certain general features of the diffracted field both near and far away from the focus. In more recent times the analysis of these authors was extended and diagrams were published which show in detail the structure of the field in this complex region; and the results were broadly confirmed by experiment, both with light and with microwaves (short radio waves).

In discussing the light distribution near focus, we shall take as our starting point the analysis of LOMMEL and STRUVE, but it will be convenient and instructive to begin from the integral representation of the field in the form employed by DEBYE.

8.8.1 Evaluation of the diffraction integral in terms of Lommel functions

Consider a spherical monochromatic wave emerging from a circular aperture and converging towards the axial focal point O. We shall consider the disturbance $U(P)$ at a typical point P in the neighbourhood of O. The point P will be specified by a position vector $\boldsymbol{R}$ relative to O, and it will be assumed that the distance $R = OP$ as well as the radius $a(\gg \lambda)$ of the aperture are small compared to the radius $f = CO$ of the wave-front W that momentarily fills the aperture (Fig. 8.40).

---

* E. LOMMEL, *Abh. Bayer. Akad.*, **15**, Abth. 2, (1885), 233; a later paper (*ibid.* **15**, Abth. 3, (1886), 531) deals with diffraction by a slit, by an opaque strip and by a straight edge.

† H. STRUVE, *Mém. de l'Acad. de St. Petersbourgh* (7), **34** (1886), 1.

‡ K. SCHWARZSCHILD, *Sitzb. München. Akad. Wiss., Math.-Phys. Kl.*, **28** (1898), 271.

§ P. DEBYE, *Ann. d. Physik.* (4), **30** (1909), 755.

If s denotes the distance from the point of observation P to a point Q on W and A/f is the amplitude at Q of the incident wave we have, by the application of the HUYGENS–FRESNEL principle,

$$U(P) = -\frac{i}{\lambda}\frac{Ae^{-ikf}}{f}\iint_W \frac{e^{iks}}{s}\,dS, \tag{1}$$

where, since only small angles are involved, the variation of the inclination factor over the wave-front has been neglected. If $\boldsymbol{q}$ denotes the unit vector in the direction OQ, we have, to a good approximation

$$s - f = -\boldsymbol{q}\,.\,\boldsymbol{R}. \tag{2}$$

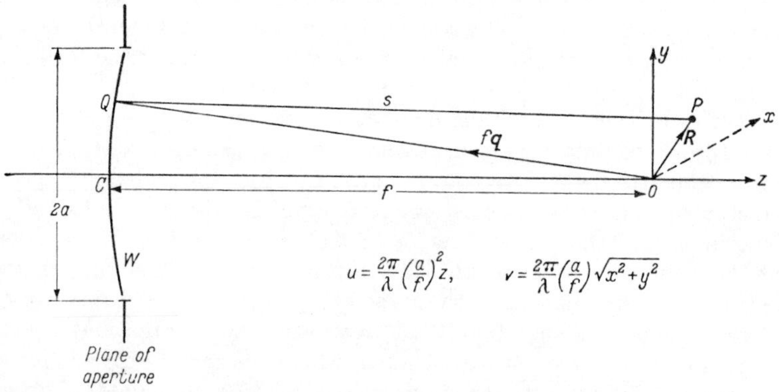

Fig. 8.40. Diffraction of a converging spherical wave at a circular aperture: Notation.

Also, the element dS of the wave-front is given by

$$dS = f^2 d\Omega, \tag{3}$$

where $d\Omega$ is the element of the solid angle that dS subtends at O. Moreover, we may replace s by f in the denominator of the integrand without introducing an appreciable error. Equation (1) then becomes

$$U(P) = -\frac{i}{\lambda}A\iint_\Omega e^{-ik\boldsymbol{q}\,.\,\boldsymbol{R}}\,d\Omega, \tag{4}$$

the integration now extending over the solid angle Ω which the aperture subtends at the focus. Equation (4) is the *Debye integral* and expresses the field as a superposition of plane waves of different directions of propagation (specified by the vectors $\boldsymbol{q}$ which fill Ω).

Before discussing the evaluation of the DEBYE integral we note the interesting fact that, being a sum of elementary solutions (plane waves), it represents a rigorous solution of the wave equation which in the limit $f \to \infty$ (aperture at infinite distance) is valid throughout the whole of space. Of course (4) is not a rigorous solution of our original problem, since no account has been taken of the nature of the screen, the exact boundary conditions being approximated to by those of the KIRCHHOFF diffraction theory. The true solution to our problem would include not only contributions of plane waves that are propagated in the directions of the incident geometrical rays, but of waves propagated in all possible directions.* If, however, the above

* This corresponds to the representation of the field in terms of a so-called angular spectrum of plane waves. Cf. § 11.4.2.

mentioned conditions are satisfied ($f \gg a \gg \lambda$, $f \gg R$), only contributions of waves that are included in (4) are appreciable.

To evaluate (4) we first express the integrand in a more explicit form. We take Cartesian axes at O, with the z direction along Oz. Let (x, y, z) be the coordinates of P and (ξ, η, ζ) those of Q. We set

$$\left.\begin{aligned} \xi &= a\rho \sin \theta, & x &= r \sin \psi, \\ \eta &= a\rho \cos \theta, & y &= r \cos \psi. \end{aligned}\right\} \tag{5}$$

Since Q lies on the spherical wave-front W,

$$\zeta = -\sqrt{f^2 - a^2\rho^2} = -f\left[1 - \frac{1}{2}\frac{a^2\rho^2}{f^2} + \dots\right]. \tag{6}$$

Then

$$\begin{aligned} \mathbf{q} \cdot \mathbf{R} &= \frac{x\xi + y\eta + z\zeta}{f} \\ &= \frac{a\rho r \cos(\theta - \psi)}{f} - z\left[1 - \frac{1}{2}\frac{a^2\rho^2}{f^2} + \dots\right]. \end{aligned} \tag{7}$$

It is useful at this stage to introduce dimensionless variables u and v, which together with ψ specify the position of P:

$$u = \frac{2\pi}{\lambda}\left(\frac{a}{f}\right)^2 z, \qquad v = \frac{2\pi}{\lambda}\left(\frac{a}{f}\right) r = \frac{2\pi}{\lambda}\frac{a}{f}\sqrt{x^2 + y^2}. \tag{8}$$

We note that the point P lies in the direct beam of light or in the geometrical shadow according as $|v/u| \lessgtr 1$.

From (7) and (8) it follows that if terms above the second power in $a\rho/f$ are neglected in comparison to unity,

$$k\mathbf{q} \cdot \mathbf{R} = v\rho \cos(\theta - \psi) - \left(\frac{f}{a}\right)^2 u + \tfrac{1}{2}u\rho^2. \tag{9}$$

Further, the element of the solid angle is

$$d\Omega = \frac{dS}{f^2} = \frac{a^2 \rho d\rho d\theta}{f^2}. \tag{10}$$

Hence (4) becomes

$$U(P) = -\frac{i}{\lambda}\frac{a^2 A}{f^2} e^{i\left(\frac{f}{a}\right)^2 u} \int_0^1 \int_0^{2\pi} e^{-i[v\rho \cos(\theta - \psi) + \frac{1}{2}u\rho^2]} \rho d\rho d\theta. \tag{11}$$

The integral with respect to θ is the same as one that we encountered in connection with FRAUNHOFER diffraction on a circular aperture (§ 8.5.2). It is equal to $2\pi J_0(v\rho)$, where $J_0(v\rho)$ is the BESSEL function of zero order. Hence (11) becomes

$$U(P) = -\frac{2\pi i a^2 A}{\lambda f^2} e^{i\left(\frac{f}{a}\right)^2 u} \int_0^1 J_0(v\rho) e^{-\frac{1}{2}iu\rho^2} \rho d\rho. \tag{12}$$

It is convenient to consider separately the real and imaginery parts of the integral. We set

$$2\int_0^1 J_0(v\rho) e^{-\frac{1}{2}iu\rho^2} \rho d\rho = C(u, v) - iS(u, v), \tag{13}$$

where

$$\left.\begin{aligned} C(u, v) &= 2\int_0^1 J_0(v\rho)\cos\left(\tfrac{1}{2}u\rho^2\right)\rho d\rho, \\ S(u, v) &= 2\int_0^1 J_0(v\rho)\sin\left(\tfrac{1}{2}u\rho^2\right)\rho d\rho. \end{aligned}\right\} \tag{14}$$

These integrals may be evaluated in terms of the *Lommel functions*

$$\left.\begin{aligned} U_n(u, v) &= \sum_{s=0}^{\infty}(-1)^s\left(\frac{u}{v}\right)^{n+2s} J_{n+2s}(v), \\ V_n(u, v) &= \sum_{s=0}^{\infty}(-1)^s\left(\frac{v}{u}\right)^{n+2s} J_{n+2s}(v), \end{aligned}\right\} \tag{15}$$

introduced by LOMMEL for this purpose.* Using the relation § 8.5 (11)

$$\frac{d}{dx}[x^{n+1}J_{n+1}(x)] = x^{n+1}J_n(x),$$

$C(u, v)$ may be written as

$$\begin{aligned} C(u, v) &= \frac{2}{v}\int_0^1 \frac{d}{d\rho}[\rho J_1(v\rho)]\cos\left(\tfrac{1}{2}u\rho^2\right)d\rho \\ &= \frac{2}{v}\left[J_1(v)\cos\tfrac{1}{2}u + u\int_0^1 \rho^2 J_1(v\rho)\sin\left(\tfrac{1}{2}u\rho^2\right)d\rho\right], \end{aligned} \tag{16}$$

on integrating by parts. Again using the relation § 8.5 (11), integrating by parts and continuing this process, we obtain

$$\begin{aligned} C(u, v) &= \frac{\cos\frac{1}{2}u}{\frac{1}{2}u}\left[\left(\frac{u}{v}\right)J_1(v) - \left(\frac{u}{v}\right)^3 J_3(v) + \ldots\right] \\ &\quad + \frac{\sin\frac{1}{2}u}{\frac{1}{2}u}\left[\left(\frac{u}{v}\right)^2 J_2(v) - \left(\frac{u}{v}\right)^4 J_4(v) + \ldots\right] \\ &= \frac{\cos\frac{1}{2}u}{\frac{1}{2}u}\cdot U_1(u, v) + \frac{\sin\frac{1}{2}u}{\frac{1}{2}u}U_2(u, v). \end{aligned} \tag{17a}$$

In a similar way we find

$$S(u, v) = \frac{\sin\frac{1}{2}u}{\frac{1}{2}u}U_1(u, v) - \frac{\cos\frac{1}{2}u}{\frac{1}{2}u}U_2(u, v). \tag{17b}$$

These formulae are valid at all points in the neighbourhood of the focus, but are only convenient for computations when $|u/v| < 1$, i.e. when the point of observation lies in the geometrical shadow. When $|u/v| > 1$, i.e. when the point of observation is in the illuminated region, it is more appropriate to use expansions involving positive

---

* For fuller discussions of these functions we refer to LOMMEL's memoirs (*loc. cit.*) and the following books: G. N. WATSON, *A Treatise on the Theory of Bessel Functions* (Cambridge University Press, 1922), pp. 537–50; A. GRAY, G. B. MATHEWS, and T. M. MACROBERT, *A Treatise on Bessel Functions* (London, Macmillan, 2nd ed. 1922), Chapter XIV; and J. WALKER, *The Analytical Theory of Light* (Cambridge University Press, 1904), p. 396.

powers of v/u. These may be derived in a similar manner by integrating by parts with respect to the trigonometric term. The first step gives

$$C(u, v) = \frac{2}{u}\int_0^1 J_0(v\rho)\frac{d}{d\rho}[\sin(\tfrac{1}{2}u\rho^2)]d\rho$$

$$= \frac{2}{u}\left[J_0(v)\sin\tfrac{1}{2}u + v\int_0^1 J_1(v\rho)\sin(\tfrac{1}{2}u\rho^2)d\rho\right], \tag{18}$$

where the relation § 8.5 (15)

$$\frac{d}{dx}[x^{-n}J_n(x)] = -x^{-n}J_{n+1}(x)$$

has been used. Integrating by parts again and using the last relation together with the well-known formula (which may be deduced from the series expansion for $J_n(x)$)

$$\lim_{x\to 0}\frac{J_n(x)}{x^n} = \frac{1}{2^n n!}, \tag{19}$$

we obtain

$$C(u, v) = \frac{\sin\frac{1}{2}u}{\frac{1}{2}u}\left[J_0(v) - \left(\frac{v}{u}\right)^2 J_2(v) + \dots\right]$$

$$-\frac{\cos\frac{1}{2}u}{\frac{1}{2}u}\left[\left(\frac{v}{u}\right)J_1(v) - \left(\frac{v}{u}\right)^3 J_3(v) + \dots\right]$$

$$+\frac{2}{u}\left[\frac{v^2}{2u} - \frac{1}{3!}\left(\frac{v^2}{2u}\right)^3 + \dots\right].$$

The series in the first two lines are two of the LOMMEL V_n functions and the series in the third line will be recognized as the expansion of $\sin(v^2/2u)$. Hence

$$C(u, v) = \frac{2}{u}\sin\frac{v^2}{2u} + \frac{\sin\frac{1}{2}u}{\frac{1}{2}u}V_0(u, v) - \frac{\cos\frac{1}{2}u}{\frac{1}{2}u}V_1(u, v). \tag{20a}$$

In a similar way we obtain for the other integral the expression

$$S(u, v) = \frac{2}{u}\cos\frac{v^2}{2u} - \frac{\cos\frac{1}{2}u}{\frac{1}{2}u}V_0(u, v) - \frac{\sin\frac{1}{2}u}{\frac{1}{2}u}V_1(u, v). \tag{20b}$$

This completes the formal solution of our problem. We shall now discuss some of the implications of these formulae.

8.8.2 The distribution of intensity

According to (12), (13), (17), and (20), the intensity $I = |U|^2$ in the neighbourhood of focus is given by the two equivalent expressions

$$I(u, v) = \left(\frac{2}{u}\right)^2 [U_1^2(u, v) + U_2^2(u, v)]\, I_0, \tag{21a}$$

and

$$I(u, v) = \left(\frac{2}{u}\right)^2 \Big[1 + V_0^2(u, v) + V_1^2(u, v)$$

$$- 2V_0(u, v)\cos\left\{\tfrac{1}{2}\left(u + \frac{v^2}{u}\right)\right\} - 2V_1(u, v)\sin\left\{\tfrac{1}{2}\left(u + \frac{v^2}{u}\right)\right\}\Big]\, I_0. \tag{21b}$$

where

$$I_0 = \left(\frac{\pi a^2 |A|}{\lambda f^2}\right)^2 \tag{22}$$

is the intensity at the geometrical focus $u = v = 0$.

It follows from (15) that $U_1(-u, v) = -U_1(u, v)$, $U_2(-u, v) = U_2(u, v)$, $V_0(-u, v) = V_0(u, v)$, $V_1(-u, v) = -V_0(u, v)$. Accordingly $I(u, v)$ remains unchanged when u is replaced by $-u$. Hence, *in the neighbourhood of the focus the intensity distribution is*

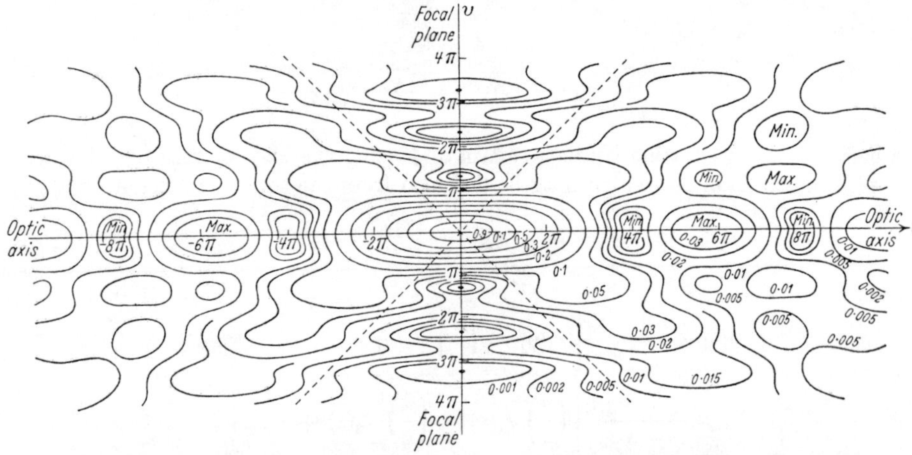

Fig. 8.41. Isophotes [contour lines of the intensity $I(u, v)$] in a meridional plane near focus of a converging spherical wave diffracted at a circular aperture. The intensity is normalized to unity at focus. The dotted lines represent the boundary of the geometrical shadow. When the figure is rotated about the u-axis, the minima on the v-axis generate the AIRY dark rings.

(Adapted from E. H. LINFOOT and E. WOLF, *Proc. Phys. Soc.*, B, **69** (1956), 823.)

symmetrical about the geometrical focal plane. Naturally, the distribution is also symmetrical about the axis $v = 0$.

From the formulae (21), LOMMEL computed the intensity distribution in a number of selected receiving planes near focus, and verified experimentally some of the predictions.* The lines of equal intensity (called *isophotes*) near focus, constructed from LOMMEL's data are shown in Fig. 8.41.†

* Related experiments were also described by M. E. HUFFORD and H. T. DAVIS, *Phys. Rev.*, **33** (1926), 589 and C. A. TAYLOR and B. J. THOMPSON, *J. Opt. Soc. Amer.*, **48** (1958), 844. Similar experiments with microwaves were carried out by M. P. BACHYNSKI and G. BEKEFI, *J. Opt. Soc. Amer.*, **47** (1957), 428, for the case of circular aperture, and by P. A. MATHEWS and A. L. CULLEN, *Proc. Inst. Elect. Engrs.*, Pt. C., **103** (1956), 449, for rectangular aperture.

† A similar, but less detailed diagram was published by M. BEREK, *Z. Phys.*, **40** (1926), 421. It is reproduced in some books with errors (incorrect position of the geometrical shadow and interchange of axes). Another version of the diagram, substantially in agreement with Fig. 8.41, was given by F. ZERNIKE and B. R. A. NIJBOER in their contribution to *La Théorie des Images Optiques*, Paris, Revue d'Optique (1949), 227. It is based on a different, but equivalent expansion (§ 9.4, eq. (12)) of the diffraction integral. Corresponding diagrams, calculated on the basis of electromagnetic theory, showing contours of the electric energy density and of the energy flow, were published by A. BOIVIN *and* E. WOLF, *Phys. Rev.*, **138** (1965), B 1561 and A. BOIVIN, J. DOW and E. WOLF, *J. Opt. Soc. Amer.*, **57** (1967), 1171.

Similar diagrams for the annular aperture were published by E. H. LINFOOT and E. WOLF, *Proc. Phys. Soc.*, B, **66** (1953), 145. Some extensions of LOMMEL's analysis to diffraction by concentric arrays of ring-shaped apertures were discussed by A. BOIVIN, *J. Opt. Soc. Amer.*, **42** (1952), 60.

Of particular interest is the tubular structure of the bright central portion of the diffraction image seen clearly in the figure and postulated on experimental grounds already in 1894 by TAYLOR.* It is this structure that is responsible for the tolerance in the setting of a receiving plane in an image-forming system.

We shall now consider several special cases of interest.

(a) *Intensity in the geometrical focal plane.* For points in the geometrical focal plane $u = 0$ and (21a) reduces to

$$I(0, v) = 4 \lim_{u \to 0} \left[\frac{U_1^2(u, v) + U_2^2(u, v)}{u^2} \right] I_0. \tag{23}$$

From the defining equation for the U_n functions it follows that

$$\lim_{u \to 0} \left[\frac{U_1(u, v)}{u} \right] = \frac{J_1(v)}{v}, \qquad \lim_{u \to 0} \left[\frac{U_2(u, v)}{u} \right] = 0, \tag{24}$$

so that

$$I(0, v) = \left[\frac{2J_1(v)}{v} \right]^2 I_0. \tag{25}$$

We thus obtain the AIRY formula § 8.5 (14) for FRAUNHOFER diffraction at a circular aperture, as was to be expected.

(b) *Intensity along the axis.* For points on the axis, $v = 0$, and the two V_n functions entering the expression (21b) reduce to

$$V_0(u, 0) = 1, \qquad V_1(u, 0) = 0.$$

Hence

$$I(u, 0) = \frac{4}{u^2} [2 - 2 \cos \tfrac{1}{2}u] I_0$$

$$= \left(\frac{\sin u/4}{u/4} \right)^2 I_0. \tag{26}$$

Thus the intensity along the axis is characterized by the function $(\sin x/x)^2$ which we discussed in § 8.5.1 in connection with FRAUNHOFER diffraction at a rectangular aperture. The first zero of intensity on the axis is given by $u/4 \equiv \pi a^2 z/2\lambda f^2 = \pm \pi$, i.e. it is at a distance $z = \pm 2f^2\lambda/a^2$ from the focus.

It is usual to regard a loss of about 20 per cent in intensity at the centre of the image patch as permissible. Since $\left(\frac{\sin u/4}{u/4} \right)^2$ decreases by this amount when the receiving plane is displaced from the central position ($u = 0$) to $u \sim 3{\cdot}2$, it follows that the focal tolerance Δz is approximately

$$\Delta z = \pm 3{\cdot}2 \frac{\lambda}{2\pi} \left(\frac{f}{a} \right)^2 \sim \pm \tfrac{1}{2} \left(\frac{f}{a} \right)^2 \lambda. \tag{27}$$

With an $f/10$ pencil for example ($f/a = 20$), and with light of wavelength $\lambda = 5.10^{-5}$ cm, the focal tolerance is about $\pm 0{\cdot}5 \,.\, 20^2 \,.\, 5.10^{-5}$ cm $= \pm 0{\cdot}1$ mm.

(c) *Intensity along the boundary of the geometrical shadow.* For points on the boundary of the geometrical shadow $u = \pm v$. Since the distribution is symmetrical with

* H. D. TAYLOR, *Mon. Not. Roy. Astr. Soc.*, **54** (1894), 67.

respect to the geometrical focal plane we may, without loss of generality, take $u = +v$. The U_n functions then reduce to

$$U_1(u, u) = \sum_{s=0}^{\infty} (-1)^s J_{2s+1}(u), \qquad U_2(u, u) = \sum_{s=0}^{\infty} (-1)^s J_{2s+2}(u). \tag{28}$$

We recall the well-known identities of JACOBI.*

$$\left.\begin{aligned} \cos(u \cos\theta) &= J_0(u) + 2\sum_{s=1}^{\infty} (-1)^s J_{2s}(u) \cos 2s\theta, \\ \sin(u\cos\theta) &= 2\sum_{s=0}^{\infty} (-1)^s J_{2s+1}(u) \cos(2s+1)\theta. \end{aligned}\right\} \tag{29}$$

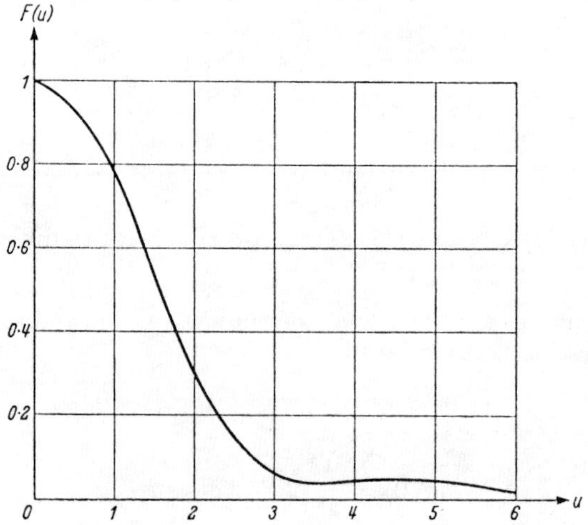

Fig. 8.42. The variation of intensity along the boundary of the geometrical shadow. The function $F(u) = \dfrac{1 - 2J_0(u)\cos u + J_0^2(u)}{u^2}$.

Setting $\theta = 0$, it follows on comparison with (28), that

$$\left.\begin{aligned} U_1(u, u) &= \tfrac{1}{2}\sin u, \\ U_2(u, u) &= \tfrac{1}{2}[J_0(u) - \cos u], \end{aligned}\right\} \tag{30}$$

and (21a) reduces to

$$I(u, u) = \frac{1 - 2J_0(u)\cos u + J_0^2(u)}{u^2} I_0. \tag{31}$$

This function is shown in Fig. 8.42.

8.8.3 The integrated intensity

It is also desirable to determine the fraction L of the (time averaged) total energy that falls within a small circle of prescribed radius r_0 about the axial point in the receiving plane u = constant. If

$$E = \pi a^2 \left(\frac{|A|}{f}\right)^2 \tag{32}$$

* See for example G. N. WATSON, *Theory of the Bessel Functions* (Cambridge University Press, 1922) p. 22.

denotes the total energy incident on to the aperture in unit time, the required fraction of energy is given by

$$L(u, v_0) = \frac{1}{E}\int_0^{r_0}\int_0^{2\pi} I(u, v) r dr d\psi$$

$$= \frac{1}{2I_0}\int_0^{v_0} I(u, v) v dv, \tag{33}$$

where

$$v_0 = \frac{2\pi}{\lambda}\left(\frac{a}{f}\right) r_0. \tag{34}$$

If LOMMEL's expressions (21) for the intensity are substituted into (33), the integral may be developed into series involving BESSEL functions. The derivation is lengthy and we shall therefore only give the final results, due to WOLF.*

Again two formally different expressions are obtained, one of which is convenient for computations when the boundary of the small circle is in the geometrical shadow, the other when it is in the direct beam of light. Suppressing the suffix zero, i.e. writing v in place of v_0, one has in the first case ($|v/u| \geqslant 1$)

$$L(u, v) = 1 - \sum_{s=0}^{\infty} \frac{(-1)^s}{2s+1}\left(\frac{u}{v}\right)^{2s} Q_{2s}(v), \tag{35a}$$

where

$$Q_{2s}(v) = \sum_{p=0}^{2s} (-1)^p [J_p(v) J_{2s-p}(v) + J_{p+1}(v) J_{2s+1-p}(v)]. \tag{36}$$

In the second case ($|v/u| \leqslant 1$)

$$L(u, v) = \left(\frac{v}{u}\right)^2 \left[1 + \sum_{s=0}^{\infty} \frac{(-1)^s}{2s+1}\left(\frac{v}{u}\right)^{2s} Q_{2s}(v)\right]$$

$$-\frac{4}{u}\left[Y_1(u, v)\cos\tfrac{1}{2}\left(u + \frac{v^2}{u}\right) + Y_2(u, v)\sin\tfrac{1}{2}\left(u + \frac{v^2}{u}\right)\right], \tag{35b}$$

where the Q's are again given by (36) and Y_1 and Y_2 are two of the functions

$$Y_n(u, v) = \sum_{s=0}^{\infty} (-1)^s (n + 2s)\left(\frac{v}{u}\right)^{n+2s} J_{n+2s}(v)$$

$$= \tfrac{1}{2}\left[\frac{v^2}{u} V_{n-1}(u, v) + u V_{n+1}(u, v)\right]. \tag{37}$$

In Fig. 8.43, the contour lines of $L(u, v)$, computed from these formulae are displayed. They may be regarded as analogous in a certain sense to the rays of geometrical optics.

We note that in the special case when the receiving plane coincides with the focal plane ($u = 0$), (35*a*) reduces to

$$L(0, v) = 1 - Q_0(v)$$

$$= 1 - J_0^2(v) - J_1^2(v), \tag{38}$$

in agreement with RAYLEIGH's formula § 8.5 (18).

Of special interest is the case when the circle over which the integral (33) is

* E. WOLF, *Proc. Roy. Soc.*, A, **204** (1951), 533. Asymptotic approximations for $L(u, v)$ were derived by J. FOCKE, *Optica Acta*, **3** (1956), 161.

extended coincides with the cross-section of the geometrical cone of rays. Then $|v/u| = 1$ and the series in (35a) and (35b) can be summed, and give*

$$L(u, u) = 1 - J_0(u) \cos u - J_1(u) \sin u. \tag{39}$$

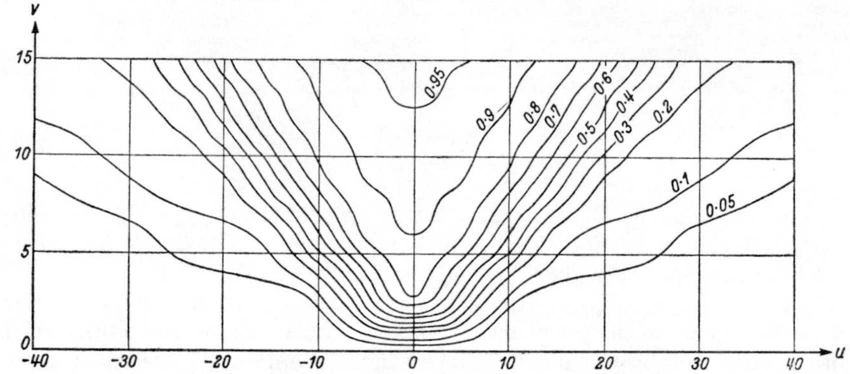

Fig. 8.43. Contour lines of $L(u, v)$, giving the fraction of the total energy which falls within small circles centred on the axis in selected receiving planes u = constant. (After E. Wolf, *Proc. Roy. Soc.*, A, **204** (1951), 542.)

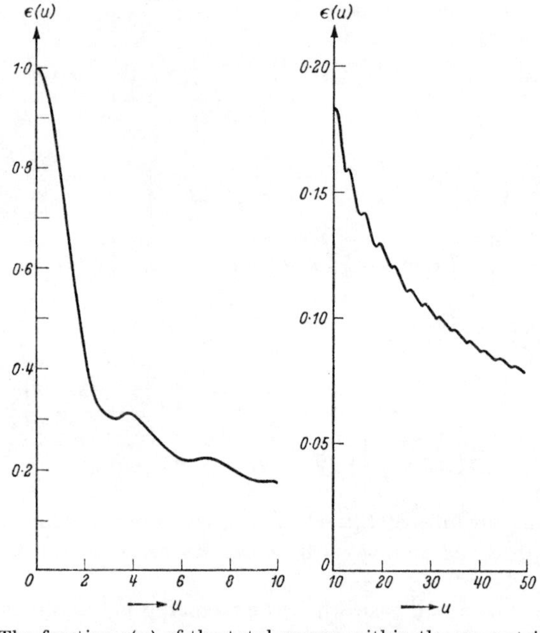

Fig. 8.44. The fraction $\varepsilon(u)$ of the total energy within the geometrical shadow. (After E. Wolf, *Proc. Roy. Soc.*, A, **204** (1951), 544.)

Hence the expression

$$\varepsilon(u) = J_0(u) \cos u + J_1(u) \sin u \tag{40}$$

gives, for the receiving plane u = constant, the fraction of the total energy in the geometrical shadow. The function $\varepsilon(u)$ is plotted in Fig. 8.44; it is not monotonically

* E. Wolf, *loc. cit.*, p. 539.

decreasing but has maxima (apart from $u = 0$) when $J_1(u) = 0$ and minima when $\sin u = 0$ $(u \neq 0)$.

8.8.4 The phase behaviour

Finally we consider the behaviour of the phase of the disturbance in the neighbourhood of the focus. According to (12) and (13) the phase is, apart from the additive term $(-\omega t)$ given by*

$$\phi(u, v) = \left(\frac{f}{a}\right)^2 u - \chi(u, v) - \frac{\pi}{2}, \qquad (\text{mod } 2\pi), \tag{41}$$

where

$$\cos \chi = \frac{C}{\sqrt{C^2 + S^2}}, \qquad \sin \chi = \frac{S}{\sqrt{C^2 + S^2}}, \tag{42}$$

the positive square root being taken in (42).

We note that, unlike the intensity distribution, the phase distribution cannot be expressed in terms of u and v alone, but has a structure which depends on the angular aperture of the geometrical pencil of rays. It also follows that each "branch" of the multivalued function $\phi(u, v)$ is continuous in u and v at all points where the intensity does not vanish, and that at points of zero intensity it is indeterminate. At the focus $u = v = 0$, one of its values is $-\pi/2$.

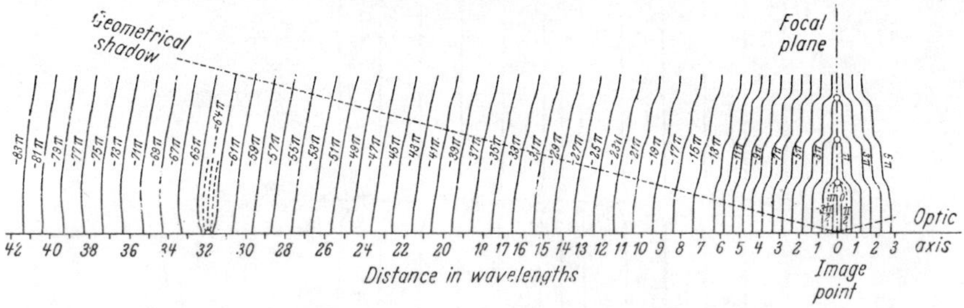

Fig. 8.45. Profiles of the co-phasal surfaces $\phi(u, v)$ = constant near the geometrical focus, calculated with $\lambda = 5 \cdot 10^{-5}$ cm, $a = 2{\cdot}5$ cm, $f = 10$ cm.
(After G. W. Farnell, *Canad. J. Phys.*, **35** (1957), 780.)†

The co-phasal surfaces (surfaces ϕ = constant) are, of course, surfaces of revolution about the u-axis. We shall now show that they possess a further symmetry, expressed by the relation

$$\phi(-u, v) + \phi(u, v) = -\pi \qquad (\text{mod } 2\pi). \tag{43}$$

From (14),

$$C(-u, v) = C(u, v), \qquad S(-u, v) = -S(u, v). \tag{44}$$

From (42) it then follows that

$$\cos \chi(-u, v) = \cos \chi(u, v), \qquad \sin \chi(-u, v) = -\sin \chi(u, v), \tag{45}$$

---

* The symbol mod 2π on the right of an equation denotes that the two sides of the equation are indeterminate to the extent of an additive constant $2m\pi$ where m is any integer.

† The values associated with the surfaces of Fig. 8.45 differ by π from those of Farnell's paper. This is so because, for the sake of consistency with our analysis, the phase at the focus is taken as $-\pi/2$ in accordance with (51), whereas in the paper by Farnell it is taken as $+\pi/2$.

so that

$$\chi(-u, v) = -\chi(u, v) \tag{46}$$

The relation (43) now follows from (46) and (41). Hence the reflection in the plane $u = 0$ of any co-phasal surface $\phi = \phi_0$ is a co-phasal surface $\phi = -\pi - \phi_0$.

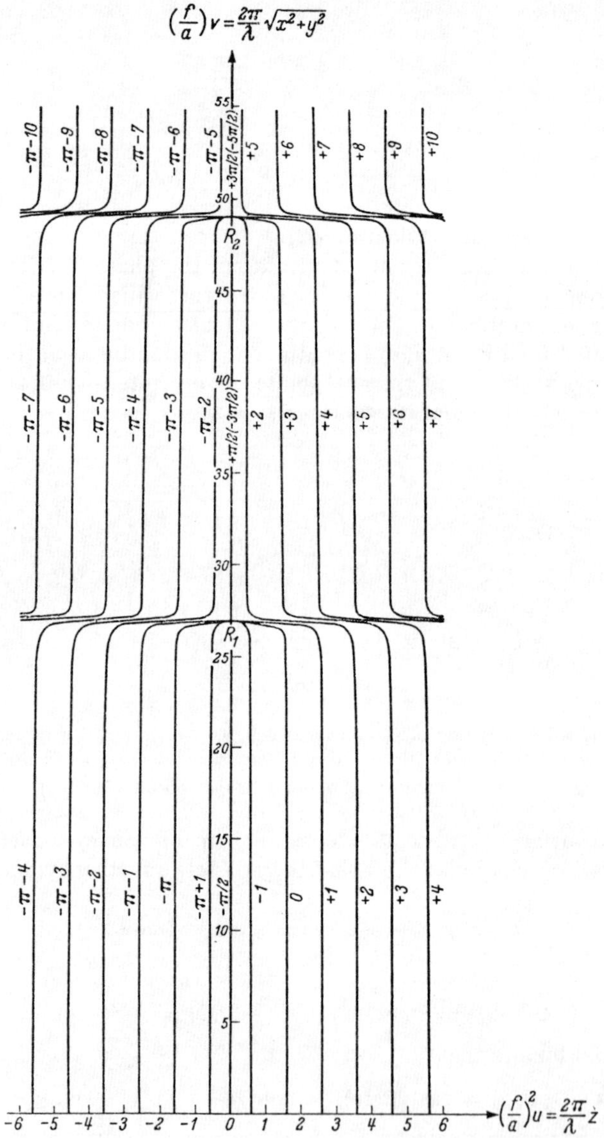

Fig. 8.46. Profiles of co-phasal surfaces in the immediate neighbourhood of the geometrical focal plane of an $f/3.5$ homocentric pencil. OR_1 and OR_2 are the radii of the first and second AIRY dark rings.

(After E. H. LINFOOT and E. WOLF, *Proc. Phys. Soc.*, B, **69** (1956), 827.)

In Fig. 8.45 the profiles of the co-phasal surfaces of an $f/2$ homocentric pencil are shown. Far away from the focus these surfaces coincide with the spherical wave-fronts of geometrical optics, but they become more and more deformed as the focal region is approached. The profiles in the immediate neighbourhood of the geometrical focal plane are shown in Fig. 8.46. It is seen that close to the focus, the surfaces are substantially plane; they are, however, spaced only $1 - a^2/4f^2$ times as far apart as they would be in a parallel beam of light of the same wavelength. Moreover (see Fig. 8.41), the intensity is not uniform over each co-phasal surface. In the immediate neighbourhood of the AIRY dark rings (denoted by R_1 and R_2 in Fig. 8.46) the co-phasal surfaces show a more complicated behaviour. Proceeding from the geometrical focus in the v direction, the phase is seen to be constant between any two successive dark rings, but changes abruptly by π on crossing each ring. That this must be so may be seen from the expression (derived easily with the help of (24), (41) and (46)) for the complex disturbance in the geometrical focal plane:

$$U(0, v) = \frac{1}{i} \frac{2J_1(v)}{v} \sqrt{I_0}. \tag{47}$$

Since this expression is purely imaginary for all values of v, the phase of $U(0, v)$ can only be equal to $-\pi/2$ or $+\pi/2$ (mod 2π), and since it changes sign on crossing each dark ring, the phase must then undergo a sudden jump by π. A discontinuity of this amount also occurs on crossing each axial point of zero intensity.

It is also of interest to study how the phase of the field changes as the point of observation moves along each ray through the focus. It is convenient to compare this variation with that of a spherical wave converging to the focus in the half-space $z < 0$ and diverging from it in the half-space $z > 0$. If the phase $\tilde{\phi}$ of this comparison wave is taken as zero at the focus we have

$$\left.\begin{aligned} \tilde{\phi}(u, v) &= -kR \qquad \text{when} \qquad u \leqslant 0 \\ &\quad +kR \qquad \text{when} \qquad u \geqslant 0 \end{aligned}\right\} \tag{48}$$

where as before $R = \sqrt{x^2 + y^2 + z^2} > 0$ is the distance of the point of observation from the focus. The difference

$$\delta(u, v) = \phi(u, v) - \tilde{\phi}(u, v) \tag{49}$$

is called the *phase anomaly*.

From (43), (48), and (49) it follows that

$$\delta(-u, v) + \delta(u, v) = -\pi \qquad (\text{mod } 2\pi) \tag{50}$$

whilst at the focus itself we have

$$\delta(0, 0) = \phi(0, 0) = -\frac{\pi}{2} \qquad (\text{mod } 2\pi). \tag{51}$$

The behaviour of the phase anomaly along selected rays through the focus of an $f/3.5$ homocentric pencil is shown in Fig. 8.47.

The figures show that on passing through the focus along any ray except the axial one, δ undergoes a rapid but continuous change in phase of π. This effect was observed a long time ago by GOUY* and has been the subject of many investigations.† Along the axis however the phase anomaly has a singular behaviour: it fluctuates periodically between the values 0 and $-\pi$.

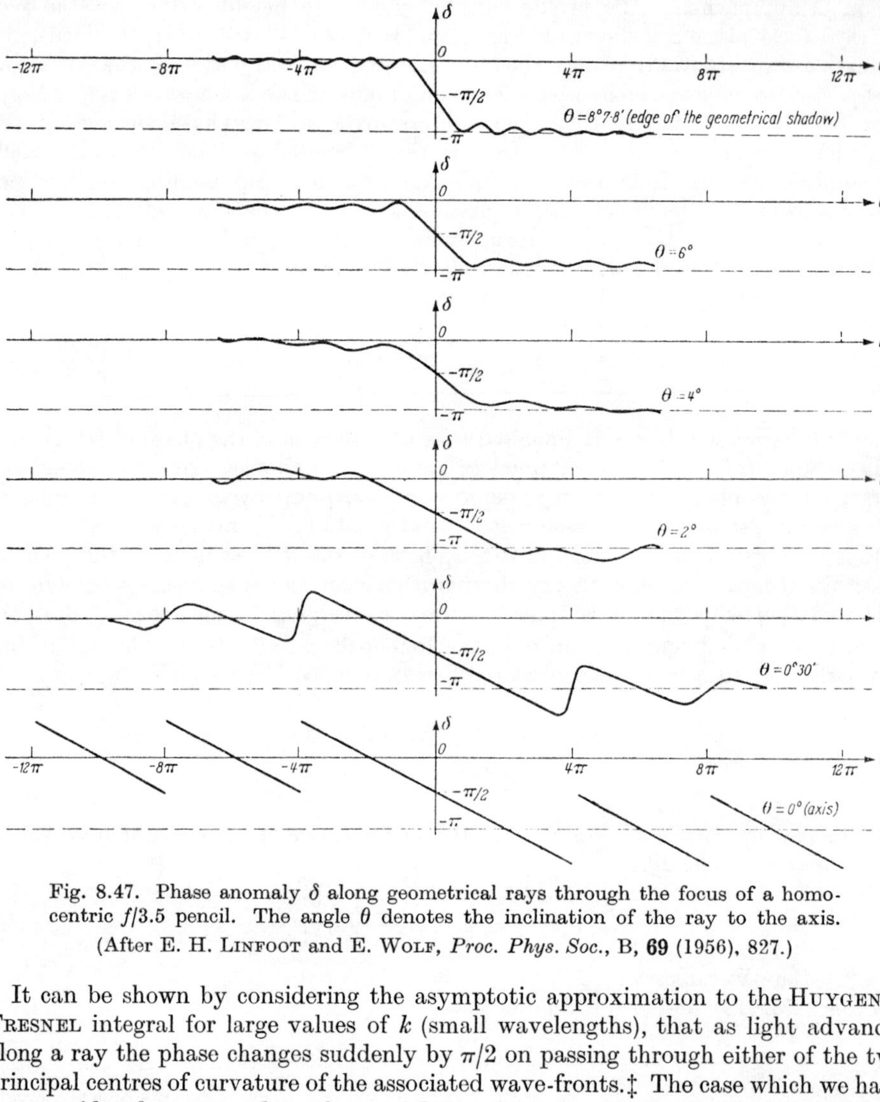

Fig. 8.47. Phase anomaly δ along geometrical rays through the focus of a homocentric $f/3.5$ pencil. The angle θ denotes the inclination of the ray to the axis. (After E. H. LINFOOT and E. WOLF, *Proc. Phys. Soc.*, B, **69** (1956), 827.)

It can be shown by considering the asymptotic approximation to the HUYGENS–FRESNEL integral for large values of k (small wavelengths), that as light advances along a ray the phase changes suddenly by $\pi/2$ on passing through either of the two principal centres of curvature of the associated wave-fronts.‡ The case which we have just considered corresponds to the special situation when the two centres of curvature coincide. Thus the change by half period is associated even with the geometrical

---

* L. G. GOUY, *Compt. Rend. Acad. Sci. Paris*, **110** (1890), 1251; *Ann Chim.* (*Phys.*) (6), **24** (1891), 145.

† For a survey of the literature see F. REICHE, *Ann. d. Physik*, (4), **29** (1909), 56, *ibid.*, 401. References to more recent investigations are given in E. H. LINFOOT and E. WOLF, *Proc. Phys. Soc.*, B, **69** (1956), 827.

‡ H. POINCARÉ, *Théorie mathématique de la lumière*, II (Paris, Georges Carré, (1892)), pp. 168–74. See also J. WALKER, *The Analytical Theory of Light* (Cambridge University Press, 1904), pp. 91–93.

optics solution.* Fig. 8.47 shows how the discontinuity of geometrical optics goes over into a continuous transition when the finiteness of the wavelength is taken into account. Finally a reference may be made to a paper by FARNELL,† which describes an experimental investigation into the structure of the phase distribution in the focal region of a microwave lens; very good agreement with the predictions of the theory was found.

8.9. THE BOUNDARY DIFFRACTION WAVE

If the edge of a diffracting aperture or obstacle is observed from points within the geometrical shadow, it appears luminous. This result was already known to THOMAS YOUNG‡ who attempted, prior to FRESNEL, to explain diffraction on a wave-theoretical basis. YOUNG believed that the incident light undergoes a kind of reflection at the edge of the diffracting body, and he regarded the diffraction pattern as arising from the interference of the incident wave and the reflected "boundary wave". However, YOUNG's views were expressed in a qualitative manner only and did not gain much recognition.

That YOUNG's theory contained an element of truth became evident after SOMMERFELD in 1894 obtained a rigorous solution for the diffraction of plane waves by a plane, semi-infinite reflecting screen (see § 11.5). This solution shows that in the geometrical shadow the light is propagated in the form of a cylindrical wave that appears to proceed from the edge of the screen, whilst in the illuminated region it is represented as superposition of the cylindrical wave and of the original incident wave.

The question arises whether also under more general conditions diffraction can be accounted for as the combined effect of an incident wave and a boundary wave. This problem had been investigated before the appearance of SOMMERFELD's paper by MAGGI,§ but his results appear to have been forgotten. It was later investigated independently and much more fully by RUBINOWICZ.|| The MAGGI–RUBINOWICZ theory was developed further by MIYAMOTO and WOLF.¶

Consider a monochromatic light wave from a point source P_0 propagated through

---

* For discussions of this point see also A. RUBINOWICZ, *Phys. Rev.*, **54** (1938), 931; and C. J. BOUWKAMP, *Physica*, **7** (1940), 485.

† G. W. FARNELL, *Canad. J. Phys.*, **36** (1958), 935.

‡ THOMAS YOUNG, *Phil. Trans. Roy. Soc.*, **20** (1802), 26.

§ G. A. MAGGI, *Annali di Matem.* (2), **16** (1888), 21. MAGGI's analysis is also discussed in a paper by F. KOTTLER, *Ann. d. Physik*, (4), **70** (1923), 413; and in B. B. BAKER and E. T. COPSON, *The Mathematical Theory of Huygens' Principle* (Oxford, Clarendon Press, 1950, 2nd ed.), p. 74.

Experimental evidence for the "existence" of the boundary wave was found by W. WIEN, *Inaug. Diss., Berlin*, 1886; E. MAEY, *Ann. d. Physik.* (9), **49** (1893), 69; and A. KALASCHNIKOW, *Journ. Russ. Phys. Chem. Ges.*, **44** (1912), *Phys. Teil*, 133. See also S. BANERJI, *Phil. Mag.* (6), **37** (1919), 112; and S. K. MITRA, *ibid.* (6), **38** (1919), 289.

|| A. RUBINOWICZ, *Ann. d. Physik.* (4), **53** (1917), 257; *ibid.* (4), **73** (1924); *ibid.*, **81** (1926), 153; *Acta Phys. Polonica*, **12** (1953), 225. See also G. N. RAMACHANDRAN, *Proc. Indian Acad. Sci. A.*, **21** (1945), 165; L. C. MARTIN, *Proc. Phys. Soc.*, **55** (1943), 104; *ibid.*, **62**B (1949), 713. Y. V. KATHAVATE, *ibid. A.*, **21** (1945), 177; R. S. INGARDEN, *Acta Phys. Polon.*, **14** (1955), 77; O. LAPORTE and J. MEIXNER, *Zeitschr. f. Phys.*, **153** (1958), 129.

A very comprehensive account of researches relating to the boundary wave is given in A. RUBINOWICZ' book *Die Beugungswelle in der Kirchhoffschen Theorie der Beugung* (Warszawa, Polska Akademia Nauk, 1957).

¶ K. MIYAMOTO and E. WOLF, *J. Opt. Soc. Amer.*, **52** (1962) 615, 626; K. MIYAMOTO, *Proc. Phys. Soc.*, **79** (1962), 617. See also E. W. MARCHAND and E. WOLF, *J. Opt. Soc. Amer.*, **52** (1962), 761; A. RUBINOWICZ, *ibid.*, **52** (1962) 717; *Acta Phys. Polonica* **21** (1962), 61, 451; *Progress in Optics*, Vol. **4**, ed. E. WOLF (Amsterdam, North Holland Publishing Company and New York, J. Wiley and Sons, 1965), p. 199.

an aperture in a plane opaque screen. As before we assume that the linear dimensions of the aperture are large compared with the wavelength but small compared to the distance of P_0 and of the point of observation P from the screen, and that the angles of incidence and of diffraction are small. Then we have in the approximations of the Kirchhoff theory (§ 8.3.2)

$$U(P) = \frac{1}{4\pi} \iint_{\mathscr{A}} \left\{ \frac{e^{ikr}}{r} \frac{\partial}{\partial n} \left(\frac{e^{iks}}{s} \right) - \left(\frac{e^{iks}}{s} \right) \frac{\partial}{\partial n} \left(\frac{e^{ikr}}{r} \right) \right\} dS, \tag{1}$$

where $\mathscr{A}$ denotes the diffracting aperture and the other symbols have the same meaning as before. We construct a closed surface bounded by (1) the opening $\mathscr{A}$, (2) the surface of a truncated cone $\mathscr{B}$ whose vertex is at P_0 and whose generators pass through the edge of the aperture, and (3) a portion $\mathscr{C}$ of a large sphere centred

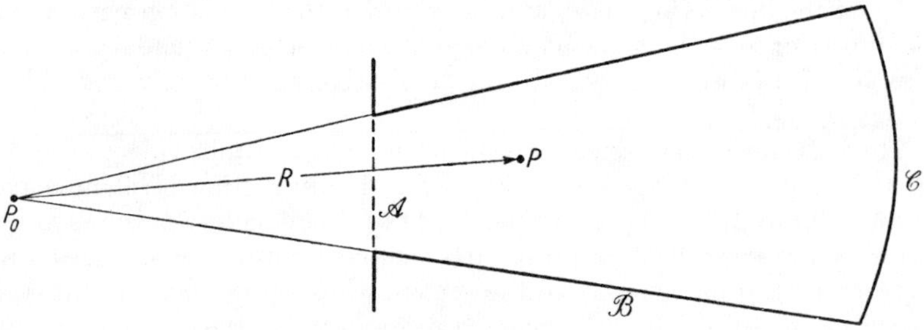

Fig. 8.48. Derivation of the boundary diffraction wave.

on P (Fig. 8.48). If R denotes the distance from P_0 to P, we have rigorously from Kirchhoff's integral theorem,

$$\frac{1}{4\pi} \iint_{\mathscr{A}+\mathscr{B}+\mathscr{C}} \left\{ \frac{e^{ikr}}{r} \frac{\partial}{\partial n} \left(\frac{e^{iks}}{s} \right) - \frac{e^{iks}}{s} \frac{\partial}{\partial n} \left(\frac{e^{ikr}}{r} \right) \right\} dS = \frac{e^{ikR}}{R} \quad \text{or} \quad 0, \tag{2}$$

according to the point P lying inside or outside the surface. Now in the same way as in § 8.3.2, the contribution from $\mathscr{C}$ can be made negligible by taking the radius of the sphere sufficiently large. We then obtain from (1) and (2),

$$U(P) = U^{(g)}(P) + U^{(d)}(P) \tag{3}$$

where

$$\left. \begin{aligned} U^{(g)}(P) &= \frac{e^{ikR}}{R} \qquad \text{when } P \text{ is in the direct beam} \\ &= 0 \qquad \text{when } P \text{ is in the geometrical shadow} \end{aligned} \right\} \tag{4}$$

and

$$U^{(d)}(P) = -\frac{1}{4\pi} \iint_{\mathscr{B}} \left\{ \frac{e^{ikr}}{r} \frac{\partial}{\partial n} \left(\frac{e^{iks}}{s} \right) - \left(\frac{e^{iks}}{s} \right) \frac{\partial}{\partial n} \left(\frac{e^{ikr}}{r} \right) \right\} dS. \tag{5}$$

$U^{(g)}$ represents the disturbance as predicted by geometrical optics, so that $U^{(d)}$ must represent the effect of diffraction. We shall now show that $U^{(d)}$ may be transformed into a line integral along the edge of the aperture.

We note first that the spheres r = constant cut orthogonally the truncated cone $\mathscr{B}$. Hence on $\mathscr{B}$,

$$\frac{\partial}{\partial n} \left(\frac{e^{ikr}}{r} \right) = 0. \tag{6}$$

Also

$$\frac{d}{dn}\left(\frac{e^{iks}}{s}\right) = \frac{d}{ds}\left(\frac{e^{iks}}{s}\right)\cos(n, s) = \left(\frac{ik}{s} - \frac{1}{s^2}\right) e^{iks}\cos(n, s). \tag{7}$$

Hence (5) reduces to

$$U^{(d)}(P) = -\frac{1}{4\pi}\iint_{\mathscr{B}} \frac{e^{ik(r+s)}}{rs}\left(ik - \frac{1}{s}\right)\cos(n, s)dS. \tag{8}$$

We can take as the element dS the area $ABB'A'$ bounded by segments of two neighbouring generators and by the arcs of circles in which the spheres r = constant

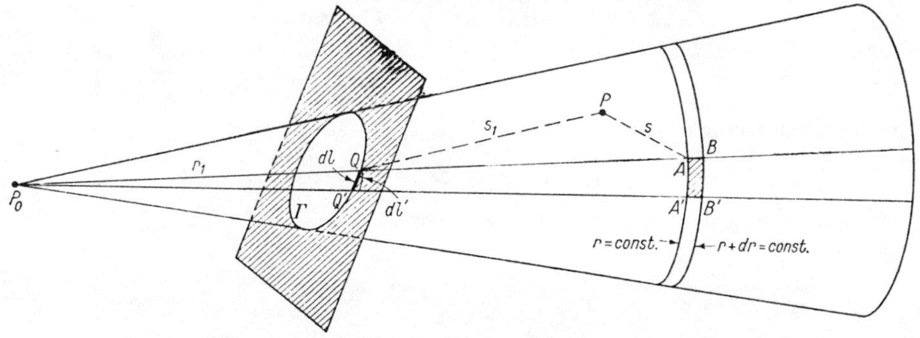

Fig. 8.49. Derivation of the boundary diffraction wave.
$P_0A = r, \quad P_0Q = r_1, \quad PA = s, \quad PQ = s_1.$

and $r + dr$ = constant intersect the cone (Fig. 8.49). If $d\phi$ is the angle between the two generators, then

$$dS = rdrd\phi. \tag{9}$$

Let Q and Q' be the points of intersection of the two generators with the edge Γ of the aperture and let dl be the length of the element of Γ between these two points. If dl' denotes the corresponding element of arc of the circle in which the sphere with radius $r_1 = P_0Q$ intersects the cone then

$$dl' = r_1 d\phi = dl\cos(dl, dl') = dl\sin(r_1, dl). \tag{10}$$

From (9) and (10)

$$dS = \frac{r}{r_1}\sin(r_1, dl)drdl. \tag{11}$$

Also, by projection on the normal to the cone at A and Q (the normals at these points being parallel),

$$s\cos(n, s) = s_1\cos(n, s_1). \tag{12}$$

On substituting from (11) and (12) into (8) it follows that

$$\begin{aligned} U^{(d)}(P) &= -\frac{1}{4\pi}\iint_{\mathscr{B}} \frac{e^{ik(r+s)}}{rs}\left(ik - \frac{1}{s}\right)\frac{s_1}{s}\cos(n, s_1)\frac{r}{r_1}\sin(r_1, dl)drdl \\ &= -\frac{1}{4\pi}\int_{\Gamma} dl\,\frac{s_1}{r_1}\cos(n, s_1)\sin(r_1, dl)\int_{r_1}^{\infty} e^{ik(r+s)}\left(\frac{ik}{s^2} - \frac{1}{s^3}\right)dr. \end{aligned} \tag{13}$$

Next we shall show that the integrand of the second integral in (13) is a total differential, namely that

$$e^{ik(r+s)}\left[\frac{ik}{s^2}-\frac{1}{s^3}\right]=\frac{d}{dr}\left\{\frac{e^{ik(r+s)}}{s[s+r-r_1+s_1\cos(s_1,r_1)]}\right\}. \tag{14}$$

We have, on carrying out the differentiation on the right of (14),

$$\frac{d}{dr}\left\{\frac{e^{ik(r+s)}}{s[s+r-r_1+s_1\cos(s_1,r_1)]}\right\}$$
$$=\frac{e^{ik(r+s)}}{s[s+r-r_1+s_1\cos(s_1,r_1)]}\left\{ik\left(1+\frac{ds}{dr}\right)-\frac{1}{s}\frac{ds}{dr}\right.$$
$$\left.-\frac{1}{(s+r-r_1+s_1\cos(s_1,r_1))}\cdot\left[1+\frac{ds}{dr}\right]\right\}. \tag{15}$$

Now from the triangle APQ

$$s^2=s_1^{\,2}+(r-r_1)^2+2s_1(r-r_1)\cos(s_1,r_1), \tag{16}$$

whence, differentiating with respect to r whilst keeping r_1 and s_1 fixed,

$$s\frac{ds}{dr}=r-r_1+s_1\cos(s_1,r_1). \tag{17}$$

On substituting from this equation into (15), the identity (14) follows. Hence

$$\int_{r_1}^{\infty}e^{ik(r+s)}\left(\frac{ik}{s^2}-\frac{1}{s^3}\right)dr=\left[\frac{e^{ik(r+s)}}{s(s+r-r_1+s_1\cos(s_1,r_1))}\right]_r^{\infty}$$
$$=-\frac{e^{ik(r_1+s_1)}}{s_1^{\,2}[1+\cos(s_1,r_1)]}, \tag{18}$$

and (13) finally becomes

$$U^{(d)}(P)=\frac{1}{4\pi}\oint_{\Gamma}\frac{e^{ik(r_1+s_1)}}{r_1s_1}\cdot\frac{\cos(n,s_1)}{[1+\cos(s_1,r_1)]}\sin(r_1,dl)dl. \tag{19}$$

This formula, together with (3) and (4), is the *Rubinowicz representation* of the KIRCHHOFF diffraction integral and may be regarded as the mathematical formulation of YOUNG's theory. It expresses the effect of diffraction in terms of the incident wave that is propagated in accordance with the laws of geometrical optics and a boundary diffraction wave that may be thought of as arising from the scattering of the incident radiation by the boundary of the aperture.

Since U is a continuous function of position, it follows from (4) that the boundary wave $U^{(d)}$ is discontinuous across the edge of the geometrical shadow so as to compensate for the discontinuity in the "geometrical wave" $U^{(g)}$. The discontinuity in $U^{(d)}$ arises from the factor $[1+\cos(s_1,r_1)]$ in the denominator in (19).

By the same argument which was given in connexion with the classification of diffraction phenomena on p. 386 and which illustrates the physical significance of the principle of the stationary phase [see Appendix III], it follows that only those points of the domain of integration contribute substantially to $U^{(d)}$ for which the phase of the integrand is stationary, i.e. for which

$$\frac{d}{dl}[k(r_1+s_1)]=0. \tag{20}$$

This relation may also be written in the form of a "reflection law",

$$\cos(r_1, dl) = -\cos(s_1 dl). \tag{21}$$

8.10. GABOR'S METHOD OF IMAGING BY RECONSTRUCTED WAVE-FRONTS (HOLOGRAPHY)

In an attempt to improve the resolving power of the electron microscope, GABOR* has proposed a two-step method of optical imagery. In the first step an object is illuminated with a coherent electron wave or a coherent light wave. The object is assumed to be such that a considerable part of the wave penetrates undisturbed through it.† A diffraction pattern, called *hologram*, which is formed by the interference of the secondary wave arising from the presence of the object with the strong background wave, is recorded on a photographic plate. If the plate, suitably processed, is replaced in the original position and is illuminated by the background wave alone, the wave that is transmitted by the plate contains information about the original object, and this can be extracted from the photograph by optical processes. In order to "reconstruct" the object from this "substitute" wave, it is only necessary to send it through a suitable image-forming system, and an image will appear in the plane conjugate to the plane in which the object was situated. We shall only be concerned with the optical principle involved‡.

8.10.1 Producing the positive hologram

Consider a monochromatic wave from a small source S, impinging on a semi-transparent object σ (Fig. 8.50(a)). Let $\mathscr{H}$ be a screen some distance behind the object and let $U = Ae^{i\psi}$ represent the complex disturbance at a typical point of $\mathscr{H}$, A being the (real) amplitude, and ψ the phase of the disturbance. We may regard U as the sum of two terms,

$$U = U^{(i)} + U^{(s)} = e^{i\psi_i}[A^{(i)} + A^{(s)}e^{i(\psi_s - \psi_i)}]. \tag{1}$$

Here $U^{(i)} = A^{(i)}e^{i\psi_i}$ denotes the *incident wave* (or *coherent background*); it is the field which would be produced at $\mathscr{H}$ in the absence of the object. The other term, $U^{(s)} = A^{(s)}e^{i\psi_s}$ represents the *secondary*, or *diffracted, wave* and it is this wave that contains information about the object. In terms of the amplitudes and phases, the amplitude of the total disturbance U may according to (1) be written as

$$A = \sqrt{UU^\star} = \sqrt{A^{(i)2} + A^{(s)2} + 2A^{(i)}A^{(s)}\cos(\psi_s - \psi_i)}. \tag{2}$$

As usual, we have suppressed the time harmonic factor $e^{-i\omega t}$, and thus have implicitly assumed that the secondary wave issuing from the object is of the same frequency as the incident wave. An object of this type is called a *Rayleigh scatterer*; to a good approximation, practically all non-fluorescent objects are of this type.

---

* D. GABOR, *Nature*, **161** (1948), 777; *Proc. Roy. Soc.*, A, **197** (1949), 454; *Proc. Phys. Soc.*, B, **64** (1951), 449.

† This means that, if the wave is expressed as the sum of the incident wave and a diffracted secondary wave, the scattering of the secondary wave is neglected. This neglection represents what is usually known as BORN's first approximation, and finds many applications in the theory of scattering of X-rays and electrons. For the limits of validity of this approximation in electron microscopy and interferometry see D. GABOR, *Rev. Mod. Phys.*, **28** (1956), 260.

‡ For a detailed treatment of this subject and for an account of various modifications and applications of this technique, see for example, J. B. DEVELIS and G. O. REYNOLDS, *Theory and Applications of Holography* (Reading, Mass., Addison-Wesley Publishing Company, 1967) and H. M. SMITH, *Principles of Holography* (New York, J. Wiley and Sons, 1969).

Suppose now that a photographic plate is placed in the $\mathscr{H}$ plane. Let α be the *transmission factor* of the plate, defined by analogy with the transmission function F of § 8.6.1 as the ratio of the complex amplitude of the wave emerging from the plate

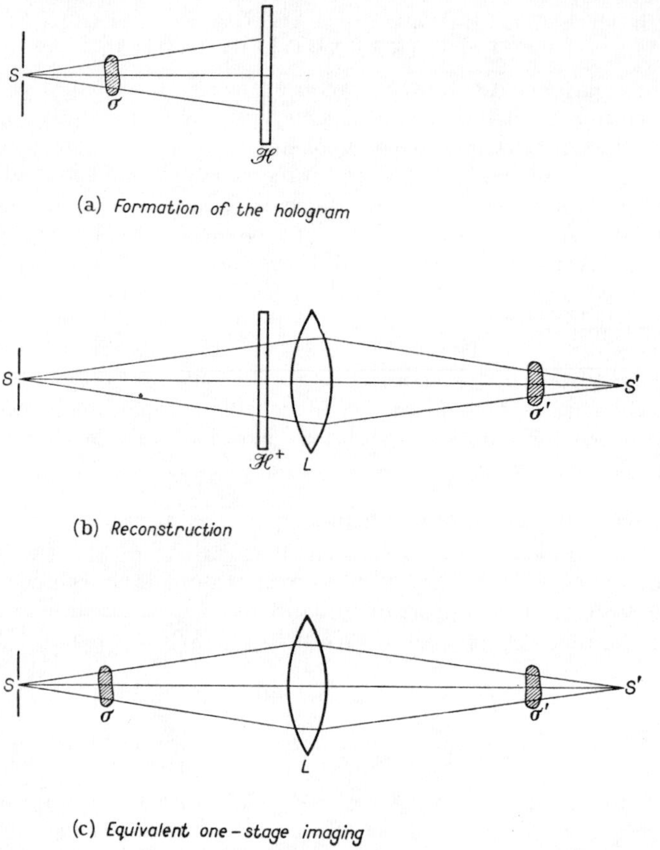

Fig. 8.50. Illustrating GABOR's method of imaging by reconstructed wave-fronts.

to that of the wave incident on the plate. The corresponding transmission factor for the intensity is $\tau = \alpha\alpha^\star$; and the quantity

$$D = -\log_{10}\tau = -\log \alpha\alpha^\star \tag{3}$$

is called the *density* of the plate. The product E of the intensity $I = A^2$ of light that reaches the plate and the time t of exposure,

$$E = It, \tag{4}$$

is called simply *exposure* (or light sum) and the curve giving D against $\log_{10}E$ is known as the *Hurter–Driffield curve*; a typical form of this curve is shown in Fig. 8.51. In the range between the points P and Q the curve is practically a straight line, and if Γ is its slope, the density of the negative is evidently given by

$$D = D_0 + \Gamma \log_{10} \frac{E}{E_0}, \tag{5}$$

where D_0 and E_0 are constants. Using (3) it follows that

$$\tau = \tau_0 \left(\frac{E}{E_0}\right)^{-\Gamma}. \tag{6}$$

For pure absorption, without phase change, α is a real number, the square root $\sqrt{\tau}$ of the intensity transmission. Thus, in this case, the amplitude transmission factor α_n of the "negative hologram" is given by a relation of the form

$$\alpha_n = (K_n A)^{-\Gamma_n}, \tag{7}$$

where K_n is proportional to the square root of the time of the exposure.

Suppose now that we take a positive print of the negative hologram. The amplitude transmission factor α_p of the positive is

$$\alpha_p = [K_p(K_n A)^{-\Gamma_n}]^{-\Gamma_p} = KA^{\Gamma}, \tag{8}$$

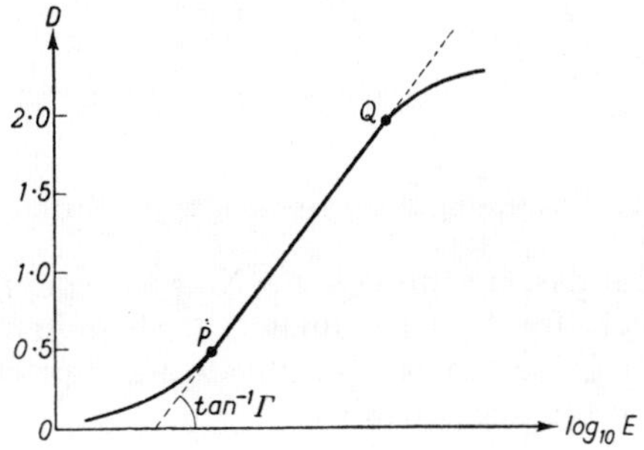

Fig. 8.51. The HURTER–DRIFFIELD curve (photographic response).

where $\Gamma = \Gamma_n \Gamma_p$ is the "overall gamma" of the negative-positive process and $K = K_p^{-\Gamma_p} K_n^{\Gamma}$.

8.10.2 The reconstruction

In the reconstruction process (Fig. 8.50(b)) the positive hologram ($\mathscr{H}^+$), whose amplitude transmission factor is given by (8), is illuminated by the coherent background $U^{(i)}$ alone. The background is obtained simply by removing the object, otherwise preserving the geometry of the original arrangement. A substitute wave U' is transmitted by the plate, and this, according to (2) and (8), is represented by

$$U' = \alpha_p U^{(i)} = KA^{(i)} e^{i\psi_i} [A^{(i)^2} + A^{(s)^2} + 2A^{(i)}A^{(s)} \cos(\psi_s - \psi_i)]^{\frac{1}{2}\Gamma}. \tag{9}$$

If we chose $\Gamma = 2$, then

$$U' = KA^{(i)^2} e^{i\psi_i} \left[A^{(i)} + \frac{A^{(s)^2}}{A^{(i)}} + A^{(s)} e^{i(\psi_s - \psi_i)} + A^{(s)} e^{-i(\psi_s - \psi_i)}\right]. \tag{10}$$

On comparing (10) and (1) it is seen that if $A^{(i)}$ is constant, i.e. if the background is uniform, the substitute wave U' contains a component, called the *reconstructed wave*, proportional to U (the first and third term in (10)). The remainder of (10) consists of two terms. One has the same phase as the background and an amplitude $A^{(s)^2}/A^{(i)}$ times that of the background. This term can be made very small by making the background sufficiently strong.* The other term has the same amplitude ($KA^{(i)2}A^{(s)}$) as the reconstructed wave, but has a phase shift of opposite sign relative to the background. We say that it represents a *conjugate wave* and we shall show that it may be regarded as being due to a fictitious object of a similar nature as the true object, but situated in a different plane.

To show this we return for the moment to the arrangement of Fig. 8.50(a) and denote by O_1 any point in the object σ, and by P any point in the $\mathscr{H}$-plane (Fig. 8.52). When the object is illuminated from the point source S, the point P may be assumed to receive light along two rays, namely along the direct ray SP (associated with the coherent background) and along the "diffracted secondary ray" O_1P (corresponding to the secondary wave). Suppose first that there is no change of phase on diffraction at O_1. Then the diffracted light at P is delayed with respect to the direct light by the path difference $O_1P - O_1A$, where A is the point on the intersection with SO_1 of the

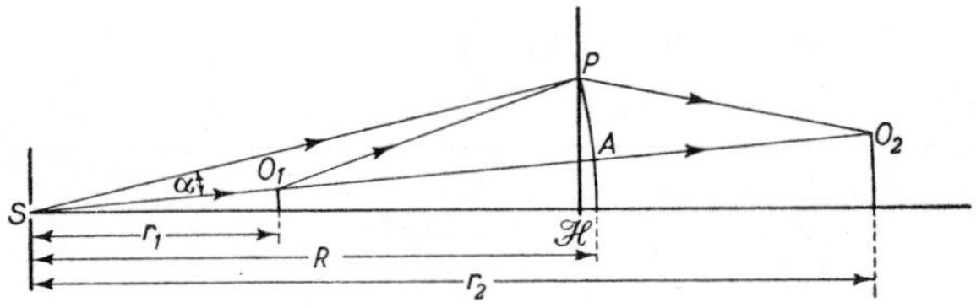

Fig. 8.52. Illustrating the position of the conjugate object.

sphere centered on S and passing through P. Now the conjugate wave is advanced relative to the direct wave by the same amount as the diffracted secondary wave is retarded with respect to it, so that the conjugate wave catches up with the direct wave at that point O_2 on the line SO_1 where

$$O_1P - O_1A = PO_2 - AO_2 \tag{11}$$

If r_1, R, and r_2 are the distances of the points O_1, A, and O_2 from S, and α is the angle between SP and the line SO_1AO_2 (see Fig. 8.52), equation (11) may be written as

$$\sqrt{(R\cos\alpha - r_1)^2 + R^2\sin^2\alpha} - (R - r_1) = \sqrt{(r_2 - R\cos\alpha)^2 + R^2\sin^2\alpha} - (r_2 - R), \tag{12}$$

Expanding both sides in powers of α and retaining leading terms only, it follows that

$$\frac{1}{r_1} + \frac{1}{r_2} = \frac{2}{R}. \tag{13}$$

If the hologram were curved to a sphere of radius R, the relation (13) would hold with respect to any point on it, and it follows that the conjugate wave could then be regarded as being due to a fictitious object which is the image of the original object in a spherical mirror, of radius R and centre S (cf. § 4.4 (16)). This result evidently also holds, as a good approximation, in the case of a plane hologram, provided that the off-axis angles of the rays are small enough. If, moreover, r_1/R is small compared to unity, then according to (13) $r_2 \sim -r_1$, so that the conjugate object and the true object are situated symmetrically about the source S.

It has been assumed so far that there is no phase change on diffraction by the object. If there is a change of phase, the preceding considerations regarding the position of the conjugate object still apply, provided that the conjugate object is assumed to

* Strong coherent background is also utilized in a method due to F. Zernike (cf. *Proc. Phys. Soc.*, **61** (1948), 158) for the display of weak secondary interference fringes.

Hologram

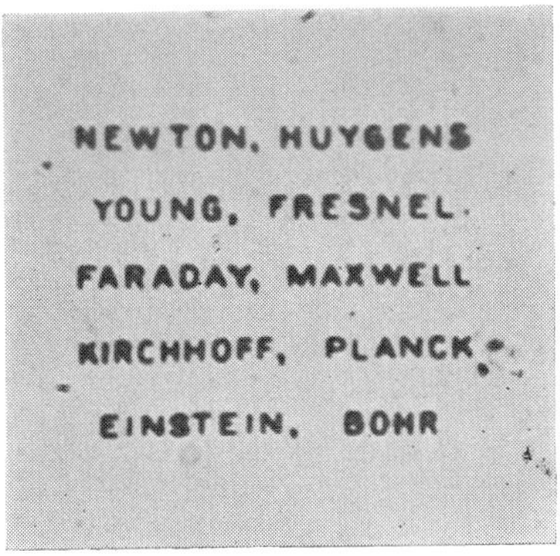

Original

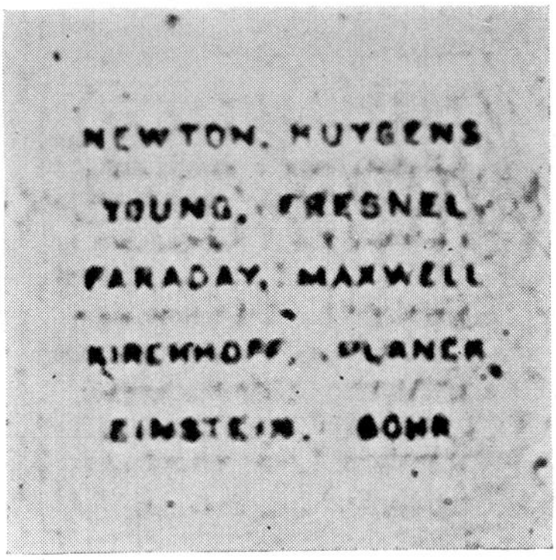

Reconstruction

Fig. 8.53. Imaging by reconstructed wave-fronts.
The object was a microphotograph of 1·5 mm diam., illuminated with light of wavelength $\lambda = 4358$ Å through a pinhole of diameter 0·2 mm, reduced by a microscope objective to 5μ nominal diameter at 50 mm from the object. Geometrical magnification 12. Effective aperture of lens used in reconstruction was 0·025. Noisy background chiefly due to imperfections of illuminating objective.

(After D. GABOR, *Proc. Roy. Soc.*, A, **197** (1949), 454.)

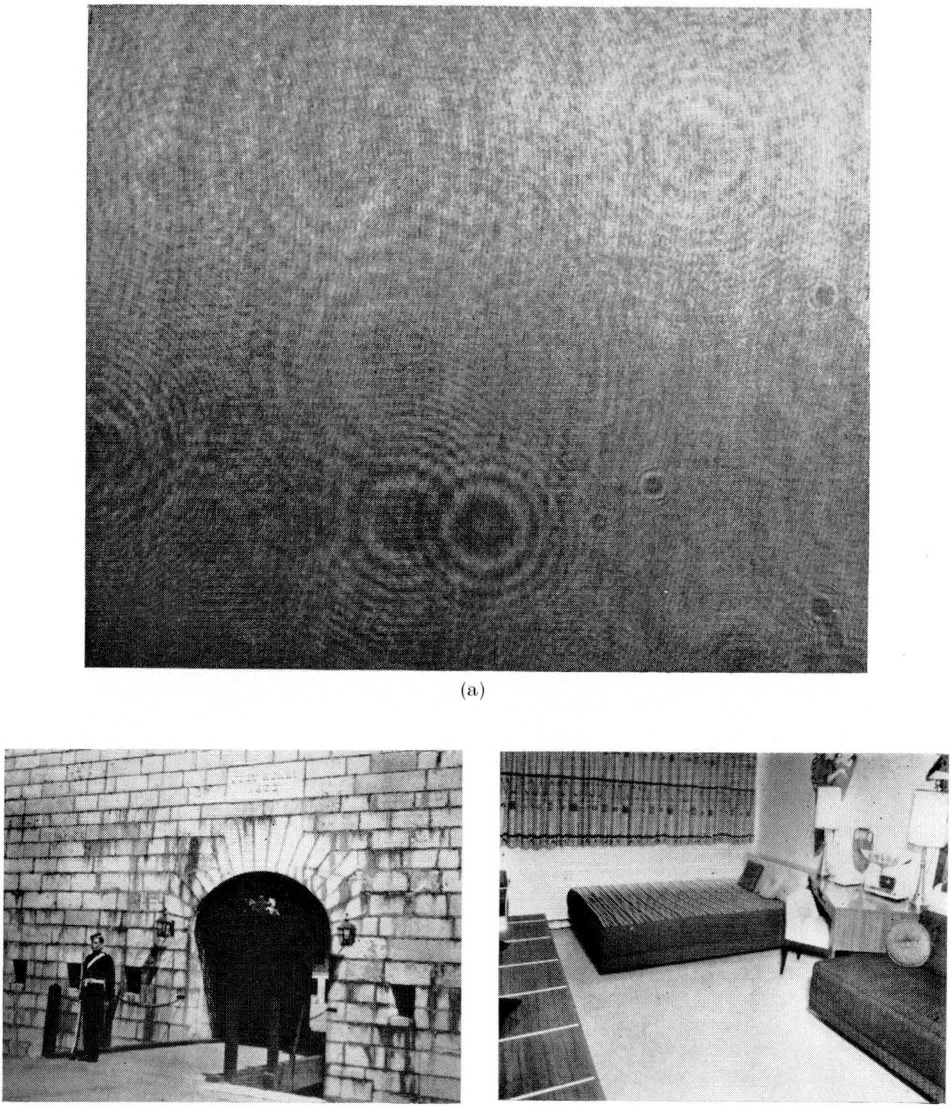

(a)

(b) (c)

Fig. 8.54. Imaging by reconstructed wavefronts. Reconstruction of two objects from a single hologram.

Hologram (a) and reconstructions (b) and (c) of two transparencies which were illuminated with diffused coherent light. The transparencies were placed 14 and 24 in. from the hologram recording plate in such positions that neither obscured the other when viewed from the position where the hologram was recorded. The observable structures seen in the hologram are mainly diffraction patterns produced by dust particles.

(After E. N. LEITH and J. UPATNIEKS, *J. Opt. Soc. Amer.*, **54** (1964), 1297.)

produce a phase change of equal amount but of opposite sign to that produced by the true object.

Returning to eq. (10), we see that, provided the background is uniform and strong in comparison with the scattered wave, the substitute wave U' is effectively the same as the original wave, apart from a contribution that may be regarded as arising from a conjugate object. Hence, *if a lens L is placed behind the positive hologram and the hologram is illuminated by the background wave alone* (Fig. 8.50 (b)), *an image σ' of the original object will be formed in a plane conjugate to that of σ,* but this image will, in general, be perturbed by a contribution due to the conjugate object. Conditions may be found under which this perturbing effect is not serious. Roughly speaking, one may expect that this effect will be small if the separation between the image of the original object and of its conjugate exceeds the focal tolerance of the image-forming pencil, this being given by the formula (27) of § 8.8.

Fig. 8.50(c) illustrates the equivalent one-stage imaging and in Fig. 8.53 the enlarged object, the hologram, and the reconstructed image are shown relating to one of the first experiments of this type.

More recently LEITH and UPATNIEKS* have considerably improved the wavefront reconstruction technique, by modifying the manner in which the coherent background is superposed onto the beam that is transmitted through the object. In their arrangement, which employs laser light, the background beam is incident on the plane of the hologram at an angle, with the help of a prism or a mirror system. The conjugate images are formed in different directions and consequently an image of high quality may be reconstructed without any disturbing effect of the other.†

The hologram contains three-dimensional information. This fact was also further demonstrated by LEITH and UPATNIEKS and is illustrated in figure 8.54.

The results confirm our conclusion that, with *coherent light,* it is possible to reconstruct an object to within a high degree of accuracy from a record of the *intensity distribution* alone, taken in any plane behind the object.

It is not essential for the success of the method that the wave-fronts from the source S should be strictly spherical. It is only necessary that the wave-fronts of the background wave in the arrangements illustrated in Figs. 8.50(a) and 8.50(b) should be of the same geometrical form. Nor is it necessary to employ the same source, or indeed radiation of the same wavelengths. It is in this connection that the potentialities of the method for electron microscopy are apparent. For one of the chief factors that limit the resolving power of the electron microscope is the spherical aberration of the objective, and as GABOR has pointed out, the effect of the spherical aberration can in principle be eliminated, or better expressed—compensated—by the reconstruction method. The hologram could be obtained with the electron beam, and the reconstruction with light. One would then have to *imitate* the spherical aberration of the electron objective by employing light waves that suffer from spherical aberration of the same amount (naturally scaled in the ratio $\lambda_{\text{light}}/\lambda_{\text{electron}}$).‡ Although, at the time of its invention, difficulties of a technical nature prevented the application

---

* E. N. LEITH and J. UPATNIEKS, *J. Opt. Soc. Amer.*, **52** (1962), 1123; **53** (1963), 1377; **54** (1964), 1295.

Some other modifications and various aspects of the method of reconstructed wavefronts are discussed in the book by J. B. DEVELIS and G. O. RENOLDS, *loc. cit.*

† The diffraction theory of this modified holographic scheme was discussed by E. WOLF and J. R. SHEWELL, *J. Math. Phys.*, **11** (1970), 2254.

‡ Light waves with a prescribed amount of spherical aberration can be produced by means of appropriate aspherical surfaces.

of GABOR's method to electron microscopy, the correctness of the basic principles of his considerations was verified by experiments with light.

As a method of two-step photography, wave-front reconstruction has an important precursor. This is a method of optical FOURIER analysis for the reconstruction of crystal structures from their diffraction patterns, first proposed by BOERSCH,* and independently by BRAGG† who called the device the "X-ray microscope." If a crystal is illuminated by a parallel beam of X-rays, the angular distribution of intensity of the diffracted light, caught on a photographic plate, is, at least for small deflexion angles, the absolute square of the FOURIER transform of the electron density distribution in the crystal, projected on a plane at right angles to the original beam.‡ The fact that only the intensities are recorded, while the phases in the experiment remain physically undefined, prevents, in general, a direct reconstruction. The necessary data must be collected from other photographs, from previous knowledge of the chemical structure, and by guess work. The work is comparatively simple when the direction of illumination is a crystal axis; in this case it can be shown that the diffracted wavelets are all in phase or in anti-phase if the illumination is strictly coherent, so that each observed diffraction maximum then corresponds to two possible phases only; this, at any rate, greatly reduces the number of combinations which must be tried.

There are, however, certain crystals in which each cell contains a heavy atom, and the amplitudes due to the diffraction by the lattice of heavy atoms are so overwhelmingly large that the resulting amplitudes can have only one sign. One can say that in this case the heavy atoms produce a coherent background. Thus, in the case of these (rather exceptional) crystals, one only needs to take the square root of the intensities of the diffracted light in order to obtain the FOURIER transform of the density distribution projected at right angles to the crystal axis, parallel to the direction of illumination.

Now an amplitude distribution proportional to the square root of the intensity can be obtained from a print of the diffraction pattern taken with an over-all gamma of unity (see eq. (8)). In the X-ray microscope this print is illuminated with a plane wave-front and the light transmitted and diffracted by the print is focused by a lens. According to the theory of FRAUNHOFER diffraction (§ 8.3) the amplitude pattern in the image plane is the FOURIER transform of the distribution in the plane of the plate and hence results from the original density distribution by two successive FOURIER transformations; from FOURIER's theorem it follows that the distribution in the focal plane is a faithful image of this (in general complex) density distribution. It is of course assumed that the aperture of the lens is large enough to admit all the diffracted rays that carry appreciable amounts of energy and that the requirement of all wavelets being in phase is satisfied. The latter condition is, in fact, only rarely fulfilled.

In the method of wave-front reconstruction, a background is artificially added. There are no conditions to be satisfied regarding symmetry or even periodicity. On the other hand this method is not suitable for X-ray analysis owing to the practical impossibility of producing a strong coherent background. However, the method may well have applications in future electron diffraction studies.

---

* H. BOERSCH, *Z. techn. Phys.*, **19** (1938), 337.

† W. L. BRAGG, *Nature*, **149** (1942), 470; see also, M. J. BUERGER, *J. Appl. Phys.*, **21** (1950), 909.

‡ See any standard book on the theory of X-ray diffraction, e.g. the books quoted on p. 406.

CHAPTER IX

THE DIFFRACTION THEORY OF ABERRATIONS

IN Chapter V we studied the effects of aberrations on the basis of geometrical optics. In that treatment the image was identified with the blurred figure formed by the points of intersection of the geometrical rays with the image plane. Since geometrical optics gives an approximate model valid in the limit of very short wavelengths, it is to be expected that the geometrical theory gradually loses its validity as the aberrations become small. For example, in the limiting case of a perfectly spherical convergent wave issuing from a circular aperture, geometrical optics predicts for the focal plane an infinite intensity at the focus and zero intensity elsewhere, whereas, as has been shown in § 8.5.2, the real image consists of a bright central area surrounded by dark and bright rings (the AIRY pattern). In the neighbourhood of the focal plane the light distribution has also been seen to be of a much more complicated nature (cf. Fig. 8.41) than geometrical optics suggests. We are thus led to the study of the effects of aberrations on the basis of diffraction theory.

The first investigations in this field are due to RAYLEIGH.* His main contribution was the formulation of a criterion (discussed in § 9.3) which, in an extended form, has come to be widely used for determining the maximum amounts of aberrations that may be tolerated in optical instruments. The subject was carried further by the researches of many writers who investigated the effects of various aberrations,† and we may mention, in particular, the more extensive treatments by STEWARD, PICHT, and BORN.‡

A very extensive diffraction treatment of image formation in the presence of aberrations is due to NIJBOER,§ carried out partly in collaboration with ZERNIKE. It is concerned with influence of small aberrations, when the departures of the wave-fronts from spherical form are only a fraction of the wavelength. The effects of large aberrations were studied on the basis of diffraction theory by VAN KAMPEN ‖ with the help of asymptotic approximations; this treatment is based on a formal extension to

---

* Lord RAYLEIGH, *Phil. Mag.*, (5), **8** (1879), 403. Reprinted in his *Scientific Papers* (Cambridge University Press, 1899), **1**, 428.

† A historical survey of diffraction theory of aberrations was given by E. WOLF in *Rep. Progr. Phys.* (London, Physical Society), **14** (1951), 95.

‡ G. C. STEWARD, *Phil. Trans. Roy. Soc.*, A, **225** (1925), 131; also his book *The Symmetrical Optical System* (Cambridge University Press, 1928). J. PICHT, *Ann. d. Physik*, (4), **77** (1925), 685. *Ibid.*, **80** (1926), 491; also his *Optische Abbildung* (Braunschweig, Vieweg, 1931). M. BORN, *Naturwissenschaften*, **20** (1932), 921; and his *Optik* (Berlin, Springer, 1933), p. 202.

§ B. R. A. NIJBOER, Thesis, University of Groningen, 1942. The main part was also published in *Physica*, **10** (1943), 679; *ibid.*, **13** (1947), 605; F. ZERNIKE and B. R. A. NIJBOER, contribution to *La Théorie des Images Optiques* (Paris, Revue d'Optique, 1949), p. 227. An extension of the theory to somewhat larger aberrations was discussed by K. NIENHUIS and B. R. A. NIJBOER, *Physica*, **14** (1948), 590. Mention must also be made of a thesis by K. NIENHUIS (University of Groningen, 1948), which is mainly concerned with the experimental study of the effects of aberrations. Some of the beautiful photographs obtained by NIENHUIS are reproduced in § 9.4.

Experimental investigations at microwave frequencies, concerning the structure of the image region in the presence of aberrations were described by M. P. BACHYNSKI and G. BEKEFI in the paper referred to on p. 440, and also in *Trans. Inst. Radio Eng.*, AP-4 (1956), 412.

‖ N. G. VAN KAMPEN, *Physica*, **14** (1949), 575; *ibid.*, **16** (1950), 817; *ibid.* **25** (1958), 437.

functions of two variables of the principle of stationary phase, which has since been formulated rigorously, first by J. FOCKE (see Appendix III, p. 752).

In the main part of this chapter we shall give an account of the NIJBOER–ZERNIKE theory and examine the structure of diffraction images affected by primary aberrations. In the final section (§ 9.5) we shall go over from point objects to extended objects and investigate imaging with coherent and incoherent illumination. Imaging with partially coherent illumination will be considered in Chapter X.

9.1. THE DIFFRACTION INTEGRAL IN THE PRESENCE OF ABERRATIONS

9.1.1 The diffraction integral

Consider a centred optical system with a point source of monochromatic light P_0 (Fig. 9.1). We take a Cartesian system of axes, with origin at the Gaussian image

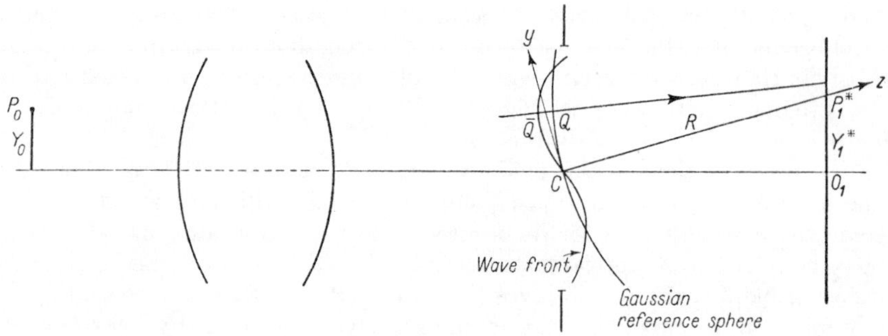

Fig. 9.1. Choice of the reference system and notation.

point $P_1^\star$ of P_0, with the z-axis along $CP_1^\star$, where C is the centre of the exit pupil. The y-axis is taken in the meridional plane (the plane containing P_0 and the axis of the system). The off-axis distances of P_0 and $P_1^\star$ will be denoted by Y_0 and $Y_1^\star$ respectively.

As in Chapter V the deformation of the wave-fronts in the region of the exit pupil will be described by the aberration function Φ. Let $\bar{Q}$ and Q be the points in which a ray in the image space intersects the wave-front through C and the Gaussian reference sphere respectively. Assuming that the refractive index of the image space is unity, Φ (taken as positive in Fig. 9.1) represents the distance $\bar{Q}Q$ measured along the ray.

Let R denote the radius $CP_1^\star$, of the Gaussian reference sphere and let s be the distance between Q and an arbitrary point P in the region of the image. The disturbance at Q is represented by $Ae^{ik(\Phi-R)}/R$, where A/R is the amplitude at Q. According to the HUYGENS–FRESNEL principle the disturbance at P is given by

$$U(P) = -\frac{i}{\lambda}\frac{Ae^{-ikR}}{R}\iint \frac{e^{ik[\Phi+s]}}{s}\,dS, \tag{1}$$

where the integration extends over the portion of the reference sphere that approximately fills the exit pupil. In (1) we have assumed that the angles involved are small, so that the variation of the inclination factor over the reference sphere may be neglected; we have also assumed that the amplitude of the wave is substantially constant over the wave-front, so that A can be taken outside the integral.

Let (ξ, η, ζ) be the coordinates of Q and (x, y, z) the coordinates of P, and let a be the radius of the exit pupil. As in § 8.8, which was concerned with the special case of an aberration-free wave ($\Phi \equiv 0$), we set

$$\left.\begin{aligned} \xi &= a\rho \sin \theta, \qquad & x &= r \sin \psi, \\ \eta &= a\rho \cos \theta, \qquad & y &= r \cos \psi, \end{aligned}\right\} \tag{2}$$

and we have, as in § 8.8 (2) and § 8.8 (9)*

$$k(s - R) = -v\rho \cos (\theta - \psi) - \tfrac{1}{2}u\rho^2 + \left(\frac{R}{a}\right)^2 u, \tag{3}$$

where u and v are the two "optical coordinates" of P,

$$u = \frac{2\pi}{\lambda}\left(\frac{a}{R}\right)^2 z, \qquad v = \frac{2\pi}{\lambda}\left(\frac{a}{R}\right)\sqrt{x^2 + y^2}. \tag{4}$$

It will now be convenient to regard Φ as function of $Y_1^{\star}$, ρ, and θ,

$$\Phi = \Phi(Y_1^{\star}, \rho, \theta). \tag{5}$$

The element of the Gaussian reference sphere is $dS = a^2\rho d\rho d\theta$, and if the angle which $CP_1^{\star}$ makes with the axis of the system is small, the range of integration may be taken as $0 \leqslant \rho \leqslant 1$, $0 \leqslant \theta < 2\pi$. Moreover, for points of observations in the region of the image, s may be replaced by R in the denominator of the integrand. Hence (1) becomes, on substitution from (3),

$$U(P) = U(u, v, \psi) = -\frac{i}{\lambda}\frac{Aa^2}{R^2} e^{i\left(\frac{R}{a}\right)^2 u} \int_0^1 \int_0^{2\pi} e^{i[k\Phi(Y_1^{\star},\rho,\theta) - v\rho\cos(\theta-\psi) - \frac{1}{2}u\rho^2]} \rho d\rho d\theta, \tag{6}$$

so that the intensity at P is

$$I(P) = |U(P)|^2 = \left(\frac{Aa^2}{\lambda R^2}\right)^2 \left| \int_0^1 \int_0^{2\pi} e^{i[k\Phi(Y_1^{\star},\rho,\theta) - v\rho\cos(\theta-\psi) - \frac{1}{2}u\rho^2]} \rho d\rho d\theta \right|^2. \tag{7}$$

It is convenient to express the intensity $I(P)$ as fraction of the intensity $I^{\star}$ which would be obtained at the Gaussian image point $P_1^{\star}$ if no aberrations were present. According to (7),

$$I^{\star} = \pi^2 \left(\frac{Aa^2}{\lambda R^2}\right)^2, \tag{8}$$

so that the *normalized intensity* is†

$$i(P) = \frac{I(P)}{I^{\star}} = \frac{1}{\pi^2} \left| \int_0^1 \int_0^{2\pi} e^{i[k\Phi(Y_1^{\star},\rho,\theta) - v\rho\cos(\theta-\psi) - \frac{1}{2}u\rho^2]} \rho d\rho d\theta \right|^2. \tag{9}$$

In the absence of aberrations the intensity is a maximum at the Gaussian image point. When aberrations are present, this will in general no longer be the case, and we may call the point of maximum intensity the *diffraction focus*.‡ Often one is

---

* R now corresponds to f of § 8.8. It is of interest to note that if equations (2) and (3) of § 8.8 are used, the diffraction integral may again be expressed in the form of an angular spectrum of plane waves (cf. J. Focke, *Optica Acta*, **3** (1956), 110).

† No confusion should arise between the symbol i used for the normalized intensity, and the same symbol used for $\sqrt{-1}$, since the former always occurs with arguments, e.g. $i(P)$, $i(u, v, \psi)$, etc.

‡ In general there may be, of course, more than one diffraction focus, but the diffraction focus is unique, if the aberrations are sufficiently small.

interested only in the maximum intensity in a particular plane of observation; this value [when normalized as in (9)] will be called the *Strehl intensity*.*

From (9) some simple results, which will be needed later, may immediately be deduced.

9.1.2. The displacement theorem. Change of reference sphere

Let Φ and Φ' be two aberration functions such that

$$\Phi' = \Phi + H\rho^2 + K\rho \sin\theta + L\rho \cos\theta + M, \tag{10}$$

where H, K, L, and M are constants of order λ. Further let $i(u, v, \psi)$ and $i'(u, v, \psi)$ be the corresponding normalized intensities. Then by (9)

$$i(u, v, \psi) = \frac{1}{\pi^2}\left|\int_0^1\int_0^{2\pi} e^{if(u,v,\psi;\,\rho,\theta)}\,\rho d\rho d\theta\right|^2, \tag{11}$$

where

$$f(u, v, \psi;\ \rho, \theta) = k\Phi - v\rho\cos(\theta - \psi) - \tfrac{1}{2}u\rho^2, \tag{12}$$

with a similar expression for i'. Now according to (10) the last expression may also be written in the form

$$\begin{aligned} f(u, v, \psi;\ \rho, \theta) &= k\Phi' - k[H\rho^2 + K\rho\sin\theta + L\rho\cos\theta + M] - v\rho\cos(\theta - \psi) - \tfrac{1}{2}u\rho^2 \\ &= k\Phi' - v'\rho\cos(\theta - \psi') - \tfrac{1}{2}u'\rho^2 - kM \\ &= f'(u', v', \psi';\ \rho, \theta) - kM, \end{aligned} \tag{13}$$

where

$$u' = u + 2kH, \qquad v'\sin\psi' = v\sin\psi + kK, \qquad v'\cos\psi' = v\cos\psi + kL. \tag{14}$$

According to (2) and (4), equations (14) represent the transformation

$$z' = z + 2\left(\frac{R}{a}\right)^2 H, \qquad x' = x + \left(\frac{R}{a}\right)K, \qquad y' = y + \left(\frac{R}{a}\right)L. \tag{15}$$

From (11) and (13) it follows that

$$i(u, v, \psi) = i'(u', v', \psi'). \tag{16}$$

We have thus established the following *displacement theorem*: *The addition to an aberration function of a term $H\rho^2 + K\rho\sin\theta + L\rho\cos\theta + M$, where H, K, L, and M are constants of order λ, results in no change in the three-dimensional intensity distribution near focus apart from a displacement of the distribution as a whole in accordance with the transformation* (15); that is, a shift of amount $2(R/a)^2H$ occurs along the principal direction CP_1^* away from the exit pupil and shifts of amounts $(R/a)K$ and $(R/a)L$ occur in the positive x- and y-directions respectively.

The additive terms on the right-hand side of (10) may be interpreted as representing a change of reference sphere. Suppose that we choose a new reference sphere, centred on a point $P'(x', y', z')$ in the image region, and of radius R' such that the new sphere is at most a few wavelengths away from the Gaussian sphere. Let a ray $\bar{Q}Q$ intersect the new reference sphere at a point N. Then the wave aberration Φ' referred to this new sphere is (see Fig. 9.2)

$$\Phi' = \bar{Q}N = \bar{Q}Q - NQ \sim \bar{Q}Q - NG, \tag{17}$$

* This concept is due to K. Strehl, *Z.f. Instrumkde.*, **22** (1902), 213, who called it "Definitionshelligkeit". In the English literature the less appropriate term "definition" is often used.

where G is the point in which the line NP' intersects the Gaussian reference sphere, the refractive index of the image space being assumed to be unity as before. Now $\bar{Q}Q = \Phi$ is the wave aberration referred to the Gaussian sphere and $NG = NP' - GP' = R' - s$, where s denotes the distance from G to P'. Hence (17) may be written as

$$\Phi' \sim \Phi + s - R' = \Phi + \frac{\lambda}{2\pi}\left[-v\rho\cos(\theta - \psi) - \tfrac{1}{2}u\rho^2 + \left(\frac{R}{a}\right)^2 u\right] + (R - R'), \quad (18)$$

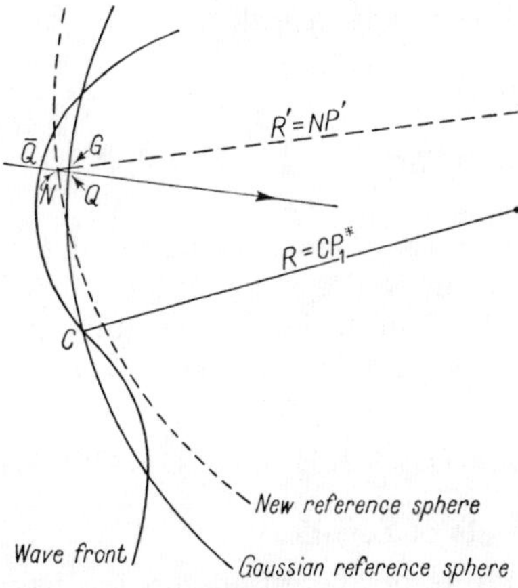

Fig. 9.2. Change of reference sphere.

where (3) was used. Here u, v, and ψ are given by (3) and (4) with x', y', z' written in place of x, y, z. The relation (18) may be written in the form (10) with

$$H = -\frac{1}{2}\left(\frac{a}{R}\right)^2 z', \quad K = -\left(\frac{a}{R}\right)x', \quad L = -\left(\frac{a}{R}\right)y', \quad M = z' + R - R'. \quad (19)$$

9.1.3 A relation between the intensity and the average deformation of wave-fronts

When the aberrations are small it is possible to express the intensity at the centre of the reference sphere in terms of the mean square value of the wave aberration. Let Φ_P be the wave aberration referred to a reference sphere centred on a point P in the image region. Then, according to equations (9) and (18), the normalized intensity at P may be expressed in the form

$$i(P) = \frac{1}{\pi^2}\left|\int_0^1\int_0^{2\pi} e^{ik\Phi_P}\rho\, d\rho\, d\theta\right|^2 = \frac{1}{\pi^2}\left|\int_0^1\int_0^{2\pi}[1 + ik\Phi_P + \tfrac{1}{2}(ik\Phi_P)^2 + \ldots]\rho\, d\rho\, d\theta\right|^2. \quad (20)$$

Let $\overline{\Phi_P^n}$ denote the average value of the nth power of Φ_P, i.e.

$$\overline{\Phi_P^n} = \frac{\displaystyle\int_0^1\int_0^{2\pi}\Phi_P^n\rho\, d\rho\, d\theta}{\displaystyle\int_0^1\int_0^{2\pi}\rho\, d\rho\, d\theta} = \frac{1}{\pi}\int_0^1\int_0^{2\pi}\Phi_P^n\rho\, d\rho\, d\theta. \quad (21)$$

If we assume that the aberrations are so small that we may neglect third and higher powers of $k\Phi_P$ in (20), the intensity at P may be written in the form

$$i(P) \sim |1 + ik\overline{\Phi_P} - \tfrac{1}{2}k^2\overline{\Phi_P^2}|^2 = 1 - \left(\frac{2\pi}{\lambda}\right)^2[\overline{\Phi_P^2} - (\overline{\Phi_P})^2]. \tag{22}$$

The quantity in the brackets on the right is the "mean-square deformation" $(\Delta\Phi)^2$ of the wave-front,

$$(\Delta\Phi_P)^2 = \frac{\displaystyle\int_0^1\int_0^{2\pi}(\Phi_P - \overline{\Phi_P})^2\rho d\rho d\theta}{\displaystyle\int_0^1\int_0^{2\pi}\rho d\rho d\theta} = \overline{\Phi_P^2} - (\overline{\Phi_P})^2, \tag{23}$$

so that (22) may be written as

$$i(P) \sim 1 - \left(\frac{2\pi}{\lambda}\right)^2(\Delta\Phi_P)^2. \tag{24}$$

This formula implies that, when the aberrations are small, the normalized intensity at the centre of the reference sphere in the region of focus is independent of the nature of the aberration and is smaller than the ideal value unity by an amount proportional to the mean-square deformation of the wave-front.

9.2. EXPANSION OF THE ABERRATION FUNCTION

9.2.1 The circle polynomials of Zernike

In our discussion of the effects of aberrations on the basis of geometrical optics (Chapter V), we expanded the aberration function Φ in a power series. In the present treatment, where integrations over the unit circle must be carried out, it is more appropriate to expand Φ in terms of a complete set of polynomials that are orthogonal over the interior of the unit circle.* Many sets of polynomials with this property can be constructed; there is, however, one such set, introduced by ZERNIKE,† which has certain simple properties of invariance. In Appendix VII it is shown how these *circle polynomials* of ZERNIKE may be derived and some of their properties are discussed; here we shall only summarize the formulae needed in the present chapter.

The circle polynomials of ZERNIKE are polynomials $V_n^l(X, Y)$ in two real variables X, Y, which, when expressed in polar coordinates ($X = \rho \sin \theta$, $Y = \rho \cos \theta$), are of the form

$$V_n^l(\rho \sin \theta, \rho \cos \theta) = R_n^l(\rho)\, e^{il\theta}, \tag{1}$$

where $l \gtrless 0$, and $n \geqslant 0$ are integers, $n \geqslant |l|$, and $n - |l|$ is even. The orthogonality and normalizing properties are expressed by the formulae

$$\iint_{X^2+Y^2\leqslant 1} V_n^{l\,*}(X, Y)V_{n'}^{l'}(X, Y)dXdY = \frac{\pi}{n+1}\,\delta_{ll'}\delta_{nn'}, \tag{2}$$

where δ_{ij} is the KRONECKER symbol and where the asterisks denotes the complex conjugate. The radial functions $R_n^l(\rho)$ are polynomials in ρ, containing the powers $\rho^n, \rho^{n-2}, \ldots \rho^{|l|}$ and, as is shown in Appendix VII, are closely related to JACOBI's

* The term "complete" implies that any reasonably well-behaved function can be expanded as a series of functions of the set. For a more precise definition of the term see, for example, R. COURANT and D. HILBERT, *Methods of Mathematical Physics*, Vol. I (1st English edition, New York, Interscience Publishers, 1953), pp. 51–54.

† F. ZERNIKE, *Physica*, **1** (1934), 689.

polynomials (terminating hypergeometric series). As is seen from (1) and (2), the radial polynomials satisfy the relations

$$\int_0^1 R_n^l(\rho) R_{n'}^l(\rho)\rho d\rho = \frac{1}{2(n+1)}\, \delta_{nn'}. \tag{3}$$

They are given by the formulae ($m = |l|$)

$$R_n^{\pm m}(\rho) = \frac{1}{\left(\dfrac{n-m}{2}\right)!\,\rho^m} \left\{\frac{d}{d(\rho^2)}\right\}^{\frac{n-m}{2}} \left\{(\rho^2)^{\frac{n+m}{2}} (\rho^2 - 1)^{\frac{n-m}{2}}\right\} \tag{4}$$

$$= \sum_{s=0}^{\frac{n-m}{2}} (-1)^s \frac{(n-s)!}{s!\left(\dfrac{n+m}{2} - s\right)!\left(\dfrac{n-m}{2} - s\right)!} \rho^{n-2s}. \tag{5}$$

The normalization has been chosen so that for all permissible values of n and m,

$$R_n^{\pm m}(1) = 1. \tag{6}$$

The radial polynomials have the generating function

$$\frac{[1 + z - \sqrt{1 - 2z(1 - 2\rho^2) + z^2}]^m}{(2z\rho)^m \sqrt{1 - 2z(1 - 2\rho^2) + z^2}} = \sum_{s=0}^{\infty} z^s R_{m+2s}^{\pm m}(\rho). \tag{7}$$

When $m = 0$, the left-hand side reduces to the generating function for the LEGENDRE polynomials* of argument $2\rho^2 - 1$, so that

$$R_{2n}^0(\rho) = P_n(2\rho^2 - 1). \tag{8}$$

In Table XXI the explicit form of the polynomials for the first few values of the indices is given.

TABLE XXI

The radial polynomials $R_n^m(\rho)$ for $m \leqslant 8$, $n \leqslant 8$

$m \backslash n$	0	1	2	3	4	5	6	7	8
0	1		$2\rho^2-1$		$6\rho^4-6\rho^2+1$		$20\rho^6-30\rho^4+12\rho^2-1$		$70\rho^8-140\rho^6+90\rho^4-20\rho^2+1$
1		ρ		$3\rho^3-2\rho$		$10\rho^5-12\rho^3+3\rho$		$35\rho^7-60\rho^5+30\rho^3-4\rho$	
2			ρ^2		$4\rho^4-3\rho^2$		$15\rho^6-20\rho^4+6\rho^2$		$56\rho^8-105\rho^6+60\rho^4-10\rho^2$
3				ρ^3		$5\rho^5-4\rho^3$		$21\rho^7-30\rho^5+10\rho^3$	
4					ρ^4		$6\rho^6-5\rho^4$		$28\rho^8-42\rho^6+15\rho^4$
5						ρ^5		$7\rho^7-6\rho^5$	
6							ρ^6		$8\rho^8-7\rho^6$
7								ρ^7	
8									ρ^8

* See, for example, R. COURANT and D. HILBERT, *loc. cit.*, p. 85.

The following relation (also proved in the Appendix) is of great importance in the NIJBOER–ZERNIKE theory:

$$\int_0^1 R_n^m(\rho) J_m(v\rho)\rho d\rho = (-1)^{\frac{n-m}{2}} \frac{J_{n+1}(v)}{v}, \tag{9}$$

where J is a BESSEL function of the first kind.

Instead of the complex polynomials V, one may use the real polynomials

$$\left.\begin{aligned} U_n^m &= \tfrac{1}{2}[V_n^m + V_n^{-m}] = R_n^m(\rho)\cos m\theta, \\ U_n^{-m} &= \frac{1}{2i}[V_n^m - V_n^{-m}] = R_n^m(\rho)\sin m\theta. \end{aligned}\right\} \tag{10}$$

In our applications we shall need the polynomials $U_n^m = R_n^m(\rho)\cos m\theta$ only. This is so because the wave distortions are symmetrical about the meridional plane $\theta = 0$ and consequently the aberration function is an even function of θ.

9.2.2 Expansion of the aberration function

Following NIJBOER, we expand the aberration function Φ in terms of ZERNIKE's circle polynomials. As in § 5.1 it follows from symmetry that the variables enter the expansion only in the combination $Y_1^{\star 2}$, ρ^2 and $Y_1^{\star}\rho\cos\theta$, so that the expansion must be of the form

$$\Phi(Y_1^{\star}, \rho, \theta) = \sum_l \sum_n \sum_m a_{lnm} Y_1^{\star 2l+m} R_n^m(\rho)\cos m\theta, \tag{11}$$

where l, n, and m are non-negative integers, $n \geqslant m$, $n - m$ is even and the a's are constants.

As we shall mainly be concerned with the diffraction image of a fixed object point ($Y_1^{\star}$ constant), it is convenient to supress the explicit dependence of Φ on $Y_1^{\star}$ and re-write (11) in the form

$$\Phi = A_{00} + \frac{1}{\sqrt{2}}\sum_{n=2}^{\infty} A_{n0}R_n^0(\rho) + \sum_{n=1}^{\infty}\sum_{m=1}^{n} A_{nm}R_n^m(\rho)\cos m\theta. \tag{12}$$

The coefficients A_{nm} are functions of $Y_1^{\star}$, and the factor $1/\sqrt{2}$ has been introduced in the second term to simplify the final formulae.

If the aberrations are sufficiently small, the normalized intensity at the Gaussian focus may be expressed in a simple form in terms of the coefficients A. We have, on substituting from (12) into § 9.1 (21) and using the orthogonality relations (3),

$$\left.\begin{aligned} \overline{\Phi} &= A_{00}, \\ \overline{\Phi^2} &= A_{00}^2 + \tfrac{1}{2}\sum_{n=1}^{\infty}\sum_{m=0}^{n}\frac{A_{nm}^2}{n+1}. \end{aligned}\right\} \tag{13}$$

The first relation implies that A_{00} represents the mean retardation of the wave-front behind the Gaussian reference sphere. The second relation is "PARSEVAL's formula" for the orthogonal set of functions $R_n^m(\rho)\cos m\theta$. On substituting from (13) into equation (22) of § 9.1, it follows that the normalized intensity at the Gaussian focus is

$$i(P_1^{\star}) = 1 - \frac{2\pi^2}{\lambda^2}\sum_{n=1}^{\infty}\sum_{m=0}^{n}\frac{A_{nm}^2}{n+1}. \tag{14}$$

A marked advantage of the expansion in terms of the circle polynomials arises in connection with the important problem of the "balancing" of aberrations of different orders against each other in such a way as to obtain maximum intensity. Suppose that the aberration is represented by a single "power series" term

$$\Phi = A'_{nm}\,\rho^n \cos{}^m\theta, \tag{15}$$

where A' is a small constant, of the order of a wavelength or less. We enquire whether it is possible to increase the intensity $i(P_1^\star)$ by introducing aberrations of lower order. More precisely, we wish to choose constants A'_{pq} in the expression

$$\Phi' = A'_{nm}\rho^n \cos{}^m\theta + \sum_{p<n}\sum_{q\leqslant p} A'_{pq}\rho^p \cos^q\theta, \tag{16}$$

so as to make the intensity at the Gaussian focus as large as possible.

With any particular choice of the constants A'_{pq}, the aberration function (16) may also be expressed in terms of the circle polynomials, in the form

$$\Phi' = \varepsilon_{nm}A_{nm}R_n^m(\rho)\cos m\theta + \sum_{p<n}\sum_{q\leqslant p}\varepsilon_{pq}A_{pq}R_p^q(\rho)\cos q\theta, \tag{17}$$

where

$$\left.\begin{aligned} \varepsilon_{nm} &= \frac{1}{\sqrt{2}} \qquad \text{when} \qquad m = 0,\, n \neq 0,\\ &= 1 \qquad \text{otherwise.}\end{aligned}\right\} \tag{18}$$

Now the coefficient of the highest power of ρ in (17), i.e. of ρ^n, is, according to (5), $n!/[\frac{1}{2}(n+m)]![\frac{1}{2}(n-m)]!$ so that on comparing the coefficients of ρ^n in (16) and (17) it follows that

$$\frac{n!}{[\frac{1}{2}(n+m)]!\,[\frac{1}{2}(n-m)]!}\,\varepsilon_{nm}A_{nm} = A'_{nm}. \tag{19}$$

Now with A'_{nm} and consequently A_{nm} fixed, it follows from (14) that the maximum intensity at $P_1^\star$ is obtained by making all the coefficients under the summation sign in (17) identically zero. The aberration function then becomes

$$\Phi' = \varepsilon_{nm}A_{nm}R_n^m(\rho)\cos m\theta, \tag{20}$$

and the intensity at $P_1^\star$ now is

$$i(P_1^\star) = 1 - \frac{2\pi^2}{\lambda^2}\,\frac{A_{nm}^2}{n+1}, \tag{21}$$

A_{nm} being given, in terms of the coefficient A'_{nm}, by (19). It is now evident, that *in the single aberration term* $A_{nm}R_n^m(\rho)\cos m\theta$ *of the expansion* (12), *a number of terms of the form* $A'_{pq}\rho^p\cos^q\theta$ *have been combined, with* $p = n, n-2, \ldots m$; $q = m, m-2, \ldots 1$ *or* 0 *in such a way, that for a given (sufficiently small) value of the coefficient of* $\rho^n\cos^m\theta$, *the normalized intensity at the Gaussian focus is a maximum.*

We illustrate this result by a simple example. Suppose that a system suffers from a small amount of sixth-order spherical aberration $[\Phi = A'_{60}\rho^6]$, and that we are in a position to introduce a controlled amount of fourth-order spherical aberration $[A'_{40}\rho^4]$ and defocusing $[A'_{20}\rho^2]$. We seek the values of the coefficients A'_{40} and A'_{20} which make the intensity at the diffraction focus as large as possible. Problems of this type were first studied by Richter,* who showed that the maximum is obtained when the two coefficients are chosen so that

$$\frac{A'_{40}}{A'_{60}} = -\frac{3}{2}, \qquad \frac{A'_{20}}{A'_{60}} = \frac{3}{5}. \tag{22}$$

A glance at Table XXI, on p. 465, shows that this is precisely the ratio of the corresponding coefficients in the polynomial $R_6^0(\rho)$:

$$R_6^0(\rho) = 20\rho^6 - 30\rho^4 + 12\rho^2 - 1. \tag{23}$$

We see that, provided the aberrations are sufficiently small, the introduction of Zernike's circle polynomials automatically solves the problem of balancing of aberrations, in the sense explained; moreover, with the help of the displacement theorem, it also enables the determination of the position of the diffraction focus.

9.3. TOLERANCE CONDITIONS FOR PRIMARY ABERRATIONS

Before considering the difficult problem of determining the intensity distribution in the diffraction image in the presence of aberrations, we consider the much simpler problem of estimating the maximum amounts of aberrations that may be tolerated in an optical system.

It is evident from the discussion of the preceding section that, when aberrations are present, the maximum intensity in the diffraction image is smaller than the intensity at the Gaussian focus (centre of the Airy pattern) in an aberration free system of the same aperture and focal length. It was shown first by Rayleigh† that when a system suffers from primary spherical aberration of such an amount that the wave-front in the exit pupil departs from the Gaussian reference sphere by less than a quarter wavelength, the intensity at the Gaussian focus is diminished by less than 20 per cent —a loss of light that can usually be tolerated. Later workers found that in the presence of other commonly occurring aberrations the quality of the image is likewise not seriously affected when the wave-front deformation is less than a quarter of a wavelength. This result has become known as *Rayleigh's quarter wavelength rule* and is a useful criterion for the amount of aberration that can be tolerated in an image-forming system. This rule is, of course, only a rough guide as to the desirable state of correction of a system, since the light distribution in the image depends not only on the maximum deformation but also on the shape of the wave-fronts (type of aberration). Moreover, the loss of light that may be tolerated depends naturally on the particular use to which the instrument is put, and more stringent tolerances have to be imposed in certain cases.

When the condition $|\Phi_{max}| = \lambda/4$ is applied to aberrations of different types, somewhat different values for the intensity at the diffraction focus are obtained. It seems more appropriate to formulate tolerance criteria which correspond to a prescribed value of the intensity at the diffraction focus. Criteria of this type were considered by Maréchal,‡ who used the relation that exists between the intensity

---

* R. Richter, *Z.f. Instrumkde.*, **45** (1925), 1.

† Lord Rayleigh, *Phil. Mag.*, (5), **8** (1879), 403. Reprinted in his *Scientific Papers* (Cambridge University Press, 1899), **1**, 432–435.

‡ A. Maréchal, *Rev. d'Optique*, **26** (1947), 257.

at the centre of the reference sphere and the root-mean-square deviation of the wave-front from spherical form.

When the aberrations are sufficiently small, the intensity at a point P in the region of the image may according to equation (24) of § 9.1 be expressed in the form

$$i(P) \sim 1 - \left(\frac{2\pi}{\lambda}\right)^2 (\Delta\Phi_P)^2. \tag{1}$$

Following MARÉCHAL *we shall regard a system to be well corrected when the normalized intensity at the diffraction focus F is greater than or equal to* 0.8. Now from (1), $i(F) \geqslant 0{\cdot}8$ when $|\Delta\Phi_F| \lesssim \lambda/14$, so that this condition is equivalent to the requirement that *the root-mean-square departure of the wave-front from a reference sphere that is centred on the diffraction focus shall not exceed the value $\lambda/14$.**

Let us now determine the position of the diffraction focus and the tolerances for primary (SEIDEL) aberrations. In the notation of the present chapter, each primary aberration represents a wave-front deformation of the form†

$$\Phi = a'_{lnm} Y_1^{\star 2l+m} \rho^n \cos^m \theta, \tag{2}$$

where $2l + m + n = 4$. It is convenient to set

$$A'_{lnm} = a'_{lnm} Y_1^{\star 2l+m}, \tag{3}$$

and (2) becomes

$$\Phi = A'_{lnm} \rho^n \cos {}^m\theta. \tag{4}$$

The constant A' can easily be expressed in terms of the SEIDEL coefficients B, C, D, E, and F, introduced in Chapter V. If we take the arbitrary constant λ_0 in equations (7) and (8) of § 5.2 equal to unity, then λ_1 denotes the magnification between the pupil planes, and if we remember that we now have $n_1 = 1$, the variables ρ and y_0 of § 5.3 (7) correspond to $a\rho/\lambda_1$ and $-\lambda_1 Y_1^\star/R$ of the present section; comparing (4) with § 5.3 (7) we obtain

---

* MARÉCHAL's condition is actually based on a somewhat different inequality. It follows from § 9.1 (20) that

$$i(\mathrm{F}) = \frac{1}{\pi^2}\left|\int_0^1\int_0^{2\pi} e^{ik\Phi_F}\rho\, d\rho\, d\theta\right|^2$$

$$\geqslant \frac{1}{\pi^2}\left|\int_0^1\int_0^{2\pi} \cos(k\Phi_F)\rho\, d\rho\, d\theta\right|^2.$$

Now if $k|\Phi_F| < \pi/2$, i.e. if $|\Phi_F| < \lambda/4$, $\cos(k\Phi_F)$ may evidently be replaced by $1 - \frac{1}{2}(k\Phi_F)^2$ in this inequality. If, moreover, the radius of the reference sphere is chosen so that $\overline{\Phi_F} = 0$, then $(\Delta\Phi_F)^2 = \overline{\Phi_F{}^2}$ and the inequality becomes

$$i(F) \gtrsim \left[1 - \frac{2\pi^2}{\lambda^2}(\Delta\Phi_F)^2\right]^2. \tag{1a}$$

MARÉCHAL's criterion is based on the inequality (1a) but the difference is evidently of no great practical consequence if the aberrations are small. For our purposes, the use of the relation (1) rather than (1a) has the advantage that it is more directly related to the extremal properties of ZERNIKE's circle polynomials.

† We attach a prime to the coefficients in the power series representation, whilst unprimed coefficients refer to the representation in terms of the ZERNIKE circle polynomials.

$$\left.\begin{aligned} A'_{040} &= -\frac{1}{4}\left(\frac{a}{\lambda_1}\right)^4 B, \\ A'_{031} &= -\left(\frac{a}{\lambda_1}\right)^3\left(\frac{\lambda_1 Y_1^\star}{R}\right) F, \\ A'_{022} &= -\left(\frac{a}{\lambda_1}\right)^2\left(\frac{\lambda_1 Y_1^\star}{R}\right)^2 C, \\ A'_{120} &= -\frac{1}{2}\left(\frac{a}{\lambda_1}\right)^2\left(\frac{\lambda_1 Y_1^\star}{R}\right)^2 D, \\ A'_{111} &= -\left(\frac{a}{\lambda_1}\right)\left(\frac{\lambda_1 Y_1^\star}{R}\right)^3 E. \end{aligned}\right\} \tag{5}$$

In the expansion in terms of the circle polynomials, a typical term represents the aberration*

$$\Phi = \varepsilon_{nm} A_{lnm} R_n^{\,m}(\rho) \cos m\theta. \tag{6}$$

The terms with indices l, m, and n such that $2l + m + n = 4$ (primary aberrations) are shown in the last column of Table XXII. It is seen that some of the SEIDEL terms are now accompanied by terms of lower degrees, and these, according to the displacement theorem (§ 9.1.2), give rise to a bodily displacement of the intensity distribution.

TABLE XXII

Representation of the primary aberrations

Type of aberration	l	n	m	Representation in form (4)	Representation in form (6)
Spherical aberration .	0	4	0	$A'_{040}\rho^4$	$\frac{1}{\sqrt{2}} A_{040} R_4^0(\rho) = \frac{1}{\sqrt{2}} A_{040}(6\rho^4 - 6\rho^2 + 1)$
Coma . . .	0	3	1	$A'_{031}\rho^3 \cos\theta$	$A_{031} R_3^1(\rho) \cos\theta = A_{031}(3\rho^3 - 2\rho) \cos\theta$
Astigmatism . .	0	2	2	$A'_{022}\rho^2 \cos^2\theta$	$A_{022} R_2^2(\rho) \cos 2\theta = A_{022}\rho^2(2\cos^2\theta - 1)$
Curvature of field .	1	2	0	$A'_{120}\rho^2$	$\frac{1}{\sqrt{2}} A_{120} R_2^0(\rho) = \frac{1}{\sqrt{2}} A_{120}(2\rho^2 - 1)$
Distortion . .	1	1	1	$A'_{111}\rho \cos\theta$	$A_{111} R_1^1(\rho) \cos\theta = A_{111}\rho \cos\theta$

Now according to the theorem of § 9.2.2, the image affected by an aberration represented by (6) has maximum intensity at the Gaussian focus. Hence from a comparison of corresponding terms in the last two columns in Table XXII, we may immediately determine the coordinates of the diffraction focus of an image affected

* The factor ε_{nm} which is equal to unity except for $m = 0$, $n \neq 0$ when it is equal to $1/\sqrt{2}$, is retained here for the sake of uniformity with the formulae of § 9.2.

by a primary aberration. We illustrate this by considering primary spherical aberration in detail. This aberration is represented by

$$\Phi = A'_{040}\rho^4. \tag{7}$$

The corresponding expression (6) is

$$\Phi = \frac{1}{\sqrt{2}} A_{040} R_4^0(\rho) = \frac{1}{\sqrt{2}} A_{040}(6\rho^4 - 6\rho^2 + 1). \tag{8}$$

Now if

$$A'_{040} = \frac{6}{\sqrt{2}} A_{040}, \tag{9}$$

then, according to the displacement theorem, the intensity distribution will be the same in both cases; but the distribution corresponding to (7) will be displaced relative to that corresponding to (8) in accordance with equation (15) of § 9.1, with

$$H = 6A_{040}/\sqrt{2} = A'_{040}, \qquad K = L = 0, \qquad M = -A_{040}/\sqrt{2} = A'_{040}/6,$$

i.e. according to the transformation

$$x' = x, \qquad y' = y, \qquad z' = z + 2\left(\frac{R}{a}\right)^2 A'_{040}. \tag{10}$$

Since the diffraction image associated with (8) has maximum intensity at the origin $x = y = z = 0$, the diffraction focus F for primary spherical aberration represented by (7) is at the point

$$x_F = y_F = 0, \qquad z_F = 2\left(\frac{R}{a}\right)^2 A'_{040}. \tag{11}$$

The point F specified by (11) has a simple geometrical interpretation. Let ΔY and ΔZ be the lateral and longitudinal spherical aberrations, measured as positive when the ray intersects the axes on the positive side of the Gaussian focus. By § 5.1 (17), with $\Phi = A'_{040}\rho^4$, $\rho = Y/a$, $D_1 \sim -R \sim -R'$, $X_0 = X = 0$, $n_1 = 1$,

$$\Delta Y = Y_1 - Y_1^\star = 4\left(\frac{R}{a}\right)\left(\frac{Y}{a}\right)^3 A'_{040}, \tag{12a}$$

and hence by elementary geometry and the use of the preceding relation

$$\Delta Z = Z \quad - Z_1^\star \sim \frac{R}{Y}\Delta Y = 4\left(\frac{R}{a}\right)^2\left(\frac{Y}{a}\right)^2 A'_{040}. \tag{12b}$$

For the marginal ray ($Y = a$) this gives $(\Delta Z)_{\max} = 4\left(\dfrac{R}{a}\right)^2 A'_{040}$, so that (11) is seen to imply that* *the diffraction focus in the presence of a small amount of primary spherical aberration is situated midway between the paraxial and marginal foci.*

Next let us determine the tolerance for primary spherical aberration. For any aberration characterized by (6) the normalized intensity at the Gaussian focus will be according to § 9.2 (14), greater than or equal to 0·8 if

$$1 - \frac{2\pi^2}{\lambda^2}\frac{A_{lnm}^2}{n+1} \geqslant 0{\cdot}8,$$

* We neglect here the small effect arising from the different choice of the z-directions in § 5.1 and in the present discussion.

i.e. provided that

$$|A_{lnm}| \lesssim \frac{\lambda\sqrt{n+1}}{10}. \tag{13}$$

In particular, for primary spherical aberration this gives

$$|A_{040}| \lesssim 0{\cdot}22\lambda,$$

or, by (9),

$$|A'_{040}| \lesssim 0{\cdot}94\lambda. \tag{14}$$

This is the required tolerance condition for primary spherical aberration and implies that the maximum deviation of the wave-front from the Gaussian reference sphere must be less than 0·94 of the wavelength.

In a strictly similar manner we may find the position of the diffraction focus and the tolerance for the other primary aberrations. In particular, the diffraction focus in the presence of a small amount of primary astigmatism is found to be at the point whose coordinates are

$$x_F = y_F = 0, \qquad z_F = \left(\frac{R}{a}\right)^2 A'_{022}. \tag{15}$$

This result has again a simple physical interpretation. According to (5) and § 5.3 (18), the radii R_t and R_s of the tangential and sagittal focal surfaces are given by (assuming $n_1 = 1$ again)

$$\frac{1}{R_t} = -\frac{4}{a^2}\left(\frac{R}{Y_1^\star}\right)^2 A'_{022}, \qquad \frac{1}{R_s} = 0, \tag{16}$$

so that the abscissa Z_t and Z_s of the two focal lines are

$$Z_t = -\frac{Y_1^{\star 2}}{2R_t} = 2\left(\frac{R}{a}\right)^2 A'_{022}, \qquad Z_s = 0. \tag{17}$$

TABLE XXIII

Position of the diffraction focus and tolerance conditions for primary aberrations

Type of aberration	Coordinates of diffraction focus F			Tolerance condition $[i(F) \geqslant 0{\cdot}8]$
	x_F	y_F	z_F	
Spherical aberration . .	0	0	$2\left(\frac{R}{a}\right)^2 A'_{040}$	$\lvert A'_{040}\rvert \lesssim 0{\cdot}94\lambda$
Coma	0	$\frac{2}{3}\left(\frac{R}{a}\right) A'_{031}$	0	$\lvert A'_{031}\rvert \lesssim 0{\cdot}60\lambda$
Astigmatism	0	0	$\left(\frac{R}{a}\right)^2 A'_{022}$	$\lvert A'_{022}\rvert \lesssim 0{\cdot}35\lambda$
Curvature of field . . .	0	0	$2\left(\frac{R}{a}\right)^2 A'_{120}$	—
Distortion	0	$\left(\frac{R}{a}\right) A'_{111}$	0	—

Hence (15) implies that *the diffraction focus in the presence of a small amount of primary astigmatism is situated midway between the tangential and sagittal focal lines.*

Since primary curvature of field and primary distortion are represented by terms of the second and first degree in ρ respectively, it follows, in accordance with the displacement theorem, that the only effect of these aberrations is a "bodily shift" of the three-dimensional distribution associated with an aberration-free image. Thus in the presence of a small amount of primary curvature of field or primary distortion, the normalized intensity i at the diffraction focus is unity, but the diffraction focus does not coincide with the Gaussian image point.

In Table XXIII the results relating to primary aberrations are summarized.

9.4. THE DIFFRACTION PATTERN ASSOCIATED WITH A SINGLE ABERRATION

We now consider the diffraction image in the presence of an aberration represented by a single term of the expansion § 9.2 (11),

$$\Phi = a_{lnm} Y_1^{\star 2l+m} R_n^m(\rho) \cos m\theta. \tag{1}$$

As before, we suppress the explicit dependence on $Y_1^\star$, and set

$$\alpha_{lnm} = \frac{2\pi}{\lambda} a_{lnm} Y_1^{\star 2l+m} = \frac{2\pi}{\lambda} \varepsilon_{nm} A_{lnm}. \tag{2}$$

We also set

$$C = -\frac{i\pi A}{\lambda}\left(\frac{a}{R}\right)^2 e^{i\left(\frac{R}{a}\right)^2 u}. \tag{3}$$

The diffraction integral § 9.1 (6) then becomes

$$U(u, v, \psi) = \frac{C}{\pi}\int_0^1\int_0^{2\pi} e^{i[-v\rho\cos(\theta-\psi) - \frac{1}{2}u\rho^2 + \alpha_{lnm}R_n^m(\rho)\cos m\theta]}\rho d\rho d\theta. \tag{4}$$

The integral (4) may be developed in an infinite series, by expanding both the terms $e^{-iv\rho\cos(\theta-\psi)}$ and $e^{i\alpha_{lnm}R_n^m(\rho)\cos m\theta}$ with the help of the JACOBI identity [§ 8.8 (29)]

$$e^{iz\cos\phi} = J_0(z) + 2\sum_{s=1}^{\infty} i^s J_s(z)\cos s\phi. \tag{5}$$

Multiplying the two expansions together we find that

$$e^{i[-v\rho\cos(\theta-\psi)+\alpha_{lnm}R_n^m(\rho)\cos m\theta]}$$
$$= 4\sum_{s=0}^{\infty}{}' \sum_{s'=0}^{\infty}{}' i^s(-i)^{s'} J_s[\alpha_{lnm}R_n^m(\rho)] J_{s'}(v\rho)\cos ms\theta\cos[s'(\theta-\psi)], \tag{6}$$

where the prime on the summation sign implies that the terms in $s = 0$ and $s' = 0$ are each to be taken with a factor $\frac{1}{2}$. We substitute this double series into (4) and integrate with respect to θ term by term. This gives

$$U(u, v, \psi) = 4C\sum_{s=0}^{\infty}{}'(-i)^{(m-1)s}\cos ms\psi\int_0^1 e^{-\frac{1}{2}iu\rho^2} J_s[\alpha_{lnm}R_n^m(\rho)] J_{ms}(v\rho)\rho d\rho, \tag{7}$$

the term $s = 0$ being again taken with the factor $\frac{1}{2}$.

As we are interested in small aberrations (α small), we develop the term $J_s[\alpha_{lnm} R_n^m(\rho)]$ under the integral sign in a power series, and re-arrange the resulting expression according to powers of α_{lnm}. This gives

$$U(u, v, \psi) = C[U_0 + i\alpha_{lnm}U_1 + (i\alpha_{lnm})^2U_2 + (i\alpha_{lnm})^3U_3 + (i\alpha_{lnm})^4U_4 + \dots], \tag{8a}$$

where

$$\left.\begin{aligned}
U_0 &= 2\int_0^1 e^{-\frac{1}{2}iu\rho^2} J_0(v\rho)\rho d\rho,\\
U_1 &= 2(-i)^m \cos m\psi \int_0^1 e^{-\frac{1}{2}iu\rho^2} R_n^m(\rho) J_m(v\rho)\rho d\rho,\\
U_2 &= \frac{1}{2!}\left\{\int_0^1 e^{-\frac{1}{2}iu\rho^2}\{R_n^m(\rho)\}^2 J_0(v\rho)\rho d\rho\right.\\
&\qquad\left. + i^{2m}\cos 2m\psi \int_0^1 e^{-\frac{1}{2}iu\rho^2}\{R_n^m(\rho)\}^2 J_{2m}(v\rho)\rho d\rho\right\},\\
U_3 &= \frac{1}{2.3!}\left\{3(-i)^m \cos m\psi \int_0^1 e^{-\frac{1}{2}iu\rho^2}\{R_n^m(\rho)\}^3 J_m(v\rho)\rho d\rho\right.\\
&\qquad\left. + (-i)^{3m}\cos 3m\psi \int_0^1 e^{-\frac{1}{2}iu\rho^2}\{R_n^m(v\rho)\}^3 J_{3m}(v\rho)\rho d\rho\right\},\\
U_4 &= \frac{1}{2^2.4!}\left\{3\int_0^1 e^{-\frac{1}{2}iu\rho^2}\{R_n^m(\rho)\}^4 J_0(v\rho)\rho d\rho\right.\\
&\qquad + 4i^{2m}\cos 2m\psi \int_0^1 e^{-\frac{1}{2}iu\rho^2}\{R_n^m(\rho)\}^4 J_{2m}(v\rho)\rho d\rho\\
&\qquad\left. + i^{4m}\cos 4m\psi \int_0^1 e^{-\frac{1}{2}iu\rho^2}\{R_n^m(\rho)\}^4 J_{4m}(v\rho)\rho d\rho\right\}.
\end{aligned}\right\} \tag{8b}$$

NIJBOER found that where both u and α_{lnm} are of the order of unity, about four terms in the expansion (8a) suffice to give the intensity to within a few per cent.

To evaluate the integrals in (8b) we may proceed as follows. We express the factor $e^{-\frac{1}{2}iu\rho^2}$ in terms of the radial polynomials, with the help of the following well-known formula due to BAUER*

$$e^{iz\cos\phi} = \left(\frac{\pi}{2z}\right)^{\frac{1}{2}} \sum_{s=0}^{\infty} i^s(2s+1)J_{s+\frac{1}{2}}(z)P_s(\cos\phi), \tag{9}$$

where the P's are the LEGENDRE polynomials. If we set $\cos\phi = 2\rho^2 - 1$ and use the relation $P_s(2\rho^2 - 1) = R_{2s}^0(\rho)$ [§ 9.2 (8)], it follows that

$$e^{-\frac{1}{2}iu\rho^2} = e^{-\frac{1}{4}iu}e^{-\frac{1}{4}iu(2\rho^2-1)} = e^{-\frac{1}{4}iu}\sqrt{\frac{2\pi}{u}}\sum_{s=0}^{\infty}(-i)^s(2s+1)J_{s+\frac{1}{2}}(\tfrac{1}{4}u)R_{2s}^0(\rho). \tag{10}$$

On substitution from (10) into (8b) integrals are obtained each of which consists of a BESSEL function multiplied by a product of the radial polynomials. Now these integrals can be evaluated by the use of the formula § 9.2 (9)

$$\int_0^1 R_n^m(\rho)J_m(v\rho)\rho d\rho = (-1)^{\frac{n-m}{2}}\frac{J_{n+1}(v)}{v}, \tag{11}$$

---

* See, for example, G. N. WATSON, *A Treatise on the Theory of Bessel Functions* (Cambridge University Press, 2nd ed., 1944), p. 368.

provided each of the products of the radial polynomials is expressed as a linear combination of the form $\sum_p A_p R_p^m(\rho)$, with m equal to the order of the Bessel function by which the product is multiplied. It is not easy to obtain a general expression for the coefficients A_p, but if m and n are not too large such linear relations may easily be established with the help of Table XXI, p. 465, as will be seen later from some examples. For the discussion of methods relating to more general cases we refer the reader to NIJBOER's thesis.

In the special case of an aberration-free wave we obtain, on substituting from (10) into (7),

$$U(u, v, \psi) = 2C\, e^{-\frac{1}{4}iu} \sqrt{\frac{2\pi}{u}} \sum_{s=0}^{\infty} (-i)^s (2s+1) J_{s+\frac{1}{2}}(\tfrac{1}{4}u) \int_0^1 R_{2s}^0(\rho) J_0(v\rho)\rho d\rho$$

$$= 2C\, e^{-\frac{1}{4}iu} \sqrt{\frac{2\pi}{u}} \sum_{s=0}^{\infty} (i)^s (2s+1) J_{s+\frac{1}{2}}(\tfrac{1}{4}u) \frac{J_{2s+1}(v)}{v}, \tag{12}$$

where (11) was used. Though formally different, this expansion is equivalent to the series developments of LOMMEL, given in § 8.8.1.

From the expansion (8) some general properties of the diffraction image may immediately be deduced. It is seen that U remains unchanged when ψ is replaced by $\psi + 2\pi\mu/m$ ($\mu = 1, 2, \ldots m$); hence *the z-axis is an m-fold axis of symmetry.* Moreover *the planes through the z-axis which makes angles $\pi\mu/m$ with the plane $x = 0$ are planes of symmetry.* When $m = 0$, there is, of course, rotational symmetry.

Next consider the symmetry with respect to the plane $z = 0$. We observe that when u is replaced by $-u$, all integrals in (8) change into their conjugates. Now, if m is odd, the coefficients by which these integrals are multiplied are all real. In this case, therefore, $U(u, v, \psi)$ is changed into its conjugate, and, in consequence, the intensity remains unaltered. Hence, *if m is odd, the intensity distribution is symmetrical with respect to the plane $z = 0$.* If, on the other hand m is even (but $m \neq 0$), those coefficients which involve a factor $\cos 2\mu m\psi$ (μ being an integer) are real, whereas the others, which involve a factor $\cos (2\mu + 1)m\psi$, are pure imaginary. Therefore, if u is replaced by $-u$ and at the same time ψ is replaced by $\psi + \pi/m$, U now changes into its conjugate. Hence *if m is even but different from zero, the intensity at any point in the plane z = constant is the same as the intensity at the point resulting from reflection in the plane $z = 0$ and an additional rotation through an angle π/m about the z-axis.* It follows that whereas, as has already been pointed out, the z-axis is in general an m-fold axis of symmetry, *it is, when m is even, a $2m$-fold axis of symmetry with respect to the pattern in the plane $z = 0$.* When $m = 0$ (spherical aberration), the diffraction image is not symmetrical with respect to the plane $z = 0$.

Finally, we observe that *when the sign of the aberration constant α_{lnm} is changed, the intensity distribution is not altered if u is replaced by $-u$ when m is even, and if ψ is replaced by $\psi + \pi$ when m is odd.*

We now briefly consider the structure of the image when the system suffers from a primary aberration of a small amount.

9.4.1 Primary spherical aberration

In this case $l = m = 0$ and $n = 4$. The aberration function is independent of θ and the three-dimensional diffraction image has rotational symmetry about the principal direction $v = 0$.

According to (8) the expansion of the diffraction integral in powers of α is

$$U(u, v, \psi) = C[U_0 + i(\alpha_{040})U_1 + (i\alpha_{040})^2 U_2 + \ldots], \tag{13}$$

where U_0 represents the disturbance in the aberration-free image and U_1, U_2, . are given by the other expressions in (8b), with $m = 0$ and $n = 4$. In particular,

$$U_1 = 2\int_0^1 e^{-\frac{1}{2}iu\rho^2} R_4^0(\rho) J_0(v\rho)\rho d\rho. \tag{14}$$

A substitution for $e^{-\frac{1}{2}iu\rho^2}$ from (10) gives

$$U_1 = 2\, e^{-\frac{1}{4}iu}\sqrt{\frac{2\pi}{u}}\sum_{s=0}^{\infty}(-i)^s(2s+1)J_{s+\frac{1}{2}}(\tfrac{1}{4}u)\int_0^1 R_{2s}^0(\rho)R_4^0(\rho)J_0(v\rho)\rho d\rho. \tag{15}$$

To evaluate the integral on the right, the product of the radial polynomials will be replaced, as already explained, by a linear combination of radial polynomials whose upper index is equal to the order of the BESSEL function (zero in this case). The linear combination is easily found by using the relation § 9.2 (8)

$$R_{2s}^0(\rho) = P_s(2\rho^2 - 1), \tag{16}$$

and certain well known relations involving the LEGENDRE polynomials. We have

$$R_{2s}^0(\rho)R_4^0(\rho) = P_s(2\rho^2 - 1)P_2(2\rho^2 - 1). \tag{17}$$

Now $P_2(t) = \frac{1}{2}(3t^2 - 1)$ so that the right-hand side of (17) may be written as

$$P_s(t)P_2(t) = \tfrac{3}{2}t^2 P_s(t) - \tfrac{1}{2}P_s(t). \tag{18}$$

Applying several times the recurrence relation*

$$tP_s(t) = \frac{1}{2s+1}[(s+1)P_{s+1}(t) + sP_{s-1}(t)], \tag{19}$$

it follows that

$$P_2(t)P_s(t) = a_s P_{s+2}(t) + b_s P_s(t) + c_s P_{s-2}(t), \tag{20}$$

where

$$a_s = \frac{3}{2}\frac{(s+2)(s+1)}{(2s+3)(2s+1)}, \quad b_s = \frac{(s+1)s}{(2s+3)(2s-1)}, \quad c_s = \frac{3}{2}\frac{s(s-1)}{(2s+1)(2s-1)}. \tag{21}$$

Hence (17) becomes

$$R_{2s}^0(\rho)R_4^0(\rho) = a_s R_{2s+4}^0(\rho) + b_s R_{2s}^0(\rho) + c_s R_{2s-4}^0(\rho). \tag{22}$$

Finally, substituting from (22) into (15) and using (11), it follows that

$$U_1 = 2\, e^{-\frac{1}{4}iu}\sqrt{\frac{2\pi}{u}}\frac{1}{v}\sum_{s=0}^{\infty}(i)^s(2s+1)J_{s+\frac{1}{2}}(\tfrac{1}{4}u)[a_s J_{2s+5}(v) + b_s J_{2s+1}(v) + c_s J_{2s-3}(v)]. \tag{23}$$

In a similar way† one may obtain series developments for U_2, U_3, ... Using these expansions one may then calculate the intensity $I = |U|^2$ at a number of points

* See, for example, E. T. WHITTAKER and G. N. WATSON, *A Course of Modern Analysis* (Cambridge University Press, fourth ed. 1940), p. 308.

† For details see F. ZERNIKE and B. R. A. NIJBOER, contribution to *La Théorie des Images Optiques* (Paris, Revue d'Optique, 1949), p. 227.

in the region of the image and construct the isophotes (lines of equal intensity). In any plane at right angle to the principal direction $v = 0$ the isophotes are, of course, circles.

Fig. 9.3 shows the isophotes in a meridional plane for primary spherical aberration $\Phi = 0{\cdot}48\lambda\rho^4$, whilst in Figs. 9.4 and 9.5 photographs are reproduced which show the appearance of images in various receiving planes in the presence of spherical aberration of somewhat larger amounts.*

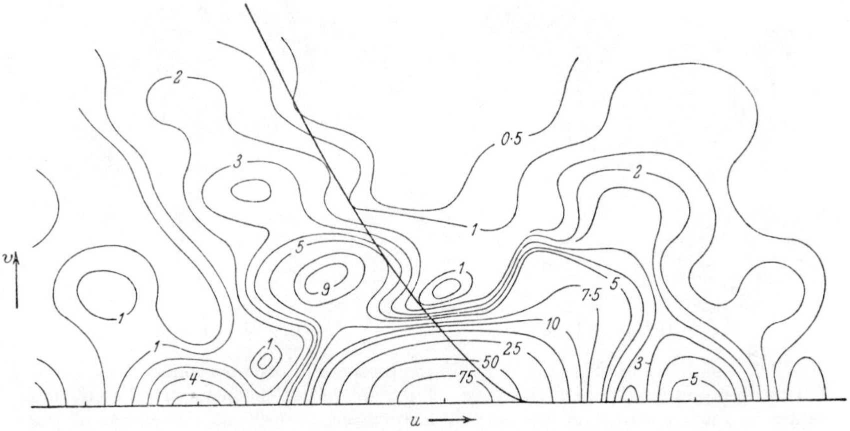

Fig. 9.3. Isophotes in a meridional plane, in presence of primary spherical aberration $\Phi = 0{\cdot}48\lambda\rho^4$. The thick line indicates the geometrical caustic. The intensity is normalized to 100 at the centre of the aberration-free image. STREHL intensity in best receiving plane: 0·95.

[After F. ZERNIKE and B. R. A. NIJBOER, contribution to *La Théorie des Images Optiques* (Paris, Revue d'Optique, 1949), p. 232.]

9.4.2 Primary coma

We now have $l = 0$, $n = 3$, $m = 1$. According to Table XXIII, p. 472, the diffraction focus is in the plane $z = 0$ and the disturbance in this plane is given by

$$U(0, v, \psi) = C[U_0(0, v, \psi) + (i\alpha_{031})U_1(0, v, \psi) + (i\alpha_{031})^2U_2(0, v, \psi) + \ldots]. \quad (24)$$

With $u = 0$, the integral U_0, defined in (8b), represents the disturbance $2J_1(v)/v$ in the focal plane of aberration-free system (AIRY pattern), and U_1 can immediately be evaluated by the use of (11). To evaluate $U_2, U_3 \ldots$ we must again express the products of the circle polynomials by the appropriate linear combination of such polynomials. In particular, it may be verified with the help of Table XXI, p. 465, that

$$(R_3^1)^2 = \tfrac{1}{4}R_0^0 + \tfrac{1}{20}R_2^0 + \tfrac{1}{4}R_4^0 + \tfrac{9}{20}R_6^0 = \tfrac{2}{5}R_2^2 + \tfrac{3}{5}R_6^2. \quad (25)$$

---

* The expansion (13) is unsuitable for computing the intensity when the aberrations are not small compared to the wavelength. Isophotes in a meridional plane for primary spherical aberration of several wavelengths were determined with the help of a mechanical integrator by A. MARÉCHAL, and are published in E. H. LINFOOT, *Recent Advances in Optics* (Oxford, Clarendon Press, 1955), pp. 60–61, and in M. FRANÇON's contribution to *Encyclopedia of Physics*, XXIV (Ed. S. FLÜGGE, Berlin, Springer, 1956), pp. 321–322; and by J. FOCKE, *Optica Acta*, **3** (1956), 110, who used asymptotic approximations. See also J. PICHT, *Ann. d. Physik*, (4), **77** (1925), 685.

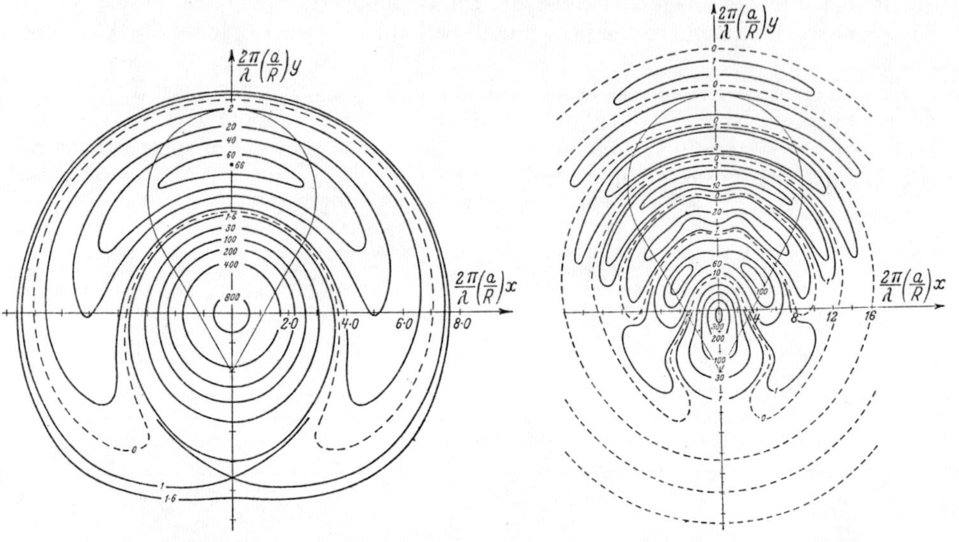

(a): $\Phi = 0{\cdot}48\lambda(\rho^3 - \frac{2}{3}\rho)\cos\theta$ (b): $\Phi = 1{\cdot}4\lambda(\rho^3 - \frac{2}{3}\rho)\cos\theta$

Fig. 9.6. Isophotes in the plane $z = 0$ in the presence of primary coma. The dotted curves represent lines of zero intensity. The boundary of the geometrical confusion figure is also shown. The intensity is normalized to 1000 at the centre of the aberration-free image. STREHL intensity: (a) 0·879; (b) 0·306.

(Fig. (a) after B. R. A. NIJBOER, Thesis, University of Groningen, 1942, p. 62; Fig. (b) after K. NIENHUIS and B. R. A. NIJBOER, *Physica* **14** (1949), 599.)

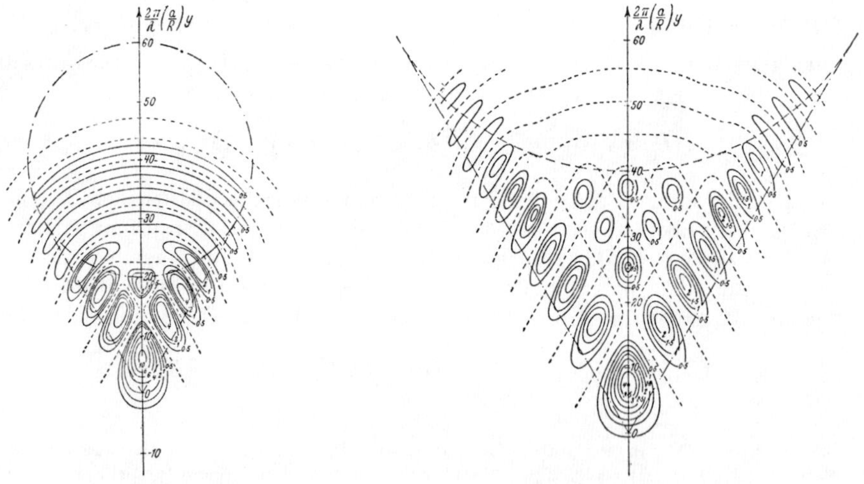

(a): $\Phi = 3{\cdot}2\lambda\rho^3\cos\theta$ (b): $\Phi = 6{\cdot}4\lambda\rho^3\cos\theta$

Fig. 9.7. Isophotes in the plane $z = 0$ in the presence of primary coma. The intensity is normalized to 100 at the centre of the aberration-free image.

(After R. KINGSLAKE, *Proc. Phys. Soc.* **61** (1948), 147.)

(a)

(b)

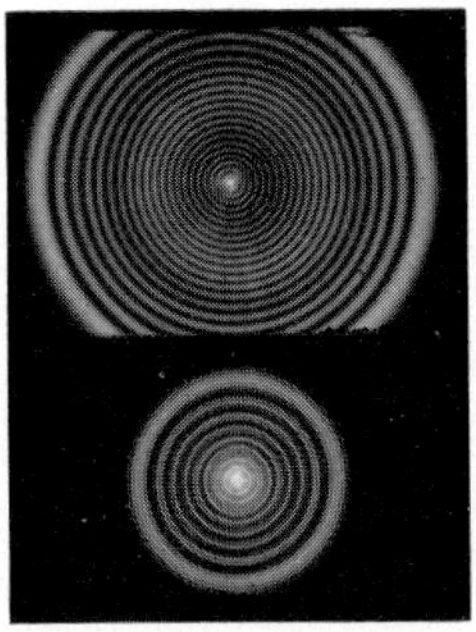

Fig. 9.4. Images in the marginal focal plane [(a)] and in the plane of the geometrical circle of least confusion [(b)] in the presence of primary spherical aberration $\Phi = 16\lambda\rho^4$.

(After K. Nienhuis, Thesis, University of Groningen, 1948, p. 56.)

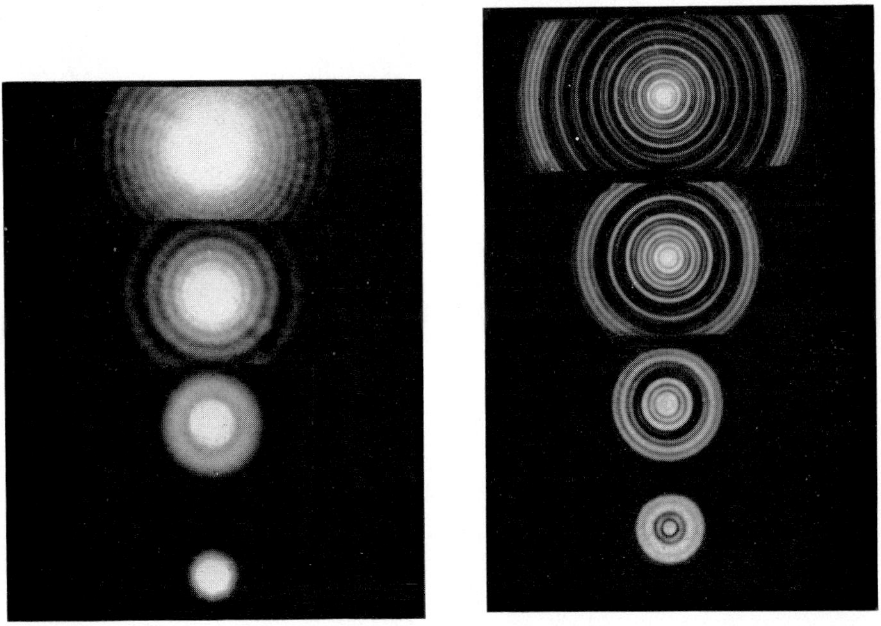

(a) (b)

Fig. 9.5. Images in the Gaussian focal plane [(a)], and in the plane of the geometrical circle of least confusion [(b)], in the presence of primary spherical aberration $\Phi = 17{\cdot}5\lambda\rho^4$, $8{\cdot}4\lambda\rho^4$, $3{\cdot}7\lambda\rho^4$, and $1{\cdot}4\lambda\rho^4$. [Scale of (b) is three times that of (a).]

(After K. Nienhuis, Thesis, University of Groningen, 1948, p. 56.)

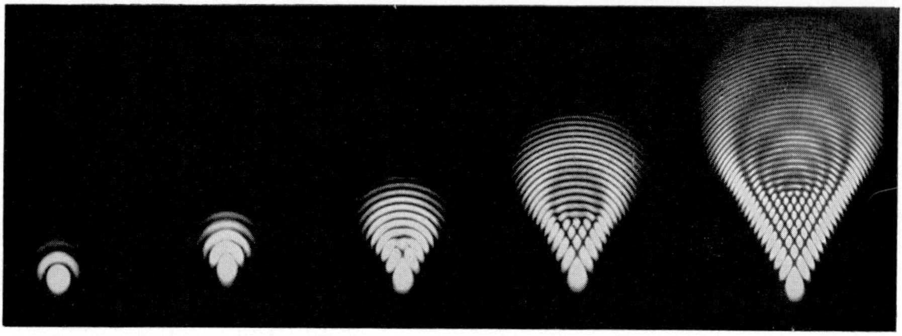

Fig. 9.8. Images in the Gaussian focal plane in the presence of coma $\Phi = 0{\cdot}3\lambda\rho^3 \cos\theta$, $\lambda\rho^3 \cos\theta$, $2{\cdot}4\lambda\rho^3 \cos\theta$, $5\lambda\rho^3 \cos\theta$, $10\lambda\rho^3 \cos\theta$.

(After K. NIENHUIS, Thesis, University of Groningen, 1948, p. 40.)

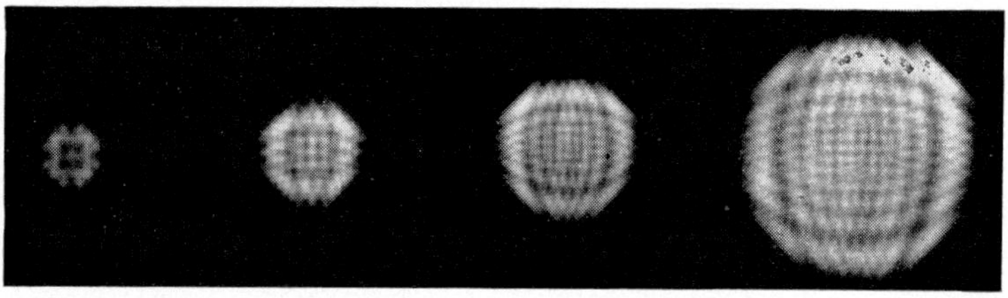

Fig. 9.10. Images in the central plane in the presence of primary astigmatism $\Phi = 1{\cdot}4\lambda\rho^2 \cos 2\theta$, $2{\cdot}7\lambda\rho^2 \cos 2\theta$, $3{\cdot}5\lambda\rho^2 \cos 2\theta$, $6{\cdot}5\lambda\rho^2 \cos 2\theta$.

(After K. NIENHUIS, Thesis, University of Groningen, 1948, p. 32.)

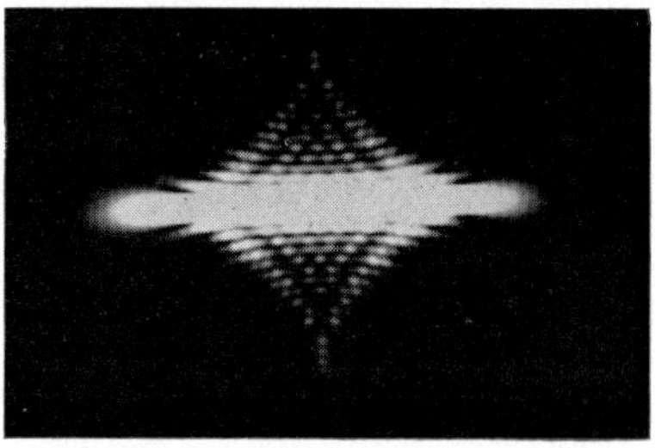

Fig. 9.11. Image in the plane containing a focal line, in the presence of primary astigmatism $\Phi = 2{\cdot}7\lambda\rho^2 \cos 2\theta$.

(After K. NIENHUIS, Thesis, University of Groningen, 1948, p. 32.)

Using these relations in the expression for U_2 in (8b) and applying (11) the integrals in U_2 are immediately evaluated and we have in all

$$\left.\begin{aligned} U_0(0, v, \psi) &= \frac{2J_1(v)}{v}, \\ U_1(0, v, \psi) &= i \cos \psi \frac{2J_4(v)}{v}, \\ U_2(0, v, \psi) &= \frac{1}{2v}\Big\{\tfrac{1}{4}J_1(v) - \tfrac{1}{20}J_3(v) + \tfrac{1}{4}J_5(v) - \tfrac{9}{20}J_7(v) \\ &\qquad - \cos 2\psi(\tfrac{2}{5}J_3(v) + \tfrac{3}{5}J_7(v))\Big\}. \end{aligned}\right\} \tag{26}$$

The isophotes for primary coma of various amounts are shown in Figs. 9.6 and 9.7. The data for Fig. 9.6 were computed from series expansions, those for Fig. 9.7 by numerical integrations. Photographs showing images affected by primary coma are given in Fig. 9.8. The figures show that when the aberration is of the order of a wavelength, the image resembles neither the Airy pattern nor the pattern predicted by geometrical optics. As the aberration is increased the true image soon becomes of the form specified by geometrical optics, but is broken up by a series of dark bands; these may be shown to arise from the interference of rays diffracted from diametrically opposite points of the aperture.

Fig. 9.6 also illustrates the general result established in § 9.2, that when a small aberration is represented in terms of a circle polynomial, the intensity pattern is displaced so as to have its maximum at the origin of the coordinates.

9.4.3 Primary astigmatism

The effect of a small amount of primary astigmatism may be investigated in a similar manner. We now have $l = 0$, $n = m = 2$, and as shown in § 9.3 the diffraction focus is midway between the two focal lines. We consider the light distribution in the *central plane*, i.e. the plane through this point, at right angles to the principal direction. When the aberration is represented in terms of the appropriate circle polynomial $[A_{022}R_2^2(\rho) \cos 2\theta]$ the central plane is the plane $u = 0$.

The disturbance in the central plane is given by

$$U(0, v, \psi) = C[U_0(0, v, \psi) + (i\alpha_{022})U_1(0, v, \psi) + (i\alpha_{022})^2U_2(0, v, \psi) \ldots], \tag{27}$$

where $U_0(0, v, \psi)$ represents as before the Airy pattern distribution and $U_1(0, v, \psi)$ may immediately be evaluated with the help of (11). To evaluate U_2 we use the identities

$$\left.\begin{aligned} (R_2^2)^2 &= \tfrac{1}{3}R_0^0 + \tfrac{1}{2}R_2^0 + \tfrac{1}{6}R_4^0, \\ &= R_4^4, \end{aligned}\right\} \tag{28}$$

and proceed as before. We obtain in all

$$\left.\begin{aligned} U_0(0, v, \psi) &= \frac{2J_1(v)}{v}, \\ U_1(0, v, \psi) &= -2 \cos 2\psi \frac{2J_3(v)}{v}, \\ U_2(0, v, \psi) &= \frac{1}{2v}\Big\{\tfrac{1}{3}J_1(v) - \tfrac{1}{2}J_3(v) + \tfrac{1}{6}J_5(v) + \cos 4\psi J_5(v)\Big\}. \end{aligned}\right\} \tag{29}$$

In Fig. 9.9 isophote-diagrams of astigmatic images are shown. Fig. 9.9 (a) was computed from the expansion (27), with terms up to and including the fourth power of α taken into account. Photographs of astigmatic images are given in Fig. 9.10 and 9.11. It is seen that when only a small amount of astigmatism is present, the isophotes in the central plane are circular near the centre but have a more complex form in the outer part of the image. When the astigmatism is increased, the image has a cushion-like appearance, and is crossed by interference fringes.

As regards the two remaining primary aberrations (curvature and distortion), we have already seen that they do not affect the structure of the three-dimensional

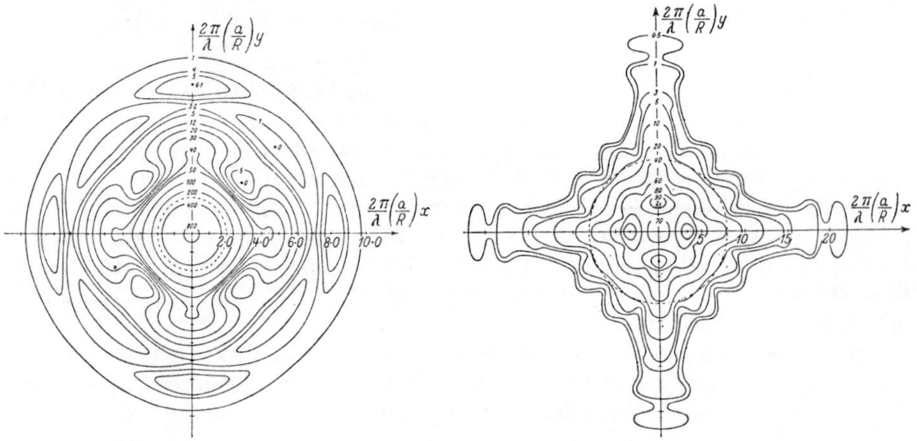

(a): $\Phi = 0{\cdot}16\lambda\rho^2 \cos 2\theta$ (b): $\Phi = 0{\cdot}64\lambda\rho^2 \cos 2\theta$

Fig. 9.9. Isophotes in the central plane in the presence of primary astigmatism. The dotted circles represent the boundary of the geometrical confusion figure. The intensity is normalized to 1000 at the centre of the aberration-free image. Strehl intensity: (a) 0·84; (b) 0·066.

(Fig. (a) after B. R. A. Nijboer, Thesis, University of Groningen, 1942, p. 55; Fig. (b) after K. Nienhuis, Thesis, University of Groningen, 1948, p. 13.)

image, but only the position of the diffraction focus. The isophote diagrams in the region of focus are therefore identical with that for an aberration-free image (Fig. 8.41), but are displaced relative to the Gaussian focus by the amounts indicated in Table XXIII, p. 472.

9.5. IMAGING OF EXTENDED OBJECTS

So far we have been concerned with images of point sources. We shall now describe some general methods, based on the techniques of Fourier transforms, relating to imaging of extended objects. These methods were developed chiefly by Duffieux,* partly in collaboration with Lansraux, and were later extended and applied to particular problems by many writers.†

We consider separately imaging with coherent and imaging with incoherent light.

* P. M. Duffieux, *L'Intégrale de Fourier et ses Applications à l'Optique* (Rennes, 1946); P. M. Duffieux and G. Lansraux, *Rev. d'Optique*, **24** (1945), 65, 151, 215.

† See, for example, A. Blanc-Lapierre, *Ann. de l'Inst. Henri Poincaré*, **13** (1953), 245; H. H. Hopkins, *Proc. Roy. Soc.*, A, **217** (1953), 408; *ibid.*, A, **231** (1955), 91; *Proc. Phys. Soc.*, B, **69** (1956), 562; K. Miyamoto, *Progress in Optics*, Vol. I, ed. E. Wolf (Amsterdam, North Holland Publishing Company and New York, J. Wiley and Sons, 1961), 41.

9.5.1 Coherent illumination

We shall specify points in the Gaussian image plane and in the plane of the exit pupil by Cartesian coordinates (x_1, y_1) and (ξ, η), respectively, referred to parallel axes with their origins at the axial points. Points in the object plane will conveniently be specified by scale normalized coordinates (x_0, y_0) which are such that if (X_0, Y_0) are the Cartesian coordinates of a typical object point and M the lateral magnification, then

$$x_0 = MX_0, \qquad y_0 = MY_0, \tag{1}$$

so that an object point and its Gaussian image point have now the same coordinate numbers.*

The imaging properties of the system may be characterized by means of a *transmission function* $K(x_0, y_0;\ x_1, y_1)$, defined as the complex amplitude, per unit area of the x_0y_0-plane, at the point (x_1, y_1) in the Gaussian image plane, due to a disturbance of unit amplitude and zero phase at the object point (x_0, y_0). The transmission function depends, of course, also on the wavelength λ of the light, but as we shall only be concerned with monochromatic light we need not consider this dependence.

Let $U_0(x_0, y_0)$ represent the complex disturbance in the plane of the object. The element at (x_0, y_0) makes a contribution $dU_1(x_1, y_1) = U_0(x_0, y_0)K(x_0, y_0;\ x_1, y_1)dx_0dy_0$ to the disturbance at the point (x_1, y_1) in the image plane. Hence the total disturbance at (x_1, y_1) is

$$U_1(x_1, y_1) = \int_{-\infty}^{+\infty}\int_{-\infty}^{+\infty} U_0(x_0, y_0)K(x_0, y_0;\ x_1, y_1)dx_0dy_0. \tag{2}$$

The integral extends only formally over an infinite domain, since U_0K is zero outside the area which does not send light into the image space of the system.

Now when we were dealing with point sources, we specified the properties of the system by the complex disturbance in the exit pupil, and this was characterized by the aberration function and by an amplitude factor, the latter being assumed to be constant in systems of moderate aperture. It is not difficult to find an expression for the transmission function in terms of these quantities. For this purpose we consider first the limiting form of (2) when the source reduces to a point source of unit strength and zero phase at the point $x_0 = x_0'$, $y_0 = y_0'$, i.e. when

$$U_0(x_0, y_0) = \delta(x_0 - x_0')\delta(y_0 - y_0'), \tag{3}$$

where δ is the Dirac delta function [cf. Appendix IV]. Then (2) gives

$$U_1(x_1, y_1) = K(x_0', y_0';\ x_1, y_1), \tag{4}$$

i.e. the transmission function K represents the disturbance due to the point source (3). Let us take the Gaussian reference sphere with centre at the Gaussian image point $x_1' = x_0'$, $y_1' = y_0'$. Let R be the radius of this reference sphere and let

$$H(x_0', y_0';\ \xi, \eta) = \frac{i}{\lambda} G(x_0', y_0';\ \xi, \eta)\frac{e^{-ikR}}{R} \tag{5}$$

be the disturbance at a typical point (ξ, η) on this sphere, due to the point source (3). Apart from an additive factor $\pi/2$, the phase of G is then the aberration function Φ

---

* (x_0, y_0) and (x_1, y_1) may be regarded as the Seidel variables of § 5.2 with the choice $l_1 = C = 1$ of the arbitrary constants.

of the system, whilst the amplitude of G is a measure of the non-uniformity in the amplitude of the image-forming wave. The factor i/λ has been introduced on the right-hand side of (5) to simplify later formulae. Now by the HUYGENS–FRESNEL principle, the disturbance in the image plane is related to the disturbance on the Gaussian reference sphere by the formula (small angles of diffraction assumed)

$$U_1(x_1, y_1) = -\frac{i}{\lambda}\iint H(x_0', y_0'; \xi, \eta)\,\frac{e^{iks}}{s}\,d\xi d\eta, \tag{6}$$

where s is the distance from the point (ξ, η) of this sphere to the point (x_1, y_1) in the Gaussian image plane, and the integral extends over the portion of the reference sphere that approximately fills the aperture. We also have according to equations (2) and (7) of § 8.8, with $x = x_1 - x_0'$, $y = y_1 - y_0'$, $z = 0$ and with R in place of f,

$$s \sim R - \frac{(x_1 - x_0')\xi + (y_1 - y_0')\eta}{R}. \tag{7}$$

From the formulae (4)–(7) we obtain

$$K(x_0, y_0;\ x_1, y_1) = \frac{1}{(\lambda R)^2}\int_{-\infty}^{+\infty}\int_{-\infty}^{+\infty} G(x_0, y_0;\ \xi, \eta)\, e^{-\frac{2\pi i}{\lambda R}[(x_1-x_0)\xi+(y_1-y_0)\eta]}\,d\xi d\eta, \tag{8}$$

G being taken to be zero at points (ξ, η) which are outside the opening. This is the required relation between the transmission function K and the "*pupil function*" G of the system.

Since K may be regarded as the disturbance in the image of a point source, it has, when considered as function of x_1, y_1, a fairly sharp maximum at or near the Gaussian image point $x_1 = x_0$, $y_1 = y_0$ and falls off rapidly, though as a rule not monotonically, with increasing distance from this point. In a well corrected system K will only be appreciable in an area whose size is of the order of the first dark ring of the AIRY pattern. Considered as function of (x_0, y_0), the transmission function varies slowly as this point explores the object surface. More precisely, the working field may be divided up into regions, each large compared to the finest detail that the system can resolve, with the property that in each such region A, K is to a good approximation a function of the displacement vector from the Gaussian image point, but not of the position of the image point itself. For example, in a well corrected system, $K(x_0, y_0;\ x_1, y_1)$ represents, apart from a constant factor, the AIRY pattern, centred on the Gaussian image point of (x_0, y_0). In such cases we may write

$$K(x_0, y_0;\ x_1, y_1) = K_A(x_1 - x_0, y_1 - y_0). \tag{9}$$

A region A with this property is said to be an *isoplanatic region* of the system. We shall restrict our discussion to objects that are so small that they fall within such an isoplanatic region.* In this case the equations (2) and (8) may be replaced by

$$U_1(x_1, y_1) = \int_{-\infty}^{+\infty}\int_{-\infty}^{+\infty} U_0(x_0, y_0)K(x_1 - x_0, y_1 - y_0)dx_0 dy_0, \tag{2a}$$

and

$$K(x_1 - x_0, y_1 - y_0) = \frac{1}{(\lambda R)^2}\int_{-\infty}^{+\infty}\int_{-\infty}^{+\infty} G(\xi, \eta)\, e^{-\frac{2\pi i}{\lambda R}[(x_1-x_0)\xi+(y_1-y_0)\eta]}\,d\xi d\eta, \tag{8a}$$

the function G now being independent of the object point.

* A thorough discussion of the conditions under which (9) holds has been published by P. DUMONTET, *Optica Acta*, **2** (1955), 53.

Let us represent U_0, U_1, and K as FOURIER integrals:

$$U_0(x_0, y_0) = \int_{-\infty}^{+\infty}\int_{-\infty}^{+\infty} \mathscr{U}_0(f, g)\, e^{-2\pi i[fx_0+gy_0]}\, dfdg, \tag{10a}$$

$$U_1(x_1, y_1) = \int_{-\infty}^{+\infty}\int_{-\infty}^{+\infty} \mathscr{U}_1(f, g)\, e^{-2\pi i[fx_1+gy_1]}\, dfdg, \tag{10b}$$

$$K(x, y) = \int_{-\infty}^{+\infty}\int_{-\infty}^{+\infty} \mathscr{K}(f, g)\, e^{-2\pi i[fx+gy]}\, dfdg. \tag{10c}$$

Then, by the FOURIER inversion formula

$$\mathscr{U}_0(f, g) = \int_{-\infty}^{+\infty}\int_{-\infty}^{+\infty} U_0(x_0, y_0)\, e^{2\pi i[fx_0+gy_0]}\, dx_0dy_0, \tag{11a}$$

$$\mathscr{U}_1(f, g) = \int_{-\infty}^{+\infty}\int_{-\infty}^{+\infty} U_1(x_1, y_1)\, e^{2\pi i[fx_1+gy_1]}\, dx_1dy_1, \tag{11b}$$

$$\mathscr{K}(f, g) = \int_{-\infty}^{+\infty}\int_{-\infty}^{+\infty} K(x, y)\, e^{2\pi i[fx+gy]}\, dxdy. \tag{11c}$$

According to equation (2a) U_1 is a convolution (also called resultant or Faltung) of U_0 and K; and on FOURIER inversion we obtain by the convolution theorem,* the simple relation

$$\mathscr{U}_1(f, g) = \mathscr{U}_0(f, g)\mathscr{K}(f, g). \tag{12}$$

This equation implies that if the disturbances in the object plane and in the image plane are each considered as a superposition of space-harmonic components of all possible "spatial frequencies" f, g, then each component of the image depends only on the corresponding component of the object, and the ratio of the components is $\mathscr{K}$. Thus the transition from the object to the image is equivalent to the action of a *linear filter*. Moreover, comparison of (10c) and (8a) shows that

$$\mathscr{K}\left(\frac{\xi}{\lambda R}, \frac{\eta}{\lambda R}\right) = G(\xi, \eta), \tag{13}$$

so that *the frequency response function* (also called *transmission factor*) $\mathscr{K}(f, g)$ *for coherent illumination is equal to the value of the pupil function* G *at the point*

$$\xi = \lambda R f, \qquad \eta = \lambda R g. \tag{14}$$

of the Gaussian reference sphere.

Since G is zero at points in the ξ, η-plane which are outside the boundary of the aperture, the spectral amplitudes belonging to frequencies above a certain value are not transmitted by the system. If the aperture is circular, and of radius a, then evidently frequency pairs such that

$$f^2 + g^2 > \left(\frac{a}{\lambda R}\right)^2 \tag{15}$$

are not transmitted. To illustrate this result consider a one-dimensional object the properties of which do not vary in the x direction. Let Δy_0 be the period belonging

---

* Cf. I. N. SNEDDON, *Fourier Transforms* (New York, McGraw-Hill, 1951), p. 23.

to the frequency g. Then by (15) the system can only transmit information about spectral components for which

$$\Delta y_0 = \frac{1}{g} > \frac{\lambda}{\sin\theta_1}, \tag{16}$$

where $\sin\theta_1 \sim a/R$ is the angular semi-aperture on the image side, assumed to be small. Now $y_0 = MY_0$, where M is the linear magnification, and if we assume that the system obeys the sine condition then (cf. § 4.5.1) $n_0 \sin\theta_0/n_1 \sin\theta_1 = M$, and (16) may be written as

$$\Delta Y_0 > \frac{\lambda_0}{n_0 \sin\theta_0}, \tag{17}$$

where $\lambda_0 = n_1\lambda$ is the vacuum wavelength and $n_0 \sin\theta_0$ is the numerical aperture of the system. Thus if the disturbance across the object plane varies sinusoidally with displacement, information about it can only be obtained in the image plane if the period exceeds the value given by the right-hand side of (17).

9.5.2 Incoherent illumination

We now consider the case where the light that emanates from the different elements of the object plane is incoherent, e.g. when the object is a primary source. Using the same coordinates as before, let $I_0(x_0, y_0)$ be the intensity at a typical point in the object plane. The intensity of the light that reaches the point (x_1, y_1) in the plane of the image from the element dx_0dy_0 centred on the object point (x_0, y_0) is $dI_1(x_1, y_1) = I_0(x_0, y_0)|K(x_0, y_0;\ x_1, y_1)|^2dx_0dy_0$, where K is again the transmission function of the system. Since the object is assumed to be incoherent, the intensities from the different elements of the object plane are additive, so that the total intensity at (x_1, y_1) is given by

$$I_1(x_1, y_1) = \int_{-\infty}^{+\infty}\int_{-\infty}^{+\infty} I_0(x_0, y_0)|K(x_0, y_0;\ x_1, y_1)|^2 dx_0dy_0. \tag{18}$$

If we again restrict ourselves to sufficiently small objects, we may replace (18) by*

$$I_1(x_1, y_1) = \int_{-\infty}^{+\infty}\int_{-\infty}^{+\infty} I_0(x_0, y_0)|K(x_1 - x_0, y_1 - y_0)|^2 dx_0dy_0. \tag{19}$$

Equation (19) shows that for imaging with incoherent illumination the intensity distribution in the image is a convolution of the intensity distribution in the object with the squared modulus of the transmission function. We represent these functions as Fourier integrals of the form (10), and denote by $\mathscr{I}_0(f, g)$, $\mathscr{I}_1(f, g)$, and $\mathscr{L}(f, g)$

---

* We note that to replace (18) by (19) it is not necessary that K should satisfy the full isoplanatic condition (9); it is sufficient that the modulus of K alone satisfies it, i.e. that throughout the region A occupied by the object, we have to a good approximation

$$|K(x_0, y_0;\ x_1, y_1)| = |K_A(x_1 - x_0, y_1 - y_0)|.$$

The analysis of Dumontet (*loc. cit.*) shows that as a rule this condition holds to a good approximation over a considerably larger region of the object plane than the relation (9). Hence the representation of optical imaging as a linear filter has a wider range of validity for incoherent than for coherent illumination. However, to be able to express the frequency response function in terms of the pupil function of the system by a relatively simple formula ((22) below), we restrict ourselves to objects that are so small that the full isoplanatic condition holds.

their "spatial spectra". Then, by the FOURIER inversion theorem, we have, in place of equations (11),

$$\mathscr{I}_0(f, g) = \int_{-\infty}^{+\infty}\int_{-\infty}^{+\infty} I_0(x_0, y_0)\, e^{2\pi i[fx_0+gy_0]}\, dx_0 dy_0, \tag{20a}$$

$$\mathscr{I}_1(f, g) = \int_{-\infty}^{+\infty}\int_{-\infty}^{+\infty} I_1(x_1, y_1)\, e^{2\pi i[fx_1+gy_1]}\, dx_1 dy_1, \tag{20b}$$

$$\mathscr{L}(f, g) = \int_{-\infty}^{+\infty}\int_{-\infty}^{+\infty} |K(x, y)|^2\, e^{2\pi i[fx+gy]}\, dxdy. \tag{20c}$$

From (19) we obtain by the convolution theorem

$$\mathscr{I}_1(f, g) = \mathscr{I}_0(f, g)\mathscr{L}(f, g). \tag{21}$$

Thus the transformation from the object to the image is again a *linear filter*, but it is now the spatial spectrum of the intensity and not of the complex amplitude that is transformed in this way. The frequency response function is the function $\mathscr{L}(f, g)$, and this, by (20c), (10c), and the convolution theorem, may be expressed in the form

$$\mathscr{L}(f, g) = \int_{-\infty}^{+\infty}\int_{-\infty}^{+\infty} \mathscr{K}(f' + f, g' + g)\, \mathscr{K}^\star(f', g')df'dg'. \tag{22}$$

The integral on the right-hand side of (22) is known as the *auto-correlation function* (of the function $\mathscr{K}$) and occurs in the analysis of many physical problems of statistical nature. We shall encounter it again later in connection with the theory of partial coherence.

We have already shown that $\mathscr{K}(f, g)$ is the value of the pupil function G at an appropriate point of the Gaussian reference sphere. If we substitute from (13) into (22) it follows that

$$\mathscr{L}\left(\frac{\xi}{\lambda R}, \frac{\eta}{\lambda R}\right) = \frac{1}{(\lambda R)^2}\int_{-\infty}^{+\infty}\int_{-\infty}^{+\infty} G(\xi' + \xi, \eta' + \eta)G^\star(\xi', \eta')d\xi' d\eta', \tag{23}$$

and we have thus established the important result that *apart from a constant factor, the frequency response function $\mathscr{L}(f, g)$ for incoherent illumination is the auto-correlation function of the pupil function of the system.*

Let $\mathscr{A}$ be the area of the exit pupil. Since the pupil function $G(\xi', \eta')$ is zero at points outside the boundary of the aperture, the domain of the $\xi'\eta'$-plane over which the integrand of (23) does not vanish is the area common to the aperture $\mathscr{A}$ and to an identical aperture displaced relative to $\mathscr{A}$ by a translation of amounts ξ and η in the negative ξ' and η' directions respectively (see Fig. 9.12). When ξ and η are so large that the two areas do not overlap, the value of the response function is evidently zero; thus, as in the coherent case, spatial frequencies only up to certain maximum values are transmitted by the system. In particular, for a circular aperture of radius a, the areas will have no region in common when $\xi^2 + \eta^2 \geqslant (2a)^2$, i.e. when

$$f^2 + g^2 > \left(\frac{2a}{\lambda R}\right)^2. \tag{24}$$

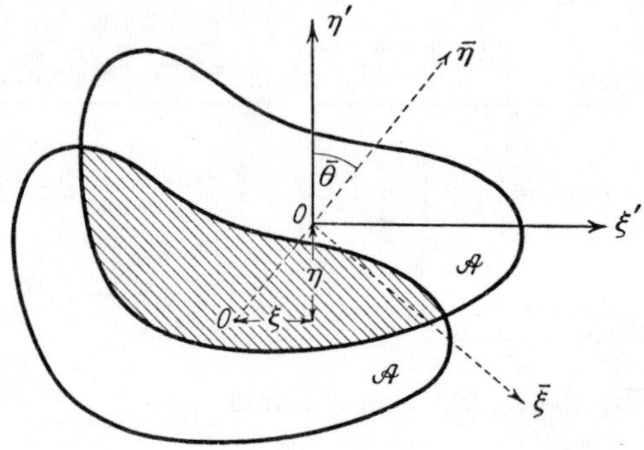

Fig. 9.12. Region of integration (shown shaded) relating to the evaluation of the response function $\mathscr{L}(f, g)$ for incoherent illumination, for the frequency pair $f = \xi/\lambda R$, $g = \eta/\lambda R$.

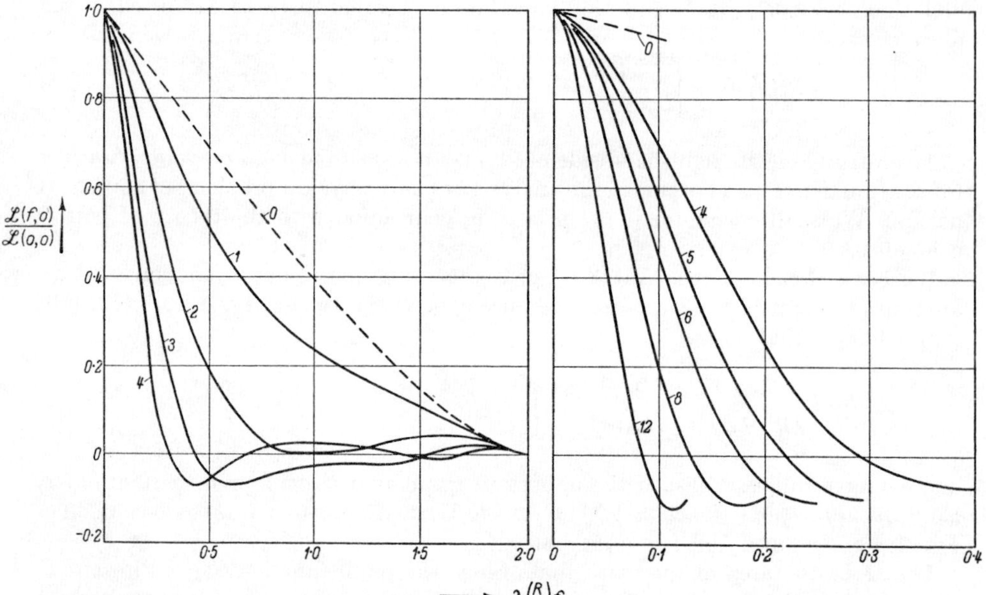

Fig. 9.13. The normalized frequency response curves for incoherent illumination of a system free of geometrical aberrations but suffering from defect of focus. $\Phi = \frac{m\lambda}{\pi}\rho^2$, $|G| = 1$. The number on each curve is the value of the parameter $m = \frac{\pi}{2\lambda}\left(\frac{a}{R}\right)^2 z$, z being the distance between the plane of observation and the Gaussian focal plane.

(After H. H. Hopkins, *Proc. Roy. Soc.*, A, **231** (1955), 98.)

By the same argument as that leading from equation (15) to (17), this implies that, with incoherent illumination, an aplanatic system can only transmit information about the spectral components whose period ΔY_0 is such that

$$\Delta Y_0 > \frac{0.5\lambda_0}{n_0 \sin \theta_0}. \tag{25}$$

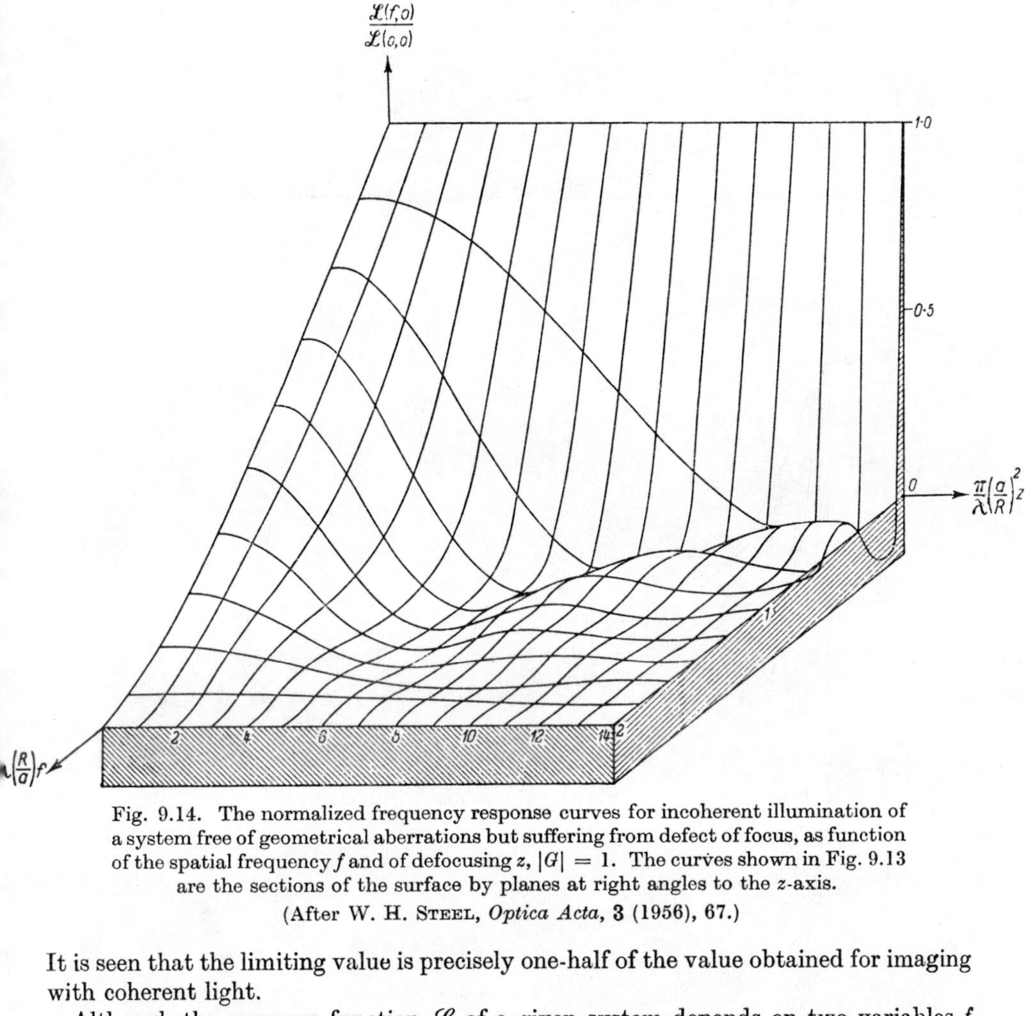

Fig. 9.14. The normalized frequency response curves for incoherent illumination of a system free of geometrical aberrations but suffering from defect of focus, as function of the spatial frequency f and of defocusing z, $|G| = 1$. The curves shown in Fig. 9.13 are the sections of the surface by planes at right angles to the z-axis.
(After W. H. Steel, *Optica Acta*, **3** (1956), 67.)

It is seen that the limiting value is precisely one-half of the value obtained for imaging with coherent light.

Although the response function $\mathscr{L}$ of a given system depends on two variables f and g, it is possible, in principle, to deduce all information about it from experiments involving one-dimensional test objects. To see this consider a frequency pair (f, g) and introduce polar coordinates such that $f = h \sin \theta$, $g = h \cos \theta$. Suppose now that the axes are rotated in their own plane through an angle $\bar{\theta}$ in the positive θ-direction. Then f and g transform into $\bar{f} = h \sin (\theta - \bar{\theta})$ and $\bar{g} = h \cos (\theta - \bar{\theta})$, but the value of $\mathscr{L}$ evidently remains unchanged. Now we may choose the angle of rotation $\bar{\theta}$ equal to $\tan^{-1} f/g$, which corresponds to taking the new η axis $(O\bar{\eta})$ along the line OO (see Fig. 9.12). Then $\bar{f} = 0$, and $\bar{g} = \sqrt{f^2 + g^2}$, and it follows that *the*

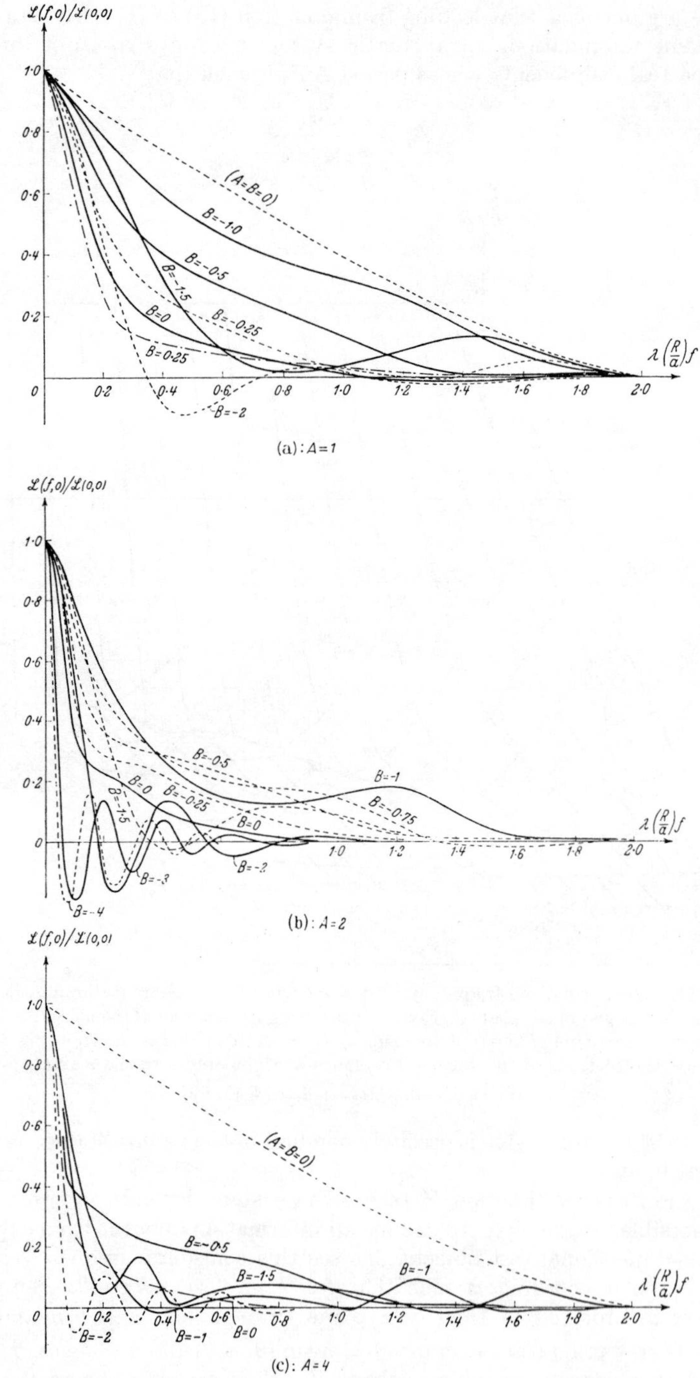

Fig. 9.15. The normalized frequency response curves for incoherent illumination, at selected focal settings of a system suffering from a small amount of primary spherical aberration $\Phi = A(\rho^4 + B\rho^2)\lambda$, $|G| = 1$. The value $B = 0$ corresponds to the paraxial focal plane and $B = -2$ corresponds to the receiving plane through the marginal focus.

(After G. Black and E. H. Linfoot, *Proc. Roy. Soc.*, A, **239** (1957), 522.)

value of the response function $\mathscr{L}$ of an optical system for the frequency pair (f, g), *is equal to that for a one-dimensional structure of frequency* $\sqrt{f^2 + g^2}$ *with its direction of periodicity inclined at the angle* $\tan^{-1} f/g$ *to the meridional plane* $\theta = 0$. This result

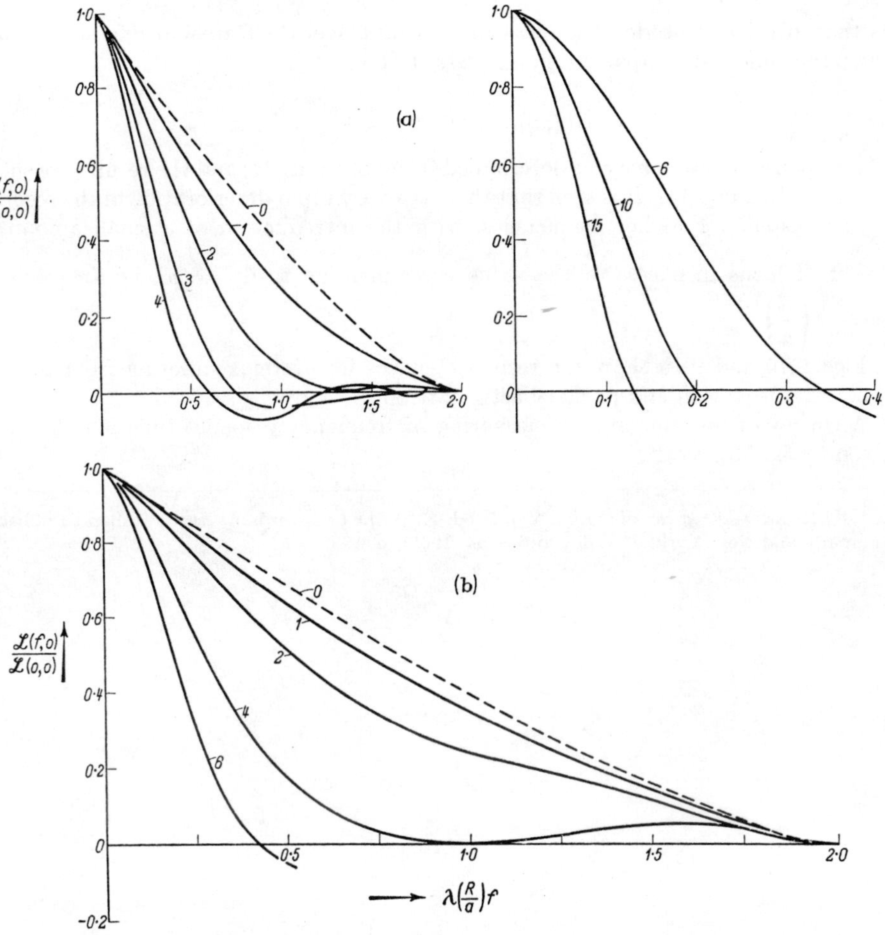

Fig. 9.16. The normalized frequency response curves for incoherent illumination, in the presence of primary astigmatism $\Phi = \dfrac{m\lambda}{\pi}\rho^2 \cos^2\theta$, $|G| = 1$, for receiving plane midway between the tangential and sagittal focal lines. Line periodic along the meridian $\theta = \pi/4$ [(a)], and along the meridians $\theta = 0$ or $\theta = \pi/2$ [(b)]. The number on each curve is the value of m.

(After M. De, *Proc. Roy. Soc.*, A, **233** (1955), 96.)

considerably simplifies the analytical evaluation of the response functions in any given case. A similar result evidently holds in connection with the response function $\mathscr{K}(f, g)$ of coherent objects but the result is of less immediate practical interest.

Let us now consider the frequency response of a centred system that is free of aberrations, but suffers from defect of focus. It follows from the discussion of § 9.1.2 that the displacement of the receiving plane by a small amount z in the positive

z-direction is formally equivalent to the introduction of a wave aberration of amount

$$\Phi(\xi, \eta) = \frac{1}{2}\left(\frac{a}{R}\right)^2 z\rho^2, \qquad \left(\rho^2 = \frac{\xi^2 + \eta^2}{a^2} \leqslant 1\right), \tag{26}$$

so that, if the amplitude of the wave is constant over the Gaussian reference sphere, the pupil function is, apart from a constant factor,

$$G(\xi, \eta) = e^{ik\Phi(\xi,\eta)} = e^{i\frac{1}{2}k\left(\frac{a}{R}\right)^2 z\rho^2}. \tag{27}$$

The response function may be determined from (23) and (27) and the results are shown in Figs. 9.13 and 9.14. It is seen that there is a very rapid deterioration of the response of the system for higher frequencies, with the introduction of a small amount of defect of focus in excess of the value corresponding to $\Phi = \frac{\lambda}{\pi}\rho^2$, i.e. in excess of $z = \frac{2\lambda}{\pi}\left(\frac{R}{a}\right)^2$.

Figs. 9.15 and 9.16 show the response curves for systems suffering from primary spherical aberration and primary astigmatism.

A survey of instruments for measuring of frequency response functions has been given by K. MURATA.*

---

* K. MURATA, *Progress in Optics*, Vol. 5, ed. E. WOLF (Amsterdam, North Holland Publishing Company and New York, J. Wiley and Sons, 1965), p. 199.

CHAPTER X

INTERFERENCE AND DIFFRACTION WITH PARTIALLY COHERENT LIGHT

10.1 INTRODUCTION

So far we have been mainly concerned with monochromatic light produced by a point source. Light from a real physical source is never strictly monochromatic, since even the sharpest spectral line has a finite width. Moreover, a physical source is not a point source, but has a finite extension, consisting of very many elementary radiators (atoms). The disturbance produced by such a source may be expressed, according to FOURIER's theorem, as the sum of strictly monochromatic and therefore infinitely long wave trains. The elementary monochromatic theory is essentially concerned with a single component of this FOURIER representation.

In a monochromatic wave field the amplitude of the vibrations at any point P is constant, while the phase varies linearly with time. This is no longer the case in a wave field produced by a real source: the amplitude and phase undergo irregular fluctuations, the rapidity of which depends essentially on the effective width $\Delta\nu$ of the spectrum. The complex amplitude remains substantially constant only during a time interval Δt which is small compared to the reciprocal of the effective spectral width $\Delta\nu$; in such a time interval the change of the relative phase of any two FOURIER components is much less than 2π and the addition of such components represents a disturbance which in this time interval behaves like a monochromatic wave with the mean frequency; however, this is not true for a longer time interval. The characteristic time $\Delta t = 1/\Delta\nu$ is of the order of the *coherence time* introduced in § 7.5.8.

Consider next the light disturbances at two points P_1 and P_2 in a wave field produced by an extended quasi-monochromatic source. For simplicity assume that the wave field is in vacuum and that P_1 and P_2 are many wavelengths away from the source. We may expect that, when P_1 and P_2 are close enough to each other, the fluctuations of the amplitudes at these points, and also the fluctuations of the phases, will not be independent. It is reasonable to suppose that, if P_1 and P_2 are so close to each other that the difference $\Delta\mathscr{S} = SP_1 - SP_2$ between the paths from each source point S is small compared to the mean wavelength $\bar{\lambda}$, then the fluctuations at P_1 and P_2 will effectively be the same; and that some correlation between the fluctuations will exist even for greater separations of P_1 and P_2, provided that for all source points the path difference $\Delta\mathscr{S}$ does not exceed the coherence length $c\Delta t \sim c/\Delta\nu = \bar{\lambda}^2/\Delta\lambda$. We are thus led to the concept of a *region of coherence* around any point P in a wave field.

In order to describe adequately a wave field produced by a finite polychromatic source it is evidently desirable to introduce some measure for the correlation that exists between the vibrations at different points P_1 and P_2 in the field. We must expect such a measure to be closely related to the sharpness of the interference fringes which would result on combining the vibrations from the two points. We must expect sharp fringes when the correlation is high (e.g. when the light at P_1 and P_2 comes from a very small source of a narrow spectral range), and no fringes at all in the absence of correlation (e.g. when P_1 and P_2 each receive light from a different physical source). We described these situations by the terms "coherent" and "incoherent" respectively. In general neither of these situations is realized and we may speak of vibrations which are *partially coherent*.

The first investigations which have a close bearing on the subject of partial coherence appear to be due to VERDET* who studied the size of the region of coherence for light from an extended primary source. Later researches by MICHELSON established the connection between the visibility of interference fringes and the intensity distribution on the surface of an extended primary source† (cf. § 7.3.6), and between the visibility and the energy distribution in a spectral line‡ (cf. § 7.5.8). MICHELSON's results were actually not interpreted in terms of correlations until much later, but his investigations have contributed to the formulation of the modern theories of partial coherence.§ The first quantitative measure of the correlation of light vibrations was introduced by VON LAUE,|| and employed in his researches on the thermodynamics of light beams. Further contributions were made by BEREK¶ who used the concept of correlation in investigations relating to image formation in the microscope.

A new stage in the development of the subject began with the publication of a paper by VAN CITTERT**, who determined the joint probability distribution for the light disturbances at any two points on a screen illuminated by an extended primary source. In a later paper†† he also determined the probability distribution for the light disturbances at any one point, but at two different instants of time. Within the accuracy of his calculations the distributions were found to be Gaussian, and he determined the appropriate correlation coefficients.

A different and simpler approach to problems of partial coherence was introduced by ZERNIKE in an important paper‡‡ published in 1938. ZERNIKE defined the "degree of coherence" of light vibrations in a manner that is directly related to experiment and established a number of valuable results relating to this quantity. While ZERNIKE's degree of coherence is for most practical purposes equivalent to the correlation factor of VAN CITTERT and is also closely related to that of VON LAUE, his methods seem particularly well suited for the treatment of practical problems of instrumental optics. The methods were simplified still further and applied to the study of image formation and resolving power by HOPKINS.§§

The investigations mentioned so far bridged the gap between two extreme cases, namely, that of complete coherence and complete incoherence, but the

---

* E. VERDET, *Ann. Scientif. l'École Normale Supérieure*, **2** (1865), 291; *Leçons d'Optique Physique* (Paris, L'Imprimerie Impériale), **1** (1869), 106.

† A. A. MICHELSON, *Phil. Mag.* (5), **30** (1890), 1; **31** (1891), 256; *Astrophys. J.*, **51** (1920), 257.

‡ A. A. MICHELSON, *Phil. Mag.* (5), **31** (1891), 338; **34** (1892), 280.

§ Cf. F. ZERNIKE, *Proc. Phys. Soc.*, **61** (1948), 158.

|| M. VON LAUE, *Ann. d. Physik* (4), **23** (1907), 1, 795.

¶ M. BEREK, *Z. Phys.*, **36** (1926), 675, 824; **37** (1926), 387; **40** (1926), 420. Experiments related to BEREK's investigations were described by C. LAKEMAN and J. TH. GROOSMULLER, *Physica* (Gravenhage), **8** (1928), 193, 199, 305.

** P. H. VAN CITTERT, *Physica*, **1** (1934), 201.

†† P. H. VAN CITTERT, *Physica*, **6** (1939), 1129; see also L. JÁNOSSY, *Nuovo Cimento*, **6** (1957), 111; *ibid.*, **12** (1959), 369.

‡‡ F. ZERNIKE, *Physica*, **5** (1938), 785.

§§ H. H. HOPKINS, *Proc. Roy. Soc.*, A, **208** (1951), 263; *ibid.* A, **217** (1953), 408.

Related investigations by D. GABOR and H. GAMO utilize the concept of partial coherence in the study of optical transmission from the standpoint of information theory. [D. GABOR, *Proc. Symp. Astr. Optics*, ed. Z. KOPAL (Amsterdam, North Holland Publishing Company, 1956), 17; *Proc. Third Symposium on Information Theory*, ed. C. CHERRY (London, Butterworths Scientific Publications, 1956), 26; H. GAMO, *J. Appl. Phys. Japan*, **25** (1956), 431; *Progress in Optics*, Vol. **3**, ed. E. WOLF (Amsterdam, North Holland Publishing Company and New York, J. Wiley and Sons, 1964), 187.

results obtained were still somewhat restricted in that they mainly applied to quasi-monochromatic light and to situations where the path differences between the interfering beams are sufficiently small. To deal with more complex situations and to formulate the theory on a rigorous basis, a further generalization was necessary. This was carried out by WOLF,* and independently by BLANC-LAPIERRE and DUMONTET† and involved the introduction of more general correlation functions. These correlation functions were found to obey rigorously two wave equations; a result which implies that not only the optical disturbance but also the correlation between disturbances is propagated in the form of waves. In the light of this result many of the theorems established previously obtained a relatively simple interpretation.

The correlation functions mentioned so far characterize the correlation between the light vibrations at *two* space-time points. Such "second order" correlation functions are entirely adequate for the analysis of the usual optical experiments involving interference and diffraction of light from steady sources.‡ For the analysis of more sophisticated experiments, higher order correlation functions—i.e. correlation functions involving the field variables in total power higher than the second—may be needed. However, so far such correlation functions have not been used to any appreciable extent.§

An attractive feature of the theory of partial coherence is the fact that it operates with quantities (namely with correlation functions and with time averaged intensities) that may, in principle, be determined from experiment. This is in contrast with the elementary optical wave theory, where the basic quantity is not measurable because of the very great rapidity of optical vibrations. In the present chapter we shall study the properties of partially coherent wave fields and we shall illustrate the results by a

---

* E. WOLF, *Proc. Roy. Soc.*, A, **230** (1955), 246. See also *ibid.*, A, **225** (1954), 96; *Nuovo Cimento*, **12** (1954), 884.

† A. BLANC-LAPIERRE and P. DUMONTET, *Rev. d'Optique*, **34** (1955), 1.

‡ More precisely from sources which give rise to a stationary field in the sense of the definition given on p. 498 below. For the analysis of non-stationary fields similar correlation functions may be employed, but they must be defined in terms of ensemble averages rather than time averages. (For a stationary field the two types of averaging will usually give the same result.) However, since the analysis of non-stationary fields requires a considerably more involved mathematical apparatus and since no experiments relating to coherence effects with non-stationary light have so far been carried out, we restrict our discussion to stationary fields only.

§ When the light originates in a thermal source, such as incandescent matter or a gas discharge, one may assume that the joint probability distribution of the field at n space-time points is, to a good approximation, Gaussian. As is well known [see, for example J. J. FREEMAN, *Principles of Noise*, New York: J. Wiley & Sons, Inc., 1958, 245–247], such distributions are completely specified by second order correlation functions, which in turn implies that all the higher order correlation functions associated with thermal light may be expressed in terms of the correlation functions of the second order. This, however, is not so for light from non-thermal sources, such as an optical maser.

Some higher order coherence effects and the appropriate correlation functions were briefly discussed by E. WOLF in *Quantum Electronics, 3rd Congress*, Vol. 1, ed. N. BLOEMBERGEN and P. GRIVET (New York, Columbia University Press; Paris, Dunod, 1964), p. 13, and L. MANDEL, *ibid.*, p. 101. See also E. WOLF, *Proc. Symp. on Optical Masers* (New York, Brooklyn Polytechnic Press and J. Wiley & Sons, Inc., 1963), p. 29, where a systematic classification of coherence effects is outlined.

Analogous quantum mechanical correlation functions were introduced by R. J. GLAUBER, in *Quantum Electronics, 3rd Congress*, Vol. 1, ed. N. BLOMBERGEN and P. GRIVET (New York, Columbia University Press; Paris, Dunod, 1964), p. 111 and *Phys. Rev.*, **130** (1963), 2529. The relation between the classical and quantum treatments is discussed by E. C. G. SUDARSHAN, *Phys. Rev. Lett.*, **10** (1963), 277 and by J. R. KLAUDER and E. C. G. SUDARSHAN, *Fundamentals of Quantum Optics* (New York, W. A. Benjamin, Inc., 1968). *See also* L. MANDEL and E. WOLF, *Rev. Mod. Phys.*, **37** (1965), 231; this article also contains a review of second and higher order coherence effects.

number of examples of practical interest. We shall only be concerned with the case of light, but our analysis has a close bearing on other fields; in particular, similar considerations apply in connection with correlation techniques for measurements of radio stars* and for the exploration of the ionosphere by radio waves.†

The mathematical techniques employed in connection with partial coherence are also very suitable for the analysis of partial polarization. Here one is concerned with phenomena which can be interpreted in terms of correlation between orthogonal components of the electromagnetic field vectors. Early investigations in this direction are due to G. G. STOKES.‡ Modern treatments which employ the concepts of correlation functions and correlation matrices are chiefly due to WIENER,§ PERRIN,‖ WOLF¶ and PANCHARATNAM.** This topic will be discussed in the concluding section (§ 10.8) of this chapter.

10.2 A COMPLEX REPRESENTATION OF REAL POLYCHROMATIC FIELDS

In discussing monochromatic wave fields we have found it useful to regard each real wave function as the real part of an associated complex wave function. In the present chapter we shall be concerned with polychromatic (i.e. non-monochromatic) fields. It will again be useful to employ a complex representation, which may be regarded as a natural generalization of that used with monochromatic fields.

Let $V^{(r)}(t)$ $(-\infty \leqslant t \leqslant \infty)$ represent a real disturbance, for example, a Cartesian component of the electric vector, at a fixed point in space, and assume that $V^{(r)}(t)$ is square integrable. It may be expressed in the form of a FOURIER integral

$$V^{(r)}(t) = \int_0^\infty a(\nu) \cos [\phi(\nu) - 2\pi\nu t] d\nu. \tag{1}$$

With $V^{(r)}$ we associate the complex function

$$V(t) = \int_0^\infty a(\nu) e^{i[\phi(\nu) - 2\pi\nu t]}\, d\nu. \tag{2}$$

Then

$$V(t) = V^{(r)}(t) + iV^{(i)}(t), \tag{3}$$

where

$$V^{(i)}(t) = \int_0^\infty a(\nu) \sin [\phi(\nu) - 2\pi\nu t] d\nu. \tag{4}$$

The functions $V^{(i)}(t)$ and $V(t)$ are uniquely specified by $V^{(r)}(t)$, $V^{(i)}$ being obtained from $V^{(r)}$ by replacing the phase $\phi(\nu)$ of each FOURIER component by $\phi(\nu) - \pi/2$. The

* Cf. R. N. BRACEWELL, *Radio Astronomy Techniques* in *Encyclopedia of Physics*, ed. S. FLÜGGE (Berlin, Springer), **54** (1959), Chapter V.

† Cf. J. A. RATCLIFFE, *Rep. Progr. Phys.* (London, Physical Society), **19** (1956), 188.

‡ G. G. STOKES, *Trans. Cambr. Phil. Soc.*, **9** (1852), 399; reprinted in his *Mathematical and Physical Papers*, Vol. III (Cambridge University Press, 1901), 233; see also P. SOLEILLET, *Ann. de Physique* (10), **12** (1929), 23.

§ N. WIENER, *J. Math. and Phys.*, **7** (1928), 109; *J. Franklin Inst.*, **207** (1929), 525; *Acta Math.*, **55** (1930), § 9.

‖ F. PERRIN, *J. Chem. Phys.*, **10** (1942), 415.

¶ E. WOLF, *Nuovo Cimento*, **12** (1954), 884; *Proc. Symp. Astr. Optics*, ed. Z. KOPAL (Amsterdam, North Holland Publishing Company, 1956), 177; *Nuovo Cimento*, **13** (1959), 1165.

** S. PANCHARATNAM, *Proc. Ind. Acad. Sci.*, A, **44** (1956), 398; *ibid.*, **57** (1963), 218, 231.

integrals (1) and (4) are said to be *allied Fourier integrals*, or *associated functions* (also called *conjugate functions*) and may be shown* to be HILBERT transforms of each other, i.e.

$$V^{(i)}(t) = \frac{1}{\pi} P \int_{-\infty}^{\infty} \frac{V^{(r)}(t')}{t'-t} dt', \qquad V^{(r)}(t) = -\frac{1}{\pi} P \int_{-\infty}^{\infty} \frac{V^{(i)}(t')}{t'-t} dt', \tag{5}$$

where P denotes the CAUCHY principal value at $t' = t$.

This complex representation is used frequently in communication theory, where V is called the *analytic signal*† belonging to $V^{(r)}$. The name derives from the fact that, provided $V^{(r)}$ satisfies certain general regularity conditions, the function $V(z)$, considered as a function of a complex variable z is analytic in the lower half of the z-plane.‡

For future use we note the transition from $V^{(r)}$ to V when $V^{(r)}$ is represented as a FOURIER integral of the form

$$V^{(r)}(t) = \int_{-\infty}^{\infty} v(\nu) e^{-2\pi i \nu t} \, d\nu. \tag{6}$$

Since $V^{(r)}$ is real,

$$v(-\nu) = v^{\star}(\nu). \tag{7}$$

Using (7), we may re-write (6) in the form (1) and we obtain on comparison

$$v(\nu) = \tfrac{1}{2} a(\nu) e^{i\phi(\nu)}, \qquad \nu \geqslant 0, \tag{8}$$

In terms of v, (2) becomes

$$V(t) = 2 \int_{0}^{\infty} v(\nu) e^{-2\pi i \nu t} \, d\nu. \tag{9}$$

Hence $V(t)$ may be derived from $V^{(r)}(t)$ by representing $V^{(r)}$ as a FOURIER integral of the form (6), suppressing the amplitudes belonging to the negative frequencies, and multiplying the amplitudes of the positive frequencies by two. For this reason V is also called *the complex half-range function* associated with $V^{(r)}$. Conversely it is evident that if the FOURIER spectrum of a complex function V contains no amplitudes belonging to negative frequencies, then the real and imaginary parts of V are associated

---

* See for example E. C. TITCHMARSH, *Introduction to the Theory of Fourier Integrals* (Oxford, Clarendon Press, 2nd ed., 1948), Chapter 5.

† The concept of an analytic signal was introduced by D. GABOR, *J. Inst. Electr. Engrs.*, **93** (1946), Pt. III, 429. See also V. I. BUNIMOVICH, *J. Tech. Phys. U.S.S.R.*, **19** (1949), 1231; J. VILLE, *Câbles et Transmission*, **2** (1948), 61; *ibid.*, **4** (1950), 9; J. R. OSWALD, *Trans. Inst. Radio Engrs.*, CT–**3** (1956), 244.

Complex functions of a real variable whose real and imaginary parts are connected by the HILBERT transform relations play an important role in many branches of physics and engineering. In Physics, the HILBERT transform relations are often called *dispersion relations* as they made their first appearance in the theory of dispersion of light by atoms [H. A. KRAMERS, *Atti Congr. Internaz. Fisici, Como* (Sept. 1927), (V), Bologna, N. Zanichelli, 1928; see also J. S. TOLL, *Phys. Rev.*, **104** (1956), 1760; J. HILGEVOORD, *Dispersion Relations and Causal Description* (Amsterdam, North Holland Publishing Company, 1960)].

‡ Cf. E. C. TITCHMARSH, *loc. cit.*, p. 128.

functions. We note the following relations which follow from (6), (7), and (9), by PARSEVAL's theorem and by the use of the relation (3):

$$\int_{-\infty}^{\infty} V^{(r)^2}(t)dt = \int_{-\infty}^{\infty} V^{(i)^2}(t)dt = \tfrac{1}{2}\int_{-\infty}^{\infty} V(t)V^{\star}(t)dt = \int_{-\infty}^{\infty} |v(\nu)|^2 d\nu = 2\int_{0}^{\infty} |v(\nu)|^2 d\nu. \quad (10)$$

In most of the applications with which we shall be concerned the spectral amplitudes will only have appreciable values in a frequency interval of width $\Delta\nu$ which is small compared to the mean frequency $\bar{\nu}$. The analytic signal then has a simple interpretation. We express V in the form

$$V(t) = A(t)e^{i[\Phi(t)-2\pi\bar{\nu}t]}, \quad (11)$$

where A ($\geqslant 0$) and Φ are real. According to (9) and (11),

$$A(t)e^{i\Phi(t)} = 2\int_{0}^{\infty} v(\nu)e^{-2\pi i(\nu-\bar{\nu})t}\, d\nu$$

$$= \int_{-\bar{\nu}}^{\infty} g(\mu)e^{-2\pi i\mu t}\, d\mu, \quad (12)$$

where

$$g(\mu) = 2v(\bar{\nu}+\mu). \quad (13)$$

Now since the spectral amplitudes were assumed to differ appreciably from zero only in the neighbourhood of $\nu = \bar{\nu}$, $|g(\mu)|$ will be appreciable only near $\mu = 0$. Hence the integral (12) represents a superposition of harmonic components of low frequencies, and since $\Delta\nu/\bar{\nu} \ll 1$, $A(t)$ and $\Phi(t)$ will vary slowly* in comparison with $\cos 2\pi\bar{\nu}t$ and $\sin 2\pi\bar{\nu}t$. Since $V^{(r)}$ and $V^{(i)}$ are the real and imaginary parts of V, we have, in terms of A and Φ,

$$\left.\begin{aligned} V^{(r)}(t) &= A(t)\cos[\Phi(t) - 2\pi\bar{\nu}t], \\ V^{(i)}(t) &= A(t)\sin[\Phi(t) - 2\pi\bar{\nu}t]. \end{aligned}\right\} \quad (14)$$

These formulae express $V^{(r)}$ and $V^{(i)}$ in the form of modulated signals of carrier frequency $\bar{\nu}$, and we see that the complex analytic signal is intimately connected with the *envelope* of the real signal†. In terms of the analytic signal V, the envelope $A(t)$ and the associated phase factor $\Phi(t)$ are given by

$$\left.\begin{aligned} A(t) &= \sqrt{V^{(r)^2} + V^{(i)^2}} = \sqrt{VV^{\star}} = |V|, \\ \Phi(t) &= 2\pi\bar{\nu}t + \tan^{-1}\frac{V^{(i)}}{V^{(r)}} = 2\pi\bar{\nu}t + \tan^{-1}\left(i\,\frac{V^{\star}-V}{V^{\star}+V}\right). \end{aligned}\right\} \quad (15)$$

We see that $A(t)$ is independent of the exact choice of $\bar{\nu}$, and that $\Phi(t)$ depends on $\bar{\nu}$ only through the additive term $2\pi\bar{\nu}t$. Evidently we could have chosen in (14) any other frequency $\bar{\nu}'$ in place of $\bar{\nu}$ without affecting the value of A; the expression for the new phase factor would differ from that given by (15) only by having $\bar{\nu}'$ written in place of $\bar{\nu}$.

In deriving (14) and (15) we have not made use of the fact that we are dealing with

* Under these circumstances one evidently has, according to (14)

$$V^{(i)}(t) \sim V^{(r)}\left(t + \frac{1}{4\bar{\nu}}\right).$$

† The envelope properties of analytic signals were studied by L. MANDEL, *J. Opt. Soc. Amer.*, **57** (1967), 613.

a narrow-band signal ($\Delta\nu/\bar{\nu} \ll 1$), so that these relations hold quite generally. However, it is only when $\Delta\nu/\bar{\nu} \ll 1$ that the concept of the envelope is useful.

We have assumed that the "disturbance" $V^{(r)}(t)$ is defined for all values of t. In practice the disturbance will exist only during a finite time interval $-T \leqslant t \leqslant T$, but this interval is as a rule so large compared to the physically significant time scales (the mean period $1/\bar{\nu}$ and the coherence time $1/\Delta\nu$) that we may idealize the situation by assuming $T \to \infty$. This idealization is mathematically desirable for reasons connected with the assumption of stationarity of the field (cf. § 10.3.1). Evidently it is then also necessary to assume that the time average of the intensity (which is proportional to $V^{(r)^2}$) tends to a finite value as the averaging interval is indefinitely increased, i.e. that

$$\lim_{T\to\infty} \frac{1}{2T}\int_{-T}^{T} V^{(r)^2}(t)dt \tag{16}$$

is finite. Now if this limit is finite and not zero then obviously $\int_{-\infty}^{\infty} V^{(r)^2}(t)dt$ diverges. Nevertheless, it is possible to utilize the techniques of FOURIER analysis.* We define the truncated functions

$$\left.\begin{aligned} V_T^{(r)}(t) &= V^{(r)}(t) \quad \text{when} \quad |t| \leqslant T,\\ &= 0 \qquad\quad\ \text{when} \quad |t| > T. \end{aligned}\right\} \tag{17}$$

Since each such truncated function may be assumed to be square integrable it may be expressed as a FOURIER integral, say

$$V_T^{(r)}(t) = \int_{-\infty}^{\infty} v_T(\nu)e^{-2\pi i\nu t}\,d\nu. \tag{18a}$$

Let $V_T^{(i)}$ be the associated function and V_T the corresponding analytic signal, i.e.

$$V_T(t) = V_T^{(r)}(t) + iV_T^{(i)}(t) = 2\int_0^{\infty} v_T(\nu)e^{-2\pi i\nu t}\,d\nu. \tag{18b}$$

Then the relations (10) hold with $V^{(r)}$ replaced by $V_T^{(r)}$, etc. Hence if we also divide each expression by $2T$, we obtain†

$$\begin{aligned} \frac{1}{2T}\int_{-\infty}^{\infty} V_T^{(r)^2}(t)\,dt &= \frac{1}{2T}\int_{-\infty}^{\infty} V_T^{(i)^2}(t)\,dt = \frac{1}{2}\frac{1}{2T}\int_{-\infty}^{\infty} V_T(t)\,V_T^*(t)\,dt \\ &= \int_{-\infty}^{\infty} G_T(\nu)\,d\nu = 2\int_0^{\infty} G_T(\nu)\,d\nu, \end{aligned} \tag{19}$$

* The problem of analysing functions of time which do not die down as t tends to infinity was encountered at the turn of the twentieth century by physicists who concerned themselves with the study of the nature of white light and noise (notably L. G. GOUY, Lord RAYLEIGH, and A. SCHUSTER). Rigorous mathematical techniques were developed chiefly by N. WIENER in his paper on generalized harmonic analysis (*Acta Math.*, **55** (1930), 117). This paper also outlines the history of the problem and includes a very full bibliography.

† Since the HILBERT transform of a truncated function is not necessarily a truncated function, $V_T^{(i)}$ and V_T do not, in general, vanish outside the range $-T \leqslant t \leqslant T$. For this reason, and also to avoid certain mathematical refinements, the limits of the time integrations in (19) and (23) are taken as $\pm\infty$ rather than $\pm T$.

where

$$G_T(\nu) = \frac{|v_T(\nu)|^2}{2T}. \tag{20}$$

It would seem natural now to proceed to the limit $T \to \infty$. Unfortunately in many cases of practical interest, the function $G_T(\nu)$, known as the *periodogram*, does not tend to a limit but fluctuates* with increasing T. However, one may overcome this difficulty by an appropriate "smoothing procedure". For example, as customary in the theory of random processes, one regards the function $V^{(r)}(t)$ to be a typical member of an ensemble of functions which characterize the statistical properties of the process. Moreover the ensembles that one normally encounters in optics are *stationary* and *ergodic*. Stationarity implies that all the ensemble averages are independent of the origin of time, whilst ergodicity implies that each ensemble average is equal to the corresponding time average involving a typical member of the ensemble. We will from now on assume that we are dealing with a stationary ergodic ensemble.† One may then show that the average of $G_T(\nu)$ taken over the ensemble of the functions $V^{(r)}(t)$ tends to a definite limit as $T \to \infty$. Thus if bar denotes the ensemble average, the "smoothed periodogram"

$$\overline{G_T(\nu)} = \frac{\overline{|v_T(\nu)|^2}}{2T} \tag{21}$$

will possess a limit‡

$$G(\nu) = \lim_{T\to\infty} \overline{G_T(\nu)} = \lim_{T\to\infty} \frac{\overline{|v_T(\nu)|^2}}{2T}. \tag{22}$$

Now if sharp brackets denote the time average,

$$\langle F(t)\rangle = \lim_{T\to\infty} \frac{1}{2T} \int_{-\infty}^{\infty} F_T(t)\, dt, \tag{23}$$

then one obtains, in the limit as $T \to \infty$, the following relations analogous to (19):

$$\langle V^{(r)^2}(t)\rangle = \langle V^{(i)^2}(t)\rangle = \tfrac{1}{2}\langle V(t)V^\star(t)\rangle = \int_{-\infty}^{\infty} G(\nu)\,d\nu = 2\int_{0}^{\infty} G(\nu)\,d\nu. \tag{24}$$

In the theory of stationary random processes, the function $G(\nu)$, defined by (22) is called the *power spectrum* of the random process, characterized by the ensemble of the functions $V^{(r)}(t)$. In our considerations, where $V^{(r)}(t)$ represents the light disturbance,

---

* See, for example, W. B. DAVENPORT and W. L. ROOT, *An Introduction to the Theory of Random Signals and Noise* (New York, McGraw–Hill, 1958), p. 107–108. See also D. MIDDLETON, *IRE Trans.*, CT-3 (1956), 299.

† For a fuller discussion of these concepts, see, for example, W. B. DAVENPORT and W. L. ROOT, *loc. cit.*; S. GOLDMAN, *Information Theory* (New York, Prentice-Hall, Inc. 1953); D. MIDDLETON, *An Introduction to Statistical Communication Theory* (New York, McGraw-Hill Co., 1960); A. M. YAGLOM, *An Introduction to the Theory of Stationary Random Functions* (Englewood Cliffs, N. J., Prentice-Hall, 1962).

‡ A rigorous proof of the existence of this limit and of some of the relations introduced heuristically in this section would lead us far into ergodic theory and cannot, therefore, be given here. Cf. J. L. DOOB, *Stochastic Processes* (New York, J. Wiley & Sons, Inc., 1953), Chapt. XI; see also S. GOLDMAN, *loc. cit.*, § 8.4, A. M. YAGLOM, *loc. cit.*, 43–51 or D. MIDDLETON, *loc. cit.*, § 3.2.

Instead of taking the ensemble average, other smoothing operations may be used [cf. S. GOLDMAN, *loc. cit.*, p. 244 or M. S. BARTLETT, *An Introduction to Stochastic Processes* (Cambridge, University Press, 1955), p. 280–284].

Another way of defining the power spectrum $G(\nu)$, which does not depend on the concept of an ensemble, is indicated in the first footnote on p. 504.

$G(\nu)d\nu$ is proportional to the contribution to the intensity from the frequency range $(\nu,\nu + d\nu)$; we shall refer to $G(\nu)$ as the *spectral density* of the light vibrations.

Since $V_T^{(r)}$ is the real part of V_T it follows that, when the operations on $V_T^{(r)}$ are linear, we may operate directly with V_T and take the real part at the end of the calculations. Moreover, just as in the monochromatic case, the relation $\langle V^{(r)^2}\rangle = \frac{1}{2}\langle VV^\star\rangle$ allows us to calculate the time average of the square of the real disturbance directly in terms of the complex disturbance which we have associated with it.

10.3 THE CORRELATION FUNCTIONS OF LIGHT BEAMS

10.3.1 Interference of two partially coherent beams. The mutual coherence function and the complex degree of coherence

We have indicated in § 10.1 that for a satisfactory treatment of problems involving light from a finite source and with a finite spectral range it is necessary to specify

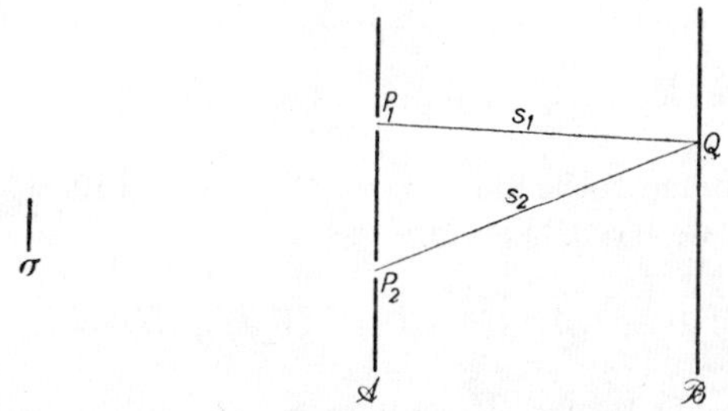

Fig. 10.1. An interference experiment with polychromatic light from an extended source σ.

the correlation that may exist between the vibrations at two arbitrary points in the wave field. A suitable measure of this correlation is suggested by the analysis of a two-beam interference experiment.

Consider the wave field produced by an extended polychromatic source σ. For the present we neglect polarization effects, so that we may regard the light disturbance as a real scalar function $V^{(r)}(P,t)$ of position and time. With $V^{(r)}(P,t)$ we associate the analytic signal $V(P,t)$. By observation it is, of course, impossible to determine how these quantities vary with time, since any detector will only record averages over time intervals during which the disturbance will have changed sign very many times. The observable intensity $I(P)$ is proportional to the mean value of $V^{(r)^2}(P,t)$, so that, apart from an inessential constant,

$$I(P) = 2\langle V^{(r)^2}(P,t)\rangle = \langle V(P,t)V^\star(P,t)\rangle, \tag{1}$$

where the relation § 10.2 (24) has been used.

Consider now two points P_1 and P_2 in the wave field. In addition to measuring the intensities $I(P_1)$ and $I(P_2)$ we may also determine experimentally the interference effects arising on superposition of the vibrations from these points. For this purpose imagine an opaque screen $\mathscr{A}$ to be placed across the field with pinholes at P_1 and P_2, and consider the intensity distribution on a second screen $\mathscr{B}$ placed some distance from $\mathscr{A}$, on the side opposite the source (Fig. 10.1). For simplicity we assume that the medium between the two screens has refractive index unity. Let s_1 and s_2 be the distances of a typical point Q on the screen $\mathscr{B}$ from P_1 and P_2. We may regard P_1

and P_2 as centres of secondary disturbances, so that the complex disturbance at Q is given by

$$V(Q,t) = K_1 V(P_1,t - t_1) + K_2 V(P_2,t - t_2). \tag{2}$$

Here t_1 and t_2 are the times needed for light to travel from P_1 to Q and from P_2 to Q respectively, i.e.

$$t_1 = \frac{s_1}{c}, \qquad t_2 = \frac{s_2}{c}, \tag{3}$$

where c is the velocity of light in vacuum. The factors K_1 and K_2 are inversely proportional to s_1 and s_2, and depend also on the size of the openings and on the geometry of the arrangement (the angles of incidence and diffraction at P_1 and P_2). Since the secondary wavelets from P_1 and P_2 are out of phase with the primary wave by a quarter of a period (cf. § 8.2, § 8.3), K_1 and K_2 are pure imaginary numbers.

It follows from (1) and (2) that the intensity at Q is given by*

$$\begin{aligned} I(Q) = K_1K_1^\star\langle V_1(t-t_1)V_1^\star(t-t_1)\rangle + K_2K_2^\star\langle V_2(t-t_2)V_2^\star(t-t_2)\rangle \\ + K_1K_2^\star\langle V_1(t-t_1)V_2^\star(t-t_2)\rangle + K_2K_1^\star\langle V_2(t-t_2)V_1^\star(t-t_1)\rangle. \end{aligned} \tag{4}$$

Now the field was assumed to be stationary. We may shift the origin of time in all these expressions, and we have, therefore,

$$\langle V_1(t-t_1)V_1^\star(t-t_1)\rangle = \langle V_1(t)V_1^\star(t)\rangle = I_1, \tag{5}$$

and similarly for the other terms. If we also use (3), and remember that K_1 and K_2 are pure imaginary numbers, (4) may be simplified to give

$$I(Q) = |K_1|^2 I_1 + |K_2|^2 I_2 + 2|K_1K_2|\Gamma_{12}^{(r)}\left(\frac{s_2 - s_1}{c}\right), \tag{6}$$

where $\Gamma_{12}^{(r)}(\tau)$ is the real part of the function

$$\Gamma_{12}(\tau) = \langle V_1(t+\tau)V_2^\star(t)\rangle. \tag{7}$$

The quantity represented by (7) is basic for the theory of partial coherence. We shall call it *the mutual coherence* of the light vibrations at P_1 and P_2, the vibrations at P_1 being considered at time τ later than at P_2; and we shall call $\Gamma_{12}(\tau)$ *the mutual coherence function*† of the wave field. When the two points coincide ($P_1 = P_2$) we obtain

$$\Gamma_{11}(\tau) = \langle V_1(t+\tau)V_1^\star(t)\rangle \tag{8}$$

and we then speak of the *self-coherence* of the light vibrations at P_1; it reduces to ordinary intensity when $\tau = 0$:

$$\Gamma_{11}(0) = I_1 \qquad \Gamma_{22}(0) = I_2.$$

---

* From now on, where convenient, we employ a shortened notation, writing $V_1(t)$ in place of $V(P_1,t)$, $\Gamma_{12}(\tau)$ in place of $\Gamma(P_1,P_2,\tau)$, etc.

† In the general theory of stationary random processes $\Gamma_{12}(\tau)$ is called the *cross-correlation function* of $V_1(t)$ and $V_2(t)$ and $\Gamma_{11}(\tau)$ the *auto-correlation function* of $V_1(t)$.

The term $|K_1|^2 I_1$ in (6) is evidently the intensity which would be observed at Q if the pinhole at P_1 alone were open ($K_2 = 0$) and the term $|K_2|^2 I_2$ has a similar interpretation. Let us denote these intensities by $I^{(1)}(Q)$ and $I^{(2)}(Q)$ respectively, i.e.,

$$I^{(1)}(Q) = |K_1|^2 I_1 = |K_1|^2 \Gamma_{11}(0), \qquad I^{(2)}(Q) = |K_2|^2 I_2 = |K_2|^2 \Gamma_{22}(0). \tag{9}$$

We also normalize $\Gamma_{12}(\tau)$:

$$\gamma_{12}(\tau) = \frac{\Gamma_{12}(\tau)}{\sqrt{\Gamma_{11}(0)}\sqrt{\Gamma_{22}(0)}} = \frac{\Gamma_{12}(\tau)}{\sqrt{I_1}\sqrt{I_2}}. \tag{10}$$

For reasons which will become apparent shortly, $\gamma_{12}(\tau)$ will be called *the complex degree of coherence* of the light vibrations. With the aid of (9) and (10), the formula (6) may finally be written in the form

$$I(Q) = I^{(1)}(Q) + I^{(2)}(Q) + 2\sqrt{I^{(1)}(Q)}\sqrt{I^{(2)}(Q)}\, \gamma_{12}^{(r)}\left(\frac{s_2 - s_1}{c}\right), \tag{11}$$

where $\gamma_{12}^{(r)}$ denotes the real part of γ_{12}.

The formula (11) is the *general interference law for stationary optical fields*. It shows that, in order to determine the intensity arising from the superposition of two beams of light, we must know the intensity of each beam and the value of the real part $\gamma_{12}^{(r)}$ of the complex degree of coherence. We shall show later how $\gamma_{12}^{(r)}$ may be calculated from data that specify the source and the transmission properties of the medium.

If the light from P_1 and P_2 does not reach Q directly, but via an intervening optical system, and if dispersion effects are negligible, (11) retains its validity provided that $s_2 - s_1$ is replaced by the path difference $P_2Q - P_1Q$. With this generalization the formula (11) also holds when the two interfering beams are derived from a primary beam, not by "wave-front division" at P_1 and P_2, but by "amplitude division" in the immediate neighbourhood of a single point P_1, for example in a Michelson interferometer. In this latter case equation (11) will involve $\gamma_{11}^{(r)}(\tau)$ in place of $\gamma_{12}^{(r)}(\tau)$.

Unlike the disturbance $V^{(r)}$, the correlation functions $\gamma_{12}^{(r)}$ and $\Gamma_{12}^{(r)}$ represent quantities which can be determined from experiment. To find the value of $\gamma_{12}^{(r)}$ for any prescribed pair of points P_1 and P_2 and for any prescribed value of τ, one places an opaque screen across the light beam, with pinholes at P_1 and P_2, as in Fig. 10.1. One then measures the intensity $I(Q)$ at a point Q behind the screen, such that $P_2Q - P_1Q = c\tau$. Next one measures the intensities $I^{(1)}(Q)$ and $I^{(2)}(Q)$ of the light from each pinhole separately. In terms of these three observed values, $\gamma_{12}^{(r)}$ is, according to (11), given by

$$\gamma_{12}^{(r)} = \frac{I(Q) - I^{(1)}(Q) - I^{(2)}(Q)}{2\sqrt{I^{(1)}(Q)}\sqrt{I^{(2)}(Q)}}. \tag{12}$$

To determine $\Gamma_{12}^{(r)}$ one must also measure the intensities $I(P_1)$ and $I(P_2)$ at each pinhole. According to (10) and (12) $\Gamma_{12}^{(r)}$ is then given by

$$\Gamma_{12}^{(r)} = \sqrt{I(P_1)}\sqrt{I(P_2)}\, \gamma_{12}^{(r)} = \frac{1}{2}\sqrt{\frac{I(P_1)I(P_2)}{I^{(1)}(Q)I^{(2)}(Q)}}\,[I(Q) - I^{(1)}(Q) - I^{(2)}(Q)]. \tag{13}$$

Returning to (10), it is not difficult to see that our normalization ensures that $|\gamma_{12}(\tau)| \leqslant 1$. To show this we introduce, as in § 10.2 (17), the truncated functions

$$\left.\begin{aligned} V_T^{(r)}(P,t) &= V^{(r)}(P,t) \quad \text{when} \quad |t| \leqslant T, \\ &= 0 \qquad\qquad\quad \text{when} \quad |t| > T \end{aligned}\right\} \tag{14}$$

and denote by $V_T(P,t)$ the associated analytic signal. By the SCHWARZ inequality*

$$\left|\int_{-\infty}^{\infty} V_T(P_1,t+\tau)V_T^{\star}(P_2,t)dt\right|^2 \leqslant \int_{-\infty}^{\infty} V_T(P_1,t+\tau)V_T^{\star}(P_1,t+\tau)dt \int_{-\infty}^{\infty} V_T(P_2,t)V_T^{\star}(P_2,t)dt. \tag{15}$$

In the first integral on the right we may replace $t+\tau$ by t. Then dividing both sides by $4T^2$, and proceeding to the limit $T \to \infty$, it follows that

$$|\Gamma_{12}(\tau)|^2 \leqslant \Gamma_{11}(0)\Gamma_{22}(0), \tag{16}$$

or, by (10),

$$|\gamma_{12}(\tau)| \leqslant 1. \tag{17}$$

The significance of γ_{12} may best be seen by expressing (11) in a somewhat different form. Let $\bar{\nu}$ be a mean frequency of the light and write

$$\gamma_{12}(\tau) = |\gamma_{12}(\tau)|e^{i[\alpha_{12}(\tau)-2\pi\bar{\nu}\tau]}, \tag{18}$$

where

$$\alpha_{12}(\tau) = 2\pi\bar{\nu}\tau + \arg \gamma_{12}(\tau). \tag{19}$$

Then (11) becomes

$$I(Q) = I^{(1)}(Q) + I^{(2)}(Q) + 2\sqrt{I^{(1)}(Q)}\sqrt{I^{(2)}(Q)}|\gamma_{12}(\tau)| \cos [\alpha_{12}(\tau) - \delta], \tag{20}$$

where the parameter τ and the phase difference δ have the values

$$\tau = \frac{s_2 - s_1}{c}, \qquad \delta = 2\pi\bar{\nu}\tau = \frac{2\pi}{\bar{\lambda}}(s_2 - s_1), \tag{21}$$

and $\bar{\lambda}$ is the mean wavelength. If $|\gamma_{12}(\tau)|$ has the extreme value unity, the intensity at Q is the same as would be obtained with strictly monochromatic light of wavelength $\bar{\lambda}$, and with the phase difference between the vibrations at P_1 and P_2 equal to $\alpha_{12}(\tau)$. In this case the vibrations at P_1 and P_2 (with the appropriate time delay τ between them) may be said to be *coherent*.† If $\gamma_{12}(\tau)$ has the other extreme value, namely zero, the last term in (20) is absent; the beams do not give rise to any interference effects and the vibrations may then be said to be *incoherent*. If $|\gamma_{12}(\tau)|$ has neither of the two extreme values, i.e. if $0 < |\gamma_{12}(\tau)| < 1$, the vibrations are said to be *partially coherent* $|\gamma_{12}(\tau)|$ representing their *degree of coherence*.‡

Whatever the value of $|\gamma_{12}|$, the intensity $I(Q)$ may also be expressed in the form

$$I(Q) = |\gamma_{12}(\tau)|\{I^{(1)}(Q) + I^{(2)}(Q) + 2\sqrt{I^{(1)}(Q)}\sqrt{I^{(2)}(Q)} \cos [\alpha_{12}(\tau) - \delta]\}$$
$$+\{1 - |\gamma_{12}(\tau)|\}\{I^{(1)}(Q) + I^{(2)}(Q)\}. \tag{22}$$

The terms in the first line may be considered to arise from *coherent* superposition of two beams of intensities $|\gamma_{12}(\tau)|I^{(1)}(Q)$ and $|\gamma_{12}(\tau)|I^{(2)}(Q)$ and of relative phase difference $\alpha_{12}(\tau) - \delta$; those in the second line from *incoherent* superposition of two beams

* See, for example, H. MARGENAU and G. M. MURPHY, *The Mathematics of Physics and Chemistry* (New York, D. van Nostrand Co., 1947), p. 131.

† General properties of coherent light have been investigated by L. MANDEL and E. WOLF, *J. Opt. Soc. Amer.*, **51** (1961), 815.

‡ Various methods for measuring the degree of coherence are discussed by M. FRANÇON and S. MALLICK in *Progress in Optics*, Vol. 6, ed. E. WOLF (Amsterdam, North-Holland Publishing Company and New York, J. Wiley, 1967), p. 71.

of intensities $[1 - |\gamma_{12}(\tau)|]I^{(1)}(Q)$ and $[1 - |\gamma_{12}(\tau)|]I^{(2)}(Q)$. Thus the light which reaches Q from both pinholes may be regarded to be a mixture of coherent and incoherent light, with intensities in the ratio

$$\frac{I_{\text{coh}}}{I_{\text{incoh}}} = \frac{|\gamma_{12}(\tau)|}{1 - |\gamma_{12}(\tau)|}, \tag{23a}$$

or

$$\frac{I_{\text{coh}}}{I_{\text{tot}}} = |\gamma_{12}|, \qquad (I_{\text{tot}} = I_{\text{coh}} + I_{\text{incoh}}). \tag{23b}$$

We have seen (equation (12)) that $\gamma_{12}^{(r)}$ may be determined from intensity measurements in an appropriate interference experiment. In § 10.4.1 we shall see that in most cases of practical interest the modulus (and in principle also the phase) of γ_{12} can likewise be determined from such experiments.

10.3.2 Spectral representation of mutual coherence

Let

$$V_T^{(r)}(P,t) = \int_{-\infty}^{\infty} v_T(P,\nu)e^{-2\pi i\nu t}\,d\nu \tag{24}$$

be the FOURIER integral representation of the truncated real function $V_T^{(r)}$. Then by the FOURIER inversion formula

$$v_T(P,\nu) = \int_{-\infty}^{\infty} V_T^{(r)}(P,t)e^{2\pi i\nu t}\,dt, \tag{25}$$

and it follows that

$$\int_{-\infty}^{\infty} V_T^{(r)}(P_1,t+\tau)V_T^{(r)}(P_2,t)dt = \int_{-\infty}^{\infty} V_T^{(r)}(P_2,t)\left[\int_{-\infty}^{\infty} v_T(P_1,\nu)e^{-2\pi i\nu(t+\tau)}\,d\nu\right]dt$$

$$= \int_{-\infty}^{\infty}\left[\int_{-\infty}^{\infty} V_T^{(r)}(P_2,t)e^{-2\pi i\nu t}\,dt\right] v_T(P_1,\nu)e^{-2\pi i\nu\tau}\,d\nu$$

$$= \int_{-\infty}^{\infty} v_T(P_1,\nu)v_T^{\star}(P_2,\nu)e^{-2\pi i\nu\tau}\,d\nu. \tag{26}$$

Next we divide both sides of (26) by $2T$ and apply to the quantity $v_T(P_1,\nu)v_T^{\star}(P_2,\nu)/2T$ a "smoothing operation" such as taking of the ensemble average (denoted by a bar) over the ensemble of the random functions $V^{(r)}$ as explained earlier in connection with eq. (20) of § 10.2. Finally proceeding to the limit $T \to \infty$ one may then expect that*

$$\langle V^{(r)}(P_1,t+\tau)V^{(r)}(P_2,t)\rangle = \int_{-\infty}^{\infty} G_{12}(\nu)e^{-2\pi i\nu\tau}\,d\nu, \tag{27}$$

where

$$G_{12}(\nu) = \lim_{T\to\infty}\left[\overline{\frac{v_T(P_1,\nu)v_T^{\star}(P_2,\nu)}{2T}}\right]. \tag{28}$$

* Similar remarks apply here as those made in footnote ‡ on p. 498.

The function $G_{12}(\nu)$ may be called the *mutual spectral density* of the light vibrations at P_1 and P_2. It is a generalization of the *spectral density* introduced earlier [§ 10.2 (22)] and reduces to it when the two points coincide. The mutual spectral density is the optical analogue of the concept of *cross-power spectrum* in the theory of stationary random processes. Equation (27) shows that the real correlation function $\langle V^{(r)}(P_1,t+\tau)V^{(r)}(P_2,t)\rangle$ and the mutual spectral density $G_{12}(\nu)$ form a FOURIER transform pair.*

Let us now pass to the complex representation. Let

$$V(P,t) = 2\int_0^\infty v(\nu)e^{-2\pi i\nu t}\,d\nu \tag{29}$$

be the analytic signal (cf. § 10.2) associated with $V^{(r)}(P,t)$. It follows by an analysis similar to that which leads from (24) to (27), that

$$\Gamma_{12}(\tau) = \langle V(P_1,t+\tau)V^\star(P_2,t)\rangle = 4\int_0^\infty G_{12}(\nu)e^{-2\pi i\nu\tau}\,d\nu. \tag{30}$$

Since Γ_{12} does not contain spectral components belonging to negative frequencies, it is an analytic signal. Hence if $\Gamma_{12}^{(r)}$ and $\Gamma_{12}^{(i)}$ denote its real and imaginary parts, i.e.

$$\Gamma_{12}(\tau) = \Gamma_{12}^{(r)}(\tau) + i\Gamma_{12}^{(i)}(\tau), \tag{31}$$

these functions are connected by the HILBERT transform relations

$$\Gamma_{12}^{(i)}(\tau) = \frac{1}{\pi}P\int_{-\infty}^{\infty}\frac{\Gamma_{12}^{(r)}(\tau')}{\tau'-\tau}\,d\tau', \qquad \Gamma_{12}^{(r)}(\tau) = -\frac{1}{\pi}P\int_{-\infty}^{\infty}\frac{\Gamma_{12}^{(i)}(\tau')}{\tau'-\tau}\,d\tau'. \tag{32}$$

It follows [cf. § 10.2 (11)–(15)] that $|\Gamma_{12}|$, considered as a function of τ, is the envelope of $\Gamma_{12}^{(r)}$; and from (30), (31), and (27) it follows that†

$$\Gamma_{12}^{(r)}(\tau) = 2\langle V^{(r)}(P_1,t+\tau)V^{(r)}(P_2,t)\rangle = 2\int_{-\infty}^{\infty} G_{12}(\nu)e^{-2\pi i\nu\tau}\,d\nu. \tag{33}$$

Furthermore $|\gamma_{12}|$ is the envelope of the real correlation factor

$$\gamma_{12}^{(r)}(\tau) = \frac{\Gamma_{12}^{(r)}(\tau)}{\sqrt{\Gamma_{11}(0)}\sqrt{\Gamma_{22}(0)}} = \frac{\langle V^{(r)}(P_1,t+\tau)V^{(r)}(P_2,t)\rangle}{\sqrt{\langle V^{(r)2}(P_1,t)\rangle}\sqrt{\langle V^{(r)2}(P_2,t)\rangle}}. \tag{34}$$

---

* In the case when $P_1 = P_2$, this result is the optical equivalent of the well-known WIENER–KHINTCHINE theorem. [N. WIENER, *Acta Math.*, **55** (1930), 117; A. KHINTCHINE, *Math. Ann.*, **109** (1934), 604.]

Instead of using a smoothing procedure one may use the Fourier inverse of (27) to define the mutual spectral density $G_{12}(\nu)$. This alternative approach is entirely adequate for the purposes of the main part of this chapter.

† It is not difficult to show that $\Gamma_{12}^{(r)}(\tau)$ is also equal to $2\langle V^{(i)}(P_1,t+\tau)V^{(i)}(P_2,t)\rangle$ and $\Gamma_{12}^{(i)}(\tau) = 2\langle V^{(i)}(P_1,t+\tau)V^{(r)}(P_2,t)\rangle = -2\langle V^{(r)}(P_1,t+\tau)V^{(i)}(P_2,t)\rangle$. [See, for example, P. ROMAN and E. WOLF, *Nuovo Cimento*, **17** (1960), 474–476 or L. MANDEL, *Progress in Optics*, Vol. 2, ed. E. WOLF (Amsterdam, North Holland Publishing Company and New York, J. Wiley and Sons, 1963), 241–242.]

Equation (30) gives the spectral representation of the mutual coherence function $\Gamma_{12}(\tau)$. Equation (33) shows that the real part of $\Gamma_{12}(\tau)$ is equal to twice the cross-correlation function of the real functions $V^{(r)}(P_1,t)$ and $V^{(r)}(P_2,t)$, and (32) gives the connection between the real and imaginary parts of $\Gamma_{12}(\tau)$.

10.4 INTERFERENCE AND DIFFRACTION WITH QUASI-MONOCHROMATIC LIGHT

We have seen that, in order to describe adequately interference with partially coherent light, it is in general necessary to know the mutual coherence function $\Gamma_{12}(\tau)$ or, what amounts to the same thing, the ordinary intensities I_1 and I_2 and the complex degree of coherence $\gamma_{12}(\tau)$. We shall now restrict ourselves to the important case of quasi-monochromatic light, i.e. light consisting of spectral components that cover a frequency range $\Delta\nu$ which is small compared to the mean frequency $\bar{\nu}$. We shall see that the theory takes a simpler form in this case. In particular, we shall find that, under a certain additional assumption which is satisfied in many applications, it is possible to employ in place of $\Gamma_{12}(\tau)$ and $\gamma_{12}(\tau)$ correlation functions which are independent of the parameter τ.

10.4.1 Interference with quasi-monochromatic light. The mutual intensity

Let us again consider the interference experiment illustrated in Fig. 10.1. According to equation (20) of § 10.3, the intensity at a point Q in the interference pattern is given by

$$I(Q) = I^{(1)}(Q) + I^{(2)}(Q) + 2\sqrt{I^{(1)}(Q)}\sqrt{I^{(2)}(Q)}|\gamma_{12}(\tau)| \cos [\alpha_{12}(\tau) - \delta], \tag{1}$$

where

$$\tau = \frac{s_2 - s_1}{c}, \qquad \delta = 2\pi\bar{\nu}\tau = \frac{2\pi}{\bar{\lambda}}(s_2 - s_1). \tag{2}$$

Suppose now that the light is quasi-monochromatic. Then it follows from equation (18), § 10.3 in the same way as in connection with equation (11) of § 10.2, that $|\gamma_{12}(\tau)|$ and $\alpha_{12}(\tau)$, considered as functions of τ, will change slowly in comparison to $\cos 2\pi\bar{\nu}\tau$ and $\sin 2\pi\bar{\nu}\tau$. Moreover, if the openings at P_1 and P_2 are sufficiently small, the intensities $I^{(1)}(Q)$ and $I^{(2)}(Q)$ of the light diffracted from each opening separately will remain sensibly constant throughout a region of the pattern in which $\cos 2\pi\bar{\nu}\tau$ and $\sin 2\pi\bar{\nu}\tau$ change sign many times. It follows that the intensity distribution in the vicinity of any point Q consist of an almost uniform background $I^{(1)}(Q) + I^{(2)}(Q)$ on which a sinusoidal intensity distribution is superimposed, with almost constant amplitude $2\sqrt{I^{(1)}(Q)}\sqrt{I^{(2)}(Q)}|\gamma_{12}(\tau)|$. The behaviour of the total intensity distribution is shown for three typical cases in Fig. 10.2. The intensity maxima and minima near Q are to a good approximation given by

$$\left.\begin{aligned} I_{\text{max}} &= I^{(1)}(Q) + I^{(2)}(Q) + 2\sqrt{I^{(1)}(Q)}\sqrt{I^{(2)}(Q)}|\gamma_{12}(\tau)|, \\ I_{\text{min}} &= I^{(1)}(Q) + I^{(2)}(Q) - 2\sqrt{I^{(1)}(Q)}\sqrt{I^{(2)}(Q)}|\gamma_{12}(\tau)|. \end{aligned}\right\} \tag{3}$$

Hence the *visibility of the fringes at Q is*

$$\mathscr{V}(Q) = \frac{I_{\text{max}} - I_{\text{min}}}{I_{\text{max}} + I_{\text{min}}} = \frac{2\sqrt{I^{(1)}(Q)}\sqrt{I^{(2)}(Q)}}{I^{(1)}(Q) + I^{(2)}(Q)}|\gamma_{12}(\tau)|. \tag{4}$$

This formula expresses the visibility of the fringes in terms of the intensity of the two beams and of their degree of coherence. If, as is often the case, the two beams are of equal intensity $[I^{(1)} = I^{(2)}]$, (4) reduces to

$$\mathscr{V}(Q) = |\gamma_{12}(\tau)|, \tag{5}$$

i.e. *the visibility of the fringes is then equal to the degree of coherence.*

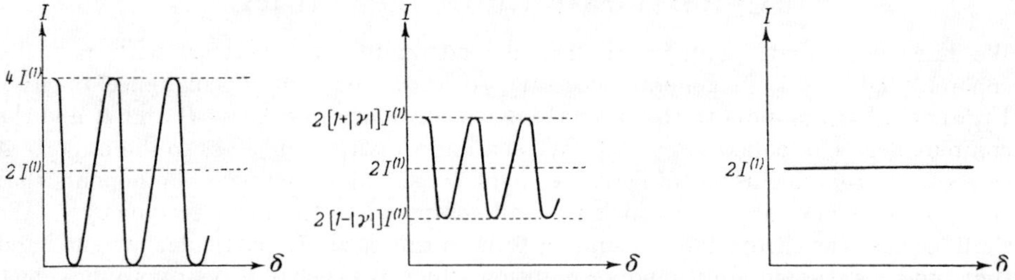

(a) Coherent superposition ($|\gamma| = 1$) (b) Partially coherent superposition ($0 < |\gamma| < 1$) (c) Incoherent superposition ($\gamma = 0$)

Fig. 10.2. Intensity distribution in the interference pattern produced by two quasi-monochromatic beams of equal intensity $I^{(1)}$ and with degree of coherence $|\gamma|$.

According to (1) and (2), the positions of the maxima of intensity near Q are given by

$$\frac{2\pi}{\bar{\lambda}}(s_2 - s_1) - \alpha_{12}(\tau) = 2m\pi, \qquad (m = 0, \pm 1, \pm 2, \ldots),$$

just as if the opening were illuminated with strictly monochromatic light of wavelength $\bar{\lambda}$ and the phase at P_1 were retarded with respect to P_2 by $\alpha_{12}(\tau)$. Now according to § 7.3 (7) a phase retardation of amount 2π corresponds to a displacement of the interference pattern in the direction parallel to P_1P_2 by an amount $a\bar{\lambda}/d$, where d is the distance between P_1 and P_2 and a is the distance between the screens $\mathscr{A}$ and $\mathscr{B}$. Hence the *quasi-monochromatic fringes are displaced relative to the fringes that would be formed with monochromatic and co-phasal illumination of P_1 and P_2 by an amount*

$$x = \frac{\bar{\lambda}}{2\pi}\frac{a}{d}\alpha_{12}(\tau) \tag{6}$$

in the direction parallel to the line joining the openings.

We see that the amplitude and the phase of the complex degree of coherence of quasi-monochromatic light beams may be determined from measurements of the visibility and the position of interference fringes. These results have a close bearing on MICHELSON's method, described in § 7.5.8, for determining the intensity distribution in spectral lines from measurements of visibility curves. It follows from equations (10) and (30) of § 10.3 that

$$\gamma_{11}(\tau) = \frac{\int_0^\infty G(\nu)e^{-2\pi i\nu\tau}\,d\nu}{\int_0^\infty G(\nu)d\nu}.$$

where $G(\nu)$ is the spectral density. Hence, by the FOURIER inversion theorem, $G(\nu)$ is proportional to the FOURIER transform of $\gamma_{11}(\tau)$. But we have just seen that the modulus of $\gamma_{11}(\tau)$ is essentially the visibility, and the phase of $\gamma_{11}(\tau)$ is simply related to the position of the fringes formed in an appropriate interference experiment. One is thus led to calculating G in exactly the same way as was done by MICHELSON. The visibility curves exhibited in Figs. 7.54 and 7.55 may be evidently interpreted as representing $|\gamma_{11}|$ as a function of the time delay between the two beams.

In practice the time delay τ introduced between the interfering beams is often very small, and it is then possible to simplify the formulae. According to equations (30), (18), and (10) of § 10.3 we have

$$|\Gamma_{12}(\tau)|e^{i\alpha_{12}(\tau)} = \sqrt{I_1}\sqrt{I_2}|\gamma_{12}(\tau)|e^{i\alpha_{12}(\tau)} = 4\int_0^\infty G_{12}(\nu)e^{-2\pi i(\nu-\bar{\nu})\tau}\,d\nu. \tag{7}$$

If $|\tau|$ is so small that $|(\nu - \bar{\nu})\tau| \ll 1$ for all the frequencies for which $|G_{12}(\nu)|$ is appreciable, i.e. if

$$|\tau| \ll \frac{1}{\Delta\nu}, \tag{8}$$

then evidently only a small error is introduced if the exponential term of the integrand in (7) is replaced by unity. The condition (8) implies, according to § 7.5 (105), that $|\tau|$ must be small compared to the coherence time of the light. When this condition is satisfied, $|\Gamma_{12}(\tau)|$, $|\gamma_{12}(\tau)|$, and $\alpha_{12}(\tau)$ differ inappreciably from $|\Gamma_{12}(0)|$, $|\gamma_{12}(0)|$, and $\alpha_{12}(0)$ respectively. It is useful to set*

$$J_{12} = \Gamma_{12}(0) = \langle V_1(t)V_2^\star(t)\rangle, \tag{9a}$$

$$\mu_{12} = \gamma_{12}(0) = \frac{\Gamma_{12}(0)}{\sqrt{\Gamma_{11}(0)}\sqrt{\Gamma_{22}(0)}} = \frac{J_{12}}{\sqrt{J_{11}}\sqrt{J_{22}}} = \frac{J_{12}}{\sqrt{I_1}\sqrt{I_2}}, \tag{9b}$$

$$\beta_{12} = \alpha_{12}(0) = \arg\gamma_{12}(0) = \arg\mu_{12}. \tag{9c}$$

Equations (18) and (10) of § 10.3 now give, subject to (8),

$$\gamma_{12}(\tau) \sim |\mu_{12}|e^{i(\beta_{12}-2\pi\bar{\nu}\tau)} = \mu_{12}e^{-2\pi i\bar{\nu}\tau}, \tag{10a}$$

$$\Gamma_{12}(\tau) \sim |J_{12}|e^{i(\beta_{12}-2\pi\bar{\nu}\tau)} = J_{12}e^{-2\pi i\bar{\nu}\tau}. \tag{10b}$$

Thus, *provided* (8) *is satisfied, we may replace* $\gamma_{12}(\tau)$ *and* $\Gamma_{12}(\tau)$ *in all our formulae by the quantities on the right-hand side of* (10a) *and* (10b) *respectively. In particular, the interference law* (1) *becomes*

$$I(Q) \sim I^{(1)}(Q) + I^{(2)}(Q) + 2\sqrt{I^{(1)}(Q)}\sqrt{I^{(2)}(Q)}|\mu_{12}|\cos(\beta_{12} - \delta), \tag{11}$$

and is valid as long as the path difference $|s_2 - s_1| = c|\tau|$, introduced between the interfering beams is small, compared to the coherence length $c/\Delta\nu$, i.e. as long as

$$|\Delta\mathscr{S}| = |s_2 - s_1| = \frac{\bar{\lambda}}{2\pi}\delta \ll \frac{\bar{\lambda}^2}{\Delta\lambda}, \tag{12}$$

where the relation $c/\Delta\nu = \bar{\lambda}^2/\Delta\lambda$ has been used.

Equation (11) is the basic formula of an elementary (quasi-monochromatic) theory of partial coherence, which forms the subject matter of the rest of this section; some applications of this theory will be considered in § 10.5. Within its range of validity (indicated by (8) or (12)) the correlation between the vibrations at any two points P_1

* We again use the shortened notation where convenient, i.e. we write J_{12} in place of $J(P_1, P_2)$, etc

and P_2 in the wave field is characterized by J_{12} rather than by $\Gamma_{12}(\tau)$, i.e. by a quantity which depends on the positions of the two points, but not on the time difference τ. It follows from (10a) that, within the accuracy of this elementary theory,

$$|\gamma_{12}(\tau)| \sim |\mu_{12}|, \tag{13}$$

so that $|\mu_{12}|$ $(0 \leqslant |\mu_{12}| \leqslant 1)$ represents the degree of coherence of the vibrations at P_1 and P_2; and we see from (11) that the phase β_{12} of μ_{12} represents their effective phase difference. μ_{12}, just like $\gamma_{12}(\tau)$ of which it is a special case, is usually called *the complex degree of coherence* (sometimes the *complex coherence factor*); and J_{12} is called the *mutual intensity*.

10.4.2 Calculation of mutual intensity and degree of coherence for light from an extended incoherent quasi-monochromatic source

(a) *The Van Cittert–Zernike theorem*

We shall now determine the mutual intensity J_{12} and the complex degree of coherence μ_{12} for points P_1 and P_2 on a screen $\mathscr{A}$ illuminated by an extended quasi-monochromatic primary source σ. For simplicity σ will be taken to be a portion of a plane parallel to $\mathscr{A}$, and we will assume that the medium between the source and the screen is homogeneous. We also assume that the linear dimensions of σ are small

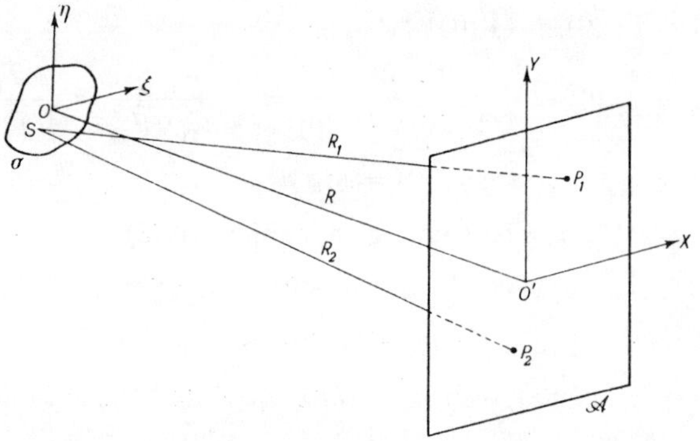

Fig. 10.3. Illustrating the VAN CITTERT–ZERNIKE theorem.

compared to the distance OO' between the source and the screen (Fig. 10.3), and that the angles between OO' and the line joining a typical source point S to P_1 and P_2 are small.

Imagine the source to be divided into elements $d\sigma_1, d\sigma_2, \ldots$ centred on points $S_1, S_2, \ldots$, of linear dimensions small compared to the mean wavelength $\bar{\lambda}$. If $V_{m1}(t)$ and $V_{m2}(t)$ are the complex disturbances at P_1 and P_2 due to the element $d\sigma_m$, the total disturbances at these points are

$$V_1(t) = \sum_m V_{m1}(t), \qquad V_2(t) = \sum_m V_{m2}(t). \tag{14}$$

Hence

$$J(P_1, P_2) = \langle V_1(t) V_2^\star(t)\rangle = \sum_m \langle V_{m1}(t) V_{m2}^\star(t)\rangle + \sum_{m \neq n}\sum \langle V_{m1}(t) V_{n2}^\star(t)\rangle. \tag{15}$$

Now the light vibrations arising from different elements of the source may be assumed to be statistically independent (mutually incoherent), and of zero mean value, so that*

$$\langle V_{m1}(t)V^{\star}_{n2}(t)\rangle = \langle V_{m1}(t)\rangle\langle V^{\star}_{n2}(t)\rangle = 0 \quad \text{when} \quad m \neq n. \tag{16}$$

If R_{m1} and R_{m2} are the distances of P_1 and P_2 from the source element $d\sigma_m$, then

$$V_{m1}(t) = A_m\left(t - \frac{R_{m1}}{v}\right)\frac{e^{-2\pi i\bar{\nu}(t - R_{m1}/v)}}{R_{m1}}, \qquad V_{m2}(t) = A_m\left(t - \frac{R_{m2}}{v}\right)\frac{e^{-2\pi i\bar{\nu}(t - R_{m2}/v)}}{R_{m2}}, \tag{17}$$

where $|A_m|$ characterizes the strength and arg A_m the phase of the radiation from the mth element,† and v is the velocity of light in the medium between the source and the screen. Hence

$$\begin{aligned}\langle V_{m1}(t)V^{\star}_{m2}(t)\rangle &= \left\langle A_m\left(t - \frac{R_{m1}}{v}\right)A^{\star}_m\left(t - \frac{R_{m2}}{v}\right)\right\rangle\frac{e^{2\pi i\bar{\nu}(R_{m1} - R_{m2})/v}}{R_{m1}R_{m2}}\\ &= \left\langle A_m(t)A^{\star}_m\left(t - \frac{R_{m2} - R_{m1}}{v}\right)\right\rangle\frac{e^{2\pi i\bar{\nu}(R_{m1} - R_{m2})/v}}{R_{m1}R_{m2}}.\end{aligned} \tag{18}$$

If the path difference $R_{m2} - R_{m1}$ is small compared to the coherence length of the light we may neglect the retardation $(R_{m2} - R_{m1})/v$ in the argument of $A^{\star}_m$, and we obtain from (15), (16), and (18)

$$J(P_1,P_2) = \sum_m \langle A_m(t)A^{\star}_m(t)\rangle\frac{e^{2\pi i\bar{\nu}(R_{m1} - R_{m2})/v}}{R_{m1}R_{m2}}. \tag{19}$$

The quantity $\langle A_m(t)A^{\star}_m(t)\rangle$ characterizes the intensity of the radiation from the source element $d\sigma_m$. In any practical case the total number of the source elements may be assumed to be so large, that we may regard the source to be effectively continuous. Denoting by $I(S)$ the intensity per unit area of the source, i.e. $I(S_m)d\sigma_m = \langle A_m(t)A^{\star}_m(t)\rangle$, (19) becomes‡

$$J(P_1,P_2) = \int_\sigma I(S)\frac{e^{i\bar{k}(R_1 - R_2)}}{R_1R_2}\,dS, \tag{20}$$

where R_1 and R_2 denote the distances between a typical source point S and the points P_1 and P_2, and $\bar{k} = 2\pi\bar{\nu}/v = 2\pi/\bar{\lambda}$ is the wave number in the medium. The complex degree of coherence $\mu(P_1,P_2)$ is, according to (20) and (9b), given by

$$\mu(P_1,P_2) = \frac{1}{\sqrt{I(P_1)}\sqrt{I(P_2)}}\int_\sigma I(S)\frac{e^{i\bar{k}(R_1 - R_2)}}{R_1R_2}\,dS, \tag{21}$$

where

$$I(P_1) = J(P_1,P_1) = \int_\sigma \frac{I(S)}{R_1^2}\,dS, \qquad I(P_2) = J(P_2,P_2) = \int_\sigma \frac{I(S)}{R_2^2}\,dS, \tag{21a}$$

are the intensities at P_1 and P_2.

---

* Incoherence always implies a finite (though not necessarily wide) spectral range, and equation (16) is, in fact, not valid for the idealized case of strictly monochromatic light. For monochromatic light one has $V_{m1}(t) = U_{m1}e^{-2\pi i\nu t}$, $V_{n2}(t) = U_{n2}e^{-2\pi i\nu t}$, where U_{m1} and U_{n2} are independent of time, so that $\langle V_{m1}(t)V^{\star}_{n2}(t)\rangle = U_{m1}U^{\star}_{n2}$ and this quantity is in general different from zero.

† In general A_m also depends on direction, but for simplicity we neglect this dependence.

‡ From now on we shall frequently use the notation dS, dP_1, . . . for surface elements centred on the points S, P_1, ...

We note that the integral (21) is the same as that which occurs in quite a different connection; namely in the calculation, on the basis of the HUYGENS–FRESNEL principle, of the complex disturbance in the diffraction pattern arising from diffraction of a spherical wave on an aperture in an opaque screen. More precisely, (21) implies that *the complex degree of coherence, which describes the correlation of vibrations at a fixed point P_2 and a variable point P_1 in a plane illuminated by an extended quasi-monochromatic primary source, is equal to the normalized complex amplitude at the corresponding point P_1 in a certain diffraction pattern, centred on P_2. This pattern would be obtained on replacing the source by a diffracting aperture of the same size and shape as the source, and on filling it with a spherical wave converging to P_2, the amplitude distribution over the wave-front in the aperture being proportional to the intensity distribution across the source.* This result was first established by VAN CITTERT* and later in a simpler way by ZERNIKE.† We shall refer to it as the *van Cittert–Zernike theorem.*

In most applications the intensity $I(S)$ may be assumed to be independent of the position of S on the surface (uniform intensity). The corresponding diffraction problem is then that of diffraction of a spherical wave of uniform amplitude by an aperture of the same size and shape as the source.

Let (ξ,η) be the coordinates of a typical source point S, referred to axes at O, and let (X_1, Y_1) and (X_2, Y_2) be the coordinates of P_1 and P_2 referred to parallel axes at O' (Fig. 10.3). Then, if R denotes the distance OO',

$$R_1^2 = (X_1 - \xi)^2 + (Y_1 - \eta)^2 + R^2,$$

so that

$$R_1 \sim R + \frac{(X_1 - \xi)^2 + (Y_1 - \eta)^2}{2R}. \tag{22}$$

Here only the leading terms in X_1/R, Y_1/R, ξ/R, and η/R have been retained. A strictly similar expression is obtained for R_2, so that

$$R_1 - R_2 \sim \frac{(X_1^2 + Y_1^2) - (X_2^2 + Y_2^2)}{2R} - \frac{(X_1 - X_2)\xi + (Y_1 - Y_2)\eta}{R}. \tag{23}$$

In the denominator of the integrands in (20) and (21), R_1 and R_2 may to a good approximation be replaced by R. We also set

$$\frac{(X_1 - X_2)}{R} = p, \qquad \frac{(Y_1 - Y_2)}{R} = q, \tag{24}$$

$$\psi = \frac{\bar{k}[(X_1^2 + Y_1^2) - (X_2^2 + Y_2^2)]}{2R}. \tag{25}$$

Then (21) reduces to

$$\mu_{12} = \frac{e^{i\psi} \iint_\sigma I(\xi,\eta) e^{-i\bar{k}(p\xi + q\eta)}\, d\xi d\eta}{\iint_\sigma I(\xi,\eta) d\xi d\eta}. \tag{26}$$

Hence *if the linear dimensions of the source and the distance between P_1 and P_2 are small compared to the distance of these points from the source, the degree of coherence $|\mu_{12}|$ is equal to the absolute value of the normalized Fourier transform of the intensity function of the source.*

* P. H. VAN CITTERT, *Physica*, **1** (1934), 201.
† F. ZERNIKE, *Physica*, **5** (1938), 785.

The quantity ψ defined by (25) has a simple interpretation. According to (23) it represents the phase difference $2\pi(OP_1 - OP_2)/\bar{\lambda}$, and may evidently be neglected when

$$OP_1 - OP_2 \ll \bar{\lambda}. \tag{27}$$

For a uniform circular source of radius ρ with its centre at O, (26) gives on integration (cf. § 8.5.2)

$$\mu_{12} = \left(\frac{2J_1(v)}{v}\right)e^{i\psi}, \tag{28}$$

where

$$\left.\begin{aligned} v &= \bar{k}\rho\sqrt{p^2+q^2} = \frac{2\pi}{\bar{\lambda}}\frac{\rho}{R}\sqrt{(X_1-X_2)^2+(Y_1-Y_2)^2}, \\ \psi &= \frac{2\pi}{\bar{\lambda}}\left[\frac{(X_1{}^2+Y_1{}^2)-(X_2{}^2+Y_2{}^2)}{2R}\right] \end{aligned}\right\} \tag{29}$$

J_1 being the Bessel function of the first kind and first order.* According to § 8.5.2, $|2J_1(v)/v|$ decreases steadily from the value unity when $v = 0$ to the value zero when $v = 3{\cdot}83$; thus as the points P_1 and P_2 are separated more and more, the degree of coherence steadily decreases and there is complete incoherence when P_1 and P_2 are separated by the distance

$$P_1P_2 = \sqrt{(X_1-X_2)^2+(Y_1-Y_2)^2} = \frac{0{\cdot}61R\bar{\lambda}}{\rho}. \tag{30}$$

A further increase in v re-introduces a small amount of coherence, but the degree of coherence remains smaller than 0·14, and there is further complete incoherence for $v = 7{\cdot}02$. Since $J_1(v)$ changes sign as v passes through each zero of $J_1(v)$, the phase $\beta_{12} = \arg \mu_{12}$ changes there by π; in consequence the position of the bright and dark fringes are interchanged after each disappearance of the fringes.

The function $|2J_1(v)/v|$ decreases steadily from the value 1 for $v = 0$ to 0·88 when $v = 1$, i.e. when

$$P_1P_2 = \frac{0{\cdot}16R\bar{\lambda}}{\rho}. \tag{31}$$

Regarding a departure of 12 per cent from the ideal value unity as the maximum permissible departure, it follows that *the diameter of the circular area that is illuminated almost coherently by a quasi-monochromatic, uniform source of angular radius* $\alpha = \rho/R$ is† $0{\cdot}16\bar{\lambda}/\alpha$. This result is useful in estimating the size of a source needed in experiments on interference and diffraction.

As an example consider the size of the "area of coherence" around an arbitrary point on a screen illuminated directly by the sun. The angular diameter 2α which the sun's disc subtends on the surface of the earth is about 0° 32′ $\sim$ 0·0093 radian. Hence, if the variation of brightness across the sun's disc is neglected, the diameter d of the area of coherence is approximately $0{\cdot}16\bar{\lambda}/0{\cdot}0047 \sim 34\bar{\lambda}$. Taking the mean wavelength $\bar{\lambda}$ as $5{\cdot}5 \,.\, 10^{-5}$ cm this gives $d \sim 0{\cdot}019$ mm.

---

* No confusion should arise from the fact that the symbol J is also used for the mutual intensity, as the latter always appears with two suffixes or with several arguments.

† As early as 1865, E. Verdet estimated that the diameter of the "circle of coherence" is somewhat smaller than $0{\cdot}5R\bar{\lambda}/\rho$. (*Ann. Scientif. de l'École Normale Supérieure*, **2** (1865), 291; also his *Leçons d'Optique Physique* (Paris, L'Imprimerie Impériale), **1** (1869), 106.)

In the present context MICHELSON's method of measuring angular diameters of stars (cf. § 7.3.6) appears in a new light. According to (5) and (13), the visibility of the fringes is equal to the degree of coherence of the light vibrations at the two outer mirrors (M_1 and M_2 in Fig. 7.16) of the MICHELSON stellar interferometer. For a uniformly bright circular star disc of angular radius α the smallest separation of the mirrors for which the degree of coherence has zero value (first fringe disappearance) is, according to (30), equal to $0{\cdot}61\bar{\lambda}/\alpha$ in agreement with § 7.3 (43). Moreover, from the measurements of both the visibility and the position of the fringes, it is in principle possible to determine not only the stellar diameter, but also the distribution of the intensity over the stellar disc. For, according to § 10.4.1, measurements of the visibility and the position of the fringes are equivalent to determining both the amplitude and the phase of the complex degree of coherence μ_{12}, and, according to (26), the intensity distribution is proportional to the inverse FOURIER transform of μ_{12}.

We mentioned in § 7.3.6 the important modification due to HANBURY BROWN and TWISS of MICHELSON's stellar interferometer. In the HANBURY BROWN–TWISS system light from the star is focused on two photo-electric detectors P_1, P_2 and information about the star is obtained from the study of the correlation in the fluctuations of their current outputs. A full analysis of the performance of this system must take into account the quantum nature of the photo-electric effect,* and requires also some knowledge of electronics, and is thus outside the scope of this book. The principle of the method may, however, be easily understood. Under ideal experimental conditions (absence of noise), the current output of each photo-electric detector is proportional to the instantaneous intensity $I(t)$ of the incident light, and the fluctuation in the current output is proportional to $\Delta I(t) = I(t) - \langle I(t)\rangle$. Hence, in the interferometer of HANBURY BROWN and TWISS, the quantity which is effectively being measured is proportional to $\Omega_{12} = \langle \Delta I_1 \Delta I_2 \rangle$. Simple statistical calculations show† that Ω_{12} is proportional to the square of the degree of coherence, so that the knowledge of Ω_{12}, just like the knowledge of $|\mu_{12}|$, yields information about the size of the star.

(b) *Hopkins' formula*

In deriving the VAN CITTERT–ZERNIKE formula (21), it was assumed that the medium between the source σ and the points P_1 and P_2 is homogeneous. It is not difficult to generalize the formula to other cases, e.g. when the medium is heterogeneous or consists of a succession of homogeneous regions of different refractive indices.

We again imagine the source to be divided into small elements $d\sigma_1$, $d\sigma_2$ . . ., centred on points S_1, S_2, . . ., of linear dimensions small compared to the mean wavelength $\bar{\lambda}$. If, as before, $V_{m1}(t)$ and $V_{m2}(t)$ represent the disturbances at P_1 and P_2 due to the element $d\sigma_m$, equations (15) and (16) still hold, but in (17) we must replace each factor $e^{i\bar{k}R_{mj}}/R_{mj}$ $(j = 1,2)$; $\bar{k} = 2\pi\bar{\nu}/v)$ by a more general function. We introduce a *transmission function* $K(S,P,\nu)$ of the medium, defined in a similar way as in § 9.5.1; it represents the complex disturbance at P, due to a monochromatic point source of frequency ν, of unit strength and of zero phase, situated at the element $d\sigma$ at S. For a homogeneous medium we have, from the HUYGENS–FRESNEL principle, that $K(S,P,\nu) = -ie^{ikR}/\lambda R$, where R denotes the distance SP, it being assumed that

* Cf. R. HANBURY BROWN and R. Q. TWISS, *Proc. Roy. Soc.*, A, **242** (1957), 300; *ibid.*, A, **243** (1957), 291. See also E. M. PURCELL, *Nature*, **178** (1956), 1449; F. D. KAHN, *Optica Acta*, **5** (1958), 93 and L. MANDEL, *Proc. Phys. Soc.*, **72** (1958), 1037; *Progress in Optics*, Vol. **2**, ed. E. WOLF (Amsterdam, North Holland Publishing Company and New York, J. Wiley and Sons, 1963), 181.

† E. WOLF, *Phil. Mag.*, **2** (1957), 351. See also J. A. RATCLIFFE, *Rep. Progr. Phys.* (London, Physical Society), **19** (1956), 233.

the angle which SP makes with the normal to $d\sigma$ is sufficiently small. It follows that in the more general case the factor $e^{i\bar{k}R_{mj}}/R_{mj}$ must be replaced by $i\bar{\lambda}K(S_m,P_j,\bar{\nu})$, and we obtain, on passing to a continuous distribution, the following relation in place of (20):

$$J(P_1,P_2) = \bar{\lambda}^2 \int_\sigma I(S)K(S,P_1,\bar{\nu})K^\star(S,P_2,\bar{\nu})dS. \tag{32}$$

According to (32) and (9b),

$$\mu(P_1,P_2) = \frac{\bar{\lambda}^2}{\sqrt{I(P_1)}\sqrt{I(P_2)}} \int_\sigma I(S)K(S,P_1,\bar{\nu})K^\star(S,P_2,\bar{\nu})dS, \tag{33}$$

where $I(P_1) = J(P_1,P_1)$ and $I(P_2) = J(P_2,P_2)$ are the intensities at P_1 and P_2 respectively.

For the purpose of later applications it will be useful to express (32) and (33) in a slightly different form. We set

$$i\bar{\lambda}K(S,P_1,\bar{\nu})\sqrt{I(S)} = U(S,P_1), \qquad i\bar{\lambda}K(S,P_2,\bar{\nu})\sqrt{I(S)} = U(S,P_2). \tag{34}$$

The formulae (32) and (33) become

$$J(P_1,P_2) = \int_\sigma U(S,P_1)U^\star(S,P_2)dS, \tag{35a}$$

$$\mu(P_1,P_2) = \frac{1}{\sqrt{I(P_1)}\sqrt{I(P_2)}} \int_\sigma U(S,P_1)U^\star(S,P_2)dS. \tag{35b}$$

We note that $U(S,P)$, defined by (34), is proportional to the disturbance which would arise at P from a strictly monochromatic point source of frequency $\bar{\nu}$, strength $\sqrt{I(S)}$ and zero phase, situated at S. Thus (35) may be interpreted as expressing the mutual intensity $J(P_1,P_2)$ and the complex degree of coherence $\mu(P_1,P_2)$ due to an extended quasi-monochromatic source, in terms of the *disturbances* produced at P_1 and P_2 by each source point of an "associated" monochromatic source.*

The expression (35b) was first suggested by HOPKINS† from heuristic considerations and is very useful in solving coherence problems of instrumental optics. The main usefulness of the formula arises from the fact that, like the VAN CITTERT–ZERNIKE theorem, it permits the calculation of the complex degree of coherence of light from an incoherent source without the explicit use of an averaging process.

10.4.3 An example

We shall illustrate the preceding considerations by the discussion of an experiment. A primary source σ_0 is imaged by a lens L_0 on to a pinhole σ_1 and the light that emerges from the pinhole is rendered parallel by a lens L_1. A second lens L_2, exactly similar to L_1, brings the interfering beams to a focus F in the focal plane $\mathscr{F}$ of the lens L_2.

---

* It must not be assumed that the *mutual intensity* and the *complex degree of coherence* of light from such a fictitious source are also given by (35). For, as already explained, the relation (16) used in the derivation of these formulae is not valid in the limiting case of monochromatic radiation; the degree of coherence of monochromatic light is, in fact, always equal to unity.

† H. H. HOPKINS, *Proc. Roy. Soc.*, A, **208** (1951), 263.

A plane mirror M is used to reduce the overall length of the instrument (Fig. 10.4). If a diffracting mask (a dark screen $\mathscr{A}$, for example a piece of uniformly blackened film), with apertures of any desired size, shape, and distribution is placed in the parallel beam between L_1 and L_2, its FRAUNHOFER diffraction pattern is formed in the focal plane $\mathscr{F}$, and, under normal use, the operator may displace the mask while viewing the diffraction pattern through a microscope.*

If σ_0 were a quasi-monochromatic point source, it would give rise to coherent illumination in the neighbourhood of its geometrical image in the plane of σ_1. The

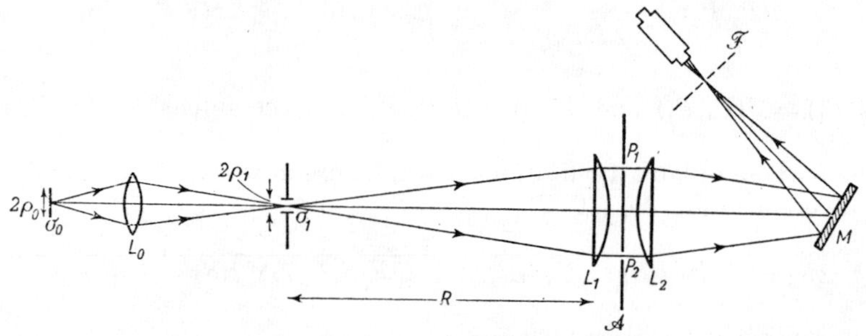

Fig. 10.4. The Diffractometer.

size of this coherently illuminated area is of the order of the effective size of the AIRY diffraction pattern σ_A, formed by the lens L_0, of the single source point. The light distribution in the plane of σ_1 due to a finite primary source may be regarded as arising from the incoherent superposition of many such patterns. If, as we assume, the image formed by L_0 of this extended source, and the pinhole σ_1, are both large compared to σ_A, the illuminated pinhole σ_1 will itself effectively act as an *incoherent source*.† According to the VAN CITTERT–ZERNIKE theorem, such a source will give rise to a correlation between vibrations at any two points on the first surface of the lens L_1 (and more generally in the plane $\mathscr{A}$); and with the usual approximations the complex degree of the coherence is given by the formula (28):

$$\mu_{12} = |\mu_{12}|e^{i\beta_{12}} = \frac{2J_1(v)}{v}\, e^{i\psi}, \tag{36a}$$

where

$$v = \frac{2\pi}{\lambda}\frac{\rho_1 d}{R}, \qquad \psi = \frac{2\pi}{\lambda}\left(\frac{r_1^{\,2} - r_2^{\,2}}{2R}\right), \tag{36b}$$

where $d = P_1P_2$, ρ_1 is the radius of σ_1, R is the distance between σ_1 and L_1, and r_1 and r_2 are the distances of P_1 and P_2 from the axis.

If the diffracting mask $\mathscr{A}$ consists of two small circular apertures centred on P_1 and P_2, the pattern observed in the focal plane $\mathscr{F}$ results from the superposition of

* This apparatus, known as the *Diffractometer*, is mainly used in connection with optical diffraction methods for solutions of problems of X-ray structure analysis. (Cf. C. A. TAYLOR, R. M. HINDE, and H. LIPSON, *Acta Cryst.*, **4** (1951), 261; A. W. HANSON, H. LIPSON, and C. A. TAYLOR, *Proc. Roy. Soc.* A, **218** (1953), 371; W. HUGHES and C. A. TAYLOR, *J. Sci. Instr.*, **30** (1953), 105.)

† This point is discussed quantitatively in § 10.5.1. See also A. T. FORRESTER, *Amer. J. Phys.*, **24** (1956), 194.

two partially coherent beams, with a degree of coherence $|\mu_{12}|$, emerging from these apertures. We shall investigate the changes in the structure of this pattern as the separation of P_1 and P_2 is gradually increased, i.e. as the degree of coherence between the two interfering beams is varied.

We assume that P_1 and P_2 are situated symmetrically about the axis. Then $\psi = 0$ and the intensities $I^{(1)}(Q)$ and $I^{(2)}(Q)$ at a point Q in the focal plane associated with either of the two beams are then equal, and are given by the FRAUNHOFER formula for

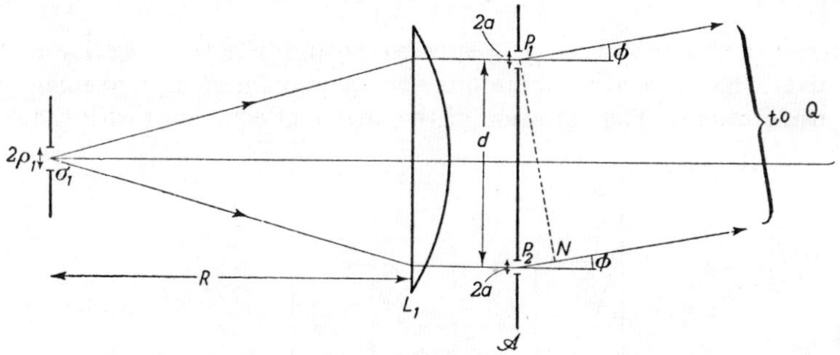

Fig. 10.5. Calculation of the intensity distribution in the focal plane of the Diffractometer.

diffraction by a circular aperture (§ 8.5 (14)). If the point Q is the focus for rays diffracted in directions that make an angle ϕ with the normal to $\mathscr{A}$, and if a is the radius of each aperture (see Fig. 10.5) then, apart from a normalizing factor,

$$I^{(1)}(Q) = I^{(2)}(Q) = \left(\frac{2J_1(u)}{u}\right)^2, \qquad u = \frac{2\pi}{\bar{\lambda}} a \sin \phi. \tag{37}$$

The phase difference δ between the beams diffracted to Q is

$$\delta = \frac{2\pi}{\bar{\lambda}} P_2N = \frac{2\pi}{\bar{\lambda}} d \sin \phi = Cuv, \qquad C = \frac{\bar{\lambda}}{2\pi} \frac{R}{\rho_1 a}, \tag{38}$$

where N is the foot of the perpendicular dropped from P_1 on to the ray diffracted at P_2. On substituting from (36), (37), and (38) into (11) we finally obtain the following expression for the intensity at the point $Q(\phi)$ in the focal plane, when the apertures at P_1 and P_2 are separated by a distance d:

$$I(\phi,d) = 2\left(\frac{2J_1(u)}{u}\right)^2\left\{1 + \left|\frac{2J_1(v)}{v}\right| \cos\left[\beta_{12}(v) - Cuv\right]\right\}, \tag{39}$$

where

$$\left.\begin{aligned} \beta_{12}(v) &= 0 \quad \text{when} \quad \frac{2J_1(v)}{v} > 0, \\ &= \pi \quad \text{when} \quad \frac{2J_1(v)}{v} < 0. \end{aligned}\right\} \tag{40}$$

In Fig. 10.6 are shown photographs of the patterns observed with such an arrangement, for various separations d. The corresponding theoretical curves, computed from the formula (39) are also shown. The chain lines represent the envelopes

$$\left.\begin{aligned} I_{\max}(\phi,d) &= 2\left(\frac{2J_1(u)}{u}\right)^2\left\{1+\left|\frac{2J_1(v)}{v}\right|\right\}, \\ I_{\min}(\phi,d) &= 2\left(\frac{2J_1(u)}{u}\right)^2\left\{1-\left|\frac{2J_1(v)}{v}\right|\right\}. \end{aligned}\right\} \tag{41}$$

It is of interest to note that when $\beta = \pi$ (cases (D) and (E)), the intensity at the centre of each pattern has a relative minimum, not a maximum, in agreement with our general considerations. The variation of the degree of coherence with the separation

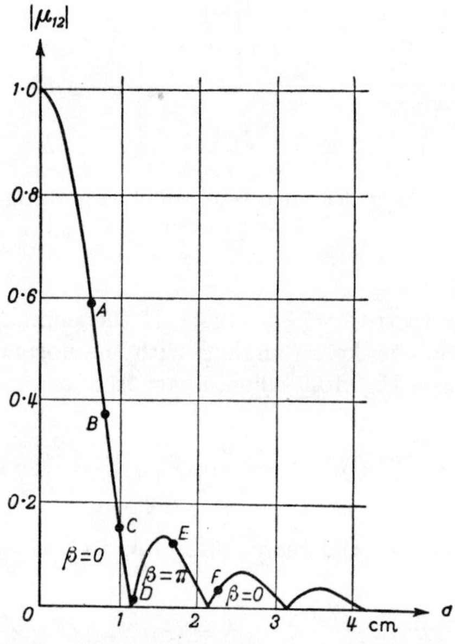

Fig. 10.7. Two-beam interference with partially coherent light. The degree of coherence as function of the separation d of the two illuminated apertures in the Diffractometer. ($\rho_1 = 0{\cdot}45 \times 10^{-2}$ cm, $R = 152$ cm, $\bar{\lambda} = 5790$ Å; incoherent illumination of σ_1 assumed.)

of the two apertures is shown in Fig. 10.7, where also the six values corresponding to the photographs of Fig. 10.6 are indicated by the appropriate letters.

10.4.4 Propagation of mutual intensity

Consider a beam of quasi-monochromatic light from an extended primary source σ, and suppose that the mutual intensity is known for all pairs of points on a fictitious surface $\mathscr{A}$ intercepting the beam. We shall show that it is then possible to determine the mutual intensity for all pairs of points on any other second surface $\mathscr{B}$ illuminated by the light from $\mathscr{A}$ either directly or via an optical system.

We assume to begin with that the medium between $\mathscr{A}$ and $\mathscr{B}$ is homogeneous and of refractive index unity. Let $U(S,Q_1)$ and $U(S,Q_2)$ be the disturbances at points Q_1

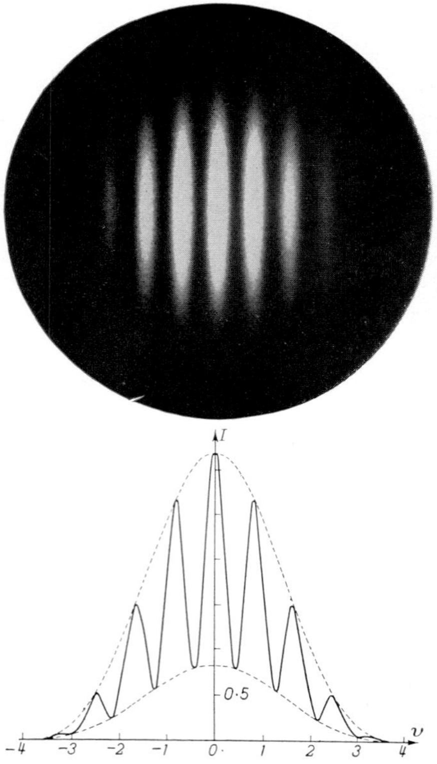

(A): $d = 0{\cdot}6$ cm
$|\mu_{12}| = 0{\cdot}593,\ \beta_{12} = 0$

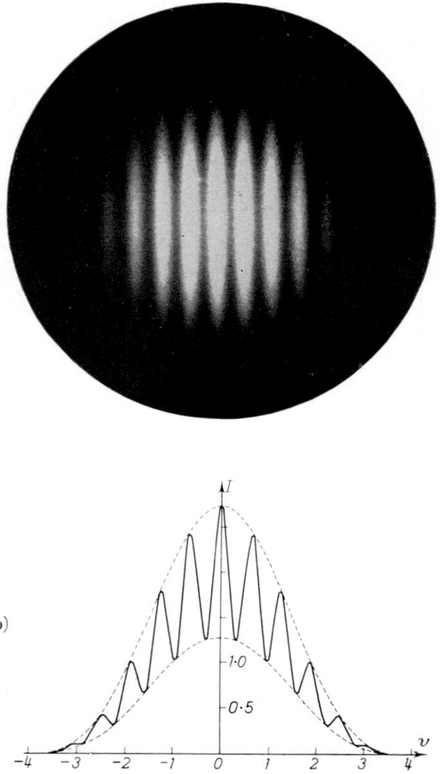

(B): $d = 0{\cdot}8$ cm
$|\mu_{12}| = 0{\cdot}361,\ \beta_{12} = 0$

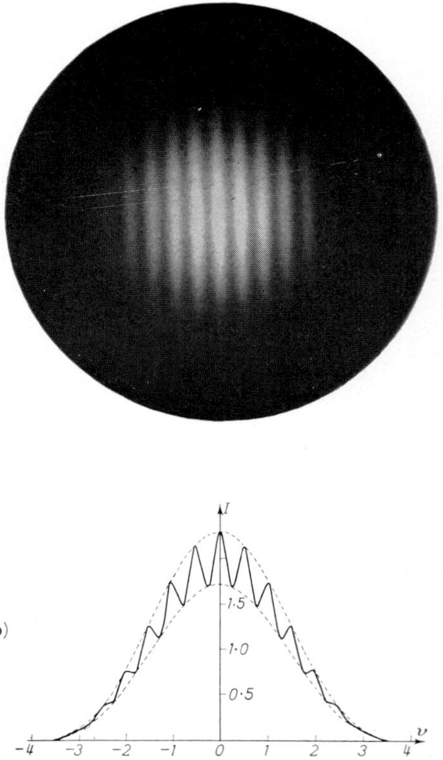

(C): $d = 1$ cm
$|\mu_{12}| = 0{\cdot}146,\ \beta_{12} = 0$

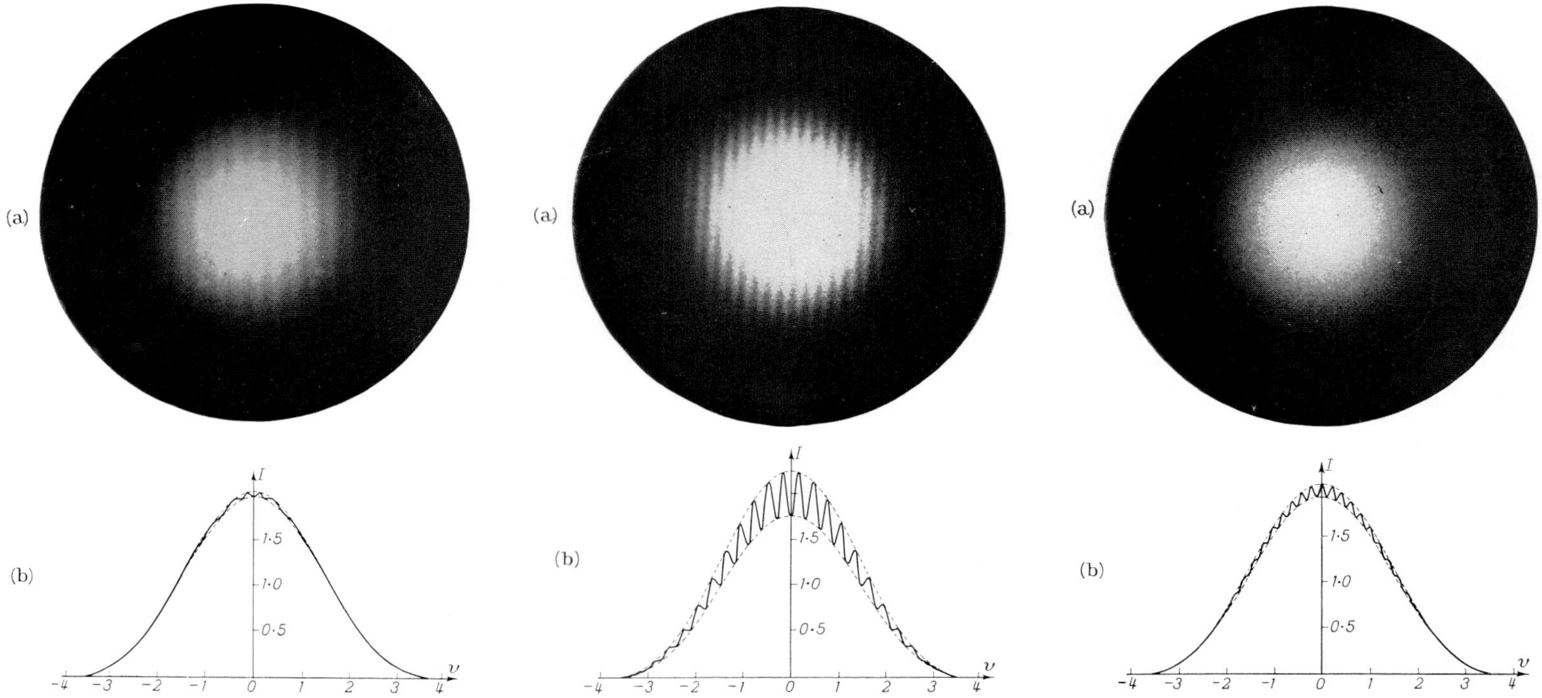

(D): $d = 1{\cdot}2$ cm
$|\mu_{12}| = 0{\cdot}015$, $\beta_{12} = \pi$

(E): $d = 1{\cdot}7$ cm
$|\mu_{12}| = 0{\cdot}123$, $\beta_{12} = \pi$

(F): $d = 2{\cdot}3$ cm
$|\mu_{12}| = 0{\cdot}035$, $\beta_{12} = 0$

Fig. 10.6. Two-beam interference with partially coherent light.
(a) Observed patterns, (b) theoretical intensity curves. Focal length of lenses L_0, L_1, and L_2 of Diffractometer: $f_0 = 20$ cm, $f_1 = f_2 = R = 152$ cm. Diameter of $L_0 = 5$ cm. Distance from L_0 to σ_1: 40 cm. Separation of L_1 and L_2: 14 cm. Distance of mirror M from $L_2 = 85$ cm. Diameter $2\rho_1$ of pinhole σ_1: $0{\cdot}9 \times 10^{-2}$ cm. Diameter $2a$ of apertures at P_1 and P_2: 0·14 cm. Mean wavelength $\bar{\lambda} = 5790$ Å.

(After B. J. Thompson and E. Wolf, *J. Opt. Soc. Amer.*, **47** (1957), 895.)

and Q_2 on $\mathscr{B}$ (Fig. 10.8) due to a typical source point S of the associated monochromatic source. Then, according to (35), the mutual intensity $J(Q_1,Q_2)$ is given by

$$J(Q_1,Q_2) = \int_\sigma U(S,Q_1)U^\star(S,Q_2)dS. \tag{42}$$

Now $U(S,Q_1)$ and $U(S,Q_2)$ may be expressed in terms of the disturbance at all points of $\mathscr{A}$ by means of the HUYGENS–FRESNEL principle:

$$U(S,Q_1) = \int_{\mathscr{A}} U(S,P_1)\frac{e^{i\bar{k}s_1}}{s_1}\Lambda_1 dP_1. \tag{43}$$

Here s_1 is the distance from a typical point P_1 on $\mathscr{A}$ to Q_1, Λ_1 is the inclination factor (denoted by K in Chapter VIII) at P_1, and $\bar{k} = 2\pi\bar{\nu}/c$ is the mean wave number. For small obliquities, $\Lambda_1 \sim -i/\bar{\lambda}$. According to (43) and a similar expression for $U(Q_2)$, we have

$$U(S,Q_1)U^\star(S,Q_2) = \int_{\mathscr{A}}\int_{\mathscr{A}} U(S,P_1)U^\star(S,P_2)\frac{e^{i\bar{k}(s_1-s_2)}}{s_1 s_2}\Lambda_1\Lambda_2^\star dP_1 dP_2, \tag{44}$$

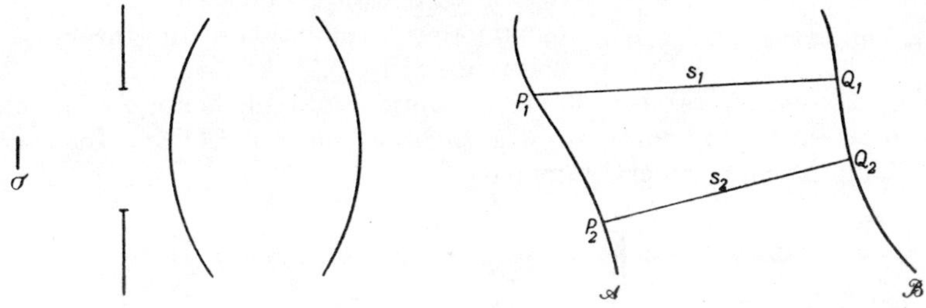

Fig. 10.8. Propagation of mutual intensity: illustrating formula (45).

where the points P_1 and P_2 take on independently all positions on the surface $\mathscr{A}$ of integration. Next we substitute from (44) into (42), and change the order of integration. The integration over σ gives precisely $J(P_1,P_2)$ and we obtain

$$J(Q_1,Q_2) = \int_{\mathscr{A}}\int_{\mathscr{A}} J(P_1,P_2)\frac{e^{i\bar{k}(s_1-s_2)}}{s_1 s_2}\Lambda_1\Lambda_2^\star dP_1 dP_2. \tag{45}$$

This is the required formula, due to ZERNIKE,* for the propagation of the mutual intensity. We have implicitly assumed in the derivation of (45) that light from every point of the surface $\mathscr{A}$ reaches the points Q_1 and Q_2. The presence of any diaphragm between the two surfaces may be taken care of by limiting the integration to those portions of the surface $\mathscr{A}$ which send light to Q_1 and Q_2, unless the diaphragm is so small that the effects of diffraction at its edges cannot be neglected. The diffraction can be taken into account by carrying out the transition from $\mathscr{A}$ to $\mathscr{B}$ in two steps, first from $\mathscr{A}$ to the plane of the diaphragm, and then from the plane of the diaphragm to the surface $\mathscr{B}$.

* F. ZERNIKE, *Physica*, **5** (1938), 791.

In the special case when the points Q_1 and Q_2 coincide, (45) reduces to the following expression for the intensity, when we also substitute for $J(P_1,P_2)$ in terms of the intensities $I(P_1)$, $I(P_2)$ and the complex degree of coherence $\mu(P_1,P_2)$:

$$I(Q) = \int_{\mathscr{A}}\int_{\mathscr{A}} \sqrt{I(P_1)}\sqrt{I(P_2)}\mu(P_1,P_2)\frac{e^{ik(s_1-s_2)}}{s_1 s_2}\Lambda_1\Lambda_2^\star dP_1 dP_2. \qquad (46)$$

This formula expresses the intensity at a point Q as the sum of contributions from each pair of elements dP_1, dP_2 of an arbitrary surface $\mathscr{A}$ intercepting the beam (see Fig. 10.9). The contribution from each pair of elements depends on the intensities at P_1 and P_2 and each contribution is weighted by the appropriate value of the complex degree of coherence factor $\mu(P_1,P_2)$. Formula (46) may be regarded as a kind of HUYGENS–FRESNEL principle for the propagation of intensity in a partially coherent field. The resemblance between the formulae just derived and those of the more elementary HUYGENS–FRESNEL theory has a deeper significance which will be brought out in our rigorous formulation of the theory of partial coherence (§ 10.7).

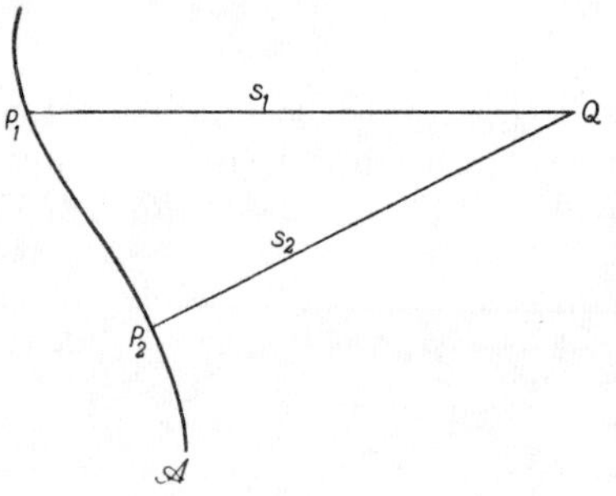

Fig. 10.9. Illustrating formula (46).

If the light from $\mathscr{A}$ reaches $\mathscr{B}$ via an optical system, factor $\Lambda e^{iks}/s$ must evidently be replaced by an appropriate transmission function $K(P,Q)$. Instead of (45) we then obtain the more general formula

$$J(Q_1,Q_2) = \int_{\mathscr{A}}\int_{\mathscr{A}} J(P_1,P_2)K(P_1,Q_1)K^\star(P_2,Q_2)dP_1 dP_2. \qquad (47)$$

10.5 SOME APPLICATIONS

10.5.1 The degree of coherence in the image of an extended incoherent quasi-monochromatic source

As a preliminary to the study of image formation with partially coherent light, it will be useful to consider the degree of coherence in the image of an extended incoherent source, formed by a centred optical system. A finite degree of correlation among vibrations in the image plane arises from the fact that, because of diffraction (and in general also because of aberrations), the light from each source point is not concentrated into a point, but spreads over a finite area. Some of these "image patterns" overlap and in consequence points in the image plane that are close enough to each other receive coherent as well as incoherent contributions.

Suppose that σ is a uniform quasi-monochromatic incoherent circular source of radius ρ emitting light with mean (vacuum) wavelength $\bar{\lambda}_0$, situated in a homogeneous object space of refractive index n. Further let D be the distance between the object plane and the plane of the entrance pupil. Corresponding quantities in the image will be denoted by primed symbols of the same type.

Let d be the distance between two points P_1 and P_2 in the entrance pupil. We assume that $\rho/\mathrm{D} \ll 1$, $d/\mathrm{D} \ll 1$ and $OP_1 - OP_2 \ll \bar{\lambda}_0$, where O is the axial point of the source*; then the complex degree of coherence $\mu(P_1,P_2)$ is, according to § 10.4 (28), given by

$$\mu(P_1,P_2) = \frac{2J_1(v)}{v}, \tag{1}$$

$$v = \frac{2\pi}{\bar{\lambda}} d \sin \alpha = \frac{2\pi n}{\bar{\lambda}_0} d \sin \alpha. \tag{2}$$

where $\alpha \sim \sin \alpha \sim \rho/D$ is the angular radius of the source as seen from the centre of the entrance pupil (Fig. 10.10).

To determine the complex degree of coherence for any pair of points in the plane of the exit pupil, we could apply the propagation law § 10.4 (47). However, we are

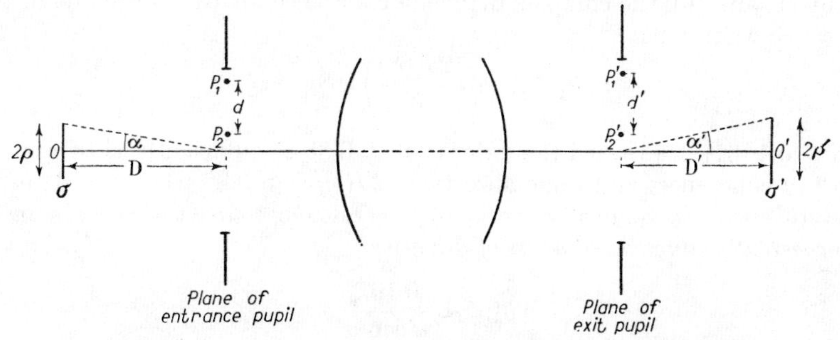

Fig. 10.10. Calculation of the degree of coherence in the image of an incoherent light source.

now concerned with the special case of propagation from a plane to its conjugate plane, and the propagation law reduces in this case to a simpler form, which may be derived directly as follows:

If $U(S,P_1)$ and $U(S,P_2)$ are the complex disturbances at P_1 and P_2 due to a source point S of the associated monochromatic source (cf. p. 513), the disturbances due to S at the conjugate points in the exit pupil are given by

$$U(S,P_1') = K_{11}U(S,P_1), \qquad U(S,P_2') = K_{22}U(S,P_2). \tag{3}$$

Here $K_{11} = K(P_1,P_1')$ is the appropriate transmission function for the propagation between the conjugate points P_1, P_1' of the pupil planes. By Hopkins' formula, § 10.4 (35b)

$$\mu(P_1',P_2') = \frac{1}{\sqrt{I(P_1')}\sqrt{I(P_2')}} \int_\sigma U(S,P_1')U^\star(S,P_2')dS. \tag{4}$$

Now the intensities $I(P_1')$ and $I(P_1)$ in the two pupil planes are related by

$$I(P_1') = \int_\sigma |U(S,P_1')|^2 dS = |K_{11}|^2 \int_\sigma |U(S,P_1)|^2 dS = |K_{11}|^2 I(P_1), \tag{5}$$

with a similar relation between $I(P_2')$ and $I(P_2)$. From (3), (4), and (5) we have

$$\mu(P_1',P_2') = \frac{K_{11}K_{22}^\star}{|K_{11}||K_{22}|} \frac{1}{\sqrt{I(P_1)}\sqrt{I(P_2)}} \int_\sigma U(S,P_1)U^\star(S,P_2)dS$$
$$= e^{i(\Phi_{11}-\Phi_{22})}\mu(P_1,P_2), \tag{6}$$

* If this last condition is not satisfied, $|\mu(P_1,P_2)|$ remains unchanged, but according to § 10.4 (28) the phase of $\mu(P_1,P_2)$ is increased by an amount $\psi = 2\pi[OP_1 - OP_2]/\bar{\lambda}_0$.

where Φ_{11} and Φ_{22} are the phases of K_{11} and K_{22} respectively. This relation implies that *the degree of coherence* $|\mu|$ *for any two points in the exit pupil is equal to the degree of coherence for the conjugate points in the entrance pupil; and the phases of the corresponding values of the complex degree of coherence for corresponding point pairs differ by the amount* $\Phi_{11} - \Phi_{22}$, *i.e. by the geometrical phase difference* $2\pi\{[P_1P_1'] - [P_2P_2']\}/\bar{\lambda}_0$.

Let

$$v' = \frac{2\pi n'}{\bar{\lambda}_0} d' \sin \alpha'. \tag{7}$$

Since P_1' is the conjugate of P_1 and P_2' the conjugate of P_2, it follows by the SMITH–HELMHOLTZ theorem (§ 4.4 (49)) that, within the accuracy of Gaussian optics, $v' = v$.* Hence, by (1) and (6), the complex degree of coherence for pairs of points in the exit pupil may be written as

$$\mu(P_1',P_2') = \left(\frac{2J_1(v')}{v'}\right) e^{i(\Phi_{11}-\Phi_{22})}. \tag{8}$$

If, as in § 10.4 (31), we regard the values $|\mu| \geqslant 0{\cdot}88$ as sufficiently close approximations to full coherence, and remember that $|2J(v)/v| \geqslant 0{\cdot}88$ when $v \leqslant 1$, it follows that *an incoherent quasi-monochromatic uniform circular source will give rise in the exit pupil to coherently illuminated areas of diameter*

$$d'_{\text{coh}} \sim \frac{0{\cdot}16\bar{\lambda}_0}{n' \sin \alpha'}, \tag{9}$$

where $2\alpha' \sim 2\rho'/D'$ *is the angle which the diameter of the image of the source subtends at the centre of the exit pupil and* $\bar{\lambda}_0/n' = \bar{\lambda}$ *is the mean wavelength of the light in the image space.*

We shall express (9) in a somewhat different form. Let r_A' denote the radius of the first dark ring in the AIRY pattern associated with the system,

$$r_A' = \frac{0{\cdot}61\bar{\lambda}_0}{n' \sin \theta'}, \tag{10}$$

where $n' \sin \theta' \sim n'a'/D'$ is the numerical aperture on the image side. Then according to (9) and (10), $d'_{\text{coh}}/r_A' \sim 0{\cdot}16 \sin \theta'/0{\cdot}61 \sin \alpha'$, so that

$$d'_{\text{coh}} \sim 0{\cdot}26a' \left(\frac{r_A'}{\rho'}\right). \tag{11}$$

This formula gives an estimate for the size of the coherently illuminated areas of the exit pupil in terms of the "physical parameters," viz. the radius r_A' of the first dark ring of the associated AIRY pattern, the radius ρ' of the geometrical image of the source, and the radius a' of the exit pupil.

The exit pupil and hence the image plane will be illuminated almost *coherently* if $d'_{\text{coh}} \geqslant 2a'$, i.e. if

$$\rho' \leqslant 0{\cdot}13r_A'. \tag{12}$$

When $d'_{\text{coh}} \ll 2a'$, i.e. when

$$\rho' \gg 0{\cdot}13r_A', \tag{13}$$

the coherently illuminated areas of the exit pupil will be small compared to the exit pupil itself so that in this case the illumination of the exit pupil is effectively *incoherent*. The complex degree of coherence for pairs of points Q_1', Q_2' in the image plane

* This means that v and v' represent a particular choice of the SEIDEL variables (§ 5.2).

will then be essentially the same as that due to an incoherent source; this source has the same size, shape, and position as the exit pupil, and the intensity distribution across this source is the same as the intensity distribution across the exit pupil. Hence according to the VAN CITTERT–ZERNIKE theorem, § 10.4 (21),

$$\mu(Q_1',Q_2') = \frac{1}{\sqrt{I(Q_1')}\sqrt{I(Q_2')}} \int_{\mathscr{A}'} I(P') \frac{e^{ik(s_1-s_2)}}{s_1 s_2} \, dP', \tag{14}$$

$$I(Q_1') = \int_{\mathscr{A}'} \frac{I(P')}{s_1^2} \, dP', \qquad I(Q_2') = \int_{\mathscr{A}'} \frac{I(P')}{s_2^2} \, dP'. \tag{15}$$

The integration is taken over the exit pupil $\mathscr{A}'$ and s_1 and s_2 denote the distances from the typical point P' in $\mathscr{A}'$ to the points Q_1' and Q_2' respectively (Fig. 10.11(a)).

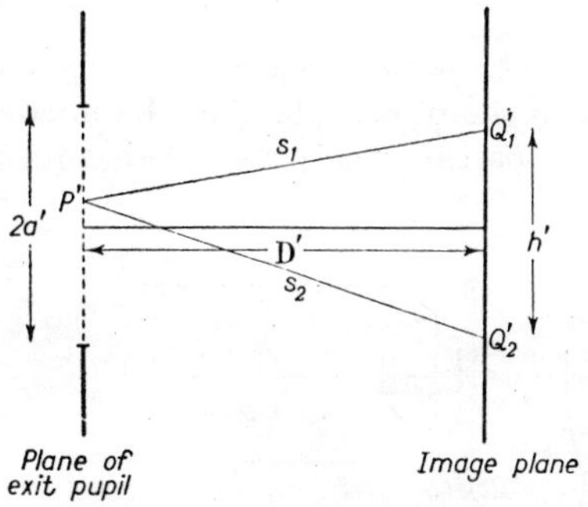

(a) Incoherent illumination of the exit pupil

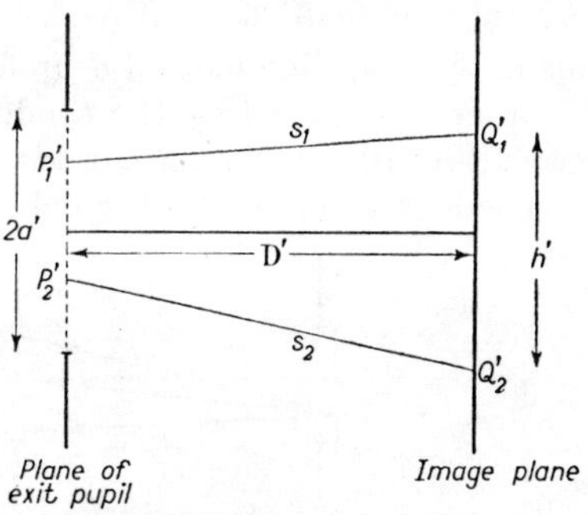

(b) Partially coherent illumination of the exit pupil

Fig. 10.11. Calculation of the complex degree of coherence in the image plane.

The intensity $I(P')$ may be calculated from the intensity $I(P)$ at the conjugate point in the entrance pupil by means of the relation (5). Since the phase of the transmission function does not appear in this relation, $\mu(Q_1',Q_2')$ *is independent of the aberrations of the system.* As a rule $I(P')$ is effectively constant; if, moreover, the points Q_1' and Q_2' are sufficiently close to each other the expression (14) then reduces to

$$\mu(Q_1',Q_2') = \frac{2J_1(u')}{u'}, \qquad u' = \frac{2\pi n'}{\lambda_0} \frac{a'}{\mathrm{D}'} h', \tag{16}$$

where h' is the distance between Q_1' and Q_2'.

In the general case, when neither condition (12) nor condition (13) holds, the exit pupil is illuminated with partially coherent light, characterized by the complex degree of coherence (8). The value of the complex degree of coherence for pairs of points in the image plane must then be calculated with the help of the propagation law § 10.4 (45) and leads to the expression

$$\mu(Q_1',Q_2') = \frac{1}{\sqrt{I(Q_1')}\sqrt{I(Q_2')}} \int_{\mathscr{A}'}\int_{\mathscr{A}'} \sqrt{I(P_1')}\sqrt{I(P_2')} \left(\frac{2J_1(v')}{v'}\right) \frac{e^{i[\Phi_{11}-\Phi_{22}+k(s_1-s_2)]}}{s_1 s_2} \Lambda_1 \Lambda_2^\star dP_1' dP_2'. \tag{17}$$

The intensities $I(Q_1')$ and $I(Q_2')$ may also be calculated from this formula, if use is made of the fact that $\mu(Q_1',Q_1') = \mu(Q_2',Q_2') = 1$. We note that, since the integrand contains the phases Φ_{11} and Φ_{22} of the transmission function, the complex degree of coherence now depends on the aberrations of the system.

10.5.2 The influence of the condenser on resolution in a microscope

In order to examine a small non-luminous object under a microscope, the object has to be illuminated. If, as is usually the case, the object is almost transparent, it is illuminated from behind, or *transilluminated*, and the light which has passed through the object is then focused on to the image plane of the microscope objective. To obtain a sufficient concentration of light, an auxiliary lens system—a condenser—is usually used. Various methods of illumination are employed. We shall briefly describe two commonly used methods, the so-called *critical illumination* and *Köhler's illumination*, and discuss the resolving power which can be achieved with them.

(a) *Critical illumination*

In this method of illumination a uniformly bright source is placed close behind the field stop and is imaged by the condenser on to the object plane of the microscope objective (Fig. 10.12). The size of the field stop aperture is adjusted so that its image by the condenser just covers the field.

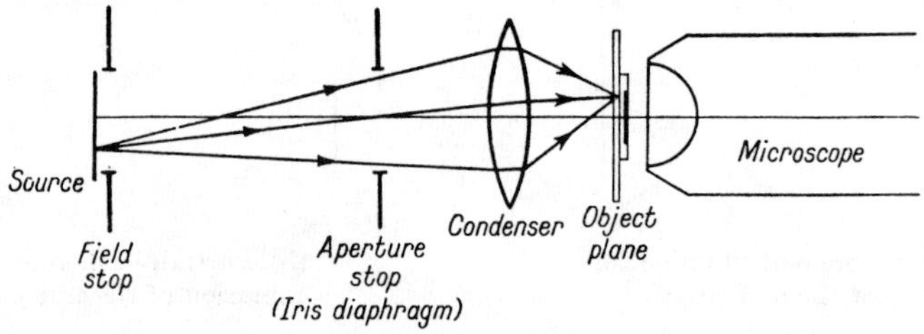

Fig. 10.12. Critical illumination.

The illuminated region in the image plane of the condenser (the object plane of the objective) is very much larger than the effective size of the AIRY pattern due to a single source point ($\rho' \gg r'_A$ in the notation of § 10.5.1). It follows from § 10.5.1 that under these circumstances *the complex degree of coherence for any pair of points in the object plane of the objective is the same as that due to an incoherent source filling the condenser aperture*; moreover, it is independent of the aberrations of the condenser. Now it is evident that the resolving power depends only on the degree of coherence (characterized by this factor) of the light incident upon the object and on the properties of the microscope objective. Hence *the aberrations of the condenser have no influence on the resolving power of a microscope*. This important result was first established in a different manner by ZERNIKE,* and shows that the widely held view, according to which a highly corrected condenser is of advantage for attaining high resolving power, is incorrect.

To estimate the effect of the size of the condenser on the resolution, consider two pinholes $P_1(X_1, Y_1)$ and $P_2(X_2, Y_2)$ in the plane of the object. With the same assumptions as before, the complex degree of coherence of the light that reaches these pinholes is given by a formula of the form (16):

$$\mu(P_1, P_2) = \frac{2J_1(u_{12})}{u_{12}}, \qquad u_{12} = \frac{2\pi}{\bar{\lambda}_0}\sqrt{(X_1 - X_2)^2 + (Y_1 - Y_2)^2}\, n'_c \sin \theta'_c, \quad (18)$$

* F. ZERNIKE, *Physica*, **5** (1938), 794.

where $n_c' \sin \theta_c'$ is the numerical aperture of the condenser on the side of the microscope objective.

Let $P(X, Y)$ be any other point in the object plane, and P' its image by the objective. If we assume that the objective is effectively free of aberrations, the intensity distribution, in the image plane of the objective, of the light arriving from P_1 alone is an AIRY pattern centred on the image P_1' of P_1. Hence if $n_0 \sin \theta_0$ is the numerical aperture of the objective, the intensity $I^{(1)}(P')$ due to light that reaches P' from P_1 alone is, apart from a constant factor, equal to

$$I^{(1)}(P') = \left(\frac{2J_1(v_1)}{v_1}\right)^2, \qquad v_1 = \frac{2\pi}{\bar{\lambda}_0} \sqrt{(X - X_1)^2 + (Y - Y_1)^2}\, n_0 \sin \theta_0. \quad (19a)$$

The intensity $I^{(2)}(P')$, due to the light reaching it from the pinhole P_2, is given by a similar expression:

$$I^{(2)}(P') = \left(\frac{2J_1(v_2)}{v_2}\right)^2, \qquad v_2 = \frac{2\pi}{\bar{\lambda}_0} \sqrt{(X - X_2)^2 + (Y - Y_2)^2}\, n_0 \sin \theta_0. \quad (19b)$$

It follows that when the two pinholes are illuminated via the condenser, the intensity $I(P')$ in the image plane of the microscope objective arises from the superposition of two partially coherent beams. The intensity of each beam is given by (19) and the complex degree of coherence of the two beams is given by (18). An expression for $I(P')$ is immediately obtained on substituting from these equations into the formula (11) of § 10.4. This gives, if we also assume that P' is very close to the geometrical images of P_1 and P_2 (more precisely that $\delta = [P_1P'] - [P_2P'] \ll \bar{\lambda}$),

$$I(P') = \left(\frac{2J_1(v_1)}{v_1}\right)^2 + \left(\frac{2J_1(v_2)}{v_2}\right)^2 + 2\left(\frac{2J_1(mv_{12})}{mv_{12}}\right)\left(\frac{2J_1(v_1)}{v_1}\right)\left(\frac{2J_1(v_2)}{v_2}\right), \quad (20)$$

where

$$m = \frac{n_c' \sin \theta_c'}{n_0 \sin \theta_0}, \qquad v_{12} = \frac{u_{12}}{m} = \frac{2\pi}{\bar{\lambda}_0} \sqrt{(X_1 - X_2)^2 + (Y_1 - Y_2)^2}\, n_0 \sin \theta_0. \quad (21)$$

Some interesting conclusions may be drawn from (20). When mv_{12} is a root other than $mv_{12} = 0$ of the equation $J_1(mv_{12}) = 0$, the product term is absent and (20) reduces to

$$I(P') = \left(\frac{2J_1(v_1)}{v_1}\right)^2 + \left(\frac{2J_1(v_2)}{v_2}\right)^2. \quad (22)$$

The distribution of the intensity in the image plane is now the same as if P_1 and P_2 were illuminated *incoherently*. In particular this will be the case when $m = 1$ and v_{12} is a non-zero root of $J_1(v_{12}) = 0$; that is, when the numerical apertures are equal and the geometrical images of the pinholes are separated by a distance equal to the radius of any dark ring of the AIRY pattern of the objective.

When the numerical aperture of the condenser is very small ($m \to 0$), then $2J_1(mv_{12})/mv_{12} \sim 1$ and (20) reduces to

$$I(P') = \left\{\frac{2J_1(v_1)}{v_1} + \frac{2J_1(v_2)}{v_2}\right\}^2. \quad (23)$$

The distribution of the intensity is now the same as with perfectly *coherent* illumination, whatever the separation of the pinholes.

Formula (20) makes it possible to study the dependence of the intensity distribution in the image plane of the microscope objective on the ratio m of the numerical apertures. In particular consider the intensity at the midpoint between P_1' and P_2'. We shall regard the pinholes to be just resolved when the intensity at the midpoint is by 26·5 per cent smaller than the intensity at either of the two points. The value 26·5 per cent corresponds to RAYLEIGH's criterion for a circular aperture in incoherent illumination (cf. § 8.6.2). We shall express this limiting separation $(P_1P_2)_{\text{lim}}$ in the same form as for incoherent [§ 8.6 (32)] and coherent [§ 8.6 (55)] illuminations:

$$(P_1P_2)_{\text{lim}} = L(m)\,\frac{\bar{\lambda}_0}{n_0 \sin \theta_0}. \tag{24}$$

The curve $L(m)$ computed from (20) on the basis of this criterion is shown in Fig. 10.13. It is seen that the best resolving power is obtained with $m \sim 1{\cdot}5$, i.e. when the

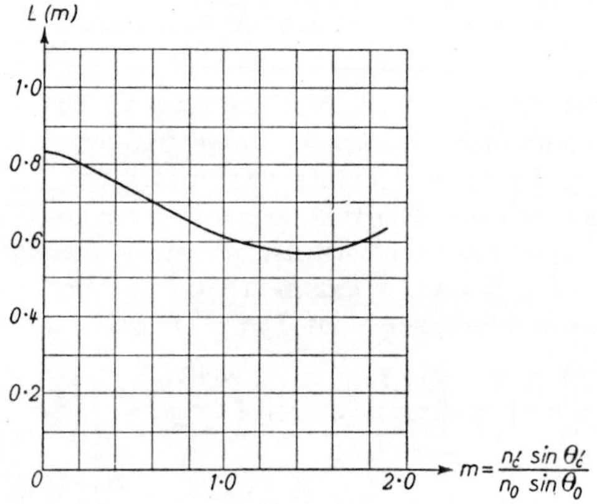

Fig. 10.13. Effect of the condenser aperture on the resolution of two pinholes of equal brightness. (After H. H. HOPKINS and P. M. BARHAM, *Proc. Phys. Soc.*, **63** (1950), 72.)

numerical aperture of the condenser is about 1·5 times that of the objective. The value of L is then slightly smaller than the value 0·61 obtained with incoherent illumination.

(b) *Köhler's illumination*

In a method of illumination due to KÖHLER,* which is illustrated in Fig. 10.14, a converging lens is placed close to the field stop and forms an image of the source σ in the focal plane of the condenser, which now contains the condenser diaphragm. The rays from each source point then emerge from the condenser as a parallel beam. This arrangement has the advantage that the irregularities in the brightness distribution on the source do not cause irregularities in the intensity of the field illumination.

To estimate the limit of resolution that is attained with KÖHLER's method of illumination we must first determine the complex degree of coherence μ for pairs of points in the object plane of the microscope objective. Let

$$U(S,P_1) = A_1 e^{i\phi_1}, \qquad U(S,P_2) = A_2 e^{i\phi_2}, \tag{25}$$

* A. KÖHLER, *Zs. f. wiss. Mikrosk.*, **10** (1893), 433; **16** (1899), 1.

be the complex disturbances at points $P_1(X_1, Y_1)$ and $P_2(X_2, Y_2)$ of the object plane of the microscope objective, due to a source point S of the monochromatic source associated with σ (cf. p. 513). Evidently

$$\phi_1 - \phi_2 = \frac{2\pi}{\bar{\lambda}_0}[p(X_1 - X_2) + q(Y_1 - Y_2)], \tag{26}$$

where p and q are the first two ray components of the two parallel rays through P_1 and P_2 from the source point S. If the condenser system suffers from aberrations, the two rays will not be strictly parallel, but as we are only considering points which

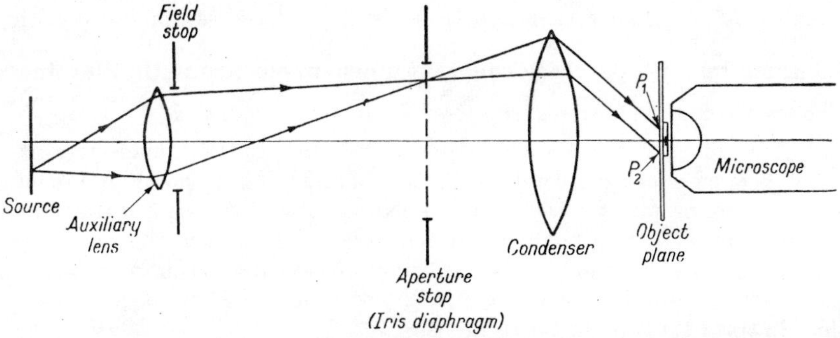

Fig. 10.14. KÖHLER's method of illumination.

are close to each other, this effect may be neglected. On substituting from (25) and (26) into HOPKINS' formula § 10.4 (35b) it follows that

$$\left.\begin{aligned} \mu(P_1,P_2) &= \frac{1}{\sqrt{I(P_1)}\sqrt{I(P_2)}} \int_\sigma A_1 A_2 e^{i\bar{k}_0[p(X_1-X_2)+q(Y_1-Y_2)]}\, dS, \\ I(P_1) &= \int_\sigma A_1^2 dS, \qquad I(P_2) = \int_\sigma A_2^2 dS. \end{aligned}\right\} \tag{27}$$

Since to every source point $S(\xi,\eta)$ there corresponds a pair of ray components (p,q), we may transform the integrals over σ into integrals over the solid angle

$$p^2 + q^2 \leqslant n_c'^2 \sin^2 \theta_c', \tag{28}$$

formed by the rays that are incident upon the object. Now within the accuracy of Gaussian optics, the relations $\xi = \xi(p,q)$, $\eta = \eta(p,q)$ are linear; in fact, as is easily seen from the formulae of § 4.3 (10), $\xi = fp$, $\eta = fq$ where f is the focal length of the condenser system. Hence the Jacobian $\partial(\xi,\eta)/\partial(p,q)$ of the transformation is constant. Outside the domain of geometrical optics, the Jacobian will in general vary over the region of integration, but this variation may be assumed to be slow compared to the variation of the exponential term and may therefore be neglected. If we also neglect the slow variation of A_1 and A_2, (27) reduces to

$$\mu(P_1,P_2) = \frac{\displaystyle\iint_\Omega e^{i\bar{k}_0[p(X_1-X_2)+q(Y_1-Y_2)]}\, dp\, dq}{\displaystyle\iint_\Omega dp\, dq}, \tag{29}$$

where Ω denotes the domain (28). Evaluation of (29) leads to the expression

$$\mu(P_1,P_2) = \frac{2J_1(u_{12})}{u_{12}}, \qquad u_{12} = \frac{2\pi}{\bar{\lambda}_0}\sqrt{(X_1-X_2)^2+(Y_1-Y_2)^2}\, n'_c \sin\theta'_c. \quad (30)$$

This formula is identical with the formula (18) for critical illumination. Hence *the complex degree of coherence of the light incident upon the object plane of a microscope is the same whether critical or Köhler's illumination is employed.* In view of this result it is somewhat unfortunate that critical illumination is often designated as "incoherent" and KÖHLER's illumination as "coherent". It follows that formula (20) holds with both types of illumination, and that Fig. 10.13 likewise applies in both cases.

10.5.3 Imaging with partially coherent quasi-monochromatic illumination*

(a) *Transmission of mutual intensity through an optical system*

In § 9.5 some general methods were described for the study of imagery of extended objects. The case of completely coherent illumination (§ 9.5.1) and completely incoherent illumination (§ 9.5.2) were studied. In the first case the transmission of the complex amplitude through the system was considered, in the second case the transmission of the intensity. We shall now investigate the more general case of partially coherent quasi-monochromatic illumination. The appropriate quantity to consider in this case is the mutual intensity.

We shall employ the same scale-normalized (SEIDEL) coordinates as in § 9.5 (1), so that the object point and its Gaussian image have the same coordinate numbers. Let $J_0(x_0,y_0; x'_0,y'_0)$ be the mutual intensity for points (x_0,y_0), (x'_0,y'_0) in the object plane. If $K(x_0,y_0; x_1,y_1)$ is the transmission function of the system (§ 9.5.1), the mutual intensity in the image plane is, according to the propagation law § 10.4 (47), given by

$$J_1(x_1,y_1; x'_1,y'_1) = \iiiint_{-\infty}^{+\infty} J_0(x_0,y_0; x'_0,y'_0)\, K(x_0,y_0; x_1,y_1)\, K^\star(x'_0,y'_0; x'_1,y'_1)\, dx_0dy_0dx'_0dy'_0. \quad (31a)$$

The integration extends only formally over an infinite domain, since J_0 is zero for all points in the object plane from which no light proceeds to the image plane.

As in § 9.5, we assume that the object is so small that it forms an isoplanatic region of the system, i.e. that for all points on it $K(x_0,y_0; x_1,y_1)$ may be replaced to a good approximation by a function depending on the differences $x_1 - x_0$ and $y_1 - y_0$ only, say $K(x_1 - x_0, y_1 - y_0)$. Equation (31a) then becomes

$$J_1(x_1,y_1; x'_1,y'_1) = \iiiint_{-\infty}^{+\infty} J_0(x_0,y_0; x'_0,y'_0)\, K(x_1 - x_0, y_1 - y_0) \times K^\star(x'_1 - x'_0, y'_1 - y'_0)\, dx_0dy_0dx'_0dy'_0. \quad (31b)$$

We represent J_0, J_1 and the product $KK^\star$ in the form of four-dimensional FOURIER integrals:

$$J_0(x_0,y_0; x'_0,y'_0) = \iiiint_{-\infty}^{+\infty} \mathscr{J}_0(f,g,;f',g')e^{-2\pi i(fx_0+gy_0+f'x_0'+g'y_0')}\, dfdgdf'dg', \quad (32a)$$

* The considerations of this section are based in parts on investigations of H. H. HOPKINS, *Proc. Roy. Soc.*, A, **217** (1953), 408 and P. DUMONTET, *Publ. Sci. Univ. d'Alger*, B, **1** (1955), 33.

$$J_1(x_1,y_1;\, x_1',y_1') = \iiiint_{-\infty}^{+\infty} \mathscr{J}_1(f,g;\, f',g')e^{-2\pi i(fx_1+gy_1+f'x_1'+g'y_1')}\, df dg df' dg', \qquad (32b)$$

$$K(x,y)K^\star(x',y') = \iiiint_{-\infty}^{+\infty} \mathscr{M}(f,g;\, f',g')e^{-2\pi i(fx+gy+f'x'+g'y')}\, df dg df' dg'. \qquad (32c)$$

Then by the FOURIER inversion formula

$$\mathscr{J}_0(f,g;\, f',g') = \iiiint_{-\infty}^{+\infty} J_0(x_0,y_0;\, x_0',y_0')e^{2\pi i(fx_0+gy_0+f'x_0'+g'y_0')}\, dx_0 dy_0 dx_0' dy_0', \qquad (33)$$

and there are strictly analogous relations for $\mathscr{J}_1$ and $\mathscr{M}$.

On applying the convolution theorem to (31b) we obtain the relation

$$\mathscr{J}_1(f,g;\, f',g') = \mathscr{J}_0(f,g;\, f',g')\mathscr{M}(f,g;\, f',g'). \qquad (34)$$

This formula implies that, if the mutual intensity in the object and image planes are represented as superposition of four-dimensional space-harmonic components of all possible spatial frequencies (f,g,f',g'), then each component in the image depends only on the corresponding component in the object, and the ratio of the components is equal to $\mathscr{M}$. Thus within the accuracy of the present approximation *the action of the optical system on the mutual intensity is equivalent to the action of a four-dimensional linear filter.* $\mathscr{M}$ is called *the frequency response function for partially coherent quasi-monochromatic illumination.*

The frequency response function $\mathscr{M}$ is related in a simple way to the pupil function of the system. If as in § 9.5 (10c), we represent K in the form of a two-dimensional FOURIER integral

$$K(x,y) = \iint_{-\infty}^{+\infty} \mathscr{K}(f,g)e^{-2\pi i(fx+gy)}\, df dg, \qquad (35)$$

and substitute for K into the inverse of (32c), we find that

$$\mathscr{M}(f,g;\, f',g') = \mathscr{K}(f,g)\mathscr{K}^\star(-f',\, -g'). \qquad (36)$$

But, according to § 9.5 (13), $\mathscr{K}(f,g)$ is equal to the value of the pupil function $G(\xi,\eta)$ of the system at the point

$$\xi = \lambda R f, \qquad \eta = \lambda R g, \qquad (37)$$

on the Gaussian reference sphere (radius R). Hence *the frequency response function for partially coherent quasi-monochromatic illumination is connected with the pupil function of the system by the formula*

$$\mathscr{M}\left(\frac{\xi}{\lambda R}, \frac{\eta}{\lambda R};\, \frac{\xi'}{\lambda R}, \frac{\eta'}{\lambda R}\right) = G(\xi,\eta)G^\star(-\,\xi',\, -\,\eta'). \qquad (38)$$

Since the pupil function is zero for points outside the area of the exit pupil, it follows that spectral components belonging to frequencies above certain values are not transmitted. If the exit pupil is a circle of radius a, then $G(\xi,\eta)G^\star(-\,\xi',-\,\eta')$

vanishes if $\xi^2 + \eta^2 > a^2$ or $\xi'^2 + \eta'^2 > a^2$. Hence spectral components of the mutual intensity belonging to frequencies (f,g,f',g') for which either

$$f^2 + g^2 > \left(\frac{a}{\bar{\lambda}R}\right)^2 \quad \text{or} \quad f'^2 + g'^2 > \left(\frac{a}{\bar{\lambda}R}\right)^2, \tag{39}$$

are not transmitted.* Here $\bar{\lambda}$ denotes the mean wavelength in the image space.

TABLE XXIV

The action of an optical system from the standpoint of its response to spatial frequencies. (Isoplanatic object region assumed)

Illumination	Basic quantity	Transition from object to image	Frequency response function
Coherent	Complex disturbance $U(x,y)$	$\mathscr{U}_1(f,g) = \mathscr{U}_0(f,g)\mathscr{K}(f,g)$	$\mathscr{K}(f,g)$
Incoherent	Intensity $I(x,y)$	$\mathscr{I}_1(f,g) = \mathscr{I}_0(f,g)\mathscr{L}(f,g)$	$\mathscr{L}(f,g) = \int\int_{-\infty}^{+\infty} \mathscr{K}(f'+f,g'+g) \times \mathscr{K}^\star(f',g')df'dg'$
Partially coherent	Mutual intensity $J(x,y;x',y')$	$\mathscr{J}_1(f,g;f',g') = \mathscr{J}_0(f,g;f',g')\mathscr{M}(f,g;f',g')$	$\mathscr{M}(f,g;f'g') = \mathscr{K}(f,g)\mathscr{K}^\star(-f',-g')$

$\mathscr{K}(f,g)$ is the FOURIER inverse of the transmission function $K(x,y)$ of the system and is related to the pupil function $G(\xi,\eta)$ by the formula $\mathscr{K}(\xi/\lambda R, \eta/\lambda R) = G(\xi,\eta)$, where R is the radius of the Gaussian reference sphere and λ the mean wavelength in the image space.

In Table XXIV, the basic formulae relating to imaging with partially coherent illumination are displayed, together with the corresponding formulae of § 9.5, relating to coherent and incoherent illumination. The formulae relating to incoherent illumination may be derived from the general formulae (34), (36), and (38) by assuming J_0 to be of the form $J_0(x_0,y_0;x_0',y_0') = I_0(x_0,y_0)\,\delta(x_0'-x_0)\,\delta(y_0'-y_0)$, where δ is the DIRAC delta function (see Appendix IV). The calculations are straightforward but somewhat lengthy and will not be given here. Formulae relating to the special case of perfectly monochromatic (and hence completely coherent) illumination may be derived somewhat more easily by noting that in this case the mutual intensity is of the form $J_0(x_0,y_0;x_0',y_0') = U_0(x_0,y_0)U_0^\star(x_0',y_0')$. The FOURIER transform $\mathscr{J}_0$ of J_0 now likewise splits into the product of two factors and the application of (34) and (36) shows that each spectral component of the complex disturbance U_0 is transmitted through the system in accordance with the formulae displayed in the first row of Table XXIV.

(b) *Images of transilluminated objects*

Suppose that a portion of the object plane is occupied by a transparent or semi-transparent object which is illuminated with partially coherent quasi-monochromatic

* If the angular aperture of the system is small and the sine condition is obeyed, we have, as in § 9.5, p. 483, $a/\bar{\lambda}R \sim n_0 \sin\theta_0/M\bar{\lambda}_0$, where $n_0 \sin\theta_0$ is the numerical aperture of the system, M the Gaussian magnification, and $\bar{\lambda}_0$ is the mean wavelength in vacuum.

light. It will be assumed that this light originates in a primary source and reaches the object plane after the passage through some illuminating system (condenser).

As in § 8.6.1, we specify the object by an appropriate transmission function $F_0(x_0,y_0)$. If $U_0^-(S; x_0,y_0)$ represents the disturbance at the point (x_0,y_0) of the object plane due to a source point S of the associated monochromatic source (cf. p. 513), the disturbance from this source point after the passage through the object is given by

$$U_0(S; x_0,y_0) = U_0^-(S; x_0,y_0)\, F(x_0,y_0). \tag{40}$$

Now according to § 10.4 (35a) the mutual intensity of the light incident on the object is given by

$$J_0^-(x_0,y_0; x_0',y_0') = \int_\sigma U_0^-(S; x_0,y_0) U_0^{-\star}(S; x_0',y_0')\, dS, \tag{41a}$$

and the mutual intensity of the light emerging from the object is

$$J_0(x_0,y_0; x_0',y_0') = \int_\sigma U_0(S; x_0,y_0) U_0^\star(S; x_0',y_0')\, dS, \tag{41b}$$

so that, because of (40),

$$J_0(x_0,y_0; x_0',y_0') = F(x_0,y_0) F^\star(x_0',y_0')\, J_0^-(x_0,y_0; x_0',y_0'). \tag{42}$$

We shall confine our attention to the important case when the mutual intensity J_0^- of the incident light depends on the four coordinates x_0, y_0, x_0', y_0' through the differences $x_0 - x_0'$, $y_0 - y_0'$ only, i.e. when J_0^- is of the form

$$J_0^-(x_0,y_0; x_0',y_0') = J_0^-(x_0 - x_0', y_0 - y_0'). \tag{43}$$

We learned in § 10.5.2 that this will be the case for both the critical and the Köhler illumination. We retain the earlier assumptions that the object is so small that it forms an isoplanatic area of the system. It then follows from (31b) that the intensity $I_1(x_1,y_1) = J_1(x_1,y_1; x_1,y_1)$ in the image plane is given by

$$I_1(x_1,y_1) = \iiiint_{-\infty}^{+\infty} J_0^-(x_0 - x_0', y_0 - y_0') F(x_0,y_0) F^\star(x_0',y_0') K(x_1 - x_0, y_1 - y_0) \times K^\star(x_1 - x_0', y_1 - y_0')\, dx_0 dy_0 dx_0' dy_0'. \tag{44}$$

We represent F and J_0^- in the form of two-dimensional Fourier integrals

$$F(x,y) = \iint_{-\infty}^{+\infty} \mathscr{F}(f,g) e^{-2\pi i(fx+gy)}\, df dg, \tag{45a}$$

$$J_0^-(x,y) = \iint_{-\infty}^{+\infty} \mathscr{J}_0^-(f,g) e^{-2\pi i(fx+gy)}\, df dg. \tag{45b}$$

If we substitute for F and $F^\star$ from (45a) into (44), use the identity $f'x_0 - f''x_0' = (f' - f'')x_1 - f'(x_1 - x_0) + f''(x_1 - x_0')$ and a similar identity involving g and y, and introduce new variables of integration $u' = x_1 - x_0$, $u'' = x_1 - x_0'$, we obtain the following expression for I_1:

$$I_1(x_1,y_1) = \iiiint_{-\infty}^{+\infty} \mathscr{T}(f',g'; f'',g'') \mathscr{F}(f',g') \mathscr{F}^\star(f'',g'') e^{-2\pi i[(f'-f'')x_1 + (g'-g'')y_1]}\, df' dg' df'' dg'', \tag{46}$$

where

$$\begin{aligned}\mathscr{T}(f',g';f'',g'') &= \iiiint_{-\infty}^{+\infty} J_0^-(u''-u',v''-v')K(u',v')K^\star(u'',v'')e^{2\pi i[(f'u'+g'v')-(f''u''+g''v'')]}\,du'dv'du''dv'' \\ &= \iiiiiint_{-\infty}^{+\infty} \mathscr{J}_0^-(f,g)K(u',v')K^\star(u'',v'')e^{2\pi i[(f+f')u'+(g+g')v'-(f+f'')u''-(g+g'')v'']}\,dfdgdu'dv'du''dv'' \\ &= \iint_{-\infty}^{+\infty} \mathscr{J}_0^-(f,g)\mathscr{K}(f+f',g+g')\mathscr{K}^\star(f+f'',g+g'')dfdg. \qquad (47)\end{aligned}$$

In passing from the second to the third lines in (47) we substituted for J_0^- from (45b) and in passing from the third to the fourth line we used the inverse of (35).

We see that in (46) the influence of the object (characterized by $\mathscr{F}$) and the combined effect of the illumination ($\mathscr{J}_0^-$) and of the system ($\mathscr{K}$) are separated. With a uniform illumination ($I_0^- =$ constant), the intensity of the light emerging from the object would be proportional to $|F|^2$, and were the imaging perfect, the intensity in the image plane would be given (apart from a constant factor) by

$$\begin{aligned}\tilde{I}_1(x_1,y_1) &= F(x_1,y_1)\,F^\star(x_1,y_1) \\ &= \iiiint_{-\infty}^{+\infty} \mathscr{F}(f',g')\mathscr{F}^\star(f'',g'')e^{-2\pi i[(f'-f'')x_1+(g'-g'')y_1]}\,df'dg'df''dg''. \qquad (48)\end{aligned}$$

Equations (46) and (48) represent the true intensity I_1 and the ideal intensity $\tilde{I}_1$ as sums of contributions from all pairs of frequencies (f',g'), (f'',g'') of the spatial spectrum of the object. Each contribution in the former case is $\mathscr{T}$ times that of the latter, and it follows that unless $\mathscr{T}$ is constant for all values f',g',f'',g'' for which both the spectral components $\mathscr{F}(f',g')$, $\mathscr{F}(f'',g'')$ differ from zero, some information about the object will be lost or falsified. The function $\mathscr{T}$ is called *the transmission cross-coefficient* of the system, working with the given transillumination.

Instead of the intensity itself, let us now consider its spatial spectrum $\mathscr{I}(f,g)$. To derive the appropriate formula we multiply both sides of (46) by $e^{2\pi i(fx_1+gy_1)}$ and integrate with respect to x_1 and y_1. If next we use the FOURIER integral theorem (or more shortly the FOURIER integral representation of the DIRAC delta function (see Appendix IV), we obtain

$$\mathscr{I}_1(f,g) = \iint_{-\infty}^{+\infty} \mathscr{T}(f'+f,g'+g,f',g')\,\mathscr{F}(f'+f,g'+g)\mathscr{F}^\star(f',g')df'dg'. \qquad (49)$$

For the ideal case represented by (48) we have

$$\tilde{\mathscr{I}}_1(f,g) = \iint_{-\infty}^{+\infty} \mathscr{F}(f'+f,g'+g)\mathscr{F}^\star(f',g')df'dg'. \qquad (50)$$

These formulae express $\mathscr{I}_1$ and $\tilde{\mathscr{I}}_1$ as the sum of contributions from each spatial frequency (f',g') of the object structure. It is seen that $\mathscr{T}$ plays a similar role as before. It characterizes the changes which arise in each contribution from the mode of illumination of the object and from the transmission characteristics of the image-forming system.

Since the response function $\mathscr{K}(f,g)$ is zero when the point $\xi = \bar{\lambda}Rf$, $\eta = \bar{\lambda}Rg$ lies outside the area of the exit pupil, it follows from (47) that $\mathscr{T}$ vanishes for sufficiently high frequencies. If the exit pupil is a circle of radius a, the product $\mathscr{K}(f+f',g+g')$ $\mathscr{K}^\star(f+f'',g+g'')$ and consequently $\mathscr{T}(f',g';f''g'')$ can only differ from zero if the two circles C' and C'', each of radius $\sqrt{f^2+g^2} = a/\bar{\lambda}R$, centred on the points $O'(-f',-g')$, $O''(-f'',-g'')$ in the f,g-plane have a domain in common (Fig. 10.15).

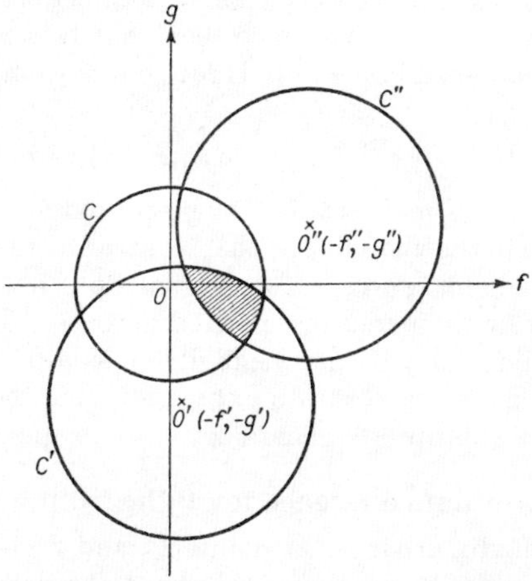

Fig. 10.15. The effective domain of integration (shown shaded) for the transmission cross-coefficient $\mathscr{T}(f',g';f'',g'')$ of an image-forming system with a circular exit pupil of radius a. The object is assumed to be transilluminated by quasi-monochromatic light of mean wavelength $\bar{\lambda}_0$ via a condenser of numerical aperture $mn_0 \sin\theta_0$, where $n_0 \sin\theta_0$ is the numerical aperture of the image-forming system. C' and C'' are circles of radius $a/\bar{\lambda}R \sim n_0 \sin\theta_0/M\bar{\lambda}_0$ centred at the points $O'(-f',-g')$ and $O''(-f'',-g'')$ respectively. C is a circle of radius $mn_0 \sin\theta_0/\bar{\lambda}_0$ centred on the origin. (R = radius of Gaussian reference sphere, M = Gaussian magnification.)

To illustrate the effect of the illumination, suppose that the illumination is either critical or KÖHLER, and that the numerical aperture $n_c' \sin\theta_c'$ of the condenser system is m times the numerical aperture $n_0 \sin\theta_0$ of the system that images the object. Then according to (18) or (30) the mutual intensity of the illuminating beam is

$$J_{\bar{0}}(x_0 - x_0', y_0 - y_0') = \left(\frac{2J_1(mv)}{mv}\right) I_{\bar{0}}, \tag{51}$$

where

$$v = \frac{2\pi}{\bar{\lambda}_0}\sqrt{(x_0' - x_0)^2 + (y_0' - y_0)^2}\, n_0 \sin\theta_0, \tag{52}$$

and $I_{\bar{0}}$ is the intensity (assumed uniform) of the incident light. Now the function on the right of (51) is a FOURIER transform of the function (cf. § 8.5.2)

$$\left.\begin{aligned} \mathscr{J}_{\bar{0}}(f,g) &= \text{constant} = \left(\frac{\bar{\lambda}_0^2}{\pi m^2 n_0^2 \sin^2\theta_0}\right) I_{\bar{0}} \quad \text{when} \quad f^2+g^2 < \frac{m^2 n_0^2 \sin^2\theta_0}{\bar{\lambda}_0^2}, \\ &= 0 \qquad\qquad\qquad\qquad\qquad\qquad\quad \text{when} \quad f^2+g^2 > \frac{m^2 n_0^2 \sin^2\theta_0}{\bar{\lambda}_0^2}. \end{aligned}\right\} \tag{53}$$

In Fig. 10.15 the circle outside which $\mathscr{J}_0^-$ vanishes is denoted by C and it follows that for given (f',g') and (f'',g'') only those points of the f,g-plane which lie within the area (shown shaded) common to the three circles C, C', and C'' contribute to the integral (47) for $\mathscr{T}$.

10.6 SOME THEOREMS RELATING TO MUTUAL COHERENCE

In § 10.4 and § 10.5 we considered interference and diffraction of quasi-monochromatic light, and confined ourselves to situations where the time delay τ was small compared to the coherence time of the light. We have seen that to a good approximation the correlation functions then depend on τ only through a harmonic term, i.e. they are of the form

$$\Gamma(P_1,P_2,\tau) \sim J(P_1,P_2)e^{-2\pi i\bar{\nu}\tau}, \qquad \gamma(P_1,P_2,\tau) \sim \mu(P_1,P_2)e^{-2\pi i\bar{\nu}\tau}.$$

The elementary theory which makes use of this approximation is adequate to take into account the decrease in the visibility at the "centre" of the pattern ($\tau = 0$), due to a finite extension of the light source. However, it does not take into account the change in the visibility with increasing path difference. To describe adequately situations in which the time delay τ is not negligible in comparison with the coherence time, it is necessary to use more accurate expressions for the correlation functions. We shall now discuss the appropriate generalization of some of our formulae.

10.6.1 Calculation of mutual coherence for light from an incoherent source

Let $V_1(t)$ and $V_2(t)$ be the disturbances at points P_1 and P_2 in a wave field produced by an extended (not necessarily quasi-monochromatic) primary source σ. To begin with we assume that the medium between σ and the points P_1 and P_2 is homogeneous.

As in § 10.4.2, we imagine the source to be divided into elements $d\sigma_1, d\sigma_2 \ldots$ centred about points $S_1, S_2, \ldots$, with linear dimensions that are small compared to the effective wavelengths. If $V_{m1}(t)$ and $V_{m2}(t)$ are the contributions to V_1 and V_2 from the element $d\sigma_m$, then

$$V_1(t) = \sum_m V_{m1}(t), \qquad V_2(t) = \sum_m V_{m2}(t), \tag{1}$$

and the mutual coherence function is given by

$$\Gamma(P_1,P_2,\tau) = \langle V_1(t+\tau)V_2^\star(t)\rangle = \sum_m \langle V_{m1}(t+\tau)V_{m2}^\star(t)\rangle. \tag{2}$$

Terms of the type $\langle V_{m1}(t+\tau)V_{n2}^\star(t)\rangle$ $(m \neq n)$ have been omitted on the right hand of (2), since the contributions from the different source elements may be assumed to be mutually incoherent.

We now proceed somewhat differently than in § 10.4. According to § 10.3 (30), each term under the summation sign in (2) may be expressed in the form

$$\langle V_{m1}(t+\tau)V_{m2}^\star(t)\rangle = 4\int_0^\infty G_m(P_1,P_2,\nu)e^{-2\pi i\nu\tau}\,d\nu \tag{3}$$

where

$$G_m(P_1,P_2,\nu) = \operatorname*{Lim}_{T\to\infty}\left[\frac{\overline{v_{mT}(P_1,\nu)v_{mT}^\star(P_2,\nu)}}{2T}\right] \tag{4}$$

is the mutual spectral density of the disturbances $V_{m1} = V_m(P_1,t)$, $V_{m2} = V_m(P_2,t)$. Now v_m represents the contribution of the appropriate frequency to the disturbance arising from the element $d\sigma_m$, and, since the medium is assumed to be homogeneous this contribution is propagated in the form of a spherical wave. Hence

$$v_{mT}(P_1,\nu) = a_{mT}(\nu)\frac{e^{ikR_{m1}}}{R_{m1}}, \qquad v_{mT}(P_2,\nu) = a_{mT}(\nu)\frac{e^{ikR_{m2}}}{R_{m2}}, \tag{5}$$

where R_{m1} and R_{m2} (assumed to be large compared to the effective wavelengths) represent the distances of P_1 and P_2 from the source point S_m, and $k = 2\pi\nu/v = 2\pi/\lambda$. The amplitude $|a_m(\nu)|$ of $a_m(\nu)$ represents the strength of the component of frequency ν from the element $d\sigma_m$, and arg $a_m(\nu)$ represents its phase. From (4) and (5),

$$G_m(P_1,P_2,\nu) = \left(\operatorname*{Lim}_{T\to\infty}\frac{\overline{|a_{mT}(\nu)|^2}}{2T}\right)\frac{e^{ik(R_{m1}-R_{m2})}}{R_{m1}R_{m2}}. \tag{6}$$

The term in the large bracket on the right represents the spectral density of the light from the source element $d\sigma_m$. As in § 10.4.2, we assume that the number of the source elements is so large that we may treat the source as being effectively continuous. Hence if $I(S_m,\nu)d\sigma_m d\nu = 4 \operatorname*{Lim}_{T\to\infty} [\overline{|a_{mT}(\nu)|^2/2T}]d\nu$, i.e. if $I(S,\nu)$ denotes the intensity per unit area of the source, per unit frequency range, then according to (2), (3), and (6),

$$\Gamma(P_1,P_2,\tau) = \sqrt{I(P_1)}\sqrt{I(P_2)}\,\gamma(P_1,P_2,\tau) = \int_0^\infty d\nu e^{-2\pi i\nu\tau}\int_\sigma I(S,\nu)\frac{e^{ik(R_1-R_2)}}{R_1R_2}\,dS, \tag{7}$$

where

$$I(P_1) = \Gamma(P_1,P_1,0) = \int_0^\infty d\nu\int_\sigma\frac{I(S,\nu)}{R_1^2}\,dS, \qquad I(P_2) = \Gamma(P_2,P_2,0) = \int_0^\infty d\nu\int_\sigma\frac{I(S,\nu)}{R_2^2}\,dS, \tag{8}$$

are the intensities at P_1 and P_2, and R_1 and R_2 are the distances of these points from the source point S. Equations (7) are generalizations of the VAN CITTERT–ZERNIKE formulae, § 10.4 (20), (21).

If the medium between the source and the points P_1 and P_2 is not homogeneous, we may proceed as in § 10.4. We only have to replace the factors e^{ikR_j}/R_j by $icK(S,P_j,\nu)/\nu$, where K is the appropriate transmission function of the medium. In place of (7) we then obtain

$$\begin{aligned}\Gamma(P_1,P_2,\tau) &= \sqrt{I(P_1)}\sqrt{I(P_2)}\,\gamma(P_1,P_2,\tau)\\ &= c^2\int_0^\infty\frac{d\nu}{\nu^2}e^{-2\pi i\nu\tau}\int_\sigma I(S,\nu)K(S,P_1,\nu)K^\star(S,P_2,\nu)\,dS,\end{aligned} \tag{9}$$

where

$$I(P_1) = c^2\int_0^\infty\frac{d\nu}{\nu^2}\int_\sigma I(S,\nu)|K(S,P_1,\nu)|^2\,dS, \qquad I(P_2) = c^2\int_0^\infty\frac{d\nu}{\nu^2}\int_\sigma I(S,\nu)|K(S,P_2,\nu)|^2\,dS. \tag{10}$$

By analogy with § 10.4, we re-write (9) or (10) in a slightly different form. We set

$$\frac{ic}{\nu}K(S,P_1,\nu)\sqrt{I(S,\nu)} = U(S,P_1,\nu), \qquad \frac{ic}{\nu}K(S,P_2,\nu)\sqrt{I(S,\nu)} = U(S,P_2,\nu). \tag{11}$$

Then (9) and (10) become

$$\Gamma(P_1,P_2,\tau) = \sqrt{I(P_1)}\sqrt{I(P_2)}\,\gamma(P_1,P_2,\tau) = \int_0^\infty d\nu e^{-2\pi i\nu\tau}\int_\sigma U(S,P_1,\nu)U^\star(S,P_2,\nu)\,dS, \quad (12)$$

where

$$I(P_1) = \int_0^\infty d\nu \int_\sigma |U(S,P_1,\nu)|^2\,dS, \qquad I(P_2) = \int_0^\infty d\nu \int_\sigma |U(S,P_2,\nu)|^2\,dS. \quad (13)$$

The formula (12), which is a generalization of HOPKINS' formula § 10.4 (35), expresses the mutual coherence function and the complex degree of coherence in terms of the light distribution arising from an associated fictitious source. For, according to (11), $U(S,P,\nu)$ may be regarded as the disturbance at P due to a monochromatic point source of frequency ν, of zero phase, and of strength proportional to $\sqrt{I(S,\nu)}$, situated at S.

10.6.2 Propagation of mutual coherence

As in § 10.4.4 let $\mathscr{A}$ be a fictitious surface which intercepts a beam of light from an extended primary source σ (Fig. 10.8). We shall show how, from knowledge of the mutual coherence for all pairs of points on $\mathscr{A}$, its value on any other surface $\mathscr{B}$ illuminated by the light from $\mathscr{A}$ can be determined. For simplicity we assume that the medium between $\mathscr{A}$ and $\mathscr{B}$ has refractive index equal to unity.

The mutual coherence for any two points Q_1 and Q_2 on $\mathscr{B}$ may be calculated by the use of (12). In this formula U represents a monochromatic disturbance and hence may be determined from the knowledge of the disturbance at all points of the surface $\mathscr{A}$ by means of the HUYGENS–FRESNEL principle:

$$U(S,Q_1,\nu) = \int_{\mathscr{A}} U(S,P_1,\nu)\frac{e^{ik_0 s_1}}{s_1}\Lambda_1 dP_1. \quad (14)$$

Here s_1 denotes the distance between a typical point P_1 on $\mathscr{A}$ and the point Q_1, and Λ_1 is the usual inclination factor. There is a strictly analogous formula for $U(S,Q_2,\nu)$. If we substitute from these formulae into (12) we obtain, after changing the order of integration, the following expression for the mutual coherence:

$$\Gamma(Q_1,Q_2,\tau) = \iint_{\mathscr{A}\mathscr{A}} \frac{dP_1 dP_2}{s_1 s_2}\int_0^\infty J(P_1,P_2,\nu)\Lambda_1(\nu)\Lambda_2^\star(\nu)e^{-2\pi i\nu\left[\tau - \frac{s_1-s_2}{c}\right]}\,d\nu, \quad (15)$$

where

$$J(P_1,P_2,\nu) = \int_\sigma U(S,P_1,\nu)U^\star(S,P_2,\nu)dS. \quad (16)$$

The repeated integration in (15) means that the points P_1 and P_2 explore the surface $\mathscr{A}$ independently. Now the inclination factors $\Lambda_1(\nu)$ and $\Lambda_2(\nu)$ depend on the frequency through a multiplicative factor ν and change slowly with ν in comparison with the other terms. If the effective spectral range of the light is sufficiently small, we may replace these factors by $\bar{\Lambda}_1 = \Lambda_1(\bar{\nu})$, $\bar{\Lambda}_2 = \Lambda_2(\bar{\nu})$, where $\bar{\nu}$ denotes the mean frequency of the light. The rest of the ν-integral is, according to (12), precisely $\Gamma(P_1, P_2, \tau - (s_1 - s_2)/c)$. We thus finally obtain the formula

$$\Gamma(Q_1,Q_2,\tau) = \iint_{\mathscr{A}\mathscr{A}} \frac{\Gamma\left(P_1,P_2,\tau - \dfrac{s_1 - s_2}{c}\right)}{s_1 s_2} \bar{\Lambda}_1 \bar{\Lambda}_2^\star \, dP_1 dP_2. \tag{17}$$

This is the required formula for the mutual coherence function at points Q_1 and Q_2 on the surface $\mathscr{B}$ in terms of the mutual intensity at all pairs of points on the surface $\mathscr{A}$.

Of particular interest is the case when the two points Q_1 and Q_2 coincide and $\tau = 0$. Denoting the common point by Q, the left-hand side of (17) reduces to the intensity $I(Q)$. Also, if we substitute for Γ on the right in terms of the intensities and the correlation factor γ, the formula reduces to

$$I(Q) = \iint_{\mathscr{A}\mathscr{A}} \frac{\sqrt{I(P_1)}\sqrt{I(P_2)}}{s_1 s_2} \gamma\left(P_1,P_2, \frac{s_2 - s_1}{c}\right) \bar{\Lambda}_1 \bar{\Lambda}_2^\star \, dP_1 dP_2. \tag{18}$$

This formula expresses the *intensity* at an arbitrary point Q as the sum of contributions from all pairs of elements of the arbitrary surface $\mathscr{A}$. Each contribution is directly proportional to the geometrical mean of the intensities at the two elements, inversely proportional to the product of their distances from Q, and is weighted by the appropriate value of the correlation factor γ.

The formulae (17) and (18) are generalizations, due to WOLF, of the propagation law of ZERNIKE (§ 10.4 (45)) and of the formula § 10.4 (46) for the intensity in a partially coherent wave field.

10.7 RIGOROUS THEORY OF PARTIAL COHERENCE*

10.7.1 Wave equations for mutual coherence

Some of the theorems relating to the correlation functions, which we derived in the preceding sections are in certain respects similar to theorems relating to the complex disturbance itself. For example, the VAN CITTERT–ZERNIKE formula § 10.4 (21) for the complex degree of coherence in a plane illuminated by an extended quasi-monochromatic primary source was seen to be identical with a formula relating to the complex disturbance in a diffraction pattern arising from the diffraction by an aperture of the same size and shape as the source. Other examples are the laws for the propagation of the mutual intensity (§ 10.4 (45)) which was seen to resemble the HUYGENS–FRESNEL principle. Now the results relating to the complex disturbance may be regarded as approximate deductions from certain rigorous theorems, namely the formulae of HELMHOLTZ and KIRCHHOFF (§ 8.3 (7), (13)), which are a consequence of the fact that the light disturbance is propagated as a wave. This analogy suggests that correlation is also propagated as a wave, and that our theorems are approximate formulations of some associated theorems of the HELMHOLTZ–KIRCHHOFF type. It is not difficult to show that this in fact is the case.

Consider now a stationary wave field in vacuum and let $V(P_1,t)$ and $V(P_2,t)$

---

* The analysis of this section is in the main part based on investigations of E. WOLF, *Proc. Roy. Soc.*, A, **230** (1955), 246; and *Proc. Phys. Soc.*, **71** (1958), 257.

represent the disturbances at points P_1 and P_2 respectively. It will be convenient to begin by expressing the mutual coherence function in the more symmetrical form

$$\Gamma(P_1,P_2,t_1,t_2) = \langle V(P_1,t_1+t)V^\star(P_2,t_2+t)\rangle$$
$$= \lim_{T\to\infty} \frac{1}{2T}\int_{-\infty}^{\infty} V_T(P_1,t+t_1)V_T^\star(P_2,t+t_2)\,dt. \tag{1}$$

Further let

$$\nabla_1^2 = \frac{\partial^2}{\partial x_1^2} + \frac{\partial^2}{\partial y_1^2} + \frac{\partial^2}{\partial z_1^2} \tag{2}$$

be the Laplacian operator with respect to the Cartesian rectangular coordinates of P_1. On applying this operator to (1), and on interchanging the order of the various operations, we obtain

$$\nabla_1^2\Gamma(P_1,P_2,t_1,t_2) = \lim_{T\to\infty}\frac{1}{2T}\int_{-\infty}^{\infty}\{[\nabla_1^2 V_T(P_1,t+t_1)]V_T^\star(P_2,t+t_2)\}\,dt. \tag{3}$$

Now the real part $V_T^{(r)}$ of V_T represents the true physical wave field (e.g. a Cartesian component of the electric vector wave) and hence satisfies the wave equation

$$\nabla_1^2 V_T^{(r)}(P_1,t+t_1) = \frac{1}{c^2}\frac{\partial^2}{\partial t_1^2}V_T^{(r)}(P_1,t+t_1). \tag{4a}$$

The imaginary part $V_T^{(i)}$ of V_T, and hence V_T itself, also satisfies the wave equation; this result follows immediately on taking the HILBERT transforms of both sides of (4a) and on using the fact that if two functions are HILBERT transforms of each other so are their derivatives. Hence*

$$\nabla_1^2 V_T(P_1,t+t_1) = \frac{1}{c^2}\frac{\partial^2}{\partial t_1^2}V_T(P_1,t+t_1), \tag{4b}$$

It follows that we may replace ∇_1^2 by $\partial^2/c^2\partial t_1^2$ on the right of (3), and we obtain after again changing the order of the operations,

$$\nabla_1^2\Gamma(P_1,P_2,t_1,t_2) = \frac{1}{c^2}\frac{\partial^2}{\partial t_1^2}\lim_{T\to\infty}\frac{1}{2T}\int_{-\infty}^{\infty}V_T(P_1,t+t_1)V_T^\star(P_2,t+t_2)\,dt,$$

i.e.

$$\nabla_1^2\Gamma(P_1,P_2,t_1,t_2) = \frac{1}{c^2}\frac{\partial^2\Gamma(P_1,P_2,t_1,t_2)}{\partial t_1^2}. \tag{5a}$$

In a strictly similar manner it also follows that

$$\nabla_2^2\Gamma(P_1,P_2,t_1,t_2) = \frac{1}{c^2}\frac{\partial^2\Gamma(P_1,P_2,t_1,t_2)}{\partial t_2^2}, \tag{5b}$$

where ∇_2^2 is the Laplacian operator with respect to the coordinates of the point P_2.

* The following alternative proof may be noted: since $V_T^{(r)}$ satisfies the wave equation, each of its spectral components $v_T(\nu)$ $(-\infty \leqslant \nu \leqslant \infty)$ satisfies the HELMHOLTZ equation. Now according to eq. (18b) of § 10.2 the spectrum of $V_T = V_T^{(r)} + iV_T^{(i)}$ is $2v_T(\nu)$ or 0 according as $\nu \gtrless 0$. Hence each spectral component of V_T also satisfies the HELMHOLTZ equation, and consequently V_T satisfies the wave equation.

Now for a stationary field Γ depends on t_1 and t_2 only through the difference $t_1 - t_2 = \tau$, and we then may write as before, $\Gamma(P_1,P_2,t_1,t_2) = \Gamma(P_1,P_2,\tau)$. Then $\partial^2/\partial t_1^2 = \partial^2/\partial t_2^2 = \partial^2/\partial \tau^2$ and we have from (5),

$$\nabla_1^2\Gamma(P_1,P_2,\tau) = \frac{1}{c^2}\frac{\partial^2\Gamma(P_1,P_2,\tau)}{\partial\tau^2}, \tag{6a}$$

$$\nabla_2^2\Gamma(P_1,P_2,\tau) = \frac{1}{c^2}\frac{\partial^2\Gamma(P_1,P_2,\tau)}{\partial\tau^2}. \tag{6b}$$

We see that *in vacuum the mutual coherence obeys two wave equations.** Each of them describes the variation of the mutual coherence when one of the points (P_2 or P_1) is fixed, while the other point and the parameter τ change. Now τ represents a time difference between the instants at which the correlation at the two points is considered, and in all experiments enters only in the combination $c\tau = \Delta\mathscr{S}$, i.e. as a path difference. The time itself has thus effectively been eliminated from our final description of the field. This is a particularly attractive feature of the theory of partial coherence, since in optical wave fields true time variations entirely escape detection. The basic entity in this theory, the mutual coherence function $\Gamma(P_1,P_2,\tau)$ is directly measurable, for example by means of the interference experiments described in § 10.3 and in § 10.4.

10.7.2 Rigorous formulation of the propagation law for mutual coherence

We again consider a stationary wave field in vacuum. Let Q_1 and Q_2 be any two points in the field and let $\mathscr{A}$ be any fictitious surface which surrounds these points. If ∇_1^2 denotes the Laplacian operator with respect to the coordinates of Q_1, we have, according to (5a),

$$\nabla_1^2\Gamma(Q_1,Q_2,t_1,t_2) = \frac{1}{c^2}\frac{\partial^2\Gamma(Q_1,Q_2,t_1,t_2)}{\partial t_1^2}. \tag{7}$$

It follows that we may apply to Γ the KIRCHHOFF integral formula (13) of § 8.3. Thus $\Gamma(Q_1,Q_2,t_1,t_2)$ may be expressed in terms of the values of $[\Gamma(P_1,Q_2,t_1,t_2)]_1$, where P_1 takes on all positions on $\mathscr{A}$ and $[\ .\ .\ .]_1$ denotes retardation with respect to the first time argument, i.e.

$$[\Gamma(P_1,Q_2,t_1,t_2)]_1 = \Gamma\left(P_1,Q_2,t_1 - \frac{s_1}{c}, t_2\right) \tag{8}$$

and s_1 is the distance between P_1 and Q_1 (Fig. 10.16). Written out explicitly, KIRCHHOFF's formula gives

$$\Gamma(Q_1,Q_2,t_1,t_2) = \frac{1}{4\pi}\iint_{\mathscr{A}}\left\{f_1[\Gamma(P_1,Q_2,t_1,t_2)]_1 + g_1\left[\frac{\partial}{\partial t_1}\Gamma(P_1,Q_2,t_1,t_2)\right]_1 + h_1\left[\frac{\partial}{\partial n_1}\Gamma(P_1,Q_2,t_1,t_2)\right]_1\right\}dP_1. \tag{9}$$

* When τ is small compared to the coherence time, we have, according to § 10.4 (10), $\Gamma(P_1,P_2,\tau) \sim J(P_1,P_2)e^{-2\pi i\bar{\nu}\tau}$. It follows from (6) that within the range of validity of the quasi-monochromatic theory, the mutual intensity J_{12} in vacuum obeys to a good approximation the HELMHOLTZ equations

$$\nabla_1^2 J(P_1,P_2) + k^2 J(P_1,P_2) = 0, \qquad \nabla_2^2 J(P_1,P_2) + k^2 J(P_1,P_2) = 0.$$

Here $\partial/\partial n_1$ denotes differentiation along the inward normal to $\mathscr{A}$ at P_1, and

$$f_1 = \frac{\partial}{\partial n_1}\left(\frac{1}{s_1}\right), \qquad g_1 = -\frac{1}{cs_1}\frac{\partial s_1}{\partial n_1}, \qquad h_1 = -\frac{1}{s_1}. \tag{10}$$

Now, according to (5b), we also have

$$\nabla_2^2\Gamma(P_1,Q_2,t_1,t_2) = \frac{1}{c^2}\frac{\partial^2\Gamma(P_1,Q_2,t_1,t_2)}{\partial t_2^2}, \tag{11}$$

where ∇_2^2 is the Laplacian operator with respect to the coordinates of Q_2. Hence $\Gamma(P_1,Q_2,t_1,t_2)$ which appears on the right of (9) may be expressed in the form of a KIRCHHOFF integral involving the values of $[\Gamma(P_1,P_2,t_1,t_2)]_2$, where P_2 takes on all possible positions on $\mathscr{A}$ and $[\ldots]_2$ denotes retardation with respect to the second time argument, e.g.

$$[\Gamma(P_1,P_2,t_1,t_2)]_2 = \Gamma\left(P_1,P_2,t_1,t_2 - \frac{s_2}{c}\right). \tag{12}$$

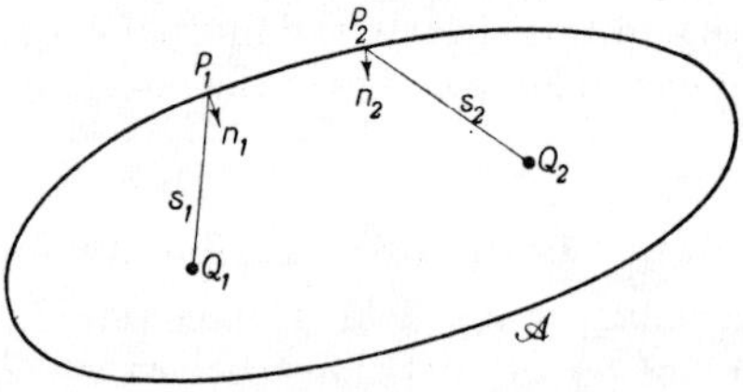

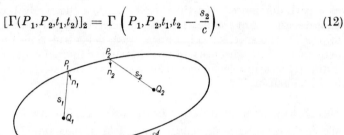

Fig. 10.16. Notation used in the rigorous formulation of the propagation law for mutual coherence.

and s_2 is the distance between P_2 and Q_2. Written out explicitly, the appropriate formula is

$$\Gamma(P_1,Q_2,t_1,t_2) = \frac{1}{4\pi}\iint_{\mathscr{A}}\left\{f_2[\Gamma(P_1,P_2,t_1,t_2)]_2 + g_2\left[\frac{\partial}{\partial t_2}\Gamma(P_1,P_2,t_1,t_2)\right]_2 + h_2\left[\frac{\partial}{\partial n_2}\Gamma(P_1,P_2,t_1,t_2)\right]_2\right\}dP_2. \tag{13}$$

Here $\partial/\partial n_2$ denotes differentiation along the inward normal at P_2, and f_2, g_2, h_2 are the same quantities as in (10) but with the suffix 1 replaced by 2. We next differentiate (13) with respect to t_1 and n_1 and obtain

$$\frac{\partial}{\partial t_1}\Gamma(P_1,Q_2,t_1,t_2) = \frac{1}{4\pi}\iint_{\mathscr{A}}\left\{f_2\left[\frac{\partial}{\partial t_1}\Gamma(P_1,P_2,t_1,t_2)\right]_2 + g_2\left[\frac{\partial^2}{\partial t_1\partial t_2}\Gamma(P_1,P_2,t_1,t_2)\right]_2 + h_2\left[\frac{\partial^2}{\partial t_1\partial n_2}\Gamma(P_1,P_2,t_1,t_2)\right]_2\right\}dP_2, \tag{14}$$

$$\frac{\partial}{\partial n_1}\Gamma(P_1,Q_2,t_1,t_2) = \frac{1}{4\pi}\iint_{\mathscr{A}}\left\{f_2\left[\frac{\partial}{\partial n_1}\Gamma(P_1,P_2,t_1,t_2)\right]_2 + g_2\left[\frac{\partial^2}{\partial n_1\partial t_2}\Gamma(P_1,P_2,t_1,t_2)\right]_2 + h_2\left[\frac{\partial^2}{\partial n_1\partial n_2}\Gamma(P_1,P_2,t_1,t_2)\right]_2\right\}dP_2. \tag{15}$$

We now substitute from (13), (14), and (15) into (9), and obtain the following expression for $\Gamma(Q_1,Q_2,t_1,t_2)$:

$$\Gamma(Q_1,Q_2,t_1,t_2)=\frac{1}{(4\pi)^2}\iint\limits_{\mathscr{A}\mathscr{A}}\Bigg\{f_1f_2[\Gamma]_{1,2}+f_1g_2\left[\frac{\partial}{\partial t_2}\Gamma\right]_{1,2}+f_1h_2\left[\frac{\partial}{\partial n_2}\Gamma\right]_{1,2}$$
$$+g_1f_2\left[\frac{\partial}{\partial t_1}\Gamma\right]_{1,2}+g_1g_2\left[\frac{\partial^2}{\partial t_1\partial t_2}\Gamma\right]_{1,2}+g_1h_2\left[\frac{\partial^2}{\partial t_1\partial n_2}\Gamma\right]_{1,2}$$
$$+h_1f_2\left[\frac{\partial}{\partial n_1}\Gamma\right]_{1,2}+h_1g_2\left[\frac{\partial^2}{\partial n_1\partial t_2}\Gamma\right]_{1,2}+h_1h_2\left[\frac{\partial^2}{\partial n_1\partial n_2}\Gamma\right]_{1,2}\Bigg\}\,dP_1\,dP_2, \tag{16}$$

where the first two arguments in Γ on the right are P_1 and P_2 and $[\,.\;.\;.\,]_{1,2}$ denotes retardation with respect to both the time arguments, e.g.

$$[\Gamma]_{1,2}=\Gamma\left(P_1,P_2,t_1-\frac{s_1}{c},t_2-\frac{s_2}{c}\right). \tag{17}$$

Finally, we make use of the assumption of stationarity which ensures that Γ depends on the two time arguments through their difference only. We write as before $\Gamma(P_1,P_2,t_1,t_2)=\Gamma(P_1,P_2,\tau)$, $\tau=t_1-t_2$. Then $\partial/\partial t_1=-\partial/\partial t_2=\partial/\partial\tau$, and (16) becomes*

$$\Gamma(Q_1,Q_2,\tau)=\frac{1}{(4\pi)^2}\iint\limits_{\mathscr{A}\mathscr{A}}\Bigg\{f_1f_2[\Gamma]-f_1g_2\left[\frac{\partial}{\partial\tau}\Gamma\right]+f_1h_2\left[\frac{\partial}{\partial n_2}\Gamma\right]$$
$$+g_1f_2\left[\frac{\partial}{\partial\tau}\Gamma\right]-g_1g_2\left[\frac{\partial^2}{\partial\tau^2}\Gamma\right]+g_1h_2\left[\frac{\partial^2}{\partial\tau\partial n_2}\Gamma\right]$$
$$+h_1f_2\left[\frac{\partial}{\partial n_1}\Gamma\right]-h_1g_2\left[\frac{\partial^2}{\partial n_1\partial\tau}\Gamma\right]+h_1h_2\left[\frac{\partial^2}{\partial n_1\partial n_2}\Gamma\right]\Bigg\}\,dP_1dP_2. \tag{18}$$

The first two arguments in Γ on the right are P_1 and P_2, and $[\,.\;.\;.\,]$ denotes "retardation" by the amount $(s_1-s_2)/c$, e.g.

$$[\Gamma]=\Gamma\left(P_1,P_2,\tau-\frac{s_1-s_2}{c}\right). \tag{19}$$

The formula (18) may be regarded as a rigorous formulation of the law for the propagation of the mutual coherence (§ 10.6 (17)). It expresses the value of the mutual coherence function for any two points Q_1 and Q_2 in terms of the values of this function and of some of its derivatives at all pairs of points on an arbitrary closed surface which surrounds both these points.

In the special case when Q_1 and Q_2 coincide and $\tau=0$, we obtain from (18), on substituting $\Gamma_{12}(\tau)=\sqrt{I_1}\sqrt{I_2}\,\gamma_{12}(\tau)$, the following expressions for the intensity:

$$I(Q)=\frac{1}{(4\pi)^2}\iint\limits_{\mathscr{A}\mathscr{A}}\Bigg(\sqrt{I_1}\sqrt{I_2}\left\{f_1f_2[\gamma]+(f_2g_1-f_1g_2)\left[\frac{\partial}{\partial\tau}\gamma\right]-g_1g_2\left[\frac{\partial^2}{\partial\tau^2}\gamma\right]\right\}$$
$$+\sqrt{I_1}\left\{f_1h_2\frac{\partial}{\partial n_2}(\sqrt{I_2}[\gamma])+g_1h_2\frac{\partial}{\partial n_2}\left(\sqrt{I_2}\left[\frac{\partial}{\partial\tau}\gamma\right]\right)\right\}$$
$$+\sqrt{I_2}\left\{f_2h_1\frac{\partial}{\partial n_1}(\sqrt{I_1}[\gamma])-g_2h_1\frac{\partial}{\partial n_1}\left(\sqrt{I_1}\left[\frac{\partial}{\partial\tau}\gamma\right]\right)\right\}$$
$$+h_1h_2\frac{\partial^2}{\partial n_1\partial n_2}(\sqrt{I_1}\sqrt{I_2}[\gamma])\Bigg)\,dP_1dP_2, \tag{20}$$

* Equation (18) applies for propagation from a closed surface $\mathscr{A}$ of arbitrary form. A much simpler formula exists for propagation from a plane surface. [See M. J. BERAN and G. B. PARRENT, *Theory of Partial Coherence* (Englewood Cliffs, N.J., 1964), § 3.3.]

where I_1 and I_2 are the intensities at P_1 and P_2 respectively, $[\gamma] = \gamma(P_1, P_2, (s_2 - s_1)/c)$. etc. The formula (20) may be regarded as a rigorous formulation of the theorem expressed by equation (18), § 10.6. It gives the intensity at an arbitrary point Q in terms of the distribution of the intensity and of the complex degree of coherence (and of some of the derivatives of these quantities) on an arbitrary surface surrounding Q.

10.7.3 The coherence time and the effective spectral width

The concept of the coherence time, which was found useful in many considerations involving polychromatic light, was introduced in § 7.5.8 from the study of the disturbance resulting from superposition of identical wave trains of finite duration. We showed from a simple example (a random sequence of periodic wave trains) that the coherence time* $\Delta\tau$ and the effective spectral width $\Delta\nu = c\Delta\lambda/\bar{\lambda}^2$ of the resulting disturbance are connected by the order of magnitude relation

$$\Delta\tau\Delta\nu \sim 1. \tag{21}$$

We also mentioned that a relation of this type holds under more general conditions, provided that $\Delta\tau$ and $\Delta\nu$ are defined as suitable averages. In this section we shall define these quantities and establish the required reciprocity relation rigorously.

Suppose that a beam of light is divided at a point P into two beams which are brought together after the introduction of a path difference $c\tau$ between them. The resulting interference effects are characterized by the self-coherence function

$$\Gamma(\tau) = \langle V(t+\tau)V^\star(t)\rangle = 4\int_0^\infty G(\nu)e^{-2\pi i\nu\tau}\,d\nu, \tag{22}$$

where $V(t)$ is the complex disturbance at P and $G(\nu)$ is the spectral density.

Since the degree of coherence of the two interfering beams is represented by $|\gamma(\tau)| = |\Gamma(\tau)|/\Gamma(0)$ it is reasonable, and mathematically convenient, to define *the coherence time* $\Delta\tau$ of the light at P as the normalized root-mean-square width (r.m.s.) of the squared modulus of $\Gamma(\tau)$, i.e.†

$$(\Delta\tau)^2 = \frac{\displaystyle\int_{-\infty}^{\infty} \tau^2|\Gamma(\tau)|^2\,d\tau}{\displaystyle\int_{-\infty}^{+\infty} |\Gamma(\tau)|^2\,d\tau}. \tag{23}$$

Next we define the *effective spectral width* $\Delta\nu$ of the light at P as the normalized r.m.s. width of the spectrum of Γ, i.e. as the normalized r.m.s. width of the square of the spectral density $G(\nu)$, taken over the range $\nu \geqslant 0$. Thus

$$(\Delta\nu)^2 = \frac{\displaystyle\int_0^\infty (\nu-\bar{\nu})^2G^2(\nu)\,d\nu}{\displaystyle\int_0^\infty G^2(\nu)\,d\nu}, \qquad \bar{\nu} = \frac{\displaystyle\int_0^\infty \nu G^2(\nu)\,d\nu}{\displaystyle\int_0^\infty G^2(\nu)\,d\nu}. \tag{24}$$

* To conform to the notation of the present chapter we now write $\Delta\tau$ in place of Δt.

† The average value $\bar{\tau} = \int_{-\infty}^{+\infty} \tau|\Gamma(\tau)|^2 d\tau \Big/ \int_{-\infty}^{+\infty} |\Gamma(\tau)|^2 d\tau$ is zero since $|\Gamma(\tau)|$ is an even function of τ.

For another definition of the coherence time, see L. Mandel, *Proc. Phys. Soc.*, **74** (1959), 233. See also L. Mandel and E. Wolf, *ibid.*, **80** (1962), 894.

To establish the required reciprocity relation we set

$$\left.\begin{aligned} \xi &= \nu - \bar{\nu}, && \text{(a)}\\ \Phi(\xi) &= 4G(\bar{\nu}+\xi) \quad \text{when} \quad \xi > -\bar{\nu}, \\ &= 0 \quad \text{when} \quad \xi < -\bar{\nu}, && \text{(b)}\\ \Psi(\tau) &= \Gamma(\tau)e^{2\pi i\bar{\nu}\tau}. && \text{(c)} \end{aligned}\right\} \tag{25}$$

We shall assume that $\Phi(\xi)$ is continuous everywhere $(-\infty < \xi < \infty)$; consequently $\Phi(-\bar{\nu}) = G(0) = 0$*. From (22) it follows that Ψ and Φ form a FOURIER transform pair,

$$\Psi(\tau) = \int_{-\infty}^{\infty} \Phi(\xi)e^{-2\pi i\xi\tau}\, d\xi, \qquad \Phi(\xi) = \int_{-\infty}^{\infty} \Psi(\tau)e^{2\pi i\xi\tau}\, d\tau. \tag{26}$$

The expressions for $\Delta\tau$ and $\Delta\nu$ become

$$(\Delta\tau)^2 = \frac{1}{N}\int_{-\infty}^{\infty} \tau^2|\Psi(\tau)|^2\, d\tau, \tag{27}$$

$$(\Delta\nu)^2 = \frac{1}{N}\int_{-\infty}^{\infty} \xi^2\Phi^2(\xi)\, d\xi, \tag{28}$$

where

$$N = \int_{-\infty}^{\infty} |\Psi(\tau)|^2\, d\tau = \int_{-\infty}^{\infty} \Phi^2(\xi)\, d\xi. \tag{29}$$

Next we express the integral (28) in terms of Ψ. We have, on using the second relation in (26),

$$\begin{aligned} (\Delta\nu)^2 &= \frac{1}{N}\int_{-\infty}^{\infty} \xi^2\Phi(\xi)\, d\xi \int_{-\infty}^{\infty} \Psi(\tau)e^{2\pi i\xi\tau}\, d\tau \\ &= \frac{1}{N}\int_{-\infty}^{\infty} \Psi(\tau)\, d\tau \left(\frac{1}{2\pi i}\right)^2 \frac{\partial^2}{\partial\tau^2} \int_{-\infty}^{\infty} \Phi(\xi)e^{2\pi i\xi\tau}\, d\xi \\ &= -\frac{1}{4\pi^2}\frac{1}{N}\int_{-\infty}^{\infty} \Psi(\tau)\frac{\partial^2}{\partial\tau^2}\Psi^\star(\tau)\, d\tau \\ &= \frac{1}{4\pi^2}\frac{1}{N}\int_{-\infty}^{\infty} \left|\frac{\partial\Psi}{\partial\tau}\right|^2 d\tau. \end{aligned} \tag{30}$$

---

* This condition is, in fact, necessary for the integral in the numerator of (23) to converge. When it is satisfied, as here assumed, the value of (23) remains unchanged when $\Gamma(\tau) = \langle V(t+\tau)V^\star(t)\rangle$ is replaced by the real correlation function $\Gamma^{(r)}(\tau) = \mathscr{R}\Gamma(\tau) = 2\langle V^{(r)}(t+\tau)V^{(r)}(t)\rangle$. [Cf. E. WOLF, *Proc. Phys. Soc.*, B, **71** (1958), 257; R. SILVERMAN, *Trans. Inst. Rad. Engrs.*, CT–**5** (1958), 84.] The more general case when $G(0) \neq 0$ was investigated by A. G. MAYER and E. A. LEONTOVICH, *Doklady Akad. Nauk SSSR*, **4** (1934), 353 and by I. KAY and R. SILVERMAN, *Information and Control*, **1** (1957), 64, 396. See also A. A. KHARKEVICH, *Spectra and Analysis*, Translated from Russian (New York, Consultants Bureau, 1960), § 12.

In passing from the second to the third line, the first relation in (26) was used, together with the relation $\Psi(-\tau) = \Psi^{\star}(\tau)$. The last line follows from the preceding one on integrating by parts and using the fact that $\Psi \rightarrow 0$ as $\tau \rightarrow \pm\infty$; this is so because the integral $\int_{-\infty}^{\infty} |\Psi(\tau)|^2 \, d\tau$ in (29) is assumed to be convergent.

It follows from (27), (29), and (30) that

$$(\Delta\tau)^2(\Delta\nu)^2 = \frac{1}{16\pi^2}\left[\frac{4\left(\int_{-\infty}^{\infty} \tau^2|\Psi(\tau)|^2\, d\tau\right)\left(\int_{-\infty}^{\infty} \left|\frac{\partial\Psi}{\partial\tau}\right|^2 d\tau\right)}{\left(\int_{-\infty}^{\infty} |\Psi(\tau)|^2\, d\tau\right)^2}\right]. \tag{31}$$

Now by a straightforward algebraical argument given in Appendix VIII, the term in the large brackets on the right side of (31) is greater than or equal to unity for any function Ψ for which the integrals exist. Hence we have established the following *reciprocity inequality** for the coherence time and the effective spectral width:

$$\Delta\tau\Delta\nu \geqslant \frac{1}{4\pi}. \tag{32}$$

We recall that when the light is quasi-monochromatic and the intensities of the two interfering beams are equal, the degree of coherence $|\gamma_{11}(\tau)| = |\Gamma_{11}(\tau)|/\Gamma_{11}(0)$ is according to § 10.4 (5) equal to the visibility $\mathscr{V}(\tau)$ of the fringes at a point corresponding to the path difference $c\tau$ between the two beams. Hence (23) may then be written in the form

$$(\Delta\tau)^2 = \frac{\int_{-\infty}^{\infty} \tau^2\mathscr{V}^2(\tau)\, d\tau}{\int_{-\infty}^{\infty} \mathscr{V}^2(\tau)\, d\tau} = \frac{\int_{0}^{\infty} \tau^2\mathscr{V}^2(\tau)\, d\tau}{\int_{0}^{\infty} \mathscr{V}^2(\tau)\, d\tau}; \tag{33}$$

thus *when the interfering beams are of equal intensity, the coherence time* $\Delta\tau$ *is equal to the normalized r.m.s. width of the square of the visibility function.*

The present definition of the coherence time is more satisfactory than that given in § 7.5.8, for we have now made no special assumptions about the nature of the elementary fields which give rise to the disturbance. In fact we now no longer require the knowledge of the detailed behaviour of the rapidly fluctuating function $V(t)$, our definition being based on the measurable correlation function $\Gamma(\tau)$. If we wish to retain the description of interference phenomena in terms of elementary wave trains,

---

* Our derivation is modelled on that given by H. WEYL and W. PAULI in connection with the HEISENBERG uncertainty relation. [H. WEYL, *The Theory of Groups and Quantum Mechanics* (London, Methuen, 1931; also Dover Publications, New York), pp. 77 and 393.]

we may regard $\Delta\tau$ as the duration of an *average* wave train; this interpretation must, however, be employed with caution.

Returning to (32) we see that the equality sign only holds when the term in the large brackets on the right-hand side of (31) is equal to unity, and this, according to Appendix VIII, is only possible when $\Psi(\tau)$ is a Gaussian function. Now the FOURIER transform of a Gaussian function is again a Gaussian function, and as this function differs from zero for all values of its argument ($-\infty < \xi < \infty$) it does not obey the second condition (25b). Thus the equality sign in (32) never applies. However, when the Gaussian function is centred on a frequency which is large compared to its r.m.s. width, the contribution to $\bar{\nu}$ and $\Delta\nu$ from the negative frequency range is negligible and it is evident that, for the high frequency spectra encountered in optics, the value of the product $\Delta\tau\Delta\nu$ cannot differ appreciably from that which corresponds to the full Gaussian curve. Thus the inequality in (32) may be replaced by the order of magnitude sign, i.e.

$$\Delta\tau\Delta\nu \sim \frac{1}{4\pi}. \tag{34}$$

The definition of the coherence time just given is appropriate when the two interfering beams are obtained from a single beam by division at a point P. The definition can be extended to situations where the two interfering beams are derived by division at two points P_1 and P_2, for example, in YOUNG's interference experiment. The generalization is straightforward. We have to employ the mutual coherence function $\Gamma_{12}(\tau) = \Gamma(P_1,P_2,\tau)$ in place of the self-coherence function $\Gamma(\tau) = \Gamma(P,P,\tau)$ and the mutual spectral density $G_{12}(\nu)$ in place of the ordinary spectral density $G(\nu)$. The only difference arises from the fact that $G_{12}(\nu)$ is now complex and that $\Gamma_{12}(\tau)$ is no longer necessarily an even function of τ and consequently $\bar{\tau}$ is not necessarily zero. The appropriate definitions are:

$$(\Delta\tau_{12})^2 = \frac{\int_{-\infty}^{\infty} (\tau - \bar{\tau}_{12})^2 |\Gamma_{12}(\tau)|^2 \, d\tau}{\int_{-\infty}^{\infty} |\Gamma_{12}(\tau)|^2 \, d\tau}, \qquad \bar{\tau}_{12} = \frac{\int_{-\infty}^{\infty} \tau |\Gamma_{12}(\tau)|^2 \, d\tau}{\int_{-\infty}^{\infty} |\Gamma_{12}(\tau)|^2 \, d\tau}, \tag{35}$$

$$(\Delta\nu_{12})^2 = \frac{\int_{0}^{\infty} (\nu - \bar{\nu}_{12})^2 |G_{12}(\nu)|^2 \, d\nu}{\int_{0}^{\infty} |G_{12}(\nu)|^2 \, d\nu}, \qquad \bar{\nu}_{12} = \frac{\int_{0}^{\infty} \nu |G_{12}(\nu)|^2 \, d\nu}{\int_{0}^{\infty} |G_{12}(\nu)|^2 \, d\nu}. \tag{36}$$

$\Delta\tau_{12}$ may be called the *mutual coherence time* and $\Delta\nu_{12}$ the *mutual effective spectral width* of the light at P_1 and P_2. By an obvious modification of the argument given in connection with $\Gamma_{11}(\tau)$ it follows that these quantities satisfy the reciprocity inequality

$$(\Delta\tau_{12})(\Delta\nu_{12}) > \frac{1}{4\pi}. \tag{37}$$

Finally, for quasi-monochromatic light we now have the following relations as generalization of (33):

$$(\Delta \tau_{12})^2 = \frac{\int_{-\infty}^{\infty} (\tau - \bar{\tau}_{12})^2 \mathscr{V}_{12}^2(\tau)\, d\tau}{\int_{-\infty}^{\infty} \mathscr{V}_{12}^2(\tau)\, d\tau}, \qquad \bar{\tau}_{12} = \frac{\int_{-\infty}^{\infty} \tau \mathscr{V}_{12}^2(\tau)\, d\tau}{\int_{-\infty}^{\infty} \mathscr{V}_{12}^2(\tau)\, d\tau}. \qquad (38)$$

Here $\mathscr{V}_{12}(\tau)$ represents the visibility of the fringes formed by the light from P_1 and P_2, when the interfering beams have the same intensity.

10.8 POLARIZATION PROPERTIES OF QUASI-MONOCHROMATIC LIGHT

In the preceding sections of this chapter we have treated the light disturbance as a scalar quantity. We shall now briefly consider some of the vectorial properties of a quasi-monochromatic light wave.

We have learned in § 1.4 that strictly monochromatic light is always *polarized*, i.e. that with increasing time the end point of the electric (and also of the magnetic) vector at each point in space moves periodically around an ellipse, which may, of course, reduce in special cases to a circle or a straight line. We have also encountered *unpolarized* light. In this case the end point may be assumed to move quite irregularly, and the light shows no preferential directional properties when resolved in different directions at right angles to the direction of propagation. Like complete coherence and complete incoherence these two cases represent two extremes. In general the variation of the field vectors is neither completely regular, nor completely irregular, and we may say that the light is *partially polarized*. Such light arises usually from unpolarized light by reflection (cf. § 1.5.3) or scattering (cf. § 13.5.2). In this section we shall investigate the main properties of a partially polarized light wave. We shall see that its observable effects depend on the intensities of any two mutually orthogonal components of the electric vector at right angles to the direction of propagation, and on the correlation which exists between them.

10.8.1 The coherency matrix of a quasi-monochromatic plane wave*

Consider a quasi-monochromatic light wave of mean frequency $\bar{\nu}$ propagated in the positive z-direction. Let

$$E_x(t) = a_1(t) e^{i[\phi_1(t) - 2\pi\bar{\nu}t]}, \qquad E_y(t) = a_2(t) e^{i[\phi_2(t) - 2\pi\bar{\nu}t]} \qquad (1)$$

represent the components, at a point O, of the electric vector in two mutually orthogonal directions at right angles to the direction of propagation. We again use the complex representation discussed in § 10.2, in which E_x and E_y are the "analytic signals" associated with the true (real) components $E_x^{(r)} = a_1(t) \cos[\phi_1(t) - 2\pi\bar{\nu}t]$, $E_y^{(r)} = a_2(t) \cos[\phi_2(t) - 2\pi\bar{\nu}t]$. If the light were strictly monochromatic, a_1, a_2, ϕ_1 and ϕ_2 would be constants. For a quasi-monochromatic wave these quantities

* The analysis given in § 10.8.1 and 10.8.2 is based on investigations of E. Wolf, *Nuovo Cimento*, **13** (1959), 1165. Some further developments are described in a paper by G. B. Parrent and P. Roman, *ibid*, **15** (1960), 370.

depend also on the time t, but, as we have seen, they change only by small relative amounts in any time interval that is small compared to the coherence time, i.e. that is small compared to the reciprocal of the effective spectral width $\Delta\nu$ of the light.

Suppose that the y-component is subjected to a retardation ε with respect to the x-component (this can be done, for example, by means of one of the compensators described in § 14.4.2), and consider the intensity $I(\theta,\varepsilon)$ of the light vibrations in the direction which makes an angle θ with the positive x-direction (Fig. 10.17). This intensity would be observed by sending the light through a polarizer (§ 14.4.1) with the appropriate orientation.

The component of the electric vector in the θ direction, after the retardation ε has been introduced, is

$$E(t;\,\theta,\varepsilon) = E_x \cos\theta + E_y e^{i\varepsilon} \sin\theta, \tag{2}$$

so that

$$\begin{aligned} I(\theta,\varepsilon) &= \langle E(t;\,\theta,\varepsilon)E^\star(t;\,\theta,\varepsilon)\rangle \\ &= J_{xx}\cos^2\theta + J_{yy}\sin^2\theta + J_{xy}e^{-i\varepsilon}\cos\theta\sin\theta + J_{yx}e^{i\varepsilon}\sin\theta\cos\theta, \end{aligned} \tag{3}$$

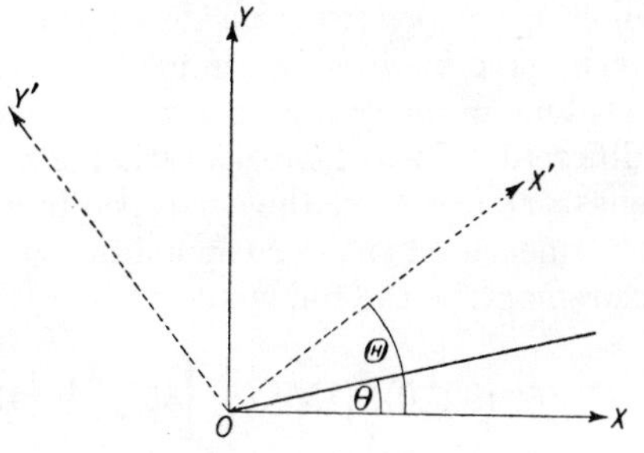

Fig. 10.17. Illustrating notation.

where $J_{xx}, \ldots$ are the elements of the matrix

$$\boldsymbol{J} = \begin{bmatrix} \langle E_x E_x^\star\rangle & \langle E_x E_y^\star\rangle \\ \langle E_y E_x^\star\rangle & \langle E_y E_y^\star\rangle \end{bmatrix} = \begin{bmatrix} \langle a_1^2\rangle & \langle a_1 a_2 e^{i(\phi_1-\phi_2)}\rangle \\ \langle a_1 a_2 e^{-i(\phi_1-\phi_2)}\rangle & \langle a_2^2\rangle \end{bmatrix}. \tag{4}$$

The diagonal elements of $\boldsymbol{J}$ are real and are seen to represent the intensities of the components in the x- and y-directions. Hence the trace $\operatorname{Tr}\boldsymbol{J}$ of the matrix, i.e. the sum of its diagonal elements, is equal to the total intensity of the light,

$$\operatorname{Tr}\boldsymbol{J} = J_{xx} + J_{yy} = \langle E_x E_x^\star\rangle + \langle E_y E_y^\star\rangle. \tag{5}$$

The non-diagonal elements are in general complex, but they are conjugates of each other. (A matrix such as this, which satisfies the relation $J_{ji} = J_{ij}^\star$ for all i and j is said to be a *Hermitian* matrix.)

We shall normalize the mixed term J_{xy} in a similar way as before (cf. § 10.4 (9b)), by setting

$$\mu_{xy} = |\mu_{xy}|e^{i\beta_{xy}} = \frac{J_{xy}}{\sqrt{J_{xx}}\sqrt{J_{yy}}}. \tag{6}$$

Then it follows by SCHWARZ' inequality, in the same way as in connection with § 10.3 (17), that

$$|\mu_{xy}| \leqslant 1. \tag{7}$$

This complex correlation factor μ_{xy} has a similar significance as the complex degree of coherence μ_{12} introduced in § 10.4.1. It is a measure of the correlation between the components of the electric vector in the x- and y-directions. The absolute value $|\mu_{xy}|$ is a measure of their "degree of coherence," and its phase β_{xy} is a measure of their "effective phase difference." We call J *the coherency matrix* of the light wave. Since J_{xx} and J_{yy} cannot be negative, (6) and (7) imply that the associated determinant is non-negative, i.e. that

$$|\mathsf{J}| = J_{xx}J_{yy} - J_{xy}J_{yx} \geqslant 0. \tag{8}$$

If we use the relation $J_{yx} = J^{\star}_{xy}$, and denote by $\mathscr{R}$ the real part, (3) becomes

$$\begin{aligned} I(\theta,\varepsilon) &= J_{xx}\cos^2\theta + J_{yy}\sin^2\theta + 2\cos\theta\sin\theta\mathscr{R}(J_{xy}e^{-i\varepsilon}) \\ &= J_{xx}\cos^2\theta + J_{yy}\sin^2\theta + 2\sqrt{J_{xx}}\sqrt{J_{yy}}\cos\theta\sin\theta|\mu_{xy}|\cos[\beta_{xy} - \varepsilon], \end{aligned} \tag{9}$$

where we substituted from (6) on going from the first to the second line. If we set $J_{xx}\cos^2\theta = I^{(1)}$, $J_{yy}\sin^2\theta = I^{(2)}$, the last formula becomes identical with the basic interference law § 10.4 (11) of quasi-monochromatic wave fields.

Like the coherence functions which we considered earlier, the elements of the coherency matrix of a given wave may be determined by means of relatively simple experiments. This may be done in many different ways. One only needs to measure the intensity for several different values of θ (orientation of the polarizer) and ε (delay introduced by the compensator), and solve the corresponding relations obtained from (3). Let $\{\theta,\varepsilon\}$ denote the measurement corresponding to a particular pair, θ, ε. A convenient set of measurements is the following:

$$\{0°,0\},\ \{45°,0\},\ \{90°,0\},\ \{135°,0\},\ \left\{45°,\frac{\pi}{2}\right\},\ \left\{135°,\frac{\pi}{2}\right\}. \tag{10}$$

It follows from (3) that, in terms of the intensities determined from these six measurements, the elements of the coherency matrix are given by

$$\left.\begin{aligned} J_{xx} &= I(0°,0), \\ J_{yy} &= I(90°,0), \\ J_{xy} &= \tfrac{1}{2}\{I(45°,0) - I(135°,0)\} + \tfrac{1}{2}i\left\{I\left(45°,\frac{\pi}{2}\right) - I\left(135°,\frac{\pi}{2}\right)\right\}, \\ J_{yx} &= \tfrac{1}{2}\{I(45°,0) - I(135°,0)\} - \tfrac{1}{2}i\left\{I\left(45°,\frac{\pi}{2}\right) - I\left(135°,\frac{\pi}{2}\right)\right\}. \end{aligned}\right\} \tag{11}$$

We see that only a polarizer is needed to determine J_{xx}, J_{yy} and the real part of J_{xy} (or J_{yx}). J_{xx} and J_{yy} may be determined from measurements with a polarizer oriented so as to transmit the components in the azimuths $\theta = 0$ and $\theta = 90°$ respectively. The real part of J_{xy} may be determined from measurements with a polarizer oriented so that it first transmits the component in the azimuth $\theta = 45°$ and then the component in the azimuth $\theta = 135°$. To determine the imaginary part of J_{xy} (or J_{yx}) we also need, according to the last two relations in (11), a compensator which introduces a phase difference of a quarter period (e.g. a quarter-wave plate, cf. § 14.4.2) between the x- and y-components; the polarizer is again used, oriented so as to pass first the component in the azimuth $\theta = 45°$ and then the component in the azimuth $\theta = 135°$. We shall learn in § 14.4.2, that the last two measurements are those made to detect right-handed and left-handed circular polarization.

It is evident from (9) that two beams of light which have the same coherency matrix are equivalent in the sense that they will yield the same (time averaged) intensity in similar experiments with a polarizer and a compensator.*

Let us now examine how the observed intensity $I(\theta,\varepsilon)$ changes for a given wave when one of the arguments (θ or ε) is kept fixed, whilst the other varies. Suppose first that we keep θ fixed and change ε. We see from (9) that the intensity varies sinusoidally between the values

$$\left.\begin{aligned} I_{\max(\varepsilon)} &= J_{xx}\cos^2\theta + J_{yy}\sin^2\theta + 2|J_{xy}|\sin\theta\cos\theta,\\ \text{and}\quad I_{\min(\varepsilon)} &= J_{xx}\cos^2\theta + J_{yy}\sin^2\theta - 2|J_{xy}|\sin\theta\cos\theta. \end{aligned}\right\} \tag{12}$$

Hence

$$\frac{I_{\max(\varepsilon)} - I_{\min(\varepsilon)}}{I_{\max(\varepsilon)} + I_{\min(\varepsilon)}} = \frac{|J_{xy}|\sin 2\theta}{J_{xx}\cos^2\theta + J_{yy}\sin^2\theta}. \tag{13}$$

Equation (13) indicates an alternative way of determining the absolute value of J_{xy} (and hence also of $|\mu_{xy}|$); it shows that this quantity may be obtained from measurements of J_{xx}, J_{yy}, $I_{\max(\varepsilon)}$ and $I_{\min(\varepsilon)}$; the phase of J_{xy} may be obtained from measurement of the value of ε at which the maxima or minima occur. For, according to (9),

$$\left.\begin{aligned} I &= I_{\max(\varepsilon)} \quad \text{when} \quad \varepsilon = \beta_{xy} \pm 2m\pi \qquad (m = 0,1,2,\ldots),\\ I &= I_{\min(\varepsilon)} \quad \text{when} \quad \varepsilon = \beta_{xy} \pm (2m+1)\pi \qquad (m = 0,1,2,\ldots). \end{aligned}\right\} \tag{14}$$

To see how the intensity changes when ε is fixed and θ is varied, it is convenient to re-write (9) in a somewhat different form. Simple calculation gives

$$I(\theta,\varepsilon) = \tfrac{1}{2}(J_{xx} + J_{yy}) + R\cos(2\theta - \alpha), \tag{15}$$

where

$$\begin{aligned} R &= \tfrac{1}{2}\sqrt{(J_{xx} - J_{yy})^2 + 4J_{xy}J_{yx}\cos^2(\beta_{xy} - \varepsilon)},\\ \tan\alpha &= \frac{2|J_{xy}|\cos(\beta_{xy} - \varepsilon)}{J_{xx} - J_{yy}}. \end{aligned} \tag{16}$$

We see from (15) that, as θ changes, the intensity again varies sinusoidally; its extremes are

$$\left.\begin{aligned} I_{\max(\theta)} &= \tfrac{1}{2}(J_{xx} + J_{yy}) + R,\\ I_{\min(\theta)} &= \tfrac{1}{2}(J_{xx} + J_{yy}) - R. \end{aligned}\right\} \tag{17}$$

On the right of (17) only R depends on ε. It takes its largest value when $\cos^2(\beta_{xy} - \varepsilon) = 1$, i.e. when ε has one of the values given by (14), and is then equal to

$$\begin{aligned} R_{\max(\varepsilon)} &= \tfrac{1}{2}\sqrt{(J_{xx} - J_{yy})^2 + 4J_{xy}J_{yx}}\\ &= \tfrac{1}{2}(J_{xx} + J_{yy})\sqrt{1 - \frac{4|\mathsf{J}|}{(J_{xx} + J_{yy})^2}}, \end{aligned} \tag{18}$$

* This statement is true only within the approximation of the quasi-monochromatic theory, for it is only within the range of validity of this theory that the expression (9) for the intensity holds. The two beams may behave quite differently when the phase delay ε introduced between the two orthogonal components is not negligible in comparison with the coherence length, measured in units of the mean wavelength $\bar{\lambda}$. For a fuller description of the observable properties of a beam it is necessary to introduce more general coherency matrices which characterize the correlations between the components at different times and also at different points. Cf. E. Wolf, *Nuovo Cimento*, **12** (1954), 884; also his contribution in *Proc. Symp. Astronom. Optics*, ed. Z. Kopal (Amsterdam, North-Holland Publishing Co., 1956), 177; P. Roman and E. Wolf, *Nuovo Cimento*, **17** (1960), 462, 477; P. Roman, *ibid.*, **20** (1961), 759; *ibid.*, **22** (1961), 1005.

where $|\boldsymbol{J}|$ is the determinant (8) of the coherency matrix. It follows that the absolute maxima and minima (with respect to both θ and ε) of the intensity are

$$\left.\begin{aligned} I_{\max(\theta,\varepsilon)} &= \tfrac{1}{2}(J_{xx}+J_{yy})\left[1+\sqrt{1-\frac{4|\boldsymbol{J}|}{(J_{xx}+J_{yy})^2}}\right], \\ I_{\min(\theta,\varepsilon)} &= \tfrac{1}{2}(J_{xx}+J_{yy})\left[1-\sqrt{1-\frac{4|\boldsymbol{J}|}{(J_{xx}+J_{yy})^2}}\right]. \end{aligned}\right\} \tag{19}$$

Hence

$$\frac{I_{\max(\theta,\varepsilon)}-I_{\min(\theta,\varepsilon)}}{I_{\max(\theta,\varepsilon)}+I_{\min(\theta,\varepsilon)}} = \sqrt{1-\frac{4|\boldsymbol{J}|}{(J_{xx}+J_{yy})^2}}. \tag{20}$$

We shall see later that this quantity has a simple physical meaning.

So far we have referred the electric vibrations to arbitrary but fixed rectangular axes OX, OY. We shall now consider how the coherency matrix transforms when a new set of axes is chosen. Suppose that we take new rectangular axes OX', OY', again in the plane perpendicular to the direction of propagation, such that OX' makes an angle Θ with OX (see Fig. 10.17). In terms of E_x, E_y, the components of the electric vector referred to the new axes are

$$\left.\begin{aligned} E_{x'} &= E_x\cos\Theta + E_y\sin\Theta, \\ E_{y'} &= -E_x\sin\Theta + E_y\cos\Theta. \end{aligned}\right\} \tag{21}$$

The elements of the transformed coherency matrix $\boldsymbol{J}'$ are

$$J_{k'l'} = \langle E_{k'}E^{\star}_{l'}\rangle, \tag{22}$$

where k' and l' each take on the values x' and y'. From (21) and (22) it follows that

$$\boldsymbol{J}' = \begin{bmatrix} J_{xx}c^2+J_{yy}s^2+(J_{xy}+J_{yx})cs & (J_{yy}-J_{xx})cs+J_{xy}c^2-J_{yx}s^2 \\ (J_{yy}-J_{xx})cs+J_{yx}c^2-J_{xy}s^2 & J_{xx}s^2+J_{yy}c^2-(J_{xy}+J_{yx})cs \end{bmatrix}, \tag{23}$$

where

$$c = \cos\Theta, \qquad s = \sin\Theta. \tag{24}$$

It is seen that the trace of the matrix is invariant under rotation of the axes. A straightforward calculation shows that its determinant is also invariant under this transformation. Both these results also follow from well-known theorems of matrix algebra.

We shall now consider the form of the coherency matrix for some cases of particular interest.

(a) *Completely unpolarized light* (*Natural light*)

Light which is most frequently encountered in nature has the property that the intensity of its components in any direction perpendicular to the direction of propagation is the same; and, moreover, this intensity is not affected by any previous retardation of one of the rectangular components relative to the other, into which the light may have been resolved. In other words

$$I(\theta,\varepsilon) = \text{constant} \tag{25}$$

for all values of θ and ε. Such light may be said to be *completely unpolarized* and is often also called *natural light*.

It is evident from (9) that $I(\theta,\varepsilon)$ is independent of ε and θ if and only if

$$\mu_{xy} = 0 \quad \text{and} \quad J_{xx} = J_{yy}. \tag{26a}$$

The first condition implies that E_x and E_y are mutually incoherent. According to (6) and the relation $J_{yx} = J^{\star}_{xy}$, (26a) may also be written as

$$J_{xy} = J_{yx} = 0, \qquad J_{xx} = J_{yy} \tag{26b}$$

and it follows that the coherency matrix of natural light of intensity $J_{xx} + J_{yy} = I_0$ is

$$\tfrac{1}{2}I_0 \begin{bmatrix} 1 & 0 \\ 0 & 1 \end{bmatrix}. \tag{27}$$

(b) *Completely polarized light*

Suppose first that the light is strictly monochromatic. Then the amplitudes a_1 and a_2, and the phase factors ϕ_1 and ϕ_2 in (1) do not depend on the time, and the coherency matrix has the form

$$\begin{bmatrix} a_1{}^2 & a_1a_2e^{i\delta} \\ a_1a_2e^{-i\delta} & a_2{}^2 \end{bmatrix} \tag{28}$$

where*

$$\delta = \phi_1 - \phi_2. \tag{29}$$

We see that in this case

$$|\mathsf{J}| = J_{xx}J_{yy} - J_{xy}J_{yx} = 0, \tag{30}$$

i.e. the determinant of the coherency matrix is zero. The complex degree of coherence of the components E_x and E_y now is

$$\mu_{xy} = \frac{J_{xy}}{\sqrt{J_{xx}}\sqrt{J_{yy}}} = e^{i\delta}, \tag{31}$$

i.e. its absolute value is unity (complete coherence) and its phase is equal to the difference between the phases of the two components.

In the special case when the light is *linearly polarized* we have (cf. § 1.4 (33)), $\delta = m\pi$ ($m = 0, \pm 1, \pm 2, \ldots$). Hence the coherency matrix of linearly polarized light is

$$\begin{bmatrix} a_1{}^2 & (-1)^m a_1a_2 \\ (-1)^m a_1a_2 & a_2{}^2 \end{bmatrix}. \tag{32}$$

The electric vector vibrates in the direction given by $E_y/E_x = (-1)^m a_2/a_1$. In particular, the matrices

$$I\begin{bmatrix} 1 & 0 \\ 0 & 0 \end{bmatrix}, \qquad I\begin{bmatrix} 0 & 0 \\ 0 & 1 \end{bmatrix}, \tag{33}$$

each represent linearly polarized light of intensity I, with the electric vector in the x-direction ($a_2 = 0$) and the y-direction ($a_1 = 0$) respectively; and the matrices

$$\tfrac{1}{2}I\cdot\begin{bmatrix} 1 & 1 \\ 1 & 1 \end{bmatrix}, \qquad \tfrac{1}{2}I\begin{bmatrix} 1 & -1 \\ -1 & 1 \end{bmatrix} \tag{34}$$

each represent linearly-polarized light of intensity I and with the electric vector in directions making angles 45° and 135° with the x-direction, respectively ($a_1 = a_2$, $m = 0$ and $a_1 = a_2$, $m = 1$).

* For the purpose of later applications of some results of § 1.4 we note that ϕ_1 and ϕ_2 correspond to $-\delta_1$ and $-\delta_2$ of § 1.4.2, so that (29) is consistent with the earlier definition § 1.4 (16), viz. $\delta = \delta_2 - \delta_1$.

For *circularly-polarized light* we have (cf. § 1.4 (35), (36)), $a_1 = a_2$, $\delta = m\pi/2$ ($m = \pm 1, \pm 3, \ldots$), so that the coherency matrix is

$$\tfrac{1}{2}I \begin{bmatrix} 1 & \pm i \\ \mp i & 1 \end{bmatrix}, \tag{35}$$

where I is the intensity of the light. By § 1.4 (38) and (40) the upper or lower sign is taken according whether the polarization is right- or left-handed.

The condition (30) may also be satisfied when the light is not monochromatic. For if a_1, a_2, ϕ_1 and ϕ_2 depend on time in such a way that the *ratio of the amplitudes* and the *difference in the phases* are time independent, so that

$$\frac{a_2(t)}{a_1(t)} = q, \qquad \delta = \phi_1(t) - \phi_2(t) = \chi, \tag{36}$$

where q and χ are constants, then

$$\begin{aligned} J_{xx} &= \langle a_1^2 \rangle, & J_{xy} &= q\langle a_1^2 \rangle e^{i\chi}, \\ J_{yx} &= q\langle a_1^2 \rangle e^{-i\chi}, & J_{yy} &= q^2\langle a_1^2 \rangle, \end{aligned} \tag{37}$$

and the condition (30) holds. The coherency matrix with the elements (37) is the same as that of strictly monochromatic light with components

$$E_x = \sqrt{\langle a_1^2 \rangle}\, e^{i[\alpha - 2\pi\bar{\nu}t]}, \qquad E_y = q\sqrt{\langle a_1^2 \rangle}\, e^{i[-\chi + \alpha - 2\pi\bar{\nu}t]}, \tag{38}$$

where α is any real constant. It follows that in experiments involving a polarizer and a compensator, the quasi-monochromatic wave which obeys the conditions (36) will behave in exactly the same way as the strictly monochromatic and hence completely polarized wave (38). (It is assumed, of course, that the phase difference introduced by the compensator is small compared to the coherence length of the light, measured in units of the mean wavelength.) The condition (30) may, therefore, be said to characterize a *completely polarized* light wave.

10.8.2 Some equivalent representations. The degree of polarization of a light wave

If several *independent* light waves which are propagated in the same direction are superposed, the coherency matrix of the resulting wave is equal to the sum of the coherency matrices of the individual waves. To prove this result let $E_x^{(n)}$, $E_y^{(n)}$ ($n = 1, 2, \ldots N$) be the components of the electric vectors (in the usual complex representation) of the individual waves. The components of the electric vector of the resulting wave are

$$E_x = \sum_{n=1}^{N} E_x^{(n)}, \qquad E_y = \sum_{n=1}^{N} E_y^{(n)}, \tag{39}$$

so that the elements of its coherency matrix are given by

$$\begin{aligned} J_{kl} = \langle E_k E_l^\star \rangle &= \sum_{n=1}^{N} \sum_{m=1}^{N} \langle E_k^{(n)} E_l^{(m)\star} \rangle \\ &= \sum_n \langle E_k^{(n)} E_l^{(n)\star} \rangle + \sum_{n \neq m} \langle E_k^{(n)} E_l^{(m)\star} \rangle. \end{aligned} \tag{40}$$

Since the waves are assumed to be independent, each term under the last summation sign is zero, and it follows that

$$J_{kl} = \sum_n J_{kl}^{(n)}, \tag{41}$$

where $J_{kl}^{(n)} = \langle E_k^{(n)} E_l^{(n)\star} \rangle$ are the elements of the coherency matrix of the nth wave. Equation (41) shows that the coherency matrix of the combined wave is equal to the sum of the coherency matrices of all the separate waves.

Conversely any wave may be regarded as the sum of independent waves, which evidently may be chosen in many different ways. One particular choice is of special significance and will now be briefly considered.

We will show that any quasi-monochromatic light wave may be regarded as the sum of a completely unpolarized and a completely polarized wave, which are independent of each other, and that this representation is unique.

To establish this result it is only necessary to show that any coherency matrix $\mathbf{J}$ can be uniquely expressed in the form

$$\mathbf{J} = \mathbf{J}^{(1)} + \mathbf{J}^{(2)}, \tag{42}$$

where, in accordance with (27) and (30),

$$\mathbf{J}^{(1)} = \begin{bmatrix} A & 0 \\ 0 & A \end{bmatrix}, \qquad \mathbf{J}^{(2)} = \begin{bmatrix} B & D \\ D^\star & C \end{bmatrix}, \tag{43}$$

with $A \geqslant 0$, $B \geqslant 0$, $C \geqslant 0$ and

$$BC - DD^\star = 0. \tag{44}$$

If J_{xx}, J_{xy}, . . . are the elements of the coherency matrix which characterizes the original wave, we must have, according to (42) and (43),

$$\left.\begin{aligned} A + B &= J_{xx}, & D &= J_{xy}, \\ D^\star &= J_{yx}, & A + C &= J_{yy}. \end{aligned}\right\} \tag{45}$$

On substituting from (45) to (44) we obtain the following equation for A:

$$(J_{xx} - A)\,(J_{yy} - A) - J_{xy}\,J_{yx} = 0; \tag{46}$$

thus A is a characteristic root (eigenvalue) of the coherency matrix $\mathbf{J}$. The two roots of (46) are

$$A = \tfrac{1}{2}(J_{xx} + J_{yy}) \pm \tfrac{1}{2}\sqrt{(J_{xx} + J_{yy})^2 - 4|\mathbf{J}|}, \tag{47}$$

where, as before, $|\mathbf{J}|$ is the determinant (8). Since $J_{yx} = J^\star_{xy}$, the product $J_{xy}J_{yx}$ is non-negative and it follows from (8) that

$$|\mathbf{J}| \leqslant J_{xx}J_{yy} \leqslant \tfrac{1}{4}(J_{xx} + J_{yy})^2,$$

so that both the roots (47) are real and non-negative. Consider first the solution with the negative sign in front of the square root. We then have

$$A = \tfrac{1}{2}(J_{xx} + J_{yy}) - \tfrac{1}{2}\sqrt{(J_{xx} + J_{yy})^2 - 4|\mathbf{J}|}, \tag{48}$$

$$\left.\begin{aligned} B &= \tfrac{1}{2}(J_{xx} - J_{yy}) + \tfrac{1}{2}\sqrt{(J_{xx} + J_{yy})^2 - 4|\mathbf{J}|}, & D &= J_{xy}, \\ D^\star = J_{yx}, \quad C &= \tfrac{1}{2}(J_{yy} - J_{xx}) + \tfrac{1}{2}\sqrt{(J_{xx} + J_{yy})^2 - 4|\mathbf{J}|}. \end{aligned}\right\} \tag{49}$$

Now

$$\sqrt{(J_{xx} + J_{yy})^2 - 4|\mathbf{J}|} = \sqrt{(J_{xx} - J_{yy})^2 + 4J_{xy}J_{yx}} \geqslant |J_{xx} - J_{yy}|.$$

Hence B and C are also non-negative, as required. The other root (47) (with the positive sign in front of the square root) leads to negative values of B and C and must therefore be rejected. We have thus obtained a unique decomposition of the required kind.

The total intensity of the wave is

$$I_{\text{tot}} = \operatorname{Tr} \mathbf{J} = J_{xx} + J_{yy}; \tag{50}$$

and the total intensity of the polarized part is

$$I_{\text{pol}} = \operatorname{Tr} \mathbf{J}^{(2)} = B + C = \sqrt{(J_{xx} + J_{yy})^2 - 4|\mathbf{J}|}. \tag{51}$$

The ratio of the intensity of the polarized portion to the total intensity is called the *degree of polarization* P of the wave; according to (50) and (51) it is given by

$$P = \frac{I_{\text{pol}}}{I_{\text{tot}}} = \sqrt{1 - \frac{4|\mathbf{J}|}{(J_{xx} + J_{yy})^2}}. \tag{52}$$

Since this expression involves only the two rotational invariants of the coherency matrix $\mathbf{J}$, the degree of polarization is independent of the particular choice of the axes OX, OY, as might have been expected. From (52) and the inequality preceding (48) it follows that

$$0 \leqslant P \leqslant 1. \tag{53}$$

When $P = 1$ there is no unpolarized component, so that the wave is then *completely polarized.* In this case $|\mathbf{J}| = 0$, so that $|\mu_{xy}| = 1$ and consequently E_x and E_y are mutually coherent. When $P = 0$ the polarized component is absent. The wave is then *completely unpolarized.* In this case $(J_{xx} + J_{yy})^2 = 4|\mathbf{J}|$, i.e.

$$(J_{xx} - J_{yy})^2 + 4J_{xy}J_{yx} = 0. \tag{54a}$$

Since $J_{yx} = J^{\star}_{xy}$ we have the sum of two squares equal to zero, and this is only possible if each vanishes separately, i.e. if

$$J_{xx} = J_{yy} \quad \text{and} \quad J_{xy} = J_{yx} = 0, \tag{54b}$$

in accordance with (26b). E_x and E_y are then mutually incoherent ($\mu_{xy} = 0$). In all other cases ($0 < P < 1$) we say that the light is *partially polarized.* Comparison of (52) and (20) shows that the quantity $(I_{\max(\theta,\varepsilon)} - I_{\min(\theta,\varepsilon)})/(I_{\max(\theta,\varepsilon)} + I_{\min(\theta,\varepsilon)})$ is precisely the degree of polarization P.

When E_x and E_y are mutually incoherent (but the light not necessarily natural), the expression for the degree of polarization takes a simple form. Since now $J_{xy} = J_{yx} = 0$, $|\mathbf{J}| = J_{xx}J_{yy}$ and (52) reduces to*

$$P = \left|\frac{J_{xx} - J_{yy}}{J_{xx} + J_{yy}}\right|. \tag{55}$$

This expression is in agreement with the formula (42) of § 1.5 employed in connection with polarization of natural light by reflection.

---

* Since every Hermitian matrix may be made diagonal by a unitary transformation, and since the values of $|\mathbf{J}|$ and $\operatorname{Tr} \mathbf{J}$ remain invariant under this transformation, the degree of polarization may always be expressed in the form

$$P = \left|\frac{A_1 - A_2}{A_1 + A_2}\right|,$$

where A_1 and A_2 are the two eigenvalues (given by (47)). However, the unitary transformation does not, in general, represent a real rotation of axes about the direction of the propagation of the wave.

It is of interest to note that the eigenvalues A_1 and A_2 are equal to the values $I_{\max(\theta,\varepsilon)}$ and $I_{\min(\theta,\varepsilon)}$ given by eq. (19).

We note some useful representations of *natural light.* The coherency matrix (27) of natural light may always be expressed in the form

$$\tfrac{1}{2}I\begin{bmatrix}1 & 0\\ 0 & 1\end{bmatrix} = \tfrac{1}{2}I\begin{bmatrix}1 & 0\\ 0 & 0\end{bmatrix} + \tfrac{1}{2}I\begin{bmatrix}0 & 0\\ 0 & 1\end{bmatrix}, \tag{56}$$

and this implies, according to (33), that a wave of natural light, of intensity I, is equivalent to two independent linearly polarized waves, each of intensity $\tfrac{1}{2}I$, with their electric vectors vibrating in two mutually perpendicular directions at right angles to the direction of propagation.

Another useful representation of natural light is

$$\tfrac{1}{2}I\begin{bmatrix}1 & 0\\ 0 & 1\end{bmatrix} = \tfrac{1}{4}I\begin{bmatrix}1 & +i\\ -i & 1\end{bmatrix} + \tfrac{1}{4}I\begin{bmatrix}1 & -i\\ +i & 1\end{bmatrix}, \tag{57}$$

and implies, according to (35), that a wave of natural light of intensity I is equivalent to two independent circularly polarized waves, one right-handed, the other left-handed, each of intensity $\tfrac{1}{2}I$.

Returning to the general case (partially polarized light), it is to be noted that unlike the degree of polarization, P, the degree of coherence $|\mu_{xy}|$ depends on the choice of the x and y directions. One may, however, readily see, that $|\mu_{xy}|$ cannot exceed P. For if in (52) we write out the determinant $|\boldsymbol{J}|$ in full and use (6) we find that

$$1 - P^2 = \frac{J_{xx}J_{yy}}{[\tfrac{1}{2}(J_{xx} + J_{yy})]^2}[1 - |\mu_{xy}|^2]. \tag{58}$$

Since the geometric mean of any two positive numbers cannot exceed their arithmetic mean, it follows that $1 - P^2 \leqslant 1 - |\mu_{xy}|^2$, i.e.

$$P \geqslant |\mu_{xy}|. \tag{59}$$

The equality sign in (59) will hold if and only if $J_{xx} = J_{yy}$, i.e. if the (time averaged) intensities associated with the x and y directions are equal. We will show now that a pair of directions always exists, for which this is the case.

If the xy-axes are rotated in their own plane, through an angle θ in the anti-clockwise sense, J_{xx} and J_{yy} are transformed into $J_{x'x'}$ and $J_{y'y'}$ respectively, where, according to (23)

$$\left.\begin{aligned}J_{x'x'} &= J_{xx}\cos^2\theta + J_{yy}\sin^2\theta + (J_{xy} + J_{yx})\cos\theta\sin\theta,\\ J_{y'y'} &= J_{xx}\sin^2\theta + J_{yy}\cos^2\theta - (J_{xy} + J_{yx})\cos\theta\sin\theta.\end{aligned}\right\} \tag{60}$$

From (60) it follows that $J_{x'x'} = J_{y'y'}$, if the axes are rotated though the angle $\theta = \Theta$, where

$$\tan 2\Theta = \frac{J_{yy} - J_{xx}}{J_{xy} + J_{yx}}. \tag{61}$$

Since $J_{yx} = J^\star_{xy}$ and J_{xx} and J_{yy} are both real, the equation (61) has always a real solution for Θ. Thus *there always exists a pair of mutually orthogonal directions for which the intensities are equal. For this pair of directions the degree of coherence* $|\mu_{xy}|$ *of the electric vibrations has its maximum value and this value is equal to the degree of polarization P of the wave.**

* The geometrical significance of this special pair of directions is disussed by E. Wolf in *Nuovo Cimento,* **13** (1959), pp. 1180–1181.

10.8.3 The Stokes parameters of a quasi-monochromatic plane wave

We have seen that, in order to characterize a quasi-monochromatic plane wave, four real quantities are in general necessary, for example J_{xx}, J_{yy} and the real and imaginary parts of J_{xy} (or J_{yx}). In his investigations relating to partially polarized light, STOKES* introduced a somewhat different four-parameter representation which is closely related to the present one. We have already encountered it, in a restricted form, in connection with monochromatic light in § 1.4.2. The general STOKES parameters are the four quantities

$$\left.\begin{aligned} s_0 &= \langle a_1^2 \rangle + \langle a_2^2 \rangle, \\ s_1 &= \langle a_1^2 \rangle - \langle a_2^2 \rangle, \\ s_2 &= 2\langle a_1 a_2 \cos \delta \rangle, \\ s_3 &= 2\langle a_1 a_2 \sin \delta \rangle, \end{aligned}\right\} \qquad (62)$$

where, as before, a_1 and a_2 are the instantaneous amplitudes of the two orthogonal components E_x, E_y of the electric vector and $\delta = \phi_1 - \phi_2$ is their phase difference. When the light is monochromatic, a_1, a_2, and δ are independent of the time, and (62) reduces to the "monochromatic STOKES parameters" defined in § 1.4 (43).

It follows from (62) and (4) that the STOKES parameters and the elements of the coherency matrix are related by the formulae

$$\left.\begin{aligned} s_0 &= J_{xx} + J_{yy}, \\ s_1 &= J_{xx} - J_{yy}, \\ s_2 &= J_{xy} + J_{yx}, \\ s_3 &= i(J_{yx} - J_{xy}). \end{aligned}\right\} \text{(63a)} \qquad \left.\begin{aligned} J_{xx} &= \tfrac{1}{2}(s_0 + s_1), \\ J_{yy} &= \tfrac{1}{2}(s_0 - s_1), \\ J_{xy} &= \tfrac{1}{2}(s_2 + is_3), \\ J_{yx} &= \tfrac{1}{2}(s_2 - is_3). \end{aligned}\right\} \text{(63b)}$$

Like the elements of the coherency matrix, the STOKES parameters of any quasi-monochromatic plane wave may be determined from simple experiments. If as before $I(\theta,\varepsilon)$ denotes the intensity of the light vibrations in the direction making an angle θ with OX, when the y-component is subjected to a retardation ε with respect to the x-component, then, according to (11) and the relations (63a),

$$\left.\begin{aligned} s_0 &= I(0°,0) + I(90°,0), \\ s_1 &= I(0°,0) - I(90°,0), \\ s_2 &= I(45°,0) - I(135°,0), \\ s_3 &= I\left(45°, \frac{\pi}{2}\right) - I\left(135°, \frac{\pi}{2}\right). \end{aligned}\right\} \qquad (64)$$

The parameter s_0 evidently represents the total intensity. The parameter s_1 is equal to the excess in intensity of light transmitted by a polarizer which accepts linear polarization in the azimuth $\theta = 0°$, over the light transmitted by a polarizer which

---

* G. G. STOKES, *Trans. Cambr. Phil. Soc.*, **9** (1852), 399. Reprinted in his *Mathematical and Physical Papers*, Vol. III (Cambridge University Press, 1901), p. 233. See also P. SOLEILLET, *Ann. de Physique* (10), **12** (1929), 23; F. PERRIN, *J. Chem. Phys.*, **10** (1942), 415; S. CHANDRASEKHAR, *Radiative Transfer* (Oxford, Clarendon Press, 1950), § 15; M. J. WALKER, *Amer. J. Phys.*, **22** (1954), 170; E. WOLF, *Nuovo Cimento*, **12** (1954), 884; S. PANCHARATNAM, *Proc. Ind. Acad. Sci.*, A, **44** (1956), 398; *ibid.*, **57** (1963), 218, 231.

The STOKES parameters are also employed in quantum mechanical treatments of polarization of elementary particles. Cf. U. FANO, *J. Opt. Soc. Amer.*, **39** (1949), 859; *ibid*, **41** (1951), 58; *Phys. Rev.*, **93** (1954), 121; D. L. FALKOFF and J. E. MACDONALD, *J. Opt. Soc. Amer.*, **41** (1951), 861; W. H. MCMASTER, *Amer. J. Phys.*, **22** (1954), 351; J. M. JAUCH and F. ROHRLICH, *The Theory of Photons and Electrons* (Cambridge, Mass., Addison-Wesley Publ. Co., 1955), § 2.8. See also N. WIENER, *Acta Math.*, **55** (1930), § 9, especially pp. 189–192.

accepts linear polarization in the azimuth $\theta = 90°$. The parameter s_2 has a similar interpretation with respect to the azimuths $\theta = 45°$ and $\theta = 135°$. Finally, the parameter s_3 is equal to the excess in intensity of light transmitted by a device which accepts right-handed circular polarization, over that transmitted by a device which accepts left-handed circular polarization.

If we use the relations (63b) our previous results may be expressed in terms of the STOKES parameters, rather than in terms of the coherency matrix. In particular, the condition (8), viz. $J_{xx}J_{yy} - J_{xy}J_{yx} \geqslant 0$, becomes

$$s_0^2 \geqslant s_1^2 + s_2^2 + s_3^2. \tag{65}$$

For monochromatic light we have, according to (30), $J_{xx}J_{yy} - J_{xy}J_{yx} = 0$ and the equality sign then holds in (65), in agreement with § 1.4 (44).

Let us now consider the decomposition of a given wave into an unpolarized and a polarized portion which are mutually independent, using the STOKES parameter representation. It follows from (41) and (63) that the STOKES parameters of a mixture of *independent* waves are sums of the respective STOKES parameters of the separate waves. From (27) and (63a) it follows that an unpolarized wave (wave of natural light) is characterized by $s_1 = s_2 = s_3 = 0$. Denoting by a single symbol $\boldsymbol{s}$ the four STOKES parameters s_0, s_1, s_2, s_3, the required decomposition of the wave characterized by $\boldsymbol{s}$ evidently is

$$\boldsymbol{s} = \boldsymbol{s}^{(1)} + \boldsymbol{s}^{(2)}, \tag{66}$$

where

$$\boldsymbol{s}^{(1)} = s_0 - \sqrt{s_1^2 + s_2^2 + s_3^2},\ 0,\ 0,\ 0, \tag{67a}$$

$$\boldsymbol{s}^{(2)} = \sqrt{s_1^2 + s_2^2 + s_3^2},\ s_1,\ s_2,\ s_3. \tag{67b}$$

$\boldsymbol{s}^{(1)}$ represents the unpolarized and $\boldsymbol{s}^{(2)}$ the polarized part. Hence, in terms of the STOKES parameters, the degree of polarization of the original wave is

$$P = \frac{I_{\text{pol}}}{I_{\text{tot}}} = \frac{\sqrt{s_1^2 + s_2^2 + s_3^2}}{s_0}, \tag{68}$$

as may also be verified by substituting from (63b) into (52). We may also easily write down expressions which give the ellipticity and the orientation of the polarization ellipse associated with the polarized part (67b). If, as in § 1.4 (28),

$$\tan \chi = \pm\, b/a \qquad (-\pi/4 < \chi \leqslant \pi/4)$$

represents the ratio of the minor and the major axes and the sense in which the ellipse is described ($\chi \gtrless 0$ according as the polarization is right- or left-handed), then according to (67b) and § 1.4 (45c),

$$\sin 2\chi = \frac{s_3}{\sqrt{s_1^2 + s_2^2 + s_3^2}}; \tag{69}$$

and the angle ψ ($0 \leqslant \psi < \pi$), which the major axis makes with OX, is according to (67b) and § 1.4 (46) given by

$$\tan 2\psi = \frac{s_2}{s_1}. \tag{70}$$

We see that the STOKES parameters, just like the coherency matrix, provide a useful tool for a systematic analysis of the state of polarization of a quasi-monochromatic wave.

CHAPTER XI

RIGOROUS DIFFRACTION THEORY

11.1 INTRODUCTION

On the basis of Maxwell's equations, together with standard boundary conditions, the scattering of electromagnetic radiation by an obstacle becomes a well-defined mathematical boundary-value problem. In the present chapter some aspects of the theory of diffraction of monochromatic waves are developed from this point of view, and in particular the rigorous solution to the classical problem of diffraction by a perfectly conducting half-plane is given in detail.

In the early theories of Young, Fresnel, and Kirchhoff, the diffracting obstacle was supposed to be perfectly "black"; that is to say, all radiation falling on it was assumed to be absorbed, and none reflected. This is an inherent source of ambiguity in that such a concept of absolute "blackness" cannot legitimately be defined with precision; it is, indeed, incompatible with electromagnetic theory.

Cases in which the diffracting body has a finite dielectric constant and finite conductivity have been examined theoretically, one of the earliest comprehensive treatments of such a case being Mie's discussion in 1908 of scattering by a sphere, which is described in Chapter XIII in connection with the optics of metals. In general, however, the assumption of finite conductivity tends to make the mathematics very complicated, and it is often desirable to accept the concept of a perfectly conducting (and therefore perfectly reflecting) body. This is clearly an idealization, but one which is compatible with electromagnetic theory; furthermore, since the conductivity of some metals (e.g. copper) is very large, it may represent a good approximation if the frequency is not too high, though it should be stressed that the approximation is never entirely adequate at optical frequencies. The simplifying assumption that the diffracting obstacle has infinite conductivity is made in most of the treatments based on a precise mathematical formulation, and the subsequent discussion is confined to this case.

The first rigorous solution of such a diffraction problem was given by Sommerfeld* in 1896, when he treated the two-dimensional case of a plane wave incident on an infinitely thin, perfectly conducting half-plane. The fame of this achievement rests partly on the skill with which the solution was constructed, and partly on the remarkable fact that it could be expressed exactly and simply in terms of the Fresnel integrals which had been such a conspicuous feature of previous approximate theories.

Many mathematicians followed Sommerfeld's lead. Early variants of his problem, dealing with line and point sources, and a noteworthy generalization to the treatment of a wedge rather than a half-plane, are associated with the names of Carslaw,† Macdonald,‡ and Bromwich.§ Other problems were attacked, and more recently new methods have been introduced, stimulated by the advance in ultra-short-wave

* A. Sommerfeld, *Math. Ann.*, **47** (1896), 317.
† H. S. Carslaw, *Proc. Lond. Math. Soc.*, **30** (1899), 121.
‡ H. M. Macdonald, *Electric Waves* (Cambridge University Press, 1902).
§ T. J. I'A. Bromwich, *Proc. Lond. Math. Soc.*, **14** (1916), 450.

radio techniques. Before proceeding to the main body of the chapter the nature of some of these investigations is very briefly indicated.

If there exists an orthogonal coordinate system, u_1, u_2, u_3, say, such that the surface of the diffracting body is identical with one of the surfaces u_i = constant, the classical technique for solving partial differential equations by separation of variables may be appropriate; this, indeed, was MIE's approach in the case of the finitely conducting sphere mentioned above. The solution of the boundary-value problem then appears, in general, as an infinite series, and its utility depends on the ease with which computation of the relevant functions can be carried out and the rapidity with which the series converges. This method has been applied to various cases apart from the sphere, notably to the circular disc or aperture.* It should be mentioned, however, that some of the work only relates to strictly scalar problems, such as those in the theory of small-amplitude sound waves; as is shown later, two-dimensional problems in electromagnetic theory are essentially of this type, but otherwise the vector nature of the electromagnetic field introduces further complications.

Another approach is based on integral equation formulations, a method apparently first considered by RAYLEIGH.† Certain problems, the simplest being that of the half-plane, yield integral equations that can be solved exactly by the method of WIENER and HOPF,‡ and the appreciation of this fact by COPSON,§ SCHWINGER, and others has led to a number of new solutions in closed form.‖ Mention should also be made, in this connection, of powerful, if somewhat complicated, variational procedures which can be used to calculate the power diffracted through an aperture.¶

For reasons of space the discussion in this chapter is largely confined to one method.** First, some aspects of considerable generality in the theory of the scattering of electromagnetic waves by perfectly conducting structures are developed. Next, a representation of any field as an integral over a spectrum of plane waves is introduced and is shown to lead to the formulation of certain diffraction problems in terms of "dual" integral equations.†† The SOMMERFELD half-plane problem is then tractable; the solution of this is obtained and examined in some detail, together with a number of ramifications. Several allied problems are discussed.

11.2 BOUNDARY CONDITIONS AND SURFACE CURRENTS

It is well known that an electromagnetic field penetrates but little into a good conductor. The idealization of infinite conductivity, when there is no penetration at all,

---

* C. J. BOUWKAMP, *Dissertation*, Groningen, 1941.
J. MEIXNER and W. ANDREJEWSKI, *Ann. d. Physik*, **7** (1950), 157.

† Lord RAYLEIGH, *Phil. Mag.*, **43** (1897), 259.

‡ For a discussion of the Wiener-Hopf method see E. C. TITCHMARSH, *Introduction to the Theory of Fourier Integrals* (Oxford, Clarendon Press, 1937), p. 339.

§ E. T. COPSON, *Quart. J. Maths.*, **17** (1946), 19.

‖ J. F. CARLSON and A. E. HEINS, *Quart. Appl. Maths.*, **4** (1947), 313; **5** (1947), 82. A. E. HEINS, *Quart. Appl. Maths.*, **6** (1948), 157, 215. H. LEVINE and J. SCHWINGER, *Phys. Rev.*, **73** (1948), 383. For a more comprehensive list of references see J. W. MILES, *J. Appl. Phys.*, **20** (1949), 760, and C. J. BOUWKAMP, *Rep. Progr. Phys.* (London, Physical Society), **17** (1954), 35.

¶ H. LEVINE and J. SCHWINGER, *Phys. Rev.*, **74** (1948), 958; **75** (1949), 1423.

** For general accounts of other methods see the article by G. WOLFSOHN in *Handbuch der Physik*, Vol. 20 (Berlin, Springer, 1928), p. 263; B. B. BAKER and E. T. COPSON, *The Mathematical Theory of Huygens' Principle* (Oxford, Clarendon Press, 1950), Chapters 4 and 5 and an article by H. HÖNL, A. W. MAUE and K. WESTPFAHL in *Handbuch d. Physik*, Vol. 25/1 (Berlin, Springer, 1961). The review by BOUWKAMP just mentioned gives in outline a most comprehensive selection of methods and formulae.

†† For a discussion of "dual" integral equations (defined on pp. 564–565 below) see E. C. TITCHMARSH, *Introduction to the Theory of Fourier Integrals* (Oxford, Clarendon Press, 1937), p. 334.

results in the concept of electric currents existing purely on the surface of the conductor, as the following argument shows.

It is a consequence of MAXWELL's curl equations (see § 1.1.3) that the tangential component of $\mathbf{E}$ is continuous in crossing an infinitely thin electric current sheet whereas that of $\mathbf{H}$ is discontinuous; more particularly, the discontinuity in $\mathbf{H}$ is in the tangential component normal to the surface current density* $\mathbf{J}$, the relative sense of the directions being indicated schematically in Fig. 11.1, and is of amount $4\pi J/c$. Furthermore, in line with the behaviour of the tangential components of $\mathbf{E}$ and $\mathbf{H}$, the normal component of $\mathbf{H}$ is continuous across the current sheet whereas that of $\mathbf{E}$ is discontinuous, the magnitude of the discontinuity being equal to 4π times the surface charge density. Hence it is clear that the field in free space exterior to a perfectly conducting body is such that on the surface of the conductor

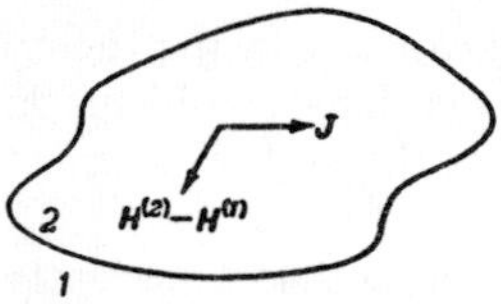

Fig. 11.1. Showing the sense of the discontinuity in $\mathbf{H}$ relative to the direction of the surface current density $\mathbf{J}$. $\mathbf{H}^{(1)}$ and $\mathbf{H}^{(2)}$ are the respective magnetic fields on sides 1 and 2 of the surface.

(a) the tangential component of $\mathbf{E}$ is zero;
(b) the tangential component of $\mathbf{H}$ is perpendicular to the surface electric current density $\mathbf{J}$, in the sense indicated above, and of amount $4\pi J/c$;
(c) the normal component of $\mathbf{H}$ is zero;
(d) the outward normal component of $\mathbf{E}$ is equal to 4π times the surface charge density.

The effect of radiation falling on a perfectly conducting body may be interpreted conveniently in terms of the induced surface currents. If $\mathbf{E}^{(i)}$ is the electric vector of the incident field and $\mathbf{E}^{(s)}$ is that of the "scattered" field due to induced currents, then the total electric vector, everywhere, is $\mathbf{E}^{(i)} + \mathbf{E}^{(s)}$. The diffraction problem may therefore be stated as follows: *given* $\mathbf{E}^{(i)}$, *to find a field* $\mathbf{E}^{(s)}$ *which could arise from a current distribution in the surface of the conductor and which is such that its tangential component on the surface is minus that of* $\mathbf{E}^{(i)}$. It is worth stressing that the boundary condition (a) above is fundamental and sufficient alone to specify the problem uniquely in the form stated.† As regards the other conditions, (b) is of value in relating the field to the induced currents, but (c) and (d) are of no particular interest.

It should be noted that an implication of $\mathbf{E}^{(s)} = -\mathbf{E}^{(i)}$ at interior points of the conductor is the existence of a unique current density, in any closed surface S, which reproduces at all points inside S the field due to sources located outside S. Likewise, from a consideration of the case when the boundary of a perfect conductor on which the radiation falls is a complete infinite plane, it follows that there is a unique current density in any plane which reproduces on one side of the plane the field due to sources situated on the other side of the plane.

In problems of diffraction by perfectly conducting screens it is desirable to make the assumption that the screens are infinitely thin; if this is not done, the mathematical difficulties become very great. Of course, the opacity of the screen is maintained, the concept, in fact, being that of a perfect conductor whose thickness tends to zero in the limit. From what has been said, the effect of such a screen can be interpreted in terms of an electric current sheet, with the difference now that the sheet is no longer a closed surface. Of particular interest is the relatively simple case when the sheet is

* The surface current was denoted by $\mathbf{j}$ in § 1.1.3.

† A discussion of proofs of uniqueness, which present some difficulties, is deferred to a later section (§ 11.9).

plane; in this case some important relations satisfied by the field $\mathbf{E}^{(s)}$, $\mathbf{H}^{(s)}$ which it radiates may be deduced immediately and are given below.

Suppose that the current sheet occupies part of the plane $y = 0$. Then, by reason of symmetry, it is clear that

$$\left.\begin{aligned} E_x^{(s)}(x, y, z) &= E_x^{(s)}(x, -y, z), & H_x^{(s)}(x, y, z) &= -H_x^{(s)}(x, -y, z), \\ E_y^{(s)}(x, y, z) &= -E_y^{(s)}(x, -y, z), & H_y^{(s)}(x, y, z) &= H_y^{(s)}(x, -y, z), \\ E_z^{(s)}(x, y, z) &= E_z^{(s)}(x, -y, z), & H_z^{(s)}(x, y, z) &= -H_z^{(s)}(x, -y, z). \end{aligned}\right\} \quad (1)$$

Moreover, if the current density in the sheet has components J_x and J_z, evidently, on $y = 0$,

$$H_x^{(s)} = \mp \frac{2\pi}{c} J_z, \qquad H_z^{(s)} = \pm \frac{2\pi}{c} J_x, \quad (2)$$

with the upper or lower sign according as to whether y reaches zero through positive or negative values respectively. As discussed in the next section, the application of these simple relations to the interesting problem of diffraction by a plane screen leads to a useful formulation which, in particular, puts in evidence an exact electromagnetic analogue of Babinet's principle.

11.3 DIFFRACTION BY A PLANE SCREEN: ELECTROMAGNETIC FORM OF BABINET'S PRINCIPLE

Suppose that an electromagnetic field $\mathbf{E}^{(i)}$, $\mathbf{H}^{(i)}$ is incident on a set of infinitely thin, perfectly conducting laminae lying in the plane $y = 0$. Let M signify those areas of the plane occupied by the metal and A the remaining "apertures", so that M and A together comprise the whole plane. Either M or A, or both M and A, may be of infinite extent.

As previously explained, a scattered field is sought which satisfies a certain boundary condition on M. Now in view of the relations § 11.2 (1), it is, in fact, only necessary to consider the scattered field in one of the half-spaces $y \geqslant 0$, $y \leqslant 0$, provided the requirement of continuity across A is explicitly recognized. Hence the problem may be formulated thus: to find, in $y \geqslant 0$ (or in $y \leqslant 0$), an electromagnetic field $\mathbf{E}^{(s)}$, $\mathbf{H}^{(s)}$, which could be generated by currents in $y = 0$, such that

$$\begin{aligned} &\text{(I)} \quad E_x^{(i)} + E_x^{(s)} = E_z^{(i)} + E_z^{(s)} = 0 \qquad &\text{on } M, \\ &\text{(II)} \quad H_x^{(s)} = H_z^{(s)} = 0 \qquad &\text{on } A. \end{aligned}$$

Here, (I) is the fundamental boundary condition for a perfect conductor, whereas (II), which follows from § 11.2 (2), is a convenient way of expressing the fact that there are no induced currents in A. If (II) is satisfied by the scattered field in $y \geqslant 0$, and § 11.2 (1) used to deduce the scattered field in $y \leqslant 0$, then continuity across A is achieved.

A form of Babinet's principle for electromagnetic waves and perfectly conducting screens which is exact* may now be easily derived. As in the classical principle (§ 8.3.2), a relation is established between the respective fields existing in the presence of the screen and the "complementary" screen obtained by interchanging the conducting laminae and the apertures; the difference lies in the fact that the field incident

* H. G. Booker, *J. Instn. Elect. Engrs.*, **93**, Pt. III A (1946), 620. L. G. H. Huxley, *The Principles and Practice of Waveguides* (Cambridge University Press, 1947), p. 284.

on the complementary screen is no longer the same as that incident on the original screen, but is derived from it by the transformation $\boldsymbol{E} \to \boldsymbol{H}$.

In the first case, then, let the field (suffix 1) defined by $\boldsymbol{E}_1^{(i)} = \boldsymbol{F}^{(i)}$ be incident, in $y > 0$, on the screen described above. From (I) and (II)

$$\text{(I}'\text{)} \quad E_{1x}^{(s)} = -F_x^{(i)}, \qquad E_{1z}^{(s)} = -F_z^{(i)} \qquad \text{on } M,$$

$$\text{(II}'\text{)} \quad H_{1x}^{(s)} = H_{1z}^{(s)} = 0 \qquad \text{on } A.$$

Secondly, let the field (suffix 2) defined by $\boldsymbol{H}_2^{(i)} = \boldsymbol{F}^{(i)}$ be incident on the complementary screen. Then, writing the boundary conditions now in terms of the total field,

$$\text{(I}''\text{)} \quad E_{2x} = E_{2z} = 0 \qquad \text{on } A,$$

$$\text{(II}''\text{)} \quad H_{2x} = F_x^{(i)}, \qquad H_{2z} = F_z^{(i)} \qquad \text{on } M.$$

Since MAXWELL's equations in free space are invariant under the transformation $\boldsymbol{E} \to \boldsymbol{H}$, $\boldsymbol{H} \to -\boldsymbol{E}$, and since there is a unique surface current density in $y = 0$ which would produce the incident field at all points in $y \leqslant 0$, it is clear from a comparison of (I′), (II′) with (II″), (I″), respectively, that

$$\boldsymbol{H}_2 = -\boldsymbol{E}_1^{(s)} \tag{1}$$

in the half-space *behind* the screen. In terms of the total field $\boldsymbol{E}_1$, (1) gives

$$\boldsymbol{E}_1 + \boldsymbol{H}_2 = \boldsymbol{F}^{(i)}, \tag{2}$$

which is the required electromagnetic form of BABINET's principle.

11.4 TWO-DIMENSIONAL DIFFRACTION BY A PLANE SCREEN

11.4.1 The scalar nature of two-dimensional electromagnetic fields

A problem which is completely independent of one Cartesian coordinate, say z, is said to be two-dimensional. As already remarked, problems of this type in electromagnetic theory are essentially of a scalar nature in that they can straightway be expressed in terms of a single dependent variable. This will now be shown.

With a time factor $\exp(-i\omega t)$ suppressed and writing $k = \omega/c$, MAXWELL's equations in free space are

$$\operatorname{curl} \boldsymbol{H} = -ik\boldsymbol{E}, \qquad \operatorname{curl} \boldsymbol{E} = ik\boldsymbol{H}.$$

Equating to zero all partial derivatives with respect to z, these may be split up into the two independent sets

$$\frac{\partial E_z}{\partial y} = ikH_x, \qquad \frac{\partial E_z}{\partial x} = -ikH_y, \qquad \frac{\partial H_y}{\partial x} - \frac{\partial H_x}{\partial y} = -ikE_z, \tag{1}$$

and

$$\frac{\partial H_z}{\partial y} = -ikE_x, \qquad \frac{\partial H_z}{\partial x} = ikE_y, \qquad \frac{\partial E_y}{\partial x} - \frac{\partial E_x}{\partial y} = ikH_z. \tag{2}$$

The first group involves only H_x, H_y, E_z, the second only E_x, E_y, H_z. Simplicity can therefore be obtained by separating any solution into a linear combination of the two solutions for which every member of one of the above sets is zero. For the sake of nomenclature we characterize the two types of field as follows:

E-polarization

$$E_x = E_y = H_z = 0,$$

$$H_x = \frac{1}{ik}\frac{\partial E_z}{\partial y}, \qquad H_y = -\frac{1}{ik}\frac{\partial E_z}{\partial x},$$

and, as is evident on substituting for H_x and H_y into the third equation of (1),

$$\frac{\partial^2 E_z}{\partial x^2} + \frac{\partial^2 E_z}{\partial y^2} + k^2 E_z = 0.$$

Here the complete field is specified in terms of E_z, which, of course, satisfies the two-dimensional form of the standard wave equation.

H-polarization

$$H_x = H_y = E_z = 0,$$

and

$$E_x = -\frac{1}{ik}\frac{\partial H_z}{\partial y}, \qquad E_y = \frac{1}{ik}\frac{\partial H_z}{\partial x}.$$

Here the complete field is specified in terms of H_z.

11.4.2 An angular spectrum of plane waves

For two-dimensional problems, to which the discussion is now confined, it has been shown that the Cartesian components of **E** and **H** satisfy the equation

$$\frac{\partial^2 V}{\partial x^2} + \frac{\partial^2 V}{\partial y^2} + k^2 V = 0. \tag{3}$$

This equation has to be solved subject to the appropriate boundary conditions.

A fundamental elementary solution of (3) is

$$e^{ikr\cos(\theta-\alpha)} = e^{ik(x\cos\alpha + y\sin\alpha)}, \tag{4}$$

r, θ $(0 \leqslant \theta \leqslant 2\pi)$ being polar coordinates related to x, y by the equations $x = r\cos\theta$, $y = r\sin\theta$. If α is real (4) represents a *homogeneous* plane wave, that is, one whose equiamplitude and equiphase planes coincide: α is the angle between the direction of propagation and the x axis (Fig. 11.2a). If, on the other hand, α is complex, (4) represents an *inhomogeneous* plane wave, that is, one whose equiamplitude and equiphase planes do not coincide. In fact, writing $\alpha = \alpha_1 + i\alpha_2$, where α_1 and α_2 are real, (4) becomes

$$e^{ikr\cosh\alpha_2\cos(\theta-\alpha_1)}\, e^{-kr\sinh\alpha_2\sin(\theta-\alpha_1)}, \tag{5}$$

from which it follows that the equiamplitude and equiphase planes are mutually perpendicular (Fig. 11.2b): the direction of phase propagation makes an angle α_1 with the x axis, the phase velocity being reduced by the factor sech α_2, and there is exponential attenuation, governed by the attenuation factor $k \sinh \alpha_2$, in the direction at right angles.

Now it can be shown* that any solution of (3) can be put in the form of an *angular spectrum of plane waves*

$$\int f(\alpha)\, e^{ikr\cos(\theta-\alpha)}\, d\alpha$$

by a suitable choice of the path of integration and the function $f(\alpha)$. Such a representation is closely linked with the representation of an arbitrary function by means of a FOURIER integral, and is likewise of great power in application. Without significant loss of generality a certain fixed path of integration can be prescribed, so that any problem becomes a matter of determining the appropriate $f(\alpha)$. We shall first express the electromagnetic field due to a plane current sheet in this way, and then show that the result leads to a formulation of the problem of diffraction by a plane screen in terms of dual integral equations.

Consider a two-dimensional current sheet in $y = 0$. As already pointed out, it is convenient to deal with E-polarization and H-polarization separately. We treat,

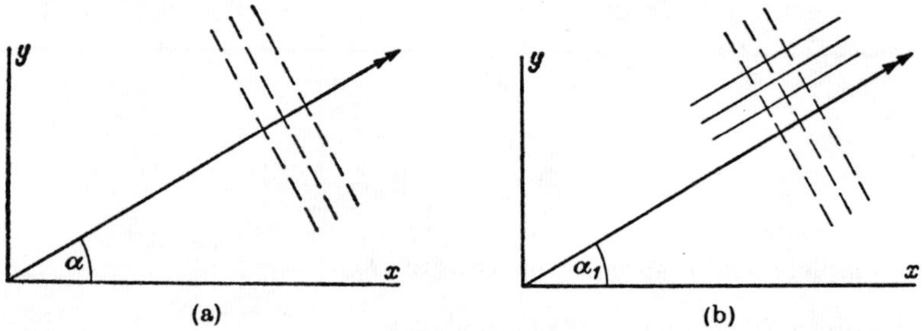

Fig. 11.2.

(a) The homogeneous plane wave (4) when α is real; the dashed lines denote both equiamplitude and equiphase planes.

(b) The inhomogeneous plane wave (4) when α is complex; the full lines denote equiamplitude planes, the dashed lines equiphase planes.

first, the former case, in which the current density has a z component only, J_z say, and begin by asking what particular distribution will radiate the E-polarized plane wave

$$\boldsymbol{E} = (0, 0, 1)e^{ikr\cos(\theta-\alpha)}, \qquad \boldsymbol{H} = (\sin\alpha, -\cos\alpha, 0)e^{ikr\cos(\theta-\alpha)} \tag{6}$$

into the half-space $y > 0$. From the first relation in § 11.2 (2) it is, in fact, immediately seen that

$$J_z(\xi) = -\frac{c}{2\pi}\, e^{ik\xi\cos\alpha}\sin\alpha \tag{7}$$

at the point $(\xi, 0)$. This could, of course, be verified by the standard method of HERTZ potentials for finding the field generated by a current distribution, though the evaluation of a quite complicated integral is then necessary.

Now, broadly speaking, any current distribution can be built up by the appropriate superposition of expressions (7) for different values of α, and the radiated field will be obtained by the corresponding superposition of the plane waves (6). More precisely, suppose the current density can be written in the form of a FOURIER integral as

$$J_z(\xi) = -\frac{c}{2\pi}\int_{-\infty}^{\infty} P(\mu)\, e^{ik\xi\mu}\, d\mu. \tag{8}$$

* E. T. WHITTAKER and G. N. WATSON, *A Course of Modern Analysis* (Cambridge University Press, 1927), p. 397.

The change of variable $\mu = \cos\alpha$ gives

$$J_z(\xi) = -\frac{c}{2\pi}\int_C \sin\alpha\, P(\cos\alpha)\, e^{ik\xi\cos\alpha}\, d\alpha, \tag{9}$$

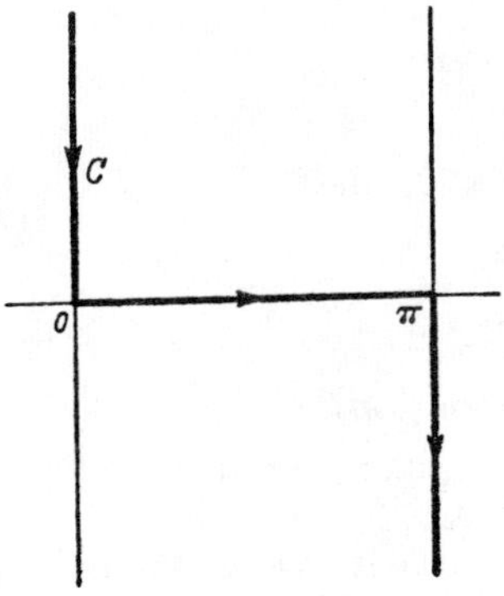

Fig. 11.3. The path C in the complex α plane.

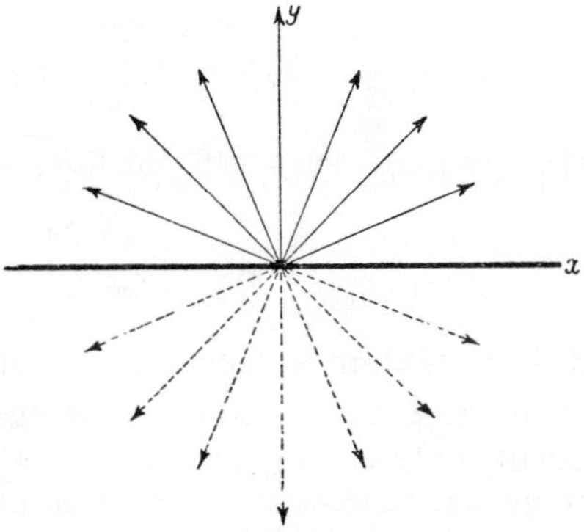

Fig. 11.4. Showing the directions of propagation of the homogeneous waves which radiate into the half-space $y > 0$ (full lines) and the half-space $y < 0$ (interrupted lines).

where C is the path in the complex α plane along which $\cos\alpha$ ranges through real values from ∞ to $-\infty$, as shown in Fig. 11.3. The resulting non-zero field components are therefore

$$E_z^{(s)} = \int_C P(\cos\alpha)\, e^{ikr\cos(\theta\mp\alpha)}\, d\alpha, \tag{10}$$

$$H_x^{(s)} = \pm\int_C \sin\alpha\, P(\cos\alpha)\, e^{ikr\cos(\theta\mp\alpha)}\, d\alpha, \tag{11}$$

$$H_y^{(s)} = -\int_C \cos\alpha\, P(\cos\alpha)\, e^{ikr\cos(\theta\mp\alpha)}\, d\alpha, \tag{12}$$

with the upper sign for $y \geqslant 0$ and the lower sign for $y \leqslant 0$.

Equations (10), (11), and (12) represent the field in the form of a plane wave spectrum determined by the function $P(\cos\alpha)$. The individual plane waves corresponding to the section of C along the real axis are homogeneous; they radiate into the regions $y > 0$ and $y < 0$, their directions of propagation embracing a range of angles π in each region, as illustrated diagrammatically in Fig. 11.4. The plane waves corresponding to the two arms of C on which $\alpha = i\beta$ and $\alpha = \pi - i\beta$ ($\beta = 0$ to ∞) are inhomogeneous; all their directions of phase propagation are along the positive or negative x axis, and they are exponentially attenuated in the direction normal to and away from the plane $y = 0$. It can easily be shown, by an examination of the Poynting vector, that on the average no energy is carried away from the plane $y = 0$ by any of these *evanescent* waves. Their presence is necessary in order to take account of structure in the current distribution which is finer than a wavelength.

For the case of H-polarization, the field due to a current density J_x in $y = 0$ would likewise be written

$$H_z^{(s)} = \pm \int_C P(\cos\alpha) e^{ikr\cos(\theta\mp\alpha)}\, d\alpha, \tag{13}$$

$$E_x^{(s)} = -\int_C \sin\alpha\, P(\cos\alpha) e^{ikr\cos(\theta\mp\alpha)}\, d\alpha, \tag{14}$$

$$E_y^{(s)} = \pm \int_C \cos\alpha\, P(\cos\alpha) e^{ikr\cos(\theta\mp\alpha)}\, d\alpha, \tag{15}$$

with the upper sign for $y \geqslant 0$, the lower sign for $y \leqslant 0$, where

$$J_x(\xi) = \frac{c}{2\pi} \int_{-\infty}^{\infty} \frac{P(\mu)}{\sqrt{(1-\mu^2)}} e^{ik\xi\mu}\, d\mu. \tag{16}$$

11.4.3 Formulation in terms of dual integral equations

The two-dimensional problem of diffraction by a plane screen can now be formulated in terms of dual integral equations.

Suppose that an electromagnetic field $\mathbf{E}^{(i)}$, $\mathbf{H}^{(i)}$ is incident on a set of infinitely thin, perfectly conducting strips lying in $y = 0$; designate by M the ranges of x within

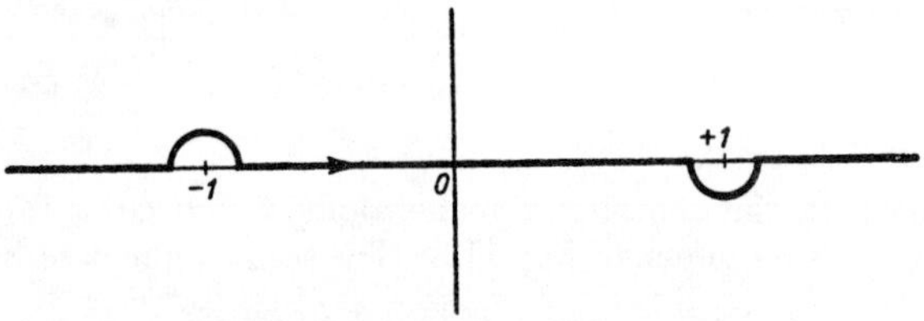

Fig. 11.5. The path of integration from $-\infty$ to ∞ in the complex μ plane.

which there is metal, by A those within which there is not. If the scattered field $\mathbf{E}^{(s)}$, $\mathbf{H}^{(s)}$ is represented as an angular spectrum of plane waves, in the form § 11.4 (10), (11), (12) or (13), (14), (15) according to the polarization, the conditions (I) and (II) of § 11.3 yield the following integral equations:

E-polarization

$$\int_{-\infty}^{\infty} \frac{P(\mu)}{\sqrt{(1-\mu^2)}} e^{ikx\mu}\, d\mu = -E_z^{(i)} \qquad \text{on } M, \tag{17}$$

$$\int_{-\infty}^{\infty} P(\mu)\, e^{ikx\mu}\, d\mu = 0 \qquad \text{on } A. \tag{18}$$

H-polarization

$$\int_{-\infty}^{\infty} P(\mu)\, e^{ikx\mu}\, d\mu = E_x^{(i)} \qquad \text{on } M, \tag{19}$$

$$\int_{-\infty}^{\infty} \frac{P(\mu)}{\sqrt{(1-\mu^2)}} e^{ikx\mu}\, d\mu = 0 \qquad \text{on } A. \tag{20}$$

A consideration of the way in which the complex α plane between $\mathscr{R}\alpha = 0$ and $\mathscr{R}\alpha = \pi$ ($\mathscr{R}$ denoting the real part) maps into the complete complex μ plane ($\mu = \cos\alpha$) shows that the path of integration along the real axis avoids the possible branch-points at $\mu = \pm 1$ as illustrated diagrammatically in Fig. 11.5. Integral equations of this type, in which a single unknown function $P(\mu)$ satisfies different equations for two distinct ranges of the parameter x, are called "dual".*

The formulation used by COPSON, SCHWINGER, and others, mentioned in § 11.1, is somewhat different from the one given above in that it only involves a single integral equation. Although it will not be wanted here, its connection with the present method should be pointed out. For the case of E-polarization, say, the solution of (16) obtained by taking its FOURIER transform could be written in the form

$$P(\mu) = -\frac{k}{c}\int_M J_z(\xi)e^{-ik\mu\xi}\,d\xi, \tag{21}$$

in agreement, of course, with eq. (8) and the fact that $J_z(\xi) = 0$ on A. Substituting this value of $P(\mu)$ into (17), and carrying out the integration with respect to μ, we have the integral equation

$$\frac{k}{c}\int_M J_z(\xi)H_0^{(1)}(k|x-\xi|)d\xi = E_z^{(i)} \qquad \text{on } M, \tag{22}$$

involving the HANKEL function $H_0^{(1)}$ of the first kind and zero order, to be solved for $J_z(\xi)$. Obviously, the left-hand side of (22) could have been derived from the direct expression of the scattered field in terms of the induced current density.

11.5 TWO-DIMENSIONAL DIFFRACTION OF A PLANE WAVE BY A HALF-PLANE

11.5.1 Solution of the dual integral equations for E-polarization

In the next few pages the diffraction of a plane wave by a semi-infinite plane sheet is treated rigorously by obtaining a simple explicit solution of the appropriate dual integral equations.

Consider, first, the E-polarized plane wave

$$E_z^{(i)} = e^{-ikr\cos(\theta-\alpha_0)} \tag{1}$$

incident on the perfectly conducting half-plane $y = 0$, $x > 0$, where it is assumed, for convenience, that α_0 is real and $0 < \alpha_0 < \pi$ (Fig. 11.6). Equations (17) and (18) of § 11.4 are now

$$\int_{-\infty}^{\infty}\frac{P(\mu)}{\sqrt{(1-\mu^2)}}e^{ikx\mu}\,d\mu = -e^{-ikx\mu_0} \qquad \text{for } x > 0, \tag{2}$$

$$\int_{-\infty}^{\infty}P(\mu)e^{ikx\mu}\,d\mu = 0 \qquad \text{for } x < 0, \tag{3}$$

where $\mu_0 = \cos\alpha_0$. We proceed to solve these equations by the use of standard techniques in contour integration.

In the integral on the left-hand side of equation (3) x is negative. Hence, by JORDAN's lemma,† provided $P(\mu) \to 0$ as $|\mu| \to \infty$ when $0 \geqslant \arg\mu \geqslant -\pi$, we can close the path of integration with an infinite semi-circle *below* the real axis without

* E. C. TITCHMARSH, *Introduction to the Theory of Fourier Integrals* (Oxford, Clarendon Press, 1937), p. 334.

† E. T. WHITTAKER and G. N. WATSON, *A Course of Modern Analysis* (Cambridge University Press, 1920), p. 115.

making any additional contribution to the integral. Thus we only require further that $P(\mu)$ should have no singularities in the half-plane below the path of integration for equation (3) to be satisfied, since the integral is then effectively round a closed contour within which the integrand is regular.

Likewise, in the integral on the left-hand side of equation (2) x is positive, and we can close the path of integration with an infinite semi-circle *above* the real axis without

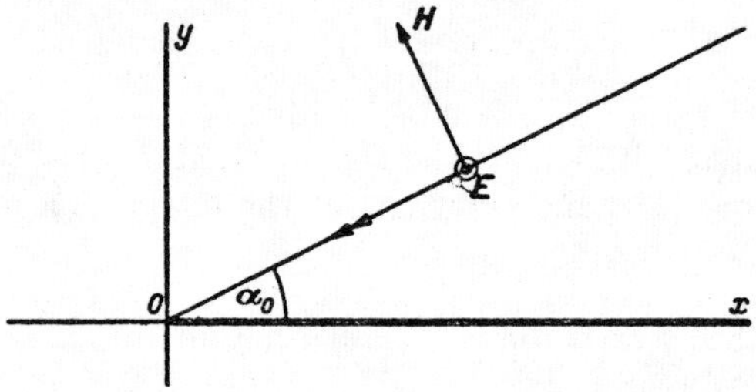

Fig. 11.6. The plane wave incident on the perfectly conducting half-plane.

making any additional contribution to the integral on the assumption that $P(\mu)/\sqrt{(1-\mu^2)} \to 0$ as $|\mu| \to \infty$ when $\pi \geqslant \arg \mu \geqslant 0$. Then if $U(\mu)$ is any function which is free of singularities in the half-plane above the path of integration and has appropriate behaviour as $|\mu| \to \infty$ therein, (2) is clearly satisfied by

$$\frac{P(\mu)}{\sqrt{(1-\mu^2)}} = -\frac{1}{2\pi i}\frac{U(\mu)}{U(-\mu_0)}\frac{1}{(\mu+\mu_0)} \tag{4}$$

if the path of integration be indented below the pole at $\mu = -\mu_0$, as shown diagrammatically in Fig. 11.7. For the only relevant singularity of the function on the right-

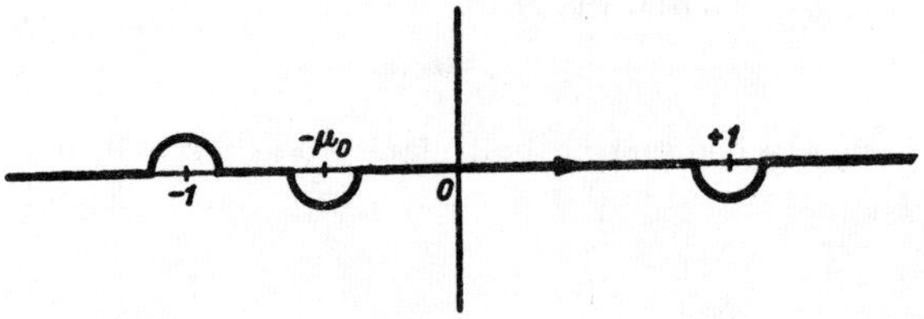

Fig. 11.7. The path of integration in the complex μ plane.

hand side of (4) is the pole at $\mu = -\mu_0$, with residue $-1/(2\pi i)$, and from CAUCHY's residue theorem this will contribute to the integral in (2) precisely the term $-\exp(-ikx\mu_0)$.

If we now rewrite (4) in the form

$$\frac{P(\mu)}{\sqrt{(1-\mu)}}(\mu+\mu_0) = -\frac{1}{2\pi i}\frac{U(\mu)}{U(-\mu_0)}\sqrt{(1+\mu)} \tag{5}$$

it can be argued that each side of (5) is a constant. For the left-hand side is free of singularities in the half-plane below the path of integration and of algebraic growth at infinity therein, whereas the right-hand side has the same characteristics in the half-plane above the path of integration. The function with which both sides are identical is thus free of singularities and of algebraic growth at infinity over the whole complex

μ plane: it must therefore be a polynomial, and since for some values of $\arg \mu$ $P(\mu) \to 0$ as $|\mu| \to \infty$, this polynomial can contain only a constant term.

The value of the constant is straightway found by putting $\mu = -\mu_0$ in the right-hand side of (5), whence

$$P(\mu) = \frac{i}{2\pi} \frac{\sqrt{(1-\mu_0)}\sqrt{(1-\mu)}}{\mu + \mu_0}, \tag{6}$$

or

$$P(\cos \alpha) = \frac{i}{\pi} \frac{\sin \tfrac{1}{2}\alpha_0 \sin \tfrac{1}{2}\alpha}{\cos \alpha + \cos \alpha_0}. \tag{7}$$

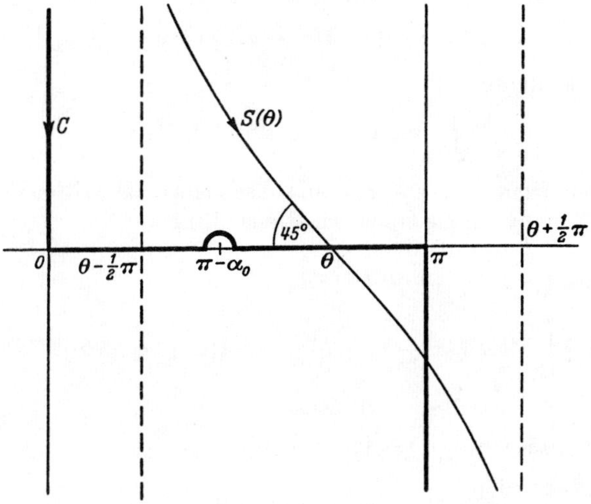

Fig. 11.8. The steepest descents path $S(\theta)$ in the complex α plane.

The significance of the symmetry of (7) in α and α_0 is mentioned at the end of § 11.7.1.

The components of the scattered field follow from § 11.4 (10), (11), (12) with the value (7) for $P(\cos \alpha)$, and hence the total field is given by

$$E_z = e^{-ikr\cos(\theta-\alpha_0)} - \frac{1}{i\pi} \int_C \frac{\sin \tfrac{1}{2}\alpha_0 \sin \tfrac{1}{2}\alpha}{\cos \alpha + \cos \alpha_0} e^{ikr\cos(\theta \mp \alpha)}\, d\alpha, \tag{8}$$

with the upper sign for $y > 0$, and the lower sign for $y < 0$. This completes the actual solution, and it only remains to cast it into a more useful form.

11.5.2 Expression of the solution in terms of Fresnel integrals

When kr is large, that is for distances greater than a wavelength or so from the origin, the evaluation of integrals of the general type

$$\int P(\cos \alpha) e^{ikr\cos(\theta-\alpha)}\, d\alpha \tag{9}$$

may be attempted by the method of steepest descent (see Appendix III). The preliminary step in this procedure is to distort the path of integration (making due allowance for the presence of any singularities in the integrand) into that of steepest descents, $S(\theta)$ say, through the saddle-point at $\alpha = \theta$. The path $S(\theta)$ is shown in Fig. 11.8; along it the new variable

$$\tau = \sqrt{2} e^{\frac{1}{4}i\pi} \sin \tfrac{1}{2}(\alpha - \theta) \tag{10}$$

goes through real values from $-\infty$ to ∞. The integral (9) then appears in the form

$$\sqrt{2}e^{-\frac{1}{4}i\pi} e^{ikr} \int_{-\infty}^{\infty} \frac{P(\cos\alpha)}{\sqrt{(1+\frac{1}{2}i\tau^2)}} e^{-kr\tau^2} d\tau, \tag{11}$$

from which asymptotic approximations for $kr \gg 1$ can be obtained.

The application of this procedure to the particular integral in (8) in fact leads, without approximation, to its expression in terms of FRESNEL integrals. This will now be shown.

Consider, first, the case $0 < \theta < \pi$. Since (7) can be put in the form

$$P(\cos\alpha) = \frac{1}{4\pi i}\{\sec\tfrac{1}{2}(\alpha-\alpha_0) - \sec\tfrac{1}{2}(\alpha+\alpha_0)\}, \tag{12}$$

it is sufficient to evaluate

$$\int_{S(\theta)} \sec\tfrac{1}{2}(\alpha-\alpha_0) e^{ikr\cos(\theta-\alpha)} d\alpha, \tag{13}$$

as the contribution from $\sec\frac{1}{2}(\alpha+\alpha_0)$ can subsequently be written down by changing the sign of α_0. Now by simple transformations (13) is

$$\begin{aligned} &\int_{S(0)} \sec\tfrac{1}{2}(\alpha-\alpha_0+\theta) e^{ikr\cos\alpha} d\alpha \\ &= \tfrac{1}{2}\int_{S(0)} \{\sec\tfrac{1}{2}(\alpha-\alpha_0+\theta) + \sec\tfrac{1}{2}(\alpha+\alpha_0-\theta)\} e^{ikr\cos\alpha} d\alpha \\ &= 2\int_{S(0)} \frac{\cos\frac{1}{2}(\alpha_0-\theta)\cos\frac{1}{2}\alpha}{\cos\alpha + \cos(\alpha_0-\theta)} e^{ikr\cos\alpha} d\alpha, \end{aligned} \tag{14}$$

and using the substitution

$$\tau = \sqrt{2}e^{\frac{1}{4}i\pi}\sin\tfrac{1}{2}\alpha$$

(14) becomes

$$-2e^{\frac{1}{4}i\pi} e^{ikr} \eta \int_{-\infty}^{\infty} \frac{e^{-kr\tau^2}}{\tau^2 - i\eta^2} d\tau, \tag{15}$$

where

$$\eta = \sqrt{2}\cos\tfrac{1}{2}(\theta-\alpha_0).$$

But

$$\int_{-\infty}^{\infty} e^{-\xi\tau^2} d\tau = \sqrt{\frac{\pi}{\xi}},$$

whence multiplication by $\exp(i\eta^2\xi)$ followed by integration over ξ from kr to infinity gives

$$e^{ikr\eta^2}\int_{-\infty}^{\infty} \frac{e^{-kr\tau^2}}{\tau^2 - i\eta^2} d\tau = \sqrt{\pi}\int_{kr}^{\infty} \frac{e^{i\eta^2\xi}}{\sqrt{\xi}} d\xi = \frac{2\sqrt{\pi}}{|\eta|}\int_{|\eta|\sqrt{(kr)}}^{\infty} e^{i\mu^2} d\mu. \tag{16}$$

Or, introducing the notation

$$F(a) = \int_a^{\infty} e^{i\mu^2} d\mu \tag{17}$$

for a form of the complex FRESNEL integral,*

$$\eta\int_{-\infty}^{\infty} \frac{e^{-kr\tau^2}}{\tau^2 - i\eta^2} d\tau = \pm 2\sqrt{\pi}e^{-ikr\eta^2} F\{\pm\eta\sqrt{(kr)}\}, \tag{18}$$

with the upper sign for $\eta > 0$, the lower sign for $\eta < 0$.

* This form of the FRESNEL integral is more convenient here than those defined in § 8.7 (12); the change in limits should be noted.

Combining these results we finally have, for $y \geqslant 0$,

$$-\frac{1}{i\pi}\int_{S(\theta)} \frac{\sin\frac{1}{2}\alpha_0 \sin\frac{1}{2}\alpha}{\cos\alpha + \cos\alpha_0} e^{ikr\cos(\theta-\alpha)}\, d\alpha$$

$$= -\frac{e^{-\frac{1}{4}i\pi}}{\sqrt{\pi}}\{e^{-ikr\cos(\theta-\alpha_0)}F[\sqrt{(2kr)}\cos\tfrac{1}{2}(\theta-\alpha_0)] \mp e^{-ikr\cos(\theta+\alpha_0)}F[\pm\sqrt{(2kr)}\cos\tfrac{1}{2}(\theta+\alpha_0)]\}, \qquad (19)$$

with the upper sign for $\theta + \alpha_0 < \pi$, the lower sign for $\theta + \alpha_0 > \pi$.

To get the complete field from (8) it only remains to take account of the simple pole at $\alpha = \pi - \alpha_0$. For $0 \leqslant \theta \leqslant \pi$, it is easily verified that in distorting the path C to the path $S(\theta)$ the sectors at infinity make no contribution, and it is clear from Fig. 11.8 that the pole is captured if and only if $\pi - \alpha_0 > \theta$. Its contribution, obtained from the residue theorem, is then

$$-e^{-ikr\cos(\theta+\alpha_0)}. \qquad (20)$$

In other words, it is the reflected wave of *geometrical optics*, the discontinuity in which across $\theta = \pi - \alpha_0$ exactly counterbalances that in the *diffraction* field (19). In fact, invoking the relation

$$F(a) + F(-a) = \sqrt{\pi}\, e^{\frac{1}{4}i\pi}, \qquad (21)$$

the complete field (8) can be written in the form

$$E_z = \frac{e^{-\frac{1}{4}i\pi}}{\sqrt{\pi}}\{e^{-ikr\cos(\theta-\alpha_0)}F[-\sqrt{(2kr)}\cos\tfrac{1}{2}(\theta-\alpha_0)] - e^{-ikr\cos(\theta+\alpha_0)}F[-\sqrt{(2kr)}\cos\tfrac{1}{2}(\theta+\alpha_0)]\}, \qquad (22)$$

which is SOMMERFELD's famous result.

When $y < 0$ the integral which has to be evaluated is

$$\int_C \frac{\sin\frac{1}{2}\alpha_0 \sin\frac{1}{2}\alpha}{\cos\alpha + \cos\alpha_0} e^{ikr\cos(\theta+\alpha)}\, d\alpha.$$

The appropriate steepest descents path is now $S(2\pi - \theta)$, and the capture of the pole at $\alpha = \pi - \alpha_0$, which occurs only for $\theta > \pi + \alpha_0$, yields minus the incident wave. The complete field is again given by (22).

To obtain the corresponding expressions for the components of $\boldsymbol{H}$ is merely a matter of differentiation, as shown in § 11.4.1. Both the Cartesian components H_x, H_y, and the polar components H_r, H_θ are of interest: in view of the fact that (22) is in terms of r, θ, it is convenient to derive the latter first, from the MAXWELL equations

$$H_r = \frac{1}{ikr}\frac{\partial E_z}{\partial\theta}, \qquad H_\theta = -\frac{1}{ik}\frac{\partial E_z}{\partial r}, \qquad (23)$$

and then to deduce the former from the relations

$$H_x = \cos\theta\, H_r - \sin\theta\, H_\theta, \qquad H_y = \sin\theta\, H_r + \cos\theta\, H_\theta.$$

The following notation is introduced in order to make the results more compact:

$$u = -\sqrt{(2kr)}\cos\tfrac{1}{2}(\theta-\alpha_0), \qquad v = -\sqrt{(2kr)}\cos\tfrac{1}{2}(\theta+\alpha_0), \qquad (24)$$

$$G(a) = e^{-ia^2}F(a). \qquad (25)$$

Note that

$$\frac{dG(a)}{da} = -1 - 2iaG(a).$$

The expression (22) then appears in the form

$$E_z = \frac{e^{-\frac{1}{4}i\pi}}{\sqrt{\pi}} e^{ikr} \{G(u) - G(v)\}, \tag{26}$$

and it follows that

$$\left.\begin{aligned} H_r &= \frac{e^{-\frac{1}{4}i\pi}}{\sqrt{\pi}} e^{ikr} \left\{\sin(\theta - \alpha_0)G(u) - \sin(\theta + \alpha_0)G(v) - i\sqrt{\left(\frac{2}{kr}\right)} \sin\tfrac{1}{2}\alpha_0 \cos\tfrac{1}{2}\theta\right\}, \\ H_\theta &= \frac{e^{-\frac{1}{4}i\pi}}{\sqrt{\pi}} e^{ikr} \left\{\cos(\theta - \alpha_0)G(u) - \cos(\theta + \alpha_0)G(v) + i\sqrt{\left(\frac{2}{kr}\right)} \sin\tfrac{1}{2}\alpha_0 \sin\tfrac{1}{2}\theta\right\}, \end{aligned}\right\} \tag{27}$$

$$\left.\begin{aligned} H_x &= -\frac{e^{-\frac{1}{4}i\pi}}{\sqrt{\pi}} e^{ikr} \left\{\sin\alpha_0[G(u) + G(v)] + i\sqrt{\left(\frac{2}{kr}\right)} \sin\tfrac{1}{2}\alpha_0 \cos\tfrac{1}{2}\theta\right\}, \\ H_y &= \frac{e^{-\frac{1}{4}i\pi}}{\sqrt{\pi}} e^{ikr} \left\{\cos\alpha_0[G(u) - G(v)] - i\sqrt{\left(\frac{2}{kr}\right)} \sin\tfrac{1}{2}\alpha_0 \sin\tfrac{1}{2}\theta\right\}. \end{aligned}\right\} \tag{28}$$

11.5.3 The nature of the solution

We now examine, in some detail, the nature of the results given in § 11.5.2. It is evident from their derivation, and can be verified directly, that $G(u) \exp(ikr)$ is itself a solution of the two-dimensional wave equation, for any value of α_0; the noteworthy point being that it has periodicity 4π, so that $G(u) - G(v)$ vanishes on $\theta = 0$, $\theta = 2\pi$, the two faces of the screen, but does not vanish on $\theta = \pi$; SOMMERFELD, indeed, arrived at his result (22) by seeking an appropriate solution of the wave equation of period 4π and combining it with its "image".* Incidentally, it follows from (28) that

$$\cos\tfrac{1}{2}\theta \frac{e^{ikr}}{\sqrt{(kr)}}, \qquad \sin\tfrac{1}{2}\theta \frac{e^{ikr}}{\sqrt{(kr)}}$$

are also solutions of the two-dimensional wave equation, a result which is well known.

The other aspect of (26) which should be examined is its behaviour as $r \to \infty$. This is a straightforward matter and is the main topic of the subsequent discussion in the present section.

A very attractive feature of the half-plane problem is that the field can be evaluated at any point from tables of the FRESNEL integrals.† Furthermore, in two cases of particular interest, namely $kr \gg 1$ and $kr \ll 1$, simple approximations (mentioned in § 8.7.2) to the FRESNEL integrals are available. The former condition is, of course, always satisfied in optical experiments, where the point of observation is likely to be millions of wavelengths from the diffracting edge; the latter condition arises in connection with the behaviour of the field in the vicinity of a sharp edge and can be studied at centimetre radio wavelengths (see § 11.5.6).

---

* A good account of this approach is given in B. B. BAKER and E. T. COPSON, *The Mathematical Theory of Huygens' Principle* (Oxford, Clarendon Press, 2nd ed., 1950), Chapter 4.

† The most convenient for this purpose seem to be those given by R. A. RANKIN, *Phil. Trans. Roy. Soc.* A, **241** (1949), 457.

$kr \gg 1$. In this case $|u|$ and $|v|$ are large compared to unity except for values of θ sufficiently close to $\pi + \alpha_0$ and $\pi - \alpha_0$, respectively. To be precise, we introduce five regions as shown in Fig. 11.9. The equations of the curves bounding regions II and IV are taken to be $u^2 = 1$ and $v^2 = 1$, respectively, so that the curves are parabolae with foci at the origin and axes $\theta = \pi + \alpha_0$, $\theta = \pi - \alpha_0$. Well within region

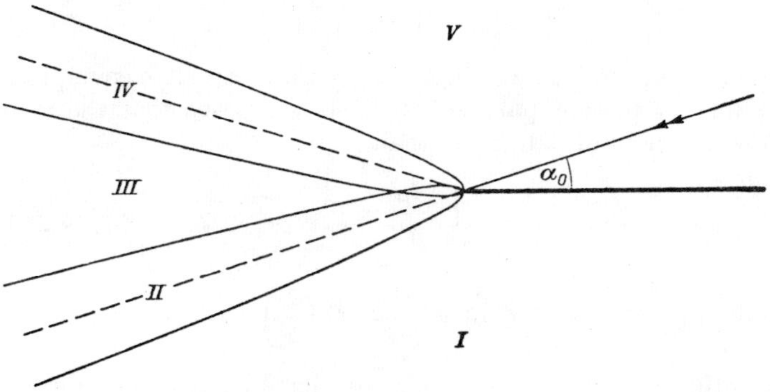

Fig. 11.9. Diffraction of a plane wave by a perfectly conducting half-plane. The five regions in terms of which the behaviour of the field may be described.

II (that is to say, inside a parabola $u^2 = \varepsilon$, where $\varepsilon \ll 1$) $|u| \ll 1$; well outside region II (that is to say, outside a parabola $u^2 = \gamma$, where $\gamma \gg 1$) $|u| \gg 1$. And similarly with $|v|$ and region IV. Furthermore, for $0 < \theta < \pi - \alpha_0$, u and v are both negative; for $\pi - \alpha_0 < \theta < \pi + \alpha_0$, u is negative but v is positive; and for $\pi + \alpha_0 < \theta < 2\pi$, u and v are both positive.

The regions I, III, and V are obviously closely connected with those which would arise from a geometrical optics treatment according to which the light travels in

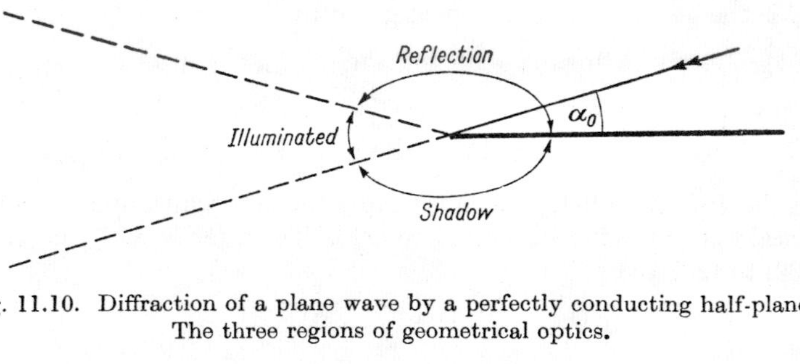

Fig. 11.10. Diffraction of a plane wave by a perfectly conducting half-plane. The three regions of geometrical optics.

straight lines; namely, as illustrated in Fig. 11.10, the *shadow* sector behind the screen where there is no field at all, the *illuminated* sector where there is the incident plane wave only, and the *reflection* sector where there is the incident plane wave together with the reflected plane wave appropriate to reflection at an *infinite* screen. In fact, broadly speaking, the regions II and IV are those in which the exact solution effects a smooth transition from the geometrical field in one sector to that in an adjacent sector. To see this in more detail we must pause to derive an asymptotic approximation to the Fresnel integral for large values of the argument.

If a is positive we write

$$G(a) = e^{-ia^2}\int_a^\infty \frac{d(e^{i\mu^2})}{2i\mu}$$

and integrate by parts twice to get

$$G(a) = \frac{i}{2a} + \frac{1}{4a^3} - \frac{3}{4}e^{-ia^2}\int_a^\infty \frac{e^{i\mu^2}}{\mu^4}\,d\mu. \tag{29}$$

A continuation of this process would, in fact, yield a complete asymptotic expansion of $G(a)$, but for our present purposes we merely observe that the modulus of the integral in the last term of (29) is less than

$$\int_a^\infty \frac{d\mu}{\mu^4} = \frac{1}{3a^3}, \tag{30}$$

whence

$$G(a) = \frac{i}{2a} + O\left(\frac{1}{a^3}\right). \tag{31}$$

It is worth noting that this result could also have been obtained from the general method outlined in § 11.5.2, which here consists in expanding the factor $(\tau^2 - i\eta^2)^{-1}$ in the integrand of (18) as a power series in τ, and then integrating term by term.

If a is negative the left-hand side of (30) diverges, but this case can easily be handled by using (21) in conjunction with the result for a positive argument. Thus

$$G(a) = \sqrt{\pi}\, e^{\frac{1}{4}i\pi}\, e^{-ia^2} + \frac{i}{2a} + O\left(\frac{1}{a^3}\right). \tag{32}$$

The fact that the asymptotic approximations (31) for a positive and (32) for a negative are different is a particular example of the STOKES phenomenon.*

Now let us write

$$E_z = E_z^{(g)} + E_z^{(d)},$$

where $E_z^{(g)}$ is the geometrical optics field given by

$$E_z^{(g)} = \begin{cases} e^{-ikr\cos(\theta-\alpha_0)} - e^{-ikr\cos(\theta+\alpha_0)} & \text{for } 0 \leqslant \theta < \pi - \alpha_0, \\ e^{-ikr\cos(\theta-\alpha_0)} & \text{for } \pi - \alpha_0 < \theta < \pi + \alpha_0, \\ 0 & \text{for } \pi + \alpha_0 < \theta \leqslant 2\pi, \end{cases} \tag{33}$$

and $E_z^{(d)}$ is the *diffraction* field, which is simply the field which must be added to that of geometrical optics to give the complete field. Then, for $kr \gg 1$, the application of (31) and (32) to (26) gives

$$E_z^{(d)} \sim \sqrt{\frac{2}{\pi}}\, e^{\frac{1}{4}i\pi} \frac{\sin\frac{1}{2}\alpha_0 \sin\frac{1}{2}\theta}{(\cos\theta + \cos\alpha_0)} \frac{e^{ikr}}{\sqrt{(kr)}} \tag{34}$$

at points not too close to regions II and IV, in the sense indicated above. It is readily seen, either from (23) or (27), that the components of $\boldsymbol{H}^{(d)}$ to the same order of approximation as (34) are $H_\theta^{(d)} = -E_z^{(d)}$ and $H_r^{(d)} = 0$. Evidently (34) implies that the diffraction field behaves as though it originates from a line-source situated along the diffracting edge whose "polar diagram" varies with angle as specified. This accords with the experimental fact that the diffracting edge, when viewed from the shadow sector for the sake of contrast, appears illuminated.

* G. G. STOKES, *Trans. Camb. Phil. Soc.*, **10** (1864), 105.

When $\cos\theta + \cos\alpha_0$ approaches zero, the approximation (34) breaks down and appeal must be made to the exact solution. Since

$$G(0) = \int_0^\infty e^{i\mu^2}\,d\mu = \tfrac{1}{2}\sqrt{\pi}e^{\frac{1}{4}i\pi}, \tag{35}$$

we see from (26) that, on $\theta = \pi + \alpha_0$,

$$E_z = \tfrac{1}{2}e^{ikr} + O\left\{\frac{1}{\sqrt{(kr)}}\right\}, \tag{36}$$

and on $\theta = \pi - \alpha_0$,

$$E_z = e^{ikr\cos(2\alpha_0)} - \tfrac{1}{2}e^{ikr} + O\left\{\frac{1}{\sqrt{(kr)}}\right\}. \tag{37}$$

Hence, near $\theta = \pi + \alpha_0$ and $\theta = \pi - \alpha_0$ the diffraction field is of the same order as the incident field. In particular, at infinity the transition between the geometrical optics fields in adjacent sectors is via their arithmetic mean.

Interference between the geometrical optics field and the diffraction field in regions where they are comparable gives rise to fringes. These are evident in Fig. 11.11, which is discussed in § 11.5.5.

$kr \ll 1$. In this case $|u|$ and $|v|$ are small compared to unity and series expansions of the FRESNEL integral are useful. Writing

$$F(a) = \int_0^\infty e^{i\mu^2}\,d\mu - \int_0^a e^{i\mu^2}\,d\mu,$$

and expanding the exponential in the integrand of the second integral, we have

$$F(a) = \tfrac{1}{2}\sqrt{\pi}\,e^{\frac{1}{4}i\pi} - a + O(a^3). \tag{38}$$

Hence, from (26) and (28), neglecting powers of kr greater than a half,

$$\left.\begin{aligned}
E_z &= 2\sqrt{\frac{2}{\pi}}\,e^{-\frac{1}{4}i\pi}\sqrt{(kr)}\sin\tfrac{1}{2}\alpha_0\sin\tfrac{1}{2}\theta,\\
H_x &= -\sin\alpha_0 - \sqrt{\frac{2}{\pi}}\,e^{-\frac{1}{4}i\pi}\sin\tfrac{1}{2}\alpha_0\cos\tfrac{1}{2}\theta\left\{\frac{i}{\sqrt{(kr)}} + (1+2\cos\alpha_0)\sqrt{(kr)}\right\},\\
H_y &= \sqrt{\frac{2}{\pi}}\,e^{-\frac{1}{4}i\pi}\sin\tfrac{1}{2}\alpha_0\sin\tfrac{1}{2}\theta\left\{-\frac{i}{\sqrt{(kr)}} + (1+2\cos\alpha_0)\sqrt{(kr)}\right\}.
\end{aligned}\right\} \tag{39}$$

It should be noticed that E_z is finite and continuous at $r = 0$, but that H_x and H_y diverge like $r^{-1/2}$, except on $\theta = \pi$ when $H_x = -\sin\alpha_0\exp(ikr\cos\alpha_0)$, and $\theta = 0$, 2π when $H_y = 0$. Such behaviour, peculiar in a physical problem, arises of course, from the idealized concept of an infinitely sharp edge. The existence of singularities in the field components in this case must be taken into account in formulating any theorem about the uniqueness of the solution (see § 11.9).

We conclude this investigation into the nature of the solution by examining the current density induced in the diffracting screen. This is $-c/4\pi$ times the difference in H_x at $\theta = 0$ and $\theta = 2\pi$; that is, from (28),

$$\frac{2\pi}{c}J_z = \sin\alpha_0\,e^{-ikx\cos\alpha_0} - \frac{e^{-\frac{1}{4}i\pi}}{\sqrt{\pi}}\,e^{ikx}\left\{2\sin\alpha_0\,G[\sqrt{(2kx)}\cos\tfrac{1}{2}\alpha_0] - i\sqrt{\left(\frac{2}{kx}\right)}\sin\tfrac{1}{2}\alpha_0\right\}. \tag{40}$$

For $\sqrt{(2kx)} \cos \frac{1}{2}\alpha_0 \gg 1$, (40) reads

$$J_z = \frac{c}{2\pi} \sin \alpha_0 \, e^{-ikx \cos \alpha_0} + O\{(kx)^{-3/2}\}. \tag{41}$$

This result is of interest in indicating the extent to which it is permissible to assume that the current density is that given by geometrical optics, a standard procedure in problems which cannot be solved exactly. Clearly the assumption is only reasonable for values of α_0 not near π, and benefits from the fact that the "correction" term in (41) tends to zero as $x \to \infty$ like $x^{-3/2}$ rather than $x^{-1/2}$.

On the other hand, for $\sqrt{(2kx)} \cos \frac{1}{2}\alpha_0 \ll 1$,

$$J_z = \frac{c}{\pi\sqrt{(2\pi)}} e^{-\frac{1}{4}i\pi} \sin \tfrac{1}{2}\alpha_0 \left\{\frac{i}{\sqrt{(kx)}} + 4 \cos^2 \tfrac{1}{2}\alpha_0 \sqrt{(kx)}\right\} e^{ikx}, \tag{42}$$

which diverges at the diffracting edge.

11.5.4 The solution for *H*-polarization

The case of H-polarization, namely when the incident field is specified by

$$H_z^{(i)} = e^{-ikr \cos (\theta - \alpha_0)}, \tag{43}$$

can be treated in the same way as that of E-polarization, the analysis being, in fact, practically identical. Alternatively, the former can be deduced from the latter by invoking the exact electromagnetic form of BABINET's principle, given in § 11.3, because the screen complementary to a half-plane is itself a half-plane. It turns out that the complete field is given by

$$H_z = \frac{e^{-\frac{1}{4}i\pi}}{\sqrt{\pi}} \{e^{-ikr \cos (\theta - \alpha_0)} F[-\sqrt{(2kr)} \cos \tfrac{1}{2}(\theta - \alpha_0)] + e^{-ikr \cos (\theta + \alpha_0)} F[-\sqrt{(2kr)} \cos \tfrac{1}{2}(\theta + \alpha_0)]\}. \tag{44}$$

This differs from the corresponding expression for E_z for an E-polarized field, (22), only in the sign of the second term.

Using the notation of § 11.5.2 the non-zero field components appear in the form

$$\left.\begin{aligned}
H_z &= \frac{e^{-\frac{1}{4}i\pi}}{\sqrt{\pi}} e^{ikr} \{G(u) + G(v)\}, \\
E_x &= \frac{e^{-\frac{1}{4}i\pi}}{\sqrt{\pi}} e^{ikr} \left\{\sin \alpha_0[G(u) - G(v)] - i\sqrt{\left(\frac{2}{kr}\right)} \cos \tfrac{1}{2}\alpha_0 \sin \tfrac{1}{2}\theta\right\}, \\
E_y &= -\frac{e^{-\frac{1}{4}i\pi}}{\sqrt{\pi}} e^{ikr} \left\{\cos \alpha_0[G(u) + G(v)] + i\sqrt{\left(\frac{2}{kr}\right)} \cos \tfrac{1}{2}\alpha_0 \cos \tfrac{1}{2}\theta\right\}.
\end{aligned}\right\} \tag{45}$$

Clearly E_x vanishes on $\theta = 0$ and $\theta = 2\pi$, and the behaviour of the field for $kr \gg 1$ can again be interpreted in terms of a diffraction field which appears to originate in a line-source along the diffracting edge for points sufficiently far from $\theta = \pi - \alpha_0$ and $\theta = \pi + \alpha_0$. As $r \to 0$, H_z remains finite and continuous, whereas E_x and E_y diverge like $r^{-1/2}$, except in so far as $E_x = 0$ for $\theta = 0$, 2π, and $E_y = -\cos \alpha_0 \exp(ikr \cos \alpha_0)$ for $\theta = \pi$.

The current density is given by

$$\frac{2\pi}{c} J_x = e^{-ikx \cos \alpha_0} - \frac{2e^{-\frac{1}{4}i\pi}}{\sqrt{\pi}} e^{ikx} G\{\sqrt{(2kx)} \cos \tfrac{1}{2}\alpha_0\}. \tag{46}$$

For $\sqrt{(2kx)} \cos \frac{1}{2}\alpha_0 \gg 1$,

$$\frac{2\pi}{c} J_x = e^{-ikx \cos \alpha_0} - \frac{e^{\frac{1}{4}i\pi}}{\sqrt{(2\pi)}} \sec \tfrac{1}{2}\alpha_0 \frac{e^{ikx}}{\sqrt{(kx)}}, \tag{47}$$

which approximates less rapidly to the current density of geometrical optics than in the case of E-polarization. For $\sqrt{(2kx)} \cos \frac{1}{2}\alpha_0 \ll 1$,

$$J_x = \frac{c}{\pi} \sqrt{\frac{2}{\pi}}\, e^{-\frac{1}{4}i\pi} \cos \tfrac{1}{2}\alpha_0 \sqrt{(kx)} e^{ikx}. \tag{48}$$

This vanishes at $x = 0$, so that, as might be expected, at the edge itself there is no current normal to the edge.

11.5.5 Some numerical calculations

A typical theoretical curve, obtained from (26), is shown in Fig. 11.11. It is for a normally incident E-polarized plane wave of amplitude unity, and is a plot of the amplitude of E_z against x at a distance of three wavelengths behind the screen ($ky = -6\pi$). It puts in evidence the diffraction fringes in the illuminated region, and the monotonic decay with deeper penetration into the shadow region.

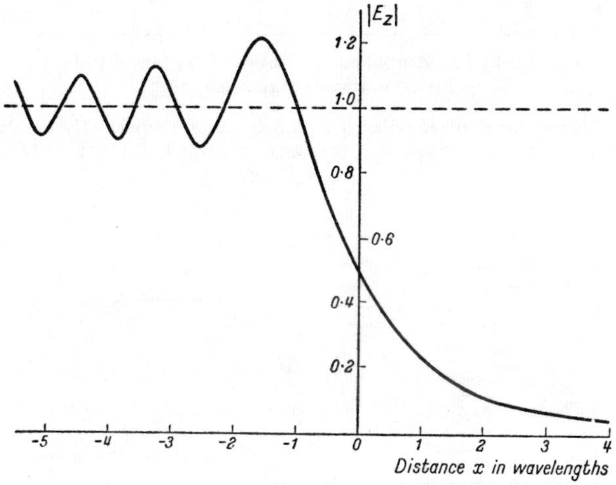

Fig. 11.11. Diffraction of a normally incident E-polarized plane wave of amplitude unity by a perfectly conducting half-plane. The variation of $|E_z|$ with x at a distance of three wavelengths behind the screen.

Some interesting calculations have been made by Braunbek and Laukien.* For a normally incident H-polarized plane wave of amplitude unity they give contours of equal amplitude (Fig. 11.12) and equal phase (Fig. 11.13) of H_z for the region within about a wavelength of the diffracting edge. They also give the lines of average energy flow (Fig. 11.14) which are orthogonal to the phase contours. That this is the case for *any* two-dimensional H-polarized field is easily proved: write $H_z = he^{i\phi}$, where h and ϕ are real; then using the relations

$$E_x = -\frac{1}{ik}\frac{\partial H_z}{\partial y}, \qquad E_y = \frac{1}{ik}\frac{\partial H_z}{\partial x},$$

* W. Braunbek and G. Laukien, *Optik*, **9** (1952), 174.

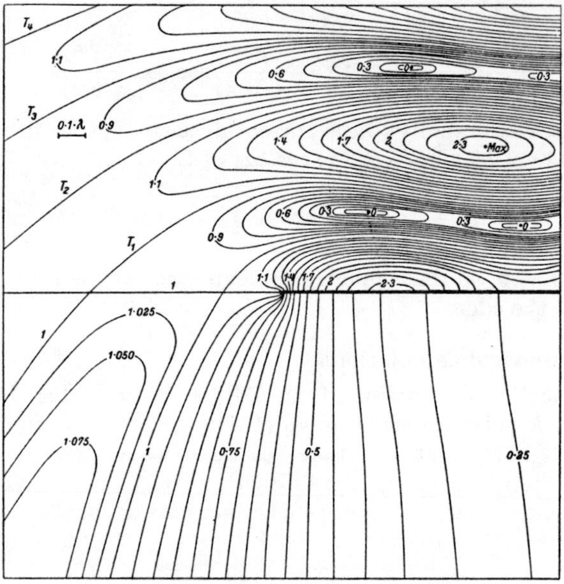

Fig. 11.12. Amplitude contours of H_z (amplitude of incident wave is taken as unity).

Diffraction of a normally incident H-polarized plane wave by a perfectly conducting half-plane [After W. Braunbek and G. Laukien, *Optik*, **9** (1952), 174.]

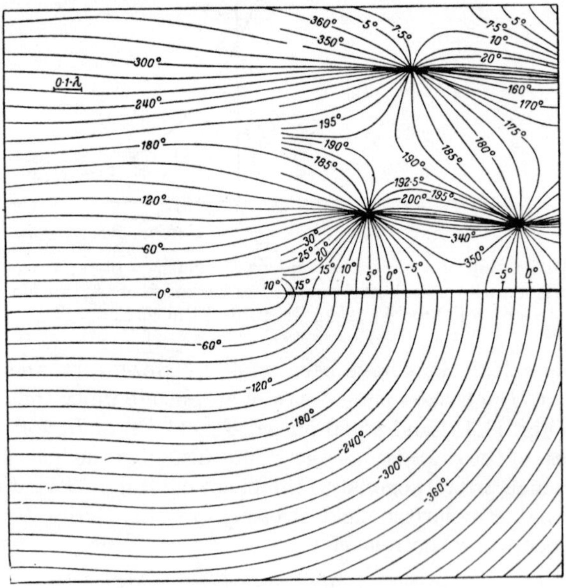

Fig. 11.13. Phase contours of H_z.

Diffraction of a normally incident H-polarized plane wave by a perfectly conducting half-plane [After W. Braunbek and G. Laukien, *Optik*, **9** (1952), 174.]

the averaging POYNTING vector [§ 1.4 (56)]

$$\frac{c}{8\pi}\mathscr{R}(\mathbf{E}\wedge\mathbf{H}^\star)=\frac{c}{8\pi}\mathscr{R}(E_yH_z^\star,\ -E_xH_z^\star,\ 0)$$

is seen to be

$$\frac{c}{8\pi}\frac{h^2}{k}\left(\frac{\partial\phi}{\partial x},\frac{\partial\phi}{\partial y},0\right), \tag{49}$$

which is orthogonal to the surfaces $\phi=$ constant. A corresponding result holds for any E-polarized field.

Fig. 11.14. Lines of average energy flow.

Diffraction of a normally incident H-polarized plane wave by a perfectly conducting half-plane. [After W. BRAUNBEK and G. LAUKIEN, *Optik*, **9** (1952), 174.]

11.5.6 Comparison with approximate theory and with experimental results

For points at a great distance from the diffracting edge in the illuminated part of region II (Fig. 11.9), where the fringes appear, the second term in each of the solutions (22) and (44) can be neglected. The intensity for both E- and H-polarization, and hence also for unpolarized light, is therefore

$$\tfrac{1}{2}\left\{\tfrac{1}{2}+\mathscr{C}\left[2\sqrt{\left(\frac{2r}{\lambda}\right)}\cos\tfrac{1}{2}(\theta-\alpha_0)\right]\right\}^2+\tfrac{1}{2}\left\{\tfrac{1}{2}+\mathscr{S}\left[2\sqrt{\left(\frac{2r}{\lambda}\right)}\cos\tfrac{1}{2}(\theta-\alpha_0)\right]\right\}^2, \tag{50}$$

where λ is the wavelength and $\mathscr{C}$, $\mathscr{S}$ are the FRESNEL "cosine" and "sine" integrals defined by § 8.7 (12). This should be compared with the analogous result § 8.7 (28) for a black half-plane on the FRESNEL–KIRCHHOFF theory. It has indeed been suggested* that the first term of the exact solution for the perfectly conducting half-plane could be regarded as giving the solution for a black half-plane.

---

* See the account in B. B. BAKER and E. T. COPSON, *The Mathematical Theory of Huygens' Principle* (Oxford, Clarendon Press, 2nd ed., 1950), p. 149 *et seq.*

Well into region I, the shadow region, the E-polarization field is given by (34), namely

$$E_z = \sqrt{\frac{2}{\pi}}\, e^{\frac{1}{4}i\pi} \frac{\sin \frac{1}{2}\alpha_0 \sin \frac{1}{2}\theta}{(\cos \alpha_0 + \cos \theta)} \frac{e^{ikr}}{\sqrt{(kr)}}. \tag{51}$$

It can likewise be shown that the H-polarization field there is

$$H_z = -\sqrt{\frac{2}{\pi}}\, e^{\frac{1}{4}i\pi} \frac{\cos \frac{1}{2}\alpha_0 \cos \frac{1}{2}\theta}{(\cos \alpha_0 + \cos \theta)} \frac{e^{ikr}}{\sqrt{(kr)}}. \tag{52}$$

The corresponding ratio of the field strengths is therefore

$$\frac{E\text{-polarization}}{H\text{-polarization}} = -\tan \tfrac{1}{2}\alpha_0 \tan \tfrac{1}{2}\theta, \tag{53}$$

and unpolarized incident light will accordingly become partly polarized on diffraction. These results are in broad agreement with optical experiments.*

Developments in microwave radio techniques provide excellent opportunities for the experimental study of the diffraction of electromagnetic waves. In particular, a diffracting screen can be used which is much nearer to the idealization of a perfectly conducting half-plane than can be realized in optical measurements, and the field in the neighbourhood of the diffracting edge can be examined. A number of measurements have been made, mainly on a wavelength of about 3 cm, which show good agreement between theory and experiment.†

11.6 THREE-DIMENSIONAL DIFFRACTION OF A PLANE WAVE BY A HALF-PLANE

In § 11.5 we solved, in effect, the problem of diffraction by a half-plane of a plane wave which was arbitrary except for the restriction that its direction of propagation was normal to the diffracting edge. It will now be shown that, by a simple device, the previous results can be extended to yield the solution for a completely arbitrary incident plane wave.

Let the incident plane wave be characterized by the phase factor

$$e^{-ikS} = e^{-ik(x \cos \alpha \cos \beta + y \sin \alpha \cos \beta + z \sin \beta)}, \tag{1}$$

where, as before, the perfectly conducting screen occupies $y = 0$, $x > 0$. The angles α and β, which specify the direction of propagation, are shown in Fig. 11.15.

Now we note that (1) is obtained from the two-dimensional form corresponding to $\beta = 0$ on first replacing k by $k \cos \beta$ and then multiplying by $\exp(-ikz \sin \beta)$. In fact, this procedure applied to any two-dimensional solution of the wave equation

$$\frac{\partial^2 V}{\partial x^2} + \frac{\partial^2 V}{\partial y^2} + k^2 V = 0 \tag{2}$$

clearly yields a solution of the three-dimensional wave equation

$$\frac{\partial^2 V}{\partial x^2} + \frac{\partial^2 V}{\partial y^2} + \frac{\partial^2 V}{\partial z^2} + k^2 V = 0 \tag{3}$$

* See WOLFSOHN's article in *Handbuch der Physik*, Vol. 20 (Berlin, Springer, 1928), 275, and J. SAVORNIN, *Ann. de Physique*, **11** (1939), 129.

† C. W. HORTON and R. B. WATSON, *J. Appl. Phys.*, **21** (1950), 16; B. N. HARDEN, *Proc. Inst. Elec. Engrs.*, **99**, Pt. III (1952), 229; R. D. KODIS, *J. Appl. Phys.*, **23** (1952), 249; R. V. ROW, *J. Appl. Phys.*, **24** (1953), 1448.

in which z enters only via the factor $\exp(-ikz \sin \beta)$. Moreover, if U, say, is such a solution of (3), then it is easily verified that two electromagnetic fields satisfying MAXWELL's equations are

$$\left.\begin{aligned} \boldsymbol{E} &= \left(-\frac{i \sin \beta}{k}\frac{\partial U}{\partial x}, -\frac{i \sin \beta}{k}\frac{\partial U}{\partial y}, \cos^2 \beta U\right), \\ \boldsymbol{H} &= \left(-\frac{i}{k}\frac{\partial U}{\partial y}, \frac{i}{k}\frac{\partial U}{\partial x}, 0\right), \end{aligned}\right\} \tag{4}$$

and

$$\left.\begin{aligned} \boldsymbol{E} &= \left(\frac{i}{k}\frac{\partial U}{\partial y}, -\frac{i}{k}\frac{\partial U}{\partial x}, 0\right), \\ \boldsymbol{H} &= \left(-\frac{i \sin \beta}{k}\frac{\partial U}{\partial x}, -\frac{i \sin \beta}{k}\frac{\partial U}{\partial y}, \cos^2 \beta U\right). \end{aligned}\right\} \tag{5}$$

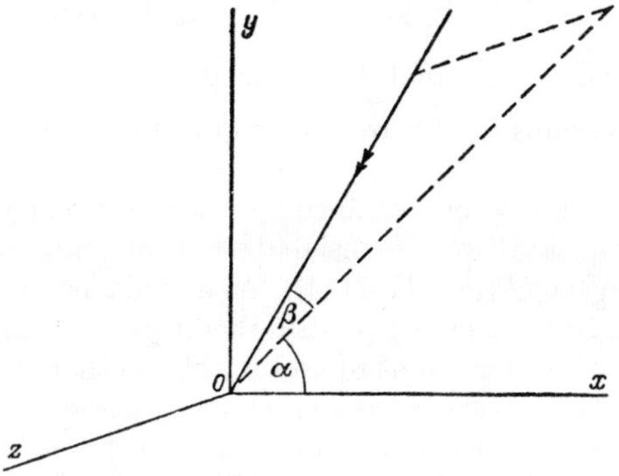

Fig. 11.15. The direction of propagation of the incident plane wave.

When $\beta = 0$, (4) gives a two-dimensional field which is E-polarized, (5) one which is H-polarized.

If we take the expression (1) for U, (4) and (5) yield, respectively, the two plane waves

$$\left.\begin{aligned} \boldsymbol{E} &= (-\cos \alpha \sin \beta, -\sin \alpha \sin \beta, \cos \beta)e^{-ikS}, \\ \boldsymbol{H} &= (-\sin \alpha, \cos \alpha, 0)e^{-ikS}, \end{aligned}\right\} \tag{6}$$

and

$$\left.\begin{aligned} \boldsymbol{E} &= (\sin \alpha, -\cos \alpha, 0)e^{-ikS}, \\ \boldsymbol{H} &= (-\cos \alpha \sin \beta, -\sin \alpha \sin \beta, \cos \beta)e^{-ikS}, \end{aligned}\right\} \tag{7}$$

where a factor $\cos \beta$ has been removed throughout. Now *any* plane wave with space variation (1) is specified by two components of $\boldsymbol{E}$ (or $\boldsymbol{H}$), because the third component would follow from $\operatorname{div} \boldsymbol{E} = 0$ (or $\operatorname{div} \boldsymbol{H} = 0$). Hence *any* plane wave can be formed by suitable superposition of (6) and (7); with the consequence that, in the diffraction problem, attention can be confined, without loss of generality, to the two cases in which they are the respective incident fields.

It should now be clear that the solution to the diffraction problem with (6) as the incident wave is given by (4) with U got from the known expression for $E_z \sec \beta$ in the two-dimensional case on first replacing k by $k \cos \beta$ and secondly multiplying by $\exp(-ikz \sin \beta)$: for the fact that $U = 0$ on $y = 0$, $x > 0$ implies also that $\partial U/\partial x = 0$

there, whence from (4) $E_x = E_z = 0$ on the screen, as required. Explicitly, we have, from § 11.5 (24) and § 11.5 (26)

$$U = \frac{e^{-\frac{1}{4}i\pi}}{\sqrt{\pi}} \sec \beta \, e^{ik(r \cos \beta - z \sin \beta)} \{G(p) - G(q)\}, \tag{8}$$

where

$$p = -\sqrt{(2kr \cos \beta)} \cos \tfrac{1}{2}(\theta - \alpha), \quad q = -\sqrt{(2kr \cos \beta)} \cos \tfrac{1}{2}(\theta + \alpha). \tag{9}$$

Thus, from (4),

$$\left.\begin{aligned} E_z &= \frac{e^{-\frac{1}{4}i\pi}}{\sqrt{\pi}} \cos \beta e^{ik(r \cos \beta - z \sin \beta)} \{G(p) - G(q)\}, \\ H_x &= -\frac{e^{-\frac{1}{4}i\pi}}{\sqrt{\pi}} e^{ik(r \cos \beta - z \sin \beta)} \left\{ \sin \alpha [G(p) + G(q)] + i\sqrt{\left(\frac{2}{kr \cos \beta}\right)} \sin \tfrac{1}{2}\alpha \cos \tfrac{1}{2}\theta \right\}, \\ H_y &= \frac{e^{-\frac{1}{4}i\pi}}{\sqrt{\pi}} e^{ik(r \cos \beta - z \sin \beta)} \left\{ \cos \alpha [G(p) - G(q)] - i\sqrt{\left(\frac{2}{kr \cos \beta}\right)} \sin \tfrac{1}{2}\alpha \sin \tfrac{1}{2}\theta \right\}, \\ E_x &= -H_y \sin \beta, \qquad E_y = H_x \sin \beta, \qquad H_z = 0. \end{aligned}\right\} \tag{10}$$

When $\beta = 0$ the expressions in (10) reduce at once to the corresponding expressions in § 11.5 (26) and (28).

Similar results can likewise be obtained for the case when the incident wave is given by (7), the appropriate two-dimensional solution being that for H-polarization, namely, the expression for H_z in § 11.5 (44). As already pointed out, the solution for a quite arbitrary incident plane wave can therefore be derived. Furthermore, a simple generalization of the argument of § 11.5.2 shows that the field radiated by *any* source distribution can be represented as a spectrum of plane waves. Thus, in principle, the solution of the diffraction problem for any source distribution can be built up from the solutions for the individual plane waves. Two cases of interest are a line-source parallel to the diffracting edge (a two-dimensional problem) and a point-source: these are treated in the next section.

11.7 DIFFRACTION OF A LOCALIZED SOURCE BY A HALF-PLANE

11.7.1 A line-current parallel to the diffracting edge

We shall consider a line-source situated at T, the point (r_0, θ_0) where $0 \leqslant \theta_0 \leqslant \pi$, which would, in free-space, radiate the E-polarized cylindrical wave

$$E_z^{(i)} = \sqrt{\frac{\pi}{2}} \, e^{\frac{1}{4}i\pi} H_0^{(1)}(kR) \sim \frac{e^{ikR}}{\sqrt{(kR)}}, \tag{1}$$

where $H_0^{(1)}$ is the HANKEL function of the first kind and zero order and R is the distance measured from T (see Fig. 11.16).

The source is, in fact, an electric current through T flowing parallel to the z axis and oscillating everywhere with the same phase. It is well known that (1) is the fundamental solution of the two-dimensional wave equation representing a diverging wave which depends only on radial distance.

In order to write (1) as an angular spectrum of plane waves we adopt, in essence, SOMMERFELD's integral representation of the HANKEL function.* Since we require

* See, for example, J. A. STRATTON, *Electromagnetic Theory* (New York, McGraw-Hill, 1941), p. 367.

the individual plane waves of the spectrum to be incident on the screen, we consider the representation in the half-space $y < r_0 \sin \theta_0$, namely

$$H_0^{(1)}(kR) = \frac{1}{\pi} \int_{S(\frac{1}{2}\pi)} e^{ikr_0 \cos(\theta_0 - \alpha)} e^{-ikr \cos(\theta - \alpha)} \, d\alpha. \tag{2}$$

The particular path $S(\frac{1}{2}\pi)$ is chosen here because (apart from the fact that it is valid for all points in $y < r_0 \sin \theta_0$) it is parallel to the path of steepest descents, which proves a convenience in the subsequent analysis. The phase factor $\exp \{ikr_0 \cos (\theta_0 - \alpha)\}$

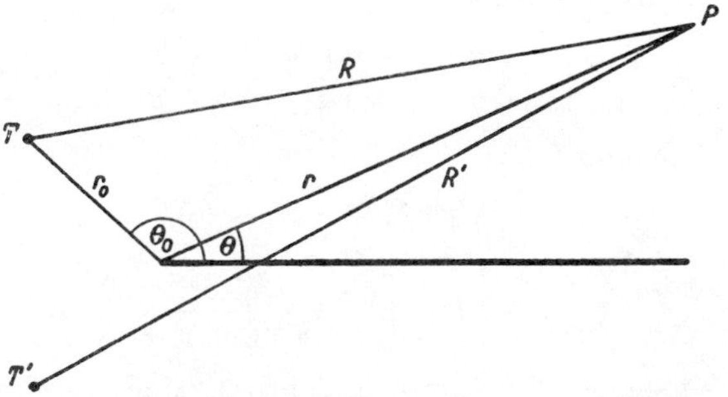

Fig. 11.16. The configuration for a line-source T in the presence of a diffracting half-plane.

in the integrand of (2) appears because the plane waves of the spectrum must all have zero phase at T.

It is apparent, therefore, that the solution of the diffraction problem for the incident field (1) is obtained from that for an incident field $\exp \{- ikr \cos (\theta - \alpha)\}$ on multiplying by the factor

$$\frac{e^{\frac{1}{4}i\pi}}{\sqrt{(2\pi)}} e^{ikr_0 \cos(\theta_0 - \alpha)} \tag{3}$$

and integrating with respect to α over the path $S(\frac{1}{2}\pi)$.

Now the solution for the incident field $\exp \{- ikr \cos (\theta - \alpha)\}$, as was shown in § 11.5, could be written

$$E_z^{(p)} = E_z^{(pg)} + E_z^{(pd)} \tag{4}$$

(the affix p being introduced here to indicate that the incident wave is plane), where $E_z^{(pg)}$ is given by § 11.5 (33) and (see § 11.5 (19))

$$E_z^{(pd)} = -\frac{i}{\pi} \int_{S(0)} \frac{\sin \frac{1}{2}\alpha \sin \frac{1}{2}(\theta + \beta)}{\cos \alpha + \cos (\theta + \beta)} e^{ikr \cos \beta} \, d\beta. \tag{5}$$

The use of this form for $E_z^{(pd)}$ (rather than that in terms of FRESNEL integrals) is adopted because, as will shortly be seen, it brings out the symmetry between α and β. The solution for the incident field (1) is thus

$$E_z = \frac{e^{\frac{1}{4}i\pi}}{\sqrt{(2\pi)}} \int_{S(\frac{1}{2}\pi)} E_z^{(pg)} e^{ikr_0 \cos(\theta_0 - \alpha)} \, d\alpha + \frac{e^{\frac{1}{4}i\pi}}{\sqrt{(2\pi)}} \int_{S(\frac{1}{2}\pi)} E_z^{(pd)} e^{ikr_0 \cos(\theta_0 - \alpha)} \, d\alpha. \tag{6}$$

In order to separate (6) into geometrical optics and diffraction terms, the path of integration for α in the second integral must be displaced to $S(\theta_0)$. In this transposition of the path the poles of $E_z^{(pd)}$, regarded as a function of α, must be taken into account: from (5) they occur where $\cos \alpha = -\cos(\theta + \beta)$ for all β on $S(0)$, and it is easy to verify that the contribution of their residues combines with the first expression in (6) to yield the geometrical optics terms

$$E_z^{(g)} = \begin{cases} \sqrt{(\tfrac{1}{2}\pi)}\, e^{\frac{1}{4}i\pi}\{H_0^{(1)}(kR) - H_0^{(1)}(kR')\} & \text{for } 0 \leqslant \theta \leqslant \pi - \theta_0, \\ \sqrt{(\tfrac{1}{2}\pi)}\, e^{\frac{1}{4}i\pi} H_0^{(1)}(kR) & \text{for } \pi - \theta_0 < \theta < \pi + \theta_0, \\ 0 & \text{for } \pi + \theta_0 < \theta \leqslant 2\pi, \end{cases} \qquad (7)$$

where R' is distance measured from T', the image of T in the plane $y = 0$ (Fig. 11.16). The diffraction term may be written

$$E_z^{(d)} = \frac{e^{-\frac{1}{4}i\pi}}{\pi\sqrt{(2\pi)}} \int_{S(0)} \int_{S(0)} \frac{\sin \frac{1}{2}(\alpha + \theta_0) \sin \frac{1}{2}(\beta + \theta)}{\cos(\alpha + \theta_0) + \cos(\beta + \theta)} e^{ik(r_0 \cos \alpha + r \cos \beta)}\, d\alpha d\beta. \qquad (8)$$

It is convenient to express (8) as

$$E_z^{(d)} = \mathscr{I}(\theta_0) - \mathscr{I}(-\theta_0) \qquad (9)$$

where

$$\mathscr{I}(\theta_0) = -\frac{e^{-\frac{1}{4}i\pi}}{4\pi\sqrt{(2\pi)}} \int_{S(0)} \int_{S(0)} \frac{e^{ik(r_0 \cos \alpha + r \cos \beta)}}{\cos \frac{1}{2}(\alpha + \beta + \theta_0 + \theta)}\, d\alpha d\beta. \qquad (10)$$

The final step is to reduce $\mathscr{I}(\theta_0)$ to a single integral. Multiplying top and bottom of the integrand in (10) by $4 \cos \frac{1}{2}(\alpha - \beta + \theta_0 + \theta)$ and discarding that part which is an odd function of β, we have

$$\mathscr{I}(\theta_0) = -\frac{e^{-\frac{1}{4}i\pi}}{8\pi\sqrt{(2\pi)}} \int_{S(0)} \int_{S(0)} \left\{ \frac{1}{\cos \frac{1}{2}(\alpha + \theta_0 + \theta) - \sin \frac{1}{2}\beta} + \frac{1}{\cos \frac{1}{2}(\alpha + \theta_0 + \theta) + \sin \frac{1}{2}\beta} \right\} \cos \tfrac{1}{2}\beta\, e^{ik(r_0 \cos \alpha + r \cos \beta)}\, d\alpha d\beta. \qquad (11)$$

In the second term of the integrand in (11) change α to $-\alpha$ and then recombine the two terms. This gives

$$\mathscr{I}(\theta_0) = -\frac{e^{-\frac{1}{4}i\pi}}{2\pi\sqrt{(2\pi)}} \times \int_{S(0)} \int_{S(0)} \frac{1}{N} \{\cos \tfrac{1}{2}(\theta_0 + \theta) \cos \tfrac{1}{2}\alpha \cos \tfrac{1}{2}\beta\, e^{ik(r_0 \cos \alpha + r \cos \beta)}\}\, d\alpha d\beta, \qquad (12)$$

where

$$N = (\cos \alpha - 1) + (\cos \beta - 1) - 4 \sin \tfrac{1}{2}(\theta_0 + \theta) \sin \tfrac{1}{2}\alpha \sin \tfrac{1}{2}\beta + 2 \cos^2 \tfrac{1}{2}(\theta_0 + \theta).$$

In (12) make the "steepest descent" substitutions

$$\xi = \sqrt{2}\, e^{\frac{1}{4}i\pi} \sin \tfrac{1}{2}\alpha, \qquad \eta = \sqrt{2}\, e^{\frac{1}{4}i\pi} \sin \tfrac{1}{2}\beta,$$

and write $R_1 = r_0 + r$, to get

$$\mathscr{I}(\theta_0) = \frac{e^{-\frac{1}{4}i\pi}}{\pi\sqrt{(2\pi)}} e^{ikR_1} \cos \tfrac{1}{2}(\theta_0 + \theta) \times \int_{-\infty}^{\infty} \int_{-\infty}^{\infty} \frac{e^{-k(r_0\xi^2 + r\eta^2)}}{\xi^2 + \eta^2 + 2 \sin \frac{1}{2}(\theta_0 + \theta)\xi\eta - 2i \cos^2 \frac{1}{2}(\theta_0 + \theta)}\, d\xi d\eta. \qquad (13)$$

Next, make the polar substitutions

$$\xi = \sqrt{(R_1/r_0)}\rho \cos \phi, \qquad \eta = \sqrt{(R_1/r)}\rho \sin \phi.$$

Then

$$\mathscr{I}(\theta_0) = \frac{e^{-\frac{1}{4}i\pi}}{\pi\sqrt{(2\pi)}} e^{ikR_1} \cos \tfrac{1}{2}(\theta_0 + \theta) \int_0^\infty \rho K(\rho)\, e^{-kR_1\rho^2}\, d\rho, \tag{14}$$

where

$$K(\rho) = \int_0^{2\pi} \left\{ \rho^2 \left[\sqrt{\frac{r}{r_0}} \cos^2\phi + \sqrt{\frac{r_0}{r}} \sin^2\phi + 2 \sin \tfrac{1}{2}(\theta_0 + \theta) \sin\phi \cos\phi \right] - 2i \frac{\sqrt{(rr_0)}}{R_1} \cos^2 \tfrac{1}{2}(\theta_0 + \theta) \right\}^{-1} d\phi. \tag{15}$$

Now $K(\rho)$ can be evaluated by the standard technique of putting $z = \exp(i\phi)$, when the path of integration becomes the unit circle and it is only necessary to calculate the residues of the enclosed poles. It is found that

$$K(\rho) = 2\pi |\sec \tfrac{1}{2}(\theta_0 + \theta)| \left\{ \rho^4 - 2i\rho^2 - \frac{4rr_0}{R_1^2} \cos^2 \tfrac{1}{2}(\theta_0 + \theta) \right\}^{-1/2}, \tag{16}$$

where the branch of the square root is that which lies (for real values of ρ) in the fourth quadrant of the complex plane. Hence (10) becomes

$$\mathscr{I}(\theta_0) = \pm \sqrt{\frac{2}{\pi}} e^{-\frac{1}{4}i\pi} e^{ikR_1} \int_0^\infty \frac{\rho e^{-kR_1\rho^2}}{\sqrt{\{[\rho^2 - i(R_1 - R')/R_1][\rho^2 - i(R_1 + R')/R_1]\}}} d\rho, \tag{17}$$

with the $\substack{\text{upper}\\\text{lower}}$ sign for $\cos \frac{1}{2}(\theta_0 + \theta) \gtrless 0$. Finally the substitution

$$\mu^2 = ikR_1\rho^2 + k(R_1 - R')$$

gives the required result, namely

$$\mathscr{I}(\theta_0) = \pm \sqrt{\frac{2}{\pi}}\, e^{-\frac{1}{4}i\pi} e^{ikR'} \int_{\sqrt{\{k(R_1 - R')\}}}^\infty \frac{e^{i\mu^2}}{\sqrt{(\mu^2 + 2kR')}} d\mu, \tag{18}$$

with the $\substack{\text{upper}\\\text{lower}}$ sign for $\cos \frac{1}{2}(\theta_0 + \theta) \gtrless 0$.

Since the incident field (1) can be written in the form

$$\sqrt{\frac{2}{\pi}}\, e^{-\frac{1}{4}i\pi} e^{ikR} \int_{-\infty}^\infty \frac{e^{i\mu^2}}{\sqrt{(\mu^2 + 2kR)}} d\mu, \tag{19}$$

the geometrical optics term (7) combines with the diffraction term (9) to give the total field

$$E_z = \sqrt{\frac{2}{\pi}}\, e^{-\frac{1}{4}i\pi} \left\{ e^{ikR} \int_m^\infty \frac{e^{i\mu^2}}{\sqrt{(\mu^2 + 2kR)}} d\mu - e^{ikR'} \int_{m'}^\infty \frac{e^{i\mu^2}}{\sqrt{(\mu^2 + 2kR')}} d\mu \right\}, \tag{20}$$

where

$$\left.\begin{aligned} m &= -2\sqrt{\left(\frac{krr_0}{R_1 + R}\right)} \cos \tfrac{1}{2}(\theta_0 - \theta) = \mp \sqrt{\{k(R_1 - R)\}}, \quad \mp \text{ for } \cos \tfrac{1}{2}(\theta_0 - \theta) \gtrless 0, \\ m' &= -2\sqrt{\left(\frac{krr_0}{R_1 + R'}\right)} \cos \tfrac{1}{2}(\theta_0 + \theta) = \mp \sqrt{\{k(R_1 - R')\}}, \quad \mp \text{ for } \cos \tfrac{1}{2}(\theta_0 + \theta) \gtrless 0. \end{aligned}\right\} \tag{21}$$

The solution was first given in essentially the form (20) by MACDONALD,* who obtained it via a transformation of an earlier solution due to CARSLAW.† It is very

* H. M. MACDONALD, *Proc. Lond. Math. Soc.*, **14** (1915), 410.

† H. S. CARSLAW, *Proc. Lond. Math. Soc.*, **30** (1899), 121.

similar in type to that of SOMMERFELD for an incident plane wave which, indeed, is immediately recovered on multiplying by $\sqrt{(kr_0)} \exp(-ikr_0)$ and letting $r_0 \to \infty$. The solution for H-polarization differs only in that the two terms in (20) are added instead of being subtracted.

If $kR_1 \gg 1$, μ may be replaced by its lower-limit value in the non-exponential factor in the integrand of (18) to give the approximate result

$$\mathscr{I}(\theta_0) = \pm \sqrt{\frac{2}{\pi}}\, e^{-\frac{1}{4}i\pi} \frac{e^{ikR'}}{\sqrt{\{k(R_1 + R')\}}} F[\sqrt{\{k(R_1 - R')\}}]; \tag{22}$$

thus, using also the corresponding approximation for $\mathscr{I}(-\theta_0)$, the diffraction field is expressed in terms of FRESNEL integrals to a degree of accuracy which is only inadequate if both the source and the point of observation are well within a wavelength of the diffracting edge.

Furthermore, if $k(R_1 - R') \gg 1$, the asymptotic approximation § 11.5 (31) may be applied to (22) to give

$$\mathscr{I}(\theta_0) = \frac{e^{\frac{1}{4}i\pi}}{2\sqrt{(2\pi)}} \sec \tfrac{1}{2}(\theta_0 + \theta) \frac{e^{ikr_0}}{\sqrt{(kr_0)}} \frac{e^{ikr}}{\sqrt{(kr)}}; \tag{23}$$

and similarly, if $k(R_1 - R) \gg 1$,

$$\mathscr{I}(-\theta_0) = \frac{e^{\frac{1}{4}i\pi}}{2\sqrt{(2\pi)}} \sec \tfrac{1}{2}(\theta_0 - \theta) \frac{e^{ikr_0}}{\sqrt{(kr_0)}} \frac{e^{ikr}}{\sqrt{(kr)}}. \tag{24}$$

The diffraction field therefore resembles the field of a certain line-source located at the diffracting edge for all points well outside the two *hyperbolae* $k(R_1 - R') = 1$ and $k(R_1 - R) = 1$, the axes of which are $\theta + \theta_0 = \pi$ and $\theta - \theta_0 = \pi$ respectively. These hyperbolae are the counterparts, for an incident cylindrical wave, of the parabolae discussed in § 11.5.3 in connection with the solution for an incident plane wave.

Finally, attention should be drawn to the fact that the solution (20) is *reciprocal* in the sense that it is unaltered by the respective interchange of r_0, θ_0, and r, θ. This, of course, is a particular example of the general theorem on reciprocity* implicit in MAXWELL'S equations: the present analysis shows that it is here associated with the fact that the spectrum function § 11.5 (7) is symmetrical in α and α_0.

11.7.2 A dipole

The most elementary source of electromagnetic waves which is localized at a point is a dipole, electric or magnetic. The problem of a dipole in the presence of a half-plane can be solved by representing the undisturbed dipole field as a (three-dimensional) spectrum of plane waves and applying the results of § 11.6 to each plane wave. The case of an electric dipole with its axis normal to the diffracting sheet has been treated in this way by SENIOR,† and the analysis is sketched in here.

The configuration is shown in Fig. 11.17 with, as before, Cartesian coordinates x, y, z and cylindrical polar coordinates r, θ, z, where the diffracting sheet occupies $y = 0, x > 0$. The dipole at T, (x_0, y_0, z_0) or (r_0, θ_0, z_0), is taken to be parallel to the y axis; T' is the image of T in the plane $y = 0$, and R, R' are the distances of the point of observation P from T and T' respectively.

---

* L. G. H. HUXLEY, *The Principles and Practice of Waveguides* (Cambridge University Press, 1947), § 7.17.

† T. B. A. SENIOR, *Quart. J. Mech. Appl. Maths.*, **6** (1953), 101. For other methods see A. E. HEINS, *Trans. Inst. Radio Engrs.*, AP–**4** (1956), 294; B. D. WOODS, *Quart. J. Mech. Appl. Maths.*, **10** (1957), 90; W. E. WILLIAMS, *Quart. J. Mech. Appl. Maths.*, **10** (1957), 210,

With a convenient choice of the magnitude of the dipole moment the undisturbed field of the dipole (cf. § 2.2) may be taken as

$$\boldsymbol{E} = \left(\frac{\partial^2\Pi}{\partial x\partial y}, \frac{\partial^2\Pi}{\partial y^2} + k^2\Pi, \frac{\partial^2\Pi}{\partial y\partial z}\right), \qquad \boldsymbol{H} = ik\left(\frac{\partial\Pi}{\partial z}, 0, -\frac{\partial\Pi}{\partial x}\right), \tag{25}$$

where

$$\Pi = \frac{e^{ikR}}{kR}. \tag{26}$$

The required resolution of (25) into plane waves is then obtained from the formula*

$$\frac{e^{ikR}}{kR} = \frac{i}{2\pi}\int_{S(\frac{1}{2}\pi)}\int_{S(\frac{1}{2}\pi)} \cos\beta\, e^{-ik\{(x-x_0)\cos\alpha\cos\beta+(y-y_0)\sin\alpha\cos\beta-(z-z_0)\sin\beta\}}\,d\alpha d\beta. \tag{27}$$

The procedure, therefore, is to write down the total field in the presence of the diffracting sheet arising from the incidence on it of each plane wave implicit in the integrand of (27), and then to carry out the integration.

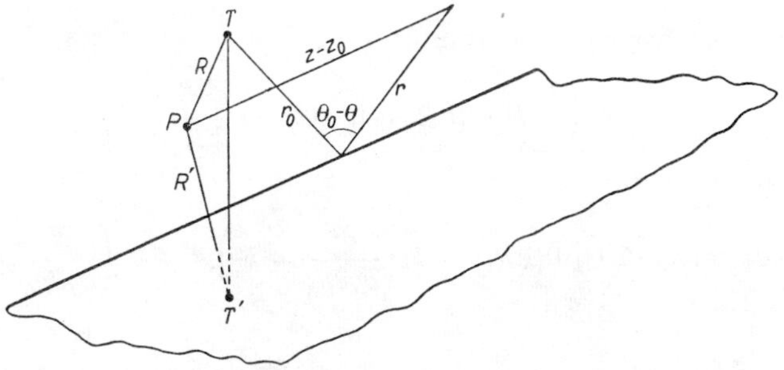

Fig. 11.17. The configuration for a dipole at T in the presence of a diffracting half-plane.

In view of the subsequent integration it is convenient to use (cf. § 11.7 (5)) the basic solutions for incident plane waves in the form § 11.5 (8) (for E-polarization) with the corresponding form for H-polarization. These expressions are modified to yield the three-dimensional solutions corresponding to the incident plane waves implicit in (27) in the manner explained in § 11.6. The component of $\boldsymbol{E}$ parallel to the dipole then appears in the form

$$E_y = E_y^{(g)} + E_y^{(d)}, \tag{28}$$

where $E_y^{(g)}$ is the field of geometrical optics and

$$E_y^{(d)} = \frac{ik}{2\pi}\{A(\theta_0) + A(-\theta_0)\} - \frac{i}{2\pi k}\left(\frac{\partial^2}{\partial y\partial y_0} - k^2\right)B(\theta_0) + \frac{i}{2\pi k}\left(\frac{\partial^2}{\partial y\partial y_0} + k^2\right)B(-\theta_0), \tag{29}$$

where

$$A(\theta_0) = \frac{ik}{2\pi}\int_{S(0)}\int_{S(\frac{1}{2}\pi)}\int_{S(0)} \cos\beta\cos\tfrac{1}{2}(\alpha+\gamma+\theta-\theta_0) \times e^{ik\{r\cos\gamma\cos\beta+r_0\cos\alpha\cos\beta+(z-z_0)\sin\beta\}}\,d\alpha d\beta d\gamma, \tag{30}$$

$$B(\theta_0) = \frac{ik}{2\pi}\int_{S(0)}\int_{S(\frac{1}{2}\pi)}\int_{S(0)} \cos\beta\sec\tfrac{1}{2}(\alpha+\gamma+\theta-\theta_0) \times e^{ik\{r\cos\gamma\cos\beta+r_0\cos\alpha\cos\beta+(z-z_0)\sin\beta\}}\,d\alpha d\beta d\gamma. \tag{31}$$

* This is a straightforward modification of a formula given by H. Weyl, *Ann. d. Physik*, **60** (1919), 481.

By means of an analysis similar to that of CARSLAW* and MACDONALD† it can be shown that

$$A(\theta_0) = \frac{\pi}{\sqrt{(rr_0)}} \cos \tfrac{1}{2}(\theta - \theta_0) H_0^{(1)}(kR_1), \tag{32}$$

where

$$R_1^2 = (r + r_0)^2 + (z - z_0)^2; \tag{33}$$

$$B(\theta_0) = \pm \pi k \int_{\pm m}^{\infty} H_1^{(1)}(kR \cosh \mu) d\mu, \qquad \pm \text{ for } \theta - \theta_0 \lessgtr \pi, \tag{34}$$

$$B(-\theta_0) = \pm \pi k \int_{\pm m'}^{\infty} H_1^{(1)}(kR' \cosh \mu) d\mu, \qquad \pm \text{ for } \theta + \theta_0 \lessgtr \pi, \tag{35}$$

where

$$m = \sinh^{-1} \left\{ 2 \frac{\sqrt{(rr_0)}}{R} \cos \tfrac{1}{2}(\theta - \theta_0) \right\}, \qquad m' = \sinh^{-1} \left\{ 2 \frac{\sqrt{(rr_0)}}{R'} \cos \tfrac{1}{2}(\theta + \theta_0) \right\}. \tag{36}$$

These results, together with the formula

$$\tfrac{1}{2} i \int_{-\infty}^{\infty} H_1^{(1)}(kR \cosh \mu) d\mu = \frac{e^{ikR}}{kR}, \tag{37}$$

enable (28) to be written

$$E_y = \frac{ik}{\sqrt{(rr_0)}} \cos \tfrac{1}{2}\theta \cos \tfrac{1}{2}\theta_0 H_0^{(1)}(kR_1) - \tfrac{1}{2} i \left(\frac{\partial^2}{\partial y \partial y_0} - k^2 \right) \mathscr{I} + \tfrac{1}{2} i \left(\frac{\partial^2}{\partial y \partial y_0} + k^2 \right) \mathscr{I}', \tag{38}$$

where

$$\mathscr{I} = \int_{-m}^{\infty} H_1^{(1)}(kR \cosh \mu) d\mu, \qquad \mathscr{I}' = \int_{-m'}^{\infty} H_1^{(1)}(kR' \cosh \mu) d\mu. \tag{39}$$

The remaining field components can likewise be expressed, exactly, in terms of $\mathscr{I}$ and $\mathscr{I}'$ as follows

$$E_x = -\frac{ik}{\sqrt{(rr_0)}} \sin \tfrac{1}{2}\theta \cos \tfrac{1}{2}\theta_0 H_0^{(1)}(kR_1) - \tfrac{1}{2} i \frac{\partial^2 \mathscr{I}}{\partial x \partial y_0} + \tfrac{1}{2} i \frac{\partial^2 \mathscr{I}'}{\partial x \partial y_0}, \tag{40}$$

$$E_z = -\tfrac{1}{2} i \frac{\partial^2 \mathscr{I}}{\partial z \partial y_0} + \tfrac{1}{2} i \frac{\partial^2 \mathscr{I}'}{\partial z \partial y_0}, \tag{41}$$

$$H_x = \frac{k(z - z_0)}{R_1 \sqrt{(rr_0)}} \cos \tfrac{1}{2}\theta \cos \tfrac{1}{2}\theta_0 H_1^{(1)}(kR_1) + \tfrac{1}{2} k \frac{\partial \mathscr{I}}{\partial z_0} + \tfrac{1}{2} k \frac{\partial \mathscr{I}'}{\partial z_0}, \tag{42}$$

$$H_y = \frac{k(z - z_0)}{R_1 \sqrt{(rr_0)}} \sin \tfrac{1}{2}\theta \cos \tfrac{1}{2}\theta_0 H_1^{(1)}(kR_1), \tag{43}$$

$$H_z = -\tfrac{1}{2} k \frac{\partial \mathscr{I}}{\partial x_0} - \tfrac{1}{2} k \frac{\partial \mathscr{I}'}{\partial x_0}. \tag{44}$$

It is interesting to note that in the complete field there is a non-zero component of ***H*** parallel to the dipole, so that the analysis cannot be formulated in terms of a single component HERTZ vector.

It is again possible, under a relatively trivial restriction, to express the solution in

---

* H. S. CARSLAW, *Proc. Lond. Math. Soc.*, **30** (1899), 121.

† H. M. MACDONALD, *Proc. Lond. Math. Soc.*, **14** (1915), 410.

terms of FRESNEL integrals. If $kR_1 \gg 1$, it is not difficult to establish the asymptotic approximation

$$\mathscr{I} = -\frac{2e^{\frac{1}{4}i\pi}}{k}\sqrt{\left\{\frac{2}{\pi R_1(R+R_1)}\right\}}\, e^{ikR}\, F\left\{-2\sqrt{\left(\frac{krr_0}{R+R_1}\right)}\cos\tfrac{1}{2}(\theta-\theta_0)\right\}, \qquad (45)$$

with a similar result for $\mathscr{I}'$.

11.8 OTHER PROBLEMS

In this section several other diffraction problems are briefly reviewed.

11.8.1 Two parallel half-planes

The problem of diffraction by two parallel half-planes which are perpendicular to the common plane through their edges is tractable by the method used in this chapter for the single half-plane.*

Consider E-polarization, with the problem precisely as in § 11.5.1 except that the diffracting obstacle now consists of *two* sheets; one (sheet 1) occupying $y = 0$, $x > 0$; the other (sheet 2), $y = -a$, $x > 0$. It is convenient to introduce the additional coordinates r', θ', measured from the edge $(0, -a)$ of sheet 2.

The scattered fields due to the induced currents in sheets 1 and 2 may be written

$$E_z^{(s1)} = \int_C P_1(\cos\alpha)e^{ikr\cos(\theta\mp\alpha)}\,d\alpha, \qquad \mp \text{ for } y \gtrless 0, \qquad (1)$$

$$E_z^{(s2)} = \int_C P_2(\cos\alpha)e^{ikr'\cos(\theta'\mp\alpha)}\,d\alpha, \qquad \mp \text{ for } y \gtrless -a, \qquad (2)$$

respectively. The continuity of $H_x^{(s1)}$ and $H_x^{(s2)}$ across the regions $y = 0$, $x < 0$, and $y = -a$, $x < 0$, respectively, is ensured by taking $P_1(\mu)$ and $P_2(\mu)$ to be free of singularities below the path of integration, which is that shown in Fig. 11.7. Furthermore, the boundary condition that E_z should vanish on the two sheets leads to the integral equations

$$\int_{-\infty}^{\infty}\frac{P_1(\mu)}{\sqrt{(1-\mu^2)}}e^{ikx\mu}\,d\mu + \int_{-\infty}^{\infty}\frac{P_2(\mu)}{\sqrt{(1-\mu^2)}}e^{ika\sqrt{(1-\mu^2)}}\,e^{ikx\mu}\,d\mu = -e^{-ikx\mu_0}, \qquad (3)$$

$$\int_{-\infty}^{\infty}\frac{P_1(\mu)}{\sqrt{(1-\mu^2)}}e^{ika\sqrt{(1-\mu^2)}}\,e^{ikx\mu}\,d\mu + \int_{-\infty}^{\infty}\frac{P_2(\mu)}{\sqrt{(1-\mu^2)}}e^{ikx\mu}\,d\mu = -e^{ika\sqrt{(1-\mu_0^2)}}\,e^{-ikx\mu_0}, \qquad (4)$$

which must hold for $x > 0$.

If we write

$$P_1(\mu) + P_2(\mu) = Q_1(\mu), \qquad P_1(\mu) - P_2(\mu) = Q_2(\mu), \qquad (5)$$

addition and subtraction, respectively, of (3) and (4) give

$$\int_{-\infty}^{\infty}\frac{Q_1(\mu)}{\sqrt{(1-\mu^2)}}\{1+e^{ika\sqrt{(1-\mu^2)}}\}\,e^{ikx\mu}\,d\mu = -\{1+e^{ika\sqrt{(1-\mu_0^2)}}\}\,e^{-ikx\mu_0}, \qquad (6)$$

$$\int_{-\infty}^{\infty}\frac{Q_2(\mu)}{\sqrt{(1-\mu^2)}}\{1-e^{ika\sqrt{(1-\mu^2)}}\}\,e^{ikx\mu}\,d\mu = -\{1-e^{ika\sqrt{(1-\mu_0^2)}}\}\,e^{-ikx\mu_0}, \qquad (7)$$

for $x > 0$.

---

* Solutions for an incident plane wave were first given, using the method introduced by SCHWINGER (mentioned in § 11.1), by A. E. HEINS, *Quart. Appl. Maths.*, **6** (1948), 157, and L. A VAINSTEIN, *Izvestiya Akad. Nauk SSSR, Ser. Fiz.* (*Bull. Acad. Sci. USSR*), **12** (1948), 144 and 166.

The form of each of the equations (6) and (7) is similar to § 11.5 (2), and to obtain solutions analogous to § 11.5 (4) the paths of integration must be closed by infinite semi-circles above the real axis. To this end, that branch of $\sqrt{(1-\mu^2)}$ is chosen which has a positive imaginary part. The required solution of (6) is then

$$\frac{Q_1(\mu)}{\sqrt{(1-\mu^2)}}\{1+e^{ika\sqrt{(1-\mu^2)}}\} = \frac{i}{2\pi}\{1+e^{ika\sqrt{(1-\mu_0^2)}}\}\frac{U(\mu)}{U(-\mu_0)}\frac{1}{\mu+\mu_0}, \tag{8}$$

where $U(\mu)$ is any function of μ which is free of singularities in the half-plane above the path of integration and tends to zero as $|\mu| \to \infty$ therein.

It remains to be seen how $Q_1(\mu)$ and $U(\mu)$ can be chosen to satisfy (8), remembering that $Q_1(\mu)$ has no singularities below the path of integration. The procedure is to factorize the coefficient of $Q_1(\mu)$ in (8) in the form

$$\frac{1+e^{ika\sqrt{(1-\mu^2)}}}{\sqrt{(1-\mu^2)}} = U_1(\mu)L_1(\mu), \tag{9}$$

where $U_1(\mu)$ is free of singularities and zeros in the half-plane above the path of integration, and is of algebraic growth at infinity therein, while $L_1(\mu)$ has similar characteristics in the half-plane below the path of integration. That such factorization is possible is known from the general theory of WIENER and HOPF,* and explicit expressions for $U_1(\mu)$ and $L_1(\mu)$ are given by HEINS.† Then clearly

$$Q_1(\mu) = \frac{i}{2\pi}\sqrt{(1-\mu_0^2)}U_1(\mu_0)\frac{1}{L_1(\mu)(\mu+\mu_0)}, \tag{10}$$

where the relation $U_1(\mu) = L_1(-\mu)$, which is implicit in (9) (apart from arbitrary constant factors), has been invoked.

Similarly, if

$$\frac{1-e^{ika\sqrt{(1-\mu^2)}}}{\sqrt{(1-\mu^2)}} = U_2(\mu)L_2(\mu) \tag{11}$$

we have

$$Q_2(\mu) = \frac{i}{2\pi}\sqrt{(1-\mu_0^2)}U_2(\mu_0)\frac{1}{L_2(\mu)(\mu+\mu_0)}. \tag{12}$$

The total scattered field is obtained by adding (1) and (2). In $y > 0$, for instance, it is therefore

$$\begin{aligned} E_z^{(s)} &= \int_C \{P_1(\cos\alpha) + P_2(\cos\alpha)\,e^{ika\sin\alpha}\}\,e^{ikr\cos(\theta-\alpha)}\,d\alpha \\ &= \tfrac{1}{2}\int_C \{Q_1(\cos\alpha)(1+e^{ika\sin\alpha}) + Q_2(\cos\alpha)(1-e^{ika\sin\alpha})\}\,e^{ikr\cos(\theta-\alpha)}\,d\alpha \\ &= \int_C P(\cos\alpha)e^{ikr\cos(\theta-\alpha)}\,d\alpha, \qquad \text{say,} \end{aligned} \tag{13}$$

where

$$\begin{aligned} P(\mu) &= \tfrac{1}{2}\sqrt{(1-\mu^2)}\{Q_1(\mu)U_1(\mu)L_1(\mu) + Q_2(\mu)U_2(\mu)L_2(\mu)\} \\ &= \frac{i}{4\pi}\frac{\sqrt{(1-\mu_0^2)}\sqrt{(1-\mu^2)}}{\mu+\mu_0}\{U_1(\mu_0)U_1(\mu) + U_2(\mu_0)U_2(\mu)\} \end{aligned} \tag{14}$$

---

* E. C. TITCHMARSH, *Introduction to the Theory of Fourier Integrals* (Oxford, Clarendon Press, 1937), p. 339.

† A. E. HEINS, *Quart. Appl. Maths.*, **6** (1948), 157.

Again the symmetry of $P(\mu)$ in μ and μ_0 should be noted. Also that, when $a = 0$, $U_1(\mu) = \sqrt{\{2/(1+\mu)\}}$, $U_2(\mu) = 0$, so that (14) reduces, as it should, to § 11.5 (6).

11.8.2 An infinite stack of parallel, staggered half-planes

In this problem* we have an infinite set of diffracting sheets, with the nth sheet occupying $y = na$, $x > nb$, where $n = 0, \pm 1, \pm 2, \ldots$

Then, for an E-polarized incident plane wave as before, the scattered field due to currents induced in the mth sheet may be written

$$E_z^{(sm)} = \int_C P_m(\cos\alpha) e^{-ikm(a\cos\alpha \pm b\sin\alpha)} e^{ikr\cos(\theta\mp\alpha)}\, d\alpha, \qquad (15)$$

with the upper sign for $y > ma$, and the lower sign for $y < ma$. All the $P_m(\mu)$ must be free of singularities below the path of integration in the $\mu = \cos\alpha$ plane.

The boundary conditions on the nth sheet yield the integral equation

$$\sum_{m=-\infty}^{\infty} \int_{-\infty}^{\infty} \frac{P_m(\mu)}{\sqrt{(1-\mu^2)}} e^{ika|n-m|\sqrt{(1-\mu^2)}} e^{ik(x-mb)\mu}\, d\mu = -e^{-ikx\mu_0} e^{-ikna\sqrt{(1-\mu_0^2)}}, \qquad (16)$$

which must hold for $x > nb$.

From the periodicity of the problem it is clear that

$$P_m(\mu) = P_0(\mu) e^{-ikm\{b\mu_0 + a\sqrt{(1-\mu_0^2)}\}} \qquad (17)$$

and on putting $n - m = q$, (16) therefore becomes

$$\sum_{q=-\infty}^{\infty} \int_{-\infty}^{\infty} \frac{P_0(\mu)}{\sqrt{(1-\mu^2)}} e^{ikq\{b\mu_0 + a\sqrt{(1-\mu_0^2)}\}} e^{ika|q|\sqrt{(1-\mu^2)}} e^{ik(qb+x)\mu}\, d\mu = -e^{-ikx\mu_0} \qquad (18)$$

for $x > 0$.

The infinite sum over q can be put in closed form, again leaving an integral equation which can be solved by the use of CAUCHY's residue theorem. As in the previous problem, it is necessary to "split" a certain function into a pair of factors, one of which is free of singularities and zeros in the upper half-plane and of algebraic growth at infinity therein, the other having similar characteristics in the lower half-plane: the factors are given explicitly, together with further details, in the papers of CARLSON and HEINS to which reference has already been made.

11.8.3 A strip

Another problem of intrinsic interest is the deceptively simple one in which the diffracting obstacle is a perfectly conducting plane strip, infinitely long with parallel edges; or the corresponding one with the complementary "screen", a slit in an infinite plane. Various methods of solution have been given,† but none yields a result in closed form. The dual integral equation approach has been used‡ to obtain the

* J. F. CARLSON and A. E. HEINS, *Quart. Appl. Maths.*, **4** (1947), 313 and **5** (1947), 82.

† Lord RAYLEIGH, *Phil. Mag.*, **43** (1897), 259 [reprinted in *Scientific Papers*, **4**, 283]; K. SCHWARZSCHILD, *Math. Ann.*, **55** (1902), 177; P. M. MORSE and P. J. RUBENSTEIN, *Phys. Rev.*, **54** (1938), 895; A. SOMMERFELD, *Optics* (New York, Academic Press, 1954), p. 273; B. B. BAKER and E. T. COPSON, *The Mathematical Theory of Huygens' Principle* (Oxford, Clarendon Press, 2nd ed., 1950), p. 177 *et seq.*; S. SKAVLEM, *Arch. Math. Naturvid.*, **51** (1951), 61; E. B. MOULLIN and F. M. PHILLIPS, *Proc. Inst. Elec. Engrs.*, **99**, Pt. IV (1952), 137; R. MÜLLER and K. WESTPFAHL, *Z. f. Phys.*, **134** (1953), 245; P. C. CLEMMOW, *Trans. Inst. Radio Engrs.*, AP-**4** (1956), 282; S. N. KARP and A. RUSSEK, *J. Appl. Phys.*, **27** (1956), 886; R. F. MILLAR, *Proc. Camb. Phil. Soc.*, **54** (1958), 479, 497.

‡ E. GROSCHWITZ and H. HÖNL, *Z. f. Phys.*, **131** (1952), 305; H. HÖNL and E. ZIMMER, *Z. f. Phys.*, **135** (1953), 196; C. J. TRANTER, *Quart. J. Mech. Appl. Maths.*, **7** (1954), 317.

first two terms of a series solution in powers of ka, where $2a$ is the width of the strip, for a normally incident plane wave, in the following manner.

For a strip occupying $y = 0$, $|x| < a$, and a normally incident H-polarized plane wave, the integral equations § 11.4 (19) and § 11.4 (20) are

$$\int_{-\infty}^{\infty} P(\mu) e^{ikx\mu}\, d\mu = 1 \qquad \text{for } |x| < a, \tag{19}$$

$$\int_{-\infty}^{\infty} \frac{P(\mu)}{\sqrt{(1-\mu^2)}} e^{ikx\mu}\, d\mu = 0 \qquad \text{for } |x| > a; \tag{20}$$

or, since the symmetry of the problem implies $P(\mu) = P(-\mu)$,

$$\int_0^{\infty} P(\mu) \cos(kx\mu) d\mu = \tfrac{1}{2} \qquad \text{for } |x| < a, \tag{21}$$

$$\int_0^{\infty} \frac{P(\mu)}{\sqrt{(1-\mu^2)}} \cos(kx\mu) d\mu = 0 \qquad \text{for } |x| > a. \tag{22}$$

A solution is sought of the form

$$\frac{P(\mu)}{\sqrt{(1-\mu^2)}} = \sum_{m=0}^{\infty} c_m \frac{J_{2m+1}(ka\mu)}{\mu}, \tag{23}$$

since (22) is satisfied by each term of the series. Substitution into (21) shows that the c_m must be found such that

$$\sum_{m=0}^{\infty} c_m \Phi_m = \tfrac{1}{2} \qquad \text{for } |x| < a, \tag{24}$$

where

$$\Phi_m = \int_0^{\infty} \frac{\sqrt{(1-\mu^2)}}{\mu} J_{2m+1}(ka\mu) \cos(kx\mu) d\mu. \tag{25}$$

Now it can be shown that, for $|ka| \ll 1$,

$$\Phi_m = i \int_0^{\infty} J_{2m+1}(ka\mu) \cos(kx\mu) d\mu + O(ka) = \frac{i \cos\left[(2m+1)\sin^{-1}\dfrac{x}{a}\right]}{ka\sqrt{(1-x^2/a^2)}} + O(ka); \tag{26}$$

whence, to the first order, (24) is

$$\frac{i}{\sqrt{(1-x^2/a^2)}} \sum_{m=0}^{\infty} c_m \cos\left[(2m+1)\sin^{-1}\frac{x}{a}\right] = \tfrac{1}{2}ka, \tag{27}$$

giving

$$c_0 = \frac{ka}{2i}, \qquad c_m = 0 \qquad \text{for } m = 1, 2, 3, \ldots \tag{28}$$

Thus, to this approximation,

$$\frac{P(\mu)}{\sqrt{(1-\mu^2)}} = \frac{ka}{2i} \frac{J_1(ka\mu)}{\mu}, \tag{29}$$

and the current density, derived from § 11.4 (16), is

$$J_x(x) = \frac{cka}{2\pi i} \int_0^{\infty} \frac{J_1(ka\mu)}{\mu} \cos(kx\mu) d\mu = \frac{cka}{2\pi i} \sqrt{(1-x^2/a^2)}. \tag{30}$$

The next approximation, involving terms in $(ka)^3$, is considerably more complicated. Identical expressions for it have been obtained independently by several authors.*

11.8.4 Further problems

There are a number of other interesting problems, allied to those which we have discussed, which are capable of solution, but we cannot do more than mention them here.

Diffraction by a two-dimensional wedge, reducing to a half-plane when the exterior wedge angle is 2π, was solved many years ago.† It involves angular spectra of period $2\pi/n$, where π/n is the exterior wedge angle.

The problem of a half-plane in the plane interface between two different homogeneous media was first considered by HANSON.‡ It is tractable by the method of this chapter and has been applied § in the theory of the propagation of radio waves over the surface of the earth.

Two investigations have been made of the effect of a half-plane under less idealized conditions. In the first,|| finite though large conductivity is introduced, which necessitates the use of approximate boundary conditions; in the second,¶ the plate is assumed to be perfectly conducting but to have a finite, though small, thickness.

11.9 UNIQUENESS OF SOLUTION

In § 11.2 we saw that the general diffraction problem with which we were concerned could be stated in the following form: given a field $\boldsymbol{E}^{(i)}$ incident on a perfectly conducting surface S, to find a field $\boldsymbol{E}^{(s)}$ which could arise from an electric current distribution in S and which is such that its tangential component on S is minus that of $\boldsymbol{E}^{(i)}$.

It is, of course, vital that the formulation should yield a unique solution,** but the demonstration that there cannot, in fact, be more than one field $\boldsymbol{E}^{(s)}$ satisfying the stated conditions is by no means straightforward, particularly when the possibilities of S being infinite and the field containing plane waves are taken into account. Only comparatively recently, it appears, has the result been satisfactorily established,†† although it has long been tacitly accepted.

A further difficulty arises in the special, though commonest, type of diffraction problem in which the obstacle may be assumed to have an infinitely sharp edge and to which the discussion in this chapter has, indeed, been confined. The reason for the

---

* R. MÜLLER and K. WESTPFAHL, *Z. f. Phys.*, **134** (1953), 245; C. J. TRANTER, *Quart. J. Mech. Appl. Maths.*, **7** (1954), 317. See also C. J. BOUWKAMP, *Rep. Progr. Phys.* (London, Physical Society), **17** (1954), 73.

† H. M. MACDONALD, *Electric Waves* (Cambridge University Press, 1902); H. S. CARSLAW, *Proc. Lond. Math. Soc.*, **18** (1919), 291.

‡ E. T. HANSON, *Phil. Trans. Roy. Soc.* A, **237** (1938), 35.

§ P. C. CLEMMOW, *Phil. Trans. Roy. Soc.* A, **246** (1953), 1.

|| T. B. A. SENIOR, *Proc. Roy. Soc.* A, **213** (1952), 436.

¶ D. S. JONES, *Proc. Roy. Soc.* A, **217** (1953), 153.

** To establish the *existence* of a solution is perhaps less important, since this question is settled when one is found in any particular case. But see C. MÜLLER, *Math. Ann.*, **123** (1951), 345, and W. K. SAUNDERS, *Proc. Nat. Acad. Sci. U.S.A.*, **38** (1952), 342.

†† F. RELLICH, *Jahr. Deut. Math. Ver.*, **53** (1943), 57. See the account in A. SOMMERFELD, *Partial Differential Equations in Physics* (New York, Academic Press, 1949), § 28. W. K. SAUNDERS, *Proc. Nat. Acad. Sci. U.S.A.*, **38** (1952), 342.

additional complication is that the solution then contains, as we have seen, a singularity at the edge, thus violating an assumption necessary to the uniqueness proof just mentioned.

It can easily be seen that, if an arbitrary edge singularity is permitted, an infinite sequence of solutions can be got by a process of differentiation.* For example, differentiation of the E-polarization solution of the half-plane problem § 11.5 (22) with respect to x gives an essentially new expression which also satisfies the wave equation and vanishes on the screen; or, again, differentiation of § 11.5 (22) with respect to y yields an expression which would apparently suffice for the H-polarization solution, but which differs from § 11.5 (44).

Each differentiation introduces a singularity of higher order at the diffracting edge. Evidently, in order to ensure uniqueness, some restriction on the nature of the singularity must be specified. The appropriate restriction, and the various ways in which it may be formulated, have been the subject of a number of papers.† The reader must be referred to these for details, but broadly speaking it may be said that the solution involving the singularity of lowest possible order is to be taken as representing the answer to the physical problem, and that this, in fact, rules out any singularities of order greater than $r^{-1/2}$ as $r \to 0$ at the diffracting edge. In particular, it can be established that a solution is unique which implies for the induced current its integrability over the diffracting surface and the vanishing at the edge of its component normal to the edge. The behaviour of the components of $\mathbf{E}$ and $\mathbf{H}$ near the edge can be deduced from these conditions.

Finally, the possibility of an infinite number of "solutions" having been seen, it may be asked why the method used in this chapter apparently yields a unique solution, which is, moreover, the correct one judged by the above-mentioned criteria. The answer is because of the assumption that the field components in the plane of the diffracting screen can be expressed as convergent FOURIER integrals: this precludes them from having singularities of too high an order.

---

* C. J. BOUWKAMP, *Physica*, **12** (1946), 467.

† J. MEIXNER, *Ann. d. Physik*, **6** (1949), 1; A. W. MAUE, *Z. f. Phys.*, **126** (1949), 601; E. T. COPSON, *Proc. Roy. Soc.* A, **202** (1950), 277; D. S. JONES, *Quart. J. Mech. Appl. Maths.*, **3** (1950), 420; A. E. HEINS and S. SILVER, *Proc. Camb. Phil. Soc.*, **51** (1955), 149; ibid., **54** (1958), 131.

CHAPTER XII

DIFFRACTION OF LIGHT BY ULTRASONIC WAVES

In Chapters I and II it was shown that the propagation of electromagnetic waves may be studied either by using Maxwell's equations, supplemented by the material equations, or by means of certain integral equations which utilize the polarization properties of the medium. In particular, either of these methods may also be applied to the study of the propagation of light through a medium whose density depends on space coordinates and on time. Though the former method has been used extensively in the past, the latter has only more recently been applied to such studies. In this chapter we shall apply the integral equation method to the problem of diffraction of light by a transparent homogeneous medium, disturbed by the passage of ultrasonic waves. It will be useful, however, to give first a qualitative description of this diffraction phenomenon and a brief summary of the theoretical work on this problem based on Maxwell's differential equations.

12.1 QUALITATIVE DESCRIPTION OF THE PHENOMENON AND SUMMARY OF THEORIES BASED ON MAXWELL'S DIFFERENTIAL EQUATIONS

12.1.1 Qualitative description of the phenomenon

Ultrasonic waves are sound waves whose frequencies are higher than those of waves normally audible to the human ear. The angular frequencies of the ultrasonic waves produced in laboratories lie from about 10^5 sec^{-1} to about 3×10^9 sec^{-1}, the former value representing the limit of audibility of the human ear. The corresponding range of wavelengths Λ of course depends on the velocity v of these waves in the medium in which they travel. For example, in water $v = 1{\cdot}2 \times 10^5$ cm/sec and the above frequency range corresponds to the wavelength range $\Lambda = 7{\cdot}5$ cm to $2{\cdot}5 \times 10^{-4}$ cm.*

In 1921 Brillouin† predicted that a liquid traversed by compression waves of short wavelengths, when irradiated by visible light, would give rise to a diffraction phenomenon similar to that due to a grating. In order to see this, consider a fluid lying between two infinite planes $y = 0$ and $y = d$, and let a plane compression wave of wavelength Λ progress through it along the positive x direction. This creates periodic stratifications of matter along the x axis, the distance between two successive planes of maximum density being Λ.

Let a monochromatic plane light wave of angular frequency ω and wavelength $\bar{\lambda}$ inside the medium be incident, with its wave-normal lying in the xy-plane and making an angle $\bar{\theta}$ with the y axis (see Fig. 12.1). Further let $\bar{\phi}$ denote the angle which a diffracted ray makes with the y axis. Since the velocity v of the compression waves is always very much smaller than the velocity of light, we may, to a first approximation,

* For methods of generation of ultrasonic waves and their many uses, see for example, L. Bergmann, *Der Ultraschall* (Zürich, Hirzel, 1954).

† L. Brillouin, *Ann. de Physique*, **17** (1921), 103.

consider the stratification of matter to be stationary. Then the directions $\bar{\phi}$ in which there is an appreciable intensity are determined by the condition that the optical path difference between the rays from two successive planes distance Λ apart shall be an integral multiple of $\bar{\lambda}$. This condition gives a relation between $\bar{\lambda}$, $\bar{\theta}$ and the directions of propagation $\bar{\phi}_l$ of the waves of various orders in the diffracted spectrum:

$$BC - AD = \Lambda(\sin \bar{\phi}_l - \sin \bar{\theta}) = l\bar{\lambda} \qquad (l = 0, \pm 1, \pm 2, \ldots), \tag{1}$$

AB and CD being portions of wave fronts associated with the refracted and diffracted rays. It will be convenient to rewrite (1) in terms of the angles θ and ϕ, and the wavelength λ outside the medium. If we use in (1) the law of refraction

$$\frac{\sin \bar{\theta}}{\sin \theta} = \frac{\sin \bar{\phi}}{\sin \phi} = \frac{\bar{\lambda}}{\lambda},$$

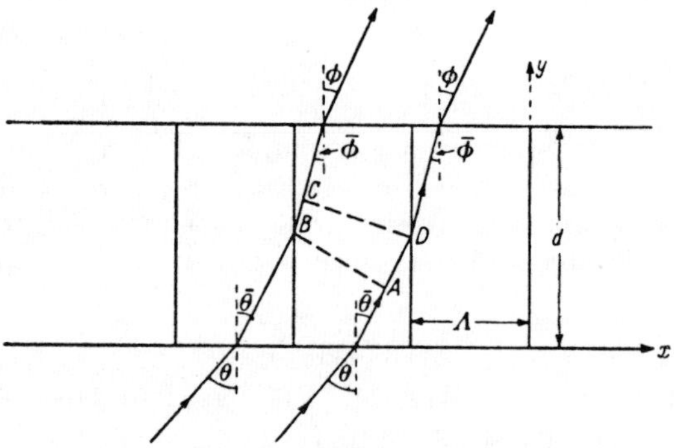

Fig. 12.1. Ultrasonic waves regarded as a diffraction grating.

we obtain

$$\Lambda(\sin \phi_l - \sin \theta) = l\lambda \qquad (l = 0, \pm 1, \pm 2, \ldots). \tag{2}$$

From (2) one has for the angular separation between successive orders

$$\sin \phi_l - \sin \phi_{l-1} \simeq \phi_l - \phi_{l-1} = \frac{\lambda}{\Lambda}.$$

Thus, for a given λ, the angular separation decreases with increasing Λ. If Λ is sufficiently large, the principal lines will be so close together that they will not be resolved in the observing instrument, and for this reason diffraction effects are not observed when ordinary sound waves are irradiated by visible light.

It was nearly a decade after BRILLOUIN's prediction that DEBYE and SEARS,* and LUCAS and BIQUARD,† independently, observed the diffraction of light by ultrasonic waves. Since then, many investigators have studied this phenomenon under a variety of experimental conditions obtained by varying one or more of the following quantities: (a) the angle of incidence θ, (b) the wavelength Λ of the ultrasonic wave, (c) the wavelength λ of the incident light, (d) the amplitude of the ultrasonic waves, (e) the width d of the ultrasonic beam.

Naturally the positions of the various orders on the screen, and their number and relative intensities, depend on one or more of these factors.‡ Fig. 12.3, p. 608, shows

* P. DEBYE and F. W. SEARS, *Proc. Nat. Acad. Sci., Wash.*, **18** (1932), 409.

† R. LUCAS and P. BIQUARD, *J. Phys. Radium*, **3** (1932), 464.

‡ For a semi-quantitative discussion of the dependence of the diffracted spectrum on these factors, see G. W. WILLARD, *J. Acoust. Soc. Amer.*, **21** (1949), 101.

in a typical case the number of orders appearing on either side of the transmitted beam at different angles of incidence θ. The usual experimental arrangement for studying diffraction spectrum is schematically drawn in Fig. 12.2.

It will be convenient at this stage to define some of the symbols and sign conventions used in this chapter.

The number density of molecules (atoms) of a medium will be denoted by $N(\boldsymbol{r},t)$. For an isotropic homogeneous medium traversed by a plane compression wave propagated in the positive x direction, $N(\boldsymbol{r},t)$ may be written in the form*

$$N(\boldsymbol{r},t) = N_0[1 + \Delta \cos (Kx - \Omega t)], \tag{3}$$

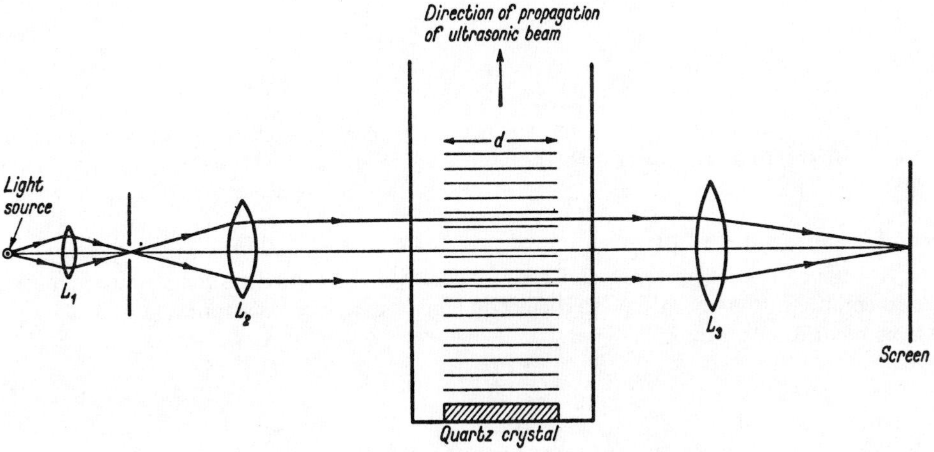

Fig. 12.2. Experimental arrangement for observing diffraction of light by ultrasonic waves.

where N_0 is the average number density of the medium, $N_0\Delta$ (usually of the order of $10^{-4}N_0$) the amplitude of the compression wave, $K = 2\pi/\Lambda$ its wave number (the magnitude of the wave-vector) and $\Omega = Kv$ the angular frequency of the ultrasonic disturbance. In such a medium the dielectric constant ε will also be a function of the space and time coordinates; this dependence may be assumed to be of the form

$$\varepsilon = \varepsilon_0 + \varepsilon_1 \cos (Kx - \Omega t). \tag{4}$$

There is, of course, a relation between ε_1 and Δ which we shall write in the form

$$\varepsilon_1 = \gamma\Delta. \tag{5}$$

If we assume the LORENTZ–LORENZ law (cf. § 2.3 (17))

$$\frac{\varepsilon - 1}{\varepsilon + 2} \cdot \frac{1}{N} = \text{constant}, \tag{6}$$

* For simplicity, we shall be concerned only with plane progressive ultrasonic waves. The diffraction of light by standing waves has also been studied experimentally (cf. BERGMANN, *loc. cit.*); the corresponding generalization of the theory is straightforward.

It should be mentioned that completely plane ultrasonic waves are hard to produce experimentally; in general, however, it is possible to regard the wave-fronts to be plane over regions of linear dimensions much greater than Λ.

differentiate the logarithm of (6) and remember that both Δ and $\varepsilon_1/\varepsilon_0$ are very much less than unity, we obtain

$$\varepsilon_1\left[\frac{1}{\varepsilon_0-1}-\frac{1}{\varepsilon_0+2}\right]-\frac{N_0\Delta}{N_0}=0,$$

or

$$\gamma=\tfrac{1}{3}(\varepsilon_0-1)(\varepsilon_0+2); \tag{7}$$

γ is of the order of magnitude of unity for most liquids.

We also set

$$\left.\begin{aligned} n&=\sqrt{\varepsilon_0}, & k&=\frac{2\pi}{\lambda}, & \bar{k}&=nk,\\ \delta&=\frac{\Delta\Lambda^2}{\lambda^2}, & \xi&=\frac{\Lambda}{\lambda}\sin\theta, & \beta&=\frac{\pi\lambda}{n\Lambda^2}. \end{aligned}\right\} \tag{8}$$

Finally the angles θ, ϕ, etc., will be measured clockwise from the positive y direction to the direction along which light advances (cf. Fig. 12.1). It may be assumed that $0 \leqslant \theta < \pi/2$.

12.1.2 Summary of theories based on Maxwell's equations

In regions free of currents and charges MAXWELL's equations for a non-magnetic, non-conducting medium, whose dielectric constant ε may be a function of space and time coordinates, are

$$\operatorname{curl}\boldsymbol{E}=-\frac{1}{c}\frac{\partial \boldsymbol{H}}{\partial t}, \qquad \operatorname{curl}\boldsymbol{H}=\frac{1}{c}\frac{\partial \boldsymbol{D}}{\partial t}, \tag{9a}$$

$$\operatorname{div}\boldsymbol{H}=0, \qquad \operatorname{div}\boldsymbol{D}=0. \tag{9b}$$

Eliminating $\boldsymbol{H}$ from (9a), and making use of the relations $\boldsymbol{D}=\varepsilon\boldsymbol{E}$, $\operatorname{div}\boldsymbol{D}=0$ and curl curl $\equiv -\nabla^2 +$ grad div, we find that

$$\frac{1}{c^2}\frac{\partial^2}{\partial t^2}(\varepsilon\boldsymbol{E})=\nabla^2\boldsymbol{E}+\operatorname{grad}(\boldsymbol{E}\,.\,\operatorname{grad}\log\varepsilon). \tag{10}$$

If we now use (4) and consider $\boldsymbol{E}$ as a superposition of plane waves of wavelengths $\bar{\lambda}\sim\lambda/n$, we find that the second term on the right-hand side of (10) is of the order $\varepsilon_1(\lambda/\Lambda)$ times the first term. Since under the usual experimental conditions both ε_1 and λ/Λ are very much less than unity, we may neglect this term in (10), and obtain

$$\frac{1}{c^2}\frac{\partial^2}{\partial t^2}(\varepsilon\boldsymbol{E})=\nabla^2\boldsymbol{E}. \tag{11}$$

Now consider the physical situation illustrated in Fig. 12.1, and let the incident monochromatic plane electromagnetic wave be linearly polarized with its electric vector perpendicular to the plane of incidence (E-polarization), i.e. along the z axis. It then follows from the preceding discussion that inside the medium the components E_x and E_y of $\boldsymbol{E}$ will be small quantities of order $\varepsilon_1(\lambda/\Lambda)$ times E_z and may, therefore, be neglected. Consequently E_x and E_y outside the medium are also negligible.

From the symmetry of the problem it is clear that E_z will be independent of the z-coordinate; hence using (4) and (11), we obtain for E_z the equation

$$\frac{\partial^2 E_z}{\partial x^2}+\frac{\partial^2 E_z}{\partial y^2}-\frac{1}{c^2}\frac{\partial^2}{\partial t^2}\{[\varepsilon_0+\tfrac{1}{2}\varepsilon_1(e^{i(Kx-\Omega t)}+e^{-i(Kx-\Omega t)})]E_z\}=0. \tag{12}$$

In order to solve (12) assume E_z to be of the form

$$E_z = \sum_l V_l(y)\, e^{i[(k\sin\theta + lK)x - (\omega + l\Omega)t]}, \tag{13}$$

where the summation is over all integral (positive, negative, and zero) values of l.

By substituting (13) into (12) and equating the coefficient of each exponential to zero, we obtain the following recurrence relations for $V_l(y)$

$$\begin{aligned} V_l''(y) + \{\varepsilon_0 c^{-2}(\omega + l\Omega)^2 - (k\sin\theta + lK)^2\}V_l(y) \\ = -\tfrac{1}{2}\varepsilon_1 c^{-2}(\omega + l\Omega)^2\{V_{l-1}(y) + V_{l+1}(y)\} \qquad (l = 0, \pm 1, \pm 2, \ldots), \end{aligned} \tag{14}$$

where a prime on $V_l(y)$ denotes differentiation with respect to y. These equations have to be solved subject to the boundary conditions*

$$\left.\begin{aligned} V_0(0) &= B, \quad \text{the amplitude of the incident light wave} \\ \text{and} \quad V_l(0) &= 0 \quad \text{for all } l \neq 0. \end{aligned}\right\} \tag{15}$$

Before solving (14) we observe that (13) represents a superposition of waves of frequencies $\omega_l = \omega + l\Omega$ $(l = 0, \pm 1, \pm 2, \ldots)$. Moreover, the x-component of the wave-vector for the wave of frequency ω_l is $k \sin \theta + lK$. Therefore the sine of the angle ϕ_l which the wave of frequency ω_l makes with the y axis beyond the scattering medium is given by

$$\begin{aligned} \sin\phi_l &= \frac{c(k\sin\theta + lK)}{\omega + l\Omega} \\ &\sim \sin\theta + l\frac{\lambda}{\Lambda}, \qquad \text{since } \frac{\Omega}{\omega} \ll 1, \end{aligned} \tag{16}$$

in agreement with (2). Further the intensity of a particular order l may be taken to be $|V_l(d)|^2$.

We first solve (14) by assuming that $|V_0| \gg |V_{\pm 1}| \gg |V_{\pm 2}| \ldots$ Remembering that one has to consider only those solutions of (14) which correspond to light waves travelling in directions along which y increases, we may put all V_l, except V_0, equal to zero in (14), make use of (15), and have in the first approximation

$$V_0^{(0)}(y) = B\, e^{i\sqrt{(\varepsilon_0 - \sin^2\theta)}ky} + O(\delta^2). \tag{17}$$

Similarly, by putting all V_l, except $V_{\pm 1}$ and V_0, equal to zero in (14), we obtain after a straightforward calculation

$$V_{\pm 1}(y) = \tfrac{1}{4}\gamma\delta\, \frac{1 - e^{-2i\beta(\xi \pm \frac{1}{2})y}}{\xi \pm \frac{1}{2}}\, V_0^{(0)}(y). \tag{18}$$

Here use has also been made of (5) and (8). If we now substitute (18) into the equations (14) for $l = 0$ and for $l = \pm 2$, we obtain a correction term to $V_0^{(0)}(y)$ and expressions for $V_{\pm 2}(y)$; both are proportional to δ^2 in this approximation. In this manner we obtain expressions for the intensities of any order l in the form of a series in ascending powers of δ. Solutions in terms of such power series were first derived by Brillouin,† who used somewhat more intricate analysis than outlined

---

* These boundary conditions are, of course, correct only if the intensities of the reflected waves are negligible; this is so in the present problem since the angle of incidence θ, for which the amplitudes of the diffracted waves are appreciable, is at most about 3° (see also § 12.2.4).

† L. Brillouin, *La Diffraction de la Lumière par des Ultrasons* (Paris, Hermann, 1933).

here, while DAVID,* following the above procedure, gave explicit expressions for the intensities of the first and second order lines on the assumption that the intensities of the higher orders are negligible. The latter formulae are quoted in § 12.2.5 [eq. (38)].

BRILLOUIN's approximation (and also DAVID's) is convenient if either $\delta \ll 1$ or $\delta/\xi \ll 1$, for then the method of successive approximations and the power series converge rapidly. When these conditions are satisfied, it is clear that only the first few orders will have appreciable intensity.

The explanation of the simultaneous appearance of many orders and approximate expressions for their intensities were first given by RAMAN and NATH.† They solved equations (14) in the following manner. Setting in (14)

$$V_l(y) = e^{\frac{1}{2}il\pi} \times e^{i\bar{k}(\cos\bar{\theta})y} \times U_l(y), \tag{19}$$

and remembering that $\Omega/\omega \sim 10^{-5}$ or less, $\bar{k} = nk$ and $k \sin\theta = \bar{k} \sin\bar{\theta}$, we obtain the following recurrence relations for $U_l(y)$

$$\begin{aligned} U_l''(y) + 2i\bar{k}\cos\bar{\theta}\, U_l'(y) - (2l\bar{k}K\sin\bar{\theta} + l^2K^2)U_l(y) \\ + \tfrac{1}{2}i(\bar{k})^2\varepsilon_1\varepsilon_0^{-1}(U_{l+1}(y) - U_{l-1}(y)) = 0 \qquad (l = 0, \pm 1, \pm 2, \ldots). \end{aligned} \tag{20}$$

Introducing a new variable

$$\chi = \tfrac{1}{2}y\bar{k}\varepsilon_1\varepsilon_0^{-1}\sec\bar{\theta},$$

equations (20) become

$$\begin{aligned} 2U_l'(\chi) + U_{l+1}(\chi) - U_{l-1}(\chi) = (\tfrac{1}{2}i\varepsilon_0^{-1}\varepsilon_1)\sec^2\bar{\theta}\, U_l''(\chi) \\ - 2i\varepsilon_0\varepsilon_1^{-1}\{2lK(\bar{k})^{-1}\sin\bar{\theta} + l^2K^2(\bar{k})^{-2}\}U_l(\chi), \end{aligned} \tag{21}$$

where now the prime on U denotes differentiation with respect to χ. Since $\lambda/d \ll 1$, the first term on the right is usually of the order $\varepsilon_1 U$, i.e. $10^{-4}U$, and may be neglected. Moreover if, following RAMAN and NATH, we put the second term also equal to zero, the resulting set of equations are the recurrence relations‡ satisfied by BESSEL functions of integral order. On making use of the boundary conditions (15), we have for the intensity of the lth order wave the expression $B^2J_l^2(\frac{1}{2}\varepsilon_1\varepsilon_0^{-1}\bar{k}d\sec\bar{\theta})$.

It will be noticed that the approximation made by RAMAN and NATH essentially consists in neglecting l^2/δ and $l\xi/\delta$ for all l. Therefore, if δ is sufficiently large compared to unity, this approximation will be a good one for the intensities of the lower orders. The BESSEL function expressions, however, overestimate the intensities of higher orders. This has been shown by EXTERMANN and WANNIER§ by numerical calculation of the intensities for three values of the parameter δ. In the work of these authors, the solution of (12) is ultimately determined by equations essentially similar to (18) and (19) of the next section.

---

* E. DAVID, *Phys. Z.*, **38** (1937), 587.

† C. V. RAMAN and N. S. N. NATH, *Proc. Ind. Acad. Sci.* A, **2** (1935), 406, 413; *ibid.*, **3** (1936), 75, 119.

It may be mentioned that RAMAN and NATH in their first two papers (*loc. cit.*), following the work of RAYLEIGH (*Proc. Roy. Soc.*, A, **79** (1907), 399) on phase gratings, found that the amplitudes of the diffracted waves can be represented by BESSEL functions. This method considers only the changes in phase of the plane light wave as it traverses the ultrasonic beam.

‡ See, for example, E. T. WHITTAKER and G. N. WATSON, *A Course of Modern Analysis* (Cambridge University Press, 4th ed., 1946), p. 360.

§ R. EXTERMANN and G. WANNIER, *Helv. Phys. Acta*, **9** (1936), 520.

Finally we mention that NATH* and AGGARWAL† have obtained solutions of (21) in the form of power series in ascending powers of $1/\delta$; these series appear to have rather limited applicability as they converge very slowly. Another treatment based on MAXWELL's equations in which the diffraction is treated as a boundary value problem was given by WAGNER.‡

12.2 DIFFRACTION OF LIGHT BY ULTRASONIC WAVES AS TREATED BY THE INTEGRAL EQUATION METHOD

It was pointed out in § 2.4 that the integral equations § 2.4 (4) for the effective electric field $\boldsymbol{E}'(\boldsymbol{r},t)$ and the accompanying formula § 2.4 (5) for $\boldsymbol{H}'$ are equivalent to MAXWELL's equations for isotropic non-magnetic substances. It was assumed there that the density of the medium is independent of time but an extension to cover the more general case of time dependence may easily be made. As before, we shall assume the medium to be non-magnetic and non-conducting.

We recall that the main content of the integral equation method is that the influence of matter on the propagation of an electromagnetic wave is equivalent to the effect of electric dipoles embedded in vacuum, the dipole moment induced in any physically infinitesimal volume element $d\boldsymbol{r}'$ of linear dimensions much smaller than λ§ being proportional to the field $\boldsymbol{E}'(\boldsymbol{r}',t)$ acting on it and to the number of molecules (atoms) in that volume. Associated with such a dipole at $\boldsymbol{r}'$ is the Hertzian vector

$$\boldsymbol{\Pi}_e = \alpha N\left(\boldsymbol{r}', t - \frac{R}{c}\right)\frac{\boldsymbol{E}'(t - R/c, \boldsymbol{r}')}{R}\, d\boldsymbol{r}',$$

from which the field at a point $\boldsymbol{r}$ and at time t may be derived by operating on it with (cf. § 2.2 (43))

$$-\frac{1}{c^2}\frac{\partial^2}{\partial t^2} + \text{grad div}.$$

Here the various symbols have the same significance as in § 2.4; thus $R = |\boldsymbol{r} - \boldsymbol{r}'|$ and the operator grad div acts on the variables $\boldsymbol{r}(x, y, z)$. Then by an argument similar to that which led to the equation § 2.4 (4), one now obtains the following integral equation for $\boldsymbol{E}'$ within the medium ||

$$\boldsymbol{E}'(\boldsymbol{r}, t) = \boldsymbol{E}^{(i)}(\boldsymbol{r}, t) + \alpha\iiint\left\{-\frac{1}{c^2}\frac{\partial^2}{\partial t^2} + \text{grad div}\right\}\left\{N(\boldsymbol{r}', t - R/c)\,\frac{\boldsymbol{E}'(\boldsymbol{r}', t - R/c)}{R}\right\} d\boldsymbol{r}'. \quad (1)$$

As in § 2.4 (4), the integration extends throughout the whole medium, except for a small domain occupied by the atom at the point of observation $\boldsymbol{r}(x, y, z)$.

This is the basic integral equation of the present theory. When it is solved for $\boldsymbol{E}'$ at all points inside the medium the field outside the medium is calculated by adding together the incident field $\boldsymbol{E}^{(i)}(\boldsymbol{r}, t)$ and the dipole field $\boldsymbol{E}^{(d)}(\boldsymbol{r}, t)$ given by the integral

* N. S. N. NATH, *Proc. Ind. Acad. Sci.*, A, **4** (1936), 222; *ibid.*, A, **8** (1938), 499.

† R. R. AGGARWAL, Ph.D. Thesis, Delhi University, India (1954).

‡ E. H. WAGNER, *Z. f. Phys.*, **141** (1955), 604, 622.

§ Such volume elements, namely elements much larger than the volume of an individual atom (molecule) but of linear dimensions small compared to λ, can be nearly always chosen for optical wavelengths.

|| Equation (1) differs from § 2.4 (4) in that the operator curl curl in the latter has been replaced by $\left(-\frac{1}{c^2}\frac{\partial^2}{\partial t^2} + \text{grad div}\right)$ and that the density N which now depends on space and time coordinates has also to be taken—just like $\boldsymbol{E}'$—at the retarded times $t - R/c$.

in (1) but extending now throughout the whole medium. It will be noted that this treatment of the propagation of light through media, in contrast to the usual method in which one sets up MAXWELL's equations for the medium and the vacuum, avoids explicit introduction of boundary conditions at the refracting surfaces but instead brings in the dimensions of the medium through the process of integration throughout the medium. Moreover, whereas in the MAXWELL equations the variation in density of the medium is taken into account through its influence on the dielectric constant ε, in the integral equations (1) the density function $N(\mathbf{r}, t)$ occurs explicitly.

Equation (1) is valid only under certain restrictive conditions. First, the polarizability α per molecule in general depends on the frequency of $\mathbf{E}'$, so that $\mathbf{E}'$ should be strictly monochromatic. However, unless one is too near the dispersion frequencies, the variation of α with the frequency of the external field is small. Hence, provided that all the component frequencies of $\mathbf{E}'$ lie close to each other, we may still use (1). Even when the incident field $\mathbf{E}^{(i)}$ is strictly monochromatic, the field $\mathbf{E}'$ acting on a molecule and producing a dipole will not necessarily be monochromatic when $N(\mathbf{r}, t)$ depends on time, on account of thermal agitation or some other cause of disorder; the spread of frequencies depends on the time variations of $N(\mathbf{r}, t)$. Hence only when the variation in time of N is slow in comparison with that of $\mathbf{E}^{(i)}$, may (1) be used with confidence*; fortunately this restriction is not severe, since in problems of scattering and diffraction of light the required condition is almost always satisfied.

Further, we have taken α to be a scalar, an assumption fully justified for atoms and for molecules having special symmetries, but holding also more generally whenever the molecules are oriented at random as already mentioned in § 2.3. Finally, the absorption of light by the medium has been assumed to be negligible here; it could be taken into account by allowing α to be complex.

We shall now apply the integral equation method to the problem of diffraction of light by a fluid traversed by ultrasonic waves, following the analysis of BHATIA and NOBLE.†

12.2.1 Integral equation for *E*-polarization

Let us again consider the physical situation described in § 12.1.1, and assume the incident light to be linearly polarized with its electric vector perpendicular to the plane of incidence; then the components of the electric vector $\mathbf{E}^{(i)}(\mathbf{r},t)$ of the incident light wave are (the real part, as usual, representing the physical quantity),

$$\left.\begin{aligned} E_x{}^{(i)} &= E_y{}^{(i)} = 0, \\ E_z{}^{(i)} &= B\, e^{i(kx \sin\theta + ky\cos\theta - \omega t)}, \end{aligned}\right\} \tag{2}$$

From the arguments given on p. 594, it may be concluded that the effective field $\mathbf{E}'(\mathbf{r},t)$ inside the medium will then be nearly parallel to the z axis, so that we may assume $E'_x = E'_y = 0$, and the vector integral equation (1) for $\mathbf{E}'$ reduces to a single integral equation for E'_z. Remembering that $N(\mathbf{r},t)$ is now given by § 12.1 (3), we may write the integral equation for E'_z in the form

$$\begin{aligned} E'_z(\mathbf{r}, t) = {} & Be^{i(kx\sin\theta + ky\cos\theta - \omega t)} + \frac{\tau_0}{4\pi}\iiint\left(-\frac{1}{c^2}\frac{\partial^2}{\partial t^2} + \frac{\partial^2}{\partial z^2}\right) \\ & \times \left[\frac{1}{R}\{1 + \tfrac{1}{2}\Delta(e^{i[Kx' - \Omega(t - R/c)]} + e^{-i[Kx' - \Omega(t-R/c)]})\} \times E'_z(\mathbf{r}', t - R/c)\right] d\mathbf{r}', \end{aligned} \tag{3}$$

* The use of the relation $\mathbf{D} = \varepsilon\mathbf{E}$ in MAXWELL's equations for heterogeneous media is in fact also justified only under similar restrictions on N.

† W. J. NOBLE, Ph.D. Thesis, University of Edinburgh (1952); A. B. BHATIA and W. J NOBLE, *Proc. Roy. Soc.*, A, **220** (1953), 356, 369.

where for convenience we have written*

$$4\pi N_0\alpha = \tau_0.$$

From § 2.3 (17)

$$\tau_0 = \frac{3(n^2-1)}{n^2+2}. \tag{4}$$

12.2.2 The trial solution of the integral equation

Since all planes perpendicular to the z axis are physically equivalent, we take as a trial solution of our integral equation (3) an expression of the form†

$$E'_z = \sum_{l,m} N_{lm}\, e^{-i(\omega_{lm}t - p_l x - q_m y)}, \tag{5}$$

where l and m are integers (positive, negative, and zero). (5) is seen to represent a doubly infinite sheaf of plane waves; this form of a possible solution is suggested by the multiple reflections and refractions to be expected in an infinite slab of stratified medium with parallel plane faces. It will be seen presently that the various unknowns N_{lm}, ω_{lm}, p_l, and q_m may be determined from the condition that (5) satisfies the integral equation (3).

To solve (3), integrals $\mathscr{J}(\omega, p, q)$ must be evaluated, defined by

$$\mathscr{J}(\omega, p, q) \times e^{\{-i(\omega t - px - qy)\}}$$
$$= \frac{1}{4\pi}\iiint_{V'}\left[-\frac{1}{c^2}\frac{\partial^2}{\partial t^2} + \frac{\partial^2}{\partial z^2}\right]\frac{e^{-i[\omega(t-R/c)-px'-qy']}}{R}\,dx'dy'dz'$$
$$= \frac{1}{4\pi}\iiint_{V'} e^{-i(\omega t - px' - qy')}\left\{\frac{\omega^2}{c^2} + \frac{\partial^2}{\partial z^2}\right\}\frac{e^{i\omega R/c}}{R}\,dx'dy'dz', \tag{6}$$

where $\omega^2 > c^2p^2$. If the point of observation x, y, z is outside the scattering medium, the volume V' extends throughout the medium ($-\infty < x' < \infty$, $0 \leqslant y' \leqslant d$, $-\infty < z' < \infty$). If the point of observation is within the scattering medium, the integral extends over the same volume, except for a small sphere of radius a (which is eventually taken to the limit $a \to 0$) around the point of observation. Setting

$$x_1 = x' - x, \qquad y_1 = y' - y, \qquad z_1 = z' - z,$$

(6) may be written in the form, after cancelling the factor $e^{-i\omega t}$ on both sides,

$$\mathscr{J}(\omega, p, q) = \mathscr{J}_1 + \mathscr{J}_2, \tag{7}$$

where

$$\mathscr{J}_1 = \frac{1}{4\pi}\iiint_{V_1}\left\{e^{i(px_1+qy_1)}\frac{\partial^2}{\partial z_1^{\,2}}\left(\frac{e^{i\omega R/c}}{R}\right)\right\}dx_1dy_1dz_1, \tag{8}$$

$$\mathscr{J}_2 = \frac{1}{4\pi}\frac{\omega^2}{c^2}\iiint_{V_1}\left\{e^{i(px_1+qy_1)}\left(\frac{e^{i\omega R/c}}{R}\right)\right\}dx_1dy_1dz_1, \tag{9}$$

the scattering medium now extending over the volume V_1: $-\infty < x_1 < \infty$ $-y \leqslant y_1 \leqslant d - y$, $-\infty < z_1 < \infty$. These integrals are evaluated in Appendix IX and give

---

* The macroscopic quantity $\tau(\mathbf{r}, t) = 4\pi\alpha N(\mathbf{r}, t)$ is sometimes called the *scattering index* of the medium.

† No confusion should arise between the amplitudes N_{lm} and the number density $N(\mathbf{r},t)$ of the molecules since from here onwards the latter quantity does not appear explicitly in our equations.

(a) when (x, y, z) lies inside the scattering medium*

$$\mathscr{J}(\omega, p, q) = \frac{1}{\sigma(\omega, p, q)} - \frac{\omega^2}{2c^2} \frac{\exp\{-ig(\omega, p, q)y\}}{g(\omega, p, q)[\omega^2 c^{-2} - p^2]^{1/2}} + \frac{\omega^2}{2c^2} \frac{\exp\{-ih(\omega, p, q)(y-d)\}}{h(\omega, p, q)[\omega^2 c^{-2} - p^2]^{1/2}}; \quad (10a)$$

(b) when (x, y, z) lies beyond the scattering medium,

$$\mathscr{J}(\omega, p, q) = \frac{\omega^2 \exp[-ig(\omega, p, q)y]}{2c^2 g(\omega, p, q)[\omega^2 c^{-2} - p^2]^{1/2}} (\exp[ig(\omega, p, q)d] - 1); \quad (10b)$$

(c) when (x, y, z) lies in front of the scattering medium (i.e. on the same side as the incident light),

$$\mathscr{J}(\omega, p, q) = \frac{\omega^2 \exp[-ih(\omega, p, q)y]}{2c^2 h(\omega, p, q)[\omega^2 c^{-2} - p^2]^{1/2}} (\exp[ih(\omega, p, q)d] - 1). \quad (10c)$$

In these expressions,

$$\left.\begin{aligned} \sigma(\omega, p, q) &= 3(p^2 + q^2 - \omega^2 c^{-2})(p^2 + q^2 + 2\omega^2 c^{-2})^{-1}, \\ g(\omega, p, q) &= q - [\omega^2 c^{-2} - p^2]^{1/2}, \\ h(\omega, p, q) &= q + [\omega^2 c^{-2} - p^2]^{1/2}. \end{aligned}\right\} \quad (11)$$

If we now substitute (5) into (3) and make use of the relations (6) and (10a), we obtain in a straightforward manner

$$\begin{aligned} &-\sum_{l,m} N_{lm} \exp[(-i)(\omega_{lm}t - p_l x - q_m y)] + B \exp[(-i)(\omega t - kx \sin\theta - ky\cos\theta)] \\ &+ \tau_0 \sum_{l,m} N_{lm} \Bigg[\frac{\exp[(-i)(\omega_{lm}t - p_l x - q_m y)]}{\sigma(\omega_{lm}, p_l, q_m)} \\ &+ \tfrac{1}{2}\Delta \sum_{+,-} \frac{\exp[(-i)\{(\omega_{lm} \pm \Omega)t - (p_l \pm K)x - q_m y\}]}{\sigma(\omega_{lm} \pm \Omega, p_l \pm K, q_m)} \\ &- \frac{\omega_{lm}^2 \exp[(-i)\{\omega_{lm}t - p_l x - [\omega_{lm}^2 c^{-2} - p_l^2]^{1/2} y\}]}{2c^2 g(\omega_{lm}, p_l, q_m)[\omega_{lm}^2 c^{-2} - p_l^2]^{1/2}} \\ &- \tfrac{1}{2}\Delta \sum_{+,-} \frac{(\omega_{lm} \pm \Omega)^2 \exp[(-i)\{(\omega_{lm} \pm \Omega)t - (p_l \pm K)x - [c^{-2}(\omega_{lm} \pm \Omega)^2 - (p_l \pm K)^2]^{1/2} y\}]}{2c^2 g(\omega_{lm} \pm \Omega, p_l \pm K, q_m)[c^{-2}(\omega_{lm} \pm \Omega)^2 - (p_l \pm K)^2]^{1/2}} \\ &+ \frac{\omega_{lm}^2 \exp[(-i)\{\omega_{lm}t - p_l x + [\omega_{lm}^2 c^{-2} - p_l^2]^{1/2} y\}] \times \exp[ih(\omega_{lm}, p_l, q_m)d]}{2c^2 h(\omega_l, p_l, q_m)[\omega_{lm}^2 c^{-2} - p_l^2]^{1/2}} \\ &+ \tfrac{1}{2}\Delta \sum_{+,-} \frac{(\omega_{lm} \pm \Omega)^2 \exp[(-i)\{(\omega_{lm} \pm \Omega)t - (p_l \pm K)x + [c^{-2}(\omega_{lm} \pm \Omega)^2 - (p_l \pm K)^2]^{1/2} y\}]}{2c^2 h(\omega_{lm} \pm \Omega, p_l \pm K, q_m)} \\ &\qquad \times \frac{\exp[ih(\omega_{lm} \pm \Omega, p_l \pm K, q_m)d]}{[c^{-2}(\omega_{lm} \pm \Omega)^2 - (p_l \pm K)^2]^{1/2}} \Bigg] \equiv 0, \end{aligned} \quad (12)$$

* Throughout § 12.2 the positive square root of an expression is to be taken, unless otherwise stated.

where the summation symbol $\sum_{+,-}$ in front of any expression is to be interpreted as follows:

$$\sum_{+,-} F(a \pm b, c \pm d) \equiv F(a + b, c + d) + F(a - b, c - d).$$

In order that (12) shall be satisfied at all times and at all points within the scattering medium, the coefficient of each exponential differing from all the others in any of the variables (x, y, t) must separately vanish. We see from (12) that ω_{lm} changes in steps of Ω and is always accompanied by a change of p_l in steps of K. The coefficients of y in the various exponentials, however, either remain unchanged (q_m) or are always the same function of the corresponding ω's and p's. Hence we may take ω_{lm} to depend only on the index l. Moreover, since we may assume without loss of generality that ω_0 is the frequency ω of the incident light, we have

$$\left.\begin{aligned} \omega_0 &= \omega, & \omega_l &= \omega + l\Omega, \\ p_0 &= k \sin\theta, & p_l &= k\sin\theta + lK \end{aligned}\right\} \quad (l = 0, \pm 1, \pm 2, \ldots). \qquad \begin{aligned}(13a)\\(13b)\end{aligned}$$

Using these relations in (12) and regrouping the various terms, (12) becomes

$$\begin{aligned} &\sum_{l,m} [N_{lm}(\tau_0/\sigma_{lm} - 1) + \tfrac{1}{2}\Delta(\tau_0/\sigma_{lm})(N_{l-1,\,m} + N_{l+1,\,m})] \times \exp[(-i)(\omega_l t - p_l x - q_m y)] \\ &\quad + \sum_l (B\delta_{l,0} - G_l)\exp[(-i)(\omega_l t - p_l x - [\omega_l^2 c^{-2} - p_l^2]^{1/2} y)] \\ &\quad + \sum_l H_l \exp[(-i)(\omega_l t - p_l x + [\omega_l^2 c^{-2} - p_l^2)^{1/2} y)] \equiv 0, \end{aligned} \qquad (14)$$

where $\delta_{l,l'}$ is the Kronecker delta symbol* (i.e. $\delta_{l,l'} = 0$ when $l \neq l'$, and $\delta_{ll} = 1$) and G_l and H_l are given by

$$G_l = \tau_0\omega_l^2 \sum_m [N_{lm} + \tfrac{1}{2}\Delta(N_{l-1,\,m} + N_{l+1,\,m})] \times \{2c^2 g_{lm}[\omega_l^2 c^{-2} - p_l^2]^{1/2}\}^{-1}, \qquad (15)$$

$$H_l = \tau_0\omega_l^2 \sum_m [N_{lm} + \tfrac{1}{2}\Delta(N_{l-1,\,m} + N_{l+1,\,m})][\exp(ih_{lm}d)]\{2c^2 h_{lm}[\omega_l^2 c^{-2} - p_l^2]^{1/2}\}^{-1}. \qquad (16)$$

Here we have also used the abbreviation (cf. (11))

$$\sigma_{lm} = \sigma(\omega_l, p_l, q_m), \qquad g_{lm} = g(\omega_l, p_l, q_m) \quad \text{and} \quad h_{lm} = h(\omega_l, p_l, q_m). \qquad (17)$$

Equating to zero the coefficient of each exponential in (14), we obtain the following sets of equations for the permissible values of q_m and the amplitudes N_{lm}:

$$N_{lm}(1 - \sigma_{lm}/\tau_0) + \tfrac{1}{2}\Delta(N_{l-1,\,m} + N_{l+1,\,m}) = 0, \qquad \text{for all } l \text{ and } m, \qquad (18)$$

$$\left.\begin{aligned} B\delta_{l,0} - G_l &= 0, \\ H_l &= 0 \end{aligned}\right\} \quad \text{for all } l. \qquad \begin{aligned}(19)\\(20)\end{aligned}$$

12.2.3 Expressions for the amplitudes of the light waves in the diffracted and reflected spectra

Before discussing the solution of the equations (18)–(20), we write down the expressions for the total light disturbance at a point (x, y, z) beyond the scattering medium. For this purpose one has to substitute (5) into the integrand on the right-hand side of (3), integrate over the scattering medium, and add to this result the incident field.

---

* No confusion should arise between the Kronecker symbol $\delta_{l,l'}$, and the parameter δ introduced in § 12.1 (8) since the former always occurs with subscripts.

One again encounters the integrals $\mathscr{J}(\omega, p, q)$ discussed in the preceding section; remembering that for a point beyond the scattering medium $\mathscr{J}(\omega, p, q)$ is given by (10b) and making use of (18), we obtain the following expression for the only non-vanishing component of the total transmitted electric field:

$$E_z = \sum_l B_l \exp[(-i)\{\omega_l t - p_l x - [\omega_l^2 c^{-2} - p_l^2]^{1/2} y\}], \tag{21}$$

where

$$B_l = \omega_l^2 \sum_m \sigma_{lm} N_{lm} \{\exp[i g_{lm} d]\} \times \{2c^2 g_{lm} [\omega_l^2 c^{-2} - p_l^2]^{1/2}\}^{-1}. \tag{22}$$

According to (21) and (22) the transmitted wave may be regarded as consisting of many plane waves, each with a different frequency and a different direction of propagation. By substituting (13) in the exponential of (21), the expressions for the frequencies ω_l and the angles ϕ_l may be easily obtained; these are the same as those given in § 12.1.2.

With the help of (10c) one may similarly write down the expressions for the amplitudes $B_l^{(r)}$ in the reflected spectrum also. $B_l^{(r)}$ are given by

$$B_l^{(r)} = -\omega_l^2 \sum_m \sigma_{lm} N_{lm} \{2c^2 h_{lm} [\omega_l^2 c^{-2} - p_l^2]^{1/2}\}^{-1}. \tag{23}$$

We shall, however, be concerned here only with points beyond the scattering medium, since in the present problem the intensities of the waves of the various orders in the reflected spectrum are generally very small.

12.2.4 Solution of the equations by a method of successive approximations

In § 12.2.4 and § 12.2.5, on the assumption that the amplitude Δ of the ultrasonic wave is small, equations (18)–(20) will be solved and approximate expressions for the intensities of the first and second order lines in the transmitted spectrum will be obtained. The case for which this approximation fails will be qualitatively discussed in § 12.2.6, and finally in § 12.2.7 equations (18)–(20) will be solved by an approximation essentially equivalent to that of Raman and Nath.

First consider the set of equations (18). The suffix m in these equations distinguishes quantities referring to different permissible values of q. Therefore, we may suppress this suffix and write (18) in the form

$$f_l(q^2) N_l(q^2) - \tfrac{1}{2}\Delta(N_{l-1}(q^2) + N_{l+1}(q^2)) = 0 \qquad (l = 0, \pm 1, \pm 2, \ldots) \tag{24}$$

with

$$f_l(q^2) = \frac{\sigma_l(q)}{\tau_0} - 1 = \frac{3(p_l^2 + q^2 - n^2\omega_l^2 c^{-2})}{(n^2 - 1)(p_l^2 + q^2 + 2\omega_l^2 c^{-2})}. \tag{25}$$

Equations (24) form an infinite set of linear homogenous equations for the amplitudes $N_l(q^2)$. The condition for the existence of a non-trivial solution, i.e. $N_l \not\equiv 0$ for all l, is that the determinant formed by the coefficients of N_l vanishes. The roots of this determinantal equation give the permissible values of q^2; let these be denoted by q_m^2 $(m = 0, \pm 1, \pm 2, \ldots)$. Corresponding to each such q^2 there are, of course, two values of q, viz. $+|q|$ and $-|q|$ and two sets of amplitudes, i.e. $N_l(+|q|) \equiv N_l^+(q^2)$ and $N_l(-|q|) \equiv N_l^-(q^2)$, $(l = 0, \pm 1, \ldots)$. Then for a given permissible value of q^2, say q_m^2, the recurrence relations (24) determine all the $N_l^+(q_m^2)$, $(l = 0, \pm 1, \ldots)$ in terms of one of them, say $N_m^+(q_m^2)$. In this way, one may obtain all the amplitudes $N_l^{\pm}(q^2)$ in terms of $N_m^{\pm}(q_m^2)$ $(m = 0, \pm 1, \pm 2, \ldots)$. These latter sets of amplitudes are to be determined from the equations (19) and (20), which are just enough in number for this purpose. (Note that in expressions (5), (12), (15), etc., the symbol $\sum_m$ implies sum over both $N^+_{\ldots,m}$ and $N^-_{\ldots,m}$ terms.)

We shall obtain here approximate solutions of (24) by using a perturbation method. Regarding Δ as a small parameter and following the usual perturbation procedure, we expand η $(= q^2)$ and N_l in powers of $\frac{1}{2}\Delta$:

$$N_l(\eta) = N_l^{(0)} + \tfrac{1}{2}\Delta N_l^{(1)} + (\tfrac{1}{2}\Delta)^2 N_l^{(2)} + \ldots, \tag{26a}$$

$$\eta = \eta^{(0)} + \tfrac{1}{2}\Delta\eta^{(1)} + (\tfrac{1}{2}\Delta)^2\eta^{(2)} + \ldots \tag{26b}$$

Making use of (26b), $f_l(\eta)$ may be written as

$$f_l(\eta) = f_l(\eta^{(0)}) + \tfrac{1}{2}\Delta\eta^{(1)}f_l'(\eta^{(0)}) + (\tfrac{1}{2}\Delta)^2[\eta^{(2)}f_l'(\eta^{(0)}) + \tfrac{1}{2}(\eta^{(1)})^2 f_l''(\eta^{(0)})] + \ldots, \tag{27}$$

where a prime on f denotes differentiation with respect to η. (Note that $f_l'(\eta)$, $f_l''(\eta)$ . . . are non-zero for every positive real value of η.) Substituting (26) and (27) in (24), we have $(l = 0, \pm 1, \pm 2, \ldots)$

$$\begin{aligned}&\{f_l(\eta^{(0)}) + \tfrac{1}{2}\Delta\eta^{(1)}f_l'(\eta^{(0)}) + (\tfrac{1}{2}\Delta)^2[\eta^{(2)}f_l'(\eta^{(0)}) + \tfrac{1}{2}(\eta^{(1)})^2 f_l''(\eta^{(0)})] + \ldots\}\\ &\times\{N_l^{(0)} + \tfrac{1}{2}\Delta N_l^{(1)} + (\tfrac{1}{2}\Delta)^2 N_l^{(2)} + \ldots\} - \tfrac{1}{2}\Delta[N_{l-1}^{(0)} + \tfrac{1}{2}\Delta N_{l-1}^{(1)}\\ &+ \ldots + N_{l+1}^{(0)} + \tfrac{1}{2}\Delta N_{l+1}^{(1)} + \ldots] = 0.\end{aligned} \tag{28}$$

Equating first to zero those terms in (28) which are independent of Δ, we obtain in the zero-order approximation

$$f_l(\eta^{(0)})N_l^{(0)} = 0 \qquad (l = 0, \pm 1, \pm 2, \ldots). \tag{29}$$

Equations (29) have as solutions,

$$\text{either}\quad f_l(\eta^{(0)}) = 0,\quad N_l^{(0)} \neq 0 \quad\text{or}\quad N_l^{(0)} = 0,\quad f_l(\eta^{(0)}) \neq 0. \tag{30}$$

Denoting by $\eta_l^{(0)}$ the value of $\eta^{(0)}$ given by $f_l(\eta^{(0)}) = 0$, we find that (30) gives

$$\eta_l^{(0)} = n^2\omega_l^2 c^{-2} - p_l^2, \qquad N_l^{(0)}(\eta_l^{(0)}) \equiv N_{ll} \neq 0, \tag{31}$$

and

$$N_{l'}^{(0)}(\eta_l^{(0)}) \equiv N_{l'l}^{(0)} = N_{ll}\delta_{l'l}. \tag{32}$$

Next equating to zero the coefficient of Δ in (28), we obtain

$$f_l(\eta_m^{(0)})N_l^{(1)}(\eta_m^{(0)}) + \eta_m^{(1)}f_l'(\eta_m^{(0)})N_l^{(0)}(\eta_m^{(0)}) = N_{l-1}^{(0)}(\eta_m^{(0)}) + N_{l+1}^{(0)}(\eta_m^{(0)}). \tag{33}$$

Putting $l = m, m+1, m-1, m+2, m-2, \ldots$ successively in (33) and making use of (32), we find

$$\eta_m^{(1)} = 0, \tag{34a}$$

$$N_{m\pm 1,m}^{(1)} = \frac{N_{mm}}{f_{m\pm 1}(\eta_m^{(0)})}, \tag{34b}$$

and

$$N_{m\pm j,m}^{(1)} = 0 \qquad \text{for } j \geqslant 2. \tag{34c}$$

Similarly, by equating to zero the coefficient of Δ^2 in (28), we obtain the following expressions for the correction to $\eta_m^{(0)}$ and for the amplitudes N_{lm} up to second order:*

---

* The formulae given here are valid only if $f_{m\pm 1}(\eta_m^{(0)})$, $f_{m\pm 2}(\eta_m^{(0)})$, . . . are all different from zero. If this is not so, the above perturbation procedure has to be modified to take this degeneracy into account. In perturbation calculations up to second order, the degeneracy comes into play when light is incident at angles $\theta = 0$ and $\theta = \sin^{-1}(\lambda/2\Lambda)$. However, we shall not pursue this matter further here, but will give the results for these two cases in § 12.2.5.

$$\eta_m^{(2)} = \frac{1}{f'_m(\eta_m^{(0)})}\left[\frac{1}{f_{m+1}(\eta_m^{(0)})} + \frac{1}{f_{m-1}(\eta_m^{(0)})}\right], \tag{35a}$$

$$N_{m\pm 2,m}^{(2)} = \frac{N_{mm}}{f_{m\pm 1}(\eta_m^{(0)})f_{m\pm 2}(\eta_m^{(0)})}, \tag{35b}$$

and

$$N_{m\pm 1,m}^{(2)} = 0, \qquad N_{m\pm j,m}^{(2)} = 0 \qquad \text{for } j \geqslant 3. \tag{35c}$$

It will be seen from the foregoing calculations that in the zero-order perturbation calculation only the quantities N_{mm} ($m = 0, \pm 1, \ldots$) are different from zero; in the first-order calculation N_{mm} and $N_{m\pm 1,m}$ differ from 0, while, in the second order, $N_{m\pm 2,m}$ are also different from zero. Likewise, by pursuing the perturbation calculation to still higher orders one obtains more and more non-diagonal amplitudes (i.e. amplitudes whose two suffixes are not equal) which are different from zero. These calculations are lengthy and cannot be given here; but it may be assumed that whenever the perturbation procedure is valid one may neglect the higher order terms.

The non-diagonal amplitudes (34b) and (35b) are completely determined once the diagonal amplitudes which form the zero-order solution of (24) are known. We determine the latter from (19) and (20). First, however, it will be instructive to examine their solutions in the simple case $\Delta = 0$, for which an exact solution of equations (18)–(20) is easily obtained. In this case the only possible non-zero amplitudes are the diagonal amplitudes $N^{\pm}_{m,m}$ ($m = 0, \pm 1, \ldots$). If we put all the non-diagonal amplitudes equal to zero in (19) and (20), we find that all the amplitudes $N^{\pm}_{m,m}$ are identically equal to zero except $N^{\pm}_{0,0}$, which are given by*

$$N^{+}_{0,0} = (2B/\sigma_{0,0}) \cos\theta[(n^2 - \sin^2\theta)^{1/2} - \cos\theta](1 + \rho^2 - 2\rho\cos 2q_0 d)^{-1/2} e^{i\psi},$$

$$N^{-}_{0,0} = N^{+}_{0,0}(g_{0,0}/h_{0,0})\, e^{2iq_0 d},$$

where

$$\rho = \left|\frac{N^{-}_{0,0}}{N^{+}_{0,0}}\right|^2, \qquad \psi = \tan^{-1}\left[\frac{\rho\sin 2q_0 d}{1 - \rho\cos 2q_0 d}\right]. \tag{36}$$

For normal incidence ($\theta = 0$), (11), (31), and (36) give

$$\rho_0 = \left|\frac{N^{-}_{0,0}}{N^{+}_{0,0}}\right|^2_{\theta=0} = \frac{(n-1)^2}{(n+1)^2}. \tag{37}$$

With the help of (36), (27), and (23), one easily obtains for the reflectivity at normal incidence on a plane parallel plate the expression

$$\left|\frac{B_0^{(r)}}{B}\right|^2 = \frac{4\rho_0 \sin^2 q_0 d}{1 + \rho_0^2 - 2\rho_0\cos 2q_0 d},$$

in agreement with § 7.6 (9), since $q_0 = nk$ for $\theta = 0$.

When Δ differs from zero but is still sufficiently small for the perturbation method to apply, the above solution for $\Delta = 0$ suggests that

$$|N_{0,0}| \gg |N_{\pm 1,\pm 1}| \gg |N_{\pm 2,\pm 2}| \ldots$$

We may, therefore, solve equations (19) and (20) for the diagonal amplitudes by successive approximations. Moreover, since for normal or near normal incidence

* There are some misprints in the expressions for $N^{+}_{0,0}$ and $N^{-}_{0,0}$ given by Bhatia and Noble (*loc. cit.*); these have been corrected here.

(θ is at most about 3° in the ultrasonic diffraction experiments) the ratio of a given N^- to the corresponding N^+ will be small (cf. (37)), we may neglect the N^- altogether and determine the N^+ from equations (19) alone.* It will be recalled that a similar approximation is also implicit in the use of the boundary conditions § 12.1 (15).

The expressions for the diagonal amplitudes $N^+_{0,0}$, $N^+_{\pm1,\pm1}$, and $N^+_{\pm2,\pm2}$ may now be obtained by using (34b), (34c), (35b), and (35c) in (19). Further, the expressions for the intensities of the first and second order lines in the transmitted spectrum can be easily written down with the help of (22). These expressions are given in the next section.

12.2.5 Expressions for the intensities of the first and second order lines for some special cases

(a) $\delta/\xi \ll 1$ *and* ξ *large compared to unity*

This is the case considered in detail in the previous section. The intensities $I_{\pm1}$ and $I_{\pm2}$ of the first and second order lines are respectively

$$I_{\pm1} = |B_{\pm1}|^2 = \tfrac{1}{4}B^2\gamma^2\delta^2 \frac{\sin^2[\beta d(\xi \pm \frac{1}{2})]}{(\xi \pm \frac{1}{2})^2}, \tag{38a}$$

and

$$I_{\pm2} = |B_{\pm2}|^2 = \tfrac{1}{64}B^2\gamma^4\delta^4 \left\{\frac{1}{2(\xi \pm 1)(\xi \pm \frac{1}{2})(\xi \pm \frac{3}{2})}\right\} \left[\frac{\sin^2[\beta d(\xi \pm \frac{1}{2})]}{(\xi \pm \frac{1}{2})} + \frac{\sin^2[\beta d(\xi \pm \frac{3}{2})]}{(\xi \pm \frac{3}{2})} - \frac{\sin^2[2\beta d(\xi \pm 1)]}{2(\xi \pm 1)}\right]. \tag{38b}$$

These equations are to be understood with either all upper signs or all lower signs.

We quote without proof the expressions for the intensities for two other cases considered by Bhatia and Noble (*loc. cit.*)

(b) $\xi \sim \frac{1}{2}$, $\delta \ll 1$

$$I_0 = \frac{1}{4}\frac{B^2\delta^2\gamma^2}{(\xi - \frac{1}{2})^2 + \frac{1}{4}\delta^2\gamma^2}\left[\frac{(\xi - \frac{1}{2})^2}{\frac{1}{4}\delta^2\gamma^2} + \cos^2[\beta d\{(\xi - \tfrac{1}{2})^2 + \tfrac{1}{4}\delta^2\gamma^2\}^{1/2}]\right], \tag{39a}$$

$$I_{-1} = \frac{1}{4}\frac{B^2\delta^2\gamma^2}{(\xi - \frac{1}{2})^2 + \frac{1}{4}\delta^2\gamma^2}\sin^2[\beta d\{(\xi - \tfrac{1}{2})^2 + \tfrac{1}{4}\delta^2\gamma^2\}^{1/2}]. \tag{39b}$$

The expressions for I_1 and I_{-2} are more complicated. For $\xi = \frac{1}{2}$, however, they take the simple form

$$I_1 = \tfrac{1}{16}\, B^2\delta^2\gamma^2\left[-\sin^2\left(\frac{\pi d\Delta\gamma}{2\lambda n}\right) + 2\sum_{+,-}\sin^2\{\beta d(1 \pm \tfrac{1}{4}\delta\gamma)\}\right] \tag{39c}$$

and

$$I_{-2} = \tfrac{1}{16}\, B^2\delta^2\gamma^2\sin^2\left(\frac{\pi d\Delta\gamma}{2\lambda n}\right). \tag{39d}$$

(c) *Normal incidence* ($\xi = 0$), $\delta \ll 1$

$$I_1 = I_{-1} = B^2\delta^2\gamma^2\sin^2[\tfrac{1}{2}\beta d(1 + \tfrac{1}{8}\delta^2\gamma^2)], \tag{40a}$$

$$I_2 = I_{-2} = \tfrac{1}{48}\, B^2\delta^4\gamma^4[-\tfrac{1}{4}\sin^2 2\beta d + \sin^2\{\tfrac{1}{2}\beta d(1 + \tfrac{1}{8}\delta^2\gamma^2)\} + \tfrac{1}{3}\sin^2\{\tfrac{3}{2}\,\beta d(1 - \tfrac{1}{24}\,\delta^2\gamma^2)\}]. \tag{40b}$$

* For a detailed discussion on these points see Bhatia and Noble (*loc. cit.*). In particular, it is shown there that the effect of the N^- on the amplitudes $B_l^{(r)}$ of the reflected spectrum is generally not negligible.

If we neglect the quantity $\delta^2\gamma^2$ occurring in the argument of the sines in (40), the resulting expressions may also be obtained from (38) by putting $\xi = 0$ in the latter.

As already mentioned, BRILLOUIN* and DAVID,† and also RYTOV,‡ derived the expressions (38) for the intensities of the first and second order lines in the manner outlined in § 12.1. AGGARWAL§ has derived the expressions (38) from the RAMAN and NATH differential equations § 12.1 (21). PHARISEAU ‖ has shown that expressions (39) for the intensities of the first and second order lines when $\xi \sim \frac{1}{2}$ can also be derived from equations (21).¶ This is, of course, as it should be, since the method based on MAXWELL's differential equations and the integral equation method of the present section are equivalent.

12.2.6 Some qualitative results

It is clear from the expressions for the intensities given in the previous section that, for values of δ and ξ such that either (a) $\delta \ll 1$ or (b) $\delta/\xi \ll 1$, only the first few lower orders will appear on each side of the transmitted beam and their intensities will diminish rapidly with increasing orders. When, however, neither (a) nor (b) is satisfied, i.e. when δ and δ/ξ are both large compared to unity, many more orders will in general appear. For this case the solution of the equations (18)–(20), and hence the calculation of the intensities of the various orders, is more difficult. By examining the conditions for the validity of the perturbation method of § 12.2.4, it may be shown** that in solving (18)–(20), one may consider only those amplitudes N_{lm} as non-zero for which both the suffixes l and m lie between the numbers $-M_1$ and M_2 defined approximately by $(0 \leqslant \xi < \delta^{2/3})$

$$\left.\begin{aligned} M_1 &\sim \xi + \delta^{2/3} + 1 \\ M_2 &\sim -\xi + \delta^{2/3} + 1 \end{aligned}\right\} \tag{41}$$

Since, in general, Δ cannot be increased much beyond 10^{-4} and the maximum value of λ/Λ for which diffraction phenomena can be observed is restricted by practical limitations of resolving power, etc., the maximum possible value of δ $(= \Delta\Lambda^2/\lambda^2)$ is about 100. Hence, even under extreme experimental conditions, one has only to solve at most about 20 linear simultaneous equations from each of the infinite sets (18), (19), (20). But even with this simplification the calculations are necessarily tedious and have not been performed.

Qualitatively, the numbers M_1 and M_2 also represent the number of orders likely to appear on the two sides of the direct transmitted beam. According to (41), the number of orders appearing on the two sides of the direct transmitted beam should become different as $\xi = (\Lambda \sin \theta)/\lambda$ increases from zero, more lines appearing on the side which can be reached by light reflected from the wave-fronts of the ultrasonic wave. PARTHASARATHY†† has studied experimentally the diffracted spectrum as a function of the angle of incidence θ, and we reproduce in Fig. 12.3 a plate from his paper. In this experiment $(\lambda/\Lambda) = 3 \times 10^{-3}$; assuming $\Delta \sim 10^{-4}$, we have from (41) for normal incidence $M_1 = M_2 = 5$. Table XXV gives the number of lines actually observed

* L. BRILLOUIN, *La Diffraction de la Lumière par des Ultrasons* (Paris, Hermann, 1933).

† E. DAVID, *Phys. Z.*, **38** (1937), 587.

‡ S. RYTOV, *Diffraction de la Lumière par les Ultra-sons* (Paris, Hermann, 1938).

§ R. R. AGGARWAL, *Proc. Ind. Acad. Sci.*, A, **31** (1950), 417.

‖ P. PHARISEAU, *Proc. Ind. Acad. Sci.*, A, **44** (1956), 165.

¶ A discussion of the experimental results in relations to the various expressions for the intensities given in this section may be found in the papers by BHATIA and NOBLE (*loc. cit.*).

** BHATIA and NOBLE (*loc. cit.*).

†† S. PARTHASARATHY, *Proc. Ind. Acad. Sci.*, A, **3** (1936), 442.

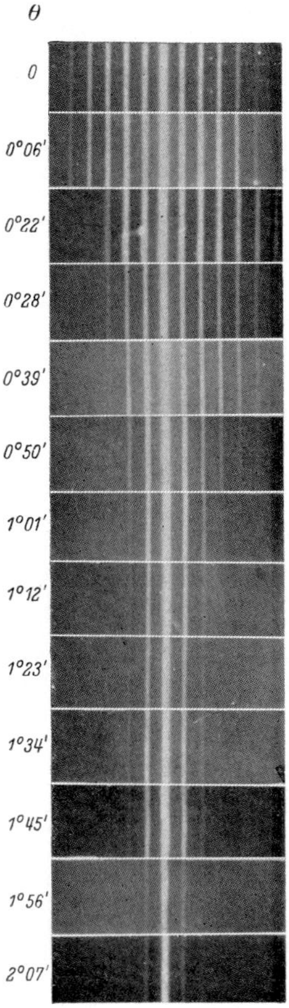

Fig. 12.3. Diffraction of light by ultrasonic waves: spectra observed with different angles θ of incidence ($\lambda/\Lambda = 3 \times 10^{-3}$, $\Delta \sim 10^{-4}$).

(After S. Parthasarathy, *Proc. Ind. Acad. Sci.*, A, **3** (1936), 442.)

to appear on each side of the direct transmitted beam at different angles of incidence. Within brackets are the corresponding theoretical numbers as given by (41). (Of course, once δ/ξ becomes much less than unity one or two orders only will appear on each side of the direct transmitted beam).

Similar experimental results have also been obtained by NOMOTO.* His curves of M_1 and M_2 against θ, though in qualitative agreement with (41), show also a slight periodic variation with ξ in the range $0 \leqslant \xi < \delta^{2/3}$. Equations (41) are based on too crude a consideration to explain this latter feature.

TABLE XXV

Diffraction of light by ultrasonic waves: number of orders observed, and (in brackets) the number predicted by theory, for different values of the angle of incidence θ

$(\lambda/\Lambda = 3 \times 10^{-3},\ \Delta \sim 10^{-4})$

θ	0	0° 06′	0° 22′	0° 39′	1° 01′	1° 23′	1° 45′	2° 07′
ξ	0	0·6	2	4	6	8	10	13
M_1	5 (5)	5 (6)	6 (7)	6 (9)	3	2	1	1
M_2	5 (5)	5 (4)	3 (3)	2 (2)	2	1	1	1

12.2.7 The Raman–Nath approximation

Finally it will be shown that the BESSEL function expressions for the intensities obtained by RAMAN and NATH may also be deduced from the solution of (18) to (20). Neglecting the small variation of the frequencies ω_l with l and remembering that $q_m{}^2 \simeq (n^2\omega_m{}^2c^{-2} - p_m{}^2)$, we can take, to a good approximation, $p_l{}^2 + q^2 + 2\omega_l{}^2c^{-2} \simeq k^2(n^2 + 2)$ in the denominator of (25). Hence (24) may be written in the form

$$\frac{q^2 - n^2k^2 + k^2\sin^2\theta}{\frac{1}{2}\Delta\gamma k^2} N_l(q) - N_{l+1}(q) - N_{l-1}(q) = -\frac{l^2 + 2l\xi}{2\gamma\delta} N_l(q)$$

$$(l = 0, \pm 1, \pm 2, \ldots). \qquad (42)$$

Now the BESSEL function expressions for the intensities were obtained by RAMAN and NATH by neglecting the terms $[(l^2 + 2l\xi)/\delta] \times U_l$ in their equations § 12.1 (21). To the same approximation we may also neglect the right-hand side of (42). (42) may then be written as

$$\frac{q^2 - b^2}{\mu^2} N_l(q) = N_{l+1}(q) + N_{l-1}(q) \qquad (l = 0, \pm 1, \ldots), \qquad (43)$$

where

$$b^2 = k^2(n^2 - \sin^2\theta) \qquad \text{and} \qquad \mu^2 = \tfrac{1}{2}\Delta\gamma k^2. \qquad (44)$$

To solve (43)† let us assume $N_l(q)$ to be of the form

$$N_l(q) = N\, e^{2\pi i l m/M}, \qquad (45)$$

* O. NOMOTO, *Proc. Phys.-Math. Soc. Japan*, **24** (1942), 380, 613.

† Equations (43) are similar to the relations which determine the normal modes of vibration of a linear chain of atoms (cf. M. BORN and TH. v. KÁRMÁN, *Phys., Z.*, **13** (1912), 297).

where M is some very large integer and m is an integer such that $0 \leqslant m < M$. (As will be seen presently the final results are independent of M.) Substituting from (45) in (43), we have for the permissible values of q^2,

$$q_m{}^2 = b^2 + 2\mu^2 \cos(2\pi m/M), \qquad 0 \leqslant m < M, \tag{46}$$

or remembering that $\mu^2 \ll b^2$,

$$\pm |q_m| = \pm [b + (\mu^2/b) \cos(2\pi m/M)]. \tag{47}$$

Thus for each m there are two values of q, namely $\pm |q_m|$; the corresponding amplitudes are given by

$$N_l(\pm q_m) \equiv N^{\pm}_{lm} = N^{\pm}\, e^{2\pi i l m/M}. \tag{48}$$

The two constants N^+ and N^- in (48) are now to be determined from (19) and (20). As in § 12.2.4, we shall neglect the amplitudes N^-_{lm} and determine the constant N^+ from (19) alone. Noting that for not too large a value of l, $[\sigma_l(q)/g_l(q)] \sim [\sigma_0(q_0{}^{(0)})/g_0(q_0{}^{(0)})]$, we may verify that the equations (19) are identically satisfied by taking

$$N^+ = \frac{1}{M}\, 2Bc^2\omega^{-2}[\omega^2c^{-2} - p_0{}^2]^{1/2}(g_{0,0}/\sigma_{0,0}) \tag{49}$$

in (48). Substituting (48) and (49) in (22) and putting $(\omega_l{}^2c^{-2} - p_l{}^2) = (\omega^2c^{-2} - p_0{}^2)$, we obtain for the amplitude B_l of the lth order diffracted wave

$$B_l \simeq \frac{B}{M} \sum_{m=0}^{M-1} e^{i[2\pi l m/M + (\mu^2 d/b) \cos(2\pi m/M)]}. \tag{50}$$

Here a phase factor independent of m has been omitted. Since by hypothesis M is a very large integer, we may replace the series in (50) by an integral. Putting $2\pi m/M = \psi'$ and $d\psi' = 2\pi/M$, we may write (50) as

$$B_l \simeq \frac{B}{2\pi} \int_0^{2\pi} e^{i[l\psi' + (\mu^2 d/b) \cos\psi']}\, d\psi', \tag{51}$$

which is independent of M. By splitting the integral in (51) into two parts (i) from 0 to $\frac{3}{2}\pi$ and (ii) from $\frac{3}{2}\pi$ to 2π, and putting in them $\psi' = \frac{1}{2}\pi - \psi$ and $\psi' = \frac{5}{2}\pi - \psi$ respectively, we obtain*

$$B_l \simeq \frac{B}{2\pi} \int_{-\pi}^{\pi} e^{i[\frac{1}{2}l\pi - l\psi + (\mu^2 d/b) \sin\psi]}\, d\psi$$
$$= B\, e^{\frac{1}{2}il\pi} J_l(\mu^2 d/b). \tag{52}$$

Hence the intensities $I_l = |B_l|^2$ are just $B^2 J_l{}^2(\mu^2 d/b)$. Moreover, it may be seen with the help of (44), § 12.1 (5) and § 12.1 (8) that the argument $(\mu^2 d/b)$ of the Bessel functions J_l is the same as that occurring in the expressions of Raman and Nath, given on p. 598.

---

* H. and B. S. Jeffreys, *Methods of Mathematical Physics* (Cambridge University Press, 1946), p. 547.

CHAPTER XIII

OPTICS OF METALS

So far we have been concerned with the propagation of light in non-conducting, isotropic media. We now turn our attention to optics of conducting media, more particularly to metals. An ordinary piece of metal is a crystalline aggregate, consisting of small crystals of random orientation. Single crystals of appreciable size are rare, but can be produced artificially; their optical properties will be studied in Chapter XIV. A mixture of randomly oriented crystallites behaves evidently as an isotropic substance, and as the theory of light propagation in a conducting isotropic medium is much simpler than in a crystal, we shall consider it here in some detail.

According to § 1.1, conductivity is connected with the appearance of Joule heat. This is an irreversible phenomenon, in which the electromagnetic energy is destroyed, or more precisely transformed into heat, and in consequence an electromagnetic wave in a conductor is attenuated. In metals, on account of their very high conductivity, this effect is so large that they are practically opaque. In spite of this, metals play an important part in optics. Strong absorption is accompanied by high reflectivity, so that metallic surfaces act as excellent mirrors. Because of the partial penetration of light into a metal, it is possible to obtain information about the absorption constants and the mechanism of absorption from observations of the reflected light, even though the depth of penetration is small.

We shall first consider the purely formal results arising from the existence of conductivity, and then briefly discuss a simple, somewhat idealized, physical model for this process, based on the classical theory of the electron. This model accounts only roughly for some of the observed effects; a more precise model can only be obtained with the help of quantum mechanics and is thus outside the scope of this book. The formal theory will be illustrated by applications to two problems of practical interest: the optics of stratified media containing an absorbing element, and the diffraction of light by a metallic sphere.

A particularly attractive mathematical feature of the theory is that the existence of conductivity may be taken into account simply by introducing a complex dielectric constant (or complex index of refraction), instead of a real one. In metals the imaginary part is preponderant.

13.1. WAVE PROPAGATION IN A CONDUCTOR

Consider a homogeneous isotropic medium of dielectric constant ε, permeability μ, and conductivity σ. Using the material equations § 1.1 (9)–(11), viz.: $\boldsymbol{j} = \sigma \boldsymbol{E}$, $\boldsymbol{D} = \varepsilon \boldsymbol{E}$, $\boldsymbol{B} = \mu \boldsymbol{H}$, Maxwell's equations take the form

$$\operatorname{curl} \boldsymbol{H} - \frac{\varepsilon}{c} \dot{\boldsymbol{E}} = \frac{4\pi}{c} \sigma \boldsymbol{E}, \tag{1}$$

$$\operatorname{curl} \boldsymbol{E} + \frac{\mu}{c} \dot{\boldsymbol{H}} = 0, \tag{2}$$

$$\operatorname{div} \boldsymbol{E} = \frac{4\pi}{\varepsilon}\rho, \tag{3}$$

$$\operatorname{div} \boldsymbol{H} = 0. \tag{4}$$

It is easy to see that for an electromagnetic disturbance, incident from outside on the conductor, one may replace (3) by div $\boldsymbol{E} = 0$. For if we take the divergence of (1) and use (3) we obtain

$$-\frac{\varepsilon}{c}\operatorname{div}\dot{\boldsymbol{E}} = \frac{4\pi\sigma}{c}\frac{4\pi}{\varepsilon}\rho$$

Also, differentiation of (3) with respect to time gives

$$\operatorname{div}\dot{\boldsymbol{E}} = \frac{4\pi}{\varepsilon}\dot{\rho}.$$

Eliminating div $\dot{\boldsymbol{E}}$ between the last two equations one obtains

$$\dot{\rho} + \frac{4\pi\sigma}{\varepsilon}\rho = 0, \tag{5}$$

giving on integration

$$\rho = \rho_0 e^{-\frac{t}{\tau}} \quad \text{where} \quad \tau = \frac{\varepsilon}{4\pi\sigma}. \tag{6}$$

Any electric charge density ρ is thus seen to fall off exponentially with time. The *relaxation time* τ is exceedingly small for any medium having appreciable conductivity. For metals, this time is very much shorter, (typically of the order of 10^{-18} sec), than the periodic time of vibration of the wave. We may, therefore, assume that ρ in a metal is always sensibly zero. Equation (3) may, therefore, be written as

$$\operatorname{div} \boldsymbol{E} = 0. \tag{7}$$

From (1) and (2) it follows by elimination of $\boldsymbol{H}$ and the use of (7) that $\boldsymbol{E}$ satisfies the wave equation

$$\nabla^2\boldsymbol{E} = \frac{\mu\varepsilon}{c^2}\ddot{\boldsymbol{E}} + \frac{4\pi\mu\sigma}{c^2}\dot{\boldsymbol{E}}. \tag{8}$$

The term in $\dot{\boldsymbol{E}}$ implies that the wave is damped, i.e. it suffers a progressive attenuation as it is propagated through the medium.

If the field is strictly monochromatic, and of angular frequency ω, i.e. if $\boldsymbol{E}$ and $\boldsymbol{H}$ are of the form $\boldsymbol{E} = \boldsymbol{E}_0 e^{-i\omega t}$, we have $\partial/\partial t \equiv -i\omega$ so that equations (1) and (2) may be re-written as

$$\operatorname{curl} \boldsymbol{H} + \frac{i\omega}{c}\left(\varepsilon + i\frac{4\pi\sigma}{\omega}\right)\boldsymbol{E} = 0, \tag{9}$$

$$\operatorname{curl} \boldsymbol{E} - \frac{i\omega\mu}{c}\boldsymbol{H} = 0, \tag{10}$$

and (8) becomes

$$\nabla^2\boldsymbol{E} + \hat{k}^2\boldsymbol{E} = 0, \tag{11}$$

where

$$\hat{k}^2 = \frac{\omega^2\mu}{c^2}\left(\varepsilon + i\frac{4\pi\sigma}{\omega}\right). \tag{12}$$

These equations are formally identical with the corresponding equations for non-conducting media if in the latter the dielectric constant ε, (which, to a good approximation, was shown to be real except for frequencies ω that are close to a resonance—cf. § 2.3.4), is replaced by

$$\hat{\varepsilon} = \varepsilon + i\,\frac{4\pi\sigma}{\omega}. \tag{13}$$

The analogy with non-conducting media becomes closer still if, in addition to the complex wave number $\hat{k}$ and the complex dielectric constant $\hat{\varepsilon}$, we also introduce a complex phase velocity $\hat{v}$ and a complex refractive index $\hat{n}$ which, in analogy with § 1.2 (8), § 1.2 (12), and § 1.3 (21), are defined by

$$\hat{v} = \frac{c}{\sqrt{\mu\hat{\varepsilon}}}, \qquad \hat{n} = \frac{c}{\hat{v}} = \sqrt{\mu\hat{\varepsilon}} = \frac{c}{\omega}\,\hat{k}. \tag{14}$$

We set

$$\hat{n} = n(1 + i\kappa), \tag{15}$$

where n and κ are real, and we call κ *the attenuation index.** The quantities n and κ may easily be expressed in terms of the material constants ε, μ, and σ. Squaring (15) we have

$$\hat{n}^2 = n^2(1 + 2i\kappa - \kappa^2). \tag{15a}$$

Also, from (14) and (13),

$$\hat{n}^2 = \mu\hat{\varepsilon} = \mu\left(\varepsilon + i\,\frac{4\pi\sigma}{\omega}\right). \tag{16}$$

Now σ, just like ε, is not a true constant of the medium, but depends on the frequency. We will see later (§ 13.3) that for sufficiently low frequencies (long wavelengths) σ is, to a good approximation, real. Assuming that ε is also real, we obtain in this case, upon equating the real and imaginary parts in (15a) and (16), the following relations:

$$n^2(1 - \kappa^2) = \mu\varepsilon, \tag{16a}$$

$$n^2\kappa = \frac{2\pi\mu\sigma}{\omega} = \frac{\mu\sigma}{\nu}. \tag{16b}$$

From these equations it follows that

$$n^2 = \frac{1}{2}\left\{\sqrt{\mu^2\varepsilon^2 + \frac{4\mu^2\sigma^2}{\nu^2}} + \mu\varepsilon\right\}, \tag{17a}$$

$$n^2\kappa^2 = \frac{1}{2}\left\{\sqrt{\mu^2\varepsilon^2 + \frac{4\mu^2\sigma^2}{\nu^2}} - \mu\varepsilon\right\}. \tag{17b}$$

The positive sign of the square roots is taken here, since n and $n\kappa$ are real, and consequently n^2 and $n^2\kappa^2$ must be positive.

Equation (11) is formally identical with the wave equation for a non-conducting medium, but the wave number is now complex. The simplest solution is that of a plane, time-harmonic wave

$$\boldsymbol{E} = \boldsymbol{E}_0\, e^{i[\hat{k}(\boldsymbol{r}.\boldsymbol{s}) - \omega t]}. \tag{18}$$

* The term "extinction coefficient" is also used.

If, in accordance with (14) and (15), we substitute for $\hat{k}$ from the relation $\hat{k} = \omega\hat{n}/c = \omega n(1 + i\kappa)/c$, (18) becomes

$$\boldsymbol{E} = \boldsymbol{E}_0 \, e^{-\frac{\omega}{c} n\kappa(\boldsymbol{r}.\boldsymbol{s})} \, e^{i\omega\left[\frac{n}{c}(\boldsymbol{r}.\boldsymbol{s}) - t\right]}.$$

The real part of this expression, viz.:

$$\boldsymbol{E} = \boldsymbol{E}_0 \, e^{-\frac{\omega}{c} n\kappa(\boldsymbol{r}.\boldsymbol{s})} \cos\left\{\omega\left[\frac{n}{c}(\boldsymbol{r}\,.\,\boldsymbol{s}) - t\right]\right\}, \tag{19}$$

which represents the electric vector, is a plane wave with wavelength $\lambda = 2\pi c/\omega n$ and with attenuation given by the exponential term. Since the energy density w of the wave is proportional to the time average of $\boldsymbol{E}^2$, it follows that w decreases in accordance with the relation

$$w = w_0 \, e^{-\chi(\boldsymbol{r}.\boldsymbol{s})}, \tag{20}$$

where

$$\chi = \frac{2\omega}{c} n\kappa = \frac{4\pi\nu}{c} n\kappa = \frac{4\pi}{\lambda_0} n\kappa = \frac{4\pi}{\lambda}\kappa, \tag{21}$$

λ_0 being the wavelength in vacuum and λ the wavelength in the medium. The constant χ is called *the absorption coefficient*.

The energy density falls to $1/e$ of its value after the wave has advanced a distance d, where

$$d = \frac{1}{\chi} = \frac{\lambda_0}{4\pi n\kappa} = \frac{\lambda}{4\pi\kappa}. \tag{22}$$

This quantity is usually a very small fraction of the wavelength (see Table XXVI).*

Returning to equations (17) we see that, when $\sigma = 0$, the first equation correctly reduces to MAXWELL's relation § 1.2 (14) $n^2 = \mu\varepsilon$, and the second gives $\kappa = 0$. For metals $\sigma \neq 0$ and is in fact so large that in (17) ε may be neglected in comparison with $2\sigma/\nu$. To get an idea of the orders of magnitude involved let it be remarked that for most metals the conductivity at frequencies up to about the infrared region of the spectrum ($\lambda \geqslant 10^{-3}$ cm) is of the order of 10^{17} sec^{-1}. Thus, for example, with $\lambda = 10^{-3}$ cm ($\nu \sim 3 \,.\, 10^{13}$ sec^{-1}), one then has $\sigma/\nu \sim 3000$. The dielectric constant ε of a metal cannot be measured directly, but as we shall see it can be deduced from optical experiments. However, as the mechanism of electric polarization in metals is not fundamentally different from that of a dielectric, it may be assumed that ε is of the same order of magnitude. Hence, provided the wavelength is not too short, one may suppose that

$$\frac{\mu\sigma}{\nu} = n^2\kappa \gg \mu\varepsilon. \tag{23}$$

Equations (17) and (22) now reduce to

$$n \sim n\kappa = \sqrt{\frac{\mu\sigma}{\nu}}. \tag{24}$$

$$d \sim \frac{\lambda_0}{4\pi}\sqrt{\frac{\nu}{\mu\sigma}} = \frac{1}{4\pi}\sqrt{\frac{c\lambda_0}{\mu\sigma}} = \frac{c}{\sqrt{8\pi\mu\sigma\omega}}. \tag{25}$$

* This phenomenon of penetration to a depth that is a small fraction of the wavelength, is well known in the conduction of alternating currents and is known to engineers as the "skin effect".

TABLE XXVI

The "penetration depth" d for copper for radiation in three familiar regions of the spectrum, calculated with the static conductivity $\sigma \sim 5{\cdot}14 \,.\, 10^{17}$ sec^{-1} *and* $\mu = 1$.

Radiation	Infra-red	Microwaves	Long radio waves
λ_0	10^{-3} cm	10 cm	1000 m $= 10^5$ cm
d	$6{\cdot}1 \,.\, 10^{-7}$ cm	$6{\cdot}1 \,.\, 10^{-5}$ cm	$6{\cdot}1 \,.\, 10^{-3}$ cm

A perfect conductor is characterized by infinitely large conductivity ($\sigma \to \infty$). Since according to (16), $\varepsilon/\sigma = (1 - \kappa^2)/\nu\kappa$, we have in this limiting case $\kappa^2 \to 1$, or by (16a), $n \to \infty$. Such a conductor would not permit the penetration of an electromagnetic wave to any depth at all and would reflect all the incident light (cf. § 13.2 below).

Whilst the refractive index of transparent substances may easily be measured from the angle of refraction, such measurements are extremely difficult to carry out for metals, because a specimen of the metal which transmits any appreciable fraction of incident light has to be exceedingly thin. Nevertheless KUNDT* succeeded in constructing metal prisms that enabled direct measurements of the real and imaginary parts of the complex refractive index to be made. Usually, however, the optical constants of metals are determined by means of katoptric rather than dioptric experiments, i.e. by studying the changes which light undergoes on reflection from a metal, rather than by means of measurements on the light transmitted through it.

13.2. REFRACTION AND REFLECTION AT A METAL SURFACE

We have seen that the basic equations relating to the propagation of a plane time-harmonic wave in a conducting medium differ from those relating to propagation in a transparent dielectric only in that the real constants ε and k are replaced by complex constants $\hat{\varepsilon}$ and $\hat{k}$. It follows that the formulae derived in Chapter I, as far as they involve only linear relations between the components of the field vectors of plane monochromatic waves, apply also in the present case. In particular, the boundary conditions for the propagation of a wave across a surface of discontinuity and hence also the formulae of § 1.5 relating to refraction and reflection remain valid.

Consider first the propagation of a plane wave from a dielectric into a conductor, both media being assumed to be of infinite extent, the surface of contact between them being the plane $z = 0$. By analogy with § 1.5 (8) the law of refraction is

$$\sin \theta_t = \frac{1}{\hat{n}} \sin \theta_i. \tag{1}$$

Since $\hat{n}$ is complex, so is θ_t, and this quantity therefore no longer has the simple significance of an angle of refraction.

Let the plane of incidence be the xz-plane. The space-dependent part of the phase of the wave in the conductor is given by $\hat{k}(\mathbf{r}.\mathbf{s}^{(t)})$ where (cf. § 1.5 (4))

$$s_x^{(t)} = \sin \theta_t, \qquad s_y^{(t)} = 0, \qquad s_z^{(t)} = \cos \theta_t. \tag{2}$$

* A. KUNDT, *Ann. d. Physik*, **34** (1888). 469.

From (1) and (2) and § 13.1 (15)

$$s_x^{(t)} = \sin\theta_t = \frac{\sin\theta_i}{n(1+i\kappa)} = \frac{1-i\kappa}{n(1+\kappa^2)}\sin\theta_i, \tag{3a}$$

$$s_z^{(t)} = \cos\theta_t = \sqrt{1-\sin^2\theta_t}$$

$$= \sqrt{1 - \frac{(1-\kappa^2)}{n^2(1+\kappa^2)^2}\sin^2\theta_i + i\frac{2\kappa}{n^2(1+\kappa^2)^2}\sin^2\theta_i}. \tag{3b}$$

It is convenient to express $s_z^{(t)}$ in the form

$$s_z^{(t)} = \cos\theta_t = qe^{i\gamma} \tag{4}$$

(q, γ real). Expressions for q and γ in terms of n, κ and $\sin\theta_i$ are immediately obtained on squaring (3b) and (4) and equating real and imaginary parts. This gives

$$\left.\begin{aligned} q^2\cos 2\gamma &= 1 - \frac{1-\kappa^2}{n^2(1+\kappa^2)^2}\sin^2\theta_i,\\ q^2\sin 2\gamma &= \frac{2\kappa}{n^2(1+\kappa^2)^2}\sin^2\theta_i. \end{aligned}\right\} \tag{5}$$

It follows that

$$\begin{aligned} \hat{k}(\mathbf{r}\,.\,\mathbf{s}^{(t)}) &= \frac{\omega}{c}n(1+i\kappa)(xs_x^{(t)} + zs_z^{(t)})\\ &= \frac{\omega}{c}n(1+i\kappa)\left[\frac{x(1-i\kappa)}{n(1+\kappa^2)}\sin\theta_i + z(q\cos\gamma + iq\sin\gamma)\right]\\ &= \frac{\omega}{c}[x\sin\theta_i + znq\,(\cos\gamma - \kappa\sin\gamma) + inzq(\kappa\cos\gamma + \sin\gamma)]. \end{aligned} \tag{6}$$

We see that the surfaces of constant amplitude are given by

$$z = \text{constant}, \tag{7}$$

and are, therefore, planes parallel to the boundary. The surfaces of constant real phase are given by

$$x\sin\theta_i + znq\,(\cos\gamma - \kappa\sin\gamma) = \text{constant}, \tag{8}$$

and are planes whose normals make an angle θ_t' with the normal to the boundary, where

$$\left.\begin{aligned} \cos\theta_t' &= \frac{nq(\cos\gamma - \kappa\sin\gamma)}{\sqrt{\sin^2\theta_i + n^2q^2(\cos\gamma - \kappa\sin\gamma)^2}},\\ \sin\theta_t' &= \frac{\sin\theta_i}{\sqrt{\sin^2\theta_i + n^2q^2(\cos\gamma - \kappa\sin\gamma)^2}}. \end{aligned}\right\} \tag{9}$$

Since the surfaces of constant amplitude and the surfaces of constant phase do not in general coincide with each other, the wave in the metal is an *inhomogeneous wave.*

If we denote the square root in (9) by n', the equation for $\sin\theta_t'$ may be written in the form $\sin\theta' = \sin\theta_i/n'$, i.e. it has the form of SNELL's law. However, n' depends now not only on the quantities that specify the medium, but also on the angle of incidence θ_i.

We may also derive expressions for the amplitude and the phase of the refracted and reflected waves by substituting for θ_t the complex value given by (1) in the FRESNEL formulae (§ 1.5.2). The explicit expressions will be given in § 13.4.1 in connection with the theory of stratified conducting media. Here we shall consider how the optical constants of the metal may be deduced from observation of the reflected wave.

Since we assumed that the first medium is a dielectric, the reflected wave is an ordinary (homogeneous) wave with a real phase factor. As in § 1.5 (21a) the amplitude components $A_\parallel$, $A_\perp$ of the incident wave and the corresponding components $R_\parallel$, $R_\perp$ of the reflected wave are related by

$$\left.\begin{aligned} R_\parallel &= \frac{\tan(\theta_i - \theta_t)}{\tan(\theta_i + \theta_t)} A_\parallel, \\ R_\perp &= -\frac{\sin(\theta_i - \theta_t)}{\sin(\theta_i + \theta_t)} A_\perp. \end{aligned}\right\} \tag{10}$$

Since θ_t is now complex, so are the ratios $R_\parallel/A_\parallel$ and $R_\perp/A_\perp$, i.e. characteristic phase changes occur on reflection; thus incident linearly polarized light will in general become elliptically polarized on reflection at the metal surface. Let $\phi_\parallel$ and $\phi_\perp$ be the phase changes, and $\rho_\parallel$ and $\rho_\perp$ the absolute values of the reflection coefficients, i.e.

$$r_\parallel = \frac{R_\parallel}{A_\parallel} = \rho_\parallel e^{i\phi_\parallel}, \qquad r_\perp = \frac{R_\perp}{A_\perp} = \rho_\perp e^{i\phi_\perp}. \tag{11}$$

Suppose that the incident light is *linearly polarized* in the azimuth α_i i.e.

$$\tan \alpha_i = \frac{A_\perp}{A_\parallel}, \tag{12}$$

and let α_r be the azimuthal angle (generally complex) of the light that is reflected. Then*

$$\tan \alpha_r = \frac{R_\perp}{R_\parallel} = -\frac{\cos(\theta_i - \theta_t)}{\cos(\theta_i + \theta_t)} \tan \alpha_i = Pe^{-i\Delta} \tan \alpha_i, \tag{13}$$

where

$$P = \frac{\rho_\perp}{\rho_\parallel}, \qquad \Delta = \phi_\parallel - \phi_\perp. \tag{14}$$

We note that α_r is real in the following two cases:

(1) For normal incidence ($\theta_i = 0$); then $P = 1$ and $\Delta = -\pi$, so that $\tan \alpha_r = -\tan \alpha_i$.

(2) For grazing incidence ($\theta_i = \pi/2$); then $P = 1$ and $\Delta = 0$, so that $\tan \alpha_r = \tan \alpha_i$.

It should be remembered that in the case of normal incidence the directions of the incident and reflected rays are opposed; thus the negative sign implies that the

---

* We write $-i\Delta$ rather than $+i\Delta$ in the exponent on the right-hand side of (13) to facilitate comparison with certain results of § 1.5.

azimuth of the linearly polarized light is unchanged in its absolute direction in space. It is also unchanged in its absolute direction when the incidence is grazing.

Between the two extreme cases just considered, there exists an angle $\bar{\theta}_i$ called *principal angle of incidence* which is such that $\Delta = -\pi/2$. At this angle of incidence, linearly polarized light is, in general, reflected as elliptically polarized light, but as may be seen from § 1.4 (31b) (with $\delta = \pi/2$), the axes of the vibration ellipse are parallel and perpendicular to the plane of incidence. If, moreover, $P \tan \alpha_i = 1$, then according to (13) $\tan \alpha_r = -i$, and the reflected light is *circularly* polarized.

Suppose that with linearly polarized incident light an additional phase difference Δ is introduced between $R_\parallel$ and $R_\perp$ by means of a suitable compensator (cf. § 14.4.2). The total phase difference is then zero, and, according to (13) the reflected light is linearly polarized in an azimuth α'_r such that

$$\tan \alpha'_r = P \tan \alpha_i. \tag{15}$$

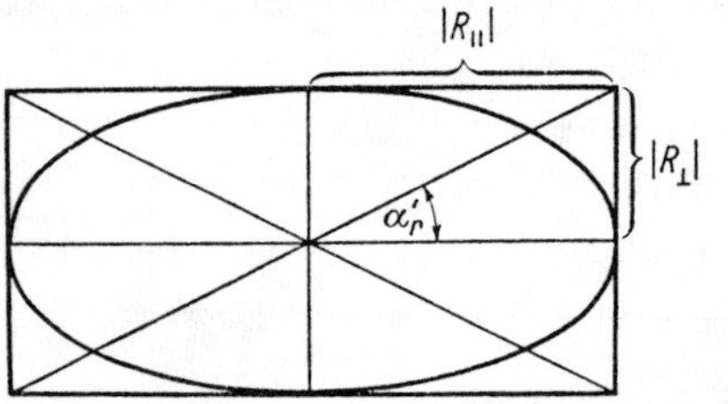

Fig. 13.1. Vibration ellipse of light reflected from a metal at the principal angle of incidence.

The angle α'_r is, for obvious reasons, called *the angle of restored polarization*, though it is usually defined only with incident light that is *linearly polarized* in the azimuth $\alpha_i = 45°$*. The values of α'_r and P relating to the principal angle of incidence $\theta_i = \bar{\theta}_i$ will be denoted by $\bar{\alpha}'_r$ and $\bar{P}$ respectively. If we imagine a rectangle to be circumscribed round the vibration ellipse of the (uncompensated) reflected light obtained from light that is incident at the principal angle, with its sides parallel and perpendicular to the plane of incidence, then the sides are in the ratio $\bar{P} \tan \alpha_i$ and the angle between a diagonal and the plane of incidence is $\bar{\alpha}'_r$ (see Fig. 13.1).

For the purpose of later calculations it is useful to introduce an angle ψ such that

$$\tan \psi = P; \tag{16}$$

the value of ψ corresponding to the principal angle of incidence will be denoted by $\bar{\psi}$.

Using (10) and (1) we can compute the quantities P ($= \tan \psi$) and Δ in terms of θ_i, if the constants n and κ of the metal are known. Fig. 13.2 (a) shows their dependence on θ_i in a typical case. In Fig. 13.2 (b) analogous curves relating to reflection from a transparent dielectric are displayed for comparison. The sudden discontinuity from $-\pi$ to 0 in the value of Δ which occurs when light is reflected from a transparent dielectric at the polarizing angle is absent when light is reflected from a metal surface. The sharp cusp when $\tan \psi$ becomes infinite is likewise absent, and the curve is replaced by a smooth curve with a comparatively broad maximum. The angle of incidence at which this maximum occurs is sometimes called the *quasi-polarizing angle;* it is nearly equal to the principal angle of incidence $\bar{\theta}_i$. It is commonly assumed that this maximum is actually at $\bar{\theta}_i$, which is almost exactly true if $n^2(1 + \kappa^2) \gg 1$, as is usually the case [cf. Table XVII]. In general the two angles are, however, different; for example, in the case of silver at the ultraviolet wavelength 3280 Å; the quantity $n^2(1 + \kappa^2)$ is small; then $\bar{\theta}_i = 47{\cdot}8°$ and $\psi = 31{\cdot}8°$, whereas $\psi_{\max} = 29{\cdot}5°$ and occurs at $\theta_i = 40°$, approximately.

Generally speaking the problem is not to find ψ and Δ from known values of n and κ, but to determine n and κ from experimental observations of the amplitude and phase of light reflected from the metal.

* Then α'_r is equal to the angle ψ introduced in (16).

As the quantities $R_{\parallel}$, $R_{\perp}$, $\phi_{\parallel}$, $\phi_{\perp}$, ψ and Δ are all functions of θ_i, and of n and κ, measurement of any two of these quantities for a specific value of the angle of incidence θ_i will in general permit the evaluation of n and κ. Since in many experiments one determines the last two of these quantities, we shall derive the fundamental expressions for n and κ in terms of ψ and Δ. From (13) and (1)

$$\frac{1 - Pe^{-i\Delta}}{1 + Pe^{-i\Delta}} = -\frac{\cos\theta_i \cos\theta_t}{\sin\theta_i \sin\theta_t} = -\frac{\sqrt{\hat{n}^2 - \sin^2\theta_i}}{\sin\theta_i \tan\theta_i}. \tag{17}$$

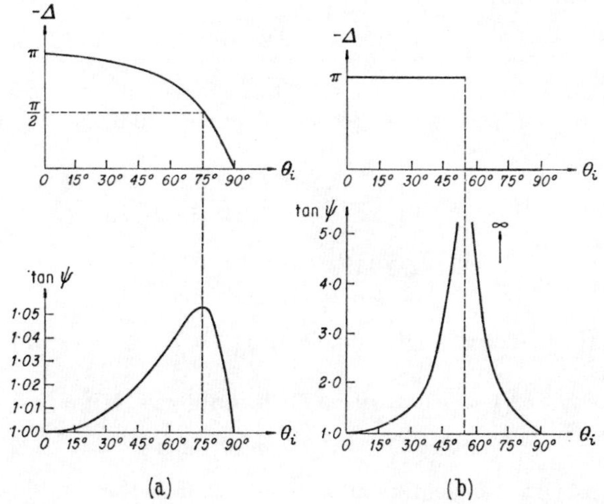

Fig. 13.2. The quantities $-\Delta = \phi_{\perp} - \phi_{\parallel}$ and $P = \tan\psi = \rho_{\perp}/\rho_{\parallel}$, which characterize the change in the state of polarization of light on reflection from a typical metal (a) and from a transparent dielectric (b).

Since $P = \tan\psi$, the left-hand side of equation (17) may be expressed in the form

$$\frac{1 - Pe^{-i\Delta}}{1 + Pe^{-i\Delta}} = \frac{1 - e^{-i\Delta}\tan\psi}{1 + e^{-i\Delta}\tan\psi} = \frac{\cos 2\psi + i \sin 2\psi \sin\Delta}{1 + \sin 2\psi \cos\Delta}. \tag{18}$$

From (17) and (18),

$$\frac{\sqrt{\hat{n}^2 - \sin^2\theta_i}}{\sin\theta_i \tan\theta_i} = -\frac{\cos 2\psi + i \sin 2\psi \sin\Delta}{1 + \sin 2\psi \cos\Delta}. \tag{19}$$

Now if, as is usually the case in the visible region,

$$n^2(1 + \kappa^2) \gg 1, \tag{20}$$

$\sin^2\theta_i$ may be neglected in comparison with $\hat{n}^2$ and we obtain

$$\frac{\hat{n}}{\sin\theta_i \tan\theta_i} = \frac{n(1 + i\kappa)}{\sin\theta_i \tan\theta_i} \sim -\frac{\cos 2\psi + i \sin 2\psi \sin\Delta}{1 + \sin 2\psi \cos\Delta}. \tag{21}$$

Equating the real parts, we obtain

$$n \sim -\frac{\sin\theta_i \tan\theta_i \cos 2\psi}{1 + \sin 2\psi \cos\Delta}. \tag{22a}$$

Equating the imaginary parts and using (22a) we find that

$$\kappa \sim \tan 2\psi \sin\Delta. \tag{22b}$$

These expressions permit the calculation of the optical constants n and κ from measurements of ψ and Δ at any angle of incidence. In the particular case of observation at the principal angle of incidence $\bar{\theta}_i$ one has $\Delta = -\pi/2$, $\psi = \bar{\psi}$ and equations (22a) and (22b) reduce to

$$n \sim -\sin\bar{\theta}_i \tan\bar{\theta}_i \cos 2\bar{\psi}, \tag{23a}$$

$$\kappa \sim -\tan 2\bar{\psi}. \tag{23b}$$

Other formulae relating to n and κ are sometimes useful. Without assuming (20) we have, on squaring (19),

$$\frac{\hat{n}^2 - \sin^2\theta_i}{\sin^2\theta_i \tan^2\theta_i} = \frac{\cos^2 2\psi - \sin^2 2\psi \sin^2\Delta + i \sin 4\psi \sin\Delta}{(1 + \sin 2\psi \cos\Delta)^2}. \tag{24}$$

If we substitute $\hat{n}^2 = n^2(1 - \kappa^2) + 2in^2\kappa$ and equate real and imaginary parts we obtain

$$n^2(1 - \kappa^2) = \sin^2\theta_i \left\{1 + \frac{\tan^2\theta_i(\cos^2 2\psi - \sin^2 2\psi \sin^2\Delta)}{(1 + \sin 2\psi \cos\Delta)^2}\right\}, \tag{25a}$$

$$2n^2\kappa = \frac{\sin^2\theta_i \tan^2\theta_i \sin 4\psi \sin\Delta}{(1 + \sin 2\psi \cos\Delta)^2}. \tag{25b}$$

In particular, at the principal angle of incidence ($\theta = \bar{\theta}_i$, $\Delta = -\pi/2$), these equations reduce to*

$$n^2(1 - \kappa^2) = \sin^2\bar{\theta}_i(1 + \tan^2\bar{\theta}_i \cos 4\bar{\psi}), \tag{26a}$$

$$2n^2\kappa = -\sin^2\bar{\theta}_i \tan^2\bar{\theta}_i \sin 4\bar{\psi}. \tag{26b}$$

The formulae (25) do not yield n and κ directly but in the combinations $n^2(1 - \kappa^2)$ and $n^2\kappa$. On reference to § 13.1 (16) we see that these quantities have a direct physical significance. With $\mu = 1$ (as is always the case at optical wavelengths), $n^2(1 - \kappa^2)$ is the dielectric constant, and $n^2\kappa$ the ratio of the conductivity and the frequency. From the magnitude of these quantities, and particularly from their variation with frequency, information may be obtained about the structure of the metal (cf. § 13.3 below).

So far our analysis has centred round the amplitudes of the components of the reflected light, but, as we shall see shortly, useful information may also be obtained from comparison of the intensity of the reflected light with that of the incident light, especially at long wavelengths. If we consider normal incidence ($\theta_i = 0$), the distinction between $R_\parallel$ and $R_\perp$ disappears, the plane of incidence then being undetermined, and we may write

$$\mathscr{R} = \left|\frac{R_\parallel}{A_\parallel}\right|^2 = \left|\frac{R_\perp}{A_\perp}\right|^2. \tag{27}$$

If we substitute from (1) and (10) (or if we replace n by $\hat{n}$ in § 1.5 (23)), we obtain

$$\mathscr{R} = \left|\frac{\hat{n} - 1}{\hat{n} + 1}\right|^2 = \frac{n^2(1 + \kappa^2) + 1 - 2n}{n^2(1 + \kappa^2) + 1 + 2n}. \tag{28}$$

* Equations such as (23) and (26), which involve measurements only at the principal angle of incidence, are simpler than the more general expressions (22) or (25), and for this reason alone many experimenters have restricted themselves to measurements at this angle. At other angles of incidence experimental accuracy may be greater. Convenient choice of the angle of incidence is discussed in P. Drude, *Ann. d. Physik*, **39** (1890), 504; J. R. Collins and R. O. Bock, *Rev. Sci. Instr.*, **14** (1943), 135; I. Simon, *J. Opt. Soc. Amer.*, **41** (1951), 336; D. G. Avery, *Proc. Phys. Soc.*, **65** (1952), 425; R. W. Ditchburn, *J. Opt. Soc. Amer.*, **45** (1955), 743.

The optical constants of many metals have been determined from measurements of reflected light. In Table XXVII values of the constants as found by various observers are given for a wavelength in the yellow region of the visible spectrum. The metals are arranged in order of their reflectivity $\mathscr{R}$. We note that in all cases $n < n\kappa$ so that according to § 13.1 (16a) $\mu\varepsilon$ and consequently (since $\mu \sim 1$ at optical wavelengths) ε is negative. At first sight it might appear that no physical significance can be attached to a negative dielectric constant. We shall see later that this is not the case and that the negative value of ε can be explained from certain simple assumptions concerning the electron mechanism of conductivity. From the table it would appear that values* $n < 1$ are associated with high reflectivity but in general this is not the case.

TABLE XXVII

The optical constants of metals, for light of wavelength $\lambda = 5893$ Å (Sodium D lines)

Condensed from H. H. LANDOLT and R. BÖRNSTEIN, *Phys. Chem. Tabellen*, (5 Aufl., Berlin, 1923; 1–3 Ergänzungsb, Berlin, 1927–1936).

Metal	n	$n\kappa$	$\mathscr{R}$	Observer	
Sodium, solid	0·044	2·42	0·97	DUNCAN	1913
Silver, massive	0·20	3·44	0·94	OPPITZ	1917
Magnesium, massive	0·37	4·42	0·93	DRUDE	1890
Potassium, molten	0·084	1·81	0·92	NATHANSON	1928
Cadmium, massive	1·13	5·01	0·84	DRUDE	1890
Aluminium, massive	1·44	5·23	0·83	DRUDE	1890
Tin, massive	1·48	5·25	0·83	DRUDE	1890
Gold, electrolytic	0·47	2·83	0·82	MEIER	1910
Mercury, liquid	1·60	4·80	0·77	LOWERY and MOORE	1932
Zinc, massive	1·93	4·66	0·75	MEIER	1910
Copper, massive	0·62	2·57	0·73	OPPITZ	1917
Gallium, single crystal	3·69	5·43	0·71	LANGE	1935
Antimony, massive	3·04	4·94	0·70	DRUDE	1890
Cobalt, massive	2·12	4·04	0·68	MINOR	1904
Nickel, electrolytic	1·58	3·42	0·66	MEIER	1910
Manganese, massive	2·41	3·88	0·64	LITTLETON	1911
Lead, massive	2·01	3·48	0·62	DRUDE	1890
Platinum, electrolytic	2·63	3·54	0·59	MEIER	1910
Rhenium, massive	3·00	3·44	0·57	LANGE	1935
Tungsten, massive	3·46	3·25	0·54	LITTLETON	1912
Bismuth, massive	1·78	2·80	0·54	MEIER	1910
Iron, evaporated	1·51	1·63	0·33	MEIER	1910

The values of n and $n\kappa$ displayed in Table XXVII cannot be expected to be in agreement with calculations based on the approximate formulae § 13.1 (17). These formulae were derived on the assumption that the conductivity σ is real and as we will see in § 13.3 this assumption is fulfilled to a good approximation only at low frequencies. It will become clear from the considerations of § 13.3 where we examine the frequency dependence of σ from an elementary model, that at the high frequencies

* When $n < 1$, the real phase velocity c/n exceeds the velocity of light in vacuum, but as explained at the end of § 1.3.3 this is not in contradiction with the theory of relativity.

($\omega \sim 3{\cdot}2 \,.\, 10^{15}\,\text{sec}^{-1}$) corresponding to the sodium D lines, to which Table XXVII refers, σ is complex, and its imaginary part is, in fact, appreciably larger than its real part. Indeed the dependence of the optical constants of metals on the wavelength, determined from experiments, shows a much more complicated behaviour than our formulae predict (see Fig. 13.3).

It appears from investigations of HAGEN and RUBENS* and subsequent workers that the reflectivity of many metals, calculated from the elementary theory that we outlined and with σ approximated by its static value, is in good agreement with results of experiments, provided the wavelength of the radiation λ is not shorter than about 10^{-3} cm. If we substitute for n and $n\kappa$ from § 13.1 (24), eq. (28) becomes (taking $\mu = 1$),

$$\mathscr{R} = \frac{2\dfrac{\sigma}{\nu} + 1 - 2\sqrt{\dfrac{\sigma}{\nu}}}{2\dfrac{\sigma}{\nu} + 1 + 2\sqrt{\dfrac{\sigma}{\nu}}}. \tag{29}$$

When ν/σ is sufficiently small, we may neglect 1 in comparison with the other terms and may develop (29) in powers of $\sqrt{\nu/\sigma}$. We then obtain

$$\mathscr{R} \sim 1 - 2\sqrt{\frac{\nu}{\sigma}} + \ldots \tag{30}$$

HAGEN and RUBENS found that, at wavelength $\lambda = 1{\cdot}2 \,.\, 10^{-3}$ cm, one has for copper $1 - \mathscr{R} = 1{\cdot}6 \,.\, 10^{-2}$, whilst on substituting for σ the static value of the conductivity, (30) gives $1 - \mathscr{R} = 1{\cdot}4 \,.\, 10^{-2}$.

As the wavelength is increased further, $\mathscr{R}$ becomes so nearly equal to unity that it is difficult to measure $1 - \mathscr{R}$ with any accuracy. HAGEN and RUBENS obtained, however, useful estimates by an indirect method. According to *Kirchhoff's law* of heat radiation† the ratio of the emissive power E_ν to the absorptive power A_ν of a body‡ depends only on the frequency ν and on the temperature T of the body and not on the nature of the body, i.e.

$$\frac{E_\nu}{A_\nu} = K(\nu, T), \tag{31}$$

where $K(\nu, T)$ is a universal function of ν and T. Evidently K is equal to the emissive power of a body whose absorptive power is unity, a so-called *black body*. Now suppose that radiation falls on a metal specimen of such a thickness that all the incident energy that is not reflected is absorbed in its interior. Then

$$A_\nu = 1 - \mathscr{R}, \tag{32}$$

* E. HAGEN and H. RUBENS, *Ann. d. Physik*, (4), **11** (1903), 873.

† See, for example, M. PLANCK, *Theory of Heat* (London, Macmillan, 1932), p. 189; or A. SOMMERFELD, *Thermodynamics and Statistical Mechanics*, edited by F. BOPP and J. MEIXNER (New York, Academic Press, 1956), p. 136.

‡ By emissive power is meant the radiant energy emitted by the body per unit time, by absorptive power the fraction which the body absorbs of the radiant energy which falls upon it.

and from (30), (31) and (32)

$$A_\nu = \frac{E_\nu}{K(\nu, T)} = 2\sqrt{\frac{\nu}{\sigma}}, \tag{33}$$

or,

$$\sqrt{\sigma}E_\nu = 2\sqrt{\nu}K(\nu, T). \tag{34}$$

The right-hand side of this equation is independent of the nature of the metal. It is a well-known function of ν and T, the function $K(\nu, T)$ being precisely known both from experiment and theory, and represented by the celebrated formula of PLANCK.*

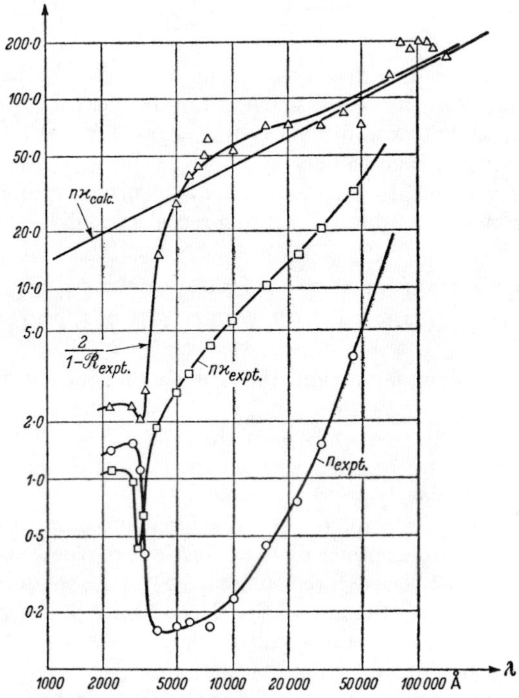

Fig. 13.3. The optical constants of silver as functions of the wavelength. Subscript "expt" refers to data obtained from experiment. The scales are logarithmic.

It follows that the validity of the formula (30) may be tested even when $\mathscr{R}$ is very close to unity by determining the conductivity σ and the emissive power E_ν as functions of the frequency and temperature and examining whether the product $\sqrt{\sigma}E_\nu$ satisfies (34). HAGEN and RUBENS confirmed that this is so at long infra-red wavelengths, using for this purpose the so-called *residual rays*. These are rays left over from a wider spectral range after repeated reflections from certain crystals, e.g. fluorite, rock-salt or sylvine. These substances have pronounced absorption maxima in the

* See, for example, M. BORN, *Atomic Physics* (London and Glasgow, Blackie and Son, 5th ed., 1951), p. 238.

spectral region $\lambda = 22{\cdot}9\mu$ to 63μ, and hence [cf. (28)] highly selective reflectivity for such wavelengths.

In Fig. 13.3 curves are given, illustrating, for the case of silver, the dependence of n and $n\kappa$ on the wavelength, as determined from experiment. For comparison the theoretical curve computed from the formula § 13.1 (24) is also shown. The scales are logarithmic so that the theoretical curve is the straight line

$$\log n \sim \log n\kappa \sim \tfrac{1}{2}\log \lambda + C,$$

where $C = \log\sqrt{\mu\sigma/c}$. From § 13.1 (24) and from (30) we may also express n and $n\kappa$ in terms of the reflectivity (for long waves):

$$n \sim n\kappa \sim \frac{2}{1-\mathscr{R}}. \tag{35}$$

The function $2/(1-\mathscr{R})$ is also displayed in the figure for comparison. We see that the experimental curve for $n\kappa$ has a sharp minimum near $\lambda = 3000$ Å and that the curve for n has a much flatter minimum near $\lambda = 5000$ Å. At about $\lambda = 3300$ Å the reflectivity of the silver is seen to be very poor.

With increasing wavelength the experimental curves approach the theoretical curve calculated from the conductivity measured electrically.

13.3. ELEMENTARY ELECTRON THEORY OF THE OPTICAL CONSTANTS OF METALS

We pointed out in the previous sections that the conductivity σ, just as the dielectric constant ε and the magnetic permeability μ, is not a true constant of the medium, but that it depends on the frequency ω of the field. We will now present a rough, simple model, (due to P. DRUDE), from which the frequency dependence of σ may be derived, at least for sufficiently low frequencies.

Let us recall first that the response of a dielectric medium to an external electromagnetic field is largely determined by the behaviour of electrons that are bound to the atomic nuclei by quasi-elastic forces (cf. § 2.3.4). In a conducting medium (such as a metal), unlike in a dielectric, not all the electrons are bound to the atoms. Some move between the molecules and are said to be *free* electrons, to distinguish them from the other electrons that are bound to the atoms, just as in a dielectric. In the absence of an external electromagnetic field, the free electrons move in a random manner and hence they do not give rise to a net current flow. When an external field is applied the free electrons acquire an additional velocity and their motion becomes more orderly, even though occasionally the electrons still collide with the (essentially stationary) atoms. This more orderly motion of the electrons gives rise to the induced current flow.

We cannot enter into a detailed discussion of this process which has to be treated by means of statistical methods of the kinetic theory of gases. The very plausible result is that the averaged total effect is the same as that of a damping force proportional and opposite in direction to the velocity of a model electron that represents the average behaviour of the whole set of electrons. The equation of motion of this model electron in an electric field $\boldsymbol{E}$ is, therefore,

$$m\ddot{\boldsymbol{r}} + m\beta\dot{\boldsymbol{r}} = e\boldsymbol{E}, \tag{1}$$

where m is the mass, e the charge of the electron and β the damping constant referred to unit mass. Unlike the equation of motion for a bound electron (§ 2.3 (33)) which contains on the right hand side an "effective field" $\boldsymbol{E}'$, Eq. (1) contains on the right hand side the macroscopic electric field $\boldsymbol{E}$, which is believed to represent more closely the field that acts on a free electron in a conductor.

In order to understand the meaning of the damping constant β in Eq. (1), consider first the case where no electric field is present. If $\boldsymbol{E} = 0$, we have

$$\ddot{\boldsymbol{r}} + \beta\dot{\boldsymbol{r}} = 0, \tag{2}$$

with the solution

$$\boldsymbol{r} = \boldsymbol{r}_0 - \frac{1}{\beta}\boldsymbol{v}_0\, e^{-\beta t}, \qquad \dot{\boldsymbol{r}} = \boldsymbol{v} = \boldsymbol{v}_0\, e^{-\beta t}; \tag{3}$$

we see that in this case the model electron starting with the velocity $\boldsymbol{v}_0$ is slowed down in an exponential way, with β as decay constant. The time $\tau = 1/\beta$ is called the *decay time*, or the *relaxation time*. It is typically of the order of 10^{-14} sec.

Let us now assume a time-harmonic field $\boldsymbol{E} = \boldsymbol{E}_0\, e^{-i\omega t}$. The solution of (1) is then the sum of two terms, one representing the decaying motion (solution of the homogeneous equation (2)) and the other representing a periodic motion

$$\boldsymbol{r} = -\frac{e}{m(\omega^2 + i\beta\omega)}\boldsymbol{E}. \tag{4}$$

This periodic motion gives rise to a current in the medium. If there are N free electrons per unit volume, the current density $\boldsymbol{j}$ is given by

$$\boldsymbol{j} = Ne\dot{\boldsymbol{r}} = \frac{Ne^2}{m(\beta - i\omega)}\boldsymbol{E}. \tag{5}$$

Comparing (5) with the constitutive relation § 1.1 (9), viz. $\boldsymbol{j} = \sigma\boldsymbol{E}$, we see that

$$\sigma = \frac{Ne^2}{m(\beta - i\omega)}. \tag{6}$$

As we already mentioned, τ is typically of the order of 10^{-14} sec, so that β is then of the order 10^{14} sec^{-1}. It is thus clear, from Eq. (6), that when $\omega \ll \beta$, σ may be approximated by its static value $\sigma_0 = Ne^2/m\beta$ which, of course, is real. On the other hand, when $\omega \gg \beta$ (which is usually the case at optical frequencies), the imaginary part of σ will become large compared to its real part. It is, therefore, evident that only for frequencies $\omega \ll \beta$ is one justified in separating the real and imaginary parts of the complex dielectric constant in the manner that leads to the formulae § 13.1 (16) and § 13.1 (17).

According to § 13.1 (16), the frequency dependence of the complex dielectric constant $\hat{\varepsilon}$ and of the complex refractive index $\hat{n}$ arises from the dependence on frequency not only of the conductivity σ (contribution from free electrons), but also of the real dielectric constant ε (contribution from bound electrons). At low enough frequencies the contribution from the bound electrons may be shown to be small compared to the contribution from the free electrons. Under these circumstances we may replace ε by unity and σ by the expression (6) in § 13.1 (16). We then obtain, if we assume the conductor to be non-magnetic ($\mu = 1$), the following expression for $\hat{\varepsilon}$:

$$\hat{\varepsilon} \equiv \hat{n}^2 = 1 - \frac{4\pi Ne^2}{m}\frac{1}{\omega(\omega - i\beta)}. \tag{7}$$

On separating the real and imaginary parts in (7), and on making use of Eq. § 3.1 (15a), viz. $\hat{n}^2 = n^2(1 + 2i\kappa - \kappa^2)$, we obtain the formulae

$$\text{Re } \hat{\varepsilon} \equiv n^2(1 - \kappa^2) = 1 - \frac{4\pi Ne^2}{m(\omega^2 + \beta^2)}, \tag{8a}$$

$$\text{Im } \hat{\varepsilon} \equiv n^2\kappa = \frac{2\pi Ne^2/\beta}{m\omega(\omega^2 + \beta^2)}. \tag{8b}$$

We may readily deduce from (8a), that if β is sufficiently small, the real part of $\hat{\varepsilon}$ is negative for low enough frequencies, but is evidently positive when ω is large. The critical value ω_c of the frequency at which the real part of $\hat{\varepsilon}$ changes sign is given by

$$\omega_c^2 = \frac{4\pi Ne^2}{m} - \beta^2. \tag{9}$$

We may re-write equations (8) in terms of this critical value and obtain

$$\text{Re } \hat{\varepsilon} \equiv n^2(1 - \kappa^2) = 1 - \frac{\omega_c^2 + \beta^2}{\omega^2 + \beta^2}, \tag{10a}$$

$$\text{Im } \hat{\varepsilon} \equiv n^2\kappa = \frac{\beta(\omega_c^2 + \beta^2)}{2\omega(\omega^2 + \beta^2)}. \tag{10b}$$

We shall now assume that ω_c^2 is much larger than β^2, so that in place of (9) we may write

$$\omega_c^2 \sim \frac{4\pi Ne^2}{m}. \tag{11}$$

If we also restrict ourselves to sufficiently high frequencies ($\omega^2 \gg \beta^2$), we obtain, in place of (10), the simpler formulae

$$\text{Re } \hat{\varepsilon} \equiv n^2(1 - \kappa^2) \sim 1 - \left(\frac{\omega_c}{\omega}\right)^2, \tag{12a}$$

$$\text{Im } \hat{\varepsilon} \equiv n^2\kappa \sim \frac{\beta}{2\omega}\left(\frac{\omega_c}{\omega}\right)^2. \tag{12b}$$

It follows from (12a) that when $\omega^2 < \omega_c^2$ (but still $\omega^2 \gg \beta^2$), the real part of $\hat{\varepsilon}$ is negative and $\kappa > 1$. The negative value of the real part of $\hat{\varepsilon}$ reflects the fact that under these circumstances the vibrations of the electrons are out of phase by a quarter of a period with the exciting field, as is evident from Eq. (5). For sufficiently low values of ω, the attenuation index κ becomes large compared to unity and the reflectivity (given for normal incidence by § 13.2 (28)), is readily seen to have a value close to unity. On the other hand, when $\omega^2 > \omega_c^2$ (but $\omega^2 \gg \beta^2$), the real part of $\hat{\varepsilon}$ is evidently positive, so that $\kappa < 1$, and when ω is sufficiently large κ becomes small compared to unity and the imaginary part of $\hat{\varepsilon}$ becomes small compared to its real part. The metal must then be expected to behave essentially as a dielectric.

The alkali metals exhibit precisely these phenomena, for in the long wavelength region they are opaque and highly reflecting, whereas at some critical wavelength in the visible or ultra-violet they become transparent, and have comparatively low absorption. Table XXVIII shows in the second row the experimentally determined wavelengths at which this transition occurs. The third row contains these critical wavelengths $\lambda_c = 2\pi c/\omega_c$ determined from the approximate formula (11), where the number of free electrons is taken to be the same as the number N of atoms in the unit volume. It is seen that the values in the two rows are different, except for sodium. The last row gives the ratio of number N_{eff} of electrons that are "effective" and the number of atoms, determined from the formula

$$\frac{N_{\text{eff}}}{N} = \frac{(\lambda_c)^2_{\text{calc}}}{(\lambda_c)^2_{\text{obs}}}. \tag{13}$$

It is seen that this number is of the order of unity though (except for sodium) considerably smaller. Thus the elementary theory gives the correct order of magnitude of the parameters, but does not describe the phenomena in detail.

TABLE XXVIII

The critical wavelengths λ_c below which the alkali metals become transparent, and above which they are opaque and highly reflecting

Metal	Lithium	Sodium	Potassium	Rubidium	Caesium
$(\lambda_c)_{\text{obs}}$	2050 Å	2100 Å	3150 Å	3600 Å	4400 Å
$(\lambda_c)_{\text{calc}}$	1500 Å	2100 Å	2900 Å	3200 Å	3600 Å
$\frac{N_{\text{eff}}}{N}$	0·54	1·00	0·85	0·79	0·67

The theory can be somewhat improved by using in place of the crude approximations (12) the more accurate formulae (10), which contain the decay constant β. However, because of the complexity of the physical processes involved in the interaction of a high frequency electromagnetic field with a metal, it is not possible to extend appreciably the range of validity of the elementary classical theory that we outlined in this section by a simple modification. A completely satisfactory theory of the optical properties of metals can only be obtained on the basis of quantum mechanics.

13.4. WAVE PROPAGATION IN A STRATIFIED CONDUCTING MEDIUM. THEORY OF METALLIC FILMS

In § 1.6 we have studied the propagation of electromagnetic waves in stratified dielectric media, that is, in dielectric media with optical properties depending on one Cartesian coordinate only. We shall now briefly discuss the extension of the theory to stratified media that contain absorbing elements. Thus we assume that in addition to ε and μ being functions of only one coordinate, there may be a finite conductivity σ which likewise is a function of this coordinate alone.

As explained at the beginning of 13.2, the formulae of Chapter I as far as they involve only linear relations between the components of the field vectors of a time harmonic wave retain their validity for conducting media, provided that the real dielectric constant ε and the real wave number k are replaced by the complex dielectric constant $\hat{\varepsilon} = \varepsilon + i4\pi\sigma/\omega$ and by the complex wave number $\hat{k} = \omega\sqrt{\mu(\varepsilon + i4\pi\sigma/\omega)}/c$ respectively. Hence we may take over the basic formulae of the theory of stratified dielectric media as developed in § 1.6, provided we make this formal change in the appropriate formulae. It follows, in particular, that a stratified absorbing medium may be characterized by a two-by-two matrix. In contrast to the case of a dielectric stratified medium, the elements of this matrix are no longer real or pure imaginary numbers but are complex numbers that contain both real and imaginary parts.

We shall illustrate the theory by studying in detail two cases of practical interest.

13.4.1 An absorbing film on a transparent substrate

Consider a plane parallel absorbing film situated between two dielectric media (Fig. 13.4). The formulae relating to the reflection and transmission of a plane mono-

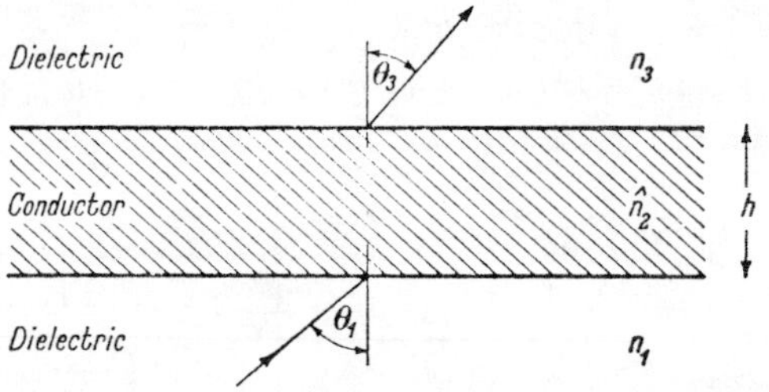

Fig. 13.4. An absorbing film situated between two dielectric media.

chromatic wave by the film are obtained from equations (55)–(58) of § 1.6 on replacing n_2 by $\hat{n}_2 = n_2(1 + i\kappa_2)$. It is convenient to set

$$\hat{n}_2 \cos \theta_2 = u_2 + iv_2, \tag{1}$$

where u_2 and v_2 are real. We can easily express u_2 and v_2 in terms of the angle of incidence and the constants which characterize the optical properties of the first and the second medium. It follows, on squaring (1) and using the law of refraction $\hat{n}_2 \sin \theta_2 = n_1 \sin \theta_1$, that

$$(u_2 + iv_2)^2 = \hat{n}_2^2 - n_1^2 \sin^2 \theta_1. \tag{2}$$

On equating real and imaginary parts this gives

$$\left.\begin{aligned} u_2^2 - v_2^2 &= n_2^2(1 - \kappa_2^2) - n_1^2 \sin^2 \theta_1, \\ u_2 v_2 &= n_2^2 \kappa_2. \end{aligned}\right\} \tag{3}$$

From (3) we find that

$$\left.\begin{aligned} 2u_2^2 &= \quad n_2^2(1 - \kappa_2^2) - n_1^2 \sin^2\theta_1 + \sqrt{[n_2^2(1 - \kappa_2^2) - n_1^2 \sin^2 \theta_1]^2 + 4n_2^4\kappa_2^2}, \\ 2v_2^2 &= -[n_2^2(1 - \kappa_2^2) - n_1^2 \sin^2 \theta_1] + \sqrt{[n_2^2(1 - \kappa_2^2) - n_1^2 \sin^2 \theta_1]^2 + 4n_2^4\kappa_2^2}. \end{aligned}\right\} \tag{4}$$

Next we must evaluate the reflection and transmission coefficients for the interfaces 1–2 and 2–3 respectively, for these coefficients enter the formulae for the reflection and transmission coefficients of the film. We consider separately the cases when the electric vector of the incident wave is perpendicular, or parallel, to the plane of incidence.

Electric vector perpendicular to the plane of incidence (TE wave)

In this case we have, on replacing $n_2 \cos \theta_2$ by $\hat{n}_2 \cos \theta_2 = u_2 + iv_2$ in § 1.6 (55),

$$r_{12} = \rho_{12}\, e^{i\phi_{12}} = \frac{n_1 \cos \theta_1 - (u_2 + iv_2)}{n_1 \cos \theta_1 + (u_2 + iv_2)}. \tag{5}$$

We shall later need explicit expressions for the amplitude ρ_{12} and the phase change ϕ_{12}. From (5) we have:

$$\rho_{12}^2 = \frac{(n_1 \cos\theta_1 - u_2)^2 + v_2^2}{(n_1 \cos\theta_1 + u_2)^2 + v_2^2}, \qquad \tan\phi_{12} = \frac{2v_2 n_1 \cos\theta_1}{u_2^2 + v_2^2 - n_1^2 \cos^2\theta_1}. \tag{6}$$

For transmission at the first interface, we have from § 1.6 (56)

$$t_{12} = \tau_{12}\, e^{i\chi_{12}} = \frac{2n_1 \cos\theta_1}{n_1 \cos\theta_1 + u_2 + iv_2}, \tag{7}$$

which gives

$$\tau_{12}^2 = \frac{(2n_1 \cos\theta_1)^2}{(n_1 \cos\theta_1 + u_2)^2 + v_2^2}, \qquad \tan\chi_{12} = -\frac{v_2}{n_1 \cos\theta_1 + u_2}. \tag{8}$$

In a strictly analogous way we obtain the following expressions relating to reflection and transmission at the second interface:

$$\rho_{23}^2 = \frac{(n_3 \cos\theta_3 - u_2)^2 + v_2^2}{(n_3 \cos\theta_3 + u_2)^2 + v_2^2}, \qquad \tan\phi_{23} = \frac{2v_2 n_3 \cos\theta_3}{u_2^2 + v_2^2 - n_3^2 \cos^2\theta_3}, \tag{9}$$

$$\tau_{23}^2 = \frac{4(u_2^2 + v_2^2)}{(n_3 \cos\theta_3 + u_2)^2 + v_2^2}, \qquad \tan\chi_{23} = \frac{v_2 n_3 \cos\theta_3}{u_2^2 + v_2^2 + u_2 n_3 \cos\theta_3}. \tag{10}$$

Since, according to the law of refraction $n_1 \sin\theta_1 = \hat{n}_2 \sin\theta_2$, $\hat{n}_2 \sin\theta_2 = n_3 \sin\theta_3$, the angle θ_3 is determined from θ_1 by means of the formula

$$n_3 \sin\theta_3 = n_1 \sin\theta_1. \tag{11}$$

Electric vector parallel to the plane of incidence (*TM wave*)

As explained in § 1.6.3 the formulae for the reflection and transmission coefficients for a TM wave can be obtained from those for a TE wave simply by replacing the quantities $p_j = n_j \cos\theta_j$ by $q_j = \cos\theta_j/n_j$, it being assumed that the media are non-magnetic. The quantities r and t now refer to the ratios of the magnetic, not the electric vectors. In particular we have from § 1.6 (55),

$$r_{12} = \rho_{12}\, e^{i\phi_{12}} = \frac{\dfrac{1}{n_1}\cos\theta_1 - \dfrac{1}{\hat{n}_2}\cos\theta_2}{\dfrac{1}{n_1}\cos\theta_1 + \dfrac{1}{\hat{n}_2}\cos\theta_2} = \frac{\hat{n}_2^2 \cos\theta_1 - n_1\hat{n}_2 \cos\theta_2}{\hat{n}_2^2 \cos\theta_1 + n_1\hat{n}_2 \cos\theta_2}$$

$$= \frac{[n_2^2(1-\kappa_2^2) + 2in_2^2\kappa_2]\cos\theta_1 - n_1(u_2 + iv_2)}{[n_2^2(1-\kappa_2^2) + 2in_2^2\kappa_2]\cos\theta_1 + n_1(u_2 + iv_2)}. \tag{12}$$

From (12) we find after a straightforward calculation

$$\left.\begin{aligned} \rho_{12}^2 &= \frac{[n_2^2(1-\kappa_2^2)\cos\theta_1 - n_1u_2]^2 + [2n_2^2\kappa_2\cos\theta_1 - n_1v_2]^2}{[n_2^2(1-\kappa_2^2)\cos\theta_1 + n_1u_2]^2 + [2n_2^2\kappa_2\cos\theta_1 + n_1v_2]^2}, \\ \tan\phi_{12} &= 2n_1n_2^2\cos\theta_1 \frac{2\kappa_2u_2 - (1-\kappa_2^2)v_2}{n_2^4(1+\kappa_2^2)^2\cos^2\theta_1 - n_1^2(u_2^2+v_2^2)}. \end{aligned}\right\} \tag{13}$$

For the ratio t_{12} we obtain from § 1.6 (56) on replacing $n_j \cos\theta_j$ by $\cos\theta_j/n_j$

$$t_{12} = \tau_{12} e^{i\chi_{12}} = \frac{\frac{2}{n_1}\cos\theta_1}{\frac{1}{n_1}\cos\theta_1 + \frac{1}{\hat{n}_2}\cos\theta_2}$$

$$= \frac{2[n_2^2(1-\kappa_2^2) + 2in_2^2\kappa_2]\cos\theta_1}{[n_2^2(1-\kappa_2^2) + 2in_2^2\kappa_2]\cos\theta_1 + n_1(u_2 + iv_2)}. \tag{14}$$

From (14) we find that

$$\left.\begin{aligned} \tau_{12}^2 &= \frac{4n_2^4(1+\kappa_2^2)^2\cos^2\theta_1}{[n_2^2(1-\kappa_2^2)\cos\theta_1 + n_1u_2]^2 + [2n_2^2\kappa_2\cos\theta_1 + n_1v_2]^2}, \\ \tan\chi_{12} &= \frac{n_1[2\kappa_2u_2 - (1-\kappa_2^2)v_2]}{n_2^2(1+\kappa_2^2)^2\cos\theta_1 + n_1[(1-\kappa_2^2)u_2 + 2\kappa_2v_2]}. \end{aligned}\right\} \tag{15}$$

In a similar way we obtain the following formulae for the reflection and transmission coefficients relating to the second interface:

$$\left.\begin{aligned} \rho_{23}^2 &= \frac{[n_2^2(1-\kappa_2^2)\cos\theta_3 - n_3u_2]^2 + [2n_2^2\kappa_2\cos\theta_3 - n_3v_2]^2}{[n_2^2(1-\kappa_2^2)\cos\theta_3 + n_3u_2]^2 + [2n_2^2\kappa_2\cos\theta_3 + n_3v_2]^2}, \\ \tan\phi_{23} &= 2n_3n_2^2\cos\theta_3\,\frac{2\kappa_2u_2 - (1-\kappa_2^2)v_2}{n_2^4(1+\kappa_2^2)^2\cos^2\theta_3 - n_3^2(u_2^2+v_2^2)}, \end{aligned}\right\} \tag{16}$$

and

$$\left.\begin{aligned} \tau_{23}^2 &= \frac{4n_3^2(u_2^2+v_2^2)}{[n_3u_2 + n_2^2(1-\kappa_2^2)\cos\theta_3]^2 + [n_3v_2 + 2n_2^2\kappa_2\cos\theta_3]^2}, \\ \tan\chi_{23} &= \frac{n_2^2[(1-\kappa_2)^2v_2 - 2\kappa_2u_2]\cos\theta_3}{n_3[u_2^2+v_2^2] + n_2^2[(1-\kappa_2)^2u_2 + 2\kappa_2v_2]\cos\theta_3}. \end{aligned}\right\} \tag{17}$$

From the knowledge of the quantities ρ_{12}, ϕ_{12}, etc., the complex reflection and transmission coefficients of the film may immediately be evaluated. It will be useful to set

$$\eta = \frac{2\pi}{\lambda_0} h, \tag{18}$$

so that

$$\beta = \frac{2\pi}{\lambda_0}\hat{n}_2 h\cos\theta_2 = (u_2 + iv_2)\eta. \tag{19}$$

The equations § 1.6 (57)–(58) now become

$$r = \rho e^{i\delta_r} = \frac{\rho_{12}\,e^{i\phi_{12}} + \rho_{23}\,e^{-2v_2\eta}\,e^{i(\phi_{23} + 2u_2\eta)}}{1 + \rho_{12}\rho_{23}\,e^{-2v_2\eta}\,e^{i(\phi_{12}+\phi_{23}+2u_2\eta)}}, \tag{20}$$

$$t = \tau e^{i\delta_t} = \frac{\tau_{12}\tau_{23}\,e^{-v_2\eta}\,e^{i[\chi_{12}+\chi_{23}+u_2\eta]}}{1 + \rho_{12}\rho_{23}\,e^{-2v_2\eta}\,e^{i[\phi_{12}+\phi_{23}+2u_2\eta]}}. \tag{21}$$

From (20) we obtain, after straightforward calculation, the following expressions for the reflectivity $\mathscr{R}$ and for the phase change δ_r on reflection:

$$\mathscr{R} = |r|^2 = \frac{\rho_{12}^2\,e^{2v_2\eta} + \rho_{23}^2\,e^{-2v_2\eta} + 2\rho_{12}\rho_{23}\cos[\phi_{23} - \phi_{12} + 2u_2\eta]}{e^{2v_2\eta} + \rho_{12}^2\rho_{23}^2\,e^{-2v_2\eta} + 2\rho_{12}\rho_{23}\cos[\phi_{12} + \phi_{23} + 2u_2\eta]}, \tag{22}$$

$$\tan \delta_r = \frac{\rho_{23}(1-\rho_{12}^2)\sin(2u_2\eta+\phi_{23})+\rho_{12}[e^{2v_2\eta}-\rho_{23}^2\,e^{-2v_2\eta}]\sin\phi_{12}}{\rho_{23}(1+\rho_{12}^2)\cos(2u_2\eta+\phi_{23})+\rho_{12}[e^{2v_2\eta}+\rho_{23}^2\,e^{-2v_2\eta}]\cos\phi_{12}}. \tag{23}$$

These formulae are valid for a TE wave as well as for a TM wave. In the former case one must substitute for ρ and ϕ the values given by (6) and (9), in the latter case those given by (13) and (16).

In a similar way we obtain from (21) the following expressions for the transmissivity $\mathscr{T}$ and for the phase change δ_t on transmission:

$$\mathscr{T} = \frac{n_3 \cos\theta_3}{n_1\cos\theta_1}|t|^2 = \frac{n_3 \cos\theta_3}{n_1\cos\theta_1}\,\frac{\tau_{12}^2\tau_{23}^2\,e^{-2v_2\eta}}{1+\rho_{12}^2\rho_{23}^2\,e^{-4v_2\eta}+2\rho_{12}\rho_{23}\,e^{-2v_2\eta}\cos[\phi_{12}+\phi_{23}+2u_2\eta]}, \tag{24}$$

$$\tan[\delta_t-\chi_{12}-\chi_{23}+u_2\eta] = \frac{e^{2v_2\eta}\sin 2u_2\eta-\rho_{12}\rho_{23}\sin(\phi_{12}+\phi_{23})}{e^{2v_2\eta}\cos 2u_2\eta+\rho_{12}\rho_{23}\cos(\phi_{12}+\phi_{23})}. \tag{25}$$

For a TM wave the factor $n_3\cos\theta_3/n_1\cos\theta_1$ must be replaced by $(\cos\theta_3/n_3)/(\cos\theta_1/n_1)$. For a TE wave the values given by (6), (8), (9), and (10) are substituted in these formulae, and for a TM wave those given by (13), (15), (16), and (17).

It may be worthwhile to recall that the phase change on reflection (δ_r) is referred to the first boundary (1–2), whereas the phase change on transmission (δ_t) is referred to the second boundary.

Equations (22)–(25) allow the computations of the four basic quantities that characterize reflection and transmission by an absorbing film of known optical proper-

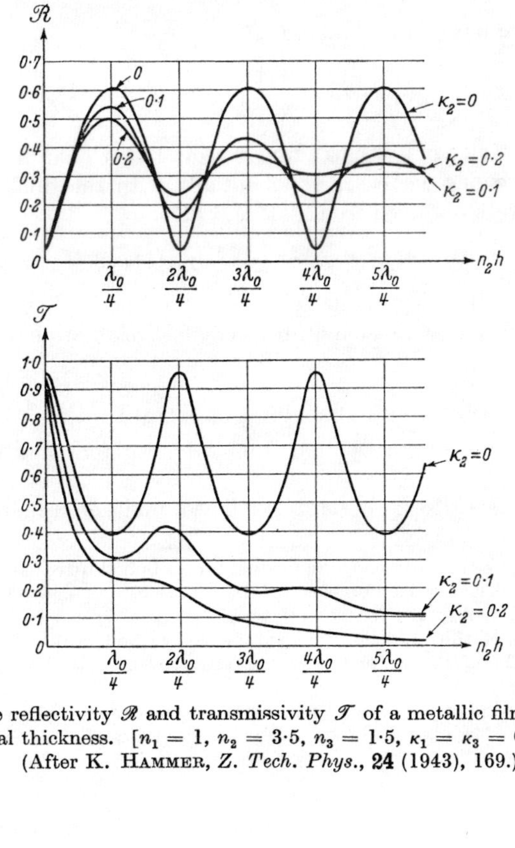

Fig. 13.5. The reflectivity $\mathscr{R}$ and transmissivity $\mathscr{T}$ of a metallic film as functions of its optical thickness. [$n_1 = 1$, $n_2 = 3{\cdot}5$, $n_3 = 1{\cdot}5$, $\kappa_1 = \kappa_3 = 0$; $\theta_1 = 0$.] (After K. Hammer, *Z. Tech. Phys.*, **24** (1943), 169.)

ties and of prescribed thickness. Figs. 13.5 and 13.6 illustrate, for some typical cases, the dependence of the reflectivity and transmissivity on the thickness of the film.

For a non-absorbing film $\mathscr{R}$ and $\mathscr{T}$ are periodic functions of the film thickness h, with a period of one wavelength. Absorption is seen to reduce the amplitude of the successive maxima and to give rise to a displacement of the maxima in the direction of smaller thickness. At optical wavelengths absorption of metals is so large that the thickness at which there is appreciable transmission is well below a quarter wavelength* (cf. Table XXVI, p. 615). With transmitted light maxima and minima are therefore not observed.

In optics metal films are chiefly used to attain high reflectivities, for example in connection with the FABRY–PEROT interferometer (§ 7.6.2). Such films used to be produced by chemical deposition but this method has in more recent times been superseded by the techniques of high vacuum evaporation.†

Finally let us briefly consider reflection and transmission with a "thick" film. If the thickness h and consequently the parameter η are sufficiently large all the terms in (22)–(25) which do not contain the multiplicative factor $\exp(2v_2\eta)$ may be neglected. For example, if $\exp(2v_2\eta) \gtrsim 100$ this neglection does not, as a rule, involve an error of more than a few per cent. For such a film, one has at normal incidence $4\pi h n_2\kappa_2/\lambda \geqslant \log_e 100 = 4{\cdot}61$, or (dropping the suffix 2),

$$\frac{h}{\lambda} \gtrsim \frac{0{\cdot}37}{n\kappa}. \tag{26}$$

For a silver film, for example, $n\kappa \sim 3{\cdot}67$ at $\lambda = 5780$ Å, and (26) gives $h \gtrsim \lambda/10 \sim 5{\cdot}8 \,.\, 10^{-6}$ cm.

For a thick film, we have from (22) and (24)

$$\mathscr{R} \sim \rho_{12}^{\ 2}, \qquad \mathscr{T} = \frac{n_3 \cos\theta_3}{n_1 \cos\theta_1} \tau_{12}^{\ 2} \tau_{23}^{\ 2}\, e^{-4v_2\eta}. \tag{27}$$

We see that the reflectivity of a "thick" film is almost that of an infinitely thick one, and that its transmissivity decreases exponentially with the thickness. The phase changes are immediately obtained from (23) and (25):

$$\delta_r \sim \phi_{12}, \qquad \delta_t \sim \chi_{12} + \chi_{23} + u_2\eta. \tag{28}$$

Formulae (27) and (28) interpret our definition of a "thick" film in somewhat more physical terms, implying that in such a film the effect of multiple beam interference is negligible.

13.4.2 A transparent film on an absorbing substrate

As a second example, consider reflection from a transparent film on an absorbing substrate (Fig. 13.6).

In this case r_{12} is real whilst r_{23} is complex. The amplitude ratio ρ_{23} and the phase

---

* Simplified formulae relating to such thin films may be obtained by expanding the numerator and denominator (22)–(25) into series in powers of the film thickness, and retaining terms in the first few powers only (cf. F. ABELÈS, *Rev. d'Optique*, **32** (1953), 257).

The optical properties of thin metallic films in the visible and infra-red spectral region are thoroughly discussed by L. N. HADLEY and D. M. DENNISON, *J. Opt. Soc. Amer.*, **37** (1947), 451; **38** (1948), 483.

† See, for example, S. TOLANSKY, *Multiple-beam Interferometry of Surfaces and Films* (Oxford University Press, 1948), p. 26; or O. S. HEAVENS, *Optical Properties of Thin Solid Films* (London, Butterworths, 1955).

change ϕ_{23} are given by (6) or (13), with suffixes 1 and 2 replaced by 2 and 3 respectively. According to § 1.6 (57) we now have

$$r = \frac{r_{12} + \rho_{23}\, e^{i(\phi_{23}+2\beta)}}{1 + r_{12}\rho_{23}\, e^{i(\phi_{23}+2\beta)}}. \tag{29}$$

This expression is identical with the formula (57) of § 1.6 if 2β is replaced by $2\beta + \phi_{23}$ and r_{23} by ρ_{23}. Thus without any calculation we may at once write down the expression

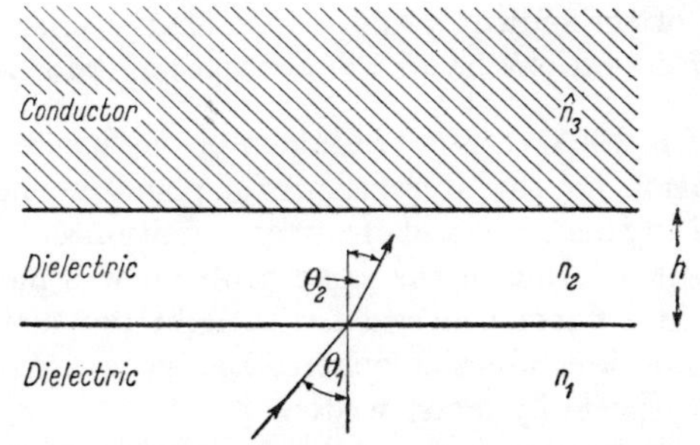

Fig. 13.6. A transparent film on an absorbing substrate.

for the reflectivity and the phase change on reflection, simply by making this substitution in equations (59) and (61) of § 1.6; we then obtain

$$\mathscr{R} = \frac{r_{12}^2 + \rho_{23}^2 + 2r_{12}\rho_{23}\cos(\phi_{23} + 2\beta)}{1 + r_{12}^2\rho_{23}^2 + 2r_{12}\rho_{23}\cos(\phi_{23} + 2\beta)}, \tag{30}$$

and

$$\tan\delta_r = \frac{\rho_{23}(1 - r_{12}^2)\sin(\phi_{23} + 2\beta)}{r_{12}(1 + \rho_{23}^2) + \rho_{23}(1 + r_{12}^2)\cos(\phi_{23} + 2\beta)}. \tag{31}$$

Thin transparent films on absorbing substrates have many practical uses. They are employed, for example, to protect metallic mirrors and to increase their reflectivity. They may also be used to reduce the reflectivity of a metal surface. We have mentioned on p. 65 that one may design a polarizer consisting of a dielectric film on a dielectric substrate for which $\mathscr{R}_\parallel = 0$ and $\mathscr{R}_\perp$ is quite large. With a metallic substrate one may have either $\mathscr{R}_\parallel = 0$ or $\mathscr{R}_\perp = 0$.*

13.5. DIFFRACTION BY A CONDUCTING SPHERE; THEORY OF MIE

Metals exhibit marked optical characteristics not only in bulk, but also when they are finely divided, as in colloidal suspensions. We may recall the brilliant ruby colours of colloidal gold, whether in liquids or glasses. These phenomena are of great interest, as refraction, absorption, and diffraction take place here side by side.

Were the metallic particles perfect conductors, one would be dealing with a problem of pure diffraction. We have, however, not discussed the subject from this standpoint in the chapters on diffraction, since it is the effects that are due to partial penetration of light into the particles that are of particular physical interest. Absorption then plays an important part and it is, therefore, more appropriate to treat the subject

* Cf. H. Schopper, *Optik*, **10** (1953), 426.

in the present chapter; the corresponding results relating to dielectric spheres are contained in our considerations as a limiting case ($\kappa \to 0$).

Of the early workers who studied the optical properties of metallic particles, mention must be made of MAXWELL GARNETT.* He considered the passage of light through a dielectric medium containing many small metallic spheres in a volume of linear dimensions of a wavelength. With the help of the LORENTZ–LORENZ formula (§ 2.3 (17)) MAXWELL GARNETT showed that such an assembly is equivalent to a medium of a certain complex refractive index $\hat{n}' = n'(1 + i\kappa')$ and he found formulae for n' and κ' in terms of the indices n and κ that characterize the metallic spheres. By means of these considerations he was able to account for some of the observed features.

In a paper published in 1908, G. MIE† obtained on the basis of the electromagnetic theory a rigorous solution for the diffraction of a plane monochromatic wave by a homogeneous sphere of any diameter and of any composition situated in a homogeneous medium. An equivalent solution of the same problem was published shortly afterwards by P. DEBYE‡ in a paper concerned with light pressure (i.e. the mechanical force exerted by light) on a conducting sphere, and since then the subject has been treated in its different aspects by many writers.§

The solution due to MIE, though derived for diffraction by a single sphere, also applies to diffraction by any number of spheres, provided that they are all of the same diameter and composition and provided also that they are randomly distributed and separated from each other by distances that are large compared to the wavelength. Under these circumstances there are no coherent phase relationships between the light that is scattered by the different spheres and the total scattered energy is then equal to the energy that is scattered by one sphere multiplied by their total number. It is particularly in this connection that MIE's solution is of great practical value and may be applied to a variety of problems: in addition to the question of colours exhibited by metallic suspensions, we may mention applications such as the study of atmospheric dust, interstellar particles or colloidal suspensions, the theory of the rainbow, the solar corona, the effects of clouds and fogs on the transmission of light, etc.

Before deriving MIE's formulae it will be helpful to explain briefly the method employed. We are concerned with finding the solution of MAXWELL's equations which describe the field arising from a plane monochromatic wave incident upon a spherical surface, across which the properties of the medium change abruptly. An appropriate system of curvilinear coordinates (spherical polar coordinates) is introduced and the field is represented as the sum of two "subfields"; one of the subfields is such that its electric vector has no radial component while the other has a magnetic vector with this property. In the spherical polar coordinates MAXWELL's equations together with the boundary conditions separate into a set of ordinary differential equations, which are then solved for the two subfields in the form of infinite series. Section 13.5.1 is concerned with the derivation of this solution and in § 13.5.2 the main consequences of the solution are discussed. The final section (§ 13.5.3) is concerned with some general results relating to the total amount of energy that is scattered and absorbed by an obstacle of arbitrary shape, and the case of a spherical obstacle is discussed in detail.

---

* J. C. MAXWELL GARNETT, *Phil. Trans. Roy. Soc.*, A. **203**, (1904), 385; *ibid.* **205** (1906), 237.

† G. MIE, *Ann. d. Physik* (4), **25** (1908), 377.

‡ P. DEBYE, *Ann. d. Physik* (4), **30** (1909), 57.

§ Of later investigations that deal with the basic theory, particular mention may be made of contributions by T. J. I'A. BROMWICH, *Phil. Trans. Roy. Soc.*, A, **220** (1920), 175 and by H. C. VAN DE HULST, *Rech. Astron. Observ. Utrecht*, XI, Pt. 1 (1946).

13.5.1 Mathematical solution of the problem

(a) *Representation of the field in terms of Debye's potentials*

We consider the diffraction of a plane, linearly polarized monochromatic wave by a sphere of radius a, immersed in a homogeneous, isotropic medium. We assume that the medium in which the sphere is embedded is a non-conductor and that both this medium as well as the sphere are non-magnetic.

Assuming, as usual, the time dependence $\exp(-i\omega t)$, the time independent parts of the electric and magnetic vectors both outside and inside the sphere satisfy MAXWELL's equations in their time-free form:

$$\left.\begin{aligned} \text{curl}\, \boldsymbol{H} &= -k_1 \boldsymbol{E}, & \text{(a)} \\ \text{curl}\, \boldsymbol{E} &= \quad k_2 \boldsymbol{H}, & \text{(b)} \end{aligned}\right\} \tag{1}$$

where

$$\left.\begin{aligned} k_1 &= \frac{i\omega}{c}\left(\varepsilon + i\frac{4\pi\sigma}{\omega}\right), & \text{(a)} \\ k_2 &= \frac{i\omega}{c}. & \text{(b)} \end{aligned}\right\} \tag{2}$$

The square of the usual wave number k (real outside and complex inside the sphere) is

$$k^2 = -k_1 k_2. \tag{3}$$

Quantities which refer to the medium surrounding the sphere will be denoted by superscript I, those referring to the sphere by superscript II. As the medium surrounding the sphere is assumed to be non-conducting, $\sigma^{(\text{I})} = 0$.

We take a rectangular system of coordinates with origin at the centre of the sphere with the z-direction in the direction of propagation of the wave and with the x-direction in the direction of its electric vector (Fig. 13.7).

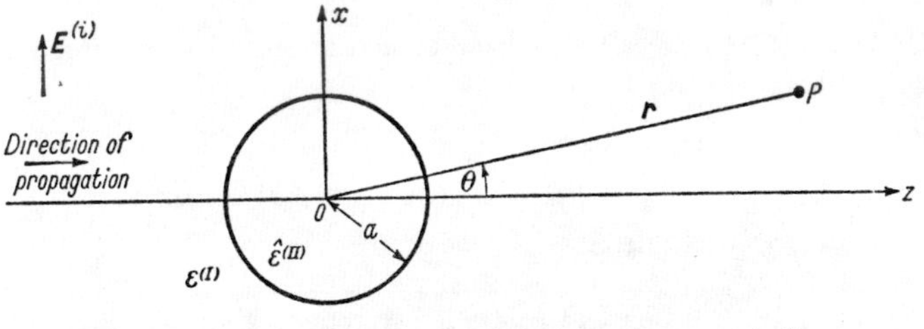

Fig. 13.7. Diffraction by a conducting sphere: notation.

If the amplitude of the electric vector of the incident wave is normalized to unity, i.e.

$$|\boldsymbol{E}^{(i)}| = |e^{ik^{(\text{I})}z}| = 1,$$

the six components of the field vectors are

$$\left.\begin{aligned} E_x^{(i)} &= e^{ik^{(\text{I})}z}, \\ H_y^{(i)} &= \frac{ik^{(\text{I})}}{k_2^{(\text{I})}} e^{ik^{(\text{I})}z}, \\ E_y^{(i)} &= E_z^{(i)} = H_x^{(i)} = H_z^{(i)} = 0. \end{aligned}\right\} \tag{4}$$

As regards boundary conditions, we only demand, in accordance with § 1.1.3, that the tangential components of $\boldsymbol{E}$ and $\boldsymbol{H}$ shall be continuous across the surface of the sphere:

$$\left.\begin{aligned} E^{(\mathrm{I})}_{\mathrm{tang}} &= E^{(\mathrm{II})}_{\mathrm{tang}}, \\ H^{(\mathrm{I})}_{\mathrm{tang}} &= H^{(\mathrm{II})}_{\mathrm{tang}}. \end{aligned}\right\} \text{ when } r = a. \tag{5}$$

The condition that the radial components of $\varepsilon \boldsymbol{E}$ and $\boldsymbol{H}$ shall also be continuous across the surface then follows from (5) and from MAXWELL's equations.

In order to satisfy the boundary conditions, we must assume that apart from the incident field $\boldsymbol{E}^{(i)}$, $\boldsymbol{H}^{(i)}$ and the field $\boldsymbol{E}^{(w)}$, $\boldsymbol{H}^{(w)}$ within the sphere there is a secondary (scattered or diffracted) field $\boldsymbol{E}^{(s)}$, $\boldsymbol{H}^{(s)}$ in the medium surrounding the sphere. Thus the total electric field in the two regions is written as

$$\left.\begin{aligned} \boldsymbol{E} &= \boldsymbol{E}^{(i)} + \boldsymbol{E}^{(s)} \qquad \text{outside the sphere} \\ &= \boldsymbol{E}^{(w)} \qquad\qquad\quad \text{within the sphere,} \end{aligned}\right\} \tag{6}$$

with similar expressions for the magnetic vector. The fields $\boldsymbol{E}^{(s)}$, $\boldsymbol{H}^{(s)}$ and $\boldsymbol{E}^{(w)}$, $\boldsymbol{H}^{(w)}$ may be considered to be analogous to the reflected field and the transmitted field respectively, for propagation involving a plane boundary (§ 1.5.1); this analogy is, however, appropriate only when the diameter of the sphere is large compared with the wavelength. Since the boundary conditions must hold for all time, all the six vectors must have the same time dependence [exp $(-i\omega t)$].

The curvilinear coordinates appropriate to the present problem are the spherical polar coordinates, r, θ, and ϕ defined by

$$\left.\begin{aligned} x &= r \sin\theta \cos\phi, \\ y &= r \sin\theta \sin\phi, \\ z &= r \cos\theta. \end{aligned}\right\} \tag{7}$$

The components of any vector $\boldsymbol{A}$ are transformed from the Cartesian system to this new system according to the rule*

$$\left.\begin{aligned} A_r &= A_x \sin\theta\cos\phi + A_y \sin\theta\sin\phi + A_z\cos\theta, \\ A_\theta &= A_x \cos\theta\cos\phi + A_y\cos\theta\sin\phi - A_z\sin\theta, \\ A_\phi &= -A_x\sin\phi + A_y\cos\phi. \end{aligned}\right\} \tag{8}$$

* See, for example, W. MAGNUS and F. OBERHETTINGER, *Formulas and Theorems for the Functions of Mathematical Physics* (New York, Chelsea Publishing Company, 1954), p. 146.

The components defined here are not those employed in the Absolute Differential Calculus of RICCI and LEVI-CIVITA. There one has two sets of different but equivalent components of a vector $\boldsymbol{A}$, the contravariant components A^i and the covariant components A_i. If $\boldsymbol{e}_1$, $\boldsymbol{e}_2$, and $\boldsymbol{e}_3$ are base vectors, in general non-orthogonal and of different lengths, the contravariant components (with respect to these base vectors) may be defined as the coefficients in the representation $\boldsymbol{A} = A^1\boldsymbol{e}_1 + A^2\boldsymbol{e}_2 + A^3\boldsymbol{e}_3$; and the covariant components may be defined as the coefficients in the representation $\boldsymbol{A} = A_1\boldsymbol{e}^1 + A_2\boldsymbol{e}^2 + A_3\boldsymbol{e}^3$, where $\boldsymbol{e}^1$, $\boldsymbol{e}^2$, and $\boldsymbol{e}^3$ are the reciprocal vectors, i.e. vectors satisfying the relations $\boldsymbol{e}^i \cdot \boldsymbol{e}_k = \delta_{ik}$ where δ_{ik} is the Kronecker symbol. In the special case when $\boldsymbol{e}_1$, $\boldsymbol{e}_2$, and $\boldsymbol{e}_3$ are orthogonal, so are also $\boldsymbol{e}^1$, $\boldsymbol{e}^2$, and $\boldsymbol{e}^3$ and the corresponding vectors in the two sets are parallel. One can then introduce one set of *natural* components $\bar{A}_i = \sqrt{A_i A^i}$ and these have a simple geometrical interpretation: they are the orthogonal projections of $\boldsymbol{A}$ on to the three directions. In the case of spherical polar coordinates, the natural components are those given by (8).

Tensor components may be treated in a similar way.

Applying these formulae to the vector curl A we obtain

$$\left.\begin{aligned}(\text{curl } A)_r &= \frac{1}{r^2 \sin\theta}\left\{\frac{\partial(rA_\phi \sin\theta)}{\partial\theta} - \frac{\partial(rA_\theta)}{\partial\phi}\right\},\\ (\text{curl } A)_\theta &= \frac{1}{r\sin\theta}\left\{\frac{\partial A_r}{\partial\phi} - \frac{\partial(rA_\phi \sin\theta)}{\partial r}\right\},\\ (\text{curl } A)_\phi &= \frac{1}{r}\left\{\frac{\partial(rA_\theta)}{\partial r} - \frac{\partial A_r}{\partial\theta}\right\}.\end{aligned}\right\} \tag{9}$$

In the spherical polar coordinates, the field equations (1) become

$$\left.\begin{aligned}&\left.\begin{aligned}-k_1E_r &= \frac{1}{r^2\sin\theta}\left\{\frac{\partial(rH_\phi\sin\theta)}{\partial\theta} - \frac{\partial(rH_\theta)}{\partial\phi}\right\}, && (\alpha)\\ -k_1E_\theta &= \frac{1}{r\sin\theta}\left\{\frac{\partial H_r}{\partial\phi} - \frac{\partial(rH_\phi\sin\theta)}{\partial r}\right\}, && (\beta)\\ -k_1E_\phi &= \frac{1}{r}\left\{\frac{\partial(rH_\theta)}{\partial r} - \frac{\partial H_r}{\partial\theta}\right\}. && (\gamma)\end{aligned}\right\}\text{(a)}\\ &\left.\begin{aligned}k_2H_r &= \frac{1}{r^2\sin\theta}\left\{\frac{\partial(rE_\phi\sin\theta)}{\partial\theta} - \frac{\partial(rE_\theta)}{\partial\phi}\right\}, && (\alpha)\\ k_2H_\theta &= \frac{1}{r\sin\theta}\left\{\frac{\partial E_r}{\partial\phi} - \frac{\partial(rE_\phi\sin\theta)}{\partial r}\right\}, && (\beta)\\ k_2H_\phi &= \frac{1}{r}\left\{\frac{\partial(rE_\theta)}{\partial r} - \frac{\partial E_r}{\partial\theta}\right\}. && (\gamma)\end{aligned}\right\}\text{(b)}\end{aligned}\right\} \tag{10}$$

The boundary conditions (5) now are

$$\left.\begin{aligned}E_\theta^{(\mathrm{I})} &= E_\theta^{(\mathrm{II})}, & E_\phi^{(\mathrm{I})} &= E_\phi^{(\mathrm{II})},\\ H_\theta^{(\mathrm{I})} &= H_\theta^{(\mathrm{II})}, & H_\phi^{(\mathrm{I})} &= H_\phi^{(\mathrm{II})}.\end{aligned}\right\}\text{ for } r = a. \tag{11}$$

The equations (10), together with the boundary conditions (11) are the basic equations of our problem.

We shall represent the solution of these equations as a superposition of two linearly independent fields $({}^eE, {}^eH)$ and $({}^mE, {}^mH)$ each satisfying equations (10) such that

$${}^eE_r = E_r, \qquad {}^eH_r = 0, \tag{12a}$$

and

$${}^mE_r = 0, \qquad {}^mH_r = H_r. \tag{12b}$$

It is not difficult to see that such a representation is consistent with our equations. With $H_r = {}^eH_r = 0$, equations (10a, β) and (10a, γ) become

$$\left.\begin{aligned}k_1\,{}^eE_\theta &= \frac{1}{r}\frac{\partial}{\partial r}(r\,{}^eH_\phi),\\ k_1\,{}^eE_\phi &= -\frac{1}{r}\frac{\partial}{\partial r}(r\,{}^eH_\theta).\end{aligned}\right\} \tag{13}$$

Substituting from these relations into (10b, β) and (10b, γ) we obtain:

$$\left.\begin{aligned}\left(\frac{\partial^2}{\partial r^2}+k^2\right)(r\,{}^eH_\theta) &= -\frac{k_1}{\sin\theta}\frac{\partial\,{}^eE_r}{\partial\phi} \qquad (\text{b},\beta)\\ \text{and}\qquad \left(\frac{\partial^2}{\partial r^2}+k^2\right)(r\,{}^eH_\phi) &= +k_1\frac{\partial\,{}^eE_r}{\partial\theta}. \qquad (\text{b},\gamma)\end{aligned}\right\} \tag{14}$$

Equations (14), together with (10a, α), constitute a system of equations for ${}^eE_r$, ${}^eH_\theta$, and ${}^eH_\phi$. Not all solutions of this system represent physical fields, however; only those do which satisfy the subsidiary conditions div ${}^e\boldsymbol{H} = 0$. We restrict ourselves to such solutions. In spherical polar coordinates, and under our assumption that ${}^eH_r = 0$, this subsidiary condition is

$$\frac{\partial}{\partial\theta}(\sin\theta\,{}^eH_\theta)+\frac{\partial}{\partial\phi}({}^eH_\phi)=0, \tag{15}$$

and ensures that the remaining equation (10b, α) is satisfied. For (10b, α) becomes, on substitution from (13)

$$0=\frac{1}{k_1{}^2r^2\sin\theta}\frac{\partial}{\partial r}\left[\frac{\partial}{\partial\theta}(r\sin\theta\,{}^eH_\theta)+\frac{\partial}{\partial\phi}(r\,{}^eH_\phi)\right],$$

and because of (15) this is identically satisfied. Strictly similar considerations apply to the complementary case with ${}^mE_r = 0$.

The solution with vanishing radial magnetic field is called *the electric wave* (or transverse magnetic wave) and that with vanishing radial electric field is called *the magnetic wave* (or transverse electric wave). We shall now show that they may each be derived from a scalar potential ${}^e\Pi$ and ${}^m\Pi$ respectively, which are known as *Debye's potentials.**

It follows, first of all from (10b, α), since ${}^eH_r = 0$, that ${}^eE_\phi$ and ${}^eE_\theta$ may be represented in terms of a gradient of a scalar,

$${}^eE_\phi=\frac{1}{r\sin\theta}\frac{\partial U}{\partial\phi},\qquad {}^eE_\theta=\frac{1}{r}\frac{\partial U}{\partial\theta}. \tag{16}$$

If now we put

$$U=\frac{\partial(r\,{}^e\Pi)}{\partial r} \tag{17}$$

then we have from (16)

$${}^eE_\theta=\frac{1}{r}\frac{\partial^2(r\,{}^e\Pi)}{\partial r\partial\theta},\qquad {}^eE_\phi=\frac{1}{r\sin\theta}\frac{\partial^2(r\,{}^e\Pi)}{\partial r\partial\phi}. \tag{18}$$

It is seen that equation (13) may be satisfied by

$$\left.\begin{aligned}{}^eH_\phi &= k_1\frac{\partial\,{}^e\Pi}{\partial\theta}=\frac{k_1}{r}\frac{\partial(r\,{}^e\Pi)}{\partial\theta},\\ {}^eH_\theta &= -\frac{k_1}{\sin\theta}\frac{\partial\,{}^e\Pi}{\partial\phi}=-\frac{k_1}{r\sin\theta}\frac{\partial(r\,{}^e\Pi)}{\partial\phi}.\end{aligned}\right\} \tag{19}$$

* If $\boldsymbol{r}$ denotes the radius vector from the origin, then ${}^e\Pi\boldsymbol{r}$ and ${}^m\Pi\boldsymbol{r}$ are *radial Hertz vectors*, i.e. HERTZ vectors (see § 2.2.2) which point everywhere in the radial direction. (Cf. A. SOMMERFELD, contribution in P. FRANK and R. v. MISES: *Riemann-Weber's Differentialgleichungen der mathematischer Physik* (Braunschweig, Vieweg, 2nd ed., 1935; also New York, Dover Publ., 1961), **2**. 790; C. J. BOUWKAMP and H. B. G. CASIMIR, *Physica*, **20** (1954), 539 and A. NISBET, *Proc. Roy. Soc.*, A, **231** (1955), 260); *Physica*, **21** (1955), 799.

If we substitute from (19) into (10a, α) we obtain

$$ {}^{e}E_r = -\frac{1}{r\sin\theta}\left\{\frac{\partial}{\partial\theta}\left(\sin\theta\frac{\partial\,{}^{e}\Pi}{\partial\theta}\right) + \frac{1}{\sin\theta}\frac{\partial^2\,{}^{e}\Pi}{\partial\phi^2}\right\}. \tag{20} $$

Substitution from (19) and (20) into (14) gives two equations, the first of which expresses the vanishing of the ϕ derivative, the second the vanishing of the θ derivative of one and the same expression. These equations may, therefore, be satisfied by equating this expression to zero, and this gives

$$ \frac{1}{r}\frac{\partial^2(r\,{}^{e}\Pi)}{\partial r^2} + \frac{1}{r^2\sin\theta}\frac{\partial}{\partial\theta}\left(\sin\theta\frac{\partial\,{}^{e}\Pi}{\partial\theta}\right) + \frac{1}{r^2\sin^2\theta}\frac{\partial^2\,{}^{e}\Pi}{\partial\phi^2} + k^2\,{}^{e}\Pi = 0. \tag{21} $$

By means of this equation, (20) may be written as

$$ {}^{e}E_r = \frac{\partial^2(r\,{}^{e}\Pi)}{\partial r^2} + k^2 r\,{}^{e}\Pi. \tag{22} $$

It can be verified by substituting from (18), (19), (20), (21) and (22) into (10) that we have obtained a solution of our set of equations.

In a similar way one may consider the magnetic wave and one finds that this wave can be derived from a potential ${}^{m}\Pi$ which satisfies the same differential equation (21) as ${}^{e}\Pi$. The complete solution of our field equations is obtained by adding the two fields; this gives

$$ \left.\begin{aligned} E_r &= {}^{e}E_r + {}^{m}E_r = \frac{\partial^2(r\,{}^{e}\Pi)}{\partial r^2} + k^2 r\,{}^{e}\Pi, && (\alpha)\\ E_\theta &= {}^{e}E_\theta + {}^{m}E_\theta = \frac{1}{r}\frac{\partial^2(r\,{}^{e}\Pi)}{\partial r\partial\theta} + k_2\frac{1}{r\sin\theta}\frac{\partial(r\,{}^{m}\Pi)}{\partial\phi}, && (\beta)\\ E_\phi &= {}^{e}E_\phi + {}^{m}E_\phi = \frac{1}{r\sin\theta}\frac{\partial^2(r\,{}^{e}\Pi)}{\partial r\partial\phi} - k_2\frac{1}{r}\frac{\partial(r\,{}^{m}\Pi)}{\partial\theta}, && (\gamma) \end{aligned}\right\} \tag{23a} $$

$$ \left.\begin{aligned} H_r &= {}^{m}H_r + {}^{e}H_r = k^2 r\,{}^{m}\Pi + \frac{\partial^2(r\,{}^{m}\Pi)}{\partial r^2}, && (\alpha)\\ H_\theta &= {}^{m}H_\theta + {}^{e}H_\theta = -k_1\frac{1}{r\sin\theta}\frac{\partial(r\,{}^{e}\Pi)}{\partial\phi} + \frac{1}{r}\frac{\partial^2(r\,{}^{m}\Pi)}{\partial r\partial\theta}, && (\beta)\\ H_\phi &= {}^{m}H_\phi + {}^{e}H_\phi = k_1\frac{1}{r}\frac{\partial(r\,{}^{e}\Pi)}{\partial\theta} + \frac{1}{r\sin\theta}\frac{\partial^2(r\,{}^{m}\Pi)}{\partial r\partial\phi}. && (\gamma) \end{aligned}\right\} \tag{23b} $$

Both potentials ${}^{e}\Pi$ and ${}^{m}\Pi$ are solutions of the differential equation (21), which is nothing but the wave equation

$$ \nabla^2\Pi + k^2\Pi = 0, $$

written in polar coordinates. In order that the components E_θ, E_ϕ, H_θ, and H_ϕ shall be continuous over the spherical surface $r = a$, it is evidently sufficient that the four quantities

$$ k_1 r\,{}^{e}\Pi, \qquad k_2 r\,{}^{m}\Pi, \qquad \frac{\partial}{\partial r}(r\,{}^{e}\Pi), \qquad \frac{\partial}{\partial r}(r\,{}^{m}\Pi) \tag{24} $$

shall also be *continuous* over this surface. Thus our boundary conditions also split into independent conditions for ${}^{e}\Pi$ and ${}^{m}\Pi$. Our diffraction problem is thus reduced to the problem of finding two mutually independent solutions of the wave equation, with prescribed boundary conditions.

(b) *Series expansions for the field components*

We first represent the solution of the wave equations as expansions with undetermined coefficients, each term representing a particular integral. We shall then determine the coefficients by using the boundary conditions.

We seek integrals of the form

$$\Pi = R(r)\Theta(\theta)\Phi(\phi). \tag{25}$$

As may easily be verified by direct substitution into (21), the functions R, Θ, and Φ must satisfy the ordinary differential equations

$$\left.\begin{aligned} \frac{d^2(rR)}{dr^2} + \left(k^2 - \frac{\alpha}{r^2}\right) rR = 0, \qquad &\text{(a)} \\ \frac{1}{\sin\theta}\frac{d}{d\theta}\left(\sin\theta\,\frac{d\Theta}{d\theta}\right) + \left(\alpha - \frac{\beta}{\sin^2\theta}\right)\Theta = 0, \qquad &\text{(b)} \\ \frac{d^2\Phi}{d\phi^2} + \beta\,\Phi = 0, \qquad &\text{(c)} \end{aligned}\right\} \tag{26}$$

where α and β are integration constants.

As the field $\boldsymbol{E}$, $\boldsymbol{H}$ is a single valued function of position, Π must likewise be single valued, and this requirement imposes certain conditions on Θ and Φ.

For each of the equations in (26) one can write down the general solution. For (c) it is

$$a\cos(\sqrt{\beta}\phi) + b\sin(\sqrt{\beta}\phi).$$

The condition of single-valuedness demands that

$$\beta = m^2, \qquad (m = \text{integer}). \tag{27}$$

Hence the single-valued solution of (26c) is

$$\Phi = a_m\cos(m\phi) + b_m\sin(m\phi). \tag{28}$$

The equation (26b) is the well-known equation for spherical harmonics. A necessary and sufficient condition for a single valued solution is that

$$\alpha = l(l+1), \qquad (l > |m|, \text{ integer}). \tag{29}$$

We substitute for β from (27) in (26b) and introduce the new variable

$$\xi = \cos\theta. \tag{30}$$

The equation then transforms to*

$$\frac{d}{d\xi}\left\{(1-\xi^2)\frac{d\Theta}{d\xi}\right\} + \left\{l(l+1) - \frac{m^2}{1-\xi^2}\right\}\Theta = 0, \tag{31}$$

the solutions of which are the associated LEGENDRE functions

$$\Theta = P_l^{(m)}(\xi) = P_l^{(m)}(\cos\theta). \tag{32}$$

* Cf. A. SOMMERFELD, *Partial Differential Equations of Physics* (New York, Academic Press, 1949), p. 127.

These functions vanish identically if $|m| > l$; for each l there are therefore $2l + 1$ such functions, namely those with

$$m = -l, \; -l+1, \; . \; . \; . \; l-1, \; l.$$

To integrate the remaining equation (26a) we set

$$kr = \rho, \qquad R(r) = \frac{1}{\sqrt{\rho}}\, Z(\rho), \tag{33}$$

and obtain the BESSEL equation*

$$\frac{d^2Z}{d\rho^2} + \frac{1}{\rho}\frac{dZ}{d\rho} + \left\{1 - \frac{(l+\frac{1}{2})^2}{\rho^2}\right\} Z = 0. \tag{34}$$

The solution of this equation is the general cylindrical function $Z = Z_{l+\frac{1}{2}}(\rho)$ of order $l + \frac{1}{2}$, so that the solution of (26a) is

$$R = \frac{1}{\sqrt{kr}}\, Z_{l+\frac{1}{2}}(kr). \tag{35}$$

Each cylindrical function may be expressed as a linear combination of two cylindrical functions of standard type, e.g. the BESSEL functions $J_{l+\frac{1}{2}}(\rho)$ and the NEUMANN functions $N_{l+\frac{1}{2}}(\rho)$. For the present purpose it is convenient to employ the functions†

$$\psi_l(\rho) = \sqrt{\frac{\pi\rho}{2}}\, J_{l+\frac{1}{2}}(\rho), \qquad \chi_l(\rho) = -\sqrt{\frac{\pi\rho}{2}}\, N_{l+\frac{1}{2}}(\rho). \tag{36}$$

The functions $\psi_l(\rho)$ are regular in every finite domain of the ρ-plane, including the origin, whereas the functions $\chi_l(\rho)$ have singularities at the origin $\rho = 0$, where they become infinite. We may, therefore, use the functions $\psi_l(\rho)$ but not the functions $\chi_l(\rho)$ for representing the wave inside the sphere.

The general integral of (26a) may then be written as

$$rR = c_l\psi_l(kr) + d_l\chi_l(kr). \tag{37}$$

In particular, with $c_l = 1$, $d_l = -i$ we have

$$rR = \zeta_l^{(1)}(kr), \tag{38}$$

where

$$\zeta_l^{(1)}(\rho) = \psi_l(\rho) - i\chi_l(\rho) = \sqrt{\frac{\pi\rho}{2}}\, H^{(1)}_{l+\frac{1}{2}}(\rho), \tag{39}$$

$H^{(1)}$ being one of the HANKEL functions.‡ The HANKEL functions are distinguished from other cylindrical functions by the property that they vanish at infinity in the

* Cf. A. SOMMERFELD; *loc. cit.*, p. 86.

† Several slightly different definitions of the ψ and χ functions exist in the literature. (Cf. G. N. WATSON, *A Treatise on the Theory of Bessel Functions* (Cambridge University Press, 2nd ed., 1944), p. 56; or A. SOMMERFELD, *loc. cit.*, pp. 113–114. On pp. 646–647 we summarize some of the formulae relating to the ψ_l functions.

‡ This formula is an immediate consequence of the well known relation between the BESSEL, NEUMANN and HANKEL functions:

$$J_p + iN_p = H_p^{(1)}.$$

There is a similar relation involving the other HANKEL function:

$$J_p - iN_p = H_p^{(2)}.$$

These two formulae are analogous to the expressions for the exponential functions $e^{i\rho}$ and $e^{-i\rho}$ in terms of $\cos\rho$ and $\sin\rho$.

complex plane. The one used here with the index 1 vanishes in the half-plane of the positive imaginary part of ρ and is thus suitable for the representation of the scattered wave.

According to (25) a particular integral $\Pi_l^{(m)}$ is obtained on multiplying together the functions given by (28), (32), and (37); we then obtain the following general solution of the wave equation:

$$
\begin{aligned}
r\Pi &= r \sum_{l=0}^{\infty} \sum_{m=-l}^{l} \Pi_l^{(m)} \\
&= \sum_{l=0}^{\infty} \sum_{m=-l}^{l} \{c_l\psi_l(kr) + d_l\chi_l(kr)\} \{P_l^{(m)}(\cos\theta)\} \{a_m \cos(m\phi) + b_m \sin(m\phi)\},
\end{aligned} \tag{40}
$$

where a_m, b_m, c_l, and d_l are arbitrary constants.

We must now determine these constants in such a way as to satisfy the boundary conditions. For this to be possible one must be able to express the potentials ${}^e\Pi^{(i)}$ and ${}^m\Pi^{(i)}$ of the incident wave in a series of the form (40). To show that this can be done, we first transform the expression (4) for the incident wave into spherical polar coordinates in accordance with (8):

$$
\left.
\begin{aligned}
E_r^{(i)} &= e^{ik^{(\mathrm{I})}r\cos\theta} \sin\theta\cos\phi, & H_r^{(i)} &= \frac{ik^{(\mathrm{I})}}{k_2^{(\mathrm{I})}} e^{ik^{(\mathrm{I})}r\cos\theta} \sin\theta\sin\phi, \\
E_\theta^{(i)} &= e^{ik^{(\mathrm{I})}r\cos\theta} \cos\theta\cos\phi, & H_\theta^{(i)} &= \frac{ik^{(\mathrm{I})}}{k_2^{(\mathrm{I})}} e^{ik^{(\mathrm{I})}r\cos\theta} \cos\theta\sin\phi, \\
E_\phi^{(i)} &= -\, e^{ik^{(\mathrm{I})}r\cos\theta} \sin\phi, & H_\phi^{(i)} &= \frac{ik^{(\mathrm{I})}}{k_2^{(\mathrm{I})}} e^{ik^{(\mathrm{I})}r\cos\theta} \cos\phi.
\end{aligned}
\right\} \tag{41}
$$

To determine the potentials ${}^e\Pi^{(i)}$ or ${}^m\Pi^{(i)}$ it is only necessary to use one of the equations (23); the first of them yields:

$$
e^{ik^{(\mathrm{I})}r\cos\theta} \sin\theta\cos\phi = \frac{\partial^2(r\, {}^e\Pi^{(i)})}{\partial r^2} + k^{(\mathrm{I})2} r\, {}^e\Pi^{(i)}. \tag{42}
$$

The first factor on the left-hand side of this equation may be expressed in the following differentiable series of LEGENDRE polynomials [BAUER's formula, § 9.4 (9)]:

$$
e^{ik^{(\mathrm{I})}r\cos\theta} = \sum_{l=0}^{\infty} i^l(2l+1) \frac{\psi_l(k^{(\mathrm{I})}r)}{k^{(\mathrm{I})}r} P_l(\cos\theta). \tag{43}
$$

We also have the identities

$$
e^{ik^{(\mathrm{I})}r\cos\theta} \sin\theta \equiv -\frac{1}{ik^{(\mathrm{I})}r} \frac{\partial}{\partial\theta} (e^{ik^{(\mathrm{I})}r\cos\theta}), \tag{44}
$$

$$
\frac{\partial}{\partial\theta} P_l(\cos\theta) \equiv -P_l^{(1)}(\cos\theta); \qquad P_0^{(1)}(\cos\theta) \equiv 0. \tag{45}
$$

Using these relations, the left-hand side of (42) may be expressed in the form

$$
e^{ik^{(\mathrm{I})}r\cos\theta} \sin\theta\cos\phi = \frac{1}{(k^{(\mathrm{I})}r)^2} \sum_{l=1}^{\infty} i^{l-1}(2l+1)\psi_l(k^{(\mathrm{I})}r) P_l^{(1)}(\cos\theta)\cos\phi. \tag{46}
$$

Accordingly we take as a trial solution of (42) a series of a similar form

$$
r\, {}^e\Pi^{(i)} = \frac{1}{k^{(\mathrm{I})2}} \sum_{l=1}^{\infty} \alpha_l \psi_l(k^{(\mathrm{I})}r) P_l^{(1)}(\cos\theta)\cos\phi. \tag{47}
$$

On substituting from (46) and (47) in equation (42) and comparing coefficients we obtain the relation

$$\alpha_l \left\{ k^{(\mathrm{I})^2} \psi_l(k^{(\mathrm{I})} r) + \frac{\partial^2 \psi_l(k^{(\mathrm{I})} r)}{\partial r^2} \right\} = i^{l-1}(2l+1) \frac{\psi_l(k^{(\mathrm{I})} r)}{r^2}. \tag{48}$$

Now from (37) (with $c_l = 1$ and $d_l = 0$) it follows that

$$\psi_l(k^{(\mathrm{I})} r) = rR \tag{49}$$

is a solution of equation (26a):

$$\frac{d^2 \psi_l}{dr^2} + \left(k^{(\mathrm{I})^2} - \frac{\alpha}{r^2} \right) \psi_l = 0, \tag{50}$$

provided [see (29)] that $\alpha = l(l+1)$. Comparing (50) with (48), we see that

$$\alpha_l = i^{l-1} \frac{2l+1}{l(l+1)}. \tag{51}$$

The calculations relating to the magnetic potential ${}^m\Pi^{(i)}$ are similar. We thus obtain the following expression for the two potentials of the incident wave:

$$\left.\begin{aligned} r\,{}^e\Pi^{(i)} &= \frac{1}{k^{(\mathrm{I})^2}} \sum_{l=1}^{\infty} i^{l-1} \frac{2l+1}{l(l+1)} \psi_l(k^{(\mathrm{I})} r) P_l^{(1)}(\cos\theta) \cos\phi, && \text{(a)} \\ r\,{}^m\Pi^{(i)} &= \frac{1}{k^{(\mathrm{I})^2}} \sum_{l=1}^{\infty} i^{l} \frac{k^{(\mathrm{I})}}{k_2^{(\mathrm{I})}} \frac{2l+1}{l(l+1)} \psi_l(k^{(\mathrm{I})} r) P_l^{(1)}(\cos\theta) \sin\phi. && \text{(b)} \end{aligned}\right\} \tag{52}$$

We have expressed both the potentials in series of the form (40) and the unknown constants can now easily be determined.

The boundary conditions (24) written more fully are

$$\left.\begin{aligned} \frac{\partial}{\partial r}\{r({}^e\Pi^{(i)} + {}^e\Pi^{(s)})\}_{r=a} &= \frac{\partial}{\partial r}\{r\,{}^e\Pi^{(w)}\}_{r=a}, && \text{(a)} \\ \frac{\partial}{\partial r}\{r({}^m\Pi^{(i)} + {}^m\Pi^{(s)})\}_{r=a} &= \frac{\partial}{\partial r}\{r\,{}^m\Pi^{(w)}\}_{r=a}, && \text{(b)} \\ k_1^{(\mathrm{I})}\{r({}^e\Pi^{(i)} + {}^e\Pi^{(s)})\}_{r=a} &= k_1^{(\mathrm{II})}\{r\,{}^e\Pi^{(w)}\}_{r=a}, && \text{(c)} \\ k_2^{(\mathrm{I})}\{r({}^m\Pi^{(i)} + {}^m\Pi^{(s)})\}_{r=a} &= k_2^{(\mathrm{II})}\{r\,{}^m\Pi^{(w)}\}_{r=a}. && \text{(d)} \end{aligned}\right\} \tag{53}$$

According to (52), these equations can only be satisfied if only those terms occur in the expansions (40) for the unknown potentials $\Pi^{(s)}$ and $\Pi^{(w)}$ for which $m = 1$ and if, moreover

$$a_1 = 0$$

for the magnetic potential, and

$$b_1 = 0$$

for the electric one.

We already noted that for the representation of $\Pi^{(w)}$ only the ψ_l functions are appropriate, since these remain regular at the origin, whilst the functions χ_l become infinite there. Hence we set

$$\left.\begin{aligned} r\,{}^e\Pi^{(w)} &= \frac{1}{k^{(\mathrm{II})^2}} \sum_{l=1}^{\infty} {}^eA_l \psi_l(k^{(\mathrm{II})} r) P_l^{(1)}(\cos\theta) \cos\phi, && \text{(a)} \\ r\,{}^m\Pi^{(w)} &= \frac{i}{k^{(\mathrm{II})} k_2^{(\mathrm{II})}} \sum_{l=1}^{\infty} {}^mA_l \psi_l(k^{(\mathrm{II})} r) P_l^{(1)}(\cos\theta) \sin\phi. && \text{(b)} \end{aligned}\right\} \tag{54}$$

We have also seen that for the scattered wave a representation in terms of the functions $\zeta_l^{(1)} = \psi_l - i\chi_l$ is appropriate, $\zeta_l^{(1)}$ being obtained from the HANKEL function $H^{(1)}$ on multiplication by $\sqrt{\pi\rho/2}$ [see (39)]. For large ρ, $H^{(1)}$ behaves as $e^{i\rho}/\sqrt{\rho}$, i.e. $\zeta_l^{(1)}$ behaves as $e^{i\rho}$ and $R = \zeta_l^{(1)}(k^{(\mathrm{I})}r)/r$ as $e^{ik^{(\mathrm{I})}r}/r$. Thus at large distances from the sphere the scattered wave is spherical, with its centre at the origin $r = 0$. We therefore set

$$\left.\begin{aligned} r\,{}^e\Pi^{(s)} &= \frac{1}{k^{(\mathrm{I})2}} \sum_{l=1}^{\infty} {}^eB_l \zeta_l^{(1)}(k^{(\mathrm{I})}r) P_l^{(1)}(\cos\theta)\cos\phi, &\text{(c)}\\ r\,{}^m\Pi^{(s)} &= \frac{i}{k^{(\mathrm{I})}k_2^{(\mathrm{I})}} \sum_{l=1}^{\infty} {}^mB_l \zeta_l^{(1)}(k^{(\mathrm{I})}r) P_l^{(1)}(\cos\theta)\sin\phi. &\text{(d)}\end{aligned}\right\} \quad (55)$$

If we now substitute the expressions (52), (54), and (55) into the boundary conditions (53) we obtain the following linear relations between the coefficients* ${}^eA_l$, ${}^mA_l$, ${}^eB_l$, and ${}^mB_l$:

$$\left.\begin{aligned} {}^eB_l \frac{1}{k^{(\mathrm{I})}} \zeta_l^{(1)\prime}(k^{(\mathrm{I})}a) + \frac{1}{k^{(\mathrm{I})}} i^{l-1} \frac{2l+1}{l(l+1)} \psi_l'(k^{(\mathrm{I})}a) &= \frac{1}{k^{(\mathrm{II})}} {}^eA_l \psi_l'(k^{(\mathrm{II})}a),\\ {}^mB_l \frac{1}{k_2^{(\mathrm{I})}} \zeta_l^{(1)\prime}(k^{(\mathrm{I})}a) + \frac{1}{k_2^{(\mathrm{I})}} i^{l-1} \frac{2l+1}{l(l+1)} \psi_l'(k^{(\mathrm{I})}a) &= \frac{1}{k_2^{(\mathrm{II})}} {}^mA_l \psi_l'(k^{(\mathrm{II})}a),\\ {}^eB_l \frac{1}{k_2^{(\mathrm{I})}} \zeta_l^{(1)}(k^{(\mathrm{I})}a) + \frac{1}{k_2^{(\mathrm{I})}} i^{l-1} \frac{2l+1}{l(l+1)} \psi_l(k^{(\mathrm{I})}a) &= \frac{1}{k_2^{(\mathrm{II})}} {}^eA_l \psi_l(k^{(\mathrm{II})}a),\\ {}^mB_l \frac{1}{k^{(\mathrm{I})}} \zeta_l^{(1)}(k^{(\mathrm{I})}a) + \frac{1}{k^{(\mathrm{I})}} i^{l-1} \frac{2l+1}{l(l+1)} \psi_l(k^{(\mathrm{I})}a) &= \frac{1}{k^{(\mathrm{II})}} {}^mA_l \psi_l(k^{(\mathrm{II})}a).\end{aligned}\right\} \quad (56)$$

We are only interested in the coefficients ${}^eB_l$ and ${}^mB_l$ which characterize the scattered wave. These may be obtained by the elimination of ${}^eA_l$ and ${}^mA_l$; this gives

$$\left.\begin{aligned} {}^eB_l &= i^{l+1} \frac{2l+1}{l(l+1)} \frac{k_2^{(\mathrm{I})}k^{(\mathrm{II})}\psi_l'(k^{(\mathrm{I})}a)\psi_l(k^{(\mathrm{II})}a) - k_2^{(\mathrm{II})}k^{(\mathrm{I})}\psi_l'(k^{(\mathrm{II})}a)\psi_l(k^{(\mathrm{I})}a)}{k_2^{(\mathrm{I})}k^{(\mathrm{II})}\zeta_l^{(1)\prime}(k^{(\mathrm{I})}a)\psi_l(k^{(\mathrm{II})}a) - k_2^{(\mathrm{II})}k^{(\mathrm{I})}\psi_l'(k^{(\mathrm{II})}a)\zeta_l^{(1)}(k^{(\mathrm{I})}a)}, &\text{(a)}\\ {}^mB_l &= i^{l+1} \frac{2l+1}{l(l+1)} \frac{k_2^{(\mathrm{I})}k^{(\mathrm{II})}\psi_l(k^{(\mathrm{I})}a)\psi_l'(k^{(\mathrm{II})}a) - k_2^{(\mathrm{II})}k^{(\mathrm{I})}\psi_l'(k^{(\mathrm{I})}a)\psi_l(k^{(\mathrm{II})}a)}{k_2^{(\mathrm{I})}k^{(\mathrm{II})}\zeta_l^{(1)}(k^{(\mathrm{I})}a)\psi_l'(k^{(\mathrm{II})}a) - k_2^{(\mathrm{II})}k^{(\mathrm{I})}\zeta_l^{(1)\prime}(k^{(\mathrm{I})}a)\psi_l(k^{(\mathrm{II})}a)}. &\text{(b)}\end{aligned}\right\} \quad (57)$$

Finally, the components of the field vectors of the scattered wave are obtained on substituting from (55) in (23). This gives

$$\left.\begin{aligned} E_r^{(s)} &= \frac{1}{k^{(\mathrm{I})2}} \frac{\cos\phi}{r^2} \sum_{l=1}^{\infty} l(l+1)\, {}^eB_l \zeta_l^{(1)}(k^{(\mathrm{I})}r) P_l^{(1)}(\cos\theta),\\ E_\theta^{(s)} &= -\frac{1}{k^{(\mathrm{I})}} \frac{\cos\phi}{r} \sum_{l=1}^{\infty} \Big\{ {}^eB_l \zeta_l^{(1)\prime}(k^{(\mathrm{I})}r) P_l^{(1)\prime}(\cos\theta)\sin\theta \\ &\qquad - i\,{}^mB_l \zeta_l^{(1)}(k^{(\mathrm{I})}r) P_l^{(1)}(\cos\theta) \frac{1}{\sin\theta} \Big\},\\ E_\phi^{(s)} &= -\frac{1}{k^{(\mathrm{I})}} \frac{\sin\phi}{r} \sum_{l=1}^{\infty} \Big\{ {}^eB_l \zeta_l^{(1)\prime}(k^{(\mathrm{I})}r) P_l^{(1)}(\cos\theta) \frac{1}{\sin\theta} \\ &\qquad - i\,{}^mB_l \zeta_l^{(1)}(k^{(\mathrm{I})}r) P_l^{(1)\prime}(\cos\theta)\sin\theta \Big\},\end{aligned}\right\} \quad (58)$$

---

* The addition of a prime to the functions ψ_l, ζ_l, and $P_l^{(1)}$ denotes differentiation with respect to their arguments.

$$\left.\begin{aligned}
H_r^{(s)} &= \frac{i}{k^{(\mathrm{I})}k_2^{(\mathrm{I})}}\frac{\sin\phi}{r^2}\sum_{l=1}^{\infty} l(l+1)\,{}^mB_l\zeta_l^{(1)}(k^{(\mathrm{I})}r)P_l^{(1)}(\cos\theta),\\
H_\theta^{(s)} &= -\frac{1}{k_2^{(\mathrm{I})}}\frac{\sin\phi}{r}\sum_{l=1}^{\infty}\Big\{{}^eB_l\zeta_l^{(1)}(k^{(\mathrm{I})}r)P_l^{(1)}(\cos\theta)\frac{1}{\sin\theta}\\
&\qquad + i\,{}^mB_l\zeta_l^{(1)\prime}(k^{(\mathrm{I})}r)P_l^{(1)\prime}(\cos\theta)\sin\theta\Big\},\\
H_\phi^{(s)} &= \frac{1}{k_2^{(\mathrm{I})}}\frac{\cos\phi}{r}\sum_{l=1}^{\infty}\Big\{{}^eB_l\zeta_l^{(1)}(k^{(\mathrm{I})}r)P_l^{(1)\prime}(\cos\theta)\sin\theta\\
&\qquad + i\,{}^mB_l\zeta_l^{(1)\prime}(k^{(\mathrm{I})}r)P_l^{(1)}(\cos\theta)\frac{1}{\sin\theta}\Big\}.
\end{aligned}\right\}\tag{58}$$

This completes the formal solution of our boundary value problem. We shall not enter into the questions of existence and convergence of the solution.

It will be helpful to recall the meaning of the various constants. Since the medium surrounding the sphere is assumed to be a non-conductor, $\sigma^{(\mathrm{I})} = 0$. If we write σ in place of $\sigma^{(\mathrm{II})}$ for the conductivity of the sphere, we have from (2)

$$\left.\begin{aligned}
&k_1^{(\mathrm{I})} = \frac{i\omega}{c}\,\varepsilon^{(\mathrm{I})} = i\frac{2\pi}{\lambda_0}\,\varepsilon^{(\mathrm{I})}, \qquad k_2^{(\mathrm{I})} = \frac{i\omega}{c} = i\frac{2\pi}{\lambda_0}, && \text{(a)}\\
&k^{(\mathrm{I})} = \sqrt{-k_1^{(\mathrm{I})}k_2^{(\mathrm{I})}} = \frac{2\pi}{\lambda_0}\sqrt{\varepsilon^{(\mathrm{I})}} = \frac{2\pi}{\lambda^{(\mathrm{I})}}, && \text{(b)}\\
&k_1^{(\mathrm{II})} = \frac{i\omega}{c}\left(\varepsilon^{(\mathrm{II})} + i\frac{4\pi\sigma}{\omega}\right) = i\frac{2\pi}{\lambda_0}\left(\varepsilon^{(\mathrm{II})} + i\frac{4\pi\sigma}{\omega}\right), \quad k_2^{(\mathrm{II})} = \frac{i\omega}{c} = i\frac{2\pi}{\lambda_0}, && \text{(c)}\\
&k^{(\mathrm{II})} = \sqrt{-k_1^{(\mathrm{II})}k_2^{(\mathrm{II})}} = \frac{2\pi}{\lambda_0}\sqrt{\varepsilon^{(\mathrm{II})} + \frac{4\pi\sigma}{\omega}}, && \text{(d)}
\end{aligned}\right\}\tag{59}$$

where λ_0 is the wavelength of the light in the vacuum and $\lambda^{(\mathrm{I})}$ is the wavelength in the medium surrounding the sphere.

For the purpose of later discussion it is also convenient to introduce the complex refractive index of the sphere relative to the surrounding medium. Denoting this index by $\hat{n}$ we have

$$\hat{n}^2 = \frac{\hat{n}^{(\mathrm{II})^2}}{n^{(\mathrm{I})^2}} = \frac{k^{(\mathrm{II})^2}}{k^{(\mathrm{I})^2}} = \frac{\varepsilon^{(\mathrm{II})}}{\varepsilon^{(\mathrm{I})}} + i\frac{4\pi\sigma}{\omega\varepsilon^{(\mathrm{I})}} = \frac{k_1^{(\mathrm{II})}}{k_1^{(\mathrm{I})}}. \tag{60}$$

We also introduce a dimensionless parameter q defined as

$$q = \frac{2\pi}{\lambda^{(\mathrm{I})}}\,a, \tag{61}$$

i.e. q is 2π times the ratio of the radius of the sphere to the wavelength of the light in the outer medium. Then, using also the relation

$$\frac{k^{(\mathrm{II})}k_2^{(\mathrm{I})}}{k^{(\mathrm{I})}k_2^{(\mathrm{II})}} = \hat{n},$$

we may express the coefficients (57) in the form

$$\left.\begin{aligned}
{}^eB_l &= i^{l+1}\frac{2l+1}{l(l+1)}\frac{\hat{n}\psi_l'(q)\psi_l(\hat{n}q) - \psi_l(q)\psi_l'(\hat{n}q)}{\hat{n}\zeta_l^{(1)\prime}(q)\psi_l(\hat{n}q) - \zeta_l^{(1)}(q)\psi_l'(\hat{n}q)}, && \text{(a)}\\
{}^mB_l &= i^{l+1}\frac{2l+1}{l(l+1)}\frac{\hat{n}\psi_l(q)\psi_l'(\hat{n}q) - \psi_l'(q)\psi_l(\hat{n}q)}{\hat{n}\zeta_l^{(1)}(q)\psi_l'(\hat{n}q) - \zeta_l^{(1)\prime}(q)\psi_l(\hat{n}q)}. && \text{(b)}
\end{aligned}\right\}\tag{62}$$

These formulae take a particularly simple form when either the dielectric constant or the conductivity of the sphere is high, and at the same time the radius of the sphere is not too small. In this case $|\hat{n}| \gg 1$, $|\hat{n}q| \gg 1$ and (62) reduce to

$$\left.\begin{aligned} {}^{e}B_l &= i^{l+1}\frac{2l+1}{l(l+1)}\frac{\psi_l'(q)}{\zeta_l^{(1)\prime}(q)}, \qquad &\text{(a)}\\ {}^{m}B_l &= i^{l+1}\frac{2l+1}{l(l+1)}\frac{\psi_l(q)}{\zeta_l^{(1)}(q)}. \qquad &\text{(b)}\end{aligned}\right\} \tag{63}$$

This approximation is of little interest for optics, but it is of importance in connection with radio waves. It is also of historical interest as the early theories were concerned with this limiting case.*

(c) *Summary of formulae relating to the associated Legendre functions and to the cylindrical functions*

For convenience of further discussion we summarize here some formulae relating to the spherical harmonics and to the cylindrical functions.

Associated Legendre functions

The LEGENDRE polynomials are the polynomials (in $\cos\theta$)

$$P_l(\cos\theta) = \sum_{m=0}^{[l/2]}(-1)^m\frac{(2l-2m)!}{2^l m!\,(l-m)!\,(l-2m)!}(\cos\theta)^{l-2m}, \tag{64}$$

and the associated LEGENDRE functions of the first kind are defined by the formula†

$$P_l^{(m)}(\cos\theta) = (\sin\theta)^m\frac{d^m P_l(\cos\theta)}{d(\cos\theta)^m}. \tag{65}$$

We shall also need the relations

$$\left.\begin{aligned} P_l^{(1)}(\cos\theta) &= \frac{l}{\sin\theta}\{P_{l-1}(\cos\theta) - \cos\theta\, P_l(\cos\theta)\},\\ P_l^{(1)}(\cos\theta) &= \frac{\cos\theta}{\sin^2\theta}\left\{P_l^{(1)}(\cos\theta) - l(l+1)\frac{P_l(\cos\theta)}{\sin\theta}\right\}.\end{aligned}\right\} \tag{66}$$

For large l one has the asymptotic approximation

$$P_l(\cos\theta) \sim \sqrt{\frac{2}{l\pi\sin\theta}}\sin\left[(l+\tfrac{1}{2})\theta + \frac{\pi}{4}\right]. \tag{67}$$

Cylindrical functions

I. *For small values of the argument x*, we have, for $\psi_l(x)$ the series expansion

$$\psi_l(x) = \frac{x^{l+1}}{1\,.\,3\,\ldots\,(2l+1)}f_l(x), \tag{68}$$

where

$$f_l(x) = 1 - \frac{2}{2l+3}\left(\frac{x}{2}\right)^2 + \ldots \tag{69}$$

---

* Cf. K. SCHWARZSCHILD, *Münch. Akad.*, *Math-phys. Kl.*, **31** (1901), 293.

† A slightly different definition is sometimes used, differing from the present one by a multiplicative factor $(-1)^m$.

For $\zeta_l^{(1)}(x)$ we have the expansion

$$\zeta_l^{(1)}(x) = -i\,\frac{1\,.\,3\,\ldots\,(2l-1)}{x^l}\,e^{ix}\,\{h_l(x) - ixg_l(x)\}, \tag{70}$$

where $h_l(x)$ and $g_l(x)$ are power series of which the first term is unity and the second term is quadratic in x. The functions $\psi_l'(x)$ and $\zeta_l^{(1)\prime}(x)$ may be expressed in an analogous form:

$$\psi_l'(x) = \frac{(l+1)x^l}{1\,.\,3\,\ldots\,(2l+1)}\,f_l^+(x), \tag{71}$$

$$\zeta_l^{(1)\prime}(x) = il\,\frac{1\,.\,3\,\ldots\,(2l-1)}{x^{l+1}}\,e^{ix}\{h_l^+(x) - ixg_l^+(x)\}, \tag{72}$$

where $f_l^+(x)$, $h_l^+(x)$, and $g_l^+(x)$ are power series of the same kind as before.

II. *For large values of the argument* x, provided that l is small compared with $|x|$, the following asymptotic formulae may be used:

$$\psi_l(x) \sim \tfrac{1}{2}\{i^{l+1}e^{-ix} + (-i)^{l+1}e^{ix}\}, \tag{73}$$

$$\zeta_l^{(1)}(x) \sim (-i)^{l+1}e^{ix}, \tag{74}$$

and

$$\psi_l'(x) \sim \tfrac{1}{2}\{i^l e^{-ix} + (-i)^l e^{ix}\}, \tag{75}$$

$$\zeta_l^{(1)\prime}(x) \sim (-i)^l e^{ix}. \tag{76}$$

For real values of x the functions $\psi_l(x)$ and $\psi_l'(x)$ are themselves real:

$$\psi_l(x) \sim \sin\left(x - \frac{l\pi}{2}\right), \tag{77}$$

$$\psi_l'(x) \sim \cos\left(x - \frac{l\pi}{2}\right). \tag{78}$$

13.5.2 Some consequences of Mie's formulae

(a) *The partial waves*

It is seen from (58) that the amplitude of the radial components $E_r^{(s)}$ and $H_r^{(s)}$ of the scattered wave fall off as the inverse square of the distance from the scattering centre, whereas the amplitudes of the other components fall off more slowly, as the inverse of this distance. At sufficiently great distances ($r \gg \lambda$), in the *radiation zone* or *wave zone*, the radial components may, therefore, be neglected in comparison with the tangential components, i.e. in this region the wave is *transverse*.

The formulae show that the scattered wave is composed of contributions involving spherical harmonics of different orders. These contributions may be called *partial waves* and their strength is determined by the absolute values of the complex coefficients ${}^eB_l$ and ${}^mB_l$. These coefficients depend on the nature of the two media, and on the ratio of the radius of the sphere to the wavelength of the incident light.

Each partial wave consists of an electrical part with amplitude ${}^eB_l$ and a magnetic part with amplitude ${}^mB_l$. The magnetic lines of

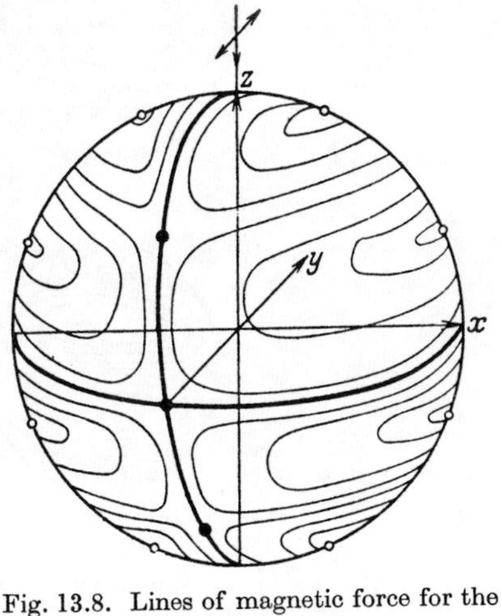

Fig. 13.8. Lines of magnetic force for the fourth electric partial wave.

force of the electric partial wave and the electric lines of force of the magnetic partial wave lie wholly on concentric spherical surfaces, since for the former $H_r^{(s)} = 0$ and for the latter $E_r^{(s)} = 0$.

Let us consider a typical partial wave, for instance the lth electric wave. We see that the corresponding components $E_\theta^{(s)}$ and $H_\phi^{(s)}$ vanish at points where either $\cos\phi$ or $P_l^{(1)\prime}(\cos\theta)\sin\theta$ is zero; likewise $E_\phi^{(s)}$ and $H_\theta^{(s)}$ vanish where either $\sin\phi$ or $P_l^{(1)}(\cos\theta)/\sin\theta$ is zero. Now within the interval $0 \leqslant \theta \leqslant \pi$, the function $P_l^{(1)\prime}(\cos\theta)$ vanishes l times and the function $P_l^{(1)}(\cos\theta)/\sin\theta$ vanishes $(l-1)$ times, but differs

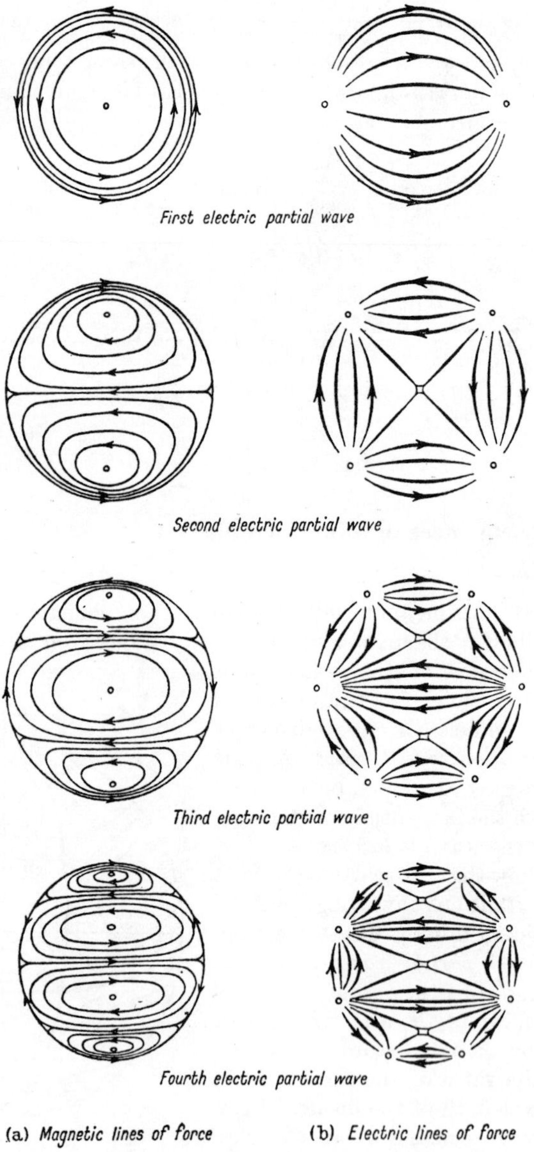

Fig. 13.9. Lines of force of the first four electric partial waves.
(After G. MIE, *Ann. d. Physik* (4), **25** (1908), 377.)

from zero when $\theta = 0$ or π. It follows that for $\phi = \pm \pi/2$ all the field-components vanish $2l$ times, i.e. $(4l - 2)$ times in all. As the magnetic lines of force must be closed curves, and as we saw they lie wholly on spherical surfaces concentric with the origin it follows that each of the $2l$ zero points on the circle $\phi = 0$ or π are the centres of families of closed magnetic loops, whereas the $2(l - 1)$ zero points lying on the circle $\phi = \pm \pi/2$ are neutral points. The lines of force avoid these neutral points just as two families of rectangular hyperbolae having common asymptotes avoid their common centre.

Fig. 13.8 shows the magnetic lines of force for the fourth electrical partial wave. The two sets of points may be clearly distinguished; the first in the xz-section (the plane of the diagram) and the second in the yz-section.

Fig. 13.9(a) shows the projections on to the yz-plane of the magnetic lines of force situated on one of the two hemispheres on either side of the yz-plane, and belonging to the first four electric partial waves. In the radiation zone the electric lines of force are orthogonal to the magnetic lines, since, according to (58), one has, for each partial wave (electric or magnetic),

$$E_\theta^{(s)} H_\theta^{(s)} + E_\phi^{(s)} H_\phi^{(s)} = 0. \tag{79}$$

Fig. 13.9(b) shows these lines in a similar projection on the xz-plane.

Similar results hold for the magnetic partial waves, except that $\cos\phi$ and $\sin\phi$ are now interchanged. The corresponding projections of the electric lines of force of the magnetic partial waves that lie on a sphere may be obtained merely by rotating the figures through an angle of 90° about the z-axis.

(b) *Limiting cases*

We shall now investigate the relative contributions of the various partial waves. The general case ($\hat{n}$ and q arbitrary) is not amenable to simple analytical treatment and we, therefore, consider in detail only two limiting cases, namely when the radius of the sphere is large compared with the wavelength ($q \gg 1$), and when it is much smaller than a wavelength ($q \ll 1$).

I: $q \gg 1$. This is the case for which our solution must be expected to give substantially the same results as the Huygens–Kirchhoff diffraction theory, or even (as $q \to \infty$) the same results as geometrical optics.

If we restrict ourselves to orders l that are very much less than q and $|\hat{n}q|$ then we may use the asymptotic approximations (73)–(76). We have

$$\frac{\psi_l(\hat{n}q)}{\psi_l'(\hat{n}q)} \sim \frac{i^{l+1}e^{-i\hat{n}q} + (-i)^{l+1}e^{i\hat{n}q}}{i^l e^{-i\hat{n}q} + (-i)^l e^{i\hat{n}q}} = \frac{\cos\left[\hat{n}q - (l+1)\dfrac{\pi}{2}\right]}{\cos\left[\hat{n}q - l\dfrac{\pi}{2}\right]} = \tan\left[\hat{n}q - l\frac{\pi}{2}\right], \tag{80}$$

and the coefficients (62) becomes

$$\left.\begin{aligned}
{}^eB_l &= (-1)^{l+1}\frac{2l+1}{l(l+1)}e^{-iq}\frac{\sin\left[q - l\dfrac{\pi}{2}\right] - \hat{n}\cos\left[q - l\dfrac{\pi}{2}\right]\tan\left[\hat{n}q - l\dfrac{\pi}{2}\right]}{1 - i\hat{n}\tan\left[\hat{n}q - l\dfrac{\pi}{2}\right]}\\
{}^mB_l &= (-1)^{l+1}\frac{2l+1}{l(l+1)}e^{-iq}\frac{\hat{n}\sin\left[q - l\dfrac{\pi}{2}\right] - \cos\left[q - l\dfrac{\pi}{2}\right]\tan\left[\hat{n}q - l\dfrac{\pi}{2}\right]}{\hat{n} - i\tan\left[\hat{n}q - l\dfrac{\pi}{2}\right]}.
\end{aligned}\right\} \tag{81}$$

We note that these coefficients are rapidly oscillating functions of both q and l, so that a small change in q or l may give rise to a large change in ${}^eB_l$ and ${}^mB_l$. We also observe that ${}^mB_{l+1}$ and ${}^eB_l$ are of the same order of magnitude, i.e. the amplitude of the electric partial wave of order l is of the same order of magnitude as that of the magnetic partial wave of next higher order.

Since the formulae (81) were derived on the assumption that l is much smaller than q, one cannot determine from these approximations the number of partial waves that make an appreciable contribution to the scattered field. DEBYE* derived asymptotic approximations valid for all orders and showed that the amplitudes of the partial waves fall off rapidly to zero as soon as $l + \frac{1}{2}$ exceeds q, so that only the first q terms need be included.

When either the dielectric constant or the conductivity of the sphere is very large ($|\hat{n}| \to \infty$), the expressions for ${}^eB_l$ and ${}^mB_l$ reduce to

$$\left.\begin{aligned} {}^eB_l &= i(-1)^{l+1}\frac{2l+1}{l(l+1)}e^{-iq}\cos\left(q - l\frac{\pi}{2}\right), \\ {}^mB_l &= (-1)^{l+1}\frac{2l+1}{l(l+1)}e^{-iq}\sin\left(q - l\frac{\pi}{2}\right), \end{aligned}\right\} \tag{82}$$

derived most simply from (63) with the help of the asymptotic approximations (73)–(76).

The approximations of this section ($q \gg 1$) may be applied in connection with the theory of the rainbow†; the parameter q, determined by the size of the raindrops, is of the order 10^4.

II. $q \ll 1$. This case is of great practical importance in connection with microscopic and submicroscopic particles in colloidal solutions. We may now use the power series expansions (68)–(72) for the cylindrical functions and obtain (restricting ourselves to terms of the leading order),

$$\left.\begin{aligned} {}^eB_l &\sim i^l \frac{q^{2l+1}}{l^2[1\,.\,3\,.\,5\,\ldots\,(2l-1)]^2}\,\frac{\hat{n}^2-1}{\hat{n}^2+\dfrac{l+1}{l}}, \\ {}^mB_l &\sim i^l \frac{q^{2l+3}}{l(l+1)(2l+1)(2l+3)[1\,.\,3\,.\,5\,\ldots\,(2l-1)]^2}(\hat{n}^2-1). \end{aligned}\right\} \tag{83}$$

When either the conductivity or the dielectric constant is very large, we have from (63)‡

$$\left.\begin{aligned} {}^eB_l &= \;\; i^l \frac{q^{2l+1}}{l^2[1\,.\,3\,.\,5\,\ldots\,(2l-1)]^2}, \\ {}^mB_l &= -\,i^l \frac{q^{2l+1}}{l(l+1)[1\,.\,3\,.\,5\,\ldots\,(2l-1)]^2}. \end{aligned}\right\} \tag{84}$$

As long as the conductivity remains finite [formulae (83)], ${}^eB_{l+1}$ and ${}^mB_l$ are proportional to the same power of q, i.e. the amplitude of the $(l+1)$th electric partial

* P. DEBYE, *Ann. d. Physik* (4), **30** (1909), 118; *Mat. Ann.*, **67** (1909), 535; *Sitzungsb. Münch. Akad. Wiss, Math. Phys. Kl.*, 5 Abh. (1910).

† Cf. B. VAN DER POL and H. BREMMER, *Phil. Mag.*, (7) **24** (1937), 857; H. BUCERIUS, *Optik*, **1** (1946), 188.

‡ A direct transition from (83) to (84) is not simple; one has to take into account that as $|\hat{n}| \to \infty$, $q \to 0$ in such a way that $q^2|\hat{n}|^2$ approaches a finite value.

wave is of the same order of magnitude as the amplitude of the lth magnetic partial wave, just as was the case when $q \gg 1$. If the radius of the sphere is so small that q^2 may be neglected in comparison with unity, only the first electric partial wave need be considered and its strength (and phase) is given by the complex amplitude

$$^eB_1 = iq^3 \frac{\hat{n}^2 - 1}{\hat{n}^2 + 2} = i \left(\frac{2\pi a}{\lambda^{(\mathrm{I})}}\right)^3 \frac{\hat{n}^2 - 1}{\hat{n}^2 + 2}. \tag{85}$$

Since ^eB_l is complex there is a phase difference between the incident primary field and the scattered secondary field.

On the other hand, in the limiting case $|\hat{n}| \to \infty$, the amplitude of the lth electric partial wave and the amplitude of the lth magnetic partial wave are of the same order of magnitude, as is seen from (84).

Let us now consider, for $|\hat{n}|$ finite, the field of the first electric partial wave, at large distances from the sphere ($r \gg \lambda$). We use in (58) the asymptotic approximations

$$\zeta_1^{(1)}(x) = -e^{ix}, \qquad \zeta_1^{(1)\prime}(x) = -ie^{ix}, \tag{86}$$

and the relations

$$P_1(\cos\theta) = \cos\theta, \qquad P_1^{(1)}(\cos\theta) = \sin\theta, \qquad P_1^{(1)\prime}(\cos\theta)\sin\theta = -\cos\theta, \tag{87}$$

where the prime denotes differentiation with respect to $\cos\theta$, and obtain,

$$\left.\begin{aligned}
E_\theta^{(s)} &= -\frac{i}{k^{(\mathrm{I})}}\cos\phi\cos\theta\, {}^eB_1 \frac{e^{ik^{(\mathrm{I})}r}}{r} = \left(\frac{2\pi}{\lambda^{(\mathrm{I})}}\right)^2 a^3 \frac{\hat{n}^2 - 1}{\hat{n}^2 + 2}\cos\phi\cos\theta \frac{e^{ik^{(\mathrm{I})}r}}{r},\\
E_\phi^{(s)} &= \frac{i}{k^{(\mathrm{I})}}\sin\phi\, {}^eB_1 \frac{e^{ik^{(\mathrm{I})}r}}{r} = -\left(\frac{2\pi}{\lambda^{(\mathrm{I})}}\right)^2 a^3 \frac{\hat{n}^2 - 1}{\hat{n}^2 + 2}\sin\phi \frac{e^{ik^{(\mathrm{I})}r}}{r},\\
H_\theta^{(s)} &= \frac{1}{k_2^{(\mathrm{I})}}\sin\phi\, {}^eB_1 \frac{e^{ik^{(\mathrm{I})}r}}{r} = \left(\frac{2\pi}{\lambda^{(\mathrm{I})}}\right)^2 \sqrt{\varepsilon^{(\mathrm{I})}}a^3 \frac{\hat{n}^2 - 1}{\hat{n}^2 + 2}\sin\phi \frac{e^{ik^{(\mathrm{I})}r}}{r},\\
H_\phi^{(s)} &= \frac{1}{k_2^{(\mathrm{I})}}\cos\phi\cos\theta\, {}^eB_1 \frac{e^{ik^{(\mathrm{I})}r}}{r} = \left(\frac{2\pi}{\lambda^{(\mathrm{I})}}\right)^2 \sqrt{\varepsilon^{(\mathrm{I})}}a^3 \frac{\hat{n}^2 - 1}{\hat{n}^2 + 2}\cos\phi\cos\theta \frac{e^{ik^{(\mathrm{I})}r}}{r}.
\end{aligned}\right\} \tag{88}$$

It is easy to see that equations (88) are identical with those for the radiation zone ($r \gg \lambda$) of an *electric dipole at O, oscillating parallel to the x-axis*, i.e. parallel to the electric vector of the primary incident field, and of moment $p = p_0\, e^{-i\omega t}$ where

$$p_0 = a^3 \left|\frac{\hat{n}^2 - 1}{\hat{n}^2 + 2}\right|. \tag{89}$$

To show this we assume first that the outer medium is vacuum ($\varepsilon^{(\mathrm{I})} = 1$) and we return to equations (53) and (54) of § 2.2. According to these equations, the radiation field in vacuum of a linear electric dipole of moment $\boldsymbol{p}_0\, e^{-i\omega t}$ is given by

$$\left.\begin{aligned}
\boldsymbol{E} &= -\left(\frac{\omega}{c}\right)^2 \frac{1}{r^3}\, \boldsymbol{r} \wedge (\boldsymbol{r} \wedge \boldsymbol{p}_0),\\
\boldsymbol{H} &= \left(\frac{\omega}{c}\right)^2 \frac{1}{r^2}\, (\boldsymbol{r} \wedge \boldsymbol{p}_0),
\end{aligned}\right\} \tag{90}$$

where the time harmonic factor $e^{-i\omega t}$ has been omitted and $\boldsymbol{r}$ is the position vector of the point of observation with respect to the origin. From (90) it follows that if $\boldsymbol{p}_0$ is the direction of the x-axis, the components of $\boldsymbol{E}$ and $\boldsymbol{H}$ are given by

$$\left.\begin{aligned} E_x &= \left(\frac{\omega}{c}\right)^2 p_0(\sin^2\theta\sin^2\phi + \cos^2\theta)\frac{e^{ikr}}{r}, \\ E_y &= -\left(\frac{\omega}{c}\right)^2 p_0(\sin^2\theta\sin\phi\cos\phi)\frac{e^{ikr}}{r}, \\ E_z &= -\left(\frac{\omega}{c}\right)^2 p_0\sin\theta\cos\theta\cos\phi\frac{e^{ikr}}{r}, \\ H_x &= \quad 0, \\ H_y &= \left(\frac{\omega}{c}\right)^2 p_0\cos\theta\frac{e^{ikr}}{r}, \\ H_z &= -\left(\frac{\omega}{c}\right)^2 p_0\sin\theta\sin\phi\frac{e^{ikr}}{r}. \end{aligned}\right\} \tag{91}$$

If now the transformation (8) to spherical polar coordinates is applied, this transforms into (88) (with $\varepsilon^{(I)} = 1$) if the expression (89) is substituted for p_0. In the more general case, when $\varepsilon^{(I)} \neq 1$, we must consider in place of (91) the corresponding expressions for the radiation field of a dipole in a dielectric. These expressions, which may be obtained in a similar way to (91), transform into (88) when transformation to spherical polar coordinates is made. We note that the refractive index of the sphere enters the expression (89) for the equivalent dipole moment only in the combination $(\hat{n}^2 - 1)/(\hat{n}^2 + 2)$; for dielectric substances (n real) we have already encountered this expression, in the theory of molecular refractivity (cf. § 2.3 (17)).

Since, according to (88) the amplitude is inversely proportional to the square of the wavelength, the intensity of the scattered light is inversely proportional to the fourth power of the wavelength. In such a case one speaks of *Rayleigh scattering*.*

The first magnetic partial wave may be similarly described in terms of a vibrating magnetic dipole. The partial waves of higher orders may be considered to be due to vibrating multipoles, but we shall not investigate this here.

(c) *Intensity and polarization of the scattered light*

Let us now return to the general case and examine briefly the intensity and the polarization of the scattered light. As we are only interested in relative values of the intensity we may take as a measure of the intensity the square of the real amplitude of the electric vector. We consider only the distant field ($r \gg \lambda$) and may therefore replace the functions $\zeta_l^{(1)}$ and $\zeta_l^{(1)\prime}$ in (58) by their asymptotic approximations. We also set

$$\left.\begin{aligned} I_{\parallel}^{(s)} &= \frac{\lambda^{(I)2}}{4\pi^2 r^2}\left|\sum_{l=1}^{\infty}(-i)^l\left({}^eB_l P_l^{(1)\prime}(\cos\theta)\sin\theta - {}^mB_l\frac{P_l^{(1)}(\cos\theta)}{\sin\theta}\right)\right|^2, \\ I_{\perp}^{(s)} &= \frac{\lambda^{(I)2}}{4\pi^2 r^2}\left|\sum_{l=1}^{\infty}(-i)^l\left({}^eB_l\frac{P_l^{(1)}(\cos\theta)}{\sin\theta} - {}^mB_l P_l^{(1)\prime}(\cos\theta)\sin\theta\right)\right|^2. \end{aligned}\right\} \tag{92}$$

* As already pointed out on p. 98, RAYLEIGH showed, in celebrated papers concerning the blue colour of the sky, that scattering of this type is also caused by spontaneous fluctuations in the density of a homogeneous medium. [Lord RAYLEIGH, *Phil. Mag.*, (4), **XLI** (1871), 274, 447; (5), **XLVII** (1899), 375. Reprinted in his *Scientific Papers*, **1** (1899), 87, 104; **4** (1903), 397.]

Then

$$|E_\theta^{(s)}|^2 = I_\parallel^{(s)} \cos^2 \phi, \qquad |E_\phi^{(s)}|^2 = I_\perp^{(s)} \sin^2 \phi. \tag{93}$$

We define the *plane of observation* as the plane that contains the direction of propagation of the incident light and the direction (θ, ϕ) of observation. According to (4) and (7), ϕ represents the angle between this plane and the direction of vibration of the electric vector of the incident wave. Since according to (93) either $E_\theta^{(s)}$ or $E_\phi^{(s)}$ vanishes when $\phi = 0$ or $\phi = \pi/2$, the scattered light is linearly polarized when the plane of observation is parallel or perpendicular to the primary vibrations. For any other direction (θ, ϕ) the light is in general elliptically polarized, since the ratio $E_\theta^{(s)}/E_\phi^{(s)}$ is complex. In the special case of RAYLEIGH scattering, represented by (88), the ratio $E_\theta^{(s)}/E_\phi^{(s)}$ is, however, always real, so that the scattered light is then linearly polarized for all directions of observation.

In practice one is usually concerned with the scattering of *natural light*. As in § 1.5 the appropriate formulae may be obtained from those for polarized light by averaging over all directions of polarization.* Denoting by a bar this average, we now have, in place of (93), since $\overline{\cos^2 \phi} = \overline{\sin^2 \phi} = \frac{1}{2}$,

$$\overline{|E_\theta^{(s)}|^2} = \tfrac{1}{2} I_\parallel^{(s)}, \qquad \overline{|E_\phi^{(s)}|^2} = \tfrac{1}{2} I_\perp. \tag{94}$$

In general neither $I_\parallel^{(s)}$ nor $I_\perp^{(s)}$ is zero, so that the scattered light is partially polarized. In analogy with § 1.5 (42) we may define the degree of polarization P of the scattered light as

$$P = \left| \frac{I_\perp^{(s)} - I_\parallel^{(s)}}{I_\perp^{(s)} + I_\parallel^{(s)}} \right|. \tag{95}$$

The unpolarized proportion of the scattered light then is

$$\left.\begin{aligned} (I_\perp^{(s)} + I_\parallel^{(s)})(1 - P) &= 2I_\parallel^{(s)} \quad \text{when} \quad I_\parallel^{(s)} < I_\perp^{(s)}, \\ &= 2I_\perp^{(s)} \quad \text{when} \quad I_\parallel^{(s)} > I_\perp^{(s)}. \end{aligned}\right\} \tag{96}$$

The dependence of the intensity and polarization of the scattered light on the direction of scattering and on the physical parameters $(\lambda, a, \hat{n})$ has been investigated on the basis of MIE's theory by many writers† and we cannot do more here than summarize briefly some of the main results obtained.

---

* Alternatively we may use the result of § 10.8.2, according to which a wave of natural light may be regarded as composed of two incoherent waves of equal amplitudes, propagated in the same direction and polarized in directions at right angles to each other. The scattering of each wave may be determined separately and the total intensity is obtained by summing the intensities of the two component waves. This leads again to (94).

† In addition to the paper by MIE already quoted we may refer to the following:

R. GANS, *Ann. d. Physik* (4), **76** (1925), 29; H. SENFTLEBEN and E. BENEDICT, *Ann. d. Physik* (4), **60** (1919), 297; H. BLUMER, *Z. f. Phys.*, **32** (1925), 119; **38** (1926), 304, 920; **39** (1926), 195; C. SCHALÉN, *Uppsala Astr. Obs. Ann.*, **1**, No. 2 (1939); *ibid.*, **1**, No. 9 (1945); G. R. PARANJPE, Y. G. NAIK, and P. B. VAIDYA, *Proc. Indian Acad.*, A, **9** (1939), 333, 352; H. HOLL, *Optik*, **1** (1946), 213; *ibid.*, **4** (1948/49), 173. A very full survey of results obtained by many writers will be found in a paper by H. C. VAN DE HULST, *Rech. Astr. Observ. Utrecht*, **11**, Pt. I (1946) and in his book *Light Scattering by Small Particles* (New York, John Wiley and Sons; London, Chapman and Hall, 1957). See also G. OSTER, *Chem. Rev.*, **43** (1948), 319.

Tables of scattering functions for spherical particles, National Bureau of Standards (Washington, D.C. 1949), Applied Mathematics Series, 4; R. O. GUMPRECHT and C. M. SLIEPCEVICH, *Light-Scattering Functions for Spherical Particles* (Ann Arbor, University of Michigan Press, 1951).

In Fig. 13.10 and 13.11 the intensity and the unpolarized proportion of the scattered light are shown as functions of the angle θ of observation for dielectric and metallic spheres of various sizes. The length of the radius vector of the outer curves is proportional to the intensity $I^{(s)} = I_{\parallel}^{(s)} + I_{\perp}^{(s)}$ and that of the inner curves, unless stated otherwise, is proportional to $I_{\perp}^{(s)}$. The units are arbitrary and different in each figure. From these *polar diagrams* as well as from other published calculations the following general conclusions may be drawn:

Excluding the case when the conductivity or the dielectric constant is very large (in which case most of the incident light is radiated in the backward direction, i.e. "reflected"), the polar diagrams, in the limit of vanishingly small spheres ($a \to 0$), are symmetrical about the plane through the centre of the sphere, at right angles to the direction of propagation of the incident light. There is an intensity maximum in the forward direction ($\theta = 0$) and in the reverse direction ($\theta = 180°$), and there is a minimum in the plane of symmetry ($\theta = 90°$). As the radius of the sphere is increased there is a departure from symmetry, more light being scattered in the forward direction than in the opposite direction. This phenomenon is often called the *Mie effect*. As the radius is increased still further practically all the scattered light appears around the forward direction $\theta = 0$; likewise for a conducting sphere there is a greater concentration of light in this direction. When the radius of the sphere is very large compared to the wavelength most of the incident light is, however, reflected, as follows from geometrical optics.

The dependence of the intensity of the scattered light on the radius of the sphere is illustrated in Table XIX. The Mie effect may be clearly seen from the comparison of

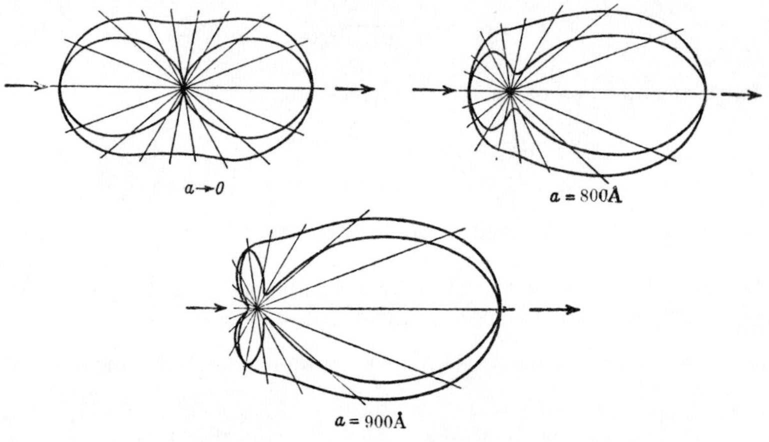

Fig. 13.10. Polar diagrams for the scattering of linearly polarized light by a spherical gold particle ($\lambda_0 = 5500$ Å, $n^{(I)} = 1{\cdot}33$, $\hat{n}^{(II)} = 0{\cdot}57 + 2{\cdot}45i$).
(After G. Mie, *Ann. d. Physik*, (4) **25** (1908), 429.)

the first and third row. The table indicates a very rapid increase in intensity with increasing size of the sphere; for true comparison the values in the table must be multiplied by the factor $\lambda^{(I)^2}/4\pi^2a^2 = 1/q^2$.

When q exceeds unity, i.e. when the diameter $2a$ of the sphere is greater than $\lambda^{(I)}/\pi$ there appear a series of maxima and minima, which at first are distributed irregularly. The appearance of a number of maxima and minima when q is large is in agreement with the Huygens–Kirchhoff theory.

Concerning polarization of the scattered light the results again are different depending on whether or not $|\hat{n}|$ is large.

For very small spheres which are highly conducting ($\sigma \to \infty$) or have very high dielectric constant ($\varepsilon \to \infty$) the polarization is largest when $\theta = 60°$ (the THOMSON

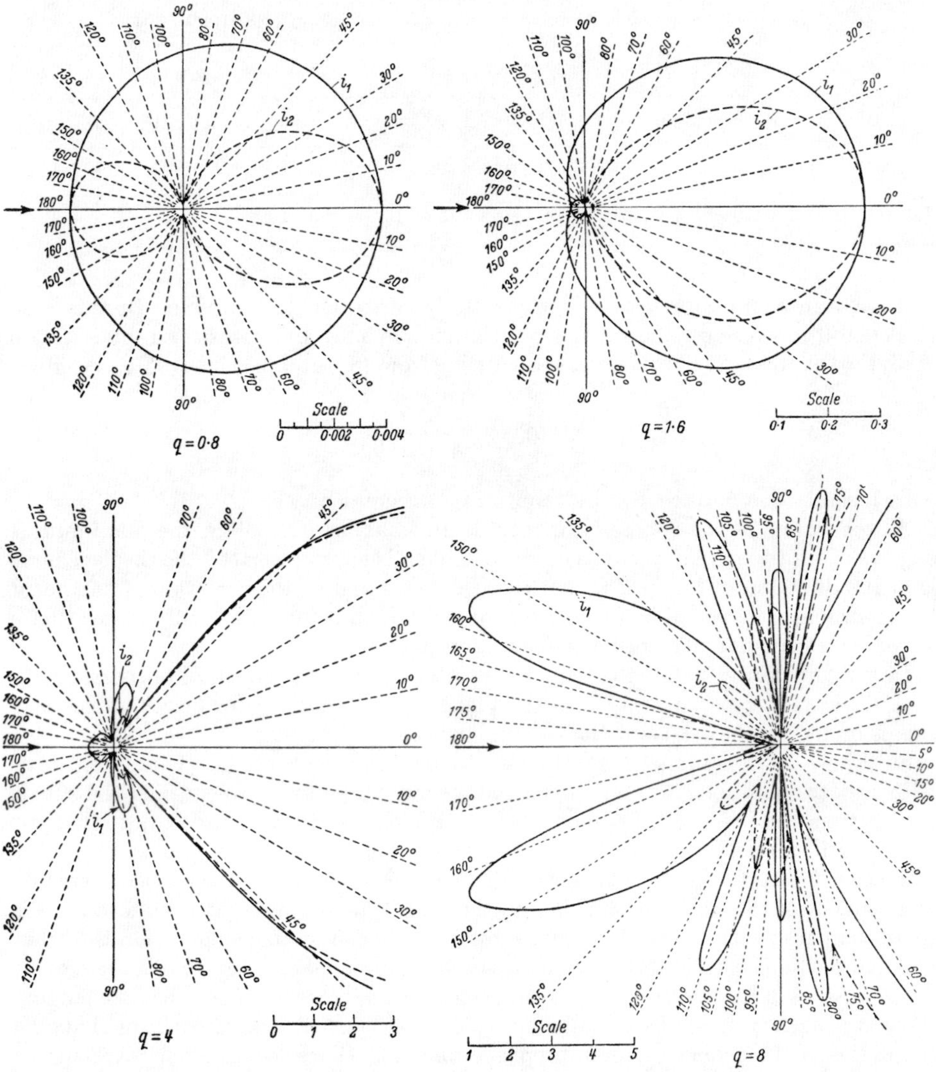

Fig. 13.11. Polar diagrams for the scattering of linearly polarized light by a dielectric sphere of refractive index $n = 1{\cdot}25$. $i_1 = q^2 I_{\perp}$, $i_2 = q^2 I_{\parallel}$.
(After H. BLUMER, *Z. f. Phys.*, **32** (1925), 119.)

angle). With increasing radius the maximum is displaced in the direction of increasing θ.

The dependence of the polarization on the angle θ of observation, for spheres of finite conductivity and finite dielectric constant, is illustrated in two typical cases in Fig. 13.10 and 13.11. When the radius of the sphere is very small ($q \to 0$), the polarization diagram is, like the intensity diagram, symmetrical about the xy-plane

TABLE XIX

The normalized intensity $4\pi^2a^2(I_\perp^{(s)} + I_\parallel^{(s)})/\lambda^{(I)^2}$ *of light scattered by dielectric spheres of refractive index* $n = 1{\cdot}25$, *as function of the parameter* $q = 2\pi a/\lambda^{(I)}$

(Compiled from calculations of H. BLUMER, *Z. f. Phys.*, **38** (1926), 304)

θ	$q = 0{\cdot}01$	$q = 0{\cdot}1$	$q = 0{\cdot}5$	$q = 1$	$q = 2$	$q = 5$	$q = 8$
0	$5{\cdot}0 \,.\, 10^{-14}$	$5{\cdot}0 \,.\, 10^{-8}$	$1{\cdot}2 \,.\, 10^{-3}$	$2{\cdot}3 \,.\, 10^{-1}$	$4{\cdot}3$	$9{\cdot}8 \,.\, 10^{2}$	$7{\cdot}5 \,.\, 10^{3}$
90°	$2{\cdot}5 \,.\, 10^{-14}$	$2{\cdot}5 \,.\, 10^{-8}$	$5{\cdot}0 \,.\, 10^{-4}$	$3{\cdot}6 \,.\, 10^{-2}$	$2{\cdot}5 \,.\, 10^{-1}$	$2{\cdot}7$	$7{\cdot}1$
180°	$5{\cdot}0 \,.\, 10^{-14}$	$4{\cdot}9 \,.\, 10^{-8}$	$7{\cdot}8 \,.\, 10^{-4}$	$1{\cdot}9 \,.\, 10^{-3}$	$2{\cdot}0 \,.\, 10^{-2}$	$1{\cdot}3$	$0{\cdot}9$

and has a maximum for $\theta = 90°$, where the polarization is complete. In this case (RAYLEIGH scattering), the degree of polarization may be represented by a single analytical expression, obtained on substituting from (88) and (93) into (95); this gives

$$P(\theta) = \frac{\sin^2 \theta}{1 + \cos^2 \theta}. \tag{97}$$

This formula was derived by RAYLEIGH, in a different way.

As the radius of the sphere is increased, up to about $a = \lambda^{(I)}/\pi$, the maximum is displaced; in the majority of the cases that have been investigated the displacement is in the direction of larger θ for dielectric spheres and in the direction of smaller θ for absorbing spheres. When the radius of the sphere is increased still further, there appears an irregular sequence of polarization maxima.

In the direction $\theta = 90°$ the light is, for $q < 1$, almost completely polarized, with its electric vector perpendicular to the plane of observation; for larger values of q this is no longer the case and the behaviour becomes irregular.

So far we have confined our attention to monochromatic light. One often deals with scattering of polychromatic light and we must therefore also consider the effects arising from the presence of components of different wavelengths. We note that the wavelength enters our formulae only through the parameter q and through the refractive index $\hat{n}$. In a sufficiently small range of wavelengths, $\hat{n}$ is practically independent of the wavelength if in (60) the term that contains the conductivity σ is small compared with the other term, i.e. for a poorly conducting sphere. On the other hand, in the limit of infinitely high conductivity, $\hat{n}$ does not enter at all. In these cases the intensities of the spectral components depend on $a/\lambda^{(I)}$ only. The effect of changing the wavelength is thus substantially equivalent to the effect that arises from changing the radius of the sphere by an appropriate amount. Since for different wavelengths the polarization maxima occur at different angles of observations, complicated colour changes are seen when observations on scattered light are made through a polarizing prism. This effect is called *polychroism*. The dependence of polarization of the scattered light on wavelength—known as *dispersion of polarization*—affords a very precise test of the theory.*

* The dispersion of polarization and polychroism was investigated by M. A. SCHIRMANN *Ann. d. Physik*, (4), **59** (1919), 493.

13.5.3 Total scattering and extinction

(a) *Some general considerations*

It is of considerable practical interest to determine the total amount of light that is scattered or absorbed by the sphere. This may be calculated by evaluating the POYNTING vector and integrating it over all directions. With the help of the orthogonality relations that exist between the associated LEGENDRE functions it is possible to express the integrals in terms of the coefficients ${}^eB_l$ and ${}^mB_l$. These calculations, which are somewhat lengthy, are carried out in full in the paper by MIE.*

The total energy lost from the incident wave, i.e. the sum of the scattered and the absorbed energy, may be determined in an alternative way, from certain general considerations that apply to an obstacle of any shape whatsoever. These considerations show that there is a close connection between the energy loss and the amplitude of the scattered wave in the forward direction ($\theta = 0$). From this result, which we shall now establish, and from MIE's formula for the scattered wave, the total energy lost by scattering and absorption from a sphere may then easily be determined.

Consider a plane monochromatic wave incident on an obstacle of arbitrary form, embedded in a dielectric medium. The field at any point in the medium surrounding the obstacle may again be represented as the sum of the incident field and the scattered field

$$\mathbf{E} = \mathbf{E}^{(i)} + \mathbf{E}^{(s)}, \qquad \mathbf{H} = \mathbf{H}^{(i)} + \mathbf{H}^{(s)}. \tag{98}$$

As usually, we omit a time factor $\exp(-i\omega t)$. The time averaged energy flow is represented by the averaged POYNTING vector, which, according to (98) and the formula (56) of § 1.4, is given by

$$\langle \mathbf{S} \rangle = \langle \mathbf{S}^{(i)} \rangle + \langle \mathbf{S}^{(s)} \rangle + \langle \mathbf{S}' \rangle, \tag{99}$$

where ($\mathscr{R}$ denoting the real part)

$$\langle \mathbf{S}^{(i)} \rangle = \frac{c}{8\pi} \mathscr{R} (\mathbf{E}^{(i)} \wedge \mathbf{H}^{(i)\star}), \tag{100a}$$

$$\langle \mathbf{S}^{(s)} \rangle = \frac{c}{8\pi} \mathscr{R} (\mathbf{E}^{(s)} \wedge \mathbf{H}^{(s)\star}), \tag{100b}$$

$$\langle \mathbf{S}' \rangle = \frac{c}{8\pi} \mathscr{R} (\mathbf{E}^{(i)} \wedge \mathbf{H}^{(s)\star} + \mathbf{E}^{(s)} \wedge \mathbf{H}^{(i)\star}). \tag{100c}$$

Consider the averaged outward flow of energy through the surface of a large sphere of radius R, centred at some point in the region occupied by the obstacle. The net flow per second is represented by the integral of the radial component $\langle \mathbf{S} \rangle_r$ of $\langle \mathbf{S} \rangle$, taken over the sphere, and is evidently zero when the obstacle is dielectric. If, however, the obstacle is a conductor, some of the incident energy is absorbed by it and the net outward flow through the surface of the sphere is equal in magnitude to the rate at which absorption takes place. Let $\mathscr{W}^{(a)}$ be the rate at which the energy is being absorbed by the obstacle. Then, from (99)

$$-\mathscr{W}^{(a)} = \mathscr{W}^{(i)} + \mathscr{W}^{(s)} + \mathscr{W}', \tag{101}$$

where $\mathscr{W}^{(i)}$, $\mathscr{W}^{(s)}$, and $\mathscr{W}'$ are the integrals of the radial components $\langle \mathbf{S}^{(i)} \rangle_r$, $\langle \mathbf{S}^{(s)} \rangle_r$, and $\langle \mathbf{S}' \rangle_r$ over the surface of the sphere. Now since the medium that surrounds the

* G. MIE, *loc. cit.*, pp. 432–436.

obstacle is assumed to be non-conducting, $\mathscr{W}^{(i)} = 0$, so that

$$\mathscr{W}^{(a)} + \mathscr{W}^{(s)} = -\mathscr{W}' = -\frac{c}{8\pi}\mathscr{R}\iint_S (\boldsymbol{E}^{(i)} \wedge \boldsymbol{H}^{(s)\star} + \boldsymbol{E}^{(s)} \wedge \boldsymbol{H}^{(i)\star}) \, . \, \boldsymbol{n} dS, \quad (102)$$

S denoting the large sphere and $\boldsymbol{n}$ the unit outward normal. Thus the expression on the right of (102) represents the rate at which energy is dissipated by heat and scattering.

Let $\boldsymbol{n}_0$ be the unit vector in the direction in which the incident wave is propagated, so that

$$\boldsymbol{E}^{(i)} = \boldsymbol{e}\, e^{ik^{(\mathrm{I})}(\boldsymbol{n}_0 . \boldsymbol{r})} \quad , \quad \boldsymbol{H}^{(i)} = \boldsymbol{h}\, e^{ik^{(\mathrm{I})}(\boldsymbol{n}_0 . \boldsymbol{r})}. \quad (103)$$

We assume that this wave is linearly polarized, so that $\boldsymbol{e}$ and $\boldsymbol{h}$ may be assumed to be real constant vectors. At a large distance from the obstacle the scattered wave may be assumed to be spherical:

$$\boldsymbol{E}^{(s)} = \boldsymbol{a}(\boldsymbol{n}) \frac{e^{ik^{(\mathrm{I})}r}}{r}, \qquad \boldsymbol{H}^{(s)} = \boldsymbol{b}(\boldsymbol{n}) \frac{e^{ik^{(\mathrm{I})}r}}{r}. \quad (104)$$

The vectors $\boldsymbol{a}(\boldsymbol{n})$ and $\boldsymbol{b}(\boldsymbol{n})$ characterize the strength of the radiation scattered in the direction $\boldsymbol{n}$. Since the incident and scattered wave obey MAXWELL's equations we have (cf. eq. (4), (5) of § 1.4)

$$\left.\begin{aligned} \boldsymbol{h} &= \sqrt{\varepsilon^{(\mathrm{I})}}\boldsymbol{n}_0 \wedge \boldsymbol{e}, & \boldsymbol{b} &= \sqrt{\varepsilon^{(\mathrm{I})}}\boldsymbol{n} \wedge \boldsymbol{a}, \\ \boldsymbol{n}_0 \, . \, \boldsymbol{e} &= \boldsymbol{n}_0 \, . \, \boldsymbol{h} = 0, & \boldsymbol{n} \, . \, \boldsymbol{a} &= \boldsymbol{n} \, . \, \boldsymbol{b} = 0, \end{aligned}\right\} \quad (105)$$

where $\varepsilon^{(\mathrm{I})}$ is the dielectric constant of the outer medium, assumed to be non-magnetic ($\mu = 1$). From these relations it follows that on the surface of the large sphere S,

$$\left.\begin{aligned} (\boldsymbol{E}^{(i)} \wedge \boldsymbol{H}^{(s)\star}) \, . \, \boldsymbol{n} &= \sqrt{\varepsilon^{(\mathrm{I})}}\boldsymbol{e} \, . \, \boldsymbol{a}^\star \, e^{ik^{(\mathrm{I})}R(\boldsymbol{n}_0 . \boldsymbol{n})} \frac{e^{-ik^{(\mathrm{I})}R}}{R}, \\ (\boldsymbol{E}^{(s)} \wedge \boldsymbol{H}^{(i)\star}) \, . \, \boldsymbol{n} &= \sqrt{\varepsilon^{(\mathrm{I})}}[(\boldsymbol{n} \, . \, \boldsymbol{n}_0)(\boldsymbol{a} \, . \, \boldsymbol{e}) - (\boldsymbol{n} \, . \, \boldsymbol{e})(\boldsymbol{n}_0 \, . \, \boldsymbol{a})] \, e^{-ik^{(\mathrm{I})}R(\boldsymbol{n}_0 . \boldsymbol{n})} \frac{e^{ik^{(\mathrm{I})}R}}{R}. \end{aligned}\right\} \quad (106)$$

We substitute these expressions in (102). To evaluate the resulting integral we use the mathematical lemma,* that when R is large, and f is an arbitrary function of $\boldsymbol{n}$,

$$\frac{1}{R}\iint_S f(\boldsymbol{n}) \, e^{-ik^{(\mathrm{I})}R(\boldsymbol{n}_0 \, . \, \boldsymbol{n})} \, dS \sim \frac{2\pi i}{k^{(\mathrm{I})}}[f(\boldsymbol{n}_0) \, e^{-ik^{(\mathrm{I})}R} - f(-\boldsymbol{n}_0) \, e^{ik^{(\mathrm{I})}R}]. \quad (107)$$

We then obtain

$$\left.\begin{aligned} \iint_S (\boldsymbol{E}^{(i)} \wedge \boldsymbol{H}^{(s)\star}) \, . \, \boldsymbol{n} dS &\sim -\frac{2\pi i}{k^{(\mathrm{I})}} \sqrt{\varepsilon^{(\mathrm{I})}}[\boldsymbol{e} \, . \, \boldsymbol{a}^\star(\boldsymbol{n}_0) - \boldsymbol{e} \, . \, \boldsymbol{a}^\star(-\boldsymbol{n}_0) \, e^{-2ik^{(\mathrm{I})}R}], \\ \iint_S (\boldsymbol{E}^{(s)} \wedge \boldsymbol{H}^{(i)\star}) \, . \, \boldsymbol{n} dS &\sim \frac{2\pi i}{k^{(\mathrm{I})}} \sqrt{\varepsilon^{(\mathrm{I})}}[\boldsymbol{e} \, . \, \boldsymbol{a}(\boldsymbol{n}_0) + \boldsymbol{e} \, . \, \boldsymbol{a}(-\boldsymbol{n}_0) \, e^{2ik^{(\mathrm{I})}R}], \end{aligned}\right\} \quad (108)$$

and the relation (102) becomes

$$\mathscr{W}^{(s)} + \mathscr{W}^{(a)} = \frac{c\sqrt{\varepsilon^{(\mathrm{I})}}}{2k^{(\mathrm{I})}} \mathscr{I}[\boldsymbol{e} \, . \, \boldsymbol{a}(\boldsymbol{n}_0)], \quad (109)$$

where $\mathscr{I}$ denotes the imaginary part.

---

* This lemma may be proved, after an obvious change of variables, by the application of the principle of stationary phase (cf. Appendix III, p. 752). See also D. S. JONES, *Proc. Camb. Phil. Soc.*, **48** (1952), 736.

The relation (109) implies that *with incident light that is linearly polarized, the rate at which the energy is dissipated is proportional to a certain amplitude component of the scattered wave; the amplitude is that which corresponds to forward scattering* ($\boldsymbol{n} = \boldsymbol{n}_0$) *and the component is in the direction of the electric vector of the incident wave.*

The ratio Q between the rate of dissipation of energy ($\mathscr{W}^{(a)} + \mathscr{W}^{(s)}$) and the rate at which the energy is incident on unit cross-sectional area of the obstacle ($|\langle \boldsymbol{S}^{(i)} \rangle|$) is called the *extinction cross-section* of the obstacle. From (100a), (103), and (105) it follows that $|\langle \boldsymbol{S}^{(i)} \rangle| = c\sqrt{\varepsilon^{(\mathrm{I})}}e^2/8\pi$ so that we have, according to (109),

$$Q = \frac{\mathscr{W}^{(s)} + \mathscr{W}^{(a)}}{|\langle \boldsymbol{S}^{(i)} \rangle|} = 2\lambda^{(\mathrm{I})} \mathscr{I}\left(\frac{\boldsymbol{e} \cdot \boldsymbol{a}(\boldsymbol{n}_0)}{e^2}\right). \tag{110}$$

This formula is due to van de Hulst.*

One may define the *scattering cross-section* $Q^{(s)}$ and the *absorption cross-section* $Q^{(a)}$ of the obstacle in a similar way:

$$Q^{(s)} = \frac{\mathscr{W}^{(s)}}{|\langle \boldsymbol{S}^{(i)} \rangle|}, \qquad Q^{(a)} = \frac{\mathscr{W}^{(a)}}{|\langle \boldsymbol{S}^{(i)} \rangle|}, \tag{111}$$

and evidently $Q = Q^{(s)} + Q^{(a)}$. For a non-absorbing obstacle $Q^{(a)} = 0$ and the extinction cross-section is then equal to the scattering cross-section.

Before applying the relation (110) to a spherical obstacle, let us consider the value of Q for an obstacle that does not transmit an appreciable fraction of the incident light. We also assume that the linear dimensions of the obstacle are large compared to the wavelength. In this case the Huygens–Kirchhoff theory applies and the

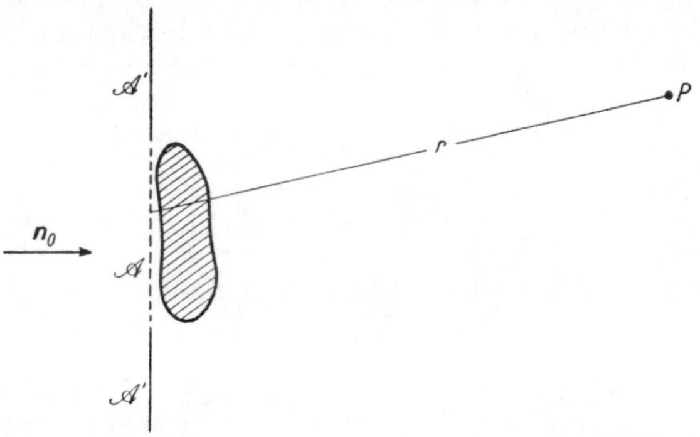

Fig. 13.12. Illustrating theorem (114), relating to the extinction cross-section of a large obstacle.

main contribution to the forward scattering arises from Fraunhofer diffraction. Let $\mathscr{A}$ be the "shadow region" and $\mathscr{A}'$ the unobstructed region of a plane wave-front of a linearly polarized wave incident upon the obstacle (Fig. 13.12), and consider the scattered field $\boldsymbol{E}^{(s)}$ at a point P at a large distance from the obstacle. According to the Huygens–Fresnel principle and the principle of Babinet (§ 8.3 (21))

$$\boldsymbol{E}^{(s)} = \frac{i}{\lambda^{(\mathrm{I})}} \boldsymbol{e} \iint_{\mathscr{A}} \frac{e^{ik^{(\mathrm{I})}r}}{r}\, dS, \tag{112}$$

* H. C. van de Hulst, *Physica*, **15** (1949), 740. The method of proof employed here is due to D. S. Jones, *Phil. Mag.*, **46** (1955), 957. There is an analogous theorem for atomic collisions (cf. E. Feenberg, *Phys. Rev.*, **40** (1932), 48; M. Lax, *Phys. Rev.*, (2), **78** (1950), 306).

if the angle of diffraction is small. If P is very far away from the obstacle in the direction of propagation of the incident wave (forward direction), r may be taken as constant and (112) gives

$$\boldsymbol{E}^{(s)}(\boldsymbol{n}_0) = \frac{i}{\lambda^{(\mathrm{I})}} D\boldsymbol{e} \frac{e^{ik^{(\mathrm{I})}r}}{r}, \tag{113}$$

where D is the geometrical cross-section of the obstacle (the area $\mathscr{A}$). Hence in this case the vector $\boldsymbol{a}(\boldsymbol{n}_0)$ in (104) is given by $iD\boldsymbol{e}/\lambda^{(\mathrm{I})}$, and (110) shows that

$$Q = 2D. \tag{114}$$

Thus *the extinction cross-section of a large opaque obstacle is equal to twice its geometrical cross-section*. This result appears somewhat paradoxical at first sight, as one might have expected that with a large obstacle the geometrical optics approximation would apply, and in this approximation the extinction cross-section is equal to D. The explanation of this apparent contradiction* is, that no matter how large the obstacle is, and no matter how far away from it the field is considered, there is always a narrow region—the neighbourhood of the edge of the geometrical shadow—where the geometrical optics approximation does not hold. In addition to the light intercepted by the obstacle (lost by reflection and absorption), with cross-section D, there is an additional contribution to the extinction, arising from the neighbourhood of the edge of the shadow and this contribution is evidently also equal to D. In order to verify the relation (114) by experiment one must collect the light over a sufficiently wide area and far enough away from the obstacle.†

Let us now apply the general formula (110) to a spherical obstacle. According to (58) the amplitude component of the scattered wave in the direction of the electric vector of the incident wave ($\phi = 0$), for forward scattering ($\theta = 0$), is given by

$$(E_\theta{}^{(s)})_{\theta=\phi=0} = -\frac{1}{k^{(\mathrm{I})}}\frac{1}{r}\sum_{l=1}^{\infty}\Big\{{}^eB_l\zeta_l{}^{(1)\prime}(k^{(\mathrm{I})}r)[P_l{}^{(1)\prime}(\cos\theta)\sin\theta]_{\theta=0} - i\,{}^mB_l\zeta_l{}^{(1)}(k^{(\mathrm{I})}r)\left[P_l{}^{(1)}(\cos\theta)\frac{1}{\sin\theta}\right]_{\theta=0}\Big\}. \tag{115}$$

The two terms involving the associated LEGENDRE functions can easily be evaluated from the expansion‡

$$P_l{}^{(m)}(x) = \frac{1}{2^m m!}\frac{(l+m)!}{(l-m)!}(1-x^2)^{m/2}\left\{1 + c_1\left(\frac{1-x}{2}\right) + c_2\left(\frac{1-x}{2}\right)^2 + \ldots\right\}, \tag{116}$$

where $c_1, c_2, \ldots$ depend on l and m only. From (116) and from the derivative of this expression we find that

---

* There is an analogous apparent paradox in quantum mechanical scattering problems, first noted by H. S. W. MASSEY and C. B. O. MOHR, *Proc. Roy. Soc.*, A, **141** (1933), 434.

† It follows from a careful analysis leading to lemma (107), that the portion of S which contributes appreciably to Q subtends at the centre of S a solid angle of order $(kR)^{-\beta}$, where $1 > \beta > 4/5$ [cf. D. S. JONES, *Proc. Camb. Phil. Soc.*, **48** (1952), 736]. For a fuller discussion of the extinction cross-section of a large obstacle and the role played by the geometrical shadow see D. SINCLAIR, *J. Opt. Soc. Amer.*, **37** (1947), 475 and L. BRILLOUIN, *J. Appl. Phys.*, **20** (1949), 1110.

‡ Cf. W. MAGNUS and F. OBERHETTINGER, *Formulas and Theorems for the Functions of Mathematical Physics* (New York, Chelsea Publishing Company, 1954), p. 54. The formula given by these authors differs from (116) by a multiplication factor $(-1)^m$. This difference arises from their use of a slightly different definition of the associated LEGENDRE polynomials (cf. footnote on p. 646).

$$\left(\frac{P_l^{(1)}(\cos\theta)}{\sin\theta}\right)_{\theta=0} = \tfrac{1}{2}l(l+1), \qquad (P_l^{(1)\prime}(\cos\theta)\sin\theta)_{\theta=0} = -\tfrac{1}{2}l(l+1). \quad (117)$$

On substituting from (117) in (115) and using the asymptotic approximations (74) and (76) for $\zeta_l^{(1)}$ and $\zeta_l^{(1)\prime}$, we obtain

$$(E_\theta^{(s)})_{\theta=\phi=0} = \frac{1}{2k^{(\mathrm{I})}}\frac{e^{ik^{(\mathrm{I})}r}}{r}\sum_{l=1}^{\infty}(-i)^l l(l+1)[{}^eB_l + {}^mB_l]. \quad (118)$$

Remembering that the incident field was taken to be of unit amplitude ($e^2 = 1$), the required quantity $\boldsymbol{e}\,.\,\boldsymbol{a}(\boldsymbol{n}_0)/e^2$ is the multiplier of $e^{ik^{(\mathrm{I})}r}/r$ in this expression; on substituting it in (110) and using the identity $\mathscr{I}(z) \equiv \mathscr{R}(-iz)$, which holds for any complex number z, we finally obtain the following expressions for the *extinction cross-section of a sphere*, in terms of the coefficients ${}^eB_l$ and ${}^mB_l$ which are given by (62):

$$Q = \frac{\lambda^{(\mathrm{I})^2}}{2\pi}\mathscr{R}\sum_{l=1}^{\infty}(-i)^{l+1}l(l+1)[{}^eB_l + {}^mB_l], \quad (119)$$

where $\mathscr{R}$ denotes the real part.

(b) *Computational results*

We shall now summarize the main computational results relating to total scattering, total absorption, and the extinction of a sphere.

We saw in § 13.5.2 that when the sphere is very much smaller than the wavelength (RAYLEIGH scattering), only the first electric partial wave needs to be taken into account. The amplitude of the scattered wave is then proportional to $1/\lambda^{(\mathrm{I})^2}$, so that the total scattering is inversely proportional to the fourth power of the wavelength. If we take into account the higher terms, which depend on the radius and the material constants, the total scattering becomes a very complicated function of the wavelength and shows selective properties.* In the case of gold, for example, even a very small sphere gives a maximum near $\lambda = 5500$ Å (see Fig. 13.13).

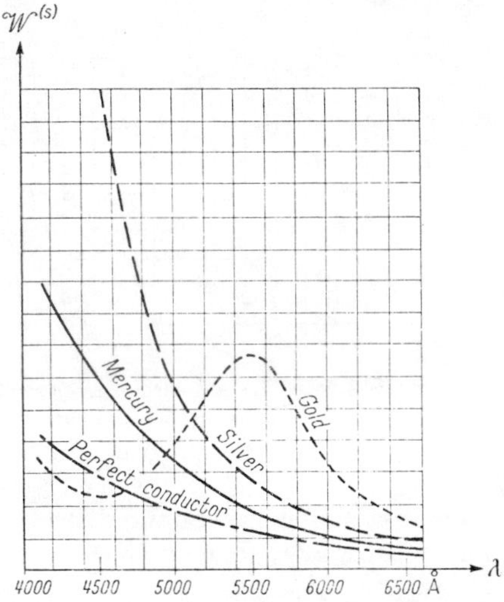

Fig. 13.13. Total scattering from very small particles ($a \to 0$) as function of the wavelength.

(After R. FEICK, *Ann. d. Physik*, (4) **77** (1925), 582.)

Such maxima may be interpreted as a kind of resonance phenomena. Let us suppose that the sphere is not under the influence of the field of the incident beam of light, but that it performs a free electromagnetic oscillation. The frequency and the damping constant of this free vibration is obtained from the theory, by omitting in

* An interesting example of selective scattering was provided by the phenomena observed in September 1950, when over a large part of Europe the sun (and the moon also) appeared deep blue. Spectrographic measurements of the "blue" and normal suns gave the extinction curve of the layer causing the phenomena, and it was concluded that the blue colour was due to selective scattering by smoke, probably consisting of oil droplets, remarkably uniform in size, which had been carried by winds in the upper atmosphere from forest fires burning in Alberta (cf. R. WILSON, *Mon. Not. Roy. Astr. Soc.*, **111** (1951), 478).

equations (56) those terms which do not contain the coefficients ${}^eB_l$ and ${}^mB_l$. The resulting equations are linear and homogeneous and have non-trivial solutions only if a consistency condition is satisfied. Each solution of the consistency equation corresponds to a damped proper vibration, and the frequency of these vibrations coincides very closely with the frequencies for which certain scattered partial waves have maxima of intensity.*

Calculations relating to scattering and extinction from spheres of finite radii, for selected values of the refractive index, have been carried out by many workers. The majority of the calculations concern dielectric spheres (n real) and spheres that are slightly absorbing. In Fig. 13.14 a typical curve, due to B. GOLDBERG, is shown for spheres of refractive index $n = 1{\cdot}33$.† This is the refractive index of water, and the results are of interest in connection with the transmission of light by mist, clouds, and fogs, in connection with the theory of the WILSON cloud chamber, etc. It is seen that the curve has a series of maxima and minima and that with increased radius the extinction cross-section tends to twice the geometrical cross-section, in agreement with equation (114). The curve has also a fine structure, i.e. small subsidiary maxima and minima. Naturally, these small fluctuations are smoothed out if the scattering is caused by many spherical particles which are not of exactly the same size.

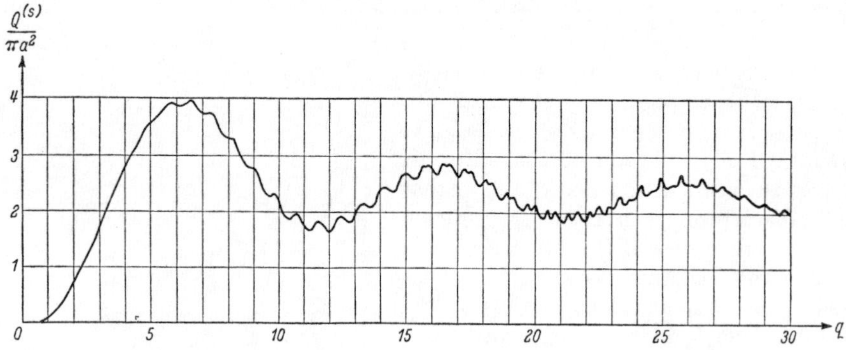

Fig. 13.14. The scattering cross-section of dielectric spheres of refractive index $n = 1{\cdot}33$ as function of the parameter $q = 2\pi a/\lambda^{(1)}$.
(After B. GOLDBERG, *J. Opt. Soc. Amer.*, **43** (1953), 1221.)

The extinction curves for dielectric spheres of other refractive indices exhibit similar behaviour. It may be shown that if n is not too different from unity all the

---

* Optical resonance was observed by R. W. WOOD (*Phil. Mag.*, (6), **3** (1902), 396), from granular films and fogs of the alkali metals.

The proper vibrations of a sphere were studied by P. DEBYE, *Ann. d. Physik*, (4), **30** (1909), 73.

† This case was also investigated by H. HOLL, *Optik*, **4** (1948), 173 and by H. G. HOUGHTON and W. R. CHALKER, *J. Opt. Soc. Amer.*, **39** (1949), 955. Similar curves relating to other values of the refractive index have been published by many writers, for example: M. D. BARNES and V. K. LA MER, *J. Col. Sci.*, **1** (1946), 79; M. D. BARNES, A. S. KENYON, E. M. ZAISER, and V. K. LA MER, *ibid.*, **2** (1947), 349; J. L. GREENSTEIN (Harvard circ., 1937, No. 422); H. C. VAN DE HULST, *Rech. Astr. Observ. Utrecht*, **11**, Pt. 1 (1946), 43–51 and his book *Light Scattering by Small Particles* (New York, J. Wiley and Sons; London, Chapman and Hall, 1957), Chapter 13; R. PENDORF, *J. Opt. Soc. Amer.*, **46** (1956), 1001.

curves have a first maximum for a value of q given by† $2q(n-1) \sim 4$, where Q may be as large as $4\pi a^2$.

For perfectly reflecting spheres† $(n \to \infty)$ the first maximum of the extinction curve is found to be at $q = 1{\cdot}2$ where $Q = 2{\cdot}29\pi a^2$ and the first minimum at $q = 1{\cdot}6$ where $Q = 2{\cdot}12\pi a^2$. Thereafter there are slight oscillations and the curve approaches $2\pi a^2$ as $q \to \infty$.

The calculations relating to absorbing spheres are much more laborious and only a few special cases have been studied in detail. In Fig. 13.15 curves relating to scattering, absorption, and extinction by small spheres of iron are shown. For larger spheres, asymptotic formulae due to JOBST,‡ based on MIE's theory and on DEBYE's asymptotic expansions of the cylinder functions may be used for calculation. Weakly

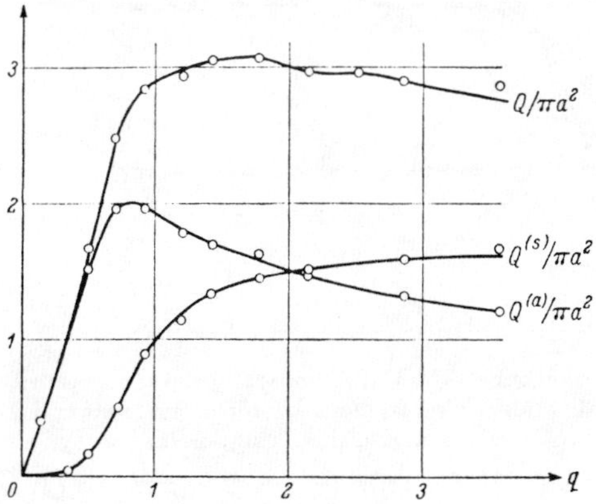

Fig. 13.15. The absorption cross-section ($Q^{(a)}$), the scattering cross-section ($Q^{(s)}$), and the extinction cross-section (Q) for iron spheres of various radii. $\hat{n} = 1{\cdot}27 + 1{\cdot}37i$, $\lambda^{(1)} = 4200$ Å.

(Based on computations of C. SCHALÉN, *Uppsala Astr. Observ. Ann.*, **1**, No. 9 (1945).)

absorbing spheres have been studied by VAN DE HULST,§ and we show results of his calculations in Fig. 13.16. In this latter case the general behaviour of the extinction curves is seen to be similar to those of dielectric spheres, but even a very small conductivity is sufficient to smooth out the small undulations completely. As the conductivity is increased still further, the first minimum disappears altogether and the extinction curve rises asymptotically from the origin to 2; the absorption curves rise asymptotically from the origin to half this value.

MIE's theory may be tested experimentally by means of observations of light scattered either by a single spherical particle, or by many particles (cloudy media, colloidal solutions). Such tests may be carried out with relative ease when the particles are large, but are rather troublesome when the diameter of each particle is of the order of a wavelength or smaller. LA MER and collaborators‖ succeeded in testing the theory

* The quantity $2q(n-1)$ represents the phase shift suffered by a ray of light that traverses the sphere along a diameter. This is a useful parameter, and if n is not too different from unity the extinction curves plotted against it are very similar to each other.

† F. W. P. GÖTZ, *Astr. Nachr.*, **255** (1935), 63 and J. L. GREENSTEIN (*loc. cit.*).

‡ G. JOBST, *Ann. d. Physik*, (4), **76** (1925), 863.

§ H. C. VAN DE HULST, *Rech. Astr. Observ. Utrecht*, **11**, Part 2 (1949), 27.

‖ M. D. BARNES and V. K. LA MER, *J. Col. Sci.*, **1** (1946), 79; M. D. BARNES, A. S. KENYON, E. M. ZAISER, and V. K. LA MER, *ibid.*, **2** (1947), 349; I. JOHNSON and V. K. LA MER, *J. Amer. Chem. Soc.*, **69** (1947), 1184.

from measurements of the angular distribution of scattered light as well as the total scattering from sulphur sols in water, of particle diameter 3000 Å to 5000 Å. Light of vacuum wavelengths in the range from 2850 Å to 10,000 Å was used and a fair agreement with the predictions of MIE's theory was found. In some cases even the minute fluctuations (fine structure) of the extinction curve (see Fig. 13.14) were observed.

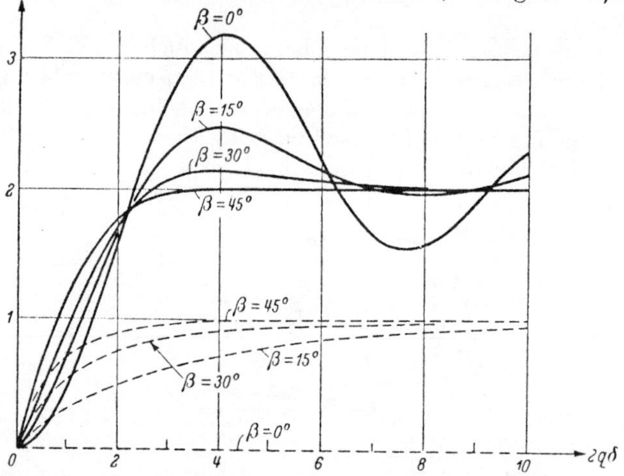

Fig. 13.16. The extinction curves $Q/\pi a^2$ (full lines) and the absorption curves $Q^{(a)}/\pi a^2$ (interrupted lines) for weakly absorbing spheres of refractive index $\hat{n} = (1 + \delta) + i\delta \tan \beta$ where δ is real and small compared to unity.

(After H. C. VAN DE HULST, *Rech. Astr. Obs. Utrecht*, **11**, Part 2 (1949), 28.)

The scattering of light by particles of shapes other than spherical has been considered by some authors, but in general the analytical properties of the corresponding wave functions are much more complicated, so that rigorous solutions are of limited practical value*. GANS† and other workers discussed the scattering of electromagnetic waves by ellipsoids with dimensions small compared to the wavelength; a rigorous solution for an ellipsoid of arbitrary size has been published by MÖGLICH.‡ The scattering from long circular conducting cylinders was studied as early as 1905 by SEITZ§ and IGNATOWSKY|| and the formulae obtained by them are similar to those of MIE relating to the sphere. Scattering by long circular dielectric cylinders and highly reflecting cylinders was investigated by SCHAEFFER and GROSSMANN.¶

---

* Approximate methods were developed by several authors. See, for example, R. W. HART and E. W. MONTROLL, *J. Appl. Phys.*, **22** (1951), 376; E. W. MONTROLL and R. W. HART, *ibid.*, **22** (1951), 1278; E. W. MONTROLL and J. M. GREENBERG, *Phys. Rev.*, **86** (1952), 889.

See also the review article by C. J. BOUWKAMP, *Rep. Progr. Phys.* (London, Physical Society), **17** (1954), 35.

† R. GANS, *Ann. d. Physik*, (4), **37** (1912), 881; *ibid.* **47** (1915), 270.

‡ F. MÖGLICH, *Ann. d. Physik*, (4), **83** (1927), 609.

§ W. SEITZ, *Ann. d. Physik*, (4), **16** (1905), 746; *ibid.*, **19** (1906), 554.

|| W. v. IGNATOWSKY, *Ann. d. Physik*, (4), **18** (1905), 495.

¶ C. SCHAEFFER and F. GROSSMANN, *Ann. d. Physik*, (4) **31** (1910), 455; see also H. C. VAN DE HULST, *Astrophys. J.*, **112** (1950), 1.

CHAPTER XIV

OPTICS OF CRYSTALS

14.1. THE DIELECTRIC TENSOR OF AN ANISOTROPIC MEDIUM

It will be remembered that our optical theory is based on two quite separate foundations; on the one hand on MAXWELL's equations § 1.1 (1) and (2), on the other hand on the material equations which in the case of an isotropic medium were given by the formulae § 1.1 (9)–(11). In dealing with crystals we must generalize these latter equations so as to take account of anisotropy. In the greater part of this chapter we assume that the medium is homogeneous, non-conducting ($\sigma = 0$), and magnetically isotropic,* but allow *electrical anisotropy*, i.e. we consider substances whose electrical excitations depend on the direction of the electric field. In general the vector $\boldsymbol{D}$ will then no longer be in the direction of the vector $\boldsymbol{E}$. In place of equation § 1.1 (10) we assume the relation between $\boldsymbol{D}$ and $\boldsymbol{E}$ to have the simplest form which can account for anisotropic behaviour, namely one in which each component of $\boldsymbol{D}$ is linearly related to the components of $\boldsymbol{E}$:

$$\left.\begin{aligned} D_x &= \varepsilon_{xx}E_x + \varepsilon_{xy}E_y + \varepsilon_{xz}E_z, \\ D_y &= \varepsilon_{yx}E_x + \varepsilon_{yy}E_y + \varepsilon_{yz}E_z, \\ D_z &= \varepsilon_{zx}E_x + \varepsilon_{zy}E_y + \varepsilon_{zz}E_z. \end{aligned}\right\} \tag{1}$$

The nine quantities $\varepsilon_{xx}, \varepsilon_{yy}, \ldots$ are constants of the medium, and constitute the *dielectric tensor*; the vector $\boldsymbol{D}$ is thus the product of this tensor with $\boldsymbol{E}$.

We shall write equation (1) in shorter form as

$$D_k = \sum_l \varepsilon_{kl} E_l, \tag{2}$$

where k stands for one of the three indices x, y, and z, and l stands for each of x, y, and z in turn in the summation. The summation sign would be omitted in formal tensor notation, the occurrence of the index l in two places in the product being understood as an instruction to sum over all l's. We shall, however, retain the summation sign, as this will help to avoid any ambiguities for readers unfamiliar with tensor calculus.

We assume that the expressions § 1.1 (31) for electric and magnetic energy densities retain their validity. Thus

$$w_e = \frac{1}{8\pi} \boldsymbol{E} \cdot \boldsymbol{D} = \frac{1}{8\pi} \sum_{kl} E_k \varepsilon_{kl} E_l. \tag{3}$$

---

* There are also magnetic crystals, but as the effect of magnetization on optical phenomena (rapid oscillations) is small, the magnetic anisotropy may be neglected. The magnetic permeability shall, however, be retained, and will be represented by a scalar μ, in order to preserve some symmetry in the formulae and to include weakly magnetic crystals; moreover, by retaining it, we facilitate the expression of the equations in systems of units in which μ is not equal to unity in free space.

and

$$w_m = \frac{1}{8\pi} \boldsymbol{B} \cdot \boldsymbol{H} = \frac{1}{8\pi} \mu H^2. \tag{4}$$

We retain also the definition § 1.1 (38) of the POYNTING vector, or the "ray-vector"

$$\boldsymbol{S} = \frac{c}{4\pi} (\boldsymbol{E} \wedge \boldsymbol{H}) \tag{5}$$

and investigate whether these definitions are consistent with the principle of the conservation of energy.

We have, as in § 1.1.4, by multiplying the first MAXWELL equation by $\boldsymbol{E}$ and the second by $\boldsymbol{H}$ and using the vector identity § 1.1 (27),

$$\begin{aligned} -c \operatorname{div} (\boldsymbol{E} \wedge \boldsymbol{H}) &= \boldsymbol{E} \cdot \dot{\boldsymbol{D}} + \boldsymbol{H} \cdot \dot{\boldsymbol{B}} \\ &= \sum_{kl} E_k \varepsilon_{kl} \dot{E}_l + \frac{1}{2} \frac{d}{dt} (\mu H^2). \end{aligned} \tag{6}$$

If we divide both sides of this equation by 4π, the second term on the right represents the rate of change of the magnetic energy per unit volume, but the first term does not represent the rate of change of the electric energy density unless

$$\frac{1}{4\pi} \sum_{kl} E_k \varepsilon_{kl} \dot{E}_l = \frac{dw_e}{dt} = \frac{1}{8\pi} \sum_{kl} \varepsilon_{kl} (E_k \dot{E}_l + E_l \dot{E}_k) \tag{7}$$

that is, unless

$$\sum_{kl} \varepsilon_{kl} (E_k \dot{E}_l - \dot{E}_k E_l) = 0.$$

The suffices k and l are dummy suffices; both run over the same values (x, y, z). Hence the expression is not altered if we interchange k and l in the second term. This leads to

$$\sum_{kl} (\varepsilon_{kl} - \varepsilon_{lk}) E_k \dot{E}_l = 0.$$

As this equation must hold whatever the value of the field, it follows that

$$\varepsilon_{kl} = \varepsilon_{lk}. \tag{8}$$

This means that the *dielectric tensor must be symmetric*; it has only six instead of nine independent components. Conversely, the condition (8) is sufficient to ensure the validity of equation (7), and we obtain the energy theorem in differential form (the "hydrodynamical continuity equation" § 1.1 (43))

$$-\operatorname{div} \boldsymbol{S} = \frac{dw}{dt}, \quad (w = w_e + w_m). \tag{9}$$

The symmetry of the tensor ε makes it possible to reduce the expression for the electric energy w_e to a form in which only the squares of the field components, and not their products, enter. Consider in a space x, y, z the surface of the second degree

$$\varepsilon_{xx} x^2 + \varepsilon_{yy} y^2 + \varepsilon_{zz} z^2 + 2\varepsilon_{yz} yz + 2\varepsilon_{xz} xz + 2\varepsilon_{xy} xy = \text{const.} \tag{10}$$

The left-hand side of (10) must be a positive definite quadratic form, because if x, y, and z are replaced by the components of $\boldsymbol{E}$ the expression becomes equal to $8\pi w_e$, and the energy w_e must be positive for any value of the field vector. Therefore equation (10) represents an ellipsoid. The ellipsoid can always be transformed to its principal axes; thus there exists a coordinate system fixed in the crystal such that the equation of the ellipsoid is

$$\varepsilon_x x^2 + \varepsilon_y y^2 + \varepsilon_z z^2 = \text{constant}. \tag{11}$$

In this system of *principal dielectric axes* the material equations and the expression for the electrical energy take the simple forms

$$D_x = \varepsilon_x E_x\,, \qquad D_y = \varepsilon_y E_y\,, \qquad D_z = \varepsilon_z E_z\,, \tag{12}$$

$$\begin{aligned} w_e &= \frac{1}{8\pi}(\varepsilon_x E_x^2 + \varepsilon_y E_y^2 + \varepsilon_z E_z^2) \\ &= \frac{1}{8\pi}\left(\frac{D_x^2}{\varepsilon_x} + \frac{D_y^2}{\varepsilon_y} + \frac{D_z^2}{\varepsilon_z}\right). \end{aligned} \tag{13}$$

ε_x, ε_y, ε_z are called the *principal dielectric constants* (or *principal permittivities*). It may be seen immediately from these formulae that $\boldsymbol{D}$ and $\boldsymbol{E}$ will have different directions, unless $\boldsymbol{E}$ coincides in direction with one of the principal axes, or the principal dielectric constants are all equal; in the latter case ($\varepsilon_x = \varepsilon_y = \varepsilon_z$) the ellipsoid degenerates into a sphere.

A note must be added here on the effect of dispersion. Just as, in the case of isotropic substances, the dielectric constant is not a constant of the material but depends on the frequency, so in an anisotropic medium the six components ε_{kl} of the dielectric tensor will also vary with frequency. As a result not only the values of the principal dielectric constants ε_x, ε_y, ε_z will vary but also the directions of the principal axes. This phenomenon is known as *dispersion of the axes*. It can arise, however, only in crystals in which the symmetry of the structure does not determine a preferential orthogonal triplet of directions, i.e. it can be observed only in monoclinic and triclinic systems (cf. § 14.3.1).*

If we restrict ourselves to monochromatic waves we may disregard dispersion; the quantities ε_{kl} are then constants depending only on the medium.

14.2. THE STRUCTURE OF A MONOCHROMATIC PLANE WAVE IN AN ANISOTROPIC MEDIUM

14.2.1 The phase velocity and the ray velocity

In a monochromatic plane wave of angular frequency $\omega = 2\pi\nu$ propagated with velocity c/n in the direction of the unit wave-normal $\boldsymbol{s}$, the vectors $\boldsymbol{E}$, $\boldsymbol{D}$, $\boldsymbol{H}$, and $\boldsymbol{B}$ are in complex notation proportional to $\exp\left[i\omega\left(\frac{n}{c}(\boldsymbol{r}.\boldsymbol{s}) - t\right)\right]$. It may be mentioned at once that in addition to the *phase* (or *wave-normal*) *velocity* c/n, we shall later have to introduce a *ray* (or *energy*) *velocity*, since, as will be seen, in an anisotropic medium the energy is in general propagated with a different velocity and in a direction different from that of the wave normal.

In such an oscillatory field the operation $\partial/\partial t$ is always equivalent to multiplication

* This dispersion phenomenon is particularly conspicuous in the infra-red: cf. Th. Liebisch and H. Rubens, *Sitzber. preuss. Akad. Wiss., Phys.-math. Kl.* (1919), 198, 876. H. Rubens, *ibid.*, 976

by $-i\omega$, while the operation $\partial/\partial x$ is equivalent to multiplication by $i\omega n s_x/c$. In particular,

$$\dot{E} = -i\omega E, \qquad \text{curl } E = i\omega \frac{n}{c} s \wedge E. \tag{1}$$

MAXWELL's equations, for a region which does not contain currents, viz.

$$\text{curl } H - \frac{1}{c}\dot{D} = 0, \qquad \text{curl } E + \frac{1}{c}\dot{B} = 0, \tag{2}$$

become

$$ns \wedge H = -D, \qquad ns \wedge E = \mu H, \tag{3}$$

where the relation $B = \mu H$ has been used. Eliminating H between equations (3) and using a well-known vector identity, we obtain

$$D = -\frac{n^2}{\mu} s \wedge (s \wedge E) = \frac{n^2}{\mu}[E - s(s \cdot E)] = \frac{n^2}{\mu} E_\perp, \tag{4}$$

where $E_\perp$ denotes the vector component of E perpendicular to s in the plane of E and s (see Fig. 14.1).

We see from equation (3) that the vector H (and hence also B) is at right angles to E, D, and s, which must therefore be *coplanar*. Further D is seen to be orthogonal to s. Thus H and D are transversal to the direction of propagation s, as before, but E is not. Fig. 14.1 shows the relative directions of these vectors, and in addition the unit

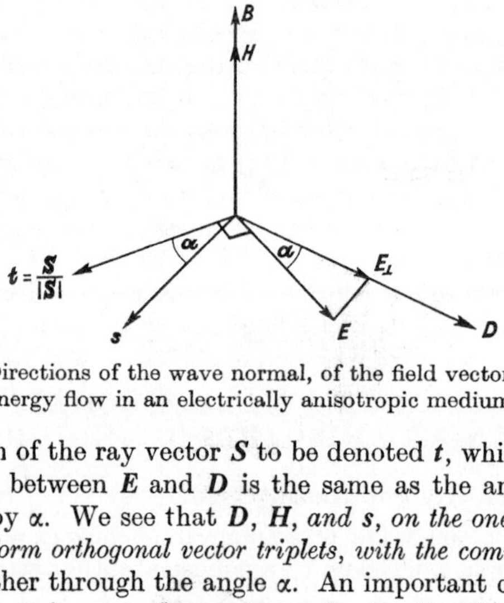

Fig. 14.1. Directions of the wave normal, of the field vectors and of the energy flow in an electrically anisotropic medium.

vector in the direction of the ray vector S to be denoted t, which is perpendicular to E and H. The angle between E and D is the same as the angle between s and t, and will be denoted by α. We see that *D, H, and s, on the one hand, and E, H, and t, on the other hand, form orthogonal vector triplets, with the common vector H*, rotated relatively to one another through the angle α. An important conclusion is that *in a crystal the energy is not in general propagated in the direction of the wave normal.* On the other hand the theorem of equal electric and magnetic energy densities retains its validity. This follows from equations (3) since*

$$\left.\begin{aligned} w_e &= \frac{1}{8\pi} E \cdot D = -\frac{n}{8\pi} E \cdot (s \wedge H), \\ w_m &= \frac{1}{8\pi} B \cdot H = \frac{n}{8\pi} (s \wedge E) \cdot H, \end{aligned}\right\} \tag{5}$$

* As we are now dealing with quadratic functions of the field variables, it is the real field vectors and not the associated complex vectors that enter equations (5) and (6) (cf. § 1.3, p. 17).

and by a well-known property of the scalar triple product the right-hand sides of these equations are equal to each other. Moreover, they are both equal to $n(\boldsymbol{E} \wedge \boldsymbol{H}).\boldsymbol{s}/8\pi$, so that the total energy density $w = w_e + w_m$ is

$$w = \frac{n}{c} \boldsymbol{S} \, . \, \boldsymbol{s}. \tag{6}$$

We must distinguish between the phase velocity and the velocity of energy transport. The former, *the phase velocity*, is in the direction of the unit vector $\boldsymbol{s}$ and its magnitude is

$$v_p = \frac{c}{n}. \tag{7}$$

The latter, *the ray velocity*, is in the same direction as the POYNTING vector $\boldsymbol{S}$, i.e. in the direction of the unit vector $\boldsymbol{t}$. Its magnitude v_r is equal to the energy that crosses in unit time an area perpendicular to the flow direction, divided by the energy per unit volume. In accordance with the energy theorem § 14.1 (9) this is given by

$$v_r = \frac{S}{w}. \tag{8}$$

From (6), (7), and (8)

$$v_p = v_r \, \boldsymbol{t}.\boldsymbol{s} = v_r \cos \alpha. \tag{9}$$

i.e. *the phase velocity is the projection of the ray velocity on to the direction of the wave normal.*

It may be noted that the ray velocity, being derived from the POYNTING vector shares with it (cf. § 1.1, p. 9–10) a certain degree of arbitrariness. It is nevertheless a useful concept although, like the phase velocity, it has no directly verifiable physical significance.

If $\boldsymbol{E}$ and $\boldsymbol{D}$ are known (as for instance when $\boldsymbol{E}$ is prescribed and $\boldsymbol{D}$ is determined from it by equation § 14.1 (1)), the refractive index n and the wave normal $\boldsymbol{s}$ are thereby determined. For in the first place, since $\boldsymbol{E}_\perp$ is the vector component of $\boldsymbol{E}$ in the direction of $\boldsymbol{D}$,

$$\boldsymbol{E}_\perp = \left(\boldsymbol{E} \, . \, \frac{\boldsymbol{D}}{D}\right) \frac{\boldsymbol{D}}{D}, \tag{10}$$

so that (4) gives

$$n^2 = \frac{\mu D}{E_\perp} = \frac{\mu D^2}{(\boldsymbol{E}.\boldsymbol{D})}. \tag{11}$$

Further, since the unit vector $\boldsymbol{s}$ is perpendicular to $\boldsymbol{D}$ and coplanar with $\boldsymbol{D}$ and $\boldsymbol{E}$, it may be expressed in the form

$$\boldsymbol{s} = \frac{\boldsymbol{E} - \boldsymbol{E}_\perp}{|\boldsymbol{E} - \boldsymbol{E}_\perp|} = \frac{\boldsymbol{E} - \dfrac{(\boldsymbol{E} \, . \, \boldsymbol{D})\boldsymbol{D}}{D^2}}{\sqrt{E^2 - \dfrac{(\boldsymbol{E} \, . \, \boldsymbol{D})^2}{D^2}}}, = \frac{D^2\boldsymbol{E} - (\boldsymbol{E} \, . \, \boldsymbol{D})\boldsymbol{D}}{\sqrt{D^2[E^2D^2 - (\boldsymbol{E} \, . \, \boldsymbol{D})^2]}}. \tag{12}$$

By analogy with the refractive index n we may also define a *ray index* or *energy index* n_r by means of the formula

$$n_r = \frac{c}{v_r}. \tag{13}$$

By (7) and (9) we have,

$$n_r = n \cos \alpha. \tag{14}$$

We shall now show that the ray index n_r and the unit vector $\boldsymbol{t}$ in the direction of the propagation of energy are given by formulae analogous to (11) and (12). We have from (14), (11), and the relation $\boldsymbol{E}.\boldsymbol{D} = ED \cos \alpha$,

$$n_r^2 = \frac{\mu(\boldsymbol{E} \cdot \boldsymbol{D})}{E^2}. \tag{15}$$

The unit vector $\boldsymbol{t}$ is perpendicular to $\boldsymbol{E}$ and coplanar with $\boldsymbol{E}$ and $\boldsymbol{D}$ and so must (apart perhaps from the sign) be given by the formulae that results in interchanging $\boldsymbol{E}$ and $\boldsymbol{D}$ in (12). Hence

$$-\boldsymbol{t} = \frac{E^2\boldsymbol{D} - (\boldsymbol{E} \cdot \boldsymbol{D})\boldsymbol{E}}{\sqrt{E^2[E^2D^2 - (\boldsymbol{E} \cdot \boldsymbol{D})^2]}}. \tag{16}$$

The negative sign on the left ensures that $\boldsymbol{s}$ and $\boldsymbol{t}$ point towards the same side of $\boldsymbol{E}$ and $\boldsymbol{D}$ as in Fig. 14.1.

Both (12) and (16) reduce to 0/0 when $\boldsymbol{E}$ and $\boldsymbol{D}$ coincide in direction, i.e. when $\boldsymbol{E}$ is in the direction of one of the principal axes of the crystal. This is to be expected since the directions of $\boldsymbol{s}$ and $\boldsymbol{t}$ are then undetermined apart from the fact that they must be perpendicular to $\boldsymbol{E}$.

We may also express the magnitude of the Poynting vector in terms of $\boldsymbol{E}$ and $\boldsymbol{D}$. According to (8), (13), and (15), remembering also that for a plane wave $w = 2w_e$

$$S = v_r w = \frac{c}{n_r}\frac{\boldsymbol{E}.\boldsymbol{D}}{4\pi} = \frac{c}{4\pi\sqrt{\mu}} E\sqrt{\boldsymbol{E} \cdot \boldsymbol{D}}, \tag{17}$$

a formula that, in the case of an isotropic medium, is seen to be in agreement with equations (8) and (9) of § 1.4.

14.2.2 Fresnel's formulae for the propagation of light in crystals

The formulae derived in § 14.2.1 are consequences of Maxwell's equations alone and therefore independent of the properties of the medium. We shall now combine these with the material equation § 14.1 (1).

We shall use a system of coordinate axes coincident with the principal dielectric axes. The relations § 14.1 (1) then reduce to the simpler form § 14.1 (12), and substitution for $\boldsymbol{D}$ into (4) gives

$$\mu\varepsilon_k E_k = n^2[E_k - s_k(\boldsymbol{E} \cdot \boldsymbol{s})] \qquad (k = x, y, z). \tag{18}$$

Equations (18), being three homogeneous linear equations in E_x, E_y, and E_z, can be satisfied by non-zero values of these components only if the associated determinant vanishes. This implies that a certain relation must be satisfied by the refractive index n, the vector $\boldsymbol{s}(s_x, s_y, s_z)$ and the principal dielectric constants $\varepsilon_x, \varepsilon_y, \varepsilon_z$. This relation may be derived by writing (18) in the form

$$E_k = \frac{n^2 s_k(\boldsymbol{E} \cdot \boldsymbol{s})}{n^2 - \mu\varepsilon_k}, \tag{19}$$

multiplying it by s_k and adding the resulting three equations; dividing the expression which then results by the common factor $\boldsymbol{E} \cdot \boldsymbol{s}$, we obtain

$$\frac{s_x^2}{n^2 - \mu\varepsilon_x} + \frac{s_y^2}{n^2 - \mu\varepsilon_y} + \frac{s_z^2}{n^2 - \mu\varepsilon_z} = \frac{1}{n^2}. \tag{20}$$

This formula may be expressed in a slightly different form. We multiply both sides of (20) by n^2 and subtract $s_x^2 + s_y^2 + s_z^2 = 1$. Next we multiply the resulting expression by $-n^2$ and find that

$$\frac{s_x^2}{\dfrac{1}{n^2} - \dfrac{1}{\mu\varepsilon_x}} + \frac{s_y^2}{\dfrac{1}{n^2} - \dfrac{1}{\mu\varepsilon_y}} + \frac{s_z^2}{\dfrac{1}{n^2} - \dfrac{1}{\mu\varepsilon_z}} = 0. \tag{21}$$

We define three *principal velocities of propagation* by the formulae*

$$v_x = \frac{c}{\sqrt{\mu\varepsilon_x}}, \qquad v_y = \frac{c}{\sqrt{\mu\varepsilon_y}}, \qquad v_z = \frac{c}{\sqrt{\mu\varepsilon_z}}. \tag{22}$$

When the expression (7) is used for the phase velocity v_p, (19) and (21) take the form

$$E_k = \frac{v_k^2}{v_k^2 - v_p^2} s_k(\boldsymbol{E} \cdot \boldsymbol{s}) \qquad (k = x, y, z), \tag{23}$$

$$\frac{s_x^2}{v_p^2 - v_x^2} + \frac{s_y^2}{v_p^2 - v_y^2} + \frac{s_z^2}{v_p^2 - v_z^2} = 0. \tag{24}$$

Equations (20), (21), and (24) are equivalent forms of *Fresnel's equation of wave normals*. This is a quadratic equation in v_p^2, as can be seen by multiplying (24) by the product of the denominators. Thus to every direction $\boldsymbol{s}$ there correspond two phase velocities v_p. (The two values $\pm v_p$ corresponding to any value v_p^2 are counted as one, since the negative value evidently belongs to the opposite direction of propagation $-\boldsymbol{s}$.) With each of the two values of v_p, the equations (23) may then be solved for the ratios $E_x{:}E_y{:}E_z$; the corresponding ratios involving the $\boldsymbol{D}$ vector may be subsequently obtained from § 14.1 (12). Since these ratios are real, the $\boldsymbol{E}$ and $\boldsymbol{D}$ fields are *linearly polarized*. Thus we have the important result that the *structure of an anisotropic medium permits two monochromatic plane waves with two different linear polarizations and two different velocities to propagate in any given direction.* It will be shown later that the two directions of the electric displacement vector $\boldsymbol{D}$ corresponding to a given direction of propagation $\boldsymbol{s}$ are perpendicular to each other.

We now show that there is an analogous formula for the ray velocity v_r. This is most easily done by showing first that there is a relation analogous to (4) in which the roles of $\boldsymbol{D}$ and $\boldsymbol{E}$ and of $\boldsymbol{s}$ and $\boldsymbol{t}$ are interchanged. It is convenient to introduce a vector $\boldsymbol{D}_\perp$ defined as the vector component of $\boldsymbol{D}$ perpendicular to $\boldsymbol{t}$, in the plane of $\boldsymbol{D}$ and $\boldsymbol{t}$. It is evidently given by

$$\boldsymbol{D}_\perp = \boldsymbol{D} - \boldsymbol{t}(\boldsymbol{D} \cdot \boldsymbol{t}). \tag{25}$$

Since the electric vector is also perpendicular to $\boldsymbol{t}$ and is coplanar with $\boldsymbol{D}$ and $\boldsymbol{t}$ (see Fig. 14.1), $\boldsymbol{D}_\perp$ is parallel to $\boldsymbol{E}$ and so may also be expressed in the form

$$\boldsymbol{D}_\perp = \left(\boldsymbol{D} \cdot \frac{\boldsymbol{E}}{E}\right)\frac{\boldsymbol{E}}{E} = \frac{n_r^2}{\mu}\boldsymbol{E}, \tag{26}$$

where (15) has been used. From (25) and (26) it follows that

$$\boldsymbol{E} = \frac{\mu}{n_r^2}[\boldsymbol{D} - \boldsymbol{t}(\boldsymbol{D}.\boldsymbol{t})] = \frac{\mu}{n_r^2}\boldsymbol{D}_\perp. \tag{27}$$

* Note that v_x, v_y, v_z are not components of a vector and are defined only with reference to the principal axes.

This equation is analogous to (4) and may be formally obtained from it by interchanging the roles of $\boldsymbol{E}$ and $\boldsymbol{D}$, n and $1/n_r$, μ and $1/\mu$ and $\boldsymbol{s}$ and $-\boldsymbol{t}$. Quite generally the following *rule of duality* follows from the basic equations:

Let the variables be arranged in two rows as follows:

$$\left.\begin{array}{llllllllllllll} \boldsymbol{E}, & \boldsymbol{D}, & \boldsymbol{s}, & \boldsymbol{t}, & c, & \mu & v_p, & n, & \varepsilon_x, & \varepsilon_y, & \varepsilon_z, & v_x, & v_y, & v_z, \\ \boldsymbol{D}, & \boldsymbol{E}, & -\boldsymbol{t}, & -\boldsymbol{s}, & \frac{1}{c}, & \frac{1}{\mu}, & \frac{1}{v_r}, & \frac{1}{n_r}, & \frac{1}{\varepsilon_x}, & \frac{1}{\varepsilon_y}, & \frac{1}{\varepsilon_z}, & \frac{1}{v_x}, & \frac{1}{v_y}, & \frac{1}{v_z}. \end{array}\right\} \tag{28}$$

If in any relation which holds between the quantities in one row each quantity is replaced by the corresponding quantity in the other row another valid relation is obtained.

Applying this rule to Fresnel's equation of wave normals (24) we immediately obtain the required *ray equation*

$$\frac{t_x^2}{\frac{1}{v_r^2}-\frac{1}{v_x^2}}+\frac{t_y^2}{\frac{1}{v_r^2}-\frac{1}{v_y^2}}+\frac{t_z^2}{\frac{1}{v_r^2}-\frac{1}{v_z^2}}=0. \tag{29}$$

We may, of course, express this equation also in forms analogous to (20) and (21). Like (24) this is again a quadratic equation and gives two possible ray velocities v_r for each ray direction $\boldsymbol{t}(t_x, t_y, t_z)$. The corresponding direction of $\boldsymbol{D}$ may be obtained on solving with the appropriate value v_r the equations dual to (23), namely

$$D_k = -\frac{v_r^2}{v_k^2 - v_r^2}\, t_k(\boldsymbol{D}\,.\,\boldsymbol{t}). \qquad (k = x, y, z). \tag{30}$$

The directions of the two $\boldsymbol{E}$ vectors (which as we saw are orthogonal to $\boldsymbol{t}$) may then be obtained by using § 1.4.1 (12).

As a rule only one of the vectors $\boldsymbol{s}$ or $\boldsymbol{t}$ is given and it is, therefore, desirable to derive relations from which the other may be directly calculated. We have from Fig. 14.1

$$\boldsymbol{E}\,.\,\boldsymbol{s} = E_\perp \tan\alpha, \qquad \boldsymbol{D}\,.\,\boldsymbol{t} = -D\sin\alpha. \tag{31}$$

But from (4), $D = n^2 E_\perp/\mu$. Hence

$$\boldsymbol{D}\,.\,\boldsymbol{t} = -\frac{n^2}{\mu}E_\perp \sin\alpha = -\frac{n^2}{\mu}\boldsymbol{E}\,.\,\boldsymbol{s}\cos\alpha = -\frac{1}{\mu}\frac{c^2}{v_p v_r}\boldsymbol{E}\,.\,\boldsymbol{s}, \tag{32}$$

where the relations (9) and (7) have been used. Substitution from (32) into (30) gives

$$D_k = \varepsilon_k E_k = \frac{1}{\mu}\frac{c^2 v_r}{v_p(v_k^2 - v_r^2)}\, t_k(\boldsymbol{E}\,.\,\boldsymbol{s}). \tag{33}$$

Comparing (33) with (23) and remembering that $\mu\varepsilon_k v_k^2 = c^2$, we obtain

$$\frac{v_p s_k}{v_k^2 - v_p^2} = \frac{v_r t_k}{v_k^2 - v_r^2}. \tag{34}$$

Solving for t_k we have

$$t_k = \frac{v_p}{v_r}\frac{v_k^2 - v_r^2}{v_k^2 - v_p^2}\, s_k, \tag{35}$$

so that

$$v_r t_k - v_p s_k = v_p s_k \frac{v_r^2 - v_p^2}{v_p^2 - v_k^2}. \tag{36}$$

Squaring and adding the three equations (36) and using the relation § 14.2 (9), namely $\boldsymbol{s} \cdot \boldsymbol{t} = v_p/v_r$, we obtain

$$v_r^2 - v_p^2 = v_p^2(v_r^2 - v_p^2)^2 \left[\left(\frac{s_x}{v_p^2 - v_x^2}\right)^2 + \left(\frac{s_y}{v_p^2 - v_y^2}\right)^2 + \left(\frac{s_z}{v_p^2 - v_z^2}\right)^2\right]. \quad (37)$$

Hence we may define

$$g^2 \equiv v_p^2(v_r^2 - v_p^2) = \frac{1}{\left(\frac{s_x}{v_p^2 - v_x^2}\right)^2 + \left(\frac{s_y}{v_p^2 - v_y^2}\right)^2 + \left(\frac{s_z^2}{v_p^2 - v_z^2}\right)^2}. \quad (38)$$

This relation expresses v_r in terms of $\boldsymbol{s}$, since v_p is already known in terms of $\boldsymbol{s}$ from Fresnel's equation (24). With v_r so determined, equation (35) then gives the unit ray vector $\boldsymbol{t}$ as a function of $\boldsymbol{s}$. Using the expression g, equation (35) may be written as

$$t_k = \frac{s_k}{v_p v_r}\left(v_p^2 + \frac{g^2}{v_p^2 - v_k^2}\right) \qquad (k = x, y, z). \quad (39)$$

Since to every $\boldsymbol{s}$ there correspond in general two phase velocities v_p, there are in general two ray directions* for each wave-normal direction. There are, however, two singular directions in certain crystals (biaxial crystals—cf. § 14.3.1) due to vanishing of the denominators in (39), to each of which there corresponds an *infinite* number of rays; there are also two singular ray directions to each of which there corresponds an infinite number of wave normal directions. These special cases give rise to an interesting phenomenon (conical refraction), which will be investigated in § 14.3.4.

14.2.3 Geometrical constructions for determining the velocities of propagation and the directions of vibration

Many results concerning the phase and the ray velocities and the directions of vibration may be illustrated by means of certain geometrical constructions.

(a) *The ellipsoid of wave normals.* By equations § 14.1 (13) the components of the vector $\boldsymbol{D}$ at a given energy density $w = 2w_e$ satisfy the relation

$$\frac{D_x^2}{\varepsilon_x} + \frac{D_y^2}{\varepsilon_y} + \frac{D_z^2}{\varepsilon_z} = C \qquad (C = 8\pi w_e = \boldsymbol{E} \cdot \boldsymbol{D}). \quad (40)$$

Let us write x, y, and z in place of $D_x/\sqrt{C}$, $D_y/\sqrt{C}$, $D_z/\sqrt{C}$, and consider these as Cartesian coordinates in space. Then

$$\frac{x^2}{\varepsilon_x} + \frac{y^2}{\varepsilon_y} + \frac{z^2}{\varepsilon_z} = 1. \quad (41)$$

This equation represents an ellipsoid, the semi-axes of which are equal to the square roots of the principal dielectric constants and coincide in directions with the principal dielectric axes. We call this ellipsoid the *ellipsoid of wave normals* in preference to the widely used but rather vague term "optical indicatrix" (also known as the index ellipsoid, or the reciprocal ellipsoid).

With the aid of the ellipsoid of wave normals we can find the two phase velocities

* The position of the rays that correspond to a given wave normal is investigated in detail for the case of a biaxial crystal in M. Born, *Optik* (Berlin, Springer, 1933), pp. 235–237.

v_p and the two directions of vibrations $\boldsymbol{D}$ which belong to a given wave-normal direction $\boldsymbol{s}$ as follows: We draw a plane through the origin at right angles to $\boldsymbol{s}$. The curve of intersection of this plane with the ellipsoid is an ellipse; the principal semi-axes of this ellipse are proportional to the reciprocals $1/v_p$ of the phase velocities, and their directions coincide with the corresponding directions of vibrations of the vector $\boldsymbol{D}$ (Fig. 14.2).

To establish this result consider the two equations that specify the ellipse:

$$xs_x + ys_y + zs_z = 0, \tag{42}$$

$$\frac{x^2}{\varepsilon_x} + \frac{y^2}{\varepsilon_y} + \frac{z^2}{\varepsilon_z} = 1. \tag{43}$$

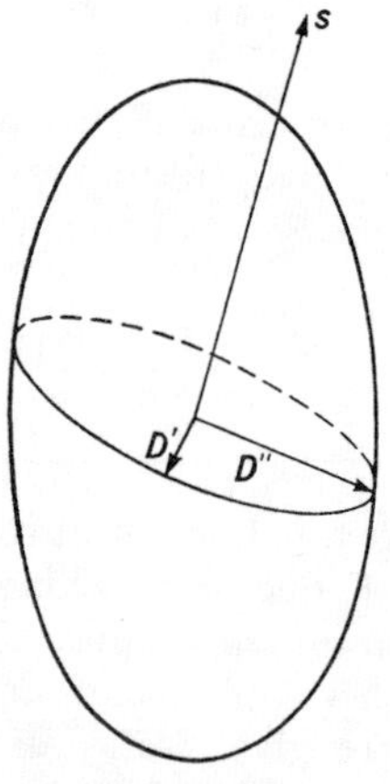

Fig. 14.2. The ellipsoid of wave normals. Construction of the directions of vibrations of the $\boldsymbol{D}$ vectors belonging to a wave normal $\boldsymbol{s}$.

Since the principal axes of an ellipse are, by definition, its shortest and longest diameters, we can determine them by finding the extrema of

$$r^2 = x^2 + y^2 + z^2, \tag{44}$$

subject to the conditions (42) and (43). This we do by means of LAGRANGE's method of undetermined multipliers.* We introduce two multipliers $2\lambda_1$ and λ_2 and construct the function

$$F = x^2 + y^2 + z^2 + 2\lambda_1(xs_x + ys_y + zs_z) + \lambda_2\left(\frac{x^2}{\varepsilon_x} + \frac{y^2}{\varepsilon_y} + \frac{z^2}{\varepsilon_z} - 1\right). \tag{45}$$

Our problem is then equivalent to finding the extremum of F subject to no subsidiary conditions. The necessary conditions for the extremum of F is that its derivatives with respect to x, y, and z should vanish, i.e. that

$$x + \lambda_1 s_x + \frac{\lambda_2 x}{\varepsilon_x} = 0, \qquad y + \lambda_1 s_y + \frac{\lambda_2 y}{\varepsilon_y} = 0, \qquad z + \lambda_1 s_z + \frac{\lambda_2 z}{\varepsilon_z} = 0. \tag{46}$$

Multiplying these equations by x, y, and z respectively and adding, we obtain, because of (42) and (43):

$$r^2 + \lambda_2 = 0. \tag{47}$$

Next we multiply the equations (46) by s_x, s_y, and s_z and add, and again use (42). We then obtain

$$\lambda_1 + \lambda_2\left(\frac{xs_x}{\varepsilon_x} + \frac{ys_y}{\varepsilon_y} + \frac{zs_z}{\varepsilon_z}\right) = 0. \tag{48}$$

Substitution for λ_1 and λ_2 from (47) and (48) into (46) gives

$$x\left(1 - \frac{r^2}{\varepsilon_x}\right) + s_x r^2\left(\frac{xs_x}{\varepsilon_x} + \frac{ys_y}{\varepsilon_y} + \frac{zs_z}{\varepsilon_z}\right) = 0, \tag{49}$$

with two similar equations. With a given $\boldsymbol{s}$, these are three homogeneous equations for x, y, and z. They are compatible only if the associated determinant vanishes; a

---

* For a full account of this method see, for example, R. COURANT, *Differential and Integral Calculus*, Vol. II (London, Blackie & Son, Ltd., 1942), pp. 188–199.

condition that gives an algebraic equation for r^2. Now it is immediately seen that the equations (49) differ only in notation from equations (18). For if we replace x by $D_x/\sqrt{C}$, x/ε_x by $E_x/\sqrt{C}$ and r^2 by $D^2/C = D^2/E \cdot D = n^2/\mu$ (in accordance with (11)), (49) becomes

$$\mu D_x = n^2 [E_x - s_x(E \cdot s)], \tag{50}$$

which, together with the two similar equations, is identical with (18).

Thus we find that the roots of the determinantal equation for $n = c/v_p$ (which as we saw is of the second degree) are proportional to the lengths r of the semi-axes of the elliptical section at right angles to $\mathbf{s}$, and, moreover, that the two possible directions $x:y:z$ of the vector $\mathbf{D}$ coincide with the directions of these axes. Since the axes of an ellipse are perpendicular to each other, we obtain the important result that *the directions of vibrations of the two vectors* $\mathbf{D}$ *corresponding to a given direction of propagation* $\mathbf{s}$ *are perpendicular to each other.* In what follows we shall denote the two $\mathbf{D}$ directions that correspond to a particular wave-normal direction $\mathbf{s}$ by $\mathbf{D}'$ and $\mathbf{D}''$; thus $\mathbf{s}$, $\mathbf{D}'$, and $\mathbf{D}''$ form an orthogonal triplet.

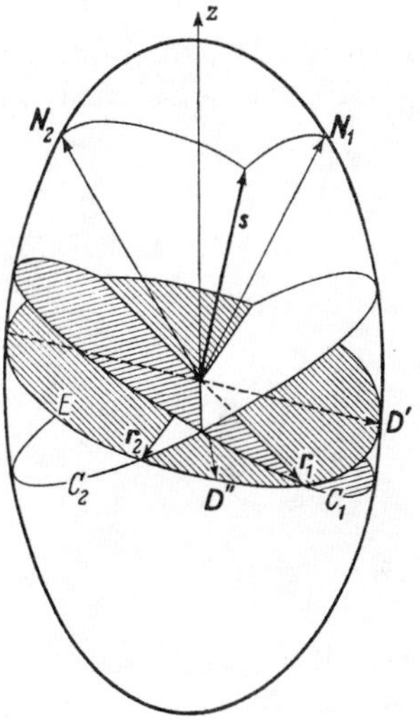

Fig. 14.3. A construction for determining the planes of vibrations ($\mathbf{s}$, $\mathbf{D}'$) and ($\mathbf{s}$, $\mathbf{D}''$).

In the special case when the direction of propagation coincides with one of the principal axes of the ellipsoid of wave normals, say the x axis, the extrema of r are, according to our construction, equal to the lengths of the other two semi-axes, i.e. to $\sqrt{\varepsilon_y}$ and $\sqrt{\varepsilon_z}$. But we saw that the extrema of r are also equal to $n/\sqrt{\mu} = c/v_p\sqrt{\mu}$. Hence *the phase velocities of waves which are propagated in the direction of the principal dielectric axis* x *are equal to* $c/\sqrt{\mu\varepsilon_y}$ *and* $c/\sqrt{\mu\varepsilon_z}$, *i.e. to the principal velocities of propagation* v_y *and* v_z *introduced formally by* (22). A corresponding result holds, of course, for propagation in the directions of the other two axes.

There is another construction by which the directions of vibrations can be determined. It is known that an ellipsoid has two circular sections C_1 and C_2 passing through the centre, and that the normals N_1 and N_2 to these sections are coplanar with the longest and shortest principal axes (z and x) of the ellipsoid. These two directions N_1 and N_2 are called the *optic axes*,* and will be considered more fully later (§ 14.3.3). Since the sections C_1 and C_2 are circular (also of the same radius), the directions N_1, N_2 have the property that there is only one velocity of propagation along them: $\mathbf{D}$ can then take *any* direction perpendicular to $\mathbf{s}$. Let E be the elliptical section through the centre, at right angles to an arbitrary unit normal $\mathbf{s}$. This plane intersects the circles C_1, C_2 in two radial vectors $\mathbf{r}_1$, $\mathbf{r}_2$ which are of equal length and must therefore make equal angles with the principal axes of E (see Figs. 14.3 and 14.4). The required directions of vibration are therefore the bisectors of the directions $\mathbf{r}_1$, $\mathbf{r}_2$. But $\mathbf{r}_1$ is perpendicular to N_1 and $\mathbf{s}$, and is, therefore, perpendicular to the plane containing N_1 and $\mathbf{s}$; similarly $\mathbf{r}_2$ is perpendicular to the plane containing N_2 and $\mathbf{s}$. If these planes intersect the ellipse E in vectors $\mathbf{r}_1'$, $\mathbf{r}_2'$ the principal axes of the ellipse

* More precisely *the optic axes of wave normals.* The corresponding sections of the ray ellipsoid (see next page) define *the optic ray axes.*

must also bisect the directions of r_1', r_2'. It follows that *the planes of vibration of the electric displacement*, i.e. the planes containing s and D' or D'', *are the internal or external bisectors of the angle between the planes* (N_1,s) *and* (N_2,s).* This construction becomes indeterminate when s coincides in direction with either N_1 or N_2, as we should expect.

(b) *The ray ellipsoid.* Rays may be treated in the same manner as wave normals if, in accordance with the duality rule (28), one starts from the *ray ellipsoid*

$$\varepsilon_x x^2 + \varepsilon_y y^2 + \varepsilon_z z^2 = 1. \tag{51}$$

In particular, the central section of this ellipsoid perpendicular to a ray direction t is an ellipse with semi-axes of lengths proportional to the two corresponding ray

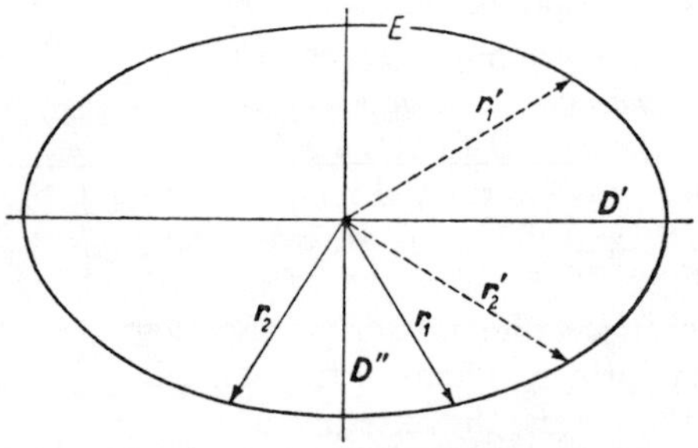

Fig. 14.4. The plane E of Fig. 14.3.

velocities v_r, oriented in the two permissible directions E' and E'' of the electric vector. Thus t, E', and E'' form an orthogonal triplet of vectors.

(c) *The normal surface and the ray surface.* Imagine that with a fixed point O inside a crystal as origin two vectors are plotted in the same direction s and of lengths that are proportional to the two corresponding phase velocities. As s takes on all possible directions the end points give rise to a surface of two shells known as *the wave-normal surface* or, for short, *the normal surface.*

Similarly, the end points of position vectors plotted from a fixed origin in all directions t, and of lengths that are proportional to the corresponding ray velocities will generate a two-sheeted surface, called *the ray surface.*

These two surfaces are more complicated than the ellipsoids that we just considered. The ray surface is of the fourth degree, and the surface of normals of the sixth degree,† as may be verified from the formulae (24) and (29). There exists an important relation between these two surfaces which we shall now derive.

We have seen that if E or D is known, the directions of s and t as well as the corresponding velocities v_p and v_r, and consequently the corresponding points (P and P' in Fig. 14.5) on the two surfaces are thereby determined. Let r and r' be the vectors representing these points:

$$r = v_r t, \qquad r' = v_p s. \tag{52}$$

* Throughout this chapter (a, b) denotes the plane containing the vectors a and b.

† The normal surface and the ray surface cannot be expected to have equations of the same degree, since they are not duals of each other. In order to construct the dual of the normal surface one would have to plot vectors of lengths $1/v_r$ (not v_r), in accordance with the duality rule (28).

We shall show that a small variation in $\boldsymbol{E}$ or $\boldsymbol{D}$ produces a change in the vector $\boldsymbol{r}$ at right angles to $\boldsymbol{r}'$.

We start from the equation (27)

$$\frac{1}{\mu}\boldsymbol{E} = \frac{1}{n_r^2}[\boldsymbol{D} - \boldsymbol{t}(\boldsymbol{D}.\boldsymbol{t})]. \tag{53}$$

Substitution for $\boldsymbol{t}$ from the first equation (52) gives, if we also set $n_r = c/v_r$:

$$\frac{c^2}{\mu}\boldsymbol{E} = \boldsymbol{r}^2\boldsymbol{D} - \boldsymbol{r}(\boldsymbol{D}.\boldsymbol{r}). \tag{54}$$

Suppose now that $\boldsymbol{E}$ is changed by a small amount $\delta\boldsymbol{E}$. If $\delta\boldsymbol{D}$ and $\delta\boldsymbol{r}$ are the corresponding changes in $\boldsymbol{D}$ and $\boldsymbol{r}$ respectively, we have, according to (54),

$$\frac{c^2}{\mu}\delta\boldsymbol{E} = 2(\boldsymbol{r}.\delta\boldsymbol{r})\boldsymbol{D} + \boldsymbol{r}^2\delta\boldsymbol{D} - \delta\boldsymbol{r}(\boldsymbol{D}.\boldsymbol{r}) - \boldsymbol{r}(\delta\boldsymbol{r}.\boldsymbol{D}) - \boldsymbol{r}(\boldsymbol{r}.\delta\boldsymbol{D}). \tag{55}$$

If we multiply both sides of this equation scalarly by $\boldsymbol{D}$ and use the relation

$$\boldsymbol{D}.\delta\boldsymbol{E} = \varepsilon_x E_x \delta E_x + \varepsilon_y E_y \delta E_y + \varepsilon_z E_z \delta E_z = \boldsymbol{E}.\delta\boldsymbol{D}, \tag{56}$$

we obtain

$$\frac{c^2}{\mu}\boldsymbol{E}.\delta\boldsymbol{D} = \delta\boldsymbol{D}.[\boldsymbol{r}^2\boldsymbol{D} - \boldsymbol{r}(\boldsymbol{D}.\boldsymbol{r})] + 2\delta\boldsymbol{r}.[\boldsymbol{r}\boldsymbol{D}^2 - \boldsymbol{D}(\boldsymbol{D}.\boldsymbol{r})]. \tag{57}$$

The terms having $\delta\boldsymbol{D}$ as a factor cancel because of (54) and the term multiplying $\delta\boldsymbol{r}$ may be written as $2\delta\boldsymbol{r}.[(\boldsymbol{D}\wedge\boldsymbol{r})\wedge\boldsymbol{D}]$. Hence, since $\boldsymbol{r} = v_r\boldsymbol{t}$,

$$\delta\boldsymbol{r}.[(\boldsymbol{D}\wedge\boldsymbol{t})\wedge\boldsymbol{D}] = 0. \tag{58}$$

Now the vector $\boldsymbol{D}\wedge\boldsymbol{t}$ is perpendicular to both $\boldsymbol{D}$ and $\boldsymbol{t}$ and so $(\boldsymbol{D}\wedge\boldsymbol{t})\wedge\boldsymbol{D}$ is in the plane of $\boldsymbol{D}$ and $\boldsymbol{t}$ and at right angles to $\boldsymbol{D}$; it is therefore parallel to $\boldsymbol{s}$ (see § 14.1). Hence

$$\boldsymbol{s}.\delta\boldsymbol{r} = 0, \tag{59}$$

i.e. $\delta\boldsymbol{r}$ is perpendicular to $\boldsymbol{s}$, thus proving our statement. This result implies that *the tangent plane of the ray surface is always perpendicular to the corresponding wave normal.* Fig. 14.5 illustrates this relation in plane section. Since the perpendicular distance from the origin to this plane is equal to $v_r\,\boldsymbol{t}.\boldsymbol{s} = v_r \cos\alpha = v_p$ by (9), it follows that *the normal surface is the pedal surface of the ray surface* and conversely *the ray surface is the envelope of planes drawn through points on the normal surface, at right angles to the radius vectors from the origin to these points.* If we know the form of either of these surfaces, the other may be determined from this relationship.

We may interpret this result in more physical terms. Consider, not a single wave, but a group formed by plane waves of the same frequency with slightly different directions of propagation. The wave normals $\boldsymbol{s}$ of the component waves fill a solid angle around a

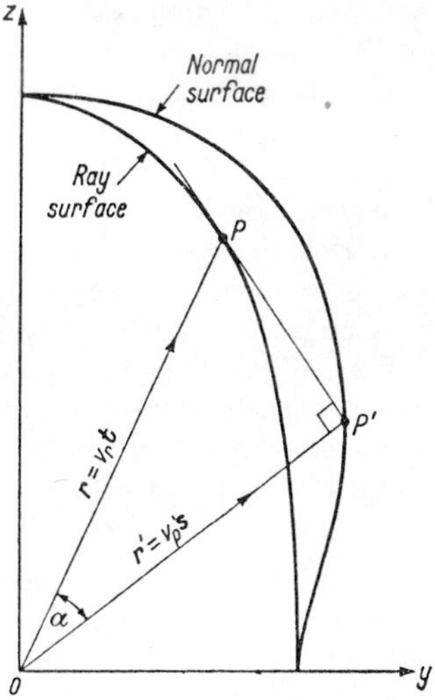

Fig. 14.5. The relation between the normal surface and the ray surface.

"mean wave normal" $\boldsymbol{s}_0$, and we assume that the amplitudes of only those waves are appreciable whose normals are close to $\boldsymbol{s}_0$. Suppose that at time $t = 0$ all the waves are in phase at a point O; the disturbance is then a maximum at this point. Let us now examine how this maximum is propagated.

Consider all the wave-fronts which pass through O at time $t = 0$. After a unit time a wave-front W, propagated with velocity v_p in the direction $\boldsymbol{s}$, will have reached a position W' such that the foot of the perpendicular dropped on to it from O has the position vector $v_p\boldsymbol{s}$; thus W' is a plane perpendicular to the appropriate radius vector of the normal surface. The amplitude of the group will be largest in a region where the waves reinforce each other, i.e. where this plane intersects the planes with neighbouring wave normals. But this is precisely the region around the envelope of these planes, i.e. in the neighbourhood of the corresponding point $v_r\boldsymbol{t}$ on the ray surface. This consideration confirms that the energy carried by the group is propagated with velocity v_r in the direction of the unit vector $\boldsymbol{t}$.

14.3. OPTICAL PROPERTIES OF UNIAXIAL AND BIAXIAL CRYSTALS

14.3.1 The optical classification of crystals

Transparent crystals fall into three distinct groups as regards their optical properties:

Group I. *Crystals in which three crystallographically-equivalent, mutually-orthogonal directions may be chosen.* These are crystals of the so-called *cubic* system. The equivalent directions evidently coincide with the principal dielectric axes and one has $\varepsilon_x = \varepsilon_y = \varepsilon_z$ ($= \varepsilon$ say); then $\boldsymbol{D} = \varepsilon\boldsymbol{E}$ and the crystal is *optically isotropic* and equivalent to an amorphous body.

Group II. *Crystals not belonging to group I in which two or more crystallographically-equivalent directions may be chosen in one plane.* These are crystals of the trigonal, tetragonal, and hexagonal systems, the plane containing the equivalent directions being perpendicular to the axis of threefold, fourfold, or six-fold symmetry. One dielectric principal axis must coincide with this distinguished direction, whilst for the other two one may choose any orthogonal line pair perpendicular to it. If the distinguished direction is taken as the z axis one then has $\varepsilon_x = \varepsilon_y \neq \varepsilon_z$. Such crystals are said to be optically *uniaxial*.

Group III. *Crystals in which no two crystallographically-equivalent directions may be chosen.* These are the crystals belonging to the so-called orthorhombic, monoclinic and triclinic systems. Here $\varepsilon_x \neq \varepsilon_y \neq \varepsilon_z$ and the directions of the dielectric axes may or may not be determined by symmetry (see Table XX) and may therefore be wavelength dependent. Crystals of this group are said to be optically *biaxial*.

That all crystals fall into these three types as regards their optical properties, may be clearly seen by considering one of the associated ellipsoids, e.g. the ellipsoid of wave normals. Evidently the ellipsoid must be unchanged by the symmetry operations that leave the crystal structure unaltered.* Now there are only *two steps in the degeneration of an ellipsoid* into a sphere: an ellipsoid has either (a) all axes of

---

* For instance, crystals of the monoclinic class are characterized by either a twofold axis parallel to, or a mirror plane perpendicular to, one of the crystal axes, or both. In either case it is clear that the ellipsoid must have one of its axes parallel to this crystal axis, in order to remain unchanged by the symmetry operation.

Of the many accounts of crystal classes and symmetry operations, we may cite C. W. Bunn, *Chemical Crystallography* (Oxford, Clarendon Press, 1945), Chapter 2.

unequal lengths or (b) two axes equal and one unequal (spheroid, i.e. ellipsoid of revolution), or (c) all axes equal (sphere); these correspond to the three groups [in the order (III), (II), and (I)] which we just considered. The terms uniaxial and biaxial refer to the number of optic axes which the ellipsoid has, i.e. the number of diameters with the property that a plane section at right angles to them through the centre of the ellipsoid is a circle. A general ellipsoid has two such diameters (biaxial crystals), a spheroid has one (uniaxial crystals), and a sphere an infinity (isotropic crystals).

Table XX gives a survey of all the possible cases. Principal dielectric axes which may be colour-dependent are shown by two thin lines at a small angle to each other (signifying positions for two wavelengths), fixed axes by thick lines, and freely rotatable (or indeterminate) axes by broken lines ending on a circle or a sphere.

TABLE XX

Crystal system	Dielectric axes		Ellipsoid of wave normals	Optical classification
Triclinic	C C C		General ellipsoid	Biaxial
Monoclinic	C C F		" "	"
Ortho-rhombic	F F F		" "	"
Trigonal Tetragonal Hexagonal	F R R		Spheroid	Uniaxial
Cubic	R R R		Sphere	Isotropic

C = Axis with colour dispersion
F = Axis fixed in direction
R = Freely rotatable, or indeterminate, axis

14.3.2 Light propagation in uniaxial crystals

We start from FRESNEL's equation of wave normals § 14.2 (24) and write it in the form

$$s_x^2(v_p^2 - v_y^2)(v_p^2 - v_z^2) + s_y^2(v_p^2 - v_z^2)(v_p^2 - v_x^2) + s_z^2(v_p^2 - v_x^2)(v_p^2 - v_y^2) = 0. \quad (1)$$

For an optically uniaxial crystal with the optic axis in the z direction, $v_x = v_y$. Writing* v_o in place of this common velocity and v_e in place of v_z, (1) reduces to

$$(v_p^2 - v_o^2)[(s_x^2 + s_y^2)(v_p^2 - v_e^2) + s_z^2(v_p^2 - v_o^2)] = 0. \quad (2)$$

Let ϑ denote the angle which the wave normal $\boldsymbol{s}$ makes with the z axis; then

$$s_x^2 + s_y^2 = \sin^2\vartheta, \qquad s_z^2 = \cos^2\vartheta,$$

and (2) becomes

$$(v_p^2 - v_o^2)[(v_p^2 - v_e^2)\sin^2\vartheta + (v_p^2 - v_o^2)\cos^2\vartheta] = 0. \quad (3)$$

* Suffixes o and e stand here for ordinary and extraordinary, a terminology which will be clarified later.

The two roots of this equation (v'_p and v''_p say) are given by

$$\left.\begin{aligned} v'^2_p &= v_o^2, \\ \text{and} \qquad v''^2_p &= v_o^2 \cos^2 \vartheta + v_e^2 \sin^2 \vartheta. \end{aligned}\right\} \tag{4}$$

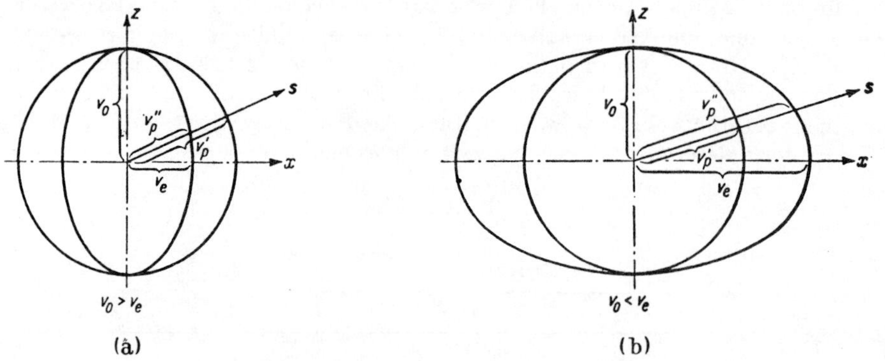

Fig. 14.6. The normal surfaces of a uniaxial crystal.
(a) Positive uniaxial crystal and (b) negative uniaxial crystal.

Equations (4) show that *the two shells of the normal surface are a sphere of radius* $v'_p = v_o$ *and a surface of revolution (of the fourth order), an ovaloid.* Thus one of the two waves that corresponds to any particular wave-normal direction is an *ordinary wave,* with a velocity independent of the direction of propagation, the other an *extraordinary wave* with velocity depending on the angle between the direction of the wave normal and the optic axis. The two velocities are only equal when $\vartheta = 0$, i.e. when the wave normal is in the direction of the optic axis.

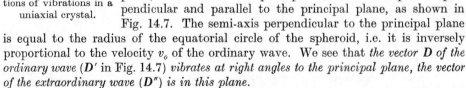

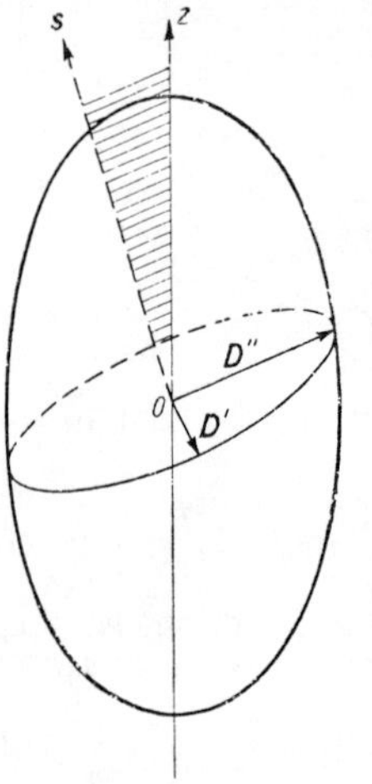

Fig. 14.7. The directions of vibrations in a uniaxial crystal.

When $v_o > v_e$ (see Fig. 14.6(a)), the ordinary wave travels faster than the extraordinary wave (except for $\vartheta = 0$ when they are equal); such a crystal is said to be a *positive* uniaxial crystal (e.g. quartz). If $v_o < v_e$ (see Fig. 14.6(b)), the ordinary wave travels more slowly than the extraordinary wave, and we speak of a *negative* uniaxial crystal (e.g. felspar).

The directions of vibration may conveniently be found with the help of the ellipsoid of wave normals, which now has two of its principal axes equal. The plane containing the wave normal $\boldsymbol{s}$ and the optic axis OZ is called the *principal plane* (shaded in Fig. 14.7). The ellipsoid is symmetrical about this plane. It follows that the elliptical section through O by the plane perpendicular to $\boldsymbol{s}$ is symmetrical about the principal plane, and therefore the principal axes of the ellipse are perpendicular and parallel to the principal plane, as shown in Fig. 14.7. The semi-axis perpendicular to the principal plane is equal to the radius of the equatorial circle of the spheroid, i.e. it is inversely proportional to the velocity v_o of the ordinary wave. We see that *the vector* $\mathbf{D}$ *of the ordinary wave* ($\mathbf{D}'$ in Fig. 14.7) *vibrates at right angles to the principal plane, the vector of the extraordinary wave* ($\mathbf{D}''$) *is in this plane.*

Optical phenomena in uniaxial crystals played a considerable part in the history of optics in connection with the question whether a "light vector" vibrates at right angles or parallel to the plane of polarization (defined as the plane of incidence for reflection from a plane air-dielectric interface at an angle of incidence at which any incident wave becomes linearly polarized, i.e. the $(\boldsymbol{H},\boldsymbol{s})$ plane in the language of electromagnetic theory; cf. p. 28 and p. 43). Today it is no longer relevant to discuss this question in detail,* since we know that there is no single physical entity that can be identified with a "light vector".

14.3.3 Light propagation in biaxial crystals

We shall now investigate the main consequences of the basic equation (1) in the general case of a biaxial crystal. It will help us to visualize the normal surface, if we consider first the sections of this surface by the three coordinate planes $x = 0$, $y = 0$,

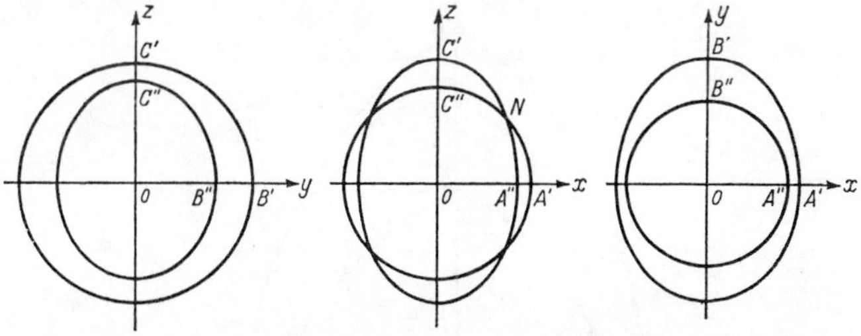

Fig. 14.8. Sections of the normal surface of a biaxial crystal.

and $z = 0$ of our reference system (principal dielectric axes). In order to fix the form of the intersection curves uniquely the three axes will be labelled so that

$$\varepsilon_x < \varepsilon_y < \varepsilon_z, \quad (v_x > v_y > v_z). \tag{5}$$

If we set $s_x = 0$ in equation (1), the equation breaks into two factors, giving

$$\left.\begin{aligned} v_p'^2 &= v_x^2, \\ v_p''^2 &= v_z^2 s_y^2 + v_y^2 s_z^2. \end{aligned}\right\} \tag{6}$$

We set $v_p s_y = y$, $v_p s_z = z$; then $v_p^2 = y^2 + z^2$ and the equations become

$$y^2 + z^2 = v_x^2, \qquad (y^2 + z^2)^2 = v_z^2 y^2 + v_y^2 z^2. \tag{6a}$$

Thus the section of the normal surface by the coordinate plane $x = 0$ is a circle and an oval. The section by each of the other two coordinate planes likewise consists of a circle and an oval, the only difference being in the relative positions of the two curves. With the choice of axes indicated by (5), the circle lies completely outside the oval in the yz-plane and lies completely inside it on the xy-plane; in the zx-plane the circle and the oval intersect in four points (Fig. 14.8). An octant of the normal surface is shown in perspective in Fig. 14.9 where each of the curves $A'B'C'N$ and $A''B''C''N$ is spanned by a smooth surface. In general two surfaces intersect in a curve, but in this case the two surfaces have only four points in common, the point N and the corresponding points in the other quadrants. The two lines joining the origin with each of these points are the two optic axes of wave normals. That there are no other such points and, therefore, no other optic axes of wave normals, follows

* For an account of the historical background see E. T. Whittaker, *History of the Theories of Aether and Electricity*, Vol. I: The Classical Theories (London, T. Nelson, 1951), p. 116.

from the geometrical theorem relating to the number of central circular sections of an ellipsoid, mentioned on p. 675. But we will now confirm this by direct calculations and establish at the same time an inequality that we shall need later.

Let

$$v_x^2 = v_y^2 + q_x, \qquad v_z^2 = v_y^2 - q_z, \qquad v_p^2 = v_y^2 + q, \tag{7}$$

where q_x and q_z are positive because of (5). With this substitution equation (1) becomes

$$s_x^2 q(q + q_z) + s_y^2(q + q_z)(q - q_x) + s_z^2(q - q_x)q = 0, \tag{8a}$$

or

$$q^2 + [s_x^2 q_z + s_y^2(q_z - q_x) - s_z^2 q_x]q - s_y^2 q_x q_z = 0. \tag{8b}$$

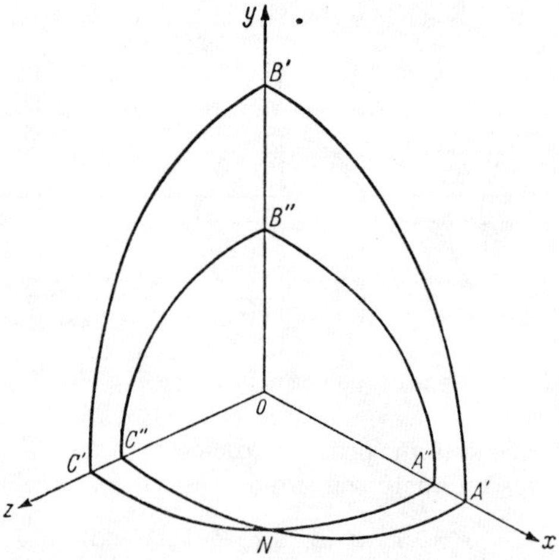

Fig. 14.9. The normal surface of a biaxial crystal.

Since the constant term $- s_y^2 q_x q_z$ cannot be positive, the roots of this equation must be real. If we denote them by q' and q'', then

$$q'q'' = - s_y^2 q_x q_z \leqslant 0.$$

Hence q' and q'' must have opposite signs. Let $q' \geqslant 0$ and $q'' \leqslant 0$.

If $q > q_x$ or $q < - q_z$ the terms on the left-hand side of (8a) would be positive. Hence q must lie in the range $- q_z \leqslant q \leqslant q_x$ and so

$$- q_z \leqslant q'' \leqslant 0 \leqslant q' \leqslant q_x. \tag{9}$$

The two roots q' and q'' can therefore only be equal if they both are zero; and for this to be the case, one must, according to (8b) have simultaneously

$$s_y^2 = 0, \qquad s_x^2 q_z = s_z^2 q_x. \tag{10}$$

Thus we obtain two directions $\boldsymbol{s}$ for each of which the corresponding two velocities are equal, confirming that there are two optical axes of wave normals. As already noted earlier, these axes lie in the xz-plane. If β is the angle which one of the axes makes with the z direction, then $s_x = \sin \beta$, $s_z = \cos \beta$ where, according to (10),

$$\tan \beta = \frac{s_x}{s_z} = \pm \sqrt{\frac{q_x}{q_z}} = \pm \sqrt{\frac{v_x^2 - v_y^2}{v_y^2 - v_z^2}}; \tag{11}$$

thus the optic axes are situated symmetrically with respect to the z-axis.

In terms of the velocities, the inequalities (9) become

$$v_z^2 \leqslant v_p''^2 \leqslant v_y^2 \leqslant v_p'^2 \leqslant v_x^2, \tag{12}$$

a relation that we shall need later. We see that the two phase velocities are real whatever the direction of the wave normal; this conclusion is, of course, also evident from the geometrical construction for the phase velocities by means of the ellipsoid of wave normals.

The expression for the two phase velocities corresponding to a given wave-normal direction $\mathbf{s}$ takes a very simple form if $\mathbf{s}$ is specified in terms of the angles ϑ_1 and ϑ_2 which it makes with the two optic axes of wave normals. Since the direction cosines of the optic axes are $\pm \sin \beta$, 0, $\cos \beta$, the angles ϑ_1 and ϑ_2 are given by the equations

$$\left.\begin{aligned} \cos \vartheta_1 &= \quad s_x \sin \beta + s_z \cos \beta, \\ \cos \vartheta_2 &= - s_x \sin \beta + s_z \cos \beta. \end{aligned}\right\} \tag{13}$$

In terms of s_x and s_z, the roots of (8b) are

$$q = -\tfrac{1}{2}P \pm \tfrac{1}{2}\sqrt{\Delta}, \tag{14}$$

where, if we also use the identity $s_x^2 + s_y^2 + s_z^2 = 1$,

$$P = s_x^2 q_x - s_z^2 q_z + q_z - q_x, \tag{15}$$

$$\begin{aligned} \Delta &= P^2 + 4s_y^2 q_x q_z \\ &= (q_x + q_z)^2 - 2(q_x + q_z)(s_x^2 q_x + s_z^2 q_z) + (s_z^2 q_z - s_x^2 q_x)^2. \end{aligned} \tag{16}$$

Now from (13) and (11),

$$\cos \vartheta_1 \cos \vartheta_2 = \frac{q_z s_z^2 - q_x s_x^2}{q_x + q_z}, \qquad \cos^2 \vartheta_1 + \cos^2 \vartheta_2 = 2\,\frac{q_x s_x^2 + q_z s_z^2}{q_x + q_z}. \tag{17}$$

Hence in terms of ϑ_1 and ϑ_2, (15) and (16) become

$$P = q_z - q_x - (q_z + q_x) \cos \vartheta_1 \cos \vartheta_2, \tag{15a}$$

$$\Delta = [(q_z + q_x) \sin \vartheta_1 \sin \vartheta_2]^2, \tag{16a}$$

and we finally obtain, on substituting from (15a) and (16a) into (14), and replacing q, q_x, and q_z by the expressions (7):

$$v_p^2 = \tfrac{1}{2}[v_x^2 + v_z^2 + (v_x^2 - v_z^2) \cos (\vartheta_1 \pm \vartheta_2)]. \tag{18}$$

Although v_y does not appear in this equation, it is contained implicitly in ϑ_1 and ϑ_2, since these angles depend on β and this is a function of all three principal velocities.

In the special case of a uniaxial crystal ϑ_1 and ϑ_2 are equal; it may readily be verified that equation (18) then correctly reduces to (4).

Similar analysis applies to the rays. Starting from the ray equation § 14.2 (29) we find that the section of the ray surface with the coordinate plane $x = 0$ consists of the two curves

$$v_r'^2 = v_x^2, \qquad \frac{1}{v_r''^2} = \frac{t_y^2}{v_z^2} + \frac{t_z^2}{v_y^2}. \tag{19}$$

If we set $v_r t_y = y$, $v_r t_z = z$ then $v_r^2 = y^2 + z^2$ and (19) becomes

$$y^2 + z^2 = v_x^2, \qquad \frac{y^2}{v_z^2} + \frac{z^2}{v_y^2} = 1, \tag{19a}$$

i.e. the section of the ray surface by the plane $x = 0$ is a circle and an ellipse (not a more general type of an oval as in (6a)). Each of the sections by the other two coordinate planes is likewise a circle and an ellipse. Because of the inequalities (5), the circle encloses the ellipse in the yz-plane, whilst in the xy-plane the ellipse encloses the circle; and in the xz-plane the circle and the ellipse intersect in four points. These points specify the position of the *optic ray axes* R_1, R_2; their directions are given by the dual of (10):

$$t_y^2 = 0, \qquad t_x^2 \left(\frac{1}{v_y^2} - \frac{1}{v_z^2}\right) = t_z^2 \left(\frac{1}{v_x^2} - \frac{1}{v_y^2}\right). \tag{20}$$

The angle γ between either of these axes and the z axis is therefore given by

$$\tan \gamma = \frac{t_x}{t_z} = \pm \frac{v_z}{v_x} \sqrt{\frac{v_x^2 - v_y^2}{v_y^2 - v_z^2}} = \pm \frac{v_z}{v_x} \tan \beta. \tag{21}$$

If $v_z < v_x$ the optic ray axis makes a smaller angle with the z axis than the axis of normals.

14.3.4 Refraction in crystals

(a) *Double refraction.* Consider a plane wave incident from vacuum on a plane surface Σ of an anisotropic medium. It will give rise to a transmitted and reflected field. We shall briefly consider the nature of the transmitted field, using substantially the same argument as in the case of an isotropic medium (§ 1.5.1). We shall confine our attention to finding the direction of propagation of the disturbance within the crystal; expressions for the amplitude ratios (corresponding to the FRESNEL formulae) will not be investigated.*

Let $\boldsymbol{s}$ be the unit wave normal of the incident wave and $\boldsymbol{s}'$ that of the transmitted wave. We shall see shortly that in general *two* waves are transmitted so that there are two possible values of $\boldsymbol{s}'$. The field vectors of the incident wave and of the transmitted waves are functions of $(t - \boldsymbol{r}.\boldsymbol{s}/c)$ and $(t - \boldsymbol{r}.\boldsymbol{s}'/v')$ respectively. The continuity of the field across the boundary demands that for any point $\boldsymbol{r}$ on the plane Σ and for all times t,

$$t - \frac{\boldsymbol{r}\,.\,\boldsymbol{s}}{c} = t - \frac{\boldsymbol{r}\,.\,\boldsymbol{s}'}{v'},$$

i.e.

$$\boldsymbol{r}\,.\left(\frac{\boldsymbol{s}'}{v'} - \frac{\boldsymbol{s}}{c}\right) = 0. \tag{22}$$

Hence the vector $\boldsymbol{s}'/v' - \boldsymbol{s}/c$ must be perpendicular to the boundary.

The permissible wave normals $\boldsymbol{s}'$ may be determined as follows: With any point O on Σ as origin we plot vectors in all directions $\boldsymbol{s}'$, each of the length $1/v'$, where v' is the phase velocity corresponding to each $\boldsymbol{s}'$ in accordance with FRESNEL equation § 14.2 (24). The locus of the end points is a two-sheeted surface which differs from the normal surface in that each radius vector is of length $1/v'$ instead of v'. We call this surface the *inverse surface of wave normals*; it is the dual of the ray surface and, therefore, like the ray surface itself, is of the fourth degree. Since the required vector $\boldsymbol{s}'/v'$ must be such that $\boldsymbol{s}'/v' - \boldsymbol{s}/c$ is perpendicular to Σ, its end point Q' must be on the normal to Σ through the end point P of the vector $\boldsymbol{s}/c$. In general, the

* These are discussed, for example, in G. SZIVESSY, *Hdb. d. Phys.*, Vol. 20 (Berlin, Springer, 1928), 715.

normal to Σ cuts the inverse surface in four points, two of which lie on the same side of the boundary as the crystal. Hence there are two such points (Q' and Q'' in Fig. 14.10), and therefore two possible wave-normal directions, so that *in general each incident wave will give rise to two refracted waves*; to each of these waves there corresponds a ray direction and a ray velocity describing the propagation of the energy within the crystal.* This is the phenomenon of *birefringence* or *double refraction*. It is illustrated by the well-known effect that two images are observed when a small object is viewed through a slab of calcite.

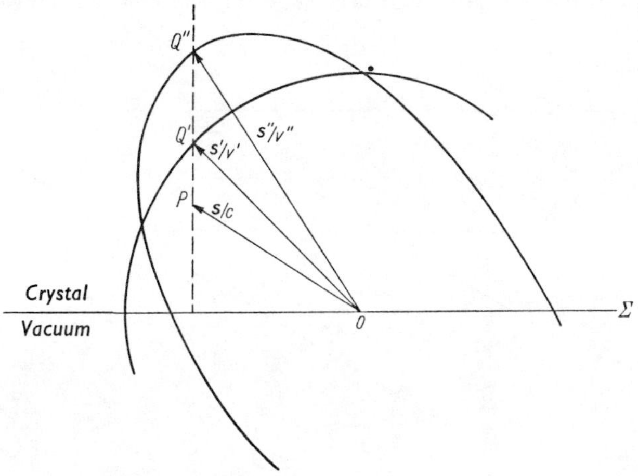

Fig. 14.10. Double refraction: construction for permissible wave normals.

Taking the boundary plane Σ as the plane $z = 0$, (22) becomes

$$\frac{xs_x + ys_y}{c} = \frac{xs'_x + ys'_y}{v'} = \frac{xs''_x + ys''_y}{v''}, \tag{23}$$

which must be satisfied for all values of x and y. This implies, firstly, that $s'_x/s_x = s'_y/s_y$ and $s''_x/s_x = s''_y/s_y$, so that *the two refracted rays lie in the plane of incidence*. Further, if θ_i, θ'_t, and θ''_t are the angles which the incident and the two transmitted waves make with the axis, we have from (23)

$$\frac{\sin\theta_i}{\sin\theta'_t} = \frac{c}{v'}, \qquad \frac{\sin\theta_i}{\sin\theta''_t} = \frac{c}{v''}. \tag{24}$$

Thus each of the transmitted waves is seen to obey the same *law of refraction* as in the case of isotropic media. However, the velocity v now depends on θ_t, so that the determination of the direction of propagation in the crystal is more complicated. In a uniaxial crystal one sheet of the inverse wave-normal surface is a sphere, so

* The origin of the double refraction may also be illustrated by generalizing HUYGENS' construction (§ 3.3.3) to anisotropic media. A proper formulation of this approach is, however, by no means as simple as is usually given in textbooks, since HUYGENS' construction operates with wavelets proceeding from point sources, and not with independent plane waves, the laws of propagation of which are as a rule taken over without justification. Some of the difficulties inherent in this approach were discussed by M. G. LAMÉ in his *Leçons sur la théorie mathématique de l'élasticité des corps solides* (Paris, Gauthier-Villars, 2nd ed., 1866), and V. VOLTERRA, *Acta Math.*, **16** (1892), 153.

that the phase velocity of *one* of the transmitted waves is then independent of θ_i, this being an *ordinary wave*.

In the special case of normal incidence ($\theta_i = 0$) one has $\theta_t' = \theta_t'' = 0$, so that the two wave-normals in the crystal coincide and are in the direction of the normal to Σ. Another special case, and one of a great theoretical interest, is that of a wave propagated in the direction of one of the optic axes of a biaxial crystal. It gives rise to a phenomenon known as conical refraction, which we shall now consider.

(b) *Conical refraction*. We mentioned earlier that when $\boldsymbol{s}$ coincides with one of the optic axes of the wave normals in a biaxial crystal, the relation between $\boldsymbol{s}$ and $\boldsymbol{t}$ has

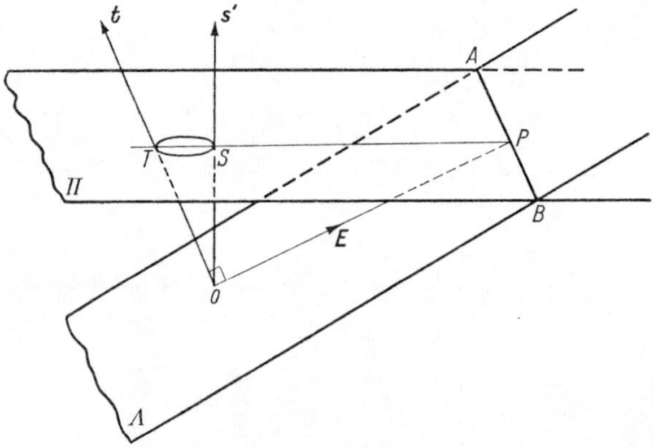

Fig. 14.11. Illustrating the position of rays that correspond to an optic axis of wave normals in a biaxial crystal.

a singularity. We must investigate the nature of this singularity before discussing the refraction phenomena associated with propagation in this special direction.

We have shown in § 14.2.3 that the electric displacement vectors $\boldsymbol{D}$ associated with a wave-normal direction $\boldsymbol{s}$ are parallel to the principal axes of the elliptical section through the ellipsoid of wave normals at right angles to $\boldsymbol{s}$. Now in the special case when the vector $\boldsymbol{s}$ is in the direction of the optic axis of wave normals, the section is circular, so that all directions of $\boldsymbol{D}$ at right angles to $\boldsymbol{s}$ are then permissible; consequently an infinity of directions of the electric vector $\boldsymbol{E}$ (calculated for each $\boldsymbol{D}$ from § 14.1 (1)) and an infinity of ray directions $\boldsymbol{t}$ (determined as in Fig. 14.1) are possible in this case. We shall now show that all these $\boldsymbol{t}$ vectors lie on the surface of a cone.

Let $\boldsymbol{s}'(s_x', 0, s_z')$ be the unit vector along one of the optic axes of wave normals, referred to the principal dielectric axes, subject to the inequality (5). Then s_x' and s_z' are related by (11) and the permissible $\boldsymbol{D}$ vectors, being perpendicular to $\boldsymbol{s}'$, satisfy the relation,

$$s_x' D_x + s_z' D_z = 0.$$

In terms of the components of the corresponding $\boldsymbol{E}$ vectors, this relation may be written as

$$s_x' \varepsilon_x E_x + s_z' \varepsilon_z E_z = 0. \tag{25}$$

In Fig. 14.11 let Π be any plane perpendicular to $\boldsymbol{s}'$ and let the line in which the electric vector $\boldsymbol{E}$ is localized cut it at P. Since according to (25) all the $\boldsymbol{E}$ vectors must lie in a plane Λ perpendicular to the vector with components $(s_x'\varepsilon_x, 0, s_z'\varepsilon_z)$, all the possible points P must lie on the straight line AB in which the planes Π and Λ intersect. Now the ray vector $\boldsymbol{t}$ is coplanar with $\boldsymbol{E}$ and $\boldsymbol{s}'$ and perpendicular to $\boldsymbol{E}$. Let $\boldsymbol{t}$ and $\boldsymbol{s}'$ cut the plane Π in the points T and S respectively. Then, by similar triangles,

$$TS\,.\,SP = OS^2 = \text{constant}. \tag{26}$$

Since the locus of P is the straight line AB, the locus of T is the inverse of AB, that is*, a circle passing through the centre of inversion S and with tangent at S parallel to AB. Hence *there is an infinity of rays corresponding to the optic axis of wave normals, and these rays form a surface of a cone.* This cone is not a circular one, since the centre of the circle is not the foot of the perpendicular from O to the plane Π.

If $\boldsymbol{E}$, $\boldsymbol{s}'$, and $\boldsymbol{t}$ all are in the xz-plane, the direction of $\boldsymbol{t}$ must be the direction $(s_x'\varepsilon_x, 0, s_z'\varepsilon_z)$ to which $\boldsymbol{E}$ is always perpendicular. If this direction makes an angle ϕ with the z axis, the aperture angle χ of the cone in this plane is given by

$$\tan\chi = \tan(\beta - \phi) = \frac{\dfrac{s_x'}{s_z'} - \dfrac{s_x'\varepsilon_x}{s_z'\varepsilon_z}}{1 + \dfrac{\varepsilon_x s_x'^2}{\varepsilon_z s_z'^2}} = \frac{1}{v_y^2}\sqrt{(v_x^2 - v_y^2)(v_y^2 - v_z^2)}, \tag{27}$$

where (11) has been used. Usually $(v_x^2 - v_y^2)/v_y^2 \ll 1$, $(v_y^2 - v_z^2)/v_y^2 \ll 1$, so that the cone is very nearly circular and of angle χ.

We may show in a strictly similar manner that *there is an infinity of wave normals corresponding to the optic ray axis, and these wave normals form the surface of a cone.* The aperture angle ψ of this cone is given by the dual of (27), i.e. by

$$\tan\psi = \frac{1}{v_x v_z}\sqrt{(v_x^2 - v_z^2)(v_y^2 - v_z^2)} = \frac{v_y^2}{v_x v_z}\tan\chi. \tag{28}$$

The situation may conveniently be illustrated in terms of the normal surface and the ray surface, using the property, derived in § 14.2.3, that the normal surface is the pedal surface of the ray surface. The intersection of these surfaces with the xz-plane is shown in Fig. 14.12. The surface of normals intersects this plane in a circle of radius $v_p' = v_y$, and in an oval with the polar radius v_p'', whilst the ray surface intersects it in the same circle $v_r' = v_y$ and an ellipse v_r''. If the circle and the oval intersect at N, the line ON is in the direction of the optic axis of wave normals, and the plane through N perpendicular to ON must touch the ray surface at all points in which the cone of permitted ray directions cuts this surface. The ray surface thus has the unusual property that certain tangent planes touch it at an infinity of points.†

The ray optic axis is represented by OR, where R is the point of intersection of the two sheets of the ray surface; the two sheets intersect in such a way that there is an infinite number of tangent planes at R, with their normals lying on a cone. The normals from O to these planes form the cone of wave normal directions corresponding to the ray direction OR. The aperture angles χ and ψ of the two cones are also shown

* See, for example, D. M. Y. Sommerville, *Analytical Conics* (London, Bell and Sons, 1941), p. 92.

† The properties of this surface are discussed in detail in G. Salmon, *Analytical Geometry of Three Dimensions.* Revised by R. A. P. Rogers, Vol. 2 (London, Longmans, Green & Co., 1915, 5th ed.), Chapter IV.

in the figure. For reasons that will become evident shortly, the cone belonging to N, i.e. formed by rays such as OA, is called the *cone of internal conical refraction*; the cone belonging to R, i.e. formed by wave normals such as OB, is called *the cone of external conical refraction*.

Consider now a plate of a biaxial crystal, e.g. aragonite, cut so that its two parallel faces are perpendicular to the optic axis of wave normals. If this be illuminated by a narrow beam of monochromatic light incident normally on one of the faces of the plate, the energy will spread out in the plate in a hollow cone, the cone of internal conical refraction, and on emerging at the other side it will form a hollow cylinder, as shown in Fig. 14.13. Thus we should expect to see on a screen parallel to the crystal

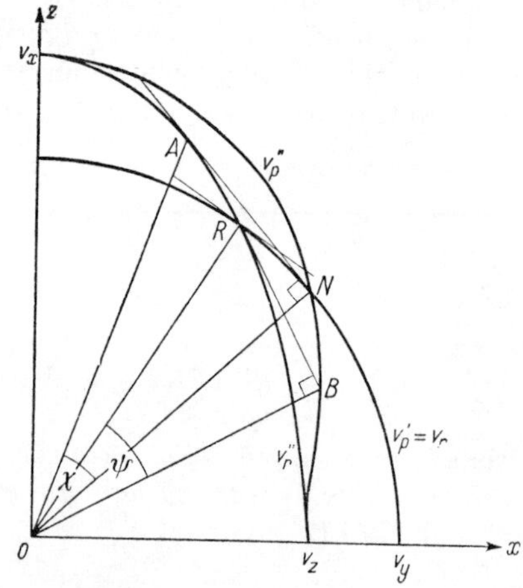

Fig. 14.12. Conical refraction: construction of the cones.

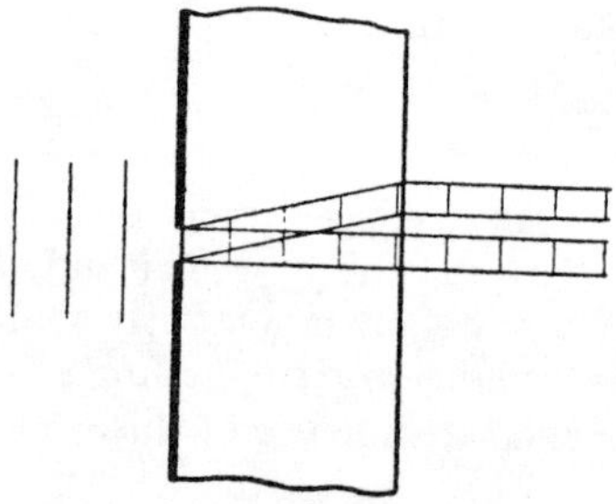

Fig. 14.13. Internal conical refraction.

face, a bright *circular ring*. This remarkable phenomenon was predicted in 1832 by Sir WILLIAM ROWAN HAMILTON, and confirmed a year later by LLOYD, who investigated it in aragonite on HAMILTON's instigation. The success of this experiment represented one of the most striking confirmations of FRESNEL's wave theory of light and contributed greatly to its general acceptance (see Historical Introduction, p. xxiv).

In practice the demonstration of the phenomenon of conical refraction is less simple than we have just indicated, for naturally it is impossible to obtain an accurately parallel beam of monochromatic light. In an experiment one always has to use beams of finite angular aperture, and in this case, as POGGENDORFF* and HAIDINGER† were the first to show, one observes two bright circles separated by a fine dark circle, as illustrated in Fig. 14.14. In LLOYD's first experiments this structure escaped observation, as the apertures limiting the width of his beam were too large and the two bright circles were therefore blurred into one. It remained unexplained for a long time after its discovery, until VOIGT‡ gave it an interpretation which may be summarized as follows:

* J. C. POGGENDORFF, *Pogg. Ann.*, **48** (1839), 461.
† W. v. HAIDINGER, *Wiener Ber.* (2), **16** (1854), 129; *Pogg. Ann.* (4), **96** (1855), 486.
‡ W. VOIGT, *Phys. Z.* **6** (1905), 672, 818.

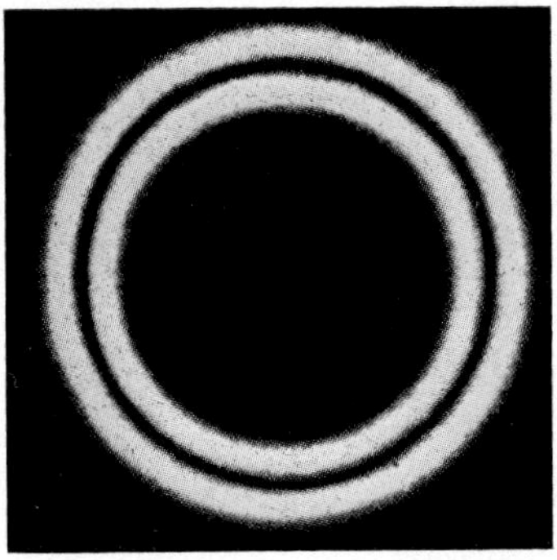

Fig. 14.14. Light distribution arising from conical refraction.

We have to consider the propagation of waves whose normals are slightly inclined to the optic axis. Each of the wave normals will give rise to two rays inside the crystal, and we should expect that the directions of these rays will differ only slightly from the directions of the generators of the cone of internal conical refraction. In order to find how the transmitted rays are distributed, we must consider part of the ray surface near the circle of contact with the tangent plane AN in Fig. 14.12. This part

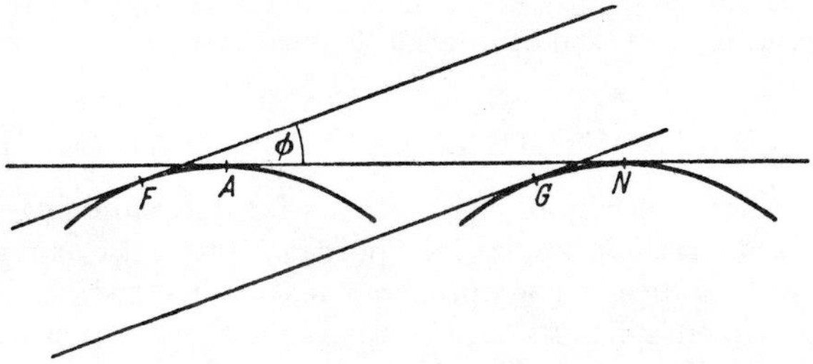

Fig. 14.15. Illustrating the position of rays belonging to wave normals that are inclined at small angles to an optic axis of wave normals.

of the surface may be likened to part of an inflated inner tube of a motor-car tyre, and the tangent plane to a flat board lying on the tyre. Fig. 14.15 shows the section with the xz-plane. Now the points on the ray surface which represent the directions of the two rays corresponding to the wave normal direction $\boldsymbol{s}$ are the points of contact of the two tangent planes to the ray surface perpendicular to $\boldsymbol{s}$ as shown in Fig. 14.5. When the wave normal ON is slightly displaced from the optic axis, the tangent plane splits into two parallel planes, one of which rolls over the surface so that its point of contact moves away from the centre of the circle of contact to F, while the second (which cannot be represented in our model because it would have to cut through the inner tube) rolls so that its point of contact moves towards the centre to G. Fig. 14.15 shows this for a displacement of the wave normal in the xz-plane, but the same will happen for a displacement in any direction.

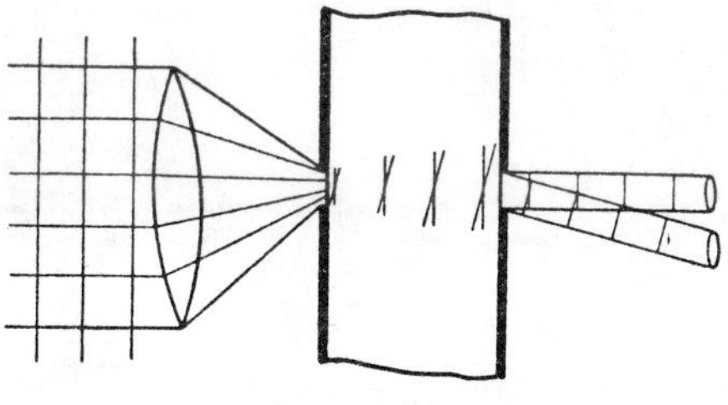

Fig. 14.16. External conical refraction.

From these remarks it is seen that all incident rays with wave normals inclined at a small angle ϕ to the optic axis will give rise to pairs of rays inclined at angles $\frac{1}{2}\chi + a\phi$ and $\frac{1}{2}\chi - a\phi$ to the central axis of the cone of inner conical refraction, where a is some constant. Thus all the energy associated with the angular range ϕ and $\phi + d\phi$ in the incident beam will appear in two cones of angular semi-apertures $\frac{1}{2}\chi \pm a\phi$ and angular separation $a d\phi$. But the corresponding energy in the incident beam is proportional to $\phi d\phi$, up to the maximum value of ϕ. It follows that, at angles $\frac{1}{2}\chi \pm a\phi$, the intensity in the cone of rays is proportional to ϕ, and in particular that it is zero at $\phi = 0$. Thus we must expect two bright circles to appear, with a dark circle between them, as is observed.

External conical refraction demonstrates the fact established earlier on that there is a whole cone of wave-normal directions corresponding to a given ray direction. This phenomenon is observed with a crystal plate cut so that its faces are perpendicular to the optic ray axis. Small apertures are placed on each face exactly opposite to each other, and one of them is illuminated with convergent light, as shown in Fig. 14.16.

Only those rays will reach the second aperture which have their directions very close to the direction of the optic ray axis, so that the waves reaching the second aperture all have their normals near the cone of the outer conical refraction. A cone of light will therefore emerge from the crystal. The angular aperture of this cone will be greater than the true angle ψ of the outer conical refraction, because of refraction from emergence from the crystal. Again, two concentric circles of light are observed on a screen parallel to the crystal face, and the explanation of this double circle is similar to that given for the case of internal conical refraction.

14.4. MEASUREMENTS IN CRYSTAL OPTICS

In this section we shall describe briefly methods for determining the character of a crystal (i.e. whether uniaxial or biaxial), the position of its optic axes and the values of its principal refractive indices. The optic axes may be located, as we shall see, from observation of interference fringes on crystal plates; the structure of the pattern indicates clearly the intersection of the optic axes with the faces of the plate. The principal refractive indices may be determined with the help of crystal prisms, from measurements of the angles of deviation or of total reflection.

As a preliminary we must discuss the production and analysis of polarized light.

14.4.1 The Nicol prism

One of the most commonly used instruments for the production of linearly polarized light is the *Nicol prism.** It consists of a natural rhomb of calcite (uniaxial crystal) which is cut into two equal parts along a diagonal plane (represented by AC in Fig. 14.17), and with the two parts cemented together with Canada balsam. The rhomb is about three times as long as it is wide, with the angles at B and D of its principal section equal to 71°; the end faces AD and BC are ground off to reduce this angle to 68°.

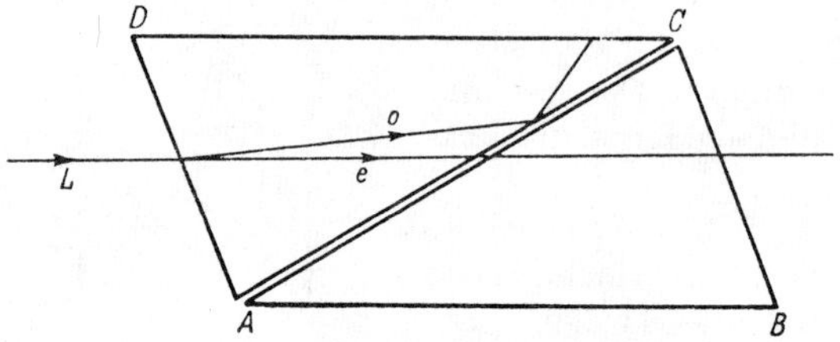

Fig. 14.17. The Nicol prism.

A ray of light incident in the direction L parallel to the long edge is split into an ordinary ray and an extraordinary ray. For the first ray the Canada balsam is of lower, for the second of higher, optical density (calcite crystal: $n_0 = 1{\cdot}66$, $n_e = 1{\cdot}49$; Canada balsam $n = 1{\cdot}53$), and it may easily be verified from the formulae of § 1.5.4 that on the Canada balsam interfaces conditions for total reflection are satisfied with respect to the ordinary ray; this ray is totally reflected towards the face DC which is blackened and so absorbs it. The extraordinary ray passes through the prism with practically no lateral displacement and is linearly polarized with its $\boldsymbol{D}$ vector in the principal section (see § 14.3.2). Thus *the Nicol prism produces linearly polarized light, whose direction of vibration is known.*

* W. Nicol, *Edinburgh New Philos. Journ.*, **6** (1829), 83.

If the incident ray is inclined to the edge of the rhomb the NICOL prism will still act as a polarizer, provided that for the ordinary ray the angle of incidence on the Canada balsam is not smaller than the critical angle; this limits the angle of the cone of incident rays in air for which the prism is effective to about* 30°.

Fairly pure linearly polarized light may also be obtained by passing natural light through a sheet of an absorbing crystalline material whose absorption coefficients for the two directions of vibrations are appreciably different. We shall consider such "polaroid sheets" in § 14.6.3.

An arrangement such as a NICOL prism, which produces linearly polarized light from light of other states of polarization, is called a *polarizer*. Such an arrangement may also be used as an *analyzer*, i.e. as a detector of linearly polarized light and of its direction of vibration. To detect linearly polarized light with a NICOL prism we only have to rotate the prism about its longitudinal axis and note whether there is a position when no light is passed through it. If there is such a position the light is linearly polarized and the direction of vibration of its $\boldsymbol{D}$ vector is perpendicular to the principal section.

14.4.2 Compensators

Crystalline material may also be used to investigate elliptically polarized light, i.e. the directions of its axes and the ratio of their lengths. For this purpose a crystal plate of suitable material and thickness is employed, and by means of it a path difference is introduced between vibrations in two mutually orthogonal directions. In particular the path difference may be made such as to convert elliptically polarized light into light that is linearly polarized; from the analysis of the linearly polarized light the required information about the elliptically polarized light is then obtained. Such a device is called a *compensator*, as its function is to compensate a phase difference.†

(a) *The quarter-wave plate.* Consider a plane parallel crystal plate of thickness h. Let the z axis be normal to the plate and the x and y axes in the directions of the corresponding $\boldsymbol{D}$-vibrations. We may assume that the plate is turned round the normal until the x and y axes are parallel to the principal axes of the vibrational ellipse of the incident light. The components of the $\boldsymbol{D}$ vector of the incident light are then represented by

$$\left.\begin{aligned} D_x^{(i)} &= a \cos \omega t, \\ D_y^{(i)} &= b \sin \omega t. \end{aligned}\right\} \tag{1}$$

On traversing the plate the two components will suffer different changes in phase because the velocities of the two rays are different. Neglecting reflection losses, the components of the $\boldsymbol{D}$ vector of the light emerging from the plate are given by

$$\left.\begin{aligned} D_x^{(t)} &= a \cos (\omega t + \delta'), \\ D_y^{(t)} &= b \sin (\omega t + \delta''), \end{aligned}\right\} \tag{2}$$

* There are several modifications of the Nicol prism which can be used over a larger angular field. For the description of these see, for example, L. C. MARTIN, *An Introduction to Applied Optics*, Vol. 1 (London, Pitman, 1930), p. 204, or R. W. WOOD, *Physical Optics* (New York, Macmillan, 3rd ed., 1934), pp. 337–8.

† For a detailed survey of compensators see H. G. JERRARD, *J. Opt. Soc. Amer.*, **38** (1948), 35. A survey of methods available for the analysis of polarized light is given in a paper by M. RICHARTZ and HSIEN-YÜ HSÜ, *ibid.*, **39** (1949), 136.

A systematic procedure for the theoretical analysis of fully or partially polarized light is afforded by the coherency matrices or by the STOKES parameters, discussed in § 10.8.

where

$$\delta' = \frac{2\pi}{\lambda} n'h, \qquad \delta'' = \frac{2\pi}{\lambda} n''h, \tag{3}$$

λ being the vacuum wavelength.* Hence the phase difference introduced by the plate is

$$\delta'' - \delta' = \frac{2\pi}{\lambda} (n'' - n')h. \tag{4}$$

In particular, if the emergent light is to be *linearly polarized*, one must have $\delta'' - \delta' = \pm \tfrac{1}{2}\pi$, or more generally $\delta'' - \delta' = (2m + 1)\pi/2$, where m is any integer, so that the plate must be of thickness

$$h = \left| \frac{2m + 1}{n'' - n'} \right| \frac{\lambda}{4}. \tag{5}$$

With this compensator the direction of the linear vibration of the transmitted light is given by

$$\frac{D_y^{(t)}}{D_x^{(t)}} = \pm \frac{b}{a}. \tag{6}$$

A compensator which introduces a phase difference $|\delta'' - \delta'| = \pi/2$, i.e. one for which the difference in the two optical thicknesses is a quarter of a wavelength, is called a *quarter-wave plate*†. It may conveniently be made of a foil of mica (biaxial crystal), split to the thickness $h = \lambda/4|n'' - n'|$.

Elliptically polarized light may be analyzed with it in the following way: the light is passed through a quarter-wave plate and then through a NICOL prism and both are rotated independently until the field seen through the NICOL prism becomes completely dark. This is the position where the axes of the mica plate are parallel to the axes of the vibrational ellipse of the incident light, and where, according to (6), the NICOL prism is set so as to extinguish linearly polarized light with its $\boldsymbol{D}$ vector inclined at an angle $\tan^{-1} b/a$ to the x axis. Thus the orientation of the axes of the ellipse and their ratio are found.

If the incident light is circularly polarized, then $b = +a$ or $b = -a$ and the $\boldsymbol{D}$ vector of the transmitted light is linearly polarized in a direction which makes an angle of 45° or 135° respectively with the axis OX. The former corresponds to left-handed, the latter to right-handed polarization.

Since the thickness of the quarter-wave plate depends on λ, exact compensation is only possible for monochromatic light of one particular wavelength. In order to achieve compensation for light of any given wavelength a wedge or a combination of wedges must be used in place of a single parallel-sided plate. We shall next consider some compensators of this type.

(b) *Babinet's compensator.* The compensator due to BABINET‡ allows the realization of all phase differences (including zero). It consists of two wedges of quartz (positive uniaxial crystal), with equal acute angles. The wedges are placed against each other

---

* In contrast to § 1.3 we write λ rather than λ_0 here, as suffix zero refers throughout this section to the ordinary ray.

† G. AIRY, *Trans. Camb. Phil. Soc.*, **4** (1833), 313.

‡ J. BABINET, *C.R. Acad. Sci., Paris*, **29** (1849), 514. J. JAMIN, *Ann. Chim. (Phys.)*, (3), **29** (1850), 274.

as shown in Fig. 14.18, and can be displaced along their plane of contact, thus forming a parallel plate of variable thickness. In one wedge the optic axis is parallel, in the other at right angles, to the edge.

Let n_o and n_e be the ordinary and extraordinary refractive indices of quartz and h_1 and h_2 the thicknesses of the two wedges at some particular point. On passing through both wedges, the phase difference between the two rays is

$$\delta = \frac{2\pi}{\lambda}(n_e - n_o)(h_1 - h_2). \tag{7}$$

The two contributions enter here with opposite signs, since the ray whose vector vibrates at right angles to the principal axis (the ordinary ray) is the faster one, and consequently the component vibrating in a direction parallel to the edge will be ahead of the other component in one wedge and lag behind it in the other. Now h_1 and h_2 and consequently also δ change continuously as the point of incidence moves across the plate, and δ is zero in the middle. In consequence there will be a series of lines along which the transmitted light is linearly polarized and can thus be extinguished by a suitably oriented NICOL prism.

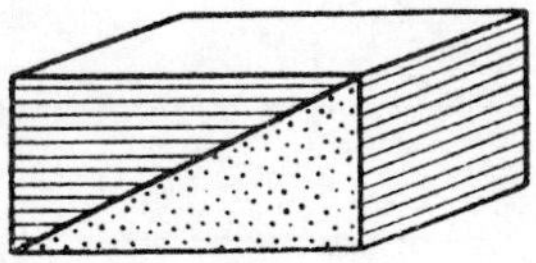
Fig. 14.18. BABINET's compensator.

Taking the x axis parallel to the edge, the elliptically polarized incident light may be represented by

$$\left.\begin{aligned} D_x^{(i)} &= a_1 \cos(\omega t + \delta_0'), \\ D_y^{(i)} &= a_2 \cos(\omega t + \delta_0''). \end{aligned}\right\} \tag{8}$$

The light transmitted by the compensator is represented by

$$\left.\begin{aligned} D_x^{(t)} &= a_1 \cos(\omega t + \delta'), \\ D_y^{(t)} &= a_2 \cos(\omega t + \delta''), \end{aligned}\right\} \tag{9}$$

where

$$\left.\begin{aligned} \delta' &= \delta_0' + \frac{2\pi}{\lambda} n_o(h_1 - h_2), \\ \delta'' &= \delta_0'' + \frac{2\pi}{\lambda} n_e(h_1 - h_2). \end{aligned}\right\} \tag{10}$$

For the emergent light to be linearly polarized one must have $\delta'' - \delta' = m\pi$ $(m = 0, \pm 1, \pm 2, \ldots)$, i.e.

$$\delta_0'' - \delta_0' = -\frac{2\pi}{\lambda}(n_e - n_o)(h_1 - h_2) + m\pi. \tag{11}$$

The direction of the linear vibrations is given by

$$\frac{D_y^{(t)}}{D_x^{(t)}} = \pm \frac{a_2}{a_1}. \tag{12}$$

Suppose first that a NICOL prism is placed in front of the compensator so that the compensator receives linearly polarized light. Then $\delta_0'' = \delta_0'$, and if the compensator is followed by an analyzer that is crossed with the polarizing NICOL prism, dark bands appear running parallel to the edge through points for which the right-hand side of

(11) is a multiple of π. These dark bands determine the *zero position*. If next elliptically polarized light is examined by means of the compensator and the analyzing prism alone, the displacement of the dark bands from the zero position immediately determines the phase difference $\delta_0'' - \delta_0'$ of the incident light; and the amplitude ratio of the components of $\boldsymbol{D}^{(i)}$ parallel and perpendicular to the edge can be determined from the orientation of the analyzer, using (12). From these data the position of the principal axes of the vibrational ellipse and their ratio can be found by application of the formulae of § 1.4.2.

(c) *Soleil's compensator*. For some purposes it is necessary to produce a phase difference (positive, negative, or zero) which is constant over the whole field of view. This may be achieved by a compensator due to SOLEIL.* It contains two quartz wedges A and A' which form, as in BABINET's compensator, a plane parallel plate, but with the difference that the optic axes in *both* wedges are now parallel to the edges. The lower wedge is cemented to a plane parallel quartz plate B whose optic axis is at right angles to the edge (see Fig. 14.19). The effective path difference which this compensator introduces between the two rays is evidently $h_B - (h_A + h_A')$. In the zero position, where this difference vanishes, the whole field of view can be obscured by a suitably oriented analyzer. The effective path difference can be altered by shifting the upper wedge, but for each position the path difference remains constant over the whole field. The analysis of elliptically polarized light is similar to that with BABINET's compensator.

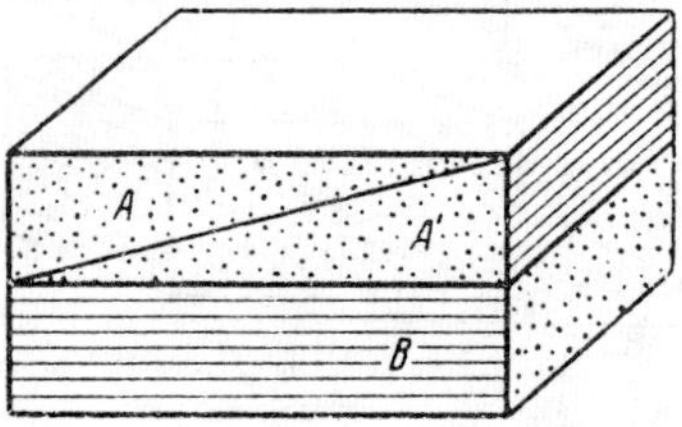

Fig. 14.19. SOLEIL's compensator.

(d) *Berek's compensator*. A compensator useful in biological microscopy (e.g. for measuring path differences in threads of birefringent material) is due to BEREK.† It consists of a slab of a uniaxial crystalline medium with its optic axes perpendicular to the faces of the slab. Phase differences between the ordinary and extraordinary rays are introduced by tilting the slab. It is mounted so that it can be turned by a graduated wheel which may be calibrated to show directly the path differences introduced.

14.4.3 Interference with crystal plates

As remarked earlier, the directions of optic axes may be determined by means of interference phenomena on crystal plates. The beautiful interference effects which are then observed, and which give a striking demonstration of the agreement between theory and experiment in crystal optics, deserve to be studied in their own rights.

Consider first a beam of linearly polarized light emerging from a polarizer and incident normally on to a plane parallel crystal plate of thickness h. On entering the plate, each ray is divided into two rays with different velocities of propagation, and with their $\boldsymbol{D}$ vectors vibrating in two mutually orthogonal directions at right angles to the direction of the plate normal. They emerge from the plate with a certain phase difference δ. If an analyzing NICOL prism is placed behind the plate, components of the two vibrations in a certain direction are then singled out and may be brought to interference in the focal plane of a lens placed behind the analyzer.

* H. SOLEIL, *C.R. Acad. Sci., Paris*, **21** (1845), 426; **24** (1847), 973; **26** (1848), 162. J. DUBOSCQ and H. SOLEIL, *ibid.*, **31** (1850), 248.

† M. BEREK, *Zbl. Miner. Geol. Paläont.* (1913), 388, 427, 464, 580.

In Fig. 14.20, the plane of the drawing is parallel to the plate. $\boldsymbol{D}'$ and $\boldsymbol{D}''$ represent the two mutually orthogonal directions of vibrations in the crystal, and OP and OA are the directions of vibrations that are passed by the polarizer and the analyzer respectively. Let ϕ be the angle that OP makes with $\boldsymbol{D}'$ and χ the angle between OA and OP. The amplitude of the light incident on the plate is represented by the vector OE (parallel to OP); its components in the directions of $\boldsymbol{D}'$ and $\boldsymbol{D}''$ are

$$OB = E \cos \phi, \qquad OC = E \sin \phi. \tag{13}$$

The analyzer transmits only the components parallel to OA which, as may be seen from the figure and from (13), have amplitudes

$$OF = E \cos \phi \cos (\phi - \chi), \qquad OG = E \sin \phi \sin (\phi - \chi). \tag{14}$$

On leaving the plate, the two components differ in phase by the amount

$$\delta = \frac{2\pi}{\lambda} (n'' - n')h. \tag{15}$$

According to § 7.2 (15), the intensity obtained from the interference of two monochromatic waves with phase difference δ is given by

$$I = I_1 + I_2 + 2\sqrt{I_1 I_2} \cos \delta,$$

where I_1 and I_2 are the intensities (squared amplitudes) of the two waves. With the amplitudes given by (14) we have

$$I = E^2 \left\{\cos^2 \chi - \sin 2\phi \sin 2 (\phi - \chi) \sin^2 \frac{\delta}{2}\right\}, \tag{16}$$

where the identity $\cos \delta = 1 - 2 \sin^2 \frac{\delta}{2}$ has been used.

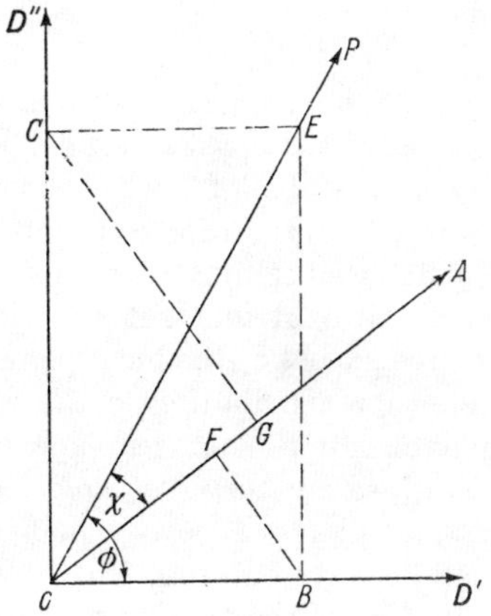

Fig. 14.20. Construction of the vibration components transmitted by a polarizer and an analyzer.

If the plate were removed ($\delta = 0$), the intensity would be $I = E^2 \cos^2 \chi$; thus the second term in (16) represents the effect of the crystal plate.

We now consider two important special cases:

(a) *Analyzer and polarizer parallel* ($\chi = 0$). In this case (16) reduces to

$$I_{\parallel} = E^2 \left(1 - \sin^2 2\phi \sin^2 \frac{\delta}{2}\right). \tag{17}$$

There is *maximum transmission* when

$$\phi = 0, \qquad \frac{\pi}{2}, \qquad \pi, \ldots, \tag{18}$$

i.e. *when the direction of vibrations passed by the analyzer coincides with one of the directions of vibrations in the plate.* The positions (18) are separated by positions of *minima of transmission*, given by $\sin 2\phi = \pm 1$, i.e. by

$$\phi = \frac{\pi}{4}, \qquad \frac{3\pi}{4}, \qquad \frac{5\pi}{4}, \ldots, \tag{19}$$

the minima being

$$I_{\parallel \min} = E^2 \left(1 - \sin^2 \frac{\delta}{2}\right) = E^2 \cos^2 \frac{\delta}{2}. \tag{20}$$

The minima are not completely dark unless δ is an odd multiple of π, i.e. unless the plate thickness has one of a number of permissible values, depending on the wavelength of the light.

(b) *Analyzer and polarizer perpendicular* ($\chi = \pi/2$). In this case (16) gives

$$I_{\perp} = E^2 \sin^2 2\phi \sin^2 \frac{\delta}{2}. \tag{21}$$

Comparison with (17) shows that the interference phenomena are now *complementary*. There are *minima of complete darkness* when ϕ has one of the values given by (18) and relative *maxima* for the intermediate positions (19) whose values are

$$I_{\perp\,\max} = E^2 \sin^2 \frac{\delta}{2}. \tag{22}$$

These phenomena may be used to produce colours that are exactly complementary; one only has to pass a parallel beam of white light through two NICOL prisms separated by a crystal plate, with the prisms being first parallel ($\chi = 0$) and then crossed ($\chi = \pi/2$). In order to obtain a uniform field it is essential to have the beam sufficiently well collimated to ensure that no significant phase differences are introduced.

If the light incident on the first polarizer originates in an extended incoherent source placed in the focal plane of a lens, each point in the source will give rise to an intensity distribution in the conjugate point, independently of all the other source points. In the image plane one then observes a light distribution which can be described by *curves of equal intensity*, forming a so-called *interference figure* of the crystal plate. To every point in the image plane there corresponds a direction of parallel rays entering and leaving the crystal and we must, therefore, consider the variation of the phase difference δ with this direction. It will be sufficient to consider only the case when the polarizer and analyzer are crossed (i.e. $\chi = \pi/2$); for when they are parallel the pattern is complementary, whilst other cases give less marked interference figures.

The intensity corresponding to a given direction of incidence depends on ϕ and δ and it is useful to consider the effects of varying each of these quantities separately. The curves along which ϕ is constant are called *isogyres* (or brushes), those along which δ is constant are called *isochromates.* The isogyres depend on the orientation of the optic axes in the plate and are independent of the thickness of the plate and the wavelength, unless there is dispersion of the axes. The isochromates depend on the direction of the wave normals and on the thickness of the plate, and are so named because, if white light is used, they are lines of equal colour. The curves of these families along which the intensity is zero are called *principal isogyres* and *principal isochromates* and are according to (21) given by $\sin 2\phi = 0$ and $\sin \frac{1}{2}\delta = 0$ respectively. On these curves the state of polarization of the light is the same as before the passage through the crystal. This is so because on the principal isogyres the direction of vibrations passed by the analyzer coincides with one of the directions of vibrations in the crystal, and on the principal isochromates the phase difference between the two emerging beams is an integral multiple of 2π. The two systems of curves are superimposed, but may be studied separately.

Before investigating the form of these curves we must consider how the phase difference δ depends on the angle of incidence. Let SA, AB', AB'' represent the wave normals to the incident and the two refracted waves at A and let θ_1, θ_2', θ_2'' be the angles of incidence and the two angles of refraction respectively (see Fig. 14.21).

Further let λ be the wavelength in the first medium (air) and $\lambda' = \lambda/n'$, $\lambda'' = \lambda/n''$ the wavelengths of the two refracted waves. The rays will emerge from the plate parallel to each other and to the incident wave normal (see Fig. 14.21) and with a phase difference

$$\delta = 2\pi \left[\frac{AB''}{\lambda''} + \frac{B''C}{\lambda} - \frac{AB'}{\lambda'}\right], \tag{23}$$

where

$$AB' = \frac{h}{\cos \theta_2'}, \qquad AB'' = \frac{h}{\cos \theta_2''}, \tag{24}$$

and

$$B''C = B''B' \sin \theta_1 = h \sin \theta_1 (\tan \theta_2' - \tan \theta_2''). \tag{25}$$

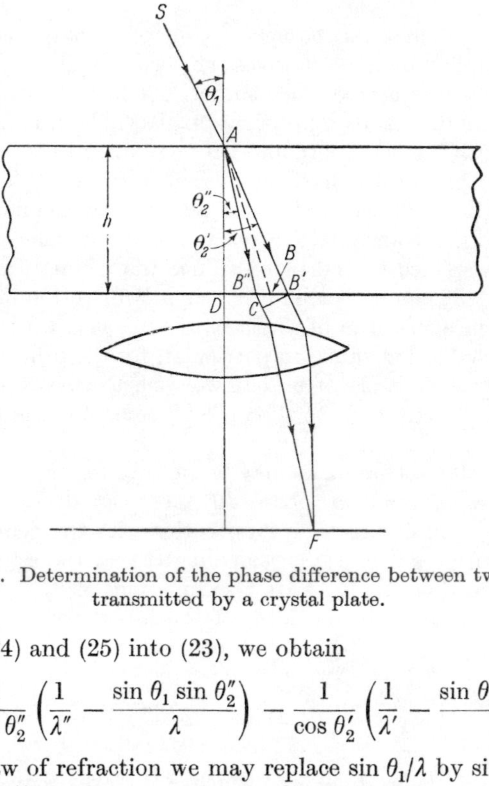

Fig. 14.21. Determination of the phase difference between two waves transmitted by a crystal plate.

Substituting from (24) and (25) into (23), we obtain

$$\delta = 2\pi h \left[\frac{1}{\cos \theta_2''}\left(\frac{1}{\lambda''} - \frac{\sin \theta_1 \sin \theta_2''}{\lambda}\right) - \frac{1}{\cos \theta_2'}\left(\frac{1}{\lambda'} - \frac{\sin \theta_1 \sin \theta_2'}{\lambda}\right)\right]. \tag{26}$$

Making use of the law of refraction we may replace $\sin \theta_1/\lambda$ by $\sin \theta_2''/\lambda''$ in the first bracket and by $\sin \theta_2'/\lambda'$ in the second bracket, giving

$$\delta = 2\pi h \left[\frac{\cos \theta_2''}{\lambda''} - \frac{\cos \theta_2'}{\lambda'}\right] = \frac{2\pi h}{\lambda}(n'' \cos \theta_2'' - n' \cos \theta_2'). \tag{27}$$

As the difference $n'' - n'$ is always small compared with n' and n'', it is permissible to use in place of (27) an approximate expression. We have, to first order,

$$\begin{aligned} n'' \cos \theta_2'' - n' \cos \theta_2' &= (n'' - n') \frac{d}{dn}(n \cos \theta_2) \\ &= (n'' - n')\left[\cos \theta_2 - n \sin \theta_2 \frac{d\theta_2}{dn}\right], \end{aligned} \tag{28}$$

where n is an average value of n' and n'' and θ_2 the corresponding average of θ_2' and θ_2''. We also have, on differentiation of the law of refraction $\sin\theta_1 = n \sin\theta_2$, keeping θ_1 fixed,

$$0 = \sin\theta_2 + n \cos\theta_2 \frac{d\theta_2}{dn}. \tag{29}$$

Hence (28) may be written as

$$n'' \cos\theta_2'' - n' \cos\theta_2' = \frac{1}{\cos\theta_2}(n'' - n'), \tag{30}$$

and (27) becomes, on substituting from (30),

$$\delta = \frac{2\pi h}{\lambda \cos\theta_2}(n'' - n'). \tag{31}$$

The quantity $h/\cos\theta_2$ represents the mean geometrical path of the two rays in the plate, and this multiplied by $n'' - n'$ gives the corresponding optical path difference.

Returning to the case of an extended source, one has to consider the transmission of waves with different directions of propagation. It will be assumed that the directions make small angles with the plate normal. Let us represent each of the incident waves by its wave normal through the fixed point A in Fig. 14.21. The points F in which the waves are brought to a focus by the lens are in a one-to-one correspondence with the points B where the transmitted wave normals AB (the average of AB' and AB'') strike the lower face of the plate. Since the inclination of AB to the plate normal AD is small, the points F will form a slightly distorted image—a projection—of the points B. Hence the form of the isochromates is essentially given by the loci of points B for which δ is constant. In particular, for the principal isochromates this constant is an integral multiple of 2π. If we wish to survey the effect of varying the plate thickness, we only have to shift the plane containing the points B parallel to itself.

It follows that all the isochromates may be surveyed by constructing around some point A *the surfaces of constant phase difference* $\delta(h, \theta_2) =$ constant, called also *isochromatic surfaces*, and finding their intersection with the planes $h =$ constant. In determining these surfaces it must be remembered that the refractive indices n' and n'' entering the expression (31) for δ are also functions of θ_2.

We shall consider separately the form of the interference figures obtained from uniaxial and biaxial crystal plates. It will be convenient to specify the points B by the polar distance

$$\rho = AB = \frac{h}{\cos\theta_2}, \tag{32}$$

and by the angle ϑ (or angles ϑ_1 and ϑ_2) which AB makes with the directions of the optic axis (or axes) of wave normals of the crystalline medium (Fig. 14.22).

14.4.4 Interference figures from uniaxial crystal plates

In a uniaxial crystal the phase velocities corresponding to a wave-normal direction that makes an angle ϑ with the optic axis are, according to § 14.3 (4), related by

$$v_p'^2 - v_p''^2 = (v_o^2 - v_e^2) \sin^2\vartheta. \tag{33}$$

Since $v_p = c/n$, and similarly for the other velocities, (33) gives

$$\frac{1}{n'^2} - \frac{1}{n''^2} = \left(\frac{1}{n_o^2} - \frac{1}{n_e^2}\right) \sin^2\vartheta. \tag{34}$$

The difference between these refractive indices is usually small compared with their values, so that (34) may be written in the approximate form

$$n'' - n' = (n_e - n_o) \sin^2\vartheta. \tag{35}$$

Substituting from this relation into (31) and using (32) we obtain

$$\delta = \frac{2\pi\rho}{\lambda} (n_e - n_o) \sin^2 \vartheta. \tag{36}$$

Hence the surfaces of constant phase difference are given by

$$\rho \sin^2 \vartheta = C, \quad (C = \text{constant}). \tag{37}$$

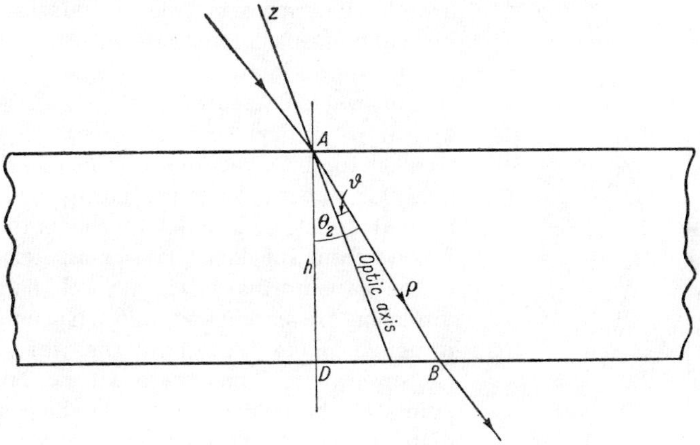

Fig. 14.22. Illustrating the theory of interference figures from crystal plates.

To visualize the form of these surfaces we take Cartesian axes with the z axis along the optic axis. Then

$$\rho^2 = x^2 + y^2 + z^2,$$
$$\rho^2 \sin^2 \vartheta = x^2 + y^2, \tag{38}$$

and, according to (37), the surfaces of constant phase difference are given by

$$(x^2 + y^2)^2 = C^2(x^2 + y^2 + z^2). \tag{39}$$

These surfaces can be generated by rotating the curves

$$x^4 = C^2(x^2 + z^2), \tag{40}$$

of which a typical one is shown in Fig. 14.23, around the z axis. At large distance from the origin ($z^2 \gg x^2$) the curve approaches asymptotically the parabolae

$$x^2 = \pm Cz. \tag{41}$$

In the neighbourhood of the x axis

$$C^2 = \frac{x^4}{x^2 + z^2} = \frac{x^2}{1 + \frac{z^2}{x^2}} = x^2 \left(1 - \frac{z^2}{x^2} + \dots\right) \tag{42}$$

so that the curve there is approximated by the hyperbola

$$x^2 - z^2 = C^2. \tag{43}$$

The parabolae and the hyperbola are shown as dotted curves in Fig. 14.23.

We can now determine all the isochromates by simply taking the sections of the surface (39) with planes at different distances h from the origin, these planes representing the exit face of the crystal plate. It is evident from Fig. 14.23 that isochromates of very different forms can arise, depending on the orientation of this face of the crystal relative to the optic axis. If the face of the plate is perpendicular to the optic axis, the isochromates are evidently circular; if the normal to the face makes a small angle with the optic axis, the curves are closed, approximating to ellipses; but if the normal makes a large angle with the optic axis the curves approximate to hyperbolae.

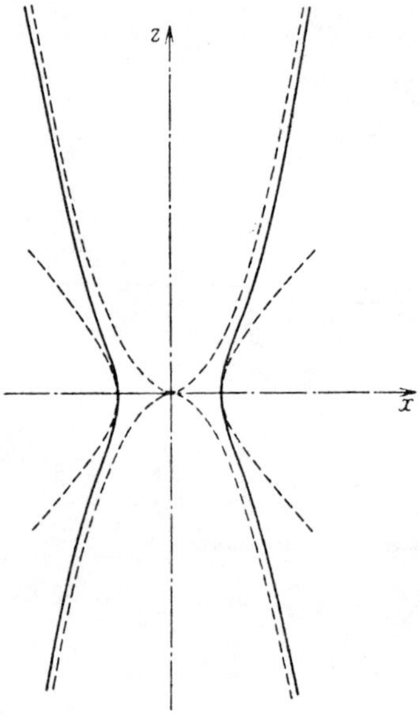

Fig. 14.23. Meridional curve of a surface of constant phase difference for an optically uniaxial crystal.

The *principal isogyres* (curves on which $\sin 2\phi = 0$), appear as a dark blurred cross whose arms are parallel to the directions of the polarizer and the analyzer, and whose centre corresponds to a wave-normal direction parallel to the optic axis. This follows from the fact that for any given direction of the wave normal in the crystal the directions of vibration are parallel to, and perpendicular to, the principal plane containing the wave normal and the optic axis. Thus there will be darkness at all points of the field of view for which the principal plane is parallel to the direction of vibration passed by either the polarizing or the analyzing Nicol prism.

Fig. 14.24 shows a typical interference figure from a uniaxial crystal; the principal isogyres and isochromates can be clearly seen.

Interference figures from plane parallel crystal plates have an important practical application in connection with *polariscopes*; these are instruments used for the detection of small proportions of polarized light in a beam which is largely unpolarized. An example is Savart's plate.* It consists of two quartz plates, each cut with its optic axis at 45° to the plate normal, and cemented together so that the planes containing the plate normals and the optic axes are perpendicular to each other. Thus the direction of ordinary vibrations in one plate is the direction of extraordinary vibration in the other, and the phase differences have opposite signs; they cancel each other exactly when the wave passes normally through the plates. This arrangement ensures that the phase difference varies very little with wavelength for nearly normal incidence, so that white light fringes may be obtained. If Savart's plate is placed between crossed Nicol prisms with its principal planes at 45° to the directions of the Nicol prisms, the interference pattern consists of light and black, almost straight, fringes. If the polarizing Nicol prism is removed, and partially polarized light passed through the system, the polarized component will give rise to fringes that are superimposed on a uniform background produced by the unpolarized

* Cf. *Ann. d. Physik*, **49** (1840), 292.

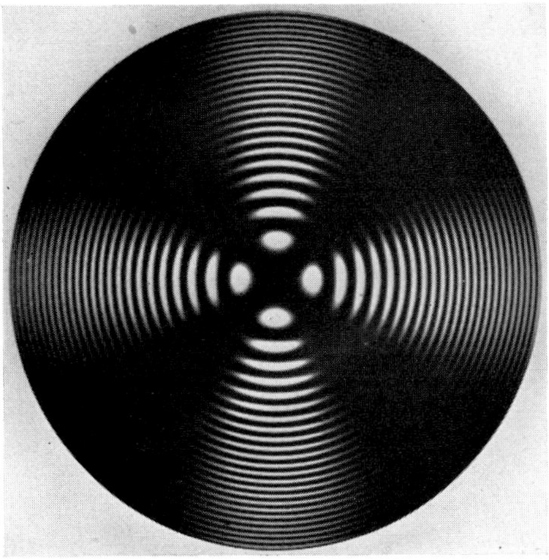

Fig. 14.24. Interference figure from fluorspar; cut perpendicular to the optic axis, between crossed Nicol prisms.

Fig. 14.26. Interference figure from Brazil topaz.

component, but the fringes can be detected even when the proportion of the polarized light is very low. The fringes have the greatest contrast when the plane of vibration of the incident light is at 45° to the principal planes of the Savart plate.

14.4.5 Interference figures from biaxial crystal plates

For a plane parallel biaxial crystal plate we have, in place of (33), the more general relation

$$v_p'^2 - v_p''^2 = (v_x^2 - v_z^2) \sin \vartheta_1 \sin \vartheta_2, \tag{44}$$

which follows from § 14.3 (18). Here ϑ_1 and ϑ_2 are the angles which the wave normal direction AB makes with the two optic axes of wave normals. Since $v_p = c/n$, $v_x = c/n_x$, etc., (44) gives

$$\left(\frac{1}{n'^2} - \frac{1}{n''^2}\right) = \left(\frac{1}{n_x^2} - \frac{1}{n_z^2}\right) \sin \vartheta_1 \sin \vartheta_2, \tag{45}$$

or approximately, since the differences between the refractive indices are small compared with their values,

$$n'' - n' = (n_z - n_x) \sin \vartheta_1 \sin \vartheta_2. \tag{46}$$

Substituting from (46) into (31) and again setting $h/\cos \vartheta_2 = \rho$, we obtain for the phase difference δ the expression

$$\delta = \frac{2\pi\rho}{\lambda} (n_z - n_x) \sin \vartheta_1 \sin \vartheta_2. \tag{47}$$

We see that the surfaces of constant phase difference are now given by

$$\rho \sin \vartheta_1 \sin \vartheta_2 = C. \qquad (C = \text{constant}). \tag{48}$$

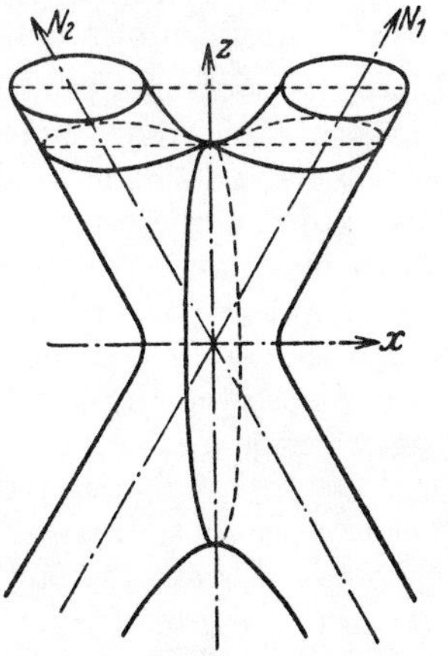

Fig. 14.25. Surface of constant phase difference for an optically biaxial crystal.

In the direction of each optic axis ($\vartheta_1 = 0$ or $\vartheta_2 = 0$) ρ tends to infinity, so that the surfaces approximate asymptotically to cylinders surrounding the optic axes. When ϑ_1 is small, ϑ_2 is approximately equal to the angle 2β between the two optic axes, and (48) then becomes

$$\rho \sin \vartheta_1 = \frac{C}{\sin 2\beta}. \tag{49}$$

But $\rho \sin \vartheta_1$ is the distance of a point on the surface from the optic axis $\vartheta_1 = 0$; similarly for $\rho \sin \vartheta_2$. Hence the "asymptotic cylinders" are circular. The general form of a surface of constant phase difference is shown in Fig. 14.25. It is evident that near the optic axes the isochromates are closed curves, approximating to ellipses, surrounding the two points in the focal plane that correspond to the optic axes.

The *principal isogyres* are obtained, as before, by finding all directions of the wave normal such that the directions of vibrations in the crystal coincide with the directions passed by the Nicol prisms. We may use the construction of § 14.2.3 (a), where it was shown that the planes of vibration bisect the angles between the planes ($\mathbf{N}_1$, $\mathbf{s}$) and ($\mathbf{N}_2$, $\mathbf{s}$), $\mathbf{N}_1$ and $\mathbf{N}_2$ being the directions of the optic axes. Thus if the crystal has its z axis vertical and the directions of vibrations which are transmitted by the Nicol prisms are parallel to the x and y axes, the principal isogyres will lie in the xz-plane and the yz-plane, and the interference figure will thus show a dark cross with one pair of arms passing through the points corresponding to the optic axes.

More generally, if the crystal is in any other orientation relative to the directions of the Nicol prisms, the principal isogyres will have the form of rectangular hyperbolae with asymptotes in the directions of vibrations transmitted by the Nicol prisms, and with arms passing through the points corresponding to the optic axes. If, with both the Nicol prisms fixed, the crystal plate is rotated in its plane, the pattern of isogyres will change, but the isochromates will, apart from rotation, remain the same; for the isochromates are defined by conditions which do not depend on the directions of the Nicol prisms. A typical interference figure from biaxial crystal plate is shown in Fig. 14.26.

14.4.6 Location of optic axes and determination of the principal refractive indices of a crystalline medium

Since the isochromates form closed curves around the optic axis (or axes), the observation of interference figures affords a ready method of determining whether a crystal is uniaxial or biaxial, and of locating the axes. The interference figures may be observed in a microscope fitted with two Nicol prisms (called *polarizing microscope*), either by removing the eyepiece and focusing the eye on the back focal plane of the objective lens (thus reproducing the conditions of Fig. 14.21) or by inserting an additional lens in the body of the microscope so that the back focal plane of the objective can be observed through the eyepiece. The second method has the advantage that a larger image of the interference figure is seen, and measurements may be made on the interference figure by the use of a calibrated scale in the eyepiece. Thus the angle between the optic axes of a biaxial crystal may be measured (naturally one must take into account the fact that the light is refracted as it leaves the crystal). Even very small crystal fragments, such as may occur in thin mineralogical sections, are sufficient for the location of the optic axes and measurements of their inclination by these means.

The principal refractive indices, n_x, n_y, n_z of a crystal may be determined either from measurements of the angle of deviation or of total reflection in a prism, or by immersion in a series of liquids of graded refractive index.

The prism method is more convenient for uniaxial crystals than for biaxial crystals. A prism is cut with its refracting edge parallel to the optic axis of wave normal. Then the ordinary and extraordinary waves have their $\boldsymbol{D}$ vectors respectively perpendicular and parallel to this edge. The two refractive indices can be found from the deviations of the two rays which emerge from the prism when an unpolarized beam is incident on one of its faces. By means of a Nicol prism one can distinguish between the ordinary and extraordinary rays.

The immersion method depends on the fact that a transparent body is not visible when immersed in a liquid of equal refractive index. Since a crystal has two refractive indices for any given direction of propagation, it will be visible in any liquid if it is observed with unpolarized light. However, if polarized light is used with its $\boldsymbol{D}$ vector in one of the directions of vibrations in the crystal, the crystal will be invisible in a liquid of the appropriate refractive index, n' or n''. If the directions of the principal dielectric axes are known, the crystal can be oriented so that light travels parallel to each axis in turn, and n' and n'' can be made equal to pairs of n_x, n_y, and n_z in turn.

If the directions of the principal dielectric axes are unknown, one can obtain a fair estimate of the refractive indices by immersing a large number of crystals, of the type under consideration, with random orientations, in a series of liquids of graded refractive indices. Each crystal will become invisible at two different values of the refractive index, for two directions of vibrations of the incident light. If these two refractive indices are n' and n'', we have, by the inequalities § 14.3 (12)

$$n_x \leqslant n' \leqslant n_y \leqslant n'' \leqslant n_z. \tag{50}$$

Thus n_x equals the lower limit of the values of n', n_z equals the upper limit of the values of n'', and n_y is equal both to the upper limit of n' and to the lower limit of n''; these limits should coincide if a sufficient number of measurements have been taken.

If the crystals are uniaxial, then every crystal will give n_o for one of its refractive indices, and the other will range between n_o and n_e.

14.5. STRESS BIREFRINGENCE AND FORM BIREFRINGENCE

14.5.1 Stress birefringence

When a transparent isotropic material is subject to mechanical stresses it may become optically anisotropic. This phenomenon, known as *stress birefringence* or the *photo-elastic effect* was first noted by BREWSTER* and finds useful practical applications. We shall only briefly indicate how optical methods may be used to obtain information about the state of stress in an initially isotropic material. As a preliminary we must consider the relations that exist between the elastic and the optical constants of matter.

The state of stress and the state of strain in an elastic solid body are characterized by second-order tensors, the stress tensor P_{kl} and the strain tensor r_{kl}, the components of which are linearly related to each other. These two tensors are always symmetric, but their principal axes are generally different from those of the dielectric tensor, which, as we saw in § 14.1, determines the optical properties of the body.

When a stress is applied to the body, the dielectric tensor is modified, and it may be assumed in the first approximation that the *changes* in the components of the dielectric tensor are linearly related to the six stress components, and hence also to the six strain components. Thus we are led to the introduction of two new sets of coefficients, *the stress-optical constants* and *the strain-optical constants* which characterize these relationships.

If we refer coordinates to the principal dielectric axes of the unstressed material, the ellipsoid of wave normals has the equation

$$\frac{x^2}{\varepsilon_x} + \frac{y^2}{\varepsilon_y} + \frac{z^2}{\varepsilon_z} = 1. \tag{1}$$

On applying a stress whose components are P_{xx}, P_{xy}, . . ., this ellipsoid is changed into another one whose equation may be written as

$$a_{xx}x^2 + a_{yy}y^2 + a_{zz}z^2 + a_{yz}yz + a_{zx}zx + a_{xy}xy = 1. \tag{2}$$

By our assumptions, each coefficient a_{kl} differs from the corresponding coefficient in (1) by a linear function in the P's. We thus have six relations, of which two typical ones are

$$\left.\begin{aligned} a_{xx} - \frac{1}{\varepsilon_x} &= q_{11}P_{xx} + q_{12}P_{yy} + q_{13}P_{zz} + q_{14}P_{yz} + q_{15}P_{zx} + q_{16}P_{xy}, \\ a_{yz} &= q_{41}P_{xx} + q_{42}P_{yy} + q_{43}P_{zz} + q_{44}P_{yz} + q_{45}P_{zx} + q_{46}P_{xy}. \end{aligned}\right\} \tag{3}$$

* D. BREWSTER, *Phil. Trans.* (1815), 60; (1816), 156. *Trans. Roy. Soc. Ed.*, **8** (1818), 369.

In this notation each numeral 1–6 in the subscripts refers to a pair of axes, thus: $1 = xx$, $2 = yy$, $3 = zz$, $4 = yz$, $5 = zx$, $6 = xy$.

There is a similar set of equations relating the coefficients a_{kl} to the strain components. The optical effect of the stress may also be described in terms of the deformation of the ray ellipsoid, and thus two further sets of linear equations with 36 coefficients are obtained. These coefficients are related to those of the ellipsoid of wave normals, since the two ellipsoids always have their principal axes in the same directions, and the semi-axes of one are reciprocals of the other.

The relations (3) take a simpler form if there are symmetry elements present in the structure. For crystals of the *cubic system* the three principal axes x, y, and z are equivalent, and in consequence the following relations hold between the stress-optical coefficients*

$$\left.\begin{aligned} q_{11} &= q_{22} = q_{33}, \\ q_{12} &= q_{21} = q_{23} = q_{32} = q_{31} = q_{13}, \\ q_{44} &= q_{55} = q_{66}, \end{aligned}\right\} \tag{4}$$

all the remaining coefficients being zero.

For *isotropic materials*, the relations (3) must remain unaltered for any change of axes. This is only possible if the stress-optical constants satisfy the conditions for cubic symmetry, and in addition the relation

$$2q_{44} = q_{11} - q_{12} \tag{5}$$

holds. Thus in this case there are only two independent constants. Since all systems of axes are now equivalent, we may use any set of axes and, in particular, the principal axes of the stress tensor; then $P_{yz} = P_{zy} = P_{xy} = 0$ and we have in place of (3) the simpler relations

$$\left.\begin{aligned} a_{xx} - \frac{1}{\varepsilon} &= q_{11}P_{xx} + q_{12}P_{yy} + q_{12}P_{zz}, \\ a_{yy} - \frac{1}{\varepsilon} &= q_{12}P_{xx} + q_{11}P_{yy} + q_{12}P_{zz}, \\ a_{zz} - \frac{1}{\varepsilon} &= q_{12}P_{xx} + q_{12}P_{yy} + q_{11}P_{zz}, \\ a_{yz} &= a_{zx} = a_{xy} = 0. \end{aligned}\right\} \tag{6}$$

Thus the principal axes of the stress tensor and those of the ellipsoid of wave normals are the same in this case, as one would expect from symmetry considerations.

Although cubic crystals such as rocksalt are optically isotropic when unstrained, they nevertheless behave differently, when strained, from truly isotropic materials like glass. The effect of stress may be conveniently observed by viewing the body between crossed NICOL prisms (or other polarizing devices such as polaroid sheets— cf. § 14.6.3). Consider a sheet of material of thickness h and let light be incident normally on it. Suppose that the sheet is stressed so that two of the principal axes, say x and y, of the stress tensor and, therefore, of the dielectric tensor, lie in the plane

---

* The relations which exist between the stress-optical coefficients of each of the crystal systems are discussed by F. POCKELS, *Ann. d. Physik*, **37** (1889), 158. Also his *Lehrbuch der Kristalloptik*, (Leipzig, 1906), pp. 469–74. They are summarized in G. SZIVESSY, *Handbuch der Physik*, Vol. 21 (Berlin, Springer, 1929), 840.

of the sheet and make angles ϕ and $\phi + \pi/2$ with the directions of the polarizer and the analyzer, as in § 14.4.3. The section of the ellipsoid of normals by the xy-plane is the ellipse

$$a_{xx}x^2 + a_{yy}y^2 = 1, \tag{7}$$

where a_{xx} and a_{yy} are given by (6). The refractive indices n' and n'' for the two waves propagated in the sheet are given by

$$n' = \frac{1}{\sqrt{a_{xx}}}, \qquad n'' = \frac{1}{\sqrt{a_{yy}}}. \tag{8}$$

Hence

$$\begin{aligned} \frac{1}{n'^2} - \frac{1}{n''^2} &= a_{xx} - a_{yy} \\ &= (q_{11} - q_{12})(P_{xx} - P_{yy}). \end{aligned} \tag{9}$$

Now n' and n'' will in practice usually differ only slightly from n_x, so that, approximately

$$n'' - n' = \tfrac{1}{2}\, n_x^3(q_{11} - q_{12})(P_{xx} - P_{yy}). \tag{10}$$

Substitution into § 14.4 (31) gives the phase difference δ between the two waves emerging from the sheet:

$$\delta = \frac{\pi h}{\lambda}\, n_x^3(q_{11} - q_{12})(P_{xx} - P_{yy}). \tag{11}$$

Thus the phase difference is proportional to $P_{xx} - P_{yy}$, which represents twice the shearing stress over planes inclined at 45° to the x and y directions. There is in this case only *one* relevant stress-optical constant, namely $q_{11} - q_{12}$.

It follows that if a stressed sheet of glass or transparent plastic is observed between crossed NICOL prisms, bright and dark fringes will be seen, and these fringes are contours of equal shearing stress. Such a "stress pattern" is shown in Fig. 14.27. The fringes are seen at maximum intensity in any given region of the pattern only if the principal axes of the stress system in this region are at 45° to the directions of vibrations transmitted by the NICOL prisms. When the principal axes of the stress are parallel to the directions transmitted by the NICOL prisms, the fringes disappear and the field of view becomes black; thus the directions of the axes of the stress system can be determined by rotating the crossed NICOL prisms, while the magnitude of the shearing stress can be obtained from the order of the fringes. This method is used to investigate stresses in engineering structures; a model of the structure is made from a suitable plastic material and the effect of the stresses is directly observed in the manner just described. In this way lengthy calculations are often avoided.*

14.5.2 Form birefringence

The birefringent properties of crystals may be explained in terms of the anisotropic electrical properties of molecules of which the crystals are composed. Birefringence may, however, arise from anisotropy on a scale much larger than molecular, namely when there is an ordered arrangement of similar particles of optically isotropic material whose size is large compared with the dimensions of molecules, but small compared with the wavelength of light. We then speak of *form birefringence.*

---

* For a full account of this method see for instance E. G. COKER and L. N. G. FILON, *A Treatise on Photoelasticity* (Cambridge University Press, 1931), or M. M. FROCHT, *Photoelasticity,* Vol. I (1941), Vol. II (1948), (New York, John Wiley).

From optical measurements information may often be obtained about the sub-microscopic particles that give rise to form birefringence. We shall explain the principle of the method by considering the somewhat idealized case of a regular assembly of particles that have the form of thin parallel plates. Let t_1 be the thickness of each plate and t_2 the widths of the spaces between them (Fig. 14.28). Further let ε_1 be the dielectric constant of each plate and ε_2 the dielectric constant of the medium in which they are immersed.

Suppose that a plane monochromatic wave is incident on the assembly and assume first that its electric vector is perpendicular to the plates. If the linear dimensions of the faces of the plates are assumed to be large but the thicknesses t_1 and t_2 small compared to the wavelength, the field in the plates and in the spaces may be considered

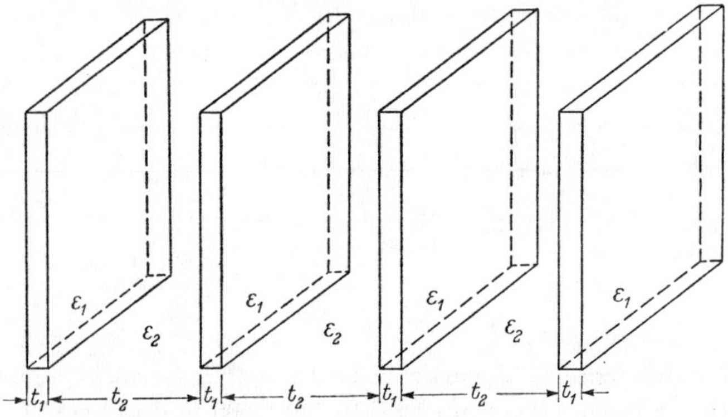

Fig. 14.28. A regular assembly of thin parallel plates.

to be uniform. Further, according to § 1.1.3, the normal component of the electric displacement must be continuous across a surface at which the properties of the medium change abruptly. Hence the electric displacement must have the same value $\boldsymbol{D}$ inside the plates and in the spaces. If $\boldsymbol{E}_1$ and $\boldsymbol{E}_2$ are the corresponding electric fields,

$$\boldsymbol{E}_1 = \frac{\boldsymbol{D}}{\varepsilon_1}, \qquad \boldsymbol{E}_2 = \frac{\boldsymbol{D}}{\varepsilon_2}, \tag{12}$$

and the mean field $\boldsymbol{E}$ averaged over the total volume is

$$\boldsymbol{E} = \frac{t_1 \dfrac{\boldsymbol{D}}{\varepsilon_1} + t_2 \dfrac{\boldsymbol{D}}{\varepsilon_2}}{t_1 + t_2}. \tag{13}$$

The effective dielectric constant $\varepsilon_\perp$ is, therefore,

$$\varepsilon_\perp = \frac{D}{E} = \frac{(t_1 + t_2)\varepsilon_1\varepsilon_2}{t_1\varepsilon_2 + t_2\varepsilon_1} = \frac{\varepsilon_1\varepsilon_2}{f_1\varepsilon_2 + f_2\varepsilon_1}, \tag{14}$$

where $f_1 = t_1/(t_1 + t_2)$, $f_2 = t_2/(t_1 + t_2) = 1 - f_1$ are the fractions of the total volume occupied by the plates and by the surrounding medium respectively.

Suppose next that the incident field has its electric vector parallel to the plates. According to § 1.1.3, the tangential component of the electric vector is continuous across a discontinuity surface, so that in this case the electric field will have the same value $\boldsymbol{E}$ inside the plates and in the spaces. The electric displacements in the two regions are

$$\boldsymbol{D}_1 = \varepsilon_1\boldsymbol{E}, \qquad \boldsymbol{D}_2 = \varepsilon_2\boldsymbol{E}, \tag{15}$$

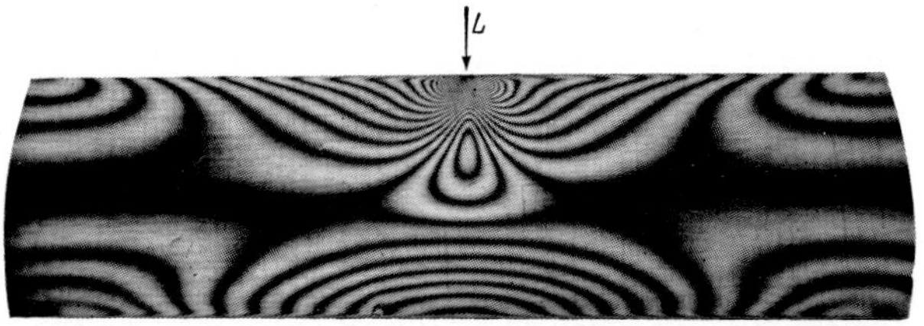

Fig. 14.27. The stress pattern of a beam under the action of a concentrated load. The position of the load is indicated by L; the ends (not shown) are clamped.

so that the mean electric displacement $\boldsymbol{D}$ is

$$\boldsymbol{D} = \frac{t_1\varepsilon_1\boldsymbol{E} + t_2\varepsilon_2\boldsymbol{E}}{t_1 + t_2}. \tag{16}$$

Hence the effective dielectric constant is now given by

$$\varepsilon_{\parallel} = \frac{D}{E} = \frac{t_1\varepsilon_1 + t_2\varepsilon_2}{t_1 + t_2} = f_1\varepsilon_1 + f_2\varepsilon_2. \tag{17}$$

Since the effective dielectric constant is the same for all directions parallel to the plates, but different for directions normal to the plates, the assembly behaves as a uniaxial crystal with its optic axis perpendicular to the plane of the plates. The difference $\varepsilon_{\parallel} - \varepsilon_{\perp}$ is always positive, since according to (14) and (17),

$$\varepsilon_{\parallel} - \varepsilon_{\perp} = \frac{f_1 f_2(\varepsilon_1 - \varepsilon_2)^2}{f_1\varepsilon_2 + f_2\varepsilon_1} \geqslant 0. \tag{18}$$

The ordinary wave has its electric vector perpendicular to the optic axis, i.e. parallel to the plane of the plates. Equation (18) implies that the assembly always behaves like a *negative uniaxial crystal* (cf. § 14.3.2). In terms of refractive indices the last equation may be written as

$$n_e^2 - n_o^2 = -\frac{f_1 f_2(n_1^2 - n_2^2)^2}{f_1 n_2^2 + f_2 n_1^2}. \tag{19}$$

For assemblies of particles of less idealized forms the calculations are naturally more complicated.*

A case of considerable practical interest is that of an assembly of parallel and similar thin cylindrical rods. It was shown by WIENER† that if the rods occupy a small fraction of the total volume ($f_1 \ll 1$), one has, in place of (19),

$$n_e^2 - n_o^2 = \frac{f_1 f_2(n_1^2 - n_2^2)^2}{(1 + f_1)n_2^2 + f_2 n_1^2}. \tag{20}$$

Such an assembly therefore behaves as a *positive uniaxial crystal*, with its optic axis parallel to the axes of the rods.

Observations on form birefringence are useful in biological microscopy. The sign of the observed difference indicates whether the shape of the particles is nearer to that of a rod or a plate, and if n_1 and n_2 are known, it may be possible to estimate from equation (19) or (20) the fraction of the volume occupied by the particles. To distinguish between form birefringence and intrinsic birefringence in the material of the particles, the refractive index n_2 of the medium is varied; the form birefringence will disappear when $n_2 = n_1$, while intrinsic birefringence will be unaffected by variation of n_2. If both forms of birefringence are present, a graph of $|n_e^2 - n_o^2|$ plotted against n_2 will show a minimum, but not a zero, at $n_2 = n_1$.

---

* For a general treatment see O. WIENER, *Abh. Sächs. Ges. Akad. Wiss., Math.-Phys. Kl. No.* 6. **32** (1912), 575. Formulae relating to ellipsoidal particles are also given by W. L. BRAGG and A. B, PIPPARD, *Acta Cryst.*, **6** (1953), 865.

† *Loc. cit.* p. 581. The principal dielectric constants of a rectangular array of parallel cylinders were also calculated by Lord RAYLEIGH, *Phil. Mag.*, (5), **34** (1892), 481, and equation (20) is in agreement with his results even for values of f_1 that are not very small compared to unity, provided that the difference between the refractive indices n_1 and n_2 is small.

The ordering of the particles that give rise to form birefringence may be permanent or semi-permanent as for example in tobacco-mosaic virus crystals,* or it may be a temporary ordering of similar particles suspended in a liquid. A suspension of similar particles in a liquid appears optically isotropic if the particles are randomly oriented, as will usually be the case if the liquid is stationary; if, however, the liquid is made to flow, there will be a tendency for the particles to align themselves in a particular direction, and the assembly will then behave as a crystal. This effect may be observed by placing the suspension between two coaxial cylinders, one rotating and the other fixed, and by viewing the liquid between crossed NICOL prisms, with light propagated in a direction parallel to the axes of the cylinders.† This phenomenon, first observed by MAXWELL,‡ is of assistance in investigations of flow of liquids past obstacles, giving information about the direction and magnitude of the velocity gradient.

14.6. ABSORBING CRYSTALS

14.6.1 Light propagation in an absorbing anisotropic medium

The crystalline media which we have so far considered have been characterized, with regard to their optical properties, by the dielectric tensor ε_{kl}. To describe media which are not only anisotropic, but are also absorbing, we must in addition introduce the *conductivity tensor* σ_{kl}. The directions of the principal axes of the two tensors will not be the same in general and, in consequence, the theory of propagation of light in such media is rather complicated. However, the principal axes of the two tensors coincide in direction for crystals of the higher symmetry classes (of at least orthorhombic symmetry), and we shall confine our attention to crystals of this type, as they illustrate all the essential features of the general theory. We then only have to replace the real dielectric constants ε_x, ε_y, and ε_z, by complex ones, $\hat{\varepsilon}_x$, $\hat{\varepsilon}_y$, and $\hat{\varepsilon}_z$. We shall see that all the earlier formulae of optics of crystals are formally retained, provided that all quantities which depend on $\hat{\varepsilon}_x$, $\hat{\varepsilon}_y$, and $\hat{\varepsilon}_z$ are assumed to be complex.

We start from MAXWELL's equations for a conducting medium, viz.

$$\operatorname{curl} \boldsymbol{H} = \frac{1}{c}\dot{\boldsymbol{D}} + \frac{4\pi}{c}\boldsymbol{j}, \qquad \operatorname{curl} \boldsymbol{E} = -\frac{1}{c}\dot{\boldsymbol{B}}, \tag{1}$$

and consider the propagation of a plane, damped wave. In the complex notation the vectors $\boldsymbol{E}$, $\boldsymbol{D}$, $\boldsymbol{B}$, $\boldsymbol{H}$, and $\boldsymbol{j}$ are then each proportional to

$$\exp\left[i\omega\left(\frac{\hat{n}}{c}(\boldsymbol{r}.\boldsymbol{s}) - t\right)\right].$$

In analogy to § 14.2 (3) we obtain the equations

$$\hat{n}\boldsymbol{s} \wedge \boldsymbol{H} = -\boldsymbol{D} + \frac{4\pi}{i\omega}\boldsymbol{j}, \qquad \hat{n}\boldsymbol{s} \wedge \boldsymbol{E} = \boldsymbol{B}. \tag{2}$$

---

* Cf. M. H. F. WILKINS, A. R. STOKES, W. E. SEEDS, and G. OSTER, *Nature*, **166** (1950), 127–9.

† The determination of the orientation of the particles from such measurements is discussed by P. BOEDER, *Z. f. Phys.*, **75** (1932), 258, and by J. T. EDSALL in *Advances in Colloidal Science*, Vol. I, edited by E. O. KRAEMER (New York, Interscience Publishers Inc., 1942), p. 269.

‡ J. C. MAXWELL, *Proc. Roy. Soc.*, **22** (1873), 46. Also his *Collected Papers*, Vol. 2 (Cambridge University Press, 1890), 379.

Setting $\boldsymbol{B} = \mu \boldsymbol{H}$ and eliminating $\boldsymbol{H}$ between these two equations, we find

$$\mu \left[\boldsymbol{D} + \frac{4\pi i}{\omega} \boldsymbol{j}\right] = \hat{n}^2[\boldsymbol{E} - \boldsymbol{s}(\boldsymbol{s} \cdot \boldsymbol{E})]. \tag{3}$$

If we take the coordinate axes in the direction of the principal dielectric tensor (which by our assumption coincide with those of the conductivity tensor), we have

$$\left.\begin{aligned} D_x &= \varepsilon_x E_x, & D_y &= \varepsilon_y E_y, & D_z &= \varepsilon_z E_z, \\ j_x &= \sigma_x E_x, & j_y &= \sigma_y E_y, & j_z &= \sigma_z E_z. \end{aligned}\right\} \tag{4}$$

Substituting from (4) into (3) and introducing the complex dielectric constants

$$\hat{\varepsilon}_k = \varepsilon_k + \frac{4\pi i}{\omega} \sigma_k, \qquad (k = x, y, z) \tag{5}$$

(3) becomes

$$\mu \hat{\varepsilon}_k E_k = \hat{n}^2[E_k - s_k(\boldsymbol{E} \cdot \boldsymbol{s})]. \tag{6}$$

This relation is formally identical with the relation § 14.2 (18), the real constants ε_k and n being replaced by complex constants $\hat{\varepsilon}_k$ and $\hat{n}_k$. Re-writing (6) in the form

$$E_k = \frac{\hat{n}^2 s_k(\boldsymbol{E} \cdot \boldsymbol{s})}{\hat{n}^2 - \mu \hat{\varepsilon}_k}, \tag{7}$$

we obtain by the same argument as in § 14.2 (or simply by using the formal substitution $\varepsilon_k \rightarrow \hat{\varepsilon}_k$, $n \rightarrow \hat{n}$), *Fresnel's equation*

$$\frac{s_x^2}{\dfrac{1}{\hat{n}^2} - \dfrac{1}{\mu \hat{\varepsilon}_x}} + \frac{s_y^2}{\dfrac{1}{\hat{n}^2} - \dfrac{1}{\mu \hat{\varepsilon}_y}} + \frac{s_z^2}{\dfrac{1}{\hat{n}^2} - \dfrac{1}{\mu \hat{\varepsilon}_z}} = 0. \tag{8}$$

Introducing the complex velocities

$$\hat{v}_p = \frac{c}{\sqrt{\mu \hat{\varepsilon}}} = \frac{c}{\hat{n}}, \qquad \hat{v}_x = \frac{c}{\sqrt{\mu \hat{\varepsilon}_x}} = \frac{c}{\hat{n}_x}, \tag{9}$$

etc., we may write FRESNEL's equation again in the form § 14.2 (24)

$$\frac{s_x^2}{\hat{v}_p^2 - \hat{v}_x^2} + \frac{s_y^2}{\hat{v}_p^2 - \hat{v}_y^2} + \frac{s_z^2}{\hat{v}_p^2 - \hat{v}_z^2} = 0. \tag{10}$$

These relations are strictly analogous to those relating to non-absorbing crystals; their physical interpretation is, however, somewhat different. From (8) or (10) we again obtain a quadratic equation for $\hat{n}^2(\boldsymbol{s})$, i.e. we find two refractive indices and two principal vibrations $\boldsymbol{D}'$ and $\boldsymbol{D}''$ corresponding to each direction of propagation $\boldsymbol{s}$. From (7) we see that the ratios $D_x : D_y : D_z$ are complex, so that the principal vibrations are in general no longer linear, but elliptical. A further difference is that the electric displacement vectors are no longer perpendicular to the wave-normal $\boldsymbol{s}$. For we have from the first equation (2), on scalar multiplication by $\boldsymbol{s}$,

$$\boldsymbol{s} \cdot \boldsymbol{D} = \frac{4\pi}{i\omega} \boldsymbol{s} \cdot \boldsymbol{j} = \frac{4\pi}{i\omega} \left[\frac{\sigma_x}{\varepsilon_x} s_x D_x + \frac{\sigma_y}{\varepsilon_y} s_y D_y + \frac{\sigma_z}{\varepsilon_z} s_z D_z\right], \tag{11}$$

and the right-hand side of (11) will in general not be zero. However, if the ratios $4\pi\sigma_k/\omega\varepsilon_k$ are small compared with unity, the component of $\boldsymbol{D}$ in the direction of $\boldsymbol{s}$ will be small compared with $\boldsymbol{D}$ itself.

Further analysis is considerably simplified if the absorption is assumed to be small, i.e. if metals are excluded and only substances that are to some extent transparent are considered. We shall restrict our discussion to this case. Formally, weak absorption implies that the second power of the attenuation index κ may be neglected in comparison with unity. We may therefore write, on using (9),

$$\left.\begin{aligned} \hat{n} &= n(1 + i\kappa), & \hat{n}^2 &= n^2(1 + 2i\kappa), \\ \hat{v}_p &= \frac{c}{n(1 + i\kappa)} = v_p(1 - i\kappa), & \hat{v}_p{}^2 &= v_p{}^2(1 - 2i\kappa), \end{aligned}\right\} \tag{12}$$

with similar expression with subscripts x, y, and z, e.g.

$$\hat{n}_x = n_x(1 + i\kappa_x), \qquad \hat{v}_x = c/\hat{n}_x = v_x(1 - i\kappa_x) \quad \text{etc.}$$

Also

$$\mu\hat{\varepsilon}_k = \hat{n}_k{}^2 = n_k{}^2\,(1 + 2i\kappa_k) = \mu\varepsilon_k(1 + 2i\kappa_k) \qquad (k = x, y, z) \tag{13}$$

and comparison with (5) gives

$$\kappa_k = \frac{2\pi}{\omega}\frac{\sigma_k}{\varepsilon_k}. \tag{14}$$

Returning to FRESNEL's equation (10) we separate the real and imaginary parts. A typical term of (10) is

$$\frac{s_x{}^2}{\hat{v}_p{}^2 - \hat{v}_x{}^2} = \frac{s_x{}^2}{v_p{}^2 - v_x{}^2 - 2i[\kappa v_p{}^2 - \kappa_x v_x{}^2]} = \frac{s_x{}^2}{v_p{}^2 - v_x{}^2}\left[1 + 2i\,\frac{\kappa v_p{}^2 - \kappa_x v_x{}^2}{v_p{}^2 - v_x{}^2}\right], \tag{15}$$

and it follows that the real part of (10) is FRESNEL's equation in the old form § 14.2 (24). The imaginary part gives the equation

$$\begin{aligned} \kappa v_p{}^2 &\left\{\frac{s_x{}^2}{(v_p{}^2 - v_x{}^2)^2} + \frac{s_y{}^2}{(v_p{}^2 - v_y{}^2)^2} + \frac{s_z{}^2}{(v_p{}^2 - v_z{}^2)^2}\right\} \\ &= \frac{\kappa_x v_x{}^2 s_x{}^2}{(v_p{}^2 - v_x{}^2)^2} + \frac{\kappa_y v_y{}^2 s_y{}^2}{(v_p{}^2 - v_y{}^2)^2} + \frac{\kappa_z v_z{}^2 s_z{}^2}{(v_p{}^2 - v_z{}^2)^2}. \end{aligned} \tag{16}$$

For any given wave-normal direction $\boldsymbol{s}$, FRESNEL's equation in general gives two phase velocities v_p as before. Some of the energy carried by the two waves is now absorbed, and (16) gives approximate values for the two attenuation indices.

We may express the formula for κ in a different form. From (7), (4), (9), and (13),

$$\begin{aligned} D_k &= \frac{c^2}{\mu}\frac{\varepsilon_k}{\hat{\varepsilon}_k}\frac{s_k(\boldsymbol{E}\,.\,\boldsymbol{s})}{\hat{v}_k{}^2 - \hat{v}_p{}^2} \\ &= -\frac{c^2}{\mu}\frac{s_k(\boldsymbol{E}\,.\,\boldsymbol{s})}{v_p{}^2 - v_k{}^2}\left[1 + 2i\,\frac{(\kappa - \kappa_k)v_p{}^2}{v_p{}^2 - v_k{}^2}\right] \qquad (k = x, y, z), \end{aligned} \tag{17}$$

where again only terms up to the first power in the attenuation indices have been retained. Now the imaginary term in (17) involves the *difference* in the κ's and may in many cases be neglected. This means that one *neglects the ellipticity of the vibrations*. In this approximation the directions of the two $\boldsymbol{D}$ vectors belonging to any given wave-normal $\boldsymbol{s}$ are the same as for a non-absorbing crystal which has the same (real) principal dielectric constants. We then have

$$\frac{s_k{}^2}{(v_p{}^2 - v_k{}^2)^2} = \left(\frac{\mu}{c^2}\right)^2\frac{D_k{}^2}{(\boldsymbol{s}\,.\,\boldsymbol{E})^2}, \tag{18}$$

and (16) may be written as

$$\kappa v_p^2 = \frac{\kappa_x v_x^2 D_x^2 + \kappa_y v_y^2 D_y^2 + \kappa_z v_z^2 D_z^2}{D^2}. \tag{19}$$

This formula may evidently break down as v_p approaches one of the principal velocities v_k, since the imaginary part in (17), which we neglected, involves the difference $v_p^2 - v_k^2$ in the denominator (see remarks at the end of § 14.6.3).

Since the two coefficients κ' and κ'' belonging to a given wave-normal direction $\mathbf{s}$ are in general different, the two waves are absorbed with different strengths. These two coefficients may be frequency-dependent and vary in different ways with the frequency, so that if white light be incident upon the crystal, the crystal will in general appear coloured, and the colour will depend on the direction of vibration of the incident light. This phenomenon is known as *pleochroism*; in the case of a uniaxial crystal one also speaks of *dichroism*; in the case of a biaxial crystal one speaks of *trichroism*.*

For *a uniaxial crystal* [$\hat{\varepsilon}_x = \hat{\varepsilon}_y = \hat{\varepsilon}_o$, $\hat{\varepsilon}_z = \hat{\varepsilon}_e$; also $\kappa_x = \kappa_y = \kappa_o$, $\kappa_z = \kappa_e$] the relations take a simpler form. As in the case of a non-absorbing uniaxial crystal, FRESNEL's equation breaks into two factors (cf. § 14.3.2), giving

$$\left.\begin{aligned} \hat{v}_p'^2 &= \hat{v}_o^2 \\ \hat{v}_p''^2 &= \hat{v}_o^2 \cos^2\vartheta + \hat{v}_e^2 \sin^2\vartheta, \end{aligned}\right\} \tag{20}$$

where, as before, ϑ is the angle which the wave normal $\mathbf{s}$ makes with the optic axis. The first equation gives, on separating real and imaginary parts,

$$v_p' = v_o, \qquad \kappa_p' = \kappa_o, \tag{21a}$$

and the second gives

$$v_p''^2 = v_o^2 \cos^2\vartheta + v_e^2 \sin^2\vartheta, \qquad \kappa_p'' v_p''^2 = \kappa_o v_o^2 \cos^2\vartheta + \kappa_e v_e^2 \sin^2\vartheta, \tag{21b}$$

where again we have neglected the terms involving the second power of the attenuation indices. We see that the absorption of the ordinary wave is the same for all directions of propagation.

For a *biaxial crystal* the relations are much more complicated and we shall confine our attention to special cases of interest. As in § 14.3.3 we consider first those directions of propagation for which $s_x = 0$. FRESNEL's equation (10) then gives, in analogy with § 14.3 (6), the equations

$$\left.\begin{aligned} \hat{v}_p'^2 &= \hat{v}_x^2 \\ \hat{v}_p''^2 &= \hat{v}_z^2 s_y^2 + \hat{v}_y^2 s_z^2. \end{aligned}\right\} \tag{22}$$

Separating real and imaginary parts, we obtain the two velocities and the two attenuation indices corresponding to the wave-normal direction $\mathbf{s}(0, s_y, s_z)$:

$$v_p'^2 = v_x^2, \qquad \kappa_p' = \kappa_x, \tag{23a}$$

and

$$v_p''^2 = v_z^2 s_y^2 + v_y^2 s_z^2, \qquad \kappa_p'' v_p''^2 = \kappa_z v_z^2 s_y^2 + \kappa_y v_y^2 s_z^2. \tag{23b}$$

Similar relations hold, of course, for directions of propagation at right angles to the y and z directions.

* This terminology derives from the fact that for a uniaxial crystal there are two characteristic colours and for a biaxial crystal there are three. Some authors, however, mean by dichroic material, any material whose absorption coefficient depends on the state of polarization of the incident light.

In general there are no real values s_y and s_z for which the two roots $\hat{v}'_p$ and $\hat{v}''_p$ are equal. Directions may be found for which the real phase velocities v'_p and v''_p are equal, but the corresponding attenuation indices (κ' and κ'') will in general be different.

Next we consider propagation in directions that are not very different from that of an optic axis of wave normals. In order to apply (19) we must determine the directions of $\boldsymbol{D}'$ and $\boldsymbol{D}''$. This may be done by using the result established in § 14.2.3, according to which the two planes of vibrations ($\boldsymbol{D}'$, $\boldsymbol{s}$) and ($\boldsymbol{D}''$, $\boldsymbol{s}$) bisect the angles between the planes ($\boldsymbol{N}_1$, $\boldsymbol{s}$) and ($\boldsymbol{N}_2$, $\boldsymbol{s}$), $\boldsymbol{N}_1$ and $\boldsymbol{N}_2$ being the axes of wave normals. Let ψ be the angle between the plane ($\boldsymbol{N}_1$, $\boldsymbol{s}$) and the xz-plane (which contains both the optic axes). Since the plane ($\boldsymbol{N}_2$, $\boldsymbol{s}$) is nearly parallel to the xz-plane, it follows from the theorem just

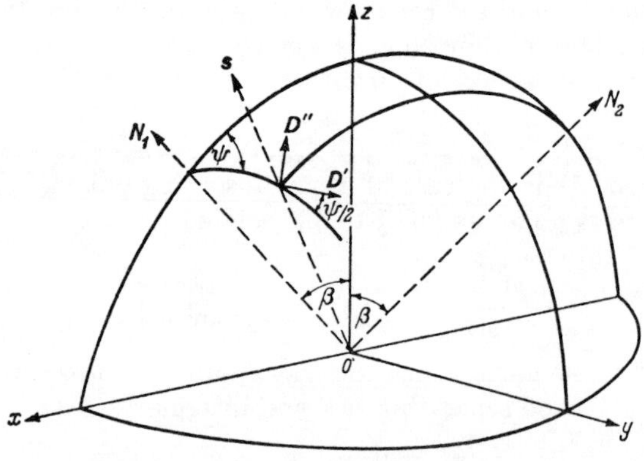

Fig. 14.29. Illustrating the theory of absorbing crystals.

quoted that the angle between $\boldsymbol{D}'$ and the xz-plane is nearly $\psi/2$ (see Fig. 14.29). The component of $\boldsymbol{D}'$ in the xz-plane is therefore $D' \cos(\psi/2)$. To obtain the x component we must project this vector on to the x axis. Since, approximately, $\boldsymbol{s}$ and $\boldsymbol{N}_1$ coincide, the projection angle is nearly equal to the angle between the optic axis $\boldsymbol{N}_1$ and the z axis and consequently (see Fig. 14.30) $D_x' = D' \cos\beta \cos(\psi/2)$. In a similar way we obtain the other components. Hence

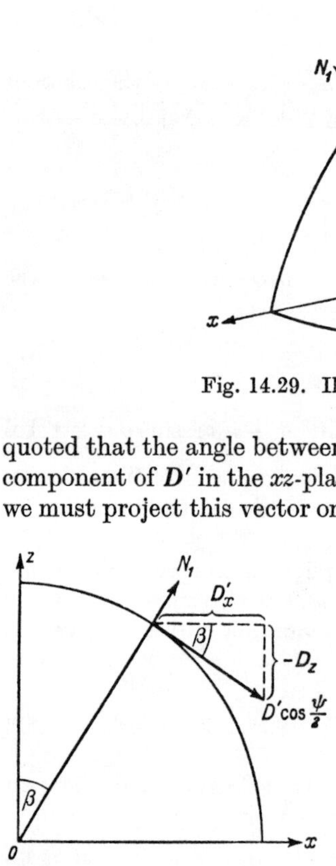

Fig. 14.30. Illustrating the theory of absorbing crystals.

$$\left.\begin{aligned} D'_x &= D' \cos\frac{\psi}{2} \cos\beta, \\ D'_y &= D' \sin\frac{\psi}{2}, \\ D'_z &= -\,D' \cos\frac{\psi}{2} \sin\beta. \end{aligned}\right\} \qquad (24a)$$

The vector $\boldsymbol{D}''$ is orthogonal to $\boldsymbol{s}$ and $\boldsymbol{D}'$; its components may be immediately obtained by replacing $\psi/2$ by $\psi/2 + \pi/2$ in (24a), giving

$$D''_x = -\,D'' \sin\frac{\psi}{2}\cos\beta, \qquad D''_y = D'' \cos\frac{\psi}{2}, \qquad D''_z = D'' \sin\frac{\psi}{2} \sin\beta. \qquad (24b)$$

We substitute from (24a) and (24b) into (19), and also use the approximation $v'_p = v''_p = v_y$ ($v_x > v_y > v_z$), which is justified since we are restricting ourselves to directions not too far from that of the optic axis. We then obtain the required attenuation indices κ' and κ'':

$$\left.\begin{aligned}\kappa' v_y{}^2 &= (\kappa_x v_x{}^2 \cos^2\beta + \kappa_z v_z{}^2 \sin^2\beta)\cos^2\frac{\psi}{2} + \kappa_y v_y{}^2 \sin^2\frac{\psi}{2},\\ \kappa'' v_y{}^2 &= (\kappa_x v_x{}^2 \cos^2\beta + \kappa_z v_z{}^2 \sin^2\beta)\sin^2\frac{\psi}{2} + \kappa_y v_y{}^2 \cos^2\frac{\psi}{2}.\end{aligned}\right\} \tag{25}$$

In view of the approximation involved in (19) these formulae cannot be expected to remain valid when $\boldsymbol{s}$ is in the immediate neighbourhood of the optic axis. In the limiting case when the wave normal is along the optic axis, the angle ψ is undetermined; to obtain the corresponding attenuation indices we return to FRESNEL's equation. For any $\boldsymbol{s}$ direction in the xz-plane ($s_y = 0$) we obtain, as in equations (23), by equating real and imaginary parts

$$\left.\begin{aligned}v_p'^2 &= v_y{}^2, & \kappa_p' &= \kappa_y,\\ v_p''^2 &= v_x{}^2 s_z{}^2 + v_z{}^2 s_x{}^2, & \kappa_p'' v_p''^2 &= \kappa_x v_x{}^2 s_z{}^2 + \kappa_z v_z{}^2 s_x{}^2.\end{aligned}\right\} \tag{26}$$

In particular, for the optic axis $s_x = \sin\beta$, $s_z = \cos\beta$, where β is the angle given by § 14.3 (11), between either of the optic axes and the z axis. Also $v_p' = v_p'' = v_y$ and the equations on the right of (26) become, if we also write $\kappa_\perp$ in place of κ' and $\kappa_\parallel$ in place of κ'':

$$\left.\begin{aligned}\kappa_\perp &= \kappa_y\\ \kappa_\parallel v_y{}^2 &= \kappa_x v_x{}^2 \cos^2\beta + \kappa_z v_z{}^2 \sin^2\beta.\end{aligned}\right\} \tag{27}$$

$\kappa_\perp$ is the attenuation index for a $\boldsymbol{D}$ wave polarized at right angles to the plane of the optic axes, and $\kappa_\parallel$ is the index for a $\boldsymbol{D}$ wave polarized in the plane of the axes. Thus the absorption of a wave propagated in the direction of the optic axis depends on its direction of vibration.

It is convenient to express κ' and κ'' in terms of the indices $\kappa_\parallel$ and $\kappa_\perp$ and the azimuthal polarization angle ψ, by substituting from (27) into (25). This gives

$$\left.\begin{aligned}\kappa' &= \kappa_\parallel \cos^2\frac{\psi}{2} + \kappa_\perp \sin^2\frac{\psi}{2} = \frac{\kappa_\parallel + \kappa_\perp}{2} + \frac{\kappa_\parallel - \kappa_\perp}{2}\cos\psi,\\ \kappa'' &= \kappa_\parallel \sin^2\frac{\psi}{2} + \kappa_\perp \cos^2\frac{\psi}{2} = \frac{\kappa_\parallel + \kappa_\perp}{2} - \frac{\kappa_\parallel - \kappa_\perp}{2}\cos\psi.\end{aligned}\right\} \tag{28}$$

14.6.2 Interference figures from absorbing crystal plates

We shall now briefly consider interference effects with absorbing crystal plates which are cut at right angles to an optic axis of wave normals. The theory is not very different from that relating to non-absorbing crystal plates; the only important distinction is that the two interfering rays are absorbed with different strengths, and as a result the visibility of the fringes is decreased. Other conclusions, and in particular the expression for the phase difference, remain unchanged in our approximation, since the geometrical laws of propagation are the same as before.

On travelling a distance l in an absorbing medium, the amplitude of a plane wave is, according to § 13.1 (19), reduced by a factor $\exp\{-\omega n\kappa l/c\}$. Hence with the same arrangement and notation as in § 14.4.3 (cf. Fig. 14.20) the amplitudes of the principal vibrations on emergence from the plate are given by [cf. § 14.4 (13)]

$$OB = Ee^{-\frac{\omega\kappa'}{v'}l}\cos\phi, \qquad OC = Ee^{-\frac{\omega\kappa''}{v''}l}\sin\phi. \tag{29}$$

Here $l = h/\cos\theta_2$, h being the thickness of the plate and θ_2 the angle which the wave normal in the plate makes with the axis, it being assumed that the two paths in the plate are the same for both waves; this is approximately so, if we restrict ourselves to wave-normal directions close to the optic axis. In the same approximation we may take $v' = v'' = v_y$ in (29). It is convenient to set

$$u = \frac{\omega l}{v_y} \sim \frac{\omega l}{v'} \sim \frac{\omega l}{v''}. \tag{30}$$

With this substitution,

$$OB = Ee^{-\kappa' u}\cos\phi, \quad OC = Ee^{-\kappa'' u}\sin\phi. \tag{31}$$

In place of § 14.4 (14) we now obtain for the amplitudes of the waves, after passing through the polarizer and the analyzer, the expressions (cf. Fig. 14.20)

$$OF = Ee^{-\kappa' u}\cos\phi\cos(\phi-\chi), \qquad OG = Ee^{-\kappa'' u}\sin\phi\sin(\phi-\chi). \tag{32}$$

The total intensity of the light brought to interference is

$$I = I_1 + I_2 + 2\sqrt{I_1 I_2}\cos\delta, \tag{33}$$

where, apart from unessential proportional factors, $I_1 = OF^2$, $I_2 = OG^2$ and the phase difference δ is calculated as before.

We now examine some special cases of interest.

(a) *Uniaxial crystals*. In this case we have

$$\left.\begin{aligned} v_x = v_y = v_o, \qquad & v_z = v_e, \\ \kappa_x = \kappa_y = \kappa_o, \qquad & \kappa_z = \kappa_e. \end{aligned}\right\} \tag{34}$$

The direction of vibration of the extraordinary ray is in the principal plane, i.e. in the plane containing the wave normal and the optic axis. We may, therefore, identify the angle between the $\boldsymbol{D}$ vector of the extraordinary wave and the direction OP of the polarizer with the angle ϕ in equation (32); to retain agreement with (21a) and (21b) we must, however, interchange κ' and κ''. With the Nicol prisms *crossed* ($\chi = \pi/2$) we have from (32)

$$OF = Ee^{-\kappa' u}\cos\phi\sin\phi, \qquad OG = -Ee^{-\kappa'' u}\sin\phi\cos\phi, \tag{35}$$

and (33) gives

$$I = \frac{E^2}{4}\sin^2 2\phi\,\{e^{-2\kappa' u} + e^{-2\kappa'' u} - 2\,e^{-(\kappa'+\kappa'')u}\cos\delta\}. \tag{36a}$$

For the optic axis, $\kappa' = \kappa''$ and $\delta = 0$, and (36a) reduces to

$$I_0 = 0; \tag{36b}$$

hence at the centre of the pattern (the point corresponding to the optic axis) there is darkness. The field of view is crossed by a dark isogyre, given by $\sin 2\phi = 0$, i.e. by $\phi = 0$ and $\phi = \pi/2$. Thus the isogyre has the form of a cross with its arms parallel to the directions of the polarizer and the analyzer (cf. § 14.4.4). The minima and maxima of intensity are

$$\left.\begin{aligned} I_{\min} &= \frac{E^2}{4}\sin^2 2\phi\{e^{-2\kappa' u} + e^{-2\kappa'' u} - 2e^{-(\kappa'+\kappa'')u}\}, \\ I_{\max} &= \frac{E^2}{4}\sin^2 2\phi\{e^{-2\kappa' u} + e^{-2\kappa'' u} + 2e^{-(\kappa'+\kappa'')u}\}, \end{aligned}\right\} \tag{37}$$

so that the visibility $\mathscr{V}$ of the fringes is

$$\mathscr{V} = \frac{I_{\max} - I_{\min}}{I_{\max} + I_{\min}} = \frac{2e^{-(\kappa'+\kappa'')u}}{e^{-2\kappa' u} + e^{-2\kappa'' u}} = \frac{1}{\cosh\{(\kappa' - \kappa'')u\}}. \tag{38}$$

Thus the visibility is the greater the smaller is the difference between κ' and κ''. Since for wave-normal directions near the optic axis (ϑ small) κ' and κ'' are nearly equal, the fringes should be clearly visible in this region, provided that sufficient light is transmitted by the plate. If κ_o is small compared with κ_e (e.g. for magnesium platinum cyanide) there will be relatively little absorption near the optic axis and the central fringes will be bright. If κ_o is large compared to κ_e (e.g. for tourmaline), the absorption will be relatively weakest near the optic axis and the central fringes will be dark. In both cases the visibility decreases from the centre towards the edge of the field. If κ_o and κ_e are nearly equal, the interference figure will be similar to that of a non-absorbing crystal, apart from a decrease in visibility with increasing distance from the centre.

(b) *Biaxial crystals.* We again restrict ourselves to the case of crossed Nicol prisms ($\chi = \pi/2$) and assume that the faces of the plate are perpendicular to one of the optic axes (N_1). Only waves whose directions of propagation are close to this axis will be considered.

Let N_1, N_2, and Q be the points of intersection of the two optic axes and of the wave normal with the exit face of the plate. Further let ψ be the angle which the line N_1Q makes with N_1N_2 and α the angle between the plane of the optic axes and the plane of vibrations transmitted by the polarizer P.

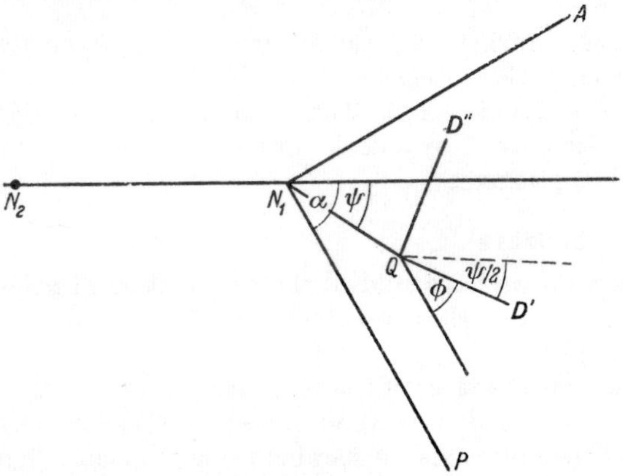

Fig. 14.31. Illustrating the theory of interference on absorbing biaxial crystal plates.

If $N_1Q \ll N_1N_2$ the angle which the direction of vibration $\boldsymbol{D}'$ makes with N_1N_2 is approximately $\psi/2$ so that the angle ϕ between $\boldsymbol{D}'$ and the direction OP is approximately (see Fig. 14.31)

$$\phi = \alpha - \frac{\psi}{2}. \tag{39}$$

The amplitudes (32) of the waves brought to interference, therefore, are

$$OF = E\cos\left(\alpha - \frac{\psi}{2}\right)\sin\left(\alpha - \frac{\psi}{2}\right)e^{-\kappa' u}, \quad OG = -E\sin\left(\alpha - \frac{\psi}{2}\right)\cos\left(\alpha - \frac{\psi}{2}\right)e^{-\kappa'' u}. \tag{40}$$

Hence the intensity is

$$I = \frac{E^2}{4} \sin^2 (2\alpha - \psi)\{e^{-2\kappa' u} + e^{-2\kappa'' u} - 2e^{-(\kappa'+\kappa'')u} \cos \delta\}. \tag{41a}$$

For a wave propagated in the direction of the optic axis, ψ is undetermined. In this special case we resolve the vibrations into components parallel and perpendicular to the plane containing the optic axes, i.e. we set $\psi = 0$ and write $\kappa_{\parallel}$ and $\kappa_{\perp}$ in place of κ' and κ'' in (41a) (cf. (28)). Also $\delta = 0$, so that we obtain, in place of (41a),

$$I_0 = \frac{E^2}{4} \sin^2 2\alpha\{e^{-\kappa_{\parallel} u} - e^{-\kappa_{\perp} u}\}^2. \tag{41b}$$

Equation (41a) shows that the intensity is zero along a line on which $\sin (2\alpha - \psi) = 0$, this being a principal isogyre. Whilst in the case of a non-absorbing crystal the isogyre passed through the point corresponding to the optic axis, now this will no longer be the case, unless, as seen from (41b), $\alpha = 0$ or $\alpha = \pi/2$, i.e. unless the plane of the polarizer is parallel or at right angles to the plane containing the two optic axes.

The dark isochromates, given by $\cos \delta = 1$, i.e. by $\delta = 0, \pm 2\pi, \pm 4\pi, \ldots$ are rings surrounding the point which corresponds to the optic axis. The visibility of the fringes, again given by (38), is appreciable only when κ' and κ'' are nearly equal. According to (28) this is so when ψ is nearly $\pi/2$ or $-\pi/2$.

Let us consider how the intensity changes with ψ when δ is kept constant. If we substitute from (28), equation (41a) may be written in the form

$$I = \frac{E^2}{2} e^{-(\kappa_{\parallel}+\kappa_{\perp})u} \sin^2 (2\alpha - \psi)\{\cosh [(\kappa_{\parallel} - \kappa_{\perp})\, u \cos \psi] - \cos \delta\}. \tag{42}$$

The term in the brace brackets is greatest when $|\cos \psi|$ is greatest and least when $|\cos \psi|$ is least. Hence the intensity is maximum when $\psi = 0$ and $\psi = \pi$, and minimum when $\psi = \pi/2$ and $-\pi/2$. Thus in addition to the dark isogyre $\psi = 2\alpha$, the field of view is also crossed by a dark "brush" with its arms perpendicular to the plane containing the optic axes.*

14.6.3 Dichroic polarizers

In the preceding section we have studied the propagation of polarized light through an absorbing crystal plate. Here we shall consider the effect of such a plate on natural (unpolarized) light.

We again assume that the faces of the plate are perpendicular to the optic axis (or to one such axis if it is biaxial) of wave normals, and consider propagation in a direction close to that of the optic axis. A parallel beam of natural light may be regarded as consisting of two mutually incoherent beams of equal amplitudes, polarized in any two mutually orthogonal directions perpendicular to the direction of propagation (cf. § 10.8 (56)). Let us choose as the directions of vibration of the two partial beams the directions of the principal vibrations in the crystal. If E is the amplitude of either of the partial beams on entering the plate, their amplitudes after travelling a distance l in the plates are

$$E' = Ee^{-\kappa' u}, \qquad E'' = Ee^{-\kappa'' u}, \tag{43}$$

* From equation (44) which, by (28), may be written as

$$I = 2E^2 e^{-(\kappa_{\parallel}+\kappa_{\perp})u} \cosh \{(\kappa_{\parallel} - \kappa_{\perp})\, u \cos \psi)\},$$

it is seen that the dark brush also appears in the absence of a polarizer and analyzer, when natural light is passed through the plate.

where as before $u = \omega l/v_y$, it being assumed that the light is quasi-monochromatic and of mean frequency ω. The total intensity is then given by

$$I = I' + I'', \tag{44}$$

where

$$I' = I_0 e^{-2\kappa' u}, \qquad I'' = I_0 e^{-2\kappa'' u}, \tag{45}$$

and $I_0 = E^2$. For the direction of the optic axes, κ' and κ'' must be replaced by $\kappa_\parallel$ and $\kappa_\perp$.

We see that after the light has travelled a distance l in the medium, the amplitudes of the two components are in the ratio $\exp\{-(\kappa' - \kappa'')u\}$, so that the light has become

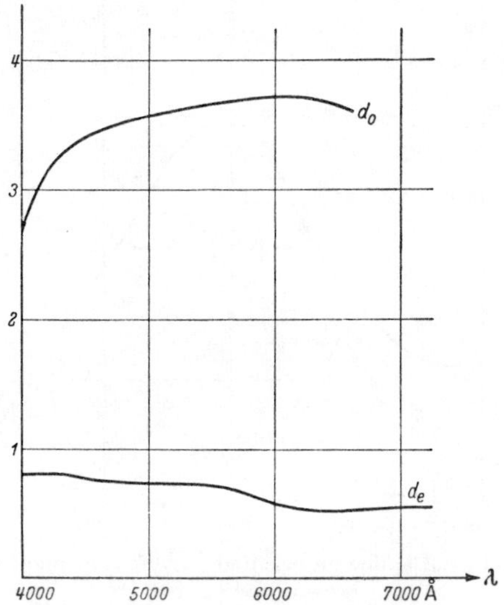

Fig. 14.32. Dichroism of black tourmaline plate about 0·2 mm thick, cut 24° from parallelism with the optic axis, for light incident at right angles to the optic axis.

$$d = 4\pi(\log_{10} e)(n\kappa l/\lambda) \sim 5{\cdot}5 n\kappa l/\lambda.$$

[After E. H. LAND and C. D. WEST, contribution in *Colloid Chemistry*, Vol. 6, ed. J. Alexander (New York, Reinhold Publishing Corporation, 1946), 167.]

partially polarized. If the two attenuation indices are very different, a relatively thin sheet of the material will be sufficient to transform the incident unpolarized beam into one that is nearly linearly polarized; i.e. the plate acts as a *polarizer*. An example of a natural crystal which will act as polarizer of this type is tourmaline, which suppresses the ordinary ray much more strongly than the extraordinary ray. However, for most wavelengths it also absorbs a considerable part of the extraordinary ray (see Fig. 14.32), so that it is not suitable for practical applications. It is possible, chiefly as the result of researches of LAND (c. 1932) and his co-workers, to produce synthetic dichroic materials which act as excellent polarizers. These materials, known commercially as *polaroids*, are not single crystals, but are sheets of organic polymers with long-chain molecules, brought into almost complete alignment by stretching or by some other treatment,* and sometimes dyed. A polymer which is particularly

* For a fuller discussion of dichroic polarizers see the article referred to in the captions to Figs. 14.32 and 14.33 or E. LAND, *J. Opt. Soc. Amer.*, **41** (1951), 957.

suitable for this purpose is the synthetic product polyvinyl alcohol $(—CH_2—CHOH—)_x$. In Fig. 14.33, the dichroism of iodine on oriented polyvinyl alcohol is shown as function of the wavelength.

The ratio of the absorption coefficients of a good dichroic polarizer may be as high as 100:1. It may transmit about 80 per cent of light polarized in one direction and less than 1 per cent of light polarized in the direction at right angles to it. Large sheets of such polarizing materials are obtainable; machines giving pieces thirty inches wide and indefinitely long are in use in commercial production. In this respect

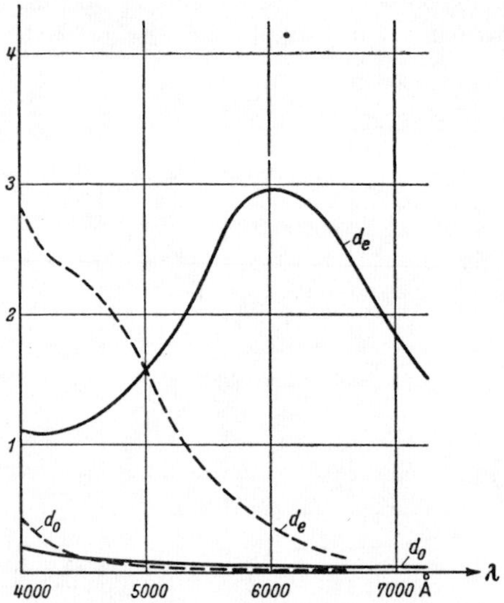

Fig. 14.33. Dichroism of iodine on oriented polyvinyl alcohol, blue (full line) and brown (broken line).

$$d = 4\pi(\log_{10} e)(n\kappa l/\lambda) \sim 5{\cdot}5 n\kappa l/\lambda.$$

[After E. H. LAND and C. D. WEST, contribution in *Colloid Chemistry*, Vol. 6, ed. J. Alexander (New York, Reinhold Publishing Corporation, 1946), 177.]

these "polaroid sheets" are superior to the NICOL prisms the dimensions of which are limited by the scarcity of large pieces of calcite of good optical quality.

Finally, let us recall that in the main part of this chapter we have assumed that the ellipticity of the principal light vibrations in the crystal may be neglected. This assumption is certainly not always justified; it may be shown, for example, that in every absorbing crystal there exist four directions, lying in pairs near the optic axes, for which the polarization is circular. However, with weaker and weaker absorption the regions of appreciable ellipticity are more and more restricted to the neighbourhood of these special directions, which themselves tend to coincidence with the optic axis. Taking this into account, the general character of the phenomenon is retained and we, therefore, do not need to enter into a more detailed discussion. Naturally our analysis does not reveal some of the finer aspects of the optical theory of absorbing crystals; for their discussion we must refer elsewhere.*

* See, for example, G. SZIVESSY, *Handbuch der Physik*, Vol. 20 (Berlin, Springer, 1928), 861–904

APPENDICES

APPENDIX I

THE CALCULUS OF VARIATIONS

It is a general feature of equations of classical physics that they can be derived from variational principles. Two early examples are Fermat's principle in optics (1657) and Maupertuis' principle in mechanics (1744). The equations of elasticity, hydrodynamics, and electrodynamics can also be represented in this way.

However, when one deals with field equations, involving as a rule four or more independent variables $x, y, z, t \dots$, one makes little use, owing to the great complexity of partial differential equations, of the property that the solution expresses stationary values of certain integrals. The only essential advantage of the variational approach in such cases is connected with the derivation of conservation laws—e.g. for energy. The situation is quite different in problems involving one independent variable (time in mechanics, or length of a ray in geometrical optics). Then one deals with a set of ordinary differential equations and it turns out that a study of the behaviour of the solution is greatly facilitated by a variational approach. This approach is in fact a straightforward generalization of ordinary geometrical optics in every detail. Its modern representation owes much to David Hilbert, on whose unpublished lectures, given at Göttingen in about 1903, we base the considerations of the following sections. The theory is presented here for a three-dimensional space (x, y, z) only, but can easily be extended to more dimensions.

1. Euler's equations as necessary conditions for an extremum

Let $F(u, v, x, y, z)$ be a given function with continuous partial derivatives up to the second order in all the five variables. Further let C be any curve $x = x(z)$, $y = y(z)$ in the x,y,z space. The derivatives of x and y will also be assumed to be continuous up to second order. If we set

$$u = x', \qquad v = y',$$

(the prime denoting differentiation with respect to z) the integral

$$I = \int_{z_1}^{z_2} F(x', y', x, y, z)dz \tag{1}$$

is a function of the curve C, i.e. of the two functions $x(z)$, $y(z)$, in other words a *functional*. The fundamental problem of the calculus of variation is:

To determine a curve C between two given points $P_1[x_1 = x(z_1), y_1 = y(z_1), z_1]$ and $P_2[x_2 = x(z_2), y_2 = y(z_2), z_2]$ for which the integral is an extremum (minimum or maximum).

The necessary conditions which such a curve C, called an *extremal*, must satisfy may be determined by the simple process of linear variation. For this purpose we choose a function $\xi(z)$ with a continuous first order derivative, which vanishes at the end points,

$$\xi(z_1) = \xi(z_2) = 0, \tag{2}$$

and form the "varied" curve C' by replacing the x-coordinate of the extremal by $x + \varepsilon\xi$, where ε is a small parameter. Equation (1) then becomes a function of ε,

$$I(\varepsilon) = \int_{z_1}^{z_2} F(x' + \varepsilon\xi', y', x + \varepsilon\xi, y, z)dz. \tag{3}$$

The quantity

$$(\delta I)_x = \left(\frac{\partial I}{\partial \varepsilon}\right)_{\varepsilon=0} = \int_{z_1}^{z_2} \left(\frac{\partial F}{\partial x'}\,\xi' + \frac{\partial F}{\partial x}\,\xi\right) dz \tag{4}$$

is called *the first variation* with respect to x. The vanishing of the first variation

$$(\delta I)_x = 0 \tag{5}$$

is clearly a necessary condition for an extremum.

Now if we integrate the first term in (4) by parts and use the boundary conditions (2), we obtain

$$(\delta I)_x = \int_{z_1}^{z_2} \left(F_x - \frac{d}{dz}\,F_{x'}\right) \xi\, dz, \tag{6a}$$

where F_x stands for $\partial F/\partial x$, etc. Similarly, replacing y by $y + \varepsilon\eta$, we find that

$$(\delta I)_y = \int_{z_1}^{z_2} \left(F_y - \frac{d}{dz}\,F_{y'}\right) \eta\, dz. \tag{6b}$$

Since ξ and η may be chosen arbitrarily in the interval $z_1 \leqslant z \leqslant z_2$, it follows from (6a) and (6b), that the conditions $(\delta I)_x = 0$ and $(\delta I)_y = 0$ may be expressed in the form of two differential equations, known as *Euler's equations*,

$$F_x - \frac{d}{dz}\,F_{x'} = 0, \tag{7a}$$

$$F_y - \frac{d}{dz}\,F_{y'} = 0. \tag{7b}$$

(7a) and (7b) are two differential equations of the second order for $x(z)$ and $y(z)$. The leading terms, i.e. those with derivatives of the highest order, will be written out fully:

$$F_{x'x'}x'' + F_{x'y'}y'' + \ldots = 0, \tag{8a}$$

$$F_{y'x'}x'' + F_{y'y'}y'' + \ldots = 0. \tag{8b}$$

These equations can be solved for x'' and y'' provided the associated determinant does not vanish, i.e. provided that

$$F_{uu}F_{vv} - F_{uv}^2 \neq 0. \tag{9}$$

We shall assume that this condition holds throughout the five-dimensional region with which we are concerned.

The solutions of two differential equations of the second order contain four arbitrary constants of integration so that the extremals form a four-parameter family of curves (∞^4 extremals).

2. Hilbert's independence integral and the Hamilton–Jacobi equation

In order to discuss the properties of these extremals it is convenient to consider another related problem. The variables u and v will be regarded as functions in the x, y, z space

$$u = u(x, y, z), \qquad v = v(x, y, z). \tag{10}$$

Then $F[u(x, y, z), v(x, y, z), x, y, z]$ as well as its partial derivatives F_u, F_v are functions of x, y, z. Let us now choose a curve $C[x = x(z),\ y = y(z)]$, and form the integral

$$S = \int_{z_1}^{z_2} \{F + (x' - u)F_u + (y' - v)F_v\}\, dz. \tag{11}$$

The problem is this: *to find such functions u, v which secure that S is independent of the choice of the curve C* : S is then a function of the end points P_1 and P_2 only, where P_1 has the coordinates $x_1 = x(z_1)$, $y_1 = y(z_1)$, z_1, and P_2 the coordinates $x_2 = x(z_2)$, $y = y(z_2)$, z_2. S will be called *Hilbert's independence integral.*

To determine u and v we first re-write (11) in the form

$$S = \int_{P_1}^{P_2} (U dx + V dy + W dz), \tag{12}$$

where

$$U = F_u, \qquad V = F_v, \qquad W = F - uF_u - vF_v. \tag{13}$$

Now it is well known that the necessary and sufficient conditions for (12) to be independent of the curve are the vanishing of the components of the curl of the vector A whose components are U, V, W, i.e.

$$\frac{\partial W}{\partial y} - \frac{\partial V}{\partial z} = 0, \qquad \frac{\partial U}{\partial z} - \frac{\partial W}{\partial x} = 0, \qquad \frac{\partial V}{\partial x} - \frac{\partial U}{\partial y} = 0. \tag{14}$$

This is a set of three partial differential equations for $u(x, y, z)$ and $v(x, y, z)$; the equations are however not quite independent as the identity

$$\frac{\partial}{\partial x}\left(\frac{\partial W}{\partial y} - \frac{\partial V}{\partial z}\right) + \frac{\partial}{\partial y}\left(\frac{\partial U}{\partial z} - \frac{\partial W}{\partial x}\right) + \frac{\partial}{\partial z}\left(\frac{\partial V}{\partial x} - \frac{\partial U}{\partial y}\right) = 0 \tag{15}$$

holds for any U, V, and W (div curl $A = 0$ for any vector A). If the equations (14) are satisfied, then $U dx + V dy + W dz$ is a total differential,

$$dS = U dx + V dy + W dz, \tag{16}$$

and S is then a function of $P_1(x_1, y_1, z_1)$ and $P_2(x_2, y_2, z_2)$ alone. Writing for short x, y, z in place of x_2, y_2, z_2 we have

$$U = \frac{\partial S}{\partial x}, \qquad V = \frac{\partial S}{\partial y}, \qquad W = \frac{\partial S}{\partial z}. \tag{17}$$

We now take an arbitrary surface $T(x, y, z) = 0$ and at each point P_1 of the surface we choose the vector (U, V, W) normal to the surface. Then by (16) $dS = 0$, and therefore

$$S(x, y, z) = S_1 \tag{18}$$

is constant on the surface. From two of the (compatible) equations (13) we can solve for u and v as functions on the surface,

$$u = u(U, V, x, y, z), \qquad v = v(U, V, x, y, z). \tag{19}$$

If we next solve the differential equations (14) with these boundary values we obtain a special solution of our problem, namely a solution which has a constant value S_1 of S on the selected surface $T(x, y, z) = 0$. This solution can be determined from a single partial differential equation for the function S. For on substituting from (19) into the remaining equation of (13) we obtain

$$W = W(U, V, x, y, z), \tag{20}$$

which, on using (17), becomes

$$\frac{\partial S}{\partial z} = W\left(\frac{\partial S}{\partial x}, \frac{\partial S}{\partial y}, x, y, z\right); \tag{21}$$

this is known as *the Hamilton–Jacobi equation* of the problem.

A function $S(x, y, z)$ which has a given constant value on the surface $T(x, y, z) = 0$ and satisfies (21) represents a solution of the problem. The two functions u, v which make the integral independent of the path can be found by solving any two of the (compatible) equations

$$F_u = \frac{\partial S}{\partial x}, \qquad F_v = \frac{\partial S}{\partial y}, \qquad F - uF_u - vF_v = \frac{\partial S}{\partial z}, \tag{22}$$

obtained by combining (13) and (17).

3. The field of extremals

We now establish *the connection between the two problems* treated in Sections 1 and 2. It is this:

If $u(x, y, z)$ and $v(x, y, z)$ are two functions which make the Hilbert integral S defined by (11) *independent of the path, then the differential equations*

$$x' = u(x, y, z), \qquad y' = v(x, y, z), \tag{23}$$

have as solutions a two-parameter set (∞^2) *of extremals, namely those which satisfy the condition that they are "transversal" to the surfaces $S(x, y, z) = S_1$.* By "transversality" we mean here that the relation

$$Udx + Vdy + Wdz = 0, \tag{24}$$

is satisfied: it states that the vector (U, V, W), defined by (13) in terms of u and v, is orthogonal to any element dx, dy, dz of the surface.

Consider a region in the x,y,z space and associate with each point of this region a vector (u,v), which is continuous and has continuous first order partial derivatives. The set of these vectors defined over the given region is called a *field*. In the present context one speaks of a *field of extremals* and $u(x, y, z)$, $v(x, y, z)$ are then called *slope functions* of the field.

The following converse of the theorem just enunciated also holds: *If a field of ∞^2 extremals is constructed in such a way that it is transversal to a given surface $T(x, y, z) = 0$, and if u and v are its slope functions, defined by* (23), *then the Hilbert integral S, given by* (11), *is independent of the path.*

Before proving these theorems we note the following corollary:

Let the curve C in (11) be an extremal of the field; the HILBERT integral (11) then reduces to the variational integral $I = \int_{z_1}^{z_2} Fdz$. Hence the value of this integral taken between each pair of "corresponding" points on the surfaces $S(x, y, z) = S_1$ and $S(x, y, z) = S_2$ (i.e. between points on the same extremal transversal to S_1 and S_2) is the same for all such pairs of points (see Fig. 1). The surfaces $S(x, y, z) =$ constant and the ∞^2 transversals may be regarded as generalizations of wave-fronts and rays of geometrical optics.

To prove the first theorem we consider a fixed curve C which satisfies (23) and is transversal to the surface $S(x, y, z) = S_1$, and apply a linear variation to it, i.e. we replace x by $x + a\xi$ and y by $y + b\eta$ where a and b are small parameters and ξ and η are arbitrary but fixed differentiable functions of z, which vanish for z_1 and z_2. Now since, by hypothesis, S is independent of the path,

$$\left(\frac{\partial S}{\partial a}\right)_0 = 0, \qquad \left(\frac{\partial S}{\partial b}\right)_0 = 0, \tag{25}$$

where the suffix 0 indicates that we substitute $a = b = 0$ after differentiation.

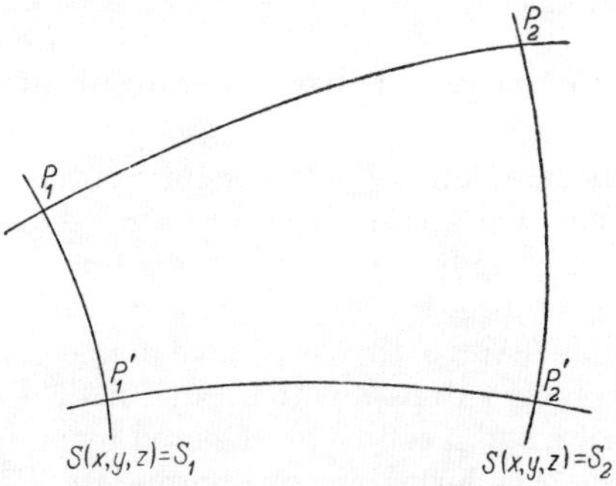

Fig. 1. Illustrating a generalization of the concept of wave-fronts and rays of geometrical optics. The variational integral (1) has a constant value for all extremals such as P_1P_2, $P_1'P_2'$, . . which are transversal to the surfaces S_1 and S_2.

Differentiation of $F[u(x, y, z), v(x, y, z), x, y, z]$ gives

$$\frac{\partial F}{\partial a} = (F_u u_x + F_v v_x + F_x)\xi. \tag{26}$$

From (11) we obtain, with the help of (26),

$$\left(\frac{\partial S}{\partial a}\right)_0 = \int_{z_1}^{z_2} \left\{ F_u u_x + F_v v_x + F_x)\xi + (\xi' - u_x\xi)F_u - v_x\xi F_v + (x' - u)\frac{\partial F_u}{\partial a} + (y' - v)\frac{\partial F_v}{\partial a} \right\}_0 dz. \tag{27}$$

The terms in the second line vanish, since the curve is assumed to satisfy (23). In the first line several terms cancel and we obtain

$$\left(\frac{\partial S}{\partial a}\right)_0 = \int_{z_1}^{z_2} (F_x\xi + F_u\xi')dz, \tag{28}$$

or, on integrating by parts,

$$\left(\frac{\partial S}{\partial a}\right)_0 = \int_{z_1}^{z_2} \left(F_x - \frac{d}{dz}F_u\right)\xi\, dz, \tag{28a}$$

and similarly

$$\left(\frac{\partial S}{\partial b}\right)_0 = \int_{z_1}^{z_2} \left(F_y - \frac{d}{dz}F_v\right)\eta\, dz. \tag{28b}$$

The right-hand side of equations (28a) and (28b) are the first variations of I [cf. (6a) and (6b)], and (25) shows that they vanish. Hence the curve C satisfies EULER's equations, i.e. it is an extremal, and the theorem is proved.

To establish the converse theorem we construct a field f_1 of ∞^2 extremals transversal to a given surface $T(x, y, z) = 0$, and compare it with another field f_2 of ∞^2 extremals. The latter field is constructed in the following way. We solve the HAMILTON–JACOBI equation (21) with the boundary condition that $S(x, y, z)$ shall be constant on $T(x, y, z) = 0$. If u and v are determined from (22), the solution can be represented by the integral (11). Then, according to the theorem just established, the equations (23) define a field f_2 of extremals which are transversal to $T = 0$. These two fields f_1 and f_2 must, however, be identical, since they satisfy the same differential equations and the same boundary conditions on $T = 0$. Hence, for the given field f_1, the integral S is independent of the path.

4. Determination of all extremals from the solution of the Hamilton–Jacobi equation

So far we have only considered an ∞^1 set of solutions $S(x, y, z) =$ constant of the HAMILTON–JACOBI equation, which correspond to an ∞^2 set of transversal extremals. To obtain all ∞^4 extremals we must consider a larger set of solutions S, namely ∞^4; these can be obtained by rotating the surface $T = 0$ round a point, and taking, in each case, $S(x, y, z) = S_1$. Assume that we have found the "complete" solution $S(x, y, z, \alpha, \beta)$ of (21), involving two parameters α and β. The function S, corresponding to any pair of values of α and β can be represented by an integral of the form (11) which is independent of the path by making an appropriate choice of the two functions u and v: $u = u(x, y, z, \alpha, \beta)$, $v = v(x, y, z, \alpha, \beta)$. Hence not only S but also $\partial S/\partial\alpha$ and $\partial S/\partial\beta$ are independent of the path. Now since $F = F[u(x, y, z, \alpha, \beta), v(x, y, z, \alpha, \beta), x, y, z]$

$$\frac{\partial F}{\partial \alpha} = F_u u_\alpha + F_v v_\alpha, \tag{29}$$

and we obtain from (11)

$$\left.\begin{aligned} S_\alpha &= \int_{z_1}^{z_2} \left\{(x' - u)\frac{\partial F_u}{\partial \alpha} + (y' - v)\frac{\partial F_v}{\partial \alpha}\right\} dz, \\ S_\beta &= \int_{z_1}^{z_2} \left\{(x' - u)\frac{\partial F_u}{\partial \beta} + (y' - v)\frac{\partial F_v}{\partial \beta}\right\} dz, \end{aligned}\right\} \tag{30}$$

where

$$\left.\begin{aligned} \frac{\partial F_u}{\partial \alpha} &= F_{uu} u_\alpha + F_{uv} v_\alpha, \\ \frac{\partial F_v}{\partial \alpha} &= F_{vu} u_\alpha + F_{vv} v_\alpha, \end{aligned}\right\} \tag{31}$$

with similar expressions for the other two derivatives.

Since the integrals (30) are independent of the path, it follows that the expressions

$$\left.\begin{aligned} dS_\alpha &= \left[(x' - u)\frac{\partial F_u}{\partial \alpha} + (y' - v)\frac{\partial F_v}{\partial \alpha}\right] dz, \\ dS_\beta &= \left[(x' - u)\frac{\partial F_u}{\partial \beta} + (y' - v)\frac{\partial F_v}{\partial \beta}\right] dz, \end{aligned}\right\} \tag{32}$$

are total differentials; S_α and S_β are thus functions of the end points P_1 and P_2 only, so that surfaces defined by constant values S_α and S_β exist, say

$$\frac{\partial S(x, y, z, \alpha, \beta)}{\partial \alpha} = A, \qquad \frac{\partial S(x, y, z, \alpha, \beta)}{\partial \beta} = B, \tag{33}$$

A, B being constants. These equations must therefore represent the solutions of the differential equations

$$x' = u(x, y, z, \alpha, \beta), \qquad y' = v(x, y, z, \alpha, \beta). \tag{34}$$

For on the surfaces defined by the equations (33), $dS_\alpha = 0$, $dS_\beta = 0$, and by (32) this implies (34), provided that the associated determinant does not vanish. Now, according to equation (31) the determinant may be written in the form

$$\begin{vmatrix} \dfrac{\partial F_u}{\partial \alpha} & \dfrac{\partial F_v}{\partial \alpha} \\[2ex] \dfrac{\partial F_u}{\partial \beta} & \dfrac{\partial F_v}{\partial \beta} \end{vmatrix} = \begin{vmatrix} F_{uu} & F_{uv} \\ F_{vu} & F_{vv} \end{vmatrix} \begin{vmatrix} u_\alpha & u_\beta \\ v_\alpha & v_\beta \end{vmatrix}. \tag{35}$$

The first factor on the right-hand side is the expression (9), which was assumed to differ from zero. The second term can only vanish at points where the equations (34) have no solutions for α and β, in other words at points where the field does not cover the space uniquely. If we exclude such cases, (34) represents ∞^2 sets of differential equations, each with ∞^2 solutions; all these together must therefore be identical with the totality of ∞^4 extremals. We have proved that the solutions

$$x = x(z, \alpha, \beta, A, B), \qquad y = y(z, \alpha, \beta, A, B), \tag{36}$$

of (33) are just this total set of ∞^4 extremals; hence the whole set of ∞^4 extremals can be obtained from a complete solution $S(x, y, z, \alpha, \beta)$ of the Hamilton–Jacobi equation by differentiation and elimination only, according to (33) and (36).

5. Hamilton's canonical equations

Each Euler's equation (7) is a differential equation of the second order. It is often convenient to replace these two second-order equations by four differential equations of the first order. This can be done in many ways. The most symmetrical way gives the so-called Hamilton canonical equations, obtained as follows:

The equations (13) are regarded as a Legendre transformation (see p. 135), which replace the variables u, v by U, V (retaining x, y, z), and the function $F(u, v, x, y, z)$ by $W(U, V, x, y, z)$. We can write the last equation in (13) as

$$W = F - uU - vV, \tag{37}$$

so that

$$dW = dF - udU - vdV - Udu - Vdv.$$

Now

$$\begin{aligned} dF &= F_u du + F_v dv + F_x dx + F_y dy + F_z dz \\ &= Udu + Vdv + F_x dx + F_y dy + F_z dz, \end{aligned}$$

and therefore

$$dW = -udU - vdV + F_x dx + F_y dy + F_z dz.$$

Since W is to be regarded as a function of U, V, x, y, z, it follows that

$$W_U = -u, \qquad W_V = -v, \qquad W_x = F_x, \qquad W_y = F_y, \qquad W_z = F_z. \tag{38}$$

If we now consider a curve $x = x(z)$, $y = y(z)$ which satisfies the equations

$$x' = u(x, y, z), \qquad y' = v(x, y, z), \tag{39}$$

these two equations, together with EULER's equations (7), may be written in the form

$$\left.\begin{aligned} x' &= -W_U, & y' &= -W_V, \\ U' &= W_x, & V' &= W_y. \end{aligned}\right\} \tag{40}$$

Equations (40) are four differential equations of the first order for x, y, U, V as functions of z, and are called *Hamilton's canonical equations*. They may be regarded as the EULER equations of the variational integral expressed in terms of the function $W(U, V, x, y, z)$. If we substitute from (37) and (39) into (1), the integral goes over into

$$I = \int_{z_1}^{z_2} \{W(U, V, x, y, z) + x'U + y'V\}\, dz. \tag{41}$$

If U, V, x, y are regarded here as four unknown functions of z and the EULER equation for each of them is formed, then (40) is immediately obtained.

6. The special case when the independent variable does not appear explicitly in the integrand

The case when F does not explicitly depend on z deserves special consideration.

We have in general for $F(x', y', x, y, z)$ that

$$\frac{dF}{dz} = F_{x'} x'' + F_{y'} y'' + F_x x' + F_y y' + F_z.$$

Assume now that $F_z = 0$, and substitute for F_x, F_y from the EULER equations (7). This gives

$$\frac{dF}{dz} = F_{x'} x'' + F_{y'} y'' + x' \frac{d}{dz} F_{x'} + y' \frac{d}{dz} F_{y'} = \frac{d}{dz} (x'F_{x'} + y'F_{y'}).$$

Hence

$$\frac{d}{dz} (F - x'F_x - y'F_{y'}) = 0,$$

so that

$$F - x'F_{x'} - y'F_{y'} = \text{constant}. \tag{42}$$

This expression, which is just the quantity W on the extremal, is independent of z; W is a constant of integration. The same result can be seen directly from the canonical equations (40), for if F does not explicitly depend on z, W is also independent of z, $W_z = 0$, and

$$\frac{dW}{dz} = W_U U' + W_V V' + W_x x' + W_y y';$$

this expression vanishes by virtue of (40).

In this case one may reduce the variational problem from three to two dimensions. Consider y as a function of x; then $y' = (dy/dx)x'$, and $F(x', y', x, y) = F(x', dy/dx, x, y)$. In the same way $F_{x'}$, $F_{y'}$, and also $F - x'F_{x'} - y'F_{y'}$ can be regarded as functions of x', dy/dx, x, y. Now the equation

$$F - x'F_x - y'F_{y'} = W \tag{43}$$

can be solved with respect to x':

$$\left.\begin{aligned} x' &= \Phi\left(\frac{dy}{dx}, x, y, W\right), \\ y' &= \frac{dy}{dx}\Phi\left(\frac{dy}{dx}, x, y, W\right). \end{aligned}\right\} \tag{44}$$

Let us consider all curves for which W has some given value; then the integral (1) can be replaced by the difference of two integrals, or

$$J = \int_{z_1}^{z_2} (F - W)dz = \int_{z_1}^{z_2} (x'F_{x'} + y'F_{y'})dz = \int_{x_1}^{x_2} \left(F_{x'} + \frac{dy}{dx} F_{y'}\right) dx.$$

If now the abbreviation

$$F_{x'} + \frac{dy}{dx} F_{y'} = f \tag{45}$$

is used, and x', y' are eliminated with the help of (44), f becomes a function of $dy/dx, x, y, W$, and

$$J = \int_{x_1}^{x_2} f\left(\frac{dy}{dx}, x, y, W\right) dx. \tag{46}$$

Thus we have, for each value of W, a variational integral of one dimension less. [This reduction corresponds in mechanics to the transition from HAMILTON's principle to MAUPERTUIS' principle: see (88) below.] If $y(x)$ is found from the EULER equation corresponding to (46), the complete extremal is obtained on integrating (44).

7. Discontinuities

It may happen that the function $F(u, v, x, y, z)$ is not everywhere continuous. The most important case (which frequently occurs in optics) is that in which F has a finite discontinuity along a surface $\sigma(x, y, z) = 0$ for all values of u, v.

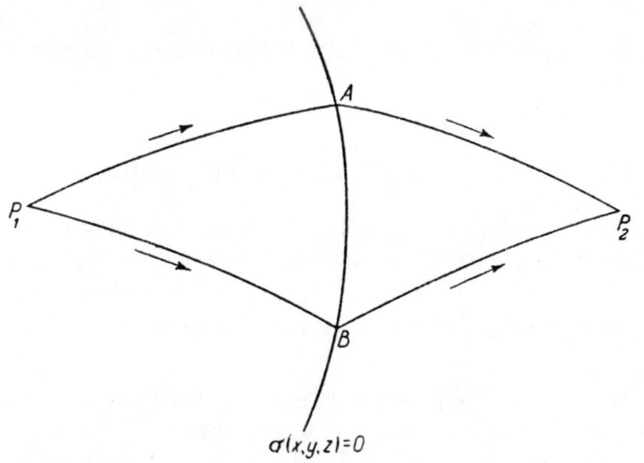

Fig. 2. Illustrating the variational analogue of the optical law of refraction.

It is obvious that outside the surface the extremals are again the solutions of the EULER equations (7); but if an extremal crosses the surface, it will have a discontinuity of direction (refraction). Let us distinguish the space to the left and right of the surface by the indices 1 and 2 (Fig. 2).

In order to find the "law of refraction" we have to establish the condition which ensures that the HILBERT integral S, defined by (11), extended from a point P_1 on the left of the surface to a point P_2 on the right of it, is independent of the path. Consider two paths P_1AP_2 and P_1BP_2, A and B being points on the surface. Then (with obvious notation) we demand that $S(P_1AP_2) = S(P_1BP_2)$ or

$$S_1(P_1A) + S_2(AP_2) = S_1(P_1B) + S_2(BP_2). \tag{47}$$

Now on the closed path P_1ABP_1 on the left of the surface,

$$S_1(P_1A) + S_1(AB) + S_1(BP_1) = 0, \tag{48}$$

and on the closed path on the right

$$S_2(P_2B) + S_2(BA) + S_2(AP_2) = 0. \tag{49}$$

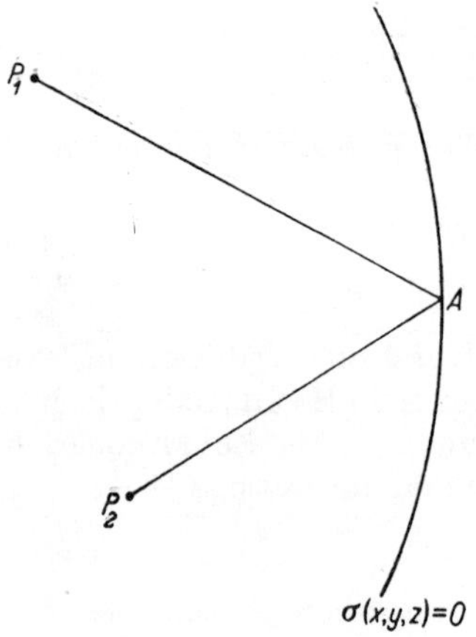

Fig. 3. Illustrating the variational analogue of the optical law of reflection.

If we add equations (48) and (49), and use (47) and the relation $S(XY) = -S(YX)$, it follows that

$$S_1(AB) = S_2(AB). \tag{50}$$

The integral (11) taken along any path on the surface $\sigma = 0$ has, therefore, the same value, whether one takes as u, v the values u_1, v_1 on the left, or the values u_2, v_2 on the right. Thus the integrands must be equal in the two cases and the *law of refraction* is equivalent to the assertion that the expression

$$F + (x' - u)F_u + (y' - v)F_v \tag{51}$$

is continuous on the surface $\sigma = 0$. According to (13) this condition may be expressed in the form

$$(Ux' + Vy' + W)_1 = (Ux' + Vy' + W)_2, \tag{52}$$

where x', y' are the derivatives of $x(z)$, $y(z)$ for any curve on the surface. This can also be expressed by saying that the vector $U_2 - U_1$, $V_2 - V_1$, $W_2 - W_1$ is normal to the discontinuity surface:

$$(U_2 - U_1)dx + (V_2 - V_1)dy + (W_2 - W_1)dz = 0. \tag{53}$$

Very similar to the problem of refracted extremals is that of reflected extremals. One has to connect two points P_1 and P_2, situated in a region where F is a continuous function of x, y, z, by a curve P_1AP_2 which has a discontinuity in direction at a point A on a given surface $\sigma(x, y, z) = 0$, the points P_1 and P_2 being on the same side of σ (Fig. 3).

It is obvious that the parts P_1A and AP_2 must be extremals, and it follows, by a consideration like that leading to the law of refraction [(53)], that the condition for the validity of the *independence theorem* for an incident field (suffix 1) and a reflected field (suffix 2) is the *law of reflection*,

$$(U_1 + U_2)dx + (V_1 + V_2)dy + (W_1 + W_2)dz = 0. \tag{54}$$

The independence theorem also holds for any fields with a finite number of refracting or reflecting discontinuities. It will be shown in the next section that in all these cases, whether the extremal which connects P_1 and P_2 is continuous or has (directional) discontinuities, it will give rise to a minimum of the integral (1), provided that the function F satisfies certain simple conditions along this curve.

8. Weierstrass' and Legendre's conditions (sufficiency conditions for an extremum)

So far no distinction has been made between maxima and minima; the extremals considered (smooth or with a "kink") may even correspond to stationary cases which are not true extrema. We shall now derive conditions necessary for a real minimum.

Let $\bar{x}(z)$, $\bar{y}(z)$ be a fixed extremal $\bar{C}$ embedded in a field $u(x, y, z)$, $v(x, y, z)$, and let

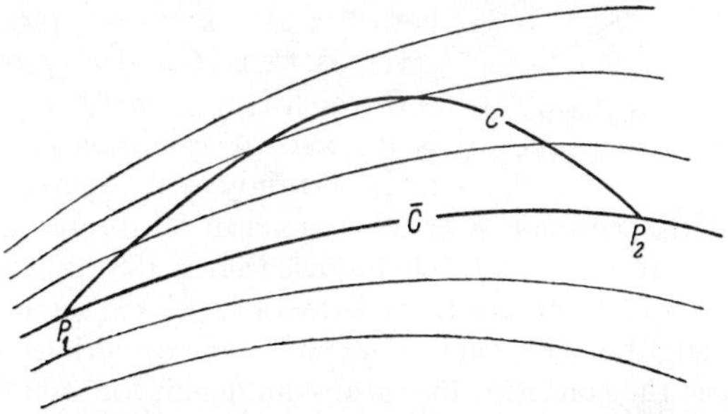

Fig. 4. Illustrating the definition of the WEIERSTRASS $\mathscr{E}$-function.

$x(z)$, $y(z)$ be any neighbouring curve C also completely embedded in the field and with the same end points P_1 and P_2 as $\bar{C}$ (Fig. 4). The extremum will be a real minimum if

$$\int_C F(x', y', x, y, z)dz - \int_{\bar{C}} F(\bar{x}', \bar{y}', \bar{x}, \bar{y}, z)dz > 0. \tag{55}$$

According to Sections 2 and 3 we may replace the second integral by one extended not over $\bar{C}$ but over C, namely by

$$\int_C \{F(u, v, x, y, z) + (x' - u)F_u + (y' - v)F_v\}dz;$$

this integral is independent of the path and reduces to $\int_{\bar{C}} Fdz$, if the path is taken to coincide with $\bar{C}$. Hence (55) becomes

$$\int_C Fdz - \int_{\bar{C}} Fdz = \int_C \mathscr{E}(x', y', u, v, x, y, z)dz > 0, \tag{56}$$

where

$$\mathscr{E}(x', y', u, v, x, y, z) = F(x', y', x, y, z) - F(u, v, x, y, z) \\ - (x' - u)F_u - (y' - v)F_v, \tag{57}$$

and the arguments in F_u, F_v are the same as in $F(u, v, x, y, z)$. The function defined by (57) is called *the $\mathscr{E}$-function (or the excess function) of Weierstrass*; the arguments x, y, z, x', y' refer to a point on the curve C and to its direction, while u, v refer to the direction of the extremal of the field, which passes through the point x, y, z.

It is seen that $\mathscr{E}$ vanishes on any portion of C which coincides with a field extremal. Now we can choose the field as the set of all ∞^2 extremals going through P_1. Then we construct a special curve C which is such that between P_1 and a point A the curve coincides with a field extremal, from A to a point B on the given extremal it is a straight line, and from B to P_2 it coincides with the given extremal (Fig. 5).

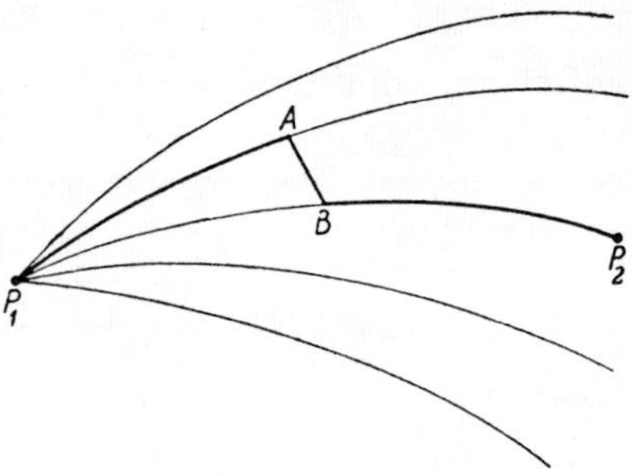

Fig. 5. Derivation of WEIERSTRASS' condition for a strong minimum.

Then $\mathscr{E}$ vanishes on the parts P_1A and BP_2, and there remains

$$\int_A^B \mathscr{E} dz > 0.$$

By letting A approach B, it is seen that this inequality is only possible if

$$\mathscr{E}(x', y', \bar{x}', \bar{y}', \bar{x}, \bar{y}, \bar{z}) > 0; \qquad (58)$$

here $\bar{x}$, $\bar{y}$, $\bar{z}$ refer to a typical point (B) on the given extremal $\bar{C}$ and x', y' refer to the direction AB which is quite arbitrary. The formula (58) is *Weierstrass' condition for a strong minimum*; it is certainly a necessary condition. But on the assumption that the function F is continuous in all its five arguments (hence $\mathscr{E}$ is continuous in its seven arguments), it follows that if (58) is satisfied for all points on the given extremal and for arbitrary directions x', y', then the inequality (56) must hold for any neighbouring curve C of arbitrary directions in a certain region surrounding $\bar{C}$. Hence the condition (58) is also sufficient for a strong minimum. This minimum is, of course, only relative, for there may be several extremals having the property of giving a minimum value to the integral (1) in comparison with all neighbouring curves; which of them gives the absolute minimum cannot be decided in this way.

The inequality (58) has a simple geometrical interpretation. For a fixed point $\bar{x}$, $\bar{y}$, $\bar{z}$ in space, F is a function of $\bar{x}'$, $\bar{y}'$ alone; this function $F(\bar{x}', \bar{y}')$ can be represented by a surface in the three-dimensional $\bar{x}'$, $\bar{y}'$, F-space. (In Fig. 6 the two-dimensional cross-section $\bar{x}'F$ is drawn.) Then

$$\mathscr{E}(x', y', \bar{x}', \bar{y}') = F(x', y') - [F(\bar{x}', \bar{y}') + (x' - \bar{x}')F_{\bar{x}'} + (y' - \bar{y}')F_{\bar{y}'}] \qquad (59)$$

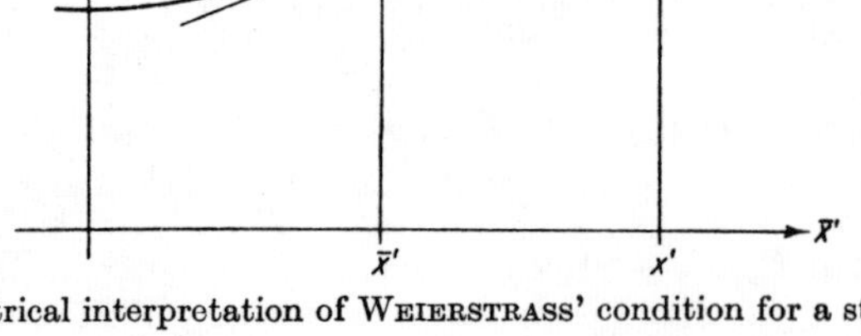

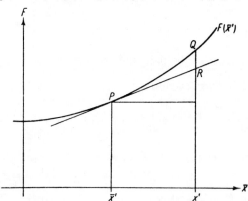

Fig. 6. A geometrical interpretation of WEIERSTRASS' condition for a strong minimum.

is obviously the distance QR along the ordinate in x', y', between the point Q on the surface $F = F(\bar{x}', \bar{y}')$ and the point R (see Fig. 6) where the tangential plane at $P(\bar{x}', \bar{y}')$ intersects this ordinate. Hence $\mathscr{E}(x', y', \bar{x}', \bar{y}') > 0$ if the surface F is above the tangential plane at P. If this holds for all x', y' there is a *strong minimum*.

If, however, (58) holds only for small intervals of $\xi = x' - \bar{x}'$, $\eta = y' - \bar{y}'$, there is a weak minimum; in this case we may expand $\mathscr{E}$ in powers of ξ, η and obtain (if again we omit the arguments x, y, z)

$$\mathscr{E}(x', y', \bar{x}', \bar{y}') = F(x', y') - F(\bar{x}', \bar{y}') - \xi F_{\bar{x}'} - \eta F_{\bar{y}'} + \tfrac{1}{2}[F_{\bar{x}'\bar{x}'}\xi^2 + 2F_{\bar{x}'\bar{y}'}\xi\eta + F_{\bar{y}'\bar{y}'}\eta^2] + \ldots$$

For small values ξ and η the quadratic terms are decisive and must evidently be positive for a minimum. Thus we obtain *Legendre's condition* (necessary and sufficient) *for a weak minimum**:

$$F_{\bar{x}'\bar{x}'} > 0, \qquad F_{\bar{x}'\bar{x}'}F_{\bar{y}'\bar{y}'} - F_{\bar{x}'\bar{y}'}{}^2 > 0. \tag{60}$$

9. Minimum of the variational integral when one end point is constrained to a surface

The $\mathscr{E}$-function provides a simple solution of the problem of finding the minimum of the variational integral (1) with respect to all curves that have one end point P_1 in common and the other constrained to a given surface $\sigma(x, y, z) = 0$.

The curve must obviously be one of the ∞^2 extremals through P_1, the problem is "which of them?" Now amongst these ∞^2 extremals there is just one† that is transversal to the surface $\sigma = 0$ and it is easily seen that this represents the solution of

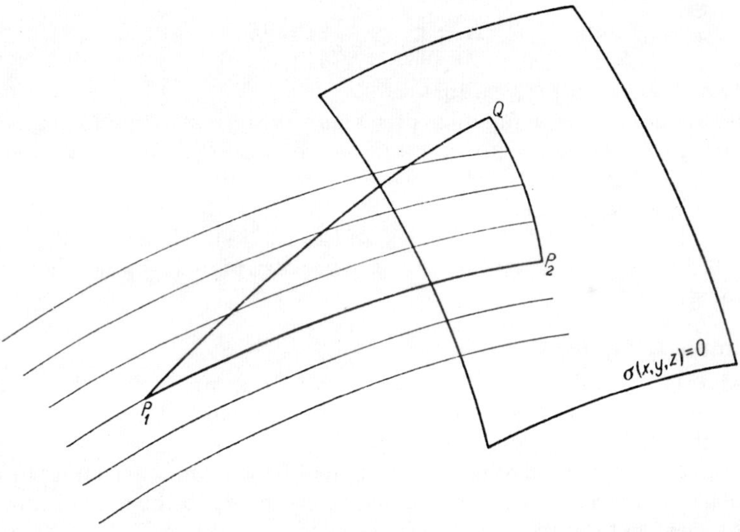

Fig. 7. Determination of the minimum of the variational integral with respect to all curves which have one end point fixed and the other constrained to a surface.

the problem. To show this let P_2 be the point in which this extremal intersects the surface $\sigma = 0$, and surround it by the field of all extremals that are transversal to the surface. Let P_1Q be any extremal through P_1 and Q its point of intersection with the surface (Fig. 7). Then the HILBERT integral $S(P_2, Q)$ vanishes. Hence the integral

* This condition has an immediate generalization when one deals not with two but with a larger number of variables (n say). In order to have a minimum, the quadratic form in the n variables must be positive definite; this implies that the associated determinant and its principal minors must all be positive.

† It is assumed that P_1 is near enough to the surface, so that the case of several such extremals is excluded.

S taken over the path P_1QP_2 is equal to the variational integral $I(P_1, Q)$. The difference $I(P_1, Q) - I(P_1, P_2)$ can now be expressed in terms of the $\mathscr{E}$-function just in the same way as before, and if (58) is satisfied, this difference is positive as long as Q differs from P_1.

10. Jacobi's criterion for a minimum

If an extremal can be embedded in a field and the LEGENDRE condition is satisfied for all points of it between P_1 and P_2, then the integral I, defined by (1), is certainly a (weak) minimum. It remains to find a criterion for the existence of the field.

Let all ∞^2 extremals that pass through P_1 be given by

$$x = x(z, \alpha, \beta), \qquad y = y(z, \alpha, \beta), \tag{61}$$

and let the given extremal C be characterized by the values $\alpha = 0$, $\beta = 0$:

$$x = x(z, 0, 0), \qquad y = y(z, 0, 0). \tag{62}$$

The curves (61) form a field as long as there is one such curve through a given point $P(x, y)$ arbitrarily close to C, i.e. as long as the equations (61) have unique solutions for α, β as functions of x, y; the condition for this is that

$$\Delta = \begin{vmatrix} x_\alpha & x_\beta \\ y_\alpha & y_\beta \end{vmatrix} \neq 0. \tag{63}$$

This is *Jacobi's criterion for a minimum.*

The determinant Δ is a function of z along the given extremal (62). The first point $\bar{P}$ where $\Delta = 0$ is called the *conjugate point of* P_1; for every interval P_1P_2 such that P_2 lies between P_1 and $\bar{P}$, there is a real minimum.

In $\bar{P}$ the given extremal is intersected by a neighbouring extremal (infinitesimally close to it); it is a point of the envelope of the set (61). Hence the *limit of the field is determined by the envelope of the set of extremals* (61). In optics these envelopes are the caustic surfaces.

11. Example I: Optics

The general theory will now be illustrated by a few examples. The first concerns the shortest line in ordinary geometry and the line of shortest optical length in geometrical optics.

Euclidean geometry is based on the theorem of PYTHAGORAS according to which the line element ds is related to its projections dx, dy, dz on to the axes of a rectangular system by means of the relation

$$(ds)^2 = (dx)^2 + (dy)^2 + (dz)^2. \tag{64}$$

The shortest lines between two points are then given by the minima of the variational integral

$$s = \int_{P_1}^{P_2} ds = \int_{z_1}^{z_2} \sqrt{x'^2 + y'^2 + 1}\, dz. \tag{65}$$

Geometrical optics can be based on a generalization of this integral, namely on FERMAT's principle of the shortest optical path (see § 3.3.2),

$$\int_{P_1}^{P_2} n ds = \int_{z_1}^{z_2} n(x, y, z)\sqrt{x'^2 + y'^2 + 1}\, dz, \tag{66}$$

where $n(x, y, z)$ is the index of refraction. We shall treat only the optical case, as (65) is the special case of (66) for $n = 1$.

We now have

$$F(x', y', x, y, z) = n(x, y, z)\sqrt{x'^2 + y'^2 + 1}. \tag{67}$$

Since $ds = \sqrt{x'^2 + y'^2 + 1}\, dz$

$$\left.\begin{aligned} U &= F_{x'} = \frac{nx'}{\sqrt{x'^2 + y'^2 + 1}} = n\frac{dx}{ds} = ns_x, \\ V &= F_{y'} = \frac{ny'}{\sqrt{x'^2 + y'^2 + 1}} = n\frac{dy}{ds} = ns_y, \\ W &= F - F_{x'}x' - F_{y'}y' = \frac{n}{\sqrt{x'^2 + y'^2 + 1}} = n\frac{dz}{ds} = ns_z. \end{aligned}\right\} \tag{68}$$

where s_x, s_y, and s_z are the components of the unit vector $\boldsymbol{s}$ tangential to the curve $x = x(z)$, $y = y(z)$. The law of refraction on a discontinuity surface of $n(x, y, z)$ may according to (53) be expressed in the form

$$(n_2\boldsymbol{s}_2 - n_1\boldsymbol{s}_1) \cdot d\boldsymbol{l} = 0, \tag{69a}$$

where $d\boldsymbol{l}(dx, dy, dz)$ is any line element of the surface. This equation implies that $\boldsymbol{s}_1$, $\boldsymbol{s}_2$ and the surface normal are coplanar, and that the angles θ_1 and θ_2 which $\boldsymbol{s}_1$ and $\boldsymbol{s}_2$ make with the surface normal are related by

$$n_2 \sin\theta_2 = n_1 \sin\theta_1, \tag{69b}$$

in agreement with the law of refraction (§ 3.2 (19)).

The Euler equations (7) associated with (66) are

$$\frac{\partial n}{\partial x}\sqrt{x'^2 + y'^2 + 1} - \frac{d}{dz}\frac{nx'}{\sqrt{x'^2 + y'^2 + 1}} = 0,$$

$$\frac{\partial n}{\partial y}\sqrt{x'^2 + y'^2 + 1} - \frac{d}{dz}\frac{ny'}{\sqrt{x'^2 + y'^2 + 1}} = 0,$$

or

$$\frac{d}{ds}\left(n\frac{dx}{ds}\right) = \frac{\partial n}{\partial x}, \quad \frac{d}{ds}\left(n\frac{dy}{ds}\right) = \frac{\partial n}{\partial y}. \tag{70a}$$

The corresponding equation in z, namely

$$\frac{d}{ds}\left(n\frac{dz}{ds}\right) = \frac{\partial n}{\partial z}, \tag{70b}$$

is an identity, as it follows from

$$\left(\frac{dx}{ds}\right)^2 + \left(\frac{dy}{ds}\right)^2 + \left(\frac{dz}{ds}\right)^2 = 1. \tag{71}$$

To derive (70b) one may proceed as follows: First one differentiates (71) with respect to s. Next one multiplies (71) by dn/ds, the differentiated equation by n and one adds the two equations. Finally, one uses (70a).* The three scalar differential equations (70) are in agreement with the vector equation § 3.2(2) for the light rays.

Since U, V, and W now represent the components of the ray vector [cf. § 4.1(4)] it is seen from (12), that the S-function belonging to the geometrical optics field is just the *Hamilton point characteristic function* [cf. § 4.1 (1)]. Moreover, by following the procedure leading to (21) one finds that the HAMILTON–JACOBI equation of the present variational problem is the *eikonal equation.*

To investigate the LEGENDRE condition (60) one has to determine the derivatives $F_{x'x'}$, etc. One has, in the present case,

$$F_{x'x'} = \quad n \frac{1 + y'^2}{(1 + x'^2 + y'^2)^{3/2}},$$

$$F_{y'y'} = \quad n \frac{1 + x'^2}{(1 + x'^2 + y'^2)^{3/2}},$$

$$F_{x'y'} = -n \frac{x'y'}{(1 + x'^2 + y'^2)^{3/2}},$$

so that

$$F_{x'x'} \cdot F_{y'y'} - F_{x'y'}{}^2 = \frac{n^2}{(1 + x'^2 + y'^2)^2} > 0. \tag{72}$$

Every extremal therefore gives a weak minimum (p. 732) if the JACOBI condition (63) is satisfied. But as the function F, for given x, y, z, i.e. for a given n, is convex downwards for all values of x', y', it follows from the geometrical interpretation of WEIERSTRASS' condition that the minimum is strong.

It remains to consider JACOBI's criterion. For $n =$ const., i.e. in ordinary Euclidean geometry, the extremals are obviously straight lines; since a bundle of straight lines through a point P_1 never has an envelope, each straight line gives a strong minimum of the distance between any two points of it. In geometrical optics on the other hand, where n in general depends on x, y, z (continuously or discontinuously), the bundle of rays from P_1 give rise to envelopes (caustic surfaces). To determine the nature of the extremum these surfaces have to be examined separately in every particular case.

12. Example II: Mechanics of material points

As a second example we consider *the mechanics of systems of material points.* Here the independent variable is the time t, and the unknown functions are the Lagrangian coordinates $q_\alpha(\alpha = 1, 2, \ldots n)$ and their derivatives the velocities $u_\alpha = \dot{q}_\alpha$.

The variational problem is given by *Hamilton's principle*

$$\int_{t_1}^{t_2} L(u_1, u_2, \ldots q_1, q_2, \ldots t)dt = \text{extremum}, \tag{73}$$

* Equation (70b) can be obtained in a more symmetrical way, which, however, introduces a superfluous EULER equation. One regards x, y, and z as functions of a parameter λ; one then has three EULER equations connected by an identity and the parameter λ is finally identified with s.

where L is the Lagrangian. In ordinary non-relativistic mechanics one has $L = T - \Phi$, where T is the kinetic energy, a quadratic form of the u_α, and Φ the potential energy; but (73) holds also in more general cases, when a magnetic force is acting and when the relativistic variation of mass is taken into account.

The function $F(u, v, x, y, z)$ is here replaced by $L(u, q, t)$; hence [cf. (13)] U, V now correspond to the momenta

$$p_\alpha = \frac{\partial L}{\partial u_\alpha} \tag{74}$$

and W to $-H$, where H is the *Hamiltonian*

$$H = \sum_\alpha u_\alpha \frac{\partial L}{\partial u_\alpha} - L = \sum_\alpha u_\alpha p_\alpha - L. \tag{75}$$

If $L = T - \Phi$, (75) becomes

$$H = \sum_\alpha u_\alpha \frac{\partial L}{\partial u_\alpha} - T + \Phi.$$

From EULER's theorem on homogeneous functions* it follows that

$$2T = \Sigma u_\alpha \frac{\partial T}{\partial u_\alpha} \left(= \Sigma u_\alpha \frac{\partial L}{\partial u_\alpha}\right),$$

and H reduces to the total energy

$$H = T + \Phi. \tag{76}$$

The EULER equations (7) become LAGRANGE's equation of motion

$$\frac{d}{dt}\left(\frac{\partial L}{\partial u_\alpha}\right) - \frac{\partial L}{\partial q_\alpha} = 0; \tag{77}$$

and the canonical equations (40) now are

$$\frac{dq_\alpha}{dt} = \frac{\partial H}{\partial p_\alpha}, \qquad \frac{dp_\alpha}{dt} = -\frac{\partial H}{\partial q_\alpha}, \tag{78}$$

where H is to be regarded as a function of the p_α, q_α and t. If H is independent of the time t, formula (42) expresses the law of conservation of energy

$$H = \sum_\alpha u_\alpha p_\alpha - L = \text{constant} = E. \tag{79}$$

In this case the HAMILTON–JACOBI differential equation (21) becomes

$$\frac{\partial S}{\partial t} + H\left(\frac{\partial S}{\partial q_\alpha}, q_\alpha\right) = 0. \tag{80}$$

---

* See, for example, R. COURANT, *Differential and Integral Calculus*, Vol. II (Glasgow, Blackie and Son, 1936), p. 109.

This gives on integration,

$$S = -Et + S_1(q_\alpha), \tag{81}$$

where, because of (80), S_1 satisfies the equation

$$H\left(\frac{\partial S_1}{\partial q_\alpha}, q_\alpha\right) = E. \tag{82}$$

From the solution of the HAMILTON–JACOBI equation one obtains the momenta in accordance with (17):

$$p_\alpha = \frac{\partial S}{\partial q_\alpha} = \frac{\partial S_1}{\partial q_\alpha}. \tag{83}$$

It follows that the line integral

$$\int_{P_1}^{P_2} \sum_\alpha p_\alpha dq_\alpha \tag{84}$$

is independent of the path joining P_1 and P_2 and is, therefore, zero when taken over a closed path in a simply connected region,

$$\oint p_\alpha dq_\alpha = 0. \tag{85}$$

If the functions p_α are multivalued it may happen that the integral over a closed path is not zero, but is a multiple of a certain constant period. This result is a generalization of the optical invariant of LAGRANGE (§ 3.3.1) and is one of *Poincaré's invariants*. It may also be expressed in the form

$$\frac{\partial p_\alpha}{\partial q_\beta} - \frac{\partial p_\beta}{\partial q_\alpha} = 0. \tag{86}$$

If H is independent of time, t may be eliminated from the minimum principle according to the general procedure expressed by (45) and (46). One then has

$$J = \int_{t_1}^{t_2} (L + E)dt = \int_{t_1}^{t_2} \sum_\alpha \dot{q}_\alpha p_\alpha dt = \int_{P_1}^{P_2} \sum_\alpha p_\alpha dq_\alpha = \text{extremum}. \tag{87}$$

This is *Maupertuis' principle of least action*, generalized for arbitrary L; it has to be understood in the following way: the equation (79) allows the elimination of the time derivatives $u_\alpha = \dot{q}_\alpha$, expressing them in terms of purely geometrical derivatives, say dq_α/dq_1 ($\alpha = 2, \ldots n$), using q_1 as an independent variable. The equation (87) represents a purely geometrical principle describing the orbits, not the motions. The latter are then to be found from (78).

If $L = T - \Phi$, then $E = T + \Phi$ and one has MAUPERTUIS' original expression

$$J = 2\int_{t_1}^{t_2} T dt, \tag{88}$$

which is to be understood in the same way. An example for this reduction of the problem of the motion to the problem of the orbit by the transition from HAMILTON's to MAUPERTUIS' principle is the treatment of electron optics given in Section 2 of Appendix II.

The LEGENDRE condition can only be investigated for a given L. If $L = T - \Phi$ and T is a quadratic form of the u_α, the condition is obviously equivalent to the postulate that T is positive definite. Then WEIERSTRASS' condition is also satisfied and one has a strong minimum as long as JACOBI's condition holds. The latter leads to the investigation of dynamical foci and caustics, but is of little importance in practice.

The LEGENDRE condition for the relativistic electron with the Lagrangian (cf. Appendix II),

$$L = -mc^2\sqrt{1-\left(\frac{v}{c}\right)^2} + e\left(\frac{\mathbf{v}}{c}\,.\,A - \phi\right) \tag{89}$$

leads to the quadratic form

$$\frac{m}{\left[1-\left(\frac{v}{c}\right)^2\right]^{3/2}}\left\{\boldsymbol{\rho}^2\left(1-\frac{v^2}{c^2}\right)+\left(\frac{\boldsymbol{\rho}\,.\,\mathbf{v}}{c}\right)^2\right\} \tag{90}$$

in the components of the vector $\boldsymbol{\rho}(\xi, \eta, \zeta)$ and is, therefore, always satisfied.

APPENDIX II

LIGHT OPTICS, ELECTRON OPTICS AND WAVE MECHANICS

IN 1831 WILLIAM ROWAN HAMILTON discovered the analogy between the trajectory of material particles in potential fields and the path of light rays in media with continuously variable refractive index. By virtue of its great mathematical beauty the "Hamiltonian Analogy" survived in the textbooks of dynamics for almost a hundred years, but did not inspire any practical applications until 1925 when H. BUSCH first explained the focusing effect of electric and magnetic fields on electron beams in optical terms. Almost at the same time E. SCHRÖDINGER took the Hamiltonian Analogy a step further by passing from geometrical optics to wave optics of particles with his wave equation, in which he incorporated the wavelength of particles, first conceived by LOUIS DE BROGLIE in 1923.

Practical electron optics developed rapidly from 1928 onwards. By this time the Hamiltonian Analogy was widely known and inspired the invention of electron-optical counterparts of light-optical instruments, such as the electron microscope. Though the mathematical analogy is general, the two techniques are not exactly parallel. Some electron optical instruments such as cathode-ray tubes and systems with curved optic axes have no important counterparts in light optics. In the available space only those problems of electron optics will be considered whose light optical analogues were developed at length in the previous chapters of this work, so that the results can be transferred almost *in toto*, with few modifications. It may be noted that this applies in particular to the most recondite chapter of electron optics: the wave theory of lens aberrations.

1. The Hamiltonian analogy in elementary form

We will show first that the determination of the trajectory of a charged particle can be reduced to an optical problem by the introduction of a suitable refractive index, variable from point to point.

Consider a particle of charge e and of mass m, to which we will for the sake of simplicity refer as an electron, moving in a steady electrostatic potential field $\phi(x, y, z)$. By NEWTON's law of motion

$$\frac{d\boldsymbol{p}}{dt} = e\boldsymbol{E} = -e \operatorname{grad} \phi, \tag{1}$$

where $\boldsymbol{p}$ is the momentum vector. This law is valid at all velocities $\boldsymbol{v}$ if NEWTON's definition of the momentum $\boldsymbol{p} = m\boldsymbol{v}$ is replaced by EINSTEIN's

$$\boldsymbol{p} = \frac{m\boldsymbol{v}}{\sqrt{1-\beta^2}}, \qquad \left(\beta = \frac{v}{c}\right), \tag{2}$$

c being the vacuum velocity of light.

It is convenient to split the equation of motion (1) into two equations, of which one is the equation of the trajectory, while the other specifies the "time table" according to which the electron moves along the trajectory. For this purpose we write $\boldsymbol{v} = v\boldsymbol{s}$, $\boldsymbol{p} = p\boldsymbol{s}$, $\boldsymbol{s}$ being the unit vector in the direction of motion. Then

$$\frac{d\boldsymbol{p}}{dt} = \frac{dp}{dt}\boldsymbol{s} + p\frac{d\boldsymbol{s}}{dt} = \frac{dp}{dt}\boldsymbol{s} + p\frac{d\boldsymbol{s}}{ds}\frac{ds}{dt} = \frac{dp}{dt}\boldsymbol{s} + pv\frac{d\boldsymbol{s}}{ds}.$$

Now it is well known from differential geometry that ds/ds is a vector in the direction of the unit principal normal $\mathbf{\nu}$, whose absolute value is equal to the curvature $1/\rho$ of the trajectory. Hence

$$\frac{d\mathbf{p}}{dt} = \frac{dp}{dt}\mathbf{s} + \frac{pv}{\rho}\mathbf{\nu}.$$

From this relation and from (1) it follows that the instantaneous centre of curvature is in the plane which passes through the tangent $\mathbf{s}$ and the electric vector $\mathbf{E} = -\operatorname{grad}\phi$. Resolving grad ϕ in the two directions it follows that

$$\frac{dp}{dt}\mathbf{s} + \frac{pv}{\rho}\mathbf{\nu} = -e[(\mathbf{s}\,.\operatorname{grad}\phi)\mathbf{s} + (\mathbf{\nu}\,.\operatorname{grad}\phi)\mathbf{\nu}]. \tag{3}$$

Equating the first terms on the two sides we obtain a scalar equation, which can be called the "time table," because ultimately it leads to the position of the electron on the trajectory as function of time. After multiplying it by $v = ds/dt$, it can be integrated and gives

$$\frac{mc^2}{\sqrt{1-\beta^2}} = -e\phi + \text{constant}. \tag{4}$$

This is EINSTEIN's energy integral. For slow-moving particles ($\beta \ll 1$), it goes over into NEWTON's integral, $\frac{1}{2}mv^2 + e\phi = \text{constant}$.

We now restrict the problem, for convenience, by considering only electrons with the same integration constant, i.e. with the same total energy. This is the case if all electrons have started at a certain potential surface ϕ_0 with zero velocity. In many practical problems this surface can be identified with the cathode. Writing

$$V = \phi - \phi_0,$$

i.e. measuring the potential V from this level, the energy integral now appears in the form

$$mc^2\left[\frac{1}{\sqrt{1-\beta^2}} - 1\right] = -eV. \tag{4a}$$

Combining this with (2) we can express both in the form of a useful double equation

$$\left(1 - \frac{eV}{mc^2}\right)^2 = 1 + \left(\frac{p}{mc}\right)^2 = \frac{1}{1-\beta^2}. \tag{5}$$

Thus the scalar value p of the momentum of these particles is determined as a function of the position x, y, z by

$$mc^2\left[\sqrt{1 + \left(\frac{p}{mc}\right)^2} - 1\right] = -eV(x, y, z). \tag{6}$$

Consider now the second part of (3), i.e. the component at right angles to the direction of motion:

$$\frac{pv}{\rho} = -e[\mathbf{\nu}\,.\operatorname{grad} V]. \tag{7}$$

Expressing v and V in terms of p by means of (2) and (6) we obtain the simple law

$$\frac{1}{\rho} = \frac{\mathbf{\nu}\,.\operatorname{grad} p}{p} = \mathbf{\nu}\,.\operatorname{grad}(\log p). \tag{8}$$

Equation (8) *is identical with equation* (14) *of* § 3.2 *for the curvature of rays in a medium of refractive index n*, that is proportional to p, and we thus obtain a formal analogy between the paths of electrons and of light.

It must be emphasized that the absolute value p of the momentum is a function of position alone only for electrons of a fixed total energy; for electrons of a different energy it is another function, given by (6). Thus the refractive index is dependent on the electron energy. This too has its analogue in light optics, in the dependence of the refractive index on the colour of light. It will be shown later that this is a very appropriate analogy, as in both cases the refractive index turns out to be a function of the wavelength.

For slow electrons p is proportional to the velocity, and this in turn is proportional to $\sqrt{V}$. The relativistic equations which we have used have the advantage that they clearly indicate that it is the momentum, and not the velocity which is the characteristic quantity. Moreover these results immediately suggest a generalization for the case of a general static field, electric and magnetic. It is well known in relativity theory that in the presence of a magnetic field one must replace the mechanical momentum $\mathbf{p}$, which now we will write $\mathbf{p}_m$, by the "total" momentum

$$\mathbf{p}_{\text{tot}} = \mathbf{p}_m + e\mathbf{A}, \tag{9}$$

where $\mathbf{A}$ is the vector potential.* This suggests that the refractive index, which in the electrostatic case was the component of $\mathbf{p}_m$ in the direction of the motion, will have to be replaced in the general case by the component of $\mathbf{p}_{\text{tot}}$ in the same direction. This guess proves to be correct, but it will now be preferable to place electron optics in electromagnetic fields on a more solid and general foundation.

2. The Hamiltonian analogy in variational form

The laws of geometrical optics may be derived from Fermat's principle (§ 3.3.2), according to which the path of light between two points P_1 and P_2 makes the optical length a minimum,

$$\int_{P_1}^{P_2} n ds = \text{min.} \tag{10}$$

It may be recalled that this *strong* formulation of Fermat's principle is valid only if the two end points are sufficiently close together, that is to say if there is no image of either P_1 or P_2 between these points on the ray connecting them. If P_2 is the image of P_1, equation (10) does not determine a ray but an infinitesimal pencil of rays connecting the two points, each with the same optical length. If P_1 and P_2 are drawn farther apart, so that an image appears between them, the *weak* formulation

$$\delta \int_{P_1}^{P_2} n ds = 0 \tag{10a}$$

again determines a ray, but this represents not a maximum, as is sometimes erroneously stated, but a stationary value of the integral, which is neither a minimum, nor a maximum.

Some consequences of Fermat's principle were discussed in Section 11 of Appendix I, on the calculus of variations. In Section 12 of Appendix I it was shown that the motion of a system of material points can be described in a similar variational form,

---

* It is remarkable that this fundamental result was discovered by K. Schwarzschild (*Königl. Gess. Wiss. Göttingen, Math. Phys. Kl.*, **3** (1903), 126), three years before the advent of relativity.

by HAMILTON's principle, which for the special case of a single material point is expressed by the condition

$$\int_{P_1,t_1}^{P_2,t_2} L(\dot{x}, \dot{y}, \dot{z}, x, y, z)dt = \text{min.} \tag{11}$$

This formulation is valid in any system of coordinates, but for illustration Cartesian coordinates, x, y, z will suffice. $\dot{x}, \dot{y}, \dot{z}$ are the velocity components, L is the Lagrangian function. Departure and arrival are specified in four dimensional space-time, and are assumed to be fixed in the variational process. It was shown in Appendix I how the equations of motion in the Lagrangian and in the Hamiltonian form can be deduced as consequences of the principle (11) [App. I, eqs. (77), (78)].

For a relativistic electron, with charge e and rest mass m, the Lagrangian function is

$$L = -mc^2\sqrt{1-\beta^2} - e\left(\phi - \frac{1}{c}\dot{\boldsymbol{r}}\ \boldsymbol{A}\right). \tag{12}$$

Here $\dot{\boldsymbol{r}} = \boldsymbol{v}$ is the velocity vector, ϕ is the electrostatic potential, and A the magnetic or vector potential.* The Lagrangian (12) can be verified by substituting it into the Lagrangian equations of motion (Appendix I (77)) and making use of the electromagnetic relations § 2.1 (7) and § 2.1 (5),

$$\boldsymbol{E} = -\frac{1}{c}\dot{\boldsymbol{A}} - \text{grad}\,\phi, \qquad \boldsymbol{B} = \text{curl}\,\boldsymbol{A}$$

The Lagrangian equations then appear in the form

$$\frac{d\boldsymbol{p}}{dt} = e\left(\boldsymbol{E} + \frac{1}{c}\boldsymbol{v}\wedge\boldsymbol{B}\right), \tag{13}$$

which is the NEWTON–LORENTZ form of the equations of motion.

HAMILTON's principle expressed by (11) is too general for the purpose of electron optics. It contains the time, which is without interest if the fields are stationary. Moreover it represents ∞^5 extremals; ∞^4 because at given total energy the points P_1, P_2 can be freely chosen on any two given surfaces, and a further infinity because the total energy has been left undetermined. In order to reduce the cumbersome dimensionality similar steps are taken in electron optics as in light optics.

First we fix the energy constant, that is to say we restrict the discussion to monochromatic light or monoenergetic electrons. As shown in Appendix I (87) this allows a reduction by one in the dimensionality of the problem. HAMILTON's principle (11) is now replaced by the principle of least action,

$$J = \int_{P_1,t_1}^{P_2,t_2} (L+E)dt = \text{min.} \tag{14}$$

where E is the total energy

$$E = \Sigma\dot{x}p_x - L.$$

In (14) the time t appears only in a formal way, because on substituting for E one obtains

$$J = \int_{P_1,t_1}^{P_2,t_2} \Sigma p_x\dot{x}\,dt = \int_{P_1}^{P_2} \Sigma p_x\,dx = \int_{P_1}^{P_2} \boldsymbol{p}\,.\,d\boldsymbol{r} = \text{min.} \tag{15}$$

* The Lagrangian (12) without the relativistic corrections was found by K. SCHWARZSCHILD, *loc. cit.*, in 1903.

In the second and third expression we have dropped t_1, t_2, because we are dealing with time-invariant fields only (otherwise the total energy would not remain constant), and in such fields the starting time t_1 is of no consequence, while the transit time $t_2 - t_1$ is fully determined by the path and by the energy constant.

Thus the principle of least action (14) is a complete analogue of FERMAT's principle (10). The study of electron motion is reduced to an optical problem, if one defines the *electron-optical refractive index* as the *component of the momentum in the direction of the trajectory*. For a purely electrostatic field this result is seen to be equivalent to that found from more elementary considerations in the previous section. In this case the momentum is purely mechanical, and parallel to the trajectory, its value being given by (2):

$$p_m = \frac{mv}{\sqrt{1-\beta^2}}.$$

In the presence of a magnetic field we must use the general definition of momentum components (Appendix I, (74)), as derivatives of the Lagrangian with respect to the velocity components. For a single particle with the Lagrangian (12) these are

$$p_x = \frac{\partial L}{\partial \dot{x}} = \frac{m\dot{x}}{\sqrt{1-\beta^2}} + \frac{e}{c} A_x,$$

etc. In vector form

$$\mathbf{p} = \frac{m\mathbf{v}}{\sqrt{1-\beta^2}} + \frac{e}{c} \mathbf{A}. \tag{16}$$

Thus, apart from an arbitrary constant factor, the general electron-optical refractive index is

$$n = \frac{mv}{\sqrt{1-\beta^2}} + \frac{e}{c} \mathbf{A}.\mathbf{s}, \tag{17}$$

where $\mathbf{A}.\mathbf{s}$ is the component of the vector potential in the direction of motion. Again it is understood that this must be interpreted as a function of position for electrons of a given total energy.

There appears to exist an important difference between the general case and the special case of purely electric fields. In an electric field the refractive index is proportional to the mechanical momentum, which is a measurable physical quantity. In the general expression (17) the second term is a component of the vector potential, which is not a physical quantity but a function, such that its *curl* is equal to the magnetic induction $\mathbf{B}$. This makes it clear that the general electron-optical refractive index itself is not a physical quantity, but a Lagrangian function.* But this is equally true of the special refractive index in a purely electric field. In both cases we could add to it the component, in the direction of motion, of the gradient of any arbitrary function of position, without altering any of the physical consequences.

There exists a more important difference between the special and the general case, which is best explained if now we follow further the simplifying procedure of optics, and reduce the ∞^4 manifoldness of trajectory fields to ∞^2. We do this by selecting

* This has been particularly emphasized by W. EHRENBERG and R. E. SIDAY, *Proc. Phys. Soc.*, B, **62** (1949), 8, who have also pointed out the interesting fact that unless the magnetic field vanishes everywhere it is not possible to normalize $\mathbf{A}$ by a gauge transformation in such a way that it vanishes with the magnetic field.

those trajectories which have started on some surface $\mathscr{S}(x, y, z) = \mathscr{S}_0$ at right angles to this surface. In geometrical optics this surface can be considered as a wave-front, and it can then be shown that the pencil of trajectories will be everywhere at right angles to a family of surfaces $\mathscr{S}(x, y, z) =$ constant; this is the theorem of MALUS and DUPIN (§ 3.3.3).

A certain property of "transversality" exists also in the general case, as shown in Sections 2 and 3 of Appendix I, but its meaning is less simple than in the MALUS–DUPIN theorem. What remains at right angles to a family of surfaces $\mathscr{S} =$ const. is not the unit vector $\boldsymbol{s}$ in the direction of the trajectories, but the momentum $\boldsymbol{p}$. As the vector potential is defined only up to a gauge transformation, there exists an infinity of families of transversal surfaces, but in the presence of a magnetic field these cannot be made transversal to the trajectories by any gauge-normalizing procedure.

In the language of geometry, two-dimensional pencils of curves which are at right angles to a family of surfaces form a "normal congruence," otherwise they form a "skew congruence" (cf. § 3.2.3). In light optics and in electrostatic electron optics the trajectories can be ordered in normal congruences, and the surfaces transversal to these are identified with the "wave-fronts." In magnetic fields the two-dimensional pencils of trajectories, usually called "beams," form skew congruences, and the concept of transversal wave-fronts cannot be applied to them. This is a rather essential difference between electron optics and light optics.

3. Wave mechanics of free electrons

The dual nature of light was first conjectured by EINSTEIN in 1905. Light propagates as if it were an electromagnetic wave, but it interacts with matter as if its energy were concentrated in photons, each with an energy quantum. Soon afterwards EINSTEIN's conjecture was brilliantly verified by observations on photoelectrical and photochemical processes.

The dual conception of material particles is due to LOUIS DE BROGLIE, who showed in 1923 that if there is a wavelength to be associated, in a relativistically invariant way, with a particle having a mechanical momentum p_m, this could only be of the form

$$\lambda = \frac{h}{p_m}, \tag{18}$$

where h is a universal constant with the dimension of an action, which DE BROGLIE identified with PLANCK's constant.

Soon afterwards HEISENBERG, BORN, and JORDAN developed, quite independently of DE BROGLIE, the first complete mathematical formulation of quantum mechanics, but their methods are less convenient for the discussion of free particles than SCHRÖDINGER's wave mechanics, which will now be briefly outlined.

SCHRÖDINGER combined DE BROGLIE's ideas with HAMILTON's, and was led to the problem of a wave-description of particle motion, which stands in the same relation to the dynamics of mass points as wave optics is to geometrical optics. His historical approach, which as we now know is correct only to a certain point, involved of course a certain amount of guessing because, though geometrical optics is logically contained in wave optics, the converse is not true.

Assume that there exists a wave field, whose intensity indicates the density of electrons in a similar way as the intensity of the electromagnetic field indicates the photon density. Moreover, assume that this is a scalar field, whose amplitude is represented by some scalar function $\Psi(x, y, z, t)$; in order to account for its presumed wave-like properties, assume that it satisfies a wave equation

$$\nabla^2\Psi = \frac{1}{u^2}\ddot{\Psi},$$

u being the wave velocity—in general a function of position. This, of course, is a highly restrictive hypothesis, because the ordinary wave equation with constant velocity of propogation can be generalized in many different ways, of which this is only one, and the simplest.

Substituting for Ψ a "monochromatic" wave

$$\Psi(x, y, z, t) = \psi(x, y, z)e^{-i\omega t}$$

we obtain a time-independent equation

$$\begin{aligned}\nabla^2\psi &= -\left(\frac{\omega}{u}\right)^2\psi \\ &= -\left(\frac{2\pi}{\lambda}\right)^2\psi,\end{aligned} \tag{19}$$

in which only the wavelength λ occurs, but not the wave velocity. Assume now that λ is identical with the DE BROGLIE wavelength, if one substitutes for the momentum p_m the value which a particle would possess at the point x, y, z according to classical mechanics. This can be calculated from (6). For simplicity consider a slow electron in an electrostatic field. In this case DE BROGLIE's relation (18) combined with (6) gives

$$\lambda = \frac{h}{\sqrt{2m(E - e\phi)}}. \tag{20}$$

Substituting this into (19) we obtain

$$\nabla^2\psi + \frac{8\pi^2 m}{h^2}(E - e\phi)\psi = 0. \tag{21}$$

This is SCHRÖDINGER's wave equation for a free particle in a scalar potential field. As it is a time-independent equation, we can interpret it as describing the stationary. e.g. periodic motion of a particle in a field of force. But we can equally apply it to the stationary beams with which we deal in electron optics, in which many particles appear, one after the other, but all under identical conditions. In either case, it appears reasonable to assume, in accordance with BORN's statistical interpretation, that the absolute square of the amplitude $|\psi|^2 = \psi\psi^\star$ is proportional to the particle density at the point x, y, z, measured over long times; or, what is the same in this case, that it is proportional to the probability that a particle will be observed at this spot at any instant.

The first verification of (21) was carried out by SCHRÖDINGER, who showed that it accounted for the atomic spectra if one assumed that the electron was bound in a field of force which was the same as in the old atom model of RUTHERFORD and BOHR. More relevant for the present purposes was its verification in the case of free electrons, in the discovery of electron diffraction by DAVISSON and GERMER, and independently by G. P. THOMSON in 1927–1928.

For slow electrons with a kinetic energy equivalent to eV electron volts, DE BROGLIE's relation (18) gives a wavelength of approximately

$$\lambda = \sqrt{\frac{150}{V}} \text{ Ångstrom units.}$$

Thus the wavelength of electrons which can be easily handled in laboratory experiments is of the order of fractions of an Ångstrom, of the same order of magnitude as X-ray wavelengths. Hence the wave behaviour of free electrons can be most easily demonstrated in experiments similar to X-ray diffraction in crystal lattices.

The terminology of X-ray analysis, which was also applied to electrons, is somewhat different from that of light optics. What is called X-ray or electron diffraction is really the interference of coherent secondary wavelets emitted by the more or less regularly arranged atoms in a lattice. Diffraction of electrons in the optical sense, by relatively large material obstacles, whose atomic structure does not come into play, results in extremely small diffraction angles, and was first observed by H. BOERSCH, in 1940 in an electron microscope.*

4. The application of optical principles to electron optics

This elementary and incomplete sketch of wave mechanics is sufficient for almost all practical purposes of electron optics. It is not even necessary to make use of the extension of wave mechanics to magnetic fields, because in electron optics it is never necessary to go beyond KIRCHHOFF's approximation in diffraction problems. The assumptions on which this approximation is based are fully justified in electron optics. With the exception of the fields inside and in the immediate neighbourhood of atomic nuclei, there exist no electric fields which are so abrupt that they change appreciably within a wavelength of the electrons used in electron optics. This is *a fortiori* true for magnetic fields, and the specifically magnetic diffraction effects, which can be inferred from the KLEIN–GORDON or DIRAC wave equation, are far too small to be detected in experiments with free electrons. Thus, with a certain caution which will be explained below, one can safely apply KIRCHHOFF's diffraction theory (§ 8.3) which, in its somewhat extended form, may be summarized as follows:

I. Construct the wave-front from the source to the obstacle or object by measuring equal optical lengths along the rays as calculated by the rules of geometrical optics.

II. If the obstacle is a dark screen, consider the free part of the wavefront as undisturbed, in phase and in amplitude, and the rest as cut off. In the case of a partially transmitting object, trace the rays through it by the laws of geometrical optics, with the changes in phase and in intensity as determined by its complex refractive index. This is justified in all practical cases because the objects which one considers, e.g. in a microscope, have such small depth that diffraction effects inside them can be safely neglected.

III. In order to calculate the diffraction effects at some remote point, for instance in an optical image, calculate, by the rules of geometrical optics, the transmission between this point and each element of the wave-front emerging from the object or obstacle, and sum the complex amplitudes, taking into account the inclination factors.

KIRCHHOFF's method, as outlined, can be transferred without modification to electron optics if only electric fields are present, including the strong microscopic

* H. BOERSCH, *Naturwissenschaften*, **28** (1940), 709. BOERSCH's diffraction photographs are reproduced in W. GLASER, *Grundlagen der Elektronenoptik* (Wien, Springer, 1952), p. 548.

fields due to the atomic structure of solid matter, which are predominantly electric. If however magnetic fields are present, the wave-fronts can no longer be determined by plotting equal numbers of DE BROGLIE wavelengths along the trajectories. As has been pointed out at the end of Section 3 of this Appendix, the wave-fronts in magnetic fields are no longer orthogonal to the trajectories, but to the lines of total momentum or "$\boldsymbol{p}$-lines," which have no simple geometric-optical interpretation. In principle one could determine the wave-fronts by starting from some given wave-front, and plotting optical lengths along the $\boldsymbol{p}$-lines. Practically this method is almost useless, because it gives the phases, but not the amplitudes. In practice one is able to avoid this complication, by the artifice of imagining the magnetic lenses replaced by equivalent electric lenses, which produce the same image field, apart from a rotation of the image as a whole, and a certain rotational distortion of the image. By ignoring these specifically magnetic effects in the propagation process, and adding them only at the end, one can safely apply KIRCHHOFF's method, and obtain results which are sufficient for all practical purposes.

Though electron optics has the same general mathematical foundation as light optics, its practical structure is very different. Practical light optics has developed on the technical basis of grinding and polishing suitably shaped surfaces on transparent or reflecting solid media. In electron optics there is only one medium: the electromagnetic field. This has its limitations when it comes to the correction of aberrations, especially as space charges and space currents cannot be used for practical reasons. It was discovered early in the development of electron optics that it was impossible to construct diverging lenses and systems free from spherical or from chromatic aberrations, so long as one considered rotationally symmetrical fields only. This led to the development of systems without rotational symmetry, in which these limitations do not hold. and also to systems with curved optic axes, which are almost unknown in light optics. Hence the knowledge of optical instruments and design principles is useful in electron optics only in so far as it may pose problems and encourage inventions; the realizations are bound to be widely divergent. On the other hand optical thinking, based on the fundamental analogy of the two fields is likely to be as useful in the future as it was in the first quarter-century of electron optics.

APPENDIX III

ASYMPTOTIC APPROXIMATIONS TO INTEGRALS

THE purpose of this Appendix is to provide the mathematical background to methods of some generality, referred to in the main text, for obtaining asymptotic approximations to certain types of integral that frequently occur in optics.

1. The method of steepest descent

THIS is a method* for obtaining, for large values of k, asymptotic approximations to complex integrals of the type

$$\int g(z)e^{kf(z)}\,dz, \tag{1}$$

where $g(z)$ and $f(z)$ are independent of k.

For present purposes k is to be considered real and positive, and is assumed so in the later discussion. In fact, however, the results to be given are in general also true for complex values of k, and it is desirable to make a few prefatory remarks about asymptotic expansions of a function of a complex variable, ζ, say.

First, a definition of what is meant by an asymptotic expansion, due to POINCARÉ,† may be stated as follows: if

$$F(\zeta) = \sum_{m=0}^{n} \frac{a_m}{\zeta^m} + R_n(\zeta), \tag{2}$$

where, for all n, $\zeta^n R_n(\zeta) \to 0$ as $\zeta \to \infty$ for arg ζ within a given interval, $a_0, a_1, \ldots$ a_n being constants, then one writes

$$F(\zeta) \sim a_0 + \frac{a_1}{\zeta} + \frac{a_2}{\zeta^2} + \ldots, \tag{3}$$

and the right-hand side of (3) is called an asymptotic expansion of $F(\zeta)$ for the given range of arg ζ.

If $F(\zeta)$ is the quotient of two functions, say $G(\zeta)$ and $H(\zeta)$, one can write

$$G(\zeta) \sim H(\zeta)\left(a_0 + \frac{a_1}{\zeta} + \frac{a_2}{\zeta^2} + \ldots\right). \tag{4}$$

In fact it should be stressed that further on in the discussion the expansions considered are in the form of the right-hand side of (4), $H(\zeta)$ being exp $(a\zeta)$, where a is some constant.

Some of the main properties of asymptotic expansions are now briefly set out.‡ If the series in (3) terminates, or converges for sufficiently large $|\zeta|$, it is asymptotic; but it often fails to converge for any value of $|\zeta|$. In general, for a given $F(\zeta)$ a particular expansion holds only for a specified range of arg ζ; if it holds for all arg ζ

---

* Essentially due to P. DEBYE, *Math. Ann.*, **67** (1909), 535.

† H. POINCARÉ, *Acta Math.*, **8** (1886), 295.

‡ See, for example, H. and B. S. JEFFREYS, *Methods of Mathematical Physics* (Cambridge University Press, 1946), Chapter 17; or A. ERDÉLYI, *Asymptotic Expansions* (New York, Dover Publications, 1956).

then it converges. Again, for a given $F(\zeta)$ in the appropriate range of arg ζ the asymptotic expansion is unique, in the sense that the coefficients in (3) are unique; on the other hand the asymptotic expansion of any one function also belongs to an infinity of other functions, that of $F(\zeta) + e^{-\zeta}$, for example, being the same as that of $F(\zeta)$ for $-\frac{1}{2}\pi < \arg \zeta < \frac{1}{2}\pi$. The asymptotic expansion of the product of two functions is obtained by multiplying out their respective asymptotic expansions. Finally (3) may be integrated term by term to yield unconditionally the asymptotic expansion of the integral of $F(\zeta)$; it also may be differentiated term by term to yield, if it exists, the asymptotic expansion of the differential of $F(\zeta)$.

In general, for a given (sufficiently large) value of $|\zeta|$, the moduli of the terms of (3) start by decreasing successively to a minimum and then subsequently increase. Roughly speaking, if the expansion is summed to any term prior to the smallest, the error is of the order of the first omitted term.* Evidently, the larger $|\zeta|$ is, the greater is the available accuracy. In physical applications it is often sufficient to use the first term only; to borrow an illustration from electromagnetic theory, the radiation field of a finite source distribution is the first term of the asymptotic expansion of the complete field in inverse powers of distance from the source.

The basis of the method for obtaining the asymptotic expansion of (1) in inverse powers of k is to relate it to integrals of the form†

$$\int_0^\infty h(\mu) e^{-k\mu^2} d\mu. \tag{5}$$

For the asymptotic development of (5) can be easily derived; it results from expanding $h(\mu)$ as an ascending power series in μ and carrying out the integration term by term.

This last remark is, in fact, the substance of *Watson's lemma*,‡ which may be stated in the following form: Let

$$h(\mu) = \frac{1}{\mu^\alpha} \sum_{s=0}^{\infty} c_s \mu^{\beta s}, \tag{6}$$

with radius of convergence ρ, where β is real and positive and the real part of α is less than 1. Let a real number d exist such that $\mu^\alpha e^{-d\mu^2} h(\mu)$ is bounded for all real values of μ greater than ρ. Then (with Γ denoting the gamma function)

$$\frac{1}{2k^{\frac{1}{2}(1-\alpha)}} \left\{ c_0 \Gamma\left(\frac{-\alpha+1}{2}\right) + c_1 \Gamma\left(\frac{\beta-\alpha+1}{2}\right) \frac{1}{k^{\frac{1}{2}\beta}} + c_2 \Gamma\left(\frac{2\beta-\alpha+1}{2}\right) \frac{1}{k^{\beta}} \right.$$
$$\left. + c_3 \Gamma\left(\frac{3\beta-\alpha+1}{2}\right) \frac{1}{k^{\frac{3}{2}\beta}} + \ldots \right\} \tag{7}$$

is the asymptotic expansion of (5).

In cases of physical interest it is likely that α will be zero and β unity. Often too, as already suggested, the first term of (7) provides an adequate approximation.

---

* For a method of estimating the error see H. and B. S. Jeffreys, *loc. cit.*, Chapter 17.

† The choice of μ^2 in the exponential of the integrand rather than the first (or some other) power of μ is of no great significance; an even power is convenient because it oftens happens that the path of integration in (1) begins and ends at infinity in such a way that (1) can be transformed to (5) with the lower limit replaced by $-\infty$.

‡ G. N. Watson, *Theory of Bessel Functions* (Cambridge University Press, 2nd ed., 1948), Chapter 8. See also H. and B. S. Jeffreys, *loc. cit.*, Chapter 17.

In order to be able to change the variable of integration so that (1) is represented by one or more integrals of the type (5) it is necessary that the path of integration should be made up of sections along each of which the imaginary part of $f(z)$ is constant and the real part of $f(z)$ decreases monotonically to $-\infty$. If this is not the case, the first step is to distort the path appropriately. The path distortion is, of course, governed by the fundamental rules of integration in the complex plane; here it will merely be shown how it may be possible to close a path, by means of sections having the required properties, with the supposition that the evaluation of any remaining contour integral is to be attempted by standard techniques.

Some quite general remarks can be made about paths along which the imaginary part of $f(z)$ is constant, irrespective of the nature of the particular function. These will suffice to indicate how two points in the complex plane are joined in the desired manner, though the method may perhaps fail if $f(z)$ has singularities. In the present connection an especially important role is played by the points at which

$$\frac{df}{dz} = 0. \tag{8}$$

These are called *saddle-points*, because they are the points where the real and imaginary parts of $f(z)$ are stationary with respect to position in the complex plane, without being absolute maxima or minima.*

Now if

$$f(z) = u(x, y) + iv(x, y) \tag{9}$$

it is easy to show from the CAUCHY–RIEMANN relations that along any path $v(x, y) =$ constant the rate of change of $u(x, y)$ can only vanish at a saddle-point. In other words, on a path $v(x, y) =$ constant which does not traverse a saddle-point, $u(x, y)$ is strictly monotonic throughout, and on a path $v(x, y) =$ constant which goes through one or more saddle-points $u(x, y)$ is strictly monotonic between adjacent saddle-points and from the terminal saddle-points to the respective ends of the path. In fact, from any point, the direction in which $u(x, y)$ decreases most rapidly is one along which $v(x, y)$ is constant, and in this sense such paths are paths of *steepest descent*. Provided $f(z)$ is single-valued, or is made so by the use of branch-cuts, it follows that, starting from any point (x_0, y_0), a path $v(x, y) = v(x_0, y_0)$ can be chosen which continues always in the direction of decreasing $u(x, y)$ and terminates either at infinity or at a singularity. It is readily seen that, apart from singularities, the only points at which a path $v(x, y) =$ constant can go in more than one direction are where $df/dz = 0$; thus, once the path $v(x, y) = v(x_0, y_0)$ has been started correctly, with $u(x, y)$ decreasing, no question of an alternative route arises unless a saddle-point is encountered, in which case at least one direction of decreasing $u(x, y)$ will be available.

Now suppose that the end-points of the given path of integration in (1) are A and B, and that it is possible to find paths of steepest descent from A and B, respectively, to infinity. If both these paths terminate at infinity in the same section of convergence of the integral, then the procedure is complete; but if the terminations at infinity are in different sections of convergence, then one must be joined to the other by a path $v(x, y) =$ constant along which the rate of change of $u(x, y)$ changes sign only once (at a saddle-point), or if necessary by several such paths via intermediary

---

* The real and imaginary parts of an analytic function have no absolute maxima or minima in the complex plane.

regions at infinity. An asymptotic expansion, in the extended sense of (4), can be obtained for each of the different paths involved, but the true asymptotic expansion of the original integral is simply that for the path on which $u(x, y)$ attains its greatest value.* Similar remarks apply when the paths of steepest descent terminate in singularities.

It should be noted that there are cases outside the precise scope of the method just outlined which may nevertheless be treated by a slight modification. As a preliminary it is remarked that replacement of the upper limit in (5) by any positive number (independent of k) leaves the asymptotic expansion (7) unaltered. Thus the case can be treated involving a path of steepest descent along which $u(x, y)$ tends down to a finite value instead of $-\infty$. Again, it is possible to use a path $v(x, y) =$ constant traversing more than one point at which the rate of change of $u(x, y)$ changes sign.

Perhaps the commonest situation where the method of steepest descent is applicable is that in which the path of integration $v(x, y) =$ constant runs from a saddle-point to infinity with $u(x, y)$ decreasing monotonically all the way.† A well-known formula for the first term of the asymptotic expansion is then readily derived. For suppose the saddle-point is at z_0, and change the variable of integration in (1) by the transformation

$$f(z) = f(z_0) - \mu^2. \tag{10}$$

Then (1) becomes

$$-2e^{kf(z_0)}\int_0^\infty \frac{g(z)}{f'(z)}\,\mu e^{-k\mu^2}\,d\mu. \tag{11}$$

To get the first term in the asymptotic expansion of (11) the value of $\mu g(z)/f'(z)$ at $\mu = 0$ ($z = z_0$) is required. Provided $f''(z_0)$ is not zero and $g(z_0)$ is not infinite this is easily seen to be $g(z_0)/\sqrt{\{-2f''(z_0)\}}$, where the sign of the square root must be determined by an examination of each particular case. The required approximation is therefore

$$\sqrt{\left\{-\frac{\pi}{2f''(z_0)}\right\}}\, g(z_0)\,\frac{e^{kf(z_0)}}{\sqrt{k}}. \tag{12}$$

A few remarks should be added regarding the cases in which the approximation (12) needs supplementing or is invalid. The full asymptotic expansion is, of course, obtained by expanding $\mu g(z)/f'(z)$ as a power series in μ and carrying out the integration in (11) term by term. As indicated in (7), the power of k appearing in the first term of the asymptotic expansion is determined by the power of μ with which the series expansion of $\mu g(z)/f'(z)$ begins. If the first term of this series expansion is $A\mu^{-p}$, where p must be less than unity for the integral to converge, then the first term of the asymptotic expansion is

$$-A\,\Gamma\left(\frac{-p+1}{2}\right)\frac{e^{kf(z_0)}}{k^{\frac{1}{2}(1-p)}}. \tag{13}$$

---

* It is, of course, possible that the greatest value of $u(x, y)$ will be attained on more than one of the paths.

† Strictly the commonest case is that in which the path begins and ends at infinity with $u(x, y)$ running monotonically from $-\infty$ up to a maximum at a saddle-point, and then monotonically from the maximum down to $-\infty$. Formula (12) holds in this case if multiplied by 2, but for the subsequent discussion it is more convenient to take a path starting at the saddle-point. Because the case in which it is arranged that the path of integration starts from or traverses a saddle-point is so common, the method of steepest descent is sometimes called the *saddle-point method*.

Thus if $g(z_0)$ is infinite or $f''(z_0)$ is zero, (12) is invalid and must be replaced by (13). In these cases the factor multiplying $\exp\{kf(z_0)\}$ tends to zero less rapidly than $k^{-1/2}$, as k tends to infinity. If $g(z_0) = 0$ or $f''(z_0)$ is infinite (12) is not incorrect, but then merely states that the factor multiplying $\exp\{kf(z_0)\}$ tends to zero more rapidly than $k^{-1/2}$, as k tends to infinity.

If the path of integration is one of steepest descent running to infinity as before but starting from a point which is not a saddle-point, it is easy to see that the non-exponential part of the first term of the asymptotic expansion is, in general, proportional to k^{-1}, in contrast to the factor $k^{-1/2}$ of (12). Again, though, if $g(z)/f'(z)$ is singular or zero at the end point, the power of k is different, and depends on the order of the singularity or zero.

What has been said so far covers fairly comprehensively the results yielded by the method of steepest descent for asymptotic expansions in the strict mathematical sense, in which k is allowed to become indefinitely great, and other parameters are supposed to have prescribed values independent of k. But it has been shown that the form of the asymptotic approximation depends on particular conditions, for example, whether the path of steepest descent starts at a saddle-point or not. That is, it may depend on parameters other than k in the sense that it changes abruptly when these parameters take certain critical values.* Thus for any *given* value of k, no matter how large, if the values of other parameters are *sufficiently* close to those for which the asymptotic form would be different, the formulae mentioned above clearly fail to provide good numerical approximations. Hence there is a practical requirement for further expressions which will give smooth transitions from one asymptotic form to another. These are naturally more elaborate functions of k than (13), but it is worth mentioning three cases which can be satisfactorily treated. In the brief outline which follows it is tacitly assumed in each case that the only complexity is the specific one under discussion.

First, suppose that $f''(z_0)$ approaches zero. Then (12) is only a good approximation for increasingly large values of k, and an expression is required to effect the transition from (12) to the different form which is appropriate when $f''(z_0) = 0$. Since $f''(z_0)$ is nearly zero there must be a second saddle-point, z_1 say, near z_0. Then the required expression can be obtained by making the transformation

$$f(z) = \alpha - \beta\mu + \tfrac{1}{3}\mu^3,$$

where

$$\alpha = \tfrac{1}{2}[f(z_1) + f(z_0)], \qquad \tfrac{2}{3}\beta^{\frac{3}{2}} = \tfrac{1}{2}[f(z_1) - f(z_0)].$$

This transformation gives z as a regular function of μ in the vicinity of z_0 and z_1, and leads to an asymptotic development expressed in terms of the Airy integral and its first derivative with argument $k\beta$.†

Next, suppose that the expansion of $\mu g(z)/f'(z)$ as a power series in μ has a radius of convergence which approaches zero because $g(z)$ has a simple pole which gets close to the saddle-point. Again, (12) is only a good approximation for increasingly large values of k, and an expression is required to effect the transition from (12) to

* This is essentially a matter of *non-uniform* convergence. To take a trivial illustration, as $k \to \infty$, $1/(1 + ka) \sim 1/ka$ for $a \neq 0$, and ~ 1 for $a = 0$.

† The rôle of the Airy integral in uniform approximations of this type has long been recognized (see, for example, H. and B. S. Jeffreys, *loc. cit.*, Chapter 17), but only relatively recently has a satisfactory method, as outlined, been fully explored. (See C. Chester, B. Friedman, and F. Ursell, *Proc. Camb. Phil. Soc.*, **53** (1957), 599.)

the different form which is appropriate when the pole is at the saddle-point. This case has been treated by various authors.* The idea is to write the non-exponential part of the integrand as the sum of two terms, one containing the pole only, the other being regular. The latter can then be handled in the usual way and the former yields a FRESNEL or error integral, with, in general, a complex argument.

Finally, suppose that the starting point of a path of steepest descent is not a saddle-point, but approaches very close to one. In this case it is clear that the error integral can again be used to link the two asymptotic forms.†

2. The method of stationary phase

This is an alternative, if very similar, method to that of steepest descent. Though perhaps less general and less immediately convincing analytically, it often has the advantage of closer contact with the physical problem. The integral to be considered is more appropriately written

$$\int g(z)e^{ikf(z)}\,dz \tag{14}$$

than in the form (1), and in practice the exponential commonly represents a travelling wave.

Keeping to the notation of (9), the paths of integration used are $v(x, y) = \text{constant}$; but in (14) this means that the amplitude part of the exponential is constant along the path, whilst the phase part varies most rapidly, a reversal of the situation in the method of steepest descent. It can still be established that the only significant contribution to the integral arises from portions of the path in the vicinity of saddle-points or end-points, but the physical interpretation of the mechanism by which this comes about is now in terms of "phase interference" (cf. § 8.3, p. 386) rather than amplitude decay.

The method of stationary phase was first introduced explicitly by Lord KELVIN.‡ A rigorous mathematical treatment which would justify the statements made above was subsequently given by WATSON§; this treatment is based on the fact that, if $0 < m < 1$, a is a positive constant and $F(x)$ has limited total fluctuation in the range $x \geqslant 0$, then

$$k^m \int_0^a x^{m-1}\, F(x)e^{ikx}\, dx \to F(0)\ \Gamma(m)\, e^{\frac{1}{2}im\pi}$$

as $k \to \infty$.‖

WATSON's discussion, however, is rather restricted in scope. In particular, it does not seem capable of producing the complete asymptotic expansion. This is given in some detail by FOCKE¶ for the case in which $f(z)$ is a real function and the path of integration is confined to the real axis. FOCKE makes use of a neutralizing function, a method suggested earlier by VAN DER CORPUT.**

---

* W. PAULI, *Phys. Rev.*, **54** (1938), 925. H. OTT, *Ann. d. Physik* (5), **43** (1943), 393. P. C. CLEMMOW, *Quart. J. Mech. Appl. Maths.*, **3** (1950), 241; *Proc. Roy. Soc.*, A, **205** (1951), 286. B. L. VAN DER WAERDEN, *Appl. Sci. Res., Hague*, B, **2** (1952), 33.

† P. C. CLEMMOW, *Quart. J. Mech. Appl. Maths.*, **3** (1950), 241.

‡ W. THOMSON, *Proc. Roy. Soc.*, A, **42** (1887), 80.

§ G. N. WATSON, *Proc. Camb. Phil. Soc.*, **19** (1918), 49.

‖ A result given by T. J. I'A. BROMWICH, *An Introduction to the Theory of Infinite Series* (London, Macmillan, 1908), p. 447.

¶ J. FOCKE, *Ber. Sächs. Ges. (Akad.) Wiss.*, **101** (1954), Heft 3.

** J. G. VAN DER CORPUT, *Indag. Math.*, **10** (1948), 201.

The deductions from the method of stationary phase follow much the same pattern as those from the method of steepest descent. For example, in the case when the path of integration in (14) starts from a saddle-point at z_0 and runs to infinity along $v(x, y) = \text{constant}$ without encountering another saddle-point, the approximation corresponding to (12) is

$$\sqrt{\left\{-\frac{\pi}{2f''(z_0)}\right\}}\, g(z_0)e^{-\frac{1}{4}i\pi}\, \frac{e^{ikf(z_0)}}{\sqrt{k}}. \tag{15}$$

But one aspect in which there is some distinction between the methods should be noted. With a steepest descent path which starts at a saddle-point and does not go to infinity, the contribution of the point at the end of the path to the strict asymptotic expansion is zero in comparison with the saddle-point contribution by virtue of the extra exponential factor it contains. On the other hand, with a stationary phase path of the same type the contribution of the point at the end of the path is, in general, of the order of that of the saddle-point merely divided by $k^{1/2}$; it is excluded, therefore, from the asymptotic approximation only if the first term of the asymptotic expansion is alone retained.

To sum up, the methods of steepest descent and stationary phase, stripped of their mathematical expression, depend on choosing a path of integration in such a way that the integrand, by virtue of its exponential factor, contributes negligibly to the integral except in the vicinity of certain *critical points*, these being either saddle-points or end-points of the path of integration.

3. Double integrals

It was shown in § 8.3 and § 9.1 that the problem of a field diffracted by an aperture leads to double integrals of the form

$$\iint g(x, y)e^{ikf(x,y)}\, dx\, dy, \tag{16}$$

where $g(x, y)$ and $f(x, y)$ are independent of k, and the domain of integration is determined by the aperture. Clearly (16) is analogous to (14), and approximations for large values of k are likewise provided by asymptotic expansions.

The theory of asymptotic expansions of double integrals is naturally more complicated than that of single integrals. The techniques of integration in the complex plane are only readily applicable to single integrals, so for double integrals it seems as though the attack must be based on an approach somewhat different from that developed in the preceding pages of this Appendix.

The case in which $f(x, y)$ is a real function is comprehensively discussed by Focke,* who uses the concept of a neutralizing function, in the paper already referred to in connection with the application of this idea to single integrals. The analysis again shows that contributions to the asymptotic expansion come only from regions in the vicinity of certain *critical points*, and that different types of critical point give rise to different powers of k in the leading terms of their respective contributions.†

There are three types of critical point. The leading terms of their respective

* J. Focke (*loc. cit.*). See also D. S. Jones and M. Kline, *J. Math. Phys.*, **37** (1958), 1; N. Chako, *C.R. Acad. Sci. Paris*, **247** (1958), 436, 580, 637; M. Kline and I. W. Kay, *Electromagnetic Theory and Geometrical Optics* (New York, Interscience Publishers, 1965), Chapt. XII.

† Critical points for double integrals were briefly mentioned by J. G. van der Corput, *Indag. Math.*, **10** (1948), 201; and their qualitative significance for optics was pointed out by N. G. van Kampen, *Physica*, **14** (1949), 575.

contributions to the asymptotic expansion are now briefly discussed, excluding cases in which a critical point is of more than one type.

A *critical point of the first kind* is a point within the domain of integration at which

$$\frac{\partial f}{\partial x} = \frac{\partial f}{\partial y} = 0. \tag{17}$$

Then near the critical point, (x_0, y_0) say,

$$f(x, y) = f(x_0, y_0) + \tfrac{1}{2}\alpha(x - x_0)^2 + \tfrac{1}{2}\beta(y - y_0)^2 + \gamma(x - x_0)(y - y_0) + \ldots, \tag{18}$$

where $\alpha = \partial^2 f/\partial x^2$, $\beta = \partial^2 f/\partial y^2$, $\gamma = \partial^2 f/\partial x\, \partial y$, the partial derivatives all being evaluated at (x_0, y_0). Now choose certain new variables of integration ξ, η which are such that

$$f(x, y) = f(x_0, y_0) + \tfrac{1}{2}\alpha\xi^2 + \tfrac{1}{2}\beta\eta^2 + \gamma\xi\eta. \tag{19}$$

Then the required asymptotic approximation to (16) is

$$g(x_0, y_0)e^{ikf(x_0, y_0)}\int_{-\infty}^{\infty}\int_{-\infty}^{\infty} e^{\frac{1}{2}ik(\alpha\xi^2+\beta\eta^2+2\gamma\xi\eta)}\, d\xi\, d\eta = \frac{2\pi i\sigma}{\sqrt{|\alpha\beta - \gamma^2|}}\, g(x_0, y_0)\, \frac{e^{ikf(x_0,y_0)}}{k} \tag{20}$$

where the positive square root* is taken and

$$\sigma = \begin{cases} +1 \text{ for } \alpha\beta > \gamma^2,\ \alpha > 0, \\ -1 \text{ for } \alpha\beta > \gamma^2,\ \alpha < 0, \\ -i \text{ for } \alpha\beta < \gamma^2. \end{cases} \tag{21}$$

The expression (20) is the analogue for double integrals of the asymptotic approximation (15) for single integrals.

Critical points of the second kind are points on the curve bounding the domain of integration at which $\partial f/\partial s = 0$, where ds is an element of arc of the bounding curve. The power of k in the non-exponential part of the leading term of the corresponding contribution to the asymptotic expansion is $k^{-3/2}$, in contrast to the factor k^{-1} of (20).

Finally, *critical points of the third kind* are corner points on the curve bounding the domain of integration, that is, points at which the slope of the curve is discontinuous. In this case the corresponding factor is k^{-2}.

---

* The term under the square root sign has a simple geometrical interpretation. Consider the surface $z = f(x,y)$. Let R_1 and R_2 be the principal radii of curvature and $K = 1/R_1R_2$ the Gaussian curvature at a typical point of the surface. Then [see, for example, G. Salmon, *Analytic Geometry of Three Dimensions*, Vol. I, revised by R. A. P. Rogers (London, Longmans, Green & Co., 5th ed., 1912), p. 411],

$$K = \frac{f_{xx}f_{yy} - f_{xy}^2}{(1 + f_x^2 + f_y^2)^2},$$

where $f_x = \partial f/\partial x$, $f_{xx} = \partial^2 f/\partial x^2$, etc. At a critical point of the first kind $f_x = f_y = 0$, $f_{xx} = \alpha$, etc., and this expression reduces to

$$K = \alpha\beta - \gamma^2.$$

APPENDIX IV

THE DIRAC DELTA FUNCTION

THE purpose of this Appendix is to summarize the main properties of the delta function,* which we found useful for representing point sources, point charges, etc. This function which is used extensively in quantum mechanics as well as in classical applied mathematics, may be defined by the equations

$$\left.\begin{aligned} \delta(x) &= 0, \quad \text{when } x \neq 0, & \text{(a)} \\ \int_{-\infty}^{+\infty} \delta(x)dx &= 1. & \text{(b)} \end{aligned}\right\} \tag{1}$$

Evidently $\delta(x)$ is not a function in the ordinary mathematical sense,† since if a function is zero everywhere except at one point, and the integral of this function exists, the value of the integral is necessarily also equal to zero. It is more appropriate to regard $\delta(x)$ as a quantity with a certain symbolic meaning.

Consider a set of functions $\delta(x, \mu)$ which, with increasing μ, differ appreciably from zero only over a smaller and smaller x-interval around the origin and which are such that for all values of μ

$$\int_{-\infty}^{+\infty} \delta(x, \mu)dx = 1. \tag{2}$$

Examples are the functions [see Fig. 8]

$$\delta(x, \mu) = \frac{\mu}{\sqrt{\pi}} e^{-\mu^2 x^2}. \tag{3}$$

It is tempting to try to interpret the DIRAC delta function as a limit of such a set, for $\mu \to \infty$, but it must be noted that the limit of $\delta(x, \mu)$ need not exist for all x. However

$$\lim_{\mu \to \infty} \int_{-\infty}^{+\infty} \delta(x, \mu)dx \tag{4}$$

does exist and is equal to unity. *We interpret any operation involving $\delta(x)$ as implying that this operation is to be performed with a function $\delta(x, \mu)$ of a suitable chosen set such as* (3), *and that the limit $\mu \to \infty$ is taken at the end of the calculation.* With this interpretation (1b) evidently holds. The exact choice of the functions $\delta(x, \mu)$ is not important, provided that their oscillations (if any) near the origin are not too violent.

---

* Also known as the *impulse function*. It was brought into prominence by P. DIRAC, *The Principles of Quantum Mechanics* (Oxford, Clarendon Press, 1930), but was known to mathematicians and physicists much earlier, chiefly through the writings of O. HEAVISIDE. Cf. B. VAN DER POL and H. BREMMER, *Operational Calculus based on the two-sided Laplace Integral* (Cambridge University Press, 1950), pp. 62–66.

† The theory of the delta function may be made mathematically rigorous by using the notion of distributions, as developed by L. SCHWARTZ in his *Théorie des distributions* (Paris, Hermann et Cie., Vol. I (1950), Vol. II (1951)). A simplified version of SCHWARTZ' theory was developed by G. TEMPLE, *Proc. Roy. Soc.*, A, **228** (1955), 175, and is described fully in M. J. LIGHTHILL, *An Introduction to Fourier Analysis and Generalised Functions* (Cambridge University Press, 1958).

An important property of the DIRAC delta function is the so-called *sifting property*, expressed by the relation

$$\int_{-\infty}^{+\infty} f(x)\delta(x-a)dx = f(a). \tag{5}$$

Here $f(x)$ is any continuous function of x. The validity of (5) is seen immediately if $\delta(x-a)$ is replaced by $\delta(x-a, \mu)$ and the behaviour of the integral is examined for large values of μ. Evidently when μ is large

$$\int_{-\infty}^{+\infty} f(x)\delta(x-a, \mu)dx \tag{6}$$

depends essentially on the values of $f(x)$ in the immediate neighbourhood of the point $x = a$ only, and the error of replacing $f(x)$ by $f(a)$ may be made negligible by taking μ sufficiently large. Using (1b), equation (5) then follows. This result implies that the

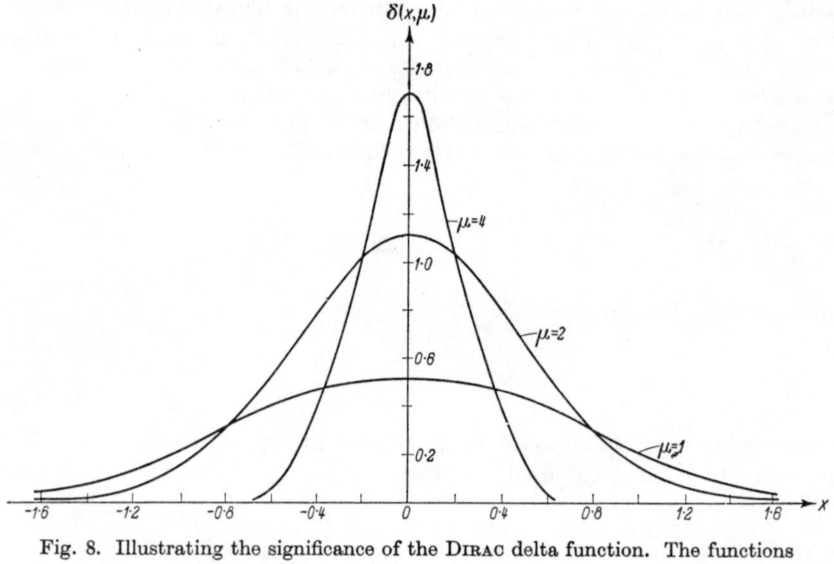

Fig. 8. Illustrating the significance of the DIRAC delta function. The functions

$$\delta(x, \mu) = \frac{\mu}{\sqrt{\pi}} e^{-\mu^2 x^2} \text{ for } \mu = 1, 2, 4.$$

The total area under each curve is unity.

process of multiplying a continuous function by $\delta(x-a)$ and integrating over all values of x is equivalent to the process of substituting a for the argument of the function. Actually, for this result to hold, the range of integration need not be taken from $-\infty$ to $+\infty$. It is only necessary that the domain of integration contains the point $x = a$ in its interior. The result is also written symbolically as

$$f(x)\delta(x-a) = f(a)\delta(x-a), \tag{7}$$

the meaning of such a relation being that the two sides give the same results when used as factors in an integral. In particular, with $f(x) = x$, $a = 0$, (7) gives

$$x\delta(x) = 0. \tag{8}$$

With a similar interpretation the following relations may easily be verified:

$$\delta(-x) = \delta(x), \tag{9}$$

$$\delta(ax) = \frac{1}{|a|}\,\delta(x), \tag{10}$$

$$\delta(x^2 - a^2) = \frac{1}{2|a|}\,\{\delta(x-a) + \delta(x+a)\}, \tag{11}$$

$$\int_{-\infty}^{+\infty} \delta(a-x)\delta(x-b)dx = \delta(a-b). \tag{12}$$

To verify the relation (10) for example, we compare the integrals of $f(x)\delta(ax)$ and of $f(x)\,\frac{1}{|a|}\,\delta(x)$. We have

$$\int_{-\infty}^{+\infty} f(x)\delta(ax)dx = \pm\int_{-\infty}^{+\infty} f\left(\frac{y}{a}\right)\,\delta(y)\,\frac{1}{a}\,dy = \frac{1}{|a|}f(0),$$

where the upper or lower sign is taken in front of the second integral according as $a \gtrless 0$. We also have

$$\int_{-\infty}^{+\infty} f(x)\,\frac{1}{|a|}\,\delta(x)dx = \frac{1}{|a|}f(0)$$

because of (5). The integrals are seen to be equal and this is the meaning of (10). Similarly (12) implies that if the two sides are multiplied by a continuous function of a or b and are integrated over all values of a or b respectively, an identity is obtained.

Let us next consider what interpretation may be given to the *derivatives* of the δ-function. We have, using the "approximation functions" $\delta(x, \mu)$ and integrating by parts

$$\int_{-\infty}^{+\infty} f(x)\delta'(x,\mu)dx = f(\infty)\delta(\infty,\mu) - f(-\infty)\delta(-\infty,\mu) - \int_{-\infty}^{+\infty} f'(x)\delta(x,\mu)dx.$$

Proceeding to the limit as $\mu \to \infty$, the first two terms on the right disappear and we obtain

$$\int_{-\infty}^{+\infty} f(x)\delta'(x)dx = -f'(0). \tag{13}$$

Repeating this process we find that

$$\int_{-\infty}^{+\infty} f(x)\delta^{(n)}(x)dx = (-1)^n f^{(n)}(0). \tag{14}$$

The following relations may easily be verified:

$$\delta'(-x) = -\delta'(x), \tag{15}$$

$$x\delta'(x) = -\delta(x). \tag{16}$$

It is often convenient (see, for example Appendix VI) to express the DIRAC delta function in terms of the *Heaviside unit function* (called also the *step function*) $U(x)$, this function being defined as

$$\left.\begin{aligned} U(x) &= 0, \qquad \text{when } x < 0, \\ &= 1, \qquad \text{when } x > 0. \end{aligned}\right\} \tag{17}$$

If as before, prime denotes the derivative with respect to the argument x, we obtain formally, on integrating by parts (with $x_1 > 0$, $x_2 > 0$),

$$\begin{aligned}\int_{-x_1}^{x_2} f(x)U'(x)dx &= \Big[f(x)U(x)\Big]_{-x_1}^{x_2} - \int_{-x_1}^{x_2} f'(x)U(x)dx \\ &= f(x_2) - \int_0^{x_2} f'(x)dx \\ &= f(x_2) - f(x_2) + f(0) \\ &= f(0).\end{aligned}$$

If we set $x = y - a$, $f(x) = f(y - a) = F(y)$ and proceed to the limit $x_1 \to \infty$, $x_2 \to \infty$, this becomes

$$\int_{-\infty}^{+\infty} F(y)U'(y - a)dy = F(a),$$

so that U' has the sifting property. In particular, with $F \equiv 1$, $a = 0$ this relation becomes

$$\int_{-\infty}^{+\infty} U'(y)dy = 1,$$

and shows that U' satisfies a relation of the form (1b). Moreover $U'(x) = 0$ when $x \neq 0$. Hence we may identify the derivative of the unit function with the delta function:

$$\delta(x) = \frac{d}{dx} U(x). \tag{18}$$

The delta function may also be introduced with the help of the FOURIER integral theorem

$$f(a) = \frac{1}{2\pi} \int_{-\infty}^{+\infty} dk \int_{-\infty}^{+\infty} f(x)e^{-ik(x-a)}\, dx. \tag{19}$$

If we set

$$K(x - a, \mu) = \frac{1}{2\pi} \int_{-\mu}^{\mu} e^{-ik(x-a)}\, dk = \frac{\sin \mu(x - a)}{\pi(x - a)} \tag{20}$$

and invert the order of integration, then (19) may be formally written as

$$f(a) = \int_{-\infty}^{+\infty} f(x)K(x - a)dx, \tag{21}$$

where $K(x - a)$ is regarded as the limit of $K(x - a, \mu)$ when $\mu \to \infty$. Strictly, this limit does not exist in the ordinary sense when $x - a \neq 0$,* but (21) has a similar symbolic meaning as the integrals discussed before, i.e. it should be interpreted as meaning that

$$f(a) = \lim_{\mu \to \infty} \int_{-\infty}^{+\infty} f(x)K(x - a, \mu)dx. \tag{22}$$

---

* The limit exists and has the value zero if interpreted in the sense of a CESÀRO limit [cf. B. VAN DER POL and H. BREMMER, *loc. cit.*, pp. 100–104].

Thus K has the sifting property. If we set $f(x) = 1$ in (21) we see that the integral of $K(x)$ taken over all x is equal to unity. We thus have another representation of the DIRAC delta function, viz.

$$\delta(x) = \frac{1}{2\pi} \int_{-\infty}^{+\infty} e^{-ikx}\, dk, \tag{23}$$

i.e. $\delta(x)$ may be regarded as the FOURIER transform of unity. There is a reciprocal relation, which follows from (21) on setting $f(x) = e^{ikx}$, $a = 0$,

$$\int_{-\infty}^{+\infty} \delta(x) e^{ikx}\, dx = 1. \tag{24}$$

So far we have considered a space of one dimension only, but the definition may easily be extended to spaces of several dimensions. In particular, consider a space of three dimensions. Then the function

$$\delta(x, y, z) = \delta(x)\delta(y)\delta(z), \tag{25}$$

often also denoted by $\delta(\boldsymbol{r})$, where $\boldsymbol{r}$ is the vector with components x, y, z, evidently satisfies relations analogous to (1), viz.

$$\left.\begin{aligned} &\delta(x, y, z) = 0, \text{ when } x \neq 0, y \neq 0, z \neq 0 && \text{(a)} \\ &\iiint_{-\infty}^{+\infty} \delta(x, y, z) dxdydz = 1. && \text{(b)} \end{aligned}\right\} \tag{26}$$

The sifting property is now expressed by the relation

$$\iiint_{-\infty}^{+\infty} f(x, y, z)\delta(x - a, y - b, z - c) dxdydz = f(a, b, c), \tag{27}$$

and $\delta(x, y, z)$ satisfies the FOURIER reciprocity relations

$$\delta(x, y, z) = \frac{1}{(2\pi)^3} \iiint_{-\infty}^{+\infty} e^{-i(k_x x + k_y y + k_z z)}\, dk_x dk_y dk_z, \tag{28}$$

$$\iiint_{-\infty}^{+\infty} \delta(x, y, z) e^{i(k_x x + k_y y + k_z z)}\, dxdydz = 1. \tag{29}$$

APPENDIX V

A MATHEMATICAL LEMMA USED IN THE RIGOROUS DERIVATION OF THE LORENTZ–LORENZ LAW (§ 2.4.2)

In this Appendix we shall establish a lemma used in § 2.4, according to which

$$\text{curl curl} \int_\sigma^\Sigma \boldsymbol{Q}(\boldsymbol{r}')G(R)dV' \to \int_\sigma^\Sigma \text{curl curl } \boldsymbol{Q}(\boldsymbol{r}')G(R)dV' + \frac{8\pi}{3}\boldsymbol{Q}(\boldsymbol{r}), \tag{1}$$

as $a \to 0$. Here $\boldsymbol{Q}(\boldsymbol{r})$ is an arbitrary vector function of position and $G(R) = e^{ikR}/R$. The integrals are taken throughout the volume bounded externally by a surface Σ and internally by a sphere σ of radius a, which is centered at a point P specified by the position vector $\boldsymbol{r}(x, y, z)$. R denotes the distance $|\boldsymbol{r} - \boldsymbol{r}'|$, where the vector $\boldsymbol{r}'(x', y', z')$ specifies the position of a typical volume element dV'.

Let $\boldsymbol{A}$ be an arbitrary vector function of position. The components of curl curl $\boldsymbol{A}$ are

$$(\text{curl curl } \boldsymbol{A})_x = \frac{\partial^2 A_y}{\partial y \partial x} + \frac{\partial^2 A_z}{\partial z \partial x} - \frac{\partial^2 A_x}{\partial y^2} - \frac{\partial^2 A_x}{\partial z^2},$$

etc., so that

$$\left(\text{curl curl} \int_\sigma^\Sigma \boldsymbol{Q} G \, dV'\right)_x$$

$$= \frac{\partial^2}{\partial y \partial x} \int_\sigma^\Sigma Q_y G dV' + \frac{\partial^2}{\partial z \partial x} \int_\sigma^\Sigma Q_z G dV' - \left(\frac{\partial^2}{\partial y^2} + \frac{\partial^2}{\partial z^2}\right) \int_\sigma^\Sigma Q_x G dV'. \tag{2}$$

Now we have for any differentiable scalar function $F(\boldsymbol{r}, \boldsymbol{r}')$,

$$\frac{\partial}{\partial x} \int_\sigma^\Sigma F dV' = \int_\sigma^\Sigma \frac{\partial F}{\partial x} dV' + \lim_{\delta x \to 0} \frac{1}{\delta x}\left[\int_{\sigma'}^\Sigma F dV' - \int_\sigma^\Sigma F dV'\right], \tag{3}$$

where σ' denotes a small sphere of radius a, centred on the point $T(x + \delta x, y, z)$. To evaluate the limit in (3), we note that the difference of the two integrals represents contributions from the two regions shown shaded in Fig. 9. The volume element may be expressed in the form $\delta V' = -\delta S' \, . \, \delta x \, . \, \rho_x$, where $\delta S'$ is the surface element and ρ_x is the x-component of the unit radial vector $\boldsymbol{\rho}$ pointing away from the point P. Hence

$$\lim_{\delta x \to 0} \frac{1}{\delta x}\left[\int_{\sigma'}^\Sigma F dV' - \int_\sigma^\Sigma F dV'\right] = -\int_\sigma F \rho_x \, dS'. \tag{4}$$

From (3) and (4) we have, if we set $F = Q_j(\boldsymbol{r})G(R)$ where Q_j ($j = x, y$ or z) is any of the Cartesian components of $\boldsymbol{Q}$,

$$\frac{\partial}{\partial x} \int_\sigma^\Sigma (Q_j G) dV' = \int_\sigma^\Sigma \frac{\partial}{\partial x}(Q_j G) dV' - \int_\sigma (Q_j G)\rho_x \, dS'. \tag{5}$$

Next we consider the partial derivatives of the second order. We differentiate (5) with respect to x and use (5) again. This gives

$$\frac{\partial^2}{\partial x^2}\int_\sigma^\Sigma (Q_j G)dV' = \int_\sigma^\Sigma \frac{\partial^2}{\partial x^2}(Q_j G)dV' - \int_\sigma \frac{\partial}{\partial x}(Q_j G)\rho_x\, dS' - \frac{\partial}{\partial x}\int_\sigma (Q_j G)\rho_x\, dS'. \quad (6)$$

Since

$$\left.\begin{aligned}\frac{\partial G}{\partial x} &= \frac{dG}{dR}\frac{\partial R}{\partial x} = -\rho_x \frac{d}{dR}\frac{e^{ikR}}{R} = \rho_x\left(\frac{1}{R} - ik\right)\frac{e^{ikR}}{R},\\ \text{and}\qquad dS' &= a^2 d\Omega,\end{aligned}\right\} \quad (7)$$

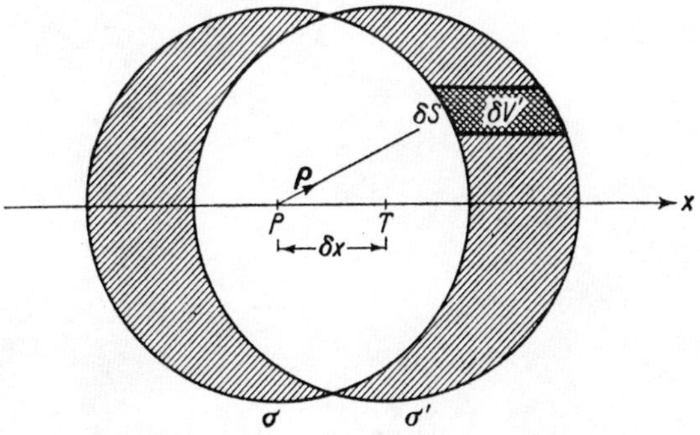

Fig. 9. Evaluation of

$$\lim_{\delta x\to 0}\frac{1}{\delta x}\left[\int_{\sigma'}^\Sigma F dV' - \int_\sigma^\Sigma F dV'\right].$$

The centre of σ is at $P(x, y, z)$, the centre of σ' is at $T(x + \delta x, y, z)$.

where $d\Omega$ is an element of the solid angle, it follows that

$$\int_\sigma \frac{\partial}{\partial x}(Q_j G)\rho_x dS' = \int_\sigma \rho_x Q_j \frac{\partial G}{\partial x} dS' =$$
$$\int_\Omega \rho_x^2 Q_j\, e^{ika}(1 - iak)d\Omega \to \frac{4\pi}{3} Q_j(\mathbf{r}) \quad \text{as} \quad a \to 0, \quad (8)$$

Ω denoting the surface of the unit sphere. The last integral in (6) tends to zero with a, so that we have, as $a \to 0$,

$$\frac{\partial^2}{\partial x^2}\int_\sigma^\Sigma (Q_j G)dV' \to \int_\sigma^\Sigma \frac{\partial^2}{\partial x^2}(Q_j G)dV' - \frac{4\pi}{3} Q_j(\mathbf{r}). \quad (9)$$

The mixed second order partial derivatives may be evaluated in the same way, and we have, for example,

$$\frac{\partial^2}{\partial y \partial x}\int_\sigma^\Sigma (Q_j G)dV' \to \int_\sigma^\Sigma \frac{\partial^2}{\partial y\partial x}(Q_j G)dV'. \quad (10)$$

The term $-\frac{4\pi}{3} Q_j(\mathbf{r})$ is now absent, since the integral corresponding to (8) is

$$\int_\sigma \frac{\partial}{\partial x}(Q_j G)\rho_y dS' = \int_\Omega \rho_x \rho_y Q_j (1 - iak)d\Omega$$

and this tends to zero with a.

On substituting from equations of the form (9) and (10) into (2) one finds that, as $a \to 0$,

$$\left[\text{curl curl} \int_{\sigma}^{\Sigma} (QG) dV'\right]_x \to \left[\int_{\sigma}^{\Sigma} \text{curl curl } (QG) dV'\right]_x + \frac{8\pi}{3} Q_x. \tag{11}$$

Similar expressions are obtained for the y and z components. Combining these three expressions into vector form, the formula (1) follows.

APPENDIX VI

PROPAGATION OF DISCONTINUITIES IN AN ELECTROMAGNETIC FIELD (§ 3.1.1)

It was mentioned in § 3.1.1 that the eikonal equation of geometrical optics is identical with an equation which describes rigorously the propagation of discontinuities in an electromagnetic field. More generally, the four equations § 3.1 (11a)–(14a) governing the behaviour of the electromagnetic field associated with the geometrical light rays may be shown to be identical with equations which connect the field vectors on a moving discontinuity surface. It is the purpose of this Appendix to demonstrate this mathematical equivalence.

1. Relations connecting discontinuous changes in field vectors

In § 1.1.3 we considered discontinuities in field vectors which arise from abrupt changes in the material parameters ε and μ, for example at a surface of a lens. Discontinuous fields may also arise from entirely different reasons, namely because a source suddenly begins to radiate. The field then spreads into the space surrounding the source and with increasing time fills a larger and larger region. On the boundary of this region the field has a discontinuity, the field vectors being in general finite inside this region and zero outside it. We shall first establish certain general relations which hold on any surface at which the field is discontinuous. For simplicity we assume that at any instant of time $t > 0$ there is only one such surface; the extension to several discontinuity surfaces (which may arise, for example, from reflections at obstacles present in the medium) is straightforward.

Let $F(x, y, z, t) = 0$ be any surface at which at least one of the field vectors is discontinuous. If this surface is fixed in space, F is, of course, independent of t. The points on either side of the surface may be distinguished by the inequalities $F < 0$ and $F > 0$ respectively. Let $\mathbf{E}$ be the electric vector and let

$$\left.\begin{aligned} \mathbf{E}(x, y, z, t) &= \mathbf{E}^{(1)}(x, y, z, t), \qquad \text{when } F(x, y, z, t) < 0, \\ &= \mathbf{E}^{(2)}(x, y, z, t), \qquad \text{when } F(x, y, z, t) > 0. \end{aligned}\right\} \tag{1}$$

$\mathbf{E}$ may then be written in the form

$$\mathbf{E} = \mathbf{E}^{(1)}\, U(-F) + \mathbf{E}^{(2)}\, U(F), \tag{2}$$

where U is the Heaviside unit function [cf. eq. (17), Appendix IV].

Using the representation (2), we derive expressions for the quantities curl $\mathbf{E}$, div $\mathbf{E}$, $\partial \mathbf{E}/\partial t$, etc., which enter Maxwell's equations. To differentiate a sum or a product that contains a discontinuous factor, we apply the ordinary rules of differentiation and use relation (18) of Appendix IV,

$$\frac{d}{dx}\, U(x) = \delta(x), \tag{3}$$

where δ is the Dirac delta function. Thus from (2) we have, for example,

$$\begin{aligned} \operatorname{curl} \mathbf{E} = U(-F) \operatorname{curl} \mathbf{E}^{(1)} &+ U(F) \operatorname{curl} \mathbf{E}^{(2)} \\ &+ [\operatorname{grad} U(-F)] \wedge \mathbf{E}^{(1)} + [\operatorname{grad} U(F)] \wedge \mathbf{E}^{(2)}. \end{aligned} \tag{4}$$

Now

$$\operatorname{grad} U(-F) = -\operatorname{grad} U(F) = -\frac{dU(F)}{dF}\operatorname{grad} F = -\delta(F)\operatorname{grad} F, \tag{5}$$

and from (4) and (5) it follows that

$$\operatorname{curl} \boldsymbol{E} = U(-F)\operatorname{curl} \boldsymbol{E}^{(1)} + U(F)\operatorname{curl} \boldsymbol{E}^{(2)} + \delta(F)\operatorname{grad} F \wedge (\Delta \boldsymbol{E}), \tag{6}$$

where

$$\Delta \boldsymbol{E} = \boldsymbol{E}^{(2)} - \boldsymbol{E}^{(1)}. \tag{7}$$

In a similar way we find

$$\operatorname{div} \boldsymbol{E} = U(-F)\operatorname{div} \boldsymbol{E}^{(1)} + U(F)\operatorname{div} \boldsymbol{E}^{(2)} + \delta(F)\operatorname{grad} F \cdot \Delta \boldsymbol{E}, \tag{8}$$

$$\frac{\partial \boldsymbol{E}}{\partial t} = U(-F)\frac{\partial \boldsymbol{E}^{(1)}}{\partial t} + U(F)\frac{\partial \boldsymbol{E}^{(2)}}{\partial t} + \delta(F)\frac{\partial F}{\partial t}\Delta \boldsymbol{E}. \tag{9}$$

Currents and charges may be represented in a similar form as the field vectors, but if the material parameters ε and μ are discontinuous at the surface $F = 0$, there may be additional terms representing the current charge density $\hat{j}$ and the surface charge density $\hat{\rho}$. Their contribution is given by equations (17a) and (18a) of § 1.1, and we have in all

$$\boldsymbol{j} = \boldsymbol{j}^{(1)}U(-F) + \boldsymbol{j}^{(2)}U(F) + \hat{\boldsymbol{j}}\,|\operatorname{grad} F|\,\delta(F), \tag{10}$$

$$\rho = \rho^{(1)}U(-F) + \rho^{(2)}U(F) + \hat{\rho}\,|\operatorname{grad} F|\,\delta(F). \tag{11}$$

We now substitute from equations (6), (8), (9), (10), (11) and from similar expressions involving the other field vectors into MAXWELL's equations § 1.1 (1)–(4). The terms with superscript (1) cancel out, and so do the terms with superscript (2), since the fields on either side of the discontinuity surface satisfy separately MAXWELL's equations. The remaining terms give the following *relations connecting the discontinuous changes in the field vectors*

$$\operatorname{grad} F \wedge \Delta \boldsymbol{H} - \frac{1}{c}\frac{\partial F}{\partial t}\Delta \boldsymbol{D} = \frac{4\pi}{c}\hat{\boldsymbol{j}}\,|\operatorname{grad} F|, \tag{12}$$

$$\operatorname{grad} F \wedge \Delta \boldsymbol{E} + \frac{1}{c}\frac{\partial F}{\partial t}\Delta \boldsymbol{B} = 0, \tag{13}$$

$$\operatorname{grad} F \cdot \Delta \boldsymbol{D} = 4\pi\hat{\rho}\,|\operatorname{grad} F|, \tag{14}$$

$$\operatorname{grad} F \cdot \Delta \boldsymbol{B} = 0. \tag{15}$$

It is of interest to note that these equations may be formally obtained from MAXWELL's equations by replacing the vectors $\boldsymbol{E}, \boldsymbol{H}, \boldsymbol{D}, \boldsymbol{B}$ by the differences $\Delta\boldsymbol{E}, \Delta\boldsymbol{H}, \Delta\boldsymbol{D}, \Delta\boldsymbol{B}$; $\boldsymbol{j}$ and ρ by $\hat{\boldsymbol{j}}$ and $\hat{\rho}$; and the differential operators $\frac{\partial}{\partial x}, \frac{\partial}{\partial y}, \frac{\partial}{\partial z}, \frac{\partial}{\partial t}$ by the multiplicative operators $\frac{1}{|\operatorname{grad} F|}\frac{\partial F}{\partial x}, \frac{1}{|\operatorname{grad} F|}\frac{\partial F}{\partial y}, \frac{1}{|\operatorname{grad} F|}\frac{\partial F}{\partial z}, \frac{1}{|\operatorname{grad} F|}\frac{\partial F}{\partial t}$.

Let $\boldsymbol{n}_{12}$ be the unit vector normal to the discontinuity surface and pointing from the region $F < 0$ (suffix 1), into the region $F > 0$ (suffix 2),

$$\boldsymbol{n}_{12} = \frac{\operatorname{grad} F}{|\operatorname{grad} F|}. \tag{16}$$

We also introduce the speed v with which the discontinuity surface advances. To a small displacement $\delta\mathbf{r}(\delta x, \delta y, \delta z)$ from a point on the discontinuity surface $F(x, y, z, t) = 0$ to a point on a neighbouring discontinuity surface, there corresponds a change δt in time, such that

$$\operatorname{grad} F \cdot \delta\mathbf{r} + \frac{\partial F}{\partial t}\,\delta t = 0. \tag{17}$$

In particular, for a displacement along the normal, $\delta\mathbf{r} = \delta s\, \mathbf{n}_{12}$, so that the speed v is given by

$$v = \frac{ds}{dt} = -\frac{1}{|\operatorname{grad} F|}\frac{\partial F}{\partial t}. \tag{18}$$

It follows that the relations (12)–(15) may also be written in the form*

$$\mathbf{n}_{12} \wedge \Delta\mathbf{H} + \frac{v}{c}\,\Delta\mathbf{D} = \frac{4\pi}{c}\hat{\mathbf{j}}, \tag{12a}$$

$$\mathbf{n}_{12} \wedge \Delta\mathbf{E} - \frac{v}{c}\,\Delta\mathbf{B} = 0, \tag{13a}$$

$$\mathbf{n}_{12} \cdot \Delta\mathbf{D} = 4\pi\hat{\rho} \tag{14a}$$

$$\mathbf{n}_{12} \cdot \Delta\mathbf{B} = 0. \tag{15a}$$

If the discontinuities in the field vectors arise from abrupt changes in the material parameters ε and μ on a surface $F(x, y, z) = 0$ which is fixed in space, then $v = 0$ and (12a)–(15a) reduce to equations (25), (23), (19), and (15) of § 1.1.

2. The field on a moving discontinuity surface

Consider a moving discontinuity surface which arises from the presence of a source which suddenly begins to radiate. We represent the surface in the form

$$F(x, y, z, t) \equiv \mathscr{S}(x, y, z) - ct = 0, \tag{19}$$

where c is the vacuum velocity of light. The field vectors on this discontinuity surface will be denoted by small letters,

$$\mathbf{e}(x, y, z) = \mathbf{E}\left[x, y, z, \frac{1}{c}\mathscr{S}(x, y, z)\right], \tag{20}$$

with similar expressions for the other field vectors. Now in the region outside the moving surface (say $F(x, y, z, t) > 0$) there is no field, so that according to (1) and

---

* These equations are a generalization of a set of discontinuity relations which appear to have been known already to O. Heaviside. See also A. E. H. Love, *Proc. Lond. Math. Soc.*, **1** (1904), 56; H. Bateman, *Partial Differential Equations of Mathematical Physics* (Cambridge University Press, 1932), p. 196; T. Levi-Civita, *Caractéristiques des Systèmes Différentiels et Propagation des Ondes* (Paris, Librairie Félix Alcan, 1932), § 10; R. K. Luneburg, *Mathematical Theory of Optics*, mimeographed lecture notes, Brown University, 1944; printed version published by University of California Press, Berkeley and Los Angeles, 1964, § 6 and § 7; M. Kline, *Comm. Pure and Appl. Math.*, **4** (1951), 239; A. Rubinowicz, *Acta Phys. Polonica*, **14** (1955), 209; M. Kline and I. W. Kay, *Electromagnetic Theory and Geometrical Optics* (New York, Interscience Publishers, 1965), pp. 37–51. The method of derivation used here is due to H. Bremmer, *Comm. Pure and Appl. Math.*, **4** (1951), 419.

(7) $\Delta \boldsymbol{E} = -\boldsymbol{E}^{(1)} = -\boldsymbol{e}$, etc. The material equations § 1.1 (10), (11) give $\boldsymbol{d} = \varepsilon \boldsymbol{e}$, $\boldsymbol{b} = \mu \boldsymbol{h}$, and if we also set $\hat{\boldsymbol{j}} = 0$, $\hat{\rho} = 0$, the relations (12)–(15) give

$$\text{grad}\, \mathscr{S} \wedge \boldsymbol{h} + \varepsilon \boldsymbol{e} = 0, \tag{21}$$

$$\text{grad}\, \mathscr{S} \wedge \boldsymbol{e} - \mu \boldsymbol{h} = 0, \tag{22}$$

$$\text{grad}\, \mathscr{S} \cdot \boldsymbol{e} = 0, \tag{23}$$

$$\text{grad}\, \mathscr{S} \cdot \boldsymbol{h} = 0. \tag{24}$$

These equations are formally identical with the basic equations [(11a)–(14a), § 3.1] of geometrical optics. Hence *the field vectors on a moving discontinuity surface obey rigorously the same equations as the field vectors associated with the geometrical optics approximation to a time-harmonic field, the moving discontinuity surface corresponding to the geometrical wave-fronts.**

Evidently the moving discontinuity surface must obey the eikonal equation

$$\left(\frac{\partial \mathscr{S}}{\partial x}\right)^2 + \left(\frac{\partial \mathscr{S}}{\partial y}\right)^2 + \left(\frac{\partial \mathscr{S}}{\partial z}\right)^2 = n^2, \tag{25}$$

where $n^2 = \varepsilon\mu$. This equation follows as before. It is the consistency condition between the equations (21) and (22) and follows from them on eliminating $\boldsymbol{h}$ or $\boldsymbol{e}$ and using (23) or (24). According to (18), (19), and (25), the discontinuity surface is propagated with the velocity $v = c/n$.

* It may also be shown that $\boldsymbol{e}$ and $\boldsymbol{h}$ obey the same transport equations [eq. (41) and (42), § 3.1] as the complex amplitude vectors of the geometrical optics field. This result was first established by R. K. Luneburg, *loc. cit.*, 1944, p. 46; *loc. cit.*, 1964, p. 44. See also E. T. Copson, *Comm. Pure and Appl. Math.*, **4** (1951), 427; M. Kline and I. W. Kay, *loc. cit.*, p. 162.

APPENDIX VII

THE CIRCLE POLYNOMIALS OF ZERNIKE (§ 9.2.1)

IN this Appendix, the circle polynomials discussed briefly in § 9.2.1 will be considered more fully. These polynomials were introduced and first investigated by ZERNIKE,* in his important paper on the knife edge test and the phase contrast method, and were studied further by him and BRINKMAN,† and by NIJBOER.‡ They were later derived from the requirement of orthogonality and invariance alone by BHATIA and WOLF§; we shall follow substantially their treatment.

1. Some general considerations

It is not difficult to show that there exists an infinity of complete sets of polynomials in two real variables x, y which are orthogonal for the interior of the unit circle, i.e. which satisfy the orthogonality condition

$$\iint\limits_{x^2+y^2\leqslant 1} V^{\star}_{(\alpha)}(x, y)\, V_{(\beta)}(x, y)dxdy = A_{\alpha\beta}\, \delta_{\alpha\beta}. \tag{1}$$

Here $V_{(\alpha)}$ and $V_{(\beta)}$ denote two typical polynomials of the set, the asterisk denotes complex conjugate, δ is the KRONECKER symbol, and $A_{\alpha\beta}$ are normalization constants, to be chosen later. The circle polynomials of ZERNIKE are distinguished from the other sets by certain simple invariance properties which can best be explained from group theoretical considerations. It is however possible to avoid the abstract formalism of group theory with the help of a kind of normalization. One considers first of all such sets as are "invariant in form" with respect to rotations of axes about the origin. By such invariance we mean that when any rotation

$$\left.\begin{aligned} x' &= x \cos \phi + y \sin \phi, \\ y' &= -x \sin \phi + y \cos \phi, \end{aligned}\right\} \tag{2}$$

is applied, each polynomial $V(x, y)$ is transformed into a polynomial of the same form, i.e. V satisfies the following relation under the transformation (2):

$$V(x, y) = G(\phi) V(x', y'), \tag{3}$$

where $G(\phi)$ is a continuous function with period 2π of the angle of rotation ϕ, and $G(0) = 1$.

Now the application of two successive rotations through angles ϕ_1 and ϕ_2 is equivalent to a single rotation through an angle $\phi_1 + \phi_2$. Hence it follows from (3) that G must satisfy the functional equation

$$G(\phi_1)G(\phi_2) = G(\phi_1 + \phi_2). \tag{4}$$

---

* F. ZERNIKE, *Physica*, **1** (1934), 689.

† F. ZERNIKE and H. C. BRINKMAN, *Verh. Akad. Wet. Amst.* (Proc. Sec. Sci.), **38** (1935), 11.

‡ B. R. A. NIJBOER, *Thesis* (University of Groningen, 1942).

§ A. B. BHATIA and E. WOLF, *Proc. Camb. Phil. Soc.*, **50** (1954), 40.

The general solution with the period 2π of this equation is well known, and is*

$$G(\phi) = e^{il\phi}. \tag{5}$$

Here l is any integer, positive, negative, or zero. On substituting from (5) into (3), setting $x' = \rho$, $y' = 0$, and using (2) it follows that V must be of the form

$$V(\rho \cos \phi, \rho \sin \phi) = R(\rho)e^{il\phi}, \tag{6}$$

where $R(\rho) = V(\rho, 0)$ is a function of ρ alone. Next we expand $e^{il\phi}$ in powers of $\cos \phi$ and $\sin \phi$. Suppose that V is a polynomial of degree n in the variables $x = \rho \cos \phi$, $y = \rho \sin \phi$; it then follows from (6) that $R(\rho)$ is a polynomial in ρ of degree n and contains no power of ρ of degree lower than $|l|$. Moreover $R(\rho)$ is evidently an even or an odd polynomial according as l is even or odd. The set of *the Zernike circle polynomials* is distinguished from all other such sets by the property that it *contains a polynomial for each pair of the permissible values of n (degree) and l (angular dependence)*, i.e. for integral values of n and l, such that $n \geqslant 0$, $l \gtreqless 0$, $n \geqslant |l|$, and $n - |l|$ is even. We denote a typical polynomial of this set by

$$V_n^l(\rho \cos \phi, \rho \sin \phi) = R_n^l(\rho)e^{il\phi}. \tag{7}$$

It follows from (1) and (7) that the *radial polynomials* $R_n^l(\rho)$ satisfy the relation

$$\int_0^1 R_n^l(\rho)R_{n'}^l(\rho)\rho d\rho = a_n^l\delta_{nn'}, \tag{8}$$

where

$$a_n = \frac{A_n^l}{2\pi}. \tag{9}$$

For any given value l, the lower index n can only take the values $|l|$, $|l| + 2$, $|l| + 4$, ... The corresponding sequence $R_{|l|}^l(\rho)$, $R_{|l|+2}^l(\rho)$, $R_{|l|+4}^l(\rho)$ . . . may be obtained by orthogonalizing the powers

$$\rho^{|l|}, \quad \rho^{|l|+2}, \quad \rho^{|l|+4}, \ldots \tag{10}$$

with the weighting factor ρ over the interval $0 \leqslant \rho \leqslant 1$. Moreover, since only the absolute values of l occur in (10),

$$R_n^{-l}(\rho) = \beta_n^l R_n^l(\rho), \tag{11}$$

where β_n^l is a constant depending only on the normalization of the two polynomials R_n^{-l} and R_n^l. In particular we may normalize in such a way that $\beta_n^l = 1$ for all l and n, and then

$$V_n^{\pm m}(\rho \cos \phi, \rho \sin \phi) = R_n^m(\rho)e^{\pm im\phi}, \tag{12}$$

where $m = |l|$ is a non-negative integer.

The set of the circle polynomials contains $\frac{1}{2}(n + 1)(n + 2)$ linearly independent polynomials of degree $\leqslant n$. Hence every monomial x^iy^j ($i \geqslant 0, j \geqslant 0$ integers) and, consequently every polynomial in x, y may be expressed as a linear combination of a finite number of the circle polynomials V_n^l. By WEIERSTRASS' approximation theorem† it then follows that the set is *complete*.

---

* See, for example, M. BORN, *Natural Philosophy of Cause and Chance* (Oxford, Clarendon Press, 1949, p. 153; Dover Publications, New York, 1964), p. 153.

† See, for example, R. COURANT and D. HILBERT, *Methods of Mathematical Physics*, Vol. 1 (New York, Interscience Publishers, 1st English edition, 1953), p. 65.

2. Explicit expressions for the radial polynomials $R_n^{\pm m}(\rho)$

Since $R_n^{\pm m}(\rho)$ is a polynomial in ρ of degree n and contains no power of ρ lower than m and is an even or odd polynomial according as n is even or odd, it follows that R may be expressed in the form

$$R_n^{\pm m}(\rho) = t^{\frac{m}{2}} Q_{\frac{n-m}{2}}(t), \tag{13}$$

where $t = \rho^2$ and $Q_{\frac{n-m}{2}}(t)$ is a polynomial in t of degree $\frac{1}{2}(n-m)$. According to (8), the Q polynomials must satisfy the relations

$$\left.\begin{aligned} \tfrac{1}{2}\int_0^1 t^m Q_k(t) Q_{k'}(t) dt &= a_n^{\pm m}\delta_{kk'}, \\ \text{where} \qquad k = \tfrac{1}{2}(n-m), \qquad k' &= \tfrac{1}{2}(n'-m). \end{aligned}\right\} \tag{14}$$

It follows that the polynomials $Q_0(t)$, $Q_1(t)$, . . . $Q_k(t)$ . . . may be obtained by orthogonalizing the sequence of natural powers

$$1, t, t^2, \ldots\ldots t^k, \ldots \tag{15}$$

with the weighting factor $w(t) = t^m$ over the range $0 \leqslant t \leqslant 1$. Now the well-known JACOBI (or hypergeometric) polynomials*

$$G_k(p, q, t) = \frac{(q-1)!}{(q+k-1)!} t^{1-q}(1-t)^{q-p} \frac{d^k}{dt^k}[t^{q-1+k}(1-t)^{p-q+k}] \tag{16}$$

$$= \frac{k!(q-1)!}{(p+k-1)!} \sum_{s=0}^{k} (-1)^s \frac{(p+k+s-1)!}{(k-s)!s!(q+s-1)!} t^s, \tag{17}$$

($k \geqslant 0, q > 0, p - q > -1$) may be defined as functions obtained by orthogonalizing (15) with the more general weighting function

$$w(t) = t^{q-1}(1-t)^{p-q}$$

over the range $0 \leqslant t \leqslant 1$. Their orthogonality and normalization properties are given by†

$$\int_0^1 t^{q-1}(1-t)^{p-q} G_k(p, q, t) G_{k'}(p, q, t) dt = b_k(p, q)\delta_{kk'}, \tag{18}$$

where

$$b_k(p, q) = \frac{k![(q-1)!]^2[p-q+k]!}{[q-1+k]![p-1+k]![p+2k]}. \tag{19}$$

(With this choice of b_k, $G_k(p, q, 0) = 1$ for all k.) On comparing (18) and (14) it follows that‡

$$Q_k(t) = \sqrt{\frac{2a_n^{\pm m}}{b_k(m+1, m+1)}}\, G_k(m+1, m+1, t). \tag{20}$$

---

* R. COURANT and D. HILBERT, *loc. cit.*, Vol. I, p. 90.

† E. KEMBLE, *The Fundamental Principles of Quantum Mechanics* (New York, McGraw-Hill, 1937), p. 594.

‡ The sign of the square root on the right hand side of eq. (20) is determined from eq. (26) below.

From (13) and (20) we obtain the following expression for the radial polynomials in terms of JACOBI polynomials:

$$R_n^{\pm m}(\rho) = \sqrt{\frac{2a_n^{\pm m}}{b_k(m+1, m+1)}}\, \rho^m G_k(m+1, m+1, \rho^2) \quad [k = \tfrac{1}{2}(n-m)]. \tag{21}$$

Following ZERNIKE, we choose the normalization so that for all n and m

$$R_n^{\pm m}(1) = 1. \tag{22}$$

Then from (21) and (22)

$$\sqrt{\frac{b_k(m+1, m+1)}{2a_n^{\pm m}}} = G_k(m+1, m+1, 1). \tag{23}$$

The value of $G_k(m+1, m+1, 1)$ can be obtained from the generating function for the JACOBI polynomials.* We have

$$\frac{[z - 1 + \sqrt{1 - 2z(1-2\rho^2) + z^2}]^m}{(2z\rho^2)^m \sqrt{1 - 2z(1-2\rho^2) + z^2}} = \sum_{s=0}^{\infty} \binom{m+s}{s} G_s(m+1, m+1, \rho^2) z^s. \tag{24}$$

For $\rho = 1$, the left-hand side reduces to $(1+z)^{-1}$ and we obtain on expanding it in a power series and on comparing it with the right-hand side:

$$G_s(m+1, m+1, 1) = \frac{(-1)^s}{\binom{m+s}{s}}. \tag{25}$$

From (25) and (23) it follows that

$$\sqrt{\frac{2a_n^{\pm m}}{b_k(m+1, m+1)}} = (-1)^{\frac{n-m}{2}} \binom{\frac{1}{2}(n+m)}{\frac{1}{2}(n-m)}, \tag{26}$$

and using (16), (17), and (26) we finally obtain from (21) the following expressions for the radial polynomials:

$$R_n^{\pm m}(\rho) = \frac{1}{\left(\frac{n-m}{2}\right)!\, \rho^m} \left\{\frac{d}{d(\rho^2)}\right\}^{\frac{n-m}{2}} \left\{(\rho^2)^{\frac{n+m}{2}} (\rho^2 - 1)^{\frac{n-m}{2}}\right\} \tag{27}$$

$$= \sum_{s=0}^{\frac{1}{2}(n-m)} (-1)^s \frac{(n-s)!}{s!\left(\frac{n+m}{2} - s\right)!\left(\frac{n-m}{2} - s\right)!} \rho^{n-2s}. \tag{28}$$

Explicit expressions for the first few polynomials are given in Table XXI on p. 465.

The normalization constant $a_n^{\pm m}$ is obtained from (26) and (19):

$$a_n^{\pm m} = \frac{1}{2n+2}. \tag{29}$$

---

* Cf. R. COURANT and D. HILBERT, *loc. cit.*, Vol. I, p. 91.

To obtain the generating function for the radial polynomials, we write s in place of $k = (n - m)/2$ and $m + 2s$ in place of n in (21) and (26) and substitute into (24). We then obtain

$$\frac{[1 + z - \sqrt{1 + 2z(1 - 2\rho^2) + z^2}]^m}{(2z\rho)^m \sqrt{1 + 2z(1 - 2\rho^2) + z^2}} = \sum_{s=0}^{\infty} z^s R_{m+2s}^{\pm m}(\rho). \tag{30}$$

Finally we shall evaluate the integral

$$\int_0^1 R_n^m(\rho) J_m(v\rho)\rho d\rho$$

which, as we saw in Chapter 9, plays an important part in the ZERNIKE–NIJBOER diffraction theory of aberrations. We substitute for $R_n^m(\rho)$ the expression (27) and for the BESSEL function J_m its series expansion.* The resulting expression may be written in the form

$$\int_0^1 R_n^m(\rho) J_m(v\rho)\rho d\rho$$

$$= \frac{1}{2\left(\dfrac{n-m}{2}\right)!} \sum_{s=0}^{\infty} \frac{(-1)^s}{s!\,(s+m)!} \left(\frac{v}{2}\right)^{m+2s} f\left(s, \frac{n-m}{2}, \frac{n+m}{2}, \frac{n-m}{2}\right), \tag{31}$$

where

$$f(s, p, q, r) = \int_0^1 u^s \left(\frac{d}{du}\right)^p \{u^q (u-1)^r\} du, \tag{32}$$

p, q, r, s being non-negative integers. Integrating (32) by parts it follows that

$$f(s, p, q, r) = \left\{u^s \left(\frac{d}{du}\right)^{p-1} [u^q(u-1)^r]\right\}_0^1 - s\int_0^1 u^{s-1} \left(\frac{d}{du}\right)^{p-1} \{u^q(u-1)^r\} du. \tag{33}$$

Now if $r \geqslant p$ and $s + q - p \geqslant 0$, the first term on the right vanishes, so that

$$f(s, p, q, r) = -s f(s-1, p-1, q, r). \tag{34}$$

We consider separately the case $s \geqslant p$ and $s < p$.

When $s \geqslant p$, we have, on applying (34) p times,

$$f(s, p, q, r) = (-1)^p s(s-1)(s-2) \ldots (s-p+1) f(s-p, 0, q, r)$$

$$= \frac{(-1)^{p+r} s!}{(s-p)!} \int_0^1 u^{s+q-p}(1-u)^r du. \tag{35}$$

The integral in (35) is the EULER integral of the first kind (beta function) and its value is† $(s + q - p)!\, r!/(s + q + r - p + 1)!$ Hence, for $s \geqslant p$,

$$f(s, p, q, r) = (-1)^{p+r} \frac{s!\,(s+q-p)!\,r!}{(s-p)!\,(s+q+r-p+1)!}. \tag{36}$$

---

* R. COURANT and D. HILBERT, *loc. cit.*, Vol. I, p. 484.

† R. COURANT and D. HILBERT, *loc. cit.*, Vol. I, p. 483.

Next consider the case $s < p$. Applying (34) s times it follows that

$$
\begin{aligned}
f(s, p, q, r) &= (-1)^s s(s-1)\ldots f(0, p-s, q, r) \\
&= (-1)^s s!\left\{\left(\frac{d}{du}\right)^{p-s-1}[u^q(u-1)^r]\right\}_0^1 \\
&= 0.
\end{aligned}
\tag{37}
$$

We now substitute from (36) and (37) into (31) and introduce a new variable l such that $s = \frac{1}{2}(n-m)+l$. We then obtain

$$
\int_0^1 R_n^m(\rho)J_m(v\rho)\rho d\rho = \frac{(-1)^{\frac{3(n-m)}{2}}}{v}\sum_{l=0}^{\infty}\frac{(-1)^l}{l!\,(n+l+1)!}\left(\frac{v}{2}\right)^{n+2l+1}. \tag{38}
$$

The series on the right will be recognized as the expansion of $J_{n+1}(v)$. Since $n-m$ is even the factor $(-1)^{\frac{3(n-m)}{2}}$ may be replaced by $(-1)^{\frac{n-m}{2}}$, and we finally obtain

$$
\int_0^1 R_n^m(\rho)J_m(v\rho)\rho d\rho = (-1)^{\frac{n-m}{2}}\frac{J_{n+1}(v)}{v}. \tag{39}
$$

APPENDIX VIII

PROOF OF AN INEQUALITY (§ 10.7.3)

LET $f(\tau)$ and $g(\tau)$ be any two functions, generally complex, of the real variable τ and let λ be a real parameter. Then

$$\int_{-\infty}^{+\infty} |f + \lambda g^\star|^2 d\tau = \int_{-\infty}^{+\infty} (f + \lambda g^\star)(f^\star + \lambda g) d\tau \geqslant 0, \tag{1}$$

or

$$\int_{-\infty}^{+\infty} ff^\star d\tau + \lambda \int_{-\infty}^{+\infty} (fg + f^\star g^\star) d\tau + \lambda^2 \int_{-\infty}^{+\infty} gg^\star d\tau \geqslant 0. \tag{2}$$

The minimum of this quadratic expression in λ is obtained by differentiating:

$$\int_{-\infty}^{+\infty} (fg + f^\star g^\star) d\tau + 2\lambda \int_{-\infty}^{+\infty} gg^\star d\tau = 0. \tag{3}$$

The root $\lambda = \lambda_{\text{min}}$ of this expression is

$$\lambda_{\text{min}} = -\frac{1}{2} \frac{\displaystyle\int_{-\infty}^{+\infty} (fg + f^\star g^\star) d\tau}{\displaystyle\int_{-\infty}^{+\infty} gg^\star d\tau}. \tag{4}$$

If this value is substituted into (2) we obtain

$$4 \left(\int_{-\infty}^{+\infty} ff^\star d\tau \right) \left(\int_{-\infty}^{+\infty} gg^\star d\tau \right) \geqslant \left(\int_{-\infty}^{+\infty} (fg + f^\star g^\star) d\tau \right)^2. \tag{5}$$

Let

$$f = \tau \psi(\tau), \qquad g = \frac{d\psi^\star(\tau)}{d\tau}. \tag{6}$$

Then

$$fg + f^\star g^\star = \tau \left(\psi \frac{d\psi^\star}{d\tau} + \psi^\star \frac{d\psi}{d\tau} \right) = \tau \frac{d}{d\tau} (\psi \psi^\star), \tag{7}$$

and (5) becomes, if we integrate by parts on the right and assume that* $\tau \psi \psi^\star \to 0$ as $\tau \to \pm \infty$,

$$4 \left(\int_{-\infty}^{+\infty} \tau^2 \psi \psi^\star d\tau \right) \left(\int_{-\infty}^{+\infty} \frac{d\psi}{d\tau} \frac{d\psi^\star}{d\tau} d\tau \right) \geqslant \left(\int_{-\infty}^{+\infty} \psi \psi^\star d\tau \right)^2. \tag{8}$$

This is the required inequality.

The equals sign in (8) can only hold if it holds in (1); this is only possible if $f \equiv -\lambda g^\star$, or, using (6), if

$$\frac{d\psi}{d\tau} = -\frac{1}{\lambda} \tau \psi. \tag{9}$$

---

* This condition is in fact satisfied whenever the integrals on the left in (8) converge [cf. H. WEYL, *The Theory of Groups and Quantum Mechanics*, translated from German (London, Methuen, 1931; also, New York, Dover Publications, Inc.), pp. 393–394].

The general solution of this differential equation is

$$\psi(\tau) = Ae^{-\frac{\tau^2}{2\lambda}}, \tag{10}$$

where A is a constant. Only solutions with $\lambda \geqslant 0$ apply, since otherwise $\psi(\tau)$ would not vanish at infinity. Hence (8) *becomes an equality if and only if* ψ *is a Gaussian function.*

APPENDIX IX

EVALUATION OF TWO INTEGRALS (§ 12.2.2)*

IN this Appendix we shall evaluate the two integrals of § 12.2 (8), (9)

$$\mathscr{J}_1 = \frac{1}{4\pi} \iiint\limits_{V_1} \left\{ e^{i(px_1+qy_1)} \frac{\partial^2}{\partial z_1^2} \left(\frac{e^{i\omega R/c}}{R} \right) \right\} dx_1 dy_1 dz_1, \tag{1}$$

$$\mathscr{J}_2 = \frac{1}{4\pi} \frac{\omega^2}{c^2} \iiint\limits_{V_1} \left\{ e^{i(px_1+qy_1)} \left(\frac{e^{i\omega R/c}}{R} \right) \right\} dx_1 dy_1 dz_1, \tag{2}$$

where

$$R = +\sqrt{x_1^2 + y_1^2 + z_1^2} \tag{3}$$

and $\omega^2 > c^2 p^2$. We are interested in two cases, namely:

(*a*) $0 < y < d$, where y and d are constants and the volume V_1 of integration is the slab $-\infty < x_1 < \infty$, $-y \leqslant y_1 \leqslant d - y$, $-\infty < z_1 < \infty$, except for a small sphere of vanishingly small radius around the origin $x_1 = y_1 = z_1 = 0$.

(*b*) $y > d$ or $y < 0$, the volume V_1 of integration now being the full region $-\infty < x_1 < \infty$, $-y \leqslant y_1 \leqslant d - y$, $-\infty < z_1 < \infty$.

To evaluate $\mathscr{J}_1$ we apply GAUSS's theorem, in the form

$$\iiint\limits_{V_1} \operatorname{div} \boldsymbol{G}\, dV_1 = \iint\limits_{S_1} \boldsymbol{G} \,.\, \boldsymbol{n}\, dS_1, \tag{4}$$

where $\boldsymbol{G}$ is an arbitrary vector function of position and $\boldsymbol{n}(n_x, n_y, n_z)$ is the unit outward normal to the surface S_1 bounding the volume V_1. We take

$$G_{x_1} = G_{y_1} = 0, \qquad G_{z_1} = \frac{1}{4\pi} \frac{\partial}{\partial z_1} \left[\frac{e^{i(px_1+qy_1+\omega R/c)}}{R} \right] \tag{5}$$

and obtain from (1) and (4)

$$\mathscr{J}_1 = \frac{1}{4\pi} \iint\limits_{S_1} n_{z_1} \frac{\partial}{\partial z_1} \left(\frac{e^{i(px_1+qy_1+\omega R/c)}}{R} \right) dS_1. \tag{6}$$

Since n_{z_1} is zero at each face ($y_1 = -y$ and $y_1 = d - y$) of the slab, the integral $\mathscr{J}_1$ is zero in cases (*b*)†. In case (*a*), we must also include the contribution from the small sphere σ_1 around the origin. If a is the radius of the sphere, we have $n_z = -z_1/a$ and the contribution is

$$\frac{1}{4\pi} \iint\limits_{\sigma_1} \left(\frac{z_1}{a} \right)^2 e^{i(px_1+qy_1+\omega a/c)} \left(\frac{1}{a^2} - \frac{i\omega}{ac} \right) d\sigma_1 \to \tfrac{1}{3} \quad \text{as } a \to 0. \tag{7}$$

---

* After C. G. DARWIN, *Trans. Camb. Phil. Soc.*, **23** (1924), §§ 6 and 8.

† Strictly, the surface S_1 should be closed, so that we should also consider the contributions from the remote edges. These contributions may, however, be rejected on physical grounds, since they would require infinitely long time to reach the point where the effect is being considered.

Hence we have in all

$$\left.\begin{aligned}\mathscr{J}_1 &= \tfrac{1}{3} \quad \text{when} \quad 0 < y < d, \\ &= 0 \quad \text{when} \quad y < 0 \text{ or } y > d.\end{aligned}\right\} \tag{8}$$

Before calculating $\mathscr{J}_2$ we note that because the integrand contains only a singularity of order $1/R$, integration throughout the vanishingly small sphere around the origin will not give a contribution. Hence in case (*a*), just as in case (*b*) we may now integrate throughout the full volume bounded by the slab:

$$\mathscr{J}_2 = \frac{\omega^2}{4\pi c^2} \int_{-y}^{d-y} e^{iqy_1} \mathscr{L}(y_1) dy_1, \tag{9}$$

where

$$\mathscr{L}(y_1) = \int_{-\infty}^{+\infty} \int_{-\infty}^{+\infty} \frac{e^{i(px_1 + \omega R/c)}}{R} dx_1 dz_1. \tag{10}$$

In (10) we take new variables ρ and χ defined by

$$px_1 + \frac{\omega}{c} R = \rho \sqrt{\left(\frac{\omega}{c}\right)^2 - p^2}, \tag{11a}$$

$$z_1 = \sqrt{\rho^2 - y_1^2} \sin \chi, \tag{11b}$$

in each case the positive square root being taken. On the $x_1 z_1$-plane, the curves of constant ρ are ellipses, with χ as the eccentric angle. For $\rho = |y_1|$ the ellipse degenerates into a point. Thus the whole $x_1 z_1$-plane is covered if ρ and χ run through the values $|y_1| \leqslant \rho \leqslant \infty$, $0 \leqslant \chi \leqslant 2\pi$. Hence

$$\mathscr{L}(y_1) = \int_0^{2\pi} \int_{|y_1|}^{\infty} \frac{e^{i\rho\sqrt{\frac{\omega^2}{c^2} - p^2}}}{R} \frac{\partial(x_1, z_1)}{\partial(\rho, \chi)} d\rho d\chi, \tag{12}$$

where

$$\frac{\partial(x_1, z_1)}{\partial(\rho, \chi)} = \frac{\partial x_1}{\partial \rho} \frac{\partial z_1}{\partial \chi} - \frac{\partial x_1}{\partial \chi} \frac{\partial z_1}{\partial \rho} \tag{13}$$

is the Jacobian of the transformation. From (11a) and (3)

$$\frac{\partial x_1}{\partial \rho} = \frac{\sqrt{\left(\frac{\omega}{c}\right)^2 - p^2}}{p + \frac{\omega}{c}\frac{x_1}{R}}, \qquad \frac{\partial x_1}{\partial \chi} = 0, \tag{14a}$$

and from (11b)

$$\frac{\partial z_1}{\partial \rho} = \frac{\rho}{\sqrt{\rho^2 - y_1^2}} \sin \chi, \qquad \frac{\partial z_1}{\partial \chi} = \sqrt{\rho^2 - y_1^2} \cos \chi = \sqrt{\rho^2 - y_1^2 - z_1^2}. \tag{14b}$$

Now we also have the identity

$$\left(pR + \frac{\omega}{c} x_1\right)^2 - \left(px_1 + \frac{\omega}{c} R\right)^2 = \left[\left(\frac{\omega}{c}\right)^2 - p^2\right][x_1^2 - R^2],$$

or using (3) and (11a),

$$\left(pR + \frac{\omega}{c} x_1\right) = \sqrt{\left(\frac{\omega}{c}\right)^2 - p^2} \ . \ \sqrt{\rho^2 - y_1^2 - z_1^2}. \tag{15}$$

From (13), (14), and (15) it follows that

$$\frac{\partial(x_1, z_1)}{\partial(\rho, \chi)} = R. \tag{16}$$

We substitute from (16) into (12). The integration with respect to χ gives immediately the value 2π. The integration with respect to ρ is also straightforward and we obtain

$$\mathscr{L}(y) = \frac{2\pi i\, e^{i|y_1|\sqrt{\left(\frac{\omega}{c}\right)^2 - p^2}}}{\sqrt{\left(\frac{\omega}{c}\right)^2 - p^2}}, \tag{17}$$

where, with the same physical justification as before, we have rejected an oscillatory contribution from infinity.

We now substitute from (17) into (9) and evaluate the resulting integral. This gives:

$$\left.\begin{aligned}
\mathscr{I}_2 &= \left(\frac{\omega}{c}\right)^2 \frac{e^{-igy}}{2g\sqrt{\left(\frac{\omega}{c}\right)^2 - p^2}}\{e^{igd} - 1\} && \text{when } y > d,\\
&= \left(\frac{\omega}{c}\right)^2 \frac{e^{-ihy}}{2h\sqrt{\left(\frac{\omega}{c}\right)^2 - p^2}}\{e^{ihd} - 1\} && \text{when } y < 0,\\
&= \left(\frac{\omega}{c}\right)^2 \left\{\frac{1}{gh} - \frac{e^{-igy}}{2g\sqrt{\left(\frac{\omega}{c}\right)^2 - p^2}} + \frac{e^{-ih(y-d)}}{2h\sqrt{\left(\frac{\omega}{c}\right)^2 - p^2}}\right\} && \text{when } 0 < y < d,
\end{aligned}\right\} \tag{18}$$

where

$$g = q - \sqrt{\left(\frac{\omega}{c}\right)^2 - p^2}, \qquad h = q + \sqrt{\left(\frac{\omega}{c}\right)^2 - p^2}. \tag{19}$$

AUTHOR INDEX

SUBJECT INDEX